This book provides a pedagogical introduction to the perturbative and non-perturbative aspects of quantum chromodynamics (QCD).

Introducing the basic theory and recent advances in QCD, it also reviews the historical development of the subject up to the present day, covering pre-QCD ideas of strong interactions such as the quark and parton models, the notion of colours, current algebra and the *S*-matrix approach. The author then discusses tools of quantum field theory, the symmetry and quantization of gauge theory, techniques of dimensional regularization and renormalization, QED high-precision tests, deep inelastic scattering and hard processes in hadron collisions, hadron jets, and inclusive processes in e + e− annihilations. Other topics include power corrections and the technologies of the Shifman–Vainshtein–Zakharov (SVZ) operator product expansion, renormalizations and phenomena beyond the SVZ expansion. The final parts of the book are devoted to modern non-perturbative approaches to QCD, such as lattice and effective theories, and the phenomenological aspects of QCD spectral sum rules.

The book will be a valuable reference for graduate students and researchers in high-energy particle and nuclear physics, both theoretical and experimental.

This title, first published in 2005, has been reissued as an Open Access publication on Cambridge Core.

STEPHAN NARISON graduated from the University of Antananarivo, Madagascar and received his Doctorat d'Etat from the University of Marseille. He is currently Director of Research in theoretical physics at the French Centre National de la Recherche Scientifique (CNRS), at the Laboratoire de Physique Mathématique et Théorique de l'Université Montpellier II. He has conducted research in laboratories and university departments throughout the world. Starting his research in the high-precision tests of QED, his main area of research is in non-perturbative aspects of QCD, using QCD spectral sum rules to study the properties of hadrons and low-energy phenomena in terms of the fundamental parameters from QCD first principles. He has worked in this field for more than two decades and has actively participated in its development. Professor Narison has had numerous publications in leading journals, as well as contributing to several books on high-energy physics. He is also the founder and chairman of the QCD Montpellier International Conference Series.

# CAMBRIDGE MONOGRAPHS ON PARTICLE PHYSICS NUCLEAR PHYSICS AND COSMOLOGY
## 17



1. K. Winter (ed.): *Neutrino Physics*
2. J. F. Donoghue, E. Golowich and B. R. Holstein: *Dynamics of the Standard Model*
3. E. Leader and E. Predazzi: *An Introduction to Gauge Theories and Modern Particle Physics, Volume I: Electroweak Interactions, the 'New Particles' and the Parton Model*
4. E. Leader and E. Predazzi: *An Introduction to Gauge Theories and Modern Particle Physics, Volume 2: CP-Violation, QCD and Hard Processes*
5. C. Grupen: *Particle Detectors*
6. H. Grosse and A. Martin: *Particle Physics and the Schrödinger Equation*
7. B. Andersson: *The Lund Model*
8. R. K. Ellis, W. J. Stirling and B. R. Webber: *QCD and Collider Physics*
9. I. I. Bigi and A. I. Sanda: *CP Violation*
10. A. V. Manohar and M. B. Wise: *Heavy Quark Physics*
11. R. K. Bock, H. Grote, R. Frühwirth and M. Regler: *Data Analysis Techniques for High-Energy Physics, Second edition*
12. D. Green: *The Physics of Particle Detectors*
13. V. N. Gribov and J. Nyiri: *Quantum Electrodynamics*
14. K. Winter (ed.): *Neutrino Physics, Second edition*
15. E. Leader: *Spin in Particle Physics*
16. J. D. Walecka: *Electron Scattering for Nuclear and Nucleon Structure*
17. S. Narison: *QCD as a Theory of Hadrons*
18. J. F. Letessier and J. Rafelski: *Hadrons and Quark–Gluon Plasma*
19. A. Donnachie, H. G. Dosch, P. V. Landshoff and O. Nachtmann: *Pomeron Physics and QCD*
20. A. Hofmann: *The Physics of Synchrotron Radiation*
21. J. B. Kogut and M. A. Stephanov: *The Phases of Quantum Chromodynamics*

# QCD AS A THEORY OF HADRONS

## From Partons to Confinement

STEPHAN NARISON

*Laboratoire de Physique Mathématique et Théorique*
*Université de Montpellier II*

CAMBRIDGE
UNIVERSITY PRESS

CAMBRIDGE
UNIVERSITY PRESS

Shaftesbury Road, Cambridge CB2 8EA, United Kingdom

One Liberty Plaza, 20th Floor, New York, NY 10006, USA

477 Williamstown Road, Port Melbourne, VIC 3207, Australia

314–321, 3rd Floor, Plot 3, Splendor Forum, Jasola District Centre, New Delhi – 110025, India

103 Penang Road, #05–06/07, Visioncrest Commercial, Singapore 238467

Cambridge University Press is part of Cambridge University Press & Assessment, a department of the University of Cambridge.

We share the University's mission to contribute to society through the pursuit of education, learning and research at the highest international levels of excellence.

www.cambridge.org
Information on this title: www.cambridge.org/9781009290319

DOI: 10.1017/9781009290296



First published 2005
Reissued as OA 2022

*A catalogue record for this publication is available from the British Library.*

ISBN 978-1-009-29031-9 Hardback
ISBN 978-1-009-29033-3 Paperback

To Larry and Rindra

# Contents

# About Stephan Narison

He is, at present, a Directeur de Recherche at the French 'Centre National de la Recherche Scientifique' (CNRS) in theoretical physics (section of high-energy elementary particle physics) at the 'Laboratoire de Physique Mathématique et Théorique de l'Université de Montpellier II' (France). He is the founder and chairman of the Series of Montpellier International Conference in Quantum ChromoDynamics (QCD) since 1985 which has been sponsored from 1996 to 2001 by the European Commission of Brussels.

He graduated at the Lycée Gallieni and University of Antananarivo (Madagascar) in 1972. After his master's degree, he was a teacher in different colleges of Antananarivo (Ambatonakanga, Esca and St Michel). In 1974, he obtained a fellowship from the 'Centre International des Etudiants et Stagiaires' of the European Commision of Brussels for preparing his Doctorat d'Etat at the University of Marseille (France). He was offered a 2-year postdoctoral position at the Abdus Salam Center for Theoretical Physics (former International Center for Theoretical Physics) in Trieste (Italy) from 1979 to 1981, a Scientific Associate position for 1 year at LAPP-Annecy (France) and a 2-year CERN (Geneva) fellowship in the Theory division in 1982. He obtained his permanent position in Montpellier in 1984. Since then, he has visited different world high-energy physics laboratories for the purpose of joint collaborations or by simple invitations. These include the traditional West European Universities and Institutes [Universities of Barcelona, Madrid, Valencia (Spain); University of Heidelberg (as a Von-Humboldt fellow), Munich (Germany), Vienna (Austria); CERN-Geneva, University of Bern (Switzerland); ICTP-Trieste, University of Pisa (Italy)], the Universities and Institutes of East European countries [University of Kracow (Poland), INR-Moscow (Russia)], the American Universities and Institutes [LBL (California), SLAC (Stanford), Brookhaven (Upton)] and the more exotic Asian Universities and Institutes [KEK-Tsukuba (Japan), KIAS-Seoul (Korea), NCTS-Hsinchu (Taiwan)]. He also participates actively in the creation of a new Theoretical Physics Institute in his home country. Finally, he is regularly invited to present contributions in the different large-scale, high-energy physics conferences (EPS, IHEP, ...) and specialized workshops.

His first research activity, which made him known in the field, was the estimate of the hadronic contributions to the anomalous magnetic moment of leptons (subject of his 3ème cycle thesis). Since then, his main research activity has been in the non-perturbative aspects of QCD using the method of QCD spectral sum rules for studying the properties of hadrons and low-energy phenomena. He has worked in this field since the date of its invention in 1979 and participates actively in its theoretical developments and new applications.

He is a member of the European Physical Society, a correspondant member of the 'Academie Nationale Malgache', a member of the New York Academy of Sciences, nominated in the Who's Who biography by the American Biographical Institute (ABI) (USA) and by the International Biographical Center of Cambridge (IBC) (UK), nominated among the 2000 exceptional men of the twentieth and twenty-first centuries by the ABI and the IBC. He has also been the President-Foundator of the 'Association Culturelle Malgache de Montpellier' (France) since 1993.

# Outline of the book

This book provides:

- A pedagogical introduction to the perturbative and non-perturbative aspects of Quantum Chromo Dynamics (QCD), which is expected to be accessible by pre-Ph.D. students who want to learn this field.
- A status of the modern developments in the field.
- An update of the different results presented in the older though successful review [2] and book [3], taking into account the developments of the field within these past 10 years.
- An extension and improvements of the presentation used in these previous review and book, where the QSSR results are compared with those from other non-perturbative approaches.

The book is divided into ten parts:

- In the first part, one starts from a general introduction to particle physics and historical survey on the developments of strong interactions prior to QCD. Then, we discuss the main ideas and basic tools of the field.
- In the second part, we present the gauge theory aspect of QCD.
- In the third part, we discuss in details the most popular techniques of dimensional regularization and renormalization and discuss some of its applications both in QCD and QED.
- In the fourth part, we present different QCD hard deep inelastic processes at hadron colliders, and discuss different unpolarized and polarized structure functions.
- In the fifth part, we present the QCD hard processes in $e^+e^-$ processes and discuss jets, fragmentation functions and totally inclusive processes.
- In the sixth part we summarize QCD tests and $\alpha_s$ measurements.
- In the seventh part, we discuss power corrections and mainly the theoretical basis and technologies of the Shifman–Vainshtein–Zakharov operating product expansion (OPE).
- In the eighth part, we present a compilation of different QCD two-point functions obtained from perturbative calculations and the SVZ-expansion. These expressions are basic ingredients for various phenomenological applications.
- In the ninth part, we present different aspects of modern non-perturbative approaches to QCD.
- In the tenth part, we present extensive phenomenological aspects of QCD spectral sum rules.
- The Appendices collect different useful conventions and formulae for QCD practitioners.
- The Contents, References and Index are useful for a quick guide for readers of the book.

# Preface

Quantum Chromodynamics (QCD) continues to be an active field of research, which one can see from the number of publications in the field, as well as from the number of presentations at different QCD dedicated conferences, such as the regular QCD-Montpellier Conference Series. This continuous activity is due to the relative difficulty in tackling its non-perturbative aspects, although its asymptotic freedom property has facilated perturbative calculations of different hard and jet processes. Therefore, we think it is still useful to write a book on QCD in which, besides the usual pedagogical introduction to the field, some reviews of its modern developments, which have not yet been 'compiled' into a book, will be presented. Elementary introductions at the level of pre-Ph.D. in different specialized topics of QCD will be discussed, which may be useful for a future deeper research and for a guide in a given subject.

We start the book with a general elementary introduction to strong interactions, parton and quark models, . . . , and present the basic tools for understanding QCD as a gauge field theory (renormalization, operator product expansion, . . . ). After, we present the usual hard processes (deep inelastic scattering, jets, . . . ) calculable in perturbative QCD, and discuss the resummation (renormalons, . . . ) of the perturbative series. Later, we discuss the different modern non-perturbative aspects of QCD (lattice, effective theories, . . . ). Among these different methods, we discuss extensively, the method and the phenomenology of the QCD spectral sum rules (QSSR) method introduced in 1979 by Shifman–Vainshtein and Zakharov (hereafter referred to as SVZ) [1]. Indeed, we have been impressed by its ability to explain low-energy phenomena such as the hadron masses, couplings and decays in terms of the first few fundamental parameters of QCD (QCD coupling, quark masses, quark and gluon condensates), and vice versa, we have been fascinated by the success of the method to extract the QCD universal parameters from experiments. In this respect, some parts of this book have been updated, improved, extended and included a latex version of the former review [2]:

Techniques of dimensional regularization and renormalization for the two-point functions of QCD and QED, S.N., *Phys. Rep.* 84 (1982) 263

and of the book [3]:

QCD Spectral Sum Rules Lecture notes in Physics, Vol. 26 (1989) World Scientific Publ. Co. Singapore.

However, the discussions in this book cannot replace the previous ones (hereafter referred to as QSSR1), as some detailed analyses carried out in the older review and book are not reported and repeated here. In this present book, we limit ourselves to review the most recent results and new developments in the field, without going into some technical details, and, in this sense, this book is a useful supplement to the former. Various misprints in QSSR1 have also been corrected.

As we have already mentioned, and as in the previous review and book, we have written this book for a large audience, not necessarily working in the field (elementary introduction to QCD, . . . ). However, experts will also appreciate this book, as they will find the most relevant and the latest results obtained so far with the QSSR method. They can also find compilations of non-trivial QCD expressions of the two-point correlators obtained within the Operator Product Expansion (OPE), and technical points relevant to the method itself (mixing of operators under renormalizations, validity of the SVZ expansion . . . ). Experimentalists will find in this book a 'quick review' of most of important results obtained from QSSR.

However, because of the large *horizontal* spectrum of the QSSR applications in different branches of low-energy physics, including nuclear matters, which we (unfortunately) cannot cover in this book, we shall limit ourselves to the well-controlled and simplest applications of the methods, namely the light and heavy quark systems and to a lesser extent the gluonia and hybrid meson channels. At present, these examples are quite well understood and will, therefore, serve as *prototype* applications of QSSR in high-energy physics and quantum field theory. Some other applications of QSSR, such as in the QCD string tension, in the composite models of electroweak interactions (QHD sum rules) and in supersymmetric QCD, were already discussed in QSSR1 and will not be discussed in detail here, since there has been no noticeable recent developments in these fields of applications, since the publication of QSSR1. We shall not discuss the uses of QSSR for nuclear matters, either, since the complexity of these phenomena still needs to be better understood. However, the enthusiasm of nuclear physicists for using this method in the baryonic sector might be restrained, owing to the delicateness of the corresponding analysis, which in my opinion has not yet been improved since the original work, in which the obstacle is due to the optimal choice of the nucleon operators. At the present stage, one can only consider the analysis done in the baryon sector to be very qualitative.

Following (actively) the developments of QCD through those of QSSR since its birth in 1979, my feeling à la Feynman (Omni magazine 1979), advocated in QSSR1 about this field remains unchanged (as already quoted in QSSR1):

*. . . A few years ago, I was very skeptical . . . I was expecting mist and now it looks like ridges and valleys after all . . . ,*

while the *great* success of QSSR in the understanding of the complexity of low-energy non-perturbative phenomena and hadron physics, is well illustrated by the Malagasy saying:

*'Vary iray no nafafy ka vary zato no miakatra!.'*

which means: with one grain of rice sowed, one can gather by the thousand!, or in other words, the method has started quite modestly and, with time, it has become more and more underground. Indeed, at present, QSSR (*used correctly*) is one of the most powerful methods for understanding (*analytically*) the low-energy dynamics of hadrons using the few fundamental parameters (coupling, masses and condensates) coming from QCD first principles.

# Acknowledgements

This book is a result of my attendance (as a chairman) at the QCD–Montpellier International Conference series from which I have learned a lot about the developments of QCD. It also comes from long-term contributions on QCD spectral sum rules and I thank former collaborators and colleagues who have contributed to the developments of this important field. Most of the figures of this book comes from the efficient work of Arlette Coudert from Cern. This work has been completed when I visited different institutes during the last 5 years (Cern–Geneva, Kek-Tsukuba, Ncts HsinChu, Ntu Taipei, Antananarivo University, MPI-Munich, Heidelberg University) and I wish to thank them for their hospitality. Finally, I am grateful for the patience of my family and friends who have sacrificed life-years during the long write-up of these materials, and I wish to thank them for their generous support.

# Part I

## General introduction

# 1

# A short flash on particle physics

Since ancient times, we have been curious to know the origin and the nature of the universe.[1] Numerous ancient philosophers and scientists have tried to answer these fundamental questions. It is only at the present time of the twentieth millennium that we can provide a partial answer to these questions, as some significant progress has been accomplished in both particle physics and astrophysics, which are two areas of research in two apparently opposite scale directions (see Fig. 1.1).[2]

On the one hand, this progress is due to our ability to explore the heart of matter, with powerful accelerators (where the accelerated particle has a velocity near to the velocity of light), which reveal their infinitely small, deepest structure (see Fig. 1.2).

As an example, we show in Figs. 1.3 and 1.4, the large electron-positron (LEP) accelerator and the reaction inside the detector after the collision of the electron and the anti-electron (positron). Notice that at LEP, the energy of the electron is in the range of 90–180 GeV which is about (5–10) $\times 10^6$ times the energy of our home TV screen. On the other hand, powerful telescopes (see Fig. 1.5) explore the enormous structure of the universe, and may reach the time of its origin. At present, these apparently two opposite (in scale) areas of research are found to have a common feature as the conditions required for exploring the smallest structure of matters (quarks) reproduce the periods which followed the big-bang (see Fig. 1.6), from which one may understand the origin of the universe.[3]

In this book, we shall concentrate on one aspect of particle physics, called Quantum ChromoDynamics, which is a part of the so-called Standard Model (SM). We know that, at the beginning of the study of nuclear physics, it was observed that, in addition, to the well-known Newton gravitation and electromagnetism (Maxwell) forces, nature is governed by two other new forces, the weak interactions responsible of the $\beta$ decay and the strong Yukawa force which binds the nucleons inside the nucleus (see Fig. 1.7).

In particle physics, only the last three forces play an important rôle as gravitation couples too weakly and cannot be directly detectable in particle physics experiments. At the particle

[1] This short review is based on the review talk in [4].
[2] Figures in this chapter come from [5].
[3] For a recent review on interfaces between these two fields, see e.g. [6].

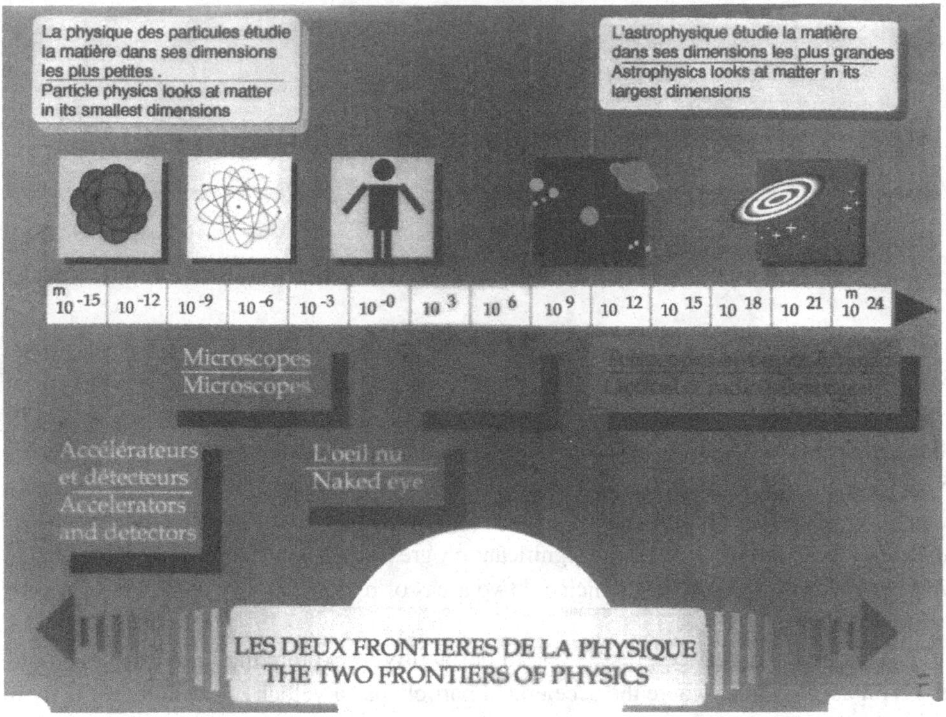


Fig. 1.1. A schematic view of the scales of the universe and related research branches. Our human body is taken as a reference scale (ref. CERN Z 11).

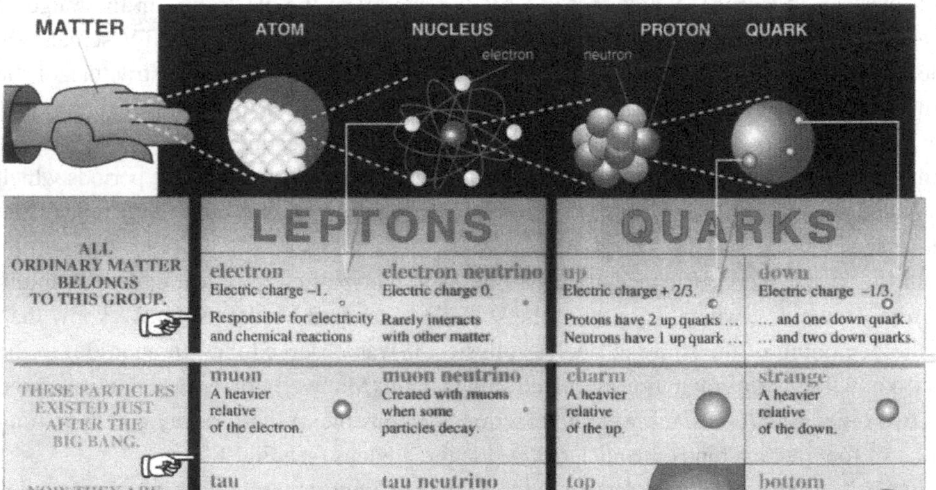

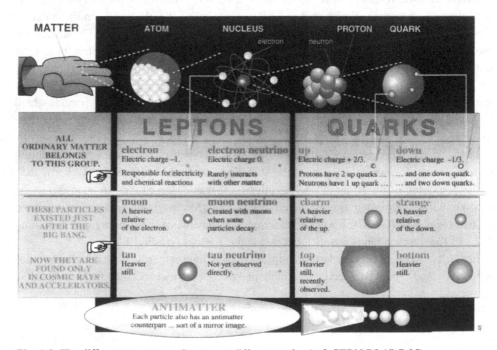

Fig. 1.2. The different structures of matter at different scales (ref. CERN DI-17-7-95).

Fig. 1.3. An aerial view of CERN-Geneva, showing the undeground LEP ring, 27 km in circumference, where also the LHC (large hadron collider) will run soon. In order to see the real size of the ring, one can see Geneva airport in the front part of the photo (ref. CERN X 973-1-87).

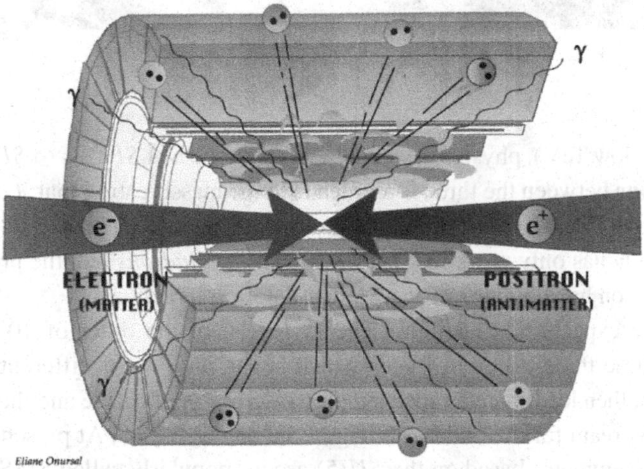


Fig. 1.4. A schematic view of the detector and particles produced after the collision (ref. cern DI-64-I-91).

Fig. 1.5. A photo of the Hubble telescope.

physics scale (below TeV), physics is well described by the SM $SU(3)_c \otimes SU(2)_L \otimes U(1)$, and the distinction between the three forces leads to the classification that: *Leptons* $(e^-, \nu_e)$ and $(\mu^-, \nu_\mu)$ pairs couple only to weak and electromagnetic $SU(2)_L \otimes U(1)$ forces (the neutral neutrino $\nu_l$ has only weak interactions), whereas *Hadrons* like the proton, neutron, pion and rho meson have mainly strong $SU(3)_c$ colour interactions.

However, one expects that at higher energy levels, of the order of $10^{15}$ GeV to the Planck scale, these three different forces which apparently are of different origins unify with gravitation, then leading to a much simpler description of nature and the realization of the old Einstein dream for the understanding of the universe laws. At present, the minimal version of supersymmetry based on the $SU(5)$ group (popularly called MSSM) is the best candidate for such a unified theory. Indeed, using the renormalization group evolution of the different couplings in the MSSM, one realizes that to second order in perturbation theory, these three couplings indeed cross with high precision at the unification scale of $10^{15}$ GeV as shown in Fig. 1.8. This result is encouraging although we still fail to find the correct

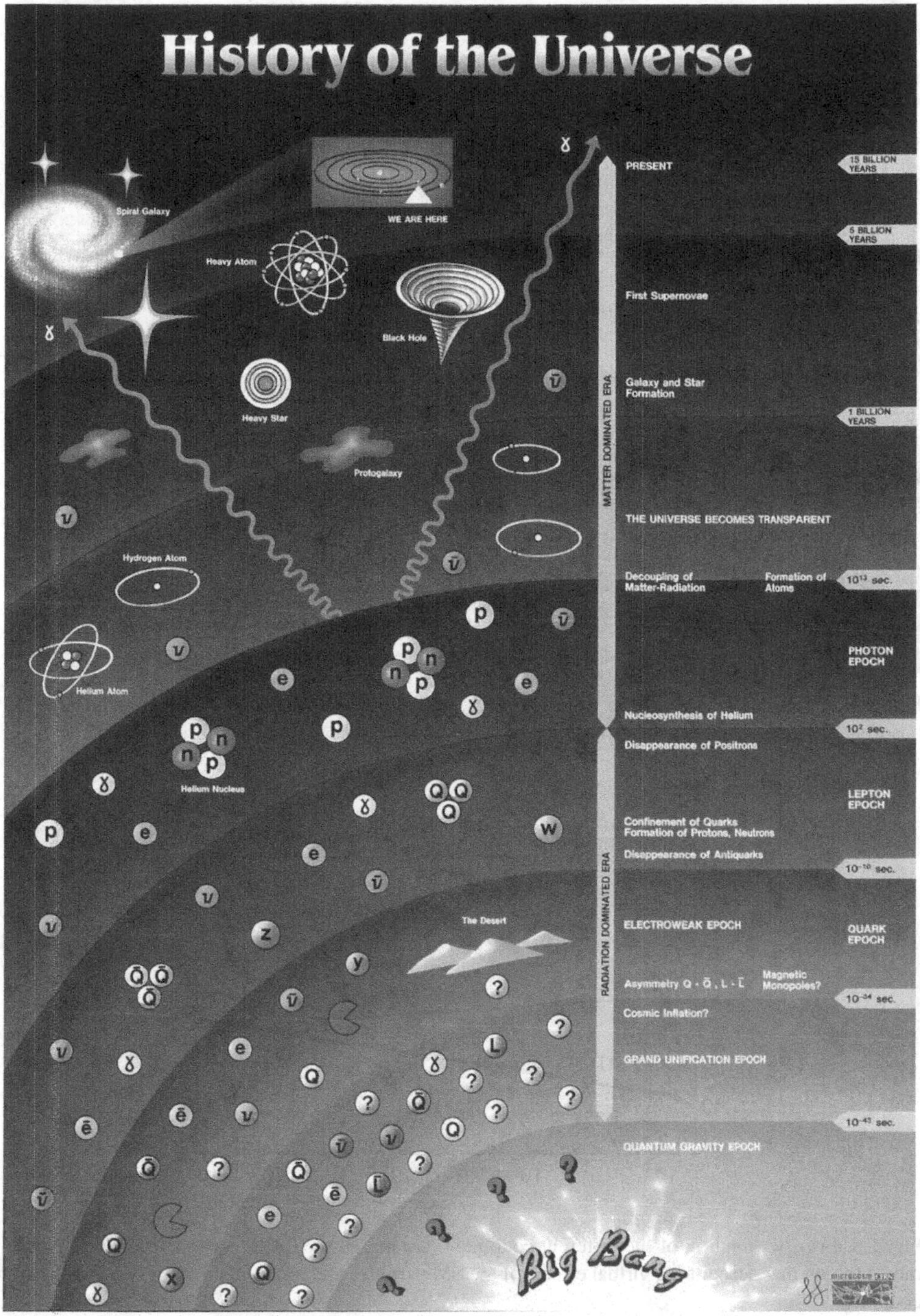


Fig. 1.6. A schematic view of the history of the universe from the big bang to the present day (ref. cern DI-2-8-91).

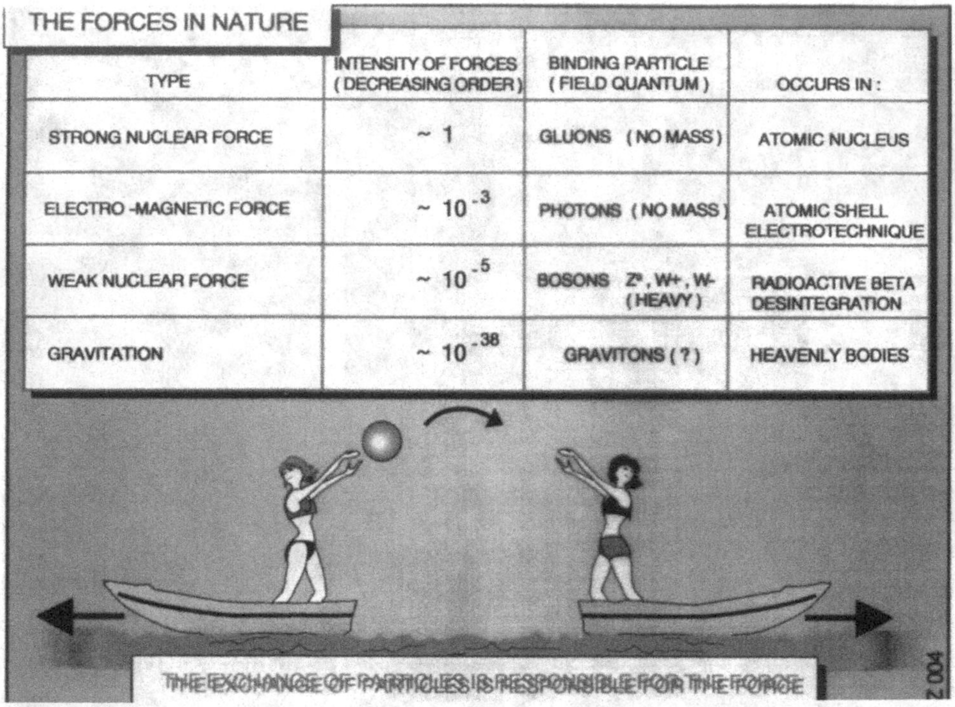


Fig. 1.7. A schematic view of the different forces in nature, and their associated vehicles (gauge bosons). The reference force is a strong interaction of strength $10^{-12}$ cm (ref. cern Z 004).

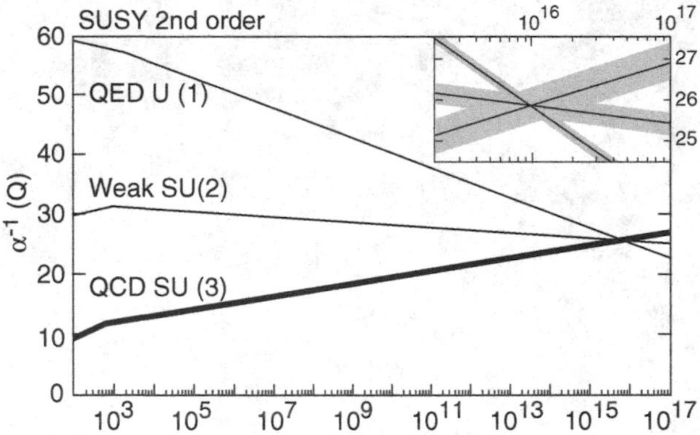


Fig. 1.8. Energy evolution of the different coupling constants of the QCD, weak and QED standard model taking into account the virtual effects of the SM particles and normalized to the MSSM $SU(5)$ coupling.

theory including gravitation. Many interesting attempts and proposals are available on the market.

The aim of this book is to present the developments of our understanding of strong interactions, and to concentrate on the exposition of its modern theory, called Quantum ChromoDynamics (QCD). Indeed, progress on strong interactions is important and necessary for making progress in the understanding of the physics beyond the SM.

# 2

# The pre-QCD era

## 2.1 The quark model

- We know that hadrons have mainly strong interactions. However, the number of observed hadrons increases drastically in comparison with that of leptons. The classification of hadrons into multiplets has been facilitated by the discovery of *internal symmetries*, which play an important rôle for obtaining relations among masses, magnetic moments and couplings of the hadrons. The classification under the $SU(3)_F$ group (named *flavour* at present) [7] has been successful, where hadrons are characterized under their isospin $I$, hypercharge $Y$, baryon number $B$ and strangeness $S$. Therefore, the pions are placed in the same pseudoscalar octet as the $K$, $\bar{K}$ and $\eta$, while the vector mesons $\rho$, $\omega$, $\phi$ fill another octet,... The splitting of hadron masses was expected, due to $SU(3)_F$ breaking that originated from strong interaction forces, whereas the $SU(2)$ isospin subgroup was found to be almost symmetric. This led to the concept of *charge independence*, which has played an important rôle in nuclear physics, where the proton and neutron form an $SU(2)$ doublet.
- However, none of the fundamental representations $SU(3)_F$ were realized by the observed hadrons, which led Gell-Mann and Zweig [8,9] to postulate that the observed hadrons, like the atoms, are not elementary, but are built by more elementary *quark*[1] constituents $q$ having three flavours *up, down and strange*. Their charge $Q$ in units of the one of the electron are:

$$Q_u = 2/3\,, \qquad Q_d = Q_s = -1/3\,. \tag{2.1}$$

In this picture, the mesons are bound states of quark–anti-quark, while the baryons are made by three quarks. The quarks internal quantum numbers are given in Table 2.1.

The $SU(3)_F$ decomposition into products of $\underline{3}$ and $\underline{3}^*$ representations gives for mesons:

$$\bar{q}q \;:\; \underline{3}^* \otimes \underline{3} = \underline{1} \oplus \underline{8} \tag{2.2}$$

and for baryons:

$$qqq \;:\; \underline{3} \otimes \underline{3} \otimes \underline{3} = \underline{1} \oplus \underline{8} \oplus \underline{8} \oplus \underline{10}\,, \tag{2.3}$$

[1] The name *quark* did not exist in the English dictionary, and may have been inspired from the following poetry *Finnengan's wake* of J. Joyce:

> "Three quarks for Muster mark!
> Sure he has'not got much of bark
> and sure any he has it's all beside the mark."

However, *quark* is a well-known German word as it means curdy milk, but more commonly it means a mess.

Table 2.1. *Additive quark-quantum numbers*

| Quark | $u$ | $d$ | $s$ |
|---|---|---|---|
| Charge $Q$ | $\frac{2}{3}$ | $-\frac{1}{3}$ | $-\frac{1}{3}$ |
| Third component of isospin $I_3$ | $\frac{1}{2}$ | $-\frac{1}{2}$ | 0 |
| Hypercharge $Y$ | $\frac{1}{3}$ | $\frac{1}{3}$ | $-\frac{2}{3}$ |
| Baryon number | $\frac{1}{3}$ | $\frac{1}{3}$ | $\frac{1}{3}$ |
| Strangeness | 0 | 0 | $-1$ |

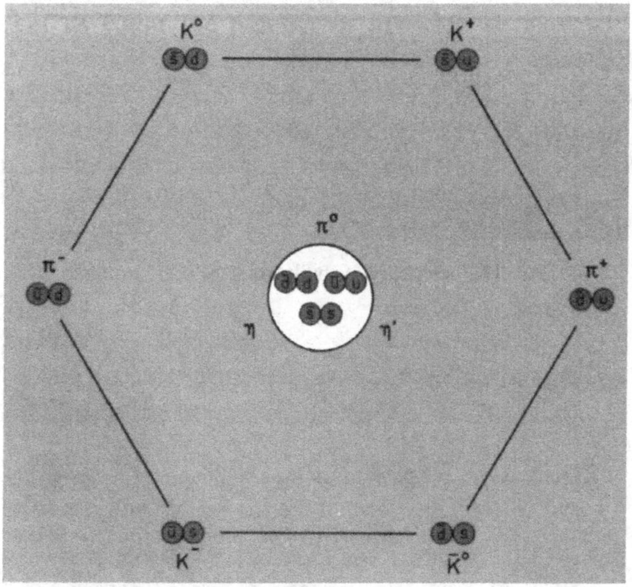


Fig. 2.1. The nine mesons built from the $u, d, s$ quarks.

from which one can built a simple but complete Periodic Table of Hadrons. These classifications are given in Figs. 2.1 to 2.3. In this sense, the quark model was a modern version of the Sakata [10] model.

- Masses and mass-splittings of hadrons have been explained by using Gell-Mann–Okubo-like mass formulae [11], and by introducing the so-called *constituent* quark masses with the values [12]:

$$M_q \approx 300 \text{ MeV} , \tag{2.4}$$

and by assuming the quark-mass differences:

$$M_d - M_u \approx 4 \text{ MeV} , \qquad M_s - M_d \approx 150 \text{ MeV} . \tag{2.5}$$

- The compositeness hypothesis for the hadrons has been supported by the measurement of the proton magnetic moment which has a value of about 2.8 in units of $\mu_p = e\hbar/2M_p$, while it is expected to be unity from a point-like spin $1/2$ object.

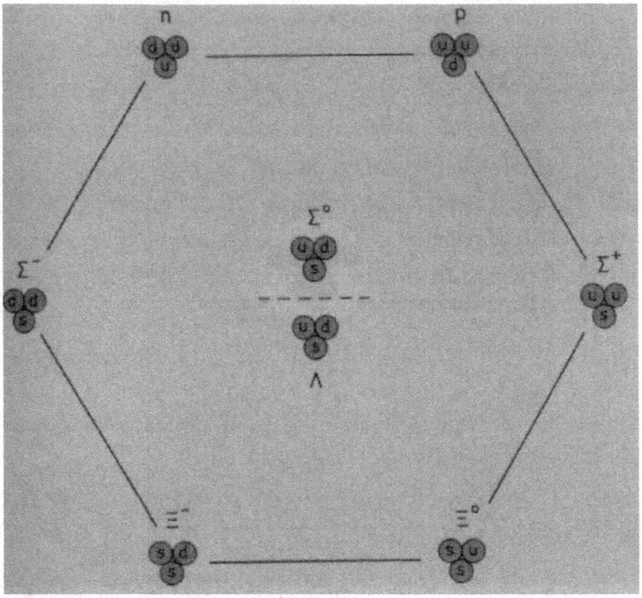


Fig. 2.2. The octet baryons built from the $u$, $d$, $s$ quarks.

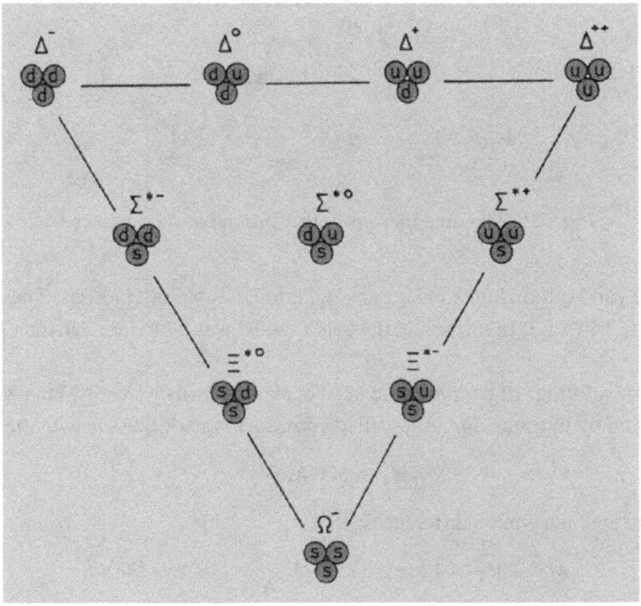


Fig. 2.3. The ten spin $3/2$ baryons built from the $u$, $d$, $s$ quarks.

## 2.2 Current algebras

Reviews on current algebras can be seen in [13]. In the following, we shall discuss some main features of the approach.

### 2.2.1 Currents conservation

- Although we have more forces in nature, *electromagnetism* plays a capital rôle. The theory of electron (muon) interacting with the photon field is the only one where the concepts of *quantum field theory* work in a satisfactory manner. Indeed, within Quantum ElectroDynamics (QED), one has been able to perform higher order approximate calculations which are confirmed by experimental measurements at an impressive, high level of accuracy (anomalous magnetic moment of the leptons, ...). Although more complicated, due to the presence of strong interactions, the study of the electromagnetic interaction of hadrons has been facilitated by the property of the *electromagnetic current conservation* leading to the *concept of universality*, which allows us to put, for example, at the same footing, an $e^-$, a $\pi^-$ and a $p^-$, and to show, for instance, that the *physical charges* of these three particles remains the same after renormalizations. Moreover, current conservation allows the use of *soft photon theorems* in order to relate the cross-section to the static properties of the hadrons (charge, magnetic moments, ...). It is also one of the basis of the popular *Vector Meson Dominance Model* (VDM) [14]. As a consequence of the current conservation, the corresponding charge is a constant of motion, such that the only non-vanishing matrix elements of this charge are between equal-mass states.
- In the case of weak current, current conservation gives a well-defined meaning to the idea of universal weak coupling which has been successfully tested experimentally in the case of non-strange weak vector currents. However, difficulty arises when one tries to explain strangeness-violating transition such as the ratio of the $K^+ \to \pi^0 e^+ \nu_e$ over the $\pi^+ \to \pi^0 e^+ \nu_e$. It can only be explained by the introduction of the Cabibbo angle $\theta_c$ [15] allowing the mixing of the strange quark with the down quark, with the experimental value $\sin\theta_c = 0.220 \pm 0.003$ [16]. In this case, the idea of weak universality appears also to work in the process involving the strange quark.
- Inspired again by the quark model, Gell-Mann [7] suggested that the vector and axial charges satisfy a $SU(3) \otimes SU(3)$ algebra. This picture naturally leads to the existence of larger multiplets of particles having the same spins but with both parities, which has been confirmed by the data. The rôle of *partially conserved axial current* (PCAC) was found to be related to the existence of the light (compared with the $\rho$ and $p$) pseudoscalar particle, the $\pi$, which has been understood, later on, from the spontaneous Nambu–Goldstone [17] nature of the symmetry breaking. More precisely, the exact current conservation of the axial current is realized when the pion is massless. Again inspired by the soft photon theorem which is a consequence of the conservation of the electromagnetic current, one can also derive *soft pion theorems* obtained from phenomenological Lagrangians satisfying the non-linear realizations of chiral symmetry.

### 2.2.2 Currents and charges

The next development is the construction of hadron currents built from quark fields in much the same way as one can write a current for lepton fields. The quark electromagnetic and charged weak currents can be written as:

$$J^{\mu}_{\text{em}} = \frac{2}{3}\bar{u}\gamma^{\mu}u - \frac{1}{3}\bar{d}\gamma^{\mu}d - \frac{1}{3}\bar{s}\gamma^{\mu}s + \cdots,$$
$$J^{\mu}_{\text{weak}} = \bar{u}\gamma^{\mu}(1-\gamma_5)d + \cdots, \tag{2.6}$$

where we ignore to a first approximation the mixing among quark fields due to the Cabibbo angle. In the massless quark limit ($m_j = 0$), the free quark Lagrangian density $\mathcal{L}_q(x)$:

$$\mathcal{L}_q(x) = i\sum_{j=1}^{n}\bar{\psi}_j\gamma_\mu\psi_j\,, \tag{2.7}$$

possesses a $SU(n)_L \times SU(n)_R$ global chiral symmetry and is invariant under the global chiral transformation:

$$\psi_i(x) \to \exp(-i\theta^A T_A)\psi_i(x)\,,$$
$$\psi_i(x) \to \exp(-i\theta^A T_A\gamma_5)\psi_i(x)\,, \tag{2.8}$$

where $T^A(A \equiv 1, \ldots, n^2-1)$ are the infinitesimal generators of the $SU(n)$ group acting on the quark-flavour components. The associated Noether currents are the vector and axial-vector currents:

$$V^A_\mu(x) = \bar{\psi}_i\gamma_\mu\, T^A_{ij}\psi_i(x)\,,$$
$$A^A_\mu(x) = \bar{\psi}_i\gamma_\mu\gamma_5\, T^A_{ij}\psi_i(x)\,, \tag{2.9}$$

which are the ones of the algebra of currents of Gell-Mann [69,13] ($n = 3$ in the original paper). The corresponding charges which are the generators of $SU(n)_L \times SU(n)_R$ are:

$$Q^A = \int d^3x\; V^A_0(x)\,,$$
$$Q^A_5 = \int d^3x\; A^A_0(x)\,. \tag{2.10}$$

The charges in Eq. (2.10) are conserved in the massless quark limit, and obey the commutation relations (simplified notations):

$$[Q^\alpha, Q^\beta] = if_{\alpha\beta\gamma}Q^\gamma\,,$$
$$\left[Q^\alpha_5, Q^\beta_5\right] = if_{\alpha\beta\gamma}Q^\gamma\,,$$
$$\left[Q^\alpha, Q^\beta_5\right] = if_{\alpha\beta\gamma}Q^\gamma_5\,, \tag{2.11}$$

i.e. $Q_V$ and $Q_A$ generate a closed algebra. They also imply:

$$[Q^\alpha, V^\beta] = if_{\alpha\beta\gamma}V^\gamma\,,$$
$$[Q^\alpha, A^\beta] = if_{\alpha\beta\gamma}A^\gamma\,,$$
$$\left[Q^\alpha_5, V^\beta\right] = if_{\alpha\beta\gamma}A^\gamma\,,$$
$$\left[Q^\alpha_5, A^\beta\right] = if_{\alpha\beta\gamma}V^\gamma\,. \tag{2.12}$$

### 2.2.3 Chiral symmetry and pion PCAC

In the Nambu–Goldstone [17] realization of chiral symmetry, the axial charge does not annihilate the vacuum, which is the basis of the successes of current algebra and pion PCAC [13]. In this scheme, the chiral flavour group $G \equiv SU(n)_L \times SU(n)_R$ is broken spontaneously by the light quark $(u, d, s)$ vacuum condensates down to a subgroup $H \equiv SU(n)_{L+R}$, where the vacua are symmetrical:

$$\langle \bar{\psi}_u \psi_u \rangle = \langle \bar{\psi}_d \psi_d \rangle = \langle \bar{\psi}_s \psi_s \rangle \ . \tag{2.13}$$

The Goldstone theorem states that this spontaneous breaking mechanism is accompanied by $n^2 - 1$ massless Goldstone $P$ (pions) bosons, which are associated with each unbroken generator of the coset space $G/H$. For $n = 3$, these Goldstone bosons can be identified with the eight lightest mesons of the Gell-Mann eightfoldway ($\pi^+$, $\pi^-$, $\pi^0$, $\eta$, $K^+$, $K^-$, $K^0$, $\bar{K}^0$). On the other hand, the vector charge is assumed to annihilate the vacuum and the corresponding symmetry is achieved à la Wigner–Weyl [18]. In the vector case, the particles are classified in irreducible representations of $SU(n)_{L+R}$ and form parity doublets. In addition to the electromagnetic mass which the Goldstone bosons can acquire [19], they get a mass mainly from an explicit breaking ($m_i \neq 0$) of the $SU(n)_L \times SU(n)_R$ global symmetry. In this case, the divergence of the axial-vector current does not vanish and reads (in the case of the $u$, $d$ quarks):

$$\partial_\mu A^\mu(x)^i_j = (m_i + m_j)\bar{\psi}_i(i\gamma_5)\psi_j \ , \tag{2.14}$$

to which are associated the quasi-Goldstone parameters defined as:

$$\langle 0|\partial_\mu A^\mu(x)^i_j|\pi\rangle = \sqrt{2} f_\pi m_\pi^2 \vec{\pi} \ , \tag{2.15}$$

where $\vec{\pi}$ is the pion field and $f_\pi = 92.4$ MeV is the pion decay constant which controls the $\pi \to \mu\nu$ decay width. In this case, the divergence of the vector current reads:

$$\partial_\mu V^\mu(x)^i_j = (m_i - m_j)\bar{\psi}_i(i)\psi_j \ , \tag{2.16}$$

to which is presumably associated the $a_0(980)$ scalar meson (the best experimental candidate).

Current algebra also tells us that the two-point correlator associated with Eq. (1.15) is related to the axial-current one via a current algebra Ward identity [20,13], up to equal-time commutator terms (in the following we shall suppress flavour indices):

$$\begin{aligned} q_\mu q_\nu \Pi_5^{\mu\nu} &= \Psi_5(q^2) - q^\nu \int d^4x e^{iqx} \delta(x_0) \langle 0[A^0(x), (A^\nu(0))^\dagger]0\rangle \\ &+ i \int d^4x e^{iqx} \delta(x_0) \langle 0[\partial_\mu A^\mu(x), (A^0(0))^\dagger]0\rangle \ , \end{aligned} \tag{2.17}$$

with:

$$\begin{aligned} \Psi_5(q^2) &= i \int d^4x e^{iqx} \langle 0|\mathbf{T}\partial_\mu A^\mu(x)(\partial_\mu A^\mu(0))^\dagger|0\rangle \ , \\ \Pi_5^{\mu\nu}(q^2) &= i \int d^4x e^{iqx} \langle 0|\mathbf{T}A^\mu(x)(A^\nu(0))^\dagger|0\rangle \ . \end{aligned} \tag{2.18}$$

At $q = 0$, the previous identity reduces to:

$$\Psi_5(0) = -i(m_u + m_d)\langle 0[\bar{\psi}_d(0) i\gamma_5 \psi_u(0), Q_5^\dagger]0\rangle \,, \tag{2.19}$$

where $Q_5$ is the axial-charge generator. In the Nambu–Goldstone realization of chiral symmetry, one has:

$$Q_5|0\rangle \neq 0 \,. \tag{2.20}$$

Therefore, we get:

$$\Psi_5(0) = -(m_u + m_d)\langle \bar{\psi}_d \psi_d + \bar{\psi}_u \psi_u\rangle \,. \tag{2.21}$$

Using Eq. (2.15) in the definition of $\Psi_5(q^2)$ and equating this with Eq. (2.19), we have the well-known pion PCAC (Gell-Mann *et al.* [21]) relation at $q = 0$ (recall that $f_\pi = 92.4$ MeV):

$$-(m_u + m_d)\langle \bar{\psi}_d \psi_d + \bar{\psi}_u \psi_u\rangle = 2m_\pi^2 f_\pi^2 \,. \tag{2.22}$$

### 2.2.4 Soft pion theorem and the Goldberger–Treiman relation

Let's consider the matrix element of the axial-vector current between two nucleon states shown in Fig. 2.4.

Using invariance properties, it can be parametrized as:

$$\langle N(p_2)|A_\mu|N(p_1)\rangle = \bar{u}(p_2)[\gamma_\mu g_A(q^2) + q_\mu g_P(q^2)]\gamma_5 u(p_1) \,, \tag{2.23}$$

where $q = p_2 - p_1$ is the momentum transfer between the nucleon states, and where experimentally $g_A(0) = 1.26$. The matrix element of the current divergence reads:

$$\mathcal{A} \equiv \langle N(p_2)|\partial^\mu A_\mu|N(p_1)\rangle = \bar{u}(p_2)[2M_N g_A(q^2) + q^2 g_P(q^2)](i\gamma_5)u(p_1) \,, \tag{2.24}$$

where the relation for the Dirac spinors:

$$q_\mu \bar{u}(p_2)(i\gamma^\mu\gamma_5)u(p_1) = 2M_N \bar{u}(p_2)\gamma_5 u(p_1) \,, \tag{2.25}$$

has been used. The PCAC hypothesis in Eq. (2.15) yields in the massless pion (chiral) limit:

$$2M_N g_A(q^2) + q^2 g_P(q^2) = 0 \,. \tag{2.26}$$

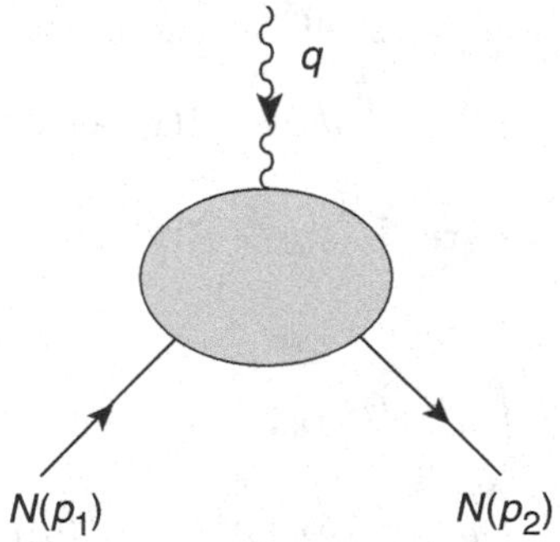


Fig. 2.4. Axial-vector scattering with nucleon.

where the divergence of the axial-vector current is zero. If $g(q^2)$ has no singularity at $q^2 = 0$, then Eq. (2.26), would imply either $M_N = 0$ or $g_A = 0$. However, none of these requirements are true. Therefore, $g_P$ should have a pole at $q^2 = 0$:

$$\lim_{q^2\to 0} g_P(q^2) = -\frac{2M_N g_A}{q^2} . \tag{2.27}$$

The matrix element in Eq. (2.24) between a one pion state and the vacuum is the same as if there were a term in $A_\mu(x)$ of the form $\sqrt{2}f_\pi \partial_\mu \vec{\pi}(x)$. Therefore, in the chiral limit, the matrix element has a pole, and reads:

$$\langle N(p_2)|A_\mu|N(p_1)\rangle = \sqrt{2}f_\pi q_\mu \langle N(p_2)|\vec{\pi}|N(p_1)\rangle = \frac{2f_\pi q_\mu}{-q^2} g_{\pi NN}(q^2)\bar{u}(k_2)(i\gamma_5)u(k_1) , \tag{2.28}$$

where $g_{\pi NN}(q^2)$ is the $\pi NN$ vertex function. Its *physical coupling* is defined at $q^2 = m_\pi^2$ at has the experimental value of $13.50 \pm 0.15$ [16]. Solving these last two equations, one can derive the *Golberger–Treiman relation* (GT) [22] in the chiral limit:

$$f_\pi g_{\pi NN}(0) = M_N g_A(0) . \tag{2.29}$$

In the case of massive quarks, one can write the matrix element in Eq. (2.24) as:

$$\mathcal{A} = \sqrt{2}f_\pi m_\pi^2 \langle N(p_2)|\vec{\pi}|N(p_1)\rangle = \frac{2f_\pi m_\pi^2}{-q^2 + m_\pi^2} g_{\pi NN}(q^2)\bar{u}(k_2)(i\gamma_5)u(k_1) . \tag{2.30}$$

By identifying Eqs. (2.24) and (2.30), and setting $q^2 = 0$, one would obtain the previous GT relation in Eq. (2.29), which one can identify with the physical coupling assuming that the coupling is a smooth function of $q^2$ from 0 to $m_\pi^2$, which is valid as there is no one-pion pole in this function. One should remark that only $g_P(q^2)$ has a pion pole term, and it is of the form:

$$g_P(q^2) = \frac{\sqrt{2}f_\pi}{m_\pi^2 - q^2}\sqrt{2}g_{\pi NN} , \tag{2.31}$$

such that at $q^2 = m_\pi^2$, Eqs. (2.30) and (2.24) leads to a trivial equality.

### 2.2.5 The Adler–Weisberger sum rule and soft pion theorems

In the case of the Golberger–Treiman relation, we have used a one-pion soft theorem for estimating the pion-nucleon-nucleon matrix element. Here, we shall be concerned by low-energy theorems for pion-nucleon scattering amplitudes involving two soft pions. The process is depicted in Fig. 2.5.

The amplitude can be written as:

$$\langle \pi_i(q_2)N(p_2)|\pi_j(q_1)N(p_1)\rangle = i(2\pi)^4\delta^4(p_1 + q_1 - p_2 - q_2)T_{ij} , \tag{2.32}$$

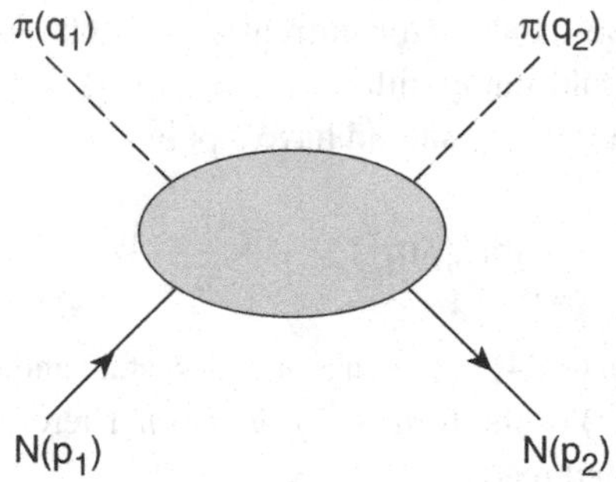


Fig. 2.5. Forward pion-nucleon scattering process.

which can be decomposed in terms of two invariants (isospin-even and -odd):

$$T_{ij} = \delta_{ij} T^{(+)} + \frac{1}{2}[\tau_i, \tau_j] T^{(-)} , \tag{2.33}$$

where $i, j$ are isospin indices. Using standard reduction formula discussed in the next section, one can apply the soft pion theorem, which gives:

$$\begin{aligned} T_{ij} &= i\left(-q_2^2 + m_\pi^2\right)\langle N(p_2)|\vec{\pi}^i(0)\vec{\pi}^j(q_1)N(p_1)\rangle \\ &= \frac{q_2^\mu\left(-q_2^2 + m_\pi^2\right)}{\sqrt{2} f_\pi m_\pi^2}\langle N(p_2)|A_\mu^i(0)|\pi^j(q_1)N(p_1)\rangle . \end{aligned} \tag{2.34}$$

For $q_2 \to 0$, we can take $T^{(-)} = 0$ since it is odd under crossing. Also, the non-singular part of the amplitude vanishes (*Adler's consistency condition*) [23]:

$$T^{(+)}\left(\nu = 0,\ \nu_B = 0,\ q_1^2 = m_\pi^2,\ q_2^2 = 0\right) = 0 , \tag{2.35}$$

where:

$$\nu \equiv q_1(p_1 + p_2)/2 , \qquad \nu_B = -q_1 \cdot q_2/2 , \tag{2.36}$$

are kinematic variables. Similarly, when $q_1^2 \to 0$, one obtains:

$$T^{(+)}\left(\nu = 0,\ \nu_B = 0,\ q_1^2 = 0,\ q_2^2 = m_\pi^2\right) = 0 . \tag{2.37}$$

Applying two times the soft pion theorems, one can reduce the amplitude as:

$$T_{ij} = i\left(q_1^2 - m_\pi^2\right)\left(q_2^2 - m_\pi^2\right)\frac{1}{2m_\pi^4 f_\pi^2}\int d^4x\, e^{iq_1x}\langle N(p_2)|\mathcal{T}\partial^\mu A_\mu(x)\partial^\mu A_\mu(0)|N(p_1)\rangle . \tag{2.38}$$

Using the current algebra Ward identity:

$$\begin{aligned} q_1^\mu q_2^\nu \int d^4x\, e^{iq_1x}\mathcal{T} A_\mu^i(x) A_\nu^j(0) = \int d^4x\, e^{iq_1x}\big[&\mathcal{T}\partial^\mu A_\mu^i(x)\partial^\mu A_\mu^j(0) \\ &- iq_1^\mu\delta(x_0)\left[A_0^j(0), A_\mu^i(x)\right] + \delta(x_0)\left[A_0^i(0), \partial^\mu A_\mu^j(x)\right]\big] , \end{aligned} \tag{2.39}$$

one can see after sandwiching between two nucleon states that the first term is the nucleon matrix element of a time-ordered product of two-pion operators; the second term can be evaluated from the current algebra commutation relation:

$$\delta(x_0)\big[A_0^i(0), A_\mu^j(x)\big] = -i\delta(x)\epsilon^{ijk} V_{k,\mu}(x) , \tag{2.40}$$

while the last term gives the *pion-sigma term*, which is symmetric in $i$, $j$, and then this $t$-channel state must have isospin 0 or 2 since the pion has isospin 1. However, since the nucleon has isospin $1/2$, only $I = 0$ state can contribute, and therefore:

$$\sigma^{ij} = \delta^{ij}\sigma_N . \tag{2.41}$$

In the low-energy limit, the following soft-pion theorems are obtained:

$$\lim_{\nu\to 0} \nu^{-1} T^{(-)}(\nu, 0, 0, 0) = \left(1 - g_A^2\right)/f_\pi^2 , \tag{2.42}$$

and

$$\lim_{\nu\to 0} \nu^{-1} T^{(+)}(0, 0, 0, 0) = -\sigma_N/f_\pi^2 . \tag{2.43}$$

It is also expected and assumed that $T^{(-)}$, which is odd under the change $\nu \to -\nu$, obeys an unsubtracted dispersion relation in the variable $\nu$:

$$\frac{T^{(-)}(\nu, q^2 = 0)}{\nu} = \frac{2}{\pi}\int_{\nu_0}^{\infty} \frac{d\nu'}{\nu'^2 - \nu^2}\, \mathrm{Im}T^{(-)}(\nu', 0) . \tag{2.44}$$

Its imaginary part can be related to the $\pi N$ cross-section if one assumes a smoothness assumption:

$$\mathrm{Im}T^{(-)}(\nu, 0) \simeq \mathrm{Im}T^{(-)}\left(\nu, m_\pi^2\right) = \nu\big[\sigma_{\rm tot}^{\pi^+ p}(\nu) - \sigma_{\rm tot}^{\pi^- p}(\nu)\big] . \tag{2.45}$$

Using the previous GT relation in Eq. (2.29) for eliminating $f_\pi$ in Eq. (2.42), the dispersion relation gives the *Adler–Weisberger relation* [24]:

$$1 - \frac{1}{g_A^2} = \frac{2M_N^2}{\pi g_{\pi NN}^2}\int_{\nu_0}^{\infty} \frac{d\nu}{\nu}\big[\sigma_{\rm tot}^{\pi^+ p}(\nu) - \sigma_{\rm tot}^{\pi^- p}(\nu)\big] , \tag{2.46}$$

which is an interesting low-energy sum rule.

### 2.2.6 Soft pion theorem for $\rho \to \pi^+\pi^-$ and the KSFR relation

We discuss here a further use of soft pion theorems. We consider the process in the chiral limit where the pions are massless:

$$\rho^0 \to \pi^+\pi^- . \tag{2.47}$$

It is described by the amplitude:

$$T_{\nu\mu}^{ij} = i\int d^4x\, \exp(iqx)\langle 0|\mathcal{T} A_\nu^i(x) A_\mu^j(0)|\rho(p)\rangle , \tag{2.48}$$

where $i,\ j$ are isospin indices. Taking its divergence, one obtains:

$$q^{\nu}T^{ij}_{\nu\mu} = \left\{ U^{ij}_{\mu} \equiv -\int d^4x\ \exp(iqx)\langle 0|\mathcal{T}\partial^{\nu}A^{i}_{\nu}(x)A^{j}_{\mu}(0)|\rho(p)\rangle \right\}$$
$$-\int d^4x\ \exp(iqx)\delta(x_0)\langle 0|\left[A^{i}_{0}(x),\ A^{j}_{\mu}(0)\right]|\rho(p)\rangle\ . \qquad (2.49)$$

Using the commutation relation given previously, one can deduce the Ward identity:

$$q^{\nu}T^{ij}_{\nu\mu} = U^{ij}_{\mu} - if^{ijk}\langle 0|V_{\mu,k}|\rho(p)\rangle\ , \qquad (2.50)$$

where $V_{\mu}$ is the vector isovector current. In the massless pion limit, the axial current is conserved such that $U^{ij}_{\mu}$ vanishes. The coupling of the *neural* $\rho$-meson to the isovector current is introduced as (from now, we shall suppress the isospin indices):

$$\langle 0|V_{\mu}|\rho(p)\rangle = \frac{M^2_{\rho}}{2\gamma_{\rho}}\epsilon_{\mu}\ . \qquad (2.51)$$

where, experimentally, $\gamma_{\rho} = 2.55$, with the normalization:

$$\Gamma_{\rho\to e^+e^-} \simeq \frac{2}{3}\pi\alpha^2\frac{M_{\rho}}{2\gamma^2_{\rho}}\ . \qquad (2.52)$$

$\epsilon^{\mu}$ is the polarization of the $\rho$ meson which ensures the conservation of the vector current. Contracting again with the pion momentum $q'$, one obtains:

$$q^{\nu}q'^{\mu}T_{\nu\mu} = (\epsilon\cdot q')\frac{M^2_{\rho}}{2\gamma_{\rho}}\ . \qquad (2.53)$$

Introducing the $\rho\pi\pi$ coupling as:

$$\langle\pi(q'),\pi(q)|\rho(p)\rangle = \epsilon^{\nu}(q'-q)_{\nu}g_{\rho\pi\pi}\ , \qquad (2.54)$$

and taking the limit $q'\to q\to 0$, one obtains the soft pion relation:

$$\frac{M^2_{\rho}}{2\gamma_{\rho}} = 4f^2_{\pi}g_{\rho\pi\pi}\ . \qquad (2.55)$$

If one assumes $\rho$-universality from the vector meson dominance model [14], one has:

$$\frac{M^2_{\rho}}{2\gamma_{\rho}} = \frac{M^2_{\rho}}{g_{\rho\pi\pi}}\ . \qquad (2.56)$$

The two equations give the Kawarabayashi–Suzuki–Ryazuddin–Fayazuddin (KSFR) relations [25]:

$$g^2_{\rho\pi\pi} = \frac{M^2_{\rho}}{4f^2_{\pi}}\ , \qquad (2.57)$$

or:

$$f^2_{\pi} = \frac{M^2_{\rho}}{16\gamma^2_{\rho}}\ , \qquad (2.58)$$

which are useful in different phenomenological applications. One can check from the data that the predictions given by these two relations are unexpectedly good despite the crude approximation used for deriving them.

### 2.2.7 Weinberg current algebra sum rules

Another important consequence of the commutation relation of currents are the different current algebra dispersion sum rules, based on the assumption that the $SU(3) \otimes SU(3)$ symmetry is realized asymptotically. Though conceptually difficult to digest, this *asymptotically free* hypothesis has been very successful in different applications [13] (Weinberg and Das–Mathur–Okubo (DMO) sum rules [26,27], Adler–Weisberger sum rule [24] discussed previously, . . .). Here, we shall discuss briefly the Weinberg and DMO sum rules. They are based on the assumed asymptotic behaviour of the absorptive amplitudes, with the assumption that the $SU(2)_L \times SU(2)_R$ chiral symmetry is asymptotically realized in nature. Weinberg has derived two superconvergent sum rules, well-known as Weinberg sum rules (WSR) [26]. In order to show this result, it is appropriate to study the two-point correlator:

$$\begin{aligned} W^{\mu\nu}_{LR} &\equiv i \int d^4x e^{iqx} \langle 0|\mathcal{T} J^\mu_L(x) \left(J^\nu_R(0)\right)^\dagger |0\rangle \\ &= -(g^{\mu\nu}q^2 - q^\mu q^\nu)\Pi^{(1)}_{LR} + q^\mu q^\nu \Pi^{(0)}_{LR} , \end{aligned} \tag{2.59}$$

where $J^\mu_L$ and $J^\mu_R$ are left- and right-handed charged currents, which read in terms of the quark fields:

$$J^\mu_L \equiv \bar{u}\gamma^\mu(1-\gamma_5)d , \qquad J^\mu_R \equiv \bar{u}\gamma^\mu(1+\gamma_5)d . \tag{2.60}$$

$\Pi^{(1)}_{LR}$ and $\Pi^{(0)}_{LR}$ are respectively the transverse and longitudinal parts of the correlator. In the asymptotic limit ($q^2 \to \infty$) or in the chiral limit ($m_{u,d} \to 0$), where the $SU(2)_L \times SU(2)_R$ chiral symmetry is realized, $W^{\mu\nu}_{LR}$ tends to zero. Using the Källen–Lehmann representation of the two-point correlator:

$$\left(\Pi^{(J)}_{LR} \equiv \left(\Pi ij^{(J)}\right)_{LR}\right)\left(q^2, m_i^2, m_j^2\right) = \int_0^\infty \frac{dt}{t-q^2-i\epsilon}\frac{1}{\pi}\,\mathrm{Im}\Pi^{(J)}_{LR}(t) + \cdots , \tag{2.61}$$

where $\cdots$ represent subtraction points, which are polynomial in the $q^2$-variable, one can transform the previous property of $W^{\mu\nu}_{LR}$ into superconvergent sum rules for its absorptive parts [26]:

$$\begin{aligned} &\int_0^\infty dt\ \mathrm{Im}\left(\Pi^{(1)}_{LR} + \Pi^{(0)}_{LR}\right) \approx 0 , \\ &\int_0^\infty dt\ t\ \mathrm{Im}\Pi^{(1)}_{LR} \approx 0 , \end{aligned} \tag{2.62}$$

where the first WSR comes from the $q^\mu q^\nu$ component of $W^{\mu\nu}_{LR}$ and the second WSR comes from its $g^{\mu\nu}$ part. These WSR express in a clear way, the *global duality* between the long-range (spectral function measurable at low-energy) and the high-energy (asymptotic theory) parts of the hadronic correlators. This quark-hadron duality is one of the basic idea behind QCD spectral sum rules, which we shall discuss in detail in the next part of the book.

In order to parametrize the spectral functions, we use a narrow-width approximation and assume that the $\pi$, $A_1$ and $\rho$ dominate the spectral functions. In this way, one can derive the constraints:

$$\frac{M_\rho^2}{2\gamma_\rho^2} - \frac{M_{A_1}^2}{2\gamma_{A_1}^2} - 2f_\pi^2 \approx 0\,,$$
$$\frac{M_\rho^4}{2\gamma_\rho^2} - \frac{M_{A_1}^4}{2\gamma_{A_1}^2} \approx 0\,, \tag{2.63}$$

where $f_\pi = 92.4$ MeV is the pion decay constant governing the $\pi \to \mu\nu$ decay; $\gamma_V$ is the $V$-meson coupling to the corresponding *charged* current:

$$\langle 0|V^\mu|\rho\rangle = \sqrt{2}\frac{M_\rho^2}{2\gamma_\rho}\epsilon^\mu\,, \tag{2.64}$$

where experimentally $\gamma_\rho \simeq 2.55$. Notice the extra $\sqrt{2}$ factor coming from the different normalizations of the charged and neutral current discussed in the analysis of the $\rho^0 \to \pi^+\pi^-$ decay. From the above crude assumptions, one can predict by solving the two WSR equations and by using the experimental values of the $\rho$ and $\pi$ parameters:

$$M_{A_1} \simeq 1.1 \text{ GeV}\,, \tag{2.65}$$

which is in good agreement with the present data [16]. If, in addition, one uses the relation between $f_\pi$, $\gamma_\rho$ and $M_\rho$ (approximate KSFR relation [25]) discussed previously:

$$f_\pi^2 \simeq \frac{M_\rho^2}{16\gamma_\rho^2}\,, \tag{2.66}$$

deduced, from $\rho \to \pi\pi$ decays, using soft pion techniques, one arrives at the successful Weinberg mass formula:

$$M_{A_1} \simeq \sqrt{2}M_\rho\,, \tag{2.67}$$

although one should notice that the data from hadronic experiments give a slightly higher value [16].

### 2.2.8 The DMO sum rules in the $SU(3)_F$ symmetry limit

*Electromagnetic current*

Weinberg-inspired sum rules have been also derived from the asymptotic realization of the flavour symmetry. The Das–Mathur–Okubo (DMO) sum rules [27] can be studied from the two-point correlator:

$$\Pi_i^{\mu\nu}(q^2) \equiv \int d^4x e^{iqx}\langle 0|\mathcal{T} V_i^\mu(x)\left(V_i^\nu(0)\right)^\dagger|0\rangle$$
$$= -(g^{\mu\nu}q^2 - q^\mu q^\nu)\Pi_i^{(1)}(q^2)\,, \tag{2.68}$$

where $V_i^\mu(x) \equiv \bar{\psi}_i \gamma^\mu \psi_i$ ($i \equiv u, d, s, \dots$) are the flavour components of the electromagnetic current:

$$J^\mu_{EM}(x) = \frac{2}{3} V^\mu_u - \frac{1}{3} V^\mu_d + \frac{2}{3} V^\mu_c - \frac{1}{3} V^\mu_s + \cdots \tag{2.69}$$

In the asymptotic limit ($q^2 \to \infty$) or in the chiral limit ($m_i \to 0$), one can derive the DMO sum rule [27]:

$$\int_0^\infty dt[\mathrm{Im}\Pi_3(t) - \mathrm{Im}\Pi_8(t)] \equiv \int_0^\infty dt \ \mathrm{Im}(\Pi_u + \Pi_d - 2\Pi_s)(t) = 0\,, \tag{2.70}$$

which corresponds to the difference between the isovector and isoscalar spectral functions associated with the $SU(3)_F$ symmetry. Saturating the spectral functions by the lowest mass resonances, one can derive the well-known successful phenomenological relation among vector mesons:

$$M_\rho \Gamma_{\rho \to e^+e^-} - 3(M_\omega \Gamma_{\omega \to e^+e^-} + M_\varphi \Gamma_{\varphi \to e^+e^-}) \simeq 0\,. \tag{2.71}$$

One can also re-write the DMO sum rules in terms of the total cross-section for $e^+e^- \to$ hadrons by using the optical theorem:

$$\sigma(e^+e^- \to \text{hadrons}) = \frac{4\pi^2\alpha}{t} e^2 \frac{1}{\pi} \mathrm{Im}\Pi(t)\,. \tag{2.72}$$

This relation is useful for testing the breaking of $SU(3)_F$, as we shall see later on, because we have complete data for the total cross-section.

*Charged current*

In the case of the charged vector or axial current:

$$V^\mu(x)^i_j = \bar{\psi}_i \gamma^\mu \psi_j\,, \qquad A^\mu(x)^i_j = \bar{\psi}_i \gamma^\mu \gamma^5 \psi_j\,, \tag{2.73}$$

the DMO sum rules read in the chiral limit:

$$\int_0^\infty dt \ \mathrm{Im}\Pi^{(1)}(t)^d_u = \int_0^\infty dt \ \mathrm{Im}\Pi^{(1)}(t)^s_u\,, \tag{2.74}$$

where the spectral function can be measured in the $\tau \to \nu_\tau +$ hadrons decays. By saturating the spectral function with the lowest resonances, one can deduce the constraint:

$$\frac{M_\rho^2}{\gamma_\rho^2} \approx \frac{M_{K^*}^2}{\gamma_{K^*}^2}\,. \tag{2.75}$$

Using $\gamma_\rho = 2.55$, $M_\rho = 0.776$ GeV and $M_{K^*} = 0.892$ GeV, it gives:

$$\gamma_{K^*} = 2.93\,, \tag{2.76}$$

which is already an interesting constraint as compared with the data from $\tau$ decay [16]. On can notice that, as in the case of the WSR, the DMO sum rules give constraints between the low-energy behaviour of the spectral functions and their asymptotic one.

### 2.2.9 $\pi^+$-$\pi^0$ mass difference

Hadronic contributions to the electromagnetic $\pi^+$-$\pi^0$ mass difference have been derived by Das *et al.* [19] by assuming a good realization of the $SU(2)_L \times SU(2)_R$ chiral symmetry at short distance. In this way, by integrating the virtual photon with momentum $q^2$, they derive the result, in the chiral limit:

$$m_{\pi^+}^2 - m_{\pi^0}^2 \simeq -i\frac{6\pi\alpha}{f_\pi^2}\int \frac{d^4q}{(2\pi)^4}\frac{1}{q^2}\int_0^\infty \frac{dt}{q^2+t-i\epsilon}\frac{t}{\pi}\mathrm{Im}\Pi_{LR}^{(1)}$$
$$\simeq -\frac{3\alpha}{4\pi f_\pi^2}\int_0^\infty dt\left(t\ln\frac{t}{\nu^2}\right)\frac{1}{2\pi}\mathrm{Im}\Pi_{LR}^{(1)} \quad (2.77)$$

where the spectral functions enter the second WSR and $\nu$ is an arbitrary UV cut-off. Using a lowest resonance saturation of the spectral functions in the narrow width appproximation (NWA), and the constraints provided by the first and second sum rules, which guarantee the convergence of the integral, one can derive the relation:

$$m_{\pi^+}^2 - m_{\pi^0}^2 \simeq \frac{3\alpha}{4\pi}\frac{M_{A_1}^2 M_\rho^2}{M_{A_1}^2 - M_\rho^2}\ln\frac{M_{A_1}^2}{M_\rho^2}\,. \quad (2.78)$$

Using the WSR relation $M_{A_1}^2 = 2M_\rho^2$, one can deduce the result of [19]:

$$m_{\pi^+} - m_{\pi^0} \simeq \frac{3\alpha}{4\pi}\frac{M_\rho^2 \ln 2}{m_\pi}\,, \quad (2.79)$$

which is in good agreement with the data $m_{\pi^+} - m_{\pi^0} = 4.5936(5)$ MeV [16]. The improvements of these prototype current algebra sum rules in the QCD context have been done in [28–34] and will be discussed in details in the following sections.

## 2.3 Parton model and Bjorken scaling

Different deep-inelastic scattering experiments such as the unpolarized electroproduction process $ep \to eX$ ($X$ being the sum of inclusive produced hadrons) at high-energy virtual photon with momentum $Q$, have been used to explore the quark structure of the proton. This unpolarized process can be characterized by two measurable *structure functions* $W_{1,2}$, which parametrize the hadronic tensor and contains all strong interaction information about the response of the target nucleon to electromagnetic probes:

$$\frac{d\sigma}{dQ^2 d\nu} = \frac{\pi\alpha^2}{4M_p E^2 \sin^4\theta E E'}\left\{2\sin^2\frac{\theta}{2}W_1(Q^2,\nu) + \cos^2\frac{\theta}{2}W_2(Q^2,\nu)\right\}\,. \quad (2.80)$$

As shown in Fig. 2.6, they depend on the usual kinematic variables $-q^2 \equiv Q^2$ and $\nu$:

$$\nu \equiv p\cdot q = M_p(E - E')\,, \quad (2.81)$$

where $\nu/M_p$ is the energy transfer in the proton rest frame; $p$ and $M_p$ are the proton momentum and mass; $E$ and $E'$ are the energies of the incident and scattered electrons in the proton rest frame, and $\theta$ is the scattering angle ($Q^2 = 4EE'\sin^2\frac{\theta}{2}$). For a point-like

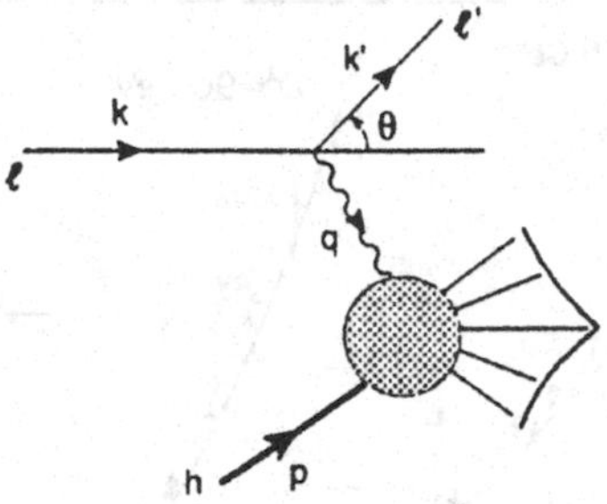


Fig. 2.6. $ep \to e$+hadrons process.

proton, the structure functions are $\delta$-functions:

$$W_1(Q^2, \nu) = \frac{Q^2}{4M_p^2}\delta\left(\nu - \frac{Q^2}{2}\right), \qquad W_2(Q^2, \nu) = \delta\left(\nu - \frac{Q^2}{2}\right). \tag{2.82}$$

It has been observed that, at large $Q^2$, contributions from pointlike spin 1/2 objects inside the proton still remain, while prominent contributions of resonances at low $Q^2$ die out quickly when $Q^2$ increases. A rough estimate of the *proton structure functions* can be done by assuming that the proton consists with pointlike spin 1/2 quark constituents (called *wee partons* by Feynman [35]), each one carrying a given fraction $\xi_i$ of the proton momentum. Defining by $f_i(\xi_i)$ the probability that a parton $i$ has momentum fraction $\xi_i$, and by $W_j^{(i)}$ the parton contribution to the structure function, then the proton structure function becomes an incoherent sum of the one of the partons, and reads:

$$\begin{aligned} W_1(Q^2, \nu) &= \sum_i \int_0^1 d\xi_i\, f_i(\xi_i) W_1^{(i)}(Q^2, \nu) = \frac{1}{2}\sum_i e_i^2 f_i(x) \equiv F_1(x) \\ W_2(Q^2, \nu) &= \sum_i \int_0^1 d\xi_i\, f_i(\xi_i) W_2^{(i)}(Q^2, \nu) = \frac{M_p^2}{\nu} x \sum_i e_i^2 f_i(x) \equiv \frac{M_p^2}{\nu} F_2(x), \end{aligned} \tag{2.83}$$

where $e_i$ is the electric charge and:

$$x \equiv \frac{Q^2}{2\nu}. \tag{2.84}$$

This simple parton description of the proton, where the structure function depends only on the kinematic variable $x$, is known as *Bjorken scaling* [36]. As a consequence of the spin-1/2 assumption of the constituent quarks, one also obtains the *Callan–Gross relation* [37]:

$$F_2(x) = 2x F_1(x). \tag{2.85}$$

These two *QCD sum rules* are well-satisfied by the data as shown in the Figs. 2.7 and 2.8,[2] which then surprisingly suggest the existence of free point-like partons inside the proton, in apparent contradiction with the confinement postulate.

[2] Small logarithmic deviations from the parton model prediction are also seen, and are well explained in QCD (as we shall see later on) after leading logs-resummation using the Altarelli–Parisi equation [38].

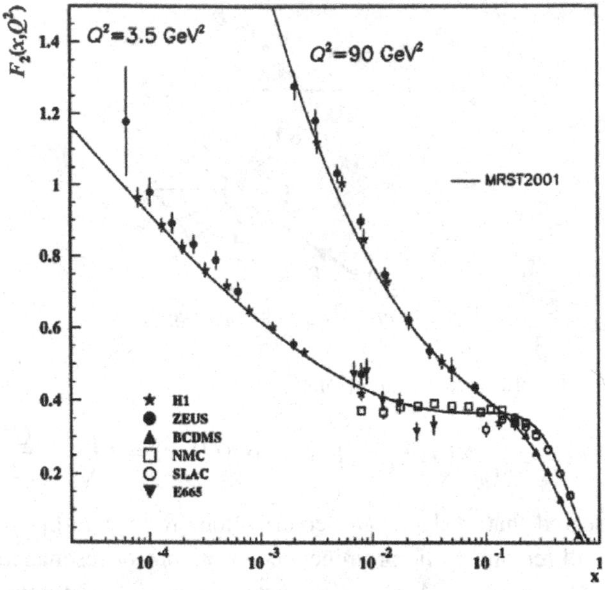


Fig. 2.7. The proton structure function $F_2$ versus $x$ at two values of $Q^2$, exhibiting scaling at the pivot point $x \approx 0.14$.

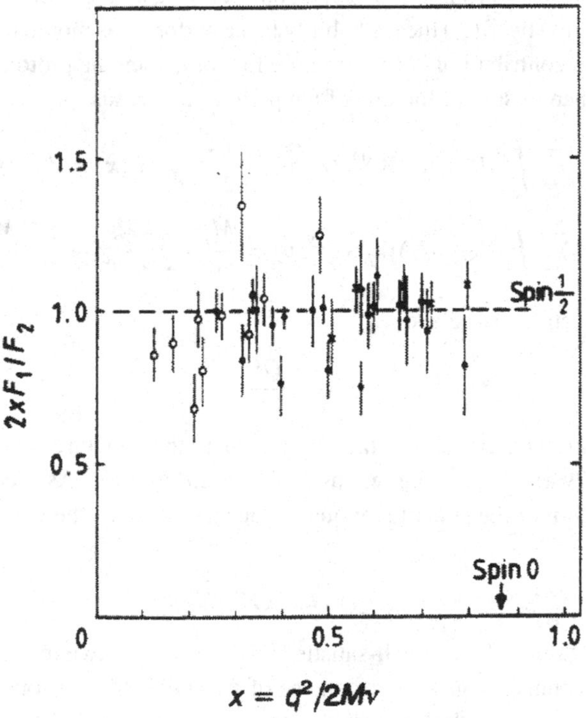


Fig. 2.8. The ratio $2xF_1/F_2$ versus $x$ for $Q^2$ values between 1.5 and 16 GeV$^2$.

## 2.4 The $S$-matrix approach and the Veneziano model

### 2.4.1 The $S$-matrix approach

An alternative to the quark model was the so-called $S$-matrix *(bootstrap)* approach which was very popular in the 1960s–1970s. It is based from a general Lagrangian, which should be constrained from general principles (relativistic covariance, substitution rule, unitarity and analyticity), and which limits the choice of the $S$-matrix. One of the main consequences of this approach is the *Regge poles* theory [39], which gives a general classification of hadrons *(Regge trajectories)* and predictions for high-energy data in terms of low-energy parameters from the study of resonances. This approach can be illustrated by the scattering process:

$$A + B \to C + D \tag{2.86}$$

and the crossed processes:

$$A + \bar{C} \to \bar{B} + D\,, \qquad A + \bar{D} \to \bar{B} + C\,, \tag{2.87}$$

characterized by the two kinematic variables $s$ and $t$. The amplitude can be written in a dispersive form:

$$A(s,t) = \frac{1}{\pi}\int \frac{\mathrm{Im}A(s',t')}{s'-s}ds', \tag{2.88}$$

where one assumes that it converges for sufficiently large $t$, while it can be written as a sum of poles:

$$A(s,t) = \beta(t)\sum_{n=0}^{\infty}\frac{s^n}{\alpha(t)-n} \tag{2.89}$$

in the variable $t$ at the solutions of the equations $\alpha(t_0) = 0$, $\alpha(t_n) = n$. Regge asymptotic law gives rise for fixed $t$ to:

$$\lim_{s\to\infty} \mathrm{Im}A(s,t) \sim \beta(t)s^{\alpha(t)}\,, \tag{2.90}$$

where one can see a direct relation between the $s$ and $t$-channels description of the scattering. This relation can also be seen more conveniently from the finite energy sum rule:

$$\int_0^L ds\; s^n\, \mathrm{Im}A(s,t) = \frac{L^{\alpha(t)+n+1}}{\alpha(t)+n+1}\,. \tag{2.91}$$

### 2.4.2 The Veneziano model and duality

The duality relation (crossing) between the $s$-channel resonance and $t$-channel Regge poles suggests the *duality bootstrap*. This has been achieved by the Veneziano approach [40], where a complete (though approximate) description of scattering can be obtained in terms of the $s$-channel resonances only. Inserting the resonance contributions from the particles contained in the trajectory $\alpha(s)$ and in its daughters, one obtains:

$$A(s,t) = \sum_n \frac{c_n(t)}{\alpha(s)-n} = \sum_n \frac{c_n(s)}{\alpha(t)-n}, \tag{2.92}$$

where the last equality is due to the duality constraints. $c_n(t)$ is a polynomial of order $n$ in $t$. The contribution of highest spin $j = n$ comes from the $\alpha(s) = n$ intercept in the leading trajectory, while the ones of lower spin come from the presence of lower 'daughter' trajectories. The solution to this equation is given by the well-known *Veneziano beta-function amplitude*:

$$A(s,t) = \frac{\Gamma[-\alpha(s)]\Gamma[-\alpha(t)]}{\Gamma[-\alpha(s)-\alpha(t)]}. \tag{2.93}$$

The Veneziano dual-resonance model for the scattering amplitude can be summarized by the following conditions:

- Only infinitely *narrow* resonances appear, and the only singularities are poles on the real axis.
- There is an exact crossing symmetry.
- There is an asymptotic Regge behaviour with linear trajectories with universal slope.

However, one should notice that straight line trajectories are very far from the expectation from a field theoretical argument which suggests a Yukawa-like potential. Instead, they follow from a harmonic oscillator potential, and seem to be supported by the data.

### 2.4.3 Duality diagrams

The previous discussion can be visualized using duality diagrams introduced in [41]. It consists to represent the quark content of non-exotic (ordinary) hadrons as:

- Ordinary baryons composed of three quarks will be represented by three quark lines oriented in the same directions.
- Mesons composed by quark–anti-quark will be represented by quark lines going in opposite directions.

The process is represented by the topological structure of the graph:

- Planar diagrams can be drawn without crossing quark lines, which coïncide with the ones suggested by duality and ordinary hadrons, and which give a non-vanishing contribution to the imaginary part of the amplitude.
- Non-planar diagrams are the other possibility, but do not contribute to the imaginary part as they are real.

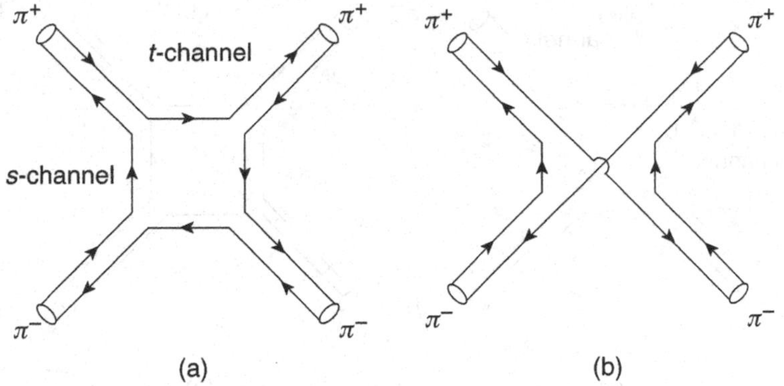


Fig. 2.9. Duality diagrams for $\pi^+\pi^-$ scattering: (a) planar $(s, t)$ graph; (b) non-planar $(s, \bar{s})$ or $(t, \bar{s})$ graph.

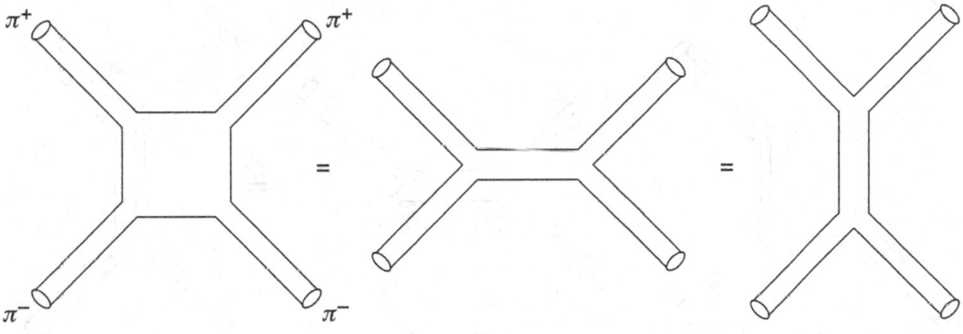


Fig. 2.10. Dual-resonance diagram for $\pi$-$\pi$ scattering.

In order to illustrate these rules, we can consider the scattering process:

$$\begin{aligned} \pi^+\pi^- &\to \pi^+\pi^- \qquad (s - \text{channel}), \\ \pi^+\pi^- &\to \pi^+\pi^- \qquad (t - \text{channel}), \\ \pi^+\pi^+ &\to \pi^+\pi^+ \qquad (\bar{s} - \text{channel}), \end{aligned} \tag{2.94}$$

shown in Fig. 2.9.

From the previous discussion, only the planar diagram contributes to the imaginary part of the amplitude. Duality invokes that a sum of resonances (or Regge poles) exchanged in the $s$ channel is equivalent to the sum of Regge poles (or resonances) exchanged in the $t$ channel, which is shown in Fig. 2.10. Similar planar diagrams can be drawn for $\pi$-$N$ scattering as shown in Fig. 2.11. In the case of $N$-$N$ (or in general baryon–antibaryon) scattering, one has the dual-resonance diagram (Fig. 2.12).

It shows that the planar graph represents exchange of non-exotic objects in the $s$ channel, but exchange of exotics in the $t$ channel. This feature signals that without exotics, the approach cannot consistently explain the hadronic world.

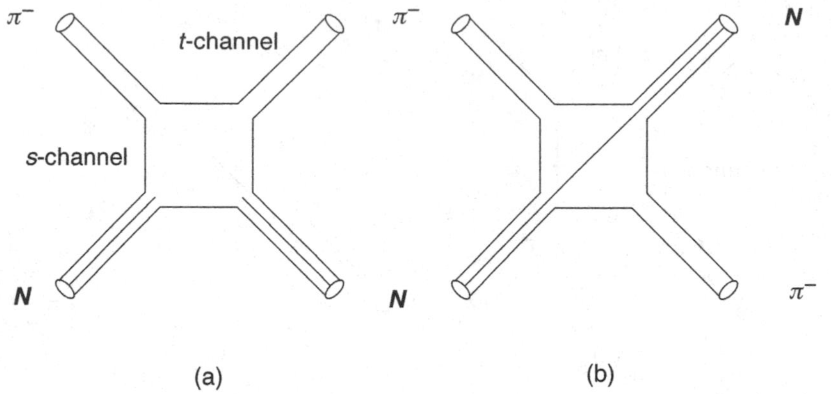


Fig. 2.11. Planar diagrams for $\pi$-$N$ scattering: (a) (($s, t$) or ($\bar{s}, t$); (b) ($s, \bar{s}$).

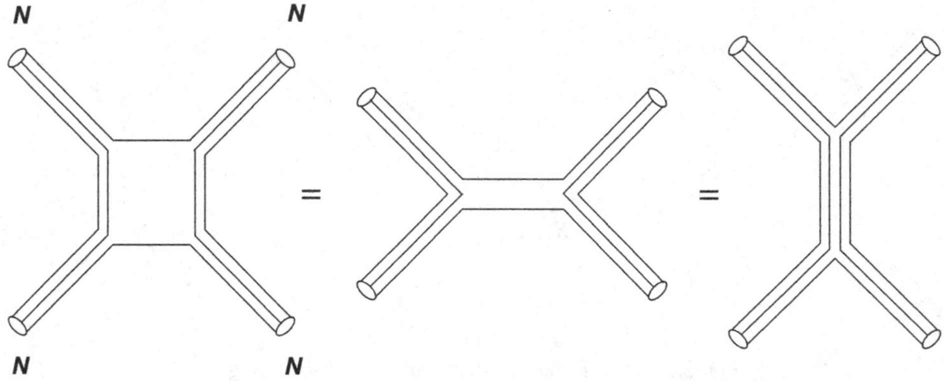


Fig. 2.12. Dual-resonance diagram for $N$-$N$ scattering.

One has expected that the previous approach based on superconvergence and duality, and implemented by the dual-resonance model suggested by Veneziano [40] will bring new insights in the developments of the theory of strong interactions. Alas, after the discovery of QCD, such theories became unsuccessful, although we know, at present, that the Veneziano model (actually it can be viewed as a string model) revives as the basics of superstring theories with which one wishes to unify the three electromagnetic, weak and strong forces with gravitation.

# 3

# The QCD story

We shall limit ourselves here to a qualitative survey of Quantum ChromoDynamics (QCD), with the aim to present in a short and simple way the main idea behind the theory. Many more complete and detailed reviews and books on QCD [2,3,42–52] and on Quantum Field Theory [53] exist in the literature, which the interested readers may consult, while recent results and developments in QCD both from theory and experiments may, for example, be found in many conferences, for instance, in the proceedings of the QCD–Montpellier Series of Conferences published regularly in *Nucl. Phys. B* (Proc. Suppl.) by Elsevier Publ. Co.

## 3.1 QCD and the notion of quarks

- QCD is by now expected (and widely accepted) to be the field theory describing the strong interactions of quarks $q$ [8,9] (elementary constituents of the matter) having *three colours (blue, red, yellow)* which are glued together inside the nucleus by eight coloured *(chromo)* gluons which provide a vehicle for the Yukawa strong nuclear forces. However, the quark scheme is not only a pure mathematical concept for classifying the hadronic world. There is indirect evidence of the existence of quarks through the observation of two-jet events, such as the one from:

$$Z^0 \rightarrow \text{hadrons}\,, \tag{3.1}$$

as shown in Fig. 3.1.

- QCD originated from the natural development of the quark model of the early 1960s, where, as we have discussed in the previous chapter, hadrons were classified under the representations of an $SU(3)_F$ (now called a *flavour group*), the so-called *eightfoldway* of Gell-Mann and Ne'eman [7], where *ordinary* mesons and baryons of this $SU(3)_F$ classification are respectively bound states $\bar{q}q$ and $qqq$ of the *light* quarks *up (u), down (d)* and *strange (s)*. The masses of these quarks,[1] which are given in the next section, are much lower than the value of the QCD scale $\Lambda \approx 300$ MeV, and have the values at the scale 2 GeV[2] [54]:

$$m_u \simeq 3.5\ \text{MeV}\,, \qquad m_d \simeq 6.3\ \text{MeV}\,, \qquad m_s \simeq 119\ \text{MeV}\,, \tag{3.2}$$

where one can notice that $m_s/m_d \simeq 20$ is a huge number obtained originally [21,55–57] from *current algebra* approaches [13]. These values have to be contrasted with the so-called *constituent* quark

[1] As quarks are not directly observed, the definitions of their masses are only theoretical. For light quarks, I will quote the values of running (or current) masses evaluated at a certain scale.
[2] The original choice of scale is 1 GeV. We take 2 GeV in order to follow current practice.

Table 3.1. *Quantum numbers of the new quarks*

| Quark | $c$ | $b$ | $t$ |
|---|---|---|---|
| Charge $Q$ | $\frac{2}{3}$ | $-\frac{1}{3}$ | $\frac{2}{3}$ |
| $C$-charm | 1 | 0 | 0 |
| $B$-bottomness or beauty | 0 | −1 | 0 |
| $T$-topness or topless | 0 | 0 | 1 |

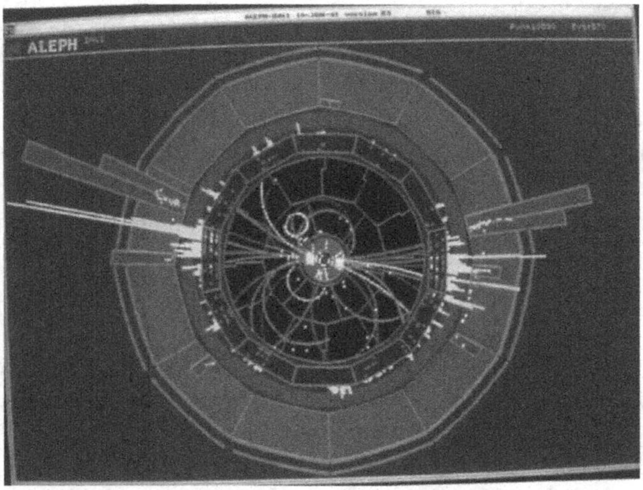


Fig. 3.1. Two-jet events from hadronic $Z^0$ decay.

values $M_q \approx 300$ MeV, used in the previous chapter for the case of the quark and potential model approaches [12], for explaining the mass-splittings of hadrons using the Gell-Mann–Okubo-like mass formulae [11].

- Including the previous three light quarks, at present *six quark flavours* have been found and classified according to their charge $Q$ in units of the electron. They are:

$$Q = 2/3 \; : \; (u, c, t)$$
$$Q = -1/3 \; : \; (d, s, b)\,, \tag{3.3}$$

where, for instance, in these triplet representations, the neutral currents of electroweak interactions are flavour conserving. The new quarks $c$, $b$, $t$ carry new quantum numbers as shown in Table 3.1.

- The *charm quark* was proposed in [58], in which the name *charm* was adopted by Bjorken and Glashow [58]. The discovery of the charm quark through the finding of the $\bar{c}c$ bound state $J/\psi$ meson [59] at 3.1 GeV, indicates that its mass is about 1/2 of the one of the meson.[3] Its discovery has been crucial for avoiding the flavour changing neutral current responsible for the excess of $Z^0$

[3] For heavy quarks ($m_q \gg \Lambda$), the mass is defined as the on-shell mass (pole mass) analogous to the one of the electron (see next section).

exchange contributions in the $K^0$-$\bar{K}^0$ oscillations, and for the huge $K_L \to \mu^+\mu^-$ and $K^\pm \to \pi^\pm \nu\bar{\nu}$ experimentally unacceptable rates. The need for charm in this mechanism was indeed advocated a long time ago by Glashow–Iliopoulos–Maiani *(GIM suppression mechanism)* [60]. Then, after the charm discovery, the two generations of quarks $(u, d)$ and $(c, s)$ for the electroweak $SU(2)_L \times U(1)$ standard model (SM) of Glashow–Weinberg–Salam [61] were completed, and could be compared with the two lepton doublets $(e, \nu_e)$ and $(\mu, \nu_\mu)$. These two quark doublets mix through the Cabibbo mixing angle $\theta_c$ introduced a long time ago [15], and has the experimental value $\sin\theta_c = 0.220 \pm 0.003$ from, for example, the $K \to \pi^0 e^+ \nu_e$ process [16].

- The discovery in 1974 of the third $\tau$ charged lepton [62], having a mass 1.8 GeV, was the first sign of the third generation, which was confirmed later on by the discovery of the $\Upsilon$, which is a $\bar{b}b$ bound state [63] in 1977, with a $b$-mass $M_b \approx 4.6$ GeV, expected to be about 1/2 of the one of the $\Upsilon$. More recently, the third family has been completed by the discovery of the $t$ quark in 1995 [64] from the analysis of the lepton + jet and dilepton channels originated from $\bar{t}t \to W^- b W^+ b$ processes at the collider experiments. This gives a top mass $M_t \simeq (174.3 \pm 3.2 \pm 4.0)$ GeV [16]. The $b$ and $t$ quarks have been predicted by Kobayashi and Maskawa [65], and the names *bottom* and *top* were first used by Harari [66]. At present, we have found three families of leptons:

$$\begin{pmatrix} \nu_e \\ e \end{pmatrix} \begin{pmatrix} \nu_\mu \\ \mu \end{pmatrix} \begin{pmatrix} \nu_\tau \\ \tau \end{pmatrix} \tag{3.4}$$

and analogous three families of quarks:

$$\begin{pmatrix} u \\ d \end{pmatrix} \begin{pmatrix} c \\ s \end{pmatrix} \begin{pmatrix} t \\ b \end{pmatrix}. \tag{3.5}$$

Quark families mix under a $3 \times 3$ unitary matrix, which is a generalization of the previous $2 \times 2$ Cabibbo unitary matrix, and which is called the CKM (Cabibbo–Kobayashi–Maskawa) mixing matrix [67]. This matrix has three real parameters (mixing angles) and one $CP$ violating phase (see Appendix A3), which cannot be absorbed by a redefinition of the quark fields. LEP studies [68] of the $Z^0$ width also indicate that it is unlikely to have more than three (almost) massless neutrinos, such that, most probably, we only have these three generations in nature.

## 3.2 The notion of colours

- Historically [69], the introduction of colours has been motivated by the failure of the quark model to expain the peculiar nature of the pion-nucleon $\Delta^{++}$ baryon, which has a total zero angular momentum $J = 3/2$. In order to fulfill this property, one has to put its three $u$-quark constituents with spins aligned up. This requirement is not allowed by Dirac statistics as the quarks are supposed to be a spin 1/2 particle. This *wrong statistic* problem is solved when one gives three colours to the quarks,[4] such that the $\Delta^{++}$ can be represented as:

$$|\Delta^{++}, J = 3/2\rangle = \frac{1}{\sqrt{6}}\epsilon^{\alpha\beta\gamma} |u_\alpha \uparrow, u_\beta \uparrow, u_\gamma \uparrow, \rangle, \tag{3.6}$$

with an antisymmetric wave function ($\alpha$, $\beta$, $\gamma$ are colour indices).

[4] A possible solution, where quarks obey parastatistics of rank three, has been proposed by Greenberg [70], which can be satisfied by the attribution, by Gell-Mann *et al.* [69], of the new internal colour quantum number to the quarks.

- It is also known that quantum anomaly spoils the renormalizability of the $SU(2) \times U(1)$ Standard Model of Electroweak interactions. Its disappearance can only be achieved if the quark number of colour is 3.

## 3.3 The confinement hypothesis

- However, the theory is amusing as one has to avoid the existence of coloured states, i.e., they should have infinite energy, such that all asymptotic states should be colourless. This leads to the *confinement hypothesis* implying the non-observability of free quarks. There is indeed an indication of such a property from a lattice measurement of heavy quark-antiquark bound state potential, where it is found to be Coulomic at short distances and increases linearly at long distances (see also Section 3.8):

$$V_{\bar{Q}Q} \sim C_{\rm F}\frac{\alpha_s(r)}{r} + \sigma r \tag{3.7}$$

with $C_{\rm F} = 4/3$ and $\sigma$ is the QCD string tension. The linear rising term renders the separation of the $\bar{Q}Q$ pair energetically impossible.
- The confinement assumption also implies that QCD should be a *local field theory* that leads to local observables described by local operators or currents built with gluons and/or quark fields. This locality property is one of the basis of the current algebras that we have outlined in the previous chapter.
- Confinement is also essential for explaining the short-range nature of the nuclear forces, while massless gluons exchange is a long-range process. This is because nucleons are colour singlet states which cannot exchange colour octet gluons but only coloureds states.
- Some qualitative ideas on the nature of confinement lead to the picture that quarks are bound by strings or chromelectric flux tubes. Indeed, if a $\bar{Q}Q$ pair is created at one space-time point in a given process, and the quark and antiquark start to move away from each other in the centre of mass of the system, then it soon becomes energetically possible to create additional pairs smoothly distributed in rapidity between the two leading charges, which neutralize colour and allow the final state to be reorganized into two jets of coloured hadrons, which communicate in the central region by a number of *wee hadrons*. This phenomena is very similar to the case of broken magnet, where an attempt to isolate a magnetic monopole by stretching a dipole, leads to the breaking of the magnet into two new monopoles at the breaking point. with small energy. Alas, nobody has succeeded yet in proving this scenario, which remains a great challenge due to the peculiar IR properties (*infrared slavery*) of the theory. At present, the *confinement hypothesis* can still be considered a *postulate*.

## 3.4 Indirect evidence of quarks

Prior QCD, constituent quark models have been used for predicting some processes. The calculations assume that one can simply produce free quarks, which, a priori, is in contradiction with the *confinement postulate*. (Indirect) evidence[5] of quarks have been observed at LEP from two hadronic jet events in the decay of the weak boson $Z^0$ through the intermediate process $Z^0 \to \bar{q}q$, where the quarks hadronize later on. However, one should remember

[5] Some direct searches based on the expectation to observe spin 1/2 quark were not successful.

that, in these hard processes, experimentalists only detect hadrons (pions, kaons, ...), but neither quarks nor gluons. It is impressive that these hard processes can be nicely explained by perturbative QCD [52].

### 3.5 Evidence for colours

- In an analogous way, the existence of gluons has been seen in three hadronic jet decays of the $Z^0$ through the process $Z^0 \to \bar{q}qg$.
- The number of colours has also been tested from different experiments. Classical examples are:

♣ The $e^+e^- \to$ hadrons total cross-section $R_{e^+e^-}$ normalized to the $e^+e^- \to \mu^+\mu^-$ cross-section is expected to be equal to the number of colours $N_c$ times the sum of the square of the quark charge, if one assumes the production of free $\bar{q}q$ pairs (*parton model*) before hadronization (see Fig. 3.2):

$$R_{e^+e^-} \equiv \frac{\sigma(e^+e^- \to \gamma,\ Z^0 \to \text{hadrons})}{\sigma(e^+e^- \to \mu^+\mu^-)} \approx N_c \sum_{u,d,s...} Q_i^2 \,. \tag{3.8}$$

This fact has been observed in $e^+e^-$ experiments for sufficiently large energy beyond the resonances structure as shown in Fig. 3.3.

◇ Similarly, the decay rate of the weak $Z^0$ boson shown in Fig. 3.4 is also controlled by $N_c$. Its hadronic branching ratio reads:

$$R_Z \equiv \frac{\Gamma(Z^0 \to \text{hadrons})}{\Gamma(Z^0 \to e^+e^-)} \approx \frac{N_c}{(v_e^2 + a_e^2)} \sum_{u,d,s...} (v_i^2 + a_i^2) \,, \tag{3.9}$$

where $v_i$ and $a_i$ are the electroweak vector and axial-vector couplings of the $\bar{q}q$ or $e^+e^-$ pairs to the $Z^0$.

Experimentally, one has [16]:

$$R_Z = 20.77 \pm 0.08 \,. \tag{3.10}$$

♡ The inclusive heavy lepton $\tau$ semi-hadronic rate $R_\tau$ normalized to its semi-leptonic one, shown in Fig. 3.5, is expected to be equal to the colour number 3 from the parton model:

$$R_\tau \equiv \frac{\Gamma(\tau \to \nu_\tau + \text{hadrons})}{\Gamma(\tau \to \nu_\tau + l + \bar{\nu}_l)} \approx N_c \,. \tag{3.11}$$

Experimentally, one has:

$$R_\tau = \frac{1 - \sum_{e,\mu} B_r(\tau \to \nu_\tau + l\bar{\nu}_l)}{B_r(\tau \to \nu_\tau + l\bar{\nu}_l)} = 3.647 \pm 0.05 \tag{3.12}$$

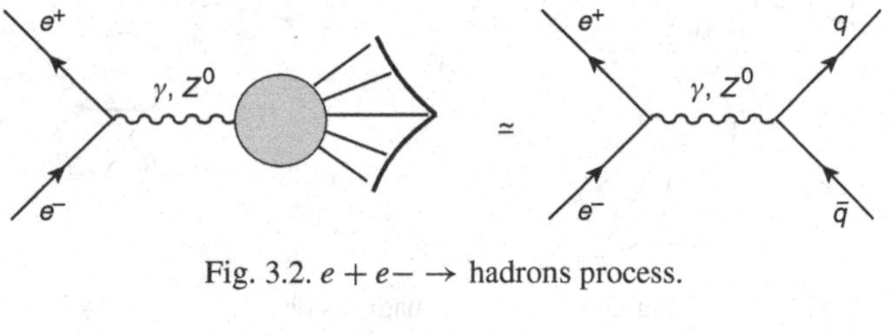


Fig. 3.2. $e + e- \to$ hadrons process.

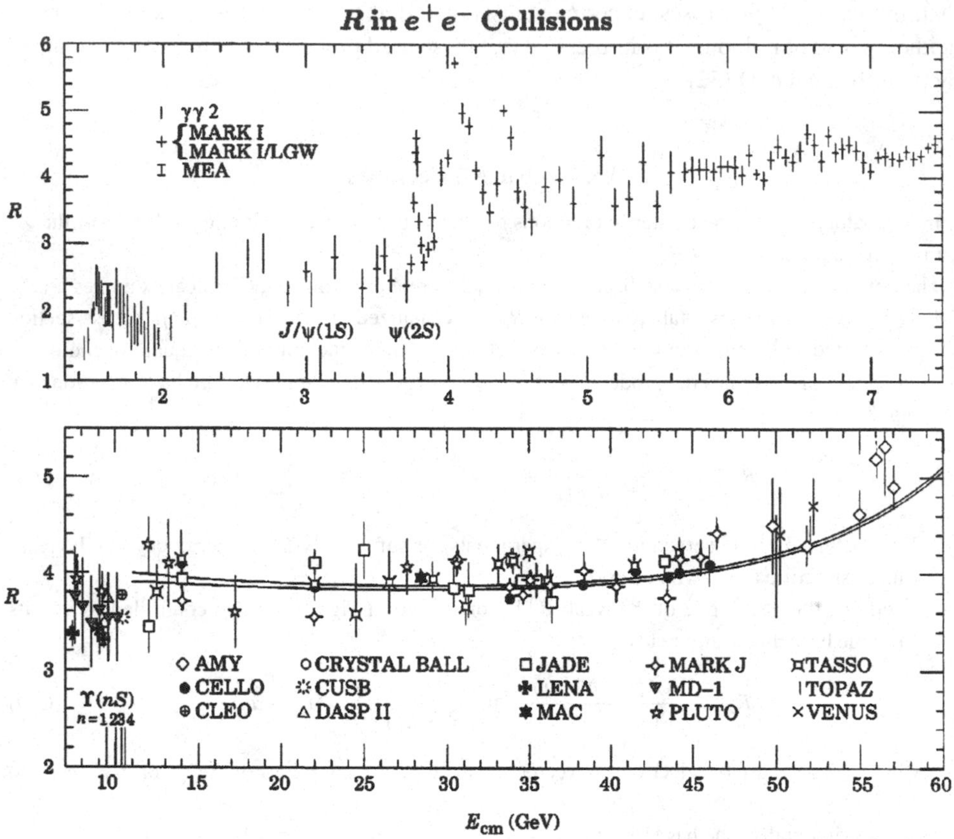


Fig. 3.3. $e + e- \to$ hadrons data. The continuous lines are QCD fit.

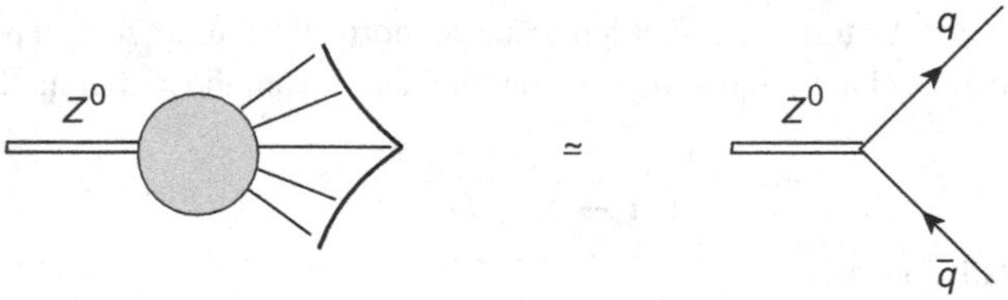

Fig. 3.4. $Z^0 \to$ hadrons decay.

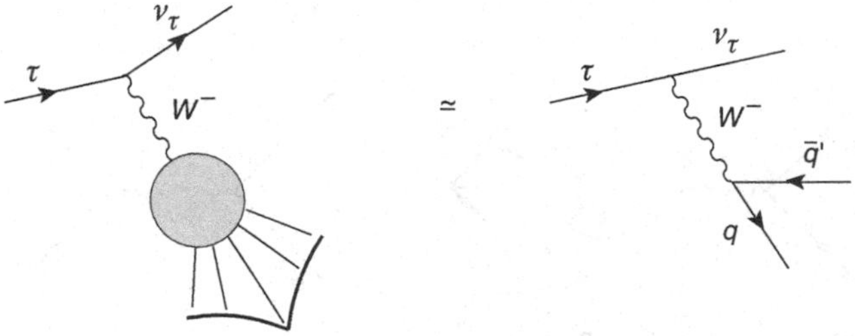

Fig. 3.5. $\tau \to \nu_\tau +$ hadrons decay.

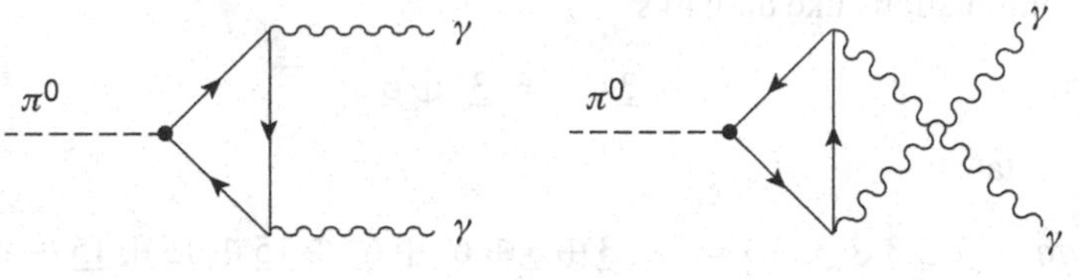


Fig. 3.6. $\pi^0 \to \gamma\gamma$ decay from the quark triangle.

where $B_r$ is the leptonic branching ratio. We shall see later on that the QCD radiative corrections explain the 20% discrepancy between the parton model prediction and the data.

♠ The decay rate of the neutral pion into two photons which occurs through the quark triangle loop (Abelian anomaly) shown in Fig. 3.6 is controlled by the square of the colour [71]:

$$\Gamma(\pi^0 \to \gamma\gamma) = \left[N_c\left(Q_u^2 - Q_d^2\right)\right]^2 \left(\frac{\alpha^2}{64\pi^3}\right) \frac{m_\pi^3}{f_\pi^2} = 7.7 \text{ eV} , \tag{3.13}$$

where $f_\pi = 92.4$ MeV is the pion decay constant controlling the decay $\pi^- \to \mu\nu$. It was shown a long time before QCD that this prediction is not affected by quantum corrections [72]. This prediction is in remarkable agreement with the data of $(7.7 \pm 0.6)$ eV [16].

## 3.6 The $SU(3)_c$ colour group

The previous properties:

- Quarks with three colours
- Quarks and anti-quarks are different objects
- Exact colour symmetry (hadrons have no colour multiplicity)

are sufficient to select the $SU(3)_c$ symmetric colour group for desribing the theory of strong interactions, instead of the other Lie group candidates $SO(3)$ and its isomorphic $SU(2) \simeq Sp(1)$, which have real representations, and then cannot distinguish the particle from its anti-particle. In this $SU(3)_c$ unitary group, quarks (anti-quarks) then belong to the fundamental presentation $\underline{3}$ (resp $\underline{3}^*$), whereas gluons are in the adjoint $\underline{8}$. The previous *Gell-Mann eightfoldway* [7] quark model classification, can now be viewed in a modern way, where *hadrons should be colour-singlet* states. The $SU(3)_c$ decomposition into products of $\underline{3}$ and $\underline{3}^*$ representations gives for mesons:

$$\bar{q}q \;:\; \underline{3}^* \otimes \underline{3} = \underline{1} \oplus \underline{8} \tag{3.14}$$

and for baryons:

$$qqq \;:\; \underline{3} \otimes \underline{3} \otimes \underline{3} = \underline{1} \oplus \underline{8} \oplus \underline{8} \oplus \underline{10} \,, \tag{3.15}$$

which guarantee the colour-singlet configurations of hadrons required by the *confinement postulate*. and which are satisfied by the experimentally observed hadrons. On the contrary, some exotic combinations like diquarks:

$$qq \; : \; \underline{3} \otimes \underline{3} = \underline{3}^* \oplus \underline{6}\,, \tag{3.16}$$

and four-quark states:

$$qqqq \; : \; \underline{3} \otimes \underline{3} \otimes \underline{3} \otimes \underline{3} = 3 \oplus \underline{3} \oplus \underline{3} \oplus 6^* \oplus \underline{6^*} \oplus \underline{15} \oplus 15 \oplus \underline{15} \oplus \underline{15'} \tag{3.17}$$

do not satisfy the colour-singlet confinement constraints, and can induce coloured states in the spectrum [73].

### 3.7 Asymptotic freedom

Gell-Mann postulated that, at short distances, the commutation relations of the *local* hadronic currents imply that the quark fields entering them are free particles (*asymptotic freedom*). These assumptions led to the success of the different current algebra superconvergent sum rules and to the Bjorken scaling. However, such assumptions a priori contradict the previous confinement postulate. As we shall see, QCD can satisfy simultaneously the two conditions thanks to the property of the QCD gauge coupling $g$, which is the only parameter that controls the QCD Lagrangian in the massless quarks limit (as we shall see in the next chapter). 't Hooft observed [74] that the slope of the first coefficient ($N_c$ and $n_f$ are respectively the colour and flavour numbers):

$$\beta_1 = -\frac{1}{2}\left(11\frac{N_c}{3} - \frac{2}{3}n_f\right) \tag{3.18}$$

of the $\beta$-function [75,76] is negative at the origin of the coupling constant for a $SU(3)_c$ *Yang–Mills gauge theory*, while, independently, Gross, Wilczek and Politzer [77] discovered that for non-Abelian gauge theories, the origin of the coupling constant is an UV stable fixed point in the deep Euclidian region. This *asymptotic freedom*[6] property thus states, after solving the renormalization group equation (RGE) (resummation of all leading logs corrections), as we shall see in Section 11.7, that at large momenta $Q$, the *running* QCD coupling falls off as:

$$\alpha_s \equiv \frac{g^2}{4\pi} \simeq \frac{\pi}{-\beta_1 \ln Q/\Lambda}\,,$$

where $\Lambda$ is the characteristic QCD scale, which indicates that below its value, the perturbative approximation breaks down. The situation in QCD is the opposite of the familiar QED described by the $U(1)$ Abelian theory, in which the *effective charge* $\alpha$ increases slowly for increasing $Q^2$ because the corresponding $\beta$ function is positive ($\beta_1 = 2/3$).[7] At the electron mass, $\alpha$ has the value $1/137$, while it is $1/129$ at an $Z^0$ mass of a distance of $1/500$ fm (It becomes infinite (so-called *Landau pole* [79]) at an energy much higher than

[6] For historical reviews on the discovery of asymptotic freedom, see the talks given by David Gross and Gerard 't Hooft at the QCD 98 Montpellier Euroconference [78].

[7] More discussions on QED will be given later.

the mass of the universe). An intuitive understanding of this decrease of the QED effective coupling at long distance is provided by the dielectric screening due to the cloud of virtual $e^+e^-$ pairs created in the vacuum, through quantum effects, surrounding the electric charge. For QCD, and more generally, for non-Abelian theories, one then expects an anti-screening effect generated by the gauge self-interactions of gluons, which spread out the QCD colour charge, and makes the Yang–Mills vacuum like a paramagnetic substance implying an anti-screening charge through relativistic invariance. This anti-screening or the asymptotic freedom property are only true for non-Abelian theories [80]. This remarkable asymptotic freedom property of QCD then permits a simple treatment of the different *QCD hard processes*, which can be approximated by perturbative series in the strong coupling $\alpha_s$ at large momenta. This feature also confirms the success of the parton model in describing (to lowest order of the $\alpha_s$-series expansion in the perturbative QCD language), the examples of QCD processes $R_{e^+e^-}$, $R_Z$, $R_\tau$ and DIS mentioned previously, but also implies that for $Q^2 \to \infty$, quarks become *free particles*.

### 3.8 Quantum mechanics and non-relativistic aspects of QCD

We have learned from previous sections that quarks are free at very short distances but tightly bounded at long distances. For an heavy $\bar{Q}Q$ bound state, the QCD potential is Coulomic at short distances and increases linearly at long distances. This behaviour is typical for quantum mechanical systems bound together by a potential which is not singular at short distance and increases infinitely with distance at large distances. This is, for instance, the case of the harmonic oscillator where its potential reads:

$$V(r) = \frac{1}{2}m\omega^2 r^2 \tag{3.19}$$

The corresponding Green's function of the system is:

$$G(\vec{x}, \vec{x}', t) = \left(\frac{m\omega}{2\pi\hbar \sin\omega t}\right)^{3/2} \exp\left\{\frac{im\omega}{2\hbar \sin\omega t}((\vec{x}^2 + \vec{x}'^2)\cos\omega t - 2\vec{x}\vec{x}')\right\}, \tag{3.20}$$

which, for small $t$ ($\omega t \ll 1$), is well approximated by the function for the free particle:

$$G_0(\vec{x}, \vec{x}', t) = \left(\frac{m}{2\pi\hbar}\right)^{3/2} \exp\left\{\frac{im\omega}{2\hbar t}(\vec{x} - \vec{x}')^2\right\}. \tag{3.21}$$

Therefore, it is not so surprising that non-relativistic potential models of quarks [12, 81–94] were able to describe some characteristic features of the systems, and successfully explain the complex hadron spectra made with heavy quarks. However, a purely quantum mechanical description of the theory is not fully satisfactory, as it does not incorporate Lorentz invariance. We shall come back to this subject in a future section.

# 4
# Field theory ingredients

In this chapter, we shall collect some of the field theory ingredients which will often be encountered in this book. More detailed discussions and derivations can be found in classic textbooks on quantum field theories [53] and some of the QCD books in [42–46].

## 4.1 Wick's theorem

Let us consider free boson or fermion fields $\varphi_i(x)$ of a particle $i$, which one can express in terms of the creation $a^\dagger$ and annihilation $a$ operators, and the corresponding ones $b^\dagger$ and $b$ for the anti-particles, where $a$ and $b$ may (or may not) coincide:

$$\varphi_i(x) = \sum_n c_i^{(n)}(x) a_n + \sum \bar{c}_i^{(n)}(x) b_n^\dagger . \tag{4.1}$$

For a fermion field ($u$ and $v$ are Dirac spinors):

$$\psi(x) = \int d\bar{k} [a(k)u(k)\, e^{-ikx} + b^\dagger(k)v(k)\, e^{ikx}] , \tag{4.2}$$

and for a boson:

$$\phi(x) = \int d\bar{k} [a(k)\, e^{-ikx} + b^\dagger(k)\, e^{ikx}] . \tag{4.3}$$

where the phase space measure is:

$$d\bar{k} \equiv \frac{d^3k}{(2\pi)^3 2E_k} = \frac{d^4k}{(2\pi)^4} 2\pi\delta(k^2 - m^2)\theta(k^0) . \tag{4.4}$$

The Wick or *normal ordered* product [95]:

$$: \varphi_1(x_1)\varphi_2(x_2) : \tag{4.5}$$

is obtained by placing all creator operators to the left of all annihilation operators, and by taking care on the (anti)-commuting relations if the fields are (fermions) bosons.

Therefore:

$$: \varphi_1(x_1)\varphi_2(x_2) : \equiv \sum_{n,n'} \left[ c_1^{(n)}(x_1)c_2^{(n')}(x_2)a_n a_{n'} + \bar{c}_1^{(n)}(x_1)\bar{c}_2^{(n')}(x_2)b_n^\dagger b_{n'}^\dagger \right.$$
$$\left. + \bar{c}_1^{(n)}(x_1)c_2^{(n')}(x_2)b_n^\dagger a_{n'} + (-1)^\delta c_1^{(n)}(x_1)\bar{c}_2^{(n')}(x_2)b_{n'}^\dagger a_n \right] , \tag{4.6}$$

where $\delta = 1(0)$ for fermions (bosons). This results can be easily generalized to more factors of fields.

## 4.2 Time-ordered product

A time-ordered product is obtained by rearranging the fields or operators in the natural sequence of time. At a time $t' > t$, we first create a particle at a time $t$ with $\varphi^\dagger$ and annihilate later on at a time $t'$ with $\varphi$. This can be encoded by the amplitude:

$$\theta(t' - t)\varphi(t', \vec{x}')\varphi^\dagger(t, \vec{x}) . \tag{4.7}$$

If, for $t' < t$, an antiparticle is produced by $\varphi(x')$, then it is annihilated by $\varphi^\dagger(x)$ at the time $t$, with the amplitude:

$$\theta(t - t')\varphi^\dagger(t, \vec{x})\varphi(t', \vec{x}') . \tag{4.8}$$

The sum of the two equations gives the *time-ordered product*:

$$\mathcal{T}\varphi(x')\varphi^\dagger(x) = \theta(t' - t)\varphi(x')\varphi^\dagger(x) + (-1)^\delta\theta(t - t')\varphi^\dagger(x)\varphi(x') , \tag{4.9}$$

where $\delta = 1(0)$ for fermion (boson), where one should also note that fermion-boson operators are taken to commute. The $T$-product is arranged from right to left with increasing times, and then the appropriate name. One can also express it in terms of the Wick product:

$$\mathcal{T}\varphi(x')\varphi^\dagger(x) =: \varphi(x')\varphi^\dagger(x) : + \langle 0|\mathcal{T}\varphi(x')\varphi^\dagger(x)|0\rangle . \tag{4.10}$$

The above results can be generalized to the $\mathcal{T}$ products of n operators/fields:

$$\mathcal{T}\varphi_1(x_1)\cdots\varphi_n(x_n) = (-1)^\delta\varphi_{i_1}(x_{i_1})\cdots\varphi_{i_n}(x_{i_n}) , \tag{4.11}$$

where in the RHS the times are ordered ($t_{i_1} > t_{i_2} > \cdots > t_{i_n}$) and $\delta$ is the number of transposition of indices of the fermion operators/fields necessary for obtaining the required form in the RHS. It can be written as:

$$\begin{aligned}
&\mathcal{T}\varphi_1(x_1)\cdots\varphi_n(x_n) = \mathcal{T}\varphi_1(x_1)\cdots\varphi_n(x_{n-1})\varphi_n(x_n) \\
&= : \varphi_1(x_1)\cdots\varphi_n(x_{n-1}) : \varphi_n(x_n) \\
&\quad + \langle 0|\mathcal{T}\varphi(x_1)\varphi(x_2)|0\rangle : \varphi_1(x_3)\cdots\varphi_n(x_{n-1}) : \varphi_n(x_n) + \text{ perm.} \\
&\quad + \langle 0|\mathcal{T}\varphi_1(x_1)\varphi_2(x_2)|0\rangle\langle 0|\mathcal{T}\varphi_3(x_3)\varphi_4(x_4)|0\rangle : \varphi_1(x_5)\cdots\varphi_n(x_{n-1}) : \varphi_n(x_n) + \text{ perm.} \\
&\quad + \cdots ,
\end{aligned} \tag{4.12}$$

where $\cdots$ stands for:

$$\langle 0|\mathcal{T}\varphi_1(x_1)\varphi_2(x_2)|0\rangle \cdots \langle 0|\mathcal{T}\varphi_{n-1}(x_{n-1})\varphi_n(x_n)|0\rangle + \text{permutations}\,, \tag{4.13}$$

if $n$ is even, and for:

$$\langle 0|\mathcal{T}\varphi_1(x_1)\varphi_2(x_2)|0\rangle \cdots \langle 0|\mathcal{T}\varphi_{n-2}(x_{n-2})\varphi_{n-3}(x_{n-3})|0\rangle\varphi_n(x_n) + \text{permutations}\,, \tag{4.14}$$

if $n$ is odd. The vacuum expectation values or contractions give rise to the field propagators.

## 4.3 The $S$-matrix

### 4.3.1 Generalities

In field theory, one measures $S$-matrix elements, which is the probability amplitudes for transition between states which contain definite numbers of particles for $t$ ranging from $-\infty$ to $+\infty$. They are usually named 'in' and 'out' states $|\alpha, \text{in}\rangle$ and $|\beta, \text{out}\rangle$, where $\alpha,\ \beta$ characterize particles momenta and quantum numbers. The $S$-matrix can be obtained from the interaction Lagrangian:

$$S = \mathcal{T}\exp\left[i\int d^4x\ \mathcal{L}_{\text{I}}\right] \tag{4.15}$$

which one can expand as:

$$S = 1 + i\int d^4x\ \mathcal{L}_{\text{I}}(x) + \cdots \frac{i^n}{n!}\int d^4x_1\cdots d^4x_n\ \mathcal{T}\mathcal{L}_{\text{I}}(x_1)\cdots\mathcal{L}_{\text{I}}(x_n)\,. \tag{4.16}$$

The $S$-matrix is relativistically invariant:

$$S = \mathcal{U}(a,\Lambda)S\mathcal{U}^{-1}(a,\Lambda)\,, \tag{4.17}$$

where $\mathcal{U}(a,\Lambda)$ is a transformation under the Poincarè group. It is also unitary:

$$S^\dagger S = 1\,. \tag{4.18}$$

It can be related to the transition amplitude:

$$\langle\beta, \text{out}|T|\alpha, \text{in}\rangle \tag{4.19}$$

which gives the probability that the incoming state $|\alpha\rangle$ will evolve in time to the outcoming state $|\beta\rangle$ as:

$$\mathcal{S} = 1 + iT \tag{4.20}$$

### 4.3.2 Applications: cross-section and decay rate

We can illustrate the discussion by considering the scattering process:

$$(p_1, J_1) + (p_2, J_2) \rightarrow (k_1, j_1) + \cdots (k_n, j_n)\,. \tag{4.21}$$

of two initial particles with momenta $p_1$ and $p_2$ and spin $J_1$ and $J_2$, and $n$ final states with momenta $p_n$ and spin $j_n$. The *unpolarized* cross-section of this process can be written as:

$$\sigma = \sum \frac{W}{FD}, \tag{4.22}$$

where $\sum$ represents an averaging over initial particle polarizations; $W$ is the transition probability per unit of time and unit of volume, $F$ is the incident particle flux and $D$ the target-particle density. In the laboratory frame of incident particle 1 on a target particle 2, one has the kinematic variables:

$$\begin{aligned} \lambda(s, m_1^2, m_2^2) &= [s - (m_1 + m_2)^2][s - (m_1 - m_2)^2] , \\ F &= 2E_1|v_1 - v_2| = \lambda^{1/2}(s)/m_2 , \\ D &= 2E_2 , \qquad (4.23) \\ s &= (p_1 + p_2)^2 , \qquad (4.24) \end{aligned}$$

where $v_i$ $(i = 1, 2)$ is the velocity of the particle $i$ $(v_2 = 0)$. The transition probability per unit of time and unit of volume is:

$$W = \frac{1}{(2\pi)^4\delta^4(0)} \int \prod_{j=1}^{n} \frac{d^3k_j}{(2\pi)^3 2E_j} |\langle f|(S-1)|i\rangle|^2 , \tag{4.25}$$

where the sum over the helicities of different particles is understood. We have used the normalization of state $|p, \lambda\rangle$ having helicity $\lambda$ and momentum $p$:

$$\langle p', \lambda' | p, \lambda\rangle = (2\pi)^3 2E\delta^3(p' - p)\delta_{\lambda'\lambda} . \tag{4.26}$$

Using trivial substitutions, one can deduce the well-known cross-section:

$$\begin{aligned} \sigma &= \frac{1}{2\lambda^{1/2}(s, m_1^2, m_2^2)} \frac{\mathcal{N}}{(2J_1+1)(2J_2+1)} \\ &\times \int (2\pi)^4 \delta^4(P_f - P_i) |\mathcal{M}(i \to f)|^2 \prod_{j=1}^{n} \frac{d^3k_j}{(2\pi)^3 2E_j} , \end{aligned} \tag{4.27}$$

where we have introduced the reduced amplitude transition $\mathcal{M}$:

$$i\langle f|T|i\rangle \equiv \delta^4(P_f - P_i)|\mathcal{M}(i \to f) . \tag{4.28}$$

Here:

$$P_f \equiv \sum_1^n p_f ; \qquad P_i = p_1 + p_2 , \tag{4.29}$$

and the statistical factor is:

$$\mathcal{N} = \prod_i \frac{1}{n_i!} , \tag{4.30}$$

if one has $n_i$ identical particles in the final state. Analogously, the decay rate reads for a particle of mass $M$ at rest is:

$$\Gamma(i \to f) = \frac{\mathcal{N}}{2M} \int (2\pi)^4 \delta^4(P_f - P_i) |\mathcal{M}(i \to f)|^2 \prod_{j=1}^{n} \frac{d^3k_j}{(2\pi)^3 2E_j} , \tag{4.31}$$

Repeated uses of the *reduction formula* will show that the transition matrix can be related to the Green's function of the relevant particles.

## 4.4 Reduction formula

Let's consider the simplest case for the elastic scattering of two scalar particles. The $S$-matrix of this process is:

$$\langle k_1 k_2 | S | p_1 p_2 \rangle \tag{4.32}$$

In terms of annihilation and creation operators, which satisfy the commutation relations:

$$[a(p), a^\dagger(p')] = (2\pi)^3 2E \delta^3(p' - p) , \quad [a(p), a(p')] = 0 , \tag{4.33}$$

the scalar field reads:

$$\phi(x) = \int \frac{d^3k_j}{(2\pi)^3 2E_j} [a(p)e^{-ipx} + a^\dagger(p)e^{ipx}] , \tag{4.34}$$

which can be inverted:

$$a(p) = i \int d^3x \, e^{ipx} \overleftrightarrow{\partial_0} \phi(x) , \tag{4.35}$$

where:

$$f \overleftrightarrow{\partial_0} g \equiv f(\partial_0 g) - (\partial_0 f) g . \tag{4.36}$$

Then, after some algebra:

$$\begin{aligned} \langle k_1 k_2 \, out | p_1 p_2 \, in \rangle &= \langle k_1 | a_{out}(k_2) | p_1 p_2 \, in \rangle \\ &= i \lim_{x_0 \to +\infty} \int d^3x \, e^{ik_2 x} \overleftrightarrow{\partial_0} \langle k_1 | \phi(x) | p_1 p_2 \, in \rangle \\ &= \langle k_1 | a_{in}(k_2) | p_1 p_2 \, in \rangle \\ &\quad + i \int d^4x \, \partial_0 [e^{ik_2 x} \overleftrightarrow{\partial_0} \langle k_1 | \phi(x) | p_1 p_2 \, in \rangle] . \end{aligned} \tag{4.37}$$

Using:

$$\begin{aligned} \partial_0 (f \partial_0 g) &= f \partial_0^2 g \\ \partial_0^2 e^{ik_2 x} &= (\nabla^2 - m^2) e^{ik_2 x} , \end{aligned} \tag{4.38}$$

one can replace the last term of the previous equation by:

$$i \int d^4x\, e^{ik_2x} \left(\partial_x^2 + m^2\right) \langle k_1|\phi(x)|p_1 p_2\ in\rangle\ . \tag{4.39}$$

Using repeatedly the above manipulations, one obtains the Fourier transform of the vacuum expectation value (VEV) of the I-product of four fields:

$$\begin{aligned}\langle k_1 k_2\ out|p_1 p_2\ in\rangle &= \langle k_1 k_2\ in|p_1 p_2\ in\rangle \\ &\quad + i^4 \int d^4x_1 d^4x_2 d^4y_1 d^4y_2\, e^{i[k_1x_1+k_2x_2-p_1y_1-p_2y_2]} \\ &\quad \times \left(\partial_{x_1}^2 + m^2\right)\left(\partial_{x_2}^2 + m^2\right)\left(\partial_{y_1}^2 + m^2\right)\left(\partial_{y_2}^2 + m^2\right) \langle k_1|\phi(x)|p_1 p_2\ in\rangle \\ &\quad \times \langle 0|T[\phi(x_1)\phi(x_2)\phi(y_1)\phi(y_2)]|0\rangle\ ,\end{aligned} \tag{4.40}$$

where:

$$\begin{aligned}\langle k_1 k_2\ out|p_1 p_2\ in\rangle - \langle k_1 k_2\ in|p_1 p_2\ in\rangle &= \langle k_1 k_2|(S-1)|p_1 p_2\rangle \\ &= (2\pi)^4 i\delta(k_1 + k_2 - p_1 - p_2)\langle k_1 k_2|T|p_1 p_2\rangle\ ,\end{aligned} \tag{4.41}$$

and we have used the shorthand notation:

$$\partial_x \equiv \frac{\partial}{\partial x_\mu}\ . \tag{4.42}$$

Similar manipulations can be extended to spinor fields.

## 4.5 Path integral in quantum mechanics

The path integral method, used long time ago by Feynman [96], has been revived by Fadeev and Popov and De Witt [97] in its application to non-Abelian theory, and by 't Hooft [98] when he derives the Feynman rules for massive gauge theories, particularly for the Standard Model of the Electroweak interactions. Detailed derivation of this method are described in modern textbooks. We shall briefly outline the method here, but starting from some examples in quantum mechanics.

### *4.5.1 Transition matrix of quantum mechanics in one dimension*

The Hermitian operator 'coordinates' $Q_a$ and a conjuguate 'momenta' $P_b$, satisfy the canonical commutation relations:

$$[Q_a, P_a] = i\delta_{ab}\ , \qquad [Q_a, Q_b] = [P_a, P_b] = 0 \tag{4.43}$$

to which correspond the eigenvectors $|q\rangle$ and $|p\rangle$ and the eigenvalues $q_a$ and $p_b$. In the Heisenberg picture, $Q$ and $P$ have a time dependence leading to:

$$|q; t\rangle = \exp(iHt)|q\rangle\ , \qquad |p; t\rangle = \exp(iHt)|p\rangle\ , \tag{4.44}$$

for the eigenstates, which satisfy the orthonormality and completeness conditions:

$$\langle q';t|q;t\rangle = \delta(q'-q)\,, \qquad \langle p';t|p;t\rangle = \delta(p'-p)\,,$$
$$\int \prod_a dq_a |q;t\rangle\langle q;t| = 1 = \int \prod_a dp_a |p;t\rangle\langle p;t| \tag{4.45}$$

and:

$$\langle q;t|p;t\rangle = \prod_a \frac{1}{\sqrt{2\pi}} \exp(iq_a p_a)\,. \tag{4.46}$$

One should remember that, in the previous notation, the state $|q;t\rangle$ in the *Heisenberg picture* coincides with the one of the *Schrödinger picture* $|q(t)\rangle$ at a given $t$. Now, we wish to calculate the scalar product:

$$\langle q';t'|q;t\rangle\,, \tag{4.47}$$

which corresponds to the probability amplitude for measurements at time $t'$ to give the state $|q';t'\rangle$, if we found that at the time $t$ our system is in a definite state $|q;t\rangle$. This is an easy task if the time $t'$ and $t$ are infinitely close to each other ($t \equiv \tau;\ t' \equiv t + d\tau$ and $d\tau \to 0$) since from Eq. (4.44):

$$\langle q';\tau + d\tau|q;\tau\rangle = \langle q';\tau|\exp(-iHd\tau)|q;\tau\rangle\,. \tag{4.48}$$

Expanding $|q;\tau\rangle$ in terms of the $P$ eigenstates $|p;\tau\rangle$ by using Eq. (4.46), one can write:

$$\begin{aligned}\langle q';\tau + d\tau|q;\tau\rangle &= \int \prod_a dp_a \langle q';\tau|\exp[-iH(Q(\tau),P(\tau))d\tau]|p;\tau\rangle\langle p;\tau|q;\tau\rangle \\ &= \int \prod_a \frac{dp_a}{2\pi} \exp\left[-iH(q',p)d\tau + i\sum_a (q'_a - q_a)p_a\right]\,,\end{aligned} \tag{4.49}$$

where each $p_a$ is integrated from $-\infty$ to $+\infty$. One can generalize this procedure by breaking the interval $t' - t$ into $N+1$ sets of infinitesimal intervals, as shown in Fig. 4.1, and sum

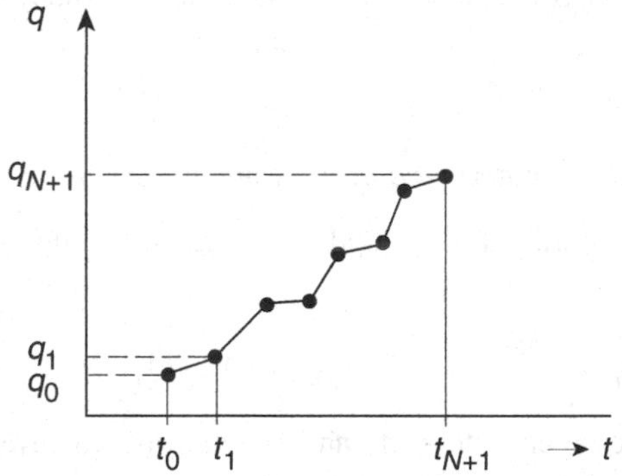


Fig. 4.1. Subdivision of the time interval $t' - t$ into $N+1$ sets of infinitesimal intervals.

over a complete set of states $|q;\tau_k\rangle$ at each time $\tau_k$. Then,

$$\langle q';t'|q;t\rangle = \int \left[\prod_{k=1}^{N}\prod_a dq_{k,a}\right]\left[\prod_{k=0}^{N}\prod_a \frac{dp_{k,a}}{2\pi}\right] \times \exp\left[i\sum_{k=1}^{N+1}\left\{-H(q_k,p_{k-1})d\tau + \sum_a (q_{k,a}-q_{k-1,a})p_{k-1,a}\right\}\right], \quad (4.50)$$

with $q_0 \equiv q$ and $q_{N+1} \equiv q'$. In the limit $\tau \to 0$, and then $N \to \infty$, one can assume that $q_a$ and $p_a$ are (to leading order in the $\tau$-expansion) independent of $\tau$, such that the argument of the exponential becomes an integral over $\tau$. Making the *formal* substitutions:

$$(q_{a,k} - q_{a,k-1}) \to \dot{q}_a d\tau\ ; \qquad \sum_{k=1}^{N+1} \to \int_t^{t'}\ ; \qquad \int\left[\prod_{k=1}^{N}\prod_a dq_{k,a}\right] \to \int \prod_a dq_a(\tau)\,, \quad (4.51)$$

one, then, obtains the *path integral* ($\equiv$ integration over all paths taken by $q(\tau)$ from $t$ to $t'$):

$$\langle q';t'|q;t\rangle = \int_{\substack{q_a\,\equiv\, q_a(t)\\ q'_a\,\equiv\, q_a(t')}} \prod_{\tau,a} dq_a(\tau)\prod_{\tau,b}\frac{dp_b(\tau)}{2\pi} \times \exp\left[i\int_t^{t'} d\tau\left\{-H\left(q(\tau),p(\tau)\right) + \sum_a \dot{q}_a(\tau)p_a(\tau)\right\}\right]. \quad (4.52)$$

One can perform the $p$ integration. In order to read off the oscillating function in the exponential, it is convenient to work in the Euclidian space by formally treating $id\tau$ to be *real*. Then, the integral has a definite norm. For a Hamiltonian of the form:

$$H(P,Q) = \frac{P^2}{2m} + V(Q)\,, \quad (4.53)$$

where $V(Q)$ is the potential, we have to perform a Gaussian integral:

$$\int_{-\infty}^{+\infty}\frac{dx}{2\pi}\exp[-ax^2+bx] = \frac{1}{\sqrt{4\pi a}}\exp[b^2/4a]\,. \quad (4.54)$$

Then, one can deduce from Eq. (4.52):

$$\langle q';t'|q;t\rangle = \int_{\substack{q_a\,\equiv\, q_a(t)\\ q'_a\,\equiv\, q_a(t')}} \prod_{\tau,a} dq_a(\tau) \times \exp\left[i\int_t^{t'} L(\tau)d\tau\right], \quad (4.55)$$

where:

$$L \equiv \frac{m}{2}\dot{q}^2 - V(q)\,, \quad (4.56)$$

is the Lagrangian.

### 4.5.2 The Green's functions

One can extend the previous discussion of matrix transition to the analysis of a Green's function which is the time-ordered products of different (local) operators. This can be illustrated by the example of a quantum mechanical two-point function, which is the matrix element of the time-ordered product between ground states:

$$G(t,t') = \langle 0|\mathcal{T}\, Q(t_1)Q(t_2)|0\rangle\ , \quad (t_1 > t_2)\ , \tag{4.57}$$

where $|0\rangle$ denotes the ground state. By inserting complete sets of states, it can be written as:

$$G(t,t') = \int dq dq' \langle 0|q';t'\rangle\langle q';t'|\mathcal{T}\, Q(t_1)Q(t_2)|q;t\rangle\langle q;t|0\rangle\ , \quad (t_1 > t_2)\ . \tag{4.58}$$

Introducing the wave function of the ground state:

$$\langle 0|q;t\rangle = \phi_0(q)\exp[-iE_0 t] \equiv \phi_0(q,t)\ , \tag{4.59}$$

and using an analogue of the derivation of Eq. (4.52) for the matrix element:

$$\begin{aligned}\langle q';t'|\mathcal{T}\, Q(t_1)Q(t_2)|q;t\rangle = \int &\langle q'|\exp[-iH(t'-t_1)]|q_1\rangle\langle q_1|Q(t_1)\exp[-iH(t_1-t_2)]|q_2\rangle \\ &\times\langle q_2|Q(t_2)\exp[-iH(t_2-t)]|q;t\rangle dq_1 dq_2\ ,\end{aligned} \tag{4.60}$$

one obtains in the Schrödinger picture:

$$\begin{aligned}G(t_1,t_2) = &\int_{\substack{q_a\,\equiv\, q_a(t)\\ q'_a\,\equiv\, q'_a(t')}} \prod_{\tau,a} dq_a(\tau)\prod_{\tau,b}\frac{dp_b(\tau)}{2\pi}\phi_0(q',t')\phi_0^*(q,t)q_1(t_1)q_2(t_2) \\ &\times\exp\left[i\int_t^{t'} d\tau\left\{-H\left(q(\tau),p(\tau)\right)+\sum_a \dot{q}_a(\tau)p_a(\tau)\right\}\right]\ .\end{aligned} \tag{4.61}$$

Now, we can remove the wave functions by introducing a complete set of states. Then:

$$\begin{aligned}\langle q';t'|Q';T'\rangle &= \langle q'|\exp[-iH(t'-T')]|Q'\rangle = \sum_n \langle q'|n\rangle\langle n|\exp\left[-iH(t'-T')\right]|Q'\rangle \\ &= \sum_n \phi_n^*(q')\phi_n(Q')\exp[-iE_n(t'-T')]\ ,\end{aligned} \tag{4.62}$$

where $E_n$ and $\phi_n$ are the energy and wave functions of the state $|n\rangle$. The contribution of the ground states can be isolated by taking the limit $t' \to i\infty$ and using the fact that $E_n > E_0$ for $n \neq 0$. In this way, one gets:

$$\lim_{t'\to -i\infty} \langle q';t'|Q';T'\rangle = \phi_0^*(q')\phi_0(Q')\exp[-E_0|t'|]\exp[iE_0T']\ . \tag{4.63}$$

Similarly:

$$\lim_{t\to +i\infty} \langle Q;T|q;t\rangle = \phi_0(q)\phi_0^*(Q)\exp[-E_0|t|]\exp[\,-iE_0T]\ , \tag{4.64}$$

from which one can deduce:

$$\mathcal{N} \equiv \lim_{\substack{t' \to -i\infty \\ t \to +i\infty}} \langle q'; t' | q; t \rangle = \phi_0^*(q')\phi_0(q) \exp[-E_0(|t| + |t'|)] . \tag{4.65}$$

Therefore, one can derive after some straightforward algebra:

$$\begin{aligned} \lim_{\substack{t' \to -i\infty \\ t \to +i\infty}} \langle q'; t' | \mathcal{T} Q(t_1) Q(t_2) | q; t \rangle = & \int dQ\, dQ' \langle q'; t' | Q'; T' \rangle \\ & \times \langle Q'; T' | \mathcal{T} Q(t_1) Q(t_2) | Q; T \rangle \langle Q; T | q; t \rangle \\ = & \, \phi_0^*(q')\phi_0(q) \exp[-E_0(|t| + |t'|)] G(t_1, t_2) . \end{aligned} \tag{4.66}$$

Combining Eqs. (4.65) and (4.66), one can deduce the result:

$$\begin{aligned} G(t_1, t_2) = & \frac{1}{\mathcal{N}} \int_{\substack{q_a \equiv q_a(t) \\ q'_a \equiv q'_a(t')}} \prod_{\tau, a} dq_a(\tau) \prod_{\tau, b} \frac{dp_b(\tau)}{2\pi} q_1(t_1) q_2(t_2) \\ & \times \exp\left[ i \int_t^{t'} d\tau \left\{ -H(q(\tau), p(\tau)) + \sum_a \dot{q}_a(\tau) p_a(\tau) \right\} \right] . \end{aligned} \tag{4.67}$$

This result for the two-point function can be generalized to $n$-point Green's function. This Green's function can be generated as:

$$G^{(n)}(\tau_1, \tau_2, \ldots \tau_n) = (-i)^n \left. \frac{\delta^n Z[J]}{\delta J(\tau_1) \cdots \delta J(\tau_n)} \right|_{J=0} . \tag{4.68}$$

by the *generating functional*:

$$Z[J] = \lim_{\substack{t' \to -i\infty \\ t \to +i\infty}} \frac{1}{\mathcal{N}} \int \mathcal{D}q \, \exp\left\{ i \int_t^{t'} d\tau \left[ -\frac{m}{2} \dot{q}^2 - V(q) + J(\tau) q(\tau) \right] \right\} , \tag{4.69}$$

which corresponds to the transition amplitude from a ground state at $\tau$ to the ground state at $\tau'$ in the presence of an external source $J(\tau)$, with the normalization $Z[0] = 1$. We have introduced the symbolic notation $\mathcal{D}$ for the integration measure:

$$\mathcal{D}q \equiv \prod_{\tau, a} dq_a(\tau) . \tag{4.70}$$

In order to elucidate the meaning of the previous expression, we recall that, by definition, a functional is an application of the space of smooth functions $f(x)$ into complex numbers:

$$J(x) \longmapsto Z[J] , \tag{4.71}$$

while a functional derivative is defined as:

$$\frac{\delta Z[J]}{\delta J(y)} = \lim_{\epsilon \to 0} \frac{Z[J + \epsilon \delta_y] - Z[J]}{\epsilon} , \tag{4.72}$$

where $\delta_y = \delta(x-y)$ is the $\delta$-function at $y$. In the case of the functional integral:

$$Z[J] = \int dx\, K(x)J(x)\,, \tag{4.73}$$

the functional derivative is:

$$\frac{\delta Z[J]}{\delta J(y)} = K(y)\,, \tag{4.74}$$

which, after performing a Taylor expansion of the kernel $K(x)$, leads to:

$$K_n(x_1,\ldots x_n) = \frac{\delta^n Z[J]}{\delta J(x_1)\cdots\delta J(x_n)}\,. \tag{4.75}$$

### 4.5.3 Euclidean Green's function

The unphysical limits $t' \to -i\infty,\ t \to i\infty$ can be interpreted in terms of the *Euclidean* Green's functions:

$$\begin{aligned} S^{(n)}(\tau_1,\ldots,\tau_n) &= i^n G^{(n)}(-i\tau_1,\ldots,-i\tau_n) \\ &= \left.\frac{\delta^n Z_{\mathrm{E}}[J]}{\delta J(\tau_1)\cdots\delta J(\tau_n)}\right|_{J=0}\,. \end{aligned} \tag{4.76}$$

where $Z_{\mathrm{E}}[J]$ can be deduced from $Z[J]$ by the formal change $\tau" \to i\tau$. In the Euclidean region, the path integral is well-defined, as it converges because the potential is bounded from below ($(m/2)\dot{q}^2 + V(q) > 0$), such that the exponential in Eq. (4.69) will give a damping factor.

## 4.6 Path integral in quantum field theory

### 4.6.1 Scalar field quantization

For simplicity let's consider a classical field $\phi(x)$ and the corresponding Lagrangian density $\mathcal{L}(\phi, \partial_\mu\phi)$ to which corresponds the action:

$$\mathcal{S} = \int d^4x\, \mathcal{L}(\phi, \partial_\mu\phi)\,. \tag{4.77}$$

The field $\phi$ satisfies the Euler–Lagrange equation of motion:

$$\partial_\mu \frac{\delta\mathcal{L}}{\delta(\partial_\mu\phi)} - \frac{\delta\mathcal{L}}{\delta\phi} = 0\,. \tag{4.78}$$

We denote by $\pi(x)$ its conjuguate momentum:

$$\pi(x) = \frac{\delta\mathcal{L}}{\delta\left(\partial_0\phi\right)}\,, \tag{4.79}$$

which obeys the equal-time canonical commutation relations:

$$[\pi(\vec{x},t),\phi(\vec{x}',t)] = -i\delta^3(\vec{x}-\vec{x}')\,, \tag{4.80}$$

while the Hamiltonian density is defined as:

$$\mathcal{H} = \int d^4x[\pi(x)\partial_0\phi(x) - \mathcal{L}(x)] \tag{4.81}$$

Therefore, in order to use the previous results of quantum mechanics, one can consider a field theory as a quantum mechanical system with infinite degrees of freedom. Therefore, one can make the substitutions:

$$\begin{aligned}
\mathcal{D}q\mathcal{D}p &\to \mathcal{D}\phi(x)\mathcal{D}\pi(x)\,,\\
L(q_i,\dot{q}_i) &\to \int d^3x\ \mathcal{L}(\phi,\partial_\mu\phi)\,;\\
H(q_i,p_i) &\to \int d^3x\ \mathcal{H}(\phi,\pi)\,.
\end{aligned} \tag{4.82}$$

Using the fact that the ground state in field theory is the vacuum state, the generating functional $Z[J]$ is the vacuum-to-vacuum transition amplitude in the presence of an external source $J(x)$, and read in the *Euclidian space*:[1]

$$Z[J] = \int \mathcal{D}\phi\ \exp\left\{\int d^4x[\mathcal{L}(\phi(x)) + J(x)\phi(x)]\right\}\,, \tag{4.83}$$

up to an inessential normalization factor; Here, $x$ is the Euclidian coordinate ($\tau \to it$). In field theory, we are interested in the *connected* Green's function, which is:

$$S^{(n)}(x_1,\dots x_n) = \left\{\frac{1}{Z[J]}\frac{\delta^n Z[J]}{\delta J(x_1)\cdots\delta J(x_n)}\right\}\Bigg|_{J=0}\,, \tag{4.84}$$

where an extra factor of $1/Z[J]$ has been inserted in order to remove the disconnected part of the Green's function.

### 4.6.2 Application to $\lambda\phi^4$ theory

We can illustrate this result by working with the Lagrangian of $\lambda\phi^4$ theory:

$$\mathcal{L} = \mathcal{L}_{\text{free}} + \mathcal{L}_{\text{I}} \tag{4.85}$$

with:

$$\begin{aligned}
\mathcal{L}_{\text{free}} &= \frac{1}{2}(\partial_\mu\phi)(\partial^\mu\phi) - \frac{1}{2}\mu^2\phi\,,\\
\mathcal{L}_{\text{I}} &= -\frac{\lambda}{4!}\phi^4\,.
\end{aligned} \tag{4.86}$$

[1] We shall work in the Euclidian space in this subsection, where, as mentioned previously, the integral has a definite norm, and is well defined. This space is useful for a path integral formulation of non-perturbative QCD. However, the derivation of Feynman rule can still be done in the Minkowski space.

In *Euclidian space*, the generating functional reads:

$$Z[J] = \int \mathcal{D}\phi \exp\left\{-\int d^4x\left[\frac{1}{2}\left(\frac{\partial\phi}{\partial\tau}\right)^2 + \frac{1}{2}(\nabla\phi)^2 + \frac{1}{2}\mu^2\phi^2 + \frac{\lambda}{4!}\phi^4 + J\phi\right]\right\}, \tag{4.87}$$

which one can rewrite as:

$$Z[J] = \left[\exp\int d^4x\ \mathcal{L}_{\mathrm{I}}\left(\frac{\delta}{\delta J}\right)\right] Z_0[J] \tag{4.88}$$

The free-field generating functional

$$Z_0[J] = \int \mathcal{D}\phi \exp\left[\int d^4x\ (\mathcal{L}_{\mathrm{free}} + J\phi)\right]. \tag{4.89}$$

can be written in the form:

$$Z_0[J] = \int \mathcal{D}\phi \exp\left\{-\frac{1}{2}\int d^4x d^4y\ \phi(x)\mathbf{K}(x, y)\phi(y) + \int d^4z J(z)\phi(z)\right\}, \tag{4.90}$$

where:

$$\mathbf{K}(x, y) = \left(-\frac{\partial^2}{\partial\tau^2} - \nabla^2 + \mu^2\right) \tag{4.91}$$

and the identity:

$$-\left(\frac{\partial\phi}{\partial\tau}\right)^2 - (\nabla\phi)^2 = \phi\left(\frac{\partial^2}{\partial\tau^2} - \nabla^2\right)\phi \tag{4.92}$$

has been used because their divergence is a total four-divergence. Integrating the Gaussian integral:

$$\lim_{N\to\infty}\int \mathcal{D}\phi_1\cdots\mathcal{D}\phi_N\ \exp\left[-\frac{1}{2}\sum_{i,j}\phi_i K_{ij}\phi_j + \sum_k J_k\phi_k\right]$$
$$\sim \frac{1}{\sqrt{\det\mathbf{K}}}\exp\left[\frac{1}{2}\sum_{i,j} J_i(\mathbf{K}^{-1})_{ij}J_j\right], \tag{4.93}$$

one obtains:

$$Z_0[J] = \exp\left[\frac{1}{2}\int d^4x d^4y\ J(x)\Delta(x, y)J(y)\right], \tag{4.94}$$

where $\Delta(x, y)$ is the inverse of $\mathbf{K}(x, y)$:

$$\int d^4y\ \mathbf{K}(x, y)\Delta(y, z) = \delta^4(x - z), \tag{4.95}$$

which leads to the desired expression of the *scalar propagator*:

$$\Delta(x, y) = \int \frac{d^4k}{(2\pi)^4} \frac{\exp[ik(x-y)]}{k^2+\mu^2} . \tag{4.96}$$

Perturbative expansion in powers of the interaction Lagrangian $\mathcal{L}_I$ generates the Feynman rules for different vertices. However, in order to keep only the connected Green's function, one should expand $\ln Z$ as defined in Eq. (4.84).

### 4.6.3 Fermion field quantization

The quantization of the fermion field can also be done by expressing the transition amplitude as a sum over possible lines connecting the initial and final states. For a classical fermion (anti-fermion) fields $\psi$, $(\bar\psi)$ and sources $\eta$, $\bar\eta$, the generating functional reads:

$$Z[\eta, \bar\eta] = \int \mathcal{D}\psi(x)\mathcal{D}\bar\psi(x) \exp\left\{i \int d^4x \, [\mathcal{L}(\psi\bar\psi) + \bar\psi\eta + \bar\eta\psi]\right\} , \tag{4.97}$$

where the functional integral must be taken over anti-commuting $c$ number functions which are elements of the Grassmann algebra:

$$\begin{aligned} &\theta(x), \theta(x') = \theta(x), \bar\theta(x') = \bar\theta(x), \bar\theta(x') = 0 , \\ &[\theta(x)]^2 = 0 , \end{aligned} \tag{4.98}$$

where $\theta \equiv \psi$ or $\eta$. The fermion Lagrangian is:

$$\mathcal{L} \equiv \mathcal{L}_{\text{free}} + \mathcal{L}_{\text{I}} , \tag{4.99}$$

with:

$$\begin{aligned} \mathcal{L}_{\text{free}}(x) &= \bar\psi(x)(i\gamma_\mu\partial^\mu - m)\psi(x) , \\ \mathcal{L}_{\text{I}}(x) &= \bar\psi(x)\gamma_\mu\psi(x)A^\mu(x) . \end{aligned} \tag{4.100}$$

Since the fermion fields always enter the Lagrangian quadratically, the previous functional is a generalized Gaussian integral. Therefore, one can write:

$$Z[J] = \int \mathcal{D}\psi(x)\mathcal{D}\bar\psi(x) \exp\left\{\int d^4x \, \bar\psi A \psi\right\} = \det A , \tag{4.101}$$

where $Z$ is the vacuum-to vacuum amplitude and the (connected) Feynman diagram generated by $\ln Z$ will be a set of single-closed fermion loops.

### 4.6.4 Gauge field quantization

Due to gauge invariance, gauge theories represent systems with constrained dynamic variables. Their quantization is more involved than the one of scalar field theory or of free fermion discussed previously, and so we shall leave it for discussion in the next section.

# Part II

## QCD gauge theory

# 5

# Lagrangian and gauge invariance

## 5.1 Introduction

After Einstein's identification of the invariance group of space and time in 1905, symmetry principles received an enthusiastic welcome in physics, with the hope that these principles could express the simplicity of nature in its deepest level. Since 1927 [99,100], it has been recognized that Quantum ElectroDynamics (QED) has a *local symmetry* under the transformations in which the electron field has a phase change that can vary point to point in space–time, and the electromagnetic vector potential undergoes a corresponding transformation. This kind of transformation is called a $U(1)$ gauge symmetry due to the fact that a simple phase change can be thought as a multiplication by a $1 \times 1$ unitary matrix. Largely motivated by the challenge of giving a field-theoretical framework to the concept of isospin invariance, Yang and Mills [101] in 1954 extended the idea of QED to the $SU(2)$ group of symmetry. However, it appears here that the symmetry would have to be approximate because gauge invariance requires massless vector bosons like the photon, and it seems obvious that strong interactions of pions were not mediated by massless but by the massive $\rho$ mesons. In 1961, there was the idea of dynamic breaking, i.e., the Hamiltonian and commutation relations of a quantum theory could possess an exact symmetry and the symmetry of the Hamiltonian might not turn to be a symmetry of the vacuum. This way of breaking the symmetry would necessarily imply the existence of massless-spin zero particle, the Nambu–Goldstone boson [17] discussed previously. Later on, Higgs and others [102] showed that if the broken symmetry is a local gauge symmetry, as in the case of QED, the Nambu–Goldstone bosons could formally exist, but can be eliminated by a gauge transformation, so that they are not physical particles. These Nambu–Goldstone bosons appear as helicity states of massive vector particles. These ideas were the starting point for building the $SU(2)_L \times U(1)$ electroweak theory by Weinberg and Salam [61] as an improvement of the model proposed earlier by Glashow [61]. The spontaneous breaking of the electroweak group into $U(1)$ via a non-vanishing expectation value of the Higgs scalar field gives masses to the $W^{\pm}$ and $Z^0$ but leaves the photon massless. At present time, one even expects that nature has a richer symmetry (*supersymmetry*), which treats in the same manner the fermions and the bosons. However, we do not have yet any *direct evidence* of a such symmetry. In the following, we shall restrict ourselves to the discussion of the symmetry of

QED and QCD described respectively by a $U(1)$ Abelian and $SU(3)_c$ non-Abelian gauge groups.

## 5.2 The notion of gauge invariance

In quantum mechanics, a multiplication of a state vector by a constant phase factor, $e^{i\alpha}$, does not induce any observable consequences. Now if you take a wave function with two very distant peaks, and multiply one by a phase factor, then you have to multiply the other by the same phase. This is the *local gauge invariance*, i.e., independence under a space–time-dependent phase factor $\exp(i\alpha(x))$, postulated by Weyl [100]. However, this requirement does not even hold for free non-relativistic particles. Indeed, if $\psi(\vec{x}, t)$ is the solution of the Schrödinger equation:

$$-\frac{\hbar^2}{2m}\vec{\nabla}^2\psi(\vec{x}, t) = i\hbar\partial_t\psi(\vec{x}, t)\,, \tag{5.1}$$

the quantity $\exp(i\alpha(x))\,\psi(\vec{x}, t)$ is not, in general, a solution of it. Then, gauge invariance necessarily implies that a particle should interact with fields. If indeed, we consider the Schrödinger equation in a magnetic field with a vector potential $\vec{A}(\vec{x}, t)$, then the equation becomes:

$$\frac{\hbar^2}{2m}\left(i\vec{\nabla} - \frac{e}{c}\vec{A}(\vec{x}, t)\right)^2\psi(\vec{x}, t) = i\hbar\partial_t\psi(\vec{x}, t)\,. \tag{5.2}$$

In this case, where the vector potential changes.[1]

$$\vec{A}(\vec{x}, t) \to \vec{A}(\vec{x}, t) - \frac{c}{e}\vec{\nabla}\alpha(\vec{x})\,, \tag{5.3}$$

one can see that both $\psi(\vec{x}, t)$ and $\exp(i\alpha(x))\,\psi(\vec{x}, t)$ are solutions of the Schrödinger equation. From this example, we learn that a local gauge invariance of the wave function necessary needs a coupling of the particle to a vector field. Such an invariance will be satisfied by the gauge theory Lagrangian that we shall discuss below.

## 5.3 The QED Lagrangian as a prototype

The previous discussion can be illustrated in field theory by the simple Lagrangian of QED. In so doing, one can consider the Lagrangian describing a free Dirac electron field having a mass $m$:

$$\mathcal{L}_{\text{free}} = \bar{\psi}(x)(i\partial_\mu\gamma^\mu - m)\psi(x)\,. \tag{5.4}$$

Under a $U(1)$ global phase transformation, one has:

$$\psi(x) \to \exp(-i\theta\mathbf{1})\,\psi(x)\,, \tag{5.5}$$

[1] Fortunately, this gauge transformation does not influence the magnetic field.

where $\theta$ is an arbitrary constant. Now, if one considers the case where $\theta$ depends one the space–time coordinate, one can notice that the Lagrangian is no longer invariant under the phase transformation as the derivative of the field has induced an extra-term:

$$\partial^\mu \psi(x) \to \exp(-i\theta \mathbf{1})\,(\partial^\mu - i\partial^\mu\theta)\,\psi(x)\,, \tag{5.6}$$

which means that, for the theory to be consistent, the same phase convention should be taken at all space–time points. However, this is not natural. *Gauge symmetry* requires that the $U(1)$ phase invariance should hold *locally*. This can be achieved, like for the case of quantum mechanics above, by adding a new spin-1 contribution which can cancel the previous extra term:

$$A_\mu(x) \to A_\mu(x) - \frac{1}{e}\partial_\mu\theta(x)\,, \tag{5.7}$$

and by defining the covariant derivative:

$$D_\mu\psi(x) \equiv (\partial_\mu + ieA_\mu(x))\psi(x)\,, \tag{5.8}$$

which transforms like the field itself:

$$\partial^\mu\psi(x) \to \exp(i\theta\mathbf{1})\,D^\mu\psi(x)\,. \tag{5.9}$$

Therefore, the Lagrangian:

$$\mathcal{L} = \bar\psi(x)(iD_\mu\gamma^\mu - m)\psi(x) = \mathcal{L}_{\text{free}} - eA_\mu(x)\bar\psi(x)\gamma^\mu\psi(x)\,, \tag{5.10}$$

is invariant under the *local* $U(1)$ transformation. As in the case of quantum mechanics, the gauge principle necessary needs a coupling of the electron field to the vector field, which is given by the second term of the Lagrangian. A complete QED Lagrangian can be achieved by adding the kinetic term of the electromagnetic field and a gauge term:

$$\mathcal{L}_\gamma = -\frac{1}{4}F_{\mu\nu}F^{\mu\nu} - \frac{1}{2\alpha_G}\partial^\mu A_\mu\partial_\mu A^\mu\,, \tag{5.11}$$

which expresses that $A_\mu$ can propagate. Here, $F_{\mu\nu} \equiv \partial_\mu A_\nu - \partial_\nu A_\mu$ is the electromagnetic field strength; $\alpha_G$ is the gauge parameter which is 0(resp. 1) in the Landau(resp. Feynman) gauge. On the other hand, a possible $m^2A^\mu A_\mu$ mass term violates gauge invariance, which then implies that the photon is massless. We have then shown that, with the alone gauge principle, one can rederive the QED Lagrangian, which leads to a very impressive quantum field theory which applications have been tested to a very high degree of accuracy (see next section).

## 5.4 The QCD Lagrangian

The case of QCD is very similar to the one of QED though more involved due to the non-Abelian structure of its $SU(3)_c$ gauge group. The QCD Lagrangian density reads:

$$\begin{aligned}\mathcal{L}_{\mathrm{QCD}}(x) = & -\frac{1}{4}G_a^{\mu\nu}G_{\mu\nu}^a + i\sum_{j=1}^{n}\bar{\psi}_j^\alpha\gamma^\mu(D_\mu)_{\alpha\beta}\psi_j^\beta - \sum_{j=1}^{n}m_j\bar{\psi}_j^\alpha\psi_{j,\alpha} \\ & -\frac{1}{2\alpha_G}\partial^\mu A_\mu^a\partial_\mu A_a^\mu - \partial_\mu\bar{\varphi}_a D^\mu\varphi^a\ ,\end{aligned} \tag{5.12}$$

where $G_{\mu\nu}^a \equiv \partial_\mu A_\nu^a - \partial_\nu A_\mu^a + g\ f_{abc}A_\mu^b A_\nu^c\ (a \equiv 1, 2, \ldots, 8)$ are the Yang–Mills field strengths constructed from the gluon fields $A_\mu^a(x)$ [101]. $\psi_j$ is the field of the quark flavour $j$ while $\varphi^a(x)$ are eight anti-commuting scalar fields in the $\underline{8}$ of $SU(3)$. $(D_\mu)_{\alpha\beta} \equiv \delta_{\alpha\beta}\partial_\mu - ig\sum_a \frac{1}{2}\lambda_{\alpha\beta}^a A_\mu^a$ are the covariant derivatives acting on the quark colour component $\alpha, \beta \equiv$ red, blue and yellow; $\lambda_{\alpha\beta}^a$ are the eight $3\times 3$ colour matrices and $f_{abc}$ are *real* structure constants which close the $SU(3)$ Lie algebra:[2]

$$[T_a, T_b] = i\ f_{abc}\ T_c\ , \tag{5.13}$$

where $(T^a)_{\alpha\beta} = \frac{1}{2}\lambda_{\alpha\beta}^a$ in the fundamental colour $\underline{3}$ representation, while $(T_a)_{bc} = -i\ f_{abc}$ in the adjoint $\underline{8}$ representation of gluon basis. The last two terms in the Lagrangian are respectively the gauge-fixing term necessary for a covariant quantization in the gluon sector [$\alpha_G = 1(0)$ in the Feynman (Landau) gauge] and the Faddeev–Popov ghost term [97] necessary to eliminate unphysical particles from the theory.

One can rewrite the above Lagrangian in a more explicit form:

$$\mathcal{L}_{\mathrm{QCD}} = \mathcal{L}_{\mathrm{free}} + \mathcal{L}_{\mathrm{I}}^{qg} + \mathcal{L}_{\mathrm{I}}^{gg} + \mathcal{L}_{\mathrm{I}}^{FPg}, \tag{5.14}$$

where:

$$\mathcal{L}_{\mathrm{free}} = \mathcal{L}_{\mathrm{free}}^{g} + \mathcal{L}_{\mathrm{free}}^{q} + \mathcal{L}_{\mathrm{free}}^{FP}\ , \tag{5.15}$$

is the free-field Lagrangian containing the kinetic terms of the different fields, with:

$$\begin{aligned}\mathcal{L}_{\mathrm{free}}^{g} &= -\frac{1}{4}\left(\partial^\mu A_a^\nu - \partial^\nu A_a^\mu\right)\left(\partial^\mu A_a^\nu - \partial^\nu A_a^\mu\right) - \frac{1}{2\alpha_G}\partial^\mu A_\mu^a\partial_\mu A_a^\mu \\ \mathcal{L}_{\mathrm{free}}^{q} &= i\sum_{j=1}^{n}\bar{\psi}_j^\alpha\gamma^\mu(\partial_\mu)_{\alpha\beta}\psi_j^\beta - \sum_{j=1}^{n}m_j\bar{\psi}_j^\alpha\psi_{j,\alpha} \\ \mathcal{L}_{\mathrm{free}}^{FP} &= -\partial_\mu\bar{\varphi}_a\partial^\mu\varphi^a\ .\end{aligned} \tag{5.16}$$

[2] More general and useful properties of the $\lambda$ matrices are given in Appendix B.

The interaction Lagrangian of the gluon fields respectively with the quarks, gluons and Faddeev–Popov ghosts reads:

$$\mathcal{L}_{\mathrm{I}}^{qg} = gA_a^{\mu}\sum_{j=1}^{n}\bar{\psi}_j^{\alpha}\gamma^{\mu}\left(\frac{\lambda_a}{2}\right)_{\alpha\beta}\psi_j^{\beta} ,$$
$$\mathcal{L}_{\mathrm{I}}^{gg} = -\frac{g}{2}f^{abc}\left(\partial^{\mu}A_a^{\nu}-\partial^{\nu}A_a^{\mu}\right)A_{b,\mu}A_{c,\nu} - \frac{g^2}{4}f^{abc}f_{ade}A_b^{\mu}A_c^{\nu}A_{\mu}^{d}A_{\nu}^{e} ,$$
$$\mathcal{L}_{\mathrm{I}}^{FPg} = gf_{abc}(\partial_{\mu}\bar{\varphi}^{a})\varphi^{b}A_{\mu}^{c} . \tag{5.17}$$

The new piece compared with the usual Abelian QED Lagrangian is the the appearance of the gluon self-interaction $\mathcal{L}_{gg}^{I}$, which is a specific feature of the non-Abelian group $SU(3)_c$. Because of this new piece, the Faddeev–Popov ghosts fields are introduced (as mentioned above) for a proper quantization of the theory, which can be done formally using path integral techniques. This method is discussed in details in various textbooks and will be briefly sketched in the next section.

## 5.5 Local invariance and BRST transformation

$\mathcal{L}_{\mathrm{QCD}}(x)$ is locally invariant under the BRST transformation [103]:

$$\begin{aligned} A_{\mu}(x) &\to A_{\mu}(x)+\omega D_{\mu}\,\varphi ,\\ \psi_i(x) &\to \exp(-ig\omega\vec{T}\cdot\vec{\varphi})\,\psi_i ,\\ \bar{\varphi} &\to \bar{\varphi}+\frac{\omega}{\alpha_G}\partial_{\mu}A^{\mu} ,\\ \phi &\to \phi-\frac{1}{2}g\omega\vec{\varphi}\times\vec{\varphi} , \end{aligned} \tag{5.18}$$

where $\omega(x)$ is an arbitrary parameter. In order to see the usefulness of the BRST transformations for generating the Slavnov–Taylor–Ward identities [104,20], let's consider the gluon propagator:

$$iD_{\mu\nu}^{ab}(k) = \int d^4x\, e^{ikx}\langle 0|\mathcal{T}A_{\mu}^{a}(x)A_{\nu}^{b}(0)|0\rangle . \tag{5.19}$$

We shall prove that order by order in perturbation theory, the non-transverse part of the gluon propagator remains the same as for the free propagator:

$$k^{\mu}k^{\nu}iD_{\mu\nu}^{ab}(k) = -i\alpha_G\delta^{ab} . \tag{5.20}$$

In so doing, we start with the trivial identity:

$$\langle 0|\partial^{\mu}A_{\mu}^{a}(x)\bar{\varphi}^{b}(0)|0\rangle = 0 . \tag{5.21}$$

The BRST invariance implies:

$$\langle 0|\partial^{\mu}A_{\mu}^{a}(x)\bar{\varphi}^{b}(0)|0\rangle = \langle 0|\partial^{\mu}\left(A_{\mu}^{a}\right)'(x)(\bar{\varphi}^{b})'(0)|0\rangle , \tag{5.22}$$

where the new fields:

$$A'^a_\mu = A^a_\mu + \omega D_\mu \varphi^a \,,$$
$$\bar{\varphi}'^a = \bar{\varphi}^a + \frac{\omega}{\alpha_G} \partial^\mu A^a_\mu \,, \tag{5.23}$$

have been introduced. Then, one can deduce:

$$\frac{\omega}{\alpha_G} \langle 0 | \partial^\mu A^a_\mu(x) \partial^\nu A^b_\nu(0) | 0 \rangle = 0 \,. \tag{5.24}$$

By taking its Fourier transform, one obtains the Ward identity written in Eq. (2.39). Using the canonical commutation relation:

$$\left[\Pi^a_\mu(x), A^b_\nu(0)\right] \delta(x_0) = -i g_{\mu\nu} \delta^{ab} \delta^4(x) \,, \tag{5.25}$$

where:

$$\Pi^a_\mu(x) = -G^a_{0\mu}(x) - \frac{1}{\alpha_G} g_{0\mu} \partial^\nu A^a_\nu(x) \,, \tag{5.26}$$

we obtain:

$$\left[A^a_0(x), \partial^\nu A^b_\nu(0)\right] \delta(x_0) = -i \alpha_G \delta^4(x) \,. \tag{5.27}$$

Using Eq. (5.27) into the Ward identity in Eq. (2.39), one can deduce the result in Eq. (5.20).

# 6

# Quantization using path integral

A quantization of a theory can be done either by considering the quark and gluon fields as operators with canonical commutation relations or by introducing Feynman integrals in functional spaces. The second procedure is very convenient for gauge theories, and especially for the non-perturbative approaches. However, although this second method preserves Lorentz invariance, it is not clear that the $S$-matrix calculated in this way is unitary. On the contrary, Lorentz invariance is obscure from the canonical commutation relations, while unitarity is obvious.

## 6.1 Path integral technique for QCD

The expression of the path integral can be obtained following the derivation discussed in a previous chapter. However, the quantization of the gauge fields is more peculiar as the source term $A^a_\mu J^\mu_a$ is not gauge invariant and hence the generating functional itself. A gauge invariant functional can be obtained (detailed derivations are given in many textbooks). This can be achieved by first introducing an invariant measure (Faddeev–Popov ansatz). One considers that the action is invariant under the gauge tranformation:

$$A_\mu \to A^\theta_\mu \,, \tag{6.1}$$

where:

$$\frac{\lambda}{2} A^\theta_\mu = U(\theta)\left[\frac{\lambda}{2} A_\mu + \frac{1}{ig} U^{-1}(\theta)\partial_\mu U(\theta)\right] U^{-1}(\theta)\,, \tag{6.2}$$

and:

$$U(\theta) = \exp\left[-i\frac{\lambda}{2}\theta(x)\right] . \tag{6.3}$$

Then, one makes an expansion for small $\theta$, which leads to:

$$\left(A^\theta_\mu\right)^a = A^a_\mu + f^{abc}\theta_b A_{\mu,c} - \frac{1}{g}\partial_\mu \theta^a \,. \tag{6.4}$$

The proper invariant measure becomes:

$$\mathcal{D}A \to \mathcal{D}A\ \Delta_G[A] \prod_{a,x} \delta\left[G^\mu A^a_\mu(x) - B^a(x)\right] . \tag{6.5}$$

$G^\mu$ has been introduced as:

$$G^\mu A^a_\mu = B^a , \tag{6.6}$$

as a generalization of the (Lorentz) gauge fixing condition $\partial_\mu A^\mu = 0$; $B^a$ is an arbitrary space–time function, independent of the gauge field; $\Delta_G$ can be obtained from the volume normalization condition:

$$\begin{aligned} 1 &= \Delta_G[A] \int \mathcal{D}\theta^a \prod_{a,x} \delta\left[G^\mu A^{\theta_a}_\mu(x) - B^a(x)\right] , \\ &= \frac{\Delta_G[A]}{\det M_G} , \end{aligned} \tag{6.7}$$

where:[1]

$$[M_G(x,y)]^{ab} = \frac{\delta\left[G^\mu \left(A^\theta_\mu\right)^a(x)\right]}{\delta\theta^b(y)} . \tag{6.8}$$

By integrating over $B^a(x)$ by the suitable choice of weight:

$$\exp\left[-\frac{i}{2\alpha_G}\int d^4x [B(x)]^2\right] , \tag{6.9}$$

where $\alpha_G$ is the gauge-fixing term, the generating functional reads:

$$\begin{aligned} Z[J] &= \int \mathcal{D}A \det M_G \prod_{a,x} \delta\left[G^\mu A^a_\mu(x) - B^a(x)\right] \\ &\quad \times \exp\left\{ i \int d^4x \left[ \mathcal{L}_{\text{kin}} - \frac{1}{2\alpha_G}\left(G^\mu A^a_\mu\right)^2 + A^a_\mu J^\mu_a \right]\right\} , \end{aligned} \tag{6.10}$$

where:

$$\mathcal{L}_{\text{kin}} = -\frac{1}{4} G^{\mu\nu} G_{\mu\nu} , \tag{6.11}$$

is the gluon kinetic term; the last term in the exponent is the external source term. In a *covariant gauge*, the matrix $M_G$ reads:

$$[M_G(x,y)]_{ab} = \left[\delta_{ab}\left(\frac{\partial}{\partial x_\mu}\right)^2 - g f_{abc}\partial^\mu A^c_\mu\right]\delta^4(x-y) , \tag{6.12}$$

which depends on the gauge field $A^a_\mu$ such that a simple perturbative expansion of the previous generating functional is not allowed. In this case, one needs to exponentiate det $M_G$

[1] One should note that in the case of axial ($n.A_0 = 0$, $n \equiv$ a space-like constant vector) and in a temporal gauge ($A_0 = 0$), det $M_G$ is a constant like in the case of an Abelian theory, where the canonical quantization can be easily done. This is not the case of the covariant gauge as we shall see later on.

and consider it as a part of the effective Lagrangian. This can be done by introducing the Faddeev–Popov fictious ghost anti-commuting fields $\varphi$ and $\bar{\varphi}$:

$$\det M_G = \int \mathcal{D}\varphi\mathcal{D}\bar{\varphi} \exp\left[-i\int d^4x d^4y\, \bar{\varphi}^a(x)\,(M_G(x,y))_{ab}\,\varphi^b(x)\right] . \tag{6.13}$$

Therefore, the complete QCD generating functional is:

$$Z[\psi, A, \varphi] = \int \mathcal{D}A\mathcal{D}\psi\mathcal{D}\bar{\psi}\mathcal{D}\varphi\mathcal{D}\bar{\varphi}\ \exp\left\{i\int d^4x[\mathcal{L}_{\text{QCD}} + \mathcal{L}_{\text{source}}]\right\}, \tag{6.14}$$

where:

$$\mathcal{L}_{\text{source}} = A_\mu J^\mu + \bar{\chi}\varphi + \bar{\varphi}\chi + \bar{\psi}\eta + \bar{\eta}\psi , \tag{6.15}$$

with $\chi,\ \bar{\chi}$ and $\eta,\ \bar{\eta}$ are respectively source functions for the ghost and fermion fields. The generating functional can now be written in the familiar form as in the case of scalar fields in Eq. (4.85), as the Lagrangian can be decomposed into:

$$\mathcal{L}_{\text{QCD}} = \mathcal{L}_{\text{free}} + \mathcal{L}_{\text{I}} , \tag{6.16}$$

where, as one can see in Eq. (5.15), that $\mathcal{L}_{\text{free}}$ has three parts. Therefore, the generating functional reads:

$$Z[\psi, A, \varphi] = \exp\left\{i\int d^4x \mathcal{L}_I\left(\frac{\delta}{i\delta J^a_\mu}, \frac{\delta}{i\delta\eta}, \frac{\delta}{i\delta\bar{\eta}}, \frac{\delta}{i\delta\chi}, \frac{\delta}{i\delta\bar{\chi}}\right)\right\} Z_0[J,\ \chi,\ \bar{\chi},\ \eta,\ \bar{\eta}] , \tag{6.17}$$

where $Z_0$ is the generating function for free fields:

$$Z_0 \equiv Z_0^g[J] Z_0^{FP}[\chi,\ \bar{\chi}] Z_0^q[\eta,\ \bar{\eta}] , \tag{6.18}$$

with:

$$\begin{aligned} Z_0^g[J] &= \int \mathcal{D}A\ \exp\left\{i\int d^4x\left(\mathcal{L}^g_{\text{free}} + AJ\right)\right\} , \\ Z_0^q[\eta,\ \bar{\eta}] &= \int \mathcal{D}\psi\mathcal{D}\bar{\psi}\ \exp\left\{i\int d^4x\left(\mathcal{L}^q_{\text{free}} + \bar{\psi}\eta + \bar{\eta}\psi\right)\right\} , \\ Z_0^{FP}[\chi,\ \bar{\chi}] &= \int \mathcal{D}\varphi\mathcal{D}\bar{\varphi}\ \exp\left\{i\int d^4x\left(\mathcal{L}^{FP}_{\text{free}} + \bar{\chi}\varphi + \bar{\varphi}\chi\right)\right\} . \end{aligned} \tag{6.19}$$

## 6.2 Feynman rules from the path integral

### 6.2.1 Free-field propagators

The propagator for a free field $\phi$ is defined as:

$$D(x, y) \equiv \langle 0|T\phi(x)\phi(0)|0\rangle = (-i)^2 \left.\frac{\delta^2 \ln Z_0}{\delta J(x) J(y)}\right|_{J=0} \tag{6.20}$$

Following closely the derivation of the scalar propagator in the case of scalar $\lambda\phi^4$ theory discussed in previous chapter, and by using the generalized Gaussian integral, one can rewrite:

$$\begin{aligned} Z_0^g[J] &= \exp\left\{i\int d^4x d^4y\; J^{a\mu}(x)D^{ab}_{\mu\nu}(x-y)J^{b\nu}(y)\right\}, \\ Z_0^q[\eta,\bar\eta] &= \exp\left\{i\int d^4x d^4y\; \bar\eta(x)S(x-y)\eta(y)\right\}, \\ Z_0^{FP}[\chi,\bar\chi] &= \exp\left\{i\int d^4x d^4y\; \bar\chi(x)D^{ab}(x-y)\chi(y)\right\}, \end{aligned} \tag{6.21}$$

where $D^{ab}_{\mu\nu}$, $S$, $D^{ab}$ are respectively the gluon, fermion and Faddeev–Popov ghost propagators, which obey respectively the conditions:

$$\begin{aligned} \int d^4y\; K^{ac}_{\mu\nu}(x-y)D^{cb}_{\nu\lambda}(y-z) &= g_{\mu\lambda}\delta^{ab}\delta^4(x-z), \\ \int d^4y\; K^{ac}(x-y)D^{cb}(y-z) &= \delta^{ab}\delta^4(x-z), \\ \int d^4y\; \Omega(x-y)S(y-z) &= \delta^4(x-z), \end{aligned} \tag{6.22}$$

where:

$$\begin{aligned} K^{ab}_{\mu\nu} &= \delta^{ab}\left[-g_{\mu\nu}\left(\frac{\partial}{\partial x_\mu}\right)^2 + \left(1-\frac{1}{\alpha_G}\right)\partial_\mu\partial_\nu\right], \\ K^{ab} &= \delta^{ab}\left(\frac{\partial}{\partial x_\mu}\right)^2, \\ \Omega &= -i\gamma^\mu\partial_\mu + m . \end{aligned} \tag{6.23}$$

Solving these equations give the Feynman rules (visualized in Appendix E) in the momentum space after Fourier transform:

$$\begin{aligned} D^{ab}_{\mu\nu}(x) &= (-i)\delta^{ab}\int \frac{d^4k}{(2\pi)^4}\frac{e^{-ikx}}{k^2+i\epsilon'}\left(g_{\mu\nu}-(1-\alpha_G)\frac{k_\mu k_\nu}{k^2}\right), \\ D^{ab}(x) &= (-i)\delta^{ab}\int \frac{d^4k}{(2\pi)^4}\frac{e^{-ikx}}{k^2+i\epsilon'}, \\ S(x) &= (i)\int \frac{d^4k}{(2\pi)^4}\frac{e^{-ikx}}{\hat k - m + i\epsilon'} . \end{aligned} \tag{6.24}$$

### 6.2.2 Vertices

Perturbative series can be generated by expanding the exponential in Eq. (6.17):

$$Z[J,\ldots] = \left\{1 + i\int d^4x\; \mathcal{L}^I\left(\frac{\delta}{i\delta J^a_\mu},\ldots\right)+\cdots\right\} Z_0[J,\ldots] . \tag{6.25}$$

Let, for instance, the three-gluon vertex at order $g$:

$$\Gamma^{a_1a_2a_3}_{3\mu_1\mu_2\mu_3}(x_1,x_2,x_3) = (-i)^2 \frac{\delta^3}{\delta J_1 \delta J_2 \delta J_3} \int d^4x\, \mathcal{L}_I^{3g}\left(\frac{\delta}{i\delta J^{a\mu}}\right) Z_0^g[J]\bigg|_{J=0} . \quad (6.26)$$

From the three-gluon terms of the Lagrangian $\mathcal{L}_{gg}$, one can deduce:

$$\begin{aligned} Z_{3g}[J] &\equiv \int d^4x\, \mathcal{L}_I^{3g}\left(\frac{\delta}{i\delta J^{a\mu}}\right) Z_0^G[J]\bigg|_{J=0} \\ &= \int d^4x \left(\frac{-g}{2}\right) f^{abc}\left(\partial_\mu \frac{\delta}{i\delta J^{a\nu}} - \partial_\nu \frac{\delta}{i\delta J^{a\mu}}\right) \frac{\delta}{i\delta J^{b\mu}} \frac{\delta}{i\delta J^{c\nu}} Z_0^g[J] \quad (6.27) \end{aligned}$$

After some algebra, one obtains:

$$\begin{aligned} Z_{3g}[J] = -i\frac{g}{2} f^{abc} \int d^4x d^4y_1 d^4y_2 d^4y_3 \big[\partial_\mu D^{aa_1}_{\nu\lambda_1}(x-y_1) - \partial_\nu D^{aa_1}_{\mu\lambda_1}(x-y_1)\big] \\ \times D^{ba_2\mu}_{\lambda_2}(x-y_2) D^{ca_3\nu}_{\lambda_3}(x-y_3) J^{a_1\lambda_1}(y_1) J^{a_2\lambda_2}(y_2) J^{a_3\lambda_3}(y_3) Z_0^g[J] , \quad (6.28) \end{aligned}$$

which gives:

$$\begin{aligned} \Gamma^{a_1a_2a_3}_{3\mu_1\mu_2\mu_3}(x_1,x_2,x_3) = g f^{abc} \int d^4x \big[\partial_\mu D^{aa_1}_{\nu\mu_1}(x-x_1) - \partial_\nu D^{aa_1}_{\mu\mu_1}(x-x_1)\big] \\ \times D^{ba_2\mu}_{\mu_2}(x-x_2) D^{ca_3\nu}_{\mu_3}(x-x_3) + \text{permutations} . \quad (6.29) \end{aligned}$$

Taking the Fourier transform, one then deduces:

$$\begin{aligned} \Gamma^{a_1a_2a_3}_{3\mu_1\mu_2\mu_3}(x_1,x_2,x_3) = \int \frac{d^4k_1}{(2\pi)^4} \frac{d^4k_2}{(2\pi)^4} \exp\left[i \sum_{i=1}^{3} k_i x_i\right] \prod_{i=1}^{3} D_{\mu_i\lambda_i} \\ \times g f^{a_1a_2a_3}\big[(k_1-k_2)^{\lambda_3} g^{\lambda_1\lambda_2} + (k_2-k_3)^{\lambda_1} g^{\lambda_2\lambda_3} + (k_3-k_1)^{\lambda_2} g^{\lambda_3\lambda_1}\big] , \end{aligned} \quad (6.30)$$

with:

$$k_1 + k_2 + k_3 = 0 \quad (6.31)$$

and:

$$D_{\mu\nu}(k) \equiv \frac{1}{k^2}\left[g_{\mu\nu} - (1-\alpha_G)\frac{k_\mu k_\nu}{k^2}\right] . \quad (6.32)$$

Equation (6.30) gives the Feynman rule for the three-gluon vertex to order $g$, which is given in Appendix E. One can extend the previous analysis to derive the different Feynman rules listed in Appendix E.

## 6.3 Quantization of QED

QED is a particular aspect of the more general non-Abelian case discussed previously. Under the $U(1)$ gauge transformation, one has instead of Eq. (6.4):

$$A^{\theta}_{\mu}(x) = A_{\mu}(x) - \frac{1}{g}\partial_{\mu}\theta(x) , \tag{6.33}$$

and the response matrix $M_G$ in Eq. (6.8) will be independent of $A_{\mu}$ for any choice of the gauge, and then the Faddeev–Popov factor det $M_G$ plays no physical role and can be dropped in the generating functional in Eq. (6.10).

## 6.4 Qualitative feature of quantization

For a qualitative physical picture of the quantization procedure, one can notice that the gluon fields $A^{\mu}_{a}$ have four Lorentz degrees of freedom, while a massless spin-1 gluon has only two physical polarizations. In QED, the gauge-fixing term is enough for making a consistent quantization, as the $U(1)$ gauge symmetry guarantees that the extra degrees of freedom do not generate physical amplitudes, and the physical results is independent of the gauge parameter $\alpha_G$. In QCD, life is more complicated. For instance, if one tries to evaluate the cross-section of the scattering process $\bar{q}q \to gg \to \bar{q}q$, one notices that, due to the propagation of the longitudinal and scalar gluon polarizations along the internal gluon lines, the contribution of the higher-order diagrams shown in Fig. 6.1, violates unitarity. In QED, the analogous process $e^+e^- \to \gamma\gamma \to e^+e^-$ does not have these drawbacks as unphysical contributions from the longitudinal and scalar components of the photons vanish

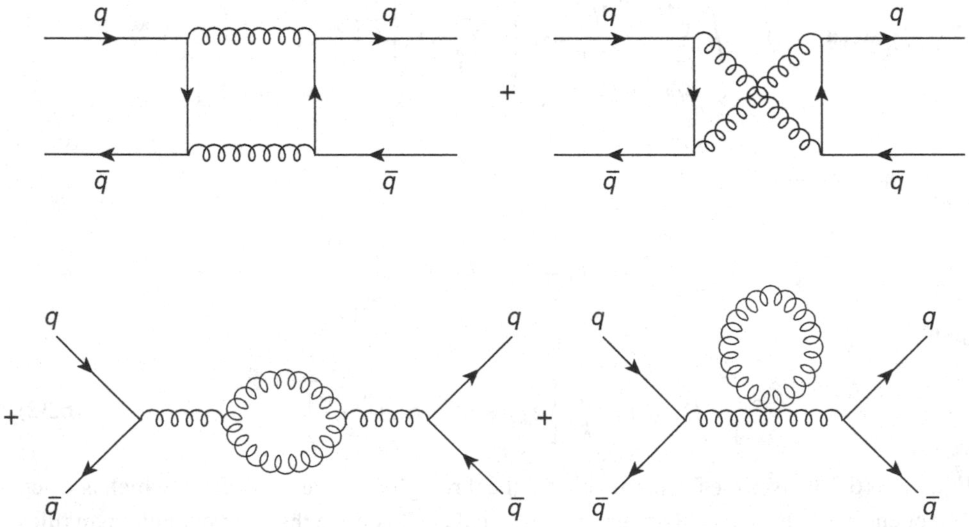


Fig. 6.1. Gluon contributions to the $\bar{q}q \to \bar{q}q$ process.

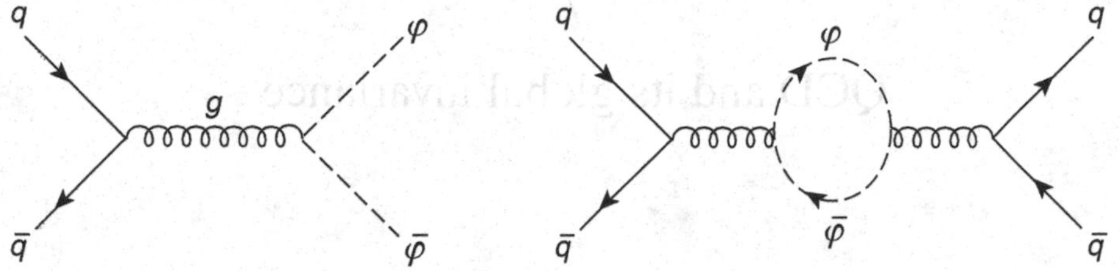


Fig. 6.2. Ghost contributions to the $\bar{q}q \rightarrow \bar{q}q$ process.

due to gauge invariance and to the conservation of the electromgnetic current. In order to recover such a property in QCD, one can introduce unphysical scalar fields with negative norms (ghosts) which eliminate the contributions of such unwanted terms from the diagrams depicted in Fig. 6.2.

More generally, the introduction of the Faddeev–Popov ghosts, in addition to the gauge-fixing term, guarantees a consistent quantization of the theory.

# 7
# QCD and its global invariance

## 7.1 $U(1)$ global invariance

$\mathcal{L}_{QCD}(x)$ is invariant under the $U(1)_B$ global transformation :

$$\psi_i(x) \rightarrow \exp(-i\theta\mathbf{1})\psi_i(x) , \tag{7.1}$$

to which corresponds the conserved baryonic current:

$$J^\mu(x) = \sum_i \bar{\psi}_i \gamma^\mu \psi_i(x) , \tag{7.2}$$

and the baryonic charge generator of the $U(1)_B$ group:

$$B = \int d^3x J^0(\vec{x}, t) . \tag{7.3}$$

For massless quarks, $\mathcal{L}_{QCD}(x)$ is also invariant under the axial $U(1)_A$ transformation:

$$\psi_i \rightarrow (-i\theta\mathbf{1}\gamma_5)\psi_i , \tag{7.4}$$

acting on quark-flavour components. The corresponding current:

$$J_5^\mu(x) = \sum_i \bar{\psi}_i \gamma^\mu \gamma_5 \psi_i(x) , \tag{7.5}$$

has an anomalous divergence:

$$\partial\mu J_5^\mu(x) = \frac{g^2}{4\pi^2}\frac{n}{8}\epsilon_{\mu\nu\rho\sigma} G_a^{\mu\nu} G_a^{\rho\sigma} , \tag{7.6}$$

where the rate of the change of the associated axial charge:

$$\dot{Q}_5 = \int d^3x\ \partial_0 J_5^0(\vec{x}, t) , \tag{7.7}$$

is zero in the absence of instanton-type solutions [105].

## 7.2 $SU(n)_L \times SU(n)_R$ global chiral symmetry

As we have already dicussed in Part 1, and we shall partly repeat here, $\mathcal{L}_{\rm QCD}(x)$ also possesses a $SU(n)_L \times SU(n)_R$ global chiral symmetry. In the massless quark limit ($m_j = 0$), it is invariant under the global chiral transformation:

$$\psi_i(x) \to \exp(-i\theta^A T_A)\psi_i(x)\,,$$
$$\psi_i(x) \to \exp(-i\theta^A T_A\gamma_5)\psi_i(x)\,, \tag{7.8}$$

where $T^A(A \equiv 1, \ldots, n^2 - 1)$ are the infinitesimal generators of the $SU(n)$ group acting on the quark-flavour components. The associated Noether currents are the vector and axial-vector currents:

$$V_\mu^A(x) = \bar{\psi}_i\gamma_\mu\, T_{ij}^A\psi_i(x)\,,$$
$$A_\mu^A(x) = \bar{\psi}_i\gamma_\mu\gamma_5\, T_{ij}^A\psi_i(x)\,, \tag{7.9}$$

which are the ones of the algebra of currents of Gell-Mann [69,13]. The corresponding charges, which are the generators of $SU(n)_L \times SU(n)_R$ are:

$$Q_L^A = \int d^3x\,\left(V_0^A - A_0^A\right)\,,$$
$$Q_R^A = \int d^3x\,\left(V_0^A + A_0^A\right). \tag{7.10}$$

The charges are conserved in the massless quark limit, and obeys the commutation relation:

$$\left[Q_L^\alpha, Q_L^\beta\right] = i f_{\alpha\beta\gamma} Q_L^\gamma\,,$$
$$\left[Q_R^\alpha, Q_R^\beta\right] = i f_{\alpha\beta\gamma} Q_R^\gamma\,,$$
$$\left[Q_L^\alpha, Q_R^\alpha\right] = 0\,, \tag{7.11}$$

where $\alpha,\ \beta,\ \gamma = 1, \ldots n$. In the Nambu–Goldstone [17] realization of chiral symmetry, the axial charge does not annihilate the vacuum, which is the basis of the successes of current algebra and pion PCAC [13]. In this scheme, the chiral flavour group $G \equiv SU(n)_L \times SU(n)_R$ is broken spontaneously by the light quark $(u, d, s)$ vacuum condensates down to a subgroup $H \equiv SU(n)_{L+R}$, where the vacua are symmetrical:

$$\langle\bar{\psi}_u\psi_u\rangle = \langle\bar{\psi}_d\psi_d\rangle = \langle\bar{\psi}_s\psi_s\rangle\,. \tag{7.12}$$

The Goldstone theorem states that this spontaneous breaking mechanism is accompanied by $n^2 - 1$ massless Goldstone $P$ (pions) bosons, which are associated with each unbroken generator of the coset space $G/H$. For $n = 3$, these Goldstone bosons can be identified with the eight lightest mesons of the Gell-Mann eightfoldway ($\pi^+,\ \pi^-,\ \pi^0,\ \eta,\ K^+,\ K^-,\ K^0,\ \bar{K}^0$). On the other hand, the vector charge is assumed to annihilate the vacuum and the corresponding symmetry is achieved à la Wigner–Weyl [18]. In the vector case, the particles are classified in irreducible representations of $SU(n)_{L+R}$ and form parity doublets.

# Part III

## $\overline{MS}$ scheme for QCD and QED

# Introduction

As in QED, the evaluation of QCD Feynman diagrams leads (in many cases) to divergent results. Finite physical answers need a regularization and a renormalization of the QCD parameters (vertices, coupling, masses...). However, the renormalization programme of QED [106] cannot be extended to QCD in a naïve way, as contrary to leptons which we can (freely) observe, quarks are off-shell, such that the standard Pauli–Villars regularization [107] and on-shell renormalization successful in QED cannot be used here. There exists different versions of off-shell renormalization schemes, which can be applied to non-Abelian gauge theories. Among them, we shall review the most elegant and powerful one, which is the: *Dimensional Regularization and Renormalization, the so-called* $\overline{MS}$ *scheme* originally proposed by 't Hooft and Veltman, Bollini and Giambiagi and by Ashmore [108,109,123].[1]

The most important feature of the method is the concept of analytic continuation of the dimension of space–time to complex $n$ ($n = 4$ for low-energy space–time). This regularization procedure has the great advantage of preserving the local invariance of the underlying Lagrangian, and allows one to treat, in a gauge-invariant way, divergent Feynman integrals to all orders of perturbation theory.

In the $\epsilon$-regularization procedure, the UV and IR divergencies are transformed into poles in $\epsilon$, where the integrals are performed in $4 - \epsilon$ space–time dimensions). In general the UV poles are of the form:

$$\sum_{p=1} \frac{Z^{(p)}}{\epsilon^p} , \tag{7.1}$$

and will appear as counterterms in the initial Lagrangian. However, these counterterms are not arbitrary as they should obey constraints imposed by the Slavnov–Taylor identities [103,104]. In the case of renormalizable theory like QCD, the $Z^{(p)}$ must be constants or polynomial in the fermion (boson) mass after the introduction of the renormalized parameters.

Finally, the most relevant term entering in the renormalization group programme is the $Z^{(1)}/\epsilon$, while the other $Z^{(p)}$ for $p \geq 2$ are related to each other via the differential equation of the renormalization group equation.

[1] For reviews see e.g. [110–112] .

In the following, we shall discuss successively the dimensional regularization and the renormalization procedure. We shall also compare this $\overline{MS}$ scheme with some other schemes proposed in the literature and discuss the link between these different schemes.

# 8

# Dimensional regularization

We shall discuss here the procedure how these divergences can be removed in QCD. Our discussion will be based on the previous QSSR1 book and review in [2,3].

## 8.1 On some other types of regularization

### *8.1.1 Pauli–Villars regularization*

In QED, one regulates an UV divergent integral using a Pauli–Villars [107] regularization (PVR), by replacing the propagator as:

$$\frac{1}{q^2 - m^2} \rightarrow \frac{1}{q^2 - m^2} - \frac{1}{q^2 - \Lambda_{\mathrm{UV}}^2}, \tag{8.1}$$

where $\Lambda_{\mathrm{UV}}$ is a UV cut-off. PVR respects translational and Lorentz invariance, and, in QED, the gauge invariance. However, the renormalization programme of QED [106] cannot be extended trivially to QCD. PVR, which is successful in QED, is not often convenient. For instance, using PVR, the proof of unitarity for massless Yang–Mills theory is quite cumbersome. For massive Yang–Mills such as the Electroweak Standard Model [61], PVR does not maintain gauge invariance [113].

### *8.1.2 Analytic regularization*

Like the case of PVR, the *analytic regularization* proposed in the literature [114], does not also maintain gauge invariance. It consists by replacing the propagator as:

$$\frac{1}{q^2 - m^2} \rightarrow \frac{1}{(q^2 - m^2)^{\alpha}}, \tag{8.2}$$

where $\alpha$ is a complex number with Re $\alpha > 1$, which ensures the convergence of the integral. The original propagator is recovered for $\alpha \rightarrow 1$.

### 8.1.3 Lattice regularization

Another type of regularization is *the lattice regularization* [115] dedicated to lattice calculations of hadron parameters but not suitable for analytic gauge theories as it breaks translation and Lorentz invariance. It is based on the fact that the space–time is discretized and made of small cells of size $a$ (lattice spacing). Due to the lattice structure of space–time, the short-distance contribution to the space–time is eliminated and then leads to a convergent integral.

## 8.2 Dimensional regularization

In QCD continuum theory or/and in the Standard Model, one uses instead the method of dimensional regularization and renormalization (so-called *MS* scheme [108–112,123]) which is proven to preserve gauge invariance to all orders of perturbation theory. Its most important feature is the concept of analytical continuation of the dimension of space–time to complex $n(n = 4$ for low-energy space–time). In practice, this means that Dirac algebra, Fierz rearrangments and the momentum integration are done in $n$ dimensions, and then analytically continued to four dimensions.[1] As mentioned in the introduction, in this approach, the IR and UV divergences are transformed into poles in $\epsilon \equiv n - 4$, as we shall see in the following explicit example of the two-point correlator of the pseudoscalar current. However, there are different variants of dimensional regularization, where the difference is due to the definitions of the Dirac matrices used in $n$ dimensions, and in particular, on the one of $\gamma_5$ which is more delicate when one works in $n > 4$ space–time dimensions. Among possible others, there are the so-called *naïve dimensional regularization (NDR)* and *'t Hooft-Veltman (HV)* [108,117] schemes, which we shall briefly sketch below.

### 8.2.1 Naïve dimensional regularization

In this case, only the $n$-dimension metric tensor satisfying the properties is introduced:

$$g_{\mu\nu} = g_{\nu\mu} , \qquad g_{\mu\rho} g^{\rho}_{\nu} = g_{\mu\nu} , \qquad g^{\mu}_{\mu} = n , \tag{8.3}$$

while the $\gamma$ matrices obey the same rules as in four dimensions (see Appendix D.5):

$$Tr\,\mathbf{1} = 4 , \qquad \{\gamma_\mu, \gamma_\nu\} = 2g_{\mu\nu} , \qquad \{\gamma_\mu, \gamma_5\} = 0 . \tag{8.4}$$

where $\gamma_5$ anti-commutes with the other Dirac matrices. NDR is very convenient and widely used in the literature because of its easy implementation in a software program. The definition of $\gamma_5$ in four dimensions given in Eqs. (D.10) and (D.11) has been proven to maintain chiral symmetry to all orders of QCD perturbation series [118]. However, care must be taken when odd parity fermion loops appear in the calculation due to the presence of the parity-violating term $Tr\,(\gamma_5\gamma_\mu\gamma_\nu\gamma_\rho\gamma_\sigma)$ [117,119], as in fact, one does not know how to deal with such a term in $n$ dimensions.

[1] Useful packages for doing these $n$-dimension calculations are given in the Appendices D and F.

### *8.2.2 Dimensional reduction for supersymmetry*

Dimensional reduction [120] is a variant of dimensional regularization, and is convenient for supersymmetric theories,[2] because the conventional dimensional regularization does not preserve supersymmetry. Indeed, in $n$ dimensions, the numbers of bosonic and fermionic degrees of freedom increases, the Fierz rearrangements need more covariants, while one also has to worry about the supersymmetric anomalies and Ward identities. This is obvious in the superfield language since the integral:

$$\int d^n x (D^\alpha D_\alpha)(\bar{D}^{\dot{\alpha}} \bar{D}_{\dot{\alpha}}) \mathcal{L}(x, \theta, \bar{\theta}) , \tag{8.5}$$

is $\theta$ independent only for $n = 4$ [122]. Here, $\theta$ is a four-component anti-commuting variable, $D^\alpha$ is a covariant derivative and $x$ is the space-time variable. The *dimensional reduction* technique is based on analytically continuing the number of co-ordinates and momenta, *but not* the number of components of the fields. In other words, the Dirac algebra should be done in four dimensions but the momentum integration has to be done in $n = 4 - \epsilon$ in order to regulate UV divergences. In particular, the average of the momentum integral should be done in $n$ dimensions:

$$\int \frac{d^n k}{(2\pi)^n} k_\mu k_\nu f(k^2, m^2) = \frac{1}{n} g_{\mu\nu} \int \frac{d^n k}{(2\pi)^n} k^2 f(k^2, m^2) . \tag{8.6}$$

More specifically, the tensor metric $\tilde{g}_{\mu\nu}$ is defined in the same way as in Eq. (8.3), except that:

$$\tilde{g}^\mu_\mu = 4 , \qquad \tilde{g}^\rho_\mu g^\rho_\nu = g_{\mu\nu} , \tag{8.7}$$

where the last equality is needed for preserving gauge invariance for $n < 4$.

### *8.2.3 't Hooft-Veltman regularization*

The HV rule can be satisfied by introducing a new metric $\hat{g}$ in addition to the previous $g$ $n$-dimensional and $\tilde{g}$ four-dimensional metrics. In 4–$\epsilon$ space–time, one has the same properties as in Eq. (8.3), except that:

$$\hat{g}^\mu_\mu = -\epsilon . \tag{8.8}$$

The difference with dimensional reduction is that, instead of the rule in Eq. (8.7), one has:

$$\tilde{g}^\rho_\mu g^\rho_\nu = \tilde{g}_{\mu\nu} , \tag{8.9}$$

which does not lead to inconsistencies, while, one also has:

$$\hat{g}^\rho_\mu g^\rho_\nu = \hat{g}_{\mu\nu} , \qquad \hat{g}_{\mu\rho} \tilde{g}^\rho_\nu = 0 . \tag{8.10}$$

[2] For a review on supersymmetry, see e.g. [121].

The $n$-dimensional Dirac matrices are now split into 4- and $-\epsilon$-dimensional parts:

$$\gamma_\mu = \tilde{\gamma}_\mu + \hat{\gamma}_\mu \,, \tag{8.11}$$

where $\gamma$, $\tilde{\gamma}$ and $\hat{\gamma}$ satisfy the usual commutation rule analogue to the one in Eq. (8.4), but, in addition, one has the novel properties:

$$\{\hat{\gamma}_\mu, \tilde{\gamma}_\nu\} = 0 \,, \qquad \hat{\gamma}_\mu \tilde{\gamma}^\mu = 0 \,, \qquad \tilde{g}^\mu_\nu \hat{\gamma}_\mu = 0 \,, \qquad \hat{g}^\mu_\nu \tilde{\gamma}_\mu = 0 \,. \tag{8.12}$$

The $\gamma_5$ matrix can be be introduced [117] which anti-commutes with $\tilde{\gamma}$ but commutes with $\hat{\gamma}$:

$$\gamma_5^2 = 1 \,, \qquad \{\gamma_5, \tilde{\gamma}_\mu\} = 0 \,, \qquad [\gamma_5, \hat{\gamma}_\mu] = 0 \,. \tag{8.13}$$

As $\gamma_5$ does not have simple commutation rules, it is important to check that chiral Ward identities are respected at each step of the calculation, where anomalous genuine terms have to be cancelled by the counterterms of the Lagrangian [119,117]. For instance, in practice, one has:

$$\frac{1}{2}(1+\gamma_5)\gamma_\mu(1-\gamma_5) = \tilde{\gamma}_\mu(1-\gamma_5) \,. \tag{8.14}$$

Equivalently, one can represent the $\gamma_5$ matrix as:

$$\gamma_5 = \frac{i}{4!}\epsilon_n^{\mu\nu\rho\sigma}\gamma_\mu\gamma_\nu\gamma_\rho\gamma_\sigma \,, \tag{8.15}$$

where for $n \geq 4$:

$$\epsilon_n^{\mu\nu\rho\sigma} = \epsilon^{\mu\nu\rho\sigma} \quad \text{for} \quad \mu\nu\rho\sigma = 0, \ldots, 3 \,, \qquad \epsilon_n^{\mu\nu\rho\sigma} = 0 \quad \text{for} \quad \mu\nu\rho\sigma > 3 \,. \tag{8.16}$$

It is clear that contrary to the NDR scheme, the HV scheme is more cumbersome, in particular, when one tries to implement it in the computer. Neverthless, it is the only dimensional regularization scheme which has been demonstrated to be consistent [119,117].

### 8.2.4 Momentum integrals in $n$ dimensions

Let us consider the typical one-loop integral:

$$I(m,r) = \int \frac{d^n k}{(2\pi)^n}\frac{(k^2)^r}{[k^2 - \mathbf{R}^2]^m} \,. \tag{8.17}$$

It is convenient to rotate (Wick's rotation) the path of integration in the complex $k_0$ plane $[k \equiv (k_0, \vec{k})]$ by $+\pi/2$ without crossing the two poles:

$$k_0 = \pm\sqrt{|\vec{k}|^2 + \mathbf{R}^2} \,. \tag{8.18}$$

Therefore, the $k_0$ integration has the limits $-i\infty$ to $+i\infty$. Going to the Euclidian space, one can define:

$$k_0 \equiv i\bar{k}_0 \,, \quad \vec{k} \equiv \vec{\bar{k}} \ \text{ and } \ \bar{k} \equiv (\bar{k}_0, \vec{\bar{k}}) \,, \tag{8.19}$$

such that the $\tilde{k}_0$ integral goes from $-\infty$ to $+\infty$. It is easy to find:

$$I(m,r) = (-1)^{r-m} i \int \frac{d^n\tilde{k}}{(2\pi)^n} \frac{(\tilde{k}^2)^r}{[\tilde{k}^2 - \mathbf{R}^2]^m} \,. \tag{8.20}$$

Going over polar co-ordinates, one has:

$$\int d^n\tilde{k} = \int_0^\infty \rho^{n-1} d_\rho \int_0^\pi d\theta_{n-1} (\sin\theta_{n-1})^{n-2} \cdots \int_0^\pi d\theta_2 (\sin\theta_2) \int_0^{2\pi} d\theta_1 \,, \tag{8.21}$$

where $\rho$ is the length of the vector $\tilde{k}$. In this way, the integrand of $I(m,r)$ only depends on $\rho$, and one can perform the angular integration using the formula:

$$\int_0^\pi d\theta\, (\sin\theta)^m = \sqrt{\pi} \frac{\Gamma\left((m+1)/2\right)}{\Gamma\left((m+2)/2\right)} \,, \tag{8.22}$$

where $\Gamma$ is the gamma function defined and having the properties in Appendix F. Then, one obtains:

$$I(m,r) = (-1)^{r-m}\, i \frac{2(\pi)^{n/2}}{\Gamma(n/2)} \int_0^\infty d\rho\, \rho^{n-1} \frac{(\rho^2)^r}{[\rho^2 - \mathbf{R}^2]^m} \,, \tag{8.23}$$

which leads to the basic formula:

$$\begin{aligned} I(m,r) &\equiv \int \frac{d^n k}{(2\pi)^n} \frac{(k^2)^r}{[k^2 - R^2]^m} \\ &= \frac{i}{(16\pi^2)^{n/4}} (-1)^{r-m} (\mathbf{R}^2)^{r-m+n/2} \frac{\Gamma(r+n/2)\Gamma(m-r-n/2)}{\Gamma(n/2)\Gamma(m)} \,. \end{aligned} \tag{8.24}$$

Using the symmetry of the integration, it is easy to show that:

$$\int \frac{d^n k}{(2\pi)^n} \frac{k_\mu k_\nu}{[k^2 - \mathbf{R}^2]^m} = \frac{1}{n} g_{\mu\nu} \int \frac{d^n k}{(2\pi)^n} \frac{k^2}{[k^2 - \mathbf{R}^2]^m} \,. \tag{8.25}$$

In the same way:

$$k_\mu k_\nu k_\rho k_\sigma \rightarrow \frac{1}{n(n+2)} (k^2)^2 (g_{\mu\nu} g_{\rho\sigma} + g_{\mu\rho} g_{\nu\sigma} + g_{\mu\sigma} g_{\nu\rho}) \,. \tag{8.26}$$

In the case where $r$ is odd:

$$\int \frac{d^n k}{(2\pi)^n} \frac{k_{\mu_1} \cdots k_{\mu_r}}{[k^2 - \mathbf{R}^2]^m} = 0 \,. \tag{8.27}$$

Finally, it is important to notice that tadpole type integral vanishes identically in dimensional regularization:

$$\int \frac{d^n k}{(2\pi)^n} (k^2)^{\beta-1} = 0 \quad \text{for} \quad \beta = 0, 1, 2, \ldots \tag{8.28}$$

We shall also see that the divergent part of $I(m,r)$ can be tranformed into $\epsilon \equiv 4 - n$ poles thanks to the properties of the $\Gamma$ function.

### 8.2.5 Example of the pseudoscalar two-point correlator

Let us consider the pseudoscalar correlator:

$$\Psi_5(q^2) \equiv i \int d^4x\, e^{iqx} \langle 0|\mathcal{T} J_P(x) (J_P(0))^\dagger |0\rangle , \tag{8.29}$$

where:

$$J_P = (m_i + m_j)\bar{\psi}_i(i\gamma_5)\psi_j , \tag{8.30}$$

is the light quark pseudoscalar current; $m_i$ is the mass of the quark $\psi_i$. In order to simplify the discussion, we shall work to lowest order of perturbative QCD and work with massless quarks in the fermion loop given in the following diagram (Fig. 8.1):[3]

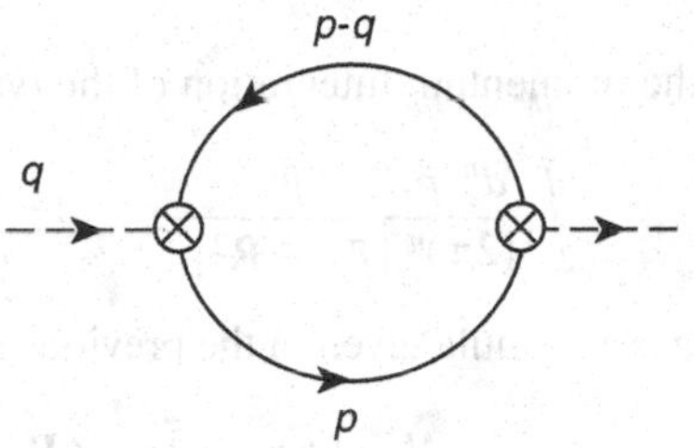


(8.31)

Using Feynman rules given in Appendix E, it reads:

$$i\Psi_5(q^2) = (m_i + m_j)^2(-1)N \int \frac{d^4p}{(2\pi)^4} \mathbf{Tr}\left\{(i\gamma_5)\frac{i}{\hat{p}+i\epsilon'}(i\gamma_5)\frac{i}{\hat{p}-\hat{q}+i\epsilon'}\right\} , \tag{8.32}$$

where one can notice that for large $k^2$, one has a divergent integral:

$$I = \int \frac{d^4k}{k^2} = \infty . \tag{8.33}$$

One can use either PVR, but it is more convenient to use dimensional regularization. In so doing, one works in $n \equiv 4 - \epsilon$ space–time dimensions, such that the previous expression becomes:

$$\nu^\epsilon \Psi_5(q^2) = (m_i + m_j)^2(-i)N \int \frac{d^np}{(2\pi)^n} Tr\left\{(i\gamma_5)\frac{1}{\hat{p}+i\epsilon'}(i\gamma_5)\frac{1}{\hat{p}-\hat{q}+i\epsilon'}\right\} . \tag{8.34}$$

The arbitrary scale $\nu$ has been introduced for dealing with dimensionless quantities in $4 - \epsilon$ dimensions.

One can parametrize the quark propagators à la Feynman (see Appendix E):

$$\frac{1}{ab} = \int_0^1 \frac{dx}{[(a-b)x+b]^2} \equiv \int_0^1 \frac{dx}{[(p-l)^2 - \mathbf{R}^2]^2} , \tag{8.35}$$

[3] The case of massive quarks will be discussed later on in Section 11.14.

where:

$$\begin{aligned} a &= (p-q)^2 + i\epsilon' , \\ b &= p^2 + i\epsilon' , \\ l &= qx , \\ \mathbf{R}^2 &= -q^2 x(1-x) - i\epsilon' . \end{aligned} \tag{8.36}$$

One uses the properties of the Dirac matrices in $n$ dimensions given previously:

$$Tr\ \hat{p}\gamma_5(\hat{p}-\hat{q})\gamma_5 = -4p(p-q) , \tag{8.37}$$

and does the shift:

$$p \to \tilde{p} + l . \tag{8.38}$$

Therefore, one arrives at the momentum integration of the type:

$$\int \frac{d^n\, \tilde{p}}{(2\pi)^n} \frac{\tilde{p}^k}{[\tilde{p}^2 - \mathbf{R}^2]^2} , \tag{8.39}$$

which one can evaluate using the formula given in the previous section. It is easy to obtain:

$$\begin{aligned} \nu^\epsilon \Psi_5^B(q^2) = (m_i + m_j)^2 \frac{N}{4\pi^2} \int_0^1 dx \Gamma\left(\frac{\epsilon}{2}\right) \left(\frac{\mathbf{R}^2 - i\epsilon'}{\nu^2}\right)^{-\epsilon/2} \\ \times \left(3 + \frac{\epsilon}{2}\right) q^2 x(1-x) . \end{aligned} \tag{8.40}$$

where $\gamma_E = 0.5772\ldots$ is the Euler constant. The loop UV divergence appears as a pole at $\epsilon = 0$ of the $\Gamma$-function:

$$\lim_{\epsilon\to 0} \Gamma\left(\frac{\epsilon}{2}\right) \simeq \frac{2}{\epsilon} + \ln 4\pi - \gamma_E + \mathcal{O}(\epsilon) , \tag{8.41}$$

which, as you may have noticed simplify the calculation, which is remarkable when one does a higher order calculation. For large value of $q^2$, one then obtains to leading order:

$$\nu^\epsilon \Psi_5^B\left(q^2\right) = (m_i + m_j)^2 q^2 \frac{N}{8\pi^2} \left\{ \frac{2}{\epsilon} + \ln 4\pi - \gamma_E - \ln\left(\frac{-q^2}{\nu^2}\right) \right\} \tag{8.42}$$

As we have discussed in the introduction, the UV (and IR) divergences originated from the $\Gamma$-function, are transformed into poles in $\epsilon \equiv n - 4$, and are, more generally, of the form:

$$\sum_{p=1} \frac{Z^{(p)}}{\epsilon^p} , \tag{8.43}$$

in the so-called [123] *Minimal Subtraction (MS) scheme*.

Later on, it has been remarked [124] that the combination in Eq. (8.41) appears always in the stage of the calculation. Therefore, the authors in [124] find that it is natural to also subtract the constant terms $\ln 4\pi - \gamma_E$ together with the $\epsilon$-pole:

$$\frac{2}{\epsilon} \to \frac{2}{\hat{\epsilon}} \equiv \frac{2}{\epsilon} + \ln 4\pi - \gamma_E \tag{8.44}$$

This is the modified version of the *MS* scheme, and called: $\overline{MS}$ *scheme*, which will be used in the forthcoming discussions of this book. These divergences will appear as counterterms in the initial Langragian constrained by the Slavnov–Taylor identities [104]. One should notice that for renormalizable theories the $Z^{(p)}$ are local, i.e. constants or polynomials in the inverse of the square of some momentum. These features will be discussed in the forthcoming section.

# 9

# The $\overline{MS}$ renormalization scheme

### 9.1 Renormalizability and power counting rules

The notion of a superficial degree of divergences, based on the power counting rule of a given Feynman diagram, is often used for studying the renormalizability of the interactions in the Lagrangian. For instance, if we consider the previous two-point correlator $\Psi_5(q^2)$, we can see, for $n$-dimensions space–time, that, to lowest order, it behaves for large $p^2$ as:

$$\Psi_5(q^2) \sim \lim_{p\to\infty} p^{n-2} , \tag{9.1}$$

and its degree of divergence is:

$$d = n - 2 . \tag{9.2}$$

More generally, for an arbitrary Green's function $\mathcal{G}$, the superficial degree of divergence reads:

$$d = nl + \sum_v \delta_v - 2n_B - n_F , \tag{9.3}$$

where:

$$\begin{aligned} n &= \text{space–time dimensions}, \\ l &= \text{number of loops (independent integrals)}, \\ \delta_v &= \text{number of momentum factors at the vertex } v, && (9.4) \\ n_B &= \text{number of internal boson lines (we consider a theory with massless bosons)}, && (9.5) \\ n_F &= \text{number of internal fermion lines}. && (9.6) \end{aligned}$$

For a given interaction Lagrangian term $\mathcal{L}_{\rm I}$, which one can write symbolically as:

$$\mathcal{L}_{\rm I} \sim g(\partial)^\delta(\phi)^b(\psi)^f , \tag{9.7}$$

where $\phi$ and $\psi$ are the bosonic and fermion fields, one can define the *index of divergence* of the interaction Lagrangian as:

$$r = \left(\frac{n-2}{2}\right)b + \left(\frac{n-1}{2}\right)f + \delta - n , \tag{9.8}$$

where:

$$\begin{aligned}\delta &= \text{number of space–time derivatives in } \mathcal{L}_{\rm I},\\ b &= \text{number of boson fields in } \mathcal{L}_{\rm I},\\ f &= \text{number of fermion lines in } \mathcal{L}_{\rm I}.\end{aligned} \tag{9.9}$$

Actually, using the fact that the action:

$$\mathcal{S} = \int \mathcal{L}_{\rm I}\, d^n x \tag{9.10}$$

is dimensionless, one can deduce from a dimensional analysis that:

$$n = \dim[g] + \left(\frac{n-2}{2}\right) b + \left(\frac{n-1}{2}\right) f + \delta\,, \tag{9.11}$$

such that:

$$r = -\dim[g]\,. \tag{9.12}$$

One can define respectively by:

$$\begin{aligned}v &= \text{number of vertices corresponding to } \mathcal{L}_{\rm I}^{\rm i} \text{ in the Green's function } \mathcal{G},\\ N_B &= \text{number of external boson lines in } \mathcal{G},\\ N_F &= \text{number of external fermion lines in } \mathcal{G},\end{aligned} \tag{9.13}$$

which obey the relations:

$$\begin{aligned}2n_B + N_B = vb; \qquad 2n_F + N_F = vf\,,\\ l = n_B + n_F - v + 1\,, \quad \sum_v \delta_v = v\delta\,.\end{aligned} \tag{9.14}$$

Eliminating for instance the internal fields through Eq. (9.14), one can rewrite Eq. (9.3) as:

$$d = rv - \left(\frac{n-2}{2}\right) N_B - \left(\frac{n-1}{2}\right) N_F + n\,, \tag{9.15}$$

where $r$ is the index divergence given above. This result can be generalized to any numbers of interaction Lagrangians by the substitution:

$$rv \to \sum_i r_i v_i \tag{9.16}$$

From these definitions, one can classify the different theories as:

- If one of the $r_i$ *is positive*, the divergences cannot be removed by any finite numbers of renormalization constants and interaction parameters. Then the theory is *not renormalizable*.
- If all $r_i \leq 0$, then there is a possibility to remove the divergences by finite numbers of renormalization constants and interaction parameters. The theory is *a candidate for a renormalizable theory*.
- If $r_i < 0$ for all $i$, then the theory is *super renormalizable* since the number of types of divergent diagrams, and the number of diagrams are finite.

- If $r_i = 0$, the theory is *renormalizable in a narrow sense*, which is the case of QCD. As QCD has a dimensionless coupling, then comes the conclusion from Eq. (9.12).

## 9.2 The QCD Lagrangian counterterms

As we have seen before, one can remove the UV divergences of a renormalizable theory by finite numbers of counterterms to any orders of perturbation theory. In QCD, the counterterms of the Langrangian are:

$$\begin{aligned}
\Delta \mathcal{L}_{\text{QCD}} = {} & \Delta_{3YM} \frac{1}{4} (\partial_\mu A_\nu - \partial_\nu A_\mu)(\partial^\mu A^\nu - \partial^\mu A^\nu) \\
& + \Delta_{1YM} \frac{1}{2} (\partial_\mu A_\nu - \partial_\nu A_\mu)\, g \vec{A}^\nu \times \vec{A}^\mu \\
& + \Delta_5 \frac{1}{4} g^2 (\vec{A}_\mu \times \vec{A}_\nu)(\vec{A}^\mu \times \vec{A}^\nu) \\
& - \Delta_{2F}\, i \sum_j \bar{\psi}_j \gamma^\mu \partial_\mu \psi_j + \Delta_4 \sum_j m_j \bar{\psi}_j \psi_j \\
& - \Delta_{1F}\, g \bar{\psi} \frac{\lambda}{2} \gamma^\mu \psi \vec{A}_\mu \\
& + \Delta_6 \frac{1}{2\alpha_G} (\partial_\mu \vec{A}^\mu)^2 + \tilde{\Delta}_3 (\partial_\mu \bar{\varphi})^2 + \tilde{\Delta}_1\, g\, \partial_\mu \bar{\varphi} A^\mu \times \psi\,,
\end{aligned} \tag{9.17}$$

which are all we need for removing the UV divergences of the theory. We have used the notation:

$$\vec{A}_\mu \times \vec{A}_\nu \equiv f_{abc} A^b_\mu A^c_\nu\,. \tag{9.18}$$

It is possible to rescale the fields in such a way that $\mathcal{L}_{\text{QCD}}$ has the form in Eq. (5.12) but in terms of 'bare' quantities. This manipulation is correlated to the introduction of renormalization constants and then to the choices of renormalization schemes.

## 9.3 Dimensional renormalization

In QED, it is natural to use the *on-shell renormalization scheme*:

$$\Psi_5(q^2)_R = \Psi_5(q^2) - \Psi_5(q^2 = 0)\,, \tag{9.19}$$

for defining a renormalized Green's function, as the photon and electron are observed, and then are on their mass-shells (for a electron self-energy diagram, on can, for example, do the subtraction at $p^2 = m_e^2$), which is not the case of QCD, as quarks are off-shell due to confinement. Therefore, there is a freedom to choose the renormalizaton schemes. We shall discuss these different renormalization schemes and their relations in the following sections. t'Hooft [123] has introduced the $MS$ (renormalization) scheme, which is specific for dimensional regularization. In this scheme, one only has to eliminate the $1/\epsilon$ poles [or in the $\overline{MS}$ scheme, the $1/\hat{\epsilon}$ poles defined in Eq. (8.44)] of the Green's functions. The renormalization constants are mass-independent and will appear as counterterms in the initial Lagrangian constrained by the Slavnov–Taylor identities [104].

Table 9.1. *Dimensions of the couplings and fields in n dimensions*

| Name | Notation | Dimension |
|---|---|---|
| gauge coupling | $g$ | $\frac{1}{2}(4-n)$ |
| quark mass | $m_i$ | 1 |
| covariant gauge parameter | $\alpha_G$ | 0 |
| fermion field | $\psi_j(x)$ | $\frac{1}{2}(n-1)$ |
| gluon field | $A^a_\mu(x)$ | $\frac{1}{2}(n-2)$ |
| Faddeev–Popov field | $\varphi^a(x)$ | $\frac{1}{2}(n-2)$ |

## 9.4 Renormalization constants

Taking into account the dimension obtained in the $4-\epsilon$ world (see Table 9.1) via the mass scale $\nu$, one has relations between renormalized and bare parameters:

$$\begin{aligned} g^R &= \nu^{-\epsilon/2} g^B Z_\alpha^{-1/2} \\ g^2/4\pi &\equiv \alpha_s , \\ m_j^R &= m_j^B Z_m^{-1} , \\ \alpha_G^R &= \alpha_G^B Z_G^{-1} , \\ \left(\psi_j^\alpha\right)^R &= \nu^{\epsilon/2} \left(\psi_j^\alpha\right)^B (Z_{2F})^{-1/2} , \\ \left(A_\mu^a\right)_R &= \nu^{\epsilon/2} \left(A_\mu^a\right)_B (Z_{3YM})^{-1/2} , \\ (\varphi^a)_R &= \nu^{\epsilon/2} (\varphi^a)_B (\tilde{Z}_3)^{-1/2} , \end{aligned} \tag{9.20}$$

where $Z_i \equiv 1 - \Delta_i$. One can introduce the renormalization constant for the quark-gluon-quark vertex as:

$$(g\bar{\psi} A \psi)_R = (g_B \bar{\psi}_B A_B \psi_B) \nu^\epsilon Z_{1F}^{-1} , \tag{9.21}$$

which corresponds to the Feynman diagrams (Fig. 9.1).

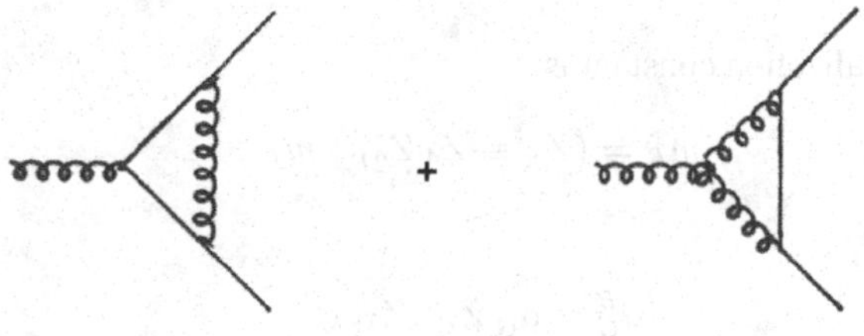

Analogously, one can introduce the three-gluon renormalization constant ($Z_{1YM}$) corresponding to the vertex (Fig. 9.2).

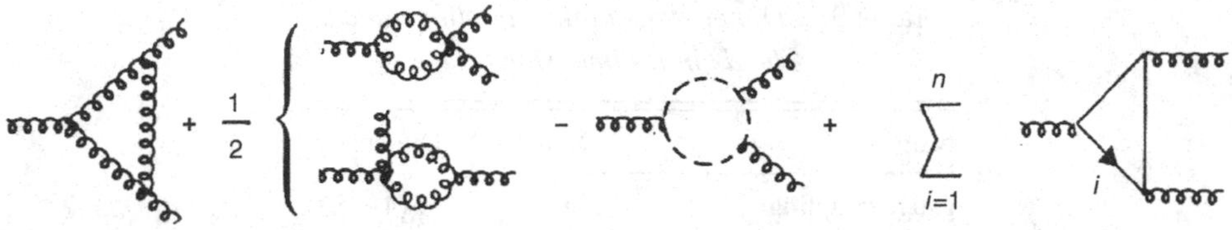


and, ($\tilde{Z}_1$) ghost self-energy and ($\tilde{Z}_3$) ghost-gluon-ghost vertex one (Fig. 9.3).

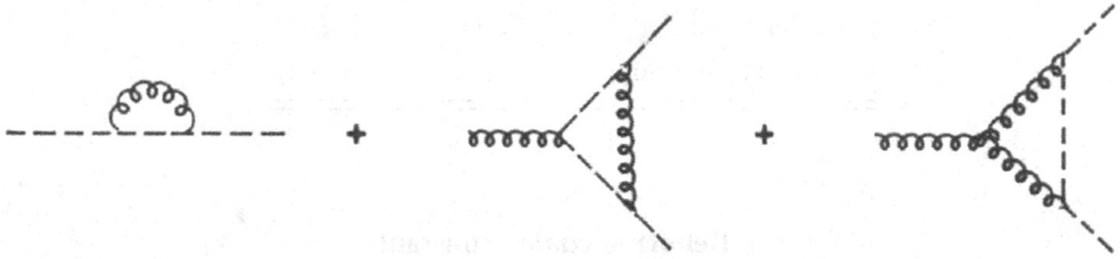


and ($Z_5$) four-gluon vertices one. Then, one can deduce:

$$\begin{aligned} g_B^{YM} &= Z_{1YM}\, Z_{3YM}^{-3/2}\, g_R \,, \\ \tilde{g}_B &= \tilde{Z}_1^{-1}\, Z_{3YM}^{-1/2}\, g_R \,, \\ g_B^F &= Z_{1F}\, Z_{3YM}^{-1/2}\, Z_{2F}^{-1}\, g_R \,, \\ \left(g_B^{(5)}\right)^2 &= Z_5\, Z_{3YM}^{-2}\, g_R^2 \,, \end{aligned} \tag{9.22}$$

which are related to each other by BRS [103] invariance:

$$g_B^{YM} = \ldots\ldots = g_{(5)}^B \,, \tag{9.23}$$

leading to the Slavnov–Taylor [104] identities:

$$\begin{aligned} Z_{3YM}/Z_{1YM} &= \tilde{Z}_3/\tilde{Z}_1 = Z_{2F}/Z_{1F} \,, \\ Z_5 &= Z_{1YM}^2/Z_{3YM} \,. \end{aligned} \tag{9.24}$$

This is the analogue of the QED relation:

$$Z_{1F} = Z_2 \,. \tag{9.25}$$

The mass renormalization constant is:

$$m_B = \left(Z_m \equiv Z_4\, Z_{2F}^{-1}\right)\, m_R \,, \tag{9.26}$$

and the gauge one is:

$$\alpha_G^B = \alpha_G^R\, Z_G^{-1}\, Z_{3YM} \,. \tag{9.27}$$

$Z_{3YM}$ comes from the evaluation of the gluon propagator (Fig. 9.4).

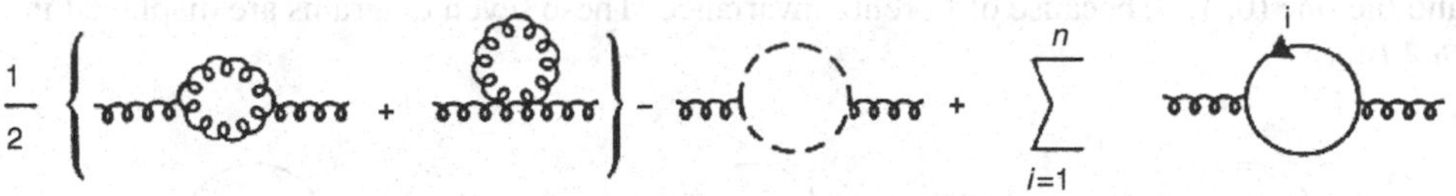

$Z_{2F}$ and $Z_m$ come from the quark self-energy diagram, which can be parametrized as:

$$\Sigma = m_B \Sigma_1 + (\hat{p} - m_B)\Sigma_2 \,, \tag{9.28}$$

and leads to:

$$Z_{2F} \equiv \frac{1}{1 - \Sigma_2|_{\text{pole}}} \,, \qquad Z_m = 1 - \Sigma_1|_{\text{pole}} \,, \tag{9.29}$$

More generally, for a Green's function with $N_G$, $N_{FP}$ and $N_F$ external gluons, ghost and fermion fields, one can associate the renormalization constants:

$$Z_\Gamma = \left(Z_{3YM}^{1/2}\right)^{-N_G} \left(Z_3^{1/2}\right)^{-N_{FP}} \left(Z_{2F}^{1/2}\right)^{-N_F} . \tag{9.30}$$

Expressions of these renormalization constants are known from standard diagram techniques (see Table 11.1).

## 9.5 Check of the renormalizability of QCD

We are now in a position to check the renormalizability of QCD. We want to see if the counterterms presented in Eq. (9.17) are sufficient for removing all divergences in Feynman integrals to all orders.

If one looks at the superficial degree of divergences for the Feynman diagrams given in Eq. (9.15), and using the fact in Eq. (9.12), we can see for QCD in four dimensions:

$$d = 4 - N_B - \frac{3}{2} N_F, \tag{9.31}$$

for $N_B$ and $N_F$ external lines of bosons and fermions. Here, $N_B$ includes gluons $N_G$ and Faddeev–Popov $N_{FP}$ ghosts. Remarking that the coupling in the ghost-gluon-ghost vertex behaves like $k_\mu$ (see Appendix E), the number of boson fields become:

$$N_B = N_G + N_{FP} + \frac{1}{2} N_{FP} \,. \tag{9.32}$$

It is easy to see that the condition $d \geq 0$ for a superficially divergent integral is obtained for seven different cases of the set $(N_F,\ N_G,\ N_{FP})$ discarding the case (0, 0, 0) (vacuum) and the one (0, 1, 0) because of Lorentz invariance. These seven diagrams are displayed in Fig. (9.5):

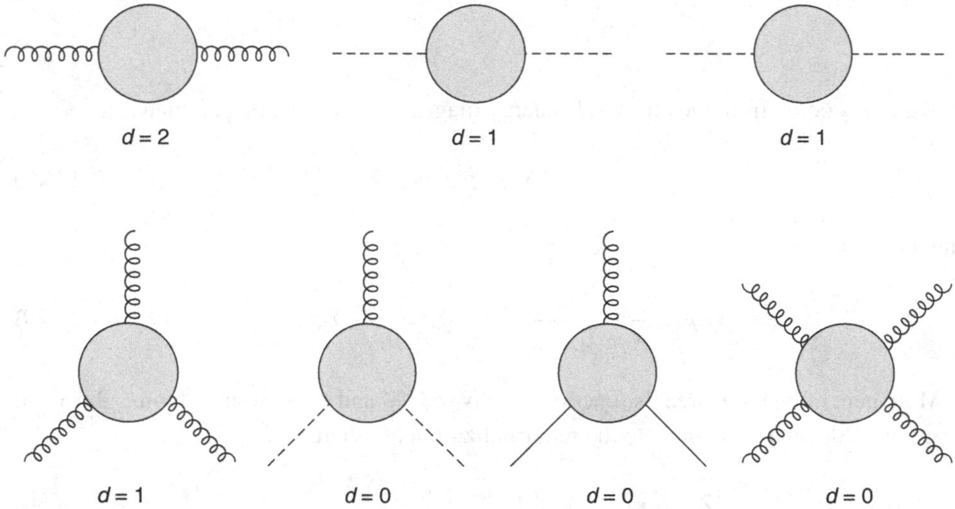


and have the same structure as the counterterms. It is an easy exercise to show that these divergences can all be absorbed by the counterterms. One should also notice that owing to gauge and Lorentz invariances, the apparent degree of divergence 2, 1, 1, 1 of the self-energies of gluons, ghost, fermions, and of the three-gluon vertex become logarithmic. These features have explicitly shown the renormalizability of QCD, which is maintained to all orders of perturbative QCD [113,108,125,104,103].

# 10

# Renormalization of operators using the background field method

In the following chapters, in order to probe the hadron properties, we will have always to deal with *local* hadronic currents or/and operators built from quark or/and gluon fields, but not only with Green's functions. Therefore, it is of prime importance to study the renormalization of such operators. Renormalization of composite operators has been studied [126,127] using background field technology and some further examples have been studied explicitly in perturbation theory [128].

## 10.1 Outline of the background field approach

The basic idea of the method is to write the gauge field appearing in the classical action as $A + Q$, where $A$ is the background field and $Q$ the quantum field which is the variable of integration in the functional integral.[1] The background field gauge is chosen, which maintains the gauge invariance in terms of the $A$ field, but breaks the one of the $Q$ field. This background field gauge invariance is further assured by coupling external sources only to the $Q$ field, which allows one to perform quantum calculations without losing the gauge invariance of the background field. More explicitly, let us consider the generating functional in Yang–Mills theory:[2]

$$Z[J] = \int \mathcal{D}Q\mathcal{D}\phi\mathcal{D}\bar{\phi}\ \exp\left(i\int d^4x[\mathcal{L}_{\rm YM} + \mathcal{L}_{\rm FP} + \mathcal{L}_{\rm source} + \mathcal{L}_{\rm gauge}]\right) . \quad (10.1)$$

where $\mathcal{L}_{\rm YM}$, $\mathcal{L}_{\rm FP}$ are the QCD and Faddeev–Popov Lagrangians defined in Eq. (5.14) without the fermion fields, and $\mathcal{L}_{\rm source}$ is defined in Eq. (6.15). The gauge fixing term is:

$$\mathcal{L}_{\rm gauge} = -\frac{1}{2\alpha_G}(G^a)^2 , \quad (10.2)$$

where $(G^a)$ is, for example, $G^a = \partial_\mu Q^a_\mu$. Doing the shift:

$$Q^\mu_a(x) \to Q^\mu_a(x) + A^\mu_a(x) , \quad (10.3)$$

[1] We shall follow closely the discussion in [127].

[2] Fermions do not play a role in this approach as they can be treated in the usual way.

where $A^{\mu}_a(x)$ is the background field, the functional integral becomes:

$$\tilde{Z}[J, A] = \int \mathcal{D}Q\mathcal{D}\phi\mathcal{D}\bar{\phi} \exp\left(i \int d^4x[\tilde{\mathcal{L}}_{\rm YM} + \tilde{\mathcal{L}}_{\rm source} + \tilde{\mathcal{L}}_{\rm gauge}]\right) , \tag{10.4}$$

where:

$$\begin{aligned}
\tilde{\mathcal{L}}_{\rm YM} &= -\frac{1}{4}\left(\partial^{\mu}(A+Q)^{\nu}_a - \partial^{\nu}(A+Q)^{\mu}_a\right)\left(\partial^{\mu}A^{\nu}_a - \partial^{\nu}A^{\mu}_a\right) , \\
\mathcal{L}_{\rm FP} &= -\partial_{\mu}\bar{\varphi}_a\left(\partial^{\mu}\delta_{ab} - gf_{abc}(A+Q)^{\mu}_c\right)\varphi^a , \\
\tilde{\mathcal{L}}_{\rm source} &= Q_{\mu}J^{\mu} + \bar{\chi}\varphi + \bar{\varphi}\chi ,
\end{aligned} \tag{10.5}$$

where the term $A_{\mu}J^{\mu}$ in the source has been omitted as $A_{\mu}$ is an external field to which one does not need to attach a source. The gauge fixing term (*background gauge*) can be chosen as:

$$\tilde{\mathcal{L}}_{\rm gauge} = -\frac{1}{2\alpha_G}(G^a)^2 , \tag{10.6}$$

where:

$$G^a = \partial^{\mu}Q^a_{\mu} + gf^{abc}A^{\mu}_bQ^{\mu}_c . \tag{10.7}$$

Like in the conventional approach, one can define the connected Green functions:

$$\tilde{W}[J] = -i\ln\tilde{Z}[J] , \tag{10.8}$$

and the effective action:

$$\tilde{\Gamma}[\tilde{Q}] = \tilde{W}[J] - \int d^4x\ J^a_{\mu}\tilde{Q}^a_{\mu} , \tag{10.9}$$

where:

$$\tilde{Q}^a_{\mu} = \delta\tilde{W}/\delta J^a_{\mu} . \tag{10.10}$$

Using the change of variable:

$$Q^a_{\mu} \to Q^a_{\mu} - f^{abc}\theta^b Q^c_{\mu} , \tag{10.11}$$

it is easy to show that $\tilde{Z}[J]$ and hence $\tilde{W}[J]$ are invariant under the infinitesimal transformations:

$$\begin{aligned}
\delta A^a_{\mu} &= gf_{abc}\theta_b A^c_{\mu} - \partial_{\mu}\theta_a , \\
\delta J^a_{\mu} &= gf_{abc}\theta_b J^c_{\mu} .
\end{aligned} \tag{10.12}$$

Then, it follows that $\tilde{\Gamma}[\tilde{Q}, A]$ is invariant under the infinitesimal transformations:

$$\delta A^a_{\mu} = gf_{abc}\theta_b A^c_{\mu} - \partial_{\mu}\theta_a , \tag{10.13}$$

and:

$$\delta\tilde{Q}^a_{\mu} = gf_{abc}\theta_b\tilde{Q}^c_{\mu} , \tag{10.14}$$

in the background field gauge. In particular, $\tilde{\Gamma}[0, A]$ should be an explicitly gauge-invariant functional of $A$, since Eq. (10.13) is an ordinary gauge transformation of the background field. The quantity $\tilde{\Gamma}[\tilde{0}, A]$ is the gauge-invariant effective action which one computes in the background field method. One can show that:

$$\tilde{\Gamma}[\tilde{0}, A] = \Gamma[\bar{Q}]|_{\bar{Q}=A} , \tag{10.15}$$

where the latter is the usual action calculated in an unconventional gauge depending on $A$. Therefore, $\tilde{\Gamma}[\tilde{0}, A]$ can be used to generate the $S$-matrix of a gauge theory in the same way as the usual effective action is used. Feynman rules in the background gauge formalism can be generated from $\tilde{\mathcal{L}}_{\text{gauge}}$ in Eq. (10.6). Since the effective action only involves 1PI diagrams, vertices with only one outgoing quantum line will never contribute. Furthermore, the propagator of the $A$ field is not defined, which does not matter as it is a classical field which never appears in the loop. Compared with ordinary Feynman rules the only difference is the appearance of the $A$ field in external legs, which one denotes by a blob. These Feynman rules are given in Appendix E.

## 10.2 On the UV divergences and $\beta$-function calculation

The UV divergences of $\tilde{\Gamma}[\tilde{0}, A]$ can be absorbed by the renormalizations $Z_A$, $Z_g$, $Z_{\alpha_G}$ of the $A$ field, the coupling constant and the gauge parameter, as it is a sum of a 1PI diagrams with $A$-field external legs and $Q$ fields inside the loops. The renormalization of the gauge parameter can be avoided by working in the Landau gauge $\alpha_G = 0$. Because explicit gauge invariance is retained in the background field method, the renormalization constants $Z_A$ and $Z_G$ are related, and the infinities must take the gauge-invariant form of a divergent constant times the product of field strength $G^a_{\mu\nu} G^{\mu\nu}_a$. Let's now consider the bare field strength:

$$G^{a,B}_{\mu\nu} = Z_A^{1/2}\left[\partial_\mu A^a_\nu - \partial_\nu A^a_\mu + g f_{abc} Z_g Z_A^{1/2} A^b_\mu A^c_\nu\right] , \tag{10.16}$$

where we have used the fact that $A_\mu$ is a classical field for renormalizing $A^b_\mu A^c_\nu$. It will only take the form constant times $G^a_{\mu\nu}$ if:

$$Z_g Z^{1/2} = 1 , \tag{10.17}$$

which is a relation analogous to the one in QED. Equation (10.17) simplifies the computation of the $\beta$-function as illustrated in the explicit calculation of [127]. In the following, we give another application of the method to the renormalization of some composite operators.

## 10.3 Renormalization of composite operators

The first thing to do is to classify these operators into three classes:

**Class I:** gauge-invariant and *do not vanish* after using the equation of motion.
**Class II:** gauge-invariant but *vanish* after using the equation of motion.
**Class III:** gauge-dependent operators.

Therefore any composite *renormalized* operators can be written as:

$$\mathcal{O} = Z_I O_I^B + Z_{II} O_{II}^B + Z_{III} O_{III}^B \,. \tag{10.18}$$

The great advantage of the background field techniques is that for graphs with external quark and background fields, one only needs gauge-invariant counterterms, i.e:

$$Z_{III} = 0 \,, \tag{10.19}$$

which is a consequence of the background field gauge invariance under quantization and renormalization. We shall now study some useful examples.

### 10.3.1 The vector and axial-vector currents

A classics example of composite operator is the local electromagnetic or neutral vector current:

$$V^\mu(x) = \bar\psi \gamma^\mu \psi(x) \,, \tag{10.20}$$

which is conserved to all orders of perturbation theory:

$$\partial_\mu V^\mu(x) = 0 \,, \tag{10.21}$$

and does not require any renormalization. The axial-vector current:

$$A^\mu_{ij}(x) = \bar\psi_i \gamma^\mu \gamma_5 \psi_j(x) \,, \tag{10.22}$$

is partially conserved for $SU(n)_L \times SU(n)_R$:

$$\partial_\mu A^\mu_{ij}(x) = (m_i + m_j)\bar\psi_i \,(i\gamma_5)\, \psi_j(x) \,. \tag{10.23}$$

It can be seen that for the divergence of the axial current, the mass renormalization compensates that of the operator $\bar\psi\gamma_5\psi$, such that at the end it does not get renormalized. We shall see, in the following, that for the $U(1)_A$ current, it needs to be renormalized.

### 10.3.2 Renormalization of $G_{\mu\nu}G^{\mu\nu}$

Let us illustrate the approach by studying the renormalization of the $G^a_{\mu\nu}G^{\mu\nu}_a$ gluon operator in the presence of massive quarks. For that, we have to take all *bare* $(B)$ operators of dimension-four:

$$\begin{aligned} O_1^B &= -\frac{i}{4} GG \,, \\ O_2^B &= -\sum_j \bar\psi_j (\hat D + i m_j)\psi_j \,, \\ O_3^B &= i \sum_j m_j \bar\psi_j \psi_j \,, \end{aligned} \tag{10.24}$$

where $\hat{D}$ is the covariant derivative. The renormalized $O_1^R$ operator is, in general, a combination of these three bare operators:

$$O_1^R = Z_{11} O_1^B + Z_{12} O_2^B + Z_{13} O_3^B \,. \tag{10.25}$$

The renormalization constants $Z_{ij}$ are mass-independent in the $\overline{MS}$ scheme, where one can notice that $Z_{11}$ and $Z_{12}$ can already be obtained in the massless limit. In order to evaluate the $Z_{ij}$, one inserts the zero momentum $O_1$ operator into the gluon and quark propagators:

$$\begin{aligned} \langle A_a^\mu \; O_1 \; A_b^\nu \rangle &= Z_\alpha^{-1} Z_{11} \langle A_a^\mu \; O_1 \; A_b^\nu \rangle + Z_{12} \langle A_a^\mu \; O_2 \; A_b^\nu \rangle \,, \\ \langle \bar\psi \; O_1 \; \psi \rangle &= Z_{11} \langle \bar\psi \; O_1 \; \psi \rangle + Z_{2F} Z_{12} \langle \bar\psi \; O_2 \; \psi \rangle + Z_{2F} Z_{13} \langle \bar\psi \; O_3 \; \psi \rangle \,. \end{aligned} \tag{10.26}$$

In practice, the insertions of $O_1^B$ and $O_2^B$ into the gluon propagator corresponds to the Feynman rules:

$$\begin{aligned} O_1 &\to -i\delta_{ab}(p^2 g_{\mu\nu} - p_\mu p_\nu) \,, \\ O_2 &\to i\hat{p} \,, \end{aligned} \tag{10.27}$$

and one has to calculate respectively:

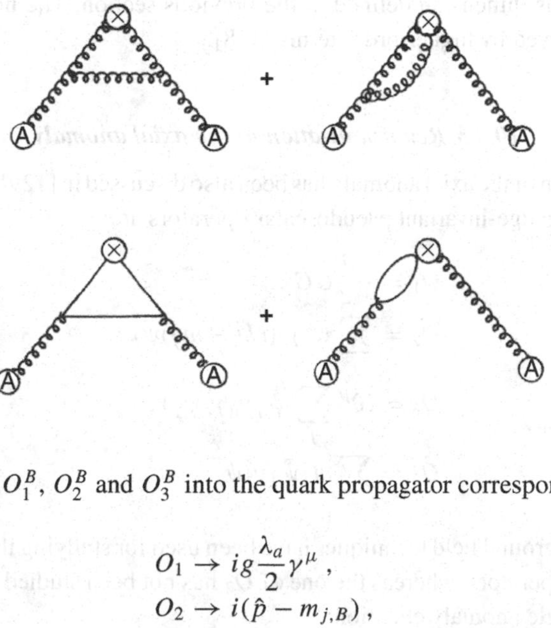

(10.28)

The insertions of $O_1^B$, $O_2^B$ and $O_3^B$ into the quark propagator correspond respectively to:

$$\begin{aligned} O_1 &\to ig\frac{\lambda_a}{2}\gamma^\mu \,, \\ O_2 &\to i(\hat{p} - m_{j,B}) \,, \\ O_3 &\to i m_{j,B} \,, \end{aligned} \tag{10.29}$$

which can be represented by the following diagrams (Fig. 10.1):

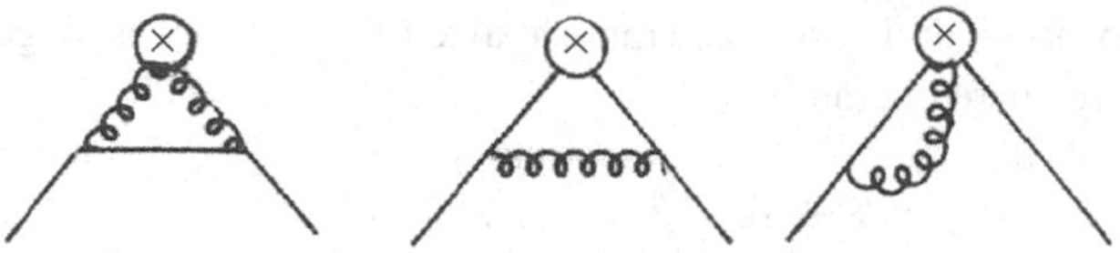

(10.30)

Evaluations of the previous diagrams give in the Landau gauge [128]:

$$Z_{11}^{(2)} = Z_{\alpha}^{(2)}, \qquad Z_{12}^{(2)} = 0, \qquad Z_{13}^{(2)} = -\frac{\gamma_1}{\epsilon}\left(\frac{\alpha_s}{\pi}\right), \tag{10.31}$$

as $Z_{2F}^{(2)} = 1$ in the Landau gauge (the index (2) means second order in $\alpha_s$). Therefore, one can deduce:

$$(GG)_R = \left(1 + \left(\frac{\alpha_s}{\pi}\right)\frac{\beta_1}{\epsilon}\right)(GG)_B + 4\frac{\gamma_1}{\epsilon}\left(\frac{\alpha_s}{\pi}\right)\sum_j m_{j,B}(\bar{\psi}_j\psi_j)_B\,, \tag{10.32}$$

i.e., $GG$ is not multiplicatively renormalizable. However, one can deduce from this expression the finite non-renormalized combination:

$$\theta_\mu^\mu = \frac{1}{4}\beta(\alpha_s)GG + \sum_j \gamma_m(\alpha_s)m_j\bar{\psi}_j\psi_j\,, \tag{10.33}$$

which is the trace of the energy-momentum tensor; $\beta(\alpha_s)$ and $\gamma_m(\alpha_s)$ are the $\beta$ function and the mass-anomalous dimension defined in the previous section. The non-renormalization of $\theta_\mu^\mu$ is also preserved by higher-order terms [128].

### *10.3.3 Renormalization of the axial anomaly*

The renormalization of the axial anomaly has been also discussed in [129]. Here, the different lowest dimension gauge-invariant pseudoscalar operators are:

$$\begin{aligned}
O_1 &= -\frac{i}{4}G\tilde{G}\,,\\
O_2 &= \sum_j \bar{\psi}_j\gamma_5(i\hat{D} - m_j)\psi_j\,,\\
O_3 &= i\partial^\mu\sum_j(\bar{\psi}_j\gamma_\mu\gamma_5\psi_j)\,,\\
O_4 &= \sum_j m_j\bar{\psi}_j\gamma_5\psi_j\,.
\end{aligned} \tag{10.34}$$

Previously background field techniques have been used for studying the renormalizations of these different operators, whereas the one of $O_2$ has not been studied because it does not appear in the triangle anomaly equation:

$$\partial^\mu\left(J_{\mu 5} \equiv \sum_j \bar{\psi}_j\gamma_\mu\gamma_5\psi_j\right) = 2i\sum_j m_j\bar{\psi}_j\gamma_5\psi_j - \left(2Q \equiv \frac{\alpha_s}{4\pi}n_f G\tilde{G}\right), \tag{10.35}$$

where $n_f$ is the number of quark flavours. After renormalizations, the divergence of the flavour singlet current and the gluon topological charge density mix as follows:

$$J^R_{\mu 5} = Z J^B_{\mu 5} ,$$
$$Q^R = Q^B - \frac{1}{2n_f}(1 - Z)\partial^\mu J^B_{\mu 5} , \tag{10.36}$$

where the renormalization constant is in $n$-dimension space–time ($n \equiv 4 - \epsilon$), and reads:

$$Z = 1 + \left(\frac{\alpha_s}{\pi}\right)^2 \frac{3}{4}\frac{4}{3} n_f \frac{1}{\epsilon} . \tag{10.37}$$

### 10.3.4 Renormalizations of higher-dimension operators

The renormalization of dimension-five and -six operators have been studied in [130,131] and reviewed in detail in [3]. In the chiral limit, one can built the RGI mixed quark-gluon $d = 5$ operator for $N$ colours and $n_f$ flavours:

$$\langle \bar{O}_5 \rangle = \alpha_s^{-(\gamma_5/\beta_1)} \left\langle g\bar{\psi}\sigma^{\mu\nu}\frac{\lambda_a}{2}\psi G^a_{\mu\nu}\right\rangle , \tag{10.38}$$

with:

$$\gamma_5 = -\frac{(N^2 - 5)}{4N} , \qquad \beta_1 = -\frac{1}{6}(11N - 2n_f) . \tag{10.39}$$

The triple gluon condensate does not mix under renormalization, and one can form the renormalization group invariant (RGI) operator:

$$\langle \bar{O}_G \rangle = \alpha_s^{-(\gamma_G/\beta_1)} \langle g f_{abc} G^a G^b G^c \rangle \; : \qquad \gamma_G = \frac{2 + 7N}{6} . \tag{10.40}$$

The renormalization of the four-quark operators involves, in general, the mixing of different operators, such that the four-quark condensate:

$$\langle O_2 \rangle = \langle g^2 \bar{\psi}\psi\bar{\psi}\psi \rangle , \tag{10.41}$$

retained in the QSSR analysis within the vacuum saturation cannot be made RGI but possesses an intrinsic $\mu$ dependence. This $\mu$ dependence is only absent in the large $N_c$-limit, where only the diagonal renormalization constant $Z_{2,2}$ (notation in [130]) contributes. Therefore, only in this limit, one can form a RGI condensate:

$$\langle \bar{O}_2 \rangle = \alpha_s^{-(\gamma_2/\beta_1)} \langle O_\psi \rangle \; : \qquad \gamma_2 = \frac{143N}{33} . \tag{10.42}$$

We shall see later on, the importance of these operators in the context of QSSR.

# 11
# The renormalization group

Renormalization invariance states that physical observables must be independent of the renormalization scheme chosen in their theoretical evaluation. The differential approach to renormalization invariance was pioneered by Stueckelberg–Peterman [75] and by Gell-Mann–Low [76], where it has been pointed out that the QED coupling constant is momentum dependent due to the definition of the renormalized charge. Such a consideration led to write a differential equation for the photon propagator. Later on, the study of the scaling behaviour in field theory (experimental observation of the Bjorken scaling [36] in deep inelastic scattering) gave rise to the Callan–Symanzik equation (CSE) [132], which is a very powerful technique for expressing the renormalization invariance constraints on the short-distance behaviour of the Green functions. The CSE takes into account the fact that scaling cannot be strictly implemented because of the necessity of a mass scale in the theory. In the $\epsilon$-regularization, such a mass scale renders the coupling constants dimensionless (see Table 9.1). A generalization of the uses of the CSE to arbitrary Green functions has been proposed [123,171]. The central idea was to treat $g,\ m_i,\ \alpha_G$ as coupling constants of various interaction terms in the Lagrangian.

The meaning of the *renormalization group* can be seen from a simple example. Let us consider a field $\phi$. One can renormalize it in two different renormalization schemes which we call $R_1$ and $R_2$. Then, the renormalized field in terms of the bare one is:

$$\phi_{R_1} = Z(R_1)\phi_B\ , \qquad \phi_{R_2} = Z(R_2)\phi_B\ , \tag{11.1}$$

where: $Z(R_i)$ is the renormalization constant for each scheme $R_i$, and $\phi_B$ is the *bare* field. As the bare field is by definition independent of the scheme, we can then, deduce:

$$\phi_{R_1} = Z(R_1, R_2)\phi_{R_2}\ , \tag{11.2}$$

with:

$$Z(R_1, R_2) \equiv Z(R_1)/Z(R_2)\ , \tag{11.3}$$

which should be finite as do the renormalized fields, despite the fact that the renormalization constants $Z(R_i)$ are divergent. Analogous reasoning can be applied for other parameters of the Lagrangian. The operation which relates quantities of two different renormalization schemes can be interpreted as a transformation from $R_1$ to $R_2$. The set of all these transformations is called the *renormalization group*. One can use the invariance of physical quantities under this group in order to study the asymptotic behaviour of the Green's functions. This can be done as shown below using the renormalization group equation.

## 11.1 The renormalization group equation

The $\epsilon$-regularized Green function reads:

$$\Gamma^R(\nu, p_1, \ldots, p_N; g, \alpha_G, m_i) = Z_\Gamma \Gamma^B(\nu, p_1, \ldots, p_N; g, \alpha_G, m_i) . \tag{11.4}$$

The $\nu$-independence of $\Gamma_B$ implies the zero of the total derivative:

$$\nu \frac{d\Gamma_B}{d\nu} = 0 , \tag{11.5}$$

which is equivalent to:

$$\left\{ \nu \frac{\partial}{\partial \nu} + \nu \frac{d\alpha_s}{d\nu} \frac{\partial}{\partial \alpha_s} + \sum_j \frac{\nu}{m_j} \frac{dm_j}{d\nu} m_j \frac{\partial}{\partial m_j} + \nu \frac{d\alpha_G}{d\nu} \frac{\partial}{\partial \alpha_G} - \frac{1}{Z_\Gamma} \nu \frac{dZ_\Gamma}{d\nu} \right\} \Gamma^R = 0 . \tag{11.6}$$

By introducing the universal $\beta$ function and anomalous dimensions $\gamma_i$:

$$\alpha_s \beta(\alpha_s) = \nu \frac{d\alpha_s}{d\nu}\bigg|_{g_B, m_B \text{ fixed}} ,$$

$$\gamma_m = -\frac{\nu}{m_i^R} \frac{dm_i^R}{d\nu}\bigg|_{g_B, m_B \text{ fixed}} ,$$

$$\gamma_i = \frac{\nu}{Z_i} \frac{dZ_i}{d\nu}\bigg|_{g_B, m_B \text{ fixed}} , \tag{11.7}$$

one can transform Eq. (11.6) into the renormalization group equation (RGE):

$$\left\{ \nu \frac{\partial}{\partial \nu} + \beta(\alpha_s)\alpha_s \frac{\partial}{\partial \alpha_s} - \sum_j \gamma_m(\alpha_s) m_j \frac{\partial}{\partial m_j} + \beta_G \frac{\partial}{\partial \alpha_G} - \gamma_\Gamma \right\} \Gamma^R = 0 . \tag{11.8}$$

For $N_G$, $N_{NP}$ and $N_F$ external gluon, ghost and fermion lines:

$$\gamma_\Gamma = -\frac{1}{2} [N_G \gamma_{3YM} + N_F \gamma_{2F} + N_{FP} \tilde{\gamma}_3] . \tag{11.9}$$

The expressions of the previous universal parameters can be easily deduced from their definitions as we shall show below.

## 11.2 The $\beta$ function and the mass anomalous dimension

Noticing that, in the $\overline{MS}$ scheme, $\beta(\alpha_s)$ is mass-independent, one can, therefore, write [110,111]:

$$\alpha_s \beta(\alpha_s, \epsilon) = \nu \frac{d\alpha_s^R}{d\nu} = \nu \frac{d}{d\nu} \left( \alpha_s^B \nu^{-\epsilon} Z_\alpha^{-1} \right) = -\epsilon \alpha_s^R - \alpha_s^R \frac{1}{Z_\alpha} \nu \frac{dZ_\alpha}{d\nu} . \tag{11.10}$$

The fact that $Z_\alpha$ is $\nu$-independent allows us to also write:

$$\left\{\alpha_s^R \beta(\alpha_s,\epsilon) + \epsilon\alpha_s^R + \left(\alpha_s^R\right)^2 \beta(\alpha_s,\epsilon)\frac{\partial}{\partial\alpha_s^R}\right\} Z_\alpha = 0 \, . \tag{11.11}$$

Using the expression of the $Z_\alpha$ in terms of the $1/\epsilon$ poles into the previous differential equation, one gets from the finite terms:

$$\alpha_s^R \beta(\alpha_s,\epsilon) = -\epsilon\alpha_s^R + \left(\text{finite term} \equiv \alpha_s^R \beta(\alpha_s)\right) \, . \tag{11.12}$$

Using this relation into the $1/\epsilon$ term, one can deduce:

$$\beta(\alpha_s) = \alpha_s^R \frac{\partial Z_\alpha}{\partial \alpha_s^R} \, , \tag{11.13}$$

i.e., $\beta(\alpha_s)$ is nothing else than the coefficient of the $1/\epsilon$-term of $Z_\alpha$. The different coefficients of $\beta$ are given in Table 11.1, showing that $\beta$ is negative for $n \leq 11$ where $n$ is the number of flavours. We shall see in the discussion of the running coupling that this negativity is important for an asymptotically free theory. We apply the same reasoning for obtaining the quark mass anomalous dimension defined as:

$$\gamma_m(\alpha_s) = -\frac{\nu}{m^R}\frac{dm^R}{d\nu}\bigg|_{g^B, m^B \text{ fixed}} \equiv \frac{\nu}{Z_m}\frac{dZ_m}{d\nu} \, . \tag{11.14}$$

where $B$ and $R$ refer to renormalized and bare quantities. Using the fact that in the $\overline{MS}$ scheme, $Z_m$ is only function of $\nu$ and $\alpha_s$, one gets:

$$\nu\frac{dZ_m}{d\nu} \equiv \left\{\nu\frac{\partial}{\partial\nu} + \beta(\alpha_s,\epsilon)\alpha_s\frac{\partial}{\partial\alpha_s}\right\} Z_m \, . \tag{11.15}$$

To lowest order of $\alpha_s$, noting that the only dependence on $Z_m$ is from $\alpha_s$, and using the previous expression of the $\beta$ function in Eq. (11.10), the previous differential equation can be written as:

$$\nu\frac{dZ_m}{d\nu} = \left\{-\epsilon\alpha_s\frac{\partial}{\partial\alpha_s} + \alpha_s\beta(\alpha_s)\frac{\partial}{\partial\alpha_s}\right\} Z_m \, . \tag{11.16}$$

Using the expression of $Z_m$, which is generically given by:

$$Z_m = 1 + \sum_n \frac{1}{\hat{\epsilon}^n} Z_m^{(n)} \, , \tag{11.17}$$

one can obtain that the mass anomalous dimension is given by the opposite of the $1/\epsilon$ pole coefficient in our sign convention ($d = 4 - \epsilon$). Analogous reasoning applies to the other anomalous dimensions, i.e., they are the opposite of the $1/\hat{\epsilon}$-coefficient. Their expressions are given in Table 11.1. The coefficients of the quark mass anomalous dimension and $\beta$ functions have been calculated in the $\overline{MS}$ scheme by: [133] ($\gamma_2$), [134] ($\beta_2$), [135] ($\gamma_3$ and $\beta_3$) and [136] ($\gamma_4$ and $\beta_4$).

## 11.3 Gauge invariance of $\beta(\alpha_s)$ and $\gamma_m$ in the $\overline{MS}$ scheme

One can also prove the gauge invariance of $\beta$ and $\gamma_m$. This property leads to a great simplicity in their evaluation, as one can perform the calculation in a given gauge like the Feynman gauge $\alpha_G = 1$. For completing the proof, we start from a dimensionless Green's function $\Gamma$ associated to a gauge-invariant amplitude. Using the fact that the bare Green's function is independent of the renormalization scale $\nu$ and of the gauge $\alpha_G$, one has the RGE:

$$\left\{\nu\frac{\partial}{\partial\nu} + \beta(\alpha_s)\alpha_s\frac{\partial}{\partial\alpha_s} - \gamma_m(\alpha_s)m\frac{\partial}{\partial m} + \beta_G\frac{\partial}{\partial\alpha_G}\right\}\Gamma^R = 0\,. \tag{11.18}$$

The fact that it is gauge invariant gives:

$$\left(\frac{\partial}{\partial\alpha_G} + \alpha_s\rho\frac{\partial}{\partial\alpha_s} + \sigma m\frac{\partial}{\partial m}\right)\Gamma^R = 0\,, \tag{11.19}$$

with:

$$\alpha_s\rho \equiv \left.\frac{d\alpha_s}{d\alpha_G}\right|_{g^B,\ \epsilon\ \text{fixed}} \quad \text{and} \quad \sigma \equiv \left.\frac{1}{m}\frac{dm}{d\alpha_G}\right|_{g^B,\ \epsilon\ \text{fixed}}\,. \tag{11.20}$$

We apply the commutators of the operators in Eqs. (11.18) and (11.19) into $\Gamma^R$:

$$[\{\ldots\},(\ldots)]\Gamma^R = 0\,. \tag{11.21}$$

Eliminating $\partial\Gamma^R/\partial\alpha_G$ with the help of Eq. (11.19), one obtains a third independent RGE:

$$\left\{\left[D\bar\beta - \bar\beta\frac{\partial(\alpha_s\rho)}{\partial\alpha_s}\right]\alpha_s\frac{\partial}{\partial\alpha_s} + \left[D\bar\gamma_m - \bar\beta\alpha_s\frac{\partial\sigma}{\partial\alpha_s}\right]m\frac{\partial}{\partial m}\right\}\Gamma^R(\alpha_s,\ \alpha_G,\ m) = 0\,, \tag{11.22}$$

where:

$$D \equiv \frac{\partial}{\partial\alpha_G} + \alpha_s\rho\frac{\partial}{\partial\alpha_s}\,, \qquad \bar\beta \equiv \beta - \rho\beta_G\,, \qquad \bar\gamma_m \equiv \gamma_m - \sigma\beta_G\,. \tag{11.23}$$

However, $\Gamma^R$ depends only on the two conditions in Eqs. (11.18) and (11.19). Therefore the third equation should be trivially satisfied:

$$\begin{aligned} D\bar\beta - \bar\beta\frac{\partial(\alpha_s\rho)}{\partial\alpha_s} &= 0 \\ D\bar\gamma_m - \bar\beta\alpha_s\frac{\partial\sigma}{\partial\alpha_s} &= 0\,. \end{aligned} \tag{11.24}$$

Therefore, the RGE becomes:

$$\left\{\nu\frac{\partial}{\partial\nu} + \bar\beta(\alpha_s)\alpha_s\frac{\partial}{\partial\alpha_s} - \bar\gamma_m(\alpha_s)m\frac{\partial}{\partial m}\right\}\Gamma^R = 0\,, \tag{11.25}$$

which shows that the physical consequences of the RGE are gauge invariant. Recalling that in the $\overline{MS}$ scheme:

$$g_B = \nu^{\epsilon/2}g_R\left(1 + \sum_n\frac{a_n}{\hat\epsilon^n}\right) \equiv \nu^{\epsilon/2}g_R Z_\alpha^{1/2}\,, \tag{11.26}$$

and using the previous definition of $\rho$, one gets:

$$\rho = -\frac{1}{Z_\alpha}\frac{dZ_\alpha}{d\alpha_G}\bigg|_{g^B,\ \epsilon \text{ fixed}} = -\frac{1}{Z_\alpha}\left\{\frac{\partial a_1}{\partial \alpha_G}\frac{1}{\epsilon} + \frac{\partial a_2}{\partial \alpha_G}\frac{1}{\epsilon^2} + \cdots\right\}. \tag{11.27}$$

Then:

$$\rho\left(1 + \frac{a_1}{\epsilon}\right) = -\frac{\partial a_1}{\partial \alpha_G}\frac{1}{\epsilon} + \mathcal{O}\left(\frac{1}{\epsilon^2}\right), \tag{11.28}$$

which is only satisfied *if and only if* $\rho = 0$ because $\rho$ is independent of $\epsilon$ (see its definition and its relation with $\bar\beta$ and $\beta$). One should notice that it is also due to the fact that in the $\overline{MS}$ scheme, $Z_\alpha$ has no constant term other than 1 (the $\ln 4\pi - \gamma$ term being already absorbed into $1/\hat\epsilon$). Inserting $\rho = 0$ into Eq. (11.24), one gets the desired result:

$$\frac{\partial \beta}{\partial \alpha_G} = 0, \tag{11.29}$$

showing that $\beta$ is gauge independent. With similar proofs, one also obtains $\sigma = 0$, leading to the gauge independence of $\gamma_m$.

## 11.4 Solutions of the RGE

One can now solve the RGE. If $D$ is the dimension of $\Gamma$ in units of mass and if one scales the momenta $p_1, \ldots, p_N$ by a dimensionless factor $\lambda$, the Euler theorem on homogeneous function gives:

$$\left\{\lambda\frac{\partial}{\partial\lambda} + \sum_j m_j\frac{\partial}{\partial m_j} + \nu\frac{\partial}{\partial\nu} - D\right\}\Gamma^R(\lambda p_1, \ldots, \lambda p_N; \alpha_s, \alpha_G, m_j, \nu) = 0. \tag{11.30}$$

Introducing for convenience the dimensionless variables:

$$t \equiv \ln\lambda \qquad x_j \equiv m_j/\nu, \tag{11.31}$$

one arrives at the desired form of the RGE:

$$\left\{-\frac{\partial}{\partial t} + \beta(\alpha_s)\alpha_s\frac{\partial}{\partial\alpha_s} - \sum_j(1+\gamma_m(\alpha_s))x_j\frac{\partial}{\partial x_j} + \beta_G\frac{\partial}{\partial\alpha_G} + D - \gamma_\Gamma\right\}$$
$$\times\,\Gamma^R(e^t p_1, \ldots, e^t p_N;\ \alpha_s,\ \alpha_G,\ x_j,\ \nu) = 0, \tag{11.32}$$

with the solution:

$$\Gamma^R(e^t p_1, \ldots, e^t p_N;\ \alpha_s,\ \alpha_G,\ x_j,\ \nu)$$
$$= \lambda^D\Gamma^R(p_1, \ldots, p_N;\ \bar\alpha_s, \overline{\alpha_G}, \bar x_j,\ t=0)\exp\left\{-\int_0^t dt'\gamma_\Gamma[\bar\alpha_s(t', \alpha_s)]\right\}. \tag{11.33}$$

Table 11.1. *Anomalous dimension* $\gamma_i = \frac{\nu}{Z_i}\frac{dZ_i}{d\nu} \equiv$ *coefficient of* $-1/\hat{\epsilon}$ *and coefficients of the* $\beta$ *function in the* $\overline{MS}$ *scheme for* $SU(N)_c \times SU(n)_F$

| | |
|---|---|
| Fermion field | $\gamma_{2F} = \left(\frac{\alpha_a}{\pi}\right)\frac{N^2-1}{2N}\frac{\alpha_G}{2} + \mathcal{O}\left(\frac{\alpha_s}{\pi}\right)^2$ |
| Gluon field | $\gamma_{3YM} = -\left(\frac{\alpha_s}{\pi}\right)\left\{\frac{N}{4}\left(\frac{13}{3} - \alpha_G\right) - \frac{2}{3}\left(\frac{1}{2}\right)n\right\}$ |
| Ghost field | $\tilde{\gamma}_3 = -\left(\frac{\alpha_s}{\pi}\right)\frac{N}{8}(3 - \alpha_G)$ |
| Mass | $\gamma_m = [\gamma_1 \equiv 2]\left(\frac{\alpha_s}{\pi}\right) + \left[\gamma_2 \equiv \frac{1}{6}\left(\frac{101}{2} - \frac{5n}{3}\right)\right]\left(\frac{\alpha_s}{\pi}\right)^2$ $+\left[\gamma_3 \equiv \frac{1}{96}\left[3747 - \left(160\zeta_3 - \frac{2216}{9}\right)n - \frac{140}{27}n^2\right]\right]\left(\frac{\alpha_s}{\pi}\right)^3$ $+\left[\gamma_4 \equiv \frac{1}{128}\left[\frac{4603055}{162} + \frac{135680}{27}\zeta_5 - 8800\zeta_5\right.\right.$ $+\left(-\frac{91723}{27} - \frac{34192}{9}\zeta_3 + 880\zeta_4 + \frac{18400}{9}\zeta_5\right)n$ $\left.\left.+\left(\frac{5242}{243} + \frac{800}{9}\zeta_3 - \frac{160}{3}\zeta_4\right)n^2 + \left(-\frac{332}{243} + \frac{64}{27}\zeta_3\right)n^3\right]\right]\left(\frac{\alpha_s}{\pi}\right)^4$ for $N = 3$; $\zeta_3 = 1.2020569\ldots, \zeta_4 = 1.0823232\ldots, \zeta_5 = 1.0369277\ldots$ |
| Coupling constant | $\beta(\alpha_s) \equiv \frac{\nu}{\alpha_s}\frac{d\alpha_s}{d\nu} = -\frac{\nu}{Z_\alpha}\frac{dZ_\alpha}{d\nu}$ $= \left[\beta_1 = -\frac{1}{2}\left(11 - \frac{2}{3}n\right)\right]\left(\frac{\alpha_s}{\pi}\right) + \left[\beta_2 = -\frac{1}{4}\left(51 - \frac{19}{3}n\right)\right]\left(\frac{\alpha_s}{\pi}\right)^2$ $+\left[\beta_3 = -\frac{1}{64}\left(2857 - \frac{5033}{9}n + \frac{325}{27}n^2\right)\right]\left(\frac{\alpha_s}{\pi}\right)^3$ $+\left[\beta_4 = -\frac{1}{128}\left[\left(\frac{149753}{6} + 3564\zeta_3\right) - \left(\frac{1078361}{162} + \frac{6508}{27}\zeta_3\right)n\right.\right.$ $\left.\left.+\left(\frac{50065}{162} + \frac{6472}{81}\zeta_3\right)n^2 + \frac{1093}{729}n^3\right]\right]\left(\frac{\alpha_s}{\pi}\right)^4$ for $N = 3$ |
| Gauge | $\beta_G = \nu\frac{d\alpha_G}{d\nu} = -\alpha_G\gamma_{3YM}$ |
| Three-gluon | $\gamma_{1YM} = -\left[\left(\frac{17}{6} - \frac{3}{2}\alpha_G\right)\frac{N}{4} - \frac{2}{3}\frac{1}{2}n\right]\left(\frac{\alpha_s}{\pi}\right)$ |
| Ghost-gluon-ghost | $\tilde{\gamma}_1 = \alpha_G\frac{N}{4}\left(\frac{\alpha_s}{\pi}\right)$ |
| Fermion-gluon-fermion | $\gamma_{1F} = \frac{1}{2}\left[(3 + \alpha_G)\frac{N}{4} - \alpha_G\frac{N^2-1}{2N}\right]$ |

where $\bar{\alpha}_s$, $\bar{\alpha}_G$ and $\bar{x}_j$ are respectively the running QCD coupling, gauge and mass, solutions of the differential equations:

$$\frac{d\bar{\alpha}_s}{dt} = \bar{\alpha}_s\beta(\bar{\alpha}_s) \;:\; \bar{\alpha}_s(0, \alpha_s) = \alpha_s^R(\nu)\,,$$
$$\frac{d\bar{\alpha}_G}{dt} = \beta_G(\bar{\alpha}_s) \;:\; \bar{\alpha}_G(0, \alpha_s) = \alpha_G(\nu)\,, \tag{11.34}$$

and:

$$\frac{d\bar{x}_i}{dt} = -[1 + \gamma_m(\bar{\alpha}_s)]\bar{x}_i(t) \;:\; \bar{x}_i(0, \alpha_s) = x_i^R(\nu)\,. \tag{11.35}$$

Their explicit expressions will be given later on. One should notice that the Green function has acquired an extra dimension induced by the exponential factor, which explains the name *anomalous dimension*.

## 11.5 Weinberg's theorem

In connection with the power counting theorem, one can derive a theorem on the asymptotic behaviour of the Green's function at large external momenta. This theorem is known as *Weinberg's theorem* [137].

It states that if non-exceptional momenta[1] are parametrized as:

$$p_{i_l} = \lambda k_{i_l} \quad : \quad l = 1, m \,, \tag{11.36}$$

the renormalized Feynman amplitude of a Feynman diagram $G$ behaves as:

$$\Gamma^R(p_1, \ldots, p_n) \sim \lambda^\alpha \ln \lambda^\beta \,, \tag{11.37}$$

when $\lambda \to \infty$ and $k_i$ kept fixed. Here $\beta$ is undetermined, while:

$$\alpha = \max d(H) \tag{11.38}$$

where $d(H)$ is the superficial degree of divergence of the subdiagram $H$ consisting of continuous path of lines connected to the external lines with momenta $p_{i_1}, \ldots, p_{i_m}$. For a renormalizable theory like QCD, the constant $d(H)$ can be obtained from Eq. (9.15) by taking $r = 0$. In other word, the Weinberg theorem tells us that the asymptotic limit in the deep Euclidean region $\lambda \to \infty$ is given by the naïve power counting times a logarithmic factor.

## 11.6 The RGE for the two-point function in the $\overline{MS}$ scheme

In order to illustrate this discussion, let us consider the generic two-point correlator:

$$\Pi(q^2) \equiv i \int d^4x \, e^{iqx} \, \langle 0|\mathcal{T} J(x)_H \, (J_H(0))^\dagger \, |0\rangle \,, \tag{11.39}$$

where $J_H(x)$ is the hadronic current of quark and/or gluon fields. In $n = 4 - \epsilon$ dimension, $\Pi(q^2)$ acquires an extra $\nu^{-\epsilon}$ dimension. The renormalized two-point correlator is [28,110,111]:

$$\Pi_R(q^2, \, \alpha_s, \, m_i, \, \nu) \equiv \Pi_B\left(q^2, \, \alpha_s^B, \, m_i^B, \, \epsilon\right) - \nu^{-\epsilon} C\left(q^2, \, \alpha_s^B, \, m_i^B, \, \epsilon\right) \,, \tag{11.40}$$

[1] A momentum configuration $(p_1, \ldots, p_n)$ of momenta are non-exceptional if *no* non-trivial partial sum $p_{i_1} + p_{i_2} + \cdots p_{i_m}$ where, ($i_j$ take any of the label $1, \ldots n$) vanishes. On the contrary, an example of vanishing trivial sum is $p_1 + p_2 + \cdots + p_n = 0$, which is due to the energy-momentum conservation.

where in the $\overline{MS}$ scheme, $C$ is the $\epsilon$-pole terms:

$$C\left(q^2,\ \alpha_s^B,\ m_i^B,\ \epsilon\right) = \sum_k \frac{1}{\epsilon^k} C_k(q^2,\ \alpha_s,\ m_j)\,, \tag{11.41}$$

where, as usual, $C_k$ are constants or polynomials in $m_j^2/q^2$. Using the fact that $\Pi_B$ is independent of $\nu$, implies the differential equation:

$$\left\{\nu\frac{\partial}{\partial\nu} + \beta(\alpha_s)\alpha_s\frac{\partial}{\partial\alpha_s} - \sum_j \gamma_m(\alpha_s)m_j\frac{\partial}{\partial m_j}\right\}\Pi^R(q^2, \alpha_s,\ m_i,\ \nu)$$
$$= -\nu\frac{d}{d\nu}\left(\nu^{-\epsilon}\sum_k \frac{1}{\epsilon^k}C_k\right). \tag{11.42}$$

Rewriting:

$$\nu\frac{d}{d\nu}\left(\nu^{-\epsilon}\sum_k \frac{1}{\epsilon^k}C_k\right) = \left\{\nu\frac{\partial}{\partial\nu} + \nu\frac{d\alpha_s}{d\nu}\frac{\partial}{\partial\alpha_s} - \sum_j \gamma_m(\alpha_s)m_j\frac{\partial}{\partial m_j}\right\}\nu^{-\epsilon}\sum_k \frac{1}{\epsilon^k}C_k\,, \tag{11.43}$$

using:

$$\nu\frac{d\alpha_s}{d\nu} = -\epsilon\alpha_s + \alpha_s\beta(\alpha_s)\,, \tag{11.44}$$

and the fact that the equation is finite for $\epsilon \to 0$, one gets:

$$\lim_{\epsilon\to 0}:\ \nu\frac{d}{d\nu}\left(\nu^{-\epsilon}\sum_k \frac{1}{\epsilon^k}C_k\right) = -\frac{\partial}{\partial\alpha_s}(\alpha_s C)\,, \tag{11.45}$$

and the set of recursive equations for $k \geq 1$:

$$\left\{\alpha_s\beta(\alpha_s) - \sum_i \gamma_m m_i\frac{\partial}{\partial m_i}\right\}C_k = \frac{\partial}{\partial\alpha_s}(\alpha_s C_{k+1})\,. \tag{11.46}$$

The dimensionless condition of $\Pi$ reads:

$$\left\{\nu\frac{\partial}{\partial\nu} + \lambda\frac{\partial}{\partial\lambda} + \sum_j m_j\frac{\partial}{\partial m_j}\right\}\Pi(\lambda^2,\ \nu^2,\ \alpha_s, m_i,\ \nu) = 0\,, \tag{11.47}$$

where $t \equiv \ln\lambda$. Therefore, one arrives at the RGE for the two-point function:

$$\left\{-\frac{\partial}{\partial t} + \beta(\alpha_s)\alpha_s\frac{\partial}{\partial\alpha_s} - \sum_j (1 + \gamma_m(\alpha_s))x_j\frac{\partial}{\partial x_j}\right\}\Pi(t, \alpha_s, x_i) = \frac{\partial}{\partial\alpha_s}(\alpha_s C) \equiv D\,, \tag{11.48}$$

with the solution:

$$\Pi(t, \alpha_s, x_i) = \Pi\,(t = 0, \bar{\alpha}_s(t), \bar{x}_i(t)) - \int_0^t dt' D[t - t',\ \bar{\alpha}_s(t'),\ \bar{x}_i(t')]\ , \quad (11.49)$$

where $\bar{\alpha}_s$ and $\bar{x}_i$ are running parameters solutions of the differential equations given in Eq. (11.34), and which will be given explicitly in the following.

## 11.7 Running coupling

### *11.7.1 Lowest order expression and the definition of the QCD scale $\Lambda$*

Solving the differential equation in Eq. (11.34), the expression of the running coupling, to one-loop accuracy is:

$$a_s^{(0)}(t, \alpha_s) = \frac{a_s(\nu)}{1 - \beta_1 a_s(\nu) t}\ , \quad (11.50)$$

where:

$$a_s \equiv \frac{\alpha_s}{\pi}\ , \qquad t \equiv \frac{1}{2} \ln \frac{-q^2}{\nu^2}\ , \quad (11.51)$$

and $\beta_1$ is the first coefficient of the $\beta$ function given in Table 11.1. It shows that for $t \to +\infty$, $a_s^{(0)} \to 0$ for $\beta_1 < 0$, which is satisfied for the number of quark flavours $n_f \leq 11$. In this case, the theory is *asymptotically free* and the use of perturbation theory is legitimate. The point $\alpha_s = 0$ is an UV fixed point as shown in Fig. 11.1 because the $\beta$-function has a negative slope at the origin.

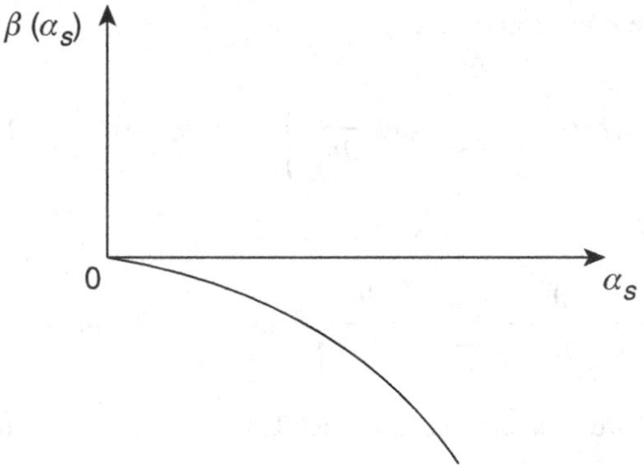


We can also re-write the solution as:

$$t = \int \frac{dz}{z} \frac{1}{\beta(z)} \equiv \varphi(z) + \text{constant} \quad (11.52)$$

where the constant term is a renormalization group invariant (RGI) quantity, which one identifies as:

$$t-\varphi(z)\equiv\frac{1}{2}\ln\nu^2+\frac{1}{\beta_1 a_s(\nu)}=\text{constant}\equiv\frac{1}{2}\ln\Lambda^2\,, \tag{11.53}$$

where $\Lambda$ is a RGI *but* renormalization scheme-dependent quantity. Therefore, the *running coupling*, in terms of $\Lambda$ to one-loop accuracy, reads:

$$a_s^{(0)}(q^2)=\frac{1}{-\beta_1}\frac{1}{\frac{1}{2}\ln\frac{-q^2}{\Lambda^2}}\,. \tag{11.54}$$

### *11.7.2 Renormalization group invariance of the first two coefficients of $\beta$*

Before discussing the high-order expression of the coupling, let us discuss the renormalization group invariance of the first two coefficients of the $\beta$ function. Let $\beta^a$ and $\beta^b$ the $\beta$ functions related to two different values of the subtraction $\nu_a$ and $\nu_b$ of the $\overline{MS}$ scheme. Using Eq. (11.52), we have:

$$t_b\equiv\frac{1}{2}\ln\frac{-q^2}{\nu_b^2}=\int_{\alpha_s(\nu_a)}^{\bar\alpha_s(t_b,\alpha_s(\nu_b^2))}\frac{dz}{z}\frac{1}{\beta^a(z)}\equiv\varphi(z)\,. \tag{11.55}$$

Applying the operator $\nu_b\partial/\partial\nu_b$ to both sides of Eq. (11.55), and using the fact that $\bar\alpha_s(t_b,\alpha_s(\nu_b^2))$ obeys the differential equation:

$$\left\{\nu_b\frac{\partial}{\partial\nu_b}+\beta^b\alpha_s(\nu_b)\frac{\partial}{\partial\alpha_s(\nu_b)}\right\}\bar\alpha_s(t_b,\,\alpha_s(\nu_b))=0\,, \tag{11.56}$$

one obtains:

$$-1=-\left(\frac{1}{\alpha_s(\nu_a)\beta^a}\right)\beta^b\alpha_s(\nu_b)\frac{\partial\alpha_s(\nu_a)}{\partial\alpha_s(\nu_b)}\Longrightarrow\beta^a=\beta^b\left(\frac{\alpha_s(\nu_b)}{\alpha_s(\nu_a)}\right)\left(\frac{\partial\alpha_s(\nu_b)}{\partial\alpha_s(\nu_a)}\right). \tag{11.57}$$

Using the $\alpha_s$ expansion:

$$\beta^a=\beta_1^a\left(\frac{\alpha_s}{\pi}\right)(\nu_a)+\beta_2^a\left(\frac{\alpha_s}{\pi}\right)^2(\nu_a)+\cdots\,,$$
$$\beta^b=\beta_1^b\left(\frac{\alpha_s}{\pi}\right)(\nu_b)+\beta_2^b\left(\frac{\alpha_s}{\pi}\right)^2(\nu_b)+\cdots\,, \tag{11.58}$$

and the relation:

$$\alpha_s(\nu_a)=\alpha_s(\nu_b)+c\alpha_s^2(\nu_b)\,, \tag{11.59}$$

where $c$ is an arbitrary constant depending on the subtraction scale, one can easily deduce:

$$\beta_1^a=\beta_1^b\quad\text{and}\quad\beta_2^a=\beta_2^b\,, \tag{11.60}$$

which achieves the proof of the RGI invariance of $\beta_1$ and $\beta_2$. The higher-order terms of the $\beta$ function will be affected by the coefficient $c$ and hence on the subtraction scale.

### 11.7.3 Higher order expression

The previous result can be extended to higher orders. To order $\alpha_s^2$, one can write the solution of Eq. (11.52) as:

$$\begin{aligned} t &= \int \frac{dz}{z^2} \frac{\pi}{\beta_1 (1 + (\beta_2/\beta_1)(z/\pi))} + \text{constant} \\ &= \frac{\pi}{\beta_1} \left\{ -\frac{1}{z} + \frac{\beta_2}{\beta_1} \frac{1}{\pi} \ln \left( \frac{1 + (\beta_2/\beta_1)(z/\pi)}{z} \right) \right\} + \text{constant} , \end{aligned} \tag{11.61}$$

where the constant is a RGI quantity which has been fixed to be $\ln \Lambda$ to lowest order. At the two-loop level, it is convenient to fix it as in [138]:

$$\text{constant} \equiv \ln \Lambda(1 \text{ loop}) - \frac{\beta_2}{\beta_1^2} \ln \left( -\frac{\beta_1}{2\pi} \right) . \tag{11.62}$$

Therefore, we get the RGI quantity to two loops:

$$\ln \nu + \frac{1}{\beta_1 a_s} - \frac{\beta_2}{\beta_1^2} \ln \left( \frac{1 + (\beta_2/\beta_1)(a_s)}{a_s \pi} \right) = \ln \Lambda(\text{two loops}) - \frac{\beta_2}{\beta_1^2} \ln \left( -\frac{\beta_1}{2\pi} \right) . \tag{11.63}$$

Expanding Eq. (11.61), and inserting the expression of the running $\alpha_s$ to one loop, we deduce:

$$a_s(q^2)^{(2)} = a_s^{(0)} \left\{ 1 - a_s^{(0)} \frac{\beta_2}{\beta_1} \ln \ln \frac{\nu^2}{\Lambda^2} \right\} . \tag{11.64}$$

It is not difficult to show that, to order $\alpha_s^2$, one can relate the one- and two-loop values of $\Lambda$ as:

$$\Lambda(\text{two loops}) = \left( -\frac{\beta_1}{2\pi} \right)^{\beta_2/\beta_1^2} \Lambda(1 \text{ loop}) \tag{11.65}$$

To three-loop accuracy the running coupling can be parametrized as:

$$\begin{aligned} a_s(\nu) = a_s^{(0)} \Bigg\{ & 1 - a_s^{(0)} \frac{\beta_2}{\beta_1} \ln \ln \frac{\nu^2}{\Lambda^2} \\ & + \left(a_s^{(0)}\right)^2 \left[ \frac{\beta_2^2}{\beta_1^2} \ln^2 \ln \frac{\nu^2}{\Lambda^2} - \frac{\beta_2^2}{\beta_1^2} \ln \ln \frac{\nu^2}{\Lambda^2} - \frac{\beta_2^2}{\beta_1^2} + \frac{\beta_3}{\beta_1} \right] + \mathcal{O}(a_s^3) \Bigg\} , \end{aligned} \tag{11.66}$$

with $\beta_i$ are the $\mathcal{O}(a_s^i)$ coefficients of the $\beta$ function in the $\overline{MS}$ scheme for $n_f$ flavours (see Table 2.2), which, for three flavours, read:

$$\beta_1 = -9/2 , \qquad \beta_2 = -8 , \qquad \beta_3 = -20.1198 . \tag{11.67}$$

$\Lambda$ is a renormalization group invariant scale but is renormalization scheme dependent. The running coupling $\alpha_s$ has been measured from LEP, $\tau$ decays[2] and deep-inelastic scattering data. We shall discuss these determinations in the next chapter. The present *world average* is [16,139]:

$$\alpha_s(M_Z) = 0.1181 \pm 0.0027 \,. \tag{11.68}$$

## 11.8 Decoupling theorem

The decoupling theorem of Appelquist and Carazzone [140] states that the effect of heavy particles (fermion, boson) of mass $M_H^2 \gg -q^2$ can be ignored below their thresholds. However, in the $MS$ and $\overline{MS}$ schemes, these heavy particles could contribute to the universal $\beta$ and $\gamma$ functions as they are mass independent, and therefore the $MS$ and $\overline{MS}$ schemes do not a priori satisfy this theorem. In order to satisfy this theorem, one should modify the scheme. References [141–143] have proposed to absorb into the renormalization constant, not only the $1/\hat{\epsilon}$ pole but also terms of the type $\ln^n M_H/\nu$ coming from heavy fermion or boson loops ($\nu$ being the scale of the $\overline{MS}$ scheme). In such an effective theory, one can relate the QCD scale of $n$ light quarks to the one with $n$ light plus one heavy flavour. To one loop, this relation is:

$$\Lambda_{n+1} = \Lambda_n \left(\frac{M_H^2}{p^2}\right)^{\frac{1}{3\beta_1}} \,. \tag{11.69}$$

At the heavy quark threshold $p^2 = 4M_H^2$, one can see that the heavy quark effect tends to decrease slightly the value of $\Lambda$. One can see more explictly such effects in Table 11.2.

## 11.9 Input values of $\alpha_s$ and matching conditions

We shall discuss below, how this decoupling is used in the practical evaluation of the running coupling. In so doing, we run the value of $\alpha_s(M_Z)$ in the range given in Table 11.2, to lower scales by taking appropriately the threshold effects due to heavy quark productions. We run this value until $M_b = 4.6$–$4.7$ GeV, using the two-loop relation:

$$\frac{\alpha_s}{\pi} = a_s^{(0)}\left(1 - a_s^{(0)}\frac{\beta_2}{\beta_1}\ln\ln(-q^2/\Lambda^2)\right) \tag{11.70}$$

and for $n_f$ flavours, we note that:

$$\beta_1 = -\frac{11}{2} + \frac{n_f}{3} \quad \text{and} \quad \beta_2 = -\frac{51}{4} + \frac{19}{12}n_f \,. \tag{11.71}$$

[2] This process gives so far the most precise measurement of $\alpha_s$ at $M_Z$ as a modest accuracy at the $\tau$-mass becomes a precise value at the $Z$-mass because the errors decrease faster than the running of $\alpha_s$. Also, here, compared with some other determinations, we have relatively the best theoretical control including the perturbative corrections to order $\alpha_s^4$, the non-perturbative condensates and the resummation of the asymptotic series.

Table 11.2. *Value of $\alpha_s$ and $\Lambda$ to two-loops at different scales and flavours*

| $\alpha_s(M_Z)$ | $\Lambda_5$[MeV] | $\alpha_s(M_b)$ | $\Lambda_4$[MeV] | $\alpha_s(M_c)$ | $\Lambda_3$[MeV] | $\alpha_s(M_\tau)$ |
|---|---|---|---|---|---|---|
| 0.112 | 160 | 0.198 | 240 | 0.312 | 290 | 0.277 |
| 0.118 | 225 | 0.218 | 325 | 0.372 | 375 | 0.319 |
| 0.124 | 310 | 0.241 | 432 | 0.463 | 480 | 0.378 |
| 0.127 | 360 | 0.254 | 495 | 0.528 | 540 | 0.417 |

Following references [144,145], we do the matching condition $\alpha_s^{(5)} = \alpha_s^{(4)}$ at this $b$-mass, in order to extract $\alpha_s$ for four flavours. We continue iteratively this procedure for completing Table 11.2, which is one of the basic inputs of numerous phenomenological analyses discussed in this book. We use here the value of the perturbative pole mass to two-loops: $M_b = 4.62$ GeV and $M_c = 1.42$ GeV which we shall discuss later on. Notice that doing a similar procedure at the three-loop level, we reproduce the value of $\alpha_s$ given in [139]. In this case, one can use the three-loop relation at the subtraction scale $M_H$ [146]:

$$\alpha_s^{(n_f-1)} = \alpha_s^{(n_f)}\left[1 - 0.291667a_s^2 - [5.32389 - (n_f - 1)0.26247]a_s^3\right], \qquad (11.72)$$

where: $a_s \equiv \alpha_s^{(n_f)}/\pi$.

## 11.10 Running gauge

The running gauge $\bar{\alpha}_G$ is the solution of the differential equation in Eq. (11.35). To leading order in $\alpha_s$, it reads [110]:

$$\bar{\alpha}_G(-q^2) = \frac{\hat{\alpha}_G}{\left[\frac{1}{2}\ln(-q^2/\Lambda)\right]^{\delta/-\beta_1}}\left\{1 + \frac{N}{4\delta}\frac{\hat{\alpha}_G}{\left[\frac{1}{2}\ln(-q^2/\Lambda)\right]^{\delta/-\beta_1}}\right\}^{-1}, \qquad (11.73)$$

where for $SU(N)_c \times SU(n)_F$:

$$\delta = \frac{13}{12}N - \frac{n}{3}. \qquad (11.74)$$

$\hat{\alpha}_G$ is a renormalization group invariant parameter defined to one loop as:

$$\hat{\alpha}_G = \frac{\alpha_G(\nu)}{1 - \frac{N}{4\delta}\alpha_G(\nu)}\left(\frac{1}{-\beta_1 a_s(\nu)}\right)^{\delta/-\beta_1}. \qquad (11.75)$$

It is interesting to notice that for $n \leq 9$, the running gauge tends to the Landau gauge ($\alpha_G = 0$) for $-q^2 \to \infty$. One also obtains:

$$\bar{\alpha}_G(q^2) = \alpha_G(\nu), \qquad (11.76)$$

for $\alpha_G = 0$ (Landau gauge) to all orders and for $\alpha_G = 4\delta$ (peculiar gauge) to lowest order in $\alpha_s$.

## 11.11 Running masses

The running masses are solutions of the differential equation in Eq. (11.35). Analogously to $\Lambda$, one can also introduce an invariant mass $\hat{m}_i$ [28]. The expression of the running quark mass in terms of the invariant mass $\hat{m}_i$ is [28]:

$$\bar{m}_i(\nu) = \hat{m}_i \left(-\beta_1 a_s(\nu)\right)^{-\gamma_1/\beta_1} \left\{ 1 + \frac{\beta_2}{\beta_1}\left(\frac{\gamma_1}{\beta_1} - \frac{\gamma_2}{\beta_2}\right) a_s(\nu) \right.$$
$$\left. + \frac{1}{2}\left[\frac{\beta_2^2}{\beta_1^2}\left(\frac{\gamma_1}{\beta_1} - \frac{\gamma_2}{\beta_2}\right)^2 - \frac{\beta_2^2}{\beta_1^2}\left(\frac{\gamma_1}{\beta_1} - \frac{\gamma_2}{\beta_2}\right) + \frac{\beta_3}{\beta_1}\left(\frac{\gamma_1}{\beta_1} - \frac{\gamma_3}{\beta_3}\right)\right] a_s^2(\nu) + \mathcal{O}\left(a_s^3\right) \right\}, \tag{11.77}$$

where $\gamma_i$ are the $\mathcal{O}(a_s^i)$ coefficients of the quark-mass anomalous dimension (see Table 11.1). For three flavours, we have:

$$\gamma_1 = 2\,, \qquad \gamma_2 = 91/12\,, \qquad \gamma_3 = 24.8404\,. \tag{11.78}$$

As we shall see later on, QSSR is, at present, the most appropriate theoretical method for extracting the absolute values of the light quark masses. A long list of these determinations is given in the recent review [54] (see also [57] and the chapter on quark masses in this book), where the QSSR results are compared with the ones from chiral perturbation theory and lattice calculations. We only quote below the results:

$$\bar{m}_d(2\text{ GeV}) = (6.5 \pm 1.2)\text{ MeV}, \qquad \bar{m}_u(2\text{ GeV}) = (3.6 \pm 0.6)\text{ MeV}\,, \tag{11.79}$$

and:

$$\bar{m}_s(2\text{ GeV}) = (117.4 \pm 23.4)\text{ MeV}\,, \tag{11.80}$$

and the bounds from the positivity of the spectral functions:

$$90\text{ MeV} \leq \bar{m}_s(2\text{ GeV}) \leq 168\text{ MeV}\,. \tag{11.81}$$

The running masses of the $c$ and $b$ quarks have been also extracted directly from the $J/\psi$ and $\Upsilon$ sum rules. To two-loop (order $\alpha_s$) accuracy, one obtains [149]:

$$\bar{m}_c(M_c) = \left(1.23^{+0.02}_{-0.04} \pm 0.03\right)\text{ GeV} \qquad \bar{m}_b(M_b) = \left(4.23^{+0.03}_{-0.04} \pm 0.02\right)\text{ MeV}\,. \tag{11.82}$$

From the $D$ and $B$ meson systems, one obtains to order $\alpha_s^2$ [150]:

$$\bar{m}_c(M_c) = \left(1.10 \pm 0.04\right)\text{ GeV} \qquad \bar{m}_b(M_b) = \left(4.05 \pm 0.06\right)\text{ MeV}\,, \tag{11.83}$$

which agree with the former within the errors though the central values are slightly lower. These results can be compared with different results based on non-relativistic and some other approaches [16].

### 11.12 The perturbative pole mass

The notion of perturbative pole mass can be useful in the phenomenology of the heavy quark systems. However, unlike in QED, where the pole mass is well-defined, due to the observation of the lepton, this definition is ambiguous in QCD due to confinement. Attempts to define the pole mass within perturbation theory have been done in the literature [141,133,148]. By analogy with QED, one can define the pole mass as the pole of the quark propagator. For definiteness, on can start with the bare quark propagator:

$$S_F(p) = \frac{1}{\hat{p} - M_B - i\epsilon}, \tag{11.84}$$

After interaction, one has:

$$S_F(p) = \left(\frac{1}{1-\Sigma_2}\right)\frac{1}{\hat{p} - M_B\left[1 + \frac{\Sigma_1}{1-\Sigma_2}\right]} \tag{11.85}$$

which shows explicitly the wave function and the mass renormalization constants in Eq. (9.20). An explicit evaluation of $\Sigma_{1,2}$ in the $\overline{MS}$ scheme gives:

$$\Sigma_1^B = (g_B\nu^{-\epsilon/2})^2 \frac{C_F}{(16\pi^2)^{1-\epsilon/4}}\int_0^1 dx \times\left[\Gamma(\epsilon/2)\left(\frac{\mathbf{R}^2}{\nu^2}\right)^{-\epsilon/2}[2(2-x) - \epsilon(1-x) + (1-\alpha_G)(1-2x)] + (1-\alpha_G)2x(1-x)\frac{p^2}{M_B^2 - p^2x}\right] \tag{11.86}$$

$$\Sigma_2^B = (g_B\nu^{-\epsilon/2})^2 \frac{C_F}{(16\pi^2)^{1-\epsilon/4}}\int_0^1 dx \times\left[\Gamma(\epsilon/2)\left(\frac{\mathbf{R}^2}{\nu^2}\right)^{-\epsilon/2}[-2x + \epsilon(1-x) + (1-\alpha_G)2(1-x)] + (1-\alpha_G)2x(1-x)\frac{p^2}{M_B^2 - p^2x}\right], \tag{11.87}$$

where:

$$\mathbf{R}^2 = (1-x)\left(M_B^2 - p^2x\right) - i\epsilon'. \tag{11.88}$$

$\alpha_G$ is the covariant gauge parameter and $C_F = (N^2-1)/(2N)$ for $SU(N)_c$. These parametric integrals lead to:

$$\Sigma_1^B = \left(\frac{\alpha_s}{\pi}\right) C_F \frac{1}{2}\left\{\frac{3}{\hat\epsilon} + \frac{5}{2} - \frac{3}{2}\ln\frac{M_B^2 - p^2}{\nu^2}\right.$$
$$+\left(\frac{1}{2}\right)\frac{M_B^2}{-p^2}\left[1 - \left(4 + \frac{M_B^2}{-p^2}\right)\ln\left(1 - \frac{p^2}{M_B^2}\right)\right]$$
$$\left.+(1-\alpha_G)\left[-\frac{1}{2} - \frac{1}{2}\frac{M_B^2}{-p^2} + \frac{1}{2}\frac{M_B^2}{-p^2}\left(1 + \frac{M_B^2}{-p^2}\right)\ln\left(1 - \frac{p^2}{M_B^2}\right)\right]\right\}, \tag{11.89}$$

$$\Sigma_2^B = \left(\frac{\alpha_s}{\pi}\right) C_F \frac{1}{4}[-1 + (1-\alpha_G)]\left\{\frac{2}{\hat\epsilon} + 1 - \ln\frac{M_B^2 - p^2}{\nu^2}\right.$$
$$\left.+\left(\frac{M_B^2}{-p^2}\right)^2 \ln\left(1 - \frac{p^2}{M_B^2}\right) - \frac{M_B^2}{-p^2}\right\}, \tag{11.90}$$

with:

$$1/\hat\epsilon \equiv 1/\epsilon + \frac{1}{2}(\ln 4\pi - \gamma), \tag{11.91}$$

which shows that $\Sigma_2^B$ vanishes to order $\alpha_s$ in the Landau gauge $\alpha_G = 0$. Their asymptotic expressions are:

$$\Sigma_1^B\big|_{p^2 \gg M^2} = \left(\frac{\alpha_s}{\pi}\right) C_F \frac{1}{2}\left\{\frac{3}{\hat\epsilon} + \frac{5}{2} - \frac{3}{2}\ln\frac{-p^2}{\nu^2} + \frac{1}{2}(1-\alpha_G) + \mathcal{O}\left(\frac{M^2}{-p^2}\ln\frac{-p^2}{M^2}\right)\right\},$$
$$\Sigma_2^B\big|_{p^2 \gg M^2} = \left(\frac{\alpha_s}{\pi}\right) C_F \frac{1}{4}[-1 + (1-\alpha_G)]\left\{\frac{2}{\hat\epsilon} + 1 - \ln\frac{-p^2}{\nu^2} + \mathcal{O}\left(\frac{M^2}{-p^2}\ln\frac{-p^2}{M^2}\right)\right\}, \tag{11.92}$$

and:

$$\Sigma_1^B\big|_{p^2 \ll M^2} = \left(\frac{\alpha_s}{\pi}\right) C_F \frac{1}{2}\left\{\frac{3}{\hat\epsilon} - \frac{3}{2}\ln\frac{M_B^2}{\nu^2} + \frac{3}{4} + \frac{5}{6}\left(\frac{-p^2}{M_B^2}\right)\right.$$
$$\left.+(1-\alpha_G)\left[-\frac{1}{4} - \frac{1}{12}\left(\frac{-p^2}{M_B^2}\right)\right]\right\},$$
$$\Sigma_2^B\big|_{p^2 \ll M^2} = \left(\frac{\alpha_s}{\pi}\right) C_F \frac{1}{4}[-1 + (1-\alpha_G)]\left\{\frac{2}{\hat\epsilon} - \ln\frac{M^2}{\nu^2} + \frac{1}{2} - \frac{2}{3}\left(\frac{-p^2}{M_B^2}\right)\right\}, \tag{11.93}$$

At $p^2 = M^2 = \nu^2$, one gets:

$$\Sigma_1^B\big|_{p^2 = M^2 = \nu^2} = \left(\frac{\alpha_s}{\pi}\right) C_F \frac{1}{2}\left[\frac{3}{\hat\epsilon} + 2\right], \tag{11.94}$$

which is gauge independent. It is related to the pole mass, which is defined at the pole $p^2 = M^2$ of the full quark propagator through Eq. (11.85). In terms of the running mass, the pole mass reads:

$$M_{\text{pole}} = \bar{m}(p^2)\left\{1 + \frac{\Sigma_1(p^2 = M^2)}{1 - \Sigma_2(p^2 = M^2)}\right\}, \tag{11.95}$$

Therefore, the previous expressions gives [148]:

$$M_{\text{pole}} = \bar{m}(p^2)\left\{1 + \left(\frac{4}{3} + \ln\frac{p^2}{M^2}\right)\left(\frac{\alpha_s}{\pi}\right)\right\}, \tag{11.96}$$

which is gauge and renormalization scheme independent. The IR finiteness of the result to order $\alpha_s^2$ has been explicitly shown in [133]. The independence of $M_{\text{pole}}$ on the choice of the regularization scheme has been demonstrated in [148]. The extension of the previous result to order $\alpha_s^2$ is [151]:

$$\begin{aligned} M_{\text{pole}} = \bar{m}(p^2)\Bigg[1 &+ \left(\frac{4}{3} + \ln\frac{p^2}{M^2}\right)\left(\frac{\alpha_s}{\pi}\right) \\ &+ \left[K_Q + \left(\frac{221}{24} - \frac{13}{36}n\right)\ln\frac{p^2}{M^2} + \left(\frac{15}{8} - \frac{n}{12}\right)\ln^2\frac{p^2}{M^2}\right]\left(\frac{\alpha_s}{\pi}\right)^2\Bigg], \end{aligned} \tag{11.97}$$

where, in the RHS, $M$ is the pole mass and:

$$K_Q = 17.1514 - 1.04137n + \frac{4}{3}\sum_{i\neq Q}\Delta\left(r \equiv \frac{m_i}{M_Q}\right). \tag{11.98}$$

For $0 \leq r \leq 1$, $\Delta(r)$ can be approximated, within an accuracy of 1% by:

$$\Delta(r) \simeq \frac{\pi^2}{8}r - 0.597r^2 + 0.230r^3, \tag{11.99}$$

while, its values in the following limiting cases are:

$$\begin{aligned} \Delta(r \to 0) &\simeq \frac{3}{4}\zeta(2)r + \mathcal{O}(r^2), \\ \Delta(r \to \infty) &\simeq \frac{1}{4}\ln^2 r + \frac{13}{24}\ln r + \frac{1}{4}\zeta(2) + \frac{151}{288} + \mathcal{O}(r^{-2}\ln r), \\ \Delta(r = 1) &\simeq \frac{3}{4}\zeta(2) - \frac{3}{8}. \end{aligned} \tag{11.100}$$

As, one can notice, the behaviour of $\Delta(r \to \infty)$ is quite bad, such that in the effective field theory where the heavy quark mass tends to infinity, one should write a well-defined relation in this limit. This can be achieved by introducing the coupling and light quark masses in the effective field theory in terms of the corresponding quantities in the full theory [144]:

$$\alpha_s^{\mathrm{eff}}(\nu) = \alpha_s(\nu) C(\alpha_s(\nu), x)$$
$$m^{\mathrm{eff}}(\nu) = m(\nu) H(\alpha_s(\nu), x) , \qquad (11.101)$$

where $x \equiv \ln(\bar{m}_h^2/\nu^2)$ and:

$$C(\alpha_s, x) = 1 + \sum_{k\geq 1} C_k \left(\frac{\alpha_s}{\pi}\right)^k , \qquad C_k(x) = \sum_{0\leq i\leq k} C_{ik} x^i ,$$
$$H(\alpha_s, x) = 1 + \sum_{k\geq 1} H_k \left(\frac{\alpha_s}{\pi}\right)^k , \qquad H_k(x) = \sum_{0\leq i\leq k} H_{ik} x^i , \qquad (11.102)$$

with:

$$C_1 = \frac{x}{6} , \qquad C_2 = \frac{11}{72} + \frac{11}{24} x + \frac{x^2}{36} ,$$
$$H_1 = 0 , \qquad H_2 = \frac{89}{32} + \frac{5}{36} x + \frac{x^2}{12} , \qquad (11.103)$$

and by expressing $\alpha_s^{\mathrm{eff}}$ in terms of the pole mass:

$$\alpha_s^{\mathrm{eff}} = \alpha_s \left\{ 1 + \frac{X}{6}\left(\frac{\alpha_s}{\pi}\right) + \left(-\frac{7}{24} + \frac{19X}{24} + \frac{X^2}{36}\right)\left(\frac{\alpha_s}{\pi}\right)^2 \right\} , \qquad (11.104)$$

where $X \equiv \ln(M_h^2/\nu^2)$. In this way, the previous expression becomes:

$$M_{\mathrm{pole}} = \bar{m}(p^2) \left[ 1 + \left(\frac{4}{3} + \ln \frac{p^2}{\bar{m}^2}\right)\left(\frac{\alpha_s}{\pi}\right) + \left[ K_Q(\bar{m}_f/\bar{m}) + \left(\frac{173}{24} - \frac{13}{36} n\right) \ln \frac{p^2}{\bar{m}^2} + \left(\frac{15}{8} - \frac{n}{12}\right) \ln^2 \frac{p^2}{\bar{m}^2} \right] \left(\frac{\alpha_s}{\pi}\right)^2 \right] , \qquad (11.105)$$

where $\bar{m}$ is the running mass of the finite mass heavy quark, $n$ is the number of finite mass quark flavours and the summation in $K_Q$ through $\Delta(\bar{m}_f/\bar{m})$ runs over the $n-1$ lightest quarks. For instance, in the case of the bottom quark mass, one uses $n = 5$, and deduce:

$$M_b = \bar{m}_b(p^2) \left[ 1 + \left(\frac{4}{3} + \ln \frac{p^2}{\bar{m}_b^2}\right)\left(\frac{\alpha_s^{\mathrm{eff}}}{\pi}\right) + \left[ K_Q(\bar{m}_f/\bar{m}_b) + \frac{389}{72} \ln \frac{p^2}{\bar{m}_b^2} + \frac{35}{24} \ln^2 \frac{p^2}{\bar{m}^2} \right] \left(\frac{\alpha_s}{\pi}\right)^2 \right] , \qquad (11.106)$$

where, by neglecting the $u$ and $d$ quark masses:

$$K_Q(\bar{m}_f/\bar{m}_b) = 9.278 + \frac{4}{3} \sum_{f\equiv s,c} \Delta(\bar{m}_f/\bar{m}_b) . \qquad (11.107)$$

Finally, a recent order $\alpha_s^3$ evaluation leads to [152]:

$$\bar{m}(M_{\text{pole}}) = M_{\text{pole}}\left[1 - \frac{4}{3}\left(\frac{\alpha_s}{\pi}\right) + [-14.3323 - 1.0414n]\left(\frac{\alpha_s}{\pi}\right)^2 + [-198.7068 + 26.9239n - 0.65269n^2]\left(\frac{\alpha_s}{\pi}\right)^3\right]. \quad (11.108)$$

However, one should be careful when using the previous mass in the OPE, as, in order to be consistent, one should use the same truncations in the mass definition and in the hadronic correlator to be analysed. For this reason, the re-summed result obtained to leading order in $(\beta_1\alpha_s)$ term within the large $n_f$-limit [154], should also be used with care. Using the previous relation with the pole and running mass as well as a direct estimate of the two-loop order $\alpha_s$ running mass from the $\psi$ and $\Upsilon$-sum rules, one obtains the value of the pole mass to two-loop accuracy [149]:[3]

$$M_c^{PT2} = (1.42 \pm 0.03)\ \text{GeV}\ , \qquad M_b^{PT2} = (4.62 \pm 0.02)\ \text{GeV}\ . \quad (11.109)$$

It is informative to compare these values with those of the pole masses from non-relativistic sum rules to two loops [149]:

$$M_c^{NR} = \left(1.45^{+0.04}_{-0.03} \pm 0.03\right)\ \text{GeV}\ , \qquad M_b^{NR} = \left(4.69^{+0.02}_{-0.01} \pm 0.02\right)\ \text{GeV}\ , \quad (11.110)$$

and, recently, to three loops of order $\alpha_s^2$ [4] including a resummation of the Coulombic corrections [156]:

$$M_b^{NR} = (4.60 \pm 0.02)\ \text{GeV}\ , \quad (11.111)$$

in good agreement with the former results.

If one uses the value of the running mass obtained to three-loop accuracy [156], and the three-loop relation between the pole and the running mass, one obtains:[5]

$$M_b^{PT3} \simeq (4.7 \pm 0.07 \pm 0.02)\ \text{GeV}\ , \quad (11.112)$$

which, although slightly higher, is in agreement within the errors with the two-loop result.

Recent extension of the sum rules analysis [157,159] have led to more accurate values of the pole mass. The one using the relation between the pole and the 1S meson mass gives [159]:

$$M_b^{PT3} \simeq (4.71 \pm 0.03)\ \text{GeV}\ , \quad (11.113)$$

in agreement with the two-loop $\alpha_s$ result given in Eq. (11.109).

One can also compare the previous values with the *dressed mass*:

$$M_b^{nr} = (4.94 \pm 0.10 \pm 0.03)\ \ \text{GeV}\ , \quad (11.114)$$

[3] We shall discuss these different points in more details in the chapter on quark masses.

[4] This result can be considered to be an improvment of the Voloshin value of 4.8 GeV [155].

[5] This value is slightly lower than the one given in [149], as the value of the running mass used there is higher. However, the results agree within the errors.

obtained from a non-relativistic Balmer formula based on a $\bar{b}b$ Coulomb potential and including higher order $\alpha_s^4$-corrections [94], or the mass obtained from the fit of the spectra within potential models [12]:

$$M_b^{\rm pot} \simeq (4.8 \sim 4.9)\ {\rm GeV}. \tag{11.115}$$

This non-relativistic mass is slightly higher than the one from the sum rules. One can remark that the mass difference is :

$$M_b^{nr} - M_b^{PT} \approx (100 \sim 200)\ {\rm MeV}\ . \tag{11.116}$$

The interpretation of this mass difference is not very well understood. If one has in mind that the non-relativistic pole mass contains a non-perturbative part, which can be of the same origin as the one induced by the truncation of the perturbative series at large order, then one might eventually consider this value as a phenomenological estimate of the renormalon contribution, which is comparable in strength with the estimate of about 100–133 MeV from the summation of higher-order corrections of large-order perturbation theory [154].

An extension of the previous analysis of the $J/\psi$ and $\Upsilon$-systems to the case of the $D$, $B$ and $D^*$, $B^*$ mesons leads to the value to order $\alpha_s$ [149]:

$$M_b^{PT2} = (4.63 \pm 0.08)\ {\rm GeV}\ , \tag{11.117}$$

in good agreement with the previous results, but less accurate. This result has been confirmed by recent estimates to order $\alpha_s^2$ [150]:

$$M_c^{PT3} = (1.47 \pm 0.06)\ {\rm GeV}\ , \qquad M_b^{PT3} = (4.69 \pm 0.06)\ {\rm GeV}\ , \tag{11.118}$$

### *11.12.1 The b and c pole mass difference*

One can also use the previous results, in order to deduce the mass difference between the $b$ and $c$ (non)-relativistic pole masses:

$$M_b(M_b) - M_c(M_c) = (3.22 \pm 0.03)\ {\rm GeV}\ , \tag{11.119}$$

in good agreement (within the errors) with potential model expectations [12,16], and with the heavy quark symmetry (HQET) result from the $B$ and $D$ mass difference [164] (see also Chapter 44):

$$M_b(M_b) - M_c(M_c) \simeq (\bar{M}_B - \bar{M}_D) \left\{1 - \frac{\lambda_1}{2\bar{M}_B\bar{M}_D} + \mathcal{O}\left(\frac{1}{M_Q^3}\right)\right\} \simeq (3.4 \pm 0.04)\ , \tag{11.120}$$

where one has used the QSSR estimate of the heavy quark kinetic term inside the meson [165,166]:

$$\lambda_1 \simeq -(0.5 \pm 0.2)\ {\rm GeV}^2. \tag{11.121}$$

A direct comparison of this mass difference with the one from the analysis of the inclusive $B$-decays needs however a better understanding of the mass definition and of the value of the scale entering into these decay processes. If one chooses to evaluate these pole masses at the scale $\nu = M_b$, which might be a natural scale for this process, one obtains to two-loop accuracy:

$$M_c(\nu = M_b) = (1.08 \pm 0.04)\ \text{GeV}, \tag{11.122}$$

which leads to the mass difference:

$$M_b - M_c|_{\nu=M_b} = (3.54 \pm 0.05)\ \text{GeV}\ . \tag{11.123}$$

## 11.13 Alternative definitions to the pole mass

It has been argued that the pole masses can be affected by non-perturbative terms induced by the resummation of the QCD perturbative series [154] (see chapter on power corrections) and alternative definitions free from such ambiguities have been proposed (residual mass [158] (see also [160]) and 1S mass [159]). Assuming that the QCD potential has no linear power corrections, the residual or potential-subtracted (PS) mass is related to the pole mass as:

$$M_{\rm PS} = M_{\rm pole} + \frac{1}{2}\int_{|\vec{q}|<\mu} \frac{d^3\vec{q}}{(2\pi)^3} V(\vec{q})\ . \tag{11.124}$$

The 1S mass is defined as half of the perturbative component to the $^3S_1$ $\bar{Q}Q$ ground state, which is half of its static energy $\langle 2M_{\rm pole} + V\rangle$.[6] The running and short distance pole mass defined at a given order of PT series will be used in the following discussions in this book.

## 11.14 $\overline{MS}$ scheme and RGE for the pseudoscalar two-point correlator

In order to illustrate the discussions in the previous sections, let us consider the two-point correlator:

$$\Psi_5(q^2) \equiv i \int d^4x\ e^{iqx}\ \langle 0|\mathcal{T} J_P(x) (J_P(0))^\dagger\ |0\rangle\ , \tag{11.125}$$

where:

$$J_P = (m_i + m_j)\bar{\psi}_i(i\gamma_5)\psi_j\ , \tag{11.126}$$

is the light quark pseudoscalar current.

[6] These definitions might still be affected by a dimension-two term advocated in [162,161,438], which might limit their accuracy [163].

### 11.14.1 Lowest order perturbative calculation

We shall be concerned with Fig. 8.1 discussed in Section 8.2.5 for massless quarks (Fig. 11.2):

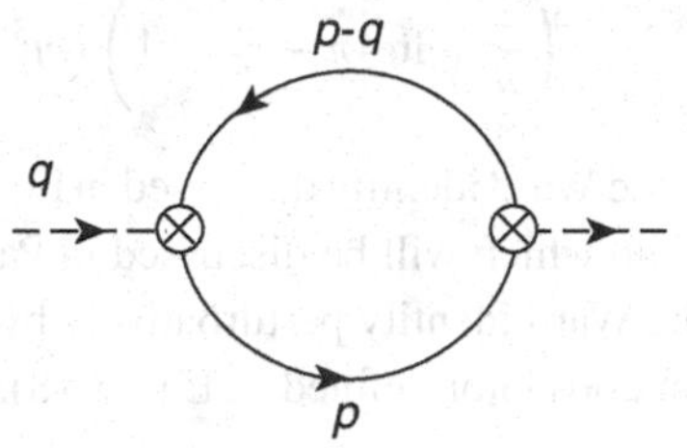


Using Feynman rules, it reads:

$$i\nu^\epsilon \Psi_5(q^2) = (m_i + m_j)^2(-1)N \int \frac{d^n p}{(2\pi)^n} \times \mathbf{Tr}\left\{(i\gamma_5)\frac{i}{\hat{p} - m_i + i\epsilon'}(i\gamma_5)\frac{i}{\hat{p} - \hat{q} - m_j + i\epsilon'}\right\}. \quad (11.127)$$

Parametrizing the quark propagators à la Feynman (Appendix E) and using the properties of the Dirac matrices (Appendix D) and momentum integrals (Appendix F) in $n$-dimensions, one obtains for the bare correlator:

$$\nu^\epsilon \Psi_5^B(q^2) = (m_i + m_j)^2 \frac{N}{4\pi^2} \int_0^1 dx \left(\frac{2}{\epsilon} + \ln 4\pi - \gamma\right)\left(\frac{\mathbf{R}^2 - i\epsilon'}{\nu^2}\right)^{-\epsilon/2} \times \left\{\left(3 + \frac{\epsilon}{2}\right) q^2 x(1-x) - 2\left(1 + \frac{\epsilon}{4}\right)\left(m_i^2 x + m_j^2(1-x) + m_i m_j\right)\right\}, \quad (11.128)$$

where:

$$\mathbf{R}^2 \equiv -q^2 x(1-x) + m_i^2 x + m_j^2(1-x), \quad (11.129)$$

and $\gamma = 0.5772\ldots$ is the Euler constant. Two limiting cases are particularly interesting:

$$\nu^\epsilon \Psi_5^B\left(q^2 \gg m_{i,j}^2\right) = (m_i + m_j)^2 q^2 \frac{N}{8\pi^2}\left[\left(\frac{2}{\epsilon} + \ln 4\pi - \gamma - \ln\left(\frac{-q^2}{\nu^2}\right)\right) \times \left[1 + 2\frac{(m_i^2 + m_j^2 - m_i m_j)}{-q^2}\right] + 2 + \frac{\epsilon}{4}\ln^2\left(\frac{-q^2}{\nu^2}\right) - \frac{\epsilon}{2}(\ln 4\pi - \gamma + 2)\ln\left(\frac{-q^2}{\nu^2}\right)\right], \quad (11.130)$$

and:

$$\nu^\epsilon \Psi_5^B(q^2=0) = (m_i+m_j)\frac{N}{4\pi^2}\left[\left(m_i^3 \ln\frac{m_i^2}{\nu^2} + m_j^3 \ln\frac{m_j^2}{\nu^2}\right) - \left(\frac{2}{\epsilon} + \ln 4\pi - \gamma - 1\right)(m_i^3+m_j^3)\right]. \tag{11.131}$$

The case $q = 0$ is useful for the Ward identity discussed in Eq. (2.17) and for the definition of the scale-invariant condensate which will be discussed in Part VII.

One can explicitly check the Ward identity perturbatively by evaluating the longitudinal part of the axial-vector current correlator defined in Eq. (2.18). One obtains:

$$q_\mu q_\nu \Pi_5^{\mu\nu} = \frac{N}{8\pi^2}q^2 \int_0^1 dx\, \left[m_i^2 x + m_j^2(1-x) + m_i m_j\right]\left(\frac{\mathbf{R}^2 - i\epsilon'}{\nu^2}\right)^{-\epsilon/2} \Gamma(\epsilon/2)\,, \tag{11.132}$$

which by comparison gives:

$$q_\mu q_\nu \Pi_5^{\mu\nu} = \Psi_5(q^2) - (m_i+m_j)\frac{N}{4\pi^2}\left(m_i^3 \ln\frac{m_i^2}{\nu^2} + m_j^3 \ln\frac{m_j^2}{\nu^2}\right). \tag{11.133}$$

Finally, one can extract the spectral function by using:

$$\ln \mathbf{R}^2 = \ln|\mathbf{R}^2| - i\pi\theta(-\mathbf{R}^2)\,. \tag{11.134}$$

Therefore, one can deduce:

$$\begin{aligned}\mathrm{Im}\Psi_5(t) &= \mathrm{Im}\left(q_\mu q_\nu \Pi_5^{\mu\nu}\right)\\ &= \frac{N}{8\pi^2}(m_i+m_j)^2 t\left(1 - \frac{(m_i-m_j)^2}{t}\right)\\ &\quad\times \lambda^{1/2}\left(1, \frac{m_i^2}{t}, \frac{m_j^2}{t}\right)\theta[t-(m_i+m_j)^2]\,.\end{aligned} \tag{11.135}$$

### 11.14.2 Two-loop perturbative calculation in the $\overline{MS}$ scheme

For a pedagogical illustration, we consider a massless quark inside the quark loop. The corresponding two-loop perturbative contribution comes from Fig. 11.3.

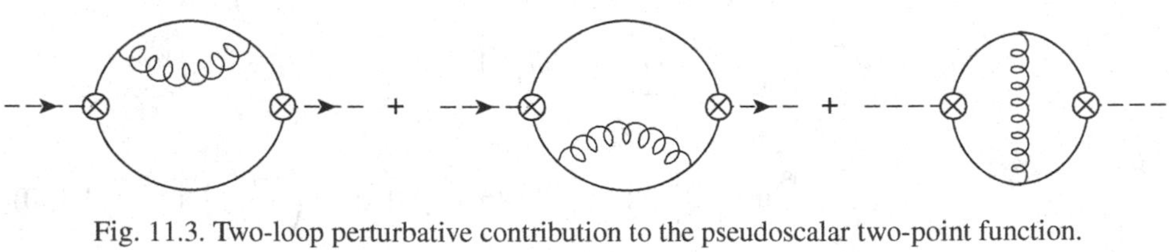

Fig. 11.3. Two-loop perturbative contribution to the pseudoscalar two-point function.

A routine application of the previous rules leads to [167]:

$$\begin{aligned}\Psi_5^B(q^2) = \nu^\epsilon \frac{3}{8\pi^2}\left(m_i^B + m_j^B\right)^2 q^2 \Bigg[ & \frac{2}{\epsilon} + \ln 4\pi - \gamma + 2 - \ln\left(\frac{-q^2}{\nu^2}\right) \\ & - \frac{\epsilon}{2}(\ln 4\pi - \gamma + 2)\ln\left(\frac{-q^2}{\nu^2}\right) + \frac{\epsilon}{4}\ln^2\left(\frac{-q^2}{\nu^2}\right) \\ & + \left(\frac{g^B \nu^{-\epsilon/2}}{4\pi^2}\right)^2 \left[\frac{4}{\epsilon^2} + \frac{4}{\epsilon}(\ln 4\pi - \gamma) + \frac{29}{3\epsilon} + \mathcal{O}(1)\right]\left(\frac{-q^2}{\nu^2}\right)^{-\epsilon}\Bigg]. \end{aligned} \tag{11.136}$$

Introducing the renormalized parameter (we shall omit the index $R$):

$$\begin{aligned} g^B \nu^{-\epsilon/2} &= g\left[1 + \mathcal{O}\left(\frac{\alpha_s}{\pi}\right)\right], \\ m_i^B &= m_i \left[1 - \frac{2}{\epsilon}\left(\frac{\alpha_s}{\pi}\right)\right], \end{aligned} \tag{11.137}$$

one can deduce [167]:

$$\begin{aligned}\Psi_5(q^2) = \frac{3}{8\pi^2}(m_i + m_j)^2 q^2 \Bigg[ & \frac{2}{\epsilon} + \ln 4\pi - \gamma + 2 - \ln\left(\frac{-q^2}{\nu^2}\right) \\ & + \left(\frac{\alpha_s}{\pi}\right)\Bigg[-\frac{4}{\epsilon^2} + \frac{5}{3\epsilon} + \ln^2\left(\frac{-q^2}{\nu^2}\right) \\ & - \left(\frac{17}{3} + 2(\ln 4\pi - \gamma)\right)\ln\left(\frac{-q^2}{\nu^2}\right)\Bigg]\Bigg]. \end{aligned} \tag{11.138}$$

This expression tells us that the lowest order term proportional to $\epsilon$ induce via the mass renormalization a non-zero finite term. It also shows how the non-local:

$$\frac{1}{\epsilon}\ln\left(\frac{-q^2}{\nu^2}\right) \tag{11.139}$$

pole has disappeared after renormalization. The disappearance of this term is a double check of the calculation as well. Finally, one can also use the RGE for checking the ln-coefficient. This can be done by working with the RGE of the two-point correlator given in Section 11.6. In so doing, we consider the coefficient of the $1/\epsilon$-terms:

$$D = D_0 + \left(\frac{\alpha_s}{\pi}\right) D_1 , \tag{11.140}$$

with:

$$\begin{aligned} D_0 &= -\frac{3}{8\pi^2}(x_i + x_j)^2 2e^{2t} , \\ D_1 &= -\frac{3}{8\pi^2}(x_i + x_j)^2 \frac{10}{3} e^{2t} . \end{aligned} \tag{11.141}$$

where $x_i \equiv m_i/\nu$ is a *dimensionless* mass and $t \equiv -1/2\ln(-q^2/\nu^2)$. Expressing $\Psi_5$ in terms of $x_i$, one has:

$$\Psi_5(t,\alpha_s,x_{i,j}) = -\frac{3}{8\pi^2}(x_i+x_j)^2 e^{2t} q^4 \times\left[-2t+\ln 4\pi-\gamma+2+\left(\frac{\alpha_s}{\pi}\right)\left(4at^2+2bt+c\right)\right], \quad (11.142)$$

where $a,\ b,\ c$ have to be determined. Using the RGE, one obtains the constraint:

$$D_0 = -\frac{3}{8\pi^2}(x_i+x_j)^2 2e^{2t}\ ,$$
$$D_1 = -\frac{3}{8\pi^2}(x_i+x_j)^2 e_{2t} \times[-8at-2b-2\gamma_1(\ln 4\pi-\gamma+2)+2\gamma_1 2t]\ , \quad (11.143)$$

where $\gamma_1 = 2$ is the mass anomalous dimension. The fact that $D_1$ cannot depend on $t$ implies:

$$-4a+2\gamma_1 = 0 \implies a = 1\ . \quad (11.144)$$

The relation between $C_1$ and $D$ given in Eq. (11.48) implies:

$$C_1^{(0)} = D_0\ . \quad (11.145)$$

$C_1^{(1)}$ is not fixed by the RGE but we know it from the previous calculation:

$$C_1^{(1)} = \frac{3}{8\pi^2}(x_i+x_j)^2 e^{2t}\frac{5}{3}\ , \quad (11.146)$$

while we deduce from Eq. (11.48):

$$2C_1^{(1)} = D_1\ . \quad (11.147)$$

The recursive relation implies:

$$C_2^{(1)} = \frac{3}{8\pi^2}(x_i+x_j)^2 e^{2t} 2\gamma_1\ . \quad (11.148)$$

Inserting the previous expressions into the one of $D_1$, one can deduce:

$$-2b-2\gamma_1(\ln 4\pi-\gamma+2) = \frac{10}{3}\ . \quad (11.149)$$

One can see that the RGE and an explicit evaluation of the $1/\epsilon$-coefficient to order $\alpha_s$ allows one to fix the coefficients of the $1/\epsilon^2$, $\ln^2$ and ln at that order. This impressive result allows to have a double check of the direct calculation.

# 12
# Other renormalization schemes

In previous chapters, we have concentrated our discussions on the modified minimal subtraction $\overline{MS}$ scheme, which is the most convenient one for QCD. However, it is known that there is a freedom for choosing a renormalization scheme. Among different existing off-shell renormalization schemes discussed in the literature, we choose to discuss the following schemes which have been widely used in the 1980s. We shall also discuss their connections by comparing the renormalized QCD coupling in these different schemes.

## 12.1 The $MS$ scheme

The $MS$ scheme is the original minimal subtraction scheme proposed for dimensional renormalization. We have already discussed the difference between the $MS$ and $\overline{MS}$ schemes, which one can illustrate by the comparison of the renormalized coupling in the two schemes:

$$\nu^{-\epsilon}\alpha_s^B = \alpha_s^R\left\{1+\left(\frac{\alpha_s}{\pi}\right)\left(\frac{\beta_1}{\epsilon}+\delta\right)+\mathcal{O}\left(\frac{\alpha_s}{\pi}\right)^2\right\}, \tag{12.1}$$

where $\delta$ is an arbitrary constant characteristic of the scheme used. In the $\overline{MS}$ scheme:

$$\delta_{\overline{MS}} = \frac{\beta_1}{2}[\ln 4\pi - \gamma], \tag{12.2}$$

and the running couplings in the two schemes are related as:

$$\bar{\alpha}_s^{\overline{MS}} = \bar{\alpha}_s^{MS}\left[1+\left(\frac{\bar{\alpha}_s}{\pi}\right)\delta + \mathcal{O}(\alpha_s^2)\right]. \tag{12.3}$$

This leading order relation can be translated by the relation between the scale $\Lambda$ in the two schemes:

$$\Lambda_{\overline{MS}} \simeq \Lambda_{MS}\exp(\delta/\beta_1), \tag{12.4}$$

i.e., one obtains to this order:

$$\Lambda_{\overline{MS}} \simeq 2.66\ \Lambda_{MS}. \tag{12.5}$$

Table 12.1. *Value of $\delta(\alpha_G, n_f)$ in the MS and MOM schemes*

| Scheme | | $\delta(\alpha_G, n_f)$ |
|---|---|---|
| $MS$ | | 0 |
| $\overline{MS}$ | | $\delta_{\overline{MS}} \equiv (\beta_1/2)[\ln 4\pi - \gamma]$ |
| MOM | Three-gluon | $\delta_{\overline{MS}} - \frac{11}{2} - \frac{23}{48}J - \alpha_G \frac{9}{16}(1-J) + \frac{\alpha_G^2}{8}(3-J) - \frac{\alpha_G^3}{16} + \frac{n_f}{3}\left(1 + \frac{2}{3}J\right)$ |
| | Quark-gluon | $\delta_{\overline{MS}} - \frac{1}{16}\left(89 - \frac{85}{9}J\right) - \alpha_G \frac{25}{24}\left(1 - \frac{2}{3}J\right) + \frac{\alpha_G^2}{16}(3-J) + \frac{5n_f}{18}$ |
| | Ghost-gluon | $\delta_{\overline{MS}} - \frac{5}{48}\left(41 + \frac{3}{2}J\right) - \frac{\alpha_G}{8}(9-2J) - \frac{\alpha_G^2}{16}\left(3 - \frac{J}{2}\right) + \frac{5n_f}{18}$ |

## 12.2 The momentum subtraction scheme

In the momentum subtraction scheme (MOM scheme), the renormalized two-point (or in general Green's) function is defined as [168–170]:

$$\Psi_5(q^2)_R = \Psi_5(q^2) - \Psi_5(q^2 = -\mu^2,\ m^2)\,, \tag{12.6}$$

where $\mu$ is the subtraction point in the Euclidean region. The choice of $\mu$ is arbitrary. It is often chosen at the symmetric point of the three-gluon vertex with which one defines the renormalized QCD coupling. However, the choice of the vertex is also arbitrary, as one can choose the quark-gluon-quark or ghost-gluon-ghost vertex for defining the renormalized coupling. In this scheme, the renormalization constants and universal parameters are mass-dependent, which is not convenient when one works with massive particles. However, due to the Appelquist–Carazzone decoupling theorem, one may ignore the effect of the heavy quarks having a mass larger than the momentum scale of the analysis. If one expresses the renormalized QCD coupling $\alpha_s$ in terms of the bare coupling $\alpha_s^B$ in $4-\epsilon$ dimensions, one has:

$$\nu^{-\epsilon}\alpha_s^B = \alpha_s\left[1 + \left(\delta(\alpha_G, n_f) + \frac{\beta_1}{\epsilon}\right)\alpha_s + \mathcal{O}(\alpha_s^2)\right]\,, \tag{12.7}$$

where $\delta(\alpha_G, n_f)$ is a finite term which depends on how $\alpha_s$ is renormalized, and are given in Table 12.1, where $\alpha_G$ is the gauge parameter; $\beta_1 = -(1/2)(11 - 2n_f/3)$ for $SU(3)_c \times SU(n)_f$, and:

$$J \equiv -2\int_0^1 dx \frac{\ln x}{x^2 - x + 1} = 2.3439072\ldots.\,. \tag{12.8}$$

Therefore, one can derive the lowest order relation between the MOM and $MS$ schemes in the case of massless quarks [170]:

$$\Lambda_{\text{mom}} = \Lambda_{\text{MS}} \exp\left\{\frac{\delta(\alpha_G, n_f)}{\beta_1}\right\}\,. \tag{12.9}$$

In the usual case of three-gluon vertex, and for some particular values of the gauge parameter, one has:

$$\delta(0, 3) = -8.46\,, \qquad \delta(1, 3) = -7.68\,, \tag{12.10}$$

which leads to the numerically lowest order relation:

$$\Lambda_{\text{mom}} \simeq \Lambda_{\overline{\text{MS}}} \begin{pmatrix} 6.55 & \text{for } \alpha_G = 0: \text{ Landau gauge} \\ 5.51 & \text{for } \alpha_G = 1: \text{ Feynman gauge} \end{pmatrix} \tag{12.11}$$

## 12.3 The Weinberg renormalization scheme

The Weinberg scheme [171] is variant of the MOM scheme. In this scheme the renormalized two-point function reads:

$$\Psi_5(q^2, m^2)_R = \Psi_5(q^2, m^2) - \Psi_5(q^2 = -\mu^2, m^2 = 0)\,, \tag{12.12}$$

and is renormalized at an off-shell space-like point $q^2 = -\mu^2$ and putting the particle masses to be zero. It coincides with the MOM scheme, in the case of massless theories. One can see that, in this scheme, the renormalization constants are also quark-mass dependent. It has been shown by [172] that the Weinberg scheme violates the Slavnov–Taylor identities due to the arbitrariness of the subtraction point at a specific vertex, the gauge dependence of the coupling and to the definition of the tensorial structure of the vertex at the subtraction point.

## 12.4 The BLM scheme

The BLM (Brodsky–Lepage–Mackenzie) scheme has been introduced in [173] and has been based on the analogy with QED where only the light fermion vacuum polarizations (VP) contribute to the renormalization of the strong coupling constant. In QED, the running effective charge can be defined as (see the next chapter on QED):

$$\alpha(Q) = \frac{\alpha}{1 + e^2\Pi_{\text{em}}(Q)}\,, \tag{12.13}$$

where to lowest order in $\alpha$, and using an on-shell renormalization:

$$\Pi_{\text{em}}(Q) = -\frac{1}{4\pi^2}\left(\frac{2}{3}\ln\frac{Q}{m_e} - \frac{5}{3}\right)\,. \tag{12.14}$$

The scheme states that an observable $\mathcal{O}$ which has the perturbative expansion:

$$\mathcal{O} = C_0\alpha(Q)\left[1 + C_1\frac{\alpha(Q)}{\pi} + \cdots\right] \tag{12.15}$$

can be replaced by:

$$\mathcal{O} = C_0\alpha(Q_0^*)\left[1 + C_1^*\frac{\alpha(Q_1^*)}{\pi} + \cdots\right] \tag{12.16}$$

where all VP corrections are absorbed into the effective coupling by an appropriate and unique choice of scales $Q_0^*$, $Q_1^*, \ldots$. Since the number $n_f$ of light flavour dependences usually enters the VP to this order, then, both $Q_i^*$ and $C_i^*$ are independent of $n_f$, while, in general, the scales $Q_i^*$ can depend on the ratio of invariants. Taking the example of the anomalous magnetic moment of the leptons, which can be expressed as (see QED section):

$$a_e = \frac{\alpha}{2\pi}\left[1 - 0.657\frac{\alpha}{2\pi}\right], \tag{12.17}$$

and the VP contribution to the muon anomaly gives:

$$A_{VP}\frac{\alpha}{\pi}a_\mu^0 = \left[\frac{2}{3}\ln\frac{m_\mu}{m_e} - \frac{25}{18}\right]\frac{\alpha}{\pi}a_\mu^0 . \tag{12.18}$$

For the muon, one can expect that, at a scale $Q^* \sim m_\mu$, the exact result can be expressed as:

$$a_\mu = \frac{\alpha(Q^*)}{2\pi}, \tag{12.19}$$

where the running coupling is defined in Eq. (12.13), such that Eq. (12.18) and Eq. (12.19) must be equal. In this way, one obtains:

$$Q^* = m_\mu e^{5/12} . \tag{12.20}$$

In this procedure, the electron and the muon anomaly have the same expression to this order, as we replace:

$$a_\mu = \frac{\alpha}{2\pi}\left[1 + \frac{\alpha}{\pi}(A_{VP} + C_1) + \cdots\right], \tag{12.21}$$

by:

$$a_\mu = \frac{\alpha(Q^*)}{2\pi}\left[1 + \frac{\alpha(Q^*)}{\pi}C_1 + \cdots\right], \tag{12.22}$$

where:

$$\alpha(Q^*) \simeq \frac{\alpha}{1 - (\alpha/\pi)A_{VP}}, \tag{12.23}$$

and:

$$C_1 = -0.657 . \tag{12.24}$$

In the case of QCD, a similar approach can be made. The observable can be written as:

$$\mathcal{M} = C_0\alpha_{\overline{MS}}(Q)[1 + (\alpha_{\overline{MS}}(Q)/\pi)(n_f A_{VP} + B)] . \tag{12.25}$$

One can change the coupling by:

$$\alpha_{\overline{MS}}(Q^*) = \alpha_{\overline{MS}}(Q)\left[1 - \beta_1(\alpha_{\overline{MS}}(Q)/\pi)\ln\frac{Q^*}{Q} + \cdots\right]^{-1} . \tag{12.26}$$

and express the observable as:

$$\mathcal{M} = C_0 \alpha_{\overline{MS}}(Q^*)[1 + (\alpha_{\overline{MS}}(Q^*)/\pi)C_1^* + \cdots] . \tag{12.27}$$

Then, one can deduce:

$$\begin{aligned} Q^* &= Q \exp(3A_{VP}) , \\ C_1^* &= \frac{33}{2} A_{VP} + B , \end{aligned} \tag{12.28}$$

where the term $\frac{33}{2} A_{VP}$ in $C_1^*$ serves to remove the part of $B$ which renormalizes the leading-order coupling.

The ratio of these gluonic corrections to the light quark ones is fixed by the $\beta$ function. In some of the examples given by BLM, the value of $Q^*$ appears to be lower than the original scale of the process, which might be inconvenient for the convergence of the QCD series. Moreover, the scheme dependence of the result in Eq. (12.28) has been pointed out in [174], while an extension of the BLM result beyond NLO shows an ambiguity in the prescription [175]. Recent interest in the resummation of perturbative QCD series using large value of $\beta$ in the naïve Abelization of QCD (see Renormalons section) has revived the use of the BLM scheme despite these previous drawbacks of the procedure.

## 12.5 The PMS optimization scheme

The principle of minimal sensitivity (PMS) scheme has been introduced by Stevenson [176] in QCD. It consists on the fact that physical quantities should be insensitive to a small variation of unphysical parameters, and is based on variational approach. It is more instructive to illustrate the method by the classical example of the $e^+e^- \to$ hadrons total cross-section, which is known to high-accuracy in perturbative QCD. To order $\alpha_s^2$, the corresponding Adler $D$-function reads:

$$D(q^2) \equiv -q^2 \int_0^\infty \frac{dt}{(t-q^2)^2} R(t) \simeq \sum_i Q_i^2 \{1 + [D_2 \equiv a_s (1 + a_s F_3)] + \cdots\} , \tag{12.29}$$

where:

$$R \equiv \frac{\sigma(e^+e^- \to \text{hadrons})}{\sigma(e^+e^- \to \mu^+\mu^-)} . \tag{12.30}$$

$Q_i$ is the quark charge in units of $e$; $F_3$ is renormalization scheme dependent; $a_s \equiv \bar{\alpha}_s/\pi$ is the QCD running coupling. The $\nu$ (subtraction scale) dependence of the dimensional renormalization scheme can be introduced via:

$$\tau \equiv -\beta_1 \ln(\nu/\Lambda) . \tag{12.31}$$

Using the differential equation obeyed by the running coupling:

$$-\beta_1 \frac{\partial a_s}{\partial \tau} = a_s \beta(a_s) = \beta_1 a_s^2 \left(1 + \frac{\beta_2}{\beta_1} a_s\right) , \tag{12.32}$$

one obtains:

$$\frac{\partial D_2}{\partial \tau} = -a_s^2\left(1+\frac{\beta_2}{\beta_1}a_s\right)(1+2F_3a_s)+a_s^2\frac{\partial F_3}{\partial \tau}\,. \tag{12.33}$$

Using the fact that $D_2$ is independent of $\tau$, the $a_s^2$ term in Eq. (12.33) must vanish, which leads to:

$$F_3(\tau) = \tau - \tau_0 + F_3(\tau_0)\,. \tag{12.34}$$

The optimization criterion imposes that the remainder term of $\partial D_2/\partial\tau$ also vanishes at a critical value $\tau \equiv \tau_c$. The optimal value of $F_3$ corresponds to:

$$\frac{\beta_2}{\beta_1} + 2F_3^{\rm opt}\left(1+\frac{\beta_2}{\beta_1}a_s(\tau_c)\right) = 0\,, \tag{12.35}$$

where the rôle of $a_s\beta_2$ can be increased by computing the next order terms. From this result, one can deduce the optimal value of $D_2$:

$$D_2^{\rm opt} = a_s(\tau_c)\left[1-\frac{(\beta_2/\beta_1)a_s(\tau_c)}{2[1+(\beta_2/\beta_1)a_s(\tau_c)]}\right]. \tag{12.36}$$

The last step of the analysis is to find $a_s(\tau_c)$. This can be done by integrating Eq. (12.32). One obtains to two loops:

$$\hat{K}_2(a_s) \equiv \tau = \int_{a_s}^{\infty}\frac{dx}{x^2[1+(\beta_2/\beta_1)a_s(x)]} = \frac{1}{a_s}+\frac{\beta_2}{\beta_1}\ln\left(\frac{(\beta_2/\beta_1)a_s}{1+(\beta_2/\beta_1)a_s}\right), \tag{12.37}$$

where the upper limit of integration is equivalent to the choice of $\Lambda$ in $\tau = -\beta_1\ln(\nu/\Lambda)$. Using Eq. (11.53) by including next leading corrections, one can derive the relation:

$$\Lambda_{\rm opt} = \Lambda_{MS}\left(-2\frac{\beta_2}{\beta_1}\right)^{\beta_2/\beta_1{}^2}. \tag{12.38}$$

Rewriting Eq. (12.34) as:

$$F_3 = \hat{K}_2(a_s) - \rho_1(Q)\,, \tag{12.39}$$

where:

$$\rho_1(Q) \equiv \tau_0 - F_3(\tau_0)\,, \tag{12.40}$$

is a constant term independent of the unphysical variable $\tau$ at fixed $Q$, where at $Q^2 \equiv -q^2 = \nu^2$, it reads:

$$\rho_1(Q^2=\nu^2) = -\beta_1\ln(Q/\Lambda) - F_3\,. \tag{12.41}$$

It is also a renormalization scheme invariant quantity, as the scheme dependence of $\Lambda$ cancels the one of $F_3$. Substituting the value of $F_3$ from Eq. (12.39) into Eq. (12.35), one gets the following transcendental equation for $a_s(\tau_c)$:

$$\hat{K}_2(a_s(\tau_c)) + \frac{1}{2}\frac{\beta_2}{\beta_1}\left(1+\frac{\beta_2}{\beta_1}(a_s(\tau_c))\right)^{-1} = \rho_1(Q) , \tag{12.42}$$

where the solution $a_s(\tau_c)$ is the one to be used in $D_2^{\text{opt}}$. As $\rho_1$ behaves like $1/a_s$, it needs to be large for a good description of the process. The PMS scheme has been quite popular in the period of 1980–1990.

At present, the interest in the method has decreased. This is probably related to the fact that it does not yet incorporate the power corrections which plays a non-negligible rôle in the extraction of the QCD coupling from different processes. However, an extension of the method including these non-perturbative corrections, although small, should be more attractive.

## 12.6 The effective charge scheme

Like the PMS, this scheme is also conceptually based on the construction of scheme-invariant quantities from combinations of scheme-dependent coefficients [177]. In order to illustrate the discussion, let's start from the $D$ function defined in Eq. (12.29), which we rewrite as:[1]

$$D_n \simeq \sum_i Q_i^2 \left\{1 + a_s d_0 \left(1 + \sum_{i=1}^{n-1} d_i a_s^i\right) + \cdots\right\} , \tag{12.43}$$

where all higher order corrections and scheme dependence of the process are absorbed into the definition of the coupling constant. The ECH approach imposes the condition that all coefficients $d_i = 0$ for all $i \geq 2$. Writing the $\beta$ function as:

$$\beta(\alpha_s) = -\beta_1 a_s \left(1 + \sum_{i=1}^{n-1} c_i a_s^i\right) , \tag{12.44}$$

and:

$$D_n^{\text{ECH}} = D_n(a_s) + \delta D_n^{\text{ECH}} , \tag{12.45}$$

these conditions imply for the remaining corrections to the physical quantities [178]:

$$\begin{aligned} \delta D_2^{\text{ECH}} &= d_0 d_1 (c_1 + d_2) \\ \delta D_3^{\text{ECH}} &= d_0 d_1 \left(c_2 - \frac{1}{2} c_1 d_1 - 2 d_1^2 + 3 d_2 + d_2\right) . \end{aligned} \tag{12.46}$$

These conditions are realized provided that the expansion of the $\beta$ function in terms of $a_s$ makes sense, which translates into the renormalization scheme-independent constraint:

$$c_1 a_s \equiv \frac{\beta_2}{\beta_1} a_s < 1 , \tag{12.47}$$

[1] We neglect in this discussion the small contribution due to the light by the light-scattering diagram (see next chapter).

which for four flavours corresponds to $Q > 1.62\lambda$. However, it is interesting to see the modification of this constraint when non-perturbative terms are included in the QCD series. In [178], relations between the corrections to $D_n$ in the PMS and ECH scheme have been also derived with the result:

$$\delta D_2^{\rm PMS} = \delta D_2^{\rm ECH} + \frac{d_0 c_1^2}{4}$$
$$\delta D_3^{\rm PMS} = \delta D_3^{\rm ECH} , \qquad (12.48)$$

as well as an extension of the analysis to $n = 4$.

# 13

# $\overline{MS}$ scheme for QED

Related discussions can e.g. be found in [147,110,111]. QED which is an Abelian theory is much simpler than the non-Abelian QCD. As we have mentioned in the introduction, QED works well with the on-shell renormalization scheme and has been experimentally tested with a high degree of accuracy ($g-2, \ldots$), such that it is a priori useless to introduce a new scheme for studying it. However, it is interesting to know the relations of different observables in the on-shell and $\overline{MS}$ schemes.

## 13.1 The QED Lagrangian

As we have already mentioned previously, the QED Lagrangian is quite simple, as we do not need ghost fields for its quantization due to its Abelian character. Its expression is given in Eq. (5.10), which we repeat below:

$$\mathcal{L}_{\mathrm{QED}} \equiv \mathcal{L}_\gamma + \mathcal{L}_l + \mathcal{L}_{\mathrm{I}} , \tag{13.1}$$

with:

$$\begin{aligned}
\mathcal{L}_\gamma &= -\frac{1}{4} F^{\mu\nu}(x) F_{\mu\nu}(x) - \frac{1}{2\alpha_G} \partial_\mu A^\mu(x) \partial_\nu A^\nu(x) , \\
\mathcal{L}_l &= \bar{\psi}(x)(i\partial_\mu\gamma^\mu - m)\psi(x) , \\
\mathcal{L}_{\mathrm{I}} &= -eA_\mu(x)\bar{\psi}(x)\gamma^\mu\psi(x)
\end{aligned} \tag{13.2}$$

where $F_{\mu\nu}(x) = \partial_\mu A_\nu(x) - \partial_\nu A_\mu(x)$ is the photon field strengths of the photon field $A_\mu(x)$; $\alpha_G$ is the gauge parameter with $\alpha_G = 0$ (Landau gauge) and $\alpha_G = 1$ (Feynman gauge); $\psi(x)$ and $m$ is the lepton field and mass; $e$ is the electric charge.

## 13.2 Renormalization constants and RGE

The renormalization constants of the fields are defined analogously to the case of QCD in Eqs. (9.20, 9.21, 9.22), and will not be repeated here. The QED Ward identity, analogous to the one in Eq. (9.24), gives:

$$Z_{1F} = Z_{2F} , \tag{13.3}$$

which implies:

$$Z_\alpha = Z_{3YM}^{-1} \equiv Z_3^{-1} , \tag{13.4}$$

while the gauge invariance of the QED Lagrangian implies that there is no renormalization counterterm for gauge term of the Lagrangian. Then:

$$Z_G = Z_3 \tag{13.5}$$

The RGE of QED has been originally introduced by [75,76] for improving the QED perturbation series. It has the form in Eq. (11.32) with the solution in Eq. (11.33).

## 13.3 $\beta$ function, running coupling and anomalous dimensions

The $\beta$ function is known to order $\alpha^3$. It reads [179]:

$$\begin{aligned}\beta(\alpha) &\equiv \frac{\nu}{Z_3}\frac{dZ_3}{d\nu} \\ &= \left(\beta_1 = \frac{2}{3}\right)\left(\frac{\alpha}{\pi}\right) + \left(\beta_2 = \frac{1}{2}\right)\left(\frac{\alpha}{\pi}\right)^2 + \left(\beta_3 = -\frac{121}{144}\right)\left(\frac{\alpha}{\pi}\right)^3 ,\end{aligned} \tag{13.6}$$

where, in the case of massless fermion, $\beta_1$ and $\beta_2$ are invariant under the renormalization schemes. It is important to notice the crucial difference between QCD and QED, as here the $\beta$ function is positive, which means that the origin of the coupling is no longer an UV fixed point and the QED coupling increases with $q^2$. The running QED coupling is solution of the differential equation analogue to the one in Eq. (11.34). Its solution to one loop is:

$$\bar{\alpha}(q^2) = \frac{\alpha(\nu)}{1 - (\alpha(\nu)/\pi)\,\beta_1 \frac{1}{2}\ln(-q^2/\nu^2)} \equiv \frac{\pi}{\beta_1 \frac{1}{2}\ln\left(-q^2/\Lambda_{em}^2\right)} , \tag{13.7}$$

where in the last identity, we have introduced the RGI parameter $\Lambda_{em}$ in analogy of QCD.

The anomalous dimensions are:

$$\begin{aligned}\gamma_F &\equiv \frac{\nu}{Z_{2F}}\frac{dZ_{2F}}{d\nu} = \frac{\alpha_G}{2}\left(\frac{\alpha}{\pi}\right) + \mathcal{O}(\alpha^2) , \\ \gamma_m &\equiv \frac{\nu}{Z_m}\frac{dZ_m}{d\nu} = \left(\gamma_{\rm em} \equiv \frac{3}{2}\right)\left(\frac{\alpha}{\pi}\right) + \mathcal{O}(\alpha^2) , \\ \gamma_\Gamma &\equiv \frac{\nu}{Z_\Gamma}\frac{dZ_\Gamma}{d\nu} = -\frac{1}{2}[n_\gamma \beta(\alpha) + n_F \gamma_F] ,\end{aligned} \tag{13.8}$$

where $n_\gamma$ and $n_F$ are respectively the number of external photon and fermion fields.

## 13.4 Effective charge and link between the $\overline{MS}$ and on-shell scheme

One can relate the electric charge in the QED on-shell scheme to the one in the $\overline{MS}$ scheme, by using the fact that the QED effective charge is invariant under the choice of renormalizations:[1]

$$\alpha_{\text{eff}}(q^2) = \frac{\alpha_B}{1 + e_B^2 \Pi_{\text{em}}^B} = \frac{\alpha_R}{1 + e_R^2 \Pi_{\text{em}}^R}, \tag{13.9}$$

where the indices $R$ and $B$ denote renormalized and bare quantities; $\alpha = e^2/4\pi$ is the fine structure constant; $\Pi_{\text{em}}$ is the electromagnetic vacuum polarization defined as:

$$\Pi_{\text{em}}^{\mu\nu}(q) = i \int d^4x \, e^{iqx} \langle 0|\mathcal{T} J^\mu(x) J^\nu(0)^\dagger|0\rangle = -(g^{\mu\nu} q^2 - q^\mu q^\nu)\Pi_{\text{em}}(q^2), \tag{13.10}$$

where $J^\mu = \bar{l}\gamma^\mu l$ is the local current built from the lepton field. Using the Feynman rule given in Appendix E and the Dirac algebra in $n$ dimensions in Appendix D, its expression can be easily obtained, if one follows closely the derivation of the pseudoscalar two-point function discussed previously. The *bare* two-point correlator reads to lowest order:

$$\Pi_{\text{em}}^B(q^2) = \frac{1}{12\pi^2}\left\{\frac{2}{\epsilon} + (\ln 4\pi) - \gamma - 3\int_0^1 dx \, 2x(1-x) \ln \frac{-q^2 x(1-x) + m^2 - i\epsilon'}{\nu^2}\right\}. \tag{13.11}$$

From this expression, one can deduce to leading order in $m^2/q^2$:

$$\Pi_{\text{em}}^B(-q^2 \gg m^2) = \frac{1}{12\pi^2}\left\{\frac{2}{\epsilon} + (\ln 4\pi) - \gamma - \ln \frac{-q^2}{\nu^2} + \frac{5}{3} - 6\frac{m^2}{-q^2} + 6\left(\frac{m^2}{-q^2}\right)^2 \ln \frac{-q^2}{m^2}\right\}, \tag{13.12}$$

and:

$$\Pi_{\text{em}}^B(q^2 = 0) = \frac{1}{12\pi^2}\left\{\frac{2}{\epsilon} + (\ln 4\pi) - \gamma - \ln \frac{m^2}{\nu^2}\right\}. \tag{13.13}$$

The renormalized *vacuum polarization* can be easily obtained:

$$\begin{aligned} \Pi_{\text{em}}^{\text{o.s.}} &\simeq \frac{1}{12\pi^2}\left\{-\ln \frac{-q^2}{m^2} + \frac{5}{3} + \cdots\right\} \\ \Pi_{\text{em}}^{\overline{MS}} &= \frac{1}{12\pi^2}\left\{-\ln \frac{-q^2}{\nu^2} + \frac{5}{3} + \cdots\right\} \end{aligned} \tag{13.14}$$

Using the fact that $\Pi_{\text{em}}^{\text{on-shell}}(q^2 = 0) = 0$ in Eq. (13.9), one can relate the charge in the on-shell and $\overline{MS}$ scheme as.[2]

$$\alpha(\nu) = \alpha_{\text{o.s.}}\left\{1 + \left(\frac{\alpha_{\text{o.s.}}}{\pi}\right)\frac{1}{3}\ln \frac{\nu^2}{m^2}\right\} \tag{13.15}$$

[1] The last equality comes from the fact that the charge renormalization constant $Z_\alpha$ and the photon field renormalization constant $Z_3$ are related to each others from Eq. (13.4).

[2] For a more elegant notation, we shall not put index for the $\overline{MS}$ coupling.

Using this relation into Eq. (13.7), one can deduce the running coupling in the $\overline{MS}$ scheme in terms of the one in the on-shell scheme evaluated at $\nu = m$:

$$\bar{\alpha}(q^2) = \frac{\alpha_{\text{o.s.}}}{1 - (\alpha_{\text{o.s.}}/\pi)\,\beta_1 \frac{1}{2}\ln(-q^2/m^2)} \,. \tag{13.16}$$

Identifying this result with the one in Eq. (13.7), one can deduce for one fermion:

$$\Lambda_{\text{em}} = m\,\exp\left(\frac{\pi}{\beta_1 \alpha_{\text{o.s.}}}\right) . \tag{13.17}$$

This result can be easily generalized to $n$ fermions of mass $m_i$:

$$\Lambda_{\text{em}} = \left(\prod_{i=1}^{n} m_i\right) \exp\left(\frac{\pi}{n\beta_1 \alpha_{\text{o.s.}}}\right) . \tag{13.18}$$

For the three observed charged leptons $e,\ \mu, \tau$, this leads to:

$$\Lambda_{\text{em}} = 5.2 \times 10^{93}\ \text{GeV}\,, \tag{13.19}$$

which is an astronomical number. This is the scale at which one expects that the QED series expansion breaks down, and is commonly called as the *Landau pole*. Using its definition in Eq. (13.9), the effective QED charge can be expressed in terms of the running charge. It reads, in the $\overline{MS}$ scheme:

$$\alpha_{\text{eff}}(q^2) \simeq \bar{\alpha}(q^2)\left\{1 - \left(\frac{\bar{\alpha}}{\pi}\right)\frac{1}{3}\left(\frac{5}{3}\right) + \mathcal{O}\left(\frac{m^2}{q^2}\right)\right\} . \tag{13.20}$$

Analogous relation from Eq. (13.9) can also be obtained in the on-shell scheme. The identification of the two relations for $\alpha_{\text{eff}}$ leads to the relation between the running coupling in the $\overline{MS}$ and on-shell schemes:

$$\bar{\alpha}(q^2) = \bar{\alpha}_{\text{o.s.}}(q^2)\left\{1 + \left(\frac{\bar{\alpha}_{\text{o.s.}}}{\pi}\right)\frac{1}{3}\ln\frac{-q^2}{m^2} + \mathcal{O}\left(\frac{\bar{\alpha}_{\text{o.s.}}}{\pi}\right)^2\right\} . \tag{13.21}$$

This relation is useful in the analysis of electroweak processes where the Green's functions are often evaluated using the $\overline{MS}$ scheme.

# 14

# High-precision low-energy QED tests

As shown in the previous chapter, one expects that QED works well until an astronomical value of $\Lambda_{\text{em}}$, which is due to the non-asymptotically free property of the theory, where the effective charge grows with the energy. Therefore, contrary to QCD, one expects that QED is best tested at low energies which is the interesting experimental region.[1]

## 14.1 The lepton anomaly

Indeed, one of the most impressive and traditional test of QED is given by the measurements of the leptons ($e,\ \mu$) anomalous magnetic moments (lepton anomaly) $a_\mu$, where one notices that from the electron to the muon, the running charge has increased from $\bar{\alpha}(t=0)$ to $\bar{\alpha}(t=\ln(m_\mu/m_e))$. The anomalous magnetic moment of charged leptons and for on-shell photon ($q^2=0$) is defined as:

$$a_l \equiv \frac{1}{2}(g_l - 2) \equiv F_2(0)\,, \tag{14.1}$$

where $F_2(0)$ is the Pauli form factor related to the lepton-photon-lepton vertex as:

$$\bar{u}(p')\Gamma_\mu(p^2 = p'^2 = m^2)u(p) = \bar{u}(p')\gamma_\mu u(p)F_1(q^2) - \frac{1}{2m}\bar{u}(p')\sigma_{\mu\nu}q^\nu u(p)F_2(q^2)\,. \tag{14.2}$$

The full vertex and the lowest order QED contribution are given by Fig. 14.1.

The lowest order contribution is:

$$a_l^{(2)} = \frac{1}{2}\left(\frac{\alpha}{\pi}\right)\,. \tag{14.3}$$

### *14.1.1 The electron anomaly and measurement of fine structure constant $\alpha$*

In the case of the electron, the anomalous magnetic moment has been measured with a high accuracy [181]:

$$\begin{aligned} a_{e^-}^{\text{exp}} &= 115\,965\,218\,84(43) \times 10^{-13}\,, \\ a_{e^+}^{\text{exp}} &= 115\,965\,218\,79(43) \times 10^{-13}\,. \end{aligned} \tag{14.4}$$

[1] For some other low-energy tests of the fermion substructure beyond the standard model, see e.g. [180].

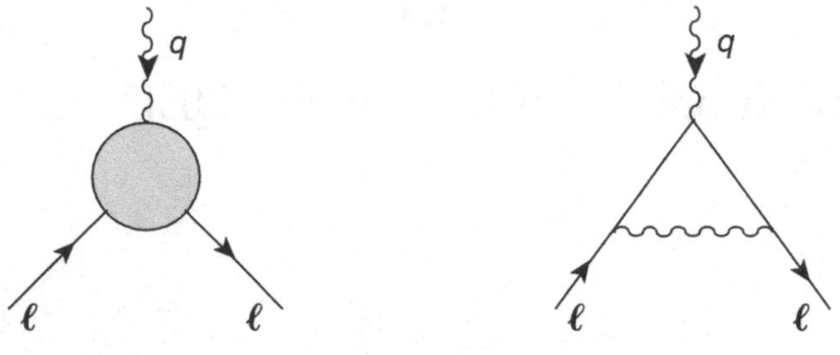


Fig. 14.1. Full vertex and lowest order QED contribution to $a_l$.

A comparison of this value with the theoretical prediction [182,183,184]:

$$a_e^{\rm SM} = \frac{1}{2}\left(\frac{\alpha}{\pi}\right) - 0.328\,478\,444\,00\left(\frac{\alpha}{\pi}\right)^2 + 1.181\,234\,017\left(\frac{\alpha}{\pi}\right)^3 \\ - 1.509\,8(384)\left(\frac{\alpha}{\pi}\right)^4 + 1.66(3)\times 10^{-12}(\text{hadronic} + \text{electroweak loops}) \quad (14.5)$$

provides a very precise measurement of the QED charge (fine structure constant) at the scale of the electron mass [185]:

$$\alpha^{-1}(a_e) = 137.035\,999\,58(52)\,, \quad (14.6)$$

which is more precise than the one from the quantum Hall effect [182]:

$$\alpha^{-1}(\text{Hall}) = 137.036\,003\,00(270)\,. \quad (14.7)$$

A resolution of the two discrepancies can provide a bound on new physics which is however not very strong as the new physics scale is constrained to be only larger than 100 GeV, assuming a generic effect of the order of $m_e^2/\Lambda^2$.

### *14.1.2 The muon anomaly and the rôle of the hadronic contributions*

In the case of the muon, higher order QED contributions are known up order $\alpha^5$. Typical higher order QED diagrams are shown in Fig. 14.2.

The total QED contribution reads [184,186]:

$$a_\mu^{\rm QED} = \frac{1}{2}\left(\frac{\alpha}{\pi}\right) + 0.765\,857\,376(27)\left(\frac{\alpha}{\pi}\right)^2 + 24.050\,508\,98(44)\left(\frac{\alpha}{\pi}\right)^3 \\ + 126.07(41)\left(\frac{\alpha}{\pi}\right)^4 + 930(170)\left(\frac{\alpha}{\pi}\right)^5 \\ = 116\,584\,705.7(2.9)\times 10^{-11}\,. \quad (14.8)$$

The electroweak contributions are known. In the standard model, the lowest order contributions come from Fig. 14.3.

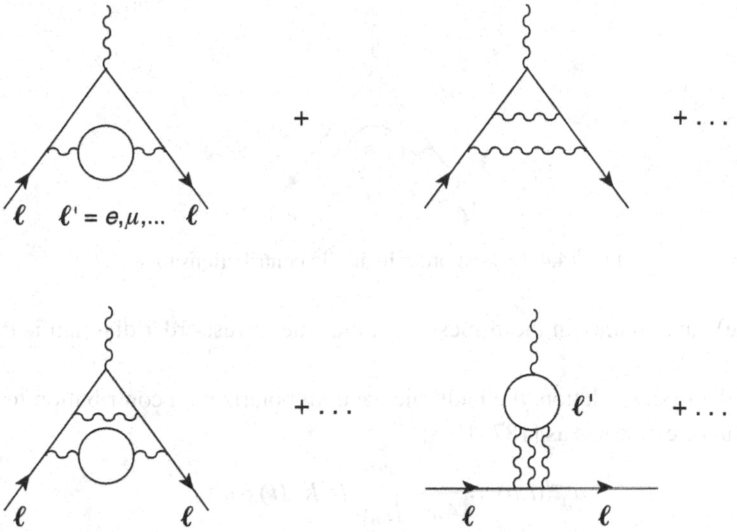


Fig. 14.2. Typical higher order QED contributions to $a_l$.

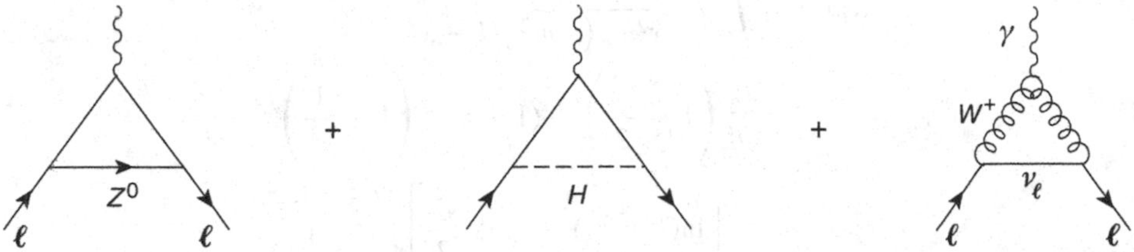


Fig. 14.3. Lowest order electroweak contribution to $a_l$.

It reads [186]:

$$a_\mu^{\rm EW}(1-\text{loop}) = \frac{G_\mu m_\mu^2}{8\sqrt{2}}\frac{5}{3\pi^2}\left[1+\frac{1}{5}(1-4\sin^2\theta_W)^2+\mathcal{O}\left(\frac{m_\mu}{M}\right)^2\right] \simeq 195\times 10^{-11}\,, \tag{14.9}$$

where the $G_\mu = 1.16637(1)\times 10^{-5}$ GeV$^2$, $\sin^2\theta_W = 0.223$ and $M$ denotes $M_{\rm W}$ or $M_{\rm Higgs}$. The full two-loop contribution including hadronic electroweak loops and a leading log-resummation is [186]:

$$a_\mu^{\rm EW}(2-\text{loop}) = -43(4)\times 10^{-11} \tag{14.10}$$

and gives the total contribution:

$$a_\mu^{\rm EW} = 152(4)\times 10^{-11}\,. \tag{14.11}$$

### 14.1.3 The lowest order hadronic contributions

The main theoretical errors in the determination of the muon anomalous magnetic moment are due to the strong interaction (hadronic loop) contributions, in the region

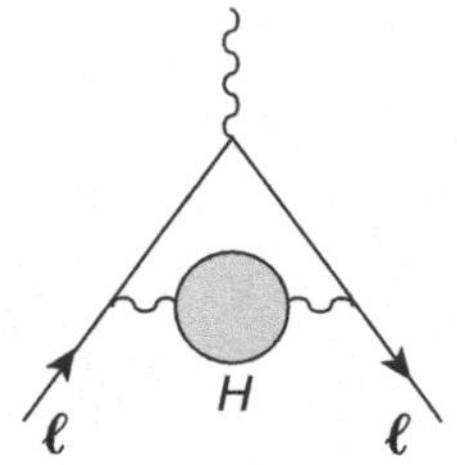


Fig. 14.4. Lowest order hadronic contributions to $a_l$.

below 2 GeV, and mainly in the $\rho$ meson region. The lowest order diagram is depicted in Fig. 14.4.

Using a dispersion relation, the hadronic vacuum polarization contribution to the muon anomaly can be expressed as [187–191]:

$$a_\mu^{\rm had}(l.o) = \frac{1}{4\pi^3}\int_{4m_\pi^2}^{\infty} dt\ K_\mu(t)\,\sigma_H(t)\,. \tag{14.12}$$

$K_\mu(t)$ is the QED kernel function [192]:

$$\begin{aligned} K_\mu(t) &= \int_0^1 dx \frac{x^2(1-x)}{x^2+\left(t/m_\mu^2\right)(1-x)} \\ &= z_\mu^2\left(1-\frac{z_\mu^2}{2}\right) + (1+z_\mu)^2\left(1+\frac{1}{z_\mu^2}\right) \\ &\quad\times\left[\ln(1+z_\mu) - z_\mu + \frac{z_\mu^2}{2}\right] \\ &\quad+\left(\frac{1+z_\mu}{1-z_\mu}\right) z_\mu^2 \ln z_\mu\,, \end{aligned} \tag{14.13}$$

with:

$$z_\mu = \frac{(1-v_\mu)}{(1+v_\mu)}\,, \quad \text{and} \quad v_\mu = \sqrt{1-\frac{4m_\mu^2}{t}}\,. \tag{14.14}$$

$K_\mu(t)$ is a monotonically decreasing function of $t$. For large $t$, it behaves as:

$$K_\mu\left(t > m_\mu^2\right) \simeq \frac{m_\mu^2}{3t}\,, \tag{14.15}$$

which will be useful for the analysis in the large $t$ regime. Such properties then emphasize the importance of the low-energy contribution to $a_\mu^{\rm had}(l.o)$, where the QCD theory cannot be (strictly speaking) applied. $\sigma_H(t) \equiv \sigma(e^+e^- \to \text{hadrons})$ is the $e^+e^- \to$ hadrons total cross-section which can be related to the hadronic two-point spectral function $\text{Im}\Pi(t)_{\rm em}$ through the optical theorem:

$$R_{e^+e^-} \equiv \frac{\sigma(e^+e^- \to \text{hadrons})}{\sigma(e^+e^- \to \mu^+\mu^-)} = 12\pi\,\text{Im}\Pi(t)_{\rm em}\,, \tag{14.16}$$

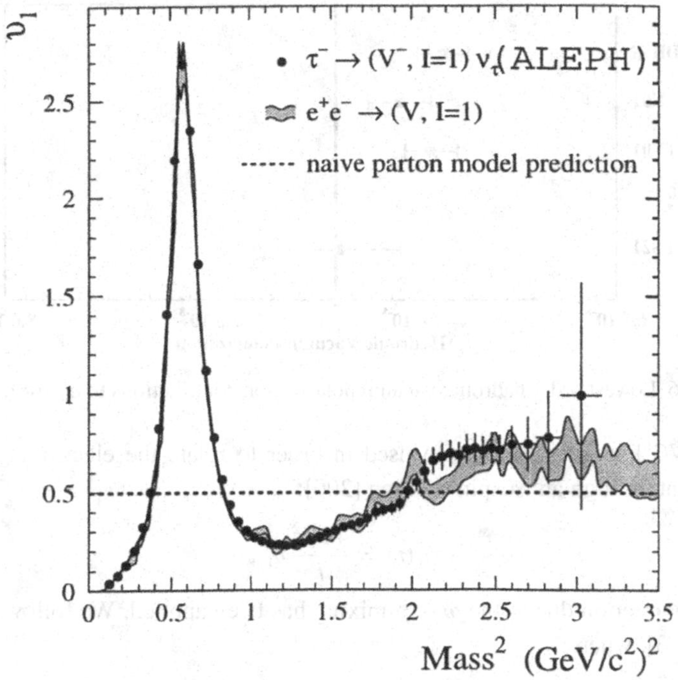


Fig. 14.5. Isovector spectral function from tau-decay and comparison with the $e^+e^-$ data.

where:

$$\sigma(e^+e^- \to \mu^+\mu^-) = \frac{4\pi\alpha^2}{3t} . \tag{14.17}$$

Here,

$$\Pi^{\mu\nu}_{\text{em}} \equiv i \int d^4x \, e^{iqx} \, \langle 0|T J^\mu_{\text{em}}(x) \left(J^\nu_{\text{em}}(x)\right)^\dagger |0\rangle$$
$$= -(g^{\mu\nu}q^2 - q^\mu q^\nu)\Pi_{\text{em}}(q^2) \tag{14.18}$$

is the correlator built from the local electromagnetic current:

$$J^\mu_{\text{em}}(x) = \frac{2}{3}\bar{u}\gamma^\mu u - \frac{1}{3}\bar{d}\gamma^\mu d - \frac{1}{3}\bar{s}\gamma^\mu s + \cdots \tag{14.19}$$

The present most precise result comes from combining the $e^+e^- \to$ hadrons compiled in [193–198,16] and the precise $\tau$-decay data [193,199]. These data are shown in Fig. 14.5.

An average of the results from $e^+e^- \to$ hadrons: $a_\mu^{\text{had}}[e^+e^-] = 7016(119) \times 10^{-11}$ and $\tau$ decay $a_\mu^{\text{had}}[\tau] = 7036(76) \times 10^{-11}$ data leads to [201]:

$$a_\mu^{\text{had}}(l.o) = 7021(76) \times 10^{-11} , \tag{14.20}$$

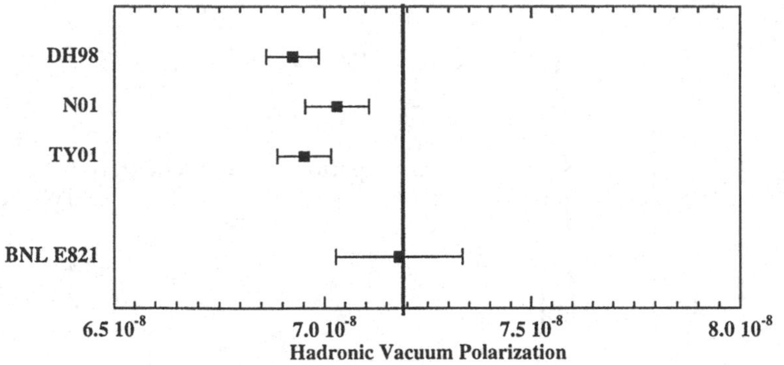


Fig. 14.6. Lowest order hadronic vacuum polarization contributions to $a_\mu$ from [207].

where the CVC hypothesis has been used in order to relate the electromagnetic to the charged current through an isospin rotation [200]:

$$\sigma_H(t) = \frac{4\pi\alpha^2}{t} v_1 , \tag{14.21}$$

and where a correction due to the $\omega - \rho$ mixing has been applied. We follow the notation of ALEPH [193], where:

$$\mathrm{Im}\Pi^{(1)}_{\bar{u}d,V} \equiv \frac{1}{2\pi} v_1 , \tag{14.22}$$

is the charged vector two-point correlator:

$$\begin{aligned} \Pi^{\mu\nu}_{\bar{u}d,V} &\equiv i \int d^4x \; e^{iqx} \langle 0|\mathcal{T} J^\mu_{\bar{u}d}(x) \left(J^\nu_{\bar{u}d}(0)\right)^\dagger |0\rangle \\ &= -(g^{\mu\nu}q^2 - q^\mu q^\nu)\Pi^{(1)}_{\bar{u}d,V}(q^2) \\ &\quad + q^\mu q^\nu \Pi^{(0)}_{\bar{u}d,V}(q^2) , \end{aligned} \tag{14.23}$$

built from the local charged current $J^\mu_{\bar{u}d,V}(x) = \bar{u}\gamma^\mu d(x)$. It is amusing to notice that the central value here coincide with the old result in [209,215]. In [202], the impressive agreement of the result of the hadronic contributions to the QED running coupling $\alpha$ and to the muonium hyperfine splitting with other determinations (see next section) is a strong support of the estimate obtained in [201]. In Fig. 14.6, we see that the most recent standard model theoretical determinations [205,201] are in good agreement with the measured value [206,207]. After the completion of this work, we became aware of recent determinations [203] using recent $e^+e^- \to$ hadrons data from CMD-2 and BES [204] and of more precise measurement of $a_\mu^{\rm exp} = 11\,659\,204(7)\,(5) \times 10^{-10}$ [208]. The estimate based on $\tau$-decay agrees with ours whereas the one from $e^+e^-$ differs by $1.4\sigma$ from ours, and leads to a discrepancy of $3\sigma$ between $a_\mu^{\rm exp}$ and $a_\mu^{\rm SM}$ of the Standard Model predictions. However, a new analysis of the scalar meson contributions (SN, hep − ph/0303004) gives an additional effect $a_\mu^S \leq 13(11)\,10^{-10}$, which reduces the discrepancy of $a_\mu^{\rm SM}$ and $a_\mu^{\rm exp}$. The difference between the results from $e^+e^-$ and $\tau$ decay still needs to be understood.

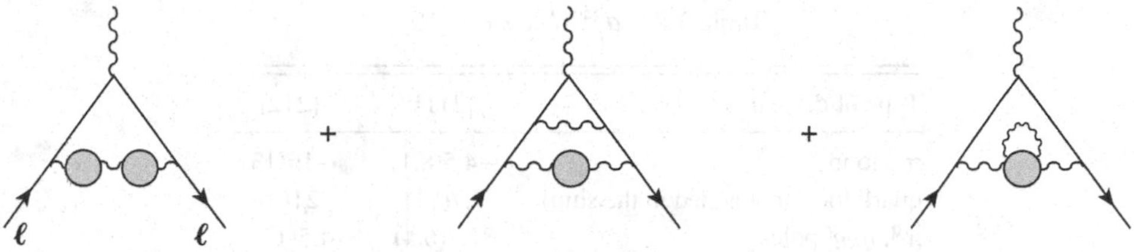


Fig. 14.7. Higher order hadronic vacuum polarization contributions to $a_l$.

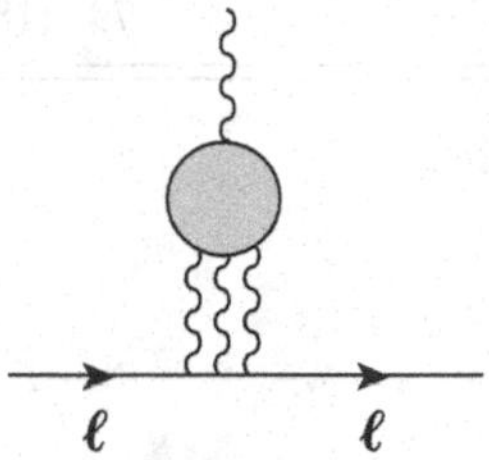


Fig. 14.8. Light-by-light scattering hadronic contribution to $a_\mu$.

### *14.1.4 The higher order hadronic contributions*

Higher order contributions were first discussed in [209]. They can be divided into two clases. The one involving the vacuum polarization is given in Fig. 14.7, and can be related to the measured $e^+e^- \to$ hadrons total cross-section, similar to the lowest order contribution. It gives [201] after rescaling the result in [209,210]:

$$a_\mu^{\rm had}(h.o.)_{\rm VP} = -101.2(6.1) \times 10^{-11} \ , \tag{14.24}$$

The second class is the light-by-light scattering diagram shown in Fig. 14.8.

Contrary to the case of vacuum polarization, this contribution is not yet related to a direct measurable quantity. In order to estimate this contribution, one has to introduce some theoretical models. The ones used at present are based on chiral perturbation [211] and ENJL model [212], where to both are added vector meson dominance and phenomenological parametrization of the pion form factors. The different contributions can be classified in diagrams Fig. 14.9, where the first two come from the quark (constituent) and boson loops, whereas the last one is due to meson pole exchanges. The first two diagrams are quite sensitive to the effects of rho-meson attached at the three off-shell photon legs which reduce the contributions by about one order of magnitude. The third diagram with pseudoscalar meson exhanges (anomaly) gives so far the most important contribution. There is a complete agreement between the two model estimates (after correcting the sign of the pseudoscalar and axial-vector contributions [213)], which may indirectly indicate that the results obtained are model independent. However, there are still some subtle issues left to be understood (is the inclusion of a quark loop a double counting?; why the inclusion of the rho-meson decreases drastically the quark and pion loop contributions?; is a single pole dominance

Table 14.1. $a_\mu^{\rm had}(h.o)_{\rm LL} \times 10^{11}$

| Type of diagrams | [211] | [212] |
|---|---|---|
| $\pi^-$ loop | −4.5(8.1) | −19(13) |
| quark loop (not added in the sum) | 9.7(11) | 21(3) |
| $\pi^0$, $\eta$, $\eta'$ poles | 82.7(6.4) | 85(13) |
| axial-vector pole | 1.74 | 2.5(1.0) |
| scalar pole | | −6.8(2.0) |
| Total | 79.9(18.2) | 61.7(23.8) |

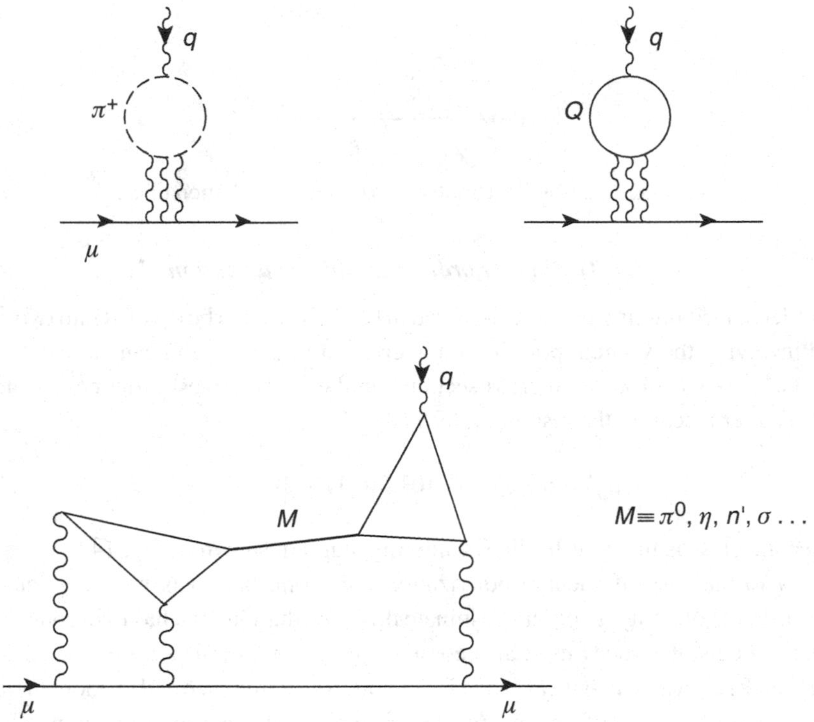


Fig. 14.9. Different light-by-light scattering hadronic contributions to $a_\mu$.

justified?, . . . ). The results in [211] and [212], after correcting the sign of the pseudoscalar and axial-vector contributions [213], are given in Table 14.1.

An arithmetic average of the central values and of the errors give:

$$a_\mu^{\rm had}(h.o)_{\rm LL} = 70.8(21.0) \times 10^{-11} . \qquad (14.25)$$

One can remark the agreement in sign and magnitude with the contribution of a quark constituent loop diagram (first used in [209]) without any hadrons [214] and YT01 in [205].

Then, we deduce:

$$a_\mu^{\rm had}(h.o) = a_\mu^{\rm had}(h.o)_{\rm VP} + a_\mu^{\rm had}(h.o)_{\rm LL} = -30.4(21.9) \times 10^{-11} \,, \qquad (14.26)$$

where one can notice a partial cancellation between the higher order vacuum polarization and the light-by-light scattering contributions.

### 14.1.5 The total theoretical contributions to $a_\mu$

Summing up different contributions, the present theoretical status in the standard model is:

$$\begin{aligned} a_\mu^{\rm SM} &= a_\mu^{\rm QED} + a_\mu^{\rm EW} + a_\mu^{\rm had} \\ &= 116\,584\,669.9(39.2) \times 10^{-11} + a_\mu^{\rm had}(l.o.) \\ &= 116\,591\,846.9(78.9) \times 10^{-11} \,. \end{aligned} \qquad (14.27)$$

This theoretical contribution can be compared with the experimental average [186] of CERN78 [216] + BNL98 [207] + BNL99 [208]:

$$a_\mu^{\rm exp}(\text{average}) = 116\,592\,023(151) \times 10^{-11} \,. \qquad (14.28)$$

which is more weighted by the new BNL precise result:

$$a_\mu^{\rm exp}(\text{BNL99}) = 116\,592\,020(160) \times 10^{-11} \,. \qquad (14.29)$$

The comparison of the theoretical and experimental numbers gives:

$$a_\mu^{\rm new} \equiv a_\mu^{\rm exp} - a_\mu^{\rm SM} = (176 \pm 170) \times 10^{-11} \,, \qquad (14.30)$$

which can indicate about $\sigma$ deviation from the SM prediction. This result can be used to give a lower bound on the presence of new physics beyond the standard model. Combining this result with the world average of $a_\mu^{\rm had}(l.o.)$, we can have the range at 90% confidence level (CL):

$$-42 \leq a_\mu^{\rm new} \times 10^{11} \leq 413 \quad (90\% \text{ CL}) \,. \qquad (14.31)$$

This range is expected to be improved in the near future both from accurate measurements of $a_\mu$ and of $e^+e^-$ data necessary for reducing the theoretical errors in the determinations of the hadronic contributions, being the major source of the theoretical uncertainties. Constraints on some models (supersymmetry, radiative muon mass, leptoquarks) beyond the standard model (SM) using this result have been discussed in [186,201]. The lower bounds on the scale of the models using this new allowed range of $a_\mu^{\rm new}$ are typically:

$$\begin{aligned} \tilde{m} &\geq 113 \text{ GeV} &&: \text{ degenerate sparticle mass} \\ M &\geq 1.7 \text{ TeV} &&: \text{ compositeness} \\ M_{S_2} &\geq 55.5 \text{ GeV} &&: \text{ Zee model singlet scalar} \\ M &\geq 1.1 \text{ TeV} &&: \text{ leptoquarks} \,. \end{aligned} \qquad (14.32)$$

### 14.1.6 The $\tau$ anomaly

The different theoretical contributions to the $\tau$ anomaly have been discussed before in [215]. Compared to the case of the electron and of the muon, an eventual precise measurement of $a_\tau$ will provide a further test of the QED calculation at short distance $t = \ln(M_\tau/m_l)$ not reached in the electron and muon case. Then, it can provide a measurement of the QED running coupling $\bar{\alpha}$ as given by the RGE discussed previously, and a test of an eventual sub-structure of the $\tau$ lepton. As can be seen in details in [215], the relative weight of the hadronic contributions has decreased compared, for example, with the weak interaction contributions. Also, because of the large value of $M_\tau$, the rôle of the $\rho$-meson has relatively decreased, which renders, almost equal, the contribution of the hadrons below and above 1 GeV. This is a positive feature which can allow a precise theoretical prediction of this observable. An update of the theoretical predictions obtained in [215] is [201] (in units of $10^{-8}$):

$$
\begin{aligned}
a_\tau^{\text{QED}} &= 117\,327.1(1.2)\,, \\
a_\tau^{\text{EW}} &\simeq 46.9(1.2)\,, \\
a_\tau^{\text{had}}(l.o) &= 353.6(4.0)\,, \\
a_\tau^{\text{had}}(h.o)_{(\text{LL})} &= 20.0(5.8)\,, \\
a_\tau^{\text{had}}(h.o) &= 27.6(5.8)\,,
\end{aligned}
\tag{14.33}
$$

which leads to:

$$
\begin{aligned}
a_\tau^{\text{SM}} &= a_\tau^{\text{had}} + a_\tau^{\text{EW}} + a_\tau^{\text{QED}} \\
&= 117\,755.2(7.2) \times 10^{-8}\,.
\end{aligned}
\tag{14.34}
$$

where we have used the present accurate value of $M_\tau = 1.77703$ GeV. This value in Eq. (14.34) can be compared with the present (inaccurate) experimental value [217]:

$$
a_\tau^{\text{exp}} = 0.004 \pm 0.027 \pm 0.023\,,
\tag{14.35}
$$

which, we hope, can be improved in the near future.

## 14.2 Other high-precision low-energy tests of QED

### 14.2.1 Lowest order hadronic contributions

In addition to the high-precision measurements of the lepton anomalies, some QED high-precision tests are also performed. As in the case of the lepton anomalies, the hadronic contributions also play an important for the QED predictions of the running QED coupling $\alpha(M_Z)$ and of the muonium hyperfine splittings $\nu$. These contributions can be expressed in a closed form as the convolution integral:[2]

$$
\mathcal{O}_{\text{had}} = \frac{1}{4\pi^3}\int_{4m_\pi^2}^{\infty} dt\; K_{\mathcal{O}}(t)\,\sigma_H(t)\,,
\tag{14.36}
$$

[2] For a recent estimate and review see e.g. Ref. [202] and references therein.

where:

$$\mathcal{O}_{\rm had} \equiv \Delta\alpha_{\rm had} \times 10^5 \quad \text{or} \quad \Delta\nu_{\rm had} \,. \tag{14.37}$$

- For the QED running coupling $\Delta\alpha_{\rm had} \times 10^5$, the kernel is:

$$K_\alpha(t) = \left(\frac{\pi}{\alpha}\right)\left(\frac{M_Z^2}{M_Z^2 - t}\right) , \tag{14.38}$$

where $\alpha^{-1}(0) = 137.036$ and $M_Z = 91.3$ GeV. It behaves for large $t$ like a constant.
- For the muonium hyperfine splitting $\Delta\nu_{\rm had}$, the kernel function is (see e.g. [220]):

$$K_\nu = -\rho_\nu \left[ (x_\mu + 2) \ln \frac{1 + v_\mu}{1 - v_\mu} - \left(x_\mu + \frac{3}{2}\right) \ln x_\mu \right] \tag{14.39}$$

where:

$$\rho_\nu = 2\nu_F \frac{m_e}{m_u} , \qquad x_\mu = \frac{t}{4m_\mu^2} \qquad v_\mu = \sqrt{1 - \frac{1}{x_\mu}} , \tag{14.40}$$

and $\nu_F$ is the Fermi energy splitting:

$$\nu_F = 445\,903\,192\,0.(511)(34) \text{ Hz}. \tag{14.41}$$

It behaves for large $t$ as:

$$K_\nu\left(t \gg m_\mu^2\right) \simeq \rho_\nu \left(\frac{m_\mu^2}{t}\right)\left(\frac{9}{2} \ln \frac{t}{m_\mu^2} + \frac{15}{4}\right) . \tag{14.42}$$

The different asymptotic behaviours of these kernel functions will influence on the relative weights of different regions contributions in the evaluation of the above integrals.

### 14.2.2 QED running coupling $\alpha(M_Z)$

Using the same data as for the anomalous magnetic moment, one can deduce the lowest order hadronic contribution [202]:

$$\Delta\alpha_{\rm had} = 2763.4(16.5) \times 10^{-5} \,. \tag{14.43}$$

We add the hadronic radiative corrections $\Delta\alpha_{\rm had} = (6.4 \pm 2.7) \times 10^{-5}$ from the radiative modes $\pi^0\gamma, \eta\gamma, \pi^+\pi^-\gamma \ldots$ coming from the largest range given in [195] and YT01 [205]. Using the renormalization group evolution of the QED coupling:

$$\alpha^{-1}(M_Z) = \alpha^{-1}(0)[1 - \Delta\alpha_{\rm QED} - \Delta\alpha_{\rm had}] \,, \tag{14.44}$$

and the available results to three-loops [218] of $\Delta\alpha_{\rm QED} = 3149.7687 \times 10^{-5}$, one can deduce:

$$\alpha^{-1}(M_Z) = 128.926(25) \,. \tag{14.45}$$

The results in Eqs. (14.43) and (14.45) are in good agreement with other determinations [205,219] as shown in Fig. 14.10, but more accurate, thanks to the combined $e^+e^-$ and $\tau$-decay data.

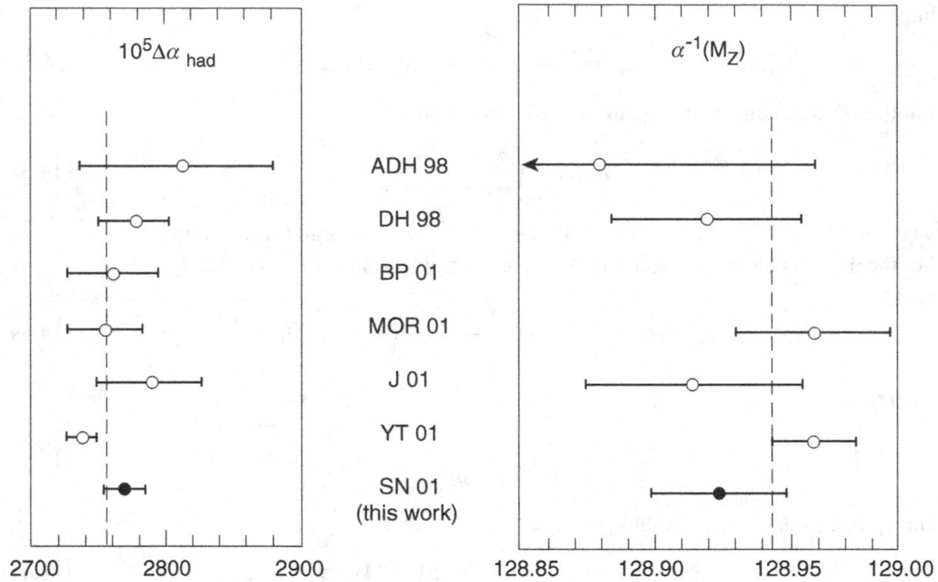


Fig. 14.10. Recent determinations of $\Delta\alpha_{\rm had}$ and $\alpha^{-1}(M_Z)$. The dashed vertical line is the mean central value. References to the authors are in [218,219,195,205].

The above results are important for a precise determination of the Higgs mass from a global fit of the electroweak data as shown in Fig. 14.11.

We expect that with this new improved estimate, the present lower bound of 114 GeV for the Higgs mass from LEP data can be improved in the near future.

### *14.2.3 Muonium hyperfine splitting*

Using again the same data input as in previous observables, the hadronic contribution to the muonium hyperfine splitting is [202]:

$$\Delta\nu_{\rm had} = (232.5 \pm 3.2)\ {\rm Hz}\ , \tag{14.46}$$

which is in excellent agreement with recent determinations shown in Table 14.2.

Here, due to the $(\ln t)/t$ behaviour of the kernel function, the contribution of the low-energy region is dominant. However, the $\rho$-meson region contribution below 1 GeV is 47% compared with the 68% in the case of $a_\mu$, while the QCD continuum is about 10% compared to 7.4% for $a_\mu$. The accuracy of our result is mainly due to the use of the $\tau$-decay data, explaining the similar accuracy of our final result with the one in [220] using new Novosibirsk data. Combined with existing QED and electroweak contributions:

$$\begin{aligned}
&\Delta\nu_{\rm QED} = 4\,270\,819(220)\ {\rm Hz}\ , \qquad \Delta\nu_{\rm weak}(l.o) = -\frac{G_F}{\sqrt{2}} m_e m_\mu \left(\frac{3}{4\pi\alpha}\right)\nu_F \simeq -65\ {\rm Hz}\ , \\
&|\Delta\nu_{\rm weak}(h.o)| \approx 0.7\ {\rm Hz}\ , \qquad \Delta\nu_{\rm had}(h.o) \simeq 7(2)\ {\rm Hz}\ ,
\end{aligned} \tag{14.47}$$

Table 14.2. *Recent determinations of* $\Delta\nu_{\rm had}$

| Authors | $\Delta\nu_{\rm had}$ [Hz] |
|---|---|
| FKM 99 [221] | $240 \pm 7$ |
| CEK 01 [220] | $233 \pm 3$ |
| SN 01 [202] | $232.5 \pm 3.2$ |

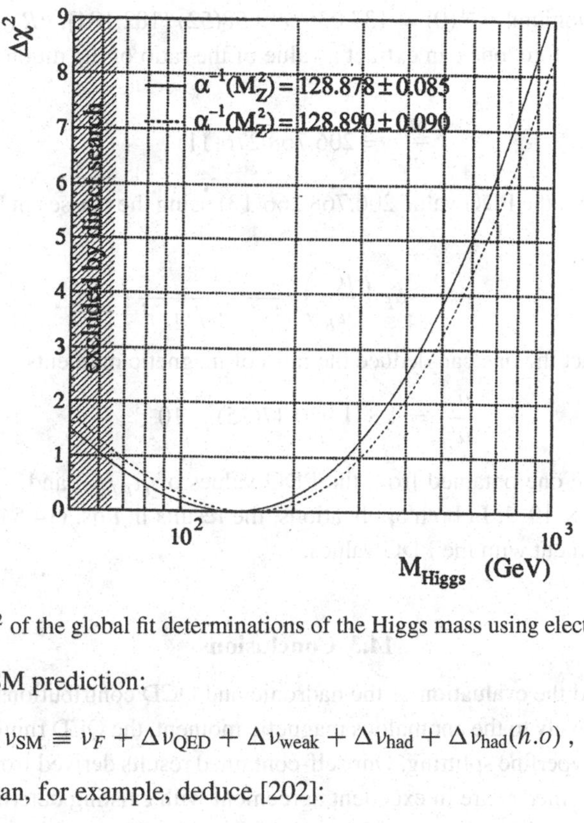


Fig. 14.11. $\chi^2$ of the global fit determinations of the Higgs mass using electroweak data.

one obtains the SM prediction:

$$\nu_{\rm SM} \equiv \nu_F + \Delta\nu_{\rm QED} + \Delta\nu_{\rm weak} + \Delta\nu_{\rm had} + \Delta\nu_{\rm had}(h.o) , \tag{14.48}$$

from which one can, for example, deduce [202]:

$$\frac{\nu_{\rm SM}}{\nu_F} = 1.000\,957\,83(5) , \tag{14.49}$$

by noting that $\nu_F$ enters as an overall factor in the theoretical contributions. Combining this result with the experimental value of $\nu$:

$$\nu_{\rm exp} = 4\,463\,302\,776(51)\ {\rm Hz} , \tag{14.50}$$

one can deduce the SM prediction:

$$\nu_F^{\rm SM} = 4\,459\,031\,783(226)\ {\rm Hz} , \tag{14.51}$$

where the error is dominated here by the QED contribution at fourth order. This result is a factor of 2 more precise than the one in [220]. One can use this result in Eq. (14.51) in the expression:

$$\nu_F = \rho_F \left(\frac{m_e}{m_\mu}\right) \frac{1}{(1+m_e/m_\mu)^3} \left(1+a_\mu\right) , \tag{14.52}$$

where $a_\mu = 1.165\,920\,3(15) \times 10^{-3}$, and:

$$\rho_F = \frac{16}{3}(Z\alpha)^2 Z^2 c R_\infty , \tag{14.53}$$

and $Z = 1$ for muonium, $\alpha^{-1}(0) = 137.035\,999\,58(52)$ [182,184], $cR_\infty = 3\,289\,841\,960\,368(25)$ kHz. Therefore, one can extract a value of the ratio of the muon over the electron mass:

$$\frac{m_\mu}{m_e} = 206.768\,276(11) , \tag{14.54}$$

to be compared with the PDG value 206.768 266(13) using the masses in MeV units. If one uses the relation:

$$\nu_F = \rho_F \left(\frac{\mu_\mu}{\mu_B^e}\right) \frac{1}{(1+m_e/m_\mu)^3} , \tag{14.55}$$

one can also extract the one can deduce the ratio of magnetic moments:

$$\frac{\mu_\mu}{\mu_B^e} = 4.841\,970\,47(25) \times 10^{-3} , \tag{14.56}$$

compared with the one obtained from the PDG values of $\mu_\mu/\mu_p$ and $\mu_p/\mu_B^e$: $\mu_\mu/\mu_B^e = 4.841\,970\,87(14) \times 10^{-3}$. In both applications, the results in Eqs. (14.54) and (14.56) are in excellent agreement with the PDG values.

## 14.3 Conclusions

We have discussed the evaluation of the hadronic and QCD contributions $a_l^{\rm had}(l.o)$, $\Delta\alpha_{\rm had}$ and $\Delta\nu_{\rm had}$ respectively to the anomalous magnetic moment, the QED running coupling and to the Muonium hyperfine splitting. Our self-contained results derived from the same input data and QCD parameters are in excellent agreement with existing determinations and are quite accurate. One of the immediate consequences of these results is the prediction of $a_\mu$, $a_\tau$ and $\alpha(M_Z)$. We have used the result for the muonium hyperfine splitting for a high precision measurement of the ratios of the muon over the electron mass and of their magnetic moments. These standard model predictions are in excellent agreement with those quoted by PDG [16] and can be used for providing strong constraints on some model building beyond the standard model.

# Part IV

Deep inelastic scatterings at hadron colliders

# 15

# OPE for deep inelastic scattering

## 15.1 Introduction

Deep-inelastic scattering (DIS) are classical QCD processes playing an important rôle in the understanding of perturbative QCD and of the nucleon structure function, where several structure functions $F_i(x, Q^2)[x$ (fraction of proton momentum) and $Q^2$ (squared of transfer momentum)] can be predicted and measured from different targets and beams and different polarizations. In the past DIS has been used for establishing the parton nature of quarks and gluons and QCD as a theory of strong interactions.

At present (as we shall see later on), DIS provide quantitative tests of QCD (measurements of quark and gluon densities in the nucleon, of $\alpha_s(Q^2), \ldots$). The theory of scaling violations for totally inclusive DIS processes are based on the operator product expansion (OPE) and renormalization group equation.

The OPE has been introduced by Wilson [222] and was proven by Zimmermann [223] in perturbation theory through the application of the BPHZ method. Let us consider the time-ordered product of two scalar fields:

$$T\phi(x)\phi(0) \tag{15.1}$$

which we can write, using the Wick's theorem studied in Section (4.1), as:

$$T\phi(x)\phi(0) = \langle 0|T\left(\phi(x)\phi(0)\right)|0\rangle + :\phi(x)\phi(0): \tag{15.2}$$

The first term in the RHS is the scalar propagator:

$$\langle 0|T\left(\phi(x)\phi(0)\right)|0\rangle = -i\Delta(x) = \int \frac{d^4p}{(2\pi)^4} e^{-ipx} \frac{1}{p^2 - m^2 + i\epsilon} \simeq \frac{i}{(2\pi)^2} \frac{1}{x^2 - i0} + \cdots, \tag{15.3}$$

where $\cdots$ means less-singular terms. It is a $c$-number (unit operator) but singular for $x \to 0$, while the operator $:\phi(x)\phi(0):$ is regular. In general, any local operators $J(x)$ and $J'(y)$ can be expanded in a series of well-defined and regular operators $\mathcal{O}_i(x)$ multiplied with the $c$-number $C_i(x)$, the Wilson coefficients containing the singularity of the product $J(x)J'(y)$

for $x = y$. This leads to the OPE or Wilson expansion:

$$J(x)J'(y) = \sum_{n=0}^{\infty} C_n(x-y)\mathcal{O}_n\left(\frac{x+y}{2}\right) \quad n = 0, 1, 2, \dots . \tag{15.4}$$

## 15.2 The OPE for free fields at short distance

As an application, let us consider the neutral vector current:

$$J_\mu(x) =: \bar\psi(x)\gamma_\mu\psi(x) : , \tag{15.5}$$

which is a normal ordered product of two quark fields. Applying the Wick theorem studied in Part 1, one can write:

$$\begin{aligned} \mathcal{T}(J_\mu(x)J_\nu(0)) &= -Tr\{\langle 0|\mathcal{T}(\psi(0)\bar\psi(x))|0\rangle\gamma_\mu\langle 0|\mathcal{T}(\psi(x)\bar\psi(0))|0\rangle\gamma_\nu\} \\ &\quad + :\bar\psi(x)\gamma_\mu\langle 0|\mathcal{T}(\psi(x)\bar\psi(0))|0\rangle\gamma_\nu\psi(0): \\ &\quad + :\bar\psi(0)\gamma_\nu\langle 0|\mathcal{T}(\psi(0)\bar\psi(x))|0\rangle\gamma_\mu\psi(x): \\ &\quad + :\bar\psi(x)\gamma_\mu\psi(x)\bar\psi(0)\gamma_\nu\psi(0): , \end{aligned} \tag{15.6}$$

where the free propagator:

$$\langle 0|\mathcal{T}(\psi(x)\bar\psi(0))|0\rangle = -iS(x) = \int \frac{d^4p}{(2\pi)^4} e^{-ipx} \frac{i}{\hat p - m + i\epsilon} , \tag{15.7}$$

is singular at *short distance* ($x \to 0$). Therefore, by inspecting Eq. (15.6), one can see that the first term is more singular than the second..., i.e. Eq. (15.6) is a typical example of an OPE. Relating the free fermion propagator to the scalar one:

$$S(x) = (i\hat\partial + m)\Delta(x) , \tag{15.8}$$

one can extract the leading singularity for $x \to 0$ from Eq. (15.3), which is quark mass independent. As the singularity behaves like $x^2$ (but not like $x$), it is on the light cone and called *light-cone singularity*. From the expression of the Fourier transform of the propagator:

$$\int dx\, e^{iqx} \frac{1}{(x - i\epsilon)^n} = 2\pi \frac{e^{i\frac{n\pi}{2}}}{\Gamma(n)} \theta(q) q^{n-1} , \tag{15.9}$$

one can see that the dominant contribution of the T-product of the two currents comes from the most singular part of the $c$-number coefficients. Therefore, near the light cone, one obtains [224]:

$$\begin{aligned} \mathcal{T}(J_\mu(x)J_\nu(0)) &= \frac{(x^2 g_{\mu\nu} - 2x_\mu x_\nu)}{\pi^4(x^2 - i\epsilon)^4} - \frac{x^\lambda}{(2\pi^2(x^2 - i\epsilon)^2} \\ &\quad \times \left[i\sigma_{\mu\lambda\nu\rho}\mathcal{O}_V^\rho(x) + \epsilon_{\mu\lambda\nu\rho}\mathcal{O}_A^\rho(x)\right] + \mathcal{O}_{\mu\nu}(x) , \end{aligned} \tag{15.10}$$

where $\mathcal{O}(x)$ are regular operators:

$$\begin{aligned}\mathcal{O}_{\mu,V}(x) &= :\bar{\psi}(x)\gamma_\mu\psi(0) - \bar{\psi}(0)\gamma_\mu\psi(x): \ ,\\ \mathcal{O}_{\mu,A}(x) &= :\bar{\psi}(x)\gamma_\mu\gamma_5\psi(0) + \bar{\psi}(0)\gamma_\mu\gamma_5\psi(x): \ ,\\ \mathcal{O}_{\mu\nu}(x) &= :\bar{\psi}(x)\gamma_\mu\psi(x)\bar{\psi}(0)\gamma_\nu\psi(0): \ ,\end{aligned} \tag{15.11}$$

and:

$$\sigma_{\mu\lambda\nu\rho} = g_{\mu\lambda}g_{\nu\rho} + g_{\mu\rho}g_{\nu\lambda} - g_{\mu\nu}g_{\nu\rho} \,. \tag{15.12}$$

We have used the relation:

$$\gamma_\mu\gamma_\lambda\gamma_\nu = (\sigma_{\mu\lambda\nu\rho} + i\epsilon_{\mu\lambda\nu\rho}\gamma_5)\gamma^\rho \,, \tag{15.13}$$

where $\epsilon_{\mu\lambda\nu\rho}$ is the totally anti-symmetric rank 4 tensor with the properties defined in Appendix D. Analogous expression can be derived for the current commutator:

$$\mathcal{T}[J_\mu(x), J_\nu(0)] \,, \tag{15.14}$$

by using:

$$\frac{1}{x^2 - i\epsilon} = \frac{\mathcal{P}}{x^2} + i\pi\delta(x^2) \,, \tag{15.15}$$

where $\mathcal{P}$ denotes principal value. Differentiating this expression, it is easy to obtain:

$$\frac{1}{(x^2 - i\epsilon)^n} - \frac{1}{(x^2 + i\epsilon)^n} = 2i\pi\frac{(-1)^{n-1}}{(n-1)!}\delta^{(n)}(x^2) \,. \tag{15.16}$$

Therefore:

$$\begin{aligned}\mathcal{T}[J_\mu(x)J_\nu(0)] - \mathcal{T}[J_\mu(x)J_\nu(0)]^\dagger &\equiv \epsilon(x_0)[J_\mu(x), J_\nu(0)]\\ &= -\frac{i}{3\pi^3}\delta^{(3)}(x^2)(x^2 g_{\mu\nu} - 2x_\mu x_\nu)\\ &\quad - \frac{1}{\pi}x^\lambda\delta^{(1)}\left[i\sigma_{\mu\lambda\nu\rho}\mathcal{O}_V^\rho(x) + \epsilon_{\mu\lambda\nu\rho}\mathcal{O}_A^\rho(x)\right]\\ &\quad + \mathcal{O}_{\mu\nu}(x) - \mathcal{O}_{\nu\mu}(x) \,,\end{aligned} \tag{15.17}$$

where:

$$\epsilon(x_0) = \frac{x_0}{|x_0|} \,, \tag{15.18}$$

is the sign function.

## 15.3 Application of the OPE for free fields: parton model and Bjorken scaling

For simplicity, we consider the unpolarized process:

$$e + p \to e + X \tag{15.19}$$

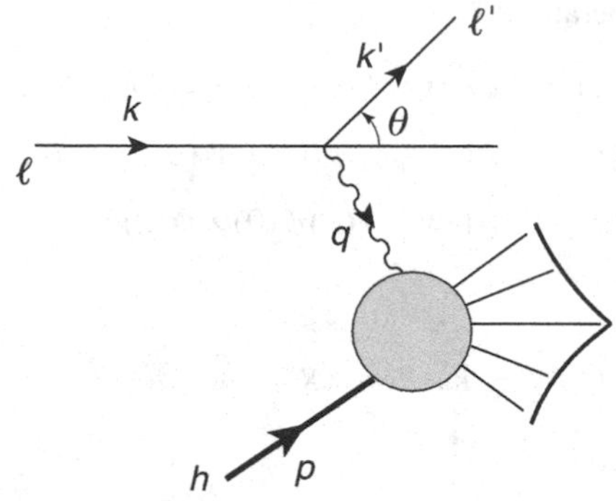


Fig. 15.1. Kinematics of the $e + p \to e + X$ process.

which we have anticipated in Section 2.3. Here, we shall derive explicitly the structure functions $W_{1,2}(Q^2, \nu)$ using OPE for free fields. The kinematics of the process is given in Fig. 15.1.

There are three independent kinematic variables:

$$s = (p+k)^2 , \qquad q^2 = (k-k')^2, \qquad W^2 = (p+q)^2 , \tag{15.20}$$

where $k$ and $k'$ are momenta of the initial and final electrons, $p$ and $q$ are respectively the proton and photon momenta. In the laboratory frame (proton rest frame) and neglecting the electron mass, one can rewrite:

$$\begin{aligned} s &= M_p(2E + M_p) , \\ q^2 &\equiv -(Q^2 > 0) = -4EE' \sin^2 \frac{\theta}{2} , \\ W^2 &= M_p^2 + 2M_p(E - E') + q^2 , \end{aligned} \tag{15.21}$$

where $E = k_0,\ E' = k_0'$ are the energies of the incident and scattered electrons in the proton rest frame, and $\theta$ is the scattering angle of the electron. The physical region is:

$$s \geq M_p^2 , \qquad q^2 \leq 0 , \qquad W^2 \geq (M_p + m_\pi)^2 , \tag{15.22}$$

$m_\pi$ being the pion mass. It is usual to introduce:

$$\nu \equiv p \cdot q = M_p(E - E') , \tag{15.23}$$

where $\nu/M_p$ is the energy transfer in the proton rest frame, in terms of which the physical region condition on $W^2$ reads:

$$2\nu + q^2 \geq m_\pi(2M_p + m_\pi) . \tag{15.24}$$

The inclusive differential cross-section of the unpolarized process can be written as:

$$E' \frac{d\sigma}{d^3k'} = \frac{1}{32(2\pi)^3} \frac{1}{k \cdot p} \sum_{\sigma,\sigma',\lambda} \sum_X (2\pi)^4 \delta^4(p_X + k' - k - p) |\langle eX|T|eN\rangle|^2 , \tag{15.25}$$

where $\sigma'$, $\sigma$, $\lambda$ are the spin components of the scattered, initial electrons and the target proton. The amplitude is:

$$\langle eX|T|eN\rangle = \bar{u}_{\sigma'}(k')(e\gamma_\mu)u_\sigma(k)\frac{1}{q^2}\langle X|(-e)J^\mu(0)|p,\lambda\rangle \ . \tag{15.26}$$

This leads to the expression of the cross-section as a convolution of the leptonic and hadronic tensors:

$$E'\frac{d\sigma}{d^3k'} = \frac{\alpha^2}{4(k\cdot p)q^4}L_{\mu\nu}W^{\mu\nu} \ , \tag{15.27}$$

where $\alpha = e^2/(4\pi)$ is the QED fine structure constant. The leptonic tensor is:

$$\begin{aligned} L^{\mu\nu} &= \frac{1}{4}Tr\{(\hat{k}+m_e)\gamma^\mu(\hat{k}'+m_e)\gamma^\nu\} \\ &= 4(k'^\mu k^\nu + k'^\nu k^\mu) + \left(2q^2+4m_e^2\right)g^{\mu\nu} \ . \end{aligned} \tag{15.28}$$

The hadronic tensor can be written as:

$$W_{\mu\nu} = \frac{1}{2\pi}\int d^4x\ e^{iqx}\frac{1}{2}\sum_\lambda\langle p;\lambda|J_\mu(x)J_\nu(0)|\lambda;p\rangle \ . \tag{15.29}$$

Using the property:

$$\int d^4x\ e^{iqx}\sum_\lambda\langle p;\lambda|J_\mu(0)J_\nu(x)|\lambda;p\rangle = 0 \ , \tag{15.30}$$

for physical process, which can be shown by using:

$$\begin{aligned} &\int d^4x\ e^{iqx}\sum_\lambda\langle p;\lambda|J_\mu(0)J_\nu(x)|\lambda;p\rangle \\ &= \sum_X(2\pi)^4\delta^4(q-p+p_X)\sum_\lambda\langle p;\lambda|J_\mu(0)|X\rangle\langle X|J_\nu(x)|\lambda;p\rangle \ . \end{aligned} \tag{15.31}$$

The assumption that $q-p+p_X = 0$ in the physical region (Eq. (15.22)) would lead to the contradiction $q_0 \geq 0$. Therefore, one obtains:

$$W_{\mu\nu} = \frac{1}{2\pi}\int d^4x\ e^{iqx}\frac{1}{2}\sum_\lambda\langle p;\lambda|[J_\mu(x),\ J_\nu(0)]|\lambda;p\rangle \ . \tag{15.32}$$

Causality requires that the commutator vanishes for $x^2 < 0$, such that the integral is only non-zero for $x^2 > 0$. Using the optical theorem, one can relate the hadronic tensor to the absorptive part of the forward Compton scattering amplitude:

$$T_{\mu\nu} = \int d^4x\ e^{iqx}\frac{1}{2}\sum_\lambda\langle p;\lambda|TJ_\mu(x)J_\nu(0)|\lambda;p\rangle \ , \tag{15.33}$$

by the relation:

$$W_{\mu\nu} = \frac{1}{\pi}\text{Im}T_{\mu\nu} \ , \tag{15.34}$$

which corresponds to the discontinuity of $T_{\mu\nu}$ across the cut along the line $q_0 \geq 0$ in the complex $q_0$ plane:

$$\mathrm{Im}T_{\mu\nu} = \frac{1}{2i}[T_{\mu\nu}(q_0 + i\epsilon) - T_{\mu\nu}(q_0 - i\epsilon)] \,. \tag{15.35}$$

Using the general Lorentz decomposition, one can express $W_{\mu\nu}$ in terms of the invariants $W_i$ (so-called structure functions) introduced in Section 2.3:

$$W_{\mu\nu} = -\left(g_{\mu\nu} - \frac{q_\mu q_\nu}{q^2}\right) W_1(Q^2, \nu) + \frac{1}{M_p^2}\left(p_\mu - \frac{p\cdot q}{q^2}q_\mu\right)\left(p_\nu - \frac{p\cdot q}{q^2}q_\nu\right) W_2(Q^2, \nu) + i\epsilon_{\mu\nu\rho\sigma}\frac{p^\rho q^\sigma}{2M_p^2} W_3(Q^2, \nu) \,. \tag{15.36}$$

For unpolarized process, only $W_{1,2}$ are relevant. Then, the differential cross-section has the form:

$$\frac{d\sigma}{dQ^2 d\nu} = \frac{\pi\alpha^2}{4M_p E^2 \sin^4\theta E E'}\left\{2\sin^2\frac{\theta}{2} W_1(Q^2, \nu) + \cos^2\frac{\theta}{2} W_2(Q^2, \nu)\right\} \,. \tag{15.37}$$

Coming back to the OPE of $W_{\mu\nu}$ given in Eq. (15.17) between two proton states, one can notice that the last term is less singular than the two former terms, such that we can neglect it to a first approximation. The first term can also be omitted as it corresponds to a disconnected diagram. Also noticing that the operators $\mathcal{O}$ are regular and finite for $x \to 0$, one can Taylor-expand the quark fields:

$$\psi(x) = \psi(0) + x^\mu[\partial_\mu \psi(x)]_{x=0} + \frac{1}{2!}[\partial_{\mu_1}\partial_{\mu_2}\psi(x)]_{x=0} + \cdots \tag{15.38}$$

and write:

$$\mathcal{O}^\rho_{V/A}(x) = \sum_0^\infty \frac{1}{n!} x^{\mu_1} \cdots x^{\mu_n} \mathcal{O}^\rho_{V/A,\mu_1\cdots\mu_n}(0) \,, \tag{15.39}$$

where:

$$\begin{aligned}
\mathcal{O}^\rho_{V,\mu_1\cdots\mu_n}(x) &= :[\partial_{\mu_1}\cdots\partial_{\mu_n}\bar\psi(x)]\gamma^\rho\psi(x) - \bar\psi(x)\gamma^\rho[\partial_{\mu_1}\cdots\partial_{\mu_n}\psi(x)]: \\
\mathcal{O}^\rho_{A,\mu_1\cdots\mu_n}(x) &= :[\partial_{\mu_1}\cdots\partial_{\mu_n}\bar\psi(x)]\gamma^\rho\gamma_5\psi(x) + \bar\psi(x)\gamma^\rho\gamma_5[\partial_{\mu_1}\cdots\partial_{\mu_n}\psi(x)]:
\end{aligned} \tag{15.40}$$

For the unpolarized process which we discuss here, the operator $\mathcal{O}^\rho_A$ will not also contribute. One can express the matrix element:

$$\langle p|\mathcal{O}^\rho_{V,\mu_1\cdots\mu_n}(0)|p\rangle = \hat{\mathcal{O}}_n p^\rho p_{\mu_1}\cdots p_{\mu_n} + \text{terms containing } g_{\mu\nu} \,. \tag{15.41}$$

where $\hat{\mathcal{O}}$ is a Lorentz invariant constant reduced matrix element which depends on $p^2 = M_p^2$ and on quark masses. We have used the fact that the matrix element only depends on $p_\mu$ and is symmetric in the indices $\mu_1,\ \mu_2,\ \cdots\ \mu_n$. The terms containing $g_{\mu\nu}$ in Eq. (15.41) are of the form $p^\rho p^2 g_{\mu_1\mu_2} p_{\mu_3}\cdots p_{\mu_n}$ and so on, which are less singular in $x^2$ because $g_{\mu_1\mu_2}$

gives rise to $x^2$, and can therefore be neglected. Therefore, the relevant part of Eq. (15.17) for our process can be written as:

$$W_{\mu\nu} = -\frac{1}{2\pi^2}\sigma_{\mu\lambda\nu\rho}p^\rho \int d^4x\, e^{iqx}x^\lambda \epsilon(x_0)\delta^{(1)}(x^2)f(p\cdot x)\,, \tag{15.42}$$

with:

$$f(z) = \sum_{n=0}^{\infty} \hat{\mathcal{O}}_n \frac{z^n}{n!}\,, \tag{15.43}$$

where one can also notice that due to the form of $\mathcal{O}^\rho_{V,\mu_1\cdots\mu_n}$, $\hat{\mathcal{O}}_n$ vanishes for $n$ even and the summation in Eq. (15.43) only runs for $n$ odd. Taking the Fourier transform:

$$f(z) = \int_{-\infty}^{+\infty} d\zeta\, e^{iz\zeta}\mathcal{F}(\zeta)\,, \tag{15.44}$$

one can rewrite:

$$W_{\mu\nu} = -\frac{i}{2\pi^2}\sigma_{\mu\lambda\nu\rho}p^\rho \frac{\partial}{\partial q_\lambda}\int_{-\infty}^{+\infty} d\zeta \mathcal{F}(\zeta)\int d^4x\, e^{i(q+p\zeta)x}\epsilon(x_0)\delta^{(1)}(x^2)\,. \tag{15.45}$$

Using:

$$\mathcal{I}_n \equiv \int d^4x\, e^{iqx}\delta^{(n)}(x^2) = \frac{i\pi^2}{4^{n-1}(n-1)!}(q^2)^{n-1}\epsilon(q_0)\theta(q^2)\,, \tag{15.46}$$

one obtains:

$$\begin{aligned} W_{\mu\nu} &= \int_{-\infty}^{+\infty} d\zeta \mathcal{F}(\zeta)\left[-\left(p\cdot q+\zeta M_p^2\right)g_{\mu\nu} + 2\zeta p_\mu p_\nu + p_\mu q_\nu + p_\nu q_\mu\right] \\ &\quad \times \epsilon(q_0+\zeta p_0)\delta\left(q^2+2\zeta p\cdot q+\zeta^2 M_p^2\right)\,. \end{aligned} \tag{15.47}$$

In the Bjorken limit:

$$p\cdot q \to \infty, \quad -q^2 \to \infty \text{ and } \quad \zeta \equiv x = -q^2/(2p\cdot q) \text{ fixed}\,, \tag{15.48}$$

one can neglect $p^2 = M_p^2$, and deduce:

$$W_{\mu\nu} = \frac{1}{2}\mathcal{F}(x)\left(-g_{\mu\nu} - \frac{q^2}{(p\cdot q)^2}p_\mu p_\nu + \frac{p_\mu q_\nu + p_\nu q_\mu}{p\cdot q}\right)\,, \tag{15.49}$$

which one can rewrite in terms of $W_{1,2}$ defined in Eq. (15.36) with:

$$\begin{aligned} W_1(\nu, Q^2) &= \frac{1}{2}\mathcal{F}(x) \equiv F_1(x)\,, \\ \frac{\nu}{M_p^2}W_2(\nu, Q^2) &= \frac{x}{2}\mathcal{F}(x) \equiv F_2(x)\,, \end{aligned} \tag{15.50}$$

as given in Eq. (2.83) in terms of the Bjorken scaling function $F_{1,2}(x)$. This result shows that the assumption of free-field light-cone structure is equivalent to that of the parton model.

## 15.4 Light-cone expansion in $\phi_6^3(x)$ theory and operator twist

For simplifying our discussions, we shall work in $\phi_6^3(x)$ theory with a mass $m$. The hadronic current is:

$$J(x) = \phi^2(x)\,, \tag{15.51}$$

and the OPE has the form given in Eq. (15.4). In the previous sections, we have used the OPE at short distance $x \to 0$, i.e. large $q$, such that we can neglect terms of order $p \cdot q$ compared with $q^2$. For instance, in this case, the tree level amplitude of a forward Compton scattering reads:

$$\frac{1}{(q+p)^2 - m^2} = \frac{1}{q^2} + \mathcal{O}\left(\frac{1}{q^4}\right)\,. \tag{15.52}$$

In deep inelastic scatterings, the light-cone region $x^2 \to 0$ corresponds to the Bjorken limit in Eq. (15.48). In this region, the tree level Compton amplitude reads:

$$\begin{aligned} F_0(q,p) &\equiv \frac{1}{(q+p)^2 - m^2} = \frac{1}{q^2}\frac{1}{1+\frac{2p\cdot q}{q^2}} \\ &\simeq \frac{1}{q^2} - \frac{2p\cdot q}{q^4} + \frac{(2p\cdot q)^2}{q^6} + \cdots + \mathcal{O}\left(\frac{1}{q^4}\right)\,, \end{aligned} \tag{15.53}$$

which expresses that the dominant term of the amplitude in the Bjorken limit is due to an infinite number of 'composite operators'. This can be seen by taking the Fourier transform of Eq. (15.53):

$$\int d^6q\, e^{-iqx} F_0(q,p) \sim \frac{1}{x^4} + i\frac{px}{x^4} - \frac{(px)^2}{8x^4} + \cdots \tag{15.54}$$

Its $k$-th term can be written as:

$$\frac{1}{x^4} x^{\mu_1} \cdots x^{\mu_k} \langle p|\mathcal{O}_{\mu_1\cdots\mu_k}|p\rangle\,. \tag{15.55}$$

In general the OPE near the light cone has the form (*light-cone expansion*):

$$J(x)J'(0) = \sum_{i,k} C_k^{(i)}(x^2) x^{\mu_1\cdots\mu_k} \mathcal{O}^{(i)}_{\mu_1\cdots\mu_k}(0)\,, \tag{15.56}$$

where the index $i$ specifies the type of composite operators. Identifying with Eq. (15.4), the coefficient functions are:

$$C_n(x) \equiv C_k^{(i)}(x^2) x^{\mu_1\ldots\mu_k}\,. \tag{15.57}$$

In free-field theory, in order to match the mass dimension of both sides of Eq. (15.56), the coefficient function should behave as:

$$C_k^{(i)}(x^2) \sim (x^2)^{-(d_{J_o}+d_{J'_o}+k-d^{(i)}_{o,k})/2}\,, \tag{15.58}$$

where $d_{J_o}$, $d_{J'_o}$ and $d_o^{(i)}$ are canonical dimensions of the current $J$, $J'$ and of the operator $\mathcal{O}^{(i)}_{\mu_1\cdots\mu_k}$. This naïve power counting is valid for free-field theory as no other mass scale is present in the OPE. The index:

$$\tau_k \equiv d^{(i)}_{o,k} - k \equiv \text{dimension} - \text{spin} \tag{15.59}$$

which governs the strength of the singularity of the coefficient function is called *the twist* of the composite operator $\mathcal{O}^{(i)}_{\mu_1\cdots\mu_k}$ [225]. $k$ is called the spin of the operator and $d$ is its dimension. The operators of lowest twist dominate in the light-cone expansion. The scalar field $\phi$, the fermion field $\psi$ and the gauge field $G_\mu\nu$ have twist one. Taking the derivative of these fields cannot reduce the twist as the derivative increases the dimension by one unit but changes the spin by 1 or 0. Therefore, the minimum twist of an operator involving $n$ fields is $n$. In the light-cone expansion the dominant operators have twist 2. In the presence of external field, the symmetric traceless tensors of rank $k$ and twist 2 are e.g. of the form:

$$\begin{aligned}
\mathcal{O}^{(i)}_{s,\mu_1\cdots\mu_k} &= \phi^* D_{\mu_1}\cdots D_{\mu_k}\phi\,, \\
\mathcal{O}^{(i)}_{f,\mu_1\cdots\mu_k} &= \frac{i^{k-1}}{k!}\{\bar\psi\gamma_{\mu_1}D_{\mu_2}\cdots D_{\mu_k}\psi + \text{permutations}\}\,, \\
\mathcal{O}^{(i)}_{g,\mu_1\cdots\mu_k} &= 2\frac{i^{k-2}}{k!}Tr\{G_{\mu_1\alpha}D_{\mu_1}\cdots D_{\mu_k}G^{\alpha}_{\mu_k} + \text{permutations}\}\,.
\end{aligned} \tag{15.60}$$

where $D_\mu$ is the covariant derivative which is half the difference of the derivative acting to the right and to the left. In the presence of an external field the scale dimension counting does not hold. In the case of a theory with an UV fixed point, the scale invariance is recovered with the anomalous dimension, and the canonical dimensions are replaced by the scale dimensions $d_J$ and $d_k^{(i)}$. Therefore, the light-cone singularity reads for $x^2 \to 0$ [222]:

$$C_k^{(i)}(x^2) \sim (x^2)^{-(d_J+d_{J'}+k-d_k^{(i)})/2}\,. \tag{15.61}$$

In QCD, this expression will only be modified by logarithmic corrections as we shall see later on.

# 16

# Unpolarized lepton-hadron scattering

## 16.1 Moment sum rules

We shall consider the previous lepton-hadron unpolarized process studied in Section 15.3 governed by the T-product of two electromagnetic currents. The general Lorentz decomposition of the hadronic tensor has the form:

$$\begin{aligned}J_\mu(x)J_\nu(0) &= (\partial_\mu \partial'_\nu - g_{\mu\nu})\mathcal{O}_L(x)\\&\quad + (g_{\mu\lambda}\partial_\rho\partial'_\nu + g_{\rho\nu}\partial_\mu\partial'_\lambda - g_{\mu\lambda}g_{\rho\nu}\partial\cdot\partial - g_{\mu\nu}\partial_\lambda\partial'_\rho)\mathcal{O}_2^{\lambda\rho}(x)\\&\quad + i\epsilon_{\mu\nu\lambda\rho}\partial^\lambda\mathcal{O}_3^\rho(x)\\&\quad + i(\epsilon_{\mu\nu\lambda\rho}\partial\cdot\partial' - \epsilon_{\mu\sigma\lambda\rho}\partial_\nu\partial'^\sigma + \epsilon_{\nu\sigma\lambda\rho}\partial_\mu\partial'^\sigma)\mathcal{O}_4^{\lambda\rho}(x)\,,\end{aligned} \tag{16.1}$$

where $\partial_\mu \equiv \partial/\partial x_\mu$ and $\mathcal{O}_i$ are suitable bilocal operators, where $\mathcal{O}_L$ corresponds to the longitudinal structure functions $W_2 - 2xW_1$ defined in Eq. (15.36). The operators $\mathcal{O}_{3,4}$ do not contribute to the unpolarized process. Using the result in Eq. (15.56), one can write an OPE for the invariants. In the QCD deep inelastic scattering region, one can neglect quark mass corrections such that we have a good realization of the $SU(n)_f$ flavour symmetry. For the case $n_f = 2$ here (isospin symmetry), the electromagnetic current corresponds to the third component of $SU(2)$ such that the product $J(x)J(0)$ and the associate composite operators $\mathcal{O}$ belong to the representations:

$$3\otimes 3 = 1\oplus 3\oplus 5\,. \tag{16.2}$$

Therefore the lowest twist ($\tau = 2$) gauge invariant operators which dominate the light-cone expansion are, the non-singlet ($\lambda_a/2$ is the $SU(n)_f$ flavour matrix):

$$\mathcal{O}^{(i)}_{NS,\mu_1\cdots\mu_k} = \frac{i^{k-1}}{k!}\left\{\bar\psi\frac{\lambda_a}{2}\gamma_{\mu_1}D_{\mu_2}\cdots D_{\mu_k}\psi + \text{permutations}\right\}\,, \tag{16.3}$$

and singlet operators which mix under renormalizations:

$$\begin{aligned}\mathcal{O}^{(i)}_{S,\mu_1\cdots\mu_k} &= \frac{i^{k-1}}{k!}\{\bar\psi\gamma_{\mu_1}D_{\mu_2}\cdots D_{\mu_k}\psi + \text{permutations}\}\,,\\ \mathcal{O}^{(i)}_{g,\mu_1\cdots\mu_k} &= 2\frac{i^{k-2}}{k!}Tr\left\{G_{\mu_1\alpha}D_{\mu_1}\cdots D_{\mu_k}G^\alpha_{\mu_k} + \text{permutations}\right\}\,.\end{aligned} \tag{16.4}$$

We have omitted terms containing $g_{\mu\nu}$, the so-called trace terms. Substituting Eq. (15.56) into Eq. (16.1), one can deduce in momentum space:

$$\begin{aligned} T_{\mu\nu} &= i\int d^4x\, e^{iqx}\langle p|T J_\mu(x)J_\nu(0)|p\rangle \\ &= -(g_{\mu\nu}q^2 - q_\mu q_\nu)\sum_{i,n}\langle p|\mathcal{O}^{(i)}_{L,\mu_1\cdots\mu_n}(0)|p\rangle C^{(i)}_{L,n}(-q^2)q^{\mu_1}\cdots q^{\mu_n}\left(\frac{-q^2}{2}\right)^{-n-1} \\ &+ (g_{\mu\lambda}q_\rho q_\nu + g_{\rho\nu}q_\mu q_\lambda - q^2 g_{\mu\lambda}g_{\rho\nu} - g_{\mu\nu}q_\lambda q_\rho) \\ &\times \sum_{i,n}\langle p|\mathcal{O}^{(i)\lambda\rho}_{2,\mu_1\cdots\mu_n}(0)|p\rangle C^{(i)}_{2,n}(-q^2)q^{\mu_1}\cdots q^{\mu_n}\left(\frac{-q^2}{2}\right)^{-n-1}, \end{aligned} \tag{16.5}$$

where we have defined the Fourier transform of the coefficient functions:

$$\begin{aligned} C^{(i)}_{L,n}(-q^2)q^{\mu_1}\cdots q^{\mu_n}\left(\frac{-q^2}{2}\right)^{-n-1} &= i\int d^4x\, e^{iqx}\, x^{\mu_1}\cdots x^{\mu_n}C^{(i)}_{L,n}(x^2), \\ C^{(i)}_{2,n+2}(-q^2)q^{\mu_1}\cdots q^{\mu_n}2\left(\frac{-q^2}{2}\right)^{-n-2} &= i\int d^4x\, e^{iqx}\, x^{\mu_1}\cdots x^{\mu_n}C^{(i)}_{2,n}(x^2), \end{aligned} \tag{16.6}$$

and we have used the simplified notation:

$$\langle p|T J_\mu(x)J_\nu(0)|p\rangle \equiv \frac{1}{2}\sum_\lambda\langle p;\lambda|T J_\mu(x)J_\nu(0)|\lambda;p\rangle\,. \tag{16.7}$$

Using the tensor structures:

$$\begin{aligned} \langle p|\mathcal{O}^{(i)}_{L,\mu_1\cdots\mu_n}(0)|p\rangle &= \hat{\mathcal{O}}_{L,n}p_{\mu_1}\cdots p_{\mu_n} + \cdots \\ \langle p|\mathcal{O}^{(i)\lambda\rho}_{2,\mu_1\cdots\mu_n}(0)|p\rangle &= \hat{\mathcal{O}}_{2,n+2}p^\lambda p^\rho p_{\mu_1}\cdots p_{\mu_n} + \cdots, \end{aligned} \tag{16.8}$$

where $\hat{\mathcal{O}}_i$ are reduced matrix elements not calculable in perturbation theory, and we have omitted terms containing $g_{\mu\nu}$, we finally deduce:

$$T_{\mu\nu} = 2\omega^n\sum_{i,n\ \text{even}} e_{\mu\nu}C^{(i)}_{L,n}(-q^2)\hat{\mathcal{O}}^{(i)}_{L,n} - d_{\mu\nu}C^{(i)}_{2,n}(-q^2)\hat{\mathcal{O}}^{(i)}_{2,n}\,, \tag{16.9}$$

with:

$$\begin{aligned} e_{\mu\nu} &\equiv g_{\mu\nu} - q_\mu q_\nu/q^2, \\ d_{\mu\nu} &\equiv g_{\mu\nu} - q^2 p_\mu p_\nu/(p\cdot q)^2 - (p_\mu q_\nu + p_\nu q_\mu)/(p\cdot q)\,, \end{aligned} \tag{16.10}$$

where $\omega^{-1} \equiv Q^2/(2p\cdot q)$ is the Bjorken variable. Because of crossing symmetry:

$$T_{\mu\nu}(\omega) = T_{\mu\nu}(-\omega)\,, \tag{16.11}$$

the sum runs only over even $n$. The unphysical relation in Eq. (16.9) ($0 \le \omega \le 1$) can be converted to a physical one $\omega \ge 1$ by using a Cauchy integral to both sides of Eq. (16.9). Since $T_{\mu\nu}$ is an analytic function in the complex $\omega$ plane with branch cuts along the real

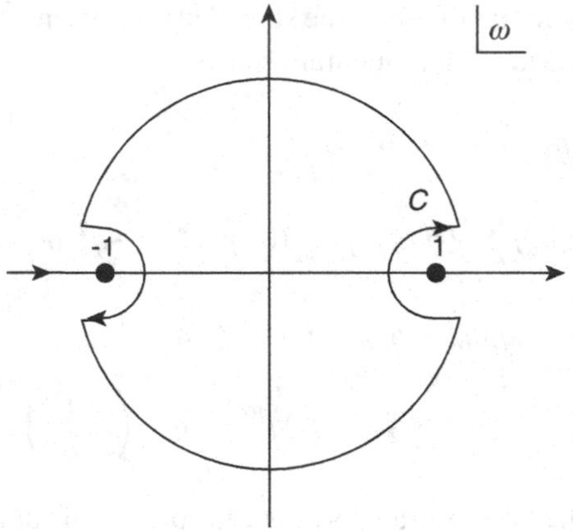


Fig. 16.1. Integration contour.

axis for $\omega \leq -1$ and $\omega \geq 1$, as shown in Fig. 16.1, it obeys the dispersion relation:

$$T_{\mu\nu} = \frac{1}{\pi}\left\{\int_{Q^2/2}^{\infty} - \int_{-\infty}^{-Q^2/2}\right\} \frac{d\nu'}{\nu' - \nu} \mathrm{Im} T_{\mu\nu}(Q^2, \nu) + \text{subtractions} . \tag{16.12}$$

Using the Cauchy integration to both sides of Eq. (16.9) along the contour in Fig. 16.1, one obtains:

$$\frac{1}{2i\pi}\oint_C \frac{T_{\mu\nu}}{\omega^n} = \frac{2}{\pi}\int_1^{\infty} \frac{d\omega}{\omega^n} \mathrm{Im} T_{\mu\nu} = 2\int_0^1 dx\, x^{n-2} W_{\mu\nu} , \tag{16.13}$$

where we have used the definitions in Eqs. (15.34) and (15.35) and the crossing symmetry in Eq. (16.11).

Noting that:

$$\oint_C d\omega\, \omega^{m-n} = \delta_{m,n-1} , \tag{16.14}$$

one can write:

$$T_{\mu\nu} = 2\sum_i e_{\mu\nu} C^{(i)}_{L,n-1}(-q^2)\hat{\mathcal{O}}^{(i)}_{L,n-1} - d_{\mu\nu} C^{(i)}_{2,n-1}(-q^2)\hat{\mathcal{O}}^{(i)}_{2,n-1} . \tag{16.15}$$

Equating Eqs. (16.15) and (16.13), one can deduce the *moment sum rules* for the structure functions [226]:

$$\begin{aligned} \mathcal{M}^{(n)}_L(Q^2) &\equiv \int_0^1 dx\, x^{n-2} F_L(x, Q^2) = \sum_i C^{(i)}_{L,n}(-q^2)\hat{\mathcal{O}}^{(i)}_{L,n} , \\ \mathcal{M}^{(n)}_2(Q^2) &\equiv \int_0^1 dx\, x^{n-2} F_2(x, Q^2) = \sum_i C^{(i)}_{2,n}(-q^2)\hat{\mathcal{O}}^{(i)}_{2,n} , \end{aligned} \tag{16.16}$$

where the structure functions $F_L \equiv F_2 - 2xF_1$ (*longitudinal structure functions*) and $F_2$ are defined through:

$$W_{\mu\nu} = \omega[e_{\mu\nu}F_L + d_{\mu\nu}F_2] , \tag{16.17}$$

and are related to the $W_{1,2}$ in Eq. (15.36) as:

$$\begin{aligned} F_L(x, Q^2) &= -W_1(\nu, Q^2) + \left(1 + \frac{\nu^2}{Q^2}\right) W_2(\nu, Q^2) , \\ F_2(x, Q^2) &= \frac{\nu}{M_p^2} W_2(\nu, Q^2) . \end{aligned} \tag{16.18}$$

The coefficient functions $C_{L,n}^{(i)}$ and $C_{2,n}^{(i)}$ in Eq. (16.16) are of short-distance nature and are calculable using perturbative QCD. The reduced matrix elements $\hat{\mathcal{O}}_{L,n}^{(i)}$ and $\hat{\mathcal{O}}_{2,n}^{(i)}$ are of long-distance nature and cannot be calculable. They can be determined experimentally, which can be done by measuring the moments in Eq. (16.16) at a fixed $Q_0^2$ and solve it for the reduced matrix elements. In practice, the moments are not very convenient as they are expressed in such a way that direct predictions of the structure functions cannot be made. Instead, one can take their inverse Mellin transform, which can be obtained by analytically continuing from integer $n$ to complex $n$ following the Carlson theorem [227].

One gets:

$$F_{L;2}(x, Q^2) = \frac{1}{2i\pi} \int_{c-i\infty}^{c+i\infty} dn\ \zeta^{1-n} C_{L;2,n}^{(i)}(Q^2) \hat{\mathcal{O}}_{L;2,n}^{(i)} , \tag{16.19}$$

where $C$ is an arbitrary real positive constant. Assuming (for simplifying the discussion) that only one operator contributes to the moment, we can suppress the index $i$. Therefore, one can deduce from the moments in Eq. (16.16):

$$\hat{\mathcal{O}}_{L;2,n} = \frac{1}{C_{L;2,n}\left(Q_0^2\right)} \int_0^1 dx\ x^{n-2} F_{L;2}\left(x, Q_0^2\right) , \tag{16.20}$$

which, when inserted into Eq. (16.19), gives after rearranging the integral:

$$F_{L;2}(x, Q^2) = \int_x^1 \frac{dy}{y} K\left(\frac{y}{x}, Q^2, Q_0^2\right) F_{L;2}\left(y, Q_0^2\right) , \tag{16.21}$$

where the kernel function is:

$$K\left(z, Q^2, Q_0^2\right) = \frac{1}{2i\pi} \int_{c-i\infty}^{c+i\infty} dn\ z^{1-n} \frac{C_{L;2,n}(Q^2)}{C_{L;2,n}\left(Q_0^2\right)} . \tag{16.22}$$

Equation (16.21) expresses that once we know the structure function at a given $Q_0^2$ for all $x$ ($0 < x < 1$), one can predict its value at another $Q^2$ using a perturbative QCD calculation of the kernel function $K$.

### 16.2 RGE for the Wilson coefficients

The $Q^2$-dependence of the structure functions is completly contained into the one of the Wilson coefficients. As the electromagnetic current is not renormalized, the anomalous dimension of the composite operators should be cancelled by the one of the Wilson coefficients. Using the discussions in Chapter 11, we can write the RGE for the Wilson coefficients:

$$\left\{\nu\frac{\partial}{\partial\nu}+\beta(\alpha_s)\alpha_s\frac{\partial}{\partial\alpha_s}-\sum_j\gamma_m(\alpha_s)m_j\frac{\partial}{\partial m_j}-\gamma_n^{(i)}\right\}C_n^{(i)}(-q^2)=0\,. \tag{16.23}$$

where $\gamma_n^{(i)}$ is the anomalous dimension of the composite operators $\hat{O}_n^{(i)}$, which can be proven to be gauge invariant such that the gauge-dependent term in the RGE is absent here. In the case of non-singlet structure functions, we have only one operator. In the case of singlet operators, we have coupled RGE due to the mixing of the two operators presented previously in Eq. (16.4). In this case, one should understand the anomalous dimension as a $2\times 2$ matrix and the Wilson coefficient as a two-component vector. The solution to the RGE is:

$$C_n^{(i)}(Q^2/\nu^2,\alpha_s,m)=C_n^{(i)}(1,\bar{\alpha}_s(t),\bar{m}(t))\exp\left[-\int_0^t dt'\gamma_n^{(i)}[\bar{g}(t')]\right], \tag{16.24}$$

where $t=1/2\log(Q^2/\nu^2)$. One can also rewrite the solution as:

$$C_n^{(i)}(Q^2)=C_n^{(i)}(1,\bar{\alpha}_s(t))\exp\left[-\int_{\alpha_s}^{\bar{\alpha}_s}dg\frac{\gamma_n^{(i)}(g)}{\beta(g)}\right], \tag{16.25}$$

where the $\beta$ function has been defined in Chapter 11 (Table 11.1):

$$\beta=\beta_1\left(\frac{\alpha_s}{\pi}\right)+\beta_2\left(\frac{\alpha_s}{\pi}\right)^2+\cdots\,. \tag{16.26}$$

### 16.3 Anomalous dimension of the non-singlet structure functions

In the following, one can safely suppress the index $i$ because in the non-singlet case, only one operator dominates the light-cone expansion. Therefore:

$$\gamma_{NS,n}^{(i)}\equiv\gamma_{NS,n}=\gamma_n^0\left(\frac{\alpha_s}{\pi}\right)+\gamma_n^1\left(\frac{\alpha_s}{\pi}\right)^2+\cdots \tag{16.27}$$

In the following, we shall explicitly discuss the evaluation of $\gamma_n^0$. It comes from the Feynman diagrams in Fig. 16.2.

Using the Feynman rules given in Appendix E for the composite operators, Fig. 16.2a gives in the massless case and in the Feynman gauge:

$$V_{ij}^{(a)}=i^5g^2\sum_{a,l}\frac{\lambda_{il}^a}{2}\frac{\lambda_{lj}^a}{2}\int\frac{d^Nk}{(2\pi)^N}\frac{\gamma^\mu\hat{k}\hat{\Delta}(\Delta\cdot k)^{n-1}\hat{k}\gamma^\nu}{k^4}\frac{(-g_{\mu\nu})}{(p-k)^2}\,. \tag{16.28}$$

The relevant contribution to the anomalous dimension is the divergent part of the coefficient of $(\Delta\cdot p)^{n-1}\hat{\Delta}$. Using standard Feynman parametrization and shift of momentum

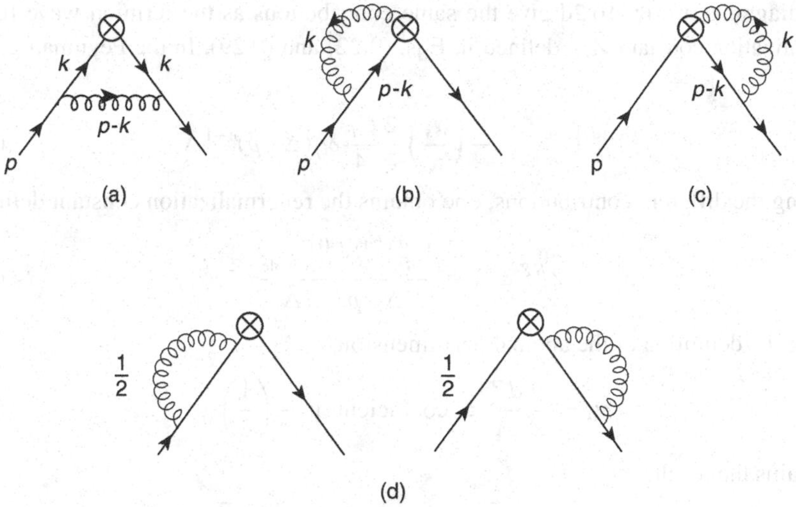


Fig. 16.2. Diagrams involved in the evaluation of $\gamma_n^0$.

(see Appendix F), the divergent part is:

$$V_{ij}^{(a)}\big|_{\epsilon \text{ pole}} = ig^2\delta_{ij}C_F \int_0^1 dx(1-x)\int \frac{d^N k}{(2\pi)^N}\frac{\mathcal{N}}{[k^2+p^2x(1-x)]^3}\,, \tag{16.29}$$

where:

$$\mathcal{N} = -\frac{2k^2}{N}\gamma^\alpha\gamma^\beta\hat{\Delta}\gamma_\beta\gamma_\alpha x^{n-1}\hat{\Delta}(\Delta\cdot p)^{n-1}\,. \tag{16.30}$$

Therefore:

$$V_{ij}^{(a)}\big|_{\epsilon \text{ pole}} = \left(\frac{\alpha_s}{\pi}\right)\frac{2}{\hat{\epsilon}}\frac{C_F}{4}\frac{2}{n(n+1)}\hat{\Delta}(\Delta\cdot p)^{n-1}\,, \tag{16.31}$$

where $C_F = (N_c^2-1)/2N_c$ for $SU(N)_c$ and:

$$\frac{2}{\hat{\epsilon}} \equiv \frac{2}{\epsilon} + \log 4\pi - \gamma_E\,. \tag{16.32}$$

Figures 16.2b and c give the same result. It reads:

$$V_{ij}^{(b)} = V_{ij}^{(c)} = -i^3g^2C_F\delta_{ij}\int\frac{d^Nk}{(2\pi)^N}\frac{\Delta^\mu\hat{\Delta}\left[\sum_{n=0}^{n-2}(\Delta\cdot p)^l[\Delta\cdot(p+k)]^{n-l-2}\right](\hat{p}\hat{k})\gamma_\mu}{k^2(k+p)^2}\,. \tag{16.33}$$

The pole part of the coefficient of $(\Delta\cdot p)^{n-1}\hat{\Delta}$ is:

$$\begin{aligned} V_{ij}^{(b)}\big|_{\epsilon \text{ pole}} &= 2ig^2C_F\delta_{ij}\hat{\Delta}\int_0^1 dx\int\frac{d^Nk}{(2\pi)^N}\frac{\sum_{n=0}^{n-2}(\Delta\cdot p)^l(\Delta\cdot k + x\Delta k)^{n-l-1}}{[k^2+p^2x(1-x)]^2} \\ &= -\left(\frac{\alpha_s}{\pi}\right)\frac{2}{\hat{\epsilon}}\frac{C_F}{2}\delta_{ij}(\Delta\cdot p)^{n-1}\hat{\Delta}\left\{\int_0^1 dx\sum_{l=1}^{n-1}x^l = \sum_{l=2}^{n}\frac{1}{l}\right\}\,. \end{aligned} \tag{16.34}$$

The diagrams in Fig. 16.2d give the same contributions as the fermion wave function renormalization constant $Z_{2F}$ defined in Eqs. (9.22) and (9.29). In the Feynman gauge, it gives:

$$V_{ij}^{(d)}\Big|_{\epsilon \text{ pole}} = -\left(\frac{\alpha_s}{\pi}\right)\frac{2}{\hat{\epsilon}}\frac{C_F}{4}\delta_{ij}(\Delta\cdot p)^{n-1}\hat{\Delta} \tag{16.35}$$

Adding the different contributions, one obtains the renormalization constant defined as:

$$Z_n^{NS} \equiv 1 + \frac{V_{ij}^{(a+b+c+d)}\Big|_{\epsilon \text{ pole}}}{(\Delta\cdot p)^{n-1}\hat{\Delta}} . \tag{16.36}$$

Using the definition of the anomalous dimension:

$$\gamma_n = \frac{\nu}{Z}\frac{dZ}{d\nu} \equiv \text{coefficient of } -\left(\frac{1}{\hat{\epsilon}}\right) , \tag{16.37}$$

one obtains the result:

$$\gamma_n^0 = \frac{C_F}{2}\left[1 - \frac{2}{n(n+1)} + 4\sum_{l=2}^{n}\frac{1}{l}\right] , \tag{16.38}$$

or equivalently:

$$\gamma_n^0 = \frac{C_F}{2}\left[4S_{1,n} - 3 - \frac{2}{n(n+1)}\right] , \tag{16.39}$$

with

$$S_{1,n} \equiv \sum_{l=1}^{n}\frac{1}{l} . \tag{16.40}$$

The expression of $S_{1,n}$ can be analytically continued to complex $n$ thanks to the Carlson theorem [227] which we have used previously when taking the inverse Mellin transform. In this case, one can write:

$$S_{1,n} = n\sum_{k=1}^{\infty}\frac{1}{k(k+n)} = \psi(n+1) + \gamma_E \ : \ \psi(z) \equiv \frac{d\log\Gamma(z)}{dz} . \tag{16.41}$$

where the expression of $\gamma_n^1$ is also known [228] and corrected in [232]. At this order, the problem of even (resp. odd) structure functions arises. The corresponding anomalous dimensions are $\gamma_n^{1,\pm}$. They read:

$$\begin{aligned}
\gamma_n^{1,\pm} = {} & \frac{32}{9}S_{1,n}\left[67 + \frac{8(2n+1)}{n^2(n+1)^2}\right] - 64S_{1,n}S_{2,n} - \frac{32}{9}[S_{2,n} - S_{2,n/2}^{\pm}]\left[2S_{1,n} - \frac{1}{n(n+1)}\right] \\
& - \frac{128}{9}\tilde{S}_n^{\pm} + \frac{32}{3}S_{2,n}\left[\frac{3}{n(n+1)} - 7\right] + \frac{16}{9}S_{3,n/2}^{\pm} - 28 \\
& - 16\frac{151n^4 + 260n^3 + 96n^2 + 3n + 10}{9n^3(n+1)^3} \\
& \pm \frac{32}{9}\frac{(2n^2+2n+1)}{n^3(n+1)^3} + \frac{32n_f}{27}\left[6S_{2,n} - 10S_{1,n} + \frac{3}{4} + \frac{11n^2+5n-3}{n^2(n+1)^2}\right] ,
\end{aligned} \tag{16.42}$$

where:

$$S^{+}_{l,n/2} = S_{l,n/2}\,, \qquad S^{-}_{l,n/2} = S_{l,(n-1)/2}\,, \qquad \tilde{S}^{\pm}_{n} = -\frac{5}{8}\zeta_3 \mp \sum_{k=1}^{\infty} \frac{(-1)^k}{(k+n)^2} S_{l,n+k}\,. \tag{16.43}$$

## 16.4 Strategy for obtaining the Wilson coefficients

The main task in perturbative QCD is to calculate the Wilson coefficients. This can be simplified by the key observation that they are independent of the states which sandwich the light-cone expansion of the T-product of the electromagnetic current for the forward Compton amplitude $T_{\mu\nu}$. For instance, instead of taking proton states, one could consider quark or gluon Green's function with the insertion of the T-product of electromagnetic current. In the case of quark fields, the truncated (quark external line) Green's function reads:

$$\Gamma_{\mu\nu}(q,p)_{\text{trunc}} = i\int d^4x d^4x_1 d^4x_2\, e^{-qx+p(x_1-x_2)} \langle 0|T J_\mu(x) J_\nu(0)\psi(x_1)\bar{\psi}(x_2)|0\rangle\,, \tag{16.44}$$

where $p$ is the quark momentum. Repeating the same reasoning as in the previous section, one can write the OPE analogous to the one in Eq. (16.9):

$$\Gamma_{\mu\nu}(q,p)_{\text{trunc}} = 2\omega^n \sum_{i,n\ \text{even}} e_{\mu\nu} C^{(i)}_{L,n}(-q^2)\hat{\mathcal{O}}^{(i,\text{pert})}_{L,n} - d_{\mu\nu} C^{(i)}_{2,n}(-q^2)\hat{\mathcal{O}}^{(i,\text{pert})}_{2,n}\,, \tag{16.45}$$

where the Wilson coefficients are the same as in Eq. (16.9) but the 'composite operators' $\hat{\mathcal{O}}^{(i,\text{pert})}_{L:2,n}$ are calculable in perturbative QCD. The strategy is to calculate $\Gamma_{\mu\nu}(q,p)_{\text{trunc}}$ and $\hat{\mathcal{O}}^{(i,\text{pert})}_{L:2,n}$ in perturbation theory and then deduce the Wilson coefficients order by order of perturbative QCD.

### 16.4.1 Non-singlet part of the Bjorken sum rule

In the non-singlet part of the Bjorken sum rule, the Wilson coefficients can be expressed as:

$$C_{n,NS}(1,\alpha_s(Q^2)) = C^0_{n,NS}\left\{1 + C^1_{n,NS}\left(\frac{\alpha_s}{\pi}\right) + \cdots\right\} \tag{16.46}$$

For their evaluation, we shall consider the quark Green's functions:

$$T_{\mu\nu}(q,\psi) = i\int d^4x\, e^{iqx}\langle\psi|T J_\mu(x) J_\nu(0)|\psi\rangle\,, \tag{16.47}$$

which has the decomposition:

$$T_{\mu\nu}(q) = e_{\mu\nu} T_L + d_{\mu\nu} T_2\,, \tag{16.48}$$

where $e_{\mu\nu}$ and $e_{\mu\nu}$ have been defined in Eq. (16.10). We shall also use:

$$\hat{\mathcal{O}}^{(i,\text{pert})}_{L:2,n}\, p^{\mu_1}\cdots p^{\mu_n} = \langle\psi|\mathcal{O}^{(i),\mu_1\cdots\mu_n}_{L:2,n}|\psi\rangle\,. \tag{16.49}$$

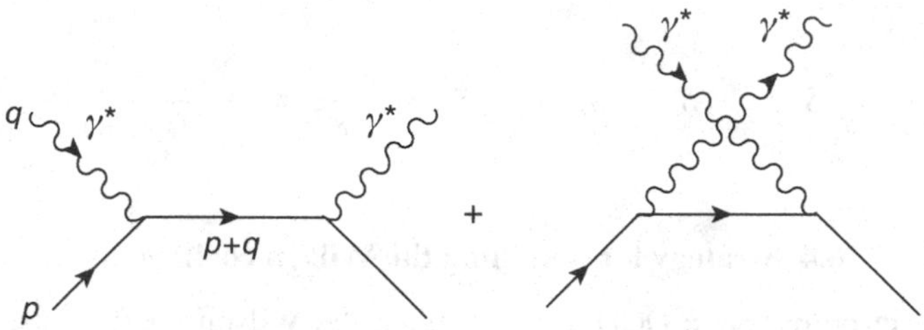


Fig. 16.3. Tree-level diagram for a photon-quark scattering.

The quark tree-level diagram shown in Fig. 16.3 leads to the amplitude:

$$T^0_{\mu\nu} = Q^2_\psi \frac{1}{2} \sum_\lambda \bar{u}_\lambda(p) \left[ \gamma_\mu \frac{1}{\hat{p}+\hat{q}-m} \gamma_\nu + \gamma_\mu \frac{1}{\hat{p}-\hat{q}-m} \gamma_\nu \right] u_\lambda(p) \,. \tag{16.50}$$

where $u_j(p)$ is the quark spinor, and $Q_\psi$ is its charge in units of $e$. Introducing the Bjorken variables, one has:

$$T^0_{\mu\nu} = Q^2_\psi d_{\mu\nu} \left( \frac{1}{x-1} - \frac{1}{x+1} \right) = 2Q^2_q d_{\mu\nu} \sum_{n=2,4\cdots} \left( \frac{1}{x} \right)^n \,, \tag{16.51}$$

where $d_{\mu\nu}$ has been defined in Eq. (16.10). Then, ones find:

$$T^0_2 = Q^2_\psi \frac{2}{x^2-1} \,, \qquad T^0_L = 0 \,. \tag{16.52}$$

These results are already known from the free-field theory discussed in the beginning of this chapter. Solving the RGE for the Wilson coefficient, one obtains the modification due to QCD at leading order:

$$C_{2,n}(Q^2) \sim \left( \log \frac{Q^2}{\Lambda^2} \right)^{\gamma^0_n/2\beta_1} \,, \tag{16.53}$$

showing that the naïve Bjorken scaling is modified by the running coupling of QCD to leading order. To second order, one has [233]:

$$C^1_{n,NS} = \frac{C_F}{4} \left( 2S^2_{1,n} + 3S_{1,n} - 2S_{2,n} - \frac{2S_{1,n}}{n(n+1)} + \frac{3}{n} + \frac{4}{n+1} + \frac{2}{n^2} - 9 \right) \,. \tag{16.54}$$

Therefore, to second order, the non-singlet moments read:

$$\begin{aligned} \mathcal{M}^{NS}_n(Q^2) &= \left( \frac{\alpha_s(Q^2_0)}{\alpha_s(Q^2)} \right)^{\gamma^0_n/\beta_1} \left( \frac{1+\beta_2/\beta_1(\alpha_s(Q^2)/\pi)}{1+\beta_2/\beta_1\left(\alpha_s\left(Q^2_0\right)/\pi\right)} \right)^{-p_n} \\ &\times \left( \frac{1+C^1_{NS,n}(\alpha_s(Q^2)/\pi)}{1+C^1_{NS,n}\left(\alpha_s\left(Q^2_0\right)/\pi\right)} \right) \mathcal{M}^{NS}_n(Q^2_0) \,, \end{aligned} \tag{16.55}$$

where:

$$p_n = \gamma_n^1/\beta_2 - \gamma_n^0/\beta_1 \,. \tag{16.56}$$

This relation is well verified experimentally and used to measure the QCD coupling $\alpha_s$.

### 16.4.2 Callan–Gross scaling violation

To leading order, the longitudinal structure function, coming from the diagram in Fig. 16.3, vanishes being defined as $F_2 - 2xF_1$. In the following, we analyze the structure function to order $\alpha_s$.

*Non-singlet part*

To order $\alpha_s$, the non-singlet part comes from the diagram in Fig. 16.4.

The analysis is simplified by noting that $T_L$ is the only amplitude multiplied by $q_\mu q_\nu$. The amplitude from the *direct* diagram is:

$$T^{ij}_{\mu\nu}\Big|_{\text{dir}} = -iC_F\delta_{ij}g^2\frac{1}{4}\sum_\sigma \bar{u}(p,\sigma) \times \int \frac{d^Nk}{(2\pi)^N}\frac{\gamma_\alpha(\hat{p}+\hat{k})\gamma^\mu(\hat{p}+\hat{k}+\hat{q})\gamma^\nu(\hat{p}+\hat{k})\gamma^\alpha}{(p+k)^4(p+k+q)^2k^2}u(p,\sigma) \,. \tag{16.57}$$

Using:

$$\sum_\sigma \bar{u}(p,\sigma)\mathcal{M}u(p,\sigma) = Tr[\hat{p}\mathcal{M}] \,, \tag{16.58}$$

and extracting term proportionnal to $q^\mu q^\nu$, one obtains after usual manipulations:

$$T_L^{NS}\Big|_{\text{dir}} = \left(\frac{\alpha_s}{\pi}\right)C_F\frac{2}{x}\int_0^1 ydy\int_0^1 dz\frac{y(1-yz)}{[y-[1-(1-y-yz)/x]]^2} \,, \tag{16.59}$$

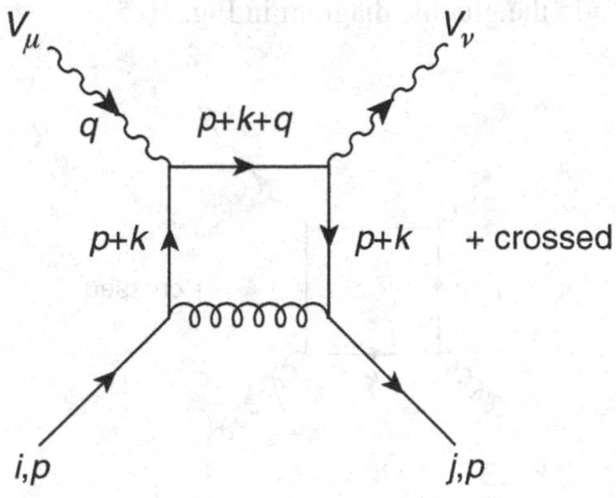


Fig. 16.4. Diagrams contributing to $F_L^{NS}$.

Expanding in powers of $1/x$ and integrating, one obtains:

$$T_L^{NS}\Big|_{\rm dir} = \left(\frac{\alpha_s}{\pi}\right) C_F \sum_{n=1}^{\infty} \frac{1}{n+1}\left(\frac{1}{x}\right)^n . \tag{16.60}$$

The *crossed* diagram doubles the even $n$ contribution and cancels the odd one. Then, one finally obtains:

$$T_L^{NS} = 2\left(\frac{\alpha_s}{\pi}\right) C_F \sum_{n={\rm even}}^{\infty} \frac{1}{n+1}\left(\frac{1}{x}\right)^n . \tag{16.61}$$

Comparing with Eqs. (16.52) and (16.16), one can deduce the *scaling violation* QCD correction to the Callan–Gross relation:

$$\mathcal{M}_{L,n}^{NS} = \delta_L^{NS}\left(\frac{\alpha_s}{\pi}\right)\frac{C_F}{n+1}\mathcal{M}_{2,n}^{NS} , \tag{16.62}$$

where for $ep$ scattering $\delta_L^{NS} = 1/6$. Taking the Mellin transforms, one can derive the non-singlet part of the structure functions:

$$F_L^{NS}(x, Q^2) = \int_x^1 dy\, C_{NS}^L(y, Q^2) F_2^{NS}\left(\frac{x}{y}, Q^2\right) , \tag{16.63}$$

where:

$$C_L^{NS}(y, Q^2) = C_F x(\alpha_s(Q^2)/\pi) + \mathcal{O}(\alpha_s^2) , \tag{16.64}$$

where the $\alpha_s^2$ correction has been evaluated in [235].

*Singlet part*

The calculation of the singlet part is similar to that for the non-singlet. To the quark diagram in Fig. 16.3, one has to add the gluonic diagram in Fig. 16.5.

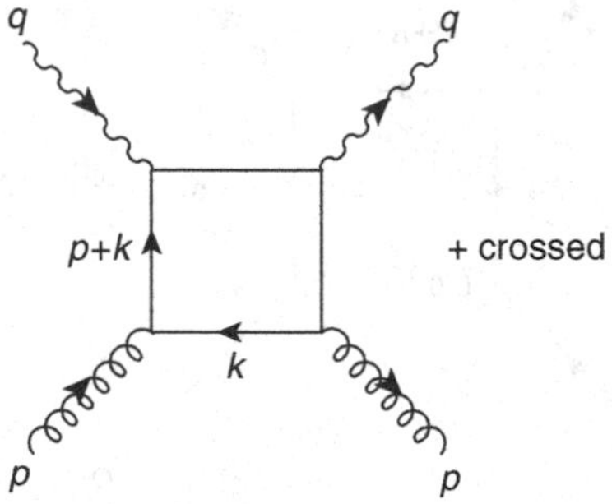


Fig. 16.5. Diagrams contributing to the gluon component of the structure function.

For electron-proton scattering, the singlet structure function can be decomposed as:

$$F_L^S(x, Q^2) = \int_x^1 dy \left\{ C_S^L(y, Q^2) F_2^S\left(\frac{x}{y}, Q^2\right) + C_G^L(y, Q^2) F_G^S\left(\frac{x}{y}, Q^2\right) \right\} , \quad (16.65)$$

where:

$$\begin{aligned} C_S^L(x, Q^2) &\equiv C_{NS}^L + C_{QS}^L , \\ C_{QS}^L(x, Q^2) &= C_{QS}^{1,L} \left(\frac{\alpha_s(Q^2)}{\pi}\right)^2 , \\ C_G^L(x, Q^2) &= 4n_f T_R x(1-x) \left(\frac{\alpha_s(Q^2)}{\pi}\right) + C_G^{1,L} \left(\frac{\alpha_s(Q^2)}{\pi}\right)^2 . \end{aligned} \quad (16.66)$$

$C_{NS}^L$ has been defined in Eq. (16.64). The coefficients $C_{QS}^{1,L}$ and $C_G^{1,L}$ have been evaluated in [235–237]. The full longitudinal structure function is the sum of the non-singlet and quark singlet components.

It is given by:

$$F_L \equiv F_2 - 2x F_3 = F_L^S + F_L^{NS} . \quad (16.67)$$

## 16.5 Singlet anomalous dimensions and moments

The singlet calculations are more involved than the case of non-singlet and longitudinal structure functions. The corresponding anomalous dimension is a $2 \times 2$ matrix because of the mixing of the operators in Eq. (16.4). Using an expansion of the anomalous dimension and Wilson coefficient function:

$$\begin{aligned} \gamma_n &= \gamma_{0n} \left(\frac{\alpha_s}{\pi}\right) + \gamma_{1n} \left(\frac{\alpha_s}{\pi}\right)^2 + \cdots \\ C_n^{(i)}(1, \alpha_s(Q^2)) &= C_{n,i}^0 \left\{1 + C_{n,i}^1 \left(\frac{\alpha_s}{\pi}\right) + \cdots\right\} \end{aligned} \quad (16.68)$$

To leading order,

$$C_n^{(i)}(1, \alpha_s(Q^2)) = C_{n,j}^0 \left(\frac{\alpha_s(Q_0^2)}{\alpha_s(Q^2)}\right)_{ij}^{\gamma_{0n}/\beta_1} \quad (16.69)$$

where the indices $i, j \equiv q, g$ indicate quark and gluon composite operators respectively. The calculation of $C_{n,i}^0$ is very analogous to the non-singlet case by considering the forward Compton amplitude sandwiched between two quark states for $C_{n,q}^0$ and two gluon states for $C_{n,g}^0$. One obtains to this order:

$$C_{n,q}^0 = \begin{cases} 1 & \text{for } \mathrm{C_{n,2}} \\ 0 & \text{for } \mathrm{C_{n,L}} \end{cases} . \quad (16.70)$$

Since the gluon does not couple to the photon to lowest order, one obtains:

$$C_{n,g}^0(Q^2) = 0 . \quad (16.71)$$

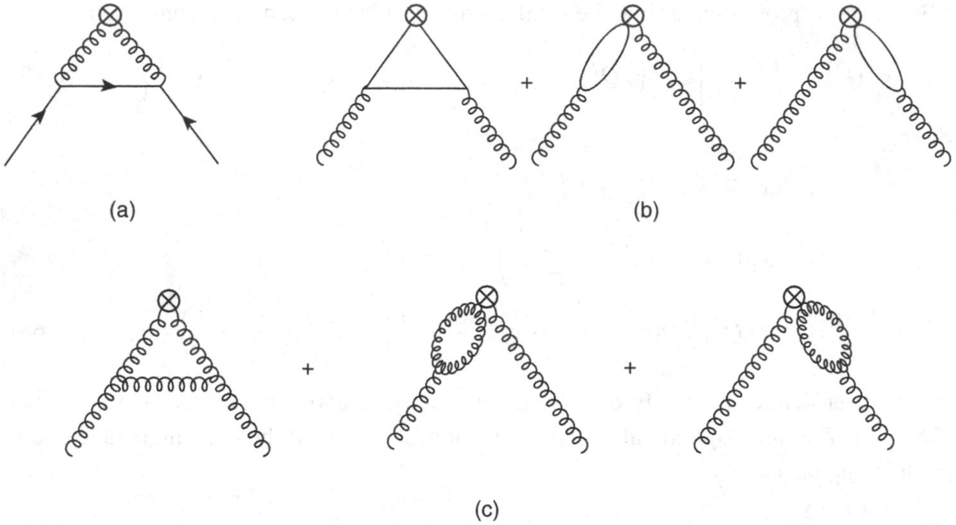


Fig. 16.6. Diagrams contributing to the singlet anomalous dimensions.

The anomalous dimension matrix reads to leading order:

$$\gamma_{0n} = \begin{pmatrix} \gamma_{0n}^{qq} & \gamma_{0n}^{qg} \\ \\ \gamma_{0n}^{gq} & \gamma_{0n}^{gg} \end{pmatrix} . \tag{16.72}$$

The diagrams contributing to the anomalous dimensions are given in Fig. 16.6, in addition to the contribution from the diagrams in Fig. 16.2. The results are [168,234]:

$$\begin{aligned}
\gamma_{0n}^{qq} &= \frac{C_F}{2}\left[1 - \frac{2}{n(n+1)} + 4\sum_{j=2}^{n}\frac{1}{j}\right] , \\
\gamma_{0n}^{qg} &= -2n_f T_R \frac{n^2+n+2}{n(n+1)(n+2)} , \\
\gamma_{0n}^{gq} &= -C_F \frac{n^2+n+2}{n(n^2-1)} , \\
\gamma_{0n}^{gg} &= 2\left[C_G\left(\frac{1}{12} - \frac{1}{n(n+1)} - \frac{1}{(n+1)(n+2)} + \sum_{j=2}^{n}\frac{1}{j}\right) + T_R\frac{n_f}{3}\right] ,
\end{aligned} \tag{16.73}$$

where $C_F = (N_c^2 - 1)/2N_c$, $T_R = 1/2$ and $C_G = N_c$ for $SU(N)_c$. To this order, the moments in Eq. (16.16) read:

$$\begin{aligned}
\mathcal{M}_{2,n}(Q^2) &= \sum_i C_{n,i}^0 \left(\frac{\alpha_s(Q_0^2)}{\alpha_s(Q^2)}\right)_{iq}^{\gamma_{0n}/\beta_1} , \\
\mathcal{M}_{L,n}(Q^2) &= 0 .
\end{aligned} \tag{16.74}$$

In order to make a comparison with experiment, it is convenient to diagonalize the anomalous dimension matrix $\gamma_{on}$. On this basis, one can write:

$$\mathcal{M}_{2,n}(Q^2) = C^0_{+,n}\left(\log\frac{Q^2}{\Lambda^2}\right)^{-\gamma^+_{0n}/2\beta_1} + C^0_{-,n}\left(\log\frac{Q^2}{\Lambda^2}\right)^{-\gamma^-_{0n}/2\beta_1}, \qquad (16.75)$$

with:

$$\gamma^\pm_{0n} = \frac{1}{2}\left[\gamma^{qq}_{0n} + \gamma^{gg}_{0n} \pm \sqrt{\left(\gamma^{qq}_{0n} + \gamma^{gg}_{0n}\right)^2 + 4\gamma^{qg}_{0n}\gamma^{gq}_{0n}}\,\right]. \qquad (16.76)$$

To the next order, the expressions of the anomalous dimensions are known and the Wilson coefficients read [233]:

$$C^1_{n,q} = C^1_{n,NS} = \frac{C_F}{4}\left(2S^2_{1,n} + 3S_{1,n} - 2S_{2,n} - \frac{2S_{1,n}}{n(n+1)} + \frac{3}{n} + \frac{4}{n+1} + \frac{2}{n^2} - 9\right),$$
$$C^1_{n,g} = T_F n_f\left(-\frac{1}{n} + \frac{1}{n^2}\frac{6}{n+1} - \frac{6}{n+2} - S_{1,n}\frac{n^2+n+2}{n(n+1)(n+2)}\right). \qquad (16.77)$$

To this order, the moments in the singlet case have more involved expressions, because of the mixing of operators. We refer the readers to, for example, the papers in [228,232,233], the review in [49] and book [46] for some expositions of this case. Finally, the expressions of few moments including three-loop corrections have been evaluated in [238].

# 17

# The Altarelli–Parisi equation

Although convenient, the use of OPE to deep inelastic scattering does not provide a transparent physical intuition of the parton model. An elegant reformulation of the moments which makes a close contact with the parton model picture is given by the Altarelli–Parisi equation [239] and the review in [48].

## 17.1 The non-singlet case

One can illustrate this equation by taking the simple example of the non-singlet structure functions, which one can write as an incoherent sum of quark parton densities $q_f(x)$. In presence of external fields, the quark parton densities acquire a $Q^2$ dependence, and the structure function can be written as:

$$F_{2,n}^{NS}(Q^2) = x \sum_f \delta_f^{NS} q_f(x, Q^2) \,, \tag{17.1}$$

where $\delta_f^{NS}$ are known coefficients. Using the QCD expression to lowest order, the corresponding moments read:

$$\mathcal{M}_{2,n}^{NS}(Q^2) = (\log(Q^2/\Lambda^2))^{\gamma_n^0/\beta_1} \,. \tag{17.2}$$

From its definition, the moments of the quark densities are:

$$\mathcal{M}_{f,n}(Q^2) \equiv \int_0^1 dx\, x^{n-1} q_f(x, Q^2) \,. \tag{17.3}$$

Using Eq. (17.2), one can derive the evolution equation:

$$\frac{\partial \mathcal{M}_{f,n}(t)}{\partial t} = -\gamma_n^0 \left(\frac{\alpha_s}{\pi}\right)(t) \mathcal{M}_{f,n}(t) \,, \tag{17.4}$$

with $t \equiv (1/2)\log(Q^2/\nu^2)$. Defining:

$$\int_0^1 dz\, z^{n-1} P_{NS}^{(0)}(z) = -\gamma_n^0 \,, \tag{17.5}$$

and taking the Mellin transform of Eq. (17.3), one can deduce the Altarelli–Parisi equation [239]:

$$\frac{\partial q_f(x,t)}{\partial t} = \frac{\alpha_s(t)}{\pi}\int_x^1 \frac{dy}{y} P^{(0)}_{NS}\left(\frac{x}{y}\right) q_f(y,t)\,, \tag{17.6}$$

which one can symbolically write as:

$$\frac{\partial q_f}{\partial t} = \frac{\alpha_s(t)}{\pi} q_f \otimes P^{(0)}_{NS}\,. \tag{17.7}$$

In its infinitesimal form, the equation can be rewritten as:

$$q_f(x,t) + dq_f(x,t) = \int_0^1 dy \int_0^1 dz\ \delta(zy-x) q_f(x,t)\left\{\delta(z-1) + \left(\frac{\alpha_s}{\pi}\right) P^{(0)}_{NS}(z)dt\right\}\,. \tag{17.8}$$

One can interpret $P^{(0)}_{NS}(z)$ (*splitting functions*) as controlling the rate of change of the parton distribution probability with respect to $t$. One can check that Eq. (17.5) is satisfied if:

$$P^{(0)}_{NS}(z) = C_F\left\{\frac{3}{2}\delta(1-z) + \frac{1+z^2}{(1-z)_+}\right\}\,, \tag{17.9}$$

where for any function $g$:

$$\int_0^1 \frac{dz}{(1-z)_+} g(z) \equiv \int_0^1 \frac{dz}{(1-z)}[g(z) - g(1)]\,. \tag{17.10}$$

## 17.2 The singlet case

In the case of singlet structure functions, analogous relations can be obtained. The Altarelli–Parisi evolution-coupled equations are:[1]

$$\frac{\partial q_f(x,t)}{\partial t} = \frac{\alpha_s(t)}{\pi}\int_x^1 \frac{dy}{y}\left\{P^{(0)}_{qq}\left(\frac{x}{y}\right) q_f(y,t) + P^{(0)}_{qg}\left(\frac{x}{y}\right) G(y,t)\right\}\,, \tag{17.11}$$

$$\frac{\partial G(x,t)}{\partial t} = \frac{\alpha_s(t)}{\pi}\int_x^1 \frac{dy}{y}\left\{P^{(0)}_{gp}\left(\frac{x}{y}\right) q_f(y,t) + P^{(0)}_{gg}\left(\frac{x}{y}\right) G(y,t)\right\}\,. \tag{17.12}$$

The splitting functions are:

$$\begin{aligned}
P^{(0)}_{qq} &= P^{(0)}_{NS}\,,\\
P^{(0)}_{qg} &= \frac{T}{n_f}(x^2 + (1-x)^2)\,,\\
P^{(0)}_{gq} &= C_F\frac{1}{x}(1 + (1-x)^2)\,,\\
P^{(0)}_{gg} &= 2C_G\left(\frac{1-x}{x} + x(1-x) + \frac{x}{(1-x)_+}\right) + \delta(1-x)\frac{(11C_G - 4T)}{6}\,,
\end{aligned} \tag{17.13}$$

[1] Recall our definition of $t$ which is $1/2$ of the one used in the original paper.

where: $T \equiv n_f/2$ , $C_F = (N^2 - 1)/(2N)$ and $C_G = N$. In the limit $C_F = C_G = 2T$ (supersymmetry of the massless QCD lagrangian where both gluons and Weyl fermions transform according to the regular representation of the group), one has the remarkable relation:

$$P^{(0)}_{qq} + P^{(0)}_{qg} = 2n_f P^{(0)}_{gq} + P^{(0)}_{gg} . \tag{17.14}$$

By taking the difference of Eq. (17.11) for $q_i$ and $q_j$, where $q_{i,j}$ are any quark or anti-quark densities, the gluon term drops out, and one recovers the previous simple result for the non-singlet (or valence) evolution equations:

$$\frac{\partial V_{ij}}{\partial t} = \frac{\alpha_s(t)}{\pi} V_{ij} \otimes P^{(0)}_{qq} \quad : \qquad V_{ij}(x,t) \equiv q_i(x,t) - q_j(x,t) . \tag{17.15}$$

Defining:

$$\Sigma(x,t) \equiv \sum_f q_f(x,t) = \sum_{\text{flavours}} [q(x,t) + \bar{q}(x,t)] , \tag{17.16}$$

one can also obtain the evolution equations in terms of two independent densities:

$$\begin{aligned} \frac{\partial \Sigma}{\partial t} &= \frac{\alpha_s(t)}{\pi} \left\{ \Sigma \otimes P^{(0)}_{qq} + G \otimes 2n_f P^{(0)}_{qg} \right\} , \\ \frac{\partial G}{\partial t} &= \frac{\alpha_s(t)}{\pi} \left\{ \Sigma \otimes P^{(0)}_{gq} + G \otimes P^{(0)}_{gg} \right\} , \end{aligned} \tag{17.17}$$

which are convenient to work with in the phenomenological analysis. The solution for a quark $i$ can be reconstructed by splitting it into its non-singlet $q_i - \Sigma/2$ and singlet $\Sigma/2$ components.

## 17.3 Some physical interpretations and factorization theorem

We have seen in Eq. (17.8) that one can interpret the (*splitting functions*) $P^{(0)}_{NS}(z)$ as controlling the rate of change of the parton distribution probability with respect to $t$. This can be understood by considering the scattering of an off-shell photon on the parton as depicted in the different diagrams in Fig. 17.1.

Diagram 17.1a shows the free quark diagram in the parton model with a certain probability $q_f(x)$ of having a fraction of the proton momentum $q_f(x)$. After a time $t$, the quark may radiate into gluons as depicted in different diagrams shown in 17.1b and 17.1c. One can show that, in the axial (physical) gauge, only diagram 17.1b contributes to the cross-section and gives a term proportional to $t$:

$$\sigma(\gamma^* q \to q + g) \simeq \frac{\alpha_s(t)^2}{e} \pi [t \; P(x) + f(x)] . \tag{17.18}$$

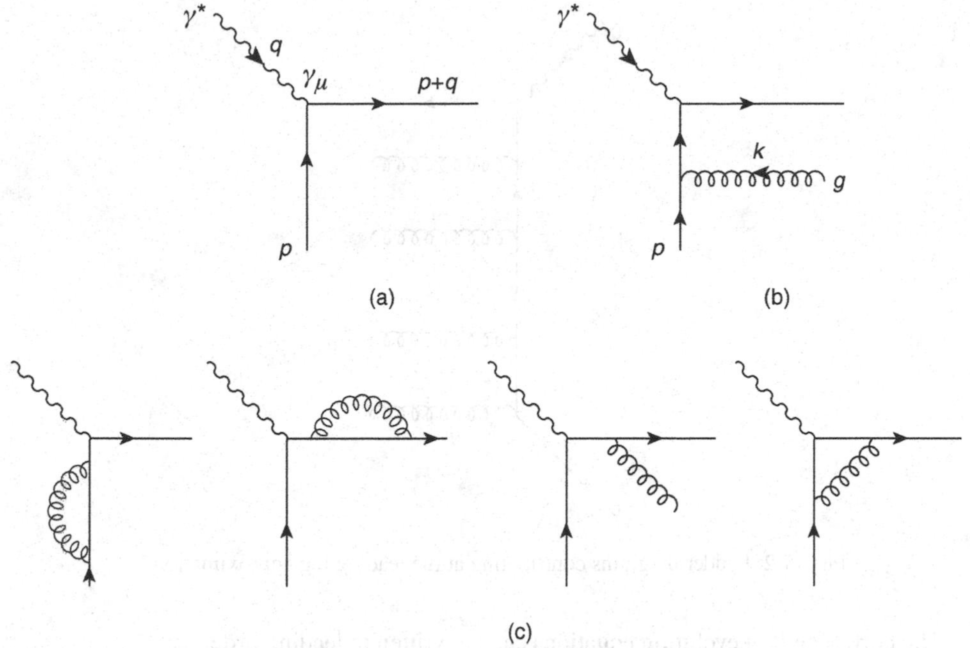


Fig. 17.1. Scattering of an off-shell photon on the parton.

$P(x)$ is well-defined perturbatively, while $f(x)$ depends on the IR regularization procedure. One can generalize the above procedure to sum up the contributions of an arbitrary number of gluons. In the leading log approximation, one can show [240] that only the ladder graphs in Fig. 17.2 contribute and lead to the factorization theorem:

$$\sigma(\gamma^* q \to q + g) \sim \left(\frac{\alpha_s(t)}{\pi}\right)^n \ln^n \frac{Q^2}{\nu^2} . \tag{17.19}$$

It is also important to recall that the splitting functions $P_{ab}^{(0)}$ are universal (anomalous dimension of the RGE) and consequently the parton densities depend only on the the target and are independent of the nature and polarization of the probe (vector, axial-vector, . . . ).

## 17.4 Polarized parton densities

The previous approach can be generalized to parton densities of definite helicity in a polarized target. The quark and gluon densities $q_{i\pm}$ and $G_\pm$, with helicity $\pm$ in a target of definite polarization, are related to the unpolarized ones as:

$$p_{A+}(x,t) + p_{A-}(x,t) = p_A(x,t) : \qquad s(p_A \equiv q_i,\ G) . \tag{17.20}$$

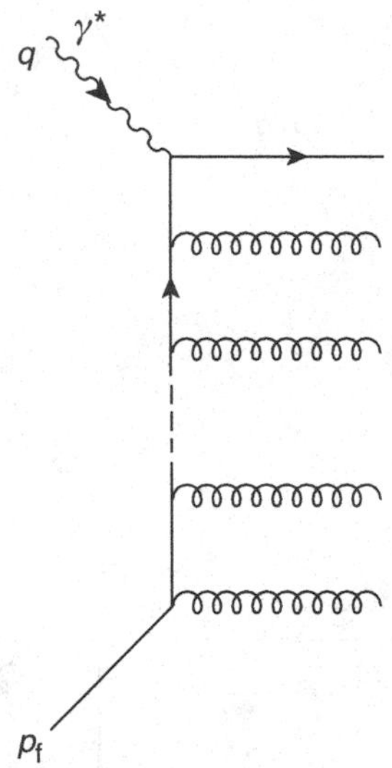


Fig. 17.2. Ladder diagrams contributing at the leading log approximation.

The corresponding evolution equations can be written to leading order as:

$$\frac{\partial p_{A\pm}}{\partial t} = \frac{\alpha_s(t)}{\pi}\left\{\sum_B \left[p_{B_+} \otimes P^{(0)}_{A_\pm B_+} + p_{B_-} \otimes P^{(0)}_{A_\pm B_-}\right]\right\} . \tag{17.21}$$

Parity and probability conservation gives:

$$\begin{aligned} P^{(0)}_{A_\pm B_-} &= P^{(0)}_{A_- B_\mp} , \\ P^{(0)}_{A_+ B_+} + P^{(0)}_{A_- B_+} &= P^{(0)}_{A_+ B_-} + P^{(0)}_{A_- B_-} , \end{aligned} \tag{17.22}$$

which imply that $(p_{A_+} + p_{A_-}) = p_A$ and $(p_{A_+} - p_{A_-}) = \Delta p_A$ evolve separately for any $A$. The evolution equation for the difference is:

$$\frac{\partial p_{A\pm}}{\partial t} = \frac{\alpha_s(t)}{\pi}\sum_B \Delta p_B \otimes \Delta P^{(0)}_{AB} . \tag{17.23}$$

The splitting function:

$$\Delta P^{(0)}_{AB} \equiv P^{(0)}_{A_+ B_+} - P^{(0)}_{A_- B_+} , \tag{17.24}$$

measures the tendency of a parton $A$ to remember the polarization of its parent $B$. From the helicity conservation at the quark gluon vertex, it follows that the non-singlet kernel is the same as in the unpolarized case:

$$\Delta P^{(0)}_{qq} \equiv P^{(0)}_{q_+ q_+} - P^{(0)}_{qq} . \tag{17.25}$$

One also finds:

$$\begin{aligned}
\Delta P_{qg}^{(0)} &= \frac{T}{n_f}\frac{1}{2}(x^2 - (1-x)^2)\,, \\
\Delta P_{gq}^{(0)} &= \frac{C_F}{2}\frac{1}{x}(1 - (1-x)^2)\,, \\
\Delta P_{gg}^{(0)} &= \frac{C_G}{2}\left((1+x^4)\left(\frac{1}{x} + \frac{1}{(1-x)_+}\right) - \frac{(1-x)^3}{x}\right) \\
&\quad + \delta(1-x)\frac{(11C_G - 4T)}{12}\,. \qquad (17.26)
\end{aligned}$$

In this case, all charge moments are well defined, as the total helicity is finite though the total number of gluons and quark pairs is infinite. To leading order, the net helicity is also conserved, such that $\Delta P_{qg}^{(0)}$ and $\Delta P_{qq}^{(0)}$ are zero.

The previous evolution equations for parton densities with definite helicities are sufficient for the prediction of scaling violations in leptoproduction on a longitudinal polarized target. Additional information is needed for a transversely polarized target.

# 18

# More on unpolarized deep inelastic scatterings

## 18.1 Target mass corrections

Target mass corrections have been introduced by Nachtmann [229], and later on in [168]. If one considers the NS part of the moments defined in Eq. (16.16), one can show [168] that they can be written as:

$$\mathcal{M}_2^{(n)}(Q^2)\big|_{\text{mass}} = \sum_{j=0}^{\infty}\left(\frac{M_N^2}{Q^2}\right)^j \frac{(n+j)!}{j!(n-2)!(n+2j)(n+2j-1)}\mathcal{M}_2^{(n)}(Q^2)\,. \quad (18.1)$$

Inverting this expression, one can express the structure function in terms of the Nachtmann variable [229]:

$$\zeta = \frac{2x}{1+\sqrt{1+4x^2M_N^2/Q^2}}\,. \quad (18.2)$$

In the region $x \to 1$, higher twist contributions can also be important and can cancel the target mass corrections [230], it is instructive to do an expansion in $x$. Keeping the leading term, one obtains (see e.g. [46]):

$$\begin{aligned} F_2^{NS}\big|_{\text{mass}}(x, Q^2) = F_2^{NS}(x, Q^2) + \frac{x^2M_N^2}{Q^2} \\ \times \left\{6x\int_0^1 \frac{dy}{y^2}\, F_2^{NS}(y, Q^2) - \frac{x\partial F_2^{NS}(x, Q^2)}{\partial x} - 4F_2^{NS}(x, Q^2)\right\} \\ +\mathcal{O}\left((3\sim 5)\frac{x^3M_N^2}{(1-x)Q^2}\right), \end{aligned} \quad (18.3)$$

where the quality of the expansion can be controlled by the size of the next term. This contribution can be compared with the higher-twist contribution in Eq. (18.6).

## 18.2 End points behaviour and the BFKL pomeron

### *18.2.1 The limit $x \to 1$*

The NLO perturbative expression of the non-singlet structure function indicates that for $x \to 1$, it behaves as [231]:

$$F^{NS} \sim (1-x)^{2[\ln(1-x)](\alpha_s/3\pi)}\,, \quad (18.4)$$

showing that perturbation theory fails. This result can be generalized to all orders by formally replacing $\alpha_s(Q^2)$ by $\alpha_s[(1-x)Q^2]$ [241]. One can also interpret this feature because, in this limit, we are in the bound state regime where the reaction of the type:

$$\gamma^* + N \rightarrow N \tag{18.5}$$

dominates. In this limit, one may also expect that non-perturbative higher twist contributions behave as [230]:

$$F_2^{HT}(x, Q^2) \sim \frac{p_T^2}{Q^2} \frac{x}{1-x} F_2(x, Q^2), \tag{18.6}$$

where $p_T$ is the transverse momentum of partons in the nucleon.

### 18.2.2 The limit $x \rightarrow 0$ for the non-singlet case

This limit has been studied extensively in hadron physics for the non-singlet scattering process.[1] It corresponds to the kinematic region where $Q^2$ is fixed and the hadronic energy $\nu$ going to infinity. This is the so-called *Regge limit*, where the cross-section of the photon scattering off the proton is proportional to the structure function:

$$\sigma(\gamma^*(Q^2)p(s)) = \frac{4\pi^2\alpha}{Q^2} F_2(x, Q^2) \ : \qquad s = Q^2/x\,, \tag{18.7}$$

and where the non-singlet amplitude can be expressed as:

$$T^{NS}(\nu \rightarrow \infty) \simeq f(Q^2)s^{\alpha_\rho(0)}, \tag{18.8}$$

due to exhange of Regge trajectories, either the $\rho$ trajectory or the one degenerate to it. $\alpha_\rho(0)$ is the *universal intercept* of the $\rho$ trajectory, which has an experimental value of about 0.5. Therefore, one can show that the structure function behaves as:

$$F_2^{NS}(x, Q^2) \sim x^{1-\alpha_\rho(0)}. \tag{18.9}$$

### 18.2.3 The limit $x \rightarrow 0$ for the singlet case and the BFKL pomeron

The singlet case is more subtle due to the coupled evolution equations from the presence of the gluon density. At present, there is no consensus on the behaviour of the structure functions at $Q_0^2 \sim$ few GeV$^2$. There are three proposals:

- **Soft pomeron**
  In this case, the structure functions are expected to behave as a constant in the $x = 0$ limit. This behaviour was first considered in [230] and completed later on. However, it has been known for a long time that a soft pomeron for off-shell processes leads to inconsistencies [243].
- **Hard pomeron**
  The previous remark then leads some people to postulate the hard pomeron exchange, where:

$$F^S(x, Q_0^2) \sim x^{-\lambda_q}, \qquad F^G(x, Q_0^2) \sim x^{-\lambda_g}. \tag{18.10}$$

[1] For a recent review, see e.g. [242].

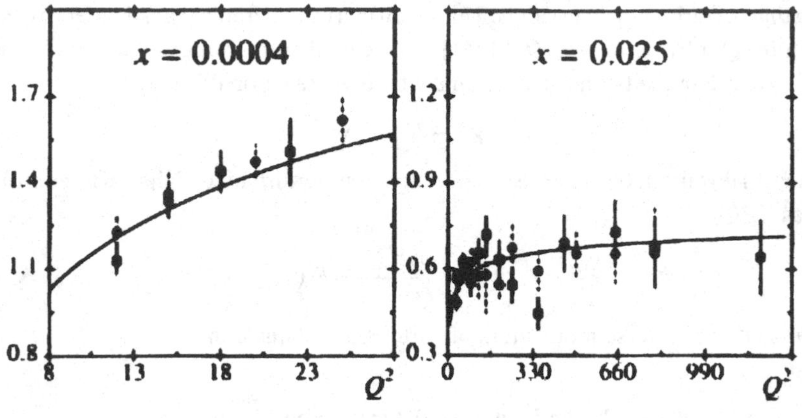


Fig. 18.1. Comparison of the measured and BFKL predictions of $F_2$ for small $x = 4 \times 10^{-4}$ and large $Q^2$. For a running value of $\alpha_s$, the HERA data are in disagreement with the BFKL result ($F_2$ should decrease with $Q^2$).

It has been proved that:

$$\lambda_q = \lambda_g \tag{18.11}$$

and are $Q^2$ independent.

- **BFKL pomeron**
The usual procedure used now is to assume a given behaviour at fixed $Q_0^2$ and then evolve the behaviour using the RGE for an arbitrary $Q^2$. Using a different approach, BFKL [244] found a different behaviour:

$$F_2(x, Q^2) \sim x^{-\omega\alpha_s(Q^2)} \; : \qquad \omega = \frac{4C_A \ln 2}{\pi} \; . \tag{18.12}$$

which is not compatible with the RGE where the exponent is constant. A comparison of this prediction with data for a given small $x$ value is given in Fig. 18.1.

A number of speculations have been suggested in order to explain this difference (two different regimes in $x$ ? $\alpha_s$ function of a soft scale of the order of $\Lambda^2$ but not of $Q^2$ ?...).

### 18.3 Experimental tests and new developments

- In the previous section, we have discussed in detail the scaling violation to the Bjorken sum rule as an illustration of the OPE approach and of the Altarelli–Parisi evolution equation. We have also concentrated the discussions on the photon scattering off a proton. A test of this prediction is given in Fig. 18.2.

  We also give the new compiled data from PDG [16] in Figs. 18.3 and 18.4.
- In [245], a model which interpolates the soft and hard pomeron parametrization and which can be used at low $Q^2$ has been proposed. It has been assumed that the soft pomeron contribution is given by an ordinary pomeron which is constant when $x \to 0$, while one has to find a parametrization

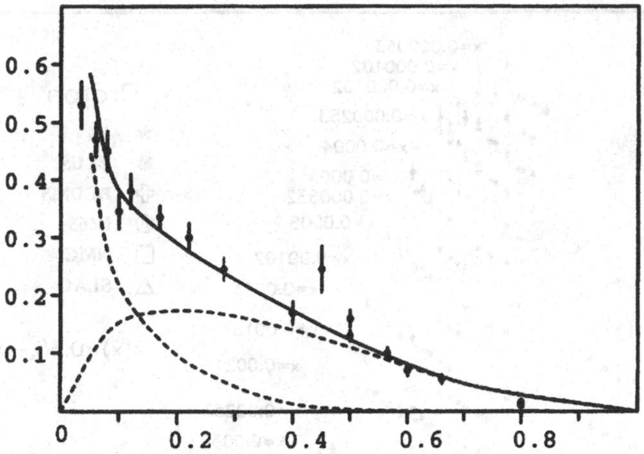


Fig. 18.2. Comparison of the measured and QCD predictions for $F_2$ where the $NS$ and $S$ components (dashed) are explicitly shown. The full curve is the sum of the two. Data points are SLAC data [246].

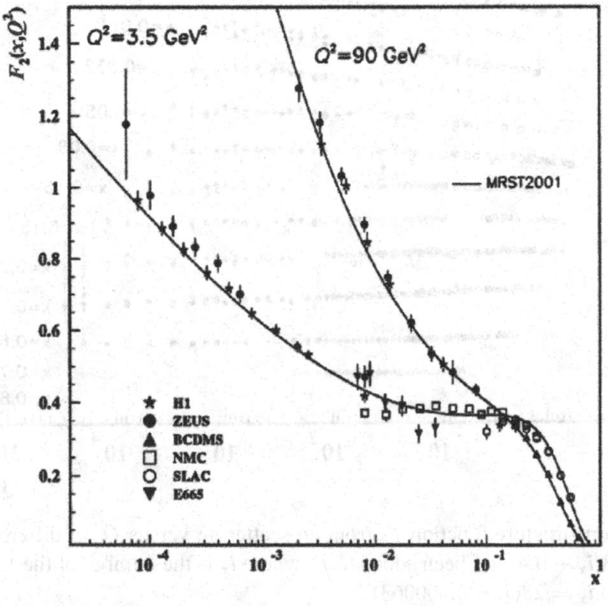


Fig. 18.3. The proton structure function $F_2$ from $ep$ scattering versus $x$ at two values of $Q^2$, exhibiting scaling at the pivot point $x \approx 0.14$.

where the cross-section does not blow up when $Q^2 \to 0$. This can be achieved by replacing the coupling by:

$$\alpha_s(Q^2) \to \tilde{\alpha}_s \equiv \frac{2\pi}{-\beta_1 \ln(Q^2 + M^2)/\Lambda^2}, \tag{18.13}$$

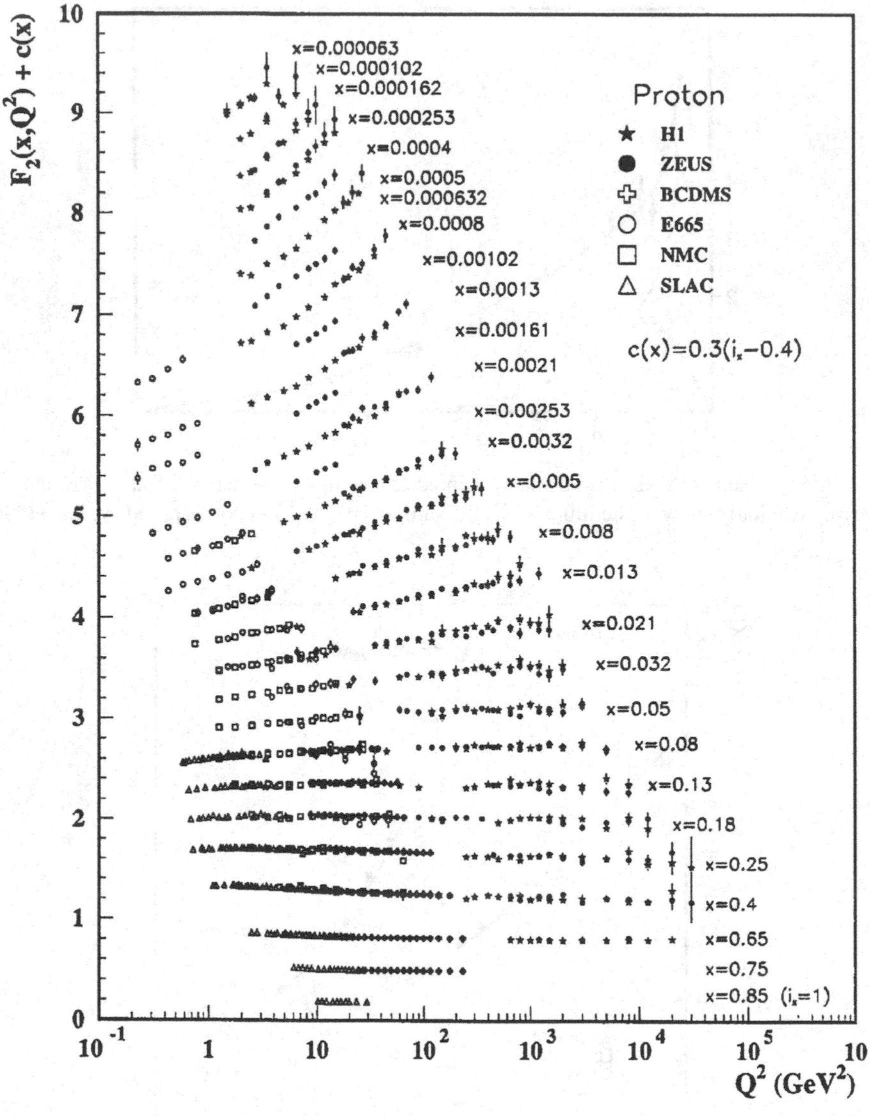


Fig. 18.4. The proton structure function $F_2$ from $ep$ scattering versus $Q^2$ at different values of $x$. A constant $c(x) = 0.3(i_x - 0.4)$ has been added to $F_2$ where $i_x$ is the number of the $x$ bin ranging from $i_x = 1 (x = 0.85)$ to $i_x = 28 (x = 0.000063)$.

and the soft pomeron term by:

$$C \to C\frac{Q^2}{Q^2 + M^2}\,, \tag{18.14}$$

where $M$ is a typical hadronic scale of the order of $M_\rho$. In this way, the structure function takes the form:

$$F_2 = \langle e_q^2 \rangle B_S \tilde{\alpha}_s^{-d^+(n=1+\lambda)} Q^{-2\lambda} s^{\lambda} + C\frac{Q^2}{Q^2 + M^2} + B_{NS} \tilde{\alpha}_s^{-d^{NS}(n=1-\lambda_{NS})} Q^{2\lambda^{NS}} s^{-\lambda^{NS}}\,, \tag{18.15}$$

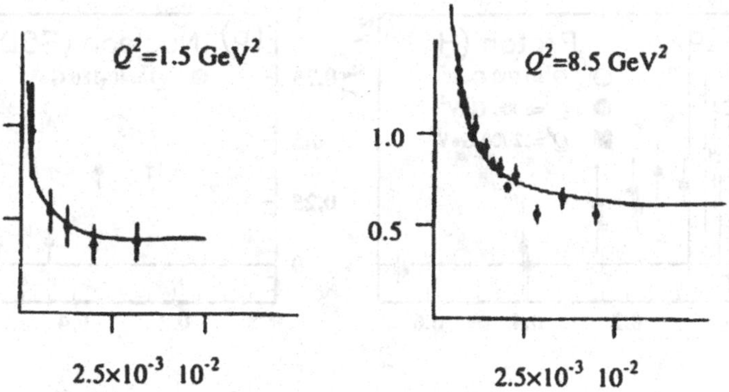


Fig. 18.5. Comparison of the measured and QCD model predictions for $F_2$ at low $x$ and small $Q^2$.

with:

$$d_+(1+\lambda_0) = 1+\lambda_0\,, \qquad d^{NS}(1-\lambda_0^{NS}) = 1-\lambda_0^{NS}\,, \tag{18.16}$$

and:

$$d(n) \equiv \gamma_n/(-2\beta_1)\,. \tag{18.17}$$

The different fits give a good description of the HERA data at low $x$ and small $Q^2$, as shown in Fig. 18.5.

The results of the fit give:

$$\lambda_0 = 0.47\,, \qquad \lambda_0^{NS} = 0.522\,. \tag{18.18}$$

which are larger than a hard pomeron fit $\lambda = 0.32 - 0.38$ but are in the range given by a soft pomeron fit $\lambda = 0.44 \pm 0.04$.

- We also know that deep inelastic scatterings and some other related sum rules have been traditionally used for extracting the QCD coupling $\alpha_s(Q^2)$ and the scale $\Lambda$ due to their sensitivity to leading order to these quantities. The determinations of $\alpha_s$ from different methods will be discussed in Section 18.4, Chapter 25 and Part VI. Various more involved systematic tests of scaling violations and modern analysis can, for example, be found in different textbooks [42–46], reviews [47–52] and also the proceedings of the QCD series of the Montpellier-Conference.

## 18.4 Neutrino scattering sum rules

For (anti)-neutrino off-proton scattering, we have the following sum rules:

- **The Adler sum rule**

$$\int_0^1 \frac{dx}{x}\left(F_2^{\bar\nu p} - F_2^{\nu p}\right) = 2\,, \tag{18.19}$$

valid for all $Q^2$, and which has no corrections because it is related to an equal-time commutator [247].

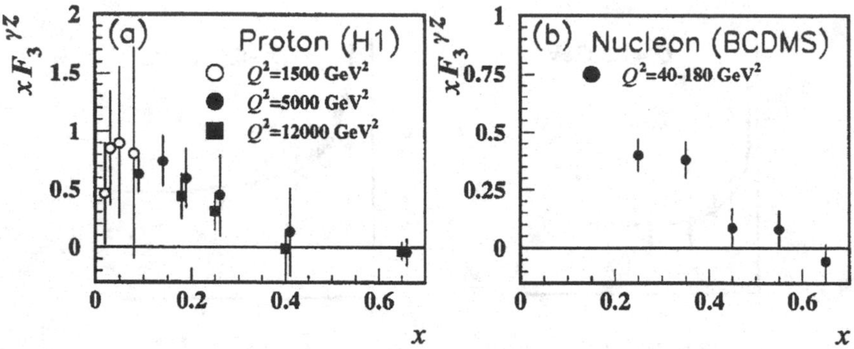


Fig. 18.6. $xF_3$ measured from electroweak scattering of (a) electrons on protons and (b) muons on carbon versus $x$ and for different $Q^2$.

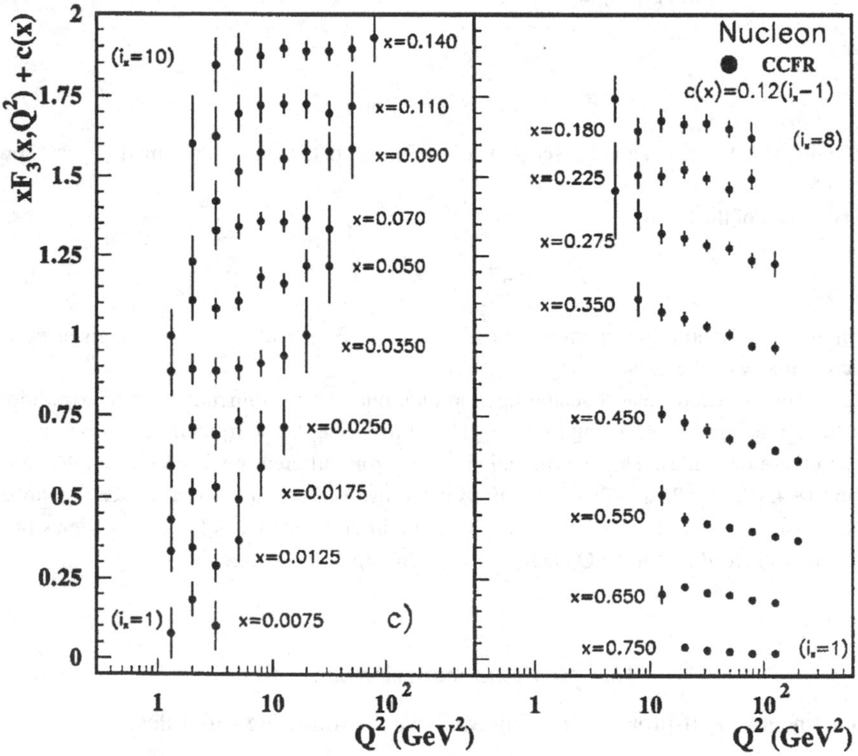


Fig. 18.7. $xF_3$ measured from $\nu - Fe$ scattering versus $Q^2$ and for different $x$.

- **The Gross–Llewellyn Smith sum rule**
  It reads [248]:

$$\int_0^1 \frac{dx}{x}\left[F_3^{\bar{\nu}p}(x, Q^2) + F_3^{\nu p}(x, Q^2)\right] = 3\left\{1 - a_s(Q^2) - 3.58\, a_s^2(Q^2) - 19.0\, a_s^3(Q^2)\right\}, \quad (18.20)$$

where higher order corrections have been evaluated by [249] and are shown in Fig. 18.6.
  Data [16] from $\nu$-$Fe$ scattering is shown in Fig. 18.7.

## 18.5 Summary of $\alpha_s$ measurements from DIS

The different analysis from DIS lead to the values of $\alpha_s$ given in Table 25.3 and Fig. 25.13 from [139]. The most recent and precise result comes from the analysis of $F_2$ by [250] using data on protons from SLAC, BCDMS, E665 and HERA. It leads to:

$$\alpha_s(M_{Z^0}) = 0.1166 \pm 0.0009\ (\text{stat}) \pm 0.0020\ (\text{syst})\ , \quad (18.21)$$

where the systematic error has been multiplied by a factor 2 as a guess of the $\mu$-dependence and effects of power corrections not fully analysed in [250]. It reaches the accuracy of the determination from, for example, the inclusive $\tau$-decay data. However, the DIS data have shown large fluctuations in recent years, and then are less satisfactory than those from $e^+e^-$ and $\tau$-decay data. The previous value $\alpha_s(M_{Z^0}) = 0.113 \pm 0.005$ from BCDMS, SLAC data [251] and soon confirmed by the CCFR result from $F_{2,3}$, become $0.119 \pm 0.002$ (stat) $\pm$ 0.003 (th) after a new energy calibration. Recent result on $F_{2\gamma}$ photon structure function is also available from LEP leading to:

$$\alpha_s(M_{Z^0}) = 0.1198 \pm 0.0028\ (\text{exp})\ {}^{+0.0034}_{-0.0046}\ (\text{th})\ . \quad (18.22)$$

These different DIS results are compared with other determinations given in Table 25.3 and Figs. 25.13 and 25.15. The overall agreement shows a great achievement of the pQCD calculations.

# 19

# Polarized deep-inelastic processes and the proton 'spin' crisis

We extend the previous unpolarized deep-inelastic scattering analysis to the case of polarized processes in the aim to study the 'spin' content of the proton and, later on (see next chapter), of the photon from $\gamma$-$\gamma$ scattering.[1] Interest on such processes has been stimulated by the EMC collaboration [253] finding that the first moment of the polarized proton structure function $g_1^p$ is unexpectedly suppressed compared with the naïve quark model prediction (OZI [254] violation), which has provoked an extensive discussion (see e.g. [255–261]) of the parton model interpretation of QCD in deep inelastic scattering processes involving the $U(1)$ axial anomaly [262–264]. We shall be concerned here with the parity-violating part of the hadronic tensor defined in Eq. (15.36) where the structure function is defined as:

$$g_1 \equiv \frac{\nu}{M_p^2} W_3 \ . \tag{19.1}$$

Data [16] are shown in Fig. 19.1.

## 19.1 The case of massless quarks

We shall discuss here the approach based on a composite operator and proper vertex. This discussion can be consulted in the reprinted paper [260] given in Section 19.4 at the end of this chapter.

## 19.2 Extension of the method to massive quarks

We extend the previous approach to the case of massive quarks [261]. In this paper, a detailed estimate of the slope of the topological susceptibility using the approach of QCD spectral sum rules in the case of massive quarks is given.[2] The result [261]:

$$\sqrt{\chi'(0)}|_{m_q \neq 0} = (33.5 \pm 3.9)\ \text{MeV}\ , \tag{19.2}$$

[1] It is a pleasure to thank Graham Shore for discussions related to this chapter.

[2] A previous estimate of the slope of the topological charge using QSSR in pure Yang–Mills theory has been done in [265] and confirmed later on by lattice calculations [266].

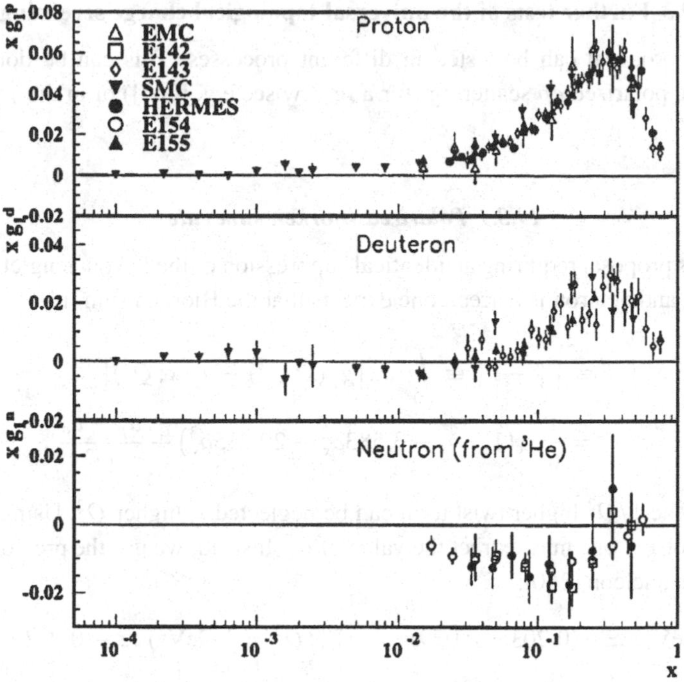


Fig. 19.1. The spin-dependent structure function $xg1$ of proton, deuteron and neutron in DIS of polarized electron/positron versus $x$ and for different $Q^2$ ranging from 0.01 to 100 GeV$^2$ (SMC) for proton and deuteron, and from 1 to 17 GeV$^2$ (E154) for neutron.

compared to the massless quark values of $(26.4 \pm 4.1)$ MeV [260] is smaller than the OZI expectation of $(43.8 \pm 5.0)$ MeV. As a result, the singlet polarized structure function:

$$a^0 = a^8 \left(\frac{\sqrt{6}}{f_\pi}\right) \sqrt{\chi'(0)|_{Q^2}} \,, \tag{19.3}$$

where:

$$G_A^{(8)} \equiv \frac{1}{2\sqrt{3}} a^8 \,, \qquad G_A^{(0)} \equiv a^0 \,, \tag{19.4}$$

has the value [261]:

$$a^0(Q^2 = 10 \text{ GeV}^2) = 0.31 \pm 0.02$$
$$\Gamma_p^1(Q^2 = 10 \text{ GeV}^2) = 0.141 \pm 0.005 \,, \tag{19.5}$$

which is about the same as the one obtained in the chiral limit, and confirms the expectation that the result is insensitve to the quark mass values. This result is in agreement with the data, which may also confirm the proposal that the proton spin suppression is a target-independent effect due to the screening of the topological charge of the QCD vacuum.

## 19.3 Further tests of the universal topological charge screening

The previous proposal can be tested in different processes. This can be done either in semi-inclusive polarized $ep$ scattering (for a review see e.g. [267]) or in the $\gamma\gamma$ polarized process.

### *19.3.1 Polarized Bjorken sum rule*

If the previous proposal requiring an identical suppression of the flavour singlet component for the proton and neutron is correct, one expects that the Bjorken sum rule:

$$\begin{aligned}\delta\Gamma_1^{p-n} &\equiv \Gamma_1^p - \Gamma_1^n \equiv \int_0^1 dx\big[g_1^p(x;Q^2) - g_1^n(x;Q^2)\big]\\ &= \frac{1}{6}g_A\big(1 - a_s - 3.583a_s^2 - 20.215a_s^3\big) + \frac{a_p - a_n}{Q^2}\,, \end{aligned} \tag{19.6}$$

should hold. The $1/Q^2$ higher twist term can be neglected at higher $Q^2$. Using the experimental value of $g_A$, one may extract the value of $\alpha_s$. Instead, we use the previous sum rule and that of the nucleon [260]:

$$\delta\Gamma_1^{p-n}(2\ \mathrm{GeV}^2) \simeq -(0.203 \pm 0.029)\,, \qquad \Gamma_1^n(Q^2 = 2\ \mathrm{GeV}^2) \simeq -(0.022 \pm 0.011)\,, \tag{19.7}$$

from which one can deduce the higher twist terms in units of $\mathrm{GeV}^2$:

$$a_p \simeq -0.117 \pm 0.145\,, \qquad a_n \simeq -0.018 \pm 0.025\,, \tag{19.8}$$

which, although consistent with zero are neverthless interesting.

### *19.3.2 Semi-inclusive polarized ep scattering*

An alternative test of the previous proposal is to perform a DIS experiment on a target other than the nucleon. This can be done by studying a semi-inclusive process in which a single hadron carrying a large target energy fraction is detected in the target fragmentation region. This is shown in the Fig. 19.2.

In terms of the *fracture function* [268] $M_i^{h/N}(x, z, t, Q^2)$ which represents the joint probability distribution for producing a parton $i$ with momentum fraction $x$ and a detected hadron $h$ carrying an energy fraction $z = p_2' \cdot q/p_2 \cdot q$ from a nucleon N ($t$ is the invariant momentum transfer), the lowest order polarized cross-section reads [267]:

$$\frac{d\Delta\sigma^{\mathrm{target}}}{dx\,dQ^2dz\,dt} = \frac{4\pi\alpha^2 y(2-y)}{Q^4}\Delta M_1^{h/N}(x, z, t, Q^2) \tag{19.9}$$

where:

$$\Delta M_1^{h/N} = \sum_i \frac{Q_i^2}{2}\Delta M_i^{h/N}\,, \tag{19.10}$$

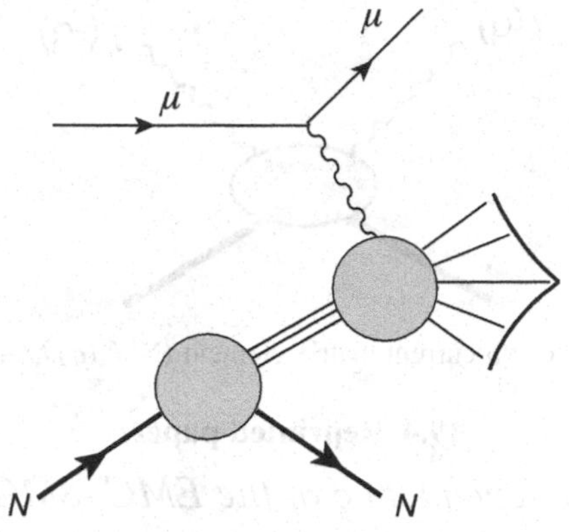


Fig. 19.2. Semi-inclusive process.

is equivalent to the inclusive structure function $g_1^N$, and where $Q_i$ is the charge of the quark $i$ in units of $e$. For large $z \to 1$, the fracture function can be modelled by, for example, a single region exchange and reads:

$$\Delta M_1^{h/N} \simeq F(t)(1-z)^{-2\alpha_{\mathcal{R}}(t)} g_1^{\mathcal{R}} \left( \frac{x}{1-z}, t, Q^2 \right) , \tag{19.11}$$

where $g_1^{\mathcal{R}}$ is the structure function of the exchanged region $\mathcal{R}$ with trajectory $\alpha_{\mathcal{R}}(t)$. Independently on the detailed model of the fracture functions, one can predict the ratio of the moments:

$$\frac{\mathcal{M}_1(ep \to e\pi^- X)}{\mathcal{M}_1(en \to e\pi^+ X)} \simeq \frac{2s+2}{2s-1} \simeq \frac{\mathcal{M}_1(ep \to eD^- X)}{\mathcal{M}_1(en \to eD^0 X)} ,$$
$$\frac{\mathcal{M}_1(ep \to eK^0 X)}{\mathcal{M}_1(en \to eK^+ X)} \simeq \frac{2s+1}{2s-1} , \tag{19.12}$$

where:

$$\mathcal{M}_1 = \int_0^{1-z} dx \Delta M_1^{h/N} \cdot (x, z, t, Q^2) , \tag{19.13}$$

and:

$$s(Q^2) = \left( \frac{C_1^S}{C_1^{NS}} \right) \left( \frac{\sqrt{2n_f}}{f_\pi} \right) \sqrt{\chi'(0)} . \tag{19.14}$$

$C_1^S$ and $C_1^{NS}$ are ratio fo the singlet and non-singlet Wilson coefficients. In the OZI limit, $s = 1$, such that one expects a large deviation from the previous value of $\chi'(0)$. In the small limit $z \to 0$, the previous ratios reduce to the first moment ratio $g_1^p/g_1^n$. In the whole range of $z$, one expects a deviation of about a factor 2.5 from the OZI prediction.

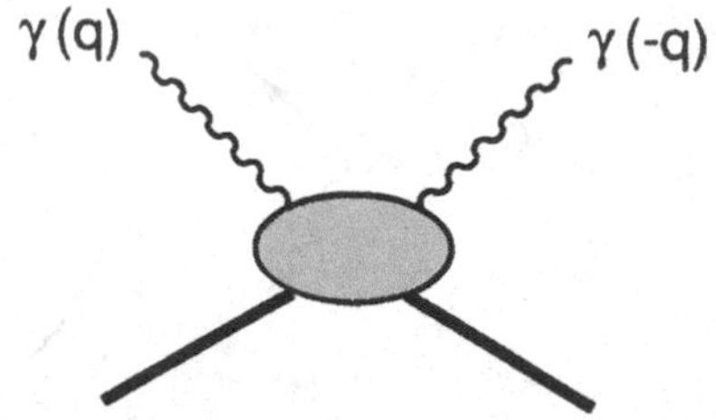


Fig. 1. The two-current matrix element $\langle N| J_\mu(q) J_\nu(-q) |N\rangle$.

### 19.4 Reprinted paper

## *Target independence of the EMC–SMC effect*

S. Narison, G. M. Snore and G. Veneziano



### 1. Introduction

The discovery by the EMC Collaboration [1] (see also Ref. [2]) of an unexpected suppression of the first moment of the polarised proton structure function $g_1^p$ has provoked an extensive discussion of the parton model interpretation of QCD in deep inelastic scattering processes involving the axial U(1) anomaly. (For reviews, see Refs. [3,4].) While it has so far proved possible with careful redefinitions and interpretations [5] to preserve the essence of the parton model description, it is becoming clear that these processes involve subtle field theoretic properties of QCD which lead beyond both the original and QCD-improved parton approximation. In this paper, we develop an alternative approach to deep inelastic scattering emphasising field theoretic concepts such as the operator product expansion (OPE), composite operator Green functions and proper vertices. This clarifies some of the difficulties encountered in the parton description and gives a new insight into the underlying reason for the EMC result. In particular, our analysis strongly suggests that the observed suppression of the first moment of $g_1^p$ is a generic QCD effect related to the anomaly and is actually *independent* of the target. Rather than revealing a special property of the proton structure, the EMC result reflects an anomalously small value of the first moment of the QCD topological susceptibility [6,7].

The essential features of this method are easily described for a general deep inelastic scattering process. The hadronic part of the scattering amplitude is given by the imaginary part of the two-current matrix element $\langle N \mid J_\mu(q) J_\nu(-q) \mid N\rangle$ illustrated in Fig. 1, where $J_\mu$ is the current coupling to the exchanged hard proton (or electroweak vector boson) and $\mid N\rangle$ denotes the target. The OPE is used to expand the large $Q^2$ limit of the product of currents as a sum of Wilson coefficients $C_i(Q^2)$ times renormalised composite operators $\mathcal{O}_i$ as follows (suppressing Lorentz indices):

$$J(q)J(-q) \underset{Q^2\to\infty}{\sim} \sum_i C_i(Q^2)\mathcal{O}_i(0). \tag{1.1}$$

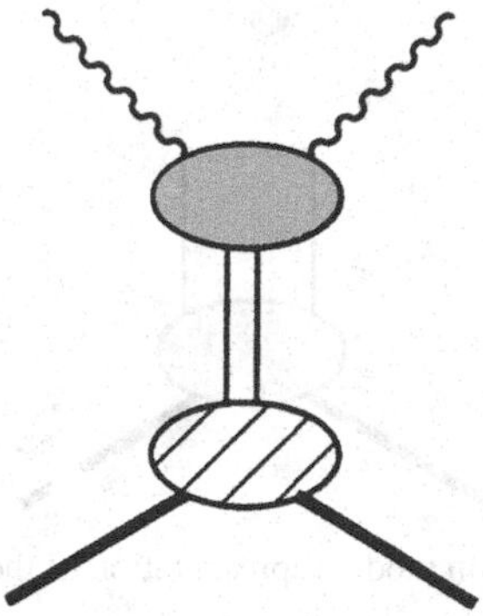

Fig. 2. Decomposition of the matrix element into a composite operator propagator (denoted by the double line) and a proper vertex (hatched).

The dominant contributions to the amplitude arise from the operators $\mathcal{O}_i$ of lowest twist. Within this set of lowest twist operators, those of spin $n$ contribute to the $n$th moment of the structure functions, i.e.

$$\int_0^1 \mathrm{d}x\, x^{n-1} F(x, Q^2) = \sum_i C_i^n(Q^2) \langle N \mid \mathcal{O}_i^n(0) \mid N \rangle. \tag{1.2}$$

The Wilson coefficients are calculable in QCD perturbation theory, so the problem reduces to evaluating the target matrix elements of the corresponding operators. We now introduce appropriately defined proper vertices $\Gamma_{\tilde{\mathcal{O}}NN}$, which are chosen to be 1PI with respect to a physically motivated basis set $\tilde{\mathcal{O}}_k$ of renormalised composite operators. The matrix elements are then decomposed into products of these vertices with zero-momentum composite operator propagators as follows:

$$\langle N \mid \mathcal{O}_i(0) \mid N \rangle = \sum_k \langle 0 \mid \mathcal{O}_i(0) \tilde{\mathcal{O}}_k(0) \mid (0) \rangle \Gamma_{\tilde{\mathcal{O}}_k NN}. \tag{1.3}$$

This is illustrated in Fig. 2. In essence, what we have done is to split the whole amplitude into the product of a "hot" (high momentum) part described by QCD perturbation theory, a "cold" part described by a (non-perturbative) composite operator propagator and finally a target-dependent proper vertex.

All the target dependence is contained in the vertex function $\Gamma_{\tilde{\mathcal{O}}NN}$. However, these are not unique – they depend on the choice of the basis $\tilde{\mathcal{O}}_k$ of composite operators. This choice is made on physical grounds based on the relevant degrees of freedom, the aim being to parametrise the amplitude in terms of a minimal, but sufficient, set of vertex functions. A good choice can often lead to an almost direct correspondence between the proper vertices and physical couplings such as, e.g., the pion–nucleon coupling $g_{\pi \mathrm{NN}}$. In particular, it will be wise to use, whenever possible, RG-invariant proper vertices.

Despite being non-perturbative, we can frequently evaluate the composite operator Green functions using a combination of exact Ward identities and dynamical approximations (see Sections 2 and 3). On the other hand, because of the target dependence, the proper vertices are not readily calculable from first principles in QCD, so we are in general left

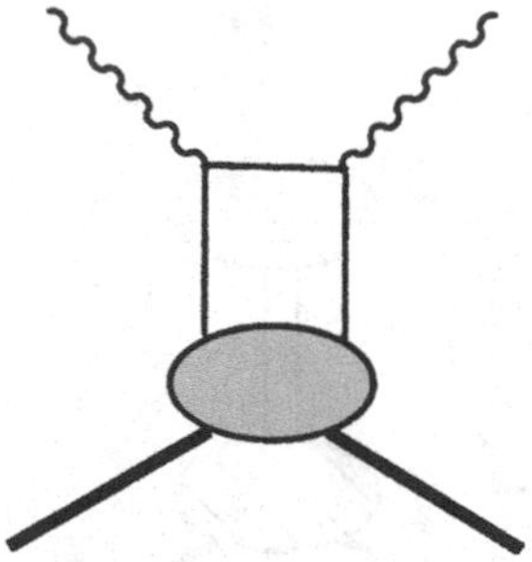

Fig. 3. The original parton model representation of the scattering amplitude.

with a parametrisation of the amplitude in terms of a (hopefully small) set of unknown vertices. These play the rôle of the non-perturbative (i.e. primordial or not-yet-evolved) parton distributions in the usual treatment. Just as for parton distributions, many different QCD processes can be related through parametrisation with the same set of vertex functions.

Now compare this approach with the parton model. In the original parton model, the amplitude is approximated by Fig. 3, describing the scattering of a large $Q^2$ photon with a parton in the target nucleon. This picture is already sufficient to reveal Bjorken scaling. It may be improved in the context of QCD by including gluonic corrections, exactly as in the OPE, as shown in Fig. 4. These give the logarithmic scaling violations characteristic of perturbative QCD. The total amplitude is therefore factorised into a perturbative scattering amplitude for the hard photon with a parton (quark or gluon) and a parton distribution function giving the probability of finding a particular parton with given fraction $x$ of the target momentum.

The question of whether the full QCD amplitude can be given a natural parton interpretation depends on the composite operators $\mathcal{O}_i$ in the Wilson expansion. For example, if the lowest twist operator for a given process is multilinear in the elementary quark and gluon fields rather than simply quadratic then the diagram of Fig. 4 is not appropriate and the process can only be described in terms of multi-parton distributions [8]. A more subtle problem arises when the operators $\mathcal{O}_i$ are non-trivially renormalised and mix with other composite operators under renormalisation. In this case, the parton interpretation is

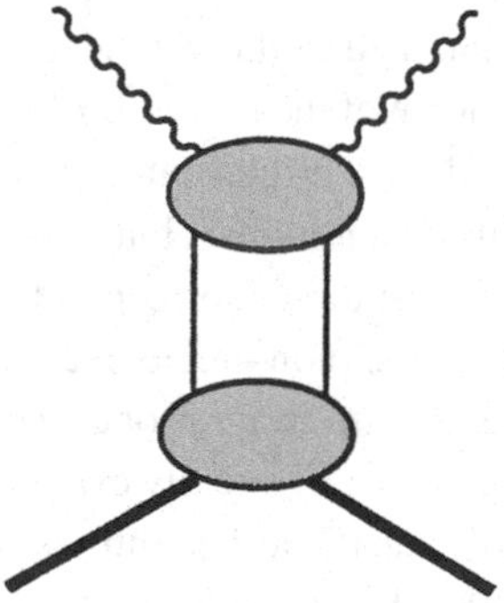

Fig. 4. The QCD-improved parton model representation.

preserved by *defining* parton distributions directly in terms of the operator matrix elements (see, e.g., Ref. [8]). This procedure becomes especially delicate [5] in the case of polarised deep inelastic scattering because of the special renormalisation properties of the relevant Wilson operator $J^0_{\mu 5\mathrm{R}}$ due to the axial U(1) anomaly.

In this paper, rather than attempt to interpret the amplitude for polarised deep inelastic scattering in terms of specially defined polarised quark and gluon distributions, we instead focus the analysis on the composite operator level. By splitting the matrix elements in the form of Eq. (1.3), we can exploit chiral Ward identities and the renormalisation group to separate out generic features of QCD manifested in the composite operator propagator from specific properties of the target. In the next section, we see how this clarifies the origin of the suppression of the first moment of $g_1^{\mathrm{p}}$ observed in polarised $\mu$p scattering.

## 2. The first moment sum rule for $g_1^{\mathrm{p}}$

Our starting point is the familiar Ellis–Jaffe [9] sum rule for the first moment of the polarised proton structure function $g_1^{\mathrm{p}}$. For $N_{\mathrm{F}} = 3$ and in the $\overline{\mathrm{MS}}$ scheme [10], this reads[1]

$$\begin{aligned}\Gamma_1^{\mathrm{p}}(Q^2) &\equiv \int_0^1 \mathrm{d}x\, g_1^{\mathrm{p}}(x;Q^2)\\ &= \frac{1}{6}\Bigg\{\left(G_{\mathrm{A}}^{(3)}(0) + \frac{1}{\sqrt{3}}G_{\mathrm{A}}^{(8)}(0)\right)\left[1 - \frac{\alpha_{\mathrm{s}}}{\pi} - 3.583\left(\frac{\alpha_{\mathrm{s}}}{\pi}\right)^2 - 20.215\left(\frac{\alpha_{\mathrm{s}}}{\pi}\right)^3\right]\\ &\quad + \frac{2}{3}G_{\mathrm{A}}^{(0)}(0;Q^2)\left[1 - \frac{1}{3}\frac{\alpha_s}{\pi} - 0.550\left(\frac{\alpha_s}{\pi}\right)^2\right]\Bigg\},\end{aligned} \tag{2.1}$$

where the $G_{\mathrm{A}}^{(a)}$ are form factors in the proton matrix elements of the axial current

$$\langle P \mid J^a_{\mu 5\mathrm{R}}(k) \mid P\rangle = G_{\mathrm{A}}^{(a)}(k^2)\bar{u}\gamma_\mu\gamma_5 u + G_{\mathrm{P}}^{(a)}(k^2)k_\mu\bar{u}\gamma_5 u, \tag{2.2}$$

and $a$ is an SU(3) flavour index. In our normalisations (see Ref. [7])

$$\begin{aligned}G_{\mathrm{A}}^{(3)} &= \tfrac{1}{2}(\Delta u - \Delta d),\\ G_{\mathrm{A}}^{(8)} &= \frac{1}{2\sqrt{3}}(\Delta u + \Delta d - 2\Delta s),\\ G_{\mathrm{A}}^{(0)} &= \Delta u + \Delta d + \Delta s \equiv \Delta\Sigma.\end{aligned} \tag{2.3}$$

We ignore heavy quarks and, for simplicity, set the light quark masses to zero in the formulae below.

The axial current occurs here since it is the lowest twist, lowest spin, odd-parity operator in the OPE of two electromagnetic currents, i.e.

$$J_\mu(q)J_\nu(-q) \underset{Q^2\to\infty}{\sim} 2\sum_{a=0,3,8}\epsilon_{\mu\nu\alpha}{}^{\beta}\frac{q^\alpha}{Q^2}C^a(Q^2)J^a_{\beta 5\mathrm{R}} + \ldots. \tag{2.4}$$

[1] We use the NLO and NNLO coefficients given in Ref. [11]. However, due to our definition (2.5) of the renormalised composite operators, the radiative corrections of the singlet are different from the corresponding terms in Ref. [11], which uses a different renormalisation of the singlet operators.

The suffix R emphasises that the current is the *renormalised* composite operator. Under renormalisation, the gluon topological density $Q_{\mathrm{R}}$ and the divergence of the flavour singlet axial current $J^0_{\mu 5\mathrm{R}}$ mix as follows [12]:

$$J^0_{\mu 5\mathrm{R}} = Z J^0_{\mu 5\mathrm{B}},$$
$$Q_{\mathrm{R}} = Q_{\mathrm{B}} - \frac{1}{2N_{\mathrm{F}}}(1-Z)\partial^\mu J^0_{\mu 5\mathrm{B}}, \tag{2.5}$$

where $J^0_{\mu 5\mathrm{B}} = \sum \bar{q}\gamma_\mu\gamma_5 q$ and $Q_{\mathrm{B}} = (\alpha_{\mathrm{s}}/8\pi)\mathrm{tr}(G^{\mu\nu}\tilde{G}_{\mu\nu})$ and we have quoted the formulae for $N_{\mathrm{F}}$ flavours. The mixing is such that the combination occurring in the axial anomaly Ward identities, e.g.

$$\langle 0 \mid \left(\partial^\mu J^0_{\mu 5\mathrm{R}} - 2N_{\mathrm{F}} Q_{\mathrm{R}}\right)\tilde{\mathcal{O}}_k \mid 0\rangle + \langle 0 \mid \delta_5 \tilde{\mathcal{O}}_k \mid 0\rangle = 0, \tag{2.6}$$

is not renormalised.

Since $J^0_{\mu 5\mathrm{R}}$ is renormalised, its matrix elements satisfy renormalisation group equations with an anomalous dimension $\gamma$, so that in particular $G^{(0)}_{\mathrm{A}}(0; Q^2)$ depends on the RG scale (which is set to $Q^2$ in Eq. (2.1)).

As we have emphasised elsewhere, $G^{(0)}_{\mathrm{A}}$ does *not*, as was initially supposed, measure the spin of the quark constituents of the proton. The RG non-invariance of $J^0_{\mu 5\mathrm{R}}$ (a consequence of the anomaly) is itself sufficient to prevent this identification. The interest in the first EMC data [1,2] on polarised $\mu p$ scattering[2], which allows the following result for $G^{(0)}_{\mathrm{A}}$ to be deduced:

$$G^{(0)}_{\mathrm{A}}(0; Q^2 = 11\,\mathrm{GeV}^2) \equiv \Delta\Sigma = 0.19 \pm 0.17, \tag{2.7}$$

is rather that this value for $G^{(0)}_{\mathrm{A}}$ represents a substantial violation of the OZI rule [13,14], according to which we would expect

$$G^{(0)}_{\mathrm{A}}(0)_{\mathrm{OZI}} = 3F - D \simeq 0.579 \pm 0.021. \tag{2.8}$$

Here, we have used [15,16]

$$F + D \simeq 1.257 \pm 0.008, \qquad F/D \simeq 0.575 \pm 0.016 \tag{2.9}$$

as fitted from hyperon and $\beta$-decays. The assumption that the OZI rule is satisfied for $G^{(0)}_{\mathrm{A}}(0)$ is equivalent to the Ellis–Jaffe sum rule prediction for the first moment of $g_1^{\mathrm{p}}$.

It follows immediately from Eq. (2.2) (assuming the absence of a massless pseudoscalar boson in the U(1) channel) that

$$G^{(0)}_{\mathrm{A}}\left(0; Q^2\right)\bar{u}\gamma_5 u = \frac{1}{2M}\langle P \mid \partial^\mu J^0_{\mu 5\mathrm{R}} \mid P\rangle, \tag{2.10}$$

[2]The combined SLAC/EMC data quoted in Ref. [1] gives

$$\Gamma_1^{\mathrm{p}}(Q^2 = 11\,\mathrm{GeV}^2) = 0.126 \pm 0.010 \pm 0.015.$$

The result for $G^{(0)}_{\mathrm{A}}$ in Eq. (2.7) is extracted from the sum rule using the values for $F$ and $D$ given below and the running coupling from tau-decay data [30] (see the remarks after Eq. (3.32)).

where $M$ is the proton mass. The anomalous chiral Ward identity then allows $G_A^{(0)}$ to be re-expressed as the forward matrix element of the renormalised gluon topological density $Q_R$, i.e.

$$G_A^{(0)}(0;Q^2)\bar{u}\gamma_5 u = \frac{1}{2M} 2N_F \langle P \mid Q_R(0) \mid P \rangle. \tag{2.11}$$

Notice that in terms of bare fields, $Q_R$ contains both gluon and quark bilinears. This, together with the explicit factor of $\alpha_s$ in the definition of the topological density, is the source of the difficully in giving a natural and unambiguous parton interpretation [5,3,4].

At this point, we apply the method described in the introduction. We choose as the composite operator basis $\tilde{O}_k$ the set of renormalised flavour singlet pseudoscalar operators, viz. $Q_R$ and $\Phi_{5R}$, where, up to a crucial normalisation factor discussed below, the corresponding bare operator is simply the singlet $i \sum \bar{q}\gamma_5 q$. We then define $\Gamma[Q_R, \Phi_{5R}; P, \overline{P}]$ to be the generating functional of proper vertces which are 1PI with respect to these composite fields only. (Here, $P$ and $\overline{P}$ denote interpolating fields for the proton – they play a purely passive rôle in the construction.) $\Gamma$ is obtained from the QCD generating functional by a Legendre transform with respect to the sources for the composite operators $Q_R$ and $\Phi_{5R}$ only. We may then write (cf. Eq. (1.3))

$$\langle P \mid Q_R(0) \mid P \rangle = \langle 0 \mid Q_R(0) Q_R(0) \mid 0 \rangle \Gamma_{Q_R P\overline{P}} + \langle 0 \mid Q_R(0) \Phi_{5R}(0) \mid 0 \rangle \Gamma_{\Phi_{5R} P\overline{P}}, \tag{2.12}$$

where the propagators are at zero momentum.

The composite operator propagator in the first term in Eq. (2.12) is the zero-momentum limit of an important quantity in QCD known as the topological susceptibility $\chi(k^2)$, viz.

$$\chi(k^2) = \int \mathrm{d}x\, e^{ik\cdot x} i \langle 0 \mid T^* Q_R(x) Q_R(0) \mid 0 \rangle. \tag{2.13}$$

The second term is clearly independent of the normalisation of the renormalised quark bilinear operator $\Phi_{5R}$. We choose to normalise this operator in such a way that the inverse two-point function $\Gamma_{\Phi_{5R}\Phi_{5R}}$, which has to vanish at $k^2 = 0$, is equal to $k^2$, the correct normalisation for a free, massless particle. With this normalisation, a straightforward but intricate argument [7] using chiral Ward identities (see Appendix A) shows that the propagator $\langle 0 \mid Q_R \Phi_{5R} \mid 0 \rangle$ at zero momentum is simply the square root of the first moment of the topological susceptibility $\chi(k^2)$. We therefore find

$$\langle P \mid Q_R(0) \mid P \rangle = \chi(0) \Gamma_{Q_R P\overline{P}} + \sqrt{\chi'(0)} \Gamma_{\Phi_{5R} P\overline{P}}. \tag{2.14}$$

The chiral Ward identities further show that for QCD with massless quarks, $\chi(0)$ actually vanishes. (This is in contrast to pure Yang–Mills theory, where $\chi(0)$ is non-zero and is related to the $\eta'$-mass in the large $N_C$ resolution of the U(1) problem [17,18].) Only the second term in Eq. (2.14) remains. Remarkably, this means that the matrix element of the renormalised *gluon* density $Q_R$ measures the coupling of the proton to the renormalised pseudoscalar *quark* operator $\Phi_{5R}$. This happens because the composite operator propagator matrix in the pseudoscalar ($Q_R$, $\Phi_{5R}$) sector is off-diagonal. We therefore arrive at our basic result [7],

$$G_A^{(0)}(0; Q^2)\bar{u}\gamma_5 u = \frac{1}{2M} 2N_F \sqrt{\chi'(0)} \Gamma_{\Phi_{5R} P\bar{P}}. \tag{2.15}$$

The renormalisation group properties of Eq. (2.15) are central to our argument. With the normalisation of $\Phi_{5R}$ chosen above, it can be shown [7] that the proper vertex $\Gamma_{\Phi_{5R} P\bar{P}}$ is RG invariant and so has no scale dependence. The scale dependence needed to match $G_A^{(0)}$ is provided entirely by the topological susceptibility which, as shown in Appendix A, satisfies the RGE

$$\left( \mu \frac{\partial}{\partial \mu} + \beta(\alpha_s)\alpha_s \frac{\partial}{\partial \alpha_s} - 2\gamma \right) \chi'(0) = 0. \tag{2.16}$$

The challenge posed by the EMC data is to understand the origin of the OZI violation in $G_A^{(0)}$. The OZI approximation applied to the RHS of Eq. (2.15) would require[3] (neglecting flavour SU(3) breaking) $\Gamma_{\Phi_{5R} P\bar{P}} \simeq \sqrt{2}\, g_{\eta_8 NN}\bar{u}\gamma_5 u$ while $\sqrt{\chi'(0)} \simeq (1/\sqrt{6}) f_\pi$.

Our proposal is that we should expect the source of the OZI violation to lie in RG non-invariant terms, i.e. in $\chi'(0)$. The reasoning is straightforward. In the absence of the U(1) anomaly, the OZI rule would be an exact property of QCD. So the OZI violation is a consequence of the anomaly. But it is the existence of the anomaly that is responsible for the non-conservation and hence non-trivial renormalisation of the axial current $J^0_{\mu 5R}$. We therefore expect to find OZI violations in quantities sensitive to the anomaly, which we identify through their RG dependence on the anomalous dimension $\gamma$. This seems reasonable since, if the OZI rule were to be good for such quantities, it would mean approximating a RG non-invariant, scale-dependent quantity by a scale-independent one. If this proposal is correct, we expect $\sqrt{\chi'(0)}$ to be significantly suppressed relative to its OZI approximation of $(1/\sqrt{6}) f_\pi$. The proper vertex $\Gamma_{\Phi_{5R} P\bar{P}}$ would behave exactly as expected according to the OZI rule. That is, the Ellis–Jaffe violating suppression of the first moment of $g_1^p$ observed by EMC would *not* be a property of the proton at all, but would simply be due to an anomalously small value of the first moment of the QCD topological susceptibility $\chi'(0)$.

In the next section, we attempt to verify this hypothesis by evaluating $\chi'(0)$ using QCD spectral sum rules.

---

[3]To understand this, we note from Ref. [7] that Eq. (2.15) is equivalently written as one form of the U(1) Goldbergen–Treiman relation, viz.

$$G_A^{(0)}(0; Q^2) = F_{\eta OZI} g_{\eta_{OZI} NN},$$

where $F_{\eta OZI}$ and $g_{\eta_{OZI} NN}$ are respectively the decay constant and nucleon coupling of a state $|\,\eta_{OZI}\rangle$. $|\,\eta_{OZI}\rangle$ is an *unphysical* state in QCD (i.e. not a mass eigenstate) which in the OZI or large $N_C$ limit, in which the anomaly is absent, can be identified as the massless U(1) Goldstone boson. Simple quark counting rules then relate $g_{\eta_{OZI} NN}$ to the $\eta_8$–nucleon coupling $g_{\eta_8 NN}$. This identification is the origin of our choice of normalisation of $\Phi_{5R}$. In the OZI limit, $\Gamma_{\Phi_{5R} NN}$ becomes the Goldstone boson–nucleon coupling.

## 3. QCD spectral sum rule estimate of $\chi'(0)$

We now present an estimate of $\chi'(0)$ in QCD with massless quarks using the method of QCD spectral sum rules (QSSR) pioneered by Shifman, Vainshtein and Zakharov [19] and reviewed recently in Ref. [20].

The correlation function $\chi(k^2)$ is defined in Eq. (2.13) and its renormalisation group equation is given in Appendix A. Including the inhomogeneous contact term [21], we have

$$\left(\mu\frac{\partial}{\partial\mu}+\beta(\alpha_s)\alpha_s\frac{\partial}{\partial\alpha_s}-2\gamma\right)\chi(k^2)=-\frac{1}{(2N_F)^2}2\beta^{(L)}k^4, \tag{3.1}$$

with the beta function

$$\beta(\alpha_s)\equiv\frac{1}{\alpha_s}\mu\frac{d}{d\mu}\alpha_s=\beta_1\frac{\alpha_s}{\pi}+\beta_2\left(\frac{\alpha_s}{\pi}\right)^2, \tag{3.2}$$

where, for QCD with $N_F$ flavours, $\beta_1=-\frac{1}{2}(11-\frac{2}{3}N_F)$ and $\beta_2=-\frac{1}{4}(51-\frac{19}{3}N_F)$, and the anomalous dimension [12]

$$\gamma\equiv\mu\frac{d}{d\mu}\log Z=-\left(\frac{\alpha_s}{\pi}\right)^2. \tag{3.3}$$

The extra RG function $\beta^{(L)}$ (so called because it appears in the longitudinal part of the Green function of two axial currents) is given by

$$\frac{1}{(2N_F)^2}\beta^{(L)}=-\frac{1}{32\pi^2}\left(\frac{\alpha_s}{\pi}\right)^2\left(1+\frac{29}{4}\frac{\alpha_s}{\pi}\right). \tag{3.4}$$

The RGE is solved in the standard way, giving

$$\begin{aligned}\chi(k^2,\alpha_s;\mu)=\exp\left(-2\int_0^t dt'\,\gamma(\alpha_s(t'))\right)\Bigg[&\chi\left(k^2,\alpha_s(t);\mu\,e^t\right)\\ &-2\int_0^t dt''\beta^{(L)}(\alpha_s(t''))\exp\left(2\int_0^{t''}dt'\,\gamma(\alpha_s(t'))\right)\Bigg],\end{aligned} \tag{3.5}$$

where $\alpha_s(t)$ is the running coupling.

The perturbative expression for the two-point correlation function in the $\overline{\text{MS}}$ scheme is [22]

$$\chi(k^2)_{\text{P.T.}}\simeq-\left(\frac{\alpha_s}{8\pi}\right)^2\frac{2}{\pi^2}k^4\log\frac{-k^2}{\mu^2}\left[1+\frac{\alpha_s}{\pi}\left(\frac{1}{2}\beta_1\log\frac{-k^2}{\mu^2}+\frac{29}{4}\right)+\ldots\right]. \tag{3.6}$$

The non-perturbative contribution from the gluon condensates (coming from the next lowest dimension operators in the OPE) is [23]

$$\chi(k^2)_{\rm N.P.} \simeq -\frac{\alpha_s}{16\pi^2}\left[\left(1+\tfrac{1}{2}\beta_1\frac{\alpha_s}{\pi}\log\frac{-k^2}{\mu^2}\right)\langle\alpha_s G^2\rangle - 2\frac{\alpha_s}{k^2}\langle gG^3\rangle\right]. \tag{3.7}$$

The RGE has been used to check the consistency of the leading log approximation in the perturbative expression and to fix the radiative correction in the gluon condensate contribution.

For the QSSR analysis of $\chi'(0)$, we use the subtracted dispersion relations

$$\frac{1}{k^2}\left[\chi(k^2)-\chi(0)\right] = \int_0^\infty \frac{dt}{t}\frac{1}{t-k^2-i\epsilon}\frac{1}{\pi}\mathrm{Im}\,\chi(t) \tag{3.8}$$

and

$$\frac{1}{k^4}\left[\chi(k^2)-\chi(0)-k^2\chi'(0)\right] = \int_0^\infty \frac{dt}{t^2}\frac{1}{t-k^2-i\epsilon}\frac{1}{\pi}\mathrm{Im}\,\chi(t). \tag{3.9}$$

Then, taking the inverse Laplace transform [20] of both sides of the dispersion relations and using the fact that $\chi(0)=0$ in massless QCD, we find[4]

$$\begin{aligned}\int_0^{t_c}&\frac{dt}{t}e^{-t\tau}\frac{1}{\pi}\mathrm{Im}\,\chi(t)\\ &\simeq\left(\frac{\overline{\alpha}_s}{8\pi}\right)^2\frac{2}{\pi^2}\tau^{-2}[1-\exp(-t_c\tau)(1+t_c\tau)]\\ &\quad\times\left[1+\frac{\overline{\alpha}_s}{4\pi}\left(29+4\beta_1(1-\gamma_E)-8\frac{\beta_2}{\beta_1}\log(-\log\tau\Lambda^2)\right)\right]\\ &\quad+\frac{\overline{\alpha}_s}{8\pi}\left(\frac{1}{2\pi}\langle\alpha_s G^2\rangle+\frac{\overline{\alpha}_s}{\pi}\tau\langle gG^3\rangle\right)\end{aligned} \tag{3.10}$$

and

$$\begin{aligned}\chi'(0)\simeq&\int_0^{t_c}\frac{dt}{t^2}e^{-t\tau}\frac{1}{\pi}\mathrm{Im}\,\chi(t)-\left(\frac{\overline{\alpha}_s}{8\pi}\right)^2\frac{2}{\pi^2}\tau^{-1}[1-\exp(-t_c\tau)]\\ &\times\left[1+\frac{\overline{\alpha}_s}{4\pi}\left(29-4\beta_1\gamma_E-8\frac{\beta_2}{\beta_1}\log(-\log\tau\Lambda^2)\right)\right]\\ &+\frac{\overline{\alpha}_s}{8\pi}\left(\frac{1}{2\pi}\tau\langle\alpha_s G^2\rangle+\frac{\overline{\alpha}_s}{2\pi}\tau^2\langle gG^3\rangle\right).\end{aligned} \tag{3.11}$$

where $\overline{\alpha}_s$ is the running coupling expressed in terms of the QCD scale $\Lambda$ from the two-loop relation:

$$\frac{\overline{\alpha}_s^{(2)}}{\pi} = \frac{\overline{\alpha}_s}{\pi}\left(1-\frac{\overline{\alpha}_s}{\pi}\frac{\beta_2}{\beta_1}\log(-\log\ \tau\Lambda^2)\right), \tag{3.12}$$

with

$$\frac{\overline{\alpha}_s}{\pi} = \frac{2}{\beta_1\ \log\ \tau\Lambda^2}. \tag{3.13}$$

[4]For the corresponding results in pure Yang–Mills theory, see Refs. [24,25].

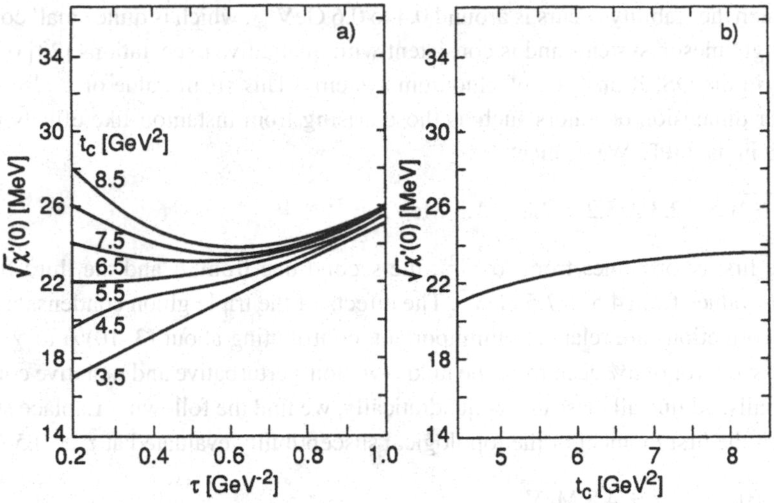


Fig. 5. (a) $\tau$-behaviour of $\sqrt{\chi'(0)}$ for different values of the continuum threshold $t_c$. (b) Behaviour of different $\tau$-minima versus $t_c$.

In these expressions, we have cut off the $t$-integration at some scale $t_c$ and used the perturbation theory approximation to Im $\chi(t)$ for $t > t_c$.

In order to extract a value for $\chi'(0)$ from these sum rules, we keep only the lowest resonance (the $\eta'$) contribution to the spectral function, i.e. we assume

$$\frac{1}{\pi}\text{Im } \chi(t) = 2\tilde{m}_{\eta'}^4 f_{\eta'}^2 \delta\left(t - \tilde{m}_{\eta'}^2\right) + \text{“QCD continuum”}\ \theta(t - t_c), \tag{3.14}$$

where $\tilde{m}_{\eta'}$ is the mass of the $\eta'$ extrapolated for massless QCD, viz.

$$\tilde{m}_{\eta'}^2 \simeq m_{\eta'}^2 - \tfrac{2}{3} m_K^2 \simeq (0.87\,\text{GeV})^2. \tag{3.15}$$

To evaluate Eqs. (3.10) and (3.11), we use

$$\Lambda \simeq 350 \pm 100\,\text{MeV} \tag{3.16}$$

for the QCD scale parameter [26],

$$\langle \alpha_s G^2 \rangle \simeq 0.06 \pm 0.02\,\text{GeV}^4 \tag{3.17}$$

from a global fit of the light mesons and charmonium data [20], and parametrise the triple gluon condensate as

$$\langle g^3 G^3 \rangle \simeq 1.5 \pm 0.5\,\text{GeV}^2 \langle \alpha_s G^2 \rangle \tag{3.18}$$

using the dilute gas instanton model [19]. We show the result in Fig. 5a for $\chi'(0)$ plotted versus $\tau$ for different values of $t_c$. In Fig. 5b we show the behaviour of the $\tau$-minima for different $t_c$. Our optimal result corresponds to the range of values of $t_c$ corresponding to the first appearance of the $\tau$-minimum until the beginning of the $t_c$ stability region. The value of $\tau$ at which the stability occurs is around 0.4 to 0.6 GeV$^{-2}$, which is quite small compared with the light meson systems and is consistent with qualitative expectations [23] of a scale hierarchy in the QSSR analysis of gluonium systems. This small value of $\tau$ also ensures that higher dimension operators such as those arising from instanton-like effects will not contribute in the OPE. We deduce

$$\sqrt{\chi'(0)} \simeq 22.3 \pm 3.2 \pm 2.8 \pm 1.3 \text{ MeV}, \tag{3.19}$$

where the first error comes from $\langle\alpha_s G^2\rangle$, the second one from $\Lambda$ and the third from the range of $t_c$-values from 4.5 to 7.5 GeV$^2$. The effects of the triple gluon condensate and the radiative corrections are relatively unimportant, contributing about (3–10)% to $\chi'(0)$. We add a guessed error of 5% each from the unknown non-perturbative and radiative correction terms. Finally, adding all these errors quadratically, we find the following Laplace sum rule estimate of the first moment of the topological susceptibility evaluated at $\tau \simeq 0.5$ GeV$^{-2}$:

$$\sqrt{\chi'(0)} \simeq 22.3 \pm 4.8 \text{ MeV}. \tag{3.20}$$

As a check on the validity of this result, we now repeat the analysis using the finite energy sum rule (FESR) local duality version of the spectral sum rules discussed in Ref. [25]. The advantage of the FESR method is that it projects out the effects of the operators of a given dimension [27] (in this case, dimension 4) in such a way that, at the order to which we are working, the FESR analogues of the sum rules (3.10) and (3.11) are not affected by higher dimension operators such as those induced by instanton-like effects.

The FESR sum rules are

$$\begin{aligned}&\int_0^{t_c} \frac{dt}{t}\frac{1}{\pi}\text{Im}\,\chi(t)\\ &\quad\simeq \left(\frac{\overline{\alpha}_s}{8\pi}\right)^2 \frac{2}{\pi^2}\frac{t_c^2}{2}\left[1+\frac{\overline{\alpha}_s}{4\pi}\left(29-4\frac{\beta_1}{2}-8\frac{\beta_2}{\beta_1}\log(-\log\,\tau\Lambda^2)\right)\right]\\ &\qquad+\frac{\overline{\alpha}_s}{8\pi}\left[\frac{1}{2\pi}\langle\alpha_s G^2\rangle\right]\end{aligned} \tag{3.21}$$

and

$$\begin{aligned}\chi'(0) \simeq &\int_0^{t_c} \frac{dt}{t^2}\frac{1}{\pi}\text{Im}\,\chi(t)\\ &-\left(\frac{\overline{\alpha}_s}{8\pi}\right)^2 \frac{2}{\pi^2}t_c\left[1+\frac{\overline{\alpha}_s}{4\pi}\left(29-4\beta_1-8\frac{\beta_2}{\beta_1}\log(-\log\,\tau\Lambda^2)\right)\right].\end{aligned} \tag{3.22}$$

Analysing Eqs. (3.21) and (3.22), we realise that the solution increases monotonically with $t_c$ so that no firm prediction can be made, although the result gives a rough indication of consistency with the previous Laplace sum rule. To overcome this problem, we repeat the analysis using only the FESR (3.22) and using as an extra input the value of the parameter $f_{\eta'}$ extracted from the first Laplace sum rule (3.10). The value of $f_{\eta'}$ is given in Appendix B. This weakens the $t_c$-dependence of the result and $t_c$-stability now appears as an inflection

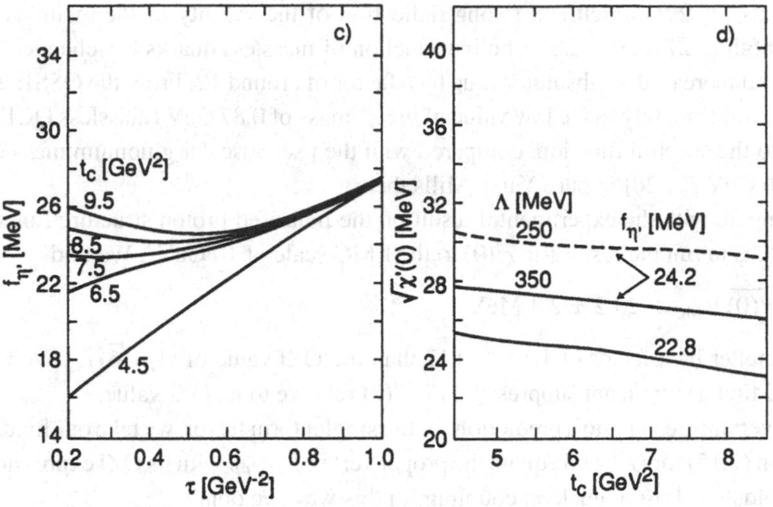


Fig. 6. (a) As Fig. 5a for the parameter $f_{\eta'}$ (b) FESR prediction of $\sqrt{\chi'(0)}$ versus $t_c$ for different values of $f_{\eta'}$.

point. We obtain the result shown in Fig. 6 for different values of $f_{\eta'}$ and $\Lambda$, from which we deduce that with $t_c \simeq 6.5$–$9.5\ \text{GeV}^2$,

$$\sqrt{\chi'(0)} \simeq 25.5 \pm 1.5 \pm 2.0 \pm 1.0\ \text{MeV}, \tag{3.23}$$

where the errors come from $f_{\eta'}$, $\Lambda$ and $t_c$ respectively. Adding the errors quadratically and including a further 5% error from the unknown higher order terms, we obtain at the scale $\tau \simeq 0.5\ \text{GeV}^{-2}$

$$\sqrt{\chi'(0)} \simeq 26.5 \pm 3.1\ \text{MeV}, \tag{3.24}$$

where we have run the result from $t_c = 8\ \text{GeV}^2$ to the scale $\tau = 0.5\ \text{GeV}^{-2}$ using the RGE solution expressed in terms of $\Lambda$, viz.

$$\chi'(0;\mu) \simeq \hat{\chi}'(0) \exp\left(\frac{8}{\beta_1^2 \log(\mu/\Lambda)}\right), \tag{3.25}$$

where $\chi'(0)$ is RG invariant. (Notice that the inhomogeneous term proportional to $\beta^{(L)}$ does not contribute to the first moment at $k^2 = 0$.) We see that the FESR result is consistent with the Laplace one.

Taking the average of the Laplace and FESR results, we obtain our final estimate of the first moment of the topological susceptibility at the scale $\tau = 0.5\ \text{GeV}^{-2}$:

$$\sqrt{\chi'(0)} \simeq 25.3 \pm 2.6\ \text{MeV}. \tag{3.26}$$

This result should be compared with that obtained [24,25] in pure $N_C = 3$ Yang–Mills theory using a similar QSSR approach:

$$\sqrt{-\chi'(0)}\ |_{YM} \simeq 7 \pm 3\ \text{MeV}. \tag{3.27}$$

It is important to notice that this pure Yang–Mills result has been confirmed by lattice calculations [28,29], which is a strong indication of the validity of the methods used in deriving both (3.27) and (3.26). The introduction of massless quarks has changed the sign of $\chi'(0)$ and increased its absolute value by a factor of around 12. From the QSSR analysis, this effect is due mainly to the low value of the $\eta'$-mass of 0.87 GeV (massless QCD) which enters into the spectral function, compared with the pseudoscalar gluonium mass of about 1.36–1.66 GeV [24,20] in pure Yang–Mills theory.

To compare with the experimental result on the polarised proton structure function, we use the RGE to run the result for $\chi'(0)$ to the EMC scale of 10 GeV$^2$. We find

$$\sqrt{\chi'(0)}\,|_{\rm EMC} \simeq 23.2 \pm 2.4 \text{ MeV}. \qquad (3.28)$$

This is smaller by a factor of $1.64 \pm 0.17$ than the OZI value of $(1/\sqrt{6})f_\pi$. We therefore do indeed find a significant suppression of $\chi'(0)$ relative to its OZI value.

To convert this result into a prediction for the singlet form factor, we take our fundamental expression (2.15) for $G_{\rm A}^{(0)}$ and equate the proper vertex $\Gamma_{\Phi_{5R}P\overline{P}}$ with its OZI expression given by the Goldstone boson–nucleon coupling. In this way, we obtain

$$G_{\rm A}^{(0)}(0) = G_{\rm A}^{(0)}(0)_{\rm OZI}\frac{\sqrt{\chi'(0)}}{(1/\sqrt{6})f_\pi}. \qquad (3.29)$$

Using the value of $G_{\rm A}^{(0)}(0)_{\rm OZI}$ in Eq. (2.8) and including an additional error of approximately 10% for the use of the OZI approximation for the proper vertex, we arrive at our final prediction:

$$G_{\rm A}^{(0)}(0; Q^2 = 10 \text{ GeV}^2) \simeq 0.353 \pm 0.052. \qquad (3.30)$$

Substituting this result[5] together with

$$G_{\rm A}^{(8)} \equiv \frac{1}{2\sqrt{3}}(3F - D), \qquad G_{\rm A}^{(3)} \equiv \tfrac{1}{2}(F + D) \qquad (3.31)$$

into the first moment sum rule (2.1), using the values of $F$ and $D$ from Eq. (2.9), and neglecting the higher twist terms (which are certainly negligible at $Q^2 = 10$ GeV$^2$), we deduce

$$\Gamma_1^{\rm p}(10 \text{ GeV}^2) \simeq 0.143 \pm 0.005. \qquad (3.32)$$

Here, we have used the coupling $\alpha_s(m_\tau) = 0.347 \pm 0.030$ extracted from tau-decay data [30]. One should notice that the radiative corrections decrease the leading order result by about 12%.

Our result, Eqs. (3.30) and (3.32), certainly goes in the right direction, i.e. that of reducing the prediction from the OZI (Ellis–Jaffe) value. At the time we obtained it, however,

[5] In terms of the quantities $\Delta u$, $\Delta d$ and $\Delta s$ defined in Eq. (2.3), we have at $Q^2 = 10$ GeV$^2$ $\Delta u = 0.84 \pm 0.01$, $\Delta d = -0.41 \pm 0.01$, $\Delta s = -0.08 \pm 0.02$.

Eq. (3.30) still appeared too high compared to the experimental result (2.7), which would have implied further OZI violations in the proper vertex. Amusingly enough, while this paper was being completed we learned of the new results from the SMC Collaboration which, combined with the earlier proton data, gives the new world average [31]:

$$\Gamma_1^{\mathrm{p}}(10\ \mathrm{GeV}^2) = 0.145 \pm 0.008 \pm 0.011 \tag{3.33}$$

from which we deduce

$$G_{\mathrm{A}}^{(0)}(0; Q^2 = 10\ \mathrm{GeV}^2) \equiv \Delta\Sigma = 0.37 \pm 0.07 \pm 0.10. \tag{3.34}$$

These results are now in excellent agreement with our predictions.

## 4. Tests of the Bjorken sum rule and estimate of higher twist effects

Recently, the SMC Collaboration at CERN [31,32] and the E142 Collaboration at SLAC [33] have produced data on the polarised neutron structure function $g_1^{\mathrm{n}}$. Since our proposal requires that the flavour singlet suppression is identical for the proton and neutron, we see no reason why the Bjorken sum rule [34],

$$\begin{aligned}\delta\Gamma_1^{\mathrm{p-n}} &\equiv \Gamma_1^{\mathrm{p}} - \Gamma_1^{\mathrm{n}}\\ &\equiv \int_0^1 \mathrm{d}x\left[g_1^{\mathrm{p}}(x; Q^2) - g_1^{\mathrm{n}}(x; Q^2)\right]\\ &\equiv \tfrac{1}{6}g_{\mathrm{A}}\left[1 - \frac{\alpha_{\mathrm{s}}}{\pi} - 3.583\left(\frac{\alpha_{\mathrm{s}}}{\pi}\right)^2 - 20.215\left(\frac{\alpha_{\mathrm{s}}}{\pi}\right)^3\right] + \frac{a_{\mathrm{p}} - a_{\mathrm{n}}}{Q^2},\end{aligned} \tag{4.1}$$

should not hold, at least up to flavour SU(2) breaking. Provided the measurements are at sufficiently high $Q^2$, the higher twist corrections related to the coefficients $a_{\mathrm{p}} - a_{\mathrm{n}}$ can be neglected. Analysis of the combined proton and deuteron data as performed in Ref. [35] gives at $Q^2 = 5\ \mathrm{GeV}^2$ [31]

$$\delta\Gamma_1^{\mathrm{p-n}} \simeq 0.203 \pm 0.029, \tag{4.2}$$

to be compared with the QCD prediction, with $\alpha_{\mathrm{s}}$ (5 GeV$^2$) $= 0.32 \pm 0.02$, of

$$\delta\Gamma_1^{\mathrm{p-n}} \simeq 0.176 \pm 0.003 + \frac{a_{\mathrm{p}} - a_{\mathrm{n}}}{Q^2}, \tag{4.3}$$

From this, one can deduce the difference of the higher twist coefficients (in units of GeV$^2$)[6],

$$a_{\mathrm{p}} - a_{\mathrm{n}} \simeq 0.135 \pm 0.145. \tag{4.4}$$

[6]Keeping the order $\alpha_{\mathrm{s}}$ term and using the estimate of the higher twist terms from QCD spectral sum rules, the authors of Ref. [36] found $\Gamma_1^{\mathrm{p-n}} \simeq 0.180 \pm 0.006$, in agreement with the data in Eq. (4.2). Our attitude here is different, as we will extract the size of the higher twist terms from the data in order to test the reliability of the previous theoretical estimate of those terms.

We can pursue an analogous analysis for the first moment of the neutron structure function, which satisfies the sum rule (cf. Eq. (2.1))

$$\begin{aligned}
&\Gamma_1^{\mathrm{n}}(Q^2)\\
&\equiv \int_0^1 \mathrm{d}x\, g_1^{\mathrm{n}}(x;Q^2)\\
&= \frac{1}{6}\left\{\left(-G_{\mathrm{A}}^{(3)}(0) + \frac{1}{\sqrt{3}}G_{\mathrm{A}}^{(8)}(0)\right)\left[1 - \frac{\alpha_{\mathrm{s}}}{\pi} - 3.583\left(\frac{\alpha_{\mathrm{s}}}{\pi}\right)^2 - 20.215\left(\frac{\alpha_{\mathrm{s}}}{\pi}\right)^3\right]\right.\\
&\quad \left. + \tfrac{2}{3}G_{\mathrm{A}}^{(0)}(0;Q^2)\left[1 - \frac{1}{3}\frac{\alpha_{\mathrm{s}}}{\pi} - 0.550\left(\frac{\alpha_{\mathrm{s}}}{\pi}\right)^2\right]\right\} + \frac{a_{\mathrm{n}}}{Q^2},
\end{aligned} \tag{4.5}$$

where we have included the higher twist contribution. Evaluating this quantity at $Q^2 = 2\ \mathrm{GeV}^2$, where the SLAC data are available, we find

$$\Gamma_1^{\mathrm{n}}(2\,\mathrm{GeV}^2) \simeq -(0.031 \pm 0.006) + \frac{a_{\mathrm{n}}}{Q^2}. \tag{4.6}$$

Comparing this with the SLAC data [33],

$$\Gamma_1^{\mathrm{n}}(2\,\mathrm{GeV}^2) \simeq -(0.022 \pm 0.011), \tag{4.7}$$

and using Eq. (4.4), we can extract the coefficients of the higher twist terms. In units of $\mathrm{GeV}^2$, we find

$$\begin{aligned}
a_{\mathrm{p}} &\simeq -0.117 \pm 0.145,\\
a_{\mathrm{n}} &\simeq 0.018 \pm 0.025.
\end{aligned} \tag{4.8}$$

These values of the higher twist terms are consistent with the previous determinations [37,38] from QCD spectral sum rules. However, these sum rules would be affected by a more general choice of the nucleon interpolating field [20] (the one used in Refs. [37,38] is not the optimal one) and by the well-known [20] large violation by a factor 2–3 of the vacuum saturation of the four-quark condensate, which is assumed in Refs. [37,38] to be satisfied to within (10–20)%. In addition, radiative corrections, which are known to be large in the baryon sum rules [20], can also be important here. More accurate data on the Bjorken and neutron sum rules, and/or a measurement of the proton sum rule at lower $Q^2$, are needed to improve the results in Eq. (4.8), which are necessary to test the validity of the QCD spectral sum rule predictions in Ref. [38].

## 5. Further discussion

In this paper, we have presented evidence that the experimentally observed suppression of the first moment of the polarised proton structure function $g_1^{\mathrm{p}}$ (the so-called EMC "proton spin" crisis) is a target-independent effect reflecting a suppression of the first moment of the QCD topological susceptibility $\chi'(0)$ relative to the OZI expectation. Not only does $G_A^{(0)}(0)$ not measure the quark spin, its suppression is not even a property of the proton structure.

It would be interesting to test this hypothesis directly by polarised deep inelastic scattering experiments on other targets not simply related to the proton by flavour symmetry. We have already studied the case of a photon target and have presented elsewhere [39] a new sum rule for the first moment of the polarised photon structure function $g_1^\gamma$ measurable in polarised $e^+e^-$ colliders. However, this turns out to be a special case because the electromagnetic U(1) anomaly contributes at leading order and so the $g_1^\gamma$ sum rule does not display the suppression mechanism described here. Another possibility is to consider semi-inclusive processes in which a particular hadron with a fraction $z$ of the incoming momentum is observed in the target fragmentation region. It was recently suggested [40] that such cross sections should be described in terms of new, non-perturbative hybrid functions $M(z, x, Q)$, called "fracture functions". To the extent that an OPE can be used, it would be possible to represent $M$ in terms of the forward matrix element of a composite operator between a suitable proton-plus-hadron state. In this case, one would again factorise $M$ into a composite propagator of the usual type and a proper vertex involving four external hadron legs. If the suppression of the polarised structure function indeed originates from the propagator, as we suggest, such a suppression should also be found at the level of the (less inclusive) fracture functions.

So far, we have only considered the first moment of $g_1^p$. Of course, we would like to extend our approach to higher moments and discuss the full $x$-dependence of the structure function. This would require knowledge of the renormalisation properties and composite operator Green functions of the higher spin axial currents and gluon densities [41], together with the associated proper vertices.

Another possible line of development would be to try to develop techniques to estimate the proper vertices themselves, rather than just the composite operator Green functions. To the extent that the quenched approximation may be trusted for the proper vertices, lattice calculations could already be suitable for the task, and QCD spectral sum rule techniques could be used in conjunction to check the validity of that approximation. We recall that, in contrast, the use of the quenched approximation directly for the matrix elements of the operator $Q$ can be shown to be completely unreliable since these are affected by low-lying poles that should disappear after dynamical quark loops are added. This is another example of how the apparent complication introduced by our splitting of matrix elements into propagators and proper vertices can ultimately pay off.

Finally, it would be interesting to attempt to apply this analysis of deep inelastic scattering using proper vertices to other QCD processes normally described in the language of the parton model rather than in terms of the OPE. Semi-inclusive deep inelastic scattering is one such example, but many other interesting possibilities can be considered, especially in the context of hadron–hadron collisions.

## Appendix A

*Chiral Ward identities and the renormalisation group*

The anomalous chiral Ward identities for Green functions of the pseudoscalar operators $Q_\mathrm{R}$ and $\Phi_{5\mathrm{R}}$ are (for zero quark masses)

$$ik_\mu \langle 0 \mid J^0_{\mu 5\mathrm{R}}(k) Q_R(-k) \mid 0\rangle - 2N_\mathrm{F}\langle 0 \mid Q_\mathrm{R}(k) Q_\mathrm{R}(-k) \mid 0\rangle = 0, \tag{A.1}$$

$$\begin{aligned} &ik_\mu \langle 0 \mid J^0_{\mu 5\mathrm{R}}(k) \Phi_{5R}(-k) \mid 0\rangle - 2N_\mathrm{F}\langle 0 \mid Q_\mathrm{R}(k) \Phi_{5\mathrm{R}}(-k) \mid 0\rangle \\ &\quad + \langle 0 \mid \delta_5 \Phi_{5\mathrm{R}}(-k) \mid 0\rangle = 0. \end{aligned} \tag{A.2}$$

So, at zero momentum, assuming there is no physical massless U(1) boson,

$$\langle 0 \mid Q_\mathrm{R}(0) Q_\mathrm{R}(0) \mid 0\rangle = 0, \tag{A.3}$$

showing that the topological susceptibility $\chi(0)$ vanishes for massless QCD, and

$$\langle 0 \mid Q_\mathrm{R}(0) \Phi_{5\mathrm{R}}(0) \mid 0\rangle = -\frac{1}{2N_\mathrm{F}} 2\langle \Phi_\mathrm{R}\rangle, \tag{A.4}$$

where $\langle \Phi_\mathrm{R}\rangle$ is the VEV of the scalar partner of $\Phi_{5\mathrm{R}}$ and is non-vanishing because of the quark condensate.

The field $\Phi_{5\mathrm{R}}$ is normalised such that the two-point proper vertex $\Gamma_{\Phi_{5\mathrm{R}}\Phi_{5\mathrm{R}}} = k^2$. This means that $\Gamma_{\Phi_{5\mathrm{R}}\Phi_{5\mathrm{R}}}$ is (minus) a component of the inverse propagator matrix in the pseudoscalar sector, i.e.

$$\Gamma_{\Phi_{5\mathrm{R}}\Phi_{5\mathrm{R}}} = \langle 0 \mid Q_\mathrm{R} Q_\mathrm{R} \mid 0\rangle (\langle 0 \mid Q_\mathrm{R} \Phi_{5\mathrm{R}} \mid 0\rangle^2 - \langle 0 \mid Q_\mathrm{R} Q_\mathrm{R} \mid 0\rangle \langle 0 \mid \Phi_{5\mathrm{R}} \Phi_{5\mathrm{R}} \mid 0\rangle)^{-1}. \tag{A.5}$$

Expanding to lowest order in $k^2$ gives

$$\Gamma_{\Phi_{5\mathrm{R}}\Phi_{5\mathrm{R}}} = \chi'(0) \langle 0 \mid Q_\mathrm{R}(0) \Phi_{5\mathrm{R}}(0) \mid 0\rangle^{-2} k^2 + \mathrm{O}(k^4), \tag{A.6}$$

where we have written $\langle 0 \mid Q_\mathrm{R}(k) Q_\mathrm{R}(-k) \mid 0\rangle = \chi'(0)k^2 + \mathrm{O}(k^4)$. We therefore deduce

$$\langle 0 \mid Q_\mathrm{R}(0) \Phi_{5\mathrm{R}}(0) \mid 0\rangle = \sqrt{\chi'(0)}, \tag{A.7}$$

as quoted in Eq. (2.15).

The renormalisation group equation for the topological susceptibility follows from the definition of the renormalised composite operators, Eq. (2.4), and the chiral Ward identities. The Ward identity for the two-current Green function is

$$ik^\mu \langle 0 \mid J^0_{\mu 5\mathrm{R}}(k) J^0_{\nu 5\mathrm{R}}(-k) \mid 0\rangle - 2N_\mathrm{F}\langle 0 \mid Q_\mathrm{R}(k) J^0_{\nu 5\mathrm{R}}(-k) \mid 0\rangle = 0. \tag{A.8}$$

Combining Eqs. (A.1), (A.8) and (2.4), we find straightforwardly

$$\langle 0 \mid Q_\mathrm{R}(k) Q_\mathrm{R}(-k) \mid 0\rangle = Z^2 \langle 0 \mid Q_\mathrm{B}(k) Q_\mathrm{B}(-k) \mid 0\rangle + \dots. \tag{A.9}$$

The dots denote the extra divergences associated with contact terms in the two-point Green functions of composite operators. Taking these into account (see Refs. [21,7] for full details) we find the full RGE for $\chi(k^2)$,

$$\left(\mu\frac{\partial}{\partial\mu} + \beta(\alpha_s)\alpha_s\frac{\partial}{\partial\alpha_s} - 2\gamma\right)\chi(k^2) = -\frac{1}{(2N_F)^2}2\beta^{(L)}(\alpha_s)k^4, \tag{A.10}$$

where $\beta^{(L)}$ is a new RG function. The inhomogeneous term does not contribute at zero momentum, however, and the required RGE (2.13) for $\chi'(0)$ follows immediately.

## Appendix B

*Decay constants and the $\eta'$*

We can estimate the parameter $f_{\eta'}$ appearing in the spectral expansion using the first Laplace QSSR, Eq. (3.10). $f_{\eta'}$ is defined by

$$\langle 0 \mid J^0_{\mu 5R}(k) \mid \eta' \rangle = ik_\mu f_{\eta'}, \tag{B.1}$$

and is RG non-invariant. On shell (see Ref. [7], Appendix D), the scale dependence is due entirely to the anomalous dimension $\gamma$ of the axial current so, using Eq. (3.3) and expressing the result in terms of the QCD scale $\Lambda$, we may write

$$f_{\eta'}(\mu) = \hat{f}_{\eta'} \exp\left(\frac{4}{\beta_1^2 \log(\mu/\Lambda)}\right), \tag{B.2}$$

where $\hat{f}_{\eta'}$ is RG invariant. From the QSSR (3.10), we find the $\tau$-stability starts at $t_c \simeq 6.5$ GeV$^2$, while the $t_c$-stability is reached for $t_c$ larger than 9.5 GeV$^2$. In this region, the radiative corrections are about 10% of the lowest order term, while the $\langle g^3G^3\rangle$ one contributes about 10%. Under such conditions, our optimal result at $\tau \simeq 0.6\,\text{GeV}^{-2}$ is (see Fig. 6a)

$$f_{\eta'} \simeq 24.1 \pm 0.6 \pm 3.4 \pm 0.3\,\text{MeV}, \tag{B.3}$$

where the first error comes from $\langle\alpha_s G^2\rangle$, the second from $\Lambda$ and the third from the range of $t_c$-values between 6.5 and 9.5 GeV$^2$. Adding a 5% error from the unknown QCD terms, adding the different errors quadratically and running to the EMC scale, we obtain

$$f_{\eta'}\mid_{\text{EMC}} \simeq 23.6 \pm 3.5\,\text{MeV}. \tag{B.4}$$

This value is strongly suppressed relative to the OZI prediction of $\sqrt{6}\, f_\pi$ for the $\eta'$-decay constant.

However, as has been shown in Refs. [42,7], this $f_{\eta'}$ is *not* the $\eta'$-decay constant measured in, e.g., the decay $\eta' \to \gamma\gamma$. In fact, the analogues of the current algebra formulae

$$f_\pi g_{\pi gg} = \frac{1}{\pi}\alpha_{\text{em}} \tag{B.5}$$

and

$$f_\pi g_{\pi NN} = m_N g_A \tag{B.6}$$

in the flavour singlet sector are [42,7]

$$F g_{\eta'\gamma\gamma} + \frac{1}{2N_F} F^2 m_{\eta'}^2 g_{G\gamma\gamma}(0) = \frac{4}{\pi}\alpha_{em} \tag{B.7}$$

and

$$F g_{\eta' NN} + \frac{1}{2N_F} F^2 m_{\eta'}^2 g_{GNN}(0) = 2m_N G_A^{(0)}(0). \tag{B.8}$$

Here, $F$ is the RG-invariant decay constant defined by

$$F = \frac{2\langle\phi_R\rangle}{m_{\eta'}} \left( \int dx\, i\langle 0 \mid T^* \phi_{5R}^0(x)\phi_{5R}^0(0) \mid 0\rangle \right)^{-1/2}, \tag{B.9}$$

where $\phi_{5R}^0 = i\Sigma\bar{q}\gamma_5 q$ and $\langle\phi\rangle = \Sigma\langle\bar{q}q\rangle$. The extra terms $g_{G\gamma\gamma}$ and $g_{GNN}$ appearing in Eqs. (B.7), (B.8) (which are properly defined as proper vertices [42,7]) may be thought of as the couplings of the gluonic component of the $\eta'$. They arise because the $\eta'$ is not a Goldstone boson in the U(1) channel and so the naive current algebra extensions of Eqs. (B.5), (B.6) are not valid. At first sight, therefore, Eqs. (B.7) and (B.8) are not predictive since $g_{G\gamma\gamma}$ and $g_{GNN}$ are unknown. However, if we follow our proposal that OZI violations are associated with RG-non-invariant quantities we can make predictions.

Taking Eq. (B.7) first, we have shown [42] that $g_{G\gamma\gamma}$ is RG invariant. Since in the OZI limit this term is absent, we therefore expect $g_{G\gamma\gamma}$ to be small, and so to a good approximation we predict

$$F g_{\eta'\gamma\gamma} = \frac{4}{\pi}\alpha_{em}. \tag{B.10}$$

Since $F$ is RG invariant, we expect it to be well approximated by its OZI value $\sqrt{6} f_\pi$. Experimentally (see Ref. [43]), the relation (B.10) is very well satisfied.

In Eq. (B.8), on the other hand, $g_{GNN}$ is not RG invariant so we do not expect this term to be small. In fact, this equation is just a rewriting of the U(1) GT formula quoted in the text, for which our proposal is successful.

An important test of our picture of the pattern of OZI breaking is therefore to evaluate the RG-invariant decay constant $F$ from first principles and check that it is close to the OZI prediction of $\sqrt{2N_F} f_\pi$. Again, we can use QCD spectral sum rules.

We require the zero-momentum limit $\Phi_5(0)$ of the two-point correlation function

$$\Phi_5(k^2) = \int dx\, e^{ik\cdot x} i\langle 0 \mid T^* \phi_{5R}^0(x)\phi_{5R}^0(0) \mid 0\rangle \tag{B.11}$$

for QCD with 3 flavours and massless quarks. However, as there is a smooth behaviour of the two-point correlator when the common light quark mass $m_R$ goes to zero, we shall work (for convenience) with the RG-invariant correlation function

$$\Psi_5(k^2) \equiv 4m_R^2 \Phi_5(k^2), \tag{B.12}$$

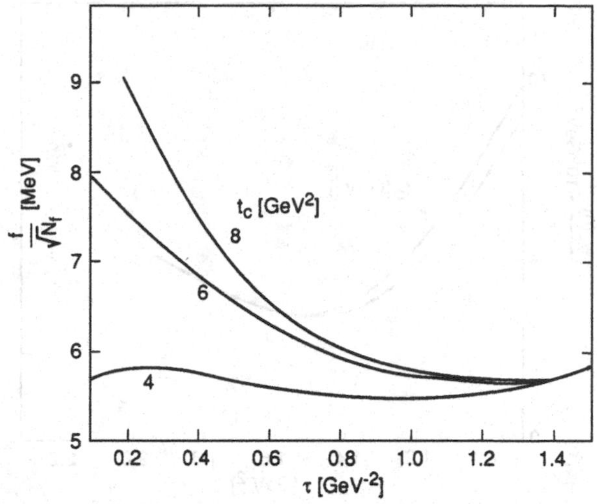


Fig. B.1. As Fig. 5a for the parameter $f$.

where $m_R$ is the average of the renormalised u and d quark masses. Now, in perturbation theory, the difference between this flavour singlet correlation function and the corresponding non-singlet one appears only at $O(\alpha_s^2)$ from the double-triangle anomaly-type diagrams. Similarly for the non-perturbative condensate terms, the difference is only of $O(\alpha_s^2)$ arising from the equivalent diagrams. Instanton-like effects appear as higher dimension operators. So, at the order we are working, we can simply use the expression for the isotriplet (pion) correlation function in QCD discussed in the literature [20].

The first Laplace sum rule to two loops reads [20]

$$\begin{aligned}
&\int_0^{t_c} \mathrm{d}t\, \mathrm{e}^{-t\tau} \frac{1}{\pi} \mathrm{Im}\Psi_5(t) \\
&\quad \simeq \frac{3N_F}{2\pi^2} \overline{m}^2(\tau) \Bigg\{ \tau^{-2}[1-\exp(-t_c\tau)(1+t_c\tau)] \\
&\qquad \times \left\{ 1 - \frac{2}{\beta_1 L}\left[ \tfrac{11}{3} + 2\gamma_E - \frac{2}{\beta_1}\left(\tilde{\gamma}_2 - \tilde{\gamma}_1 \frac{\beta_2}{\beta_1}\right) + 2\frac{\tilde{\gamma}_1 \beta_2}{\beta_1^2} \log L \right] \right\} \\
&\qquad + \left( \tfrac{1}{3}\pi \langle \alpha_s G^2 \rangle + \tfrac{896}{81}\pi^3 \rho \alpha_s \langle \overline{u}u \rangle^2 \tau \right) \Bigg\},
\end{aligned} \tag{B.13}$$

where $L = -\log \tau \Lambda^2$ and [20]

$$\begin{aligned}
&\rho \alpha_s \langle \overline{u}u \rangle^2 \simeq (3.8 \pm 2.0) \times 10^{-4}\,\mathrm{GeV}^6, \\
&\overline{m}(\tau) \equiv \tfrac{1}{2}(\overline{m}_u + \overline{m}_d)(\tau) \simeq \left(-\tfrac{1}{2}\log \tau \Lambda^2\right)^{\tilde{\gamma}_1/\beta_2} (12.1 \pm 1.0)\,\mathrm{MeV}.
\end{aligned} \tag{B.14}$$

As before, we parametrise the spectral function keeping only the lowest ($\eta'$) resonance, i.e.

$$\frac{1}{\pi}\,\mathrm{Im}\Psi_5(t) = 2\tilde{m}_{\eta'}^4 f^2 \delta\left(t - \tilde{m}_{\eta'}^2\right) + \text{“QCD continuum”}\ \theta(t - t_c), \tag{B.15}$$

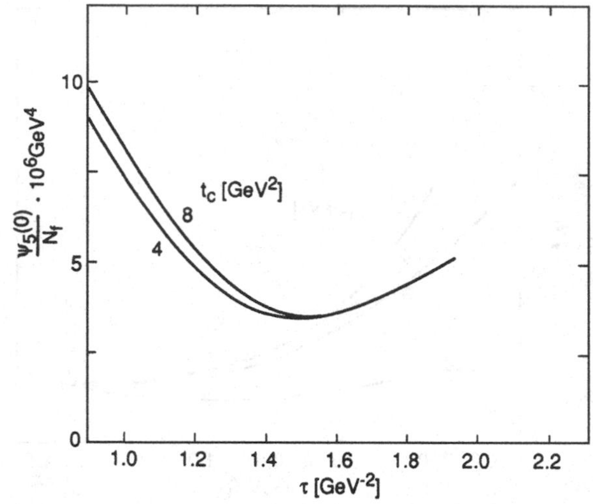


Fig. B.2. As Fig. 5a for $\Psi_5(0)$.

where the unknown parameter $f$, which is defined by

$$2m_{\mathrm{R}}\langle 0 \mid \phi^0_{5\mathrm{R}} \mid \eta'\rangle = \sqrt{2} f \tilde{m}^2_{\eta'}, \tag{B.16}$$

can be estimated from the sum rule (B.13). We study the $\tau$-and $t_c$-behaviours of $f$ in Fig. B.1. The $\tau$-stability starts for $t_c \simeq 4$ GeV$^2$, while stability in $t_c$ appears above $t_c \simeq 7$ GeV$^2$, a range which is equal to the one for the correlation function for $Q(x)$. The value for the $\tau$-stability of about 0.9 GeV$^{-2}$ is typical of light quark correlation functions. At the minimum, we obtain

$$f = \sqrt{N_{\mathrm{F}}}(5.55 \pm 0.08 \pm 0.65 \pm 0.35 \pm 0.06 \pm 0.03)\ \mathrm{MeV}, \tag{B.17}$$

where the errors come respectively from $t_c$, $\Lambda$, $\overline{m}$, $\langle \alpha_s G^2\rangle$ and $\rho\alpha_s\langle \overline{u}u\rangle^2$. Adding these errors quadratically, we deduce

$$f = \sqrt{N_{\mathrm{F}}}(5.55 \pm 0.75)\ \mathrm{MeV}. \tag{B.18}$$

With this value for $f$, we are now able to estimate $\Psi_5(0)$ itself using a second Laplace sum rule [44,25]:

$$\begin{aligned}\Psi_5(0) \simeq \int_0^{t_c} \frac{\mathrm{d}t}{t} \mathrm{e}^{-t\tau} \frac{1}{\pi} \mathrm{Im}\Psi_5(t) - \frac{3N_{\mathrm{F}}}{2\pi^2}\overline{m}^2(\tau)\Bigg(&\tau^{-1}[1-\exp(-t_c\tau)] \\ &\times\left\{1 - \frac{2}{\beta_1 L}\left[\tfrac{11}{3} + 2\gamma_{\mathrm{E}} - \frac{2}{\beta_1}\left(\tilde{\gamma}_2 - \tilde{\gamma}_1\frac{\beta_2}{\beta_1}\right) + 2\frac{\tilde{\gamma}_1\beta_2}{\beta_1^2}\log L\right]\right\} \\ &+ \left(\tfrac{1}{3}\pi\langle\alpha_s G^2\rangle + \tfrac{1}{2}\tfrac{896}{81}\pi^3\rho\alpha_s\langle\overline{u}u\rangle^2\tau\right)\Bigg),\end{aligned} \tag{B.19}$$

where $\tilde{\gamma}_1, \tilde{\gamma}_2$ are the coefficients in the anomalous dimension for the light quark mass. For three flavours, $\tilde{\gamma}_1 = 2$ and $\tilde{\gamma}_2 = \frac{91}{12}$. The sum rule analysis of this quantity shows a strong $t_c$-dependence and the $\tau$-stability only appears at unrealistic values of $t_c$ larger than 8 GeV$^2$. In order to circumvent this difficulty, we work with a combination of the sum rules (B.13) and (B.19) which has been used successfully in the past for measuring the deviation from pion and kaon PCAC to a good accuracy [44]. The combined sum rule reads

$$\Psi_5(0) \simeq \int_0^{t_c} \frac{dt}{t} e^{-t\tau}(1 - t\tau)\frac{1}{\pi} \text{Im}\, \Psi_5(t) - \frac{3N_F}{2\pi^2}\bar{m}^2(\tau)\Bigg( \tau^{-1}[t_c\tau \exp(-t_c\tau)]$$
$$\times \left\{1 - \frac{2}{\beta_1 L}\left[\tfrac{11}{3} + 2\gamma_E - \frac{2}{\beta_1}\left(\tilde{\gamma}_2 - \tilde{\gamma}_1\frac{\beta_2}{\beta_1}\right) + 2\frac{\tilde{\gamma}_1\beta_2}{\beta_1^2}\log L\right]\right\}$$
$$+ \tau\left(\tfrac{2}{3}\pi\langle\alpha_s G^2\rangle + \tfrac{3}{2}\tfrac{896}{81}\pi^3\rho\alpha_s\langle\bar{u}u\rangle^2\tau\right)\Bigg). \tag{B.20}$$

This sum rule is studied in Fig. B.2. The position of the stability is almost insensitive to the value of $t_c$ due to some cancellations amongst the perturbative terms. However, this feature also implies that the stability is obtained at values of $\tau$ larger than in the previous cases, making the result sensitive to the errors on the four-quark condensates, which affects the accuracy of the result. We deduce

$$\Psi_5(0) \simeq N_F(3.70 \pm 0.90 \pm 0.30 \pm 0.70 \pm 2.00) \times 10^{-6}\ \text{GeV}^4, \tag{B.21}$$

where the errors are due to $f$, $\Lambda$, $\langle\alpha_s G^2\rangle$ and $\rho\alpha_s\langle\bar{u}u\rangle^2$. Adding these errors quadratically, we obtain

$$\sqrt{\Psi_5(0)} \simeq \sqrt{N_F}(1.92 \pm 0.53) \times 10^{-3}\ \text{GeV}^2. \tag{B.22}$$

Using this value in Eq. (B.9) (with $\tilde{m}_{\eta'}$), after multiplying the numerator and denominator by the overall $2m_R$ factor and using Dashen's formula for $m_R\langle\phi_R\rangle$, we finally find

$$F \simeq (1.55 \pm 0.43)\sqrt{2N_F} f_\pi, \tag{B.23}$$

to be compared with the OZI prediction of $\sqrt{2N_F} f_\pi$.

This result is again in broad agreement with our expectations, although of course the errors are much too large to draw a definitive conclusion. Nevertheless, this confirmation can be taken as providing extra support for the reliability of the estimate in the text for $\chi'(0)$.

**Acknowledgements**

One of us (G.M.S.) would like to thank John Ellis and the TH Division, CERN for their hospitality during several visits.

## References


[1] J. Ashman et al., Phys. Lett. B 206 (1988) 364; Nucl. Phys. B 328 (1990) 1.
[2] G. Baum et al., Phys. Rev. Lett. 51 (1983) 1135.
[3] G. Altarelli, Lectures at the Int. School of Subnuclear physics, Erice, preprint CERN-TH.5675/90.
[4] S.D. Bass and A.W. Thomas, J. Phys. G 19 (1993) 925.
[5] G. Altarelli and G.G. Ross, Phys. Lett. B 212 (1988) 391.
[6] G.M. Shore and G. Veneziano, Phys. Lett. B 244 (1990) 75.
[7] G.M. Shore and G. Veneziano, Nucl. Phys. B 381 (1992) 23.
[8] R.G. Roberts, The structure of the proton (Cambridge U. P., Cambridge, 1990).
[9] J. Ellis and R.L. Jaffe, Phys. Rev. D 9 (1974) 1444; D 10 (1974) 1669.
[10] J. Kodaira, Nucl. Phys. B 165 (1980) 129.
[11] S.A. Larin and J.A.M Vermaseren, Phys. Lett. B 259 (1991) 213;
S.A. Larin, preprint CERN-TH.7208/94.
[12] D. Espriu and R. Tarrach, Z. Phys. C 16 (1982) 77.
[13] G. Veneziano, Lecture at the Okubofest, May 1990, preprint CERN-TH.5840/90.
[14] G. Veneziano, Mod. Phys. Lett. A 4 (1989) 1605.
[15] M. Bourquin et al., Z. Phys. C 21 (1983) 27.
[16] Z. Dziembowski and J. Franklin, J. Phys. G 17 (1991) 213;
F.E. Close and R.G. Roberts, Phys. Lett. B 302 (1993) 533.
[17] E. Witten, Nucl. Phys. B 156 (1979) 269.
[18] G. Veneziano, Nucl. Phys. B 159 (1979) 213.
[19] M.A. Shifman, A.I. Vainshtein and V.I. Zakharov, Nucl. Phys. B 147 (1979) 385, 448.
[20] S. Narison, QCD spectral sum rules, Lecture notes in physics, Vol. 26 (World Scientific, Singapore, 1989).
[21] G.M. Shore, Nucl. Phys. B 362 (1991) 85.
[22] A.L. Kataev, N.V. Krasnikov and A.A. Pivovarov, Nucl. Phys. B 198 (1982) 508.
[23] V.A. Novikov et al., Nucl. Phys. B 237 (1984) 525.
[24] S. Narison, Z. Phys. C 26 (1984) 209.
[25] S. Narison, Phys. Lett. B 255 (1991) 101.
[26] Particle Data Group, Phys. Rev. D 45 (1992) Part 2.
[27] R.A. Bertlmann, G. Launer and E. de Rafael, Nucl. Phys. B 250 (1985) 61.
[28] G. Briganti, A. Di Giacomo and H. Panagopoulos, Phys. Lett. B 253 (1991) 427.
[29] A. Di Giacomo, Lecture at the QCD 90 Workshop, Montpellier July 1990, Nucl. Phys. B (Proc. Suppl.) 23B (1991).
[30] E. Braaten, A. Pich and S. Narison, Nucl. Phys. B 373 (1992) 581;
ALEPH Collaboration, D. Buskulic et al., Phys. Lett. B 307 (1993) 209;
S. Narison, Talk given at the Fubini fest, Torino 1994, preprint CERN-TH.7188/94, and references therein.
[31] SMC Collaboration, P. Shanahan, Talk given at the 29th rencontre de Moriond on QCD and high-energy hadronic interactions, Meribel, Savoie, March 1994;
D. Adams et al., CERN-PPE/94-57 (1994), to be published.
[32] SMC Collaboration, B. Aveda et al., Phys. Lett. B 302 (1993) 533.
[33] E142 Collaboration, P.L. Anthony et al., Phys. Rev. Lett. 71 (1993) 959.
[34] J.D. Bjorken, Phys. Rev. 148 (1966) 1467; D 1 (1970) 1376.
[35] SMC Collaboration, B. Aveda et al., Phys. Lett. B 320 (1994) 400.
[36] J. Ellis and M. Karliner, Phys. Lett. B 313 (1993) 131.

[37] I.I. Balitsky, V.M. Braun and A.V. Kolesnichenko, 7 Phys. Lett. B 242 (1990) 245; B 318 (1993) 648 (E)
X. Ji and M. Unrau, MIT-CTP-2232 (1993).
[38] G.C. Ross and R.G. Roberts, Phys. Lett. B 322 (1994) 425.
[39] S. Narison, G.M. Shore and G. Veneziano, Nucl. Phys. B 391 (1993) 69.
[40] L. Trentadue and G. Veneziano, Phys. Lett. B 323 (1994) 201.
[41] S.D. Bass, Z. Phys. C 55 (1992) 653.
[42] G.M. Shore and G. Veneziano, Nucl. Phys. B 381 (1992) 3.
[43] M.S. Chanowitz, in Proc. VI Int. Workshop on Photon-Photon Collisions, Lake Tahoe, 1984, ed. R. Lander, World Scientific, Singapore, 1984.
[44] S. Narison, Phys. Lett. B 104 (1981) 485.

# 20

# Drell–Yan process

It corresponds to the sub-process, where the quark and anti-quark come from the two scattering hadrons, and annihilate into vector bosons (photon, $W^{\pm}$, $Z^0$) with large invariant mass and then produce a lepton pair. A classical example is the annihilation into photon and with the production of $e^+e^-$:

$$\bar{q}q \to e^+e^- , \tag{20.1}$$

shown in Fig. 20.1. Drell–Yan process offers the possibility to test perturbative QCD as the large scale is given by the invariant mass of the lepton pair (of the order of $M_{W,Z}$ at CERN and Tevatron energies), while the parton densities enter quadratically in this process where the final state is totally inclusive.

## 20.1 Kinematics

The kinematics of the process is characterized by the parton distribution $q_f^{h_i}(x)$ for a quark of flavour $f$ issued from the hadron $h_i$. The total momentum squared of the subprocess is:

$$Q^2 = (x_1 p_1 + x_2 p_2)^2 , \tag{20.2}$$

and coincides with the invariant mass squared of the photon. The total energy squared of the hadron is:

$$s = (p_1 + p_2)^2 . \tag{20.3}$$

For large $s$, one usually neglects the hadron mass, such that one can approximately write:

$$Q^2 \simeq x_1 x_2 s . \tag{20.4}$$

Another useful variable is:

$$x_F \equiv x_1 - x_2 , \tag{20.5}$$

and the *rapidity* $y$ defined as:

$$\tanh y = \frac{x_1 - x_2}{x_1 + x_2} \quad \text{or} \quad y = \frac{1}{2} \ln \frac{x_1}{x_2} . \tag{20.6}$$

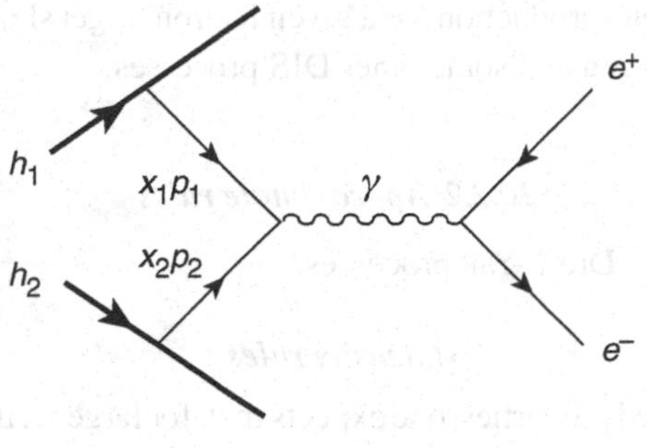


Fig. 20.1. Drell–Yan process.

Alternatively, in the hadron-hadron centre of mass where the photon momentum is:

$$q = (E; q_\parallel, q_\perp) , \tag{20.7}$$

one has:

$$x_F = 2q_\parallel/\sqrt{s} , \qquad y = \frac{1}{2}\ln\frac{E+q_\parallel}{E-q_\parallel} . \tag{20.8}$$

## 20.2 Parton model

### 20.2.1 Cross-section

In order to evaluate the production cross-section, one calculates the reduced cross-section corresponding to the subprocess in Eq. (20.1), and write the total cross-section as a convolution. Neglecting quark and electron masses, the point-like cross-section reads, to lowest order:

$$\hat\sigma_{l.o}(\bar q + q \to e^+e^-) = \frac{4\pi\alpha^2 Q_f^2}{3N_c Q^2} , \tag{20.9}$$

where $Q_f$ is the quark charge in units of $e$. The full lowest order differential cross-section reads:

$$\frac{d\sigma_{l.o}}{dQ^2} = \frac{4\pi\alpha^2}{3N_c Q^2}\sum_f Q_f^2 \int_0^1 \frac{dx_1}{x_1}\int_0^1 \frac{dx_2}{x_2}\delta(1-z)\left[q_f^{h_1}(x_1)\bar q_f^{h_2}(x_2) + \bar q_f^{h_1}(x_1) q_f^{h_2}(x_2)\right] , \tag{20.10}$$

where:

$$\tau \equiv Q^2/s \quad \text{and} \quad z \equiv \frac{\tau}{x_1x_2} . \tag{20.11}$$

$\tau$ quantifies the fraction of energy squared that goes into the lepton pair. If $\tau$ is small, then, one of the $x_i$ is small and then favours the sea quark contribution. If the $x_i$ is maximal i.e. around $1/3 \sim 1/4$, then the valence contribution will dominate. The Drell–Yan processes are important as they can provide a non-trivial test of the validity of the parton approach and of its extension in QCD through the factorization theorem. One expects that the parton densities measured in lepto-production for a given hadron target should be relevant to make predictions on the Drell–Yan and some other DIS processes.

### 20.2.2 Approximate rules

There are typical rules for Drell–Yan processes.

*Intensity rules*

From the above-mentioned properties, one expects that, for large $x_i$, the cross-section involving two valence quarks for producing the $e^+e^-$ pair, is much larger than the one involving one valence and one sea quarks. For an isoscalar target one, e.g., expects:

$$\frac{\sigma(\pi^+N(I=0))}{\sigma(\pi^-N(I=0))} \to \frac{1}{4} . \tag{20.12}$$

*Scaling*

In the region where the naïve parton model is valid, one expects that the dimensioneless quantities:

$$Q^4\frac{d\sigma}{dQ^4} , \qquad Q^4\frac{d\sigma}{dQ^2dx_F} , \qquad Q^4\frac{d\sigma}{dQ^2dy} , \tag{20.13}$$

should scale as functions of the scaling variables $\tau$, $x_F$ and $y$ independently of $Q^2$.

*Angular distribution of leptons*

For large $Q^2$, where the longitudinal structure function ($W_L$) is much smaller than the transverse ($W_T$) one, the lepton pair angular distribution originated from an off-shell photon is predominantly of the form:

$$\frac{d\sigma}{dQ^2d\cos\theta} \sim W_T(Q^2,\tau)(1+\cos^2\theta) . \tag{20.14}$$

*Atomic number*

The cross-section being proportional to the number of quarks or antiquarks in the target nucleus, each contribution adding up incoherently, one expects a linear dependence with the atomic number $A$ in the Drell–Yan region.

## 20.3 Higher order corrections to the cross-section

The different processes relevant to the NLO corrections are:

$$\begin{aligned} &q+\bar{q} \to \gamma^* \\ &q+\bar{q} \to \gamma^* + g \\ &g+q(\text{or } \bar{q}) \to \gamma^* + q(\text{or } \bar{q}) , \end{aligned} \tag{20.15}$$

where $\gamma^*$ produces the lepton pairs $e^+e^-$. They are shown in Fig. 20.2.

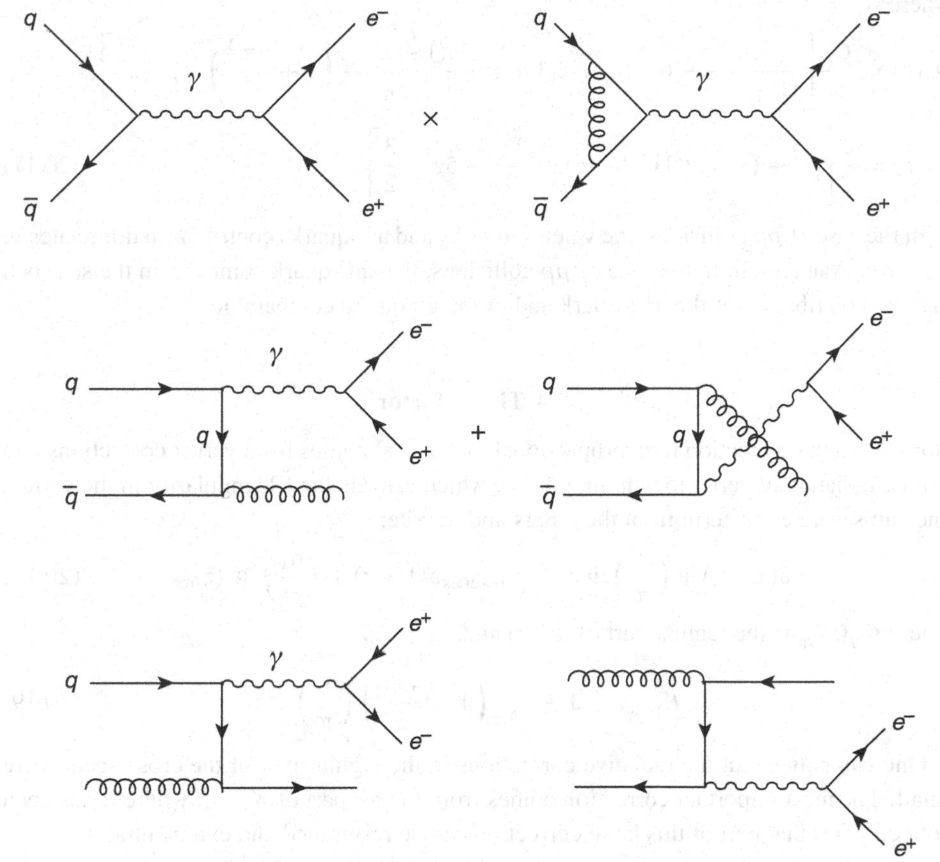


Fig. 20.2. NLO corrections to the Drell–Yan process.

Technically, the evaluation of higher order corrections is not easy because of the interplay between the IR and mass singularities. The NLO corrections have been obtained in [270], and the NNLO corrections in [271]. The interactions with the spectator quarks induce a $1/Q^2$ power corrections analogue of the higher twist term in DIS. The expression of the cross-section including the NLO corrections reads:

$$\frac{d\sigma_{l.o}}{dQ^2} = \frac{4\pi\alpha^2}{3N_c Q^2}\sum_f Q_f^2 \int_0^1 \frac{dx_1}{x_1}\int_0^1 \frac{dx_2}{x_2}\left\{\left[\delta\,(1-z) + \left(\frac{\alpha_s}{\pi}\right)\theta(1-z)\Phi_q(z)\right]\right.$$
$$\times\left[q_f^{h_1}(x_1)\bar{q}_f^{h_2}(x_2) + \bar{q}_f^{h_1}(x_1)q_f^{h_2}(x_2)\right],$$
$$\left. + \left(\frac{\alpha_s}{\pi}\right)\theta(1-z)\Phi_g(z)\left[q_f^{h_1}(x_1) + \bar{q}_f^{h_2}(x_1)\right]g^{h_2}(x_2, Q^2) + (1 \leftrightarrow 2)\right\}, \qquad (20.16)$$

where:

$$\Phi_q(z) = \frac{C_F}{2}\left[\frac{3}{(1-z)_+} - 6 - 4z + 2(1+z^2)\frac{\ln(1-z)}{(1-z)_+} + \left(1+\frac{4\pi^2}{3}\right)\delta(1-z)\right],$$

$$\Phi_g(z) = \frac{1}{2}\left[[z^2+(1-z)^2]\ln(1-z) + \frac{9z^2}{2} - 5z + \frac{3}{2}\right]. \tag{20.17}$$

In the case of $\bar{p}p$ collisions, the valence quarks and antiquarks contribution dominates in the Drell–Yan region. In the case of $pp$ collisions, the anti-quark comes from the sea such that the contribution of the anti-quark and of the gluon are comparable.

## 20.4 The *K* factor

Noting that the correction term proportional to $\delta(1-z)$ comes from vertex corrections and from a radiation of zero momentum gluons, which cancels the IR singularity in the vertex, one can separate this term from the others and rewrite:

$$\delta(1-z) + \left(\frac{\alpha_s}{\pi}\right)\Phi_q(z) \equiv K_{\text{vertex}}\delta(1-z) + \left(\frac{\alpha_s}{\pi}\right)\Phi_q(z)_{\text{reg}} \tag{20.18}$$

where $\Phi_q(z)_{\text{reg}}$ is the regular part of $\Phi_q(z)$ and:

$$K_{\text{vertex}} = 1 + \frac{C_F}{2}\left(1+\frac{4\pi^2}{3}\right)\left(\frac{\alpha_s}{\pi}\right). \tag{20.19}$$

One can notice that the radiative corrections in the regular part of the cross-section are small. The most important correction comes from the $\pi^2$ part of $K_{\text{vertex}}$, where it has been noticed [272] that part of this large correction can be resummed and exponentiates:

$$1 + C_F\frac{\pi^2}{2}\left(\frac{\alpha_s}{\pi}\right) \to K(Q^2) \equiv \exp\left(\frac{C_F}{2}\pi\alpha_s\right), \tag{20.20}$$

while the remaining correction:

$$1 + \frac{C_F}{2}\left(1+\frac{\pi^2}{3}\right)\left(\frac{\alpha_s}{\pi}\right), \tag{20.21}$$

is comfortably small. However, one should be aware of the fact that the resummation procedure is not unique. Different phenomenology of the Drell–Yan processes have been performed at Tevatron, which can be consulted from different contributions at various conferences, like the QCD-Montpellier series.

# 21

# One 'prompt photon' inclusive production

We shall be concerned with the process:

$$h_1 + h_2 \to \gamma + X . \tag{21.1}$$

This process is very similar to the one hadron inclusive process:

$$e^+e^- \to H + X \tag{21.2}$$

with the hadron $H$ replaced by a photon, which we shall study in the next part of this book. As it has been studied in hadronic collisions rather than in $e^+e^-$ [273], further difficulties and complications arise in practice. However, in contrast to quarks, the photon does not hadronize and their energies and directions can be measured with better accuracy than hadron jets. To leading order, the production cross-section is $\mathcal{O}(\alpha\alpha_s)$ which is relatively smaller than the hadron cross-section $\mathcal{O}(\alpha_s^2)$, while backgrounds due to photons initiated from $\pi^0$ and $\eta$ productions, are experimentally difficult to separate. In terms of the photon transverse momentum $p_T$ and rapidity variable, the cross-section can be written in the form [273]:

$$\frac{d\sigma}{d\mathbf{p_T}d\eta} = \frac{d\sigma^{\text{dir}}}{d\mathbf{p_T}d\eta} + \frac{d\sigma^{\text{brem}}}{d\mathbf{p_T}d\eta} , \tag{21.3}$$

where one distinguishes between the 'direct' and 'bremsstrahlung' photon productions, which are known to NLO. Assuming factorization, they read:

$$\begin{aligned}
\frac{d\sigma^{\text{dir}}}{d\mathbf{p_T}d\eta} &= \sum_{i,j=q,g} \int dx_1 dx_2 F_i^{h_1}(x_1,\mu) F_j^{h_1}(x_2,\mu) \left(\frac{\alpha_s(\nu)}{2\pi}\right) \\
&\quad \times \left(\frac{d\hat{\sigma}_{ij}}{d\mathbf{p_T}d\eta} + \frac{\alpha_s(\nu)}{2\pi} K_{ij}^{\text{dir}}(\nu,\mu,\mu_f)\right) \\
\frac{d\sigma^{\text{brem}}}{d\mathbf{p_T}d\eta} &= \sum_{i,j,k=q,g} \int dx_1 dx_2 F_i^{h_1}(x_1,\mu) F_j^{h_1}(x_2,\mu) \frac{dz}{z^2} D_{\gamma/k}(z,\mu_f) \left(\frac{\alpha_s(\nu)}{2\pi}\right)^2 \\
&\quad \times \left(\frac{d\hat{\sigma}_{ij}^k}{d\mathbf{p_T}d\eta} + \frac{\alpha_s(\nu)}{2\pi} K_{ij,k}^{\text{brem}}(\nu,\mu,\mu_f)\right) .
\end{aligned} \tag{21.4}$$

$F_i^{h_l}$ are parton densities in the initial hadrons, which depend on the factorization scale $\mu$; $D_{\gamma/k}$ is the parton to photon fragmentation function which depends on the fragmentation scale $\mu_f$, while $\nu$ is the renormalization scale. $\hat{\sigma}$ are the point-like cross-section, while the $K$ factors are higher-order QCD corrections evaluated in [274]. In principle the differential cross-section is function of the three arbitrary variables $(\mu, \mu_f, \nu)$, and the optimal physical results should present stabilities or extrema against their variations, which is not often reached. In practice, the choice $\mu_f = \nu$ or $\mu_f = \nu = \mu$ is chosen, which minimizes the arbitrariness in the analysis. Using the NLO QCD predictions, the UA6 collaboration determined $\alpha_s$ from a measurement of the cross-section difference in the $p_T$ range from about 4 to 8 GeV [275]:

$$\sigma(\bar{p}p \to \gamma X) - \sigma(pp \to \gamma X) , \tag{21.5}$$

which is free from the poorly known sea quarks and gluons distributions, with the results:

$$\alpha_s(24.3 \text{ GeV}) = 0.135 \pm 0.006 \text{ (exp)} \, ^{+0.011}_{-0.005} \text{ (th)} . \tag{21.6}$$

# Part V

## Hard processes in $e^+e^-$ collisions

# Introduction

In this part, we study different hard and jet processes in $e^+e^-$. These concern:

- one hadron inclusive production.
- $\gamma\gamma$ scatterings and the 'spin' of the photon.
- QCD jets.
- heavy quarkonia inclusive decays.
- $e^+e^- \to$ hadrons total cross-section.
- $Z \to$ hadrons inclusive decay
- $\tau \to \nu_\tau +$ hadrons semi-inclusive decays.

These processes are used as classical tests of perturbative QCD, where values of the running QCD coupling have been extracted. A pedagogical introduction to the physics of $e^+e^-$ can be found in, for example, the book of [276]. More modern QCD phenomenology in $e^+e^-$ can be found in different reviews and in the proceedings of the QCD-Montpellier series of conferences and many others.

# 22

# One hadron inclusive production

## 22.1 Process and fragmentation functions

We shall be concerned here with the one hadron production inclusive process:

$$e^+e^- \to \gamma^*(s) \to H + X\,, \tag{22.1}$$

which is the twin in the timelike region of the leptoproduction discussed previously on the target $\bar{H}$:

$$\gamma^*(-s) + \bar{H} \to X\,. \tag{22.2}$$

In the centre of mass of $\gamma^*$:

$$q = (\sqrt{s}, \mathbf{0})\,, \tag{22.3}$$

the kinematics of the process can be described by the momentum $p$ of the hadron $H$ and the fraction of beam energy $z\sqrt{s}/2$, where $0 \le z \le 1$:

$$p = (z\sqrt{s}/2, \mathbf{p})\,. \tag{22.4}$$

By formal analogy with leptoproduction, one can introduce the structure functions $\bar{F}^H_{1,2}(z, Q^2)$, such that the angular differential cross-section reads:

$$\frac{d\sigma}{dz\,d\cos\theta} = \frac{3}{4}\tilde{\sigma}^{(0)} z\left[2\bar{F}_1(z,s) + \frac{z}{2}\sin^2\theta\,\bar{F}_2(z,s)\right], \tag{22.5}$$

where in the naïve parton with spin $1/2$ quarks:

$$\tilde{\sigma}^{(0)} = \frac{4\pi\alpha^2}{3s}\,. \tag{22.6}$$

Alternatively, one can introduce the transverse and longitudinal structure functions:

$$\begin{aligned} \bar{F}_T(z,Q^2) &= 2\bar{F}_1(z,Q^2) \\ \bar{F}_L(z,Q^2) &= 2\bar{F}_1(z,Q^2) + z\bar{F}_2(z,Q^2)\,, \end{aligned} \tag{22.7}$$

with which one can express the differential cross-section:

$$\frac{d\sigma}{dz} = \tilde{\sigma}^{(0)} z\left[\bar{F}_T(z,s) + \frac{1}{2}\bar{F}_L(z,s)\right], \tag{22.8}$$

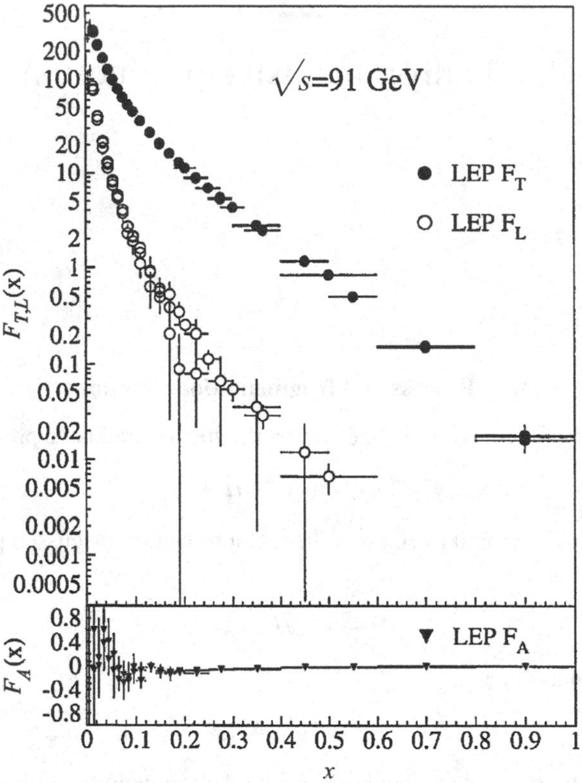


Fig. 22.1. Transverse $\bar{F}_T \equiv F_T$ and longitudinal $\bar{F}_L \equiv F_L$ fragmentation functions versus $x$ at $\sqrt{Q} =$ 91 GeV. $F_A$ is a parity-violating contribution coming from the interference between the vector and axial-vector contributions.

where in the naïve parton with spin 1/2 quarks:

$$\bar{F}_L(z, s) = 0$$
$$\bar{F}_T(z, s) = 3 \sum_i Q_i^2 \left[ D^H_{0q_i}(z) + D^H_{0\bar{q}_i}(z) \right] . \tag{22.9}$$

$D^H_{0q_i}$ is the *fragmentation* or decay function, which is the number of density of $H$ in the jet of parton $p$.

The data of these functions compiled by [16] are given in Fig. 22.1.

## 22.2 Inclusive density, correlations and hadron multiplicity

As in all inclusive processes, one can define the inclusive total cross-section:

$$\sigma_{\text{tot}} = \sum_H \sigma_H , \tag{22.10}$$

as the sum of all exclusive channels production of $H$ particles. The one particle inclusive cross-section density is:

$$\rho(p) = \frac{1}{\sigma_{\rm tot}} \frac{p^0 d\sigma}{d_3 p} , \tag{22.11}$$

for a particle of momentum $p$. Similarly for two particles 1 and 2, the inclusive density is defined as:

$$\rho(p_1, p_2) = \frac{1}{\sigma_{\rm tot}} \frac{p_1^0 p_2^0 d\sigma}{d^3 p_1 d^3 p_2} , \tag{22.12}$$

and one can define their correlations:

$$C(p_1, p_2) = \rho(p_1, p_2) - \rho(p_1)\rho(p_2) . \tag{22.13}$$

The *average hadron mutiplicity* for one inclusive particle are defined as:

$$\langle n_H \rangle = \int \frac{d^3 p}{p^0} \rho(p) , \tag{22.14}$$

and for two particles:

$$\langle n_1 n_2 \rangle = \int \frac{d^3 p_1 d^3 p_2}{p_1^0 p_2^0} \rho(p_1, p_2) . \tag{22.15}$$

In the same way, one can also define the third isospin components for the hadron $H$:

$$I_3 = \frac{1}{\sigma} \sum_H \int d^3 p I_3^H \frac{d\sigma_H}{d^3 p} . \tag{22.16}$$

## 22.3 Parton model and QCD description

To the leading order approximation, one has for each parton $p$:

$$\sum_H \int_0^1 dz I_3^H D_{0p}^H(z) = I_3^p$$
$$\sum_H \int_0^1 dz D_{0p}^H(z) = 1 , \tag{22.17}$$

where the first equation reflects the non-singlet charge conservation sum rule, while the second is the momentum conservation in the jet of parton $p$. The parton model description of a one hadron inclusive production is shown in Fig. 22.2, where the photon produces a hard parton $p$ with four momentum $k$ and with an energy fraction $y$ of the beam energy:

$$k = (y\sqrt{s}/2, \mathbf{k}) , \tag{22.18}$$

such that, independently of other partons, the produced parton fragments into hadrons. One expects that no hard interactions can take place between produced partons because their separation in rapidity is too wide at higher energies. In the limit of massless partons and

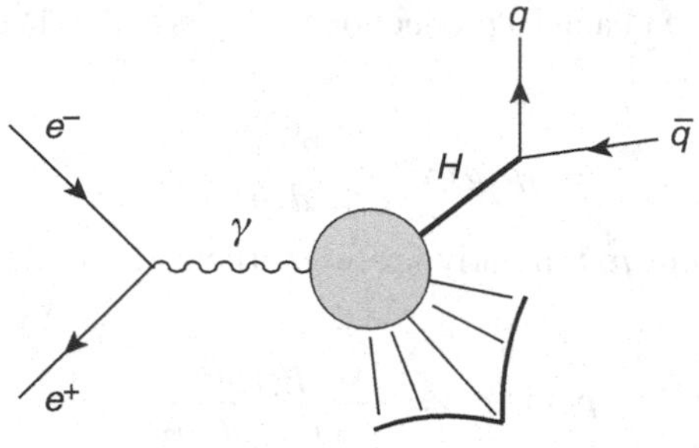


Fig. 22.2. $e^+e^- \to \gamma^* \to H+$ all in the parton model.

negligible intrinsic transverse momentum of the fragments, one has the relation between the hadron and parent parton:

$$p = (z/y)k\,. \tag{22.19}$$

The cross-section for producing a hadron $H$ with fraction $z$ of the beam energy is obtained as a convolution of the cross-section for producing a parton with energy fraction $y$ times the density of a hadrons $H$ in the parton $p$ with the fraction $z/y$ of the proton momentum:

$$z\bar{\mathcal{F}}_a^H(z,s) = \frac{1}{\tilde{\sigma}^{(0)}}\int_z^1 \frac{dy}{y}\sum_i \sigma_a^{\gamma^*\to p_i}(y,s)D_{0p_i}^H(z/y)\,, \tag{22.20}$$

where:

$$\bar{\mathcal{F}}_a^H = \left(2\bar{F}_1^H, -z\bar{F}_2^H\right). \tag{22.21}$$

In the case of the naïve parton model, the cross-section reads:

$$\sigma^{\gamma^*\to q} = 3\sum_i Q_i^2\delta(1-z)\,. \tag{22.22}$$

One can easily see that the inclusive quark production cross-section to order $\alpha_s$ is:

$$\begin{aligned}\sigma^{\gamma^*\to q}(x_q<1) &= \frac{C_F}{2}\left(\frac{\alpha_s}{\pi}\right)\int_{1-x_q}^1 dx_{\bar{q}}\frac{x_q^2+x_{\bar{q}}^2}{(1-x_q)(1-x_{\bar{q}})} = \frac{C_F}{2}\left(\frac{\alpha_s}{\pi}\right)\frac{1+x_q^2}{1-x_q}t+\cdots\\ &= \frac{1}{2}\left(\frac{\alpha_s}{\pi}\right)P_{qq}(x_q)t+\cdots\end{aligned} \tag{22.23}$$

where the log-divergence of the integral at $x_{\bar{q}} = 1$ has been re-interpreted as a factor $t \equiv (1/2)\ln(Q^2/\nu^2)$.

In the same way, the cross-section production of a gluon is:

$$\begin{aligned}\sigma^{\gamma^*\to g}(x_g) &= \frac{C_F}{2}\left(\frac{\alpha_s}{\pi}\right)\int_{1-x_g}^1 dx_q\frac{x_q^2+(2-x_{\bar{q}}-x_q)^2}{(1-x_q)(x_q+x_g-1)}\\ &= 2\frac{C_F}{2}\left(\frac{\alpha_s}{\pi}\right)\frac{1+(1-x_g)^2}{x_g}t+\cdots = 2C_F\left(\frac{\alpha_s}{\pi}\right)P_{gq}(x_g)t+\cdots\,.\end{aligned} \tag{22.24}$$

where the factor 2 indicates that the gluons can be emitted either by quark or by antiquarks. Therefore, one can deduce:

$$
\begin{aligned}
z\bar{\mathcal{F}}_a^H(z,s) = \frac{1}{\bar{\sigma}^{(0)}} 3Q^2 \int_z^1 \frac{dy}{y} \Bigg\{ \left[ \delta(1-y) + \left(\frac{\alpha_s}{\pi}\right) \left( t P_{qq}(y) + \frac{1}{2} \bar{f}_q^a(y) \right) \right] \\
\times \left[ D_{0q}^H(z/y) + D_{0\bar{q}}^H(z/y) \right] + \left(\frac{\alpha_s}{\pi}\right) \left( 2t P_{qq}(y) + \frac{1}{2} \bar{f}_g^a(y) \right) D_{0g}(z/y) \Bigg\} ,
\end{aligned}
\tag{22.25}
$$

where the sum over flavours is understood. As in the case of electroproduction for the structure functions, the fragmentation functions obey similar Altarelli–Parisi evolution equations. To order $\alpha_s$, it reads:

$$
\begin{aligned}
\frac{\partial}{\partial t} D_{q_i}(z,t) &= \left(\frac{\alpha_s}{\pi}\right) [P_{qq} \otimes D_{q_i} + P_{gq} \otimes D_g] \\
\frac{\partial}{\partial t} D_g(z,t) &= \left(\frac{\alpha_s}{\pi}\right) \left[ P_{qg} \otimes \sum_i (D_{q_i} + D_{\bar{q}_i}) + P_{gg} \otimes D_g \right] ,
\end{aligned}
\tag{22.26}
$$

where the only difference with electroproduction is the transposition $P_{qg} \leftrightarrow P_{gq}$. In terms of the singlet and non-singlet fragmentation functions:

$$
\begin{aligned}
D_{NS} &= D_{q_i} - D_{q_j} \\
D_S &= \sum_i (D_{q_i} + D_{\bar{q}_i}) ,
\end{aligned}
\tag{22.27}
$$

the evolution equations read:

$$
\begin{aligned}
\frac{\partial}{\partial t} D_S(z,t) &= \left(\frac{\alpha_s}{\pi}\right) [P_{qq} \otimes D_S + 2n_f P_{gq} \otimes D_g] \\
\frac{\partial}{\partial t} D_g(z,t) &= \left(\frac{\alpha_s}{\pi}\right) [P_{qg} \otimes D_S + P_{gg} \otimes D_g] .
\end{aligned}
\tag{22.28}
$$

Factorization of the perturbative (hard gluon radiation) and non-perturbative (hadronization) regime at a scale $\mu_f$ in the time-like region has been proved by many authors (see, however, the notion of fracture functions introduced in [268]). In this case, the inclusive cross-section of the process can be expressed as:

$$
\frac{d\sigma}{dx}(e^+e^- \to H + X) = \sum_i \int_z^1 \frac{dy}{y} C_i(y, \mu^2, \mu_f^2) D_i^H(z/y, \mu_f^2) , \tag{22.29}
$$

where $C_i$ are Wilson coefficients calculable perturbatively and correspond to the cross-section for the creation of a hard parton $i$ and a momentum fraction $y$ of the beam energy; $D_i$ is the fragmentation function (density of a hadron $H$ in a parton $i$ with fraction $z/y$ of the parton momentum). The coefficient functions vanish to lowest order for gluons and are known to higher orders. However, the previous factorization assumption may not work, and it can be more appropriate to introduce the notion of *fracture functions*. The phenomenology of fragmentation functions has been discussed in the literature using different Monte-Carlo

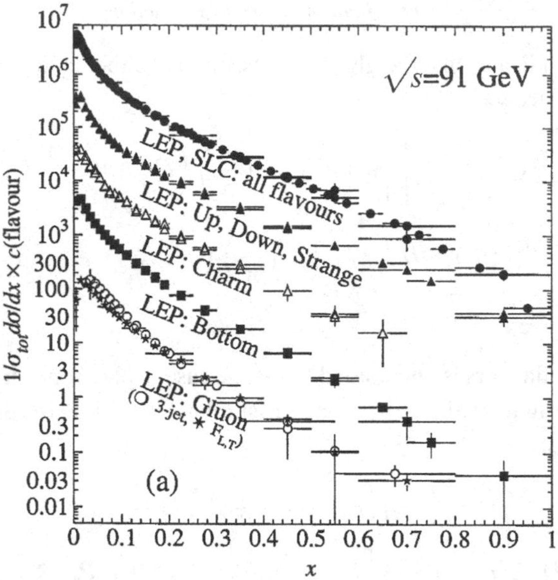


Fig. 22.3. Charged-particle and flavour-dependent $e^+e^-$ fragmentation functions versus $x$ at $\sqrt{Q} = 91$ GeV. The data are shown for the inclusive, light ($u, d, s$), $c$ and $b$ quarks, and the gluon. The distributions were scaled by c(flavour)= $10^n$, where $n$ ranges from $n = 0$ (gluon) and $n = 4$ (all flavours).

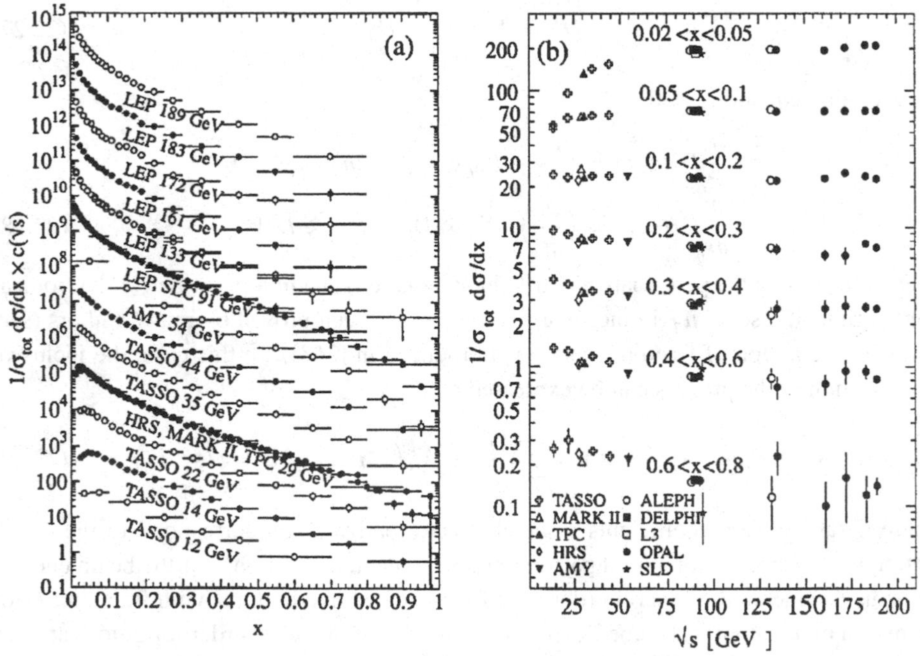


Fig. 22.4. All charged-particle $e^+e^-$ fragmentation functions (a) for different c.m. energies $\sqrt{s}$, versus $x$ and (b) for different $x$ versus $\sqrt{s}$. The data are shown for the inclusive, light ($u$, $d$, $s$), $c$ and $b$ quarks, and the gluon. For plotting (a), the distributions were scaled by $c(\sqrt{s}) = 10^i$, where $i$ ranges from $i = 0$ ($\sqrt{s} = 12$ GeV) to $i = 12$ ($\sqrt{s} = 189$ GeV).

simulation programs (see e.g. [277]). Detailed analyses of the charged hadron fragmentation functions have been performed by different LEP groups using data samples at PETRA, PEP and LEP energies from c.m. energy in the range from 14 to 92 GeV. We show the data in Figs. 22.3 and 22.4.

These analyses have been used for extracting $\alpha_s$ and some QCD power-like corrections. Combined ALEPH [278] and DELPHI [279] results give:

$$\alpha_s(M_{Z^0}) = 0.125^{+0.006}_{-0.007}\ (\text{exp})\ \pm 0.009\ (\text{theo})\ , \tag{22.30}$$

where the theoretical uncertainties are mainly due to the scale variations.

# 23

# $\gamma\gamma$ scatterings and the 'spin' of the photon

$\gamma\gamma$ collisions in $e^+e^-$ process are known to be an important source of hadrons as the cross-section $e^+e^- \to e^+e^- +$ hadrons increases logarithmically with the energy while the annihilation process $e^+e^- \to$ hadrons decreases like $1/s$. The dominant contribution comes from two on-shell photons emitted at small angles using the so-called equivalent photon approximation [280].

## 23.1 OPE and moment sum rules

The subprocess:

$$\gamma + \gamma \to \text{hadrons}\ , \tag{23.1}$$

depicted in Fig. 23.1, where one photon is far off-shell (large $Q^2$) and the other almost on shell (small $k^2$), can be considered as a deep-inelastic scattering on a photon target with the kinematic variables:

$$\nu \equiv p_2 \cdot q\ , \quad \tilde{\nu} = k \cdot q\ , \quad Q^2 \equiv -q^2\ , \quad x = Q^2/2\nu\ , \quad y = Q^2/2\tilde{\nu}\ , \tag{23.2}$$

and the DIS limit:

$$Q^2\ ,\ \nu\ ,\ \tilde{\nu} \to \infty\ , \qquad -k^2/Q^2 \ll 1\ . \tag{23.3}$$

One can also express these variables in terms of the energy $E_1'$ and scattering angle $\theta_1$ of the hard scattered electron, the energy $E$ of the incident electron, the scattered angle $\theta_2$ of the target electron and the invariant hadronic mass $W$. In this way, one has:

$$Q^2 = 4EE_1' \sin^2\frac{\theta_1}{2}\ , \qquad -k^2 \simeq EE_2'\theta_2^2\ , \tag{23.4}$$

and:

$$x = \frac{E_1' \sin^2(\theta_1/2)}{E - E_1' \cos^2(\theta_1/2)}\ , \qquad y = \frac{Q^2}{Q^2 + W^2}\ . \tag{23.5}$$

The formalism is very similar to the case of $ep$ scattering discussed previously where the gluon is now replaced by a photon. The derivation of the moment sum rules is based on the

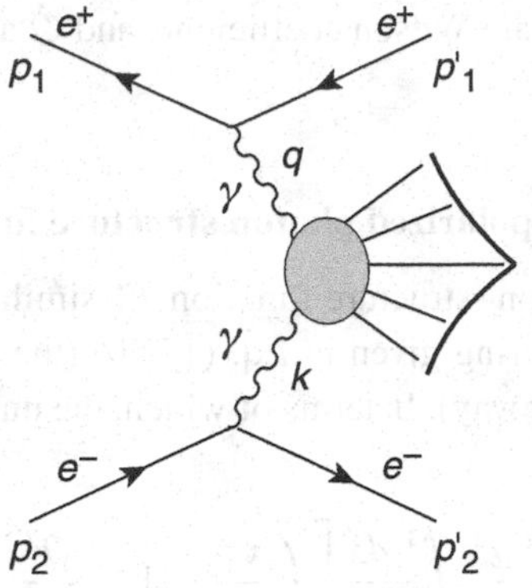


Fig. 23.1. $e^+e^- \to e^+e^- +$ hadrons process.

OPE of the T-product of two electromagnetic currents ($-q^2 \to \infty$):

$$\begin{aligned} iJ_\mu(q)J_\nu(-q) \sim & \sum_{n=2,\text{even}} \sum_h \mathcal{O}^{h,n}_{\mu_1...\mu_n}(0) \frac{2^n}{(-q^2)^{n+1}} \\ & \times \left[ C_1^{h,n}(-q^2) q^{\mu_1} \ldots q^{\mu_n} (g^{\mu\nu} q^2 - q^\mu q^\nu) \right. \\ & \left. + C_2^{h,n}(-q^2) q^{\mu_3} \ldots q^{\mu_n} (g^{\mu\mu_1} q^2 - q^\mu q^{\mu_1})(g^{\nu\mu_2} q^2 - q^\nu q^{\mu_2}) \right] \\ & + \sum_{n=1,\text{odd}} \sum_h \mathcal{O}^{h,n}_{3,\mu_1...\mu_n}(0) \frac{2^n}{(-q^2)^n} C_3^{h,n}(-q^2) q^{\mu_2} \ldots q^{\mu_n} i\epsilon^{\mu\nu\alpha\mu_1} q_\alpha \, . \end{aligned} \tag{23.6}$$

where $\mathcal{O}^{h,n}_{\mu_1...\mu_n}$ and $\mathcal{O}^{h,n}_{3,\mu_1...\mu_n}$ are set of even and odd parity, twist-2 operators (including photons) listed in Eqs. (15.60), (16.3) and (16.4). The sum $h$ runs over non-singlet, singlet, gluon and photon operators. Introducing this expression into the four-point function $J_\mu J_\nu A_\lambda A_\rho$, one obtains:

$$\begin{aligned} & \langle 0 | \mathcal{O}^{h,n}_{\mu_1...\mu_n} A_\lambda(k) A_\rho(-k) | 0 \rangle \\ & = \frac{1}{k^4} \hat{\mathcal{O}}^{h,n}(k^2) k_{\mu_3} \cdots k_{\mu_n} (k^2 g_{\lambda\mu_1} g_{\rho\mu_2} - k_\lambda k_{\mu_1} g_{\mu_2\rho} - k_\rho k_{\mu_2} g_{\mu_1\lambda} + k_{\mu_2} k_{\mu_1} g_{\lambda\rho}) \\ & \qquad (n \geq 2, \text{ even}) \, , \end{aligned}$$

and

$$\langle 0 | \mathcal{O}^{h,n}_{3,\mu_1...\mu_n} A_\lambda(k) A_\rho(-k) | 0 \rangle = \frac{1}{k^4} \hat{\mathcal{O}}_3^{h,n}(k^2) k_{\mu_2} \cdots k_{\mu_n} i\epsilon_{\lambda\rho\mu_1\alpha} k^\alpha \qquad (n \geq 2, \text{ odd}) \tag{23.7}$$

Therefore, the moments of the photon structure functions read:

$$\begin{aligned} \mathcal{M}_L^{(n)} & \equiv \int_0^1 dy \, y^{n-1} F_L^\gamma(y, Q^2, k^2) = \sum_h C_L^{h,n+1}(Q^2) \hat{\mathcal{O}}^{h,n+1}(k^2) \, , \\ \mathcal{M}_2^{(n)} & \equiv \int_0^1 dy \, y^{n-1} F_2^\gamma(y, Q^2, k^2) = \sum_h C_2^{h,n+1}(Q^2) \hat{\mathcal{O}}^{h,n+1}(k^2) \, , \\ \mathcal{M}_1^{(n)} & \equiv \int_0^1 dy \, y^{n-1} g_1^\gamma(y, Q^2, k^2) = \sum_h C_3^{h,n}(Q^2) \hat{\mathcal{O}}_3^{h,n}(k^2) \, , \end{aligned} \tag{23.8}$$

where $C_L \equiv C_2 - C_1$ and $C_3$ are Wilson coefficients and $\hat{\mathcal{O}}$ are reduced operators or form factors.

## 23.2 Unpolarized photon structure functions

One can introduce the 'electron' structure function $F_2^e$ similarly to the case of the proton structure function in $ep$ scattering given in Eq. (15.50) (the other structure functions $F_1^e$ and $F_L^e$ are defined in a similar way). In terms of which, the unpolarized cross-section reads [267]:

$$\sigma = 2\pi\alpha^2 \frac{1}{s} \int_0^\infty \frac{dQ^2}{Q^2} \int_0^1 \frac{dx}{x^2} \left[ \left( \frac{xs}{Q^2} - 1 + \frac{Q^2}{2xs} \right) F_2^e - \frac{Q^2}{2xs} F_L^e \right] . \tag{23.9}$$

The 'electron' structure function can be related to the conventional photon structure function $F_i^\gamma$ using the Altarelli–Parisi evolution equation [281]:

$$F_i^e(x, Q^2) = \frac{\alpha}{2\pi} \int_0^\infty \frac{dk^2}{k^2} \int_x^1 \frac{dy}{y} \frac{x}{y} P_{\gamma e} \left( \frac{x}{y} \right) F_i^\gamma(y, Q^2, k^2) , \tag{23.10}$$

where $i \equiv 2, L$ and:

$$P_{\gamma e} = \frac{1}{z}(1 - (1 - z)^2) , \tag{23.11}$$

is the splitting function. Using the previous evolution equation into the expression of the cross-section, one can derive the $x$-moments of the cross-section:

$$\int_0^1 dx\, x^n \frac{d^3\sigma}{dQ^2 dx dk^2} = \frac{\alpha^3}{Q^4 k^2} \int_0^1 z^n\, P_{\gamma e}(z) \int_0^1 y^{n-1} F_i^\gamma(y, Q^2, k^2) . \tag{23.12}$$

For $n = 1,3,\ldots$, the $z$ integral is finite, to which corresponds the moment sum rules of the structure functions:

$$\mathcal{M}_i^{(n)} = \int_0^1 y^{n-1} F_i^\gamma(y, Q^2, k^2) = \sum_h C_i^{h,n+1}(Q^2) \hat{\mathcal{O}}^{h,n+1}(k^2) . \tag{23.13}$$

One can notice that for $n = 0$ (total cross-section), the $z$ integration is logarithmically divergent. More explicitly, one can express the cross-section as:

$$\frac{d^2\sigma}{dQ^2 dy}(ee \to eeX) \simeq \frac{d^2\sigma}{dQ^2 dy}(e\gamma \to eeX)\Phi(E) . \tag{23.14}$$

where the photon flux factor is:

$$\Phi(E) \equiv \frac{\alpha}{2\pi} \int_0^1 z^n\, P_{\gamma e}(z) \int_0^\infty \frac{dk^2}{k^2} \approx 2\frac{\alpha}{\pi} \ln \frac{E}{E_{\min}} \ln \frac{E\theta_{2\max}}{m_e} \tag{23.15}$$

after taking the cuts:

$$-k^2_{\max} = E^2\theta^2_{2\max} , \quad -k^2_{\min} = m_e^2 , \quad E_\gamma \geq E_{\min} , \quad z_{\min} = E_{\min}/E . \tag{23.16}$$

Assuming that the photon structure function is crudely approximately constant, and using the differential cross-section:

$$\frac{d^2\sigma}{dQ^2dy}(e\gamma \to eX) = 2\pi\alpha^2 \frac{1}{Q^4}\left(1 - t + \frac{t^2}{2}\right)\frac{1}{y}F_2^\gamma(y, Q^2), \quad (23.17)$$

where $t = Q^2/ys$, one can deduce for $Q^2/xs \ll 1$:

$$\sigma = \alpha^3 \frac{4}{Q^2_{\min}} \ln\frac{E}{E_{\min}} \ln\frac{E\theta_{2\max}}{m_e} \ln\frac{2E^2}{\tilde{\nu}_{\min}} F_2^\gamma, \quad (23.18)$$

where further cuts $\tilde{\nu}_{\max} = s/2$ and $\tilde{\nu}_{\min}$ have been taken for $\tilde{\nu} \equiv k \cdot q$. One recovers the result of [280] obtained using the equivalent photon approximation.

The parton model contribution to $F_2^\gamma$ comes from the box diagram and dominates over the vector meson contribution. For large $Q^2$, the *parton model* expression reads:

$$F_2^\gamma(x, Q^2) = \left(N_c \sum_i Q_i^4\right) 8\alpha^2 x P_{q\gamma}(x) \ln Q^2 \quad (23.19)$$

where $P_{q\gamma}(x)$ is the splitting function encountered previously in the case of $ep$ scattering but the gluon is now replaced by a photon. Witten [282] pointed out that QCD corrections affect the parton model expression in Eq. (23.19), and his result has been extended to next order in [283]. The moments of the photon structure functions can be expressed in a similar way as in the case of gluons, where there is a mixing between the quark and photon operators. It reads [282]:

$$\int_0^1 dx\, x^{n-2} F_2^\gamma(x, Q^2) \sim \alpha\left[a_n \ln\frac{Q^2}{\Lambda^2} + \tilde{a}_n \ln\ln\frac{Q^2}{\Lambda^2} + b_n + \mathcal{O}\left(\frac{1}{\ln\frac{Q^2}{\Lambda^2}}\right)\right], \quad (23.20)$$

where the VDM contributions are included in the $1/\ln Q^2$ term. $a_n$, $\tilde{a}_n$ and $b_n$ have been calculated in perturbation theory by the previous authors: $a_n$ depends on the one-loop anomalous dimension and one-loop $\beta$-function; $\tilde{a}_n$ depends in addition on the two-loop $\beta$ function. In addition to the previous dependences, $b_n$ depends also on the two-loop anomalous dimensions and one-loop contribution to the Wilson coefficients, and is renormalization-scheme dependent. Extensive phenomenology of this process exists in the literature (see for example [49]).

## 23.3 Polarized process: the 'spin' of the photon

### *23.3.1 Moments and cross-section*

We will be interested here in the polarized $\gamma\gamma$ process, in which one can test the idea of the universality of the topological charge screening discussed in the previous chapter. An approach similar to the case of the unpolarized $\gamma\gamma$ process gives the results in terms of the

$g_1$ structure function as defined in Eq. (19.1):

$$g_1^e(x, Q^2) = \frac{\alpha}{2\pi} \int_0^\infty \frac{dk^2}{k^2} \int_x^1 \frac{dy}{y} \frac{x}{y} \Delta P_{\gamma e} \left(\frac{x}{y}\right) g_1^\gamma(y, Q^2, k^2) , \tag{23.21}$$

where:

$$\Delta P_{\gamma e} = 2 - z , \tag{23.22}$$

is the splitting function. The ratio of the polarized over the unpolarized cross-section is:

$$\frac{\Delta\sigma}{\sigma} = \frac{1}{2} \frac{\tilde{a}_n}{a_n} \frac{Q^2_{\min}}{s} \ln \frac{Q^2_{\max}}{Q^2_{\min}} \left[ 1 + \ln \frac{Q^2_{\max}}{\Lambda^2} \left( \ln \frac{Q^2_{\min}}{\Lambda^2} \right)^{-1} \right] , \tag{23.23}$$

where one can approximately take $a_n \simeq \tilde{a}_n$. The moment is given in Eq. (23.8). The Wilson coefficients have a $3 \times 3$ anomalous dimension matrix $\gamma_n^{hh}$ in the hadron sector and another $\gamma_n^{h\gamma}$ reflecting the mixing of the photon and singlet hadron operators. It explicitly reads:

$$\begin{aligned} \mathcal{M}_1^{(n)}(Q^2, k^2) = {} & C_3^{h,n}(1, \alpha_s(Q^2)) \mathcal{T} \exp - \int_0^t dt' \gamma_n^{hh}(\alpha_s(t')) \hat{O}_3^{h,n}(k^2, \alpha_s(\mu), \alpha) \\ & + \left[ C_3^{h,n}(1, \alpha_s(Q^2)) \mathcal{T} \exp - \int_0^t dt' \gamma_n^{h\gamma}(\alpha_s(t')) + C_3^{\gamma,n}(1, \alpha_s(Q^2)) \right] \\ & \times \hat{O}_3^{\gamma,n}(k^2, \alpha_s(\mu), \alpha) . \end{aligned} \tag{23.24}$$

To leading order, one obtains:

$$\mathcal{M}_1^{(n)}(Q^2, k^2) = \frac{\alpha}{4\pi} \tilde{a}_n \ln \frac{Q^2}{\Lambda^2} \qquad n \geq 3 \text{ odd.} \tag{23.25}$$

For $n = 1$, there is no operator $R_{\gamma,1}$, such that the lowest twist 2 operator is the axial current $J_{\mu 5}$. To, leading order, one can write

$$\begin{aligned} \mathcal{M}_1^{(n)}(Q^2, k^2) = {} & \sum_{a \neq 0} 2 \mathrm{Tr}\, (Q^2 \lambda^a) \hat{O}_3^{a,1}(k^2, \alpha_s, \alpha) \\ & + n_f^{-1} \mathrm{Tr}\, Q^2 \exp \left\{ - \int_0^t dt' \gamma(\alpha_s(t')) \hat{O}_3^{\gamma,n}(k^2, \alpha_s(\mu), \alpha) \right\} . \end{aligned} \tag{23.26}$$

### 23.3.2 The $g_1^\gamma$ sum rule and the axial anomaly

*The AVV vertex and chiral Ward identities*

Let us define the vertices:

$$\begin{aligned} \Gamma^a_{\mu\lambda\rho}(p, k_1, k_2) & \equiv \langle 0 | J^a_{\mu 5}(p) J_\lambda(k_1) J_\rho(k_2) | 0 \rangle \ : \ J^a_{\mu 5} = \bar{\psi} \gamma_\mu \gamma_5 \lambda^a \psi , \\ \Gamma^a_{5\lambda\rho}(p, k_1, k_2) & \equiv \langle 0 | J^a_5(p) J_\lambda(k_1) J_\rho(k_2) | 0 \rangle \ : \ \Phi^a_5 = i \bar{\psi} \gamma_5 \lambda^a \psi , \\ \Gamma_{Q\lambda\rho}(p, k_1, k_2) & \equiv \langle 0 | Q(p) J_\lambda(k_1) J_\rho(k_2) | 0 \rangle \ : \ Q = (\alpha_s / 8\pi) \mathrm{Tr}\, \tilde{G}_{\mu\nu} G^{\mu\nu} , \end{aligned} \tag{23.27}$$

where $\lambda^a$ are $SU(3)$ matrices. The conservation of the electromagnetic currents implies:

$$k_1^\lambda \Gamma^a_{\mu\lambda\rho}(p, k_1, k_2) = 0 = k_2^\rho \Gamma^a_{\mu\lambda\rho}(p, k_1, k_2) . \tag{23.28}$$

The vertices obey the anomalous chiral Ward identities.

$$\begin{aligned} ip^\mu \Gamma^a_{\mu\lambda\rho} - 2m\Gamma^a_{5\lambda\rho} + \frac{N_c}{4\pi^2} l_a \epsilon_{\lambda\rho\alpha\beta} k_1^\alpha k_2^\beta &= 0 \quad (a \neq 0) \\ ip^\mu \Gamma^0_{\mu\lambda\rho} - 2m\Gamma^0_{5\lambda\rho} - 2n_f \Gamma_{Q\lambda\rho} + \frac{N_c}{4\pi^2} l_a \epsilon_{\lambda\rho\alpha\beta} k_1^\alpha k_2^\beta &= 0 , \end{aligned} \tag{23.29}$$

where: $l_a = \mathrm{Tr}\, Q^2 \lambda^a$ is related to the quark charge $Q$ in units of $e$. Then, for $n_f = 3$, $l_0 = 2/3$, $l_3 = 1/6$, and $l_8 = 1/(6\sqrt{3})$. The AVV vertex function has the general Lorentz decomposition:

$$\begin{aligned} -i\langle 0|J^a_{\mu 5}(p) J_\lambda(k_1) J_\rho(k_2)|0\rangle = {} & A_1^a \epsilon_{\mu\lambda\rho\alpha} k_1^\alpha + A_2^a \epsilon_{\mu\lambda\rho\alpha} k_2^\alpha \\ & + A_3^a \epsilon_{\mu\lambda\alpha\beta} k_1^\alpha k_2^\beta k_{2\rho} + A_4^a \epsilon_{\mu\rho\alpha\beta} k_1^\alpha k_2^\beta k_{1\lambda} \\ & + A_5^a \epsilon_{\mu\lambda\alpha\beta} k_1^\alpha k_2^\beta k_{1\rho} + A_6^a \epsilon_{\mu\rho\alpha\beta} k_1^\alpha k_2^\beta k_{2\lambda} , \end{aligned} \tag{23.30}$$

where $A_i^a(p^2, k_1^2, k_2^2)$ are invariants. In the case of the $n = 1$ sum rule with $p = 0$, $k_1 = -k_2 = k$, one can deduce away from the chiral limit $m_\pi^2 \neq 0$:

$$\hat{\mathcal{O}}_3^{a,1}(k^2) = 4\pi\alpha \left(A_1^a - A_2^a\right)(0, k^2, k^2) . \tag{23.31}$$

Rewriting:

$$\left(A_1^a - A_2^a\right)(0, k^2, k^2) = \frac{N_c}{4\pi^2} \mathrm{Tr}(Q^2\lambda^a) F_a(k^2, \mu^2) , \tag{23.32}$$

one obtains for $n_f = 3$:

$$\begin{aligned} \mathcal{M}_1^{(1)}(Q^2, k^2) &\equiv \int_0^1 dy\, g_1^\gamma(y, Q^2, k^2) \\ &= \frac{\alpha}{18\pi}[3F_3(k^2) + F_8(k^2) + 8F_0(k^2, Q^2)] , \end{aligned} \tag{23.33}$$

where the singlet form factor $F_0$ has a non-trivial $Q^2$ dependence due the anomalous dimension $\gamma$.

*Non-singlet form factors*

Using the conservation of the electromagnetic current on the AVV (amputated) vertex in Eq. (23.28), one can derive:

$$A_1^a = A_3^a k_2^2 - A_5^a \frac{1}{2}\left(k_1^2 + k_2^2 - p^2\right) . \tag{23.34}$$

Assuming a smooth behaviour of the form factors in the limit $p \to 0$ and $k_1 \to -k_2 \to \pm k$, one obtains:

$$A_1^a(0, k^2, k^2) = k^2\left(A_3^a - A_5^a\right)(0, k^2, k^2)\, \mathcal{O}(k^2) , \tag{23.35}$$

by assuming in addition that there is no $1/k^2$ pole in the form factors $A_i$ $(i \equiv 3, 5)$ (and $i \equiv 4, 6$ if one assumes that a similar result holds for $A_2^a$). Defining the form factor $\mathcal{F}_a$:

$$2m\Gamma^a_{5\lambda\rho} = \mathcal{F}_a \epsilon_{\lambda\rho\alpha\beta} k_1^\alpha k_2^\beta \ , \tag{23.36}$$

and considering the previous Ward identities, one obtains:

$$\left(A_1^a - A_2^a\right)(0, k^2, k^2) = -\mathcal{F}_a(k^2) + \frac{N_c}{4\pi^2} l_a \ . \tag{23.37}$$

Identifying with the result in Eq. (23.32), one obtains:

$$F_a(k^2) = 1 - \frac{\mathcal{F}_a(k^2)}{\mathcal{F}_a(0)} \ . \tag{23.38}$$

Expressing the PVV vertex in terms of the pion field and coupling to $\gamma\gamma$, one obtains the leading-order relation:

$$\mathcal{F}_a(k^2) = \frac{1}{8\pi\alpha} f_\pi g_{\pi_a\gamma^*\gamma^*}(k^2) \ : \quad f_\pi = 92.4 \text{ MeV} \ , \tag{23.39}$$

which gives:

$$F_a(k^2) = 1 - \frac{g_{\pi_a\gamma^*\gamma^*}(k^2)}{g_{\pi_a\gamma\gamma}(0)} \ . \tag{23.40}$$

Using an OPE of the PVV vertex for large $k^2$, one obtains:

$$\langle \pi_a | J_\lambda(k) J_\rho(-k) | 0 \rangle \sim 2\epsilon_{\lambda\rho\alpha\mu} \frac{k^\alpha}{k^2} C_3^{a,1}(k^2) \langle \pi_a | J_5^{a\mu}(0) | 0 \rangle \ . \tag{23.41}$$

Therefore, one can deduce:

$$F_a(k^2) = 1 - \frac{16\pi^2}{N_c} \frac{f_\pi^2}{(-k^2)} + \cdots \tag{23.42}$$

Combining this result with the one in Eq. (23.40), one can deduce that the form factor interpolates smoothly from 0 to 1 when $k^2$ varies from zero to infinity. We can parametrize this behaviour as:

$$F_a(k^2) = \frac{-k^2}{-k^2 + M_a^2} \ , \tag{23.43}$$

where:

$$M_a^2 \simeq \left(\frac{16\pi^2}{N_c}\right) f_\pi^2 \simeq 0.67^2 \text{ GeV}^2 \ , \tag{23.44}$$

is a characteristic hadronic mass scale indicative of the non-perturbative realization of the AVV vertex in the spontaneously broken chiral symmetry phase of QCD. It can be related to the quark vacuum condensate in the QCD spectral function analysis of the vertex function.

### Singlet form factors

The situation is much more involved here due to the presence of the $U(1)$ anomaly [255,256]. Defining the form factor as:

$$\Gamma_{Q\lambda\rho} = \frac{1}{n_f}\mathcal{F}_0 \epsilon_{\lambda\rho\alpha\beta} k_1^\alpha k_2^\beta \,, \tag{23.45}$$

and using the fact that $A_1^0 - A_2^0 = \mathcal{O}(k^2)$, one can write:

$$F_0(k^2) = 1 - \frac{\mathcal{F}_0(k^2, \mu^2)}{\mathcal{F}_0(0)} \,. \tag{23.46}$$

One can introduce the OZI Nambu–Goldstone boson $\eta_0$ associated with the singlet pseudoscalar field $\Phi_5^0$ defined in Eq. (23.27) and its decay constant $f_{\eta_0}$, with $\eta^0 = 2\langle\Phi\rangle^{-1} f_{\eta_0}\Phi_5^0$. The latter being related to the first moment of the topological susceptibility:

$$f_{\eta_0} = 2n_f\sqrt{\chi'(0)} \,, \tag{23.47}$$

with:

$$\chi(p^2) = i\int d^4x\, e^{ipx}\langle 0|\mathcal{T} Q(x) Q^\dagger(0)|0\rangle \,. \tag{23.48}$$

An approach similar to the case of the non-singlet current gives:

$$F_0(k^2, \mu^2) = 1 - \frac{g_{\eta_0\gamma^*\gamma^*}(k^2)}{g_{\eta_0\gamma\gamma}(0)} \,. \tag{23.49}$$

However, the situation is more complicated as $\eta_0$ is not a physical state, while $g_{\eta_0\gamma^*\gamma^*}$ and $g_{\eta_0\gamma\gamma}$ are not RG invariants. Therefore, we approximate the $\eta_0$ by the $\eta'$ and replace the difference of the couplings by their OZI limit $g_{\eta'\gamma^*\gamma^*} - g_{\eta'\gamma\gamma}$ which is RG invariant. We also replace the anomaly coefficient using the relation:

$$f_{\eta'} g_{\eta'\gamma\gamma} = 2N_c \frac{2}{3}\frac{\alpha}{\pi} \,, \tag{23.50}$$

where:

$$f_{\eta'} = \frac{1}{M'_\eta} 2\langle\Phi\rangle \left[ i\int d^4x\, e^{ipx}\langle 0|\mathcal{T}\Phi_5^0(x)\Phi_5^0(0)|0\rangle \right]^{-1/2} \,. \tag{23.51}$$

The form factor reads [281]:

$$F_0(k^2, \mu^2) \sim \frac{f_{\eta_0}(0, \mu^2)}{f_{\eta'}} \left[ 1 - \frac{8\pi^2}{N_c n_f} f_{\eta'} f_{\eta_0}(0, \mu^2) \times \left( \mathcal{T}\exp - \int_0^t dt'\, \gamma(\alpha_s(t')) \right) \frac{1}{(-k^2)} + \cdots \right] . \tag{23.52}$$

The associated scale for interpolating the singlet form factor from 0 to 1 is:

$$M_0^2 \simeq \frac{8\pi^2}{N_c n_f} f_{\eta'} f_{\eta_0}(0, k^2) \,. \tag{23.53}$$

The explanation of the proton spin proposed in the previous section requires a small value of $f_{\eta_0}(0, Q^2 = 11\ \mathrm{GeV}^2)$ compared to its OZI value of $\sqrt{6} f_\pi$.

*Implications for the moment sum rules*

Introducing the previous behaviour of the form factors, one can deduce from Eq. (23.33):

$$\int_0^1 dy\, g_1^\gamma(y, Q^2, k^2 = 0) = 0 \,, \tag{23.54}$$

and for $M_\rho^2 \ll -k^2 \ll Q^2$:

$$\int_0^1 dy\, g_1^\gamma(y, Q^2, k^2) \simeq N_c \frac{\alpha}{\pi} Q_f^4 \left(1 - c + c\frac{f_{\eta_0}(0, Q^2)}{f_{\eta'}}\right) , \tag{23.55}$$

with:

$$c = \frac{1}{n_f}\left(\sum_f Q_f^2\right)^2 \Big/ \sum_f Q_f^4 \,, \tag{23.56}$$

where the deviation from the naïve leading-order value comes from the effect of the $U(1)$ anomaly. Related phenomenology of the $U(1)$ anomaly but on the $\eta'(\eta) \rightarrow \gamma\gamma$ decays is reviewed in [284].

# 24

# QCD jets

## 24.1 Introduction

We shall focus our discussions for jet productions in $e^+e^-$. More complete discussions can, for example, be found in [52] and the different contributions of the LEP groups at the QCD-Montpellier conference series. The aim is to study final states which do not depend on the identification of particular hadronic channels. High-energy $e^+e^-$ experiments offer a such opportunuity, although many aspects of the analysis can be extended to other processes. We shall consider the parton process:

$$e^+e^- \to \gamma^* \to \bar{q}q \,, \tag{24.1}$$

if one assumes that quarks are produced as free particles. In that case, one obtains, the angular distribution:

$$\frac{d\sigma^{(0)}}{d\cos\theta} = \frac{\pi\alpha^2 Q_q^2}{2s}(1+\cos\theta) \,, \tag{24.2}$$

which after integration gives the parton model total-cross section:

$$\sigma^{(0)} = \frac{4\pi\alpha^2 Q_q^2}{3s} \,. \tag{24.3}$$

## 24.2 IR divergences: Bloch–Nordsieck and KLN theorems

However, the process in Eq. (24.1) does not exist in practice as the production of quarks is always accompained by the emission of gluons. Formally, this feature is signalled by the appearance of the IR divergences when one evaluates the QCD radiative corrections given by diagrams in Fig. 24.1.

The IR divergence from the vertex correction is cancelled by the one from soft gluon radiation, which renders the total cross-section finite:

$$\sigma^{(1)} = \sigma^{(0)}\left[1 + \frac{3C_F}{4}\left(\frac{\alpha_s}{\pi}\right)\right] \,, \tag{24.4}$$

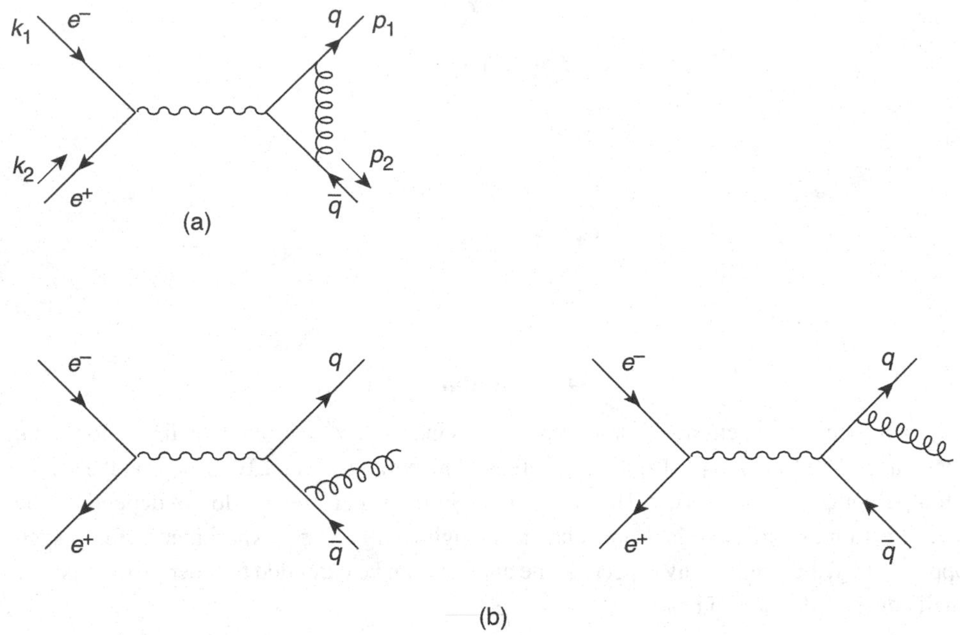


Fig. 24.1. $\alpha_s$ corrections to $e^+e^- \rightarrow \gamma^* \rightarrow \bar{q}q$. (a) vertex corrections. (b) gluon radiation.

which is the well-known *inclusive cross-section*. Therefore, only the sum of the cross-sections:

$$e^+e^- \rightarrow \gamma^* \rightarrow \bar{q}q + \bar{q}qg + \cdots \tag{24.5}$$

is expected to be finite, and this is the quantity that one measures. This cancellation of IR divergence is a general property already encountered in QED for soft and collinear photons and is known as the *Bloch–Nordsieck theorem* [285]. It states that soft divergence is absent for a totally inclusive cross-section. However, new features appear in QCD at higher orders due to the self-gluon interactions, or if one works in a covariant gauge, due to the emission of soft ghosts and the appearance of ghost loops. The theorem has been generalized to QCD by the *Kinoshita–Lee–Nauenberg (KLN) theorem* [286]. The KLN theorem states that in a theory with massless fields, transition rates are free of IR soft and collinear (mass singularities) divergences if the summation over the initial and final *degenerate states* (a massless quark accompanied by an arbitrary number of gluons cannot be distinguished from a single quark) are carried out. That is, for a single-quark state of mass $m$, we should add all final states that in the limit $m \rightarrow 0$ have the same mass, including massless gluons and quarks. In order to quantify this feature, one can mimic the IR problem in QED, where, under certain conditions, the processes: $e^+e^- \rightarrow \bar{q}q$ and $e^+e^- \rightarrow \bar{q}qg, \ldots$ are indistinguishable. This is realized if the gluon energy $k_0$ is below a certain detection threshold or if the angle formed by its three-momentum $\mathbf{k}$ with the quark momenta $\mathbf{p_i}$ is smaller than

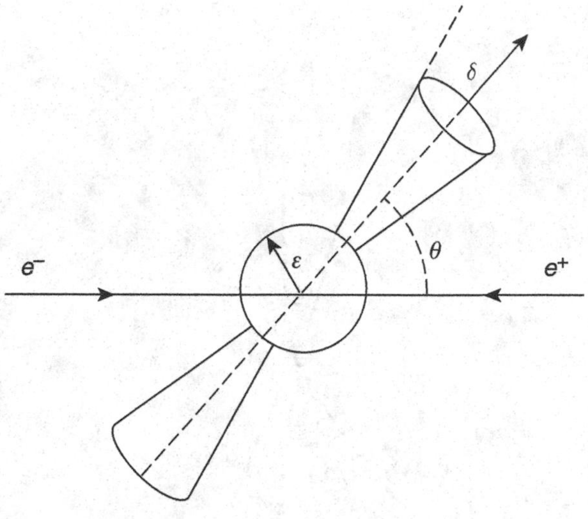


Fig. 24.2. Two 'fat' jets with possible extra soft partons (inside the sphere).

the detector resolution [287]:

$$k_0, p_{i0} \leq \epsilon\sqrt{s}$$
$$\angle(\mathbf{k}, \mathbf{p_i}), \ \angle(\mathbf{p_1}, \mathbf{p_2}) \leq \delta , \tag{24.6}$$

where $\epsilon$ and $\delta$ which characterizes the detection efficiency are defined in Fig. 24.2.

The previous conditions can be generalized for more produced numbers of quarks and gluons. If one considers the massless quark propagator in Fig. 24.2:

$$\frac{i}{\hat{p}_1 + \hat{k}} \simeq i\frac{\hat{p}_1 + \hat{k}}{2p_1 \cdot k} , \tag{24.7}$$

which indicates that for soft partons $k_0$, $p_{i0} \simeq 0$ or for collinear momenta $\mathbf{p_1} \parallel \mathbf{k}$, the denominator vanishes (*collinear mass singularities*). The conditions in Eq. (24.6) guarantee that this does not happen because:

$$p_1 \cdot k \geq \frac{1}{2}s(\epsilon\delta)^2 , \tag{24.8}$$

which after integration over final particle momenta, corresponds for the cross-section, to the singularity:

$$\sigma^{(1)}_{\text{sing}} \sim \alpha_s \ln \epsilon \ln \delta . \tag{24.9}$$

This result informs us that at higher energies this contribution becomes more and more negligible as $\alpha_s$ is smaller, such that the parton model description of the cross-section will be much better and the events are more jet-like. However, the complete analysis is more complicated because we see jets of hadrons (*hadronization*) not quark jets. This leads to the introduction of *fragmentation functions* discussed previously.

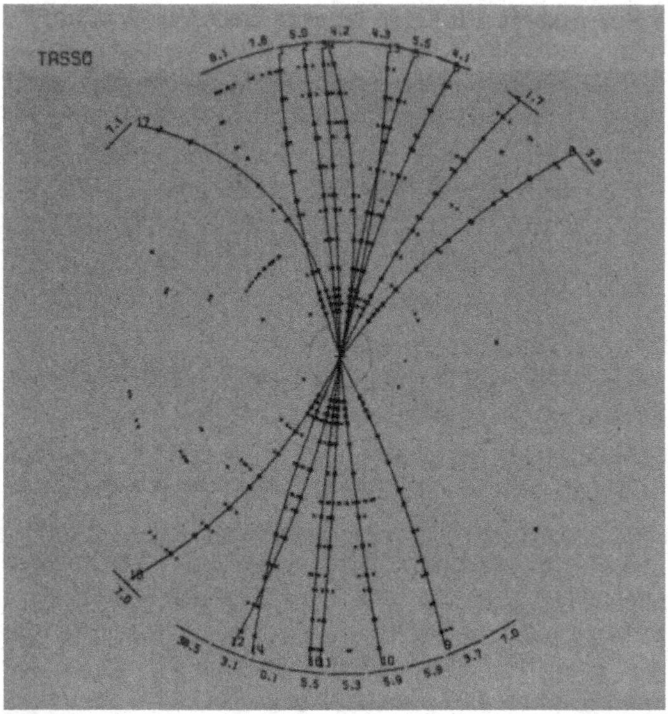


Fig. 24.3. Two-jet events seen in $e^+e^-$ at PETRA (1979).

## 24.3 Two-jet events

It is instructive to compare Fig. 24.2 with the two jet events seen inside the detector (Fig. 24.3).

Using the Sterman–Weinberg parametrization, one can explicitly show the different contributions from Fig. (24.2), where each individual contributions are IR divergent, which we regulate by attributing a mass $\lambda$ to the gluons. The contribution of the diagrams in Fig. (24.1b) for the production of a real gluon can be divided into three parts:

- A contribution of a $\bar{q}q$ jet plus a jet due to *a hard gluon inside the cone* with an energy greater than $\epsilon\sqrt{s}$ from Fig. (24.2b), which is:

$$\sigma(\text{hard})^{(b)} = \sigma^{(0)} C_F \left(\frac{\alpha_s}{\pi}\right) \left[-\ln\left(\frac{\delta\sqrt{s}}{\lambda}\right)(3 + 4\ln 2\epsilon) - 2\ln^2 2\epsilon + \frac{17}{4} - \frac{\pi^2}{3} + \mathcal{O}(\epsilon, \delta)\right]. \tag{24.10}$$

- A contribution due to two jets from $\bar{q}q$ and the one due to *a soft gluon inside the cone* with an energy smaller than $\epsilon\sqrt{s}$, which is:

$$\sigma(\text{soft})^{(b)} = \sigma^{(0)} C_F \left(\frac{\alpha_s}{\pi}\right) \left[2\ln^2\left(\frac{2\epsilon\sqrt{s}}{\lambda}\right) - \frac{\pi^2}{6} + \mathcal{O}(\epsilon, \delta)\right]. \tag{24.11}$$

- A contribution from the interference of the lowest order diagram with the vertex and self-energy corrections, which is:

$$\sigma(\text{interf})^{(c)} = \sigma^{(0)}\left[1 + C_F\left(\frac{\alpha_s}{\pi}\right)\right]\left[-2\ln^2\left(\frac{\sqrt{s}}{\lambda}\right) + 3\ln\left(\frac{\sqrt{s}}{\lambda}\right) - \frac{7}{4} + \frac{\pi^2}{6} + \mathcal{O}(\epsilon, \delta)\right]. \tag{24.12}$$

The sum of the different contributions, where all but a fraction of the total energy is emitted inside these cones, are IR finite (cancellation of soft and collinear singularities) and reads:

$$\sigma = \sigma^{(0)}\left[1 - C_F\left(\frac{\alpha_s}{\pi}\right)\left[(3 + 4\ln 2\epsilon)\ln\delta - \frac{5}{2} + \frac{\pi^2}{3} + \mathcal{O}(\epsilon, \delta)\right]\right]. \tag{24.13}$$

Therefore, the fraction of events which have all but a fraction $\epsilon$ of their energy in some pairs of cones with half-angle $\delta$ is:

$$R(2\text{jet}) = \frac{\sigma}{\sigma^{(1)}} = 1 - C_F\left(\frac{\alpha_s}{\pi}\right)\left[(3 + 4\ln 2\epsilon)\ln\delta - \frac{7}{4} + \frac{\pi^2}{3} + \mathcal{O}(\epsilon, \delta)\right]\Bigg], \tag{24.14}$$

where $\sigma^{(1)}$ is the inclusive total cross-section to order $\alpha_s$. This expression is valid if $\epsilon$ and $\delta$ are not too small such that perturbation theory is valid [288]. Alternatively, one can take another parametrization (e.g. cylindrical jet picture). Noting that the previous inclusive total cross-section in Eq. (24.4) includes the two- and three-jet events, the two-jet events can be obtained as:

$$\sigma(2\text{jet}) = \sigma^{(1)} - \sigma(\not{2}\text{jet}). \tag{24.15}$$

where $\sigma(\not{2}\text{jet})$ does not contain two-jet events. The cross-section for the process:

$$e^+e^- \to \gamma^* \to \bar{q}qg \tag{24.16}$$

can be obtained from Fig. (24.2). Defining:

$$s = (p_1 + p_2 + k)^2 \quad \text{and} \quad x_i = 2p_{0i}/\sqrt{s}, \tag{24.17}$$

one obtains:

$$\frac{1}{\sigma^{(0)}}\frac{d^2\sigma}{dx_1 dx_2} = \frac{C_F}{2}\left(\frac{\alpha_s}{\pi}\right)\frac{x_1^2 + x_2^2}{(1 - x_1)(1 - x_2)}, \tag{24.18}$$

with:

$$x_1 + x_2 \geq 1, \qquad 0 \leq x_i \leq 1. \tag{24.19}$$

Using the geometry of the $\bar{q}qg$ produced state given in Fig. 24.4, this process will not be considered as a two-jet event if the angle $\theta$ between the quark momenta is smaller than $\pi - \eta_0$, where $\eta_0$ is the resolution of the detector. Therefore, the not two-jet (three-jet)

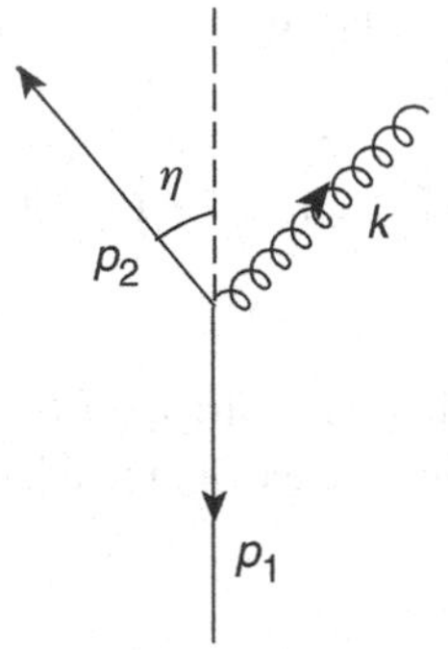


Fig. 24.4. Configuration of $\bar{q}qg$ produced state.

cross-section will be:

$$\sigma(\not{2}\text{jet}) = \int^{sup} \int dx_1 dx_2 \frac{d^2\sigma}{dx_1 dx_2} , \qquad (24.20)$$

where *sup* corresponds to the domain:

$$x_1 + x_2 = 1 + \frac{x_1 x_2}{2}(1 + \cos \eta_0) . \qquad (24.21)$$

In the limit $\eta_0 = 0$, which corresponds to a much better experimental precision, one obtains:

$$\sigma(\not{2}\text{jet}) = \sigma^{(0)} \frac{C_F}{2} \left(\frac{\alpha_s}{\pi}\right) \left[ \ln^2 \frac{4}{\eta_0^2} - 3 \ln \frac{4}{\eta_0^2} + \frac{\pi^2}{3} + \frac{7}{2} \right] , \qquad (24.22)$$

from which one can deduce the observed two-jet total cross-section:

$$\sigma(2\text{jet}) = \sigma^{(0)} \left\{ 1 - \frac{C_F}{2} \left(\frac{\alpha_s}{\pi}\right) \left[ \ln^2 \frac{4}{\eta_0^2} - 3 \ln \frac{4}{\eta_0^2} + \frac{\pi^2}{3} + 2 \right] \right\} , \qquad (24.23)$$

which depends on the resolution $\eta_0$. This result differs from that of Sterman–Weinberg, which shows the dependence of the cross-section on the parametrization of the two-jet event.

## 24.4 Three-jet events

Experimentally, three-jet events have been observed in $e^+e^-$ experiments. We show these events in Fig. 24.5.

It is interpreted in QCD as coming from quark-anti-quark plus a gluon emitted from one of the quark.

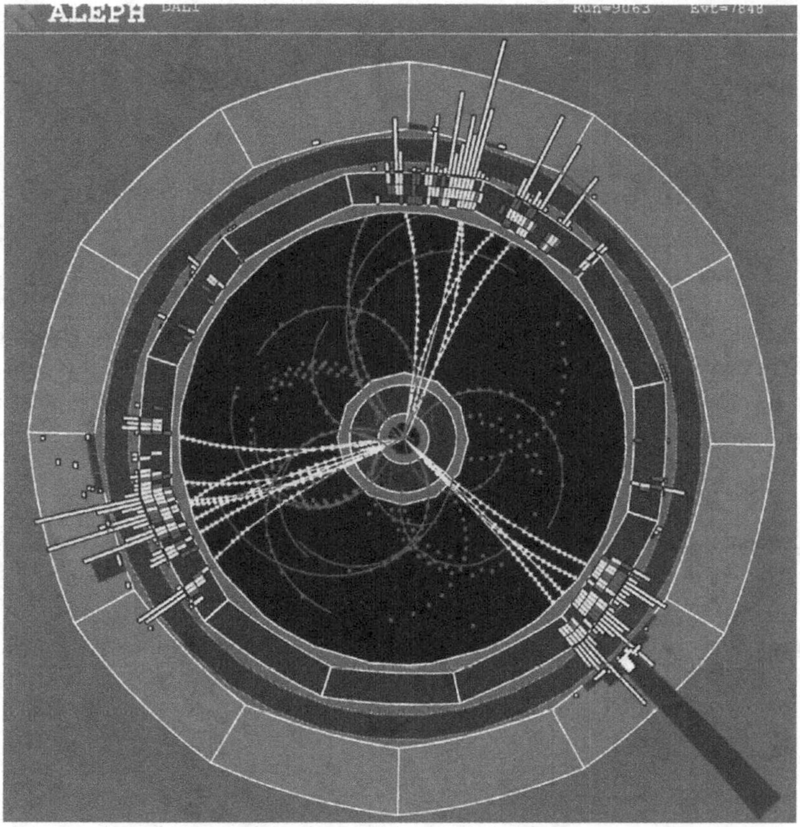


Fig. 24.5. Three-jet events seen at LEP.

The three-jet cross-section has been already evaluated in Eq. (24.20). For studying these events, it is convenient to introduce the kinematic variables:

$$x_1 = 2p_{01}/\sqrt{s}\,, \qquad x_2 = 2p_{02}/\sqrt{s}\,, \qquad x_3 = 2k_0/\sqrt{s} = 2 - x_1 - x_2\,. \quad (24.24)$$

### 24.4.1 Thrust as a jet observable

Different observables have been proposed in the literature for a qualitative description of final state topology. They are, for example, useful to define the axis or the plane of the event and therefore longitudinal and transverse momentum distributions. These variables should be linear in energy and/or momentum in order to meet the necessary condition of cancellation of IR divergence. *Thrust* and *spherocity* are two alternative IR safe quantities for a parametrization over the continuous range from the topology of a sphere to that of an ideal collinear two-jet event. *Spherocity* is defined as [290]:

$$S = \left(\frac{4}{\pi}\right)^2 \min \left(\frac{\sum_i |\mathbf{p}_{i\perp}|}{\sum_i |\mathbf{p}_i|}\right)^2 , \quad (24.25)$$

where $\mathbf{p_{i\perp}}$ is the transverse momentum with respect to the minimum direction (spherocity axis). It has the extremal values:

$$0 \leq S \leq 1 \qquad S = \begin{matrix} 1: & \text{sphere} \\ 0: & \text{line} \end{matrix} \tag{24.26}$$

The *thrust* variable is defined as [289]:

$$T = 2\text{max}\frac{\sum |\mathbf{p_{i\|}}|^2}{\sqrt{s}}, \tag{24.27}$$

where the sum runs over all particles in a hemisphere; $\mathbf{p_{i\|}}$ are the components of particle momenta along the jet axis contained in the hemisphere. The plane of the hemisphere is chosen to be perpendicular to the jet axis. The latter is found by requiring $T$ to be maximal. This can be obtained by choosing an arbitrary jet axis characterized by the polar angles $(\theta, \phi)$, and evaluates $T(\theta, \phi)$ as a function of these angles. In terms of partonic variables:

$$T = \max\{x_1, x_2, x_3\}, \tag{24.28}$$

and, in general, it has the boundaries:

$$1/2 \leq T \leq 1. \tag{24.29}$$

Integrating the cross-section in Eq. (24.18) at fixed $T$, one finds the differential cross-section:

$$(1-T)\frac{d\sigma}{dT} = \sigma^{(0)}\frac{C_F}{2}\left(\frac{\alpha_s}{\pi}\right)\left[\left(6T - 6 + \frac{4}{T}\right)\ln\left(\frac{2T-1}{1-T}\right) + 3\left(3T^2 - 8T + 4\right)\right], \tag{24.30}$$

The average value is [291]:

$$\begin{aligned}\langle 1 - T\rangle_{\bar{q}qg} &= \frac{C_F}{2}\left(\frac{\alpha_s}{\pi}\right)\left[-\frac{3}{4}\ln 3 - \frac{1}{18} + 4\int_{2/3}^{1}\frac{dT}{T}\ln\left(\frac{2T-1}{1-T}\right)\right] \\ &\simeq 1.05\left(\frac{\alpha_s}{\pi}\right).\end{aligned} \tag{24.31}$$

Another alternative definition of thrust, mostly used at LEP, is:

$$T = \max_{\mathbf{n}} \frac{\sum_i |\mathbf{p_i} \cdot \mathbf{n}|}{\sum_i |\mathbf{p_i}|}, \tag{24.32}$$

where $\mathbf{p}$ is the momenta of particles produced, while $\mathbf{n}$ is a unit vector. The thrust axis $\mathbf{n_T}$ is the direction at which the maximum is attained.

### 24.4.2 Other event-shape variables

Below we shall list some other event-shape parameters useful in the jet analysis. They are IR safe quantities, i.e. free from IR divergences, which are insensitive to the emission of soft or collinear partons at the logarithmic level.

- **Heavy (resp. light) jet mass** A plane through the origin and orthogonal to the thrust axis $\mathbf{n_T}$ divides the event into two hemispheres $H_1$ and $H_2$, from which one obtains the corresponding normalized hemisphere invariant masses:

$$M_i = \frac{1}{s}\left(\sum_k p_k\right)^2, \qquad i = 1, 2\,, \tag{24.33}$$

where $s \equiv E_{vis}$ is the square of the *total visible energy* of the events. The heavier (resp. lighter) of the two hemispheres is called heavy (resp. light) jet mass $M_h$ (resp. $M_l$).
- **The jet broadening** corresponding to the definition in Eq. (24.32), is defined as:

$$B_k = \left(\sum_{i\in H_k} |\mathbf{p_i} \times \mathbf{n_T}|\right) \Big/ \left(2\sum_i |\mathbf{p_i}|\right). \tag{24.34}$$

- **The total jet broadening** is defined as:

$$B_T = B_1 + B_2\,. \tag{24.35}$$

- **The wide jet broadening** is defined as:

$$B_W = \max(B_1, B_2)\,. \tag{24.36}$$

- The **C parameter** is defined as:

$$C = 3(\lambda_1\lambda_2 + \lambda_2\lambda_3 + \lambda_3\lambda_1)\,, \tag{24.37}$$

where $\lambda_i$ is the eigenvalue of the quantity:

$$\left(\sum_i (p_i^a p_i^b)/|\mathbf{p_i}|\right) \Big/ \sum_i |\mathbf{p_i}|\,. \tag{24.38}$$

### 24.4.3 Event-shape distributions

One can generally study any particle distributions in terms of the shape parameters:

$$\mathcal{H}_l = \sum_{k,k'} \frac{|\mathbf{p}(k)||\mathbf{p}(k')|}{s} P_l(\cos\phi_{kk'})\,, \tag{24.39}$$

where $s$ is the square of the $e^+e^-$ c.m. energy and $P_l$ is the Legendre polynomials of the angle $\phi_{kk'}$ between two final momenta; $p(k)$ is the final momenta of the particle $k$. In the massless limit, the energy–momentum conservation requires:

$$\mathcal{H}_0 = 1 \quad \text{and} \quad \mathcal{H}_1 = 0\,, \tag{24.40}$$

while collinear jets give:

$$\begin{aligned} \mathcal{H}_l &= 1 \quad (l \text{ even}) \\ &= 0 \quad (l \text{ odd})\,. \end{aligned} \tag{24.41}$$

In general:

$$0 \le \mathcal{H}_l \le \mathcal{H}_0 . \tag{24.42}$$

For a continuous distribution of momenta, $\mathcal{H}_l$ corresponds to the multiple momenta:

$$\mathcal{H}_l = \frac{4\pi}{2l+1} \sum_m |A_l^m|^2 , \tag{24.43}$$

where:

$$A_l^m = \int \rho(\Omega)\, Y_l^m(\Omega) d\Omega . \tag{24.44}$$

$\Omega$ is the solid angle and:

$$\rho(\Omega) \sim \frac{|\mathbf{p}(k)|}{\sqrt{s}} . \tag{24.45}$$

### 24.4.4 Energy-energy correlation

If $\omega$ is an angle between 0 and $\pi$, the energy-energy correlation is defined as [293]:

$$\frac{1}{\sigma^{(0)}} \frac{d\Sigma}{d\cos\omega} = \frac{2}{N_S \Delta\omega \, \sin\omega} \sum_{A=1}^{N} \sum_{\text{pairs in } \Delta\omega} E_{A_a} E_{A_b} , \tag{24.46}$$

where $A$ labels the events. In each event, $E_{A_a}$ and $E_{A_b}$ are the energies of two particles separated by an angle $\omega \pm \frac{1}{2}\Delta\omega$. For small resolution $\Delta\omega$, one can impose the conditions:

$$\begin{aligned} &\omega - \frac{1}{2}\Delta\omega \le \theta_{ab} \le \omega + \frac{1}{2}\Delta\omega \\ &\delta(\omega - \theta_{ab})\Delta\omega = \delta(\cos\omega - \cos\theta_{ab})\Delta\omega \sin\omega . \end{aligned} \tag{24.47}$$

In terms of the partonic variables, one has:

$$\begin{aligned} \cos\theta_{ab} &= \left(x_c^2 - x_a^2 - x_b^2\right)/2x_a x_b \qquad c \neq a, b \\ E_a E_b &= \frac{s}{4} x_a x_b , \end{aligned} \tag{24.48}$$

where $a, b, c$ vary from 1 to 3 and $x_3 = 2 - x_1 - x_2$, Substituting into the jet cross-section in Eq. (24.18), one can deduce:

$$\frac{1}{\sigma^{(0)}} \frac{d\Sigma}{d\cos\omega} = \frac{C_F}{2} \left(\frac{\alpha_s}{\pi}\right) \int dx_1 \int dx_2 \frac{x_1^2 + x_2^2}{(1-x_1)(1-x_2)} \sum_{a<b} x_a x_b \delta(\cos\omega - \cos\theta_{ab}) . \tag{24.49}$$

After integration, one obtains:

$$\frac{1}{\sigma^{(0)}} \frac{d\Sigma}{d\cos\omega} = \frac{C_F}{8} \left(\frac{\alpha_s}{\pi}\right) \frac{(3-2z)}{z^5(1-z)} [2(3 - 6z + 2z^2)\ln(1-z) + 3z(2-3z)] , \tag{24.50}$$

where:

$$z = (1 - \cos\omega)/2 . \tag{24.51}$$

The next order correction to this expression has been evaluated in [294].

### 24.4.5 Jade and Durham algorithms

- **Jade algorithm** Another popular jet definition is the so-called Jade algorithm [295]. For the three jet, one uses the invariant mass cut $y$:

$$s_{ij} = (p_i + p_j)^2 > y_{\rm cut} s \qquad (i, j = 1, 2, 3) , \tag{24.52}$$

  where $s$ is the squared of the sum of the measured energies of all particles of an event. In this original Jade algorithm, one can define:

$$s_{ij} = 2E_i E_j (1 - \cos\theta_{ij}) , \tag{24.53}$$

  or the *jet rates*:

$$y_{\rm cut} = s_{ij}/s , \tag{24.54}$$

  where $E_i$ and $E_j$ are the energies of the particles and $\theta_{ij}$ is the angle between them.
- **Durham algorithm** In its variant (Durham and Cambridge algorithms [296]), one defines instead:

$$s_{ij} = 2\min\left(E_i^2, E_j^2\right)(1 - \cos\theta_{ij}) \quad \text{or} \quad y_{\rm cut} = s_{ij}/s . \tag{24.55}$$

  These two definitions are the most used at LEP due to their less sensitivity to hadronization and mass effects.
- **Jet resolution parameter** $y_n$ They are defined as the particular values of $y_{\rm cut}$ at which events switch from $n-1$ to $n$-jet configuration.

The QCD expression of the fraction of the three-jet cross-section is of the form:

$$R_3 = \frac{\sigma(3\text{jet})}{\sigma^{(0)}} = \frac{C_F}{2}\left(\frac{\alpha_s}{\pi}\right)\left[2\ln^2\left(\frac{y}{1-2y}\right) + 3(1-2y)\ln\left(\frac{y}{1-2y}\right) + \frac{5}{2} - 6y - \frac{9}{2}y^2 + \text{Li}_2\left(\frac{y}{1-y}\right) - \frac{\pi^2}{3}\right] \tag{24.56}$$

where:

$$\text{Li}_2(z) \equiv -\int_0^z \frac{dx}{1-x}\ln x , \tag{24.57}$$

is the dilogarithm function with the properties given in Appendix F. The limit $y = 0$ reflects the IR singularities. The fraction of the two-jet event is:

$$R_2 = 1 - R_3 . \tag{24.58}$$

This result can be generalized for $n$-jet configuration provided that the constraint $s_{ij} > ys$ is satisfied for all $i, j = 1, \ldots, n$. In this case, the pair $i,\ j$ of particles or cluster of particles satisfying the previous cut condition is replaced or *recombined* into a single jet or a cluster $k$ with four-momentum $p_k = p_i + p_j$. This procedure is repeated until all pair $y_{ij}$ are larger than the jet resolution parameter cut $y_{cut}$, and the remaining clusters of particles are called jets. One has:

$$R_n = \left(\frac{\bar{\alpha}_s}{\pi}\right)^{n-2} \sum_{j=0} C_j(y)^{C_j^{(n)}} \left(\frac{\bar{\alpha}_s}{\pi}\right)^j , \tag{24.59}$$

with:

$$\sum_n R_n = 1 . \tag{24.60}$$

and $\bar{\alpha}_s(s)$ corresponds to the summation of the higher-order term $\alpha_s(\nu^2)^j \ln^k(s/\nu^2)$. For large $y$, the jet fractions $R_n$ with $n \geq 3$ are small, while for $y \to 0$, the IR-divergence reappears making the QCD series unreliable. Other jet algorithms in order to improve the QCD predictions at low values of $y$ have been proposed in the literature (see e.g. [297]).

## 24.5 QCD tests from jet analysis

- As we have mentioned previously, one observes jet of hadrons but not jets of quarks or/and gluons. Therefore, one has to take into account the hadronization which is quantified into the fragmentation functions. This effect is modelled through Monte-Carlo analysis and introduces theoretical uncertainties not under control. Jet analyses, like the deep inelastic processes discussed in the previous chapters, have been used to measure the value of the QCD coupling constant where complete results for different energies (91.2, 133, 161, 172, 183, 189 GeV) from LEP studies will be shown in the next chapter. At the $Z_0$ mass, the average results from LEP and SLC are [139]:

$$\alpha_s(91.2\text{ GeV}) = 0.121 \pm 0.001\text{ (exp)} \pm 0.006\text{ (th)} . \tag{24.61}$$

  Recent ALEPH result from four-jets [298] at NLO (order $\alpha_s^3$) leads to:

$$\alpha_s(91.2\text{ GeV}) = 0.1170 \pm 0.0001\text{ (stat)} \pm 0.0013\text{ (syst)} , \tag{24.62}$$

  where the analysis of the error needs to be reconsidered to being convincing.
- Three-jet events are also used to test the gluon spin, where for a spin zero gluon, the term $x_1^2 + x_2^2$ of the cross-section in Eq. (24.18) should be replaced by $x_3^2/4$. The measured distributions agree well with a spin-1 gluon and excludes the spin-0 one.
- One can also notice that the jet event-shape variables are functions of the colour group factors:

$$T_F = 1/2 , \qquad C_F = (N_c^2 - 1)/(2N_c) , \qquad C_A = N_c , \tag{24.63}$$

  which originate from the $SU(N)_c$ algebra given in Appendix B at the different vertices. Combined fit of these quantities favour the $SU(3)$ group for QCD (see different contributions at the QCD-Montpellier conference). In Fig. 24.6, one compares the scaling violation rates in the hadron spectra

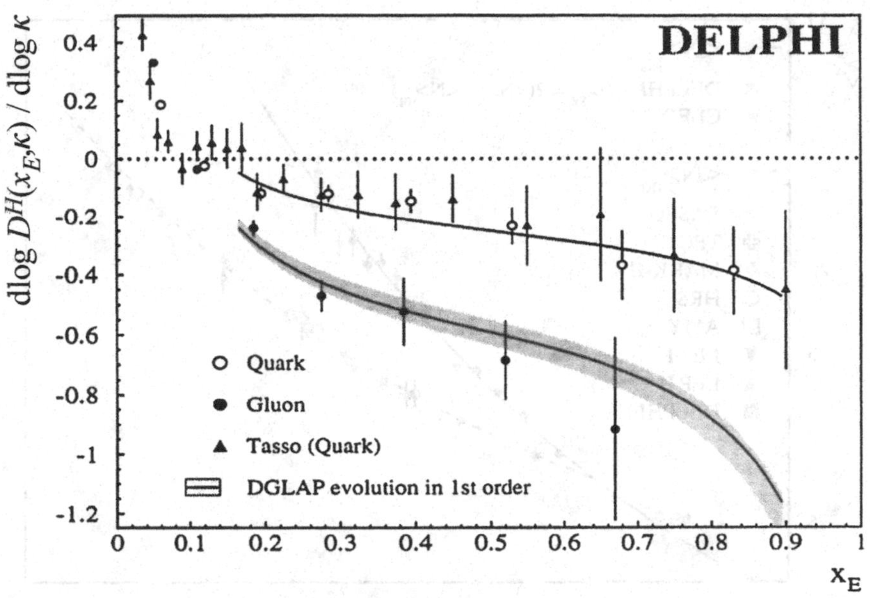


Fig. 24.6. Scaling violation rates in inclusive hadron distributions from quarks and gluon jets.

from gluon and quark jets as a function of the hardness scale $\kappa$ which caracterizes a given jet [299]. At large $x_E \sim 1$, one expects that the log-derivatives between the quark and gluon jet is close to $C_A/C_F$, which is 9/4 for a $SU(3)_c$ QCD group. As shown in the figure, experimentally, one obtains:

$$\frac{C_A}{C_F} = 2.23 \pm 0.09_{\text{stat}} \ \pm 0.06_{\text{syst}} \, . \tag{24.64}$$

In the same way, one expects that hadron multiplicity increases with the hardness of the jets proportional to the multiplicity of secondary gluons and sea quarks. This is shown in Fig. 24.7. The ratio of the slopes in the gluons and quarks jets are proportional to $C_A/C_F$, which is again verified experimentally:

$$\frac{C_A}{C_F} = 2.246 \pm 0.062_{\text{stat}} \ \pm 0.08_{\text{syst}} \ \pm 0.095_{\text{th}} \, . \tag{24.65}$$

## 24.6 Jets from heavy quarkonia decays

Quarkonia decays can also produce gluon jets:

$$\begin{aligned} 1^{--} &\to 3g \\ &\to 2g\gamma \\ 0^{-+} &\to 2g \\ &\to g\gamma \end{aligned} \tag{24.66}$$

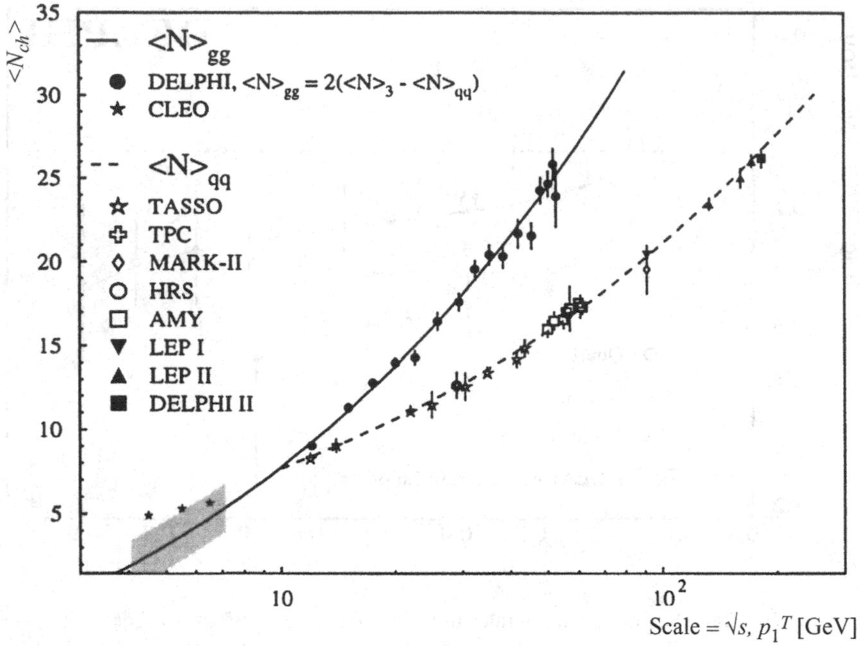


Fig. 24.7. Charged hadron multiplicity in gluons and quark jets.

via OZI violating processes. To leading order, the differential decay rate for $1^{--} \to 3g$, can be written as:

$$\frac{1}{\Gamma_{3g}^{(0)}} \frac{d\Gamma_{3g}}{dx_1 dx_2} = \frac{1}{\pi^2 - 9} \left\{ \left( \frac{1 - x_1}{x_2 x_3} \right)^2 + (x_i \leftrightarrow x_j) \right\} \qquad i, j = 1, 2, 3, \quad (24.67)$$

where $x_i = 2k_i^0/M_V$, $k_i$ is the gluon momenta and $M_V$ is the vector meson mass. $\Gamma_{3g}^{(0)}$ is the lowest order decay rate:

$$\Gamma_{3g}^{(0)} = \frac{160\alpha_s^3}{81} \frac{|^3S_1(0)|^2}{M_V^2}, \qquad (24.68)$$

where $|^3S_1(0)|$ is the wave function at the origin. In terms of the thrust variable, one has [291]:

$$\frac{1}{\Gamma_{3g}^{(0)}} \frac{d\Gamma_{3g}}{dT} = \frac{3}{\pi^2 - 9} \left\{ \frac{4(1-T)}{T^2(2-T)^3} (5T^2 - 12T + 8) \ln \frac{2(1-T)}{T} \right. \\ \left. + \frac{2(3T-2)(1-T)^2}{T^3(2-T)^2} \right\}, \qquad (24.69)$$

and the average:

$$\langle T\rangle_{3g} = \frac{3}{\pi^2 - 9}\left\{6\ln(2/3) - \frac{3}{2} + \frac{4\pi^2}{3} + 20\int_o^1 dx\,\frac{\ln x}{2+x}\right\} \simeq 0.889\,. \quad (24.70)$$

## 24.7 Jets from $ep$, $\bar{p}p$ and $pp$ collisions

QCD jets may also be produced in $ep$ or hadronic reactions and from heavy quarkonia decays. In $ep$ scattering, and to leading order in $\alpha_s$, two identified jets in addition to the beam jet from the remnants of the incoming proton, arise from photon gluon fusion and from QCD Compton processes. This process has been also used for determining the QCD coupling $\alpha_s$, where the theoretical uncertainties come from the scale variations and structure functions, while the systematic ones come from the uses of jet algorithms and hadronization models. The result from HERA is [300]

$$\alpha_s(M_{Z^0}) = 0.118 \pm 0.002\ (\text{stat}) \pm 0.008\ (\text{syst}) \pm 0.007\ (\text{th})\,. \quad (24.71)$$

Jets from hadronic collisions followed the previous strategies used in $e^+e^-$. However, one has to separate *(jet finders)* the jets from the proton remnants from the ones from reconstructed jets, which is different from the case of $e^+e^-$ where all particles are assigned to be jets. At present, one follows the jet definitions used in [301], where jets are defined by concentrations of transverse energy $E_T = |E\sin\theta|$ in cones of radius:

$$R = \sqrt{(\Delta\eta)^2 + (\Delta\phi)^2}\,, \quad (24.72)$$

where $\eta = -\ln\tan(\theta/2)$ is the pseusdorapidity, $\phi$ is the azimuthal and $\theta$ the polar angles of a particle in the calorimeter of the detector, measured with respect of the point of beam crossing. Jets study have been used by the CDF collaboration for determining $\alpha_s$, as a function of $E_T$ and for a radius $R = 0.7$, with the result [302]:

$$\alpha_s(M_{Z^0}) = 0.1178 \pm 0.0001\ (\text{stat})\ ^{+0.008}_{-0.010}\ (\text{syst})\ ^{+0.007}_{-0.005}\ (\text{th}) \pm 0.006\ (\text{pdf})\,, \quad (24.73)$$

where the theoretical error is due to the scale dependence.

Analogously, heavy quark production has been also studied at Tevatron hadron colliders, where there is a good agreement with QCD predictions for the top production, while the total rate and $E_T$ distribution of $b$ quarks produced by CDF exceeds the QCD predictions up to the largest values of $E_T$ by a factor of 3–4. According to [303], this rare discrepancy between the data and QCD predictions can be attributed to the inconsistency of the input $B$ meson fragmentation functions used in previous analysis (mismatch between the perturbative and non-perturbative contributions). This result can be tested in some other processes.

# 25

# Total inclusive hadron productions

## 25.1 Heavy quarkonia OZI-violating decays

In the previous chapter, we have studied the QCD jets from the OZI-violating decays of quarkonia, which occurs through the diagrams in Fig. 25.1.

OZI or Zweig rule [9,254] states that the decays of an heavy resonance involving disconnected diagrams such as in the previous figure are suppressed. In QCD, the rate behaves as $\alpha_s^3$ for a spin one and to order $\alpha_s^2$ for a spin zero resonance. In the case of the $\bar{b}b$ states:

$$\begin{aligned} \Upsilon &\to \text{hadrons} \sim M_\Upsilon \alpha_s^3(M_\Upsilon) \\ \eta_b &\to \text{hadrons} \sim M_{\eta_b} \alpha_s^2(M_{\eta_b}) \,. \end{aligned} \tag{25.1}$$

The rule works better for heavier and heavier resonances, which can be understood from a $1/N_c$ argument [304]. Phenomenologically, a decay of a $\bar{Q}Q$ resonance into a $\bar{Q}Q$ pair should involve a pair of open $Q$ states $\bar{Q}q$ and $\bar{q}Q$. As the open $Q$ states are too heavy, there is not enough phase space for the $\bar{Q}Q$ resonance to decay into them. An explanation of the smallness of this width was one of the successes of QCD [305]. In QCD, the evaluation of the width consists of replacing the sum over hadron states by the gluons. Let's consider the $1^-(^3S_1)$ quarkonia states described by the hadronic current:

$$J_V^\mu(x) = \bar{Q}\gamma^\mu Q \,. \tag{25.2}$$

To lowest order of QCD, one has:

$$1^- \to \text{hadrons} \simeq 1^- \to 3g \,. \tag{25.3}$$

In this way, the onium decay is very similar to the one of positronium up to an overall colour factor:

$$\Gamma(V \to \text{hadrons}) \simeq \frac{64(\pi^2-9)}{9} C_V \frac{|^3\Psi_1(0)|^2}{M_V^2} \alpha_s^3\left(M_V^2\right) , \tag{25.4}$$

where $|^3\Psi_1(0)|^2$ is the square of the onium wave function at the origin, and is proportional to the matrix element:

$$\langle V|\bar{Q}\gamma^\mu Q|0\rangle \,, \tag{25.5}$$

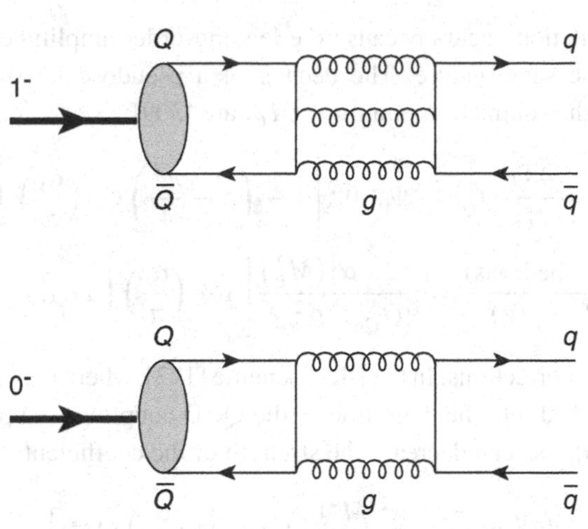


Fig. 25.1. Hadronic decays of an heavy quarkonium.

while:

$$C_V = \frac{1}{16N_c}\sum_{abc} d_{abc}^2 = 5/18 \tag{25.6}$$

is the colour factor. The wave function can be also related to the $V \to e^+e^-$ width as:

$$\Gamma(V \to e^+e^-) = \frac{16\pi Q_Q^2\alpha^2}{M_V^2}|^3\Psi_1(0)|^2 . \tag{25.7}$$

where $Q_Q$ is the heavy quark charge in units of $e$. Therefore, one obtains the branching ratio:

$$R_V \equiv \frac{\Gamma(V \to \text{hadrons})}{\Gamma(V \to e^+e^-)} = \frac{10(\pi^2-9)}{81\pi}\frac{\alpha_s^3(M_V^2)}{Q_Q^2\alpha^2} . \tag{25.8}$$

Including the next-to-leading order (NLO) corrections, one obtains in the $\overline{MS}$ scheme:

$$\begin{aligned}\Gamma(V \to e^+e^-) &= \Gamma(V \to e^+e^-)_{LO}\left[1 + 4C_F\left(\frac{\alpha_s}{\pi}\right)\right] \quad \text{in [306]}\\ \Gamma(V \to \text{hadrons}) &= \Gamma(V \to \text{hadrons})_{LO}\left[1 - (3.8 \pm 0.5)\left(\frac{\alpha_s}{\pi}\right)\right] \quad \text{in [307]} , \end{aligned}\tag{25.9}$$

and therefore (for $n_f = 4$):

$$R_V = \frac{10(\pi^2-9)}{81\pi}\frac{\alpha_s^3(M_V^2)}{Q_Q^2\alpha^2}\left[1 - (9.1 \pm 0.5)\left(\frac{\alpha_s}{\pi}\right)\right] , \tag{25.10}$$

which is a huge coefficient correction, and requires an evaluation of the non-trivial next-to-next-leading order (NNLO) contribution. The situation is much better for the ratio [308]:

$$R_\gamma \equiv \frac{\Gamma(V \to \dot{\gamma} + \text{hadrons})}{\Gamma(V \to \text{hadrons})} = \frac{36Q_Q^2}{5}\frac{\alpha}{\alpha_s(M_V^2)}\left[1 + (2.2 \pm 0.6)\left(\frac{\alpha_s}{\pi}\right)\right] , \tag{25.11}$$

where a large cancellation occurs because the leading-order amplitudes for $V \to 3g$ and $V \to gg\gamma$ are of the same nature. The decays of a pseudoscalar $0^-({}^1S_0)$ state in the $\overline{MS}$ scheme and, at the subtraction point $\nu = M_P$, are [309]:

$$\Gamma(P \to \gamma\gamma) = \frac{48\pi}{M_P^2} Q_Q^4 \alpha^2 |{}^1\Psi_0(0)|^2 \left[1 - \left(5 - \frac{\pi^2}{4}\right) C_F \left(\frac{\alpha_s}{\pi}\right)\right]$$

$$R_P \equiv \frac{\Gamma(P \to \text{hadrons})}{\Gamma(P \to \gamma\gamma)} = \frac{2}{9Q_Q^4} \frac{\alpha_s^2(M_P^2)}{\alpha^2} \left[1 + \left(\frac{\alpha_s}{\pi}\right)\left(17.13 - \frac{8}{9} n_f\right)\right] \tag{25.12}$$

and also have huge $\alpha_s$ corrections. In the BLM scheme [173], where the vacuum polarization corrections are absorbed into the definition of the QCD coupling (see previous chapter on the renormalizations), one can decrease the strength of the coefficient:

$$R_P^{BLM} \simeq \frac{2}{9Q_Q^4} \frac{\alpha_s^2(M^*)}{\alpha^2} \left[1 + 2.46 \left(\frac{\alpha_s}{\pi}\right)(M^*)\right], \tag{25.13}$$

but the scale at which the coupling is evaluated becomes too low $M^* \simeq 0.26 M_P$. Another unclear situation is the possible effect of the analytical continuation from Euclidean (QCD result) to the time-like (the process) regions (see e.g. [310] for related discussions). These processes have been used to estimate the value of $\alpha_s$ from $J/\Psi$ and $\Upsilon$ decays [311], after the inclusions of relativistic and finite mass corrections, and an estimate of higher-order corrections. The analysis gives:

$$\alpha_s(M_{Z^0}) = 0.113^{+0.007}_{-0.005}, \tag{25.14}$$

which is comparable with other results, although most probably, the error has been underestimated. The result needs to be confirmed by the inclusion of the NNLO terms.

## 25.2 Alternative extractions of $\alpha_s$ from heavy quarkonia

Alternative to these non-relativistic approaches, are the QCD spectral sum rule (QSSR) analysis which will be discussed in the following chapters. They have also been used to extract the QCD coupling $\alpha_s$ from the leptonic widths [312,155] after the resummation of Coulombic corrections. However, the result should be affected by the value of the quark mass and of the non-perturbative terms which are strongly correlated in the sum rule analysis [3,148,149,313]. The result quoted in [139] is:

$$\alpha_s(M_{Z^0}) = 0.118 \pm 0.006\,. \tag{25.15}$$

Using also QSSR, $\alpha_s$ has been extracted from the meson mass-splitting to order $\alpha_s$ [313], with the NLO result:

$$\frac{M^2_{{}^1P_1}}{M^2_{{}^3P_1}} \simeq 1 + \alpha_s(\sigma)\left[\Delta_\alpha^{13}(exact) = 0.014^{+0.008}_{-0.004}\right] + \mathcal{O}(\alpha_s^2), \tag{25.16}$$

where $\sigma^{-1} \simeq 1.3$ GeV is the sum rule scale. Using the experimental value $M^2_{{}^1P_1} \simeq 3526.1$ GeV, one can deduce:

$$\alpha_s(1.3\ \text{GeV}) = 0.64^{+0.36}_{-0.18} \Longrightarrow \alpha_s(M_{Z^0}) = 0.127 \pm 0.009\ , \qquad (25.17)$$

in fair agreement with the different predictions given in the next section, although not included in the 'world summary table' (Table VI.1). For a comparison, one can also use non-perturbative lattice calculations of the $\Upsilon$ mass splittings. The resulting value of $\alpha_s(M_{Z^0})$ ranges from $0.105 \pm 0.004$ (quenched approximation [314]) to $0.1174 \pm 0.0024$ [315] and $0.1118 \pm 0.0017$ [316] for two dynamic quarks, indicating that systematic errors are not under good control. A more conservative lattice result has been adopted to be:

$$\alpha_s(M_{Z^0}) = 0.115 \pm 0.006\ , \qquad (25.18)$$

as quoted in the 'world summary table' (Table VI.1), given in Part VI [139].

## 25.3 $e^+e^- \to$ hadrons total cross-section

The inclusive $e^+e^- \to$ hadrons production is the simplest though fundamental deep inelastic process. The data until LEP energies are shown in Figs. 25.2 and 25.3.

In the one photon approximation (below the $Z^0$ mass), the hadronic production occurs through the process shown in Fig. 25.4, in which the $\bar{q}q$ pairs interact through QCD forces, and then exchange and emit gluons in different ways.

However, we do not yet have a good understanding on the way quarks and gluons hadronize. At short distance $x \sim 1/\sqrt{t}$, one can use perturbative QCD for predicting the

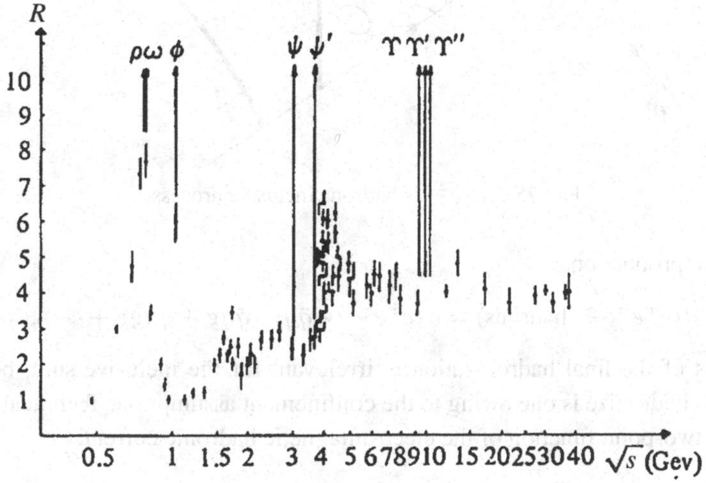


Fig. 25.2. $e^+e^- \to$ hadrons data at lower energies.

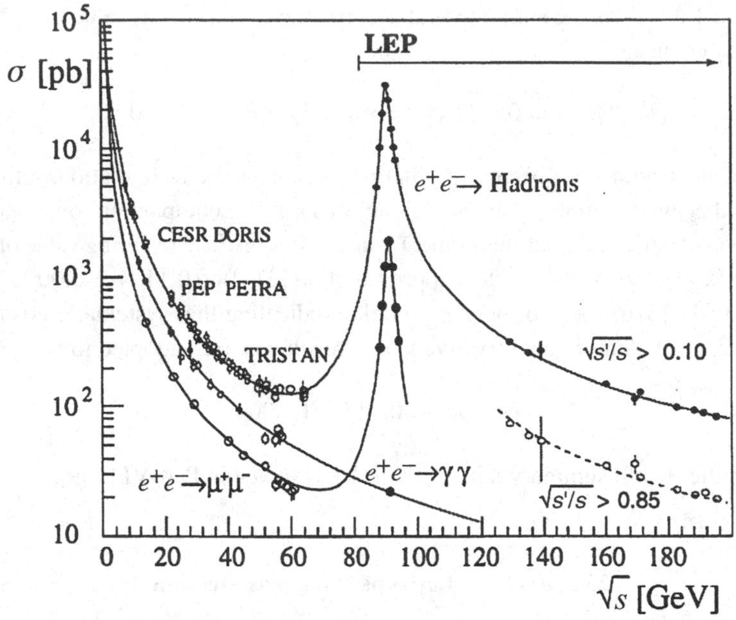


Fig. 25.3. $e^+e^- \to$ hadrons data at LEP energies.

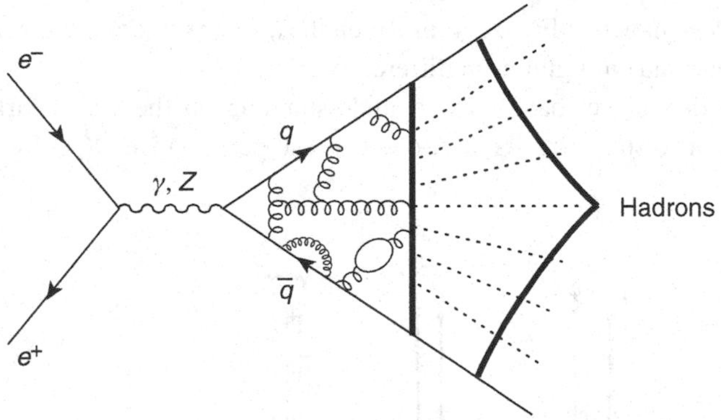


Fig. 25.4. $e^+e^- \to$ hadrons inclusive process.

total inclusive productions:

$$\sigma(e^+e^- \to \text{ hadrons}) = \sigma(e^+e^- \to \bar{q}q + \bar{q}qg + \bar{q}qgg + \cdots)\,, \tag{25.19}$$

as the details of the final hadronization is irrelevant for the inclusive sum, because the probability to hadronize is one owing to the confinement assumption. Technically, one can consider the two-point function of the electromagnetic hadronic current:

$$\Pi^{\mu\nu}_{\text{em}}(q^2) = i \int d^4x e^{iqx} \langle 0|\mathbf{T} J^\mu(x) J^{\nu\dagger}(0)|0\rangle = -(g^{\mu\nu}q^2 - q^\mu q^\nu)\Pi_{\text{em}}(q^2)\,, \tag{25.20}$$

where:

$$J_\mu(x) = \frac{2}{3}\bar{u}(x)\gamma_\mu u(x) - \frac{1}{3}\bar{d}(x)\gamma_\mu d(x) - \frac{1}{3}\bar{s}(x)\gamma_\mu s(x) + \cdots \tag{25.21}$$

is the electromagnetic current associated to the quarks $u,\ d,\ s.\dots$ Thanks to its analyticity property, $\Pi_{em}(q^2)$ obeys the well-known Källen–Lehmann dispersion relation (Hilbert representation):

$$\Pi_{em}(q^2) = \int_{t_<}^{\infty} \frac{dt}{t - q^2 - i\epsilon}\frac{1}{\pi}\mathrm{Im}\Pi_{em}(t) + \cdots , \tag{25.22}$$

where $\cdots$ represents subtraction terms, which are, in general, polynomial in $q^2$; $t_<$ is the hadronic threshold. Its imaginary (absorptive) part is related to the total cross-section:

$$\sigma(e^+e^- \to \text{ hadrons}) = \frac{4\pi^2\alpha}{q^2}e^2\frac{1}{\pi}\mathrm{Im}\Pi_{em}(q^2) , \tag{25.23}$$

where:

$$-3\theta(q)q^2\frac{1}{\pi}\mathrm{Im}\Pi_{em}(q^2) = \sum_\Gamma \langle 0|J^\mu_{em}(0)|0\rangle\langle 0|J^\dagger_{\mu,em}(0)|0\rangle(2\pi)^3\delta^{(4)}(q - p_\Gamma) . \tag{25.24}$$

Normalized to the $e^+e^- \to \mu^+\mu^-$ total cross-section:

$$\sigma(e^+e^- \to \mu^+\mu^-) = -\frac{4\pi^2\alpha^2}{3q^2} , \tag{25.25}$$

it reads:

$$R_{e^+e^-} \equiv \frac{\sigma(e^+e^- \to \text{ hadrons})}{\sigma(e^+e^- \to \mu^+\mu^-)} = 12\pi\ \mathrm{Im}\Pi_{em}(t + i\epsilon) . \tag{25.26}$$

It is also convenient, in the perturbative calculation, to relate this quantity to the Adler $D$-function [317] defined as:

$$D(Q^2) \equiv -Q^2\frac{d}{dQ^2}\Pi_{em}(Q^2) . \tag{25.27}$$

In this way, one obtains:

$$R(t) = \frac{1}{2i\pi}\int_{-t-i\epsilon}^{t+i\epsilon}\frac{dQ^2}{Q^2}D(Q^2) , \tag{25.28}$$

where it is necessary to transform the result into the physical region by taking into account the effects due to the analytic continuation of the terms of the type:

$$\ln^n(-q^2/\nu^2) \to (\ln(-q^2/\nu^2) + i\pi)^n . \tag{25.29}$$

Away from thresholds, one can neglect quark mass corrections, and obtain the perturbative series in $\alpha_s$, in the $\overline{MS}$ scheme. To order $\alpha_s^4$, one has:

$$R_{e^+e^-} = 3\left(\sum_1^{n_f} Q_i^2\right)\left[1 + F_2 a_s(t) + F_3 a_s^2(t) + F_4 a_s^3(t)\right] + F_4' a_s^3(t)\left(\sum_i Q_i\right)^2 , \tag{25.30}$$

where:

$$\begin{aligned}
F_2 &= 1 \text{ in } [318,319] \\
F_3 &= 1.9857 - 0.1153n_f \text{ in } [317,320] \\
F_4 &= -6.6368 - 1.2001n_f - 0.005n_f^2 \text{ in } [321] \\
F_4' &= -1.2395 \text{ in } [321] \, ,
\end{aligned} \tag{25.31}$$

where, for, for example, five flavours:

$$R_{e^+e^-} = \frac{11}{3}\left[1 + a_s(t) + 1.411a_s^2(t) - 12.80a_s^3(t) + \mathcal{O}\left(a_s^4\right)\right] \, . \tag{25.32}$$

The perturbative uncertainties are of the order $a_s^4$ and includes ambiguities related to the choice of renormalization scale and scheme, which leads to slightly different predictions for the truncated series. Because of the above functional forms of $R_{e^+e^-}$, relative errors in $R_{e^+e^-}$ lead to an absolute error in $\alpha_s$ of the same size:

$$\frac{\Delta R_{e^+e^-}}{R_{e^+e^-}} \sim \Delta\alpha_s \, , \tag{25.33}$$

such that precise measurement of $R_{e^+e^-}$ still leads to large errors in $\alpha_s$. Re-analysis of PETRA and TRISTAN data in the c.m. energy range from 20 to 65 GeV, gives at NNLO [322]:

$$\alpha_s(42.4 \text{ GeV}) = 0.175 \pm 0.028 \quad \Longrightarrow \quad \alpha_s(M_{Z_0}) = 0.126 \pm 0.022 \, . \tag{25.34}$$

## 25.4 $Z \to$ hadrons

On top of the $Z^0$, LEP experiments have produced a large statistical data sample that allow a precise measurement of $\alpha_s$. The hadronic $Z^0$ width can be parametrized in a similar way:

$$R_Z \equiv \frac{\Gamma(Z^0 \to \text{ hadrons})}{\Gamma(Z^0 \to e^+e^-)} = 3R_Z^{ew}\left[1 + \sum_{n\geq 1} \tilde{F}_n a_s^n(M_{Z^0}) + \mathcal{O}\left(\frac{m_f^2}{M_Z^2}\right)\right] , \tag{25.35}$$

where:

$$R_Z^{ew} = \frac{\sum_i \left(v_i^2 + a_i^2\right)}{\left(v_e^2 + a_e^2\right)}(1 + \delta_{ew}) \tag{25.36}$$

contains the underlying $Z \to \sum_i \bar{q}_i q_i$ decay amplitude, $v_e$ is the weak coupling of fermion to $Z_0$; $\delta_{ew}$ is the weak correction. The QCD correction coefficients $\tilde{F}_n$ are slightly different from $F_n$ due to the presence of both vector and axial-vector coupling. Combined LEP results lead to [68]:

$$R_Z = 20.768 \pm 0.024 \, , \tag{25.37}$$

which gives:

$$\alpha_s(M_{Z^0}) = 0.124 \pm 0.004\ (\text{exp})\ \pm 0.002\ (\text{M}_\text{H},\ \text{M}_\text{t})\ ^{+0.003}_{-0.001}\ (\text{QCD})\,, \quad (25.38)$$

where the second errors are from those from $M_t$ and from $M_H$ ranging from 100 to 1000 GeV. The last errors come from the scheme and scale dependences at NNLO.

## 25.5 Inclusive semi-hadronic $\tau$ decays

The QCD evaluation of the inclusive semi-hadronic process:

$$\tau \to \nu_\tau + \text{hadrons} \quad (25.39)$$

is diagramatically similar to the $e^+e^- \to$ hadrons process. One puts all possible gluon and $\bar{q}q$ corrections to the QCD diagram in Fig. 25.5 and computes the sum of all partonic subprocesses. As in $e^+e^-$, one considers the two-point correlator:

$$\begin{aligned}\Pi_L^{\mu\nu}(q^2) &= i\int d^4x e^{iqx}\langle 0|\mathbf{T}J_L^\mu(x)J_L^{\nu\dagger}(0)|0\rangle \\ &= -(g^{\mu\nu}q^2 - q^\mu q^\nu)\Pi_L^{(1)}(q^2) + q^\mu q^\nu)\Pi_L^{(0)}(q^2)\,, \end{aligned} \quad (25.40)$$

associated with the charged current:

$$J_L^\mu = \bar{u}\gamma^\mu(1-\gamma_5)(d\cos\theta_C + s\sin\theta_C)\,, \quad (25.41)$$

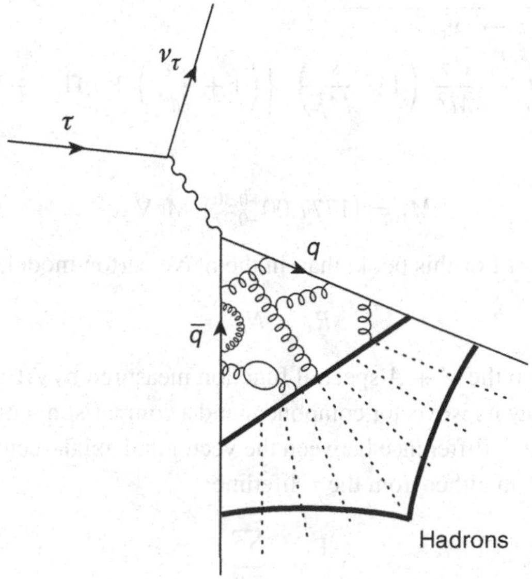


Fig. 25.5. $\tau \to \nu_\tau$ + hadrons inclusive process.

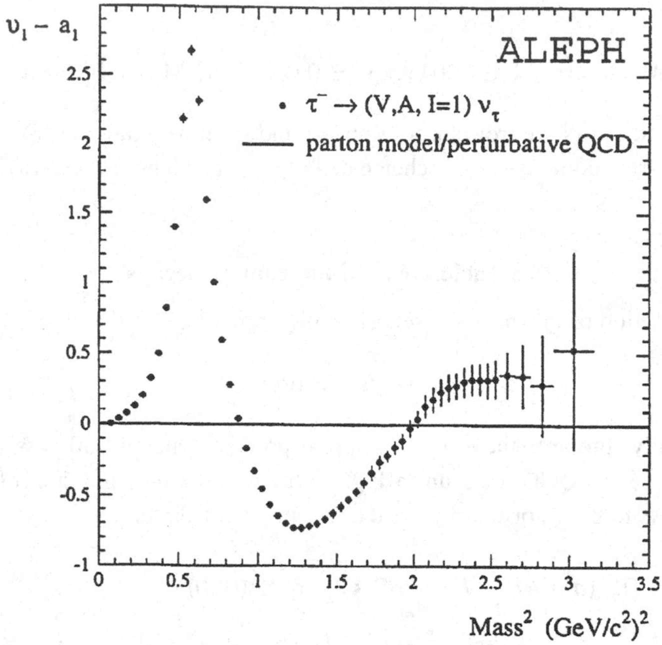


Fig. 25.6. Sum of the vector and axial-vector spectral functions from tau-decay.

where $u$, $d$ and $s$ are quark fields and $\theta_C$ is the Cabibbo angle. In terms of which, one can express the inclusive semi-hadronic branching ratio:

$$R_\tau \equiv \frac{\Gamma(\tau \to \nu_\tau + \text{hadrons})}{\Gamma(\tau \to \nu_\tau e \nu_e)}$$
$$= 12\pi \int_0^{M_\tau^2} \frac{ds}{M_\tau^2}\left(1 - \frac{s}{M_\tau^2}\right)^2 \left\{\left(1 + \frac{2s}{M_\tau^2}\right) \text{Im}\,\Pi_L^{(1)} + \text{Im}\,\Pi_L^{(0)}\right\}, \quad (25.42)$$

where [16]:

$$M_\tau = \left(1777.00^{+0.30}_{-0.27}\right)\ \text{MeV}\,. \quad (25.43)$$

We have seen, in Part I of this book, that, in the naïve parton model, one expects:

$$R_\tau = N_c\,. \quad (25.44)$$

We show in Fig. 25.6 the $V + A$ spectral function measured by ALEPH.

In Fig. 14.5, we show its isovector component and a comparison with the $e^+e^-$ data, and in Fig. 25.7, we show the difference between the vector and axial-vector spectral function.

The experimental data either from the $\tau$-lifetime:

$$R_\tau^\Gamma \equiv \frac{\Gamma_\tau - \sum_{e,\mu} \Gamma_{\tau \to \ell}}{\Gamma_{\tau \to \ell}} \quad (25.45)$$

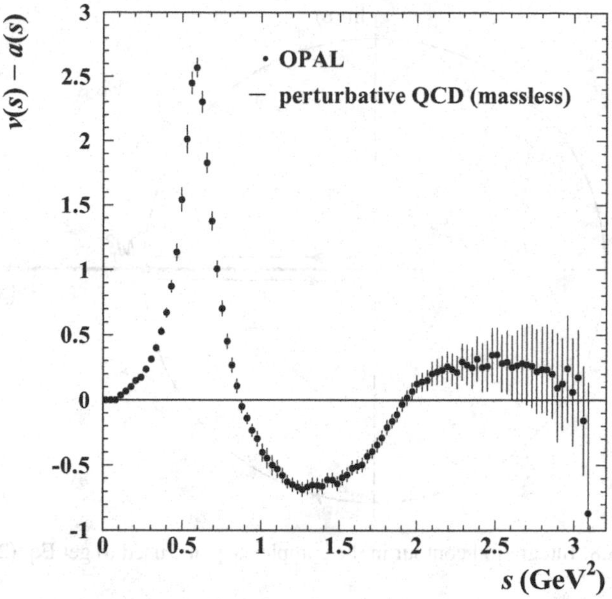


Fig. 25.7. Difference between the vector and axial-vector spectral functions from tau-decay.

or/and from the $\tau$-leptonic branching ratios:

$$R_\tau^B \equiv \frac{1 - B_e - B_\mu}{B_e} \tag{25.46}$$

have the present average [323]:

$$R_\tau = 3.649 \pm 0.014 \,. \tag{25.47}$$

This experimental value is indeed a good evidence for the existence of colour but it is 20% higher than the quark-parton model estimate, such that one (a priori) can expect that QCD perturbative and/or non-perturbative corrections can resolve this discrepancy. From the expression of the width, it is clear that $R_\tau$ in Eq. (25.45) cannot be calculated directly from QCD for $s \leq \Lambda^2$. However, exploiting the analyticity of the correlators $\Pi^{(J)}$ and the Cauchy theorem, one can express $R_\tau$ as a contour integral in the complex $s$-plane running counter-clockwise around the circle of radius $|s| = M_\tau^2$ shown in Fig. 25.8:

$$R_\tau = 6i\pi \oint_{|s|=M_\tau^2} \frac{ds}{M_\tau^2}\left(1 - \frac{s}{M_\tau^2}\right)^2 \left\{\left(1 + \frac{2s}{M_\tau^2}\right) \Pi^{(1)}_{(s)} + \Pi^{(0)}_{(s)}\right\}. \tag{25.48}$$

One should notice the existence of the double zero at $s = M_\tau^2$, which suppresses the uncertainties near the time-like axis. As $|s| = M_\tau^2 \gg \Lambda^2$, one can use the standard operator product expansion (OPE) à la SVZ [1] (as will be discussed in the following chapters) for the estimate of the correlators. In this way, one can express the QCD expression of the decay

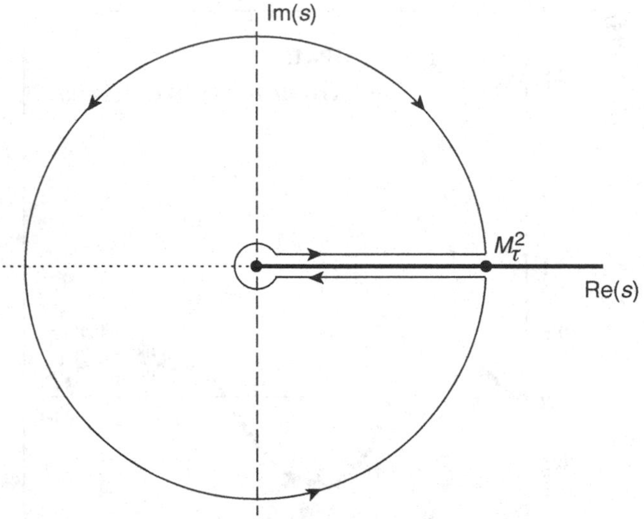


Fig. 25.8. Integration contour in the complex $s$-plane used to get Eq. (25.48).

width as [324] (hereafter referred to as BNP [325]):

$$R_\tau = 3(|V_{ud}|^2 + |V_{us}|^2)\, S_{EW}\{1 + \delta_{EW} + \delta^{(0)} + \delta_{NP}\}\,, \tag{25.49}$$

where:

$$|V_{ud}| \simeq 0.9753 \pm 0.0006, \quad \text{and} \quad |V_{us}| \simeq 0.221 \pm 0.003\,, \tag{25.50}$$

are the CKM mixing angles, while the Cabibbo angle is defined as:

$$\sin^2\theta_C \equiv \frac{|V_{us}|^2}{|V_{ud}|^2 + |V_{us}|^2}\,. \tag{25.51}$$

$S_{EW} = 1.0194$ and $\delta_{EW} = 0.0010$ are LO and NLO electroweak corrections [326,327]. Based on the SVZ-Operator Product Expansion [1],[1] these non-perturbative corrections have been estimated to be small by BNP [325]:

$$\delta_{NP} \simeq -(0.7 \pm 0.4)\% \tag{25.52}$$

A direct measurement of these effects from $\tau$ decay gives [33,328]:

$$\delta_{NP} \equiv \sum_{D\geq 4}\left(\cos^2\theta_c\delta_{ud}^{(D)} + \sin^2\theta_c\delta_{us}^{(D)}\right) \simeq -(0.5 \pm 0.7)\%\,, \tag{25.53}$$

and from the most recent analysis from $e^+e^-$ data [329] (reprinted article):[2]

$$\delta^{NP} \simeq -(2.8 \pm 0.6)10^{-2}\,, \tag{25.54}$$

[1] Here and in the rest of this section, we anticipate the discussions of the SVZ expansion and of the QCD condensates in Part VII, which the reader may consult for understanding the origin of the non-perturbative corrections.

[2] This result has been obtained by combining the fitted value of non-perturbative corrections in the vector channel with the theoretical estimate which relates the vector and axial-vector terms.

confirm the previous estimate of BNP. The smallness of these non-perturbative effects are related to the fact that within the SVZ expansion the numerical leading contribution behaves as $(\Lambda/M_\tau)^6$, while the radiative corrections are relatively large at the $\tau$ mass. These properties indeed show that $\tau$ decay is a good laboratory ( or a *lucky process* as stated by Gabriele Veneziano) for extracting $\alpha_s$. The perturbative QCD correction $\delta^{(0)}$ gives the dominant contribution, and can then be used to determine $\alpha_s$ at the $\tau$-mass scale. It reads:

$$
\begin{aligned}
\delta^{(0)}_{BNP} &= \sum_{n=1} (K_n + g_n)\, a_\tau^n\,, \\
&= a_\tau + \left(K_2 - \frac{19}{24}\beta_1\right) a_\tau^2 \\
&\quad + \left(K_3 - \frac{19}{12}K_2\beta_1 - \frac{19}{24}\beta_2 + \frac{265-24\pi^2}{288}\beta_1^2\right) a_\tau^3 + \mathcal{O}\left(a_\tau^4\right), \qquad (25.55)
\end{aligned}
$$

where, here:

$$
a_\tau \equiv \frac{\bar{\alpha}_s(M_\tau)}{\pi}\,. \qquad (25.56)
$$

$K_n$ are the coefficients appearing in the $D$-function given in Eq. (25.31), which, for $n = 3$ flavours read:

$$
K_1 \equiv F_2 = 1\,, \qquad K_2 \equiv F_3 = 1.63982\,, \qquad K_3 \equiv F_4 = 6.37101\,, \qquad (25.57)
$$

while $g_n$ are induced by the contour integral and depend on $K_{m\leq n}$ and on $\beta_{m\leq n}$. For $n = 3$ flavours, one has:

$$
g_2 = 3.5626, \qquad g_3 = 19.9949\,, \qquad (25.58)
$$

and

$$
\delta^{(0)}_{BNP} = a_\tau + 5.2023 a_\tau^2 + 26.366 a_\tau^3 + \mathcal{O}\left(a_\tau^4\right), \qquad (25.59)
$$

while a bold-guess of

$$
K_4 \approx K_3(K_3/K_2) \approx 25\,, \qquad (25.60)
$$

confirmed by the estimate [178] ($K_4 \approx 27.5$) based on PMS [176] and ECH [177] renormalization invariant schemes, from the large $\beta$ limit of QCD [330,331,154] ($K_4 \approx 24.8$) and from an experimental measurement [332] ($K_4 \approx 29 \pm 5$) gives [323]:

$$
g_4 \approx 78\,, \qquad (25.61)
$$

which indicates that $g_n$ are larger than the corresponding $K_n$ coefficients, and implies a sizeable renormalization scale dependence [333,334]. As observed in [333], these large corrections come from the running along the circle $s = M_\tau^2 \exp(i\phi)$ $(0 \leq \phi \leq 2\pi)$, which leads to the imaginary logarithm $\log(-s/M_\tau^2) = i(\phi - \pi)$, which are large in some parts of the integration range, and leads to the *small* convergence radius $a_\tau \leq 0.11$. Using this remark, [333,335] deduce that the series is more convergent if one expands it in terms of the *contour coupling* $A^{(n)}$:

$$A^{(n)}(a_\tau) = \frac{1}{2i\pi}\oint_{|s|=M_\tau^2} \frac{ds}{s}\left(1 - 2\frac{s}{M_\tau^2} + 2\frac{s^3}{M_\tau^6} - \frac{s^4}{M_\tau^8}\right) a_\tau^n(s)\,, \tag{25.62}$$

such that:

$$\delta^{(0)} = \sum_{n\geq 1} K_n A^{(n)}(a_\tau)\,, \tag{25.63}$$

where the QCD series is more convergent and the renormalization scale dependence is very small [336]. The error in the truncation of the perturbative series can be estimated from the last known term of the series [337]. In this way, one can deduce the conservative estimate [323,338]:

$$\delta^{\text{trunc}} \simeq (25 \pm 50)a_\tau^4 \tag{25.64}$$

where the factor 2 has been included in the estimate of the error. In this way, the estimate of the error is about the effect of $K_3 A^{(3)}$, which appears to be conservative enough, and is, therefore, realistic. This result agrees with the one (though slightly larger) from the renormalons effect within an optimized PMS renormalization scheme, which has been estimated to be [339] (see also [340]):

$$\delta^{\text{ren}} \simeq 0.01\,, \tag{25.65}$$

for a typical value of $\alpha_s(M_\tau) = 0.33$. It also agrees with the fit from the $e^+e^-$ data of the $1/M_\tau^2$ contribution [329,341]:

$$\delta^{1/M_\tau^2} \approx 0.01\,, \tag{25.66}$$

confirmed later on from other channels [161] (hereafter referred to as CNZ). The existence of the small $D = 2$ dimension term beyond the usual OPE expansion may be justified from the short distance linear term of the QCD potential and from monopole studies [162] as we shall see in the following chapters. The fit from $e^+e^-$ data does not allow the existence of an eventual huge contribution from the quark constituent mass advocated sometimes in the literature, due to the small value of the contribution obtained from the fit as well as to the opposite sign compared with the expected contribution from the known coefficient of the quark mass. Indeed, the result of the fit would correspond to a tachyonic mass naturally interpreted as the one of tachyonic gluon by [161]. These results indicate that those obtained from a naïve resummation of the QCD series [331,154], [342–344] may be an overestimate of the *true* error.

Table 25.1. *Values of $\alpha_s$ from $R_\tau$.*

| Pert. Theory | ALEPH | OPAL |
|---|---|---|
| FOPT | $0.322 \pm 0.005$ (exp) $\pm 0.019$ (th) | $0.324 \pm 0.006$ (exp) $\pm 0.013$ (th) |
| CIPT | $0.345 \pm 0.007$ (exp) $\pm 0.017$ (th) | $0.348 \pm 0.010$ (exp) $\pm 0.019$ (th) |
| RCFT | $\sim$ | $0.306 \pm 0.005$ (exp) $\pm 0.011$ (th) |

Estimates of some other sources of the errors can be found in [338]. For a typical value of $R_\tau = 3.56 \pm 0.03$, these errors have been classified as follows for $\alpha_s(M_Z)$:

$$
\begin{array}{ll}
0.0003 & \text{electroweak} \\
0.0010 & M_\tau \to M_Z \\
0.0005 & RS\text{-dependence} \\
0.0009 & \mu\text{-dependence} \\
0.0005 & \text{quark masses} \\
0.0009 & \text{SVZ condensates} \\
0.0014 & \text{truncation of the PT series at } \alpha_s^4 \\
 & \text{or UV renormalon and } 1/M_\tau^2 .
\end{array} \tag{25.67}
$$

Adding them in quadrature, the total *theoretical* error is expected to be:

$$\Delta\alpha_s(M_Z) \simeq 0.0023 , \tag{25.68}$$

which is reflected in Table 25.1 above. The most extensive determinations of $\alpha_s$ from $\tau$-decays are based on recent sutudies from LEP, making use of the large amount of statistical data available at LEP-1. Measurements of the vector and axial-vector differential hadronic mass distributions of $\tau$-decays have been performed by the ALEPH and OPAL collaborations [33,328], which allow a simultaneous measurement of $\alpha_s$ and the non-perturbative corrections, where, as mentioned earlier, these latter are found to be small. At NNLO corrections, the resulting values of $\alpha_s$ from recent measurements [33] are given in Table 25.1, corresponding to the different structure of the perturbative series:

- FOPT: naïve perturbative expansion in terms of $\alpha_s(M_\tau)$ given in BNP.
- CIPT: contour-improved perturbation theory where $\delta^{(0)}$ is expressed as a contour integral in the complex-$s$ plane.
- RCPT: renormalon chain-improved perturbation theory, where the leading terms of the $\beta$-functions are resummed by insertion of renormalon chains (gluon lines with fermion loop insertions) as will be seen in the next section.

The resulting mean value from the two experiments and from the different structure of perturbative series is:

$$\alpha_s(M_\tau) = 0.323 \pm 0.005 \text{ (exp) } \pm 0.030 \text{ (th)} , \tag{25.69}$$

at $M_\tau = (1777.00^{+0.30}_{-0.27})$ MeV, which shows that FOPT gives the mean theoretical values. Runned to $M_{Z^0}$, and taking account of the different threshold effects, this value gives:

$$\alpha_s(M_{Z^0}) = 0.1181 \pm 0.0007 \text{ (exp)} \pm 0.0030 \text{ (th)}, \tag{25.70}$$

which is in excellent agreement with the direct measurement at the $Z^0$ peak and with a similar error bar. This agreement of the two determinations of $\alpha_s$ in two extreme regime from $M_\tau$ to $M_{Z^0}$ provides a beautiful test of the QCD prediction of the running coupling behaving as $1/\log$, which is a very significant experimental verification of asymptotic freedom.

### 25.5.1 Running of $\alpha_s$ below the $\tau$-mass

The analysis of the running of $\alpha_s$ from the inclusive distribution of $\tau$-decays [345] and from $e^+e^- \to$ hadrons data [346,329] has been extended to a lower mass below $M_\tau$, where one does not find any deviation from the one expected from QCD. We show in Fig. 25.9 the comparison of the theoretical prediction (FOPT) and ALEPH measurements of $R_{\tau,V+A}$ for different values of the $\tau$-mass.

In Fig. 25.10, one shows the running of $\alpha_s(s_0)$ below 3 GeV$^2$ using FOPT. The other structures of PT series (CIPT and RCPT) show the same behaviour [33].

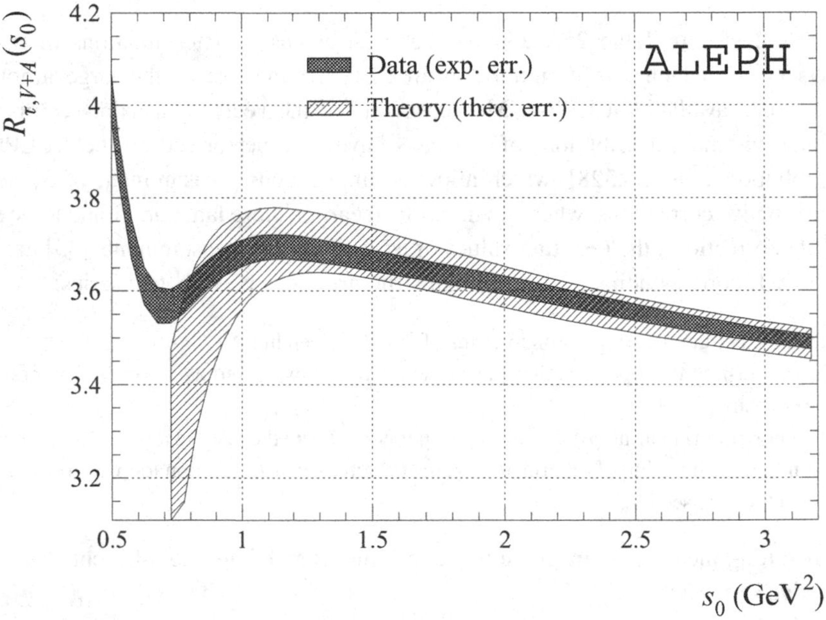


Fig. 25.9. $R_{\tau,V+A}$ as a function of the $\tau$ mass $s_0$. The theoretical predictions is from FOPT.

Table 25.2. *Values of $\alpha_s$ from different observables in $\tau$-decays.*

| Observables | $\alpha_s\left(M_\tau^2\right)$ |
|---|---|
| $R_{\tau,V} = 1.78 \pm 0.03$ | $0.35 \pm 0.05$ |
| $R_{\tau,A} = 1.67 \pm 0.03$ | $0.34 \pm 0.05$ |
| $R_{\tau,\mathrm{excl}} = 3.58 \pm 0.05$ | $0.34 \pm 0.04$ |

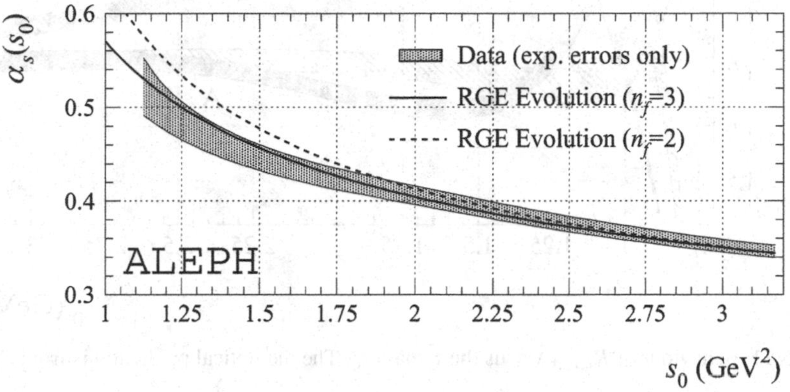


Fig. 25.10. Running of $\alpha_s$ from the theoretical predictions of $R_{\tau,V+A}$ from FOPT to four-loop RGE evolution for two and three flavours. The shaded band is the data.

## 25.6 Some other $\tau$-like processes

### *25.6.1 $\alpha_s$ from other $\tau$ widths*

One can also extend the previous analysis in order to extract the value of $\alpha_s$ from the vector, axial-vector and from the sum of the exclusive modes by applying a $SU(2)$ isospin rotation [347]. This analysis has been done in [338] using the compilation of data in [346]. The analysis and predictions are summarized in Table 25.2 from [338] where one should remark that the errors in each separate channel are larger than in the total inclusive mode, which comes from the fact that the non-perturbative contributions have larger errors in each separate vector and axial-vector channel than in the sum. Indeed, from [338], one can deduce the sum of the non-perturbative terms in each channel at the $\tau$ mass:

$$\delta^V_{NP} \simeq (2.4 \pm 1.3)10^{-2}, \qquad \delta^A_{NP} \simeq -(4.4 \pm 2.1)10^{-2}\,. \tag{25.71}$$

The larger error from the exclusive modes is mainly due to the data. These different determinations are consistent with each others. In Fig. 25.11, we show the behaviour of the vector and axial-vector components of $\tau$ decays versus an hypothetical heavy lepton of mass $\sqrt{s_0}$.

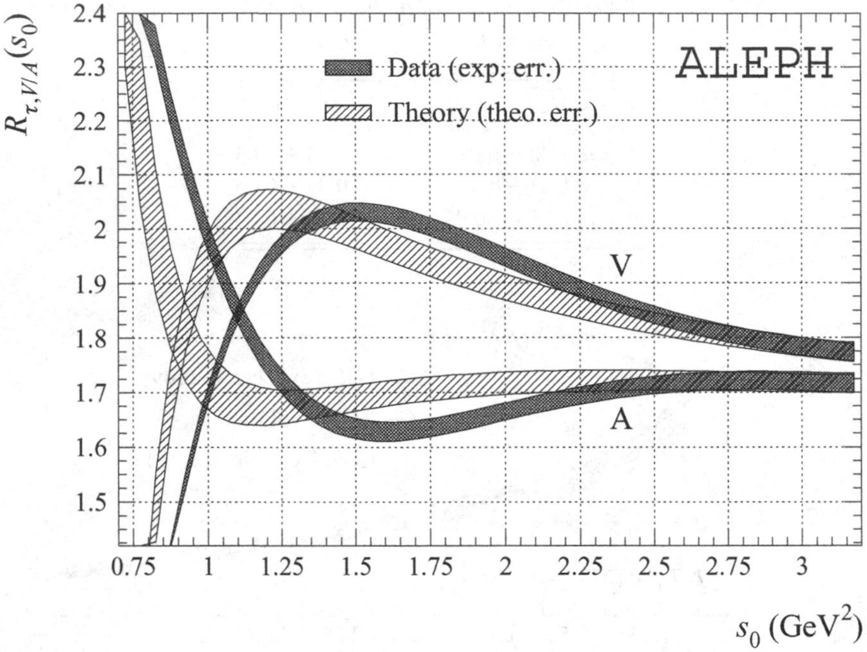


Fig. 25.11. Behaviour of $R_{\tau,V/A}$ versus the $\tau$ mass $s_0$. The theoretical prediction is from FOPT.

One can again notice that the pQCD prediction is in very good agreement with the data for a value of $s_0$ above 1 GeV$^2$, confirming that the determination of $\alpha_s$ from $\tau$ decays is robust.

### 25.6.2 $\alpha_s$ from $e^+e^- \to I = 1$ hadrons data

We have discussed that the vector component of the $\tau$ decay can provide an estimate of $\alpha_s$ although the accuracy is less than in the case of the total inclusive mode. Equivalently, one can use the $e^+e^-$ data into the vector spectral function using an isospin rotation [347] in order to estimate $\alpha_s$. In this way, the decay width reads [346]:

$$R_{\tau,V} \equiv \frac{3\cos^2\theta_c}{2\pi\alpha^2} S_{EW} \int_0^{M_\tau^2} ds \left(1 - \frac{s}{M_\tau^2}\right)^2 \left(1 + \frac{2s}{M_\tau^2}\right) \frac{s}{M_\tau^2} \sigma_{e^+e^- \to I=1} \,. \quad (25.72)$$

This quantity has been used for studying the mass dependence of the prediction on $\alpha_s$. Therefore, it can provide an independent test on the reliability of the result from $\tau$ decays and a test of the isospin symmetry. The value of $\alpha_s$ obtained in this way [346,338,329], at the observed value of the $\tau$ mass, is given in Table 25.2. From this analysis we conclude that the $e^+e^-$ data give a value of $\alpha_s$ compatible with the one from $\tau$ decay data. We show in Fig. 25.12, the behaviour of $R_{\tau,1}$ versus the value of the $\tau$ mass using FOPT.

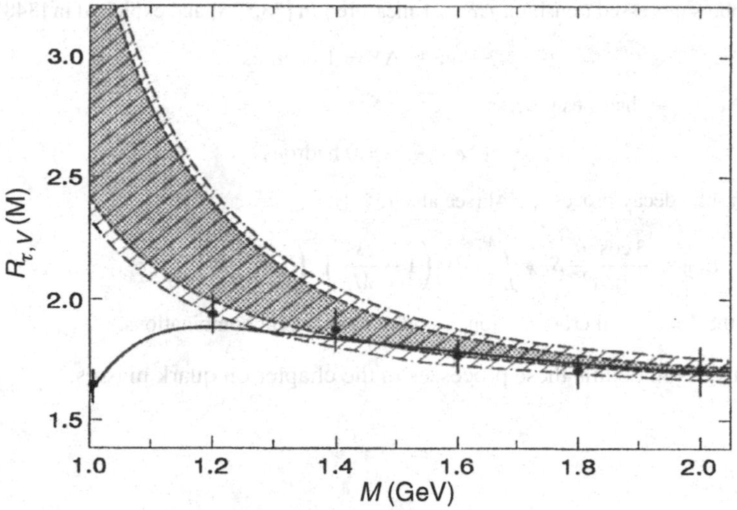


Fig. 25.12. $R_{\tau,V}$ versus the hypothetical $\tau$ mass M using $e^+e^-$ data. The shaded region is the theoretical predictions corresponding to the choice of parameters discussed in the text.

There is a good agreement between the data and the theory above 1.2 GeV. The shaded area between the two dashed curves corresponds to the theoretical predictions for $\alpha_s(M_\tau^2) = 0.33$. The bigger allowed region at low value of $M$ is due to the uncertainty in the leading non-perturbative contribution taken to be:

$$\delta_V^{D=6}(M) \simeq (2.4 \pm 1.3)10^{-2} \times \left(\frac{M_\tau}{M}\right)^6 , \tag{25.73}$$

as given in Eq. 25.71. The departure of the theoretical prediction from the data points below 1.2 GeV signals the important role of higher dimension non-perturbative contributions which we shall discuss in the part of this book dedicated to QCD spectral sum rules. Here, a reasonable fit represented by the continuous line corresponds to the choice:

$$\alpha_s\left(M_\tau^2\right) = 0.33 , \quad \delta_V^{D=6}\left(M_\tau^2\right) \simeq 2.4 \times 10^{-2} , \quad \delta_V^{D=8}\left(M_\tau^2\right) \simeq -9.5 \times 10^{-3} . \tag{25.74}$$

However, though the $D = 8$ condensate contribution is tiny at the $\tau$ mass, its effect at 1.2 GeV is 1.25 larger than the one of $D = 6$, which changes completely the shape of the QCD prediction and can raise some doubt on the validity of the OPE at this scale.

### 25.6.3 Strange quark mass from $\tau$-like processes

$\tau$-like processes have also been used for extracting the strange quark running mass defined in the previous chapter. These processes are:

- The Cabibbo suppressed transition $\Delta S = 1$ measured in [33,328] and exploited in [348]:

$$\tau \to \nu_\tau + \Delta S = 1 \text{ hadrons} \tag{25.75}$$

- The $I = 0$ $e^+e^- \to$ hadrons process:

$$e^+e^- \to I = 0 \text{ hadrons} \tag{25.76}$$

using the $\tau$-like decay process [354] (see also [355]):

$$R_{\tau,0} \equiv \frac{3\cos^2\theta_c}{2\pi\alpha^2} S_{EW} \int_0^{M_\tau^2} ds \left(1 - \frac{s}{M_\tau^2}\right)^2 \left(1 + \frac{2s}{M_\tau^2}\right) \frac{s}{M_\tau^2} \sigma_{e^+e^- \to I=0} \,. \tag{25.77}$$

involving the $I = 0$ total cross-section or/and of its different combinations.

We shall discuss in details these processes in the chapter on quark masses.

# Part VI

## Summary of QCD tests and $\alpha_s$ measurements

## VI.1 The different observables

We have discussed in the previous two parts of this book, deep inelastic scatterings in hadron colliders and different hard processes in $e^+e^-$ annihilations. These hard processes have been used for testing the underlying ideas of perturbative QCD at short distances. Among others, one has studied and measured:

- Scaling violations in different parton model sum rules.
- Structure functions.
- Spin content of the proton.
- Fragmentation functions.
- Spin of the photon.
- One hadron inclusive production.
- Jets.
- Total inclusive $e^+e^-$ cross-sections.
- Hadronic $\tau$ and $Z^0$ decays.

In all these hard processes, most of the perturbative QCD predictions based on the $SU(3)_c$ colour group and on asymptotic freedom properties have been confirmed by the data.

## VI.2 Different tests of QCD

The main outcomes of these analysis in the previous parts of the book are given in the following sections.

### VI.2.1 Deep inelastic scatterings

- A measurement of the scaling violations to parton model predictions in deep inelastic processes using different moments of the structure functions as predicted by QCD. In the unpolarized case, one has used these processes to extract the value of the QCD running coupling. In the polarized case, one has been able to emphasize the important universal rôle of the QCD anomaly for explaining the relative suppression of the first moment of the structure function compared to the OZI prediction (so-called proton spin) and a proposal for testing its effect from the measurement of the photon spin in $\gamma$-$\gamma$ scattering processes, and of some semi-inclusive processes.

- An extension of the test of the validity of perturbation theory in the low-$x$ region leading to a modification of the Altarelli–Parisi evolution equations.

### VI.2.2 QCD jets

- A confirmation of the vector nature of the gluon rather than its scalar nature from the measurement of the moment distributions in three-jet events.
- A measurement of the ratio of the colour group factor $C_A/C_F$ from the scaling violation rates in inclusive hadron distributions and charged hadron multiplicites from gluon and quark jets, which leads to

$$\frac{C_A}{C_F} = 2.24 \pm 0.11 \,, \tag{VI.1}$$

in agreement with the QCD expectation:

$$\frac{C_A}{C_F} = 9/4 = 2.25 \,. \tag{VI.2}$$

This fact confirms the $SU(3)_c$ colour group structure of QCD for describing the strong interactions, and the appearance of the different vertices involving gluon interactions. It also rules out some other group candidates (Abelian, semi-simple Lie group...).
- An extraction of the QCD running coupling $\alpha_s$.

*Inclusive $e^+e^-$, $Z \to$ hadrons and $\tau \to \nu_\tau$ hadrons processes*

- Most precise extractions of the QCD running coupling $\alpha_s$ using the high statistics LEP measurements of the $Z \to$ hadrons and $\tau \to \nu_\tau +$ hadrons decays and the best QCD approximation available today (NNLO and resummation of the asymptotic terms of the QCD series). Unlike the previous deep inelastic and jet processes, one does not need to introduce structure and/or fragmentation functions which can limit the accuracy of the predictions.

## VI.3 Summary of the $\alpha_s$ determinations

- In the massless quark limits which are a good approximation for the light quarks, QCD is a one-parameter theory gouverned by its running coupling $\alpha_s(Q^2)$ evaluated at a scale $Q$, such that all hard strong interaction processes, where one can apply perturbative QCD, should be described in terms of this single input parameters.
- A determination of the values of the running QCD coupling $\alpha_s(Q^2)$ at different energies from various processes as summarized in the table and figures from [139]. In this comparison, the coupling should be defined in the same way everywhere. The $\overline{MS}$ scheme has been adopted as the most convenient renormalization scheme for defining this coupling.

One can see that the running of the coupling shown in Fig. VI.1 from 1 to 100 GeV and at LEP2 energies in Fig. VI.2 satisfies the $1/\log$ behaviour predicted by QCD. The slope of the curve interpreted in terms of the first coefficient of the $\beta$ function lead to an alternative measurement of

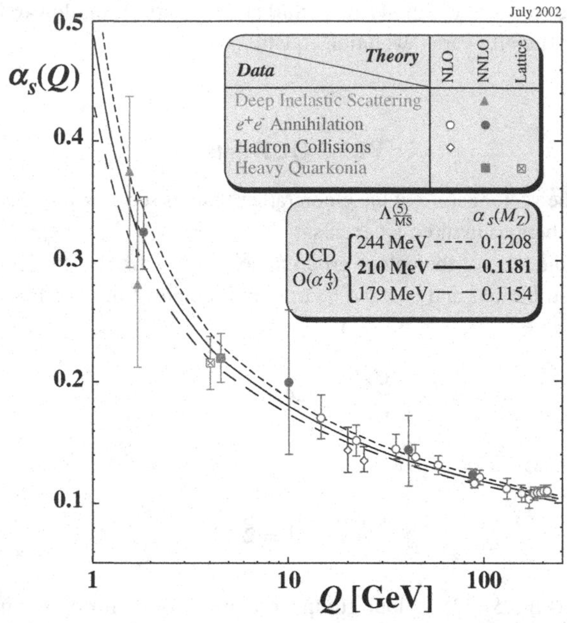


Fig. VI.1. Summary of the different $\alpha_s$ determinations at different energies from [139].

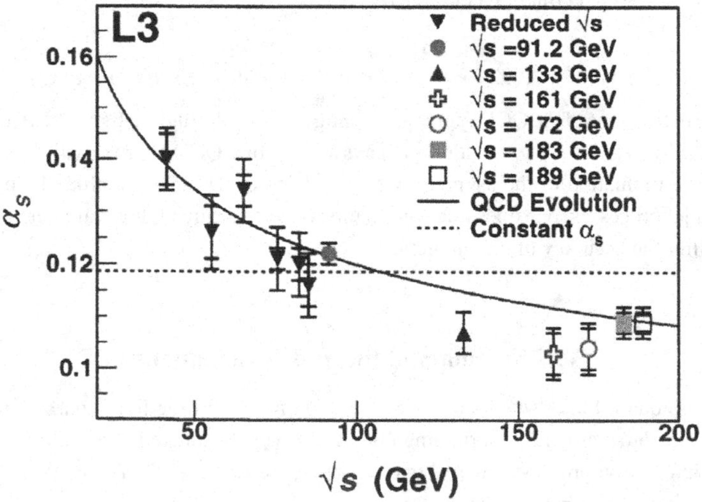


Fig. VI.2. $\alpha_s$ determinations from hadronic event shapes at LEP2 energies.

the number of colours:

$$N_c = 3.03 \pm 0.12 , \tag{VI.3}$$

which is an internal consistency check of the results between data and QCD ($N_c = 3$ in QCD!).

- Evaluated at the common scale $Q = M_{Z^0}$, the different experiments lead to consistent values of $\alpha_s$ as shown in Fig. VI.3, with the average value from the six most significant NNLO determinations

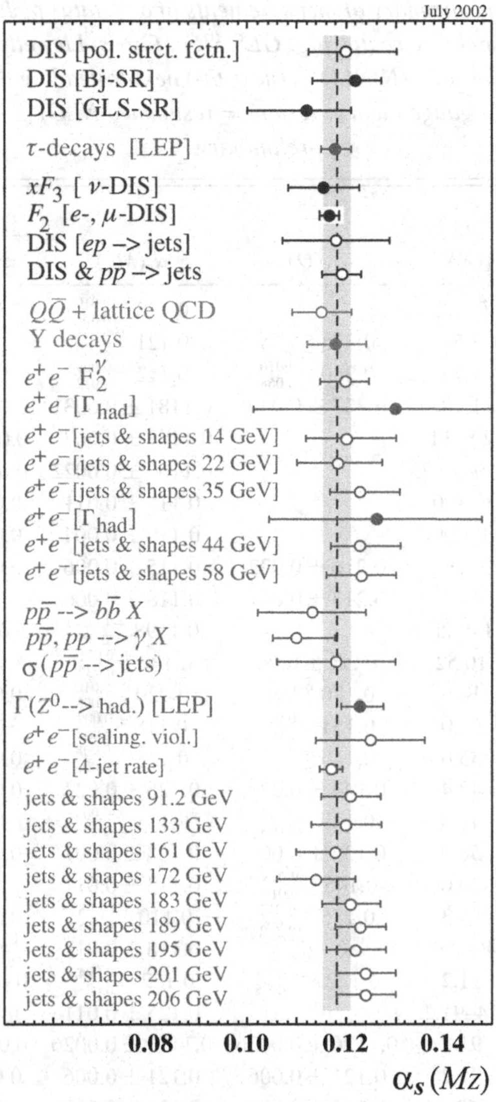


Fig. VI.3. Summary of the different $\alpha_s$ determinations at the common scale $M_{Z^0}$ from [139].

(total error less or equal than 0.008) [139]:[1]

$$\alpha_s(M_{Z^0}) = 0.1181 \pm 0.0027\ . \tag{VI.4}$$

As a result, the corresponding value of the QCD scale for five flavours is:

$$\Lambda_5 = \left(210^{+34}_{-31}\right)\ \text{MeV}\ . \tag{VI.5}$$

[1] The one coming from PDG 2000 [16] is slightly more precise than the average of different determinations from Table VI.1. This is mainly due to the inclusion of the result from [250], where the error of 0.001 has been taken literally.

Table VI.1. *World summary of measurements of $\alpha_s$ (status of July 2002) from [139]: DIS = deep inelastic scattering; GLS-SR = Gross–Llewellyn-Smith sum rule; Bj-SR = Bjorken sum rule; (N)NLO = (next-to-) next-to-leading order perturbation theory; LGT = lattice gauge theory; resum. = resummed NLO). Some entries are still preliminary.*

| Process | Q [GeV] | $\alpha_s(Q)$ | $\alpha_s(M_Z)$ | $\Delta\alpha_s(M_Z)$ exp. | $\Delta\alpha_s(M_Z)$ theor. | Theory |
|---|---|---|---|---|---|---|
| DIS [pol. strct. fctn.] | 0.7–8 | | $0.120^{+0.010}_{-0.008}$ | $^{+0.004}_{-0.005}$ | $^{+0.009}_{-0.006}$ | NLO |
| DIS [Bj-SR] | 1.58 | $0.375^{+0.062}_{-0.081}$ | $0.121^{+0.005}_{-0.009}$ | – | – | NNLO |
| DIS [GLS-SR] | 1.73 | $0.280^{+0.070}_{-0.068}$ | $0.112^{+0.009}_{-0.012}$ | $^{+0.008}_{-0.010}$ | 0.005 | NNLO |
| $\tau$-decays | 1.78 | $0.323 \pm 0.030$ | $0.1181 \pm 0.0031$ | 0.0007 | 0.0030 | NNLO |
| DIS [$\nu$; $xF_3$] | 2.8–11 | | $0.1153 \pm 0.0073$ | 0.0040 | 0.0061 | NNLO |
| DIS [$e/\mu$; $F_2$] | 1.9–15.2 | | $0.1166 \pm 0.0022$ | 0.0009 | 0.0020 | NNLO |
| DIS [e-p $\rightarrow$ jets] | 6–100 | | $0.118 \pm 0.011$ | 0.002 | 0.011 | NLO |
| DIS & $p\bar{p}$ $\rightarrow$jets | 1–400 | | $0.119 \pm 0.004$ | 0.002 | 0.003 | NLO |
| $Q\bar{Q}$ states | 4.1 | $0.216 \pm 0.022$ | $0.115 \pm 0.006$ | 0.000 | 0.006 | LGT |
| $\Upsilon$ decays | 4.75 | $0.217 \pm 0.021$ | $0.118 \pm 0.006$ | – | – | NNLO |
| $e^+e^-$ [$F_2^\gamma$] | 1.4–28 | | $0.1198^{+0.0044}_{-0.0054}$ | 0.0028 | $^{+0.0034}_{-0.0046}$ | NLO |
| $e^+e^-$ [$\sigma_{had}$] | 10.52 | $0.20 \pm 0.06$ | $0.130^{+0.021}_{-0.029}$ | $^{+0.021}_{-0.029}$ | 0.002 | NNLO |
| $e^+e^-$ [jets & shapes] | 14.0 | $0.170^{+0.021}_{-0.017}$ | $0.120^{+0.010}_{-0.008}$ | 0.002 | $^{+0.009}_{-0.008}$ | resum |
| $e^+e^-$ [jets & shapes] | 22.0 | $0.151^{+0.015}_{-0.013}$ | $0.118^{+0.009}_{-0.008}$ | 0.003 | $^{+0.009}_{-0.007}$ | resum |
| $e^+e^-$ [jets & shapes] | 35.0 | $0.145^{+0.012}_{-0.007}$ | $0.123^{+0.008}_{-0.006}$ | 0.002 | $^{+0.008}_{-0.005}$ | resum |
| $e^+e^-$ [$\sigma_{had}$] | 42.4 | $0.144 \pm 0.029$ | $0.126 \pm 0.022$ | 0.022 | 0.002 | NNLO |
| $e^+e^-$ [jets & shapes] | 44.0 | $0.139^{+0.011}_{-0.008}$ | $0.123^{+0.008}_{-0.006}$ | 0.003 | $^{+0.007}_{-0.005}$ | resum |
| $e^+e^-$ [jets & shapes] | 58.0 | $0.132 \pm 0.008$ | $0.123 \pm 0.007$ | 0.003 | 0.007 | resum |
| $p\bar{p} \rightarrow b\bar{b}X$ | 20.0 | $0.145^{+0.018}_{-0.019}$ | $0.113 \pm 0.011$ | $^{+0.007}_{-0.006}$ | $^{+0.008}_{-0.009}$ | NLO |
| $p\bar{p}$, $pp \rightarrow \gamma X$ | 24.3 | $0.135^{+0.012}_{-0.008}$ | $0.110^{+0.008}_{-0.005}$ | 0.004 | $^{+0.007}_{-0.003}$ | NLO |
| $\sigma(p\bar{p} \rightarrow$ jets) | 40–250 | | $0.118 \pm 0.012$ | $^{+0.008}_{-0.010}$ | $^{+0.009}_{-0.008}$ | NLO |
| $e^+e^-$ [$\Gamma(Z^0 \rightarrow$ had.)] | 91.2 | $0.1227^{+0.0048}_{-0.0038}$ | $0.1227^{+0.0048}_{-0.0038}$ | 0.0038 | $^{+0.0029}_{-0.0005}$ | NNLO |
| $e^+e^-$ scal. viol. | 14–91.2 | | $0.125 \pm 0.011$ | $^{+0.006}_{-0.007}$ | 0.009 | NLO |
| $e^+e^-$ four-jet rate | 91.2 | $0.1170 \pm 0.0026$ | $0.1170 \pm 0.0026$ | 0.0001 | 0.0026 | NLO |
| $e^+e^-$ [jets & shapes] | 91.2 | $0.121 \pm 0.006$ | $0.121 \pm 0.006$ | 0.001 | 0.006 | resum |
| $e^+e^-$ [jets & shapes] | 133 | $0.113 \pm 0.008$ | $0.120 \pm 0.007$ | 0.003 | 0.006 | resum |
| $e^+e^-$ [jets & shapes] | 161 | $0.109 \pm 0.007$ | $0.118 \pm 0.008$ | 0.005 | 0.006 | resum |
| $e^+e^-$ [jets & shapes] | 172 | $0.104 \pm 0.007$ | $0.114 \pm 0.008$ | 0.005 | 0.006 | resum |
| $e^+e^-$ [jets & shapes] | 183 | $0.109 \pm 0.005$ | $0.121 \pm 0.006$ | 0.002 | 0.005 | resum |
| $e^+e^-$ [jets & shapes] | 189 | $0.109 \pm 0.004$ | $0.121 \pm 0.005$ | 0.001 | 0.005 | resum |
| $e^+e^-$ [jets & shapes] | 195 | $0.109 \pm 0.005$ | $0.122 \pm 0.006$ | 0.001 | 0.006 | resum |
| $e^+e^-$ [jets & shapes] | 201 | $0.110 \pm 0.005$ | $0.124 \pm 0.006$ | 0.002 | 0.006 | resum |
| $e^+e^-$ [jets & shapes] | 206 | $0.110 \pm 0.005$ | $0.124 \pm 0.006$ | 0.001 | 0.006 | resum |

- However, one should have in mind that the different values of $\alpha_s$ for each process are not obtained within the same QCD approximations. In some processes, they are known very precisely to NNLO, while in some others they are poorly known to NLO. In addition, the theoretical uncertainties are also affected by the asymptotic behaviour of the perturbative series in powers of $\alpha_s$, and by small non-perturbative effects which should be present in different processes. We shall come back to this point in subsequent chapters.

# Part VII

Power corrections in QCD

# 26
# Introduction

The problems of power corrections have been intensively discussed during the last few years [1,3,329–344,356–398].[1] By power corrections to the parton model, one means terms of the order of:

$$\left(\frac{\Lambda}{Q}\right)^n \sim \exp\left(\frac{n\pi}{\alpha_s(Q^2)\beta_1}\right) , \tag{26.1}$$

where $Q$ is a typical momentum much larger than the QCD scale $\Lambda$, and $\beta_1 = -1/2(11 - 2n_f/3)$ in our notation. A priori, this is problematic as these contributions are exponentially small in the inverse of the running coupling, and can be related to many orders of the perturbative expansion. In order to develop a phenomenology of these power corrections, one then assumes that they are numerically important and are responsible for the breaking of asymptotic freedom at intermediate energies. This fact has been firstly indicated in the QCD spectral sum rules phenomenology [1,2,3,356–365] and in the analysis of renormalons [329–380] and instantons [382–398]. However, as noticed in [162], the idea of power corrections are not quite new but can be traced back to an old paper [416]. It has been considered an $e^+e^-$ pair at distance $r$ placed into a centre of a conducting cage of size $L$. Assuming that $L \gg r$, the potential energy of the pair can be approximated by:

$$V_{e^+e^-}(r) \simeq -\frac{\alpha}{r} + \text{const}\,\frac{\alpha r^2}{L^3} \quad \text{for} \quad L \gg r , \tag{26.2}$$

where the second term can be viewed as a power correction to the Coulomb potential. In classical electrodynamics, this correction corresponds to the interaction of a dipole with its image, or can be also derived in terms of one-photon exchange. From this example, one can derive, by analogy, the heavy quark potential of QCD, which at short distances looks as [417]:

$$V_{\bar{Q}Q}(r \to 0) \simeq \frac{C}{r} + \text{const}\,\Lambda^3 r^2 , \tag{26.3}$$

where $C$ is calculable in perturbative QCD as series in $\alpha_s$, and where one should notice the absence of a linear term at short distance. For deriving this expression, one has replaced $L$ by $1/\Lambda$, assuming that the gluon propagator is modified by IR effects at the scale $1/\Lambda$. The shift

[1] It is a pleasure to thank Valya Zakharov for reading this Part.

of the atomic levels in the cage are sensitive to a local characteristic of the non-perturbative fields, which on dimensional grounds reads [416,162]:

$$(\delta E)_{NP} \sim \frac{\langle 0|\mathbf{E}^2|0\rangle}{m_e^3} , \tag{26.4}$$

where $\langle 0|\mathbf{E}^2|0\rangle$ corresponds to the difference of the mean value of $\mathbf{E}^2$ in one photon approximation evaluated without and with the cage, and which is UV finite. Translated in QCD, one can have the picture of the $\bar{Q}Q$ bound states, in terms of the density of the colour field strengths, or more popularly known as gluon condensate $\langle 0|\alpha_s(G^a_{\mu\nu})^2|0\rangle$, first discussed in [418]:

$$\delta E_{nl} \sim C_{nl}\frac{\langle 0|\alpha_s\left(G^a_{\mu\nu}\right)^2|0\rangle}{M_Q^4} + \cdots \tag{26.5}$$

Considering now the QCD correlation function, which generically reads:

$$\Pi_H(-q^2) \equiv i\int d^4x\, e^{iqx}\langle 0|\mathcal{T}J(x)_H\,(J_H(0))^\dagger\,|0\rangle , \tag{26.6}$$

where $J_H(x)$ is the hadronic current of quark and/or gluon fields, a use of the standard OPE leads to [1]:

$$\Pi_H(-q^2) \sim \Pi_H(-q^2)_{\text{parton}}\left[1 + a_H\alpha_s(-q^2) + \frac{b_H}{q^4} + \cdots\right] \tag{26.7}$$

for massless quarks, and, where $a_H$ and $b_H$ are constants which depend on the hadron quantum numbers. One should notice the absence of the $1/Q^2$ power term in the standard OPE as we shall discuss in detail in the next chapter. We shall also see in the following chapters how the resummation of the QCD asymptotic series and its phenomenological picture can induce such a term.

# 27

# The SVZ expansion

## 27.1 The anatomy of the SVZ expansion

For definiteness, let us illustrate our discussion from the generic two-point correlator:

$$\Pi_H(q^2) \equiv i \int d^4x \, e^{iqx} \langle 0|\mathcal{T} J_H(x) (J_H(0))^\dagger |0\rangle , \tag{27.1}$$

where $J_H(x)$ is the hadronic current of quark and/or gluon fields. Here, the analysis is in principle much simpler than in the case of deep inelstic scatterings, because one has to sandwich the $T$-product of currents between the vacuum rather than between two proton states. Following SVZ [1], the breaking of ordinary perturbation theory at low $q^2$ is due to the manifestation of non-perturbative terms appearing as power corrections in the operator product expansion (OPE) of the Green function à la Wilson [222]. In this way, one can write:

$$\Pi_H(q^2, m^2) \simeq \sum_{D=0,2,4,\ldots} \frac{1}{(m^2 - q^2)^{D/2}} \sum_{dim\mathcal{O}=D} C(q^2, m^2, \nu^2) \langle \mathcal{O}(\nu) \rangle , \tag{27.2}$$

provided that $m^2 - q^2 \gg \Lambda^2$. For simplicity, $m$ is the heaviest quark mass entering into the correlator; $\nu$ is an arbitrary scale that separates the long- and short-distance dynamics; $C$ are the Wilson coefficients calculable in perturbative QCD by means of Feynman diagrams techniques; $\langle \mathcal{O} \rangle$ are the non-perturbative (non-calculable) condensates built from the quarks or/and gluon fields. Though, separately, $C$ and $\langle \mathcal{O} \rangle$ are (in principle) $\nu$-dependent, this $\nu$-dependence should (in principle) disappear in their product.

- The case $D = 0$ corresponds to the naïve perturbative contribution.
- For $D = 2$ and owing to gauge invariance which forbids the formation of a condensate, one can only have the contributions from the light quark running mass squared. Moreover, one may (or may not) also expect that the summation of perturbation theory via UV renormalon technology (see next section) could also induce such a term, while the possibility from the freezing mechanism of the coupling constant is negligibly small as it is expected to be of the order of $(\alpha_s/\pi)^2$ [338].
- For $D = 4$, the condensates that can be formed are the quark and gluon ones:

$$m\langle \bar{\psi}\psi \rangle , \qquad \langle \alpha_s G^2 \rangle , \tag{27.3}$$

where the former can be fixed by pion PCAC (see Part I) in the standard Gell-Man–Oakes–Renner (GMOR) realization of chiral symmetry.

- For $D = 5$, one can only have, in the massless quark limit, the mixed quark-gluon condensate:

$$\langle \bar{\psi} \sigma_{\mu\nu} \lambda^a / 2 G_a^{\mu\nu} \psi \rangle . \tag{27.4}$$

- For $D = 6$ one has, in the chiral limit, the triple gluon and the four-quark condensates:

$$g^3 f_{abc} \langle G^a G^b G^c \rangle , \qquad \alpha_s \langle \bar{\psi} \Gamma_1 \psi \bar{\psi} \Gamma_2 \psi \rangle , \tag{27.5}$$

where $\Gamma_i$ are generic notations for any Dirac and/or colour matrices.

The validity of the SVZ expansion has been understood formally, using renormalon technology (see next sections) and the mixing of operators under renormalizations. The SVZ expansion has been also tested in the $\lambda\phi^4$ [367,368] and QCD-like models [369] (Schwinger two-dimensional gauge theories [370], the $CP^{N-1}$ model [371], which both have instantons and $\theta$-vacua; the Gross–Neveu model [372] with dynamical chiral symmetry breaking [373] and 2-$d$ $O(N)$ free non-linear $\sigma$ model [374,367], where both have the asymptotic freedom property of QCD).

Its phenomenological confirmation can be viewed from the unexpected accurate extraction of $\alpha_s$ from $\tau$ decays and from independent measurements of the QCD condensates (see chapter on QCD condensates).

## 27.2 SVZ expansion in the $\lambda\phi^4$ model

For a simple pedagogical introduction, let us illustrate the SVZ expansion for scalar field theory.[1] The bare Lagrangian of the theory reads:

$$\mathcal{L}_\phi = \frac{1}{2}(\partial_\mu \phi_B)^2 - \frac{1}{2} m_B^2 \phi_B^2 - \frac{\lambda_B}{4!}\phi_B^4 , \tag{27.6}$$

where $\phi$ is the scalar field, $m$ and $\lambda$ are its mass and coupling. $B$ refers to bare quantity. It is known that for $m_B^2 < 0$, one has a spontaneous breaking mechanism where the field acquires a non-vanishing expectation value, which is non-analytical in the coupling, such that the model mimics non-perturbative effects. In order to further simplify our discussions, let us, however, work in the case $m_B^2 > 0$, where no condensate breaks spontaneously the symmetry and let us ignore (for the moment) renormalization effects. We shall be concerned with the propagator:

$$\mathcal{D}(q^2) = i \int d^4x e^{iqx} \langle 0|\mathbf{T}\phi(x)\phi(0)|0\rangle , \tag{27.7}$$

which we shall evaluate in two different ways. In the first one, we evaluate it using the standard perturbative expansion in $\lambda$ (Fig. 27.1).

[1] We shall ignore in this illustrative example the radiative corrections discussed in [367,368].

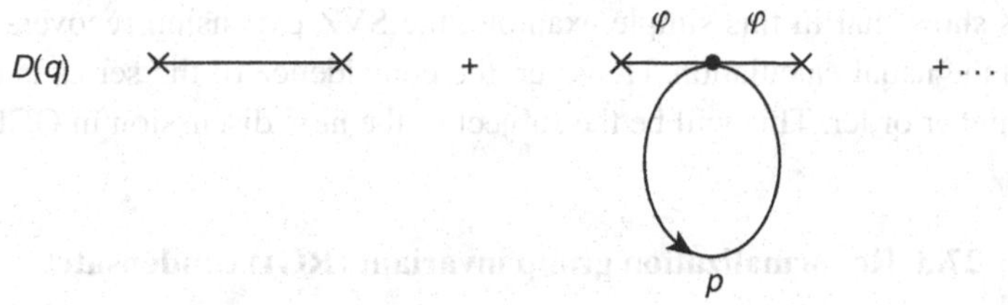


Fig. 27.1. Lowest order perturbative contribution to the scalar correlator.

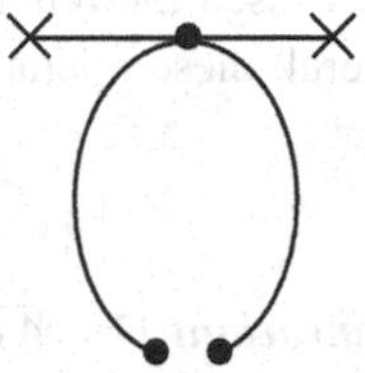

Fig. 27.2. Lowest order scalar condensate contribution to the scalar correlator.

Using, for instance, a Pauli–Villars regularization (the following conclusion is regularization invariant), one obtains for $-q^2 \gg m_B^2 \equiv m^2$:

$$\mathcal{D}(q^2) \simeq \frac{1}{q^2} + \frac{m^2}{q^4}\left\{1 - \frac{\lambda}{32\pi^2}\left(\log\frac{\Lambda^2}{m^2} - \frac{\Lambda^2}{m^2}\right)\right\}, \tag{27.8}$$

where $\Lambda$ is an arbitrary UV cut-off.

In the second method, one evaluates the propagator using the SVZ expansion for $-q^2 \gg m^2$. Therefore, it reads:

$$\mathcal{D}(q^2) \simeq C_1\mathbf{1} + C_\phi\langle\phi^2\rangle + \cdots \tag{27.9}$$

By introducing the scale $\nu$, which separates the long- and short-wavelength fluctuations, one can extract $C_1$ from the perturbative graph for $p > \nu$ (short fluctuations):

$$C_1 \simeq \frac{1}{q^2} + \frac{m^2}{q^4}\left\{1 - \frac{\lambda}{32\pi^2}\left(\log\frac{\Lambda^2}{\nu^2} - \frac{\Lambda^2 - \nu^2}{m^2}\right)\right\}. \tag{27.10}$$

The Wilson coefficient $C_\phi$ is associated to the $\phi^2$ 'condensate':

$$C_\phi \simeq \frac{\lambda}{2q^4}, \tag{27.11}$$

and comes from Fig. 27.2.

The condensate $\langle\phi^2\rangle$ corresponds to the evaluation of the tadpole-like graph for $p < \nu$ (large fluctuations), from which, one obtains:

$$\langle\phi^2\rangle \simeq \frac{1}{16\pi^2}\left(\nu^2 - m^2\log\frac{\nu^2}{m^2}\right). \tag{27.12}$$

These results show that in this simple example, the SVZ expansion recovers (to this approximation) the usual calculation. However, the coincidence of the series is not trivial if one goes to higher order. This will be the subject of the next discussion in QCD.

## 27.3 Renormalization group invariant (RGI) condensates

Now, the next step is to see if the condensates can be well-defined in perturbation theory, namely if one can form quantities which are invariant under the RGE. This discussion has already been anticipated when we discussed the renormalization of composite operators, where it has been shown that, in general, these operators can mix under renormalizations [126–131].

### 27.3.1 Scale invariant $D = 4$ condensates

- **Generalities and definitions**

In the previous section, it was demonstrated that the condensates:

$$m_j\langle\bar{\psi}_j\psi_j\rangle\,,$$
$$\langle\theta^\mu_\mu\rangle = \frac{1}{4}\langle\beta(\alpha_s)GG\rangle + \sum_j \gamma_m(\alpha_s)m_j\langle\bar{\psi}_j\psi_j\rangle\,, \tag{27.13}$$

are renormalization group invariant [126,128]. However, perturbative evaluations of the quark-mass corrections to the correlation functions give rise to IR logarithms of the form $m^4\alpha_s^n(\nu)\log^k(m/\nu)$ $(k \leq n+1)$, where $\nu$ is the $\overline{MS}$ renormalization scale [167,399]. The mass singularities arise from the region of small loop-momenta in the relevant Feynman diagrams, and therefore should be absorbed (like the IR renormalons) into the non-perturbative condensates $\langle\mathcal{O}(\nu)\rangle$. The IR logarithms are nothing more than the perturbative contributions to the $D = 4$ vacuum condensates.

However, one should notice that the calculation of the $D = 2$ quark-mass corrections does not produce any logarithms $\log^k(m/\nu)$, which is a consequence of the absence of the $D = 2$ operators in QCD.

In order to be explicit, let us consider the pseudoscalar two-point correlator defined in Eq. (8.29). At $q = 0$, one obtains from a perturbative calculation of the correlator [167]:

$$\Psi_5^R(0)|_{\rm pert} = \frac{3}{4\pi^2}(m_i + m_j)\left(m_i^3 Z_i + m_j^3 Z_j\right)\,, \tag{27.14}$$

with:

$$Z_i = 1 - \log\frac{m_i^2}{\nu^2} + \frac{2}{3}\left(\frac{\alpha_s}{\pi}\right)\left(5 - 5\log\frac{m_i^2}{\nu^2} + 3\log^2\frac{m_i^2}{\nu^2}\right)\,, \tag{27.15}$$

which improves the non-perturbative Ward identity in Eq. (2.17), and which indicates that, in order to absorb the mass singularities, one should add a perturbative piece to the quark condensate. In a similar way, the perturbative piece to the gluon condensate reads [325]:

$$\langle GG(\nu)\rangle^{\overline{MS}}_{\rm pert} = -\frac{1}{2\pi^2}\left(\frac{\alpha_s}{\pi}\right)\sum_i m_i^4(\nu)\left[9 - 8\log\frac{m_i^2}{\nu^2} + 3\log^2\frac{m_i^2}{\nu^2}\right]\,. \tag{27.16}$$

The summation of the log-terms using the RGE becomes more convenient by working with the scale invariant *non-normal ordered* condensates [325]:[2]

$$\langle\overline{\alpha_s GG}\rangle \equiv \left(1+\frac{16}{9}\alpha_s(\nu)+O\left(\alpha_s^2\right)\right)\frac{\alpha_s(\nu)}{\pi}\langle\ :GG(\nu):\ \rangle^{\overline{MS}}$$
$$-\frac{16}{9}\frac{\alpha_s(\nu)}{\pi}\left(1+\frac{19}{24}\frac{\alpha_s(\nu)}{\pi}+O\left(\alpha_s^2\right)\right)\sum_i m_i\langle\bar\psi_i\psi_i\rangle^{\overline{MS}}(\nu)$$
$$-\frac{1}{3\pi}\left(1+\frac{4}{3}\frac{\alpha_s(\nu)}{\pi}+O\left(\alpha_s^2\right)\right)\sum_i m_i^4(\nu)\,, \tag{27.17}$$

and:

$$\langle\overline{m_i\bar\psi_j\psi_j}\rangle \equiv m_i\langle\ :\bar\psi_j\psi_j:\ \rangle^{\overline{MS}}(\nu)$$
$$+\frac{3}{7\pi\alpha_s(\nu)}\left(1-\frac{53}{24}\frac{\alpha_s(\nu)}{\pi}+O\left(\alpha_s^2\right)\right)m_i(\nu)m_j^3(\nu)\,, \tag{27.18}$$

where one can notice the inverse power of $\alpha_s$ in the expression for the quark condensate. The use of these scale-invariant condensates in the OPE implies that the coefficient functions obey an homogeneous RGE, which then facilitates the summation of the $\log(Q/\nu)$ terms in the analysis.

Analogously to the invariant mass $\hat m_i$ in Eq. (11.77), a spontaneous RGI mass $\hat\mu_i$ associated to the quark vacuum condensate can also be introduced by taking into account the fact that the product $m_i\langle\bar\psi\psi\rangle$ is RGI (at least to leading order in $m_i$) [28,110,2]. Then, one obtains:

$$\overline{\langle\ :\bar\psi_i\psi_i:\ \rangle}(\nu) = -\,\hat\mu_i^3\left(-\beta_1 a_s(\nu)\right)^{\gamma_1/\beta_1}\left\{1+\frac{\beta_2}{\beta_1}\left(\frac{\gamma_1}{\beta_1}-\frac{\gamma_2}{\beta_2}\right)a_s(\nu)\,,\right.$$
$$+\frac{1}{2}\left[\frac{\beta_2^2}{\beta_1^2}\left(\frac{\gamma_1}{\beta_1}-\frac{\gamma_2}{\beta_2}\right)^2-\frac{\beta_2^2}{\beta_1^2}\left(\frac{\gamma_1}{\beta_1}-\frac{\gamma_2}{\beta_2}\right)+\frac{\beta_3}{\beta_1}\left(\frac{\gamma_1}{\beta_1}-\frac{\gamma_3}{\beta_3}\right)\right]a_s^2(\nu)$$
$$\left.+\,\mathcal{O}\left(a_s^3,\ m_i^3\right)\right\}^{-1}\,, \tag{27.19}$$

where the values of the $\beta$ functions and mass anomalous dimensions can be found in Table 11.1.

- **Values of the light quark condensates**

Assuming a GMOR realization of chiral symmetry as commonly accepted, the light quark condensate can be estimated, to leading order of the light quark mass, from the PCAC relation given in Eq. (2.22):

$$(m_u+m_d)\langle\ :\bar\psi_u\psi_u+\bar\psi_d\psi_d:\ \rangle \simeq -2m_\pi^2 f_\pi^2\,. \tag{27.20}$$

Anticipating the values of the running masses in the next chapter, one can deduce:

$$\frac{1}{2}\langle\bar uu+\bar dd\rangle(2\ \mathrm{GeV}) = -[(254\pm 15)\ \mathrm{MeV}]^3\,. \tag{27.21}$$

We shall also see, in the next chapter, that one has a large breaking from the $SU(3)_F$ flavour symmetric value of the condensates as first noticed in [400] from the pseudoscalar sum rule, where

[2] One should notice that the use of the Wick's theorem in the evaluation of the Feynman diagrams generates automatically *normal-ordered* condensates which we shall denote as $\langle\ :\mathcal{O}:\ \rangle$

a recent update estimate [354,419–421] (see also [423]) leads to the ratio of the *normal ordered* condensates:

$$\langle\ :\bar{s}s:\ \rangle/\langle\ :\bar{u}u:\ \rangle = 0.66 \pm 0.10\ , \tag{27.22}$$

in agreement within the errors with different baryon sum rules results [426–430]. Combining Eq. (27.17) with this result, one can also deduce the ratio of the *non-normal ordered* condensates:

$$\langle\ \bar{s}s\ \rangle/\langle\ \bar{u}u\ \rangle = 0.75 \pm 0.12\ . \tag{27.23}$$

- **Value of the gluon condensate**
  The gluon condensate has been orginally estimated by SVZ from charmonium sum rules [1]. We shall see in the next chapter that a re-extraction of this quantity from the $e^+e^- \to I = 1$ hadrons data and from the heavy quarkonia mass-splittings, lead to [329,313] (Sections 51.3 and 52.10):

$$\langle\alpha_s G^2\rangle = (7.1 \pm .9)10^{-2}\ \text{GeV}^4\ , \tag{27.24}$$

  as expected from various post-SVZ estimates [3,356–365], but about a factor of two higher than the original SVZ estimate.

### 27.3.2 $D = 5$ *mixed quark-gluon condensate*

The renormalization of the mixed quark-gluon condensate has been studied in [130], where it has been shown that the scale invariant, which one can form, is the combination:

$$\langle\bar{O}_5\rangle = \alpha_s^{\gamma_M/-\beta_1}\langle O_4 + x\ O_1 + y\ O_2\rangle \tag{27.25}$$

where the dimension-five gauge invariant operators are:

$$\begin{aligned}
O_1 &\equiv i m_j^2\langle\bar{\psi}_j\psi_j\rangle\ ,\\
O_2 &\equiv -\frac{i}{4}m_j\langle G^2\rangle\ ,\\
O_3 &\equiv -m_j\bar{\psi}_j(\hat{D}+im_j)\psi_j\ ,\\
O_4 &\equiv g\left\langle\bar{\psi}_j\sigma^{\mu\nu}\frac{\lambda^a}{2}\psi_j G^a_{\mu\nu}\right\rangle\ .
\end{aligned} \tag{27.26}$$

For $SU(3)_C$:

$$x = -\frac{1944}{315} \qquad y = -\frac{72}{63} \qquad \gamma_M = -\frac{1}{3}\ , \tag{27.27}$$

which indicates that working with $\langle O_4\rangle$, is only valid to leading order in the quark mass-expansion.

The value of the mixed quark-gluon condensate has been estimated from baryon sum rules [424–430] to be about 0.8 GeV$^2$, and alternatively from the heavy-light quark systems $B$ and $B^*$ [401]. Further discussions will be given in the next chapter. The QSSR value of the mixed quark-gluon condensate is:

$$g\left\langle\bar{\psi}_j\sigma^{\mu\nu}\frac{\lambda^a}{2}\psi_j G^a_{\mu\nu}\right\rangle \equiv M_0^2\ \alpha_s^{\gamma_M/-\beta_1}\langle\bar{\psi}_j\psi_j\rangle\ , \tag{27.28}$$

where:

$$M_0^2 = (0.8 \pm 0.01)\,\mathrm{GeV}^2\,. \tag{27.29}$$

For a conservative result, we shall adopt:

$$M_0^2 = (0.8 \pm 0.1)\,\mathrm{GeV}^2\,, \tag{27.30}$$

assuming a 10% uncertainty typical for the sum rule approach.

### 27.3.3 $D = 6$ gluon condensates

- **Triple gluon condensate:**
  It reads:

$$\langle \mathcal{O}_G \rangle \equiv g^3 f_{abc} \langle G^a G^b G^c \rangle\,, \tag{27.31}$$

and does not mix under renormalization with the other $D = 6$ operators. Its renormalization improved expression is [130]:

$$\langle \bar{\mathcal{O}}_G \rangle = \alpha_s^{-\gamma_G/\beta_1} \langle \mathcal{O}_G \rangle\,, \tag{27.32}$$

where for $SU(N)$:

$$\gamma_G = \frac{1}{6}(2 + 7N)\,. \tag{27.33}$$

A crude estimate of this condensate can be obtained using a dilute gas instanton approximation model (DIGA) (see next chapter). For an instanton size $\rho_c \approx 200$ MeV, one obtains [382]:[3]

$$\langle \mathcal{O}_G \rangle \approx (1.5 \pm 0.5)\,\mathrm{GeV}^2 \langle \alpha_s G^2 \rangle\,, \tag{27.34}$$

where we have assumed a 30% error. This estimate is in good agreement with the $SU(2)$ lattice estimate [402]:

$$\langle \mathcal{O}_G \rangle \approx (1.2\,\mathrm{GeV}^2) \langle \alpha_s G^2 \rangle\,, \tag{27.35}$$

although one should be careful in using this result (in particular the sign) as the latter has been obtained in the Euclidean region.

- **Gluon derivative condensate:**

$$g^2 \langle DG\,DG \rangle \equiv g^2 \langle D_\mu G_a^{\alpha\mu}\, D^\nu G^a_{\alpha\nu} \rangle\,, \tag{27.36}$$

It reduces to the four-quark condensates:

$$g^4 \left\langle \sum \bar\psi \gamma_\mu \frac{\lambda_a}{2} \psi \right\rangle^2\,, \tag{27.37}$$

after the use of the equation of motion.

[3] The effect of the anomalous dimension has not been included in the analysis but it is negligible.

### 27.3.4 $D = 6$ *four-quark condensates*

On the contrary, the four-quark condensates mix with each other [130,131]. In the chiral limit and after the use of the equation of motion, one remains only with five operators:

$$O_\Gamma \equiv \langle \bar\psi^\alpha \Gamma_1 \psi_\alpha \bar\psi^\beta \Gamma_2 \psi_\beta \rangle \,, \tag{27.38}$$

where the colour indices $(\alpha, \beta)$ run from 1 to $N$. $\Gamma_i$ is any Dirac and colour indices. One can choose the basis:

$$\begin{aligned} &\Gamma_S = 1 \,, \quad \Gamma_V = \gamma_\mu \,, \\ &\Gamma_P = \gamma_5 \,, \quad \Gamma_A = \gamma_\mu \gamma_5 \,, \\ &\Gamma_T = \sigma_{\mu\nu} \equiv \tfrac{i}{2}[\gamma_\mu, \gamma_\nu] \,. \end{aligned} \tag{27.39}$$

In the large $N$-limit, where the vacuum saturation is expected, one has:

$$\langle \bar\psi \Gamma_1 \psi \bar\psi \Gamma_2 \psi \rangle = \frac{1}{16N^2} [\mathbf{Tr}\, \Gamma_1\ \mathbf{Tr}\, \Gamma_2 - \mathbf{Tr}(\Gamma_1 \Gamma_2)] \langle \bar\psi \psi \rangle^2 \,. \tag{27.40}$$

With the inclusion of the next $1/N$-correction, one has [130]:

$$\begin{aligned} &O_S = \langle \bar\psi \psi \rangle^2 \left(1 - \tfrac{1}{4N}\right) \,, \quad O_P = \langle \bar\psi \psi \rangle^2 \left(-\tfrac{1}{4N}\right) \,, \\ &O_V = \langle \bar\psi \psi \rangle^2 \left(-\tfrac{1}{N}\right) \,, \qquad O_A = \langle \bar\psi \psi \rangle^2 \left(\tfrac{1}{N}\right) \,, \\ &O_T = \langle \bar\psi \psi \rangle^2 \left(-\tfrac{3}{4N}\right) \,, \end{aligned} \tag{27.41}$$

indicating that in the large $N$-limit, only $O_S$ is relevant. However for finite $N$, the operators mix with each other and signal the fact that the vacuum saturation is inconsistent with the RGE, as one cannot form a RGI condensate. In the solely case of $N \to \infty$, the situation improves, as one can construct the RGI condensate:

$$\langle \bar O_2 \rangle = \alpha_s^{\gamma_6/-\beta_1} \langle g^2 \bar\psi \psi \bar\psi \psi \rangle \,, \tag{27.42}$$

where:

$$\gamma_6(N \to \infty) = \frac{143}{33} N \,. \tag{27.43}$$

The size of the four-quark condensates has been estimated from the $e^+e^- \to$ hadrons data [403,404], [405–409] and from the $\tau \to$ hadrons decay width [328,33]. It has been noticed for *the first time* in [404] that the vacuum saturation assumption, used previously, underestimates the real value of the four-quark condensate by a factor:

$$\rho \simeq 2 - 3 \,, \tag{27.44}$$

while in the $e^+e^-$ analysis, there is a strong correlation between the four-quark and the $\langle \alpha_s G^2 \rangle$ condensates, as they appear in opposite signs in the OPE. This result has been also found from baryon sum rules [424,426]. A recent analysis of the $e^+e^-$ data gives [329] (reprinted paper):

$$\rho \alpha_s \langle \bar\psi \psi \rangle^2 = (5.8 \pm 0.9) 10^{-4}\ \mathrm{GeV}^6 \,, \tag{27.45}$$

a result confirmed by the ALEPH and OPAL measurement from $\tau$-decay data [328,33].

### 27.3.5 Higher dimensions gluonic condensates

The dimension-eight gluonic condensates have been discussed in [410–412]. In general, one can form eight operators:

$$
\begin{aligned}
O_1 &= \langle \mathbf{Tr}\, G^2\ \mathbf{Tr}\, G^2 \rangle\,, \\
O_2 &= \langle \mathbf{Tr}\, G_{\nu\mu} G^{\rho\mu}\ \mathbf{Tr}\, G_{\nu\tau} G^{\rho\tau} \rangle\,, \\
O_3 &= \langle \mathbf{Tr}\, G_{\nu\mu} G^{\tau\rho}\ \mathbf{Tr}\, G_{\nu\mu} G^{\tau\rho} \rangle\,, \\
O_4 &= \langle \mathbf{Tr}\, G_{\nu\mu} G^{\tau\rho}\ \mathbf{Tr}\, G^{\nu}_{\tau} G^{\mu}_{\rho} \rangle\,, \\
O_5 &= \langle \mathbf{Tr}\, G_{\nu\mu} G^{\mu\rho}\ G_{\rho\tau} G^{\tau\nu} \rangle\,, \\
O_6 &= \langle \mathbf{Tr}\, G_{\nu\mu} G^{\nu\mu} G^{\tau\rho} G_{\tau\rho} \rangle\,, \\
O_7 &= \langle \mathbf{Tr}\, G_{\nu\mu} G^{\nu\rho}\ G_{\mu\tau} G^{\rho\tau} \rangle\,, \\
O_8 &= \langle \mathbf{Tr}\, G_{\nu\mu} G^{\rho\tau}\ G^{\nu\mu} G^{\rho\tau} \rangle\,.
\end{aligned}
\tag{27.46}
$$

Using the symmetry properties of the colour indices and an explicit evaluation of the trace, one can show that one has only six independent operators and the relation for $N = 3$ [411]:

$$
\begin{aligned}
O_5 + 2O_7 &= O_2 + \frac{1}{2} O_4\,, \\
O_8 + 2O_6 &= O_3 + \frac{1}{2} O_1\,.
\end{aligned}
\tag{27.47}
$$

The use of the vacuum saturation in the large $N$-limit gives:

$$
\begin{aligned}
O_1 &= \langle G^2\rangle^2 \tfrac{1}{4}\left(1 + \tfrac{1}{3}\tfrac{1}{N^2-1}\right), & O_2 &= \langle G^2\rangle^2 \tfrac{1}{4}\left(\tfrac{1}{4} + \tfrac{1}{3}\tfrac{1}{N^2-1}\right), \\
O_3 &= \langle G^2\rangle^2 \tfrac{1}{4}\left(\tfrac{1}{6} + \tfrac{7}{6}\tfrac{1}{N^2-1}\right), & O_4 &= \langle G^2\rangle^2 \tfrac{1}{4}\left(\tfrac{1}{12} + \tfrac{1}{2}\tfrac{1}{N^2-1}\right), \\
O_5 &= \langle G^2\rangle^2 \tfrac{1}{4N}\left(\tfrac{1}{2} - \tfrac{1}{12}\tfrac{1}{N^2-1}\right), & O_6 &= \langle G^2\rangle^2 \tfrac{1}{4N}\left(\tfrac{7}{6} - \tfrac{1}{6}\tfrac{1}{N^2-1}\right), \\
O_7 &= \langle G^2\rangle^2 \tfrac{1}{4N}\left(\tfrac{1}{3} - \tfrac{1}{4}\tfrac{2}{N^2-1}\right), & O_8 &= \langle G^2\rangle^2 \tfrac{1}{4N}\left(\tfrac{1}{3} - \tfrac{1}{N^2-1}\right),
\end{aligned}
\tag{27.48}
$$

which indicates that only the first four operators are leading in $1/N$, and they do not satisfy the previous constraints. Moreover, the $1/N^2$ corrections to these leading-term are also large for $N = 3$ in the case of $O_3$ and $O_4$, and put some doubts on the validity of the $1/N$-approximation. A modified factorization has been therefore proposed [411] based on the evaluation of the typical $G^4$ one-point function:

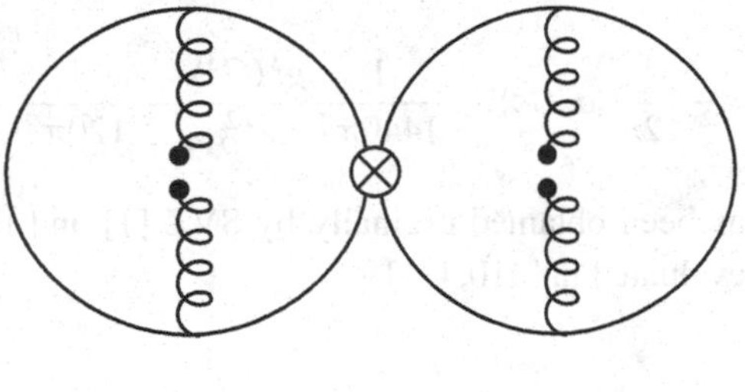

(27.49)

within a heavy-quark mass expansion and on the approximate validity of the factorization of the four-quark operator. In this way, one obtains the constraints:

$$
\begin{aligned}
O_1 &= \frac{\langle G^2\rangle^2}{4}\,, \\
O_3 &= 2O_4\,, \\
O_6 &= \frac{1}{4}\frac{1}{N}\langle G^2\rangle^2\,, \\
O_5 &= O_7\,, \\
O_8 &= 4O_7 - O_6\,,
\end{aligned} \tag{27.50}
$$

which leave one operator unconstrained.

Although theoretically interesting, these results are not very rewarding in practice. However, phenomenological fits from the $\tau$ decay [407,328,346] and $e^+e^- \to$ hadrons data [329] (Section 52.10) indicate that the estimate based on the factorization assumption gives about a factor 5 underestimate of the real value of the dimension-eight operators.

### 27.3.6 Relations among the different condensates

The heavy quark expansion has been used to derive the relations among the different condensates by studying the OPE in the inverse of the quark mass of the corresponding one-point function. More rigorously, the following results apply in the condensates of a heavy quark $Q$, although it has been extended to the light quark one in the literature, in an attempt to derive the light quark constituent mass from an effective QCD action [413].

A study of the $\langle \bar{Q}Q\rangle$ one-point function as shown in the following figure:

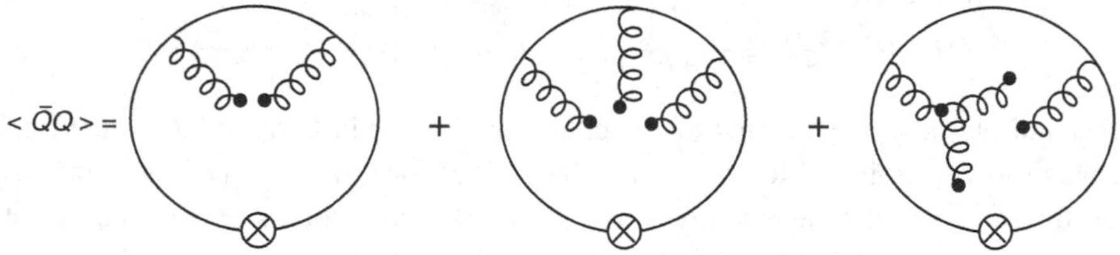


(27.51)

gives:

$$
M_Q\langle\bar{Q}Q\rangle = -\frac{1}{12\pi}\langle\alpha_s G^2\rangle - \frac{1}{1440\pi^2}\frac{g^3\langle G^3\rangle}{M_Q^2} - \frac{1}{120\pi^2}\frac{\langle DGDG\rangle}{M_Q^2}\,, \tag{27.52}
$$

where the first term has been obtained orginally by SVZ [1] and the higher dimensions corrections have been evaluated in [410,411].

A similar relation has been also derived from the one-point function of the mixed quark-gluon condensate [410,411,414]:

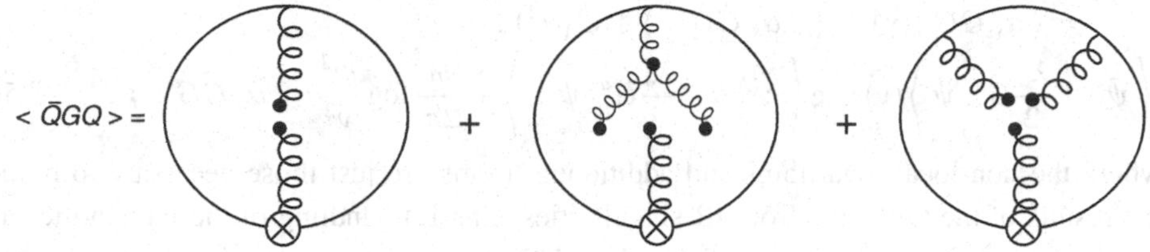


(27.53)

It reads:

$$M_Q\left\langle \bar{Q}\sigma^{\mu\nu}\frac{\lambda_a}{2}QG^a_{\mu\nu}\right\rangle = -\frac{5}{24}\left(\frac{\alpha_s}{\pi}\right)g\langle G^3\rangle, \tag{27.54}$$

where we have dropped the non-local *perturbative* term, which has no physical relevance, though its rôle is useful for absorbing the $m^k \log^n(m/\nu)$ mass singularities (subtleties related to these terms have been discussed in detail in [411]).

The previous relations show that the quark and mixed quark-gluon condensates vanish in the world with an infinitely heavy quark mass. Due to the positivity of the $G^3$-condensate, the previous relation also shows that the mixed quark-gluon condensate is positive, which is a less trivial result. Finally, a relation among the condensates has also been derived using a Cauchy–Scharwz-like inequality [415]:

$$\langle g\bar{\psi}G\psi\rangle^2 \leq 16\pi\langle\alpha_s G^2\rangle|\langle\bar{\psi}\gamma_5\psi\bar{\psi}\gamma_5\psi\rangle|\,, \tag{27.55}$$

which we should regard with a great caution as we can not control the error in deriving this formula. However, it indicates that the previous values of the condensates obtained from QSSR are self-consistent and disfavours the SVZ *standard* value of the gluon and four-quark condensates.

### 27.3.7 Non-normal ordered condensates and cancellation of mass singularities

We have discussed in previous sections the mixing among different condensates and the neccessity to form scale-invariant quantities which facilitates the log-resummation. This definition is also intimately connected with the absence of quark-mass singularities in the OPE. In order to show explicitly how these IR singularities are absorbed, it is informative to write the renormalized *non-normal-ordered* condensates in terms of the *normal-ordered* ones denoted by $\langle\ :\mathcal{O}:\ \rangle$, where the latter appear naturally in the use of the Wick's theorem for the calculation of the Feynman diagrams. To order $\alpha_s$, one has:

$$\langle\ \bar{\psi}\psi\ \rangle(\nu) = \langle\ :\bar{\psi}\psi\ :\ \rangle - \frac{3}{4\pi^2}m^3\left[\log\frac{m^2}{\nu^2} - 1\right] - \frac{1}{12\pi}\frac{1}{m}\langle\ :\alpha_s\, GG:\ \rangle\,,$$
$$\langle\ \alpha_s\, GG\ \rangle(\nu) = \langle\ :\alpha_s\, GG:\ \rangle + \mathcal{O}(\alpha_s^2)\,,$$
$$g\left\langle\ \bar{\psi}\sigma^{\mu\nu}\frac{\lambda_a}{2}G^a_{\mu\nu}\psi\ \right\rangle(\nu) = g\left\langle\ :\bar{\psi}\sigma^{\mu\nu}\frac{\lambda_a}{2}G^a_{\mu\nu}\psi:\ \right\rangle + \frac{m}{2\pi}\log\frac{m^2}{\nu^2}\langle\ :\alpha_s\, GG:\ \rangle\,, \tag{27.56}$$

where the non-local logarithms and additionnal terms are just those necessary to render the results of the OPE free from IR singularities. Careful handling of these quantities are required for correct treatments of the Green's functions in which one or more internal fermion masses are much smaller than the QCD scale. We also expect that most of the perturbative results available in the literature (e.g. supersymmetric calculations or QCD high-energy processes, ...), which are strongly affected by the change of the light quark masses (or the IR scale of the theory), should be treated in an analogous way in order to absorb such divergences.

# 28

# Technologies for evaluating the Wilson coefficients

There are nice expositions of these methods in the literature [431–434,3,362,45,399]. We shall not try to replace these discussions, but for a pedagogical reason, we shall repeat within our proper style some of these results. Let us remind ourselves that the evaluation of the Green's function of some local colourless currents is reduced to its evaluation in external gluons or/and quark fields, assuming that the field is weak, namely its momentum is much smaller than the characteristic scale of the problem. In this way, the expansion in power series à la Wilson makes sense.

## 28.1 Fock–Schwinger fixed-point technology

### *28.1.1 Fock-Schwinger gauge*

Let us now come to the methods for evaluating the Wilson coefficients appearing in the SVZ-expansion. Among others, the Fock–Schwinger method is the most convenient one in practice [431]. It corresponds to the choice of the Fock–Schwinger gauge [435,319]:

$$(x - x_0)A^a_\mu(x) = 0\,, \tag{28.1}$$

used often in QED. $A^a_\mu(x)$ is the four-potential and $x_0$ is an arbitrary choice of coordinate which plays the role of a gauge. As Eq. (28.1) breaks explicitly the translation invariance, its restoration (cancellation of the $x_0$ terms) for gauge-invariant quantities provides a double check of the validity of the calculation. Unfortunately, due to algebraic complications, most calculations have been done in the special choice $x_0 = 0$ of the gauge.

### *28.1.2 Gluon fields and condensates*

Using the identity:

$$A^\mu(x) = \frac{\partial}{\partial x^\mu}[A^\rho(x)x_\rho] - x_\rho\frac{A^\rho(x)}{\partial x^\mu}\,, \tag{28.2}$$

and from Eq. (28.1) at $x_0 = 0$:

$$x_\rho\frac{A^\rho}{\partial x^\mu} = x_\rho G^{\rho\mu} + x_\rho\frac{A^\mu(x)}{\partial x_\rho}\,, \tag{28.3}$$

one can deduce:

$$A^\mu(x) + x_\rho \frac{A^\mu(x)}{\partial x_\rho} = x_\rho G^{\rho\mu} . \tag{28.4}$$

By substituting $x \equiv \alpha z$ in the previous equation, it is easy to realize that this equation is a full derivative:

$$\frac{d}{d\alpha}[\alpha A_\mu(\alpha z)] , \tag{28.5}$$

which gives after integration:

$$A^\mu(x) = \int_0^1 d\alpha\, \alpha G^{\rho\mu}(\alpha x) x_\rho , \tag{28.6}$$

which expresses the gauge field $A^a_\mu(x)$ in terms of the gluon-strength tensor $G^a_{\mu\nu}$. One can use Eq. (28.6) by Taylor expanding $G_{\rho\mu}$ around $x^\mu = 0$:

$$A^a_\mu = \sum_{x=0}^{\infty} \frac{1}{n!(n+2)} x^\rho x^{\nu_1} \cdots x^{\nu_n} \partial_{\nu_1} \cdots \partial_{\nu_n} G^a_{\rho\mu}\big|_{x=0} . \tag{28.7}$$

By Taylor expanding $A^\mu(x)$, the gauge condition $x_\mu A^\mu(x) = 0$ becomes:

$$x_\mu \left[ A^\mu(x) + x_{\nu_1} \partial_{\nu_1} A^\mu(x)(0) + \frac{1}{2} x_{\nu_1} x_{\nu_2} \partial_{\nu_1} \partial_{\nu_2} A^\mu(x)(0) + \cdots \right] = 0 , \tag{28.8}$$

for all $x$ and leads to:

$$\begin{aligned} x_\mu A^\mu(x)(0) &= 0 , \\ x_\mu x_{\nu_1} \partial_{\nu_1} A^\mu(x)(0) &= 0 , \\ x_\mu x_{\nu_1} x_{\nu_2} \partial_{\nu_1} \partial_{\nu_2} A^\mu(x)(0) &= 0 . \end{aligned} \tag{28.9}$$

Therefore:

$$\begin{aligned} x_{\nu_1} \partial_{\nu_1} G(0) &= x_{\nu_1} [D_{\nu_1}, \ G(0)] , \\ x_{\nu_1} x_{\nu_2} \partial_{\nu_1} \partial_{\nu_2} G(0) &= x_{\nu_1} x_{\nu_2} \partial_{\nu_1} [D_{\nu_2}, \ G(0)] = x_{\nu_1} x_{\nu_2} [D_{\nu_1}, \ [D_{\nu_2}, \ G(0)]], \ldots \end{aligned} \tag{28.10}$$

and then the useful formula:

$$A^\mu(x) = \sum_{x=0}^{\infty} \frac{1}{n!(n+2)} x^\rho x^{\nu_1} \cdots x^{\nu_n} \big[D_{\nu_1}, \ [D_{\nu_2}, \ [\ldots [D_{\nu_n}, \ G^a_{\rho\mu}\big|_{x=0}] \cdots ]]\big] . \tag{28.11}$$

One can immediately form the gluon normal ordered condensate:

$$\begin{aligned} A_\mu(x) A_\nu(y) &= \frac{1}{4} x^\lambda y^\rho G_{\lambda\mu} G_{\rho\nu} + \cdots \\ &= \frac{1}{4d(d-1)} x^\lambda y^\rho [g_{\lambda\rho} g_{\mu\nu} - g_{\lambda\nu} g_{\mu\rho}] G^{\alpha\beta} G_{\alpha\beta} + \cdots , \end{aligned} \tag{28.12}$$

where $d = 4 - \epsilon$ is the space-time dimension.

### 28.1.3 Light quark fields and condensates

Analogous arguments can be used for the quark fields. One obtains the Taylor expansion:

$$\psi(x)=\sum_n \frac{1}{n!}x^{\nu_1}\cdots x^{\nu_n} D_{\nu_1}\, D_{\nu_2}\cdots D_{\nu_n}\psi|_{x=0}$$
$$\bar\psi(x)=\sum_n \frac{1}{n!}x^{\nu_1}\cdots x^{\nu_n}\bar\psi D^\dagger_{\nu_1}\, D^\dagger_{\nu_2}\cdots D^\dagger_{\nu_n}|_{x=0}\,, \tag{28.13}$$

with:

$$\bar\psi(0)\partial^\dagger_{\nu_1}=\partial_{\nu_1}\bar\psi\,. \tag{28.14}$$

From the previous expressions, one can also form the normal-ordered quark condensate:

$$\begin{aligned}\langle :\bar\psi_{i,\alpha}(x)\psi_{i,\alpha}(0):\rangle=&\frac{1}{4N}\delta_{\alpha\beta}\left[\left(\delta_{ij}+\frac{i}{4}m\,x^\mu(\gamma_\mu)_{ij}\right)\langle :\bar\psi\psi:\rangle\right.\\&-\frac{i}{16}x^2\left(\delta_{ij}+\frac{i}{6}m\,x^\mu(\gamma_\mu)_{ij}\right)\left\langle :\bar\psi\sigma^{\mu\nu}\frac{\lambda_a}{2}G^a_{\mu\nu}\psi:\right\rangle\\&\left.+\frac{i}{288}x^2\,x^\mu(\gamma_\mu)_{ij}g^2\left\langle :\bar\psi\gamma_\rho\frac{\lambda_a}{2}\psi\sum_f\bar\psi_f\gamma^\rho\frac{\lambda_a}{2}\psi_f:\right\rangle\right]\,. \end{aligned}\tag{28.15}$$

This expression tells us that one should be careful in evaluating the Wilson coefficients of high-dimension condensates as the *propagation* of the $\langle\bar\psi\psi\rangle$ condensate induces extra-contributions due to the mixed and four-quark condensates. This effect has been one of the main source of errors in the existing QSSR literature.

### 28.1.4 Mixed quark-gluon condensate

By combining the Taylor expressions of the quark and gluon fields, one can form the normal-ordered mixed quark-gluon condensate:

$$\begin{aligned}\langle :\bar\psi_i(x)A_\rho(z)\psi_j(0):\rangle=&\frac{1}{2}z^\mu\langle :\bar\psi G_{\mu\rho}\psi:\rangle+\frac{1}{2}x^\nu z^\mu\langle :\bar\psi D^\dagger_\nu G_{\mu\rho}\psi:\rangle+\cdots\\=&\frac{z^\mu}{96}\left[\left[\sigma_{\mu\rho}-\frac{m}{2}(x_\mu x_\rho-x_\rho x_\mu)+\frac{i}{2}mx^\nu\sigma_{\mu\rho}x_\nu\right]_{ij}\right.\\&\times\langle :\bar\psi\sigma_{\tau k}G^{\tau k}\psi:\rangle+\left[i\left(-\frac{2}{3}z_\mu\gamma_\rho+\frac{2}{3}z_\rho\gamma_\mu\right)+\frac{1}{2}x^\nu\gamma_\nu\sigma_{\mu\rho}\right]_{ij}\\&\left.\times g^2\left\langle :\bar\psi\gamma_\rho\frac{\lambda_a}{2}\psi\sum_f\bar\psi_f\gamma^\rho\frac{\lambda_a}{2}\psi_f:\right\rangle\right]\,.\end{aligned}\tag{28.16}$$

This expression also indicates that the *propagation* of the mixed quark-gluon condensate induces a quartic condensate. Here, one should remark that the non-local condensates used in some literature can be identified with the LHS of Eqs. (28.15) and (28.16). In this framework, the Wilson coefficients of these non-local condensates differ from the standard SVZ expansion.

### 28.1.5 Gluon propagator

For a complete calculational purpose, one also likes to have the expression of the propagators. We only quote their expressions in this gauge. The gluon propagator reads:

$$\begin{aligned} D_{\mu\nu}(q) =& \int d^4x e^{iqx} D_{\mu\nu}(x,0) \\ =& \frac{g_{\mu\nu}}{q^2} + g\frac{2}{q^4}G_{\mu\nu} + g\frac{4i}{q^6}(qD)G_{\mu\nu} - g\frac{2i}{3q^6}g_{\mu}\nu D_\alpha G_{\alpha\beta}\, q^\beta \\ &+ g\frac{2}{q^8}(qD)D_\alpha G_{\alpha\beta}g^\beta g_{\mu\nu} + g\frac{2}{q^8}(q^2D^2G_{\mu\nu} - 4(qD)^2G_{\mu\nu}) \\ &+ g^2\frac{1}{2q^8}g_{\mu\nu}(q^2G_{\alpha\beta}G^{\alpha\beta} - 4(q_\alpha G_{\alpha\beta})^2) + g^2\frac{4}{q^6}G_{\mu\alpha}G_{\alpha\nu}\,, \end{aligned} \tag{28.17}$$

where:

$$G_{\mu\nu} \equiv G^{ab}_{\mu\nu} = G^a_{\mu\nu}\lambda^b = f^{abc}G^c_{\mu\nu}(0)\,. \tag{28.18}$$

### 28.1.6 Quark propagator

The quark propagator satisfies the Dirac equation:

$$\left(i\frac{\partial}{\partial x_\mu}\gamma_\mu + gA^\mu(x)\gamma_\mu - M\right)S(x,y) = \delta^{(4)}(x-y)\,, \tag{28.19}$$

where: $A^\mu \equiv (\lambda_a/2)A^\mu_a$ and $M$ is the quark mass. Under the condition that the position of the field is much smaller than the characteristic distance $x - y$, one can have the Taylor expansion:

$$\begin{aligned} iS(x,y) =& iS^{(0)}(x,y) + g\int d^4z\, iS^{(0)}(x,z)\, i\hat{A}(z)\, iS^{(0)}(z,y) \\ &+ g^2\int d^4z'\, d^4z\, iS^{(0)}(x,z')\, i\hat{A}(z')\, iS^{(0)}(z',z)\, i\hat{A}(z)\, iS^{(0)}(z,y) \\ &+ \cdots\,, \end{aligned} \tag{28.20}$$

where $S^{(0)}(x,y)$ is the free quark propagator, and $\hat{A} \equiv A^\mu\gamma_\mu$. This expression shows explicitly how many times the quark from the point $y$ scatters 0,1,... external fields before annihilating at $x = 0$.

We shall consider the case of the heavy quark propagators in the next section due to the subtlety that the quark and gluon condensates are related to each other through Eq. (27.52).

Let us now consider the massless quark propagator in external fields. It reads in the $x$-space:

$$2\pi^2 S(x,y) = \frac{\hat{r}}{(r^2)^2} - \frac{1}{4}\frac{r^\alpha}{r^2}\tilde{G}_{\alpha\mu}(0)\gamma_\mu\gamma_5 + \left\{\frac{i}{2}\frac{\hat{r}}{(r^2)^2}\gamma_\rho x_\mu G_{\rho\mu}(0) - \frac{1}{96}\frac{\hat{r}}{(r^2)^2}(x^2y^2 - (xy)^2)G_{\mu\nu}(0)G^{\mu\nu}(0)\right\} + \text{operators of higher dimensions}\,, \tag{28.21}$$

where:

$$r = x - y \qquad G_{\mu\nu} \equiv g\frac{\lambda^a}{2}G^a_{\mu\nu} \qquad \tilde{G}_{\alpha\mu} = \frac{1}{2}\epsilon_{\alpha\mu\nu\beta}G^{\nu\beta}\,. \tag{28.22}$$

For two-point correlators without derivative currents, only the first two terms are operative in the evaluation of the gluon condensate effects, while the other terms contribute in the case of three-point functions or current with derivatives. The extension of this expression including higher-dimension gluon operators can be done. For completeness, this expression is:

$$\begin{aligned}
S(p) = \int d^4x\, e^{ipx}\, S(x,o) &= \frac{1}{\hat{p}} - \frac{p^\alpha}{p^4}g\tilde{G}_{\alpha\beta}\gamma^\beta\gamma^5 \\
&+ \frac{2}{3}g\frac{1}{p^6}[p^2 D^\alpha G_{\alpha\beta}\gamma^\beta - \hat{p}D^\alpha G_{\alpha\beta}\, p^\beta - p_\nu D^\nu D^\alpha G_{\alpha\beta}\gamma^\beta - 3ip_\nu D^\nu D^\alpha \tilde{G}_{\alpha\beta}\gamma^\beta\gamma^5] \\
&+ \frac{2}{p^8}\Big[ip^2 ip_\nu D^\nu D^\alpha G_{\alpha\beta}\gamma^\beta - i\hat{p}p_\nu D^\nu D^\alpha G_{\alpha\beta}\gamma^\beta - i\,(p_\nu D^\nu)^2\, p^\alpha G_{\alpha\beta}\gamma^\beta \\
&+ 2\,(p_\nu D^\nu)^2\, p^\alpha \tilde{G}_{\alpha\beta}\gamma^\beta\gamma^5 - \frac{1}{2}D_\nu D^\nu p^2 p^\alpha \tilde{G}_{\alpha\beta}\gamma^\beta\gamma^5\Big] \\
&+ \frac{1}{p^8}g^2[-2\hat{p}p^\alpha G_{\alpha\beta}G^{\beta\nu}p_\nu + p^2p^\alpha(G_{\alpha\beta}G^{\beta\nu} + G^{\beta\nu}G_{\alpha\beta})\gamma_\nu \\
&- p^2p^\alpha(G_{\alpha\beta}G^{\beta\nu} - G^{\beta\nu}G_{\alpha\beta})\gamma_\nu] + \cdots
\end{aligned} \tag{28.23}$$

where here $G_{\alpha\beta} = (\lambda_a/2)G^a_{\alpha\beta}$. The Wilson coefficient of the gluon condensate having dimension $D$ is proportional to $p^{-D+1}$.

## 28.2 Application of the Fock–Schwinger technology to the light quarks pseudoscalar two-point correlator

In order to illustrate the discussions in the previous sections and chapters, let us consider the two-point correlator:

$$\Psi_5(q^2) \equiv i\int d^4x\, e^{iqx}\,\langle 0|\mathcal{T}J_P(x)\,(J_P(0))^\dagger\,|0\rangle\,, \tag{28.24}$$

where:

$$J_P = (m_i + m_j)\bar{\psi}_i(i\gamma_5)\psi_j \tag{28.25}$$

is the light quark pseudoscalar current. The lowest order perturbative result for massive quarks and the two loop expression for massless quarks has been discussed for illustration in previous chapters. Here, we shall discuss explicitly the evaluation of the non-perturbative contributions.

### *28.2.1 Quark condensate* $\langle : \bar{\psi}\psi : \rangle$

In order to compute the Wilson coefficient, one can start from the Wick's theorem and leave one pair of $\langle : \bar{\psi}\psi : \rangle$ without contraction. Therefore:

$$\Psi_5(q^2) = (m_u + m_d)^2(\gamma_5)_{ij}(\gamma_5)_{kl}(-i)\int d^4x\, e^{iqx}$$
$$\times [d(x)_{\alpha j}\bar{d}(o)_{\beta k}\langle : \bar{u}(x)_{\alpha i}u(0)_{\beta l} : \rangle + u(0)_{\beta l}\bar{u}(x)_{\alpha i}\langle : \bar{d}(0)_{\beta k}d(x)_{\alpha j} : \rangle] . \tag{28.26}$$

Using the definition of the propagator:

$$\bar{\psi}^F_{\alpha i}(x)\psi^{F'}_{\beta j}(y) = i\delta_{\alpha\beta}\delta^{FF'}S_{ij}(x-y)$$
$$= i\delta_{\alpha\beta}\delta^{FF'}\int \frac{d^4p}{(2\pi)^4}\delta_{ij}(p)e^{-ip(x-y)} , \tag{28.27}$$

with:

$$S^F_{ij}(p) = \frac{1}{\hat{p} - m_F + i\epsilon'} , \tag{28.28}$$

one obtains:

$$\Psi_5(q^2) = (m_u + m_d)^2 \int d^4x \int \frac{d^4p}{(2\pi)^4} e^{-i(p-q)x}$$
$$\times [\langle : \bar{u}(x)_{\alpha i}u(0)_{\beta l} : \rangle[\gamma_5 S^d(p)\gamma_5]_{il} + \langle : \bar{d}(0)_{\beta k}d(x)_{\alpha j} : \rangle[\gamma_5 S^d(p)\gamma_5]_{kj}] . \tag{28.29}$$

In terms of Feynman diagrams, Eq. (28.29) reads:

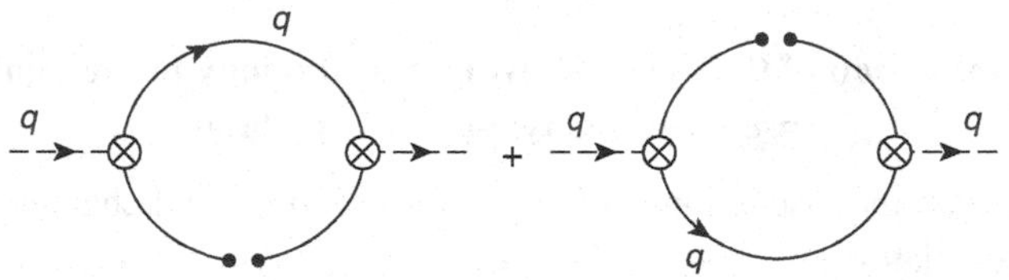


(28.30)

where • • means that the two-quark fields condense at the same point, so that a Taylor expansion in $x_\mu$ of $\langle : \bar{\psi}(x)\psi(0) : \rangle$ makes sense. Using Eq. (28.15), wherein we shall limit ourselves to the first two terms of the expansion, one obtains:

$$\Psi_5(q^2) = (m_u + m_d)^2 \frac{3}{12}$$
$$\times \left[ \langle\ :\bar{u}u:\ \rangle \left\{ \mathbf{Tr}\ \gamma_5 S^d(q)\gamma_5 - \frac{1}{4} m_u \left[ -\frac{\partial}{\partial p_\lambda} \mathbf{Tr}(\gamma_5 S^d(q)\gamma_5\gamma^\lambda) \right]_{p=q} \right\} \right.$$
$$\left. \times (u \longleftrightarrow d) \right] . \tag{28.31}$$

Using the property:

$$-\frac{\partial}{\partial p_\lambda} S(p) = S(p)\gamma_\lambda S(p) , \tag{28.32}$$

one can deduce the final result:

$$\Psi_5(q^2)|_{\bar\psi\psi} = \frac{(m_u + m_d)^2}{q^2} \left[ \left( m_d - \frac{m_u}{2} \right) \langle\ :\bar{u}u:\ \rangle + (u \longleftrightarrow d) \right] . \tag{28.33}$$

The minus sign is due to the $\gamma_5$ chirality flip which acts on the term $\partial/\partial p^\lambda$. This implies that for the scalar current, one has to change this minus sign.

### 28.2.2 Gluon condensate $\langle\ :\alpha_s G^2:\ \rangle$

The evaluation of the effect of the gluon condensate can be done by using the previous expression of the quark propagators in external fields. Diagramatically, one has to compute (Fig. 28.2):

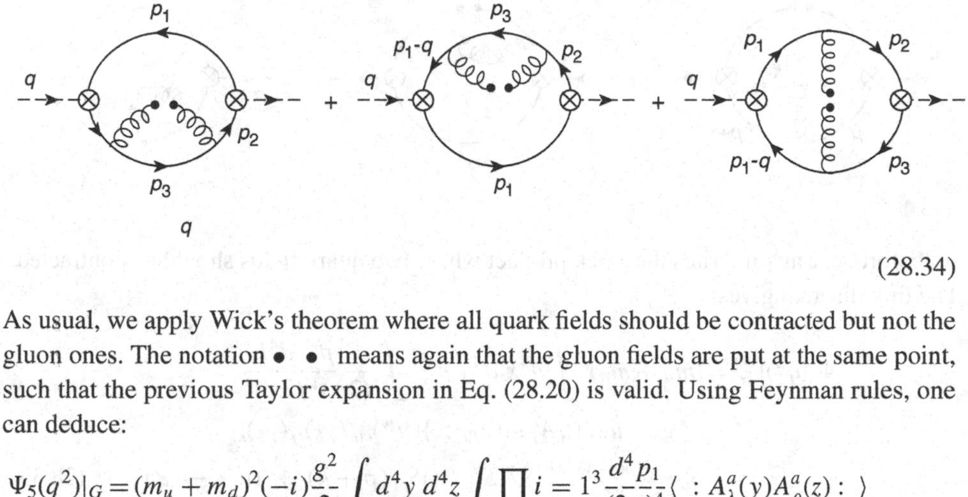


(28.34)

As usual, we apply Wick's theorem where all quark fields should be contracted but not the gluon ones. The notation • • means again that the gluon fields are put at the same point, such that the previous Taylor expansion in Eq. (28.20) is valid. Using Feynman rules, one can deduce:

$$\Psi_5(q^2)|_G = (m_u + m_d)^2(-i)\frac{g^2}{2} \int d^4y\, d^4z \int \prod i = 1^3 \frac{d^4 p_1}{(2\pi)^4} \langle\ :A^a_\lambda(y)A^a_\rho(z):\ \rangle$$
$$\times \left[ \mathbf{Tr}[\gamma_5 S(p_1+q)\gamma^\lambda S(p_3)\gamma^\rho S(p_2)\gamma_5 S(p_1)] e^{i(q+p_1-p_3)y+i(p_3-p_2)z} \right.$$
$$+ \mathbf{Tr}[\gamma_5 S(p_1)\gamma_5 S(p_2)\gamma^\rho S(p_3)\gamma^\lambda S(p_1-q)] e^{i(-p_1+q+p_3)y+i(p_2-p_3)z}$$
$$\left. + \mathbf{Tr}[\gamma_5 S(p_1)\gamma^\rho S(p_2)\gamma_5 S(p_3)\gamma^\lambda S(p_1-q)] e^{i(q-p_1+p_3)y+i(p_1-p_2)z} \right] , \tag{28.35}$$

where we have omitted the flavour indices $u$, $d$ as we shall work in the massless quark limit. Now, one takes advantage of Eq. (28.12), which is valid in the Schwinger gauge. Substituting it in Eq. (28.35), one gets:

$$\begin{aligned}\Psi_5(q^2)|_G = &(m_u+m_d)^2(-i)\frac{1}{16d(d-1)}\langle g^2 G_a^{\mu\nu}G_{\mu\nu}^a\rangle[g_{\nu\tau}g_{\lambda\rho}-g_{\nu\rho}g_{\lambda\tau}]\int\frac{d^4p_1}{(2\pi)^4}\\ &\times\left[2\frac{\partial}{\partial p_{1\nu}}\frac{\partial}{\partial p_{2\tau}}\mathbf{Tr}[\gamma_5 S(p_1+q)\gamma^\lambda S(p_3)\gamma^\rho S(p_2)\gamma_5 S(p_1)]_{p_2=p_3=p_1+q}\right.\\ &\left.+\frac{\partial}{\partial p_{3\nu}}\frac{\partial}{\partial p_{2\tau}}\mathbf{Tr}[\gamma_5 S(p_1)\gamma^\rho S(p_2)\gamma_5 S(p_3)\gamma^\lambda S(p_1-q)]_{p_2=p_1=p_3=p_1-q}\right],\end{aligned} \tag{28.36}$$

where we have used the fact that the two self-energy-like diagrams give the same contribution. Using Eq. (28.32) and the properties of Dirac matrices and Feynman integrals given in Appendices D and F and in that of QSSR1, one can deduce:

$$\Psi_5(q^2)|_G = -\frac{1}{8\pi}\frac{(m_u+m_d)^2}{q^2}\left\langle\alpha_s G_a^{\mu\nu}G_{\mu\nu}^a\right\rangle. \tag{28.37}$$

### 28.2.3 Mixed quark-gluon condensate

This contribution corresponds to the diagram:

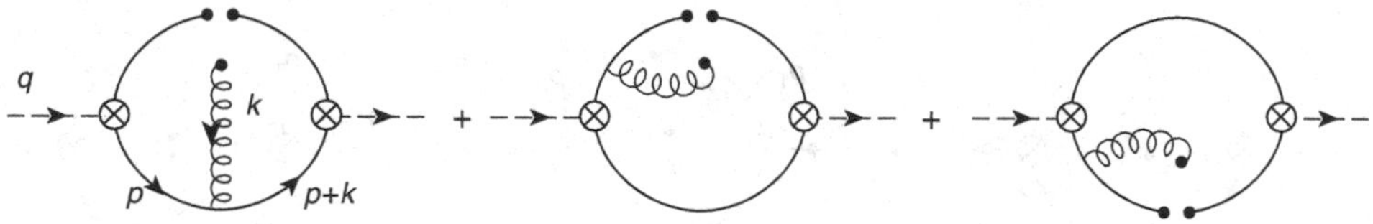


(28.38)

As before, one again writes the Wick product where two quark fields should be contracted. The first diagram gives:

$$\begin{aligned}\Psi_5(q^2)\big|_M^{(1)} = &(m_u+m_d)^2\int d^4x\, d^4y\, e^{iqx}\int\frac{d^4p}{(2\pi)^4}\frac{d^4k}{(2\pi)^4}\\ &\times g\langle\ :\bar u(x)_{\alpha i}A_\mu^a u(0)_{\beta l}:\ \rangle(\gamma^\mu)_{mn}(\gamma_5)_{ij}(\gamma_5)_{kl}\\ &\times e^{-i[p(x-y)+(p+k)y]}S_{jm}^d(p)S_{nk}^d(p+k)\ +\ (u\longleftrightarrow d)\,.\end{aligned} \tag{28.39}$$

We use now Eq. (28.16), the property in Eq. (28.32) and we do the Dirac algebra.

The self-energy-like diagram can be obtained by considering the *propagation* of the $\langle\bar\psi\psi\rangle$ condensate in a weak external field as given in Eq. (28.15). Using iteratively the property in Eq. (28.32) and doing as usual the Dirac algebra, one obtains the desired result.

The sum of the mixed quark-gluon condensate contributions is:

$$\Psi_5(q^2)|_M = -\frac{(m_u + m_d)^2}{2(q^2)^2} g[m_d\langle \bar{u}Gu\rangle + m_u\langle \bar{d}Gd\rangle] , \tag{28.40}$$

with the shorthand notation:

$$g\langle \bar{\psi}G\psi\rangle \equiv g\left\langle \; : \bar{\psi}\sigma^{\mu\nu}\frac{\lambda^a}{2}G^a_{\mu\nu}\psi : \; \right\rangle . \tag{28.41}$$

The result for the scalar current can be deduced from Eq. (28.40) by changing the overall -$(m_d + m_u)^2$ factor with $(m_d - m_u)^2$.

### 28.2.4 Four-quark condensates

Two classes of diagrams contribute to the four-quark condensates.

- **Class 1**: is that where the gluon fields once contracted give a hard momentum gluon propagator:

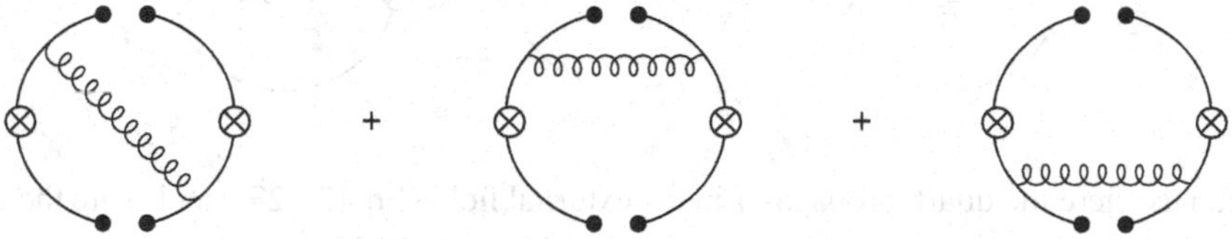

(28.42)

The computation of these diagrams can be done using standard perturbation theory, namely by writing the Wick product, contracting the gluon fields and two pairs of quark fields and by taking the vacuum expectation values (v.e.v) of the four-quark operators. Then, one obtains:

$$\Psi_5(q^2)|^{(1)}_{4\psi} = \frac{(m_u + m_d)^2}{2(q^2)^2}\pi\alpha_s\left\langle \; : \bar{u}\sigma^{\mu\nu}\gamma_5\frac{\lambda^a}{2}u - \bar{d}\sigma^{\mu\nu}\gamma_5\frac{\lambda^a}{2}d : \; \right\rangle^2 . \tag{28.43}$$

- **Class 2** is that where the momentum of the gluon propagator is zero. This contribution is represented by the diagrams:

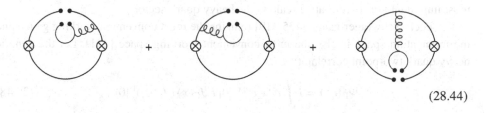

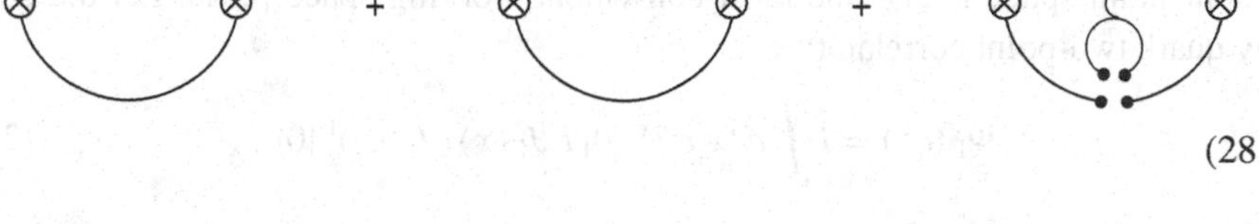

(28.44)

The first two diagrams are generated by the propagation of the $\langle \; : \bar{\psi}\psi : \; \rangle$ condensate in a weak external field as given in Eq. (28.15). The third diagram is generated by the mixed quark-gluon condensate as in Eq. (28.16). Evaluation of these diagrams leads to:

$$\Psi_5(q^2)|^{(2)}_{4\psi} = \frac{(m_u+m_d)^2}{2(q^2)^2}\frac{\pi\alpha_s}{6}\left\langle\ :\left(\bar{u}\gamma^\mu\frac{\lambda^a}{2}u+\bar{d}\gamma^\mu\frac{\lambda^a}{2}d\right)\sum_{u,d,s}\bar{\psi}\gamma_\mu\frac{\lambda^a}{2}\psi:\ \right\rangle . \qquad (28.45)$$

If one uses the vacuum saturation and the $SU(2)_F$ flavour symmetry of the quark condensates, the sum of the four-quark contributions reads:

$$\Psi_5(q^2)|_{4\psi} = \frac{(m_u+m_d)^2}{2(q^2)^2}\frac{112}{27}\rho\pi\alpha_s\langle\ :\bar{\psi}\psi:\ \rangle^2 , \qquad (28.46)$$

where $\rho$ measures the deviation from the vacuum saturation estimate.

### *28.2.5 Triple gluon condensate*

The contribution of the triple gluon condensate $g\langle\ : f_{abc}G^a_{\mu\nu}G^b_{\nu\rho}G^c_{\rho\mu}:\ \rangle$ has been evaluated in [436,432] and comes from the diagrams:

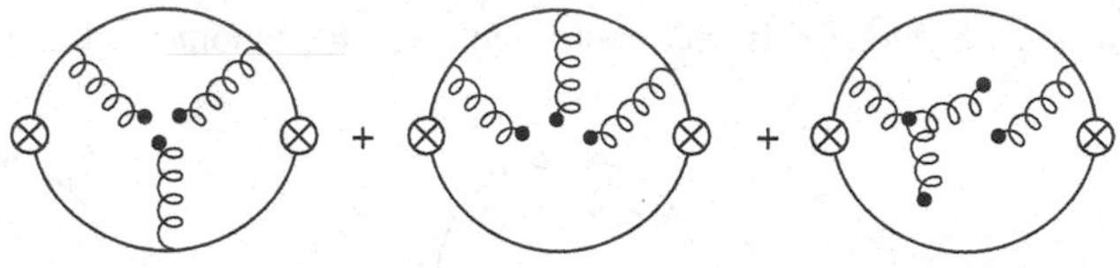

(28.47)

One can use here the quark propagator in the external field (Eq. (28.23)) and write the gluon fields in terms of the field strengths as in Eqs. (28.11) and (28.12) in order to form the triple condensate. The calculation can be done using standard perurbation theory. In the chiral limit ($m_{i,j}=0$), the effect of the triple gluon condensate vanishes for any quark-bilinear currents.

## 28.3 Fock–Schwinger technology for heavy quarks

### *28.3.1 General procedure*

The technology differs slightly from the light quark one as we can no longer neglect the quark mass $M$ which is the most important scale in the OPE. Moreover, due to the Wigner–Weyl realization of chiral symmetry for the heavy quark systems, the heavy quark condensate vanishes as $1/M$ and is correlated to the gluon condensate as in Eq. (27.52), which is the most important non-perturbative scale in the heavy quark sector.

The Fock–Schwinger gauge [435,319] remains the most convenient working gauge and the momentum space is also the most convenient working space [431]. Let the generic heavy quark two-point correlator:

$$\Psi_\Gamma(q^2) = i\int d^4x\, e^{iqx}\,\langle 0|\mathcal{T}J_\Gamma(x)(J_\Gamma(0))^\dagger|0\rangle , \qquad (28.48)$$

where:

$$J_\Gamma = \bar{\psi}_i(\Gamma)\psi_j \qquad (28.49)$$

and Γ is any Dirac matrices. The non-perturbative contributions to the correlator are typically of the form:

$$\Psi_\Gamma(q^2,\, M^2) = \langle gG \cdots G \rangle \int \frac{d^4k}{(2\pi)^4} \frac{\mathbf{Tr}(\Gamma \ldots,\, \hat{k},\, \hat{q},\, m, \ldots \Gamma \ldots)}{(k^2 - M^2)[(k+q)^2 - M^2]^n} \,. \tag{28.50}$$

The trace can be done using some algebraic programs. It is convenient to express the result as inverse powers of $k^2 - M^2$ and $(k+q)^2 - M^2$. After a Feynman parametrization, one encounters integrals of the form:

$$I_n^{\alpha\beta}(q^2, M^2) = \int_0^1 dx \frac{x^\alpha (1-x)^\beta}{[-q^2 x(1-x) + M^2]^n} \,. \tag{28.51}$$

By noting the symmetry $x \to (1-x)$, one can re-expand the previous integral in $x(1-x)$ and deduce the recursive relation:

$$I_n^{\alpha\alpha} \equiv I_n^\alpha = \frac{1}{Q^2}\left(I_{n-1}^{\alpha-1} - M^2 I_n^{\alpha-1}\right) \,. \tag{28.52}$$

This leads to the basic integral:

$$J_n = \int_0^1 \frac{dx}{[1 - x(1-x)q^2/M^2]^n} \,, \tag{28.53}$$

which reads:

$$J_n = \frac{(2n-3)!!}{(n-1)!}\left[\left(\frac{v^2-1}{2v^2}\right)^n v^{1/2} \log\frac{v+1}{v-1} + \sum_{k=1}^{n-1} \frac{(k-1)!}{(2k-1)!!}\left(\frac{v^2-1}{2v^2}\right)^{n-k}\right], \tag{28.54}$$

where:

$$v \equiv \left(1 - \frac{4M^2}{q^2}\right)^{1/2} \,. \tag{28.55}$$

### 28.3.2 $D = 4$ gluon condensate of the electromagnetic correlator

We use the Fock–Schwinger gauge in order to express the gluon fields in terms of the field strengths as in Eq. (28.12). The algorithm is very similar to the one used for the light quarks. The first two self-energy-like diagrams normalized to $\langle \alpha_s G^2 \rangle$ give [431]:

$$C_G^a = -\frac{1}{96\pi}\frac{1}{q^4}\left[2\frac{(5v^4+3)}{v^4} + \frac{(v^2-1)^2(5v^2+3)}{v^5}\log\frac{v-1}{v+1}\right]. \tag{28.56}$$

The vertex-like diagram contributes as:

$$C_G^b = \frac{1}{48\pi}\frac{1}{q^4}\left[2\frac{2(v^2+1)}{v^2} + \frac{(v^2-1)^2}{v^3}\log\frac{v-1}{v+1}\right], \tag{28.57}$$

where one can notice that each set of diagrams develops a non-transverse part:

$$q^\mu q^\nu \Pi^b_{\mu\nu} = -\frac{1}{16\pi}\frac{1}{q^4}\frac{(1-v^2)}{v^2}\left[1+\frac{(v^2+1)}{2v}\log\frac{v-1}{v+1}\right], \tag{28.58}$$

which vanishes in the sum. One can also express the sum of the transverse contribution in terms of the basic integral in Eq. (28.53):

$$C_4 \equiv C^a_G + C^b_G = \frac{1}{24\pi}\frac{1}{q^4}(-1+3J_2-2J_3), \tag{28.59}$$

which is a useful compact expression for further analysis.

### 28.3.3 $D = 6$ condensates of the electromagnetic correlator

The *light* four-quark condensates contribute through the diagram:

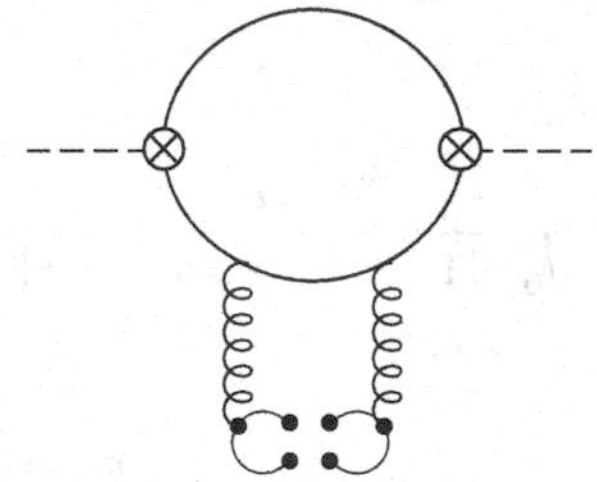

(28.60)

via the equation of motion of the gluon fields:

$$gJ^\mu_a \equiv D_\mu G^{\mu\nu}_a = -\frac{g^2}{2}\sum_{u,d,s}\bar\psi\gamma_\nu\frac{\lambda^a}{2}\psi, \tag{28.61}$$

while the triple gluon condensate contributes via the diagrams in Eq. (28.3.3):

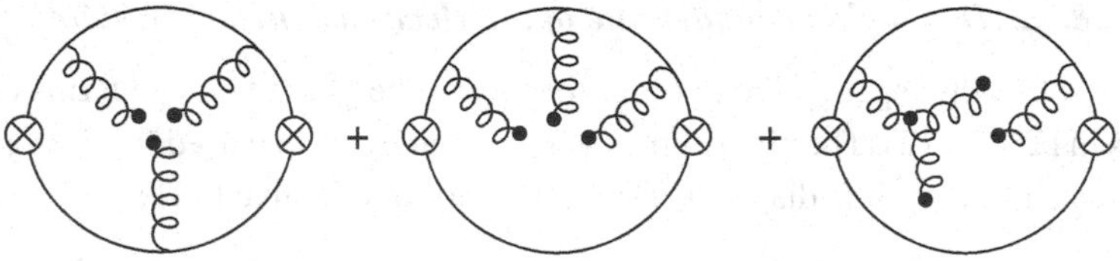

(28.62)

In order to reach the desired result, it is useful to express the v.e.v:

$$\langle : D_\alpha D_\beta G_{\mu\nu} G_{\rho\sigma} : \rangle, \quad \langle : D_\alpha G_{\mu\nu} D_\beta G_{\rho\sigma} : \rangle, \quad \left\langle : g^3 f_{abc} G^a_{\mu\nu} G^b_{\alpha\beta} G^c_{\rho\sigma} : \right\rangle. \tag{28.63}$$

In so doing, one uses the colour trace due to two and three $\lambda$ matrices, the previous gluon field equation of motion and the Bianchi identity. After a lengthy but straightforward algebraic manipulation, one can express the result in terms of the two condensates:

$$C_{3G}\langle g^3 f_{abc} G^a_{\mu\nu} G^b_{\nu\rho} G^c_{\rho\mu}\rangle \,, \qquad C_{JJ}\langle g^4 J^a_\mu J^a_\mu\rangle \,, \tag{28.64}$$

where the Wilson coefficients are [1,433]:

$$\begin{aligned} C_{3G} &= \frac{1}{72\pi^2 q^6}\left[\frac{2}{15} + 4J_2 - \frac{31}{3}J_3 + \frac{43}{5}J_4 - \frac{12}{5}J_5 + \frac{q^2}{10M^2}\right], \\ C_{JJ} &= \frac{1}{36\pi^2 q^6}\left[\frac{41}{45} + \left(\frac{2}{3} - \frac{q^2}{3M^2}\right) J_1 - J_2 - \frac{4}{9}J_3 - \frac{26}{15}J_4 + \frac{8}{5}J_5 + \frac{3q^2}{5M^2}\right]. \end{aligned} \tag{28.65}$$

### 28.3.4 Matching the heavy and light quark expansions

It is instructive to compare the coefficient functions obtained directly from a light quark expansion and from the heavy quark one by taking the limit $v = 1$. In order to be explicit, let us consider the coefficient of the gluon condensate $\langle\ :\alpha_s G^2:\ \rangle$. In the light quark-expansion, one obtains [411]:

$$\begin{aligned} G^a_G(m=0) &= 0\,, \\ G^b_G(m=0) &= \frac{1}{12\pi}\frac{1}{q^4}\langle\ :\alpha_s G^2:\ \rangle\,. \end{aligned} \tag{28.66}$$

If one takes naively the heavy quark result, one obtains from Eqs. (28.56) and (28.57):

$$\begin{aligned} G^a_G(v\to 1) &= -\frac{1}{6\pi}\frac{1}{q^4}\langle\ :\alpha_s G^2:\ \rangle\,, \\ G^b_G(v\to 1) &= \frac{1}{12\pi}\frac{1}{q^4}\langle\ :\alpha_s G^2:\ \rangle\,, \end{aligned} \tag{28.67}$$

which shows that the two limits do not coincide (!). This discrepancy can be restored by including the effect of the quark condensate which is known to be correlated to that of the gluon through Eq. (27.52).

One obtains in the two cases [411]:

$$\begin{aligned} C_\psi(m=0) &= \frac{2}{q^4}m\langle\ :\bar\psi\psi:\ \rangle\,, \\ C_\psi(v) &= \frac{8}{3}\frac{v+2}{(v+1)^2}\frac{1}{q^4}m\langle\ :\bar\psi\psi:\ \rangle\,, \end{aligned} \tag{28.68}$$

where the two results coincide for $v \to 1$. Using the relation in Eq. (27.56), one can introduce the non-normal-ordered quark and gluon condensates, where an extra gluon condensate term has been induced by the quark condensate. This term cancels the extra part in $C^a_G(v\to 1)$.

This lesson just tells us that one cannot directly take the $v = 1$ limit of the heavy quark correlator in order to get the light-quark result without paying attention to the *masked* contribution of the quark condensate, which induces a gluon condensate effect. Some other similar relations and properties hold for higher dimension condensates.

### *28.3.5 Cancellation of mass singularities*

Let us now discuss another example related to the previous subtlety of the quark and gluon condensates.

Let the example of the correlator of the vector current built from one light and one heavy quark fields:

$$J^\mu(x)^i_j = \bar{\psi}_i \gamma^\mu \psi_j \,. \tag{28.69}$$

By keeping the quark mass terms and taking the limit $-q^2 \to \infty$ after integration, one obtains for the transverse part [437]:

$$C_G^T = \frac{1}{12\pi}\left(1 - \frac{m_i}{m_j} - \frac{m_j}{m_i}\right)\frac{1}{q^4}\langle \alpha_s G^2 \rangle \tag{28.70}$$

which exhibits a dangerous mass singularity. The *normal ordered* quark condensate contribution is:

$$C_\psi = \frac{1}{q^4}\langle\ : m_i \bar{\psi}_j \psi_j + m_j \bar{\psi}_i \psi_i :\ \rangle\,, \tag{28.71}$$

Expressing it in terms of the *non-normal ordered* quark condensate as defined Eq. (27.56) and adding it to the previous gluon condensate contribution, one obtains the IR stable result:

$$C_G^T = \frac{1}{12\pi}\frac{1}{q^4}\langle\ : \alpha_s G^2 :\ \rangle\,. \tag{28.72}$$

and:

$$C_\psi = \frac{1}{q^4}\langle\, m_i \bar{\psi}_j \psi_j + m_j \bar{\psi}_i \psi_i\, \rangle(\nu)\,, \tag{28.73}$$

However, the natural question to ask is the commutativity of the operation by taking the limit $m_{i,j} = 0$ before the loop integration. A positive answer to this question can only be provided if one treats the IR integral in dimensional regularization and if one removes the $1/\epsilon$-pole at the *very end* of the calculation.

Indeed, in this calculation, one encounters integrals of the type:

$$\begin{aligned} I &\equiv \int \frac{d^n l}{(2\pi)^n}\left(\frac{q^2}{l^2 + i\epsilon}\right)^a \left(\frac{q^2}{(l+q)^2 + i\epsilon}\right)^b , \\ &= \left(\frac{-q^2}{4\pi}\right)^{n/2} \frac{\Gamma(a+b-n/2)\Gamma(n/2-a)\Gamma(n/2-b)}{\Gamma(a)\Gamma(b)\Gamma(n-a-b)} \end{aligned} \tag{28.74}$$

where the IR singularity is transformed into $1/\epsilon$-pole, which can be removed. In general, the extension of this method for the calculation of the Wilson coefficients of higher dimension condensates can be easily done provided one takes care of the mixing of the operators under renormalizations as discussed earlier.

## 28.4 The plane wave method

This method exploits the fact that the Wilson's expansion is an operator identity, namely that one can single out a given operator by sandwiching it between appropriate states. Let us consider the two-point correlator associated with the quark current:

$$J^\Gamma(x) = \bar\psi \Gamma \psi \;, \tag{28.75}$$

characterized by the Dirac matrix $\Gamma$, and which possesses the generic OPE (omitting Lorentz indices):

$$\Pi(q^2) \simeq C_1 1 + C_m \bar\psi\psi + G_g G^2 + D_G \left\{ G_{\alpha\delta} G^\alpha_\beta q^\delta q^\beta = \frac{1}{4} q^2 G^2 \right\} . \tag{28.76}$$

The first unit term corresponds to the usual perturbative calculation, which one obtains by sandwiching the correlator between the vacua. The next term is obtained by sandwiching the correlator between one-quark states and corresponds to the quark-current scattering amplitude shown in Fig. 28.1.

The Wilson coefficient $C_G$ can be obtained by sandwiching the correlator between one-gluon states. Therefore, the problem reduces to the evaluation of the forward gluon scattering amplitude on a colour-singlet current. From Lorentz invariance, this amplitude can be decomposed as:

$$\begin{aligned} T^{\mu\nu}(q,k) &\equiv i \int d^4x e^{iqx} \langle k, \mu | T J^\Gamma(x) \left(J^\Gamma(0)\right)^\dagger | k, \nu\rangle \\ &= F_1^{\mu\nu} C(q,k) + F_2^{\mu\nu} D(q,k) \;, \end{aligned} \tag{28.77}$$

where:

$$F_1^{\mu\nu} = 4(k^2 g^{\mu\nu} - k^\mu k^\nu) \equiv \langle k, \mu | G^2 | k, \nu \rangle \;, \tag{28.78}$$

and:

$$\begin{aligned} F_2^{\mu\nu} &= 2[k^2 q^\mu q^\nu - (k.q)(q^\mu k^\nu + q^\nu k^\mu) + g^{\mu\nu}(k\cdot q)^2] - q^2(k^2 g^{\mu\nu} - k^\mu k^\nu) \\ &\equiv \langle k, \mu | G_{\alpha\delta} G^{\alpha\beta} q^\delta q^\beta - \frac{1}{4} q^2 G^2 | k, \nu \rangle \;. \end{aligned} \tag{28.79}$$

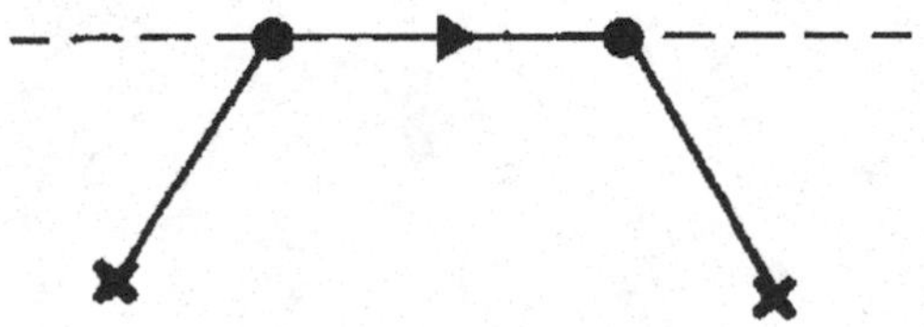

Fig. 28.1. 'Weak' quark (full line)-current (dashed line) scattering amplitude.

They correspond to the diagrams:

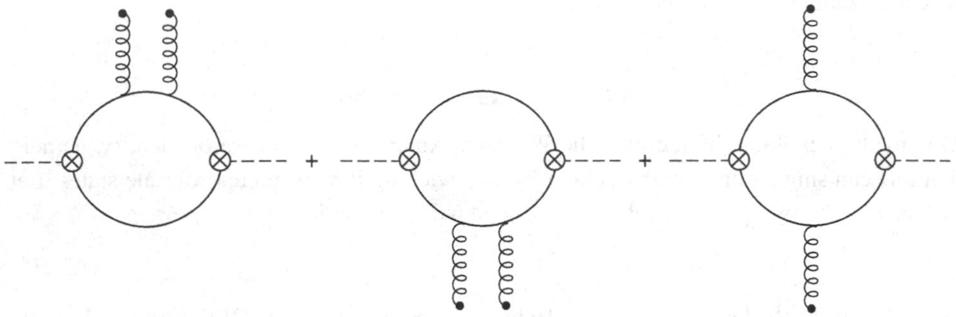

(28.80)

A comparison of Eqs (28.76) and (28.77)–(28.79) gives:

$$D_G(q^2) = C_G(q, k)|_{k_\alpha=0}. \qquad (28.81)$$

In practice, the plane wave method is convenient when one has external weak quark fields as in Fig. 28.1. In the case of many 'weak' external gluon fields, the extraction of a particular operator from various possible candidates having the same dimensions becomes very difficult. In one sense, this is the main inconvenience of this method.

## 28.5 On the calculation in a covariant gauge

The evaluation of the Wilson coefficients can be also obtained in a covariant gauge. Unlike the usual perturbative term, and the quark condensate term, which are easily obtained in this gauge, the evaluation of the Wilson coefficients of the gluon condensates is much more cumbersome in this gauge than in that of Fock–Schwinger. A published evaluation of the gluon condensate contribution in this gauge can be found in [626]. As applications of this method, we give at the end of this part a compilation of QCD two-point functions useful for further analysis.

# 29

# Renormalons

## 29.1 Introduction

The renormalon problem is related to the well-known fact [372,375] (for more complete reviews, see for example [162,154]) that the QCD series is unfortunately divergent (no finite radius of convergence) like $n!$, which is the number of diagrams of $n$th order. Indeed, a given observable can be expressed as a power series of the coupling $g$ as:

$$F(g) = \sum_n f_n g^n \ , \tag{29.1}$$

where the series are divergent:

$$f_n(n \to \infty) \sim K a^n n! n^b \ , \tag{29.2}$$

and where the $n$th order grows like $n!$, such that it is not practicable to have a quantitative meaning of Eq. (29.1). For the approximation to be meaningful, the approximation should asymptotically approach the exact result in the complex $g$-plane, such that:

$$\left| F(g)_{\text{exact}} - \sum_{n=0}^{N} f_n g^n \right| < K_{N+1} g^{N+1} \ , \tag{29.3}$$

where the truncation error at order $N$ should be bounded to the order $g^{N+1}$. If $f_n$ behaves like in Eq. (29.2), $K_N$ usually behaves as $a^N N! N^b$. The truncation error behaves similarly as the terms of the series. It first decreases until:

$$N_0 \sim \frac{1}{|a| g} \ , \tag{29.4}$$

beyond which the approximation to $F$ does not improve through the inclusion of higher-order terms. For $N_0 \gg 1$, the approximation is good up to terms of the order:

$$K_{N_0} g^{N_0} \sim e^{-1/|a|g} \ . \tag{29.5}$$

Provided $f_n \sim K_n$, the best approximation is reached when the series is truncated at its minimal term and the truncation error is given by the minimal term of the series.

One can use the well-known technique (*Borel transform*) for improving the convergence of a power series whose $n$th order grows like $n!$, by considering the related series:

$$B(z) \equiv \sum_n f_n \frac{z^n}{n!} \,. \tag{29.6}$$

If $f_n$ grows not faster than $n!$, then $B(z)$ will at least have a finite radius of convergence. Using the usual formula:

$$\int_0^\infty \exp(-z/g) z^n dz = n! g^{n+1} \,, \tag{29.7}$$

one can see that:

$$\tilde{F} \equiv g F(g) = \int_0^\infty \exp(-z/g) B(z) dz \,. \tag{29.8}$$

The relation in Eq. (29.8) is true order by order in perturbation theory, but there are arguments that it cannot be true for the full Greens functions. From Eq. (29.8), in order to calculate $F(g)$, one needs $B(z)$ only for real positive values of $z$ less than or of the order $g$, which can be obtained from the series in Eq. (29.1) if the singularities of $B(z)$ in the complex plane are all at distances from the origin much greater than $g$. Even if a few poles $z_1,\ z_2, \ldots$ have moduli of order $g$, one can calculate $B(z)$ by using power series for $(z - z_1)(z - z_2)\ldots B(z)$, where we should know the position of the poles. Singularities of $B(z)$ on the positive real axis are much worse, as they invalidate Eq. (29.8). One can distort the contour integral to avoid singularities on the positive real axis, but the ambiguities come from the question of distortion of the contour below or above the singularity? In the following, we will show that some of the singularities of the Borel transform $B(z)$ are associated with terms in the OPE (*renormalons*) and the others with solutions of the classical field equations (*instantons*).

In order to illustrate this discussion, let us assume that:

$$F(g)_{\text{exact}} = K a^n \Gamma(n + 1 + b) \tag{29.9}$$

For positive $b$, its Borel transform is:

$$\mathcal{B}[F](z) = K \frac{\Gamma(1 + b)}{(1 - az)^{1+b}} \,, \tag{29.10}$$

while for negative integer $b = -l$, one can write:

$$\mathcal{B}[F](z) = \frac{(-1)^l}{\Gamma(l)} (1 - az)^{l-1} \ln(1 - az) \; + \; \text{polynomial in } z. \tag{29.11}$$

In the case of QCD and QED, where one expects $a > 0$ (non-alternating series), one has singularities in the positive $z$ axis, such that the Borel integral does not exist. However, it may still be defined by taking the contour above or below the singularities, where it acquires an imaginary part:

$$\text{Im}\, \tilde{F}(g) = \mp \pi \frac{K}{a} e^{-1/(ag)} (ag)^b \,, \tag{29.12}$$

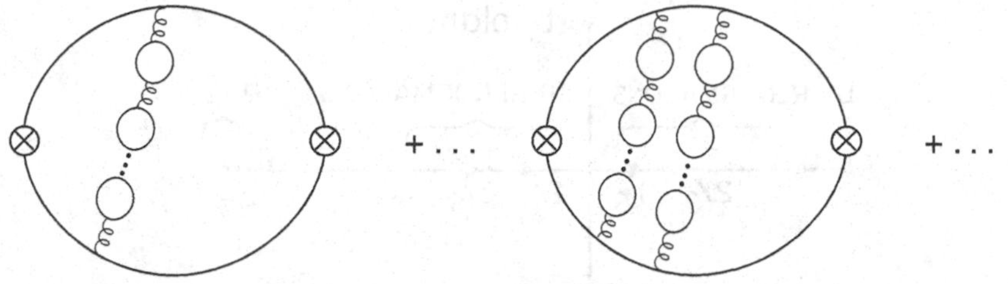

Fig. 29.1. Renormalon chains contributions to the QCD Adler $\mathcal{D}$-function.

where the sign depends on whether the integration is taken in the upper or lower complex plane. The difference in the two definitions is the so-called *ambiguity of the Borel integral.* As it behaves as an exponential in the coupling, it is of non-perturbative origin and induces power corrections.

In the following, we shall discuss for definiteness, the Adler $\mathcal{D}$-function in QCD:

$$\mathcal{D}(Q^2 = -q^2) \equiv -4\pi^2 Q^2 \frac{d\Pi(Q^2)}{dQ^2} , \tag{29.13}$$

built from the electromagnetic current $J_\mu(x) = \bar\psi\gamma_\mu\psi$ and which governs the $e^+e^- \to$ hadrons total cross-section. For the $\mathcal{D}$-function, the unnecessary $\nu$-dependence appearing in the two-point correlator $\Pi(q^2)$ from the leading-log term is not there, i.e., $\mathcal{D}$ is RGI. Therefore, its perturbative expansion reads:

$$\mathcal{D}\left(a_s \equiv \frac{\alpha_s}{\pi}\right) = \sum_n K_n a_s^n , \tag{29.14}$$

where $a_s(Q^2)$ is the running coupling and $K_n$ are pure numbers which are, however, RS-dependent.

Renormalon effects are associated to the insertion of n bubbles of quark loops into gluon lines (gluon chains) exchanged between the two quark lines in the $\mathcal{D}$-function built from the quark current as shown in Fig. 29.1.

It is well-known that they induce a n! growth into the perturbative series. This difficulty can be (in principle) cured by working with the Borel transform $\tilde{\mathcal{D}}$ of the correlator $\mathcal{D}(s)$:

$$\mathcal{D}(a_s) - \mathcal{D}(0) = \int_0^\infty db\, \tilde{\mathcal{D}}(b)\, \exp(-b/a_s) , \tag{29.15}$$

which possesses an explicit $1/n!$ suppression factor. However, life is not so simple because of the features described in the following.

## 29.2 Convergence of the Borel integral

The $b$-integral does not converge for $b \to \infty$. This can be seen from the fact that, in the chiral limit, hadrons have a non-zero mass in QCD, such that $\tilde{\mathcal{D}}$ should have singularities at $Q = M_0$, where $M_0$ is the mass of any hadrons having the quantum number of the photon

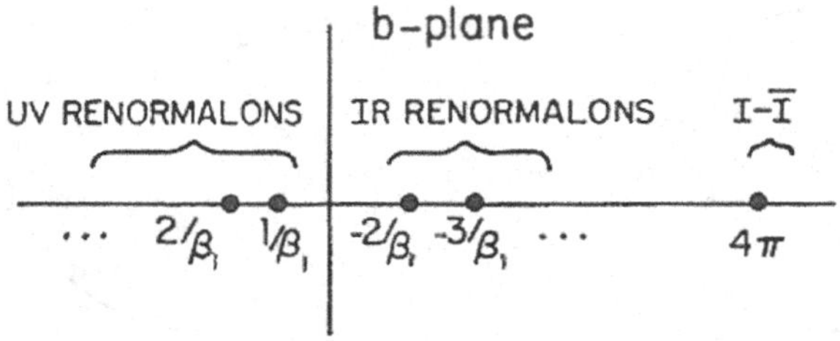


Fig. 29.2. Singularities in the Borel plane of the QCD Adler $\mathcal{D}$-function.

(or gluons). As for large $Q$, $\tilde{\mathcal{D}}$ is only function of:

$$a_s(Q^2) = \frac{1}{(-\beta_1/2)[\log Q^2/\Lambda^2 + (2n+1)i\pi]} , \tag{29.16}$$

one can see that it has an infinite number of singularities in $\alpha_s$, where $\alpha_s = 0$, corresponding to $M_0 = \infty$, is an accumulation point of these singularities. However, the singularities at $\alpha_s = \alpha_s(M_0)$ can arise through the behaviour:

$$\lim_{b\to\infty} \tilde{\mathcal{D}} \sim \exp -b\beta_1[\log Q^2/\Lambda^2 + (2n+1)i\pi] , \tag{29.17}$$

which indicates that the $b$-integral does not converge for $b \to \infty$. However, a large $b$-region corresponds to large $Q^2$ where $\tilde{\mathcal{D}}$ decreases rapidly like $\alpha_s$, such that the $\alpha_s$-singularities are very weak and justify the uses of the Borel integral for studying, without any ambiguities, the asymptotic behaviour of QCD at large $Q^2$. In general, $\tilde{\mathcal{D}}$ develops singularities at $b = kb_0 \equiv 2\pi k/(-\beta_1)$ in the real $b$-axis, where the integral is also ambiguous.

## 29.3 The Borel plane in QCD

There are three known types of singularities in the Borel plane of QCD as shown in Fig. 29.2.

- UV renormalons occur in the negative real axis ($\beta_1$ is negative in our notation.), and are harmless since the integration contour in Eq. (29.15) is along the positive $b$-axis. At the $n$th order of perturbation theory, integrand of the form:

$$\frac{d^4p}{p^6} \ln^n p^2 , \tag{29.18}$$

  gives a $n!$ factor and reflects the fact that such integrals are less convergent for large $n$.
- IR renormalons are singularities in the positive $b$-axis, which are due to the IR region of the Feynman integrals.
- Instanton–anti-instanton singularities occur because far separated instanton–anti-instanton pairs which can exist cannot be properly treated in a perturbative expansion around $\alpha_s = 0$.

## 29.4 IR renormalons

The IR renormalons correspond to the singularities at $k = +2, \ +3, \ldots$, and are generated by the low-energy behaviour of these higher-order diagrams, where fermion bubbles are inserted into the internal gluon line exchanged between the two fermion lines. In order to illustrate this feature, let us consider the one-gluon exchange diagram with a gluon of momentum $p$, where we shall focus on the low-$p$ region $R$.[1] Then:

$$\mathcal{D}(a_s(Q)) \sim \int_R \frac{d^4p}{p^2} a_s(p^2)F(p) . \tag{29.19}$$

where $F(P^2 \equiv -p^2)$ behaves as $P^2$. Using:

$$a_s(p) \simeq \frac{a_s(Q)}{1 - \beta_1 a_s(Q)\log(P^2/Q^2)} , \tag{29.20}$$

and:

$$\tilde{\mathcal{D}}(b) = \frac{1}{2i\pi}\int_{a_s-i\infty}^{a_s+i\infty} d(1/a_s)e^{b/a_s}\mathcal{D}(a_s) , \tag{29.21}$$

one obtains:

$$\tilde{\mathcal{D}}(b) \sim \int_R P^2 dP^2 \left(\frac{P^2}{Q^2}\right)^{-b/b_0} , \tag{29.22}$$

which gives the singularity near $b = -4\pi/\beta_1$:

$$\tilde{\mathcal{D}}(b) \sim \left(1 + \frac{b\beta_1}{4\pi}\right)^{-1} , \tag{29.23}$$

or to two loops, i.e., for two gluons-exchange:

$$\tilde{\mathcal{D}}(b) \sim \left(1 + \frac{b\beta_1}{4\pi}\right)^{-1+2\beta_2/\beta_1^2} , \tag{29.24}$$

for $b > 2b_0 \equiv -4\pi/\beta_1$. This indicates that the pole at $b = 2b_0$ gives rise to an IR ambiguity, if one tries to reconstruct $\mathcal{D}(a_s)$ from $\tilde{\mathcal{D}}(b)$ taken from perturbation theory. Converting the $a_s$-dependence into a $Q$-one, one can expect that the non-perturbative corrections to perturbation theory are of the size $1/Q^4$. More generally, diagrams with one chain of gluons contribute as:

$$\mathcal{D}(a_s) \sim \sum_n n\left(\frac{\alpha_s}{kb_0}\right)^n \implies \mathcal{B}(\mathcal{D}) \equiv \tilde{\mathcal{D}}(b) \sim -\frac{kb_0}{b - kb_0} , \tag{29.25}$$

for $b > kb_0$, which indicates that the pole at $b = kb_0$ gives rise to an IR ambiguity:

$$\delta\mathcal{D}(a_s) \sim \exp\left(-\frac{kb_0}{\alpha_s}\right) \sim \left(\frac{\Lambda^2}{-q^2}\right)^k , \tag{29.26}$$

if one tries to reconstruct $\mathcal{D}(a_s)$ from $\tilde{\mathcal{D}}(b)$ taken from perturbation theory. However, different prescriptions for defining $\mathcal{D}$ in perturbation theory for $b > kb_0$ can be compensated by the changes in the value of the non-perturbative condensates introduced via the SVZ

[1] IR renormalons have been studied in the $O(N)$ non-linear $\sigma$ model [374] and in QCD [376,377]. Here, we shall limit ourselves to the QCD case.

expansion, which *one must add* to perturbation theory in order to obtain a reliable result [162,342].

The absence of a $k = 1$ singularity is related to the absence of any gauge invariant operator of dimension 2. The absence of this singularity has been proved [331] from an explicit calculation in the limit of large $n_f$-number of flavours, where it has been shown that the relation:

$$\mathcal{B}\,(\mathrm{Im}\Pi)\,(b) \sim \frac{\sin(\pi b/b_0)}{(\pi b/b_0)}\tilde{\mathcal{D}}(b)\,, \tag{29.27}$$

implies that $\mathcal{B}\,(\mathrm{Im}\Pi)$ has only a zero at $k = 1$ but not the other alternative where $\tilde{\mathcal{D}}$ has a pole at this point, which follows from the simple factorization of the $Q^2$ dependence in the Borel transform of the $D$-function in the large $\beta$-limit. Then, one can conclude that no $1/Q^2$-ambiguity can be generated by the IR renormalons and they are intimately connected to the gauge invariant condensates in the SVZ-expansion. Restricting to the lowest IR renormalon pole, one can derive the perturbative contribution to the gluon condensate [378]:

$$\frac{\langle 0|\alpha_s G^2|0\rangle_{\mathrm{ren}}}{24\pi Q^4} = \sum_{n\ \mathrm{large}} \frac{3}{2\pi}\left(\frac{\alpha_s(Q^2)}{\pi}\right)^{n+1}\left(\frac{-\beta_1}{4}\right)^n . \tag{29.28}$$

One should notice that renormalons are target-blind:

$$\langle p|\alpha_s G^2|p\rangle_{\mathrm{ren}} = \langle 0|\alpha_s G^2|0\rangle_{\mathrm{ren}} . \tag{29.29}$$

They cannot also produce a non-vanishing quark condensate $\langle \bar{q}q\rangle$ as they respect the symmetries of the QCD Lagrangian, and cannot bring some insights on confinement due to their 'perturbative' origin.

However, at the one-loop level, renormalons are not the only way to probe the IR regions perturbatively. Another possibility is the introduction of the gluon mass $\lambda$ [478] as a fit parameter, while an IR perturbative contribution to the gluon condensate has been obtained in [479]:

$$\langle 0|\alpha_s G^2|0\rangle_{\mathrm{pert}} = \frac{3\alpha_s}{\pi^2}\lambda^4 \ln\lambda^2 . \tag{29.30}$$

A similar result has been obtained in a QCD-like model [369,374], which is an alternative to the renormalon contribution for massless gluon. Phenomenology using gluon mass has been developed [366], while in [162], a one-to-one correspondence between the two approaches has been proposed. Keeping only IR-sensitive contributions, a one-loop calculation with a gluon mass $\lambda$ can be translated as:

$$\begin{aligned} &C_0\alpha_s \ln\lambda^2 + C_1\alpha_s\frac{\sqrt{\lambda^2}}{Q} + C_2\alpha_s\frac{\lambda^2\ln\lambda^2}{Q^2} + \cdots \to \\ &C_0'\alpha_s \ln\Lambda^2 + C_1'\alpha_s\frac{\Lambda}{Q} + C_2'\alpha_s\frac{\Lambda^2}{Q^2} + \cdots . \end{aligned} \tag{29.31}$$

where $C_i,\ C_i'$ are coefficients.

## 29.5 UV renormalons

The UV renormalon singularities correspond to $k = -1, -2, \ldots$, and are generated by the high-energy behaviour of the virtual momenta. They lead to a Borel-summable series thanks to the asymptotic freedom property of the theory. After a Borel sum, they cannot limit the applicability of perturbation theory [377,379], although they can induce an uncertainty in the truncated perturbative series when the Borel sum is not done. Their contributions are dominated by the leading singularity at $k = -1$:

$$K_n \sim \frac{n!}{(-b_0)^n} , \tag{29.32}$$

which gives rise to an asymptotic series:

$$|K_1 a| > |K_2 a^2| \cdots > K_{N-1} a^{N-1} \sim |K_N a^N| < |K_{N+1} a^{N+1}| < \cdots , \tag{29.33}$$

where the successive terms decrease like $N \sim b_0/a$, at which the minimum value is attained, while the series explodes afterwards. The alternating sign of $K_n$ guarantees that the series is Borel summable. For a truncated series, the accuracy is limited by the size of the minimum term:

$$4\pi^2 \delta\mathcal{D}(a) \equiv |K_N a^N| \sim N!\, n^N \sim \sqrt{2\pi N} e^{-N} \sim \exp(-b_0/a) \sim \Lambda^2/-q^2 , \tag{29.34}$$

which indicates that the UV renormalon can contribute as $1/Q^2$ [161,162,297–300].

However, this result is subtraction-scale dependent [162], as a more careful analysis shows that the ambiguity scales as:

$$A\sqrt{\alpha_s(\nu)}\left(\frac{\Lambda^2 Q^2}{\mu^4}\right) , \tag{29.35}$$

where $A$ and $\Lambda$ absorb this renormalization scheme (RS)-dependence, whilst $\mu$ is an arbitrary UV cut-off. However, it can be shown that the results obtained in the limit of infinite numbers of flavours within the one-chain approximation, can be strongly affected by the UV renormalon induced by the two-, three-,...chains of gluons [342,343], such that, it is premature to deduce any reliable quantitative estimates from this approach. However, some more optimistic authors have considered a more refined version of the one-chain of gluons approximation, involving next-to-leading $\beta$ functions and RS-invariant quantities. The analysis indicates that the UV renormalon effect is much smaller [339,340] than naïvely expected [344,331], and than that of the perturbative error based on the last calculated coefficient term of the series (theorem of divergent series [337]) [338,323]. Taking into account the different existing (qualitative) estimates of UV renormalon effects [331–340], one can conclude that the estimate of the perturbative errors based on the last calculated term of the QCD series [338,323] gives a reasonable or presumably an overestimate of the true error. It is also clear that the UV renormalon contribution *cannot* be considered as a *new source* of uncertainty, but it is of the same nature as the perturbative error. An independent extraction of such a contribution is needed. The only available alternative attempt for doing this, is its phenomenological extraction from the $e^+e^- \to I = 1$ hadrons data [341,329](Section 52.10)], from QSSR. It has been noticed from the analysis of [329], that the obtained constraint is strongly correlated to the value of the gluon condensate. Postulating that a new term of dimension-two exists in the QCD series, the OPE is modified as:

$$\mathcal{D}(Q^2) = 1 + \left(\frac{\alpha_s}{\pi}\right) + \cdots + \frac{d_2}{Q^2} + \cdots \tag{29.36}$$

one obtains [329]:

$$d_2 \approx (0.03 \sim 0.05)\ \text{GeV}^2\ , \tag{29.37}$$

if one uses the value of the gluon condensate $\langle\alpha_s G^2\rangle \simeq 0.08\ \text{GeV}^4$. This term would induce an effect of about 1% in the QCD expression of the $\tau$-width [325], which is a negligible effect.

## 29.6 Some phenomenology in the large $\beta$-limit

The large $\beta$-limit corresponds to the case where one takes large numbers of quark flavours and neglect the remainder terms of the $\beta$-function:

$$\beta_1(n_f \to \infty) \simeq n_f/3\ , \tag{29.38}$$

and then corresponds to the abelianisation of QCD.

### *29.6.1 The D-function*

In this limit the $D$-function can be expressed as:

$$\mathcal{D}(Q^2) = 1 + \left(\frac{\alpha_s}{\pi}\right) \sum_{n=0} \alpha_s^n \left[ d_n \left(\frac{\beta_1}{2\pi}\right)^n + \delta_n \right], \tag{29.39}$$

where $d_0 = 1$ and $\delta_0 = 0$. The coefficient $d_n$ comes from the bubble diagrams. Its Borel transform reads:

$$\mathcal{B}(\mathcal{D})(b) = \sum_{n=0} \frac{d_n}{n!} b^n = \frac{32}{3} \left(\frac{Q^2}{\nu^2} e^C\right)^{-b} \frac{b}{1-(1-b)^2} \sum_{j=2}^{\infty} \frac{(-1)^j j}{(j^2-(1-b)^2)^2}\ , \tag{29.40}$$

where in the $\overline{MS}$ scheme $C = -5/3$. The UV renormalon poles at $b = -1, -2, \ldots$ are double poles, while the IR renormalon poles at $b = 3, 4, \ldots$ are double poles and a single pole at $b = 2$. It is informative to decompose the Borel transform into the sum of leading poles:

$$\begin{aligned} \mathcal{B}(\mathcal{D})(b) = {} & e^{-5/3} \left[ \frac{4}{9} \frac{1}{(1+u)^2} + \frac{10}{9} \frac{1}{(1+u)} \right] + e^{10/3} \frac{2}{(2-u)} \\ & e^{-10/3} \left[ -\frac{2}{9} \frac{1}{(2+u)^2} - \frac{1}{2(2+u)} \right] + \cdots \end{aligned} \tag{29.41}$$

Working in the $\overline{MS}$ scheme, the ambiguity in summing the series can be quantified as (see e.g. [154]):

$$\delta\mathcal{D}(Q^2)_{\text{ren}} = \left(\frac{4}{-\beta_1}\right)\frac{e^{10/3}}{\pi}\frac{\Lambda^4}{Q^4} \approx \frac{0.06\ \text{GeV}^4}{Q^4} . \tag{29.42}$$

This effect is smaller than the non-perturbative gluon condensate contribution:

$$\delta\mathcal{D}(Q^2)_{\text{cond}} \simeq \frac{2\pi}{3}\frac{\langle 0|\alpha_s G^2|0\rangle}{Q^4} \simeq \frac{0.14\ \text{GeV}^4}{Q^4} , \tag{29.43}$$

where we have used the most recent QSSR determination [313,329]. This result is not significant for raising doubts on the existence of the non-perturbative gluon condensate in the SVZ expansion [1], although it can contribute to its perturbative component.

### 29.6.2 Semi-hadronic inclusive $\tau$ decays

Semi-hadronic tau decays have been discussed in details in BNP [325]. We shall be interested here in its asymptotic perturbative expansion, which have been discussed by many authors [331–345]. In the large $\beta$-function limit, one can write, the branching ratio [154]:

$$R_\tau = 3(|V_{ud}|^2 + |V_{us}|^2)\left\{1 + \left(\frac{\alpha_s}{\pi}\right)\sum_{n=0}\alpha_s^n\left[d_n^\tau\left(\frac{\beta_1}{2\pi}\right)^n + \delta_n^\tau\right]\right\} , \tag{29.44}$$

where one can neglect the remainder $\delta_n^\tau$. The Borel transform is [154]:

$$\mathcal{B}(\mathcal{R}_\tau)(b) = \mathcal{B}(\mathcal{D})(b)\sin(\pi b)\left[\frac{1}{\pi b} + \frac{2}{\pi(1-b)} - \frac{2}{\pi(3-b)} + \frac{1}{\pi(4-b)}\right] , \tag{29.45}$$

where the sinus attenuates all renormalon poles except those at $b = 3, 4$. The point $b = 1$ is regular but will not be suppressed by a factor $\alpha_s$ when one uses the Cauchy contour integral for evaluating $R_\tau$.

Taking the leading renormalon poles, one can approximately have:

$$\mathcal{B}(\mathcal{R}_\tau)(b) \simeq e^{-5/3}\frac{2}{15(1+u)} + e^{-10/3}\frac{2}{135(2+u)} + e^5\left[\frac{8}{3(3-u)^2} - \frac{8}{9(3-u)}\right] + \cdots \tag{29.46}$$

Expressing the rate as in BNP:

$$R_\tau = 3(|V_{ud}|^2 + |V_{us}|^2)S_{EW}[1 + \delta_{PT} + \delta_{EW} + \delta_{NP}] , \tag{29.47}$$

one can compare the measured value of $\delta_{PT}$ with the one obtained from the large $\beta$-limit prediction. One can notice that the value of $\alpha_s(M_\tau)$ can reduce by 15% compared to the one from the truncated series but this effect is smaller than one obtained by adding the $\alpha_s^3$ correction. Another point is that the error induced by the $\Lambda^2/M_\tau^2$ term which arises when the series is truncated at the onset of UV renormalon divergence is numerically very small due to the smallness of its coefficient. Therefore, the induced uncertainty is negligible in the $\overline{MS}$ scheme.

### 29.7 Power corrections for jet shapes

The phenomenology of power corrections in jets and DIS has been developed [162,366], while numerous experimental studies have been devoted for measuring these contributions [480,481]. Renormalons are most useful in these frameworks as they can fix unambiguously the power $n$ of the corrections $(\Lambda/Q)^n$. However, in order to find relations between various corrections, models are still needed as one expects [342,162] that any number of renormalon chains gives power corrections of the same order, and there is no way to evaluate all of them. Some other reservations to be made in the renormalon approach are also the extrapolation of the small QCD coupling expansion in the UV regime down to the IR domain where the QCD coupling is of order one, and where, terms which dominate in the UV region do not necessarily dominate in the IR region.

For definiteness, let us consider the thrust variable defined in the previous chapters dedicated to jets:

$$T = {}^{\max}_{\mathbf{n}} \frac{\sum_i |\mathbf{p_i} \cdot \mathbf{n}|}{\sum_i |\mathbf{p_i}|} , \tag{29.48}$$

where $\mathbf{p}$ is the momenta of particles produced and $\mathbf{n}$ is a unit vector. From perturbation theory $T \neq 1$ due to the emission of gluons from quarks. The contribution due to a soft gluon emission can be quantified as [162]:

$$\langle 1 - T \rangle_{\text{soft}} \sim \int_0^{\Lambda} \frac{d\omega}{\omega} \frac{\omega}{Q} \alpha_s \sim \frac{\Lambda}{Q} , \tag{29.49}$$

where $d\omega/\omega$ is the standard factor for the gluon emission; $\omega/Q$ comes from the definition of $T$ while $\alpha_s$ is of the order one. Alternatively, if one attributes to the gluon an intrinsic invariant mass squared $\zeta Q^2$, and evaluate the thrust mean value, one obtains [366,154]:

$$\langle 1 - T \rangle = C_F \left(\frac{\alpha_s}{\pi}\right) [0.788 - k\sqrt{\zeta} + \cdots] , \tag{29.50}$$

where $\sqrt{\zeta} \sim \Lambda/Q$, and its coefficient depends on the definition of thrust used ($k = -7.32$ with the previous definition, while it is 4 for the definition used in [366]). One can generalize the previous result by using an *universality picture*. That can be done by keeping terms which contributes perturbatively as $\alpha_s^n \ln^k Q$ and extrapolating such terms in the IR region where, however, they no longer dominate! In this way, the $1/Q$ correction can be expressed in terms of the *universal factor* [162]:

$$E_{\text{soft}} = \int dk_\perp \gamma_{eik}(\alpha_s(k_\perp^2)) , \tag{29.51}$$

where $\gamma_{eik}$ is the so-called eikonal anomalous dimension, and the integral over the Landau pole is understood as the principal value. In this way, on gets the different relations among the event-shape variables (see the definitions in the jet chapters from Eqs. (24.32) to (24.37))[162]:

$$\langle 1-T\rangle_{1/Q} = \frac{2}{3\pi}\langle C\rangle_{1/Q}$$
$$= \left(\frac{\langle M_h^2 + M_l^2\rangle}{Q^2}\right)_{1/Q} \tag{29.52}$$

and:

$$\left(\frac{\langle M_h^2\rangle}{Q^2}\right)_{1/Q} \approx \left(\frac{\langle M_l^2\rangle}{Q^2}\right)_{1/Q} . \tag{29.53}$$

These relations are well verified experimentally [480]:

$$\frac{1}{2}\langle 1-T\rangle_{1/Q} = \frac{C}{Q}(0.511 \pm 0.009)$$
$$\frac{1}{3\pi}\langle C\rangle_{1/Q} = \frac{C}{Q}(0.482 \pm 0.008)$$
$$2\left(\frac{\langle M_h^2\rangle}{Q^2}\right)_{1/Q} = \frac{C}{Q}(0.616 \pm 0.018)\,, \tag{29.54}$$

where $C$ is a constant.

## 29.8 Power corrections in deep inelastic scattering

Power corrections in deep inelastic scattering have been developed at the single renormalon chain level [482], and an alternative derivation using Landau pole of the power corrections has been given in [162].

### 29.8.1 Drell–Yan process

The inclusive cross-section can be expressed in terms of the moments:

$$\int d\tau\, \tau^{n-1}\frac{d\sigma(Q^2,\tau)}{dQ^2} = M_n\left[1+\alpha_s C_\lambda \frac{\sqrt{\lambda}^2}{Q}\right] , \tag{29.55}$$

where $Q$ is the invariant mass of the lepton pair; $\sqrt{s}$ is the invariant mass of the $\bar{q}q$ from the initial hadrons $h_{1,2}$ and $\tau = Q^2/s$; $\lambda$ is the gluon mass. To one loop, one finds [154]:

$$C_\lambda = 0 \quad \text{for} \quad n\cdot\Lambda/\sqrt{s} \ll 1\,. \tag{29.56}$$

An understanding of this result from general arguments based on the inclusive nature of momenta has been given in [162].

### 29.8.2 Non-singlet proton structure functions $F_2$

A systematic measurement of power corrections in DIS for the moments of the non-singlet structure functions $F_2$ has been performed [481]. The moments:

$$\mathcal{M}_2^{(n)}(Q^2) \equiv \int_0^1 dx\, x^{n-2} F_2(x, Q^2)\,, \tag{29.57}$$

have been parametrized as:

$$\mathcal{M}_2^{(n)}(Q^2) = \mathcal{M}_{2,PT}^{(n)} \left(1 + C_2^{(n)} \frac{\tau^2}{Q^2} + C_4^{(n)} \frac{\lambda^4}{Q^4} \cdots \right)\,, \tag{29.58}$$

where $\mathcal{M}_{2,PT}^{(n)}$ is the perturbative QCD prediction. The $n$-dependence of the power corrections has been included into $C_{2,4}^{(n)}$. In the range of $Q^2$ values from 5 to 260 GeV$^2$, and studying different figures, the analysis leads to a non-vanishing contribution:

$$\tau^2 \simeq (0.2 \pm 0.1)\ \text{GeV}^2\,, \tag{29.59}$$

if $\lambda^4 = 0$ and $(0.25 \pm 0.2)$ GeV$^2$ if $\lambda^4 \neq 0$. This result and the $n$-dependence agree with the renormalon-based result [162].

### 29.8.3 Gross–Llewellyn Smith and polarized Bjorken sum rules

Power corrections to other DIS sum rules (Gross–Llewellyn Smith (GLS), polarized Bjorken (PBj) sum rules) have been also analysed from the renormalon approach [331]. In the large $\beta$-limit, one can approximately assume them to be the same because the perturbative contributions differ only by light-by-light scattering starting at $\alpha_s^3$. Let's remind ourselves of the GLS sum rule given in Eq. (29.60):

$$\begin{aligned}&\int_0^1 \frac{dx}{x}[F_3^{\bar{\nu}p}(x, Q^2) + F_3^{\nu p}(x, Q^2)] \\ &= 3\left\{1 - a_s(Q^2) - 3.58\, a_s^2(Q^2) - 19.0\, a_s^3(Q^2) + \delta_{GLS}\right\}\,,\end{aligned} \tag{29.60}$$

to which we add the power correction (twist-4) term $\delta_{GLS}$. In the large $\beta$-limit, one obtains in the $\overline{MS}$ scheme [154]:

$$\delta_{GLS} \simeq e^{5/3}\left(-\frac{16}{9\beta_1}\right)\frac{\Lambda^2}{Q^2} \approx \frac{0.1\ \text{GeV}^2}{Q^2}\,, \tag{29.61}$$

which is comparable in strength but differs in sign with the twist-4 QSSR estimate [483] and fit using the CCFR data [249]:

$$\delta_{HT} \approx -\frac{(0.10 \pm 0.05)\ \text{GeV}^2}{Q^2}\,. \tag{29.62}$$

However, an extraction of this power correction from the polarized Bjorken sum rule [260] leads to an inaccurate value consistent with zero as given in Eq. (19.8).

## 29.9 Power corrections to the heavy quark pole mass

We have defined in the chapter on perturbation theory the notion of pole mass, which is defined at the pole of the propagator. We have seen that this definition is not renormalized [141,147,133], independent of the regularization procedure [148] and free from IR singularities [133]. However, when this mass is related to the short distance $\overline{MS}$ running mass, one can notice its sensitivity to long distances. In the renormalon approach, this difference is given by the self-energy diagram with one-gluon chain:

$$M(p^2 = \bar{m}^2) = \bar{m}(\nu) + (-i)\int \frac{d^n k}{(2\pi)^n}\alpha_s(ke^{-5/6})\frac{\gamma^\mu(\hat{p}+\hat{k}+m)\gamma_\mu}{k^2[(p-k)^2-m^2]}\bigg|_{p^2=m^2} . \quad (29.63)$$

It shows that for $p^2 = m^2$, the integral behaves for small $k$ like $d^4k/k^3$, which implies that the series gives an IR renormalon singularity at $b = -\pi/\beta_1$. The asymptotic behaviour of the series expansion reads [331,154]:

$$M(p^2 = \bar{m}^2) = \bar{m}(\nu) + C_F\frac{e^{5/6}}{\pi}\nu\sum_n\left(\frac{-\beta_1}{\pi}\right)^n n!\alpha_s^{n+1} . \quad (29.64)$$

Writing:

$$\delta m \equiv M_{\text{pole}} \equiv M(p^2 = \bar{m}^2) - \bar{m}(\bar{m}) = \bar{m}(\bar{m})\frac{C_F}{4}\left(\frac{\alpha_s}{\pi}\right)\sum_{n=0}[d_n(-\beta_1/\pi)^n + \delta_n]\alpha_s^{n+1} , \quad (29.65)$$

its Borel transform reads, in the large $\beta$-limit, [331]:

$$\mathcal{B}[\delta m/\bar{m}] = \left(\frac{\bar{m}^2}{\nu^2}\right)^{-u} e^{5u/3}6(1-u)\frac{\Gamma(u)\Gamma(1-2u)}{\Gamma(3-u)} + \cdots , \quad (29.66)$$

where $\cdots$ indicates subtraction terms which are rapidly convergent and give negligible contributions to the coefficients $d_n$ for increasing $n$. Comparisons of the values of $d_n$ with the available calculations [151,153] show that the asymptotic series reproduce approximately the first two coefficients [331]. One can also notice that the series is rapidly dominated by the IR renormalon contributions and the series start to diverge to order $\alpha_s^3$ for the charm quark mass, and to order $\alpha_s^4$ for the bottom. An intuitive derivation of this IR effect can be obtained from the Coulomb potential. In this way, the IR correction to the heavy quark mass is [162]:

$$\frac{\delta m}{\bar{m}} = -\frac{1}{2\bar{m}}\int_{|\vec{q}|<\mu}\frac{d^3\vec{q}}{(2\pi)^3}V(\vec{q}) \simeq -C_F\alpha_s\frac{\mu}{\bar{m}} , \quad (29.67)$$

where:

$$V(\vec{q}) = -4\pi C_F\frac{\alpha_s(\vec{q})}{\vec{q}^2} . \quad (29.68)$$

It has also been noticed [158,159] that the IR singularity of the Borel transform for the pole mass in Eq. (29.66) is cancelled by that of the potential [486,162]:

$$\mathcal{B}[V(\vec{r})] = -(4\pi C_F)(\nu^2 e^{-C})^u \int \frac{d^3\vec{q}}{(2\pi)^3} \frac{e^{i\vec{q}\cdot\vec{r}}}{(\vec{q}^2)^{1+u}}$$
$$= -\frac{C_F}{r}(\nu^2 r^2 e^{-C})^u \frac{\Gamma((1/2)+u)\Gamma((1/2)-u)}{\pi\Gamma(1+2u)} . \quad (29.69)$$

This leads to the proposal of a new mass definition that is less IR sensitive than the pole mass in this approximation (see Section 11.13). In, for example, the derivations of the inclusive $B$-decays using a $1/m_b$ expansion, which behaves to leading order as $m_b^5$, it has been noticed that the use of the pole mass definition introduces an ambiguity of the order of $\Lambda/m_b$ when summing the series, which does not match with any non-perturbative parameters of the heavy quark expansion. This problem does not appear when one expresses the width in terms of the short distance $\overline{MS}$-mass, where a cancellation of the leading divergence between that of the width and of the relation between the pole and running mass occurs.

# 30

# Beyond the SVZ expansion

## 30.1 Tachyonic gluon mass

We have extensively discussed power corrections related to the IR regions, where the physical picture is simply the increase of the running coupling at large distance. Unconventional $1/Q^2$ corrections which go beyond this simple picture have been also analysed in the literature [161,162,342,341,329,344]. A lattice calculation of [487] shows that the $1/Q^2$ correction arises within a dispersive approach or from a removal of the Landau pole of the running coupling [162,488]. We have sketched this point when presenting the UV renormalons in the previous chapter. Following the presentation in [162], the leading UV renormalon gives the series expansion:

$$F = \left(\sum_n a_n \alpha_s(Q^2)\right)_{UV} = \sum_n n!(-b_0)^n \alpha_s^n(Q^2) . \tag{30.1}$$

Using its Borel transform, one has the integral representation:

$$\mathcal{B}[F] = \int dz \, \exp(-z) \, \frac{(\alpha_s b_0 z)^N}{1+\alpha_s b_0 z} , \tag{30.2}$$

where $N = 1/b_0\alpha_s$ is the value of $n$ for which the absolute value of the terms reaches its minimum. The integral is of the form:

$$\left(a_n \alpha_s^n(Q^2)\right)_{n=N} \simeq \frac{1}{2}\frac{\Lambda^2}{Q^2} , \tag{30.3}$$

where one can notice that the correction comes from the large virtual momenta $p^2 \sim Q^2 \exp(N)$, which is very different with the case of IR renormalon. However, in a theory such as lattice, which possesses an intrinsic UV cut-off, this effect can be irrelevant. Therefore, the alternative dispersive approach of the coupling can be used. The coupling can be parametrized as:

$$\frac{1}{\ln Q^2/\Lambda^2} \to \frac{1}{\ln Q^2/\Lambda^2} - \frac{\Lambda^2}{Q^2 - \Lambda^2} . \tag{30.4}$$

This modification can be justified if one argues that at finite order of perturbation theory the coupling satisfies dispersion relations with cuts at physical $s > 0$. More explicitly, one has:

$$\sum_n a_n \alpha_s(Q^2) \to \sum_n a_n \alpha_s(Q^2) - \sum_n a_n n! b_0^n \frac{\Lambda^2}{Q^2} . \tag{30.5}$$

In the case:

$$a_n = n!(-b_0)^n , \tag{30.6}$$

the power correction in the second term is still poorly defined. Taking its Borel transform, one obtains:

$$\left(\sum_n (-1)^n \frac{\Lambda^2}{Q^2}\right)_{\text{Borel}} = \frac{1}{2}\frac{\Lambda^2}{Q^2} , \tag{30.7}$$

showing that the power corrections from the procedures in Eqs (30.2) and (30.5) are the same, which may indicate that the Borel summation of the UV renormalon series and the removal of the Landau pole from dispersion relation can be intimately connected.

Another issue is the short distance ($r \ll \Lambda$) modification of the QCD potential, which becomes ($k$ is the string tension):

$$\lim_{r\to 0} V(r) = -C_F \frac{\alpha_s(r)}{r} + kr , \tag{30.8}$$

while in standard QCD, the leading power correction at short distance is $r^2$. This leads to the introduction of new small-size non-perturbative corrections and of a new picture of the QCD vacuum. In [161] one discusses this modification of the standard picture in terms of the phenomenology of the tachyonic gluon mass which is assumed to mimic the short-distance non-perturbative effects of QCD. We have seen previously that the $1/Q^2$ corrections to DIS can be explained from the IR region and is consistent with the OPE. In this picture, the constant term of the linear correction can be expressed as [162]:

$$k \approx -\frac{4}{6}\alpha_s \lambda^2 , \tag{30.9}$$

where $\lambda^2 < 0$ is the tachyonic gluon mass. In this framework, the standard picture of the OPE within the SVZ expansion gets modified due to the presence of the new $1/Q^2$ term. A systematic evaluation of this contribution using Feynman diagrams has been developed in CNZ. For the current–current two-point functions, it corresponds, to lowest order in $\alpha_s$, to the evaluation of the diagram in Fig. 30.1.

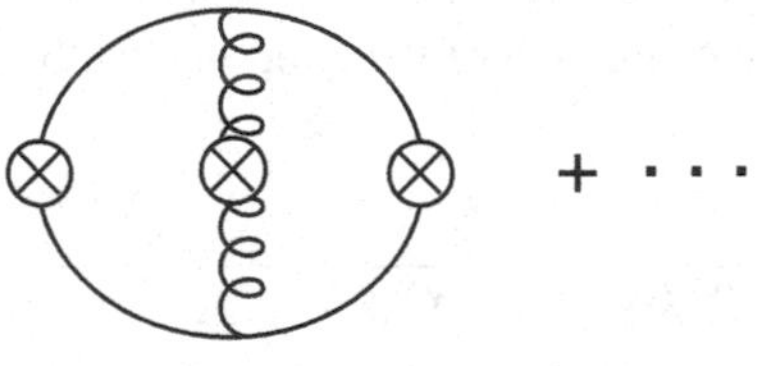


Fig. 30.1. Lowest order diagram contributing to $1/Q^2$. The cross in the gluon propagator corresponds to the tachyonic gluon mass insertion $\lambda^2$.

The value of the tachyonic gluon mass has been extracted phenomenologically using previous analysis in [341,329] from $e^+e^-$ data. Analyses of some other channels by CNZ have confirmed such findings. The pion and $\rho$ meson channels give the intersection range:

$$(\alpha_s/\pi)\lambda^2 \simeq -(0.06 \sim 0.07)\ \text{GeV}^2\ , \quad \text{or} \quad \lambda^2(1.25\text{GeV}^2) \approx -(0.34 \sim 0.52)\ \text{GeV}^2\ , \tag{30.10}$$

leading to the value of the string tension:

$$\sqrt{k} \simeq (369 \pm 14)\ \text{MeV}\ , \tag{30.11}$$

in agreement with the lattice results. The consequences of this result in some paradoxical QCD spectral sum rules channels have been also studied by CNZ, and lead to a resolution of different puzzles for the sum rule scales. One finds, for instance, for these scales:

$$M_\pi^2 \simeq 4M_\rho^2\ , \tag{30.12}$$

in agreement with the expectations of [382]. Analogous expectations in the gluonium channel has been also recovered. However, this change in the scale does not affect the predictions on the QCD parameters from the sum rules (quark mass, ...).

## 30.2 Instantons

Instanton–anti-instanton singularities occur for $b = 4\pi$ in the positive real axis [375]. They occur because far-separated instanton–anti-instanton pairs cannot be properly treated in a perturbative expansion around $\alpha_s = 0$. Due to graph counting rules, perturbation theory has a singularity at $b = 4\pi$, such that perturbation theory alone cannot give an unambiguous answer to the Borel integral for $b > 4\pi$. However, the singularity at $b = 4\pi$ in perturbation theory should also appear in the valley method for instanton-anti-instanton pairs. In addition, a proper definition of $\tilde{\mathcal{D}}(b)$ for $b > 4\pi$, including non-perturbative and non-analytic terms in $b$ should also emerge from the valley method.

In QCD, one expects an important rôle of the instantons due to the topologically non-trivial fluctuations of the gauge fields [381,264], where they are expected to explain the large mass of the $\eta'$ compared with the usual pseudoscalar mesons [262–264].

### *30.2.1 't Hooft instanton solution*

For a pedagogical introduction, let us start from the example of the Riemann integral (see e.g. [51]):

$$\mathcal{I} = \int_{-\infty}^{+\infty} F(x) e^{-\lambda(x)} dx\ , \tag{30.13}$$

where $\lambda(x)$ is some positive-definite functions. If $\lambda(x)$ has a minimum at the position $x_0$, one can approximate this function by:

$$\lambda(x) \simeq \lambda_0 + \frac{\lambda''}{2}(x - x_0)^2 + \cdots\ , \tag{30.14}$$

and obtain:

$$\mathcal{I} \simeq F(x_0)e^{-\lambda_0}\sqrt{\frac{2\pi}{\lambda''}} . \tag{30.15}$$

If instead, $\lambda(x)$ has several minimas $\lambda_{0,i}$ at positions $x_{0,i}$, one can approximately write:

$$\mathcal{I} \simeq \sum_i F(x_{0,i})e^{-\lambda_{0,i}}\sqrt{\frac{2\pi}{\lambda_i''}} . \tag{30.16}$$

A similar procedure can be done in the evaluation of a functional integral. If the action $S[\Phi]$ has a minimum for a field $\Phi_0(x)$, then this field gives a classical contribution to the functional integral analogously to Eq. (30.15):

$$\int F[\Phi]\mathcal{D}\Phi \sim F[\Phi_0]e^{-S[\Phi_0]} , \tag{30.17}$$

where corrections to this result are quantum corrections. If the field $\Phi_0(x)$ leads to a minimum of the action $S[\Phi]$, it is a solution of the Euler–Lagrange equations for that action. Hence solutions of the classical equations of motion lead to classical contributions to the functional integral in Eq. (30.17). There exist classical solutions of pure $SU(2)$ Yang–Mills theory which can be embedded in any $SU(N)$ gauge theory, which are called instantons. The *'t Hooft instanton* solution of the Yang–Mills equation is [264]:

$$G^a_{\mu\nu} = \frac{4\eta^a_{\mu\nu}\rho^2}{g[(x-x_0)^2+\rho^2]^2} , \tag{30.18}$$

where $x_0$ is the instanton position and $\eta^a_{\mu\nu}$ is the t'Hooft anti-symmetric symbol with the properties:

$$\begin{aligned} \eta^a_{\mu\nu} &= \epsilon^a_{\mu\nu} \quad \text{for} \quad \mu,\ \nu = 1,2,3 \\ \eta^a_{4\nu} &= -\delta^a_\nu \\ \eta^a_{4\mu} &= \delta^a_\mu \\ \eta^a_{44} &= 0 , \end{aligned} \tag{30.19}$$

where $\epsilon_{ijk}$ is the totally anti-symmetric tensor in three-dimensions, while $a = 1,2,3$ for the (subgroup) $SU(2)$. The *anti-instanton* solution is obtained by replacing $\eta^a_{\mu\nu}$ by its dual:

$$\bar{\eta}^a_{\mu\nu} = (-1)^{\delta_{\mu4}+\delta_{\nu4}}\eta^a_{\mu\nu} . \tag{30.20}$$

In Euclidian space–time, these solutions would correspond to particles of size $\rho$ at a position $x_0$, while in Minkowskian space–time, the solutions are not particles but can be considered as contributions to the tunnelling between different vacua. The action corresponding to the solution in Eq. (30.18), is easily obtained:

$$\mathcal{S}[G^a]_{cl} = -\frac{1}{4g}\int d^4x\ G^a_{\mu\nu}G^{\mu\nu}_a = \frac{8\pi^2}{g^2} . \tag{30.21}$$

The instanton fields are self-dual:

$$\tilde{G}^a_{\mu\nu} = \frac{1}{2}\epsilon_{\mu\nu\alpha\beta}G^a_{\alpha\beta} = G^a_{\mu\nu} \,. \tag{30.22}$$

The action of a self-dual field configuration is determined by its topological charge defined as:

$$Q = \int d^4x \left(\frac{g^2}{32\pi^2}\right) G^a_{\mu\nu}\tilde{G}^{\mu\nu}_a \,, \tag{30.23}$$

where an instanton has topological charge $+1$, and an anti-instanton $-1$. According to Eq. (30.17), the contribution of a single instanton to the vacuum expectation value of the functional $F[G]$ of $G^a_{\mu\nu}$ is:

$$F[G]e^{-S[G]} = F[G]e^{-\frac{8\pi^2}{g^2}} \,. \tag{30.24}$$

From Eqs. (30.21) and (30.24), one can deduce that instantons give genuine non-perturbative contributions, since the exponential cannot be expanded in a convergent power series of $g$ and its asymptotic expansion in $g$ is identically zero.

### 30.2.2 Instanton phenomenology

Qualitative estimates of the instanton effects based on the dilute gas approximation have been done in the literature [382–385], while an instanton liquid model has also been proposed [386]. However, the results obtained in these papers, for example, for the pseudoscalar quark currents are controversial, which come mainly from the uncontrolled use of the chiral symmetry-breaking parameters entering the analysis. Indeed, one does not know exactly if one should use the light quark current masses or the quark condensate. Moreover, the effects depend also crucially on the size of the instanton, whose value is very inaccurate. In practice, in this model, the instantons contribute as operators of dimension larger or equal than 9–11. For $Q^2 \geq 1$ GeV$^2$, no appreciable evidence of these effects has been detected in the phenomenological analysis, (even in the pseudoscalar channel, where one often claims that the effects are important!), as we shall see later on. A quantitative estimate of these effects from $e^+e^- \to I = 1$ hadrons data indeed shows that they are small [387,329], as expected from [385]. A program for measuring instanton induced hard scattering processes at HERA has been proposed [388]. In DIS, one expects to probe small-size instantons, which, in principle, are calculable, where the cross-section behaves as the square of the instanton density $D \sim e^{-\frac{2\pi}{\alpha_s}}$ times a function $F(\epsilon = \sqrt{s}/Q')$ of the total energy over the invariant mass of the particle produced:

$$\sigma \sim e^{-\frac{4\pi}{\alpha_s}F(\epsilon)} \,. \tag{30.25}$$

### 30.2.3 Dilute gas approximation

In principle, the superposition of two instanton solutions will not be a solution of the Euler–Lagrange equations, due to the non-linearity of these equations for a non-Abelian gauge theory. If one considers two far-away solutions, the superposition should be a good approximation for two instanton solutions (topological charge 2). In the dilute gas instanton approximation (DIGA), the instanton contribution can be estimated very roughly [382]. In so doing, one starts from the dilute gas density:

$$d(\rho) \approx C\left(\frac{2\pi}{\alpha_s(\rho)}\right)^6 \exp\left(\frac{-2\pi}{\alpha_s(\rho)}\right) \;:\quad C \approx 0.06 \quad \text{for QCD}\,, \tag{30.26}$$

where $\rho$ is the instanton radius. Using the previous t'Hooft instanton solution of the Yang–Mills equation the gluon condensates of $2n$ dimensions can be represented as:

$$\langle \mathcal{O}_{2n}\rangle \equiv \left\langle \left(gG^a_{\mu\nu}\right)_1 \cdots \left(gG^a_{\mu\nu}\right)_n \right\rangle = \int_0^{\rho_c} d\rho \frac{d(\rho)}{\rho^{2n+1}}\,, \tag{30.27}$$

where $\rho_c$ is the critical cut-off size of the instanton. By introducing the approximate relation:

$$\frac{2\pi}{\alpha_s(\rho)} \approx \frac{2\pi}{\alpha_s(\rho_c)} + 11\log(\rho_c/\rho)\,, \tag{30.28}$$

one obtains:

$$\begin{aligned}\langle \mathcal{O}_{2n}\rangle \approx\; & \frac{1}{(11-2n)}\frac{1}{\rho_c^{2n}}\left(\frac{2\pi}{\alpha_s(\rho_c)}\right)^6 \exp\left(-\frac{2\pi}{\alpha_s(\rho_c)}\right) \\ & \times \sum_{k=0}^{6} \frac{6!}{(6-k)!}\left(\frac{11}{2(11-2n)}\alpha_s(\rho_c)\right)^k\,, \end{aligned} \tag{30.29}$$

indicating that, for condensates of a critical dimension:

$$2n = 11\,, \tag{30.30}$$

one has a phase transition which separates the large-size instantons ($2n \leq 11$), that is, *ordinary low-dimension* condensates, with the small-size instanton (*instanton–anti-instanton or one-instanton*) effects. As emphasized in the previous derivation, the small-size instanton is very sensitive to the value of the instanton radius $\rho_c$, which renders (among many other unknown) uncertain the quantitative estimate of its effect. Some other reasons, as we shall see below are the inconsistency of the size and distance between instanton ensembles. For these different reasons, the estimate based on DIGA should only be considered at the qualitative level. Using the general expression in Eq. (30.29) for estimating, the contribution of the instanton to the NP gluon condensate $\langle g^2G^2\rangle$, and using the value of $\alpha_s(\rho_c) \approx 1$, one can deduce:

$$\langle g^2G^2\rangle^{1/4}_{\text{inst}} \geq \frac{4}{\rho_c}\,. \tag{30.31}$$

Using the previous expression of the topological charge and the self-duality relation, one obtains for $n_d$ dilute instantons in a volume $V$ greater than the instanton size, the instanton density:[1]

$$n_0 \equiv \frac{n_d}{V} = \frac{1}{V}\int_V Q(x)d^4x = \frac{1}{32\pi^2}\langle g^2G^2\rangle_{\text{inst}} . \tag{30.32}$$

Therefore the average distance $d_I$ between two instantons is:

$$d_I \equiv n_0^{-1/4} = \left(\frac{32\pi^2}{2\langle g^2G^2\rangle_{\text{inst}}}\right)^{1/4} . \tag{30.33}$$

These two equations give the ratio:

$$\frac{d_I}{\rho_c} \lesssim 0.7 , \tag{30.34}$$

which is smaller than 1. It may indicate that the dilute gas approximation is inconsistent, or it can indicate that higher unknown perturbative corrections or non-perturbative contributions (multi-instantons) to the classical result are important. Alternatively, one can integrate the tunnelling rate in order to get the phenomenological value of the instanton density [389]:

$$n_{\text{phen}} = \int_0^{\rho_c} d\rho \; n_0(\rho) , \tag{30.35}$$

which for $n_{\text{phen}} = 1 \text{ fm}^{-4}$, gives $\rho_c = 1$ fm using the SVZ value of the gluon condensate, which is rather pessimistic.[2]

### 30.2.4 The instanton liquid model

A more promising picture is the instanton liquid model [386,390]. The non-perturbative contribution to the instanton density defined previously can be estimated from the gluon condensate. The interaction of an instanton with an arbitrary external field $G^a_{\mu\nu}$ is:

$$S_{\text{int}} = \frac{2\pi^2\rho^2}{g^2}\bar{\eta}^a_{\mu\nu}U^{ab}G^b_{\mu\nu} , \tag{30.36}$$

which is a dipole interaction, and then does not contribute to the average action to first order. $U$ is an unitary matrix describing the orientation of the instanton in colour space. One can deduce [391]:

$$n(\rho) = n_0(\rho)\left[1 + \frac{\pi^4\rho^4}{2g^4}\langle G^2\rangle + \cdots\right] , \tag{30.37}$$

which has been exponentiated by [392]. In this way, and using $n_{\text{phen}} = 1$ fm, one obtains using the SVZ value of the condensate:

$$\rho_c = 1/3 \text{ fm} , \tag{30.38}$$

[1] In the classical field approach, the quantity below has no $g^2$ factor.

[2] However, the SVZ value of the gluon condensate has been underestimated by a factor of about 2 [329,313] such that the value of $\rho_c$ becomes 0.5 fm which leads to a more optimistic situation.

which is rather small. This result gives a different picture of the QCD vacuum. The instanton size being smaller than the separation between instantons implies that the vacuum is dilute. Also, the field inside the instanton is very strong:

$$G_{\mu\nu} \gg \Lambda^2 , \tag{30.39}$$

implying that the semi-classical approximation is valid. The action is large:

$$S = 8\pi^2/g^2 \sim 10 - 15 \gg 1 . \tag{30.40}$$

Also, instantons retain their individuality and are not destroyed by interactions:

$$\delta S_{\text{int}} \ll S_0 , \tag{30.41}$$

while interactions are important for the structure of the instanton ensemble:

$$(\exp|\delta S_{\text{int}}| \sim 20 \gg 1) . \tag{30.42}$$

The phenomenology of the instanton liquid model has been published in [386], which readers can consult for more details.

## 30.3 Lattice measurements of power corrections

Recent lattice measurements of the $V \pm A$ and (pseudo)scalar $(S, P)$ two-point correlators have been done in [393] in the $x$-space and have been compared with different models of power corrections (SVZ, ILM). Using the expressions of the correlators in the momentum space given in the previous section, and using the Fourier transform formulae in Table G.1 from [394] given in Appendix G, the different QCD expressions of the $V + A$ and $S + P$ correlators of interest here[3] in the $x$-space normalized to the perturbative contributions are [394]:

$$\frac{\Pi^{V+A}}{\Pi^{V+A}_{\text{pert}}} \to 1 - \frac{\alpha_s}{4\pi}\lambda^2 \cdot x^2 - \frac{\pi}{48}\langle\alpha_s\left(G^a_{\mu\nu}\right)^2\rangle x^4 \ln x^2 + \frac{2\pi^3}{81}\alpha_s\langle\bar{q}q\rangle^2 x^6 \ln x^2 , \tag{30.43}$$

where we adopt the convention $\ln x^2 < 0$. We have added to the usual SVZ-expansion the quadratic $x^2$ correction from [161]. In the $V - A$ channel, the usual SVZ expansion works quite well but for a small radius of convergence. In the $V + A$ channel, the SVZ-expansion as well as the ILM describe quite well the quantity $Q^2\Pi(Q^2)$, which is expected not to have a $1/Q^2$-term [161]:

$$\frac{Q^2 \cdot \Pi^{V+A}}{Q^2 \cdot \Pi^{V+A}_{\text{pert}}} \to 1 - \frac{\pi}{96}\langle\alpha_s\left(G^a_{\mu\nu}\right)^2\rangle x^4 + \frac{2\pi^3}{81}\alpha_s\langle\bar{q}q\rangle^2 x^6 \ln x^2. \tag{30.44}$$

This is to be contrasted to the case of $\Pi(Q^2)$, which needs also to be measured on the lattice, in order to test the existence of the $1/Q^2$ in the $V + A$ channel. However, the channel

[3] Some other correlators in the $x$-space are given in Chapter 39.

Table 30.1. *Different parameters used in the analysis of the S + P data in units of GeV^d (d is the dimension of the operator)*

| Sources | $\langle\alpha_s G^2\rangle$ | $\alpha_s\langle\bar\psi\psi\rangle^2$ | $(\alpha_s/\pi)\lambda^2$ |
|---|---|---|---|
| SET 1 (SVZ) [1] | 0.04 | $0.25^6$ | 0 |
| SET 2 [313,329] | 0.07 | $5.8\times10^{-4}$ | 0 |
| SET 3 [313,329,161] | 0.07 | $5.8\times10^{-4}$ | −0.12 |

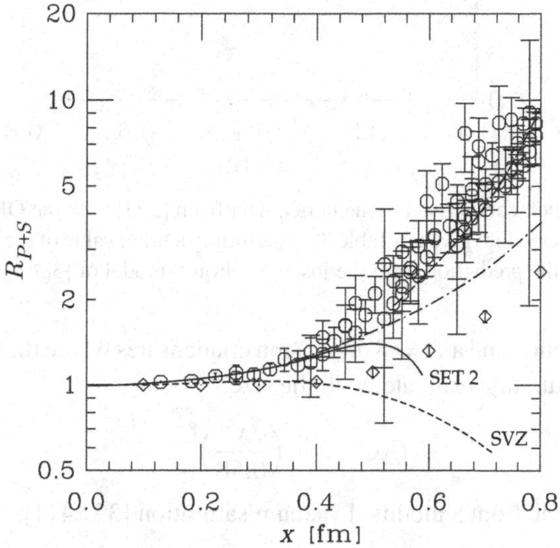


Fig. 30.2. $S+P$ channel: comparison of the lattice data from [393] with the OPE predictions for the two sets of QCD condensate values given in Table 30.1 . The dot-dashed curve is the prediction for SET 3 where the contribution of the $x^2$-term has been added to SET 2. The bold dashed curve is SET 3 + a fitted value of the $D=8$ condensate contributions. The diamond curve is the prediction from the instanton liquid model of [386].

which is crucial for the present analysis is the $(S+P)$ one. In this channel:

$$R_{P+S} \equiv \frac{1}{2}\left(\frac{\Pi^P}{\Pi^P_{\rm pert}}+\frac{\Pi^S}{\Pi^S_{\rm pert}}\right)$$
$$\to 1-\frac{\alpha_s}{2\pi}\lambda^2x^2+\frac{\pi}{96}\left\langle\alpha_s\left(G^a_{\mu\nu}\right)^2\right\rangle x^4+\frac{4\pi^3}{81}\alpha_s\langle\bar qq\rangle^2x^6\ln x^2\,. \tag{30.45}$$

As shown in Fig. 30.2, neither the SVZ-expansion nor the ILM can describe the lattice data, where we have used the sets of condensate values given in Table 30.1.

If such data are confirmed, it may indicate a strong evidence of the quadratic $1/Q^2$ power correction. We can see in Fig. 30.2, that for large $x$, the data is better fitted by including

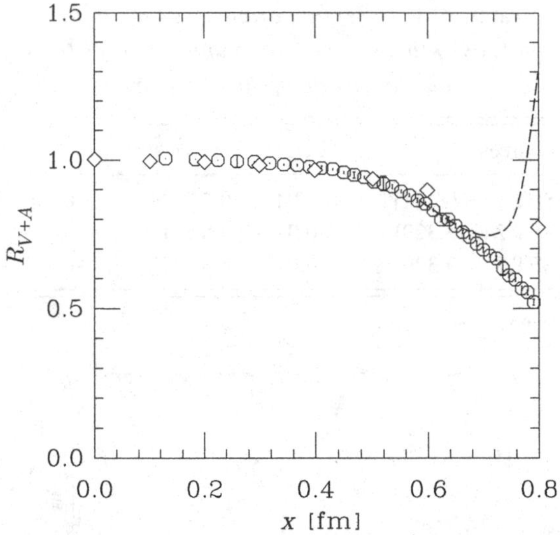


Fig. 30.3. $V + A$ channel: comparison of the lattice data from [393] with the OPE predictions for the SET 3 QCD condensates values given in Table 30.1 including a fitted value of the $D = 8$ contributions. The diamond curve is the prediction from the instanton liquid model of [386].

both the $1/Q^2$ correction and a $D = 8$ dimension condensates where the latter differ notably from the vacuum saturation estimate, with the size:

$$C_8\mathcal{O}_8 \simeq +\left(\frac{x}{0.58}\right)^8 , \tag{30.46}$$

compared with the one from a modified vacuum saturation [399,411]:

$$C_8\mathcal{O}_8|_{\text{fac}} \simeq +\frac{3395}{30855168}\langle\alpha_s G^2\rangle^2 x^8 \approx \left(\frac{x}{1.2}\right)^8 . \tag{30.47}$$

For completeness we also show in Fig. 30.3, a fit of the $V + A$ channel including the $D = 8$ condensate contributions. One can notice that like in the case of the $S + P$ channel, the value of the $D = 8$ condensates differs notably from the vacuum saturation estimate. It reads:

$$C_8\mathcal{O}_8 \simeq +\left(\frac{x}{0.7}\right)^8 , \tag{30.48}$$

compared with the one from a modified vacuum saturation [399,411]:

$$C_8\mathcal{O}_8|_{\text{fac}} \simeq +\frac{2}{3428352}\langle\alpha_s G^2\rangle^2 x^8 \approx \left(\frac{x}{2.5}\right)^8 . \tag{30.49}$$

One can conclude from the lattice measurement of the $S + P$ correlators that, if the data have to be explained by power corrrections, it can only be done by the presence of $\lambda^2$ quadratic corrections at moderate distance (less than 0.5 fm). For larger distances, one needs to add the contributions of higher eight-dimension condensates. It has been argued [395] that the $\lambda^2$ correction can be better understood within the effective Higgs-like theories which are common within the monopole mechanism of confinement, where, in the presence of a magnetically charged (effective) scalar field, the symmetry of the theory is $SU(3)_{\text{colour}} \times U(1)_{\text{magnetic}}$. Upon the spontaneous breaking of the magnetic $U(1)$ the gauge boson acquires a non-vanishing mass and its mass squared is the only parameter of dimension $d = 2$ consistent with the symmetry. Moreover, in exchanges between (colour) charged particles the gauge-boson mass appears to be the tachyonic mass as was demonstrated on the $U(1)$ example in [396,397]. Detailed analysis of various power corrections within the Higgs-like models can be found in [396–398]. Moreover, if the monopole size is indeed as small as indicated above, then the effective Higgs-like theories can apply at all distances $\sim (0.1 \div 0.5)$ fm.

# Part VIII

QCD two-point functions

# 31

# References guide to original works

In this chapter, we give a compilation of the different Wilson coefficients, as applications of the discussions in the previous chapter.[1] In order to minimize the missprint errors in the transcription of the formulae, we have used as much as possible the transfer of the formulae from the original files. These QCD two-point functions are useful for further uses in high-energy physics processes ($e^+e^- \to$ hadrons total cross-section, Higgs decays, ...), and not only for the QCD spectral sum rules analysis.

## 31.1 Electromagnetic current

- Historically, the electromagnetic spectral function has been obtained to order $\alpha$ in QED [318,319].
- In the massless quark limit, the order $\alpha_s^2$ correction has been obtained by [317], while the order $\alpha_s^3$ terms have been computed in [321]. The order $\alpha_s^4$ terms have been estimated [178] using the principle of minimal sensitivity (PMS) [176] and effective charge (ECH) approaches [177], or using $\tau$ decay data [332]. The order $\alpha_s^4 n_f^2$ has been computed recently in [438].
- The non-perturbative corrections were orginally obtained by SVZ [1]. Radiative corrections to the non-perturbative quark condensate have been calculated for the first time in [439].
- This observable is the most accurate quantity known in QCD today.

## 31.2 (Pseudo)scalar and (axial-)vector currents

- The results for the bilinear (pseudo)scalar and (axial)-vector quark correlators come essentially from [325,399,440,441,444].
- The $\alpha_s$ correction to the massless pseudoscalar correlator as well as the non-perturbative corrections were computed for the first time in [167]. The $\alpha_s^2$ term has been obtained in [445]. The $\alpha_s^3$ correction has been obtained in [446].

## 31.3 Quark mass corrections to the (pseudo)scalar and (axial)-vector quark correlators

- Quark mass corrections to the quark current–current correlators have been calculated to higher orders in [325,399,440,441], where it has been emphasized that the perturbative terms resulting from the relation between the normal and non-normal ordered quark condensates are essential for removing the mass logarithms singularities.

[1] This list of references might not be complete but only representative.

- The complete $\alpha_s$ correction to the massive (pseudo)scalar and (axial-)vector correlator has been evaluated in [399], while the $\alpha_s^2$ corrections come from [448,449].

## 31.4 Tachyonic gluon corrections to the (pseudo)scalar and (axial)-vector quark correlators

- Dimension two contributions due to tachyonic gluon mass have been obtained for the first time in [161].

## 31.5 Tensor quark correlators

- The correlator associated to the quark tensor current has been evaluated in [357,451]. It has been revised and corrected in [452].

## 31.6 Baryonic correlators

- Radiative corrections and non-perturbative effects to the light baryonic correlators have been calculated in [424–430].
- Correlators of heavy baryons have been evaluated in [453,454].

## 31.7 Four-quark correlators

- The two-point correlator associated to the four-quark current has been evaluated in [465,466] for analysing the four-quark states.
- Analogous correlators have been evaluated for the study of the $\Delta S = 2$ [467,468] and $\Delta I = 1/2$ kaon weak decays [469,470]. These results have been revised in [471].
- Similar correlators for the analysis of the $\bar{B}B$ mixing have been obtained in [472] to lowest order and including non-perturbative corrections. Radiative corrections including non-factorizable ones have been evaluated in [473]. $SU(3)$ breaking corrections are given in [474].

## 31.8 Gluonia correlators

- Radiative perturbative corrections to the bilinear gluonic correlators have been computed in [455], while the non-perturbative terms have been obtained in [382,456].
- The two-point correlator associated to three-gluonic current including non-perturbative corrections has been computed in [457].
- The off-diagonal quark-gluon two-point correlators have been calculated in [458,450,457].

## 31.9 Hybrid correlators

- The two-point correlator associated to the hybrid massless quark and gluonic current has been calculated in [459,460], where the final correct expression is given in [461]. The contribution of the tachyonic gluon acting as a new operator of dimension two has been obtained in [462].
- Two-point correlator associated to the heavy hybrid meson have been calculated in [463]. The contribution of the tachyonic gluon has been obtained in [464].

# 32

# (Pseudo)scalar correlators

We shall be concerned with the two-point correlators:

$$\Psi_5(q^2)^i_j \equiv i \int d^4x\, e^{iqx} \langle 0|\mathcal{T}\partial_\mu A^\mu(x)^j_i \left(\partial_\nu A^\nu(0)^j_i\right)^\dagger|0\rangle ,$$

$$\Psi(q^2)^i_j \equiv i \int d^4x\, e^{iqx} \langle 0|\mathcal{T}\partial_\mu V^\mu(x)^j_i \left(\partial_\nu V^\nu(0)^j_i\right)^\dagger|0\rangle , \tag{32.1}$$

associated to the pseudoscalar and scalar currents:

$$\begin{aligned}\partial_\mu A^\mu_{ij} &= (m_i + m_j)\bar{\psi}_i\,(i\gamma_5)\,\psi_j\\ \partial_\mu V^\mu_{ij} &= (m_i - m_j)\bar{\psi}_i(i)\psi_j\end{aligned} \tag{32.2}$$

Here the indices $i,\ j$ correspond to the light quark flavours $u,\ d,\ s$; $m_i$ is the mass of the quark $i$. It will be convenient to introduce the notation:

$$m_\pm = m_i \pm m_j\ . \tag{32.3}$$

The result of the scalar current can be deduced from the one of the pseudoscalar by the change $m_j$ into $-m_j$ or, equivalently, by the change $m_+$ into $m_-$ and vice-versa.

## 32.1 Exact two-loop perturbative expression in the $\overline{MS}$ scheme

The *complete* two-loop result for the pseudoscalar correlator, using the $\overline{MS}$ renormalized mass is:

$$\Psi_5(q^2)^i_j = \frac{3}{8\pi^2}(m_i + m_j)^2\left[(q^2 - (m_i - m_j)^2)\left[K + \left(\frac{\alpha_s}{\pi}\right)\frac{L}{3}\right] + M + \left(\frac{\alpha_s}{\pi}\right)\frac{N}{3}\right], \tag{32.4}$$

with:

$$\begin{aligned}
K &\equiv 1 + \frac{1}{2}l_i + \frac{1}{2}(1 + x_i)f_i\ +\ (i \longleftrightarrow j)\,,\\
L &\equiv 3K^2 + 2K + 6 - 2EI - 10x_i f_i^2\\
&\quad + m_i\big[(3K-2)\,\partial K/\partial m_i - 2\big(E + m_j^2\big)\partial I/\partial m_i\big] + (i \longleftrightarrow j)\,,\\
M &\equiv -m_i^2(1 + l_i) +\ (i \longleftrightarrow j)\,,
\end{aligned}$$

$$
\begin{aligned}
N &\equiv -\frac{1}{8}q^2(12K^2 - 4K + 5) + 3m_i^2(1 + f_i)(1 + x_j f_j) \\
&\quad - 2m_i^2\left(5 + 5l_i + 3l_i^2\right) + (i \longleftrightarrow j)\,, \\
I &\equiv [F(x_i) + F(x_j) - F(x_i x_j) - F(1)]/q^2, \\
E &\equiv \frac{1}{2}\left(m_i^2 + m_j^2 - q^2\right), \\
l_i &\equiv \log\left(\nu^2/m_i^2\right), \quad f_i \equiv \frac{\log x_i}{(1 - x_i)}\,, \\
x_{i,j} &\equiv m_{i,j}^2/\{E + E[1 - (m_i m_j/E)^2]^{1/2}\}\,, \\
F(x) &\equiv \int_o^x dy \left(\frac{\log y}{1-y}\right)^2 \log\left(\frac{x}{y}\right) = \sum_{n=1}^{\infty}[(2 - n\log x)^2 + 2]x^n/n^3\,. \qquad (32.5)
\end{aligned}
$$

This expression reproduces the massless result given previously in Section 11.14. The use of these results at $q = 0$ lead to the two-loop expression given in Eq. (27.14).

## 32.2 Three-loop expressions in the chiral limit

To order $\alpha_s^2$, the correlator reads:[1]

$$
\begin{aligned}
&(16\pi^2)\Psi_5(q^2) \\
&= -q^2 m_+^2\Bigg[\left[-12 - 6\ln\frac{\nu^2}{-q^2}\right] \\
&\quad + \left(\frac{\alpha_s}{\pi}\right)\left[-\frac{131}{2} + 24\,\zeta(3) - 34\ln\frac{\mu^2}{Q^2} - 6\ln^2\frac{\mu^2}{Q^2}\right] \\
&\quad + \left(\frac{\alpha_s}{\pi}\right)^2\Bigg[-\frac{17645}{24} + 353\,\zeta(3) + \frac{3}{2}\zeta(4) - 50\,\zeta(5) \\
&\qquad + \frac{511}{18}n_f - 8\,\zeta(3)\,n_f - \frac{10801}{24}\ln\frac{\nu^2}{-q^2} + 117\,\zeta(3)\ln\frac{\nu^2}{-q^2} \\
&\qquad + \frac{65}{4}n_f\ln\frac{\nu^2}{-q^2} - 4\,\zeta(3)\,n_f\ln\frac{\nu^2}{-q^2} - 106\ln^2\frac{\nu^2}{-q^2} + \frac{11}{3}n_f\ln^2\frac{\nu^2}{-q^2} \\
&\qquad - \frac{19}{2}\ln^3\frac{\nu^2}{-q^2} + \frac{1}{3}n_f\ln^3\frac{\nu^2}{-q^2}\Bigg]\Bigg]. \qquad (32.6)
\end{aligned}
$$

The same equation with $n_f = 3$ reads:

$$
\begin{aligned}
&(16\pi^2)\Psi_5(q^2) \\
&= -q^2 m_+^2\Bigg[\left[-12 - 6\ln\frac{\nu^2}{-q^2}\right] \\
&\quad + \left(\frac{\alpha_s}{\pi}\right)\left[-\frac{131}{2} + 24\,\zeta(3) - 34\ln\frac{\nu^2}{-q^2} - 6\ln^2\frac{\nu^2}{-q^2}\right]
\end{aligned}
$$

[1] From now, we shall omit the indices $i$ and $j$ on $\Psi_5(q^2)^i_j$.

$$+\left(\frac{\alpha_s}{\pi}\right)^2\left[-\frac{15601}{24}+329\,\zeta(3)+\frac{3}{2}\,\zeta(4)-50\,\zeta(5)-\frac{9631}{24}\ln\frac{\nu^2}{-q^2}\right.$$
$$\left.+105\,\zeta(3)\ln\frac{\nu^2}{-q^2}-95\ln^2\frac{\nu^2}{-q^2}-\frac{17}{2}\ln^3\frac{\nu^2}{-q^2}\right]\Bigg]. \quad (32.7)$$

## 32.3 Dimension-two

For a practical application, one should substract the mass singularities with the help of the Ward identity in Eqs. (2.17) and (27.14) and by working with the non-normal ordered condensate. To next-to-leading order in the quark mass terms, the IR stable result is:

$$\Psi_\pm(q^2)|^{(D=2)}=\frac{3}{8\pi^2}(m_i+m_j)^2\left[A\left(m_i^2\pm m_j^2\right)\mp Bm_im_j\right], \quad (32.8)$$

where:

$$A\equiv 2l-2+C_F\left(\frac{\alpha_s}{\pi}\right)\left[-3l^2+8l-\frac{25}{2}+6\zeta(3)\right],$$
$$B\equiv 2l-4+C_F\left(\frac{\alpha_s}{\pi}\right)[-3l^2+14l-22+6\zeta(3)], \quad (32.9)$$

with: $C_F=(N^2-1)/(2N)$ and $l=\log(-q^2/\nu^2)$; $\Psi_+$ and $\Psi_-$ are the pseudoscalar and scalar correlators.

## 32.4 Dimension-four

In terms of the non-normal ordered quark condensate, where the $m^4\log m^2$ terms have been absorbed, one obtains:

$$\Psi_\pm(q^2)\Big|_{m^4}^{(D=4)}=\frac{3}{8\pi^2}(m_i\pm m_j)^2\frac{1}{-2q^2}\left[m_i^4+4m_i^2m_j^2+m_j^4\right.$$
$$\left.+2\left(m_i^4\mp 2m_i^3m_j\mp 2m_j^3m_i+m_j^4\right)l\right]. \quad (32.10)$$

To lowest order in $\alpha_s$ and to *all orders* in the quark mass, the normal ordered quark condensate contribution reads:

$$\Psi_\pm(q^2)\Big|_{\psi}^{(D=4)}=-(m_i\pm m_j)^2\left[\frac{\langle:\bar\psi_i\psi_i:\rangle}{2m_i}\left[1-\frac{q_\pm^2}{q^2-m_i^2-m_j^2}f(z_i)\right]+(i\longleftrightarrow j)\right], \quad (32.11)$$

where:

$$f(z_i)=\frac{1}{2z_i}[1-\sqrt{1-4z_i}],$$
$$z_i\equiv\frac{m_i^2q^2}{(q^2-m_j^2+m_i^2)^2},$$

$$q_\pm \equiv q^2 - (m_i \pm m_j)^2 \,,$$
$$u \equiv \sqrt{1 - \frac{4m_i m_j}{q_-^2}} \,. \tag{32.12}$$

To order $\alpha_s$ and to leading order in the quark mass, one obtains:

$$\Psi_\pm(q^2 0|_\psi^{(D=4)} = \frac{(m_i \pm m_j)^2}{-q^2}\left[\frac{1}{2}\left[1 + C_F\left(\frac{\alpha_s}{\pi}\right)\left(-\frac{3}{2}l + \frac{11}{4}\right)\right]\langle m_i\bar{\psi}_i\psi_i + m_j\bar{\psi}_j\psi_j\rangle \right.$$
$$\left. \mp \left[1 + C_F\left(\frac{\alpha_s}{\pi}\right)\left(-\frac{3}{2}l + \frac{7}{2}\right)\right]\langle m_j\bar{\psi}_i\psi_i + m_i\bar{\psi}_j\psi_j\rangle\right] \,, \tag{32.13}$$

To all orders in the quark mass, one obtains the contribution of the normal ordered gluon condensates:

$$\Psi_\pm(q^2)|_G^{(D=4)} = -(m_i \pm m_j)^2\frac{1}{48\pi}\langle\, :\alpha_s GG:\,\rangle$$
$$\times\left[\frac{q^2}{q_\pm^4}\left[\frac{3(3+u^2)(1-u^2)}{2u^3}\log\frac{u+1}{u-1} - \frac{3u^4+4u^2+9}{u^2(1-u^2)}\right] \pm \frac{4}{m_i m_j}\right] \,, \tag{32.14}$$

where the expression of the scalar correlator can be deduced from the former by the additional change of $u$ into $1/u$. The previous expression still contains mass singularities. The introduction of *non-normal ordered* condensates as given in Eq. (27.17) leads to a cancellation of these terms. One obtains the IR stable result:

$$\Psi_\pm(q^2)|_G^{(D=4)} = (m_i \pm m_j)^2\frac{1}{-q^2}\frac{1}{8\pi}\langle\alpha_s GG\rangle$$
$$\times\left[1 + \frac{2}{3}\frac{m_i^2 + m_j^2}{q^2} \pm \frac{2m_i m_j}{q^2}(3 - 2l)\right] \,, \tag{32.15}$$

where the mass-logs have cancelled.

## 32.5 Dimension-five

To all order in the quark mass, the contribution of the normal ordered mixed quark-gluon condensate reads:

$$\Psi_\pm(q^2)|_{\text{mix}}^{(D=5)} = -(m_i \pm m_j)^2\langle\, :\bar{\psi}_i G\psi_i:\,\rangle\frac{1}{2m_i^3 q_\pm^2}\left[q^2 - m_j^2 \mp m_i m_j \right.$$
$$\left. - \frac{\left[(q^2 - m_j^2)^2 \mp m_i m_j(q^2 - m_j^2 + m_i^2) - m_i^2 m_j^2\right]}{q^2 - m_j^2 - m_i^2} f(z_i)\right]$$
$$+ (i \longleftrightarrow j) \,. \tag{32.16}$$

The leading contribution of the non-normal ordered condensate is:

$$\Psi_\pm(q^2)\big|_{\rm mix}^{(D=6)} = \mp\frac{(m_i \pm m_j)^2}{2q^2} g\langle m_i\bar\psi_j G\psi_j + m_j\bar\psi_i G\psi_i\rangle\ . \tag{32.17}$$

## 32.6 Dimension-six

The leading contributions are:

$$\begin{aligned}\Psi_\pm(q^2)\big|_{4\psi}^{(D=6)} &= \frac{(m_i \pm m_j)^2}{q^4}\pi\alpha_s\Bigg[\mp 4\langle\bar\psi_i\sigma_{\mu\nu}T_A\psi_j\bar\psi_j\sigma_{\mu\nu}T_A\psi_i\rangle \\ &+\frac{4}{3}\Big\langle 9\bar\psi_i\gamma_\mu T_A\psi_i + \bar\psi_j\gamma_\mu T_A\psi_j)\sum_{u,d,s}\bar\psi_k\gamma^\mu T^A\psi_k\Big\rangle\Bigg]\ . \end{aligned}\tag{32.18}$$

Using the vacuum saturation-like parametrization, one can write:

$$\Psi_\pm(q^2)\big|_{4\psi}^{(D=6)} = -\frac{(m_i \pm m_j)^2}{q^4}\left(\frac{4C_F}{3N}\right)\pi\alpha_s[\langle\bar\psi_i\psi_i\rangle^2 + \langle\bar\psi_j\psi_j\rangle^2 \mp 9\langle\bar\psi_i\psi_i\rangle\langle\bar\psi_j\psi_j\rangle]\ . \tag{32.19}$$

## 32.7 Exact two-loop expression of the spectral function

The complete two-loop pseudoscalar spectral function expressed in terms of the pole mass reads:

$$\begin{aligned}\frac{1}{\pi}{\rm Im}\Psi_5(t) = \frac{3}{8\pi^2}(m_i+m_j)^2\frac{\bar q^4}{t}v\Bigg[1+\frac{4}{3}\left(\frac{\alpha_s}{\pi}\right)\Bigg[\frac{3}{8}(7-v^2) \\ +\sum_i(v+v^{-1})\left[{\rm Li}_2(\alpha_i\alpha_j)-{\rm Li}_2(\alpha_i)-\log\alpha_i\log\beta_i\right] \\ +A_i\log\alpha_i + B_i\log\beta_i\Bigg]+\mathcal{O}(\alpha_s^2)\Bigg]\ ,\end{aligned}\tag{32.20}$$

where:

$$\begin{aligned}{\rm Li}_2(x) &= -\int_0^x\frac{dx}{x}\log(1-x)\ ,\\ A_i &= \frac{3}{4}\left(\frac{3m_i+m_j}{m_i+m_j}\right)-\frac{19+2v^2+3v^4}{32v}-\frac{m_i(m_i-m_j)}{\bar q^2 v(1+v)}\left(1+v+\frac{2v}{1+\alpha_i}\right)\ ,\\ B_i &= 2+2\frac{(m_i^2-m_j^2)}{\bar q^2 v}\ ,\\ \alpha_i &= \frac{m_i}{m_j}\frac{1-v}{1+v}\ ,\\ \beta_i &= \frac{(1+v)^2}{4v}\sqrt{1+\alpha_i}\ ,\end{aligned}\tag{32.21}$$

with:

$$\bar{q}^2 \equiv t - (m_i - m_j)^2 ,$$
$$v \equiv \sqrt{1 - \frac{4m_i m_j}{\bar{q}^2}} , \tag{32.22}$$

while $A_j$, $\alpha_j$, $B_j$ and $\beta_j$ can be obtained by interchanging the label $i$ and $j$. The spectral function of the scalar current $\partial^\mu(\bar{\psi}_i \gamma_\mu \psi_j)$ can be obtained from the former by changing the sign of $m_j$. For $m_i = 0$, one has $A_i = 0$ and $\beta_i = 1$, which guarantee the absence of mass singularities. In this case, the expression simplifies and reads ($m \equiv m_i$):

$$\frac{1}{\pi}\mathrm{Im}\Psi_5(t) = \frac{3}{8\pi^2}x(1-x)^2 t^2 \Bigg[1 + \frac{4}{3}\left(\frac{\alpha_s}{\pi}\right)\Bigg[\frac{9}{4} + \mathrm{Li}_2(x)$$
$$+ \log x \log(1-x) - \frac{3}{2}\log\left(\frac{1}{x} - 1\right) - \log(1-x)$$
$$+ x\log\left(\frac{1}{x} - 1\right) - \frac{x}{1-x}\log x\Bigg]\Bigg]\theta(t - m^2) , \tag{32.23}$$

where $x \equiv m^2/t$.

The order $\alpha_s^2$ corrections to the (pseudo)scalar spectral function has been also obtained recently for massive quarks [448] where the result is available in a Mathematica package Rvs.m from the url: *http://www-ttp.physik.uni-karlsruhe.de/Progdata/ttp00/ttp00-25.*

## 32.8 Heavy-light correlator

We have given in the previous section the expression of the spectral function when one of the quark mass goes to zero. In the following, we give useful lowest order expressions in $\alpha_s$ when $m \equiv m_i \ll M \equiv m_j$ for the (pseudo)scalar current. In taking the small mass $m$ limit, different cares have been taken in order to have IR stable results. Expressing the correlator as:

$$\Psi_\pm = (m \pm M)^2 \Big[\Psi_\pm^{pert} + \Psi_\pm^{\bar{\psi}\psi}\langle\bar{\psi}\psi\rangle + \Psi_\pm^{\bar{Q}Q}\langle\bar{Q}Q\rangle + \Psi_\pm^{G^2}\langle\alpha_s G^2\rangle +$$
$$\Psi_\pm^{\bar{\psi}G\psi}\langle\bar{\psi}\frac{\lambda_a}{2}\sigma^{\mu\nu}G^a_{\mu\nu}\psi\rangle + \Psi_\pm^{\bar{Q}GQ}\langle\bar{Q}\frac{\lambda_a}{2}\sigma^{\mu\nu}G^a_{\mu\nu}Q\rangle\Big] \tag{32.24}$$

One obtains[488] ($W \equiv M^2 - q^2$):

$$\Psi_\pm^{pert} = \frac{N}{8\pi^2}\Bigg[2q^2 - 3M^2 + \frac{M^4}{q^2}\ln\frac{M^2}{W} - (2M^2 - q^2)\ln\frac{\mu^2}{W}$$
$$\pm 2mM\left(2 - \frac{M^2}{q^2}\ln\frac{M^2}{W} + \ln\frac{\mu^2}{W}\right) - 2m^2\left(1 + \ln\frac{\mu^2}{W}\right)$$
$$\mp 2\frac{m^3 M}{W}\left(1 - \frac{M^2}{q^2}\ln\frac{M^2}{W} - \ln\frac{m^2}{W}\right)$$
$$+\frac{m^4}{W^2}\left[M^2 - \frac{3}{2}q^2 - \frac{M^4}{q^2}\ln\frac{M^2}{W} - (2M^2 - q^2)\ln\frac{m^2}{W}\right]\Bigg]$$

$$\Psi_{\pm}^{\bar\psi\psi} = \left[ \mp \frac{Mq^2}{W} + \frac{mq^2(2M^2-q^2)}{2W^2} \mp \frac{m^2M^3q^2}{W^3} \right]$$

$$\Psi_{\pm}^{\bar QQ} = \left[ -\frac{M}{2} \pm m \mp \frac{m^3}{W} \right] \qquad \text{for} \quad -q^2 > M^2$$

$$\Psi_{\pm}^{G^2} = \frac{1}{12\pi W}\left[ \pm \frac{Mq^2}{m} - \frac{q^4}{2W} \pm mq^4MW^2\left(q^2+6M^2\ln\frac{mM}{W}\right) - \frac{m^2q^4}{W^3}\left(q^2+7M^2+6M^2\ln\frac{mM}{W}\right)\right]$$

$$\Psi_{\pm}^{\bar\psi G\psi} = \left[ \pm \frac{Mq^6}{2W^3} - \frac{mM^2q^6}{2W^4}\right]$$

$$\Psi_{\pm}^{\bar QGQ} = \pm\frac{mq^2}{2W} \qquad \text{for} \quad -q^2 > M^2 \tag{32.25}$$

One can notice that some of these terms are IR singular and behave like $\log m$ and $1/m$. In order to have an IR stable result, one should work with the renormalized condensates defined in Eq. (27.56). In this way, one obtains:

$$\begin{aligned}
\bar\Psi_{\pm}^{pert} &= \Psi_{\pm}^{pert} + \frac{3}{\pi^2}\left[\frac{M^3}{q^2}\left[\log\frac{M^2}{\mu^2}-1\right]\Psi_{\pm}^{\bar QQ} + \frac{m^3}{q^2}\left[\log\frac{m^2}{\mu^2}-1\right]\Psi_{\pm}^{\bar\psi\psi}\right],\\
\bar\Psi_{\pm}^{\bar QQ} &= \Psi_{\pm}^{\bar QQ},\\
\bar\Psi_{\pm}^{\bar\psi\psi} &= \Psi_{\pm}^{\bar\psi\psi},\\
\bar\Psi_{\pm}^{G^2} &= \Psi_{\pm}^{G^2} + \frac{1}{12M}\Psi_{\pm}^{\bar QQ} + \frac{1}{12m}\Psi_{\pm}^{\bar\psi\psi} - \frac{M}{2q^2}\log\frac{M^2}{\mu^2}\Psi_{\pm}^{\bar QGQ} - \frac{m}{2q^2}\log\frac{m^2}{\mu^2}\Psi_{\pm}^{\bar\psi G\psi},\\
\bar\Psi_{\pm}^{\bar QGQ} &= \Psi_{\pm}^{\bar QGQ},\\
\bar\Psi_{\pm}^{\bar\psi G\psi} &= \Psi_{\pm}^{\bar\psi G\psi} = \mp\frac{M}{2}.
\end{aligned} \tag{32.26}$$

Therefore, one can deduce for small $m$:

$$\begin{aligned}
\bar\Psi_{\pm}^{pert} &= \frac{3}{16\pi^2}\left[4q_{\mp}^2 - (M^4+4M^2m^2+m^4)\frac{1}{q^2}\right.\\
&\left. -2\left[q^2 \mp -M^2 - m^2 + (M^4 \mp 2M^3m \mp 2Mm^3 + m^4)\frac{1}{q^2}\right]\log\frac{-q^2}{\mu^2}\right].\\
\bar\Psi_{\pm}^{G^2} &= -\frac{1}{8} - \frac{(M^2+m^2)}{12q^2} \mp \frac{Mm}{4q^2}\left[3 - 2\log\frac{-q^2}{\mu^2}\right],
\end{aligned} \tag{32.27}$$

where:

$$q_{\mp}^2 = q^2 - (M \mp m)^2, \tag{32.28}$$

and where all IR infinities have disappeared.

# 33

# (Axial-)vector two-point functions

The Wilson coefficients for the OPE of these correlators were first calculated by SVZ to leading order in $\alpha_s$ and in the quark mass terms. Calculations of the coefficients beyond the leading order exist in the literature. These results are collected here.

We shall be concerned with the two-point correlator for the vector $V^\mu_{ij} = \bar{\psi}_i \gamma^\mu \psi_j$ and axial-vector currents $A^\mu_{ij} = \bar{\psi}_i \gamma^\mu \gamma_5 \psi_j$:

$$\Pi^{\mu\nu}_{ij,V}(q) \equiv i \int d^4x\, e^{iqx} \langle 0|\mathcal{T} V^\mu(x)^j_i \left(V^\nu(0)^j_i\right)^\dagger |0\rangle \,,$$
$$\Pi^{\mu\nu}_{ij,A}(q) \equiv i \int d^4x\, e^{iqx} \langle 0|\mathcal{T} A^\mu(x)^j_i \left(A^\nu(0)^j_i\right)^\dagger |0\rangle \,. \qquad (33.1)$$

Here the indices $i,\ j$ correspond to the light quark flavours $u,\ d,\ s$. The vector ($V$) and axial-vector ($A$) correlators have the Lorentz decomposition:

$$\Pi^{\mu\nu}_{ij,V/A} = -(g^{\mu\nu} q^2 - q^\mu q^\nu)\Pi^{(1)}_{ij,V/A}(q^2, m_i^2, m_j^2) + q^\mu q^\nu \Pi^{(0)}_{ij,V/A}(q^2, m_i^2, m_j^2)\,, \quad (33.2)$$

where $m_i$ is the mass of the quark $i$; $\Pi^{(J)}_{ij}$ is the correlator associated to the hadrons of spin $J = 0, 1$. The (pseudo)scalar correlators $\Psi_{(5)}(q^2)^i_j$ is related to $\Pi^{(0)}_{ij}$ via the non-anomalous Ward identity in Eq. (2.17). It will be convenient to introduce the notation:

$$\Pi^{(1+0)}_{ij} \equiv \Pi^{(1)}_{ij} + \Pi^{(0)}_{ij}\,. \qquad (33.3)$$

The result of the axial-vector current can be deduced from the one of the vector by the change $m_j$ into $-m_j$ or, equivalently, by the change $m_-$ into $m_+$ and vice-versa.

### 33.1 Exact two-loop perturbative expression in the $\overline{MS}$ scheme

The complete two-loop result for the vector correlator is:

$$\begin{aligned}\Pi^{(1+0)}_{ij,V} \equiv & -\frac{1}{3}\left[1 + \left(\frac{\alpha_s}{\pi}\right)\frac{15}{4}\right] + PK \\ & + \alpha l_i + \beta l_j + 2(\alpha - \beta)(\alpha Z_i - \beta Z_j) \\ & + \frac{2}{3}\left(\frac{\alpha_s}{\pi}\right)\left[\frac{1}{2}PL + \alpha l_i(1 + 2l_i) + \beta l_j(1 + 2l_j)\right.\end{aligned}$$

$$
\begin{aligned}
&-\frac{1}{4}(1+2N_i+2N_j)\left(1+6\frac{(m_i-m_j)^2}{q^2}\right) \\
&+x_i f_i^2 + x_j f_j^2 + \frac{1}{2}(N_i-N_j)^2 \\
&-(\alpha-\beta)(G(x_i)-G(x_j)) - (3-2(\alpha+\beta))\,K^2 \\
&-\left(\frac{1}{2}+\alpha\,(2+l_i)+\beta(2+l_j)\right)\Bigg] ,
\end{aligned}
\tag{33.4}
$$

with:

$$
\begin{aligned}
\alpha &\equiv -m_i^2/q^2 , \\
\beta &\equiv -m_j^2/q^2 , \\
P &\equiv 1-\alpha-\beta-2(\alpha-\beta) , \\
N_i &\equiv \alpha(1+f_i)(1+x_j f_j) , \\
N_j &\equiv \beta(1+f_j)(1+x_i f_i) , \\
Z_i &\equiv 1+l_i+\frac{2}{3}\left(\frac{\alpha_s}{\pi}\right)(5+5l_i+3l_i^2) ,
\end{aligned}
$$

$$
G(x) \equiv xF'(x) = \int_o^x dy\left(\frac{\log y}{1-y}\right)^2 = \sum_{n=1}^{\infty}[(1-n\log x)^2+1]x^n/n^2 , \tag{33.5}
$$

where $K$ has been defined in Eq. (32.5). The log-mass terms appearing there should be cancelled once one introduces the contributions of non-normal ordered condensates.

### 33.2 Three-loop expression including the $m^2$-terms

Including the $m^2$-term to order $\alpha_s^2$, the correlator reads:

$$
\begin{aligned}
&(16\pi^2)\Pi_V^{(0+1)} \\
&= +\left[\frac{20}{3}+4\ln\frac{\nu^2}{-q^2}\right] \\
&\quad+\frac{\alpha_s}{\pi}\left[\frac{55}{3}-16\,\zeta(3)+4\ln\frac{\nu^2}{-q^2}\right] \\
&\quad+\left(\frac{\alpha_s}{\pi}\right)^2\left[\frac{41927}{216}-\frac{1658}{9}\zeta(3)+\frac{100}{3}\zeta(5)-\frac{3701}{324}n_f+\frac{76}{9}\zeta(3)\,n_f\right. \\
&\qquad+\frac{365}{6}\ln\frac{\nu^2}{-q^2}-44\,\zeta(3)\ln\frac{\nu^2}{-q^2}-\frac{11}{3}n_f\ln\frac{\nu^2}{-q^2} \\
&\qquad\left.+\frac{8}{3}\zeta(3)\,n_f\ln\frac{\nu^2}{-q^2}+\frac{11}{2}\ln^2\frac{\nu^2}{-q^2}-\frac{1}{3}n_f\ln^2\frac{\nu^2}{-q^2}\right]
\end{aligned}
$$

$$
\begin{aligned}
&+\frac{m_-^2}{-q^2}\left[-6+\frac{\alpha_s}{\pi}\left(-12-12\ln\frac{\nu^2}{-q^2}\right)\right]\\
&+\frac{m_+^2}{-q^2}\left[-6+\frac{\alpha_s}{\pi}\left(-16-12\ln\frac{\nu^2}{-q^2}\right)\right]\\
&+\left(\frac{\alpha_s}{\pi}\right)^2\frac{m_-^2}{-q^2}\left[-\frac{4681}{24}-34\,\zeta(3)+115\,\zeta(5)+\frac{55}{12}n_f+\frac{8}{3}\zeta(3)\,n_f\right.\\
&\qquad\left.-\frac{215}{2}\ln\frac{\nu^2}{-q^2}+\frac{11}{3}n_f\ln\frac{\nu^2}{-q^2}-\frac{57}{2}\ln^2\frac{\nu^2}{-q^2}+n_f\ln^2\frac{\nu^2}{-q^2}\right]\\
&+\left(\frac{\alpha_s}{\pi}\right)^2\frac{m_+^2}{-q^2}\left[-\frac{19691}{72}-\frac{124}{9}\zeta(3)+\frac{1045}{9}\zeta(5)+\frac{95}{12}n_f-\frac{253}{2}\ln\frac{\nu^2}{-q^2}\right.\\
&\qquad\left.+\frac{13}{3}n_f\ln\frac{\nu^2}{-q^2}-\frac{57}{2}\ln^2\frac{\nu^2}{-q^2}+n_f\ln^2\frac{\nu^2}{-q^2}\right]\\
&+\left(\frac{\alpha_s}{\pi}\right)^2\frac{\sum_f m_f^2}{-q^2}\left[\frac{128}{3}-32\,\zeta(3)\right]. \qquad (33.6)
\end{aligned}
$$

$$
\begin{aligned}
&(16\pi^2)\Pi_A^{(0+1)}\\
&=+\left[\frac{20}{3}+4\ln\frac{\nu^2}{-q^2}\right]\\
&+\frac{\alpha_s}{\pi}\left[\frac{55}{3}-16\,\zeta(3)+4\ln\frac{\nu^2}{-q^2}\right]\\
&+\left(\frac{\alpha_s}{\pi}\right)^2\left[\frac{34525}{216}-\frac{1430}{9}\zeta(3)+\frac{100}{3}\zeta(5)+\frac{299}{6}\ln\frac{\nu^2}{-q^2}-36\,\zeta(3)\ln\frac{\nu^2}{-q^2}\right.\\
&\qquad\left.+\frac{9}{2}\ln^2\frac{\nu^2}{-q^2}\right]\\
&+\frac{m_-^2}{-q^2}\left[-6+\frac{\alpha_s}{\pi}\left(-12-12\ln\frac{\nu^2}{-q^2}\right)\right]\\
&+\frac{m_+^2}{-q^2}\left[-6+\frac{\alpha_s}{\pi}\left(-16-12\ln\frac{\nu^2}{-q^2}\right)\right]\\
&+\frac{m_-^2}{-q^2}\left(\frac{\alpha_s}{\pi}\right)^2\left[-\frac{4351}{24}-26\,\zeta(3)+115\,\zeta(5)-\frac{193}{2}\ln\frac{\nu^2}{-q^2}-\frac{51}{2}\ln^2\frac{\nu^2}{-q^2}\right]\\
&+\frac{m_+^2}{-q^2}\left(\frac{\alpha_s}{\pi}\right)^2\left[-\frac{17981}{72}-\frac{124}{9}\zeta(3)+\frac{1045}{9}\zeta(5)-\frac{227}{2}\ln\frac{\nu^2}{-q^2}-\frac{51}{2}\ln^2\frac{\nu^2}{-q^2}\right]\\
&+\sum_f m_f^2\left(\frac{\alpha_s}{\pi}\right)^2\left[\frac{128}{3}-32\,\zeta(3)\right]. \qquad (33.7)
\end{aligned}
$$

### 33.3 Dimension-four

The dynamic operators of dimension-four are the gluon and quark condensates. Let us start by giving the contributions coming from the normal ordered condensates, which are obtained from a direct calculation of the Feynman diagrams within the Wick's theorem. One obtains:

$$\left[\Pi^{(1)}_{ij,V/A}\right]^{(D=4)}_{\psi} = -\frac{1}{3m_i q^2}\langle\ :\bar\psi_i\psi_i:\ \rangle\Bigg[1+2\frac{(m_j^2-m_i^2)}{q^2} - \frac{[q^2+m_j^2+m_i^2-2(m_j^2-m_i^2)^2/q^2]}{q^2-m_j^2+m_i^2}f(z_i)\Bigg] + (i \longleftrightarrow j)\,, \tag{33.8}$$

$$\left[\Pi^{(1)}_{ij,V/A}\right]^{(D=4)}_{G} = \frac{1}{48\pi}\langle :\alpha_s GG:\rangle\frac{1}{q^4_\pm}\left[\frac{3(1+u^2)(1-u^2)^2}{2u^5}\log\frac{u+1}{u-1} - \frac{3u^4+2u^2+3}{u^4}\right], \tag{33.9}$$

where the result for the axial-vector can be obtained by the additionnal change of $u$ into $1/u$.

Let us now use the previous results and truncate the series to the $D=4$ contributions but including radiative corrections. In so doing, we shall consider these quark and gluon operators defined previously in the $\overline{MS}$ scheme. The remaining $D=4$ operators are product of the running quark masses. In terms of the scale invariant condensates defined previously, the contributions to the correlators are:

$$\begin{aligned}
&Q^4\left[\Pi^{(1+0)}_{ij,V/A}(-Q^2)\right]^{(D=4)}\\
&= \frac{1}{12}\left[1-\frac{11}{18}\left(\frac{\alpha_s}{\pi}\right)(Q)\right]\left\langle\frac{\alpha_s}{\pi}GG\right\rangle\\
&+\left[1-\left(\frac{\alpha_s}{\pi}\right)(Q)-\frac{13}{3}\left(\frac{\alpha_s}{\pi}\right)^2(Q)\right]\langle m_i\bar\psi_i\psi_i+m_j\bar\psi_j\psi_j\rangle\\
&\pm\left[\frac{4}{3}\left(\frac{\alpha_s}{\pi}\right)(Q)+\frac{59}{6}\left(\frac{\alpha_s}{\pi}\right)^2(Q)\right]\langle m_j\bar\psi_i\psi_i+m_i\bar\psi_j\psi_j\rangle\\
&+\left[\frac{4}{27}\left(\frac{\alpha_s}{\pi}\right)(Q)+\left(-\frac{257}{486}+\frac{4}{3}\zeta(3)\right)\left(\frac{\alpha_s}{\pi}\right)^2(Q)\right]\sum_k\langle m_k\bar\psi_k\psi_k\rangle\\
&+\frac{3}{2\pi^2}\left[1+\left(\frac{76}{9}-\frac{4}{3}\zeta(3)\right)\left(\frac{\alpha_s}{\pi}\right)^2(Q)\right]m_i^2(Q)m_j^2(Q)\\
&+\frac{1}{4\pi^2}\left[-\frac{12}{7}\left(\frac{\alpha_s}{\pi}\right)^{-1}(Q)+1\right]\left[m_i^4(Q)+m_j^4(Q)\right]\\
&\mp\frac{4}{7\pi^2}m_i(Q)m_j(Q)\left[m_i^2(Q)+m_j^2(Q)\right]\\
&-\frac{1}{28\pi^2}\left[1-\left(\frac{65}{6}-16\zeta(3)\right)\left(\frac{\alpha_s}{\pi}\right)(Q)\right]\sum_k m_k^4(Q)\,,
\end{aligned} \tag{33.10}$$

and:

$$\begin{aligned}
&Q^4\left[\Pi^{(0)}_{ij,V/A}(-Q^2)\right]^{(D=4)}\\
&= \langle(m_i\mp m_j)(\bar\psi_i\psi_i\mp\bar\psi_j\psi_j)\rangle
\end{aligned}$$

$$+\frac{1}{4\pi^2}\left[-\frac{12}{7}\left(\frac{\alpha_s}{\pi}\right)^{-1}(Q)+\frac{11}{14}\right][m_i(Q)\mp m_j(Q)]\left[m_i^3(Q)\mp m_j^3(Q)\right]$$
$$\mp\frac{3}{4\pi^2}m_i(Q)m_j(Q)[m_i(Q)\mp m_j(Q)]^2\,. \tag{33.11}$$

### 33.4 Dimension-five

This contribution is due to the mixed quark-gluon condensate. It has been evaluated to all orders in the quark mass. In terms of the normal ordered condensate, it reads:

$$\begin{aligned}\left[\Pi^{(1)}_{ij,V/A}(-Q^2)\right]^{(D=5)}_{\rm mix} &= -\langle\,:\bar\psi_i G\psi_i:\,\rangle\frac{1}{3m_i^3q^4q_\pm^2q_\pm^2}\\ &\times\Big[[q^2+2m_j(m_j\mp m_i)]\big(q^2-m_j^2\big)^2-m_i^2q^2\big(q^2+m_j^2\big)\\ &-2m_jm_i^2(m_j\mp m_i)\big(2m_j^2-m_i^2\big)\\ &-\frac{P(q^2,m_i,m_j)}{q^2-m_j^2+m_i^2}f(z_i)\Big]+(i\longleftrightarrow j)\,,\end{aligned} \tag{33.12}$$

with:

$$\begin{aligned}P(q^2,m_i,m_j) &= [q^2+2m_j(m_j\mp m_i)]\big(q^2-m_j^2\big)^3\\ &-m_j^2m_i^2q^2\big(4m_j^2\mp 6m_jm_i+m_i^2\big)\\ &-m_i^2q^4\big(q^2+m_j^2\big)\\ &+2m_jm_i^2(m_j\mp m_i)\big(3m_j^4-3m_j^2m_i^2+m_i^4\big)\,.\end{aligned} \tag{33.13}$$

### 33.5 Dimension-six

Here we shall consider the contributions which do not vanish for massless quarks. Then we shall neglect the triple gluon condensate contribution $g^3\langle f_{abc}G^a_{\mu\nu}G^b_{\nu\rho}G^c_{\rho\mu}\rangle$, where the coefficient vanishes in the chiral limit. Therefore, we have:

$$\begin{aligned}&Q^6\left[\Pi^{(1+0)}_{ij,V/A}(-Q^2)\right]^{(D=6)}\\ &= -8\pi^2\left[1+\left(\frac{431}{96}-\frac{9}{8}L\right)\left(\frac{\alpha_s}{\pi}\right)(\nu)\right]\left(\frac{\alpha_s}{\pi}\right)(\nu)\left\langle\bar\psi_i\gamma_\mu\begin{pmatrix}\gamma_5\\1\end{pmatrix}T_a\psi_j\bar\psi_j\gamma_\mu\begin{pmatrix}\gamma_5\\1\end{pmatrix}T_a\psi_i(\nu)\right\rangle\\ &+\frac{5\pi^2}{4}(3+4L)\left(\frac{\alpha_s}{\pi}\right)^2(\nu)\left\langle\bar\psi_i\gamma_\mu\begin{pmatrix}1\\\gamma_5\end{pmatrix}T_a\psi_j\bar\psi_j\gamma_\mu\begin{pmatrix}1\\\gamma_5\end{pmatrix}T_a\psi_i(\nu)\right\rangle\\ &+\frac{2\pi^2}{3}(3+4L)\left(\frac{\alpha_s}{\pi}\right)^2(\nu)\left\langle\bar\psi_i\gamma_\mu\begin{pmatrix}1\\\gamma_5\end{pmatrix}\psi_j\bar\psi_j\gamma_\mu\begin{pmatrix}1\\\gamma_5\end{pmatrix}\psi_i(\nu)\right\rangle\\ &-\frac{8\pi^2}{9}\left[1+\left(\frac{107}{48}-\frac{95}{72}L\right)\left(\frac{\alpha_s}{\pi}\right)(\nu)\right]\left(\frac{\alpha_s}{\pi}\right)(\nu)\\ &\times\sum_k\langle(\bar\psi_i\gamma_\mu T_a\psi_i+\bar\psi_j\gamma_\mu T_a\psi_j)\bar\psi_k\gamma^\mu T_a\psi_k(\nu)\rangle\end{aligned}$$

$$
\begin{aligned}
&+\frac{5\pi^2}{54}(-7+6L)\left(\frac{\alpha_s}{\pi}\right)^2(\nu)\sum_k\langle(\bar\psi_i\gamma_\mu\gamma_5 T_a\psi_i+\bar\psi_j\gamma_\mu\gamma_5 T_a\psi_j)\bar\psi_k\gamma^\mu\gamma_5 T_a\psi_k(\nu)\rangle \\
&+\frac{4\pi^2}{81}(-7+6L)\left(\frac{\alpha_s}{\pi}\right)^2(\nu)\sum_k\langle(\bar\psi_i\gamma_\mu\gamma_5\psi_i+\bar\psi_j\gamma_\mu\gamma_5\psi_j)\bar\psi_k\gamma^\mu\gamma_5\psi_k(\nu)\rangle \\
&+\frac{4\pi^2}{81}(1+6L)\left(\frac{\alpha_s}{\pi}\right)^2(\nu)\sum_{k,l}\langle\bar\psi_k\gamma_\mu T_a\psi_k\bar\psi_l\gamma^\mu T_a\psi_l(\nu)\rangle\,. \qquad (33.14)
\end{aligned}
$$

where $L \equiv \log(Q^2/\nu^2)$. The upper component of $\binom{1}{\gamma_5}$ or $\binom{\gamma_5}{1}$ is for the vector($V$) correlator, while the lower one is for the axial-vector ($A$).

## 33.6 Vector spectral function to higher order

### *33.6.1 Complete two-loop perturbative expression of the spectral function*

In the case where one of the quark mass is zero, the spectral function of the vector current reads:

$$
\begin{aligned}
\mathrm{Im}\Pi^{(1)}(t) = &\frac{(2+x)}{3m^2t}\mathrm{Im}\Psi_5(t) \\
&-\frac{1}{6\pi}\left(\frac{\alpha_s}{\pi}\right)\Big[(3+x)(1-x)^3\log\frac{x}{1-x} \\
&+2x\log x+(3-x^2)(1-x)\Big]\,, \qquad (33.15)
\end{aligned}
$$

where $x \equiv m^2/t$, and becomes:

$$
\begin{aligned}
\frac{1}{\pi}\mathrm{Im}\Pi^{(1)}(t) = &\frac{1}{8\pi^2}\Bigg[(2+x)\Bigg[1+\frac{4}{3}\left(\frac{\alpha_s}{\pi}\right)\Bigg[\frac{13}{4}+2\log x+\log x\log(1-x) \\
&+\frac{3}{2}\log\frac{x}{1-x}-\log(1-x)-x\log\frac{x}{1-x}-\frac{x}{1-x}\log x\Bigg]\Bigg] \\
&+\frac{4}{3}\left(\frac{\alpha_s}{\pi}\right)\Bigg[-(3+x)(1-x)\log\frac{x}{1-x}-\frac{2x}{(1-x)^2}\log x \\
&-5-2x-\frac{2x}{1-x}\Bigg]\Bigg]\theta(t-m^2)\,. \qquad (33.16)
\end{aligned}
$$

For the case of the electromagnetic current, one has the well-known QED result, which is accurately reproduced by the Schwinger interpolating formula:

$$
\frac{1}{\pi}\mathrm{Im}\Pi^{(1)}(t) = \frac{1}{4\pi}v\left(\frac{3-v^2}{2}\right)\left[1+\frac{4}{3}\alpha_s f(v)\right]\theta(t-4m^2)\,, \qquad (33.17)
$$

where:

$$
\begin{aligned}
v &\equiv \sqrt{1-4x}\,, \\
f(v) &\equiv \frac{\pi}{2v}-\frac{(3+v)}{4}\left(\frac{\pi}{2}-\frac{3}{4\pi}\right). \qquad (33.18)
\end{aligned}
$$

### 33.6.2 Four-loop perturbative expression of the spectral function

The neutral vector spectral function can be related to the $e^+e^- \to$ hadrons total cross-section as:

$$R_{e^+e^-}(t) \equiv \frac{\sigma(e^+e^- \to \text{ hadrons}(\gamma))}{\sigma(e^+e^- \to \mu^+\mu^-(\gamma))} = 12\pi \, \text{Im}\Pi_{\text{em}}(t + i\epsilon) \,, \tag{33.19}$$

where $\text{Im}\Pi_{\text{em}}$ is associated to the conserved electromagnetic current $J^\mu_{\text{em}} \equiv \sum_i Q_i \bar{\psi}_i \gamma^\mu \psi_i$ $(i = u, \ d, \ s, \ \ldots)$. However, the perturbative calculation has been done in the Euclidian region and corresponds to the $D$-function:

$$D(Q^2) \equiv -Q^2 \frac{d}{dQ^2} \Pi_{\text{em}}(Q^2) \,, \tag{33.20}$$

which can be related to $R_{e^+e^-}$ through:

$$R(s) = \frac{1}{2i\pi} \int_{-s-i\epsilon}^{-s+i\epsilon} \frac{dQ^2}{Q^2} D(Q^2) \,, \tag{33.21}$$

where it is necessary to transform the result into the physical region by taking into account the effects due to the analytic continuation of the terms of the type:

$$\log^n(-q^2/\nu^2) \to (\log(t/\nu^2) + i\pi)^n \,. \tag{33.22}$$

The asymptotic four-loop expression reads:

$$(16\pi^2)\frac{1}{\pi}\text{Im}\Pi_{\text{em}}(t) = 3\left(\sum_i Q_i^2\right)\left[1 + \frac{\bar{\alpha}_s}{\pi} + F_3\left(\frac{\bar{\alpha}_s}{\pi}\right)^2 + F_4\left(\frac{\bar{\alpha}_s}{\pi}\right)^3\right] + \left(\sum_i Q_i\right)^2 F_4'\left(\frac{\bar{\alpha}_s}{\pi}\right)^3 \,, \tag{33.23}$$

where $\bar{\alpha}_s$ is the running coupling evaluated at the scale $t$ and:

$$\begin{aligned} F_3 &= 1.9857 - 0.1153n \,, \\ F_4 &= -6.6368 - 1.2001n - 0.0052n^2 \,, \\ F_4' &= -1.2395 \,. \end{aligned} \tag{33.24}$$

The last term comes from the light-by-light diagrams specific for the neutral electromagnetic current. The expression of the $D$-function reads:

$$\begin{aligned}(16\pi^2)D(-Q^2) = 3\left(\sum_i Q_i^2\right)&\left[1 + \left(\frac{\alpha_s}{\pi}\right) + \left[F_3 + \frac{b_1}{2}L\right]\left(\frac{\alpha_s}{\pi}\right)^2\right. \\ &\left. + \left[F_4 + \left(F_3\beta_1 + \frac{\beta_2}{2}\right)L + \frac{\beta_1^2}{4}\left(L^2 + \frac{\pi^2}{3}\right)\right]\left(\frac{\alpha_s}{\pi}\right)^3\right] \\ &+ \left(\sum_i Q_i\right)^2 F_4'\left(\frac{\alpha_s}{\pi}\right)^3 \,, \end{aligned} \tag{33.25}$$

where the values of the $\beta$-function have been given in Table 11.1 and $L \equiv \ln(Q^2/\nu^2)$.

## 33.7 Heavy-light correlator

In the following, we give useful lowest order expressions in $\alpha_s$ when $m \equiv m_i \ll M \equiv m_j$ for the (axial-)vector current. The notations are the same as in previous section but differs from $\Pi_{V/A}$ given in [488]. They are related as:

$$\Pi^{(1)}_{V/A} = \Pi_{V/A} - \frac{(M \mp m)^2}{q^2}\Psi_\mp - \frac{(M \mp m)}{q^2}\Big[\langle \bar{Q}Q \mp \bar{q}q\rangle\Big], \qquad (33.26a)$$

where $\Pi^{(1)}_{V/A}$ is the $g_{\mu\nu}$ coefficients in [488], $\Psi_\mp$ has been given in Section (32.8), and:

$$\Pi_{V/A} = \Pi_{V/A}\Big|_{pert} + \Pi_{V/A}\Big|_{\bar\psi\psi}\langle\bar\psi\psi\rangle + \Pi_{V/A}\Big|_{\bar{Q}Q}\langle\bar{Q}Q\rangle + \Pi_{V/A}\Big|_{G^2}\langle\alpha_s G^2\rangle +$$
$$\Pi_{V/A}\Big|_{\bar\psi G\psi}\langle\bar\psi\frac{\lambda_a}{2}\sigma^{\mu\nu}G^a_{\mu\nu}\psi\rangle + \Pi^{(1)}_{V/A}\Big|_{\bar{Q}GQ}\langle\bar{Q}\frac{\lambda_a}{2}\sigma^{\mu\nu}G^a_{\mu\nu}Q\rangle. \qquad (33.26b)$$

The different contributions read:

$$\Pi_{V/A}\Big|_{pert} = \frac{3}{24\pi^2}\Bigg[\frac{10}{3}q^2 + 4M^2 - 4\frac{M^4}{q^2} + 2(2M^2 - 3q^2)\frac{M^4}{q^4}\ln\frac{M^2}{W} + 2q^2\ln\frac{\mu^2}{W}$$
$$+6m^2\left(1 + 2\frac{M^2}{q^2} - 2\frac{M^4}{q^4}\ln\frac{M^2}{W}\right)$$
$$-3\frac{m^4}{W^2}\left[\frac{(2M^2 - q^2)^2}{q^2} - 2(2M^2 - 3q^2)\frac{M^4}{q^4}\ln\frac{M^2}{W} + 2q^2\ln\frac{m^2}{W}\right]\Bigg]$$
$$\Pi_{V/A}\Big|_{\bar\psi\psi} = \left[\frac{mq^4}{W^2} + \frac{2m^3q^4(4M^2 - q^2)}{3W^4}\right]$$
$$\Pi_{V/A}\Big|_{\bar{Q}Q} = \left[M - \frac{2M^3}{3q^2} + \frac{2m^2M}{q^2}\right] \quad \text{for } -q^2 > M^2$$
$$\Pi_{V/A}\Big|_{G^2} = -\frac{q^4}{12\pi W^2}\left[1 + \frac{2m^2}{W^2}\left(q^2 + 7M^2 + 6M^2\ln\frac{mM}{W}\right)\right]$$
$$\Pi_{V/A}\Big|_{\bar\psi G\psi} = -\frac{mM^2q^6}{W^4}$$
$$\Pi_{V/A}\Big|_{\bar{Q}GQ} = \mp\frac{mM^2}{3W} \quad \text{for } -q^2 > M^2. \qquad (33.27a)$$

with $W \equiv M^2 - q^2$. Introducing the renormalized condensates defined in Eq. (27.56), and using relations similar to Eq. (32.26), one can deduce:

$$\bar\Pi_{V/A}\Big|_{pert} = \frac{3}{24\pi^2}\Bigg[10q^2 + 18(M^2 + m^2) - 9(3M^4 - 4M^2m^2 + 3m^4)\frac{1}{q^2}$$
$$-6\Big[q^2 - 3(M^4 + m^4)\frac{1}{q^2}\Big]\log\frac{-q^2}{\mu^2}\Bigg],$$

$$\begin{aligned}
\bar{\Pi}_{V/A}\Big|_{\bar\psi\psi} &= \Pi_{V/A}\Big|_{\bar\psi\psi}\,,\\
\bar{\Pi}_{V/A}\Big|_{\bar{Q}Q} &= \Pi_{V/A}\Big|_{\bar{Q}Q}\,,\\
\bar{\Pi}_{V/A}\Big|_{G^2} &= \frac{1}{12} - \frac{(M^2+m^2)}{18q^2}\,,\\
\bar{\Pi}_{V/A}\Big|_{\bar\psi G\psi} &= \Pi_{V/A}\Big|_{\bar\psi G\psi}\,,
\end{aligned} \tag{33.27b}$$

from which the expressions for $\Pi^{(1)}_{V/A}$ can be easily derived.

## 33.8 Beyond the SVZ expansion: tachyonic gluon contributions to the (axial-)vector and (pseudo)scalar correlators

Here, we shall give contributions coming from the dimension $D = 2$ operators induced by a tachyonic gluon mass. This contribution has been introduced in [161], where one expects that the gluon mass phenomenologically mimics the resummation of the QCD perturbative series due to renormalons.

### 33.8.1 Vector correlator

This effect can be systematically obtained from the Feynman diagram given in Fig. 30.1. The derivation of the results is explictly given in [161]. Here, we only quote these results which are consistent if one uses normal non-ordered condensates for the $D = 4$ contribution. To first order in $\alpha_s$ and expanding in $\lambda^2$, $m^2_{1,2}$ we obtain:

$$\begin{aligned}
(16\pi^2)\Pi^{(1)}_V = &\left[\frac{20}{3} + 6\frac{m_-^2}{Q^2} - 6\frac{m_+^2}{Q^2} + 4l_{\mu Q} + 6\frac{m_-^2}{Q^2}l_{\mu Q}\right]\\
&+\frac{\alpha_s}{\pi}\left[\frac{55}{3} - 16\,\zeta(3) + \frac{107}{2}\frac{m_-^2}{Q^2} - 24\,\zeta(3)\frac{m_-^2}{Q^2} - 16\frac{m_+^2}{Q^2}\right.\\
&\left.+4l_{\mu Q} + 22\frac{m_-^2}{Q^2}l_{\mu Q} - 12\frac{m_+^2}{Q^2}l_{\mu Q} + 6\frac{m_-^2}{Q^2}l^2_{\mu Q}\right]\\
&+\frac{\alpha_s}{\pi}\frac{\lambda^2}{Q^2}\left[-\frac{128}{3} + 32\,\zeta(3) - \frac{76}{3}\frac{m_-^2}{Q^2} + 16\,\zeta(3)\frac{m_-^2}{Q^2}\right.\\
&\left.-8\frac{m_+^2}{Q^2} + 12\frac{m_-^2}{Q^2}l_{\mu Q} - 12\frac{m_+^2}{Q^2}l_{\mu Q}\right],
\end{aligned} \tag{33.28}$$

The above result is in the $\overline{MS}$ scheme and the notations are as follows: $m_\pm = m_1 \pm m_2$ and $l_{\mu Q} = \log(\mu^2/Q^2)$. Note that the terms of order $\lambda^2/Q^2$ in Eq. (33.28) are $\mu$ independent and, thus, do not depend on the way the overall UV subtraction of the vector correlator is implemented. The quark mass-logs appearing in the $\lambda^2 m^2/Q^4$ terms have been absorbed after adding the contribution of the quark condensate to the correlators and its modified renormalization group invariant combination:

$$\langle\bar\psi_i\psi_i\rangle = \frac{3m_i^3}{4\pi^2}\left[1+\ln\left(\frac{\mu^2}{m_i^2}\right)+2\frac{\alpha_s}{\pi}\left(\ln^2\left(\frac{\mu^2}{m_i^2}\right)+\frac{5}{3}\ln\left(\frac{\mu^2}{m_i^2}\right)+\frac{5}{3}\right)\right]$$
$$+\frac{m_i\lambda^2}{4\pi^2}\frac{\alpha_s}{\pi}\left(-5+6\ln\frac{\mu^2}{m_i^2}\right). \tag{33.29}$$

In the light-quark case relevant to the $\rho$-channels we can neglect the $m^2$ terms and $\Pi_\rho(M^2)$ simplifies greatly:

$$\frac{1}{\pi}\mathrm{Im}\,\Pi_\rho(s) = \frac{1}{4\pi^2}\left\{1+\left(\frac{\alpha_s}{\pi}\right)\left[1-1.05\,\frac{\lambda^2}{s}\delta(s)\right]\right\}. \tag{33.30}$$

### 33.8.2 (Pseudo)scalar correlator

In the chiral limit ($m_u \simeq m_d = 0$), the QCD expression of the absorptive part of the (pseudo)scalar correlator reads:

$$\frac{1}{\pi}\mathrm{Im}\psi_{(5)}(s) \simeq (m_i+(-)m_j)^2\frac{3}{8\pi^2}s\left[1+\left(\frac{\alpha_s}{\pi}\right)\left(-2L+\frac{17}{3}-4\frac{\lambda^2}{s}\right)\right], \tag{33.31}$$

where one should notice that the coefficient of the $\lambda^2$ term:

$$b_\pi \approx 4b_\rho\,. \tag{33.32}$$

# 34
# Tensor-quark correlator

We shall be concerned with the two-point correlator:

$$\begin{aligned}\Psi_{2,\mu\nu\rho\sigma} &\equiv i\int d^4x\ e^{iqx}\langle 0|\mathcal{T}\theta^q_{\mu\nu}(x)\theta^q_{\rho\sigma}(0)^\dagger|0\rangle \\ &= \frac{1}{2}\left(\eta_{\mu\rho}\eta_{\nu\sigma}+\eta_{\mu\sigma}\eta_{\nu\rho}-\frac{2}{3}\eta_{\mu\nu}\eta_{\rho\sigma}\right)\Psi_2(q^2)\,, \end{aligned} \tag{34.1}$$

where

$$\theta^q_{\mu\nu}(x)=i\bar{q}(x)(\gamma_\mu\bar{D}_\nu+\gamma_\nu\bar{D}_\mu)q(x) \tag{34.2}$$

is the quark component of the energy-momentum tensor $\theta^q_{\mu\nu}(x)$. Here, $\bar{D}_\mu \equiv \vec{D}_\mu - \overleftarrow{D}_\mu$ is the covariant derivative. The previous current mixes under renormalization with the gluonic current:

$$\theta^g_{\mu\nu}(x)=-G_{\alpha\mu}G^\alpha_\nu+\frac{1}{4}g_{\mu\nu}G_{\alpha\beta}G^{\alpha\beta}\,, \tag{34.3}$$

as:

$$\begin{aligned}\theta^{q,R}_{\mu\nu} &= Z_{11}\theta^{q,B}_{\mu\nu}+Z_{12}\theta^{g,B}_{\mu\nu}\\ \theta^{g,R}_{\mu\nu} &= Z_{21}\theta^{q,B}_{\mu\nu}+Z_{22}\theta^{g,B}_{\mu\nu}\,. \end{aligned} \tag{34.4}$$

The indices $B$ and $R$ refer respectively to bare and renormalized quantities. The renormalization constants have been evaluated in [450]. For the quark currents, they read in $4-\epsilon$ dimension space–time:

$$Z_{11}=1+\left(\frac{\alpha_s}{\pi}\right)\frac{1}{\hat{\epsilon}}\frac{4}{3}C_F\,,\qquad Z_{12}=-\left(\frac{\alpha_s}{\pi}\right)\frac{1}{\hat{\epsilon}}\frac{4}{3} \tag{34.5}$$

with: $\hat{\epsilon}^{-1}=\epsilon^{-1}+(\ln 4\pi-\gamma_E)/2$. To the order, we are working, only $Z_{11}$ is relevant. The corresponding anomalous dimension is:

$$\gamma_{11}=-\frac{\nu}{Z_{11}}\frac{dZ_{11}}{d\nu}=\left(\frac{16}{9}\equiv\gamma^1_{11}\right)\left(\frac{\alpha_s}{\pi}\right)\,. \tag{34.6}$$

The renormalized perturbative contribution to the correlator to order $\alpha_s$ is [452]:

$$\Psi^R_{2,\text{pert}}(q^2 \equiv -Q^2) = -\frac{3}{10\pi^2} Q^4 \log\frac{Q^2}{\nu^2}\left[1 - \left(\frac{\alpha_s}{\pi}\right)\left(\frac{473}{135} - \frac{8}{9}\log\frac{Q^2}{\nu^2}\right)\right]. \quad (34.7)$$

The bare quark and mixed condensate contributions read:

$$\Psi^R_{2,q+m}(q^2) = \frac{1}{q^2}\left[-8m^3\langle\bar\psi\psi\rangle^B + \frac{16}{3}mg\left\langle\bar\psi\sigma_{\mu\nu}\frac{\lambda_a}{2}\psi G^{\mu\nu}_a\right\rangle^B\right]. \quad (34.8)$$

The evaluation of the gluon condensate is much more cumbersome. Evaluating the Feynman integrals for arbitrary mass and expanding the result in powers of $m^2/q^2$, one obtains:

$$\Psi^B_{2,G}(q^2) = \left(\frac{\alpha_s}{\pi}\right)\langle G^2\rangle^B\left[\frac{8}{9}\left(\frac{-2}{\hat\epsilon} + \ln\frac{m^2}{\nu^2}\right) + 3\frac{m^2}{q^2} + \frac{8}{9}\left(1 - 3\frac{m^2}{q^2}\right)\ln -\frac{q^2}{m^2}\right]. \quad (34.9)$$

In order to remove the IR logarithm appearing in the bare result, one has to write the heavy- to light-quark expansions of the condensates discussed in previous chapters:

$$\langle\bar\psi\psi\rangle = -\frac{1}{12\pi}\left(\frac{\alpha_s}{\pi}\right)\langle G^2\rangle + \cdots,$$
$$\left\langle\bar\psi\sigma_{\mu\nu}\frac{\lambda_a}{2}\psi G^{\mu\nu}_a\right\rangle = \frac{m}{2}\left(-\frac{2}{\hat\epsilon} + \ln\frac{m^2}{\nu^2}\right)\left(\frac{\alpha_s}{\pi}\right)\langle G^2\rangle + \cdots. \quad (34.10)$$

In this way, one obtains the bare gluon condensate contribution:

$$\Psi^B_{2,G}(q^2) = \left(\frac{\alpha_s}{\pi}\right)\langle G^2\rangle^B\frac{1}{3}\left[\frac{8}{3}\left(1 - 3\frac{m^2}{q^2}\right)\left(\frac{-2}{\hat\epsilon} + \log -\frac{q^2}{\nu^2}\right) + 7\frac{m^2}{q^2}\right]. \quad (34.11)$$

The remaining $m^2/\hat\epsilon q^2$ pole can be eliminated by the introduction of the renormalized mixed condensate [130] discussed in previous chapters:

$$\left\langle\bar\psi\sigma_{\mu\nu}\frac{\lambda_a}{2}\psi G^{\mu\nu}_a\right\rangle^B = \left\langle\bar\psi\sigma_{\mu\nu}\frac{\lambda_a}{2}\psi G^{\mu\nu}_a\right\rangle^R - \frac{m}{\hat\epsilon}\left(\frac{\alpha_s}{\pi}\right)\langle G^2\rangle^R. \quad (34.12)$$

Then, one obtains the renormalized result:

$$\Psi^R_{2,G}(q^2) = \left(\frac{\alpha_s}{\pi}\right)\langle G^2\rangle^R\frac{1}{3}\left[\frac{8}{3}\left(1 - 3\frac{m^2}{q^2}\right)\log -\frac{q^2}{\nu^2} + 7\frac{m^2}{q^2}\right]. \quad (34.13)$$

This explicit exercise has shown how delicate is the evaluation of the Wilson coefficients of the non-perturbative condensate contributions.

The four-quark condensate contribution is:

$$\Psi_{2,4q}(q^2) = \frac{64\pi}{9q^2}\rho\left(\frac{\alpha_s}{\pi}\right)\langle\bar\psi\psi\rangle^2, \quad (34.14)$$

where $\rho$ is the deviation from the vacuum saturation estimate.

# 35

# Baryonic correlators

## 35.1 Light baryons

### *35.1.1 The decuplet*

The simplest interpolating field for the $\Delta(\frac{3}{2})$ is [424–430]:

$$\Delta_\mu = \frac{1}{\sqrt{2}} : (\psi^T C \psi)\left(g_{\mu\lambda} - \frac{1}{4}\gamma_\mu\gamma_\lambda\right)\psi : , \tag{35.1}$$

where $C$ is the charge conjugation matrix and colour indices. The corresponding correlator is:

$$\begin{aligned} S_{\mu\nu} &= i \int d^4q\, e^{iqx} \langle 0|\mathcal{T}\Delta_\mu(x)\Delta_\mu(0)^\dagger|0\rangle \\ &= (\hat{q}F_1 + F_2)\, g_{\mu\nu} + \cdots \end{aligned} \tag{35.2}$$

Using the SVZ-expansion, the form factor can be expressed as:

$$\begin{aligned} -F_1 &= q^4 A_1 \log -\frac{q^2}{\nu^2} + A_2\pi \log -\frac{q^2}{\nu^2}\langle\alpha_s G^2\rangle + m_s A_3\pi^2\langle\bar\psi\psi\rangle + A_4\frac{\pi^4}{q^2}\langle\bar\psi\psi\rangle^2 \\ -F_2 &= m_s B_1 q^4 \log -\frac{q^2}{\nu^2} + B_2\pi^2 \log -\frac{q^2}{\nu^2}\langle\bar\psi\psi\rangle q^2 + B_3\pi^2\left\langle\bar\psi\sigma^{\mu\nu}\frac{\lambda_a}{2}G^a_{\mu\nu}\psi\right\rangle , \end{aligned} \tag{35.3}$$

where $A_i$ and $B_i$ are Wilson coefficients determined from perturbative calculation of the QCD diagrams shown in Fig. 35.1. The different expressions of these Wilson coefficients compiled in [426] are given in Tables 35.1 and 35.2 to lowest order of $\alpha_s$. In the table, we also introduce the parameters controlling the $SU(3)$ breaking of the condensates:

$$\chi_3 = \frac{\langle\bar s s\rangle}{\langle\bar u u\rangle} , \quad \text{and} \quad \chi_5 = \frac{\langle\bar s\sigma^{\mu\nu}\lambda_a G^a_{\mu\nu} s\rangle}{\langle\bar u\sigma^{\mu\nu}\lambda_a G^a_{\mu\nu} u\rangle} . \tag{35.4}$$

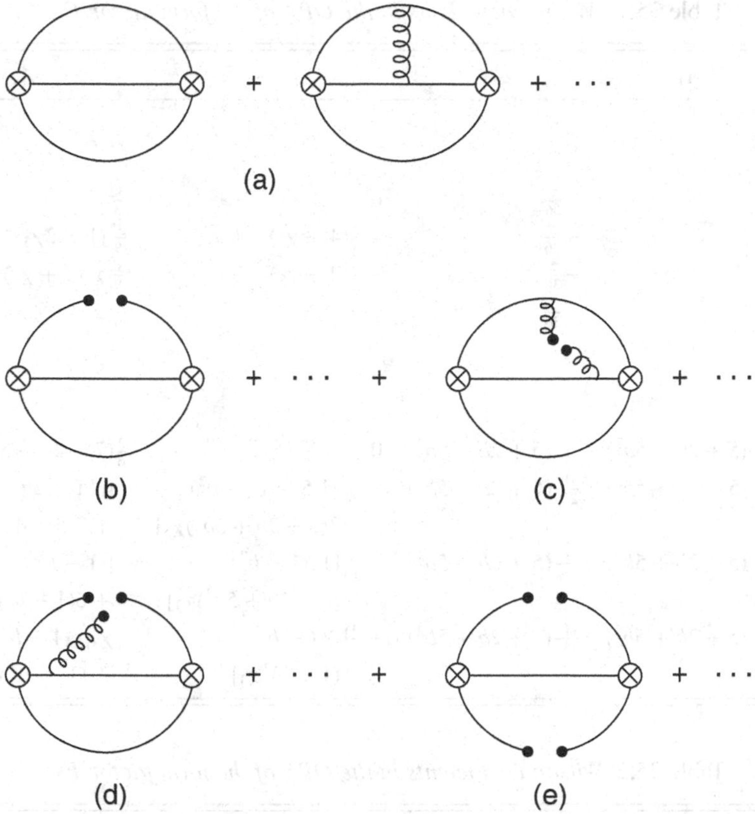


Fig. 35.1. Feynman diagrams corresponding to the OPE of the baryon correlator: (a) perturbative; (b) quark condensate; (c) gluon condensate; (d) mixed condensate; (e) four-quark condensate.

### 35.1.2 The octet

The nucleon can be in general interpolated by the lowest dimension operators:

$$N = \frac{1}{\sqrt{2}} : [(\psi C\gamma_5\psi)\,\psi + b\,(\psi C\psi)\,\gamma_5\psi] : \tag{35.5}$$

where $b$ is an arbitrary parameter. We shall discuss the different choices of $b$ in the sum rule analysis. The corresponding correlator is:

$$\begin{aligned} S_{\mu\nu} &= i\int d^4q\, e^{iqx}\langle 0|\mathcal{T}N(x)N(0)^\dagger|0\rangle \\ &= \hat{q}F_1 + F_2 + \cdots , \end{aligned} \tag{35.6}$$

Using the SVZ-expansion, the form factor can be expressed as in Eq. (35.3). The corresponding Wilson coefficients are given Tables 35.1 and 35.2.

Table 35.1. *Wilson coefficients in the OPE of the form factor $F_1$.*

| Type | $A_1$ | $A_2$ | $A_3$ | $A_4$ |
|---|---|---|---|---|
| **3/2** | | | | |
| $\Delta$ | $\frac{1}{20}$ | $-\frac{5}{36}$ | $0$ | $\frac{32}{3}$ |
| $\Sigma^*$ | $\frac{1}{20}$ | $-\frac{5}{36}$ | $-\frac{2}{3}(4-\chi_3)$ | $\frac{32}{9}(1+2\chi_3)$ |
| $\Xi^*$ | $\frac{1}{20}$ | $-\frac{5}{36}$ | $-\frac{4}{3}(2+\chi_3)$ | $\frac{32}{9}\chi_3(2+\chi_3)$ |
| $\Omega$ | $\frac{1}{20}$ | $-\frac{5}{36}$ | $-6\chi_3$ | $\frac{32}{3}\chi_3^2$ |
| **1/2** | | | | |
| $N$ | $\frac{1}{256}(5+2b+5b^2)$ | $\frac{1}{256}(5+2b+5b^2)$ | $0$ | $\frac{2}{6}(7-2b-5b^2)$ |
| $\Lambda$ | $\frac{1}{256}(5+2b+5b^2)$ | $\frac{1}{256}(5+2b+5b^2)$ | $-\frac{1}{48}[4(5-4b-b^2)$ $-3(5+2b+5b^2)\chi_3]$ | $\frac{1}{9}[(11+2b-13b^2)$ $+2(5-4b-b^2)\chi_3]$ |
| $\Sigma$ | $\frac{1}{256}(5+2b+5b^2)$ | $\frac{1}{256}(5+2b+5b^2)$ | $-\frac{1}{16}[12(1-b^2)$ $-(5+2b+5b^2)\chi_3]$ | $\frac{1}{3}[(1-b)^2$ $+6(1-b^2)\chi_3]$ |
| $\Xi$ | $\frac{1}{256}(5+2b+5b^2)$ | $\frac{1}{256}(5+2b+5b^2)$ | $-\frac{3}{8}[2(1-b^2)$ $-(1+b)^2\chi_3]$ | $\frac{1}{3}\chi_3[6(1-b^2)$ $+(1-b)^2\chi_3]$ |

Table 35.2. *Wilson coefficients in the OPE of the form factor $F_2$.*

| Type | $B_1$ | $B_2$ | $B_3$ |
|---|---|---|---|
| **3/2** | | | |
| $\Delta$ | $0$ | $-\frac{8}{3}$ | $\frac{4}{3}$ |
| $\Sigma^*$ | $\frac{1}{8}$ | $-\frac{8}{9}(2+\chi_3)$ | $\frac{4}{9}(2+\chi_5)$ |
| $\Xi^*$ | $\frac{1}{4}$ | $-\frac{8}{9}(1+2\chi_3)$ | $-\frac{4}{9}(1+2\chi_5)$ |
| $\Omega$ | $\frac{3}{8}$ | $-\frac{8}{3}\chi_3$ | $\frac{4}{3}\chi_5$ |
| **1/2** | | | |
| $N$ | $0$ | $-\frac{1}{8}(7-2b-5b^2)$ | $\frac{3}{8}(1-b^2)$ |
| $\Lambda$ | $\frac{1}{192}(11+2b-13b^2)$ | $-\frac{1}{24}[2(5-4b-b^2)$ $+(11+2b-13b^2)\chi_3]$ | $\frac{1}{32}(1-b^2)(10+4\chi_5)$ |
| $\Sigma$ | $\frac{1}{64}(1-b)^2$ | $-\frac{1}{8}[6(1-b^2)$ $+(1-b)^2\chi_3]$ | $\frac{3}{9}(1-b^2)$ |
| $\Xi$ | $\frac{3}{32}(1-b^2)$ | $-\frac{1}{8}[(1-b)^2$ $+6(1-b^2)\chi_3]$ | $\frac{3}{8}(1-b^2)\chi_5$ |

### 35.1.3 Radiative corrections

As the current gets renormalized, the previous correlators have anomalous dimensions. These anomalous dimensions are equal for the $\Delta$ and $N$ and read:

$$\gamma = -2 \times \left(\frac{2}{3}\right) . \tag{35.7}$$

Radiative corrections to these lowest-order terms have been first obtained in the chiral limit in [424] and corrected in [428,425]. For the nucleon, one has [428,425]:

$$A_1 = \frac{1}{256}(5+2b+5b^2)\left[1+\frac{71}{12}\left(\frac{\alpha_s}{\pi}\right)-\frac{1}{2}\log -\frac{q^2}{\nu^2}\right]$$
$$B_2 = -\frac{1}{8}\left[7\left[1+\left(\frac{\alpha_s}{\pi}\right)\frac{15}{14}\right]-2b\left[1+\left(\frac{\alpha_s}{\pi}\right)\frac{3}{2}\right]-5b^2\left[1+\left(\frac{\alpha_s}{\pi}\right)\frac{9}{10}\right]\right] . \tag{35.8}$$

## 35.2 Heavy baryons

Analogous correlators but for baryons containing heavy quarks have been evaluated in [453,454,731].

### 35.2.1 Spin 1/2 baryons

Let us consider the baryonic current:

$$J = r_1\,(u^t C\gamma^5 c)b + r_2(u^t Cc)\gamma^5 b + r_3\,(u^t C\gamma^5\gamma^\mu c)\gamma_\mu b\,, \tag{35.9}$$

which has the quantum numbers of the $\Lambda(bcu)$; $r_1$, $r_2$ and $r_3$ are arbitrary mixing parameters where, in terms of the $b$ parameter used in [454]:

$$r_1 = (5+b)/\,2\sqrt{6};\quad r_2 = (1+5b)/\,2\sqrt{6};\quad r_3 = (1-b)/\,2\sqrt{6}\,. \tag{35.10}$$

The choice of operators in [453] is recovered in the particular case where:

$$r_1 = 1;\quad r_2 = k;\quad r_3 = 0\,. \tag{35.11}$$

The associated two-point correlator is:

$$i\int d^4x\; e^{ip\cdot x}\langle 0|\mathbf{T}J(x)\bar{J}(0)|0\rangle = \not{p}F_1 + F_2\,. \tag{35.12}$$

The QCD expressions of the form factors $F_1$ and $F_2$ can be parametrized as:

$$F_i = F_i^{\rm Pert} + F_i^{G} + F_i^{\rm Mix}\,, \tag{35.13}$$

where:

$$\text{Im}\;F_2^{\rm Pert}(t) = \frac{1}{128\pi^3 t}\left\{(2r_3^2+r_2^2-r_1^2)\,m_b\,\left\{6\left[m_b^2t^2+(m_b^4-2m_b^2m_c^2-m_c^4)t\right.\right.\right.$$
$$\left.+\,2m_b^2m_c^4\right]\mathcal{L}_1 - 6t\left[m_b^2\,t+(m_b^2-m_c^2)^2\right]\mathcal{L}_2$$

$$- \left[t^2 + 5\left(2m_b^2 - m_c^2\right)t + m_b^4 - 5m_b^2 m_c^2 - 2m_c^4\right] \lambda_{bc}^{1/2}\}$$
$$- 2r_1 r_3\, m_c \left\{6\left[m_c^2 t^2 + \left(m_c^4 - 2m_c^2 m_b^2 - m_b^4\right)t + 2m_c^2 m_b^4\right]\mathcal{L}_1\right.$$
$$+ 6t\left[m_c^2\, t + \left(m_c^2 - m_b^2\right)^2\right]\mathcal{L}_2$$
$$\left. - \left[t^2 + 5\left(2m_c^2 - m_b^2\right)t + m_c^4 - 5m_c^2 m_b^2 - 2m_b^4\right] \lambda_{cb}\right\}\} \tag{35.14}$$

$$\text{Im}\, F_2^{\psi}(t) = \frac{\langle \bar\psi \psi\rangle}{8\pi t}\lambda_{bc}^{1/2}\left\{-\left(r_1^2 + r_2^2 + 4r_3^2\right)m_b m_c + r_1 r_3\left(m_b^2 + m_c^2 - t\right)\right\} \tag{35.15}$$

$$\text{Im}\, F_2^{G}(t) = \frac{\langle \alpha_s\, G^2\rangle}{384\pi^2 t}\left\{\left[2\frac{r_3^2}{m_b}\left(-2t + 7m_b^2 + 2m_c^2\right)\right.\right.$$
$$+ \frac{r_1^2 - r_2^2}{m_b}\left(2t + 5m_b^2 - 2m_c^2\right) + 2\frac{r_1 r_3}{m_c}\left(2t - 2m_b^2 - m_c^2\right)$$
$$\left. + 12 r_2 r_3\, m_c\right]\lambda_{bc}^{1/2}$$
$$+ 6\left[\left(r_2^2 - r_1^2\right)m_b\, t + 2r_3^2 m_b m_c^2 - r_1 r_3 m_c t - r_2 r_3 m_c\left(t - 2m_b^2\right)\right]\mathcal{L}_1$$
$$\left. - 6t\left[\left(r_2^2 - r_1^2\right)m_b + \left(r_1 + r_2\right)r_3 m_c\right]\mathcal{L}_2\right\} \tag{35.16}$$

$$\text{Im}\, F_2^{\text{Mix}}(t) = \frac{M_0^2\langle \bar\psi \psi\rangle}{64\pi t\, \lambda_{bc}^{3/2}}\left\{2\left(r_1^2 + r_2^2\right)m_b m_c\left[-t^3 + t^2\left(m_b^2 + 3m_c^2\right)\right.\right.$$
$$\left. + t\left(m_b^2 + m_c^2\right)\left(m_b^2 - 3m_c^2\right) - \left(m_b^2 - m_c^2\right)^3\right]$$
$$+ 4r_3^2 m_b m_c\left[-t^3 + t^2\left(3m_b^2 + m_c^2\right)\right.$$
$$\left. + t\left(-3m_b^4 - 6m_b^2 m_c^2 + m_c^4\right) + \left(m_b^2 - m_c^2\right)^3\right]$$
$$+ 2r_1 r_3\left[t^4 + t^3\left(-3m_b^2 - 2m_c^2\right) + 3t^2 m_b^2\left(m_b^2 - m_c^2\right)\right.$$
$$\left. + t\left(-m_b^6 + 4m_b^4 m_c^2 + 3m_b^2 m_c^4 + 2m_c^6\right) + m_c^2\left(m_b^2 - m_c^2\right)^3\right]$$
$$+ 2r_2 r_3\left[t^4 + t^3\left(-4m_b^2 - 3m_c^2\right) + 3t^2\left(2m_b^4 + m_b^2 m_c^2 + m_c^4\right)\right.$$
$$\left.\left. - t\left(m_b^2 - m_c^2\right)\left(4m_b^4 + m_b^2 m_c^2 - m_c^4\right) + m_b^2\left(m_b^2 - m_c^2\right)^3\right]\right\} \tag{35.17}$$

$$\text{Im}\, F_1^{\text{Pert}}(t) = \frac{1}{512\pi^3 t^2}\left\{\left(r_1^2 + r_2^2 + 4r_3^2\right)\left\{12\left[t^2\left(m_b^4 + m_c^4\right) - 2m_b^4 m_c^4\right]\mathcal{L}_1\right.\right.$$
$$- 12t^2\left(m_b^4 - m_c^4\right)\mathcal{L}_2$$
$$+ \left[t^3 - 7t^2\left(m_b^2 + m_c^2\right) + t\left(-7m_b^4 + 12m_b^2 m_c^2 - 7m_c^4\right)\right.$$
$$\left.\left. + m_b^6 - 7m_b^4 m_c^2 - 7m_b^2 m_c^4 + m_c^6\right]\lambda_{bc}^{1/2}\right\}$$
$$- 4r_1 r_3\, m_b m_c\left\{12\left[t^2\left(m_b^2 + m_c^2\right) - 4t m_b^2 m_c^2 + 2m_b^2 m_c^2\left(m_b^2 + m_c^2\right)\right]\mathcal{L}_1\right.$$
$$- 12t^2\left(m_b^2 - m_c^2\right)\mathcal{L}_2$$
$$\left.\left. - 2\left[2t^2 + 5t\left(m_b^2 + m_c^2\right) - m_b^4 - 10m_b^2 m_c^2 - m_c^4\right]\lambda_{cb}\right\}\right\} \tag{35.18}$$

$$\begin{aligned}
\text{Im } F_1^{\psi}(t) &= \frac{\langle\bar\psi\psi\rangle}{16\pi t^2}\lambda_{bc}^{1/2}\left\{\left(2r_3^2+r_2^2-r_1^2\right)m_c\left(t+m_b^2-m_c^2\right)\right.\\
&\quad \left.+2r_1r_3\, m_b\left(m_b^2-m_c^2-t\right)\right\} \qquad (35.19)\\
\text{Im } F_1^{G}(t) &= \frac{\langle\alpha_s G^2\rangle}{768\pi^2 t^2}\left\{\left[-4r_3^2\left(t+3m_b^2\right)-\left(r_2^2+r_1^2\right)\left(t-3m_b^2+3m_c^2\right)\right.\right.\\
&\quad +4\frac{r_1r_3}{m_bm_c}\left(2t\left(m_b^2+m_c^2\right)-2m_b^4-11m_b^2m_c^2-2m_c^4\right)\\
&\quad \left.-36r_2r_3\, m_bm_c\right]\lambda_{bc}^{1/2}\\
&\quad +12m_bm_c\left[-2r_3^2\, m_bm_c+2r_1r_3\left(t-2m_b^2-3m_c^2\right)\right.\\
&\quad \left.\left.+\ 2r_2r_3\left(t-m_b^2-2m_c^2\right)\right]\mathcal{L}_1\right\} \qquad (35.20)\\
\text{Im } F_1^{\text{Mix}}(t) &= \frac{M_0^2\langle\bar\psi\psi\rangle}{64\pi t^2\,\lambda_{bc}^{3/2}}\left\{2\left(r_1^2-r_2^2\right)m_c\left[-t^4+t^3\left(2m_b^2+5m_c^2\right)\right.\right.\\
&\quad -t^2\left(2m_b^4+3m_b^2m_c^2+9m_c^4\right)+t\left(m_b^2-m_c^2\right)\left(2m_b^4-m_b^2m_c^2-7m_c^4\right)\\
&\quad \left.-\left(m_b^2-m_c^2\right)^3\left(m_b^2-2m_c^2\right)\right]\\
&\quad +2r_3^2m_c\left[t^3\left(m_b^2-m_c^2\right)+t^2\left(-3m_b^4+4m_b^2m_c^2+3m_c^4\right)\right.\\
&\quad \left.+3t\left(m_b^2-m_c^2\right)\left(m_b^2+m_c^2\right)^2-\left(m_b^2-m_c^2\right)^3\left(m_b^2+m_c^2\right)\right]\\
&\quad +2r_1r_3\, m_b\left[-t^4+t^3\left(5m_b^2+m_c^2\right)+t^2\left(-9m_b^4-4m_b^2m_c^2+m_c^4\right)\right.\\
&\quad \left.+t\left(m_b^2-m_c^2\right)\left(7m_b^4+4m_b^2m_c^2+m_c^4\right)-2m_b^2\left(m_b^2-m_c^2\right)^3\right]\\
&\quad +2r_2r_3m_b\left[-t^4+2t^3\left(2m_b^2+m_c^2\right)-2t^2\left(3m_b^4+m_c^4\right)\right.\\
&\quad \left.\left.+2t\left(2m_b^6-3m_b^4m_c^2+m_c^6\right)-\left(m_b^2-m_c^2\right)^4\right]\right\}, \qquad (35.21)
\end{aligned}$$

with:

$$\begin{aligned}
\mathcal{L}_1(t) &= \frac{1}{2}\log\frac{1+v}{1-v}; \qquad v = \sqrt{1-\frac{4m_b^2m_c^2}{\left(t-m_b^2-m_c^2\right)^2}}\\
\lambda_{bc}^{1/2} &= \left(t-m_b^2-m_c^2\right)v; \quad \mathcal{L}_2 = \log\frac{(m_b^2+m_c^2)t+\left(m_b^2-m_c^2\right)\left(\lambda_{bc}^{1/2}-m_b^2+m_c^2\right)}{2m_bm_c\,t}.\\
&= \lambda^{1/2}\left(m_b^2,m_c^2,t\right) \qquad (35.22)
\end{aligned}$$

### 35.2.2 Spin 3/2 baryon

Let us consider the two-point correlator:

$$S_{\mu\nu}(q^2) = i\int d^4x\ \langle 0|\mathcal{T} J_\mu(x)J_\nu(0)|0\rangle \equiv g_{\mu\nu}(\hat q F_1+F_2)+\cdots \qquad (35.23)$$

built from the *simplest* interpolating $\Xi^*_Q$ operator:

$$J_\mu^{\Xi^*_Q} = \frac{1}{\sqrt{3}}\epsilon^{\alpha\beta\lambda} : 2\left(Q^T_\alpha C\gamma^\mu u_\beta\right)Q_\lambda + \left(Q^T_\alpha C\gamma_\mu Q_\beta\right)u_\lambda : , \tag{35.24}$$

where $Q$ and $u$ are respectively the heavy- and light-quark fields. The QCD expressions of the form factors for a heavy quark of mass $M$ are [453]:

$$\begin{aligned}
\mathrm{Im}F_1^{\mathrm{pert}}(x) &= \frac{M^4}{480\pi^3}\Big[60(-1+4x-4x^2-2x^3)\mathcal{L}_v \\
&\quad + \frac{v}{x^2}(3-19x+98x^2-130x^3-60x^4)\Big] , \\
\mathrm{Im}F_2^{\mathrm{pert}}(x) &= \frac{M^5}{288\pi^3x^2}[24(-2+3x-5x^3)\mathcal{L}_v + v(9+34x-10x^2-60x^3)] , \\
\mathrm{Im}F_1^{G}(x) &= -\frac{\langle\alpha_s G^2\rangle}{12\pi^2}\left[x^2(2+x)\mathcal{L}_v + \frac{v}{24}(1+26x+12x^2)\right] , \\
\mathrm{Im}F_2^{G}(x) &= -\frac{\langle\alpha_s G^2\rangle}{36\pi^2}M\left[(2+3x^2)\mathcal{L}_v + \frac{v}{12}\left(11+18x-\frac{8}{x}\right)\right] , \\
\mathrm{Im}F_1^{\psi}(x) &= -M\langle\bar\psi\psi\rangle\frac{v}{3\pi} , \quad \mathrm{Im}F_2^{\psi}(x) = -M^2\langle\bar\psi\psi\rangle\frac{v}{18\pi}\left(7+\frac{2}{x}\right) , \\
\mathrm{Im}F_1^{\mathrm{mix}}(x) &= M_0^2\frac{\langle\bar\psi\psi\rangle}{9\pi}\frac{x^2}{Mv^3}(2-11x) , \\
\mathrm{Im}F_2^{\mathrm{mix}}(x) &= M_0^2\frac{\langle\bar\psi\psi\rangle}{36\pi}\frac{1}{v^3}(2-11x+12x^2-30x^3) ,
\end{aligned} \tag{35.25}$$

with:

$$x = \frac{M^2}{t} , \quad v \equiv \sqrt{1-4x} , \quad \mathcal{L}_v = \ln\left(\frac{1+v}{1-v}\right) . \tag{35.26}$$

# 36
# Four-quark correlators

We shall be concerned here with the two-point correlators associated with the four-quark operators. These operators can describe a four-quark state but play also a crucial rôle for describing the flavour changing $\Delta S = 1$ for the $\Delta I = 1/2$ rule processes of the weak hamiltonian and the $\Delta S = 2$ and $\Delta B = 2$ for the $K - \bar{K}$ and $B - \bar{B}$ oscillations.

## 36.1 Four-quark states

The two-point function associated to the colour singlet operator:

$$\mathcal{O}^{\pm} = \frac{1}{\sqrt{2}} \sum_{\Gamma=1,\gamma_5} \bar{s}\Gamma s(\bar{u}\Gamma u \pm \bar{d}\Gamma d) \tag{36.1}$$

has been evaluated in [465] to leading order in $\alpha_s$ and including non-perturbative corrections. It is shown in Fig. 36.1, and reads:

$$\begin{aligned}\Psi_4(q^2) = q^8 \ln \frac{-q^2}{\nu^2} \Bigg\{ &-\frac{1}{40960\pi^6} + \frac{1}{1020\pi^6}\left(\frac{m_s^2}{q^2}\right) + \frac{1}{q^4}\left[\frac{m_s\langle\bar{s}s\rangle}{128\pi^4} + \frac{\langle\alpha_s G^2\rangle}{64\pi^5}\right] \\ &- \frac{1}{q^6}\left[\frac{1}{64\pi^4}\left\langle \bar{s}\sigma_{\mu\nu}\frac{\lambda^a}{2}G_a^{\mu\nu}s\right\rangle + \frac{3}{8\pi^2}(\langle\bar{u}u\rangle^2 + \langle\bar{s}s\rangle^2)\right]\Bigg\} \\ &- \left(\frac{8}{3q^2}\right) m_s\langle\bar{s}s\rangle\langle\bar{u}u\rangle^2 ,\end{aligned} \tag{36.2}$$

which is free from non-local $\frac{1}{\epsilon}\ln -q^2/\nu^2$ pole absorbed by the addition of evanescent diagrams. The two-point correlator associated to the operator:

$$\mathcal{O}^{\pm} = \frac{1}{\sqrt{2}} \sum_{\Gamma=1,\gamma_5} \bar{s}\Gamma\lambda_a s(\bar{u}\Gamma\lambda_a u \pm \bar{d}\Gamma\lambda_a d) , \tag{36.3}$$

has been analysed in citeSN4Q and can be easily deduced from the former to leading order using the Fierz transform.

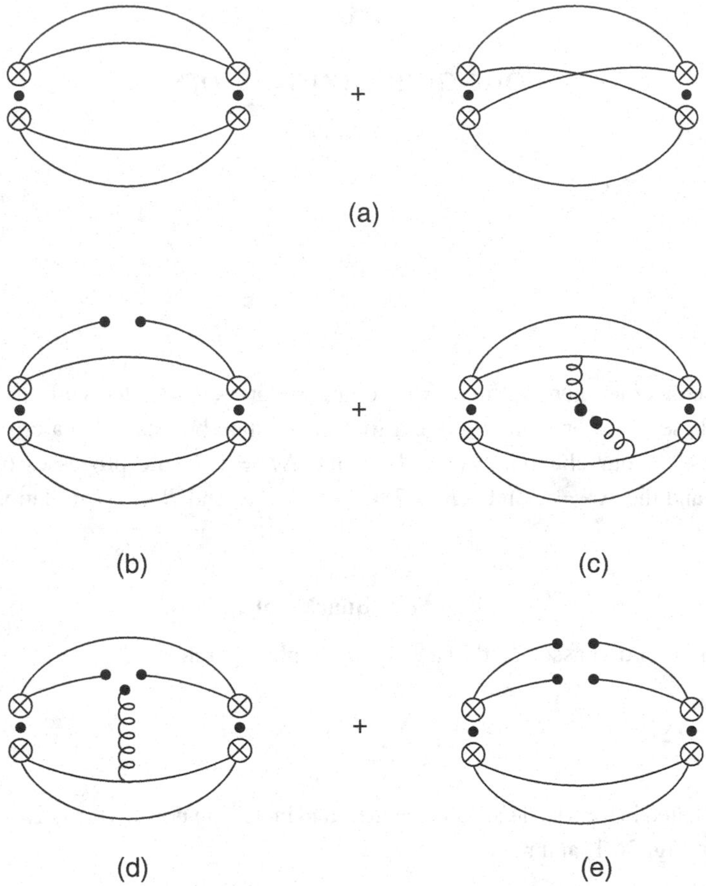


Fig. 36.1. Feynman diagrams corresponding to the OPE of the four-quark correlator: (a) perturbative; (b) quark condensate; (c) gluon condensate; (d) mixed condensate; (e) four-quark condensate.

## 36.2 $\Delta S = 1$ correlator and $\Delta I = 1/2$ rule

In these weak processes, the short-distance Hamiltonian can be described by the four-quark operators $Q_i(x)$ obtained from the operator product expansion:

$$\mathcal{H}_{\rm eff} = \frac{G_F}{\sqrt{2}} V_{ud} V_{us}^* \sum_i C_i(\mu^2) Q_i \, , \tag{36.4}$$

where $V_{uq}$ are elements of the CKM mixing matrix, while $C_i$ is the Wilson coefficient obtained from pQCD calculation. The relevant two-point function for these processes is:

$$\begin{aligned} \Psi(q^2) &\equiv i \int dx\, e^{iqx} \, \langle 0|\, T\{\, \mathcal{H}_{\rm eff}(x)\, \mathcal{H}_{\rm eff}(0)^\dagger\}|0\rangle \\ &= \left(\frac{G_F}{\sqrt{2}}\right)^2 |V_{ud} V_{us}^*|^2 \sum_{i,j} C_i(\mu^2)\, C_j^*(\mu^2)\, \Psi_{ij}(q^2)\, . \end{aligned} \tag{36.5}$$

This vacuum-to-vacuum correlator can be studied with perturbative QCD methods, allowing for a consistent combination of Wilson-coefficients $C_i(\mu^2)$ and two-point functions of the four-quark operators, $\Psi_{ij}$, in such a way that the renormalization scheme and scale dependences exactly cancel (to the computed order). The associated spectral function $\frac{1}{\pi}\mathrm{Im}\Psi^{\Delta S=1}(q^2)$ is a quantity with definite physical information. It describes in an inclusive way how the weak Hamiltonian couples the vacuum to physical states of a given invariant mass. In the following we shall analyse the four-quark correlators but build a RS combination that is useful for the physical processes. Here, we shall consider the correlators associated to the $\Delta S = 1$ operators:

$$Q_1 = 4\,(\bar{s}_L\gamma^\mu d_L)\,(\bar{u}_L\gamma_\mu u_L)\,, \qquad Q_2 = 4\big(\bar{s}_L^\alpha\gamma^\mu d_L^\beta\big)\big(\bar{u}_L^\beta\gamma_\mu u_L^\alpha\big)\,. \tag{36.6}$$

It is usual to work in the diagonal basis:

$$Q_\pm = \frac{1}{2}\,(Q_1 \pm Q_2)\,, \tag{36.7}$$

and to define the RS-invariant operators [476]:

$$\bar{Q}_\pm \equiv \left[1 + \left(\frac{\alpha_s}{\pi}\right)B_\pm\right]Q_\pm\,, \tag{36.8}$$

where in the t'Hooft–Veltman (HV) and naïve dimensional regularization (NDR) schemes (see Chapter 8):

$$B_\pm^{HV} = \frac{7}{8}\left(\pm 1 - \frac{1}{N}\right), \qquad B_\pm^{NDR} = \frac{11}{8}\left(\pm 1 - \frac{1}{N}\right). \tag{36.9}$$

In this basis, the corresponding correlator is:

$$\bar{\Psi}_{\pm\pm} = \frac{1}{2}\left[1 + 2\left(\frac{\alpha_s}{\pi}\right)B_\pm\right][\Psi_{11} \pm \Psi_{12}]\,, \tag{36.10}$$

and is RS-invariant.

$$\begin{aligned}\frac{1}{\pi}\mathrm{Im}\bar{\Psi}_{\pm\pm}(s,\mu^2) &= \theta(s)\,\frac{s^4}{(4\pi)^6}\,A_\pm\left\{1 + \left(\frac{\alpha_s}{\pi}\right)\left[\frac{3}{2}\left(\pm 1 - \frac{1}{N}\right)\ln\left|\frac{s}{\mu^2}\right|\right.\right.\\ &\qquad \left.\left. + \frac{3}{4}N \mp \frac{101}{20} + \frac{43}{10}\frac{1}{N}\right]\right\},\end{aligned} \tag{36.11}$$

with:

$$A_\pm = \frac{2}{45}N(N \pm 1)\,. \tag{36.12}$$

The coefficient of the logarithm is just equal to the leading-order anomalous dimensions $\gamma_\pm^{(1)}$ of $Q_\pm$. Introducing the $\mu^2$-dependent Wilson coefficient:

$$C_\pm(\mu^2) = \alpha_s(\mu^2)^{\gamma_\pm^{(1)}/\beta_1}\left[1 - \frac{\alpha_s(\mu^2)}{4\pi}R_\pm\right], \tag{36.13}$$

where the NLO correction $R_\pm$ can be found in [476], it is possible to form the RGI spectral functions:

$$\frac{1}{\pi}\text{Im}\hat{\Psi}_{\pm\pm}(s) = \frac{1}{\pi}\text{Im}\bar{\Psi}_{\pm\pm}(s)C_\pm^2(s) \tag{36.14}$$

For $N = 3$ the two spectral functions read:

$$\frac{1}{\pi}\text{Im}\hat{\Psi}_{++}(s) = \theta(s)\frac{8}{15}\frac{s^4}{(4\pi)^6}\alpha_s(s)^{-4/9}\left[1 - \frac{3649}{1620}\left(\frac{\alpha_s}{\pi}\right)\right],$$
$$\frac{1}{\pi}\text{Im}\hat{\Psi}_{--}(s) = \theta(s)\frac{4}{15}\frac{s^4}{(4\pi)^6}\alpha_s(s)^{8/9}\left[1 + \frac{9139}{810}\left(\frac{\alpha_s}{\pi}\right)\right]. \tag{36.15}$$

Taking $\alpha_s(s)/\pi \approx 0.1$, at the NLO we find a moderate suppression of $\text{Im}\hat{\Psi}_{++}$ by roughly 20%, whereas $\text{Im}\hat{\Psi}_{--}$ acquires a huge enhancement on the order of 100%. Because $\text{Im}\hat{\Psi}_{++}$ solely receives contributions from $\Delta I = 3/2$, and $\text{Im}\hat{\Psi}_{--}$ is a mixture of both $\Delta I = 1/2$ and $\Delta I = 3/2$, this pattern of the radiative corrections entails a strong enhancement of the $\Delta I = 1/2$ amplitude, which can provide a promising picture for the emergence of the $\Delta I = 1/2$–rule.

## 36.3 The $\Delta S = 2$ correlator

Here, we shall consider the correlator associated to the $\Delta S = 2$ operator:

$$\mathcal{O}_{\Delta S=2} = (\bar{s}_L\gamma^\mu d_L)(\bar{s}_L\gamma_\mu d_L) \tag{36.16}$$

where:

$$\psi_L \equiv \frac{1}{2}(1-\gamma_5)\psi\ . \tag{36.17}$$

We shall analyse its phenomenological application in the next chapter. The QCD expression of the spectral function reads [468]:

$$\frac{1}{\pi}\text{Im}\Psi_{\Delta S=2}(t) = \frac{1}{(16\pi^2)^3}\frac{1}{10}\left(1+\frac{1}{N}\right)t^4\alpha_s(t)^{-4/9}\Bigg\{1 - A\left(\frac{\alpha_s}{\pi}\right)$$
$$-\frac{40\bar{m}_s^2}{t} - \frac{20\pi^2}{t^2}(16\pi m_s\langle\bar{s}s\rangle - \langle\alpha_s G^2\rangle)\Bigg\}\ . \tag{36.18}$$

The coefficient of the perturbative correction is RS dependent. In [471], it has been shown that one can define a RS invariant combination $\hat{Q}_{\Delta S=2}$:

$$\hat{Q}_{\Delta S=2} \equiv \alpha_s(\nu)^{\gamma_{\Delta S=2}/\beta_1}\left[1 - \left(\frac{\alpha_s}{\pi}\right)Z\right]Q_{\Delta S=2}\ , \tag{36.19}$$

where $Z$ depends on the regularization scheme used [475]; $\gamma_{\Delta S=2}$ is the anomalous dimension of the operator $Q_{\Delta S=2}$ defined as:

$$Q_{\Delta S=2} \equiv \frac{1}{2}\left[\mathcal{O}_{\Delta S=2} + (\bar{s}_L^\alpha\gamma^\mu d_L^\beta)(\bar{s}_L^\beta\gamma_\mu d_L^\alpha)\right]. \tag{36.20}$$

It coincides with $\mathcal{O}_{\Delta S=2}$ in the HV scheme since in HV Fierz symmetry is respected for current–current operators while $\mathcal{O}_{\Delta S=2}$ renormalizes into itself. This is not the case for the NDR scheme where the $\gamma_5$ matrix is naïvely anti-commuting while the rest of the calculation is done in $n$-dimensions. Within this RS invariant combination one obtains [471]:

$$A = -\frac{3649}{1620} , \tag{36.21}$$

where the global effect reduces by about 20% the lowest-order result.

## 36.4 The $\Delta B = 2$ correlator

We shall consider the two-point correlator:

$$\psi_{\Delta B=2}(q^2) \equiv i \int d^4x \, e^{iqx} \, \langle 0|\mathcal{T}\mathcal{O}_q(x)(\mathcal{O}_q(0))^\dagger|0\rangle , \tag{36.22}$$

built from the $\Delta B = 2$ weak operator $\mathcal{O}_q$ defined as:

$$\mathcal{O}_q(x) \equiv (\bar{b}\gamma_\mu L q)(\bar{b}\gamma_\mu L q) , \tag{36.23}$$

with: $L \equiv (1-\gamma_5)/2$ and $q \equiv d, s$. This correlator has been firstly evaluated to lowest order in [472] in the case of massless light quark mass and including non-perturbative corrections. The perturbative radiative corrections including *non-factorizable* corrections have been obtained in [473]. The $SU(3)$ breaking correction has been evaluated in [474]. The lowest-order perturbative contribution for $m_s \neq 0$ to the two-point correlator is [474]:

$$\begin{aligned}
\frac{1}{\pi}\mathrm{Im}\psi^{\mathrm{pert}}_{\Delta B=2}(t) &= \theta(t-4(m_b+m_s)^2) \times \frac{t^4}{1536\pi^6} \times \int_{(\sqrt{\delta}+\sqrt{\delta'})^2}^{(1-\sqrt{\delta}-\sqrt{\delta'})^2} dz \int_{(\sqrt{\delta}+\sqrt{\delta'})^2}^{(1-\sqrt{z})^2} du \, zu \\
&\times \lambda^{1/2}(1,z,u)\lambda^{1/2}\left(1,\frac{\delta}{z},\frac{\delta'}{z}\right)\lambda^{1/2}\left(1,\frac{\delta}{u},\frac{\delta'}{u}\right) \\
&\times \left[4f\left(\frac{\delta}{z},\frac{\delta'}{z}\right)f\left(\frac{\delta}{u},\frac{\delta'}{u}\right) - 2f\left(\frac{\delta}{z},\frac{\delta'}{z}\right)g\left(\frac{\delta}{u},\frac{\delta'}{u}\right)\right. \\
&\left. - 2g\left(\frac{\delta}{z},\frac{\delta'}{z}\right)f\left(\frac{\delta}{u},\frac{\delta'}{u}\right) + \frac{(1-z-u)^2}{zu}g\left(\frac{\delta}{z},\frac{\delta'}{z}\right)g\left(\frac{\delta}{u},\frac{\delta'}{u}\right)\right] .
\end{aligned} \tag{36.24}$$

Here $\delta \equiv m_b^2/t$ and $\delta' \equiv m_s^2/t$, respectively. The functions $f(x,y)$ and $g(x,y)$ are defined by

$$\begin{aligned}
f(x,y) &\equiv 2 - x - y - (x-y)^2 , \\
g(x,y) &\equiv 1 + x + y - 2(x-y)^2 .
\end{aligned} \tag{36.25}$$

The function $\lambda(x, y, z)$ is a phase space factor,

$$\lambda(x, y, z) \equiv x^2 + y^2 + z^2 - 2xy - 2yz - 2zx \,. \tag{36.26}$$

We include the $\alpha_s$ correction from factorizable diagrams to the $m_s$ contribution by using the results for the two-point correlators of currents [477]. This can be done using the convolution formula:

$$\frac{1}{\pi}\text{Im}\psi^{\alpha_s}_{\Delta B=2}(t) = \theta(t - 4(m_b + m_s)^2) \times \frac{t^2}{6\pi^4}\int_{(\sqrt{\delta}+\sqrt{\delta'})^2}^{(1-\sqrt{\delta}-\sqrt{\delta'})^2} dz \int_{(\sqrt{\delta}+\sqrt{\delta'})^2}^{(1-\sqrt{z})^2} du\lambda^{1/2}(1, z, u)$$
$$\times \left\{\text{Im}\Pi^0_{\mu\nu}(zt)\text{Im}\Pi^{\alpha_s\mu\nu}(ut) + \text{Im}\Pi^{\alpha_s}_{\mu\nu}(zt)\text{Im}\Pi^{0\mu\nu}(ut)\right\} \tag{36.27}$$

Here $\Pi^0_{\mu\nu}(q^2)$ and $\Pi^{\alpha_s}_{\mu\nu}(q^2)$ are respectively the lowest and the next-to-leading order QCD contribution to the two-point correlator $\Pi_{\mu\nu}(q^2)$ which is defined by

$$\Pi_{\mu\nu}(q^2) \equiv i\int d^4x e^{iqx} \times \langle 0|T(\bar{b}_L(x)\gamma_\mu s_L(x))(\bar{s}_L(0)\gamma_\nu b_L(0))|0\rangle \,. \tag{36.28}$$

The quark condensate contribution reads:

$$\begin{aligned}\frac{1}{\pi}\text{Im}\psi^{\bar{s}s}_{\Delta B=2}(t) = {}& \theta(t - 4(m_b + m_s)^2)\frac{1}{384\pi^3}m_s\langle\bar{s}s\rangle \\ & \times \int_{(m_b+m_s)^2}^{(\sqrt{t}-m_b)^2} dq_1^2\sqrt{\lambda_1}\left(4 + 2q^2\frac{\partial}{\partial q^2}\right) \\ & \times \left[\sqrt{\lambda_0}\left\{\lambda_1\left(1 + \frac{m_b^2}{q^2} - \frac{q_1^2}{q^2}\right)q_1^2\right.\right. \\ & \left.\left. + f_1\left(1 - \frac{m_b^2}{q^2} + \frac{q_1^2}{q^2}\right)(q^2 - m_b^2 - q_1^2)\right\}\right].\end{aligned} \tag{36.29}$$

Here $\lambda_0$, $\lambda_1$, and $f_1$ are defined by

$$\begin{aligned}\lambda_0 &\equiv \lambda\left(1, \frac{q_1^2}{q^2}, \frac{m_b^2}{q^2}\right), \\ \lambda_1 &\equiv \lambda\left(1, \frac{m_b^2}{q_1^2}, \frac{m_s^2}{q_1^2}\right), \\ f_1 &\equiv 1 + \frac{m_b^2}{q_1^2} + \frac{m_s^2}{q_1^2} - 2\frac{(m_b^2 - m_s^2)^2}{q_1^4}\,.\end{aligned} \tag{36.30}$$

The gluon condensate contribution reads in the case $m_s = 0$ [472]:

$$\begin{aligned}\frac{1}{\pi}\text{Im}\psi^{G^2}_{\Delta B=2}(t) = {}& \frac{t^2}{(16\pi^2)^2}\frac{1}{\pi}\langle\alpha_s G^2\rangle\int_{x_0}^1 dx\int_{y_-}^{y_+} dy \\ & \times \{-(\Delta/2y^2)[\Delta - y(1-y)][2xy + (1-x)^2(1-y)]\end{aligned}$$

$$+ (\delta x/3y^3)(1-x)^2(1-y)^3[2\Delta - y(1-y)]\}$$
$$- \int_{\delta}^{(1-\sqrt{\delta})^2} dz\, z(1-\delta/z)^2 \lambda^{1/2}(1, z, \delta)\,, \tag{36.31}$$

where:

$$\Delta \equiv \delta(y/x + 1 - y) - y(1-y) \tag{36.32}$$

and the parametric integration limits are given by:

$$x_0 = \delta/(1-\sqrt{\delta})^2\,,$$
$$y_\pm = \frac{1}{2}[1 + \delta(1 - 1/x) \pm \lambda^{1/2}(1, \delta, \delta/x)]\,. \tag{36.33}$$

# 37

# Gluonia correlators

## 37.1 Pseudoscalar gluonia

We shall be concerned with the correlator:

$$\chi(k^2) = \int dx\, e^{ik.x} i\langle 0|\mathcal{T}\, Q_R(x)\, Q_R(0)|0\rangle\,, \tag{37.1}$$

where $Q_R(x)$ is the renormalized gluon topological density which mixes under renormalization with the divergence of the flavour singlet axial current $J^0_{\mu 5R}$ mix as follows [129] (see Section 10.3.3 in Part III):

$$\begin{aligned} J^0_{\mu 5R} &= Z J^0_{\mu 5B} \\ Q_R &= Q_B - \frac{1}{2n_f}(1-Z)\partial^\mu J^0_{\mu 5B}\,, \end{aligned} \tag{37.2}$$

where:

$$\begin{aligned} J^0_{\mu 5B} &= \sum \bar{q}\gamma_\mu\gamma_5 q \\ Q_B &= \frac{\alpha_s}{8\pi}\mathbf{Tr}\, G^{\mu\nu}\tilde{G}_{\mu\nu} \end{aligned} \tag{37.3}$$

and we have quoted the formulae for $n_f$ flavours. The correlation function $\chi(k^2)$ obeys the inhomogeneous RGE [260]:

$$\left(\mu\frac{\partial}{\partial\mu} + \beta(\alpha_s)\alpha_s\frac{\partial}{\partial\alpha_s} - 2\gamma\right)\chi(k^2) = -\frac{1}{(2n_f)^2}2\beta^{(L)}k^4\,. \tag{37.4}$$

The anomalous dimension is:

$$\gamma \equiv \mu\frac{d}{d\mu}\log Z = -\left(\frac{\alpha_s}{\pi}\right)^2\,. \tag{37.5}$$

The extra RG function $\beta^{(L)}$ (so called because it appears in the longitudinal part of the Green function of two axial currents) is given by

$$\frac{1}{(2n_f)^2}\beta^{(L)} = -\frac{1}{32\pi^2}\left(\frac{\alpha_s}{\pi}\right)^2\left[1+\frac{29}{4}\left(\frac{\alpha_s}{\pi}\right)\right]. \tag{37.6}$$

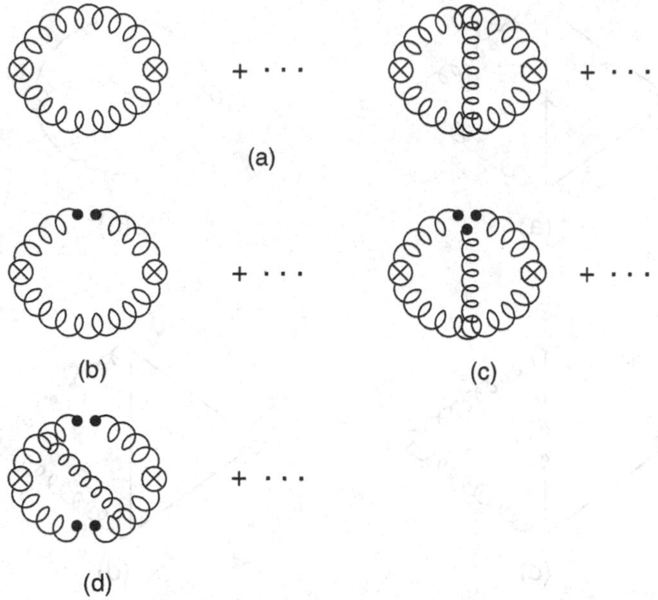

Fig. 37.1. Feynman diagrams corresponding to the OPE of the gluonium correlator: (a) perturbative; (b) two-gluon condensate; (c) three-gluon condensate; (d) four-gluon condensate.

The RGE is solved in the standard way, giving

$$\chi(k^2, \alpha_s; \mu) = e^{-2\int_0^t dt'\gamma(\alpha_s(t'))}\Big[\chi(k^2, \alpha_s(t); \mu e^t) - 2\int_0^t dt''\beta^{(L)}(\alpha_s(t''))e^{2\int_0^{t''} dt'\gamma(\alpha_s(t'))}\Big], \tag{37.7}$$

where $\alpha_s(t)$ is the running coupling. The different QCD diagrams contributing to the correlator are shown in Fig. 37.1.

The perturbative expression for the two-point correlation function in the $\overline{MS}$ scheme is [455]:

$$\chi(k^2)_{\rm P.T.} \simeq -\left(\frac{\alpha_s}{8\pi}\right)^2 \frac{2}{\pi^2}k^4 \log\frac{-k^2}{\mu^2}\left[1 + \left(\frac{\alpha_s}{\pi}\right)\left(\frac{1}{2}\beta_1 \log\frac{-k^2}{\mu^2} + \frac{29}{4}\right) + \cdots\right] \tag{37.8}$$

The non-perturbative contribution from the gluon condensates (coming from the next lowest dimension operators in the OPE) is [382]:

$$\chi(k^2)_{\rm N.P.} \simeq -\frac{\alpha_s}{16\pi^2}\left[\left(1 + \frac{1}{2}\beta_1\left(\frac{\alpha_s}{\pi}\right)\log\frac{-k^2}{\mu^2}\right)\langle\alpha_s G^2\rangle - 2\frac{\alpha_s}{k^2}\langle gG^3\rangle\right]. \tag{37.9}$$

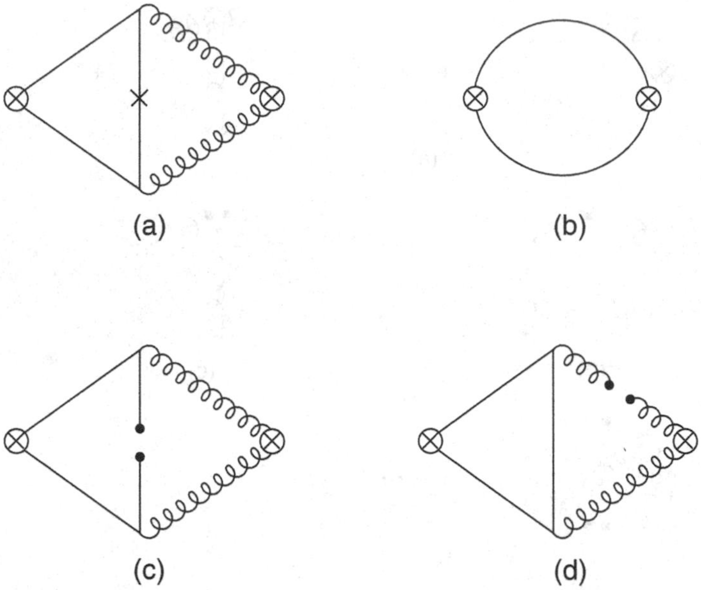


Fig. 37.2. Feynman diagrams corresponding to the OPE of the meson-gluonium correlator: (a) perturbative; (b) diagram which mixes with (a) under renormalization; (c) quark condensate; (d) gluon condensate.

### 37.2 Pseudoscalar meson-gluonium mixing

Let us consider the off-diagonal two-point correlator:

$$\Psi^{-}_{gq}(q^2) = \int dx\, e^{iq.x} i\langle 0|\mathcal{T}\, Q_R(x)\, \partial_\mu J^\mu_5(0)|0\rangle \tag{37.10}$$

shown in Fig. 37.2.

Its QCD expression reads [458]:

$$\begin{aligned}(8\pi)^2\Psi^{-}_{gq}(q^2) = \alpha_s\left(\frac{\alpha_s}{\pi}\right)\frac{3}{\pi^2}m_s^2 q^2\log -\frac{q^2}{\nu^2}\left[\log -\frac{q^2}{\nu^2} - \frac{2}{3}\left(\frac{11}{4} - 3\gamma_E\right)\right]\\ -8\alpha_s\left(\frac{\alpha_s}{\pi}\right)m_s\langle\bar{s}s\rangle\log -\frac{q^2}{\nu^2}\\ +2\left(\frac{\alpha_s}{\pi}\right)\langle\alpha_s G^2\rangle\left(\frac{m_s^2}{q^2}\right)\log -\frac{q^2}{m_s^2}\,,\end{aligned} \tag{37.11}$$

where one can notice that the mixing from the OPE vanishes in the chiral limit. However, one should notice that this mixing acts on the gluonium propagator, that is, it affects the mass splitting but not its decay width which is governed by a three-point function. Unfortunately, several authors mix these two features in the literature. This feature may justify why the lattice prediction in the world without quark can give a prediction that is almost compatible with the experimentally observed gluonium candidate.

## 37.3 Scalar gluonia

We shall be concerned with the correlator:

$$\Psi_s(q^2) \equiv 16i \int d^4x \, e^{iqx} \, \langle 0|\mathcal{T}\theta^\mu_\mu(x)\theta^\mu_\mu(0)^\dagger|0\rangle \, , \tag{37.12}$$

where $\theta_{\mu\nu}$ is the improved QCD energy-momentum tensor (neglecting heavy quarks) whose anomalous trace reads, in standard notations:

$$\theta^\mu_\mu(x) = \frac{1}{4}\beta(\alpha_s)G^2 + (1+\gamma_m(\alpha_s))\sum_{u,d,s} m_i\bar{\psi}_i\psi_i \, . \tag{37.13}$$

Its leading-order perturbative and non-perturbative expressions in $\alpha_s$ have been obtained by the authors of [382]. To two-loop accuracy in the $\overline{MS}$ scheme, its perturbative expression has been obtained by [455], while the radiative correction to the gluon condensate has been derived in [456]. Using a simplified version of the OPE:

$$\Psi_s(q^2) = \sum_{D=0,4,\cdots} C_D\langle O_D\rangle \, , \tag{37.14}$$

one obtains for three flavours and by normalizing the result with $(\beta(\alpha_s)/\alpha_s)^2$:

$$\begin{aligned}
C_0 &= -2\left(\frac{\alpha_s}{\pi}\right)^2(-q^2)^2\log-\frac{q^2}{\nu^2}\left\{1+\frac{59}{4}\left(\frac{\alpha_s}{\pi}\right)+\frac{\beta_1}{2}\left(\frac{\alpha_s}{\pi}\right)\log-\frac{q^2}{\nu^2}\right\}\\
C_4\langle O_4\rangle &= 4\alpha_s\left\{1+\frac{49}{12}\left(\frac{\alpha_s}{\pi}\right)+\frac{\beta_1}{2}\left(\frac{\alpha_s}{\pi}\right)\log-\frac{q^2}{\nu^2}\right\}\langle\alpha_s G^2\rangle\\
C_6\langle O_6\rangle &= 2\alpha_s\left\{1-\frac{29}{4}\alpha_s\log-\frac{q^2}{\nu^2}\right\}g^3 f_{abc}\langle G^aG^bG^c\rangle\\
C_8\langle O_8\rangle &= 14\left\langle\left(\alpha_s f_{abc}G^a_{\mu\rho}G^{b\rho}_\nu\right)^2\right\rangle-\left\langle\left(\alpha_s f_{abc}G^a_{\mu\nu}G^b_{\rho\lambda}\right)^2\right\rangle .
\end{aligned} \tag{37.15}$$

## 37.4 Scalar meson-gluonium mixing

Let's consider the off-diagonal two-point correlator:

$$\Psi^+_{gq}(q^2) = \int dx \, e^{iq.x} i\langle 0|\mathcal{T} J_{2g}(x) \, J^\dagger_q(0)|0\rangle \, , \tag{37.16}$$

where:

$$J_{2g} = \alpha_s G^2 \, , \qquad J_q = 2m\bar{q}q \, . \tag{37.17}$$

Its perturbative QCD expression reads [458]:

$$\Psi^+_{gq}(q^2) = \alpha_s\left(\frac{\alpha_s}{\pi}\right)\frac{3}{\pi^2}m_s^2q^2\log-\frac{q^2}{\nu^2}\left[\log-\frac{q^2}{\nu^2}-\frac{2}{3}(4-3\gamma_E)\right] . \tag{37.18}$$

The evaluation of the quark and gluon condensates is very similar to the case of the pseudoscalar channel, which the reader can easily evaluate as an exercise. The result

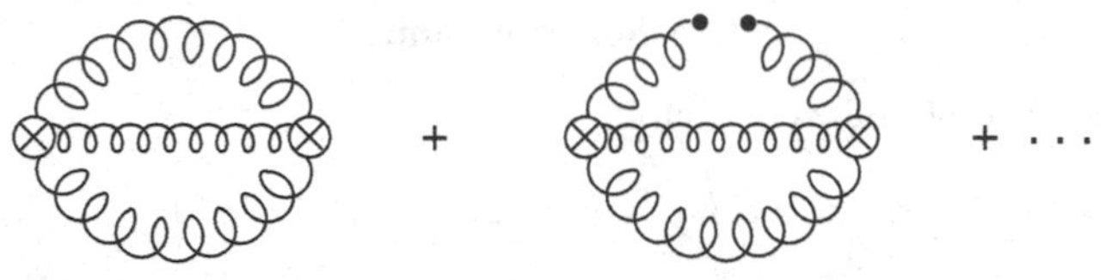

Fig. 37.3. Feynman diagrams corresponding to the OPE of the tri-gluonium correlator: (a) perturbative; (b) gluon condensate.

indicates that the mixing also vanishes in the chiral limit like in the case of the pseudoscalar channel.

### 37.5 Scalar tri-gluonium correlator

Here, one studies the correlator in Fig. 37.3 associated to the interpolating trigluon current:

$$J_{3g} = g^3 f_{abc}\langle G^a G^b G^c\rangle \tag{37.19}$$

Its QCD expression reads [457]:

$$\psi_3(q^2) = -\alpha_s^2\left\{\frac{3\alpha_s}{10\pi}q^8\log-\frac{q^2}{\nu^2} + 18\pi q^4\langle\alpha_s G^2\rangle - \frac{27}{2}\left(q^2\log-\frac{q^2}{\nu^2}\right)g^3 f_{abc}\langle G^a G^b G^c\rangle + \alpha_s\pi^3 36\times 64\,(\phi_7-\phi_5)\right\}, \tag{37.20}$$

with:

$$\phi_5 = \frac{1}{16}\mathbf{Tr}\langle G_{\nu\mu}G^{\mu\rho}G_{\rho\tau}G^{\tau\nu}\rangle$$
$$\phi_7 = \frac{1}{16}\mathbf{Tr}\langle G_{\nu\mu}G^{\nu\rho}G^{\mu\rho}G_{\mu\tau}\rangle\ . \tag{37.21}$$

### 37.6 Scalar di- and tri-gluonium mixing

We shall be concerned with the off-diagonal correlator:

$$\psi_{23}(q^2) \equiv i\int d^4x\, e^{iqx}\,\langle 0|T J_{2g}(x)J_{3g}(0)^\dagger|0\rangle\ . \tag{37.22}$$

Its QCD expression reads [457]:

$$\psi_{23}(q^2) = \alpha_s^2\left\{\log-\frac{q^2}{\nu^2}\left[\frac{9}{4\pi^3}g^2q^6 - \frac{9}{4\pi}g^2\langle G^2\rangle q^2\right] - 24\pi g\langle f_{abc}G^aG^bG^c\rangle\right\}. \tag{37.23}$$

## 37.7 Tensor gluonium

We shall be concerned with the two-point correlator:

$$\Psi^T_{\mu\nu\rho\sigma} \equiv i \int d^4x \, e^{iqx} \langle 0|T\theta^g_{\mu\nu}(x)\theta^g_{\rho\sigma}(0)^\dagger|0\rangle$$
$$= \frac{1}{2}\left(\eta_{\mu\rho}\eta_{\nu\sigma} + \eta_{\mu\sigma}\eta_{\nu\rho} - \frac{2}{3}\eta_{\mu\nu}\eta_{\rho\sigma}\right)\psi_T(q^2) , \tag{37.24}$$

where:

$$\theta^g_{\mu\nu} = -G^\alpha_\mu G_{\nu\alpha} + \frac{1}{4}g_{\mu\nu}G_{\alpha_\beta}G^{\alpha\beta} . \tag{37.25}$$

and:

$$\eta_{\mu\nu} \equiv g_{\mu\nu} - \frac{q_\mu q_\nu}{q^2} . \tag{37.26}$$

To leading order in $\alpha_s$ and including the non-perturbative condensates, the QCD expression of the correlator reads [382]:

$$\Psi_T(q^2 \equiv -Q^2) = -\frac{1}{20\pi^2}(Q^4)\log\frac{Q^2}{\nu^2} + \frac{5}{12}\frac{g^2}{Q^4}\langle 2O_1 - O_2\rangle , \tag{37.27}$$

where:

$$O_1 = (f_{abc}G_{\mu\alpha}G_{\nu\alpha})^2 \quad \text{and} \quad O_2 = (f_{abc}G_{\mu\nu}G_{\alpha\beta})^2 . \tag{37.28}$$

Using the vacuum saturation hypothesis, one can write:

$$\langle 2O_1 - O_2\rangle \simeq -\frac{3}{16}\langle G^2\rangle^2 . \tag{37.29}$$

## 37.8 Tensor meson-gluonium mixing

We shall be concerned with the off-diagonal two-point correlator:

$$\Psi^T_{gq,\mu\nu\rho\sigma} \equiv i \int d^4x \, e^{iqx} \langle 0|T\theta^g_{\mu\nu}(x)\theta^q_{\rho\sigma}(0)^\dagger|0\rangle$$
$$= \frac{1}{2}\left(\eta_{\mu\rho}\eta_{\nu\sigma} + \eta_{\mu\sigma}\eta_{\nu\rho} - \frac{2}{3}\eta_{\mu\nu}\eta_{\rho\sigma}\right)\Psi^T_{gq}(q^2) , \tag{37.30}$$

where

$$\theta^q_{\mu\nu}(x) = i\bar{q}(x)(\gamma_\mu\bar{D}_\nu + \gamma_\nu\bar{D}_\mu)q(x) . \tag{37.31}$$

Here, $\bar{D}_\mu \equiv \vec{D}_\mu - \overleftarrow{D}_\mu$ is the covariant derivative, and the other quantities have already been defined earlier. Taking into account the mixing of the currents, one obtains [452]:

$$\Psi^T_{gq}(q^2 \equiv -Q^2) \simeq \frac{q^4}{15\pi^2}\left(\frac{\alpha_s}{\pi}\right)\left(\log^2\frac{Q^2}{\nu^2} - \frac{91}{15}\log\frac{Q^2}{\nu^2}\right) - \frac{7}{36}\log\frac{Q^2}{\nu^2}\left(\frac{\alpha_s}{\pi}\right)\langle G^2\rangle . \tag{37.32}$$

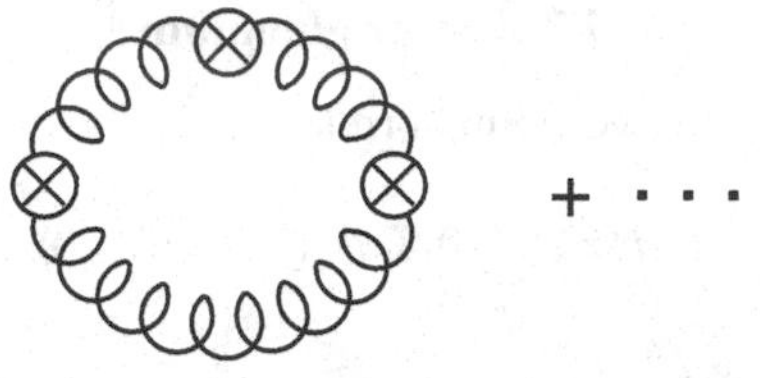

Fig. 37.4. Lowest order tachyonic gluon contribution to the gluonic correlator. The cross in the internal gluon propagator corresponds to the tachyonic gluon mass insertion $\lambda^2$.

## 37.9 Contributions beyond the OPE: tachyonic gluon mass

As we have seen previously, there are also contributions beyond the SVZ-expansion. We shall first be concerned with the two-point correlator:

$$\Psi_s(Q^2) \equiv i \int d^4x\, e^{iqx} \langle 0|\mathcal{T} J_{2g}(x)(J_{2g}(0))^\dagger|0\rangle \tag{37.33}$$

associated to the scalar gluonium current:

$$J_{2g} = \alpha_s \left(G^a_{\mu\nu}\right)^2 . \tag{37.34}$$

Its evaluation leads to:

$$\frac{1}{\pi}\mathrm{Im}\Psi_s(s) \approx (\text{parton model})\left(1 - \frac{6\lambda^2}{s} + \cdots\right). \tag{37.35}$$

The tachyonic gluon contribution comes from the diagram in Fig. 37.4.

Thus, one can expect that the $\lambda^2$ correction in this channel is relatively much larger since it is not proportional to an extra power of $\alpha_s$. Let us consider now the case of the tensor gluonium with the correlator:

$$\begin{aligned}
\psi^T_{\mu\nu\rho\sigma}(q) &\equiv i \int d^4x\, e^{iqx} \langle 0|\mathcal{T}\theta^g_{\mu\nu}(x)\theta^g_{\rho\sigma}(0)^\dagger|0\rangle \\
&= \psi^T_4\left(q_\mu q_\nu q_\rho q_\sigma - \frac{q^2}{4}(q_\mu q_\nu g_{\rho\sigma} + q_\rho q_\sigma g_{\mu\nu}) + \frac{q^4}{16}(g_{\mu\nu}g_{\rho\sigma})\right) \\
&\quad + \psi^T_2\left(\frac{q^2}{4} g_{\mu\nu}g_{\rho\sigma} - q_\mu q_\nu g_{\rho\sigma} - q_\rho q_\sigma g_{\mu\nu} + q_\mu q_\sigma g_{\nu\rho} + q_\nu q_\sigma g_{\mu\rho}\right. \\
&\quad \left. + q_\mu q_\rho g_{\nu\sigma} + q_\nu q_\rho g_{\mu\sigma}\right) \\
&\quad + \psi^T_0\left(g_{\mu\sigma}g_{\nu\rho} + g_{\mu\rho}g_{\nu\sigma} - \frac{1}{2} g_{\mu\nu}g_{\rho\sigma}\right) ,
\end{aligned} \tag{37.36}$$

where $\theta^g_{\mu\nu}$ has been defined in Eq. (37.25). A direct calculation gives the following results for the structure functions $\psi_i^T$ and their respective Borel/Laplace transforms:

$$\pi^2\psi_4^T = \frac{l_{\mu Q}}{15} + \frac{17}{450} - \lambda^2\frac{1}{3Q^2} \Longrightarrow \frac{1}{15}\left(1-5\frac{\lambda^2}{M^2}\right), \tag{37.37}$$

$$\pi^2\psi_2^T = \frac{Q^2 l_{\mu Q}}{20} + \frac{9Q^2}{200} + \lambda^2\left(\frac{l_{\mu Q}}{6} - \frac{2}{9}\right) \Longrightarrow -\frac{M^2}{20}\left(1-\frac{10}{3}\frac{\lambda^2}{M^2}\right), \tag{37.38}$$

$$\pi^2\psi_0^T = \frac{Q^4 l_{\mu Q}}{20} + \frac{9Q^4}{200} + \lambda^2 Q^2\left(\frac{l_{\mu Q}}{4} - \frac{1}{12}\right) \Longrightarrow \frac{M^4}{20}\left(1-\frac{5}{2}\frac{\lambda^2}{M^2}\right). \tag{37.39}$$

If, instead of considering $\theta^g_{\mu\nu}$, we would introduce the total energy-momentum tensor of interacting quarks and gluons $\theta_{\mu\nu}$, then various functions components of $\psi_{\mu\nu\rho\sigma}$ are related to each other because of the energy-momentum conservation. Indeed, requiring that

$$\psi^T_{\mu\nu\rho\sigma} q_\mu \equiv 0$$

we immediately obtain:

$$\psi_2^T = \frac{3}{4}Q^2\psi_4^T \quad \text{and} \quad \psi_0^T = \frac{3}{4}Q^4\psi_4^T\,,$$

and, as a consequence, the following representation of the function in Eq. 37.36:

$$\psi^T_{\mu\nu\rho\sigma}(q) = \left(\eta_{\mu\rho}\eta_{\nu\sigma} + \eta_{\mu\sigma}\eta_{\nu\rho} - \frac{2}{3}\eta_{\mu\nu}\eta_{\rho\sigma}\right)\psi^T(Q^2)\,, \tag{37.40}$$

where:

$$\psi^T(Q^2) \equiv Q^4\frac{3}{4}\psi_4^T(Q^2), \qquad \eta_{\mu\nu} \equiv g_{\mu\nu} - \frac{q_\mu q_\nu}{q^2}\,. \tag{37.41}$$

# 38

# Hybrid correlators

## 38.1 Light hybrid correlators

We shall be concerned with the two-point correlator (standard notations):

$$\Pi^{\mu\nu}_{V/A}(q^2) \equiv i \int d^4x \; e^{iqx} \; \langle 0|\mathcal{T}\mathcal{O}^{\mu}_{V/A}(x) \left(\mathcal{O}^{\nu}_{V/A}(0)\right)^{\dagger} |0\rangle$$
$$= -(g^{\mu\nu}q^2 - q^{\mu}q^{\nu})\Pi^{(1)}_{V/A}(q^2) + q^{\mu}q^{\nu}\Pi^{(0)}_{V/A}(q^2), \tag{38.1}$$

built from the hadronic local currents $\mathcal{O}^{V/A}_{\mu}(x)$:

$$\mathcal{O}^{\mu}_{V}(x) \equiv: g\bar{\psi}_i\lambda_a\gamma_\nu\psi_j G^{\mu\nu}_a : , \qquad \mathcal{O}^{\mu}_{A}(x) \equiv: g\bar{\psi}_i\lambda_a\gamma_\nu\gamma_5\psi_j G^{\mu\nu}_a : \tag{38.2}$$

which select the specific quantum numbers of the hybrid mesons; A and V refer respectively to the vector and axial-vector currents. The invariant $\Pi^{(1)}$ and $\Pi^{(0)}$ refer to the spin one and zero mesons. The correlator is represented in Fig. 38.1.

The perturbative QCD expressions of the invariants are:

$$\frac{1}{\pi}\text{Im}\Pi^{(1)}_{V/A}(t)_{\text{pert}} = \frac{\alpha_s}{60\pi^3}t^2\left\{1+\frac{\alpha_s}{\pi}\left[\frac{121}{16}-\frac{257}{360}n_f+\left(\frac{35}{36}-\frac{n_f}{6}\right)\log\frac{\nu^2}{t}\right]\right\}$$
$$\frac{1}{\pi}\text{Im}\Pi^{(0)}_{V/A}(t)_{\text{pert}} = \frac{\alpha_s}{120\pi^3}t^2\left\{1+\frac{\alpha_s}{\pi}\left[\frac{1997}{432}-\frac{167}{360}n_f+\left(\frac{35}{36}-\frac{n_f}{6}\right)\log\frac{\nu^2}{t}\right]\right\}. \tag{38.3}$$

The anomalous dimension of the current can be easily deduced to be:

$$\gamma_H = \beta_1 + \frac{32}{9}, \tag{38.4}$$

where $\beta_1 = -1/2(11 - 2n_f/3)$ is the first coefficient of the beta function. The short-distance tachyonic gluon mass effect is given by the diagram in Fig. 38.2 and reads [462]:

$$\frac{1}{\pi}\text{Im}\Pi^{(1)}_{V/A}(t)_{\lambda} = -\frac{\alpha_s}{60\pi^3}\frac{35}{4}\lambda^2 t$$
$$\frac{1}{\pi}\text{Im}\Pi^{(0)}_{V/A}(t)_{\lambda} = \frac{\alpha_s}{120\pi^3}\frac{15}{2}\lambda^2 t\,. \tag{38.5}$$

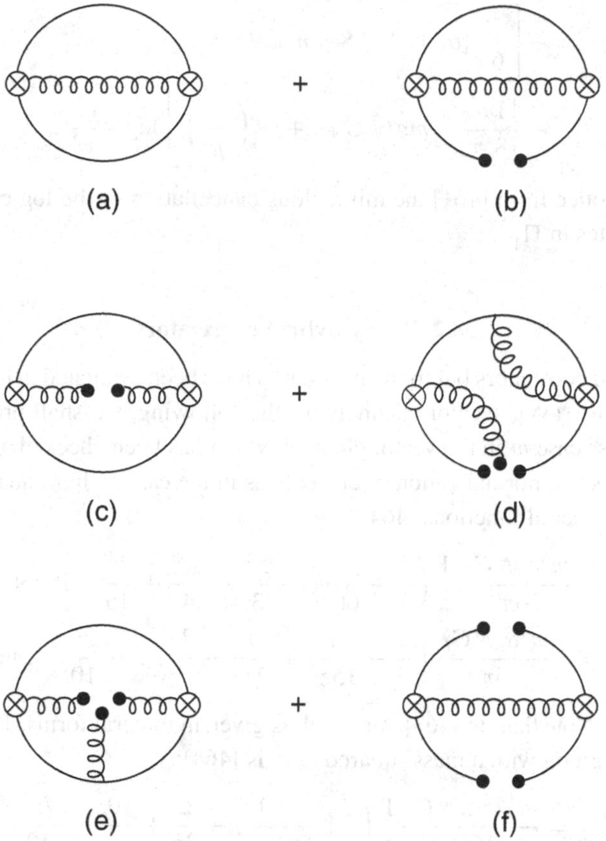


Fig. 38.1. Feynman diagrams corresponding to the OPE of the hybrid correlator: (a) perturbative; (b) quark condensate; (c) gluon condensate; (d) mixed condensate; (e) three-gluon condensate; (f) four-quark condensate.

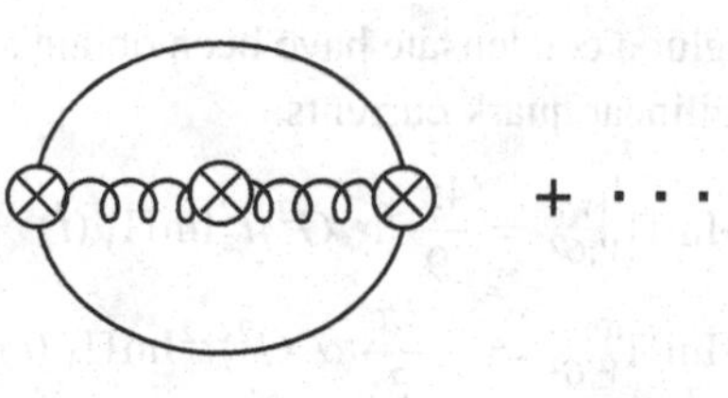


Fig. 38.2. Lowest order tachyonic gluon contribution to the hybrid correlator. The cross in the internal gluon propagator corresponds to the tachyonic gluon mass insertion $\lambda^2$.

The (corrected) contributions of the dimension-four and -six terms have been obtained by [461] and reads in the limit $m^2 = 0$:

$$\Pi_V^{(1)}(q^2)_{NP} = -\frac{1}{9\pi}[\alpha_s\langle G^2\rangle + 8\alpha_s m\langle\bar\psi\psi\rangle]\log -\frac{q^2}{\nu^2} + \frac{1}{q^2}\left[\frac{16\pi}{9}\alpha_s\langle\bar\psi\psi\rangle^2 + \frac{1}{48\pi^2}g^3\langle G^3\rangle - \frac{83}{432}\frac{\alpha_s}{\pi}mg\langle\bar\psi G\psi\rangle\right]$$

$$\Pi_A^{(0)}(q^2)_{NP} = -\left[\frac{1}{6\pi}[\alpha_s\langle G^2\rangle - 8\alpha_s m\langle\bar\psi\psi\rangle] - \frac{11}{18}\frac{\alpha_s}{\pi}\frac{1}{q^2}mg\langle\bar\psi G\psi\rangle + \mathcal{O}\left(\frac{1}{q^2}\right)\right]\log -\frac{q^2}{\nu^2}, \quad (38.6)$$

where one can notice from [461] the miraculous cancellation of the log-coefficient of the $D = 6$ condensates in $\Pi_V^{(1)}$.

## 38.2 Heavy hybrid correlators

Analogous hybrid correlators but for heavy quarks have been evaluated in [463] for unequal masses and for the (axial-)vector channels. In the following, we shall present the results for the equal mass case $m$ in the vector channel which has been checked and completed in [464]. Using the same normalization of currents as in the case of light quarks, one obtains the perturbative spectral functions [464]:

$$\begin{aligned}\frac{1}{\pi}\mathrm{Im}\Pi_{V,\mathrm{pert}}^{(1)} &= \frac{m^6\alpha_s N C_F}{16\pi^3}\frac{1}{t}\left(\frac{7}{3} + \frac{1}{60z^2} - \frac{5z}{3} - \frac{3z^2}{4} + \frac{z^3}{15} + \ln z + 2z\ln z\right)\\ \frac{1}{\pi}\mathrm{Im}\Pi_{V,\mathrm{pert}}^{(1+0)} &= -\frac{m^4\alpha_s N C_F}{16\pi^3}\left(\frac{2}{3} - \frac{1}{15z^3} + \frac{1}{2z^2} - \frac{2}{z} + z - \frac{z^2}{10} - 2\ln z\right),\end{aligned} \quad (38.7)$$

where $z = t/m^2$. Note that, in [463], the result is given in integral forms. The contributions of the tachyonic gluon with a mass squared $-\lambda^2$ is [464]:

$$\begin{aligned}\frac{1}{\pi}\mathrm{Im}\Pi_{V,\lambda}^{(1)} &= \frac{m^4\lambda^2\alpha_s N C_F}{16\pi^3}\frac{1}{t}\left(-2 - \frac{1}{12z^2} - \frac{2}{3z} + \frac{10z}{3} - \frac{7z^2}{12} - 3\ln z\right)\\ \frac{1}{\pi}\mathrm{Im}\Pi_{V,\lambda}^{(1+0)} &= -\frac{m^2\lambda^2\alpha_s N C_F}{16\pi^3}\left(-\frac{4}{3} + \frac{1}{3z^3} - \frac{4}{3z^2} + \frac{2}{z} + \frac{z}{3}\right).\end{aligned} \quad (38.8)$$

The contributions of the gluon condensate have been obtained in [463] and expressed in terms of the correlators of bilinear quark currents:

$$\begin{aligned}\frac{1}{\pi}\mathrm{Im}\Pi_{V,G^2}^{(1)} &= \frac{4\pi}{9}\langle\alpha_s G^2\rangle t^2\mathrm{Im}\Pi_V(t)\\ \frac{1}{\pi}\mathrm{Im}\Pi_{V,G^2}^{(0)} &= -\frac{2\pi}{3}\langle\alpha_s G^2\rangle t^2\mathrm{Im}\Pi_V(t),\end{aligned} \quad (38.9)$$

where:

$$\mathrm{Im}\Pi_V(t) = \frac{N}{24\pi}v(3 - v^2) \quad (38.10)$$

is the vector bilinear current spectral function and where $v^2 = 1 - 4m^2/t$ is the square of the heavy quark velocity.

# 39

# Correlators in $x$-space

In previous chapters we have discussed correlators in the momentum space. In some applications, some authors prefer to work in the $x$-space. From the pure theoretical point of view, the use of the $x$-space is no better than the use of the momentum space, which is the traditional tool of QSSR [1,3]. However, each representation has its own advantages and inconveniences. The $x$-space approach is described in detail, for example in [386]. In particular, the current correlators are measured in the most direct way on the lattice [393]. In the coordinate space, the two-point functions obey a dispersion representation:

$$\Pi(x) = \frac{1}{4\pi^2}\int_0^\infty dt\, \frac{\sqrt{t}}{x} K_1(x\sqrt{t})\, \mathrm{Im}\Pi(t)\,, \tag{39.1}$$

where $K_1(z)$ is the modified Bessel function, which behaves for small $z$ as:

$$K(z \to 0) \simeq \frac{1}{z} + \frac{z}{2}\ln z\,. \tag{39.2}$$

In the limit $x \to 0$, $\Pi(x)$ coincides with the free-field correlator. For the sake of completeness, we begin with a summary of theoretical expressions for the current correlators, both in the $Q-$ and $x-$spaces. We will focus on the $(V \pm A)$ and $(S \pm P)$ channels since the recent lattice data [393] refer to these channels.

## 39.1 (Axial-)vector correlators

In case of $(V \pm A)$ currents the correlator is defined as:

$$\Pi_{\mu\nu}(q) = i\int d^4x\, e^{iqx}\langle T J_\mu(x) J_\nu(0)^\dagger\rangle = (q_\mu q_\nu - g_{\mu\nu}q^2)\Pi(q^2)\,, \tag{39.3}$$

where $-q^2 \equiv Q^2 > 0$ in the Euclidean space–time. For the sake of definiteness we fix the flavour structure of the light-quark current $J_\mu$ as:

$$J_\mu^{V\pm A} = \bar{u}\gamma_\mu(1 \pm \gamma_5)d\,. \tag{39.4}$$

In the chiral limit one has in the $(V+A)$ case (see, e.g., [1,3] and previous chapters):

$$\Pi^{V+A}(Q^2) = \frac{1}{2\pi^2}\left\{ -\left(1+\frac{\alpha_s}{\pi}\right)\ln\frac{Q^2}{\nu^2} - \frac{\alpha_s}{\pi}\frac{\lambda^2}{Q^2} + \frac{\pi}{3}\frac{\langle\alpha_s (G^a_{\mu\nu})^2\rangle}{Q^4} + \frac{256\pi^3}{81}\frac{\alpha_s\langle\bar{q}q\rangle^2}{Q^6}\right\}. \tag{39.5}$$

The corresponding relation for the $(V-A)$ case reads as:

$$\Pi^{V-A}(Q^2) = \frac{4m_q < \bar{q}q >}{Q^4} - \frac{64\pi}{9}\frac{\alpha_s\langle\bar{q}q\rangle^2}{Q^6} + 8\pi\frac{\alpha_s M_0^2\langle\bar{q}q\rangle^2}{Q^8}, \tag{39.6}$$

where $M_0^2 \approx 0.8\,\text{GeV}^2$ parametrizes the mixed condensate as discussed in previous chapters. In the $x$-space the same correlators, upon dividing by $\Pi^{V+A}_{\text{pert}}$ where $\Pi^{V+A}_{\text{pert}}$ stands for the perturbative correlator, are obtained by applying the equations collected for convenience in the Table G.1 from [394] given in Appendix G. Therefore, one obtains [394]:

$$\frac{\Pi^{V+A}}{\Pi^{V+A}_{\text{pert}}} \to 1 - \frac{\alpha_s}{4\pi}\lambda^2\cdot x^2 - \frac{\pi}{48}\langle\alpha_s (G^a_{\mu\nu})^2\rangle x^4 \ln x^2 + \frac{2\pi^3}{81}\alpha_s\langle\bar{q}q\rangle^2 x^6 \ln x^2 . \tag{39.7}$$

Note that $\ln x^2$ is negative since we start from small $x$. In the $(V-A)$ case:

$$\frac{\Pi^{V-A}}{\Pi^{V+A}_{\text{pert}}} \to \frac{\pi^2}{2} m_q\langle\bar{q}q\rangle x^4 \ln x^2 - \frac{\pi^3}{9}\alpha_s\langle\bar{q}q\rangle^2 x^6 \ln x^2 . \tag{39.8}$$

The $x$-transform of the $Q^2\cdot\Pi(Q^2)$ is given by:

$$\frac{Q^2\cdot\Pi^{V+A}}{Q^2\cdot\Pi^{V+A}_{\text{pert}}} \to 1 - \frac{\pi}{96}\langle\alpha_s (G^a_{\mu\nu})^2\rangle x^4 + \frac{2\pi^3}{81}\alpha_s\langle\bar{q}q\rangle^2 x^6 \ln x^2. \tag{39.9}$$

Similarly:

$$\frac{Q^2\cdot\Pi^{V-A}}{Q^2\cdot\Pi^{V+A}_{\text{pert}}} \to -\frac{\pi^3}{9}\alpha_s\langle\bar{q}q\rangle^2 x^6 \ln x^2 . \tag{39.10}$$

### 39.2 (Pseudo)scalar correlators

Next, we will concentrate on the currents having the quantum numbers of the pion and of $a_0(980)$-meson. The correlator of two pseudoscalar currents is defined as

$$\Pi^P(Q^2) \equiv i\int d^4x\, e^{iqx}\langle T\{J^\pi(x)J^\pi(0)\}\rangle , \tag{39.11}$$

where

$$J^P = i(m_u + m_d)\bar{u}\gamma_5 d , \tag{39.12}$$

In the momentum space, it reads in terms of the renormalized coupling, masses and condensates:

$$\Pi^P(Q^2) \equiv i \int d^4x\, e^{iqx} \langle T\{J^\pi(x) J^\pi(0)\}\rangle$$
$$= \frac{3}{8\pi^2}(m_u+m_d)^2 \left\{ \left[1 + \left(\frac{17}{3} - \ln\frac{Q^2}{\nu^2}\right)\frac{\alpha_s}{\pi}\right] Q^2 \ln\frac{Q^2}{\nu^2} + \frac{4\alpha_s}{\pi}\lambda^2 \ln\frac{Q^2}{\nu^2} \right.$$
$$\left. + \frac{\pi}{3}\frac{\langle \alpha_s (G^a_{\mu\nu})^2\rangle}{Q^2} + \frac{896\pi^3}{81}\frac{\alpha_s\langle \bar{q}q\rangle^2}{Q^4} \right\}. \tag{39.13}$$

Here, the standard OPE terms can be found in [1,3,167] as compiled in previous chapters, while the gluon-mass correction was introduced first in [161]. It is more convenient to introduce the running QCD coupling $\bar{\alpha}_s(Q^2)$, the quark running mass $\bar{m}_i(Q^2)$ and condensate $\langle\bar{q}q\rangle(Q^2)$,[1] into the second derivative in $Q^2$ of $\Pi^P(Q^2)$ defined in Eq. (39.13), which obeys an homogeneous RGE:

$$\frac{\partial^2 \Pi^P}{(\partial Q^2)^2} = \frac{3}{8\pi^2}\frac{(\bar{m}_u+\bar{m}_d)^2}{Q^2}\left\{ 1 + \frac{11}{3}\frac{\bar{\alpha}_s}{\pi} - \frac{4\alpha_s}{\pi}\frac{\lambda^2}{Q^2} \right.$$
$$\left. + 2\frac{\pi}{3}\frac{\langle \alpha_s (G^a_{\mu\nu})^2\rangle}{Q^4} + 2\cdot 3\frac{896\pi^3}{81}\frac{\bar{\alpha}_s\langle \bar{q}q\rangle^2}{Q^6} \right\}. \tag{39.14}$$

In what follows, we shall work with the appropriate ratio where the pure perturbative corrections are absorbed into the overall normalization and concentrate on the power corrections assuming that these corrections are responsible for the observed rather sharp variations of the correlation functions. Thus, in the $x$-space we have for the pion channel [394]:

$$\frac{\Pi^P}{\Pi^P_{\text{pert}}} \to 1 - \frac{\alpha_s}{2\pi}\lambda^2 x^2 + \frac{\pi}{96}\langle \alpha_s (G^a_{\mu\nu})^2\rangle x^4 - \frac{7\pi^3}{81}\alpha_s\langle\bar{q}q\rangle^2 x^6 \ln x^2 . \tag{39.15}$$

Note that the coefficient in front of the last term in Eq. (39.15) differs both in the absolute value and sign from the corresponding expression in [386].

Similarly, in the $S$-channel, the correlator associated with the scalar current having the quantum number of the $a_0$:

$$J^S = i(m_u - m_d)\bar{u}d \tag{39.16}$$

is obtained from Eq. (39.13) by changing $m_i$ into $-m_i$ and by taking the coefficient in front of the $1/Q^6$ correction to be $-1408\pi^3/81$ instead of $896\pi^3/81$ in Eq. (39.13). This term was found first in [666].

Therefore, we have in the $x-$space:

$$\frac{\Pi^S}{\Pi^S_{\text{pert}}} \to 1 - \frac{\alpha_s}{2\pi}\lambda^2 x^2 + \frac{\pi}{96}\langle \alpha_s (G^a_{\mu\nu})^2\rangle x^4 + \frac{11\pi^3}{81}\alpha_s\langle\bar{q}q\rangle^2 x^6 \ln x^2 . \tag{39.17}$$

[1] We assume that $\alpha_s\lambda^2$ does not run like $\langle\alpha_s(G^a_{\mu\nu})^2\rangle$.

The channel which is crucial for the analysis in [393,394] is the $(S+P)$, which is less affected by some eventual direct instanton contributions than the individual $S$ and $P$ correlators. In this channel:

$$R_{P+S} \equiv \frac{1}{2}\left(\frac{\Pi^P}{\Pi^P_{\rm pert}} + \frac{\Pi^S}{\Pi^S_{\rm pert}}\right) \rightarrow 1 - \frac{\alpha_s}{2\pi}\lambda^2 x^2 + \frac{\pi}{96}\langle\alpha_s (G^a_{\mu\nu})^2\rangle x^4 + \frac{4\pi^3}{81}\alpha_s\langle\bar{q}q\rangle^2 x^6 \ln x^2 \,. \tag{39.18}$$

This expression concludes the summary of the power corrections to the current correlators.

We shall see later on that the QCD expressions of the two-point functions given in this part of the book are crucial inputs in the discussions of QCD spectral sum rules analysis and in various high-energy processes ($e^+e^- \rightarrow$ hadrons total cross-section, Higgs decays, ...).

# Part IX

QCD non-perturbative methods

# 40
# Introduction

In the previous parts of this book, we have studied different methods based on power corrections, namely renormalons, instantons and the SVZ-expansion in terms of condensates and eventual quadratic corrections, which are one important aspect of non-perturbative QCD. In this part, we shall shortly discuss the other most popular non-perturbative methods used in QCD for studying the low-energy properties of the hadrons and of QCD. These are:

- Lattice gauge theory
- Chiral perturbation theory (ChPT)
- Models of the QCD effective action
- Heavy quark effective theory (HQET)
- Potential approaches for quarkonia.

The method and phenomenology of QCD Spectral Sum Rules (QSSR) which will be discussed in more details will be devoted to a new part. Here, we do not aim to present a complete review and references but we shall limit ourselves to the outline of the general features of the different approaches. A more extensive list of non-perturbative methods, which will not be discussed here, such as the:

- Skyrme model
- Bag models
- Discretized light-cone quantization

can be found in [3]. We shall shortly discuss in the chapter dedicated to the heavy quarks, the two approaches:

- Potential and quark models
- Stochastic vacuum model
- Non-relativistic effective theories

which are discussed in details in some other reviews (see e.g. [51]). We shall complete this part by a short discussion on monopole and confinement.

# 41

# Lattice gauge theory

## 41.1 Introduction

In this chapter, we shall discuss very briefly the main idea behind the lattice approach in QCD. More detailed discussions and some introductions can be found in different textbooks on lattice gauge theories [489] and some non-specialized reviews. (see e.g., Yndurain's book [46] or Dosch's review [51]). More recent reviews on the lattice results can be found in different contributions at the annual Lattice conferences (*Nucl. Phys. B* (Proc, Suppl.)). The starting point is the *Euclidian* generating functional:

$$Z = \int \mathcal{D}\psi(x)\mathcal{D}\bar{\psi}(x)\exp\left\{-\mathcal{S} \equiv \int d^4x\ \mathcal{L}_{\rm QCD}\right\}, \tag{41.1}$$

where the QCD action $\mathcal{S}$ is positive, thus providing the convergence factor. It is convenient to write the Lagrangian in a matrix notation:

$$G_{\mu\nu} \equiv \sum_a \frac{\lambda_a}{2} G^a_{\mu\nu}, \tag{41.2}$$

where $\lambda_a$ are the generators of the $SU(3)_c$ gauge transformation group:

$$U(x) = \exp\left\{ig\frac{\lambda_a}{2}A^a(x)\right\}. \tag{41.3}$$

Therefore, it reads:

$$\mathcal{L}_{\rm QCD}(x) = \frac{1}{2}\sum_{\mu\nu} G^2_{\mu\nu}(x) + \bar{\psi}(x)(\partial^\mu\gamma_\mu + m)\psi(x) - ig\bar{\psi}\gamma_\mu A^\mu(x)\psi(x), \tag{41.4}$$

where, in this notation, the gauge transformations become:

$$\begin{aligned} A_\mu(x) &\to U^{-1}(x)A_\mu(x)U(x) + \frac{i}{g}U^{-1}(x)\partial_\mu(x)U(x) \\ G_{\mu\nu}(x) &\to U^{-1}(x)G_{\mu\nu}(x)U(x), \\ \psi(x) &\to U^{-1}(x)\psi(x). \end{aligned} \tag{41.5}$$

Next we introduce the essential ingredients for the *lattice formulation* of QCD. Here, one expects that all expressions introduced below are well-defined, and, *in principle*, can be evaluated numerically. This feature has made lattice gauge theory one of the most important non-perturbative methods for QCD. The functional integral introduced before has to be understood as the limiting value of a high-dimensional volume integral where the fields at the lattice points $i, j, \ldots$ are the integration variables. For definiteness, we shall consider a finite hypercube lattice, with lattice spacing $a$ and volume $V = (Na)^4$ with periodic boundary conditions. The physical (continuum) limit is reached for $V \to \infty$ first and after $a \to 0$. The lattice provides a *regularization* as $a$ is finite, such that UV divergences do not occur. As long as $N$ is bounded from above, IR divergences are prevented. The UV divergences will reappear as $1/a$ or/and $\log a$, when one goes to the continuum limit, where $a \to 0$.

- A point on the lattice is denoted by its coordinates in units of $a$, i.e. by the integers: $(n) \equiv (n_1, n_2, n_3, n_4)$, representing the point with coordinates $X = (an_1, an_2, an_3, an_4)$.
- The neighbour of the point $(n)$ in the $\mu$-direction is denoted by $(n + \mu)$.
- The link from point $n$ to its neighbour in the $\mu$-direction, $n + \mu$ is denoted by $(n, n + \mu)$. Its plays an essential rôle in the lattice.

## 41.2 Gluons on the lattice: the Wegner–Wilson action

- An element of the gauge group is attached to each link, while its inverse is attached to the link in the opposite direction [490,491]:

$$(n, n + \mu) \to U(n, n + \mu)\,, \qquad (n + \mu, n) \to U^{-1}(n, n + \mu)\,, \tag{41.6}$$

  where the group elements $U(n, n + \mu)$ can be expressed by the generators $\lambda_a/2$ of the group as:

$$U(n, n + \mu) = \exp\left\{ig\frac{a\lambda_a}{2}A^a_\mu a(n)\right\}\,. \tag{41.7}$$

- In a local gauge theory, an element of the gauge group is attached to each point on the lattice:

$$U(n) = \exp\left\{ig\frac{\lambda_a}{2}\Lambda^a(n)\right\}\,. \tag{41.8}$$

  The gauge transformation for the group element $U(n, n + \mu)$ is defined as:

$$U(n, n + \mu) \to U(n)U(n, n + \mu)U^{-1}(n + \mu)\,, \tag{41.9}$$

  where one may notice that there is no inhomogeneous term on the lattice version of gauge transformation.
- The continuum limit is achieved by connecting quantities attached to neighbouring lattice points through the Taylor expansion and retaining the lowest-order contribution in the lattice spacing $a$:

$$U(n + \mu) = U(x) + a\partial_\mu U(x) + \mathcal{O}(a^2)\,. \tag{41.10}$$

  Using the expansion:

$$U(n, n + \mu) = 1 + iag\frac{\lambda_a}{2}A^a(x) + \mathcal{O}(a^2)\,, \tag{41.11}$$

  the gauge tranformation in Eq. (41.9), becomes the one of the continuum limit in Eq. (41.5).

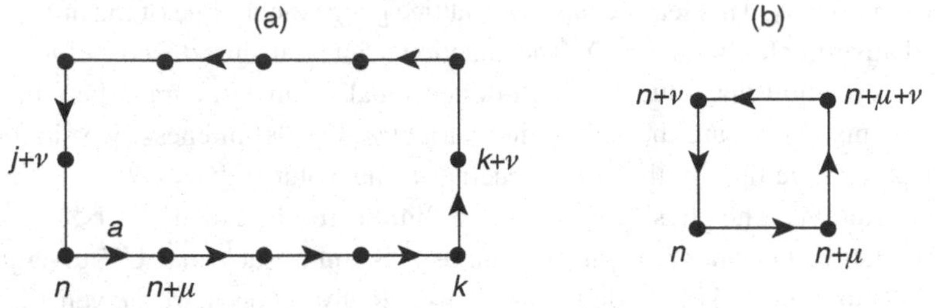


Fig. 41.1. Plaquette: a group element $U(n, n+\mu)$ is attached to each link.

- The Wegner–Wilson loop [490,491] corresponds to the product of group elements $U(n, n+\mu)$ along a closed contour $L$. It is defined as:

$$W[L] = U(n, n+\mu)U(n+\mu, n+\mu+\lambda)\cdots U(n+\nu, n)\,. \tag{41.12}$$

Using the property $UU^{-1} = 1$ and the fact that the trace is cyclic: TrABC = TrBCA = ..., it is easy to show that, under the gauge transformations in Eq. (41.9):

$$Tr W[L] \qquad \text{is gauge invariant}\,. \tag{41.13}$$

- A *Plaquette* is the simplest non-trivial Wegner–Wilson loop, which is the product of four group elements attached to a square with a sidelength $a$ and lattice points as corners (see Fig. 41.1).

$$\begin{aligned}\mathcal{P}(n, \mu, \nu) &= U(n+\mu, n+\mu+\nu)U(n+\mu+\nu, j+\nu)U(j+\nu, j)U(j, j+\mu)\\ &= U(n+\mu, n+\mu+\nu)U^{-1}(n+\nu, j+\mu+\nu)U^{-1}(j, j+\nu)U(j, j+\mu)\\ &= e^{[igaA_\nu(n+\mu)]}e^{[-igaA_\mu(n+\nu)]}e^{[-igaA_\nu(n)]}e^{[-igaA_\mu(n)]}\end{aligned} \tag{41.14}$$

Using the Campbell–Hausdorff formula:

$$e^{ax}e^{ay} \simeq e^{ax+ay+a^2[x,y]+\mathcal{O}(a^3)}\,, \tag{41.15}$$

with each pair of the previous exponentials, one obtains:

$$\begin{aligned}\mathcal{P}(n, \mu, \nu) = &\, e^{\{-iag(A_\nu(n+\mu)-A_\mu(n+\nu))+a^2g^2[A_\nu(n+\mu),A_\mu(n+\nu)]/2+i\mathcal{O}(a^3)\}}\\ &\times e^{\{-iag(A_\nu(n)-A_\mu(n))+a^2g^2[A_\nu(n),A_\mu(n)]/2+i\mathcal{O}(a^3)\}}\,,\end{aligned} \tag{41.16}$$

Using the Taylor expansion:

$$A_\mu(j+\nu) = A_\mu(n) + a\partial_\nu A_\mu(n) + \mathcal{O}(a^2)\,, \tag{41.17}$$

and applying again Eq. (41.15), one can deduce the form of the plaquette in the continuum limit:

$$\mathcal{P}(n, \mu, \nu) = e^{\{ia^2g^2[G_{\mu\nu}(n)+\mathcal{O}(a)]\}}\,, \tag{41.18}$$

with the usual definition of the field tensor:

$$G_{\mu\nu}(x) \equiv G^a_{\mu\nu}\frac{\lambda_a}{2} = \partial_\mu A_\nu(x) - \partial_\nu A_\mu(x) - ig[A_\nu, A_\mu]\,. \tag{41.19}$$

In terms of the plaquette, one can now define a positive real and gauge-invariant action on the lattice:

$$\mathcal{S}_g = -\frac{1}{g^2}\sum_n\sum_{\mu<\nu}\mathbf{Tr}\{\mathcal{P}(n,\mu,\nu)+\mathcal{P}^\dagger(n,\mu,\nu)\}\,. \tag{41.20}$$

It is customary to express the action in terms of the variable:

$$\beta \equiv \frac{2N_c}{g^2}\,. \tag{41.21}$$

In the continuum limit, one can write:

$$\begin{aligned}\mathcal{S}_g &= -\frac{1}{g^2}\sum_n\sum_{\mu<\nu}2\mathrm{Re}\,\mathbf{Tr}\exp\{ia^2gG_{\mu\nu}(n)+\mathcal{O}(a)\}\\ &= -\frac{1}{g^2}\sum_n\sum_{\mu,\nu}\mathrm{Re}\,\mathbf{Tr}\left\{1+ia^2gG_{\mu\nu}(n)-\frac{1}{2}a^4g^2G_{\mu\nu}(n)G^{\mu\nu}(n)\right\}\,,\end{aligned} \tag{41.22}$$

where the sum over $\mu,\nu$ gets a factor $1/2$ because $\mu\nu$ and $\nu\mu$ define the same plaquette. Using the fact that $\mathrm{Tr}\lambda_a = 0$, one recovers the usual continuum action given in Eq. (41.4):

$$\mathcal{S}_g = \frac{1}{2}\sum_n\sum_{\mu,\nu}a^4G^{\mu\nu}(n)G_{\mu\nu}(n)+\mathcal{O}(a^6)+\text{constant}\,. \tag{41.23}$$

The vacuum expectation value of a function of the fields $F[U(n,n+\mu)]$ is:

$$\langle F[U(n,n+\mu)]\rangle = \frac{1}{\int \mathcal{D}U\, e^{\mathcal{S}_g}}\int \mathcal{D}U\, e^{\mathcal{S}_g}F[U(n,n+\mu)]\,, \tag{41.24}$$

where the invariant measure on the group attached to the link is:

$$\mathcal{D}U \equiv \prod_{n,\nu} dU(n,n+\nu)\,. \tag{41.25}$$

For an Abelian group, the measure is:

$$dU(n,n+\nu) = d(aA_\nu(n)) \qquad \text{with}: -\pi/a \leq A_\nu(n) \leq \pi/a\,. \tag{41.26}$$

For a non-Abelian $SU(N)_c$ group, one has:

$$dU(n,n+\nu) = \sqrt{\det\left[\frac{1-\cos aA_\nu(n)}{(aA_\nu(n))^2}\right]}\prod_{b=1}^{N_c^2-1} d\big(aA_\nu^b(n)\big)\,. \tag{41.27}$$

## 41.3 Quarks on the lattice

In this section, we turn to the less understood subject of the formulation of quarks (fermions) on the lattice, where the complications are already present at the free-field level. Since fermions obey Pauli exclusion principle, they are described at the classical level by anticommuting variables forming the so-called *Grassmann algebra*, which anticommute themselves but commute with complex numbers. To each lattice points, with coordinates $(n)$, are attached $N_c \times 8$ anticommuting quantities:

$$\psi_\alpha^c(n) , \qquad \bar{\psi}_\alpha^c(n) , \tag{41.28}$$

where the spinor $\alpha$ runs from 1 to 4, the colour index $c$ from 1 to $N_c$. The field transform as:

$$\psi_\alpha^c(n) \to U_{cc'}\psi_\alpha^{c'}(n) , \qquad \bar{\psi}_\alpha^c(n) \to U_{cc'}^{-1}\psi_\alpha^{c'}(n) , \tag{41.29}$$

Therefore terms like:

$$\bar{\psi}_\alpha^c(n)\psi_\alpha^c(n) , \qquad \bar{\psi}_\alpha^c(n+\mu)U_{cc'}(n+\mu, n)\psi_\alpha^{c'}(n) , \tag{41.30}$$

are gauge invariant. It is usual to start from the free continuum Lagrangian in Eq. (41.4):

$$\mathcal{L}_{\text{free}} = \bar{\psi}(x)(\partial\gamma_\mu + m)\psi(x) , \tag{41.31}$$

which possesses a $SU(n_f)_L \times SU(n_f)_R$ global symmetry in the massless limit $m = 0$. As in previous section, one introduces a four-dimensional hypercubic lattice of $N^4$ sites. To each site $n$, one associates an independent four-component spinor variable:

$$\psi_n \equiv \psi(an) \to \psi(x) \tag{41.32}$$

characterizing the quark fields. For simplifying the lattice action, one defines the derivative symetrically:

$$\partial_\mu\psi \to \frac{1}{2a}(\psi_{n+\mu} - \psi_{n-\mu}) . \tag{41.33}$$

Therefore, the lattice action reads:

$$\mathcal{S}_{\text{free}} = \sum_{n,k} \bar{\psi}_n M_{nk} \psi_k \tag{41.34}$$

with:

$$M_{nk} = \frac{1}{2}a^3 \sum_\mu \gamma_\mu(\delta_{k,n+\mu} - \delta_{k,n-\mu}) + a^4 m\delta_{nk} . \tag{41.35}$$

Now, one can put this action into a path integral:

$$Z_{\text{free}} = \int \mathcal{D}\psi\mathcal{D}\bar{\psi}\ e^{-\mathcal{S}} , \tag{41.36}$$

where:

$$\mathcal{D}\psi \equiv \prod_k d\psi_k , \tag{41.37}$$

after a relatively long, though straightforward manipulation, one finds:

$$Z_{\text{free}} = \frac{(-1)^{2N+1}}{(2N+1)!} \det M . \tag{41.38}$$

The quark propagator can be obtained by inverting $M$, which one can do with the help of a (finite) Fourier transform:

$$(M^{-1})_{nk} = a^{-4}(2N+1)^{-4} \sum_j \tilde{M}_j^{-1} \exp\left\{\frac{2i\pi}{2N+1} \sum_\mu j_\mu (n-k)_\mu\right\}, \quad (41.39)$$

and by using the relation:

$$\sum_{j_\mu=-N}^{N} \exp \frac{2i\pi}{2N+1} j_\mu (n-k)_\mu = (2N+1)\delta_{n_\mu k_\mu} . \quad (41.40)$$

Therefore, one finds:

$$S(j) = \tilde{M}_j^{-1} = \left(m + \frac{i}{a} \sum_\mu \gamma_\mu \sin \frac{2\pi j_\mu}{2N+1}\right)^{-1} . \quad (41.41)$$

In the case of a large lattice, one has:

$$\frac{2\pi j_\mu}{2N+1} \equiv a p_\mu , \quad (41.42)$$

which leads to the $p$-space propagator:

$$S(p) = \left(m + \frac{i}{a} \sum_\mu \gamma_\mu \sin a p_\mu\right)^{-1} . \quad (41.43)$$

Replacing the sum over $j$ by integrals:

$$\frac{1}{2N+1} \sum_{j_\mu=-N}^{N} \to a \int_{-\pi/a}^{+\pi/a} \frac{dp_\mu}{2\pi} , \quad (41.44)$$

Equation (41.39) becomes:

$$(M^{-1})_{nk} = \int_{-\pi/a}^{+\pi/a} \frac{d^4 p}{(2\pi)^4} \frac{e^{i \sum_\mu p^\mu (an-ak)_\mu}}{m + (i/a) \sum_\mu \gamma_\mu \sin a p_\mu} . \quad (41.45)$$

In the continuum limit ($a \to 0,\ an \to x,\ ak \to y$), this previous equation becomes:

$$(M^{-1})_{nk} \to S(x-y) = \int_{\infty}^{+\infty} \frac{d^4 p}{(2\pi)^4} \frac{e^{i \sum_\mu p^\mu (x-y)_\mu}}{m + i \sum_\mu \gamma_\mu p_\mu} , \quad (41.46)$$

which is the Euclidian propagator. However, by analysing Eq. (41.43), for example in the case $m = 0$, one can see that, for finite $a$, it has too many poles as the denominator vanishes for $p_\mu = 0$ and $p_\mu = \pi/a$. On the hypercube lattice, one has $2^4 = 16$ poles instead of one! This *fermion doubling* is catastrophic as one loses asymptotic freedom, the existence of the $U(1)$ anomaly (the 16 fermions contribute with alternate signs to the anomaly triangle),... Several solutions to this fermion doubling problem have been proposed in the literature [489]. One of the most popular is the one proposed by Wilson [492]. It consists of adding to the Lagrangian a quadratic term:

$$\mathcal{L}_q^W = m\bar{\psi}_n\psi_n + \frac{4r}{a}\bar{\psi}_n\psi_n + \frac{1}{2a}\sum_\mu\{(r+\gamma_\mu)\psi_{n+\mu} + (r-\gamma_\mu)\psi_{n-\mu}\}\,, \quad (41.47)$$

where $r$ is arbitrary. In the large lattice limit, the corresponding $p$-space propagator is:

$$S^W(p) = \left(m + \frac{1}{a}\sum_\mu\left[i\gamma_\mu \sin ap_\mu + \frac{r}{a}(1-\cos ap_\mu)\right]\right)^{-1}\,. \quad (41.48)$$

One can notice that for small momentum, the new term is of the order of $a$ and thus drops out. When a component $p$ is near $\pi/a$, the addition increases the mass of the unwanted state by $2r/a$:

$$m + \frac{r}{a}\sum_\nu(1-\cos ap_\nu) = m + \frac{2rn_\pi}{a}\,, \quad (41.49)$$

where the sum $\nu$ runs over $ap_\nu = \pi$, and $n_\pi$ is the number of extra particles. Therefore, in the continuum limit, all extra states have infinite mass and then decouple. Only one species of *physical particle* mass $m$ survives for $ap_\mu = 0$. However, it was shown [493] that the propagator in Eq. (41.48) breaks chiral invariance. One hopes that, working with Wilson fermions, one can recover chiral symmetry in the continuum limit.

### 41.4 Quark and gluon interactions

Now, one can formulate the quark and gluon interactions on the lattice. In the case of Abelian theory:

$$\begin{aligned}\mathcal{L}_{\text{free}} + \mathcal{L}_{A\psi} &= m\bar{\psi}_n\psi_n \\ &+ \frac{1}{2a}\bar{\psi}_n\sum_\mu\gamma_\mu[U(n,n+\mu)\psi_{n+\mu} - U(n-\mu,n)\psi_{n-\mu}]\,, \end{aligned}\quad (41.50)$$

which is invariant under the gauge transformation in Eq. (41.9) of the link matrices ($g\lambda_a/2 \equiv e$ electric charge). Using the expansions:

$$\lim_{a\to 0} U(n,n+\mu) = 1 - iagA_\mu + \mathcal{O}(a^2)\,, \quad (41.51)$$

and:

$$\lim_{a\to 0}\psi(n+\mu) = \psi(n) + a\partial_\mu\psi(n) + \mathcal{O}(a^2)\,, \quad (41.52)$$

it is easy to show that the previous Lagrangian gives the correct continuum limit:

$$\lim_{a\to 0}\{\mathcal{L}_{\text{free}} + \mathcal{L}_{A\psi}\} = m\bar{\psi}(x)\psi(x) + \bar{\psi}(x)\gamma^\mu\partial_\mu\psi(x) - ie\bar{\psi}(x)\gamma^\mu A_\mu\psi(x)\,. \quad (41.53)$$

In QCD, the interaction between quarks and gluons can be introduced as in the Abelian case. For Wilson fermions, the action reads:

$$\mathcal{S}_{gq} = a^4 \sum_n \bar{\psi}_n \left( m + \frac{4r}{a} \right) \psi_n + \frac{1}{2a} \sum_{n,\mu} \bar{\psi}_n [(r + \gamma_\mu) U(n, n+\mu) \psi_{n+\mu} + (r - \gamma_\mu) U^{-1}(n-\mu, n) \psi_{n-\mu}] . \tag{41.54}$$

The continuum limit of the action can also be obtained:

$$\lim_{a \to 0} \mathcal{S}_{gq} = a^4 \Bigg\{ m \sum_n \bar{\psi}_n \psi_n + \frac{1}{2} \sum_{n,\nu} [\bar{\psi}_n \gamma_\mu \partial^m u \psi_n - \partial^\mu \bar{\psi}_n \gamma_\mu \psi_n - ig \bar{\psi}_n \gamma_\mu \psi_n A^\mu] + \frac{r}{2a} \partial_\mu [\bar{\psi}_n \gamma^\mu \psi_n] \Bigg\} . \tag{41.55}$$

The last term vanishes after summation over $n$ (integration over $x$), such that the continuum limit reproduces the usual QCD action in Eq. (41.4). Therefore, the corresponding full generating functional for Wilson fermions is:

$$Z = \int \mathcal{D}U \, \mathcal{D}\psi \, \mathcal{D}\bar{\psi} \, e^{-(\mathcal{S}_g + \mathcal{S}_{gq})} , \tag{41.56}$$

where the measures have been defined in Eqs. (41.25) and (41.37). One should notice that unlike the continuum case, the gauge-fixing term is not necessary to obtain some vacuum expectation values (except the gluon propagator or some gauge-dependent quantities), as Eq. (41.56) averages over all gauges. In order to define the Green's functions, one has to define the integration over the Grassmann variables. which obey the following general properties:

$$\int d\eta_1 d\eta_2 \cdots d\eta_n \; (\eta_1 \eta_2 \cdots \eta_n) = 1, \text{ all other integrals are zero.} \tag{41.57}$$

For instance, one has:

$$\int d\eta_1 d\eta_2 \eta_1 \eta_2 = 1 = -\int d\eta_1 d\eta_2 \eta_2 \eta_1 , \qquad \int d\eta_1 d\eta_2 \eta_2 = 0 = \int d\eta_1 \eta_2 . \tag{41.58}$$

With the previous properties, any analytic function of the Grassmann variables can be integrated. This can be done by Taylor-expanding it and then by applying Eq. (41.57). For instance, the integral over the Grassmann algebra with four generators $\bar{\eta}_j, \eta_j, \; j = 1, 2$ reads:

$$\int \prod_{j=1}^{2} d\eta_j d\bar{\eta}_j \exp\left[ \sum_{i,j=1}^{2} \bar{\eta}_i A_{ij} \eta_j \right] = \int \prod_{j=1}^{2} d\eta_j d\bar{\eta}_j [\bar{\eta}_1 A_{11} \eta_1 \bar{\eta}_2 A_{22} \eta_2 + \bar{\eta}_1 A_{12} \eta_2 \bar{\eta}_2 A_{21} \eta_1] = A_{11} A_{22} - A_{12} A_{21} = \det A . \tag{41.59}$$

Collecting the different results in the previous section, the vacuum expectation value of a function of the gauge and fermion fields is defined as:

$$\langle F[A_m u^a(p), \bar{\psi}(n)\psi(k)]\rangle = \frac{1}{\mathcal{N}}\int \mathcal{D}U\mathcal{D}\bar{\psi}^c_\alpha(n)\mathcal{D}\psi^c_\alpha(n)F[A_m u^a(p), \bar{\psi}(n)\psi(k)]e^{-\mathcal{S}_{\text{latt}}} \tag{41.60}$$

with:

$$\mathcal{N} = \int \mathcal{D}U\mathcal{D}\bar{\psi}^c_\alpha(n)\mathcal{D}\psi^c_\alpha(n)e^{-\mathcal{S}_{\text{latt}}} : \qquad \mathcal{S}_{\text{latt}} = \mathcal{S}_g + \mathcal{S}_{gq} . \tag{41.61}$$

From the lattice action, one can, for example, derive different Feynman rules on the lattice. For example, the propagators can be obtained from the quadratic terms of the fields entering into the action. The quark propagator has been already given in the previous section (see e.g. Eq. (41.48) for the Wilson fermion). In the Feynman gauge, the gluon propagator is:

$$D^{cb}_{\mu\nu}(p) = \delta^{cb}\delta_{\mu\nu}\frac{1}{2a^{-2}\sum_\rho(1-\cos ap_\rho)} . \tag{41.62}$$

Feynman rules for the vertices are more involved as the interactions are non-polynomial functions of the fields, and there are infinite numbers of vertices associated with higher powers of the lattice spacing $a$. More discussions can be found in [494].

## 41.5 Some applications of the lattice

A large spectrum of the lattice applications can be found in the different references given in the introduction of this chapter. Here, we shall limit with very few examples as an illustration of the method.

### *41.5.1 The QCD coupling and the weak coupling regime*

We have noticed that for finite $a$, QCD on the lattice is UV finite, such that we do not worry to distinguish between bare and renormalized quantities. Hower, for $a \to 0$, loop diagrams become divergent in the *weak coupling limit*, and the lattice can be considered as a regularization procedure with the cut-off $1/a \to \infty$. To leading order of pQCD, the QCD coupling reads:

$$g^2(a) = \frac{4\pi^2}{\beta_1 \log \Lambda_{\text{latt}} a} . \tag{41.63}$$

The scale $\Lambda_{\text{latt}}$ can be related to the one of the $\overline{MS}$ scheme by simply evaluating one-loop renormalization for $\alpha_s$, including constant terms using the two different schemes and by equating. The lattice calculation has been done in [495] but is quite cumbersome due to the peculiarity of the lattice regularization (Lorentz invariance, ...). For $n_f = 0$ fermions, one obtains to one loop:

$$\Lambda_{\rm latt} \simeq \frac{\Lambda^0_{\rm mom}}{83.5} \simeq \frac{\Lambda^0_{\overline{MS}}}{39} . \tag{41.64}$$

The present values are (see previous chapters and [16]):

$$\Lambda^0_{\overline{MS}} \approx 400 \text{ MeV} \Longrightarrow \Lambda_{\rm latt} \approx 10 \text{ MeV} , \tag{41.65}$$

showing that $\Lambda_{\rm latt}$ has a very small value. From Eq. (41.63), one can also derive the leading-order relation between the lattice spacing $a$ and $\Lambda_{\rm latt}$:

$$a = \Lambda^{-1}_{\rm latt} e^{\frac{4\pi^2}{\beta_1 g^2(a)}} \tag{41.66}$$

valid for small $a$ and for weak coupling:

$$a\Lambda_{\rm latt},\ g^2(a) \ll 1 . \tag{41.67}$$

In order to check if one has reached the continuum limit from the numerical analysis, one should see if the lattice results behave as predicted by the renormalization group equation.

### *41.5.2 Wilson loop, confinement and the strong coupling regime*

Here one considers the Green's function of a pair of a static infinitely heavy ($m \to \infty$) quark and anti-quark at lattice points $j$ and $j + n\mu$. A gauge-invariant function of such a state is given by:

$$J(k) = \bar\psi_k U(k, k+\nu) \cdots U(k + (n-1)\nu, k + n\nu)\psi(k + n\nu) . \tag{41.68}$$

Its propagation in the Euclidian space–time is described by the Green's function:

$$G(k, l) = \langle J(l)^\dagger J(k) \rangle , \tag{41.69}$$

where the lattice point $l$ is displaced with respect to $k$ by $r$ units in four-direction. Since, in the action, the fermionic variables $\bar\psi\psi$ occur quadratically, hence the integration is Gaussian, such that the integration over the fermion fields will not pose (in principle) any problem. It is possible to show that for $m \to \infty$, the Green's function behaves as:

$$G(k, l) \sim \left(\frac{p}{m}\right)^{2n} \langle Tr W[L] \rangle_U \ : \qquad p \equiv a^3/2 \tag{41.70}$$

where $W(L)$ is the rectangular Wilson loop with corners $k,\ k+\nu,\ l,\ l+\nu$, and $\langle \ldots \rangle_U$ corresponds to the vacuum expectation value in Eq. (41.24) over the gauge field $U$. One can sketch the derivation of this result by considering the integration over the fermion fields at the point $k$. In the integrand one has from Eq. (41.68) the term $\bar\psi(k)$. The integral will not vanish if one has an additional factor $\psi(k)$, which one can obtain by expanding the action $e^{-S_{\rm latt}}$. This expansion leads, among others, to the term: $p\bar\psi(k+\mu)(r + \gamma_\mu)U^{-1}(k, k+\mu)\psi(k)$ . After fermion integration at the point $k$, the fermion field $\bar\psi(k)$ is no longer present, but now we have a fermion field at the position $k + \mu$ and the previous factor $p \ldots$ . We thus hopped with the fermion field from $k$ to $k+\mu$, such that we may hop from $k$ to $j$, and from $j+\nu$ to $k+\nu$. For other points, we need to expand the mass term in $e^{-S_{\text{latt}}}$, which yields a factor $m$ for each points. The final factor $(p/m)^{2n}$ and the group elements $U$ attached to the links of the loop are obtained after dividing by the normalization factor $\mathcal{N}$.

On the other hand, one knows that the Wilson loop measures the response of the gauge fields to an external quark-like source passing around its perimeter. For a timelike loop, this represents the production of a quark pair at the earliest time, moving them along the world lines dictated by the sides of the loop, and then annihilating at the latest time. If the loop is a rectangle of dimensions $T$ and $R$, a transfer matrix argument suggests that for large $T$:

$$\lim_{T\to\infty} \langle W[L]\rangle_U = -\exp[-E(R)T] \tag{41.71}$$

where $E(R)$ is the static quark–anti-quark energy separated by a distance $R$. In the *strong coupling regime* $1/g^2 \to 0$, one obtains to leading order:

$$\langle W[L]\rangle_U \sim \left(\frac{1}{g}\right)^{RT/a^2} , \tag{41.72}$$

showing that in that approximation the static energy of the quarks increases linearly with the spatial distance $R$:

$$\lim_{R\to\infty} E(R) = \sigma R , \tag{41.73}$$

where $\sigma$ is called *the string tension* and characterizes long-distance physics effects. Therefore a separation of the two quarks would need infinite energy. Unfortunately, this result is also obtained for Abelian theory. Since we do not observe confinement in QED, we have to assume that there is a phase transition between the confining phase in the strong-coupling regime and the deconfined phase in the weak-coupling regime. There is no formal proof that such a transition does not exist in non-Abelian QCD. A numerical evaluation of the expectation value $\langle W[L]\rangle_U$ indicates that the area law in Eq. (41.72) is also verified for weak coupling, strongly indicating that *confinement is a consequence of the QCD-Lagrangian*. Phenomenologically, the string tension can be related to the slope of the Regge trajectory if one uses a string model for describing the hadrons [496]:

$$\alpha' = (2\pi\sigma)^{-1} , \tag{41.74}$$

where using the phenomenological value $\alpha' \simeq 1\,\text{GeV}^{-2}$, one finds:

$$\sigma \simeq (400\ \text{MeV})^2 . \tag{41.75}$$

Using the previous equations, one can notice that this quantity is proportional to the QCD coupling $g^2$, i.e. $\Lambda_{\text{lattice}}$. This is a remarkable feature as one is able to relate a long-distance ($\sigma$) to a short-distance ($\Lambda_{\text{latt}}$) quantities.

### 41.5.3 Some other applications and limitations of the lattice

Some observables like hadron masses, ... can also be obtained by calculating numerically Green's functions of interpolating fields on the lattice. In so doing, let us consider the vector current:

$$J_\mu(x) = \bar{\psi}(k)\gamma_u\psi(k)\,, \tag{41.76}$$

which has the quantum number of the $\rho$-meson. After a rotation in the Euclidian space–time, the two-point correlator reads (we omit indices for simplicity):

$$\Pi(T) \equiv \langle J(T)J(0)\rangle = \langle J(0)e^{-HT}J(0)\rangle \tag{41.77}$$

Inserting a complete set of energy eigenstates and taking the large $T$ limit, one may select the lowest ground state $\rho$-meson contribution:

$$\Pi(T) = \sum_n |\langle J(0)|n\rangle|^2 e^{-E_nT} \to |\langle J(0)|0\rangle|^2 e^{-E_0T}\ , \tag{41.78}$$

where $E_0$ is equal to the $\rho$-meson mass $M_\rho$. In this way, one can recover the whole hadron spectrum, ... However, in practice, there are many difficulties and questions which the *lattice experimentalists* should clearly answer. Besides the usual statistical and finite size (about 1% if the lattice size $L \geq 3$ fermi, and $m_\pi L \geq 6$) errors inherent to the numerical lattice calculations, which can be minimized using modern technology, there are still large uncertainties related to the uses of field theory on the lattice.

- When one approximates the functional integral by a product of Riemann integrals, when do we reach the continuum limit ? The renormalization group analysis shows that one should expect an exponential dependence of the lattice spacing on the coupling constant. This can be reached if the lattice spacing $a$ is relatively small like the coupling $g$.
- However, if the lattice spacing $a$ is small say a fraction of a fermi, the lattice should be large enough in order to accommodate a hadron of a typical size of one fermi. Therefore, the lattice should at least have $4 \times 10^4$ lattice points. Since for $SU(3)$, we have, for each lattice point, eight groups of integrations and 24 fermionic integrations, it is clear that one needs very sophisticated integraltion methods. However, even with these sophisticated integration methods, one has to do some approximations, as an exact evaluation of the fermionic integrals are not possible with most of the present computers.
- In the case of (*quenched approximation*), one ignores quark loops, thus simplifying the evaluation of the integral, but with a brutal non-inclusion of the fermion determinant into the action. This implies a modification of chiral symmetry ($\chi S$) for $m_q = 0$ as well as the disappearance of the QCD anomaly: $M_{\eta'} \approx m_\pi$. At present, some progress towards including active quark flavours has been achieved by some groups.
- Another obstacle is the small values of the light quark masses. Generally, one evaluates the Green's functions at large mass and then extrapolates the results to zero quark mass values with the help of the mass dependence expected from chiral perturbation theory (ChPT) (see next section). For a typical value of the lattice spacing $1/a \simeq 2$ GeV, and keeping the condition $m_\pi L \geq 6$, one requires $L/a \geq 90$ in order to avoid finite volume effects. At present the lattice size $L/a$ is about 32 (quenched) and about 24 (unquenched) which is far below this limit.

- There are also discretization errors specific to each lattice actions, which are $\mathcal{O}(a)$ for the Wilson (explicit breaking of $\chi S$) and domain walls (extra fifth dimension for preserving $\chi S$) actions. The errors are $\mathcal{O}(a^2)$ for the staggered (reduction of quark couplings with high-momenta gluons) and $\mathcal{O}(a\alpha_s)$ for the Clover (inclusion of the mixed quark-gluon operator) actions.
- There are also errors due to the mixing of different operators at finite $a$.
- How good is the separation of the ground state from the rest of the spectra in the large Euclidian time limit if the mass splitting between the ground state and the first radial excitation is accidentally small?

The list of difficulties which we have given is not exhaustive but lattice experts know all of them completely. These difficulties will have to be resolved before reliable lattice results on the hadron and QCD parameters, will be available. We hope that such difficulties can be solved gradually in the future. However, it is unfortunate that most non-lattice experts and especially experimentalists *blindly* use the present lattice results without asking about their reliability, although this is, however, difficult to quantify by non-experts in the field. Some lattice results will be presented in subsequent chapters as a comparison with the QCD spectral-sum rules results.

# 42
# Chiral perturbation theory

## 42.1 Introduction

In the general introduction of this book, we have discussed that, below the vector meson resonances region ($E \leq M_\rho$), the hadronic spectrum of light flavours only consists of an octet of quasi-Goldstone pseudoscalar mesons ($\pi, K, \eta$), whose interactions can be easily understood using the global symmetry of the QCD Lagrangian. In the limit of massless quarks, the QCD Lagrangian is invariant under the rotations of the left and right quark fields triplets:

$$\psi_L \equiv \frac{1}{2}(1-\gamma_5)\psi\,, \qquad \psi_R \equiv \frac{1}{2}(1+\gamma_5)\psi\,, \qquad \psi \equiv u, d, s\,. \tag{42.1}$$

These rotations generate the chiral group $SU(3)_L \times SU(3)_R$, which at the level of hadronic spectrum is broken down to the diagonal flavour $SU(3)_V$ ($V \equiv L + R$) group of the eightfoldway [7]. Thc Goldstonc bosons are associated to the spontaneous breakdown of chiral symmetry and obey low-energy theorems which are the basis of successful predictions of current algebra and pion PCAC [13]. Since there is a mass gap separating the Goldstone bosons from the rest of the hadronic spectrum, one can build an effective field theory including the symmetry of QCD where the Goldstone bosons are the only dynamic degrees of freedom [497]. This allows to a systematic analysis of the low-energy implications of the QCD symmetries which simplifies current algebra calculations and allows an investigation of higher-order corrections in the sense of perturbative field theory [498]. This approach is known as chiral perturbation theory (ChPT), which is a low-energy effective field theory of QCD, where many excellent reviews and lectures have been devoted to the subject [500–502]. Our presentation has been mainly inspired from the reviews in [500,501] and the works of Gasser–Leutwyler [499].

A well-known example of effective theories is the low-energy limit of QED ($E_\gamma \ll m_e$). In this limit the $\gamma\gamma$ scattering process can be described by the effective Euler–Heisenberg Lagrangian:

$$\mathcal{L}_{\text{eff}} = -\frac{1}{4}F_{\mu\nu}(x)F^{\mu\nu}(x) + \frac{A}{m_e^4}(F_{\mu\nu}(x)F^{\mu\nu}(x))^2 + \frac{B}{m_e^4}F_{\mu\nu}(x)F^{\nu\sigma}(x)F_{\sigma\rho}(x)F^{\rho\mu}(x) + \cdots \tag{42.2}$$

which is only based on the gauge, Lorentz and parity invariance conditions. The coefficients $A$ and $B$ are known and can be computed by integrating out the electron field from the original QED generating functional, or equivalently by computing the corresponding $\gamma\gamma$ box diagram. They read [503]:

$$A = -\frac{\alpha^2}{36}, \qquad B = \frac{7}{90}\alpha^2 . \tag{42.3}$$

However, this QED example is academic since perturbation theory in terms of the QED coupling $\alpha$ is known to work at high accuracy. In QCD, due to confinement which induces that quark and gluon are not asymptotic states, the effective approach is more useful as we know the symmetry properties of QCD, from which we can write the effective theory in terms of hadronic asymptotic states, and parametrize the unknown dynamics of the theory in terms of some few couplings.

In the following discussions, we shall limit ourselves to the presentation of the main idea behind the method and illustrate its applications for the estimate of the light quark mass ratios.

## 42.2 PCAC relation from ChPT

One can also derive the previous PCAC relation ontained in Part I of this book using ChPT. In this approach, it is convenient to formulate the strong interactions of the pseudoscalar mesons in terms of an effective low-energy QCD Lagrangian described by the octet of Goldstone fields:

$$\phi(x) = \frac{1}{\sqrt{2}}\vec{\lambda}.\vec{\varphi}(x) = \begin{pmatrix} \frac{\pi^0}{\sqrt{2}} + \frac{\eta}{\sqrt{6}} & \pi^+ & K^+ \\ \pi^- & -\frac{\pi^0}{\sqrt{2}} + \frac{\eta}{\sqrt{6}} & K^0 \\ K^- & \bar{K}^0 & -\frac{2}{\sqrt{6}}\eta \end{pmatrix}, \tag{42.4}$$

instead of in terms of the usual quark and gluon fields. The associated $3 \times 3$ unitary matrix:

$$U(\phi) = \exp(i\sqrt{2}\phi/f_\pi), \tag{42.5}$$

transforms linearly under the global chiral rotations, although $\vec{\varphi}$ transforms non-linearly. The unique lowest order (in derivative) effective Lagrangian, satisfying chiral symmetry and generating non-trivial interaction is:

$$\mathcal{L} = \frac{f^2}{4}\mathbf{Tr}\{\partial_\mu U^\dagger \partial^\mu U\}, \tag{42.6}$$

where $f$ is a constant which cannot be fixed by symmetry requirements alone. Expanding $U(\phi)$ in a power series of $\phi$, the Lagrangian reads:

$$\mathcal{L} = \frac{1}{2}\mathbf{Tr}\{\partial_\mu\phi\partial^\mu\phi\} + \frac{1}{12f^2}\mathbf{Tr}\{(\phi\partial_\mu\phi)(\phi\partial^\mu\phi)\} + \mathcal{O}\left(\frac{\phi^6}{f^4}\right), \tag{42.7}$$

where one should note that the $\phi^4$ interaction fixes the $\pi$–$\pi$ scattering amplitude [504]:

$$T(\pi^+\pi^0 \to \pi^+\pi^0) = \frac{t}{f^2} \tag{42.8}$$

where $t \equiv (p'_+ - p_+)^2$ is the usual kinematic variable. Now, one can go to a step further by introducing the couplings of external sources to the usual massless QCD Lagrangian:

$$\mathcal{L}_{\rm QCD}(x) = \mathcal{L}_{\rm QCD}^{\rm massless}(x) + \bar\psi\gamma^\mu(v_\mu + \gamma_5 a_\mu)\psi\bar\psi\gamma^\mu\,(s - i\gamma_5)\,\psi\ , \tag{42.9}$$

where $v_\mu,\ a_\mu,\ s$ and $p$ are Hermitian $3\times 3$ matrices in flavour and colour singlets. The Lagrangian $\mathcal{L}$ is now invariant under the *local* $SU(3)_L \times SU(3)_R$ gauge transformations. The generalized effective Lagrangian satisfying the local invariance reads to lowest order:

$$\mathcal{L}_{\rm eff}^{(2)} = \frac{f^2}{4}\mathbf{Tr}\{D_\mu U^\dagger D^\mu U\} + U^\dagger\chi + \chi^\dagger U\ , \tag{42.10}$$

where $D_\mu$ is the covariant derivative:

$$D_\mu U = \partial_\mu U - i(v_\mu + a_\mu)U + iU(v_\mu - a_\mu) \tag{42.11}$$

and:

$$\chi = 2B(s + ip)\ . \tag{42.12}$$

$B$ is a constant which, like $f$, cannot be fixed by symmetry requirements alone. With the choice of directions:

$$\begin{aligned} s + ip &= \mathcal{M} + \cdots \\ r_\mu &= v_\mu + a_\mu = eQA_\mu + \cdots \\ l_\mu &= v_\mu - a_\mu = eQA_\mu + \frac{e}{\sqrt{2}\sin\theta_W}(W_\mu^\dagger T_+ + \text{h.c.}) + \cdots\ , \end{aligned} \tag{42.13}$$

where $A_\mu$ and $W_\mu$ are the photon and $W^-$ bosons,

$$\mathcal{M} = \text{diag}(m_u, m_d, m_s), \qquad Q = \frac{1}{3}\text{diag}(2, -1, -1)\ , \tag{42.14}$$

and:

$$T_+ = \begin{pmatrix} 0 & V_{ud} & V_{us} \\ 0 & 0 & 0 \\ 0 & 0 & 0 \end{pmatrix}, \tag{42.15}$$

one can break chiral symmetry explicitly and select the electroweak standard model couplings. The Green functions are obtained as functional derivatives of the generating functional:

$$\exp\{iZ\} = \int \mathcal{D}\psi\mathcal{D}\bar\psi\mathcal{D}A_\mu \exp\left\{i\int d^4x\ \mathcal{L}_{\rm QCD}\right\} = \int \mathcal{D}\exp\left\{i\int d^4x\ \mathcal{L}_{\rm eff}\right\}. \tag{42.16}$$

At lowest order in momenta, the generating functional reduces to the classical action:

$$S_2 = \int d^4x \, \mathcal{L}_{\text{eff}}^{(2)}(x) \, . \tag{42.17}$$

The Noether currents can be derived by taking appropriate derivatives with respect to the external fields:

$$\begin{aligned} J_L^\mu &= \frac{\delta S_2}{\delta l_\mu} = \frac{i}{2} f^2 D_\mu U^\dagger U = \frac{f}{\sqrt{2}} D_\mu \phi - \frac{i}{2}(\phi \vec{D}^\mu \phi) + \cdots \\ J_R^\mu &= \frac{\delta S_2}{\delta r_\mu} = \frac{i}{2} f^2 D_\mu U U^\dagger = -\frac{f}{\sqrt{2}} D_\mu \phi - \frac{i}{2}(\phi \vec{D}^\mu \phi) + \cdots \, , \end{aligned} \tag{42.18}$$

which shows the indentification of the coupling $f$ with the decay constant $f_\pi = 92.4$ MeV to order $p^2$:

$$\langle 0 | J_A^\mu | \pi \rangle \equiv i\sqrt{2} f_\pi p^\mu \, . \tag{42.19}$$

In a similar way:

$$\begin{aligned} \bar{\psi}_L^i \psi_R^j &= -\frac{\delta S_2}{\delta (s - ip)^{ij}} = -\frac{f^2}{2} B \, U^{ij} \\ \bar{\psi}_R^i \psi_L^j &= -\frac{\delta S_2}{\delta (s + ip)^{ij}} = -\frac{f^2}{2} B (U^{ij})^\dagger \, , \end{aligned} \tag{42.20}$$

which implies:

$$\langle 0 | \bar{\psi}^j \psi^i | 0 \rangle = -f^2 B \delta^{ij} \, , \tag{42.21}$$

By taking $s = \mathcal{M}$ and $p = 0$, the $\chi$ term in Eq. (42.10) gives a quadratic pseudoscalar mass plus additional interactions proportional to the quark mass. Expanding in powers of $\phi$, one obtains:

$$\frac{f^2}{4} 2B \, \mathbf{Tr}\{\mathcal{M}(U + U^\dagger)\} = B \left\{ -\mathbf{Tr}\,(\mathcal{M}\phi^2) + \frac{1}{6f^2} \mathbf{Tr}\,(\mathcal{M}\phi^4) + \cdots \right\} . \tag{42.22}$$

An explicit evaluation of the trace in the quadratic mass term provides:

$$\begin{aligned} M_{\pi^\pm}^2 &= (m_u + m_d) B + \mathcal{O}\big(m_q^2\big) \, , \\ M_{\pi^0}^2 &= (m_u + m_d) B - \epsilon + \mathcal{O}\big(\epsilon^2, m_q^2\big) \, , \\ M_{K^+}^2 &= (m_u + m_s) B + \mathcal{O}\big(m_q^2\big) \, , \\ M_{K^0}^2 &= (m_d + m_s) B + \mathcal{O}\big(m_q^2\big) \, , \\ M_{\eta_8}^2 &= \frac{1}{3} (m_u + m_d + 4m_s) \, B + \epsilon + \mathcal{O}\big(\epsilon^2, m_q^2\big) \, , \end{aligned} \tag{42.23}$$

where:

$$\epsilon = \frac{B}{4} \frac{(m_u - m_d)^2}{(m_s - \hat{m})} \, , \qquad \hat{m} = \frac{1}{2}(m_u + m_d), \tag{42.24}$$

originates from the small mixing between the $\pi^0$ and $\eta_8$ fields. Previous relations explain why the masses of the multiplet break strongly explicitly the eightfoldway symmetry because $m_s \gg m_d > m_u$. Using also these results in Eqs. (42.21) and (42.23), one can deduce the pion PCAC relation given in Part I of this book, namely:

$$(m_u + m_d)\langle \bar{u}u + \bar{d}d \rangle = -2 f_\pi^2 m_\pi^2 \,. \tag{42.25}$$

However, there is no rigorous evidence on the dominance of this linear quark mass term over the quadratic one in the previous relation in Eq. (42.23) leading to the previous PCAC relation where the quark mass is a quadratic function of the pseudoscalar mass. Some alternative scenario (*so-called Generalized ChPT*), where the value of the $\langle\bar{\psi}\psi\rangle$ condensate is smaller than the 'standard' value, is discussed in the literature [505]. We might expect that lattice calculations will clarify this issue in the near future, and at present, there are some lattice indications that $M_P^2$ behaves like $m_q$ [506]. We shall see in the next section that direct extractions of the light quark masses from QCD spectral sum rules also favour the result that $m_q \sim M_P^2$.

## 42.3 Current algebra quark mass ratios

The ratios of the expressions in Eq. (42.23) imply the old current algebra mass ratios [21],[55–57]:

$$\frac{M_{\pi^\pm}^2}{(m_u + m_d)} = \frac{M_{K^+}^2}{(m_u + m_s)} = \frac{M_{K^0}^2}{(m_d + m_s)} \approx \frac{3M_{\eta_8}^2}{(m_u + m_d + 4m_s)} \,, \tag{42.26}$$

while the estimate of their absolute values needs more QCD theoretical inputs (renormalization and scale dependence). Neglecting the $m^2$ [1] and small $\mathcal{O}(\epsilon)$ corrections, one can deduce the mass ratios [55]:

$$\begin{aligned} \frac{m_u}{m_d} &\approx \frac{M_{\pi^+}^2 - M_{K^0}^2 + M_{K^+}^2}{M_{\pi^+}^2 + M_{K^0}^2 - M_{K^+}^2} \approx 0.66 \\ \frac{m_s}{m_d} &\approx \frac{-M_{\pi^+}^2 + M_{K^0}^2 + M_{K^+}^2}{M_{\pi^+}^2 + M_{K^0}^2 - M_{K^+}^2} \approx 20 \,, \end{aligned} \tag{42.27}$$

where the electromagnetic part of the $K^+ - K^0$ squared mass-difference has been subtracted by using the fact that it is the same for the $K^+$ and $\pi^+$ [507]:

$$(M_{K^0}^2 - M_{K^+}^2)_{\rm QCD} \simeq (M_{K^0}^2 - M_{K^+}^2) - (M_{\pi^0}^2 - M_{\pi^+}^2) \,. \tag{42.28}$$

Up to order $(m_d - m_u)$, one can also derive the quadratic Gell-Mann–Okubo mass relation [11]:[2]

$$3M_{\eta_8}^2 \approx 4M_K^2 - M_\pi^2 \,. \tag{42.29}$$

[1] This is not justified in the approach of [505].

[2] Analogous GMO mass formula for vector mesons might be affected by large perturbative $m_s^2$ corrections [32].

One should also note that the $\phi^4$ interaction in Eq. (42.22) gives a mass correction to the $\pi$–$\pi$ scattering amplitude given in Eq. (42.8):

$$T(\pi^+\pi^0 \to \pi^+\pi^0) = \frac{t - M_\pi^2}{f_\pi^2}\,, \tag{42.30}$$

in good agreement with the current algebra result [504].

## 42.4 Chiral perturbation theory to order $p^4$

Improvements of this lowest order effective Lagrangian with the inclusion of $p^4$- and $p^6$-terms are actively discussed in the literature [502]. To order $p^4$, three different sources contribute to the generating functional:

- The most general effective Lagrangian $\mathcal{L}_{\text{eff}}^{(4)}$ to order $p^4$ to be considered at the tree level.
- The one-loop graphs generated from the lowest order $\mathcal{L}_{\text{eff}}^{(2)}$ Lagrangian.
- The Wess–Zumino–Witten functional [508,509] induced by the non-Abelian chiral anomaly [510].

### *42.4.1 The chiral Lagrangian to order* $(p^4)$

The most general expression of the $\mathcal{O}(p^4)$ Lagrangian is:

$$\begin{aligned}
\mathcal{L}_{\text{eff}}^{(4)} &= L_1\,\mathbf{Tr}\,(D_\mu U^\dagger D^\mu U)^2 + L_2\,\mathbf{Tr} D_\mu U^\dagger D_\nu U\,\mathbf{Tr} D^\mu U^\dagger D^\nu U \\
&+ L_3\,\mathbf{Tr} D_\mu U^\dagger D^\mu U D_\nu U^\dagger D^\nu U \\
&+ L_4\,\mathbf{Tr} D_\mu U^\dagger D^\mu U\,\mathbf{Tr}\,(\chi^\dagger U + U^\dagger \chi) + L_5\,\mathbf{Tr} D_\mu U^\dagger D^\mu U\,(\chi^\dagger U + U^\dagger \chi) \\
&+ L_6\,[\mathbf{Tr}\,(\chi^\dagger \mathcal{U} + U^\dagger \chi)]^2 + L_7\,[\mathbf{Tr}\,(\chi^\dagger U - U^\dagger \chi)]^2 \\
&+ L_8\,\mathbf{Tr}(U\chi^\dagger U\chi^\dagger + U^\dagger \chi U^\dagger \chi) \\
&+ iL_9\,\mathbf{Tr}\left(F_R^{\mu\nu} D_\mu U D_\nu U^\dagger + F_L^{\mu\nu} D_\mu U^\dagger D_\nu U\right) + L_{10}\,\mathbf{Tr} U^\dagger F_R^{\mu\nu} U F_{L\mu\nu} \\
&+ H_1\,\mathbf{Tr}\left(F_R^{\mu\nu} F_{R\mu\nu} + F_L^{\mu\nu} F_{L\mu\nu}\right) + H_2\,\mathbf{Tr}\chi^\dagger \chi\,.
\end{aligned} \tag{42.31}$$

In this Lagrangian the parameters $L_i$, $i = 1, 2, 3, \ldots, 10$ are dimensionless coupling constants, which like $f_\pi$ and $B$ in the lowest order effective Lagrangian, are not fixed by chiral symmetry requirements alone. The terms proportional to the coupling constants $H_1$ and $H_2$ involve only the external fields. As a result these coupling constants cannot be fixed from low-energy observables alone. By contrast, most of the other couplings can be fixed from low-energy observables. The $L_i$ constants, like $f_\pi$ and $B$, are *in principle* calculable parameters in terms of the intrinsic $\Lambda_{\text{QCD}}$ scale only.

### *42.4.2 Chiral loops*

Here, we consider that ChPT is an effective field theory for low energies despite the fact that a simple power counting shows that loops generated by the lowest order Lagrangian are highly divergent as a consequence of the fact that the non-linear sigma model in four-dimensions is not renormalizable and then needs an infinite number of local counterterms. In order to define the loop integrals it is necessary to fix a regularization that preserves the symmetries of the Lagrangian, which can be done by using the well-known dimensional regularization technique. Since by construction, the $\mathcal{O}(p^4)$ Lagrangian $\mathcal{L}_{\rm eff}^{(4)}$ contains all possible terms which are allowed by chiral invariance, all the one-loop divergences from $\mathcal{L}_{\rm eff}^{(2)}$ can be absorbed by suitable renormalizations of the $L_i$ and $H_{1,2}$ constants. This feature can be understood by power counting where one-loop divergences can only give rise to local $\mathcal{O}(p^4)$ terms. This program has been explicitly realized by Gasser and Leutwyler in [498], and leads to the renormalized low-energy couplings:

$$L_i = L_i^{\rm R}(\nu) + \gamma_i\,\lambda^{\rm loop}, \quad i = 1, 2, 3, \ldots 10; \qquad H_i = H_i^{\rm R}(\nu) + \tilde{\gamma}_j\,\lambda^{\rm loop}, \quad j = 1, 2, \tag{42.32}$$

where for $n = 4 - \epsilon$ space-time dimension:

$$\lambda^{\rm loop} = \frac{\nu^{-\epsilon}}{16\pi^2}\left\{-\frac{1}{\epsilon} - \frac{1}{2}[\log(4\pi) + \Gamma'(1) + 1]\right\}, \quad j = 1, 2; \tag{42.33}$$

and $\gamma_i$, $x\tilde{\gamma}_j$ have the following rational values:

$$\gamma_1 = \frac{3}{32}, \quad \gamma_2 = \frac{3}{16}, \quad \gamma_3 = 0, \quad \gamma_4 = \frac{1}{8},$$
$$\gamma_5 = \frac{3}{8}, \quad \gamma_6 = \frac{11}{144}, \quad \gamma_7 = 0, \quad \gamma_8 = \frac{5}{48},$$
$$\gamma_9 = \frac{1}{4}, \quad \gamma_{10} = -\frac{1}{4}, \quad \tilde{\gamma}_1 = -\frac{1}{8}, \quad \tilde{\gamma}_2 = \frac{5}{24}. \tag{42.34}$$

The renormalized coupling constants depend as usual on the scale $\nu$ introduced by the dimensional regularization. The running in $\nu$ is governed by the coefficients $\gamma_i$ (and $\tilde{\gamma}_j$), which play the rôle of one–loop $\beta$–functions:

$$L_i^{\rm r}(\nu) = L_i^{\rm r}(\nu') + \frac{\gamma_i}{16\pi^2}\log\frac{\nu'}{\nu}. \tag{42.35}$$

The $\nu$–scale dependence cancels however in the full $\mathcal{O}(p^4)$ calculation of a given physical observable. The non-polynomial contribution to a specific physical process will in general have a logarithmic $\nu$–scale dependence (*the so called chiral logarithms*), which cancels with the $\nu$–dependence of the tree level contribution modulated by the $L_i(\nu)$–constants. A typical $\mathcal{O}(p^4)$ amplitude will then consist of a non-polynomial part, coming from the loop computation, plus a polynomial in momenta and pseudoscalar masses, which depends on the unknown constants $L_i$. The non-polynomial part (the so-called chiral logarithms) is completely predicted as a function of the lowest-order coupling $f$ and the Goldstone masses.

Finally, it is important to notice that ChPT is an expansion in powers of momenta over some typical hadronic scale, usually called the scale of chiral symmetry breaking $\Lambda_\chi$.

The variation of the loop contribution under a rescaling of $\mu$ provides a natural order-of-magnitude estimate[3] of $\Lambda_\chi$ [498,512] :

$$\Lambda_\chi \sim 4\pi f_\pi \sim 1.2\,\text{GeV} \approx M_\rho \approx M_p\,. \tag{42.36}$$

This result has been recovered from the analysis of the connection between the low- and high-energy behaviours of the pion form factor [281].

### 42.4.3 The non-Abelian chiral anomaly

Although the QCD Lagrangian with external sources is formally invariant under local chiral transformations, this is no longer true for the associated generating functional. The anomalies of the fermionic determinant break chiral symmetry at the quantum level. The anomalous change of the generating functional under an infinitesimal chiral transformation:

$$g_{L,R} = 1 + i\alpha \mp i\beta + \cdots \tag{42.37}$$

is given by [510]:

$$\delta Z[v, a, s, p] = -\frac{N_c}{16\pi^2}\int d^4x\, \text{tr}\beta(x)\,\Omega(x), \tag{42.38}$$

where:

$$\begin{aligned}\Omega(x) = \varepsilon^{\mu\nu\sigma\rho}\Big[ & v_{\mu\nu}v_{\sigma\rho} + \frac{4}{3}\nabla_\mu a_\nu \nabla_\sigma a_\rho + \frac{2}{3}i\,\{v_{\mu\nu}, a_\sigma a_\rho\} \\ & + \frac{8}{3}i\, a_\sigma v_{\mu\nu} a_\rho + \frac{4}{3}\, a_\mu a_\nu a_\sigma a_\rho\Big]\,, \qquad \varepsilon_{0123} = 1;\end{aligned} \tag{42.39}$$

and:

$$v_{\mu\nu} = \partial_\mu v_\nu - \partial_\nu v_\mu - i\,[v_\mu, v_\nu], \qquad \nabla_\mu a_\nu = \partial_\mu a_\nu - i\,[v_\mu, a_\nu] \tag{42.40}$$

This anomalous variation of $Z$ is an $\mathcal{O}(p^4)$ effect in the chiral counting. Chiral symmetry is the basic requirement to construct the effective $\chi$PT Lagrangian. Since chiral symmetry is explicitly violated by the anomaly at the fundamental QCD level, one is forced to add an effective functional with the property that its change under chiral gauge transformations reproduces Eq. (42.38). Such a functional was first constructed by Wess and Zumino [508]. An interesting topological interpretation was later found by Witten [509]. The functional in question, has the following explicit form:

$$\begin{aligned}\Gamma[U, \ell, r]_{WZW} = & -\frac{iN_c}{240\pi^2}\int d\sigma^{ijklm}\, \mathbf{Tr}\left\{\Sigma_i^L \Sigma_j^L \Sigma_k^L \Sigma_l^L \Sigma_m^L\right\} \\ & -\frac{iN_c}{48\pi^2}\int d^4x\, \varepsilon_{\mu\nu\alpha\beta}\big(W(U, \ell, r)^{\mu\nu\alpha\beta} - W(\mathbf{1}, \ell, r)^{\mu\nu\alpha\beta}\big)\,,\end{aligned} \tag{42.41}$$

[3] Since the loop amplitude increases with the number of possible Goldstone mesons in the internal lines, this estimate results in a slight dependence of $\Lambda_\chi$ on the number of light-quark flavours $N_f$ [511]: $\Lambda_\chi \sim 4\pi f_\pi/\sqrt{N_f}$.

with:

$$W(U,\ell,r)_{\mu\nu\alpha\beta} = \mathbf{Tr}\Big\{U\ell_\mu\ell_\nu\ell_\alpha U^\dagger r_\beta + \frac{1}{4}U\ell_\mu U^\dagger r_\nu U\ell_\alpha U^\dagger r_\beta + iU\partial_\mu\ell_\nu\ell_\alpha U^\dagger r_\beta$$
$$+ i\partial_\mu r_\nu U\ell_\alpha U^\dagger r_\beta - i\Sigma^L_\mu \ell_\nu U^\dagger r_\alpha U\ell_\beta + \Sigma^L_\mu U^\dagger\partial_\nu r_\alpha U\ell_\beta$$
$$- \Sigma^L_\mu\Sigma^L_\nu U^\dagger r_\alpha U\ell_\beta + \Sigma^L_\mu\ell_\nu\partial_\alpha\ell_\beta + \Sigma^L_\mu\partial_\nu\ell_\alpha\ell_\beta$$
$$- i\,\Sigma^L_\mu\ell_\nu\ell_\alpha\ell_\beta + \frac{1}{2}\Sigma^L_\mu\ell_\nu\Sigma^L_\alpha\ell_\beta - i\Sigma^L_\mu\Sigma^L_\nu\Sigma^L_\alpha\ell_\beta\Big\}$$
$$- (L \leftrightarrow R)\,, \tag{42.42}$$

where:

$$\Sigma^L_\mu = U^\dagger\partial_\mu U\,, \qquad \Sigma^R_\mu = U\partial_\mu U^\dagger\,, \tag{42.43}$$

and $(L \leftrightarrow R)$ stands for the interchanges $U \leftrightarrow U^\dagger$, $\ell_\mu \leftrightarrow r_\mu$ and $\Sigma^L_\mu \leftrightarrow \Sigma^R_\mu$. The integration in the first term of Eq. (42.41) is over a five-dimensional manifold whose boundary is four-dimensional Minkowski space. The integrand is a surface term; therefore both the first and the second terms of $\Gamma_{WZW}$ are $\mathcal{O}(p^4)$ according to the chiral counting rules.

Since the effect of anomalies is perturbatively calculable, their translation from the fundamental quark-gluon level to the effective chiral level is unaffected by hadronization problems. Despite its apparent complexity, the anomalous action [Eq. (42.41)] has no free parameters. It is responsible for the $\pi^0 \to 2\gamma$, $\eta \to 2\gamma$ decays, and the $\gamma 3\pi$, $\gamma\pi^+\pi^-\eta$ interactions among others. The five-dimensional surface term generates interactions among five or more Goldstone bosons.

## 42.5 Some low-energy phenomenology to order $p^4$

At lowest order in momenta, the predictive power of the chiral Lagrangian was quite impressive; with only two low-energy couplings, it was possible to describe all Green functions associated with the pseudoscalar-meson interactions, and to reproduce all old current algebra results [13]. The symmetry constraints become less powerful at higher orders. Ten additional constants appear in the $\mathcal{L}_4$ Lagrangian, and many more would be needed at $\mathcal{O}(p^6)$.

Higher-order terms in the chiral expansion are much more sensitive to the non-trivial aspects of the underlying QCD dynamics. With $p \ll M_K$ $(M_\pi)$, we expect $\mathcal{O}(p^4)$ corrections to the lowest-order amplitudes at the level of $p^2/\Lambda_\chi^2 \leq 20\%$ (2%). We need to include those corrections if we aim to increase the accuracy of the ChPT predictions beyond this level. Although the number of free constants in $\mathcal{L}_4$ looks quite big, only a few of them contribute to a given observable. In the absence of external fields, for instance, the Lagrangian reduces to the first three terms; elastic $\pi\pi$ and $\pi K$ scatterings are then sensitive to $L_{1,2,3}$. The two-derivative couplings $L_{4,5}$ generate mass corrections to the meson decay constants (and mass-dependent wave-function renormalizations). Pseudoscalar masses are affected by the non-derivative terms $L_{6,7,8}$; $L_9$ is mainly responsible for the charged-meson electromagnetic radius and $L_{10}$, finally, only contributes to amplitudes with at least two external vector or

Table 42.1. *Phenomenological values of the renormalized couplings* $L_i^r(M_\rho)$.

| $i$ | $L_i^r(M_\rho) \times 10^3$ | Source |
|---|---|---|
| 1 | $0.7 \pm 0.5$ | $K_{e4}$, $\pi\pi \to \pi\pi$ |
| 2 | $1.2 \pm 0.4$ | $K_{e4}$, $\pi\pi \to \pi\pi$ |
| 3 | $-3.6 \pm 1.3$ | $K_{e4}$, $\pi\pi \to \pi\pi$ |
| 4 | $-0.3 \pm 0.5$ | Zweig rule |
| 5 | $1.4 \pm 0.5$ | $F_K : F_\pi$ |
| 6 | $-0.2 \pm 0.3$ | Zweig rule |
| 7 | $-0.4 \pm 0.2$ | Gell-Mann–Okubo, $L_5$, $L_8$, sum rules |
| 8 | $0.9 \pm 0.3$ | $M_{K^0} - M_{K^+}$, $L_5$, $(m_s - \hat{m}) : (m_d - m_u)$ |
| 9 | $6.9 \pm 0.7$ | $\langle r^2 \rangle^\pi_{\rm em}$ |
| 10 | $-5.5 \pm 0.7$ | $\pi \to e\nu\gamma$ |

axial-vector fields, like the radiative semi-leptonic decay $\pi \to e\nu\gamma$. Table 42.1 summarizes the present status of the phenomenological determination of the renormalized constants $L_i$ [499,502], evaluated at a scale $\mu = M_\rho$. The values of these couplings at any other renormalization scale can be trivially obtained, through the logarithmic running given in Eq. (42.35).

### 42.5.1 Decay constants

In the isospin limit ($m_u = m_d = \hat{m}$), the $\mathcal{O}(p^4)$ calculation of the meson-decay constants gives [499]:

$$\begin{aligned} f_\pi &= f\left\{1 - 2\mu_\pi - \mu_K + \frac{4M_\pi^2}{f^2} L_5^r(\mu) + \frac{8M_K^2 + 4M_\pi^2}{f^2} L_4^r(\mu)\right\}, \\ f_K &= f\left\{1 - \frac{3}{4}\mu_\pi - \frac{3}{2}\mu_K - \frac{3}{4}\mu_{\eta_8} + \frac{4M_K^2}{f^2} L_5^r(\mu) + \frac{8M_K^2 + 4M_\pi^2}{f^2} L_4^r(\mu)\right\}, \\ f_{\eta_8} &= f\left\{1 - 3\mu_K + \frac{4M_{\eta_8}^2}{f^2} L_5^r(\mu) + \frac{8M_K^2 + 4M_\pi^2}{f^2} L_4^r(\mu)\right\}, \end{aligned} \tag{42.44}$$

where:

$$\mu_P \equiv \frac{M_P^2}{32\pi^2 f^2} \log\left(\frac{M_P^2}{\mu^2}\right). \tag{42.45}$$

The result depends on two $\mathcal{O}(p^4)$ couplings, $L_4$ and $L_5$. The $L_4$ term generates a universal shift of all meson-decay constants, $\delta f^2 = 16 L_4 B \mathbf{Tr}\mathcal{M}$, which can be eliminated taking ratios. From the experimental value [513]:

$$\frac{f_K}{f_\pi} = 1.22 \pm 0.01 \; , \tag{42.46}$$

one can then fix $L_5(\mu)$; this gives the result quoted in Table 42.1. Moreover, one gets the absolute prediction [499]:

$$\frac{f_{\eta_8}}{f_\pi} = 1.3 \pm 0.05 \; . \tag{42.47}$$

Taking into account isospin violations, one can also predict [499] a tiny difference between $f_{K^\pm}$ and $f_{K^0}$, proportional to $m_d - m_u$.

### 42.5.2 Electromagnetic form factors

At $\mathcal{O}(p^2)$ the electromagnetic coupling of the Goldstone bosons is just the minimal one, obtained through the covariant derivative. The next-order corrections generate a momentum-dependent form factor:

$$F_V^{\phi^\pm}(p^2) = 1 + \frac{1}{6}\langle r^2\rangle_V^{\phi^\pm} p^2 + \cdots \quad ; \quad F_V^{\phi^0}(p^2) = \frac{1}{6}\langle r^2\rangle_V^{\phi^0} p^2 + \cdots \tag{42.48}$$

The pion electromagnetic radius $\langle r^2\rangle_V^{\phi}$ gets local contributions from the $L_9$ term, plus logarithmic loop corrections [499]:

$$\langle r^2\rangle_V^{\pi^\pm} = \frac{12 L_9^r(\mu)}{f^2} - \frac{1}{32\pi^2 f^2}\left\{2\log\left(\frac{M_\pi^2}{\mu^2}\right) + \log\left(\frac{M_K^2}{\mu^2}\right) + 3\right\} \tag{42.49}$$

The measured electromagnetic pion radius, $\langle r^2\rangle_V^{\pi^\pm} = 0.439 \pm 0.008\,\text{fm}^2$ [514], is used as input to estimate the coupling $L_9$.

- The factor $1/(16\pi^2 f^2)$ is a characterisitc factor of a loop-expansion, where chiral logs are expected to contribute as $p^2/(16\pi^2 f^2)\log$ in physical processes.
- The form factor provides a good example of the importance of higher-order local terms in the chiral expansion [515]. If one tries to ignore the $L_9$ contribution, using instead some *physical* cut-off $p_{\max}$ to regularize the loops, one needs an unrealistic value $p_{\max} \sim 60\,\text{GeV}$, in order to reproduce the experimental value. This fact shows that the pion charge radius is dominated by the $L_9^r(\mu)$ contribution, for any reasonable value of $\mu$, which can be better understood from a $1/N_c$ (number of colour) counting rules, where for large $N_c$, $L_9$ and $f_\pi^2$ are order $N_c$, implying that the chiral loops are $1/N_c$ suppressed compared to the tree level contributions.
- The phenomenological value of dimensionless couplig $L_9$ might be understood as originating from the $f_\pi^2/4$ factor from $\mathcal{L}_{\text{eff}}^{(4)}$ divided by the chiral symmetry breaking scale $\Lambda_\chi^2$, which leads to the order of magnitude value of about $10^{-3}$; an expected value for all other $L_i$ couplings as found experimentally in Table 42.1.

The kaon electromagnetic radius reads:

$$\langle r^2\rangle_V^{K^0} = -\frac{1}{16\pi^2 f^2}\log\left(\frac{M_K}{M_\pi}\right), \tag{42.50}$$

$$\langle r^2\rangle_V^{K^\pm} = \langle r^2\rangle_V^{\pi^\pm} + \langle r^2\rangle_V^{K^0}. \tag{42.51}$$

Since neutral bosons do not couple to the photon at tree level, $\langle r^2\rangle_V^{K^0}$ only gets a loop contribution, which is moreover finite (there cannot be any divergence because there exists no counterterm to renormalize it). The predicted value:

$$\langle r^2\rangle_V^{K^0} = -0.04 \pm 0.03\,\mathrm{fm}^2, \tag{42.52}$$

is in perfect agreement with the experimental determination [516]

$$\langle r^2\rangle_V^{K^0} = -0.054 \pm 0.026\,\mathrm{fm}^2. \tag{42.53}$$

The measured $K^+$ charge radius [517]:

$$\langle r^2\rangle_V^{K^\pm} = 0.28 \pm 0.07\,\mathrm{fm}^2, \tag{42.54}$$

has a larger experimental uncertainty. Within present errors, it is in agreement with the parameter-free relation in Eq. (42.51).

### 42.5.3 $K_{l3}$ decays

The semi-leptonic decays $K^+ \to \pi^0 l^+ \nu_l$ and $K^0 \to \pi^- l^+ \nu_l$ are governed by the corresponding hadronic matrix elements of the vector current $[t \equiv (P_K - P_\pi)^2]$:

$$\langle \pi | \bar{s}\gamma^\mu u | K\rangle = C_{K\pi}\,[(P_K + P_\pi)^\mu\, f_+^{K\pi}(t) + (P_K - P_\pi)^\mu\, f_-^{K\pi}(t)], \tag{42.55}$$

where $C_{K^+\pi^0} = 1/\sqrt{2}$, $C_{K^0\pi^-} = 1$. At lowest order, the two form factors reduce to trivial constants: $f_+^{K\pi}(t) = 1$ and $f_-^{K\pi}(t) = 0$. There is however a sizeable correction to $f_+^{K^+\pi^0}(t)$, due to $\pi^0\eta$ mixing, which is proportional to $(m_d - m_u)$:

$$f_+^{K^+\pi^0}(0) = 1 + \frac{3}{4}\,\frac{m_d - m_u}{m_s - \hat{m}} = 1.017. \tag{42.56}$$

This number should be compared with the experimental ratio:

$$\frac{f_+^{K^+\pi^0}(0)}{f_+^{K^0\pi^-}(0)} = 1.028 \pm 0.010. \tag{42.57}$$

The $\mathcal{O}(p^4)$ corrections to $f_+^{K\pi}(0)$ can be expressed in a parameter-free manner in terms of the physical meson masses [499]. Including those contributions, one gets the more precise values:

$$f_+^{K^0\pi^-}(0) = 0.977, \qquad \frac{f_+^{K^+\pi^0}(0)}{f_+^{K^0\pi^-}(0)} = 1.022, \tag{42.58}$$

which are in perfect agreement with the experimental result of Eq. (42.57). The accurate ChPT calculation of these quantities allows us to extract [513] the most precise determination of the Cabibbo–Kobayashi–Maskawa matrix element $V_{us}$:

$$|V_{us}| = 0.2196 \pm 0.0023 \,. \tag{42.59}$$

At $\mathcal{O}(p^4)$, the form factors get momentum-dependent contributions. Since $L_9$ is the only unknown chiral coupling occurring in $f_+^{K\pi}(t)$ at this order, the slope $\lambda_+$ of this form factor can be fully predicted:

$$\lambda_+ \equiv \frac{1}{6}\,\langle r^2\rangle_V^{K\pi}\, M_\pi^2 = 0.031 \pm 0.003 \,. \tag{42.60}$$

This number is in excellent agreement with the experimental determinations [16], $\lambda_+ = 0.0300 \pm 0.0016$ ($K^0_{e3}$) and $\lambda_+ = 0.0286 \pm 0.0022$ ($K^\pm_{e3}$). Contrary to this case, the experimental determination of the slope of the form factor $f_0^{K\pi}$ is still controversial. It is predicted to be [499]:

$$\lambda_0 \equiv \frac{1}{6}\,\langle r^2\rangle_S^{K\pi}\, M_\pi^2 = 0.017 \pm 0.004 \,, \tag{42.61}$$

and is determined by the constant $L_5$.

### 42.5.4 Ratios of light quark masses to order $p^4$

Ratios of light quark masses to this order have been discussed in details in [57]. Here, we outline the different derivations of the results obtained there. The relations in Eq. (42.23) get modified at $\mathcal{O}(p^4)$. The additional contributions depend on the low-energy constants $L_4$, $L_5$, $L_6$, $L_7$ and $L_8$. It is possible, however, to obtain one relation between the quark and meson masses, which does not contain any of the $\mathcal{O}(p^4)$ couplings. The dimensionless ratios

$$Q_1 \equiv \frac{M_K^2}{M_\pi^2}\,, \qquad Q_2 \equiv \frac{\left(M_{K^0}^2 - M_{K^+}^2\right)_{\rm QCD}}{M_K^2 - M_\pi^2}\,, \tag{42.62}$$

get the same $\mathcal{O}(p^4)$ correction [499]:

$$Q_1 = \frac{m_s + \hat{m}}{2\hat{m}}\,\{1 + \Delta_m\}\,, \qquad Q_2 = \frac{m_d - m_u}{m_s - \hat{m}}\,\{1 + \Delta_m\}\,, \tag{42.63}$$

where

$$\Delta_m = -\mu_\pi + \mu_{\eta_8} + \frac{8}{f^2}\left(M_K^2 - M_\pi^2\right)\left[2L_8^r(\mu) - L_5^r(\mu)\right] \,. \tag{42.64}$$

Therefore, at this order, the ratio $Q_1/Q_2$ is just given by the corresponding ratio of quark masses,

$$Q^2 \equiv \frac{Q_1}{Q_2} = \frac{m_s^2 - \hat{m}^2}{m_d^2 - m_u^2}\,. \tag{42.65}$$

where $Q^2 = 22.7 \pm 0.8$ using the value of the $\eta \to \pi^+\pi^-\pi^0$ decay rate from the PDG average [16], though this value can well be in the range 22–26, to be compared with the Dashen's formula [507] value of 24.2 including next-to-leading chiral corrections [518];

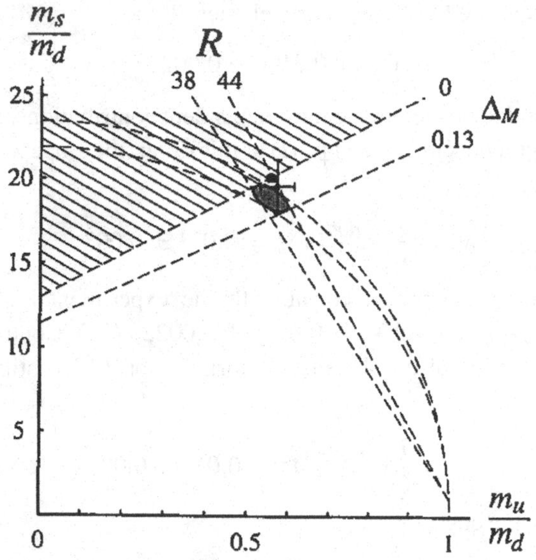


Fig. 42.1. $m_s/m_d$ versus $m_u/m_d$ from [57].

$\hat{m} \equiv (1/2)(m_u + m_d)$. To a good approximation, Eq. (42.65) constrains the quark-mass ratios to be on the ellipse,

$$\left(\frac{m_u}{m_d}\right)^2 + \frac{1}{Q^2}\left(\frac{m_s}{m_d}\right)^2 = 1\,, \tag{42.66}$$

In Fig. 42.1, one shows the range spanned by the corrections to the GMO mass formula:

$$\Delta_M :\ M_{\eta_8}^2 = (1/3)\big(4M_K^2 - M_\pi^2\big)(1 + \Delta_M)\,, \tag{42.67}$$

where to order $p^4$, one has:

$$\Delta_M \equiv \left(\frac{M_8^2 - M_\pi^2}{4M_K^2 - M_\pi^2}\right)\Delta_{\rm GMO}\,, \tag{42.68}$$

with:

$$\Delta_{\rm GMO} \equiv \frac{4M_K^2 - 3M_{\eta_8}^2 - M_\pi^2}{M_{\eta_8}^2 - M_\pi^2}\,. \tag{42.69}$$

Neglecting the mass difference $m_d - m_u$, one gets [499]

$$\begin{aligned}\Delta_{\rm GMO} =&\ \frac{-2\left(4M_K^2\mu_K - 3M_{\eta_8}^2\mu_{\eta_8} - M_\pi^2\mu_\pi\right)}{M_{\eta_8}^2 - M_\pi^2}\\ &- \frac{6}{f^2}\left(M_{\eta_8}^2 - M_\pi^2\right)\left[12L_7^r(\mu) + 6L_8^r(\mu) - L_5^r(\mu)\right]\,.\end{aligned} \tag{42.70}$$

Experimentally, correcting the masses for electromagnetic effects, one obtains:

$$\Delta_{\rm GMO} = 0.21. \tag{42.71}$$

Since $L_5$ is already known, this allows the combination $2L_7 + L_8$ to be fixed. However, in order to determine the individual quark-mass ratios from Eqs. (42.63), we would need to fix the constant $L_8$. However, there is no way to find an observable that isolates this coupling. The reason is an accidental symmetry of the Lagrangian $\mathcal{L}_2 + \mathcal{L}_4$, which remains invariant under the following simultaneous change [519] of the quark-mass matrix and some of the chiral couplings:

$$\begin{aligned} \mathcal{M}' &= \alpha\,\mathcal{M} + \beta\,(\mathcal{M}^\dagger)^{-1}\det\mathcal{M}\,, \qquad B_0' = B_0/\alpha\,, \\ L_6' &= L_6 - \zeta\,, \qquad L_7' = L_7 - \zeta\,, \qquad L_8' = L_8 + 2\zeta\,, \end{aligned} \tag{42.72}$$

where $\alpha$ and $\beta$ are arbitrary constants, and $\zeta = \beta f^2/(32\alpha B_0)$. The only information on the quark-mass matrix $\mathcal{M}$ that we used to construct the effective Lagrangian was that it transforms as $\mathcal{M} \to g_R \mathcal{M} g_L^\dagger$.

The matrix $\mathcal{M}'$ transforms in the same manner; therefore, symmetry alone does not allow us to distinguish between $\mathcal{M}$ and $\mathcal{M}'$. In order to resolve this ambiguity, additional information outside the framework of the pseudoscalar meson chiral Lagrangian has been used, by the introduction of the ratio:

$$R \equiv (m_s - \hat{m})/(m_d - m_u)\,. \tag{42.73}$$

Its value comes from the analysis of isospin breaking in the $\omega - \rho$ mixing and from the baryon spectrum [499]. At the intersection of different ranges, one deduces from Fig. 42.1:

$$\begin{aligned} \frac{m_u}{m_d} &= 0.553 \pm 0.043\,, \qquad \frac{m_s}{m_d} = 18.9 \pm 0.8\,, \\ \frac{2m_s}{(m_d + m_u)} &= 24.4 \pm 1.5\,. \end{aligned} \tag{42.74}$$

However, the possibility to have a $m_u = 0$ advocated in [519], where chiral symmetry can be still broken by, for example, instantons, appears to be unlikely as it implies too strong flavour symmetry breaking and is not supported by the QSSR results from two-point correlators of the divergences of the axial and vector currents, as will be shown in the following chapters.

# 43

# Models of the QCD effective action

## 43.1 Introduction

Our purpose is to briefly present the general features of different models of the low-energy hadronic interactions based on the effective action of QCD using a well-defined set of approximations. In this chapter, we shall follow closely the discussions in [500]. The chiral symmetry of the underlying QCD theory implies that the generating functional $\Gamma(v, a, s, p)$ of the Green's functions of quark currents:

$$e^{i\Gamma(v,a,s,p)} = \frac{1}{Z}\int \mathcal{D}G_\mu \det \not{D} \exp\left(-i\int d^4x \, \frac{1}{4}\, \vec{G}_{\mu\nu}\vec{G}^{\mu\nu}\right) , \tag{43.1}$$

with: $\not{D}$ the Dirac operator

$$\not{D} = \gamma^\mu(\partial_\mu + ig_s G_\mu) - i\gamma^\mu(v_\mu + \gamma_5 a_\mu) + i(s - i\gamma_5 p) ; \tag{43.2}$$

$G_\mu$ is the gluon field, $\vec{G}^{\mu\nu}$ the gluon field strength tensor; and $v_\mu, a_\mu, s, p$ external field sources; the normalization factor $Z$ is such that $\Gamma(0, 0) = 1$, admits a low-energy representation:

$$e^{i\Gamma(v,a,s,p)} = \frac{1}{Z}\int \mathcal{D}U \, \exp\left[i\int d^4x \, \mathcal{L}_{\rm eff}(U; v, a, s, p)\right] , \tag{43.3}$$

in terms of an effective Lagrangian $\mathcal{L}_{\rm eff}(U; v, a, s, p)$ with $U(x)$ a $3\times 3$ unitary matrix containing the octet of pseudoscalar fields $(\pi, K, \eta)$. However, the single term in $\mathcal{L}_{\rm eff}$ which is known from first principles, is the one associated with the existence of anomalies in the fermionic determinant [510]. The corresponding effective action is the Wess and Zumino [508,509] functional that we have discussed in the previous section. All possible other terms in $\mathcal{L}_{\rm eff}$, are not fixed by symmetry requirements alone. The desire is to build some effective dynamical QCD models with a mimimum set of parameters that can fix the different coupling constants of the effective chiral Lagrangian, and that are needed for making progress in the phenomenology of non-leptonic flavour dynamics. In the following, we shall list the following models:

- QCD in the large–$N_c$ limit.
- Low-lying resonances dominance models.
- The constituent chiral quark model.

- Effective action approach models.
- The extended Nambu and Jona-Lasinio Model (ENJL model.)

## 43.2 QCD in the large–$N_c$ limit

### 43.2.1 Large $N_c$ counting rules for mesons

The study of QCD in the limit of large $N_c$ was suggested by t'Hooft [520], soon after the discovery of asymptotic freedom, as an attempt to get an insight into the non-perturbative aspects of QCD. The large $N_c$ limit of QCD corresponds to the case where the number of colours is sent to infinity and the QCD coupling $\alpha_s$ sent to zero in such a way that:

$$N_c\alpha_s = \text{constant}\,. \tag{43.4}$$

Therefore the Green's function of the theory is proportional to a power of $N_c$ [520–522]. Denoting by $\mathcal{G}_{qw}$ the general connected Green's function containing $q$ quark currents and $w$ winding number densities:

$$\mathcal{G}_{qw} = \langle 0|\mathcal{T} J_1(x-1)\cdots J_q(x_q)Q(y_1)\cdots Q(y_w)|0\rangle_{\text{connect}} \tag{43.5}$$

with:

$$J_i = \bar{\psi}\Gamma_i\psi\,, \qquad Q(x) = \frac{g^2}{8\pi^2}\mathbf{Tr}\,(G_{\mu\nu}\tilde{G}^{\mu\nu})\,, \tag{43.6}$$

where $\Gamma_i$ is neutral colour matrices acting on the spin and quark flavours. For large $N_c$, the Green's function behaves as:

$$\begin{aligned}\mathcal{G}_{qw} &= \mathcal{O}\big(N_c^{2-w}\big)\,, \qquad q=0\\ &= \mathcal{O}\big(N_c^{1-w}\big)\,, \qquad q\neq 0\,.\end{aligned} \tag{43.7}$$

This counting rule holds only for generic momenta, but is modified by, for example, the exchange of an $\eta'$ pole, which at zero momentum produces an additionnal power of $N_c$ ($M^2_{\eta'} \sim 1/N_c$ in the chiral limit). This counting rule can be understood in the following way: the leading contributions to the Green's functions containing quark currents ($q \neq 0$) arise from graphs with a single quark loop (planar diagrams with the quark loop running at the edge of the diagram). These graphs are given by the functional integral over the gluon field of the product of the form $\mathbf{Tr}\left(\Gamma_{i_1}S\Gamma_{i_2}S\ldots\Gamma_{i_q}S\right)$, where $i_1,\ldots i_q$ is some permutation of $1,\ldots,q$ and where $S$ denotes the quark propagator in the presence of the gluon field. In the chiral limit, the propagator is flavour independent, and the leading contribution to the Green's function depends on the flavour of the current only through the trace $\mathbf{Tr}\left(\lambda_{i_1},\ldots,\lambda_{i_q}\right)$ where $\lambda_i$ is the flavour factor in the matrix $\Gamma_i$. From Eq. (43.7), one can deduce the large-$N_c$ behaviour of the generating functional:

$$Z(v,a,s,p,\theta) = N_c^2 f_0(\theta/N_c) + N_c f_1(v,a,s,p,\theta/N_c) + \mathcal{O}(1)\,, \tag{43.8}$$

where the functional $f_0(\alpha)$ and $f_1(v,a,s,p,\alpha)$ are independent of $N_c$. One can deduce the counting rule for one particle matrix elements [522]:

$$\langle 0|J|\text{meson}\rangle = \mathcal{O}(N_c^{1/2}), \qquad \langle 0|J|\text{glueball}\rangle = \mathcal{O}(1),$$
$$\langle 0|Q|\text{meson}\rangle = \mathcal{O}(N_c^{-1/2}), \qquad \langle 0|Q|\text{glueball}\rangle = \mathcal{O}(1). \tag{43.9}$$

Every additional meson in a vertex brings a suppression factor $1/N_c^{-1/2}$. Therefore, three-meson amplitudes are of order $1/N_c^{-1/2}$, four-meson amplitudes are of order $1/N_c, \ldots$. Loop corrections in the meson sector are suppressed by powers of $1/N_c$, and are consistent with a semiclassical expansion in powers of $\hbar$.

### 43.2.2 Chiral Lagrangian in the large $N_c$-limit

It would be a major breakthrough, if one could derive the low-energy effective Lagrangian of the interactions between Nambu–Goldstone modes in the large-$N_c$ limit of QCD. To analyse the large $N_c$ behaviour of the effective Lagrangian, it suffices to expand the matrix in terms of the meson fields and to look at the terms independent of these fields. The desired results are obtained by comparing these terms with those in Eq. (43.8). As examples, one obtains:

$$f = \mathcal{O}(N_c^{1/2}), \qquad B = \gamma = \mathcal{O}(1), \tag{43.10}$$

where $\gamma$ quantifies the $\eta - \eta'$ mixing. For the non-vanishing coupling constants, one has obtained the large $N_c$ behaviour [499]:

$$\begin{aligned} \mathcal{O}(N_c^2) &: L_7, \\ \mathcal{O}(N_c) &: L_1, L_2, L_3, L_5, L_8, L_9, L_{10}, H_1, H_2, \\ \mathcal{O}(1) &: 2L_1 - L_2, L_4, L_6. \end{aligned} \tag{43.11}$$

So far, it has only been possible to obtain *constraints* among various coupling constants in this limit; but not their *values* in terms, say, of $\Lambda_{\text{QCD}}$. A typical example is the relation:

$$2L_1 = L_2, \tag{43.12}$$

which, as first noticed by Gasser and Leutwyler [499], follows in the large-$N_c$ limit of QCD. Unfortunately, nobody can claim as yet to be able to *compute*, say $L_2$, in that limit. Often in the literature, there appear statements about 'large-$N_c$ predictions' but, in fact, they have been all derived with some extra *ad hoc* assumptions.

An interesting approach to do approximate calculations within the framework of the $1/N_c$-expansion is the one proposed by Bardeen, Buras and Gérard [524], which they have applied extensively to the calculation of non-leptonic weak matrix elements. The basic idea is to start with the factorized form of the four-quark operators in the effective weak Hamiltonian, and to do one-loop chiral perturbation theory, keeping track of the quadratic divergences which appear. If one was able to work with the *full* hadronic low-energy effective Lagrangian, it would be possible to obtain a smooth matching between the scale dependence of the Wilson coefficients, calculated at short distances, and the hadronic matrix elements calculated with the *full* hadronic low-energy effective Lagrangian. The hope with the approach proposed by [524] is that the *numerical* matching of the *quadratic* long-distance scale with the *logarithmic* short-distance scale, may turn out to be already a good first approximation to the problem one would like to solve. The technology of their approach is explained with detail in their papers.

### 43.2.3 Minimal hadronic ansatz to large $N_c$ QCD

The hadronic spectrum predicted by large $N_c$–QCD seems a priori different from the *real world*, as one expects here the presence of an infinite sum of narrow resonances with specific quantum numbers [520]. This feature can be better understood from the Coleman–Witten theorem [525] which states that if QCD at $N_c = 3$ is confined, and if confinement persists for large $N_c$, then, in this limit, the chiral $U(n_f) \times U(n_f)$ invariance of the QCD Lagrangian with $n_f$ massless flavours is spontaneously broken down to the diagonal $U(n_f)$ subgroup. Though, the real world has a much more complicated structure, one expects that the hadronic world predicted by large $N_c$ can give an approximate good prediction of this real world, when observables in terms of spectral functions are involved, as in this case, one needs only to know the *global properties of the hadronic spectrum.*

*The left–right correlation function*

Of particular interest for our purposes is the correlation function ($Q^2 \equiv -q^2 \geq 0$ for $q^2$ space-like):

$$\Pi^{\mu\nu}_{LR}(q) = 2i \int d^4x\, e^{iq\cdot x} \langle 0|\mathrm{T}(L^\mu(x)R^\nu(0)^\dagger)|0\rangle\,, \tag{43.13}$$

with colour singlet currents:

$$R^\mu\ (L^\mu) = \bar{d}(x)\gamma^\mu \frac{1}{2}(1 \pm \gamma_5)u(x)\,. \tag{43.14}$$

In the chiral limit, ($m_{u,d,s} \rightarrow 0$,) this correlation function has only a transverse component

$$\Pi^{\mu\nu}_{LR}(Q^2) = (q^\mu q^\nu - g^{\mu\nu}q^2)\Pi_{LR}(Q^2)\,. \tag{43.15}$$

The self-energy like function $\Pi_{LR}(Q^2)$ vanishes order by order in perturbative QCD (pQCD) and is an order parameter of ChSB for all values of $Q^2$; therefore it obeys an unsubtracted dispersion relation

$$\Pi_{LR}(Q^2) = \int_0^\infty dt \frac{1}{t+Q^2}\frac{1}{\pi}\mathrm{Im}\Pi_{LR}(t)\,. \tag{43.16}$$

In large $N_c$–QCD, the spectral function $\frac{1}{\pi}\mathrm{Im}\Pi_{LR}(t)$ consists of the difference of an infinite number of narrow vector and axial-vector states, together with the Goldstone pole of the pion:

$$\frac{1}{\pi}\mathrm{Im}\Pi_{LR}(t) = \sum_V f_V^2 M_V^2 \delta(t - M_V^2) - F_0^2\delta(t) - \sum_A f_A^2 M_A^2 \delta(t - M_A^2) . \quad (43.17)$$

The low $Q^2$ behaviour of $\Pi_{LR}(Q^2)$, namely the long-distance behaviour of the correlation function in Eq. (43.13), is governed by chiral perturbation theory:

$$-Q^2\Pi_{LR}(Q^2)|_{Q^2\to 0} = f_0^2 + 4L_{10}Q^2 + \mathcal{O}(Q^4) , \quad (43.18)$$

where $f_0$ is the pion coupling constant in the chiral limit, and $L_{10}$ is one of the coupling constants of the $\mathcal{O}(p^4)$ effective chiral Lagrangian. The high $Q^2$ behaviour of $\Pi_{LR}(Q^2)$, that is, the short-distance behaviour of the correlation function in Eq. (43.13), is governed by the operator product expansion (OPE) of the two local currents in Eq. (43.13) [1],

$$\lim_{Q^2\to\infty} Q^6\Pi_{LR}(Q^2) = \left[-4\pi^2\frac{\alpha_s}{\pi} + \mathcal{O}(\alpha_s^2)\right]\langle\bar\psi\psi\rangle^2 , \quad (43.19)$$

which implies the two Weinberg sum rules:

$$\int_0^\infty dt \mathrm{Im}\Pi_{LR}(t) = \sum_V f_V^2 M_V^2 - \sum_A f_A^2 M_A^2 - F_0^2 = 0 , \quad (43.20)$$

and:

$$\int_0^\infty dt t \mathrm{Im}\Pi_{LR}(t) = \sum_V f_V^2 M_V^4 - \sum_A f_A^2 M_A^4 = 0 . \quad (43.21)$$

In fact, as pointed out in [526], in large $N_c$ QCD, there exist an infinite number of Weinberg-like sum rules. In full generality, the moments of the $\Pi_{LR}$ spectral function with $n = 3, 4, \ldots$,

$$\int_0^\infty dt\, t^{n-1}\left[\frac{1}{\pi}\mathrm{Im}\Pi_V(t) - \frac{1}{\pi}\mathrm{Im}\Pi_A(t)\right] = \sum_V f_V^2 M_V^{2n} - \sum_A f_A^2 M_A^{2n} , \quad (43.22)$$

govern the short-distance expansion of the $\Pi_{LR}(Q^2)$ function;

$$\Pi_{LR}(Q^2)|_{Q^2\rangle\infty} = \left(\sum_V f_V^2 M_V^6 - \sum_A f_A^2 M_A^6\right)\frac{1}{Q^6} + \left(\sum_V f_V^2 M_V^8 - \sum_A f_A^2 M_A^8\right)\frac{1}{Q^8} + \cdots . \quad (43.23)$$

On the other hand, inverse moments of the $\Pi_{LR}$ spectral function, with the pion pole removed, (which we denote by $\mathrm{Im}\tilde\Pi_A(t)$,) determine a class of coupling constants of the low-energy effective chiral Lagrangian.

For example:

$$\int_0^\infty dt\frac{1}{t}\left[\frac{1}{\pi}\mathrm{Im}\Pi_V(t) - \frac{1}{\pi}\mathrm{Im}\tilde\Pi_A(t)\right] = \sum_V f_V^2 - \sum_A f_A^2 = -4L_{10} . \quad (43.24)$$

Moments with higher inverse powers of $t$ are associated with couplings of composite operators of higher dimension in the chiral Lagrangian. Tests of the two Weinberg sum rules in Eqs. (43.20) and (43.21) and of the $L_{10}$ sum rule in Eq. (43.24), albeit in a different context to the one we are interested in here, have also been discussed in the literature, (see e.g. [527,528], [33,34]).

*The minimal ansatz*

We shall now consider the approximation which we call the *minimal hadronic ansatz* to large $N_c$–QCD. In the case of the left–right two-point function in Eq. (43.13), this is the approximation where the hadronic spectrum consists of one vector state $V$, one axial-vector state $A$ and the Goldstone pion, with the ordering [526] $M_V < M_A$. This is the *minimal spectrum* which is required to satisfy the two Weinberg sum rules in Eqs. (43.20) and (43.21.) In this approximation, $\Pi_{LR}(Q^2)$ has a very simple form:

$$\begin{aligned}-Q^2\Pi_{LR}(Q^2) &= \frac{f_0^2}{(1+\frac{Q^2}{M_V^2})(1+\frac{Q^2}{M_A^2})}\\ &= \frac{M_A^2M_V^2}{Q^4}\frac{f_0^2}{(1+\frac{M_V^2}{Q^2})(1+\frac{M_A^2}{Q^2})}\,. \end{aligned}\tag{43.25}$$

This equation shows, explicitly, a remarkable short-distance $\rightleftharpoons$ long-distance duality [529]. Indeed, with $g_A$ defined as:

$$M_V^2 = g_A M_A^2 \quad \text{and} \quad z \equiv \frac{Q^2}{M_V^2}\,, \tag{43.26}$$

the non-local order parameters corresponding to the long-distance expansion for $z\rangle 0$, which are couplings of the effective chiral Lagrangian i.e.:

$$-Q^2\Pi_{LR}(Q^2)|_{z\rangle 0} = f_0^2\left\{1-(1+g_A)z+\left(1+g_A+g_A^2\right)z^2+\cdots\right\}, \tag{43.27}$$

are correlated to the local-order parameters of the short-distance OPE for $z\rangle\infty$ in a very simple way:

$$-Q^2\Pi_{LR}(Q^2)|_{z\rangle\infty} = f_0^2\frac{1}{g_A}\frac{1}{z^2}\left\{1-\left(1+\frac{1}{g_A}\right)\frac{1}{z}+\left(1+\frac{1}{g_A}+\frac{1}{g_A^2}\right)\frac{1}{z^2}+\cdots\right\}\,; \tag{43.28}$$

in other words, there is a one-to-one correspondence between the two expansions by changing

$$g_A \rightleftharpoons \frac{1}{g_A} \quad \text{and} \quad z^n \rightleftharpoons \frac{1}{g_A}\frac{1}{z^{n+2}}\,. \tag{43.29}$$

The moments of the $\Pi_{LR}$ spectral function, when evaluated in the *minimal hadronic ansatz* approximation, can be converted into a very simple set of finite energy sum rules

(FESR's), corresponding to the OPE in Eq. (43.28):

$$\int_0^{s_0} dt\, t^2 \frac{1}{\pi} \mathrm{Im}\Pi_{LR}(t) = -f_0^2 M_V^4 \frac{1}{g_A}\,, \tag{43.30}$$

$$\int_0^{s_0} dt\, t^3 \frac{1}{\pi} \mathrm{Im}\Pi_{LR}(t) = -f_0^2 M_V^6 \frac{1+\frac{1}{g_A}}{g_A}\,, \tag{43.31}$$

$$\int_0^{s_0} dt\, t^4 \frac{1}{\pi} \mathrm{Im}\Pi_{LR}(t) = -f_0^2 M_V^8 \frac{1+\frac{1}{g_A}+\frac{1}{g_A^2}}{g_A}\,, \tag{43.32}$$

$$\cdots \qquad\qquad \cdots\,.$$

where the upper limit of integration $s_0$ denotes the onset of the pQCD continuum which, in the chiral limit, is common to the vector and axial-vector spectral functions. It is important to realize that $s_0$ is not a free parameter. Its value is fixed by the requirement that the OPE of the correlation function of two vector currents, (or two axial-vector currents,) in the chiral limit, have no $1/Q^2$ term, which results in an implicit equation for $s_0$ [405,530]. In the *minimal hadronic ansatz* approximation the onset of the pQCD continuum, which we shall call $s_0^*$, is then fixed by the equation

$$\frac{N_c}{16\pi^2}\frac{2}{3}s_0^*\left(1+\mathcal{O}(\alpha_s)\right) = f_0^2\frac{1}{1-g_A}\,. \tag{43.33}$$

Also, the moments which correspond to the chiral expansion in Eq. (43.27) are given by another simple set of FESR's:

$$\int_0^{s_0} dt\, \frac{1}{\pi} \mathrm{Im}\tilde{\Pi}_{LR}(t) = f_0^2\,, \tag{43.34}$$

$$\int_0^{s_0} \frac{dt}{t} \frac{1}{\pi} \mathrm{Im}\tilde{\Pi}_{LR}(t) = \frac{f_0^2}{M_V^2}(1+g_A)\,, \tag{43.35}$$

$$\int_0^{s_0} \frac{dt}{t^2} \frac{1}{\pi} \mathrm{Im}\tilde{\Pi}_{LR}(t) = \frac{f_0^2}{M_V^4}\left(1+g_A+g_A^2\right)\,, \tag{43.36}$$

$$\cdots \qquad\qquad \cdots\,.$$

These duality relations have been tested by comparing moments of the *physical spectral function* $1/\pi\ \mathrm{Im}\Pi_{LR}^{\mathrm{exp}}(t)$ determined from experiment (tau-decay data) to the predictions of the *minimal hadronic ansatz* as shown in the RHS of Eqs. (43.30) to (43.32) and Eqs. (43.34) to (43.36), where one finds that the tau-decay data is consistent with the simple pattern of duality properties between short and long distances which follow from the minimal hadronic ansatz of a narrow resonances in the large $N_c$ limit of QCD.

#### 43.2.4 Baryons in the large $N_c$ limit

Many features of the baryon sector have been also understood using the $1/N_c$ expansion, where a new $SU(4)$ symmetry connects the $u\uparrow, u\downarrow, d\uparrow, u\downarrow$ states in the baryon (see e.g. the review of Manohar in [502]). A systematic computation of the $1/N_c$ corrections then becomes possible, and some results obtained previously from the quark and Skyrme models [3] can be proved to order $1/N_c$ or $1/N_c^2$. However, in the large $N_c$ limit, baryons are more difficult to study than mesons, as the number of quarks in the baryon is $N_c$. Large $N_c$ counting rules for baryons were given by Witten [522]. In particular, if one assumes that the baryon mass and axial coupling $g_A$ are of order $N_c$, one can deduce using a non-relativistic quark model:

$$g_A = \frac{N_c + 2}{3}, \tag{43.37}$$

which is equal to the well-known quark model prediction $5/3$ for $N_c = 3$. Phenomenology of multicolour QCD in the baryon sector using QCD spectral sum rules has been studied in [531] for $N_c$ flavour, and in [532] for two flavours. In the latter case (see the details of the derivation in [532,3]), the Skyrme parameter has been obtained to be:

$$e \simeq 9/N_c^{1/2}, \tag{43.38}$$

in agreement with large $N_c$-expectations.

## 43.3 Lowest meson dominance models

There has been quite a lot of progress during the last few years in understanding the rôle of resonances in ChPT. At the phenomenological level [533,534], it turns out that the observed values of the $L_i$-constants are practically saturated by the contribution from the lowest resonance exchanges between the pseudoscalar particles; and particularly by vector-exchange, whenever vector mesons can contribute. The specific form of an effective chiral invariant Lagrangian describing the couplings of vector and axial-vector particles to the (pseudo) Nambu–Goldstone modes is not uniquely fixed by chiral symmetry requirements alone. When the vector fields describing heavy vector particles are integrated out, different field theory descriptions may lead to different predictions for the $L_i$-couplings. It has been shown however that, if a few QCD short-distance constraints are imposed, the ambiguities of different formulations are then removed [535]. The most compact effective Lagrangian formulation, compatible with the short-distance constraints, has two free parameters: $f_\pi$ and $M_V$. When the vector and axial-vector fields are integrated out, it leads to specific predictions for *five* of the $L_i$ constants:

$$L_1^{(V)} = L_2^{(V)}/2 = -L_3^{(V)}/6 = L_9^{(V)}/8 = -L_{10}^{(V+A)}/6 = \frac{f_\pi^2}{16M_V^2} \simeq 0.6 \times 10^{-3}, \tag{43.39}$$

in good agreement, within errors, with experiment. [See Table 42.1]

It is fair to conclude that the old phenomenological concept of *vector meson dominance* (VMD) [14] can now be formulated in a way that is compatible with the chiral symmetry properties and the short-distance behaviour of QCD.

### 43.4 The constituent chiral quark model

This model was introduced by Georgi and Manohar [512], in an attempt to reconcile the successful features of the constituent quark model [81], with the chiral symmetry requirements of QCD. The basic assumption of the model is the idea that between the scale of chiral symmetry breaking $\Lambda_\chi$ and the confinement scale $\sim \Lambda_{\rm QCD}$ the underlying QCD theory, may admit a useful effective Lagrangian realization in terms of *constituent quark fields* $Q$; *pseudoscalar particles*; and, perhaps, *'gluons'*. The Lagrangian in question has the form:

$$
\begin{aligned}
\mathcal{L}_{\rm eff}^{\rm GM} &= i\bar{Q}\gamma_\mu(\partial_\mu + ig_s G_\mu + \Gamma_\mu)Q \\
&+ \frac{i}{2} g_A \bar{Q}\gamma_5\gamma^\mu \xi_\mu Q - M_Q\bar{Q}Q \\
&+ \frac{1}{4} f_\pi^2\, tr\, D_\mu U D^\mu U^\dagger - \frac{1}{4}\vec{G}_{\mu\nu}\vec{G}^{\mu\nu} \,. \qquad (43.40)
\end{aligned}
$$

Some explanations about the notation here are in order. Remember that under chiral rotations $(V_L, V_R)$, $U$ transforms like: $U \to V_R U V_L$. The unitary matrix $U$ is the product of the so-called left and right coset representatives: $U = \xi_R \xi_L^\dagger$ and, without lost of generality, one can always choose the gauge where $\xi_L^\dagger = \xi_R \equiv \xi$. The coset representative $\xi, (U = \xi\xi^\dagger,)$ transforms as:

$$
\xi \to V_R \xi h^\dagger = h\xi V_L^\dagger \qquad h \in SU(3)_V \,, \qquad (43.41)
$$

where $h$ denotes the rotation induced by the chiral transformation $(V_L, V_R)$ in the diagonal $SU(3)_V$. In Eq. (43.40) the constituent quark fields $Q$ transform as:

$$
Q \to hQ, \qquad h \in SU(3)_V \,. \qquad (43.42)
$$

In the presence of external sources:[1]

$$
\Gamma_\mu = \frac{1}{2}\{\xi^\dagger[\partial_\mu - i(v_\mu + a_\mu)]\xi + \xi[\partial_\mu - i(v_\mu - a_\mu)]\xi^\dagger\} \qquad (43.43)
$$

and:

$$
\xi_\mu = i\xi^\dagger D_\mu U \xi^\dagger \,. \qquad (43.44)
$$

The free parameters of the theory are $f_\pi$, $M_Q$, and $g_A$. The QCD coupling constant is assumed to have entered a regime (below $\Lambda_\chi$,) where its running is frozen and is taken to be constant.

The merit of this model is that it automatically incorporates the phenomenological successes of the constituent quark model, in a way compatible with chiral symmetry. This model indeed appears in practically all QCD low-energy models where quarks are not confined. The weak point of the model is its 'vagueness' about the gluonic sector. In the absence of a dynamic justification for the 'freezing' of the QCD running coupling constant, it is very unclear what the 'left out' gluonic interactions mean; and in fact, in most applications, they are simply ignored.

[1] The original formulation of the model of Georgi and Manohar [512] was in fact made without external fields.

## 43.5 Effective action approach models

The basic idea in this class of models is to make some kind of drastic approximation to compute the non–anomalous part of the QCD-fermionic determinant in the presence of external $v_\mu$ and $a_\mu$ fields, but with the external $s$ and $p$ fields frozen to the quark matrix:

$$s + ip = \mathcal{M} = \text{diag}(m_u, m_d, m_s)\,. \tag{43.45}$$

Although the integral over the quark fields in Eq. (42.16) can be done explicitly, we do not know how to perform analytically the remaining integration over the gluon fields. A perturbative evaluation of the gluonic contribution would obviously fail in reproducing the correct dynamics of Spontaneous Chiral Symmetry Breaking (SCSB). A possible way out is to parametrize phenomenologically the SCSB and make a weak gluon-field expansion around the resulting physical vacuum. The simplest parametrization [413] is obtained by adding to the QCD Lagrangian the chiral invariant term:

$$\Delta\mathcal{L}_{\text{QCD}} = -M_Q(\bar{q}_R U q_L + \bar{q}_L U^\dagger q_R)\,, \tag{43.46}$$

which serves to introduce the $U$ field, and a mass parameter $M_Q$, which regulates the IR behaviour of the low-energy effective action. In the presence of this term the operator $\bar{q}q$ acquires a vacuum expectation value; therefore, Eq. (43.46) is an effective way to generate the order parameter due to SCSB. Making a chiral rotation of the quark fields, $Q_L \equiv u(\phi)q_L$, $Q_R \equiv u(\phi)^\dagger q_R$, with $U = u^2$, the interaction Eq. (43.46) reduces to a mass-term for the *dressed* quarks $Q$; the parameter $M_Q$ can then be interpreted as a *constituent-quark mass*.

The derivation of the low-energy effective chiral Lagrangian within this framework has been extensively discussed by [413]. In the chiral and large-$N_C$ limits, and including the leading gluonic contributions, one gets:

$$8L_1 = 4L_2 = L_9 = \frac{N_C}{48\pi^2}\left[1 + \mathcal{O}(1/M_Q^6)\right]\,,$$
$$L_3 = L_{10} = -\frac{N_C}{96\pi^2}\left[1 + \frac{\pi^2}{5N_C}\frac{\langle\frac{\alpha_s}{\pi}GG\rangle}{M_Q^4} + \mathcal{O}(1/M_Q^6)\right]\,, \tag{43.47}$$

where the positive sign of the corrections helps for a better agreement with experiments. Due to dimensional reasons, the leading contributions to the $\mathcal{O}(p^4)$ couplings only depend on $N_C$ and geometrical factors. It is remarkable that $L_1$, $L_2$ and $L_9$ do not get any gluonic correction at this order; this result is independent of the way SCSB has been parametrized ($M_Q$ can be taken to be infinite). Table 43.1 compares the predictions obtained with only the leading term in Eq. (43.47) (i.e. neglecting the gluonic correction) with the phenomenological determination of the $L_i$ couplings. The numerical agreement is quite impressive; both the order of magnitude and the sign are correctly reproduced (notice that this is just a free-quark result!). Moreover, the gluonic corrections shift the values of $L_3$ and $L_{10}$ in the right direction, making them more negative.

Table 43.1. *Leading-order ($\alpha_s = 0$) predictions for the $L_i$'s, within the QCD-inspired model in Eq. (43.46). The phenomenological values are shown in the second row for comparison. All numbers are given in units of* $10^{-3}$

| | $L_1$ | $L_2$ | $L_3$ | $L_9$ | $L_{10}$ |
|---|---|---|---|---|---|
| $L_i^{\text{th}}(\alpha_s = 0)$ | 0.79 | 1.58 | −3.17 | 6.33 | −3.17 |
| $L_i^r(M_\rho)$ | $0.4 \pm 0.3$ | $1.4 \pm 0.3$ | $-3.5 \pm 1.1$ | $6.9 \pm 0.7$ | $-5.5 \pm 0.7$ |

The results in Eq. (43.47) obey almost all relations in (43.39). In the same way, one also obtains a relation between the quark constituent mass and the pion decay constant [413]:

$$f_\pi^2 = \frac{N_c}{16\pi^2} 4M_Q^2 \left[ \log \frac{\Lambda^2}{M_Q^2} + \frac{\pi^2}{6N_c} \frac{< \frac{\alpha_s}{\pi} GG >}{M_Q^4} + \frac{1}{360 N_c} \frac{< g^3 GGG >}{M_Q^6} + \cdots \right] . \tag{43.48}$$

The authors mention that the gluon condensate appearing here has nothing to do with the one from QCD spectral sum rules phenomenology, which is hard to digest as the quantity $\langle \alpha_s G^2 \rangle$ has a very weak scale dependence. This approach has been also extended to the estimate of four-fermion non-leptonic weak operators, which the interested readers can find in [537]. Analogous result has been derived in [538] using a variational mass expansion.

## 43.6 The Extended Nambu–Jona-Lasinio Model

There have been many suggestions in the literature proposing that Nambu and Jona-Lasinio [539]-like models are relevant models for low-energy hadron dynamics. In e.g. [540,541], one assumes that at intermediate energies below or of the order of the spontaneous chiral symmetry breaking scale $\Lambda_\chi$, the leading operators of higher dimension which, after integration of the high-frequency modes of the quark and gluon fields down to the scale $\Lambda_\chi$, become relevant in the QCD Lagrangian, are those which can be cast in the form of four-fermion operators, i.e.:

$$\mathcal{L}_{\text{QCD}} \Longrightarrow \mathcal{L}_{\text{QCD}}^{\chi} + \mathcal{L}_{\text{S,P}} + \mathcal{L}_{\text{V,A}} + \cdots , \tag{43.49}$$

where:

$$\mathcal{L}_{\text{S,P}} = \frac{1}{N_c} \frac{8\pi^2}{\Lambda_\chi^2} \mathbf{G}_S \sum_{i,j} \left( \bar{q}_R^i q_{Lj} \right) \left( \bar{q}_L^j q_{Ri} \right) , \tag{43.50}$$

and:

$$\mathcal{L}_{\text{V,A}} = -\frac{1}{N_c} \frac{8\pi^2}{\Lambda_\chi^2} \mathbf{G}_V \sum_{i,j} \left[ \left( \bar{q}_L^i \gamma^\mu q_{Lj} \right) \left( \bar{q}_L^j \gamma_\mu q_{Li} \right) + L \leftrightarrow R \right] . \tag{43.51}$$

Here $i,j$ denote $u$, $d$, and $s$ flavour indices and summation over colour degrees of freedom within each bracket is understood; $q_{L,R} \equiv \frac{1}{2}(1 \pm \gamma_5) q$. The couplings $\mathbf{G}_{S,V}$ are dimensionless functions of the UV integration cut-off $\Lambda$. They are expected to grow as $\Lambda$ approaches the critical value $\Lambda_\chi$, where spontaneous chiral symmetry breaking occurs. (This is the reason why the operators $\mathcal{L}_{\rm S,P}$ and $\mathcal{L}_{\rm V,A}$ become relevant). In QCD, and with the factor $N_c^{-1}$ pulled out, both couplings $\mathbf{G}_S$ and $\mathbf{G}_V$ are $\mathcal{O}(1)$ in the large-$N_c$ limit. These constants are in principle calculable functions of the ratio $\Lambda/\Lambda_{\rm QCD}$. In practice however, the calculation requires non-perturbative knowledge of QCD in the region where $\Lambda \simeq \Lambda_\chi$, and we shall take $\mathbf{G}_S$ and $\mathbf{G}_V$, as well as $\Lambda_\chi$, as independent unknown parameters. The $\chi$ index in $\mathcal{L}^\chi_{\rm QCD}$ means that only the low-frequency modes $\Lambda \leq \Lambda_\chi$ of the quark and gluon fields are to be considered from now onwards.

Notice that in QCD, couplings of the type $\mathcal{L}_{\rm S,P}$ and $\mathcal{L}_{\rm V,A}$ appear naturally from gluon exchange between two QCD colour currents. Using Fierz rearrangement, one has in the large-$N_c$ limit:

$$g_s^2 \sum_a \left(\bar{q}\gamma^\mu \frac{\lambda^a}{2} q\right)\left(\bar{q}\gamma_\mu \frac{\lambda^a}{2} q\right) \Rightarrow \frac{1}{N_c}\frac{8\pi^2}{\Lambda_\chi^2} 4 \frac{\alpha_s N_c}{\pi} \sum_{i,j} (\bar{q}^i_R q_{Lj})(\bar{q}^j_L q_{Ri})$$
$$- \frac{1}{N_c}\frac{8\pi^2}{\Lambda_\chi^2}\frac{\alpha_s N_c}{\pi} \sum_{i,j} \big[(\bar{q}^i_L \gamma^\mu q_{Lj})(\bar{q}^j_L \gamma_\mu q_{Li})$$
$$+ L \leftrightarrow R\big]; \tag{43.52}$$

i.e.; $\mathbf{G}_V = \mathbf{G}_S/4 = \alpha_s N_c/\pi$ in this case. The two operators $\mathcal{L}_{\rm S,P}$ and $\mathcal{L}_{\rm V,A}$ have, however, different anomalous dimensions, and it is therefore not surprising that $\mathbf{G}_S \neq 4\mathbf{G}_V$ for the corresponding physical values.

If furthermore, one assumes that the relevant gluonic effects for low-energy physics are those already absorbed in the new couplings $\mathbf{G}_S$ and $\mathbf{G}_V$, then:

$$\mathcal{L}^\chi_{\rm QCD} \Rightarrow i\bar{q}\, \not{D} q \tag{43.53}$$

in Eq. (43.49) with $\not{D}$ the Dirac operator given in Eq. (43.2), where now the gluon field $G_\mu$ plays the rôle of an external colour field source. There is no gluonic kinetic term any longer.

As is well known from the early work of Nambu and Jona-Lasinio [539], the operator $\mathcal{L}_{\rm S,P}$, for values of $\mathbf{G}_S > 1$, is at the origin of the spontaneous chiral symmetry breaking. This can best be seen following the standard procedure of introducing auxiliary field variables to convert the four-fermion coupling operators into bilinear quark operators. For this purpose, one introduces a $3 \times 3$ auxiliary field matrix $M(x)$ in flavour space; the so called collective field variables, which under chiral-$SU(3)$ transform as:

$$M \rightarrow V_R M V_L^\dagger; \tag{43.54}$$

and uses the functional integral identity:

$$\exp\left[i\int d^4x \frac{1}{N_c}\frac{8\pi^2}{\Lambda_\chi}\mathbf{G}_S \sum_{i,j} (\bar{q}^i_R q_{Lj})(\bar{q}^j_L q_{Ri})\right]$$
$$= \int \mathcal{D}M \exp\left[i\int d^4x \left\{-(\bar{q}_L M^\dagger q_R + h.c.) - N_c \frac{\Lambda_\chi^2}{8\pi^2}\frac{1}{\mathbf{G}_S} tr M M^\dagger\right\}\right]. \tag{43.55}$$

By polar decomposition:

$$M = \xi H \xi^\dagger , \tag{43.56}$$

with $\xi\xi^\dagger = U$ unitary and $H$ hermitian.

Next, we look for translational-invariant solutions, which minimize the effective action;

$$\left.\frac{\partial \Gamma_{\text{eff}}}{\partial M}\right|_{H=<H>=M_Q,\xi=1;v=a=s=p=0.} = 0 .$$

The minimum is reached when all the eigenvalues of $< H >$ are equal, i.e., $< H >= M_Q 1$; and the minimum condition leads to

$$\text{Tr}\left(x\left|\frac{1}{\not{D}}\right|x\right) = -2M_Q N_c \frac{\Lambda_\chi^2}{8\pi^2}\frac{1}{\mathbf{G}_S}\int d^4x . \tag{43.57}$$

The trace in the LHS of this equation is formally proportional to $< \bar{\psi}\psi >$. The calculation, however, requires a regularization, with $\Lambda_\chi$ the UV cut-off. We choose the proper time regularization. [See e.g. [540] for technical details.] Then:

$$< \bar{\psi}\psi >= -\frac{N_c}{16\pi^2} 4M_Q^3 \Gamma\left(-1, \frac{M_Q^2}{\Lambda_\chi}\right) ; \tag{43.58}$$

and the minimum condition in Eq. (43.57) leads to the so-called *gap equation*:

$$\frac{M_Q}{\mathbf{G}_S} = M_Q \left\{ \exp\left(-\frac{M_Q^2}{\Lambda_\chi^2}\right) - \frac{M_Q^2}{\Lambda_\chi^2}\Gamma\left(0, \frac{M_Q^2}{\Lambda_\chi^2}\right)\right\} . \tag{43.59}$$

The functions:

$$\Gamma\left(n-2, x \equiv \frac{M_Q^2}{\Lambda_\chi^2}\right) = \int_x^\infty \frac{dz}{z} e^{-z} z^{n-2} ; \quad n = 1, 2, 3, \ldots , \tag{43.60}$$

are incomplete gamma functions. Equations (43.58) and (43.59) show the existence of two phases with regards to chiral symmetry. The unbroken phase corresponds to the trivial solution $M_Q = 0$, which implies $< \bar{\psi}\psi > = 0$. The broken phase corresponds to the possibility that the coupling $\mathbf{G}_S$ increases as we decrease the UV cut-off $\Lambda$ down to $\Lambda_\chi$, allowing for solutions to Eq. (43.59) with $M_Q > 0$ and therefore $< \bar{\psi}\psi > \neq 0$ and negative. In this phase the Hermitian auxiliary field $H(x)$ develops a non-vanishing vacuum expectation value, which is at the origin of a constituent chiral quark mass term [see the RHS of Eq. (43.55)]:

$$-M_Q(\bar{q}_L U^\dagger q_R + \bar{q}_R U q_L) = -M_Q \bar{Q} Q , \tag{43.61}$$

like the one which appears in the Georgi–Manohar model [512]; and like the one proposed in the effective action approach of [413]. In the presence of the operator $\mathcal{L}_{\text{V,A}}$, we need two more auxiliary $3 \times 3$ complex field matrices $L_\mu(x)$ and $R_\mu(x)$ to rearrange the Lagrangian in

Eq. (43.49) into an equivalent Lagrangian which is only quadratic in the quark fields. Under chiral $(V_L, V_R)$ transformations these collective field variables are chosen to transform as follows:

$$L_\mu ) V_L L_\mu V_L^\dagger , \qquad R_\mu ) V_R R_\mu V_R^\dagger .$$

Then, the following functional identity follows:

$$\exp\left(-i\int d^4x \frac{1}{N_c}\frac{8\pi^2}{\Lambda_\chi^2}\mathbf{G}_V \sum_{i,j}\left[(\bar{q}_L^i\gamma^\mu q_{Lj})(\bar{q}_L^j\gamma_\mu q_{Li}) + L \leftrightarrow R\right]\right)$$
$$= \int \mathcal{D}L_\mu \mathcal{D}R_\mu \exp\left[i\int d^4x \left\{\bar{q}_L\gamma^\mu L_\mu q_L + N_c \frac{\Lambda_\chi^2}{8\pi^2}\frac{1}{\mathbf{G}_V}\frac{1}{4} trL^\mu L_\mu + L \leftrightarrow R\right\}\right] . \tag{43.62}$$

It is convenient to trade the auxiliary field matrices $L_\mu(x)$ and $R_\mu(x)$ by new vector field matrices:

$$W_\mu^{(\pm)} = \xi L_\mu \xi^\dagger \pm \xi^\dagger R_\mu \xi ,$$

which transform homogeneously under chiral transformations $(V_L, V_R)$; i.e.:

$$W_\mu^{(\pm)} ) h W_\mu^{(\pm)} h^\dagger ,$$

with $h$ the $SU(3)_V$ rotation induced by $(V_L, V_R)$. The fermionic determinant can then be obtained using standard techniques, like for example the *heat kernel* expansion we described earlier. When computing the resulting effective action, there appears a mixing term between the fields $W_\mu^{(-)}$ and $\xi_\mu$. One needs a new redefinition of the auxiliary field $W_\mu^{(-)}$:

$$W_\mu^{(-)} ) \hat{W}_\mu^{(-)} + (1 - g_A)\xi_\mu ,$$

in order to diagonalize the quadratic form in the variables $W_\mu^{(-)}$ and $\xi_\mu$. It is this mixing which is at the origin of an effective axial coupling of the constituent quarks with the Nambu–Goldstone modes:

$$\frac{1}{2} i g_A \bar{Q}\gamma^\mu\gamma_5\xi_\mu Q ,$$

a term like the axial coupling which appears in the Georgi–Manohar model. but with a specific form for the axial coupling constant $g_A$:

$$g_A = \frac{1}{1 + \mathbf{G}_V \frac{4M_Q^2}{\Lambda_\chi^2}\Gamma\left(0, \frac{M_Q}{\Lambda_\chi^2}\right)} . \tag{43.63}$$

In terms of Feynman diagrams this result can be understood as an *infinite* sum of constituent quark bubbles, with a coupling at the end to the pion field. These are the diagrams generated by the $\mathbf{G}_V$ four-fermion coupling to leading order in the $1/N_c$ expansion. The quark propagators in these diagrams are constituent quark propagators, solution of the

Schwinger–Dyson which is at the origin of the *gap equation* in Eq. (43.59). In the limit where $\mathbf{G}_V = 0, g_A = 1$; but in general [542], $g_A \neq 1$ to leading order in the $1/N_c$ expansion.

Kinetic terms for the auxiliary field variables are also generated by the functional integral over the quark fields $Q$ and $\bar{Q}$. The resulting Lagrangian, after wave-function rescaling of the auxiliary fields, has the form of a constituent chiral quark model, with scalar $S(x)$, vector $V(x)$, and axial-vector $A(x)$ field couplings:

$$
\begin{aligned}
\mathcal{L}_{\text{eff}}^{ENJL} &= i\bar{Q}\gamma^\mu\left(\partial_\mu + \Gamma_\mu - \frac{i}{\sqrt{2}f_V}V_\mu\right)Q - M_Q\bar{Q}Q \\
&+ \frac{i}{2}g_A\bar{Q}\gamma_5\gamma^\mu\left(\xi_\mu - \frac{\sqrt{2}}{f_A}A_\mu\right)Q - \frac{1}{\lambda_S}\bar{Q}S(x)Q \\
&+ \frac{1}{2}\text{tr}\left[\partial_\mu S\partial^\mu S - M_S^2 SS\right] \\
&- \frac{1}{4}\text{tr}[(\partial_\mu V_\nu - \partial_\nu V_\mu)(\partial^\mu V^\nu - \partial^\nu V^\mu) - 2M_V V_\mu V^\mu] \\
&- \frac{1}{4}\text{tr}\left[(\partial_\mu A_\nu - \partial_\nu A_\mu)(\partial^\mu A^\nu - \partial^\nu A^\mu) - 2M_A^2 A_\mu A^\mu\right] \\
&+ \frac{1}{4}f_\pi^2\text{tr}D_\mu U D^\mu U^\dagger + \mathcal{O}(p^4)\text{terms}\,,
\end{aligned}
\tag{43.64}
$$

where $\Gamma_\mu$ and $\xi_\mu$ are the same as those defined in Eqs.(43.43) and (43.44), and the coupling constants and masses are now expressed in terms of only three input parameters. As input parameters, we can either fix: $\mathbf{G}_S$, $\mathbf{G}_V$, and $\Lambda_\chi$; or the more physical parameters:

$$
M_Q\,, \qquad \Lambda\chi\,, \qquad g_A\,. \tag{43.65}
$$

The coupling constants are then:

$$
\begin{aligned}
f_\pi^2 &= \frac{N_c}{16\pi^2}4M_Q^2 g_A\Gamma(0, M_Q^2/\Lambda_\chi^2)\,, \\
f_V^2 &= \frac{N_c}{16\pi^2}\frac{2}{3}\Gamma(0, M_Q^2/\Lambda_\chi^2)\,, \\
f_A^2 &= \frac{N_c}{16\pi^2}\frac{2}{3}g_A^2[\Gamma(0, M_Q^2/\Lambda_\chi^2) - \Gamma(1, M_Q^2/\Lambda_\chi^2)]\,, \\
[3pt]\lambda_S^2 &= \frac{N_c}{16\pi^2}\frac{2}{3}[3\Gamma(0, M_Q^2/\Lambda_\chi^2) - 2\Gamma(1, M_Q^2/\Lambda_\chi^2)]\,;
\end{aligned}
\tag{43.66}
$$

and the masses:

$$
\begin{aligned}
M_V^2 &= 6M_Q^2\frac{g_A}{1-g_A}\,, \\
M_A^2 &= 6M_Q^2\frac{1}{1-g_A}\frac{1}{1-\frac{\Gamma\left(1,M_Q^2/\Lambda_\chi^2\right)}{\Gamma\left(0,M_Q^2/\Lambda_\chi^2\right)}}\,, \\
M_S^2 &= 4M_Q^2\frac{1}{1-\frac{2}{3}\frac{\Gamma\left(1,M_Q^2/\Lambda_\chi^2\right)}{\Gamma\left(0,M_Q^2/\Lambda_\chi^2\right)}}\,.
\end{aligned}
\tag{43.67}
$$

Table 43.2. *The $L_i$-coupling constants in the ENJL model of [540], with $g_A$ defined in Eq.(43.63), and $\Gamma_n \equiv \Gamma(n, M_Q^2/\Lambda_\chi^2)$. The second column gives the results corresponding to the input parameter values in Eq. (43.68). The third column gives the experimental values of Table 42.1.*

| The $L_i$ couplings of $\mathcal{O}(p^4)$ in the ENJL–model | Fit 1 | Experiment |
|---|---|---|
| $L_1 = \frac{N_c}{16\pi^2}\frac{1}{48}\left[\left(1-g_A^2\right)^2\Gamma_0 + 4g_A^2\left(1-g_A^2\right)\Gamma_1 + 2g_A^4\Gamma_2\right]$ | 0.85 | 0.7 $\pm 0.5$ |
| $L_2 = 2L_1$ | 1.7 | 1.2 $\pm 0.4$ |
| $L_3 = -\frac{N_c}{16\pi^2}\frac{1}{8}\left\{\left[\left(1-g_A^2\right)^2\Gamma_0 + 4g_A^2\left(1-g_A^2\right)\Gamma_1 + \right.\right.$ $\left.\left. -\frac{2}{3}g_A^4\left[2\Gamma_1 - 4\Gamma_2 + 3\frac{1}{\Gamma_0}(\Gamma_0-\Gamma_1)^2\right]\right]\right\}$ | −4.2 | −3.6 $\pm 1.3$ |
| $L_5 = \frac{N_c}{16\pi^2}\frac{1}{4}g_A^3[\Gamma_0 - \Gamma_1]$ | 1.6 | 1.4 $\pm 0.5$ |
| $L_8 = \frac{N_c}{16\pi^2}\frac{1}{16}g_A^2\left[\Gamma_0 - \frac{2}{3}\Gamma_1\right]$ | 0.8 | 0.9 $\pm 0.3$ |
| $L_9 = \frac{N_c}{16\pi^2}\frac{1}{6}\left[\left(1-g_A^2\right)\Gamma_0 + 2g_A^2\Gamma_1\right]$ | 7.1 | 6.9 $\pm 0.7$ |
| $L_{10} = -\frac{N_c}{16\pi^2}\frac{1}{6}\left[\left(1-g_A^2\right)\Gamma_0 + g_A^2\Gamma_1\right]$ | −5.9 | −5.5 0.7 |

In the absence of the vector and axial-vector four-fermion-like coupling i.e., when $\mathbf{G}_V = 0$: $g_A = 1$, $M_V\rangle\infty$ and $M_A\rangle\infty$. Then the vector and axial-vector interactions decouple, and the model becomes equivalent to the *Constituent Chiral Quark Model* of [512], with $g_A = 1$ and a non-trivial coupling to a scalar field.

The functional integration over the quark fields and the auxiliary $S(x)$, $V(x)$, and $A(x)$ fields results in an effective action among the Nambu–Goldstone boson particles, with all the couplings fixed by the three parameters $M_Q$, $\Lambda_\chi$, and $g_A$. The explicit results one gets for the $L_i$ constants which appear in the large-$N_c$ limit at $\mathcal{O}(p^4)$ in the chiral expansion are shown in Table 43.2. The reason why the constant $L_7$ does not appear in this table is that, phenomenologically, this constant gets a large contribution from the integration of the *heavy* singlet $\eta'$ particle. However, in the chiral limit, the mass of the $\eta'$ is induced by the axial-$U(1)$ anomaly, which only appears to next-to-leading order in the $1/N_c$ expansion. By definition, the ENJL model, as formulated here, ignores this effect. In order to take these next-to-leading effects in $1/N_c$ systematically, together with the chiral expansion, one has to resort to a $U(3)\times U(3)$ formulation of the effective theory [543]. The constants $L_4$ and $L_6$ are of next-to-leading order in the $1/N_c$ expansion; this is the reason why they do not appear in Table 43.2 either. We also show in Table 43.2 the numerical results of the fit 1 discussed in [540]. These results correspond to the set of input parameter values:

$$M_Q = 265\ MeV\,, \qquad \Lambda_\chi = 1165\ MeV\,, \qquad g_A = 0.61\,. \tag{43.68}$$

The overall picture which emerges from this simple model is quite remarkable. The main improvement with respect to the results obtained in the *effective action approach model* discussed in the previous section is on the constants $L_5$ and $L_8$, where the combined effect of the vector and scalar degrees of freedom leads to rather simple results modulated by powers of the $g_A$-constant, which agree very well with the phenomenological determinations. One of the characteristic features of the ENJL model, is that it interpolates successfully between pure VMD-type predictions and those of the constituent chiral quark model. A nice illustration is the result for $L_9$ in Table 43.2, where the first term is the one coming from vector–exchange, whereas the second one comes from the chiral quark loop integral.

There is no difficulty to reproduce the anomalous Wess–Zumino–Witten functional within the ENJL model [544] QCD two-point functions, beyond the low-energy expansion, have also been evaluated in the ENJL model [541]. This involves calculations to leading order in the $1/N_c$ expansion (i.e., an infinite number of chains of fermion bubbles; but no loops of chains) and to all orders in powers of momenta $Q^2/\Lambda_\chi^2$. As a result, vector and axial-vector correlation functions have a VMD-like form, but with slowly varying couplings and masses. For the transverse invariant functions for example, the results are:

$$\Pi_V^{(1)}(Q^2) = \frac{2f_V^2(Q^2)M_V^2(Q^2)}{M_V^2(Q^2) - Q^2}, \tag{43.69}$$

and:

$$\Pi_A^{(1)}(Q^2) = \frac{2f_\pi^2(Q^2)}{Q^2} + \frac{2f_A^2(Q^2)M_A^2(Q^2)}{M_A^2(Q^2) - Q^2}, \tag{43.70}$$

where:

$$f_V^2(Q^2) = 4\frac{N_c}{16\pi^2}\int_0^1 dx x(1-x)\Gamma\big(0, x_Q \equiv [M_{Q^2} + x(1-x)Q^2]/\Lambda_\chi^2\big). \tag{43.71}$$

The product:

$$2f_V^2(Q^2)M_V^2(Q^2) = N_c\frac{\Lambda_\chi^2}{8\pi^2}\frac{1}{\mathbf{G}_V} \tag{43.72}$$

is scale invariant. With:

$$g_A(Q^2) = \frac{1}{1 + \mathbf{G}_V\frac{4M_{Q^2}}{\Lambda_\chi^2}\int_0^1 dx\Gamma(0, x_Q)}, \tag{43.73}$$

the other couplings are fixed by:

$$f_A^2(Q^2) = g_A^2(Q^2)f_V^2(Q^2), \tag{43.74}$$

and the relations

$$\begin{aligned} f_V^2(Q^2)M_V^2(Q^2) &= f_A^2(Q^2)M_A^2(Q^2) + f_\pi^2(Q^2), \\ f_V^2(Q^2)M_V^4(Q^2) &= f_A^2(Q^2)M_A^4(Q^2), \end{aligned} \tag{43.75}$$

where the last two equalities are the $Q^2$-dependent version of the first- and second-Weinberg sum rules [26]. In the case of the scalar two-point function there appears a pole in the

$Q^2$-summed expression at:

$$M_S = 2M_Q \,. \tag{43.76}$$

The case of the other two-point functions is somewhat more involved because they mix through the four-fermion interaction terms. In principle, the ENJL model can be applied to obtain a systematic calculation of the low-energy constants of the weak non-leptonic Lagrangian ($B_K$-parameter, ...). These applications can be found in some more dedicated reviews.

# 44

# Heavy quark effective theory

## 44.1 Introduction

Over a decade, a lot of experimental informations on heavy-quark decays and masses have been obtained from $e^+e^-$ and hadron collider experiments. These have led to a detailed knowledge of the flavour sector of the standard model and to the discoveries of the $B^0 - \bar{B}^0$ mixing, rare decays induced by penguin operators, ... The experimental progress in the heavy flavour physics has been accompanied by some theoretical progress. Among other approaches, the discovery of the heavy-quark symmetry has led to the development of the heavy quark effective theory (HQET), which provides a systematic analysis of the properties of a hadron containing a heavy quark in terms of an expansion of the inverse of the heavy quark mass. Detailed discussions and references to the original works can be found in different reviews and lectures (see e.g. [545]).

## 44.2 Heavy-quark symmetry

When the mass of the heavy quark is much larger than the QCD scale $\Lambda_{QCD}$, the QCD running coupling $\alpha_s(m_Q)$ is small, implying that at this scale of the order of the Compton wavelength $\lambda_Q \sim 1/m_Q$, one can safely use perturbative QCD for describing the hadrons. In this case the $\bar{Q}Q$-bound states with the size $\lambda_Q/\alpha_s(m_Q) \ll R_{\text{had}} \sim 1$ fermi are like the hydrogen atom. However, systems composed of a heavy quark plus a light quark are more complicated because the size of the system is of the order of $R_{\text{had}}$ while the typical momenta exchanged between the heavy and light quarks is of the order of $\Lambda_{QCD}$. Therefore, the heavy quark is surrounded by strongly interacting clouds of light quarks, antiquarks and gluons. In this case, the simplification is provided by the fact that the Compton wavelength $\lambda_Q$ is much smaller than the hadron size $R_{\text{had}}$. To resolve the quantum numbers of the heavy quark would require a hard probe; the soft gluons exchanged between the heavy quark and the light constituents can only resolve distances much larger than $\lambda_Q$. Therefore, the light degrees of freedom are blind to the flavour (mass) and spin orientation of the heavy quark. They experience only its colour field, which extends over large distances because of confinement. In the rest frame of the heavy quark, it is in fact only the electric colour field that is important. Since the heavy-quark spin participates in interactions only through such relativistic effects, it decouples for $m_Q \to \infty$. It then follows that, in the limit $m_Q \to \infty$, hadronic systems which differ only in the flavour or spin quantum numbers of the heavy quark have the same configuration of their light degrees of freedom. Although this observation still does not allow us to calculate what this configuration is, it provides relations between the properties of, for example, the heavy mesons $B$, $D$, $B^*$ and $D^*$ in the ideal case where the $b$ and $c$ quark masses are infinitely heavy and the corrections to this limit are negligible. These relations result from some approximate symmetries of the effective strong interactions of heavy quarks at low energies. The configuration of light degrees of freedom in a hadron containing a single heavy quark with velocity $v$ does not change if this quark is replaced by another heavy quark with different flavour or spin, but with the same velocity. Both heavy quarks lead to the same static colour field. For $n_h$ heavy-quark flavours, there is thus an SU($2n_h$) spin-flavour symmetry group, under which the effective strong interactions are invariant. These symmetries are in close correspondence to familiar properties of atoms. The flavour symmetry is analogous to the fact that different isotopes have the same chemistry, since to good approximation the wave function of the electrons is independent of the mass of the nucleus. The electrons only see the total nuclear charge. The spin symmetry is analogous to the fact that the hyperfine levels in atoms are nearly degenerate. The nuclear spin decouples in the limit $m_e/m_N \to 0$. This heavy-quark symmetry looks quite similar to the chiral symmetry ($m \to 0$) but in the opposite way ($m_Q \to \infty$), although there is a striking difference.

Whereas chiral symmetry is a symmetry of the QCD Lagrangian in the limit of vanishing quark masses, heavy-quark symmetry is not a symmetry of the Lagrangian (not even an approximate one), but rather a symmetry of an effective theory that is a good approximation to QCD in a certain kinematic region. It is realized only in systems in which a heavy quark interacts predominantly by the exchange of soft gluons. In such systems the heavy quark is almost on-shell; its momentum fluctuates around the mass shell by an amount of order $\Lambda_{\rm QCD}$. The corresponding fluctuations in the velocity of the heavy quark vanish as $\Lambda_{\rm QCD}/m_Q \to 0$. The velocity becomes a conserved quantity and is no longer a dynamical degree of freedom [546]. Nevertheless, results derived on the basis of heavy-quark symmetry are model-independent consequences of QCD in a well-defined limit. To this end, it is however necessary to cast the QCD Lagrangian for a heavy quark:

$$\mathcal{L}_{\rm Q} = \bar{Q}\,(i\not{D} - m_Q)\,Q\,, \tag{44.1}$$

into a form suitable for taking the limit $m_Q \to \infty$.

## 44.3 Heavy quark effective theory

### *44.3.1 Introduction*

As the effects of infinitely heavy quark are irrelevant at low energies, it becomes useful to built a low-energy effective theory, where the heavy quark no longer appears. This is very similar to the Fermi's theory where weak interactions in weak processes can be approximated by a four-fermion interaction governed by the weak coupling $G_F$. The removal of the heavy particle degrees of freedom can be done in the following ways [547–549]:

- One integrates out the heavy fields in the generating functional of the Green's functions of the theory, which is possible as the heavy particles do not appear as an external source. The resulting action is nonlocal, as in full QCD the heavy particle with mass $M \simeq m_Q$ can appear in virtual processes and propagate over a short but finite distance $\Delta x \sim 1/M$.
- Thus, one needs to get a local effective Lagrangian, which can be done by rewriting the non-local effective action as an infinite series of local terms in an operator product expansion (OPE) [222], which approximately corresponds to an expansion in powers of $1/M$. In this step, the short- and long-distance physics is disentangled, and their domain is separated by a scale $\nu$ such that $\Lambda_{QCD} \ll \nu \ll m_Q$. The long-distance physics corresponds to interactions at low energies and is the same in the full and the effective theory below $\nu$.
- In a third step, one needs to add, in a perturbative way using renormalization-group techniques, short-distance effects arising from quantum corrections involving large virtual momenta (of order $M$), which have not been described correctly in the effective theory once the heavy particle has been integrated out. These short-distance effects lead to a renormalization of the coefficients of the local operators in the effective Lagrangian. An example is the effective Lagrangian for non-leptonic weak decays, in which radiative corrections from hard gluons with virtual momenta in the range between $m_W$ and some low renormalization scale $\mu$ give rise to Wilson coefficients, which renormalize the local four-fermion interactions [550–552]. The fact that the physics must be independent of the arbitrary scale $\nu$ allows us to derive renormalization-group equations, which can be employed to deal with the short-distance effects in an efficient way.

However, one should notice that the HQET approach is peculiar as it is motivated to describe the properties and decays of hadrons which do contain a heavy quark. Hence, it is not possible to remove the heavy quark completely from the effective theory, but only to integrate out the 'small components' in the full heavy-quark field, which describe the fluctuations around the mass shell.

### 44.3.2 The HQET Lagrangian

The starting point in the construction of the HQET is the observation that a heavy quark bound inside a hadron moves with the hadron's velocity $v$ and is almost on-shell. Its momentum can be written as:

$$p_Q^\mu = m_Q v^\mu + k^\mu \,, \tag{44.2}$$

where the components of the so-called *residual momentum* $k$ are much smaller than $m_Q$. Note that $v$ is a four-velocity, so that $v^2 = 1$. Interactions of the heavy quark with light degrees of freedom change the residual momentum by an amount of order $\Delta k \sim \Lambda_{QCD}$, but the corresponding changes in the heavy-quark velocity vanish as $\Lambda_{QCD}/m_Q \to 0$. In this situation, it is appropriate to introduce large- and small-component fields, $h_v$ and $H_v$, by:

$$h_v(x) = e^{im_Q v\cdot x}\, P_+\, Q(x)\,, \qquad H_v(x) = e^{im_Q v\cdot x}\, P_-\, Q(x)\,, \tag{44.3}$$

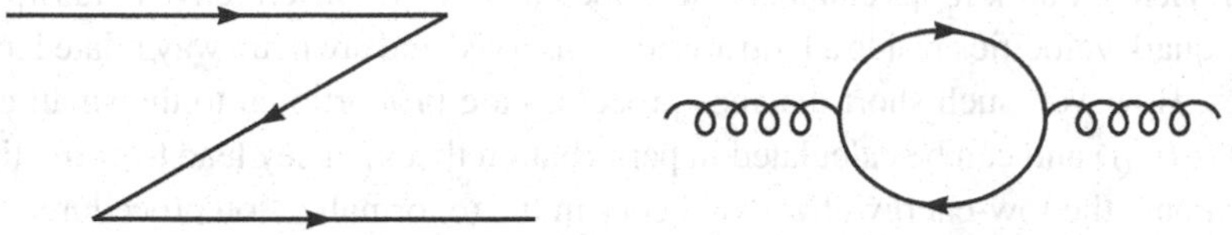

Fig. 44.1. Virtual fluctuations involving pair creation of heavy quarks. Time flows to the right.

where $P_+$ and $P_-$ are projection operators defined as:

$$P_\pm = \frac{1 \pm \slashed{v}}{2}\,. \tag{44.4}$$

It follows that

$$Q(x) = e^{-im_Q v\cdot x}\,[h_v(x) + H_v(x)]\,. \tag{44.5}$$

Because of the projection operators, the new fields satisfy $\slashed{v}h_v = h_v$ and $\slashed{v}H_v = -H_v$. In the rest frame, i.e. for $v^\mu = (1, 0, 0, 0)$, $h_v$ corresponds to the upper two components of $Q$, while $H_v$ corresponds to the lower ones. Whereas $h_v$ annihilates a heavy quark with velocity $v$, $H_v$ creates a heavy antiquark with velocity $v$.

In terms of the new fields, the QCD Lagrangian (44.1) for a heavy quark takes the form:

$$\mathcal{L}_Q = \bar{h}_v\, iv\cdot D\, h_v - \bar{H}_v\,(iv\cdot D + 2m_Q)\,H_v + \bar{h}_v\, i\slashed{D}_\perp H_v + \bar{H}_v\, i\slashed{D}_\perp h_v\,, \tag{44.6}$$

where $D^\mu_\perp = D^\mu - v^\mu\, v\cdot D$ is orthogonal to the heavy-quark velocity: $v\cdot D_\perp = 0$. In the rest frame, $D^\mu_\perp = (0, \vec{D}\,)$ contains the spatial components of the covariant derivative. From Eq. (44.6), it is apparent that $h_v$ describes massless degrees of freedom, whereas $H_v$ corresponds to fluctuations with twice the heavy-quark mass. These are the heavy degrees of freedom that will be eliminated in the construction of the effective theory. The fields are mixed by the presence of the third and fourth terms, which describe pair creation or annihilation of heavy quarks and antiquarks. As shown in the first diagram in Fig. 44.1, in a virtual process, a heavy quark propagating forward in time can turn into an antiquark propagating backward in time, and then turn back into a quark. The energy of the intermediate quantum state $hh\bar{H}$ is larger than the energy of the incoming heavy quark by at least $2m_Q$. Because of this large energy gap, the virtual quantum fluctuation can only propagate over a short distance $\Delta x \sim 1/m_Q$. On hadronic scales set by $R_{\text{had}} = 1/\Lambda_{\text{QCD}}$, the process essentially looks like a local interaction of the form:

$$\bar{h}_v\, i\slashed{D}_\perp \frac{1}{2m_Q}\, i\slashed{D}_\perp h_v\,, \tag{44.7}$$

where we have simply replaced the propagator for $H_v$ by $1/2m_Q$. A more correct treatment is to integrate out the small-component field $H_v$, thereby deriving a non-local effective action for the large-component field $h_v$, which can then be expanded in terms of local operators. Before doing this, let us mention a second type of virtual corrections involving pair creation, namely heavy-quark loops. An example is shown in the second diagram in

Fig. 44.1. Heavy-quark loops cannot be described in terms of the effective fields $h_v$ and $H_v$, since the quark velocities inside a loop are not conserved and are in no way related to hadron velocities. However, such short-distance processes are proportional to the small coupling constant $\alpha_s(m_Q)$ and can be calculated in perturbation theory. They lead to corrections that are added onto the low-energy effective theory in the renormalization procedure.

On a classical level, the heavy degrees of freedom represented by $H_v$ can be eliminated using the equation of motion. Taking the variation of the Lagrangian with respect to the field $\bar{H}_v$, we obtain:

$$(iv \cdot D + 2m_Q)\, H_v = i\not{D}_\perp h_v \,. \tag{44.8}$$

This equation can formally be solved to give:

$$H_v = \frac{1}{2m_Q + iv \cdot D}\, i\not{D}_\perp h_v \,, \tag{44.9}$$

showing that the small-component field $H_v$ is indeed of order $1/m_Q$. We can now insert this solution into Eq. (44.6) to obtain the *non-local effective Lagrangian*:

$$\mathcal{L}_{\text{eff}} = \bar{h}_v\, iv \cdot D\, h_v + \bar{h}_v\, i\not{D}_\perp \frac{1}{2m_Q + iv \cdot D}\, i\not{D}_\perp h_v \,. \tag{44.10}$$

Clearly, the second term corresponds to the first class of virtual processes shown in Fig. 44.1.

One can derive this Lagrangian in a more elegant way using the generating functional for QCD Green functions containing heavy-quark fields [553]. To this end, one starts from the field redefinition in Eq. (44.5) and couples the large-component fields $h_v$ to external sources $\rho_v$. Green functions with an arbitrary number of $h_v$ fields can be constructed by taking derivatives with respect to $\rho_v$. No sources are needed for the heavy degrees of freedom represented by $H_v$. The functional integral over these fields is Gaussian and can be performed explicitly, leading to the effective action:

$$S_{\text{eff}} = \int \mathrm{d}^4x\, \mathcal{L}_{\text{eff}} - i \ln \Delta \,, \tag{44.11}$$

with $\mathcal{L}_{\text{eff}}$ as given in Eq. (44.10). The appearance of the logarithm of the determinant:

$$\Delta = \exp\left( \frac{1}{2} \,\text{Tr}\, \ln[2m_Q + iv \cdot D - i\eta] \right) \tag{44.12}$$

is a quantum effect not present in the classical derivation presented above. However, in this case the determinant can be regulated in a gauge-invariant way, and by choosing the gauge $v \cdot A = 0$ one can show that $\ln \Delta$ is just an irrelevant constant [553,554].

Because of the phase factor in Eq. (44.5), the $x$ dependence of the effective heavy-quark field $h_v$ is weak. In momentum space, derivatives acting on $h_v$ produce powers of the residual momentum $k$, which is much smaller than $m_Q$. Hence, the non-local effective Lagrangian

$$= \frac{i}{v \cdot k} \frac{1+\not{v}}{2} \delta_{ji}$$

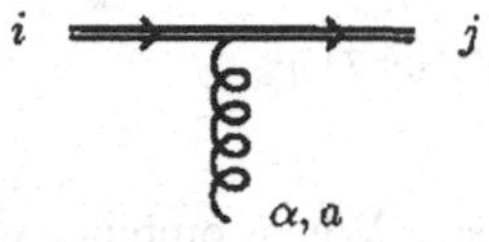

$$= i g_s v^{\alpha} (T_a)_{ji}$$

Fig. 44.2. Feynman rules of the HQET ($i$, $j$ and $a$ are colour indices). A heavy quark with velocity $v$ is represented by a double line. The residual momentum $k$ is defined in Eq. (44.2).

in Eq. (44.10) allows for a derivative expansion:

$$\mathcal{L}_{\text{eff}} = \bar{h}_v \, iv \cdot D \, h_v + \frac{1}{2m_Q} \sum_{n=0}^{\infty} \bar{h}_v \, i\not{D}_{\perp} \left( -\frac{iv \cdot D}{2m_Q} \right)^n i\not{D}_{\perp} h_v \, . \tag{44.13}$$

Taking into account that $h_v$ contains a $P_+$ projection operator, and using the identity

$$P_+ \, i\not{D}_{\perp} \, i\not{D}_{\perp} P_+ = P_+ \left[ (iD_{\perp})^2 + \frac{g}{2} \sigma_{\mu\nu} G^{\mu\nu} \right] P_+ \, , \tag{44.14}$$

where $i[D^{\mu}, D^{\nu}] = g \, G^{\mu\nu}$ is the gluon field-strength tensor, one finds that [555,556]

$$\mathcal{L}_{\text{eff}} = \bar{h}_v \, iv \cdot D \, h_v + \frac{1}{2m_Q} \bar{h}_v \, (iD_{\perp})^2 \, h_v + \frac{g_s}{4m_Q} \bar{h}_v \, \sigma_{\mu\nu} \, G^{\mu\nu} \, h_v + O\big(1/m_Q^2\big) \, . \tag{44.15}$$

In the limit $m_Q \to \infty$, only the first term remains:

$$\mathcal{L}_{\infty} = \bar{h}_v \, iv \cdot D \, h_v \, . \tag{44.16}$$

This is the effective Lagrangian of the HQET. It gives rise to the Feynman rules shown in Fig. 44.2.

### 44.3.3 Symmetries of the Lagrangian

Study of these symmetries can be, for example, found in [546]. Since there appear no Dirac matrices, interactions of the heavy quark with gluons leave its spin unchanged. Associated with this is an SU(2) symmetry group, under which $\mathcal{L}_{\infty}$ is invariant. The action of this symmetry on the heavy-quark fields becomes most transparent in the rest frame, where the generators $S^i$ of SU(2) can be chosen as:

$$S^i = \frac{1}{2} \begin{pmatrix} \sigma^i & 0 \\ 0 & \sigma^i \end{pmatrix} ; \qquad [S^i, S^j] = i\epsilon^{ijk} S^k \, . \tag{44.17}$$

Here $\sigma^i$ are the Pauli matrices. An infinitesimal SU(2) transformation $h_v \to (1 + i\vec{\epsilon} \cdot \vec{S}\,) \, h_v$ leaves the Lagrangian invariant:

$$\delta \mathcal{L}_{\infty} = \bar{h}_v \, [iv \cdot D, i\vec{\epsilon} \cdot \vec{S}\,] \, h_v = 0 \, . \tag{44.18}$$

Another symmetry of the HQET arises since the mass of the heavy quark does not appear in the effective Lagrangian. For $n_h$ heavy quarks moving at the same velocity, Eq. (44.16) can be extended by writing:

$$\mathcal{L}_\infty = \sum_{i=1}^{n_h} \bar{h}_v^i \, iv \cdot D \, h_v^i \,. \tag{44.19}$$

This is invariant under rotations in flavour space. When combined with the spin symmetry, the symmetry group is promoted to SU($2n_h$). This is the heavy-quark spin-flavour symmetry [557,546]. Its physical content is that, in the limit $m_Q \to \infty$, the strong interactions of a heavy quark become independent of its mass and spin.

Now, let us consider the operators appearing at order $1/m_Q$ in the effective Lagrangian in Eq. (44.15). They can be easily identified in the rest frame. The first operator:

$$\mathcal{O}_{\rm kin} = \frac{1}{2m_Q} \bar{h}_v \, (iD_\perp)^2 \, h_v \to -\frac{1}{2m_Q} \bar{h}_v \, (i\vec{D}\,)^2 \, h_v \,, \tag{44.20}$$

is the gauge-covariant extension of the kinetic energy arising from the residual motion of the heavy quark. The second operator is the non-Abelian analogue of the Pauli interaction, which describes the colour-magnetic coupling of the heavy-quark spin to the gluon field:

$$\mathcal{O}_{\rm mag} = \frac{g_s}{4m_Q} \bar{h}_v \, \sigma_{\mu\nu} \, G^{\mu\nu} \, h_v \to -\frac{g_s}{m_Q} \bar{h}_v \, \vec{S} \cdot \vec{B}_c \, h_v \,. \tag{44.21}$$

Here $\vec{S}$ is the spin operator defined in (44.17), and $B_c^i = -\frac{1}{2}\epsilon^{ijk} G^{jk}$ are the components of the colour-magnetic field. The chromo-magnetic interaction is a relativistic effect, which scales like $1/m_Q$. This is the origin of the heavy-quark spin symmetry.

### *44.3.4 Heavy quark wave-function renormalization in HQET*

As an illustration of the previous discussion, we consider the heavy quark wave-function renormalization using dimensional regularization in $n = 4 - \epsilon$-space–time, which we have discussed in length in previous sections. For QCD, one introduces renormalized quantities by $Q^{\rm bare} = Z_Q^{1/2} \, Q^{\rm ren}$, $A^{\rm bare} = Z_A^{1/2} A^{\rm ren}$, $\alpha_s^{\rm bare} = \mu^{2\epsilon} Z_\alpha \, \alpha_s^{\rm ren}$, etc., where $\mu$ is an arbitrary mass scale introduced to render the renormalized coupling constant dimensionless. Similarly, in the HQET one defines the renormalized heavy-quark field by $h_v^{\rm bare} = Z_h^{1/2} \, h_v^{\rm ren}$. From now on, the superscript "ren" will be omitted. In the minimal subtraction $MS$ scheme, $Z_h$ can be computed from the $1/\epsilon$ pole in the heavy-quark self energy using:

$$1 - Z_h^{-1} = \frac{1}{\epsilon} \text{pole of } \frac{\partial \Sigma(v \cdot k)}{\partial v \cdot k} \,. \tag{44.22}$$

As long as $v \cdot k < 0$, the self-energy is IR finite and real. The result is gauge-dependent, however. Evaluating the diagram shown in Fig. 44.3 in the Feynman gauge, we obtain at

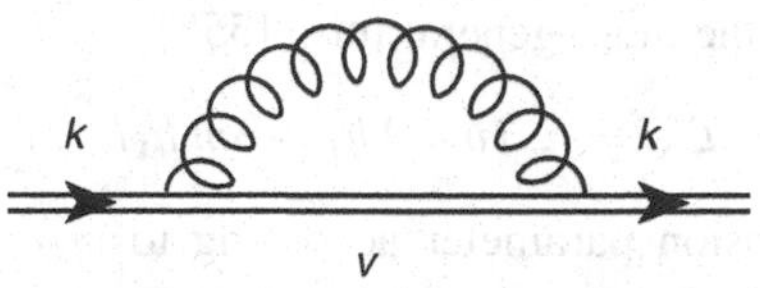


Fig. 44.3. One-loop self-energy $-i\Sigma(v\cdot k)$ of a heavy quark in the HQET.

one-loop order:

$$\begin{aligned}\Sigma(v\cdot k) &= -ig_s^2\,t_a t_a \int \frac{\mathrm{d}^n t}{(2\pi)^n}\,\frac{1}{(t^2+i\eta)[v\cdot(t+k)+i\eta]}\\ &= -2iC_F g_s^2 \int_0^\infty \mathrm{d}\lambda \int \frac{\mathrm{d}^n t}{(2\pi)^n}\,\frac{1}{[t^2+2\lambda\, v\cdot(t+k)+i\eta]^2}\\ &= \frac{C_F\alpha_s}{2\pi}\,\Gamma(\epsilon)\int_0^\infty \mathrm{d}\lambda\left(\frac{\lambda^2+\lambda\omega}{4\pi\mu^2}\right)^{-\epsilon}, \end{aligned} \tag{44.23}$$

where $C_F = 4/3$ is a colour factor, $\lambda$ is a dimensionful Feynman parameter, and $\omega = -2v\cdot k > 0$ acts as an IR cutoff. A straightforward calculation leads to:

$$\begin{aligned}\frac{\partial\Sigma(v\cdot k)}{\partial v\cdot k} &= \frac{C_F\alpha_s}{\pi}\,\Gamma(1+\epsilon)\left(\frac{\omega^2}{4\pi\mu^2}\right)^{-\epsilon}\int_0^1 \mathrm{d}z\, z^{-1+2\epsilon}\,(1-z)^{-\epsilon}\\ &= \frac{C_F\alpha_s}{\pi}\,\Gamma(2\epsilon)\,\Gamma(1-\epsilon)\left(\frac{\omega^2}{4\pi\mu^2}\right)^{-\epsilon}, \end{aligned} \tag{44.24}$$

where we have substituted $\lambda = \omega(1-z)/z$. From an expansion around $\epsilon = 0$, we obtain:

$$Z_h = 1 + \frac{C_F\alpha_s}{2\pi\epsilon}\,. \tag{44.25}$$

This result was first derived by Politzer and Wise [572].

### *44.3.5 Residual mass term and definition of the heavy quark mass*

In the derivation presented earlier in this section, $m_Q$ has been chosen to be the *mass in the Lagrangian*. Using this parameter in the phase redefinition in Eq. (44.5) we obtained the effective Lagrangian in Eq. (44.16), in which the heavy-quark mass no longer appears. However, this treatment has its subtleties. The symmetries of the HQET allow a *residual mass* $\delta m$ for the heavy quark, provided that $\delta m$ is of order $\Lambda_{\text{QCD}}$ and is the same for all heavy-quark flavours. Even if we arrange that such a mass term is not present at the tree level, it will in general be induced by quantum corrections. (This is unavoidable if the theory is regulated with a dimensionful cutoff.) Therefore, instead of Eq. (44.16) we should write the effective Lagrangian in the more general form [558]:

$$\mathcal{L}_\infty = \bar{h}_v \, iv \cdot D \, h_v - \delta m \, \bar{h}_v h_v \, . \tag{44.26}$$

If we redefine the expansion parameter according to $m_Q \to m_Q + \Delta m$, the residual mass changes in the opposite way: $\delta m \to \delta m - \Delta m$. This implies that there is a unique choice of the expansion parameter $m_Q$ such that $\delta m = 0$. Requiring $\delta m = 0$, as it is usually done implicitly in the HQET, defines a heavy-quark mass, which in perturbation theory coincides with the pole mass [133,147,148]. This, in turn, defines for each heavy hadron $H_Q$ a parameter $\bar{\Lambda}$ (sometimes called the *binding energy*) through

$$\bar{\Lambda} = (m_{H_Q} - m_Q)|_{m_Q \to \infty} \, . \tag{44.27}$$

If one prefers to work with another choice of the expansion parameter, the values of non-perturbative parameters such as $\bar{\Lambda}$ change, but at the same time one has to include the residual mass term in the HQET Lagrangian. However, like the pole mass, the previous definition might be affected by renormalons as we have discussed in previous chapters.

## 44.4 Hadron spectroscopy from HQET

The spin-flavour symmetry leads to many interesting relations between the properties of hadrons containing a heavy quark. The most direct consequences concern the spectroscopy of such states [559,560]. In the limit $m_Q \to \infty$, the spin of the heavy quark and the total angular momentum $j$ of the light degrees of freedom are separately conserved by the strong interactions. Because of heavy-quark symmetry, the dynamics is independent of the spin and mass of the heavy quark. Hadronic states can thus be classified by the quantum numbers (flavour, spin, parity, etc.) of their light degrees of freedom [561]. The spin symmetry predicts that, for fixed $j \neq 0$, there is a doublet of degenerate states with total spin $J = j \pm \frac{1}{2}$.

The flavour symmetry relates the properties of states with different heavy-quark flavour.

In general, the mass of a hadron $H_Q$ containing a heavy quark $Q$ obeys an expansion of the form:

$$m_{H_Q} = m_Q + \bar{\Lambda} + \frac{\Delta m^2}{2m_Q} + O(1/m_Q^2) \, . \tag{44.28}$$

The parameter $\bar{\Lambda}$ represents contributions arising from terms in the Lagrangian that are independent of the heavy-quark mass [558], whereas the quantity $\Delta m^2$ originates from the terms of order $1/m_Q$ in the effective Lagrangian of the HQET. For the ground-state pseudoscalar and vector mesons, one can parametrize the contributions from the kinetic energy and the chromomagnetic interaction in terms of two quantities $\lambda_1$ and $\lambda_2$, in such a way that [562]:

$$\Delta m^2 = -\lambda_1 + 2\left[J(J+1) - \tfrac{3}{2}\right] \lambda_2 \, , \tag{44.29}$$

where $J = j \pm 1/2$ is the total spin of the states. The hadronic parameters $\bar{\Lambda}$, $\lambda_1$ and $\lambda_2$ are independent of $m_Q$. They characterize the properties of the light constituents.

Consider, as a first example, the SU(3) mass splittings for heavy mesons. The heavy quark expansion predicts that:

$$m_{B_S} - m_{B_d} = \bar{\Lambda}_s - \bar{\Lambda}_d + O(1/m_b)\,,$$
$$m_{D_S} - m_{D_d} = \bar{\Lambda}_s - \bar{\Lambda}_d + O(1/m_c)\,, \tag{44.30}$$

where we have indicated that the value of the parameter $\bar{\Lambda}$ depends on the flavour of the light quark. Thus, to the extent that the charm and bottom quarks can both be considered sufficiently heavy, the mass splittings should be similar in the two systems. This prediction is confirmed experimentally, since:

$$m_{B_S} - m_{B_d} = (90 \pm 3)\ \text{MeV}\,,$$
$$m_{D_S} - m_{D_d} = (99 \pm 1)\ \text{MeV}\,. \tag{44.31}$$

As a second example, consider the spin splittings between the ground-state pseudoscalar ($J = 0$) and vector ($J = 1$) mesons, which are the members of the spin-doublet with $j = \frac{1}{2}$. From Eqs. (44.28) and (44.29), it follows that

$$m_{B*}^2 - m_B^2 = 4\lambda_2 + O(1/m_b)\,,$$
$$m_{D*}^2 - m_D^2 = 4\lambda_2 + O(1/m_c)\,. \tag{44.32}$$

The data are compatible with this prediction:

$$m_{B*}^2 - m_B^2 \approx 0.49\ \text{GeV}^2\,,$$
$$m_{D*}^2 - m_D^2 \approx 0.55\ \text{GeV}^2\,. \tag{44.33}$$

Assuming that the $B$ system is close to the heavy-quark limit, one can obtain the value:

$$\lambda_2 \approx 0.12\ \text{GeV}^2 \tag{44.34}$$

for one of the hadronic parameters in Eq. (44.29). This quantity plays an important role in the phenomenology of inclusive decays of heavy hadrons. Similar relations can also be obtained in the case of heavy baryons:

$$m_{\Lambda_b} - m_B - \frac{3\lambda_2}{2m_B} \simeq 311\ \text{MeV}\,,$$
$$m_{\Lambda_c} - m_D - \frac{3\lambda_2}{2m_D} \simeq 320\ \text{MeV}\,, \tag{44.35}$$

which are close to each other to be compared with the data. The dominant correction in Eq. (44.35) comes from the contribution of the chromo-magnetic interaction to the masses of the heavy mesons,[1] which adds a term $3\lambda_2/2m_Q$ on the right-hand side.

The mass formula in Eq. (44.28) can also be used to derive information on the heavy-quark masses from the observed hadron masses. Introducing the 'spin-averaged' meson masses

[1] Because of spin symmetry, there is no such contribution to the masses of $\Lambda_Q$ baryons.

$\bar{m}_B = \frac{1}{4}(m_B + 3m_{B^*}) \approx 5.31$ GeV and $\bar{m}_D = \frac{1}{4}(m_D + 3m_{D^*}) \approx 1.97$ GeV, we find that:

$$m_b - m_c = (\bar{m}_B - \bar{m}_D)\left\{1 - \frac{\lambda_1}{2\bar{m}_B\bar{m}_D} + O\left(1/m_Q^3\right)\right\}. \tag{44.36}$$

Using theoretical estimates for the parameter $\lambda_1$, which lie in the range (for a complete reference, see e.g. [545]):

$$\lambda_1 = -(0.3 \pm 0.2)\ \text{GeV}^2\,, \tag{44.37}$$

this relation leads to:

$$m_b - m_c = (3.39 \pm 0.03 \pm 0.03)\ \text{GeV}\,, \tag{44.38}$$

where the first error reflects the uncertainty in the value of $\lambda_1$, and the second one takes into account unknown higher-order corrections. The fact that the difference of the pole masses, $m_b - m_c$, is known rather precisely is important for the analysis of inclusive decays of heavy hadrons.

## 44.5 The $\bar{B} \to D^* l \bar{\nu}$ exclusive process

We shall be concerned here with the semi-leptonic decay process $\bar{B} \to D^* l \bar{\nu}$ shown schematically in Fig. 44.4, and which has the largest branching fraction of all $B$-meson decay modes.

The strength of the $b \to c$ transition vertex is governed by the element $V_{cb}$ of the CKM matrix, which is a fundamental parameter of the Standard Model. A primary goal of the study of semi-leptonic decays of $B$ mesons is to extract with high precision the values of $|V_{cb}|$ (as well as $|V_{ub}|$ for $b \to u$ transitions).

### *44.5.1 Semi-leptonic form factors: the Isgur–Wise function*

Heavy-quark symmetry implies relations between the weak decay form factors of heavy mesons, which are of particular interest. These relations have been derived by Isgur and Wise [557], generalizing ideas developed by Nussinov and Wetzel [563], and by Voloshin and Shifman [564,565].

Consider the elastic scattering of a $B$ meson, $\bar{B}(v) \to \bar{B}(v')$, induced by a vector current coupled to the $b$ quark. Before the action of the current, the light degrees of freedom inside the $B$ meson orbit around the heavy quark, which acts as a static source of colour. On

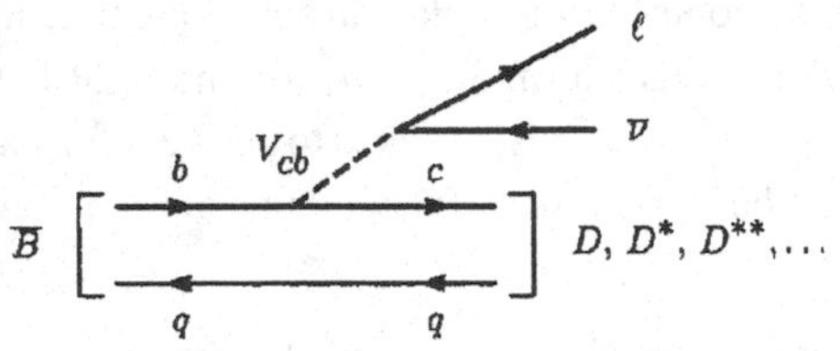


Fig. 44.4. Semi-leptonic decays of $B$ mesons.

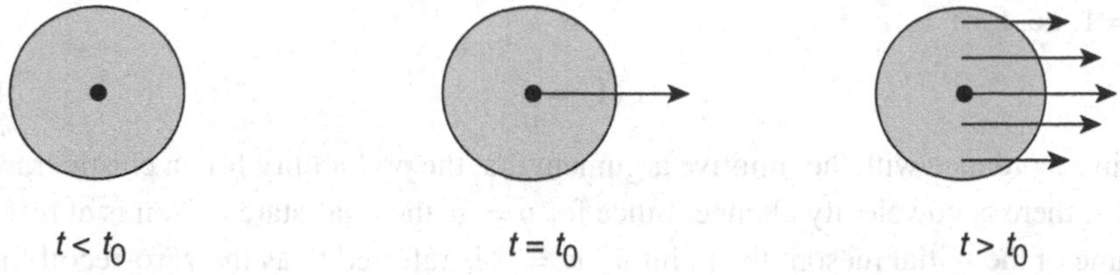


Fig. 44.5. Elastic transition induced by an external heavy-quark current.

average, the $b$ quark and the $B$ meson have the same velocity $v$. The action of the current is to replace instantaneously (at time $t = t_0$) the colour source by one moving at a velocity $v'$, as indicated in Fig. 44.5. If $v = v'$, nothing happens; the light degrees of freedom do not realize that there was a current acting on the heavy quark. If the velocities are different, however, the light constituents suddenly find themselves interacting with a moving colour source. Soft gluons have to be exchanged to rearrange them so as to form a $B$ meson moving at velocity $v'$. This rearrangement leads to a form-factor suppression, reflecting the fact that, as the velocities become more and more different, the probability for an elastic transition decreases. The important observation is that, in the limit $m_b \to \infty$, the form factor can only depend on the Lorentz boost $\gamma = v \cdot v'$ connecting the rest frames of the initial- and final-state mesons. Thus, in this limit a dimensionless probability function $\xi(v \cdot v')$ describes the transition. It is called the Isgur–Wise function [557]. In the HQET, which provides the appropriate framework for taking the limit $m_b \to \infty$, the hadronic matrix element describing the scattering process can thus be written as:

$$\frac{1}{m_B} \langle \bar{B}(v')| \bar{b}_{v'} \gamma^\mu b_v |\bar{B}(v)\rangle = \xi(v \cdot v')(v + v')^\mu . \tag{44.39}$$

Here $b_v$ and $b_{v'}$ are the velocity-dependent heavy-quark fields of the HQET. It is important that the function $\xi(v \cdot v')$ does not depend on $m_b$. The factor $1/m_B$ on the left-hand side compensates for a trivial dependence on the heavy-meson mass caused by the relativistic normalization of meson states, which is conventionally taken to be:

$$\langle \bar{B}(p')|\bar{B}(p)\rangle = 2m_B v^0 (2\pi)^3 \delta^3(\vec{p} - \vec{p}\,') . \tag{44.40}$$

Note that there is no term proportional to $(v - v')^\mu$ in Eq. (44.39). This can be seen by contracting the matrix element with $(v - v')_\mu$, which must give zero since $\not{v} b_v = b_v$ and $\bar{b}_{v'} \not{v}' = \bar{b}_{v'}$.

It is more conventional to write the above matrix element in terms of an elastic form factor $F_{\rm el}(q^2)$ depending on the momentum transfer $q^2 = (p - p')^2$:

$$\langle \bar{B}(v')| \bar{b} \gamma^\mu b |\bar{B}(v)\rangle = F_{\rm el}(q^2)(p + p')^\mu , \tag{44.41}$$

where $p^{(\prime)} = m_B v^{(\prime)}$. Comparing this with Eq. (44.39), we find that

$$F_{\rm el}(q^2) = \xi(v \cdot v') , \qquad q^2 = -2m_B^2(v \cdot v' - 1) . \tag{44.42}$$

Because of current conservation, the elastic form factor is normalized to unity at $q^2 = 0$. This condition implies the normalization of the Isgur–Wise function at the kinematic point

$v \cdot v' = 1$, i.e. for $v = v'$:

$$\xi(1) = 1\,. \tag{44.43}$$

It is in accordance with the intuitive argument that the probability for an elastic transition is unity if there is no velocity change. Since for $v = v'$ the final-state meson is at rest in the rest frame of the initial meson, the point $v \cdot v' = 1$ is referred to as the zero-recoil limit.

The heavy-quark flavour symmetry can be used to replace the $b$ quark in the final-state meson by a $c$ quark, thereby turning the $B$ meson into a $D$ meson. Then the scattering process turns into a weak decay process. In the infinite-mass limit, the replacement $b_{v'} \to c_{v'}$ is a symmetry transformation, under which the effective Lagrangian is invariant. Hence, the matrix element:

$$\frac{1}{\sqrt{m_B m_D}} \langle D(v')|\, \bar{c}_{v'} \gamma^\mu b_v \,|\bar{B}(v)\rangle = \xi(v \cdot v')\,(v + v')^\mu \tag{44.44}$$

is still determined by the same function $\xi(v \cdot v')$. This is interesting, since in general the matrix element of a flavour-changing current between two pseudoscalar mesons is described by two form factors:

$$\langle D(v')|\, \bar{c}\, \gamma^\mu b \,|\bar{B}(v)\rangle = f_+(q^2)\,(p + p')^\mu - f_-(q^2)\,(p - p')^\mu\,. \tag{44.45}$$

Comparing the above two equations, we find that:

$$\begin{aligned} f_\pm(q^2) &= \frac{m_B \pm m_D}{2\sqrt{m_B m_D}}\, \xi(v \cdot v')\,, \\ q^2 &= m_B^2 + m_D^2 - 2 m_B m_D\, v \cdot v'\,. \end{aligned} \tag{44.46}$$

Thus, the heavy-quark flavour symmetry relates two a priori independent form factors to one and the same function. Moreover, the normalization of the Isgur–Wise function at $v \cdot v' = 1$ now implies a non-trivial normalization of the form factors $f_\pm(q^2)$ at the point of maximum momentum transfer, $q^2_{\max} = (m_B - m_D)^2$:

$$f_\pm\left(q^2_{\max}\right) = \frac{m_B \pm m_D}{2\sqrt{m_B m_D}}\,. \tag{44.47}$$

The heavy-quark spin symmetry leads to additional relations among weak decay form factors. It can be used to relate matrix elements involving vector mesons to those involving pseudoscalar mesons. A vector meson with longitudinal polarization is related to a pseudoscalar meson by a rotation of the heavy-quark spin. Hence, the spin-symmetry transformation $c_{v'}^{\Uparrow} \to c_{v'}^{\Downarrow}$ relates $\bar{B} \to D$ with $\bar{B} \to D^*$ transitions. The result of this transformation is [557]:

$$\begin{aligned} \frac{1}{\sqrt{m_B m_{D^*}}} \langle D^*(v', \varepsilon)|\, \bar{c}_{v'} \gamma^\mu b_v \,|\bar{B}(v)\rangle &= i\epsilon^{\mu\nu\alpha\beta}\, \varepsilon^*_\nu\, v'_\alpha v_\beta\; \xi(v \cdot v')\,, \\ \frac{1}{\sqrt{m_B m_{D^*}}} \langle D^*(v', \varepsilon)|\, \bar{c}_{v'} \gamma^\mu \gamma_5\, b_v \,|\bar{B}(v)\rangle &= [\varepsilon^{*\mu}\,(v \cdot v' + 1) - v'^\mu\, \varepsilon^* \cdot v]\xi(v \cdot v')\,, \end{aligned} \tag{44.48}$$

where $\varepsilon$ denotes the polarization vector of the $D^*$ meson. Once again, the matrix elements are completely described in terms of the Isgur–Wise function. Now this is even more remarkable, since in general four form factors, $V(q^2)$ for the vector current, and $A_i(q^2)$, $i = 0, 1, 2$, for the axial current, are required to parametrize these matrix elements. In the heavy-quark limit, they obey the relations [566]

$$\begin{aligned}\frac{m_B + m_{D^*}}{2\sqrt{m_B m_{D^*}}}\,\xi(v\cdot v') &= V(q^2) = A_0(q^2) = A_1(q^2)\\ &= \left[1 - \frac{q^2}{(m_B+m_D)^2}\right]^{-1} A_1(q^2)\,,\\ q^2 &= m_B^2 + m_{D^*}^2 - 2m_B m_{D^*}\, v\cdot v'\,. \end{aligned} \tag{44.49}$$

Equations (44.46) and (44.49) summarize the relations imposed by heavy-quark symmetry on the weak decay form factors describing the semi-leptonic decay processes $\bar{B} \to D\,\ell\,\bar{\nu}$ and $\bar{B} \to D^*\ell\,\bar{\nu}$. These relations are model-independent consequences of QCD in the limit where $m_b, m_c \gg \Lambda_{\rm QCD}$. They play a crucial role in the determination of the CKM matrix element $|V_{cb}|$. In terms of the recoil variable $w = v\cdot v'$, the differential semi-leptonic decay rates in the heavy-quark limit become [567]:

$$\begin{aligned}\frac{{\rm d}\Gamma(\bar{B}\to D\,\ell\,\bar{\nu})}{{\rm d}w} &= \frac{G_F^2}{48\pi^3}\,|V_{cb}|^2\,(m_B+m_D)^2\,m_D^3\,(w^2-1)^{3/2}\,\xi^2(w)\,,\\ \frac{{\rm d}\Gamma(\bar{B}\to D^*\ell\,\bar{\nu})}{{\rm d}w} &= \frac{G_F^2}{48\pi^3}\,|V_{cb}|^2\,(m_B-m_{D^*})^2\,m_{D^*}^3\,\sqrt{w^2-1}\,(w+1)^2\\ &\times\left[1+\frac{4w}{w+1}\,\frac{m_B^2-2w\,m_B m_{D^*}+m_{D^*}^2}{(m_B-m_{D^*})^2}\right]\xi^2(w)\,. \end{aligned} \tag{44.50}$$

### 44.5.2 The Luke's theorem for the $1/m_Q$ corrections

These expressions receive symmetry-breaking corrections, since the masses of the heavy quarks are not infinitely large. Perturbative corrections of order $\alpha_s^n(m_Q)$ can be calculated order-by-order in perturbation theory. A more difficult task is to control the non-perturbative power corrections of order $(\Lambda_{\rm QCD}/m_Q)^n$. The HQET provides a systematic framework for analysing these corrections. For the case of weak-decay form factors the analysis of the $1/m_Q$ corrections was performed by Luke [568], where, in the zero-recoil limit, an analogue of the Ademollo–Gatto theorem [569] can be proved. This is *Luke's theorem* [568], which states that the matrix elements describing the leading $1/m_Q$ corrections to weak decay amplitudes vanish at zero recoil. This theorem is valid to all orders in perturbation theory [562,570,571], and then protects the $\bar{B} \to D^*\ell\,\bar{\nu}$ decay rate from receiving first-order $1/m_Q$ corrections at zero recoil [567]. A similar statement is not true for the decay $\bar{B} \to D\,\ell\,\bar{\nu}$. The reason is simple but somewhat subtle. Luke's theorem protects only those form factors not multiplied by kinematic factors that vanish for $v = v'$. By angular momentum conservation, the two pseudoscalar mesons in the decay $\bar{B} \to D\,\ell\,\bar{\nu}$ must be in a relative $p$ wave, and hence the amplitude is proportional to the velocity $|\vec{v}_D|$ of the $D$ meson in the $B$-meson rest frame. This leads to a factor $(w^2 - 1)$ in the decay rate. In such a situation, kinematically suppressed form factors can contribute [566]. Later, the authors in [562] have analysed the structure of $1/m_Q^2$ corrections for both meson and baryon weak decay form factors [562].

### 44.5.3 Short-distance corrections and matching conditions

We have shown previously that HQET reproduces correctly the non-perturbative part of the full theory but does not contain correctly its short-distance part. This can be understood by denoting that the heavy quark only participates to strong interactions through its interaction with gluons, where hard gluons can resolve the spin and flavour quantum numbers of a heavy quark. Their effects lead to a renormalization of the coefficients of the operators in the HQET. A new feature of such short-distance corrections is that through the running coupling constant they induce a logarithmic dependence on the heavy-quark mass [564], which can be calculated in perturbation theory, since $\alpha_s(m_Q)$ is small.

Let us for example, consider the matrix elements of the vector current $V = \bar{q}\,\gamma^\mu Q$. In QCD this current is partially conserved and needs no renormalization. Therefore, its matrix elements are free of UV divergences. Still, these matrix elements have a logarithmic dependence on $m_Q$ from the exchange of hard gluons with virtual momenta of the order of the heavy-quark mass. If one goes over to the effective theory by taking the limit $m_Q \to \infty$, these logarithms diverge. Consequently, the vector current in the effective theory does require a renormalization [572]. Its matrix elements depend on an arbitrary renormalization scale $\nu$, which separates the regions of short- and long-distance physics. If $\nu$ is chosen such that $\Lambda_{\rm QCD} \ll \nu \ll m_Q$, the effective coupling constant in the region between $\nu$ and $m_Q$ is small, and perturbation theory can be used to compute the short-distance corrections. These corrections have to be added to the matrix elements of the effective theory, which contain the long-distance physics below the scale $\nu$. The relation between matrix elements in the full and in the effective theory is:

$$\langle\, V(m_Q)\,\rangle_{\rm QCD} = C_0(m_Q, \nu)\,\langle V_0(\nu)\rangle_{\rm HQET} + \frac{C_1(m_Q, \nu)}{m_Q}\,\langle V_1(\nu)\rangle_{\rm HQET} + \cdots\,, \qquad (44.51)$$

where we have indicated that matrix elements in the full theory depend on $m_Q$, whereas matrix elements in the effective theory are mass-independent, but do depend on the renormalization scale. The Wilson coefficients $C_i(m_Q, \nu)$ are defined by this relation. Order by order in perturbation theory, they can be computed from a comparison of the matrix elements in the two theories. Since the effective theory is constructed to reproduce correctly the low-energy behaviour of the full theory, this *matching* procedure is independent of any long-distance physics, such as IR singularities, non-perturbative effects, and the nature of the external states used in the matrix elements.

The coefficient functions can be evaluated in perturbation theory using the renormalization group equation. Most of the existing calculations of short-distance corrections in the HQET can be found, for example, in [545].

### 44.5.4 Determination of $|V_{cb}|$ from HQET

For this purpose, one considers the decay rate given Eq. (44.50), where the Isgur–Wise function $\xi^2(w)$ is replaced by the function $\mathcal{F}(w)$, which takes into account corrections of the order $\alpha_s(m_Q)$ and $\Lambda_{\rm QCD}/m_Q$ to the Isgur–Wise function. The aim is to measure the quantity $|V_{cb}|\mathcal{F}(w)$ as a function of $w$, and and to extract $|V_{cb}|$ from an extrapolation of the data to the zero-recoil point $w = 1$, where the $B$ and the $D^*$ mesons have a common rest frame. At this kinematic point, heavy-quark symmetry helps us to calculate the normalization $\mathcal{F}(1)$ with small and controlled theoretical errors. Since the range of $w$ values accessible in this decay is rather small ($1 < w < 1.5$), the extrapolation can be done using an expansion around $w = 1$:

$$\mathcal{F}(w) = \mathcal{F}(1)\,[1 - \hat{\varrho}^2\,(w-1) + \hat{c}\,(w-1)^2 \ldots]\,. \tag{44.52}$$

The slope $\hat{\varrho}^2$ and the curvature $\hat{c}$, and indeed more generally the complete shape of the form factor, are tightly constrained by analyticity and unitarity requirements [573,574]. In the long run, the statistics of the experimental results close to zero recoil will be such that these theoretical constraints will not be crucial to get a precision measurement of $|V_{cb}|$. They will, however, enable strong consistency checks. Measurements of the recoil spectrum have been performed by several experimental groups. Figure 44.6 shows, as an example, the data reported some time ago by the CLEO Collaboration. The weighted average of the

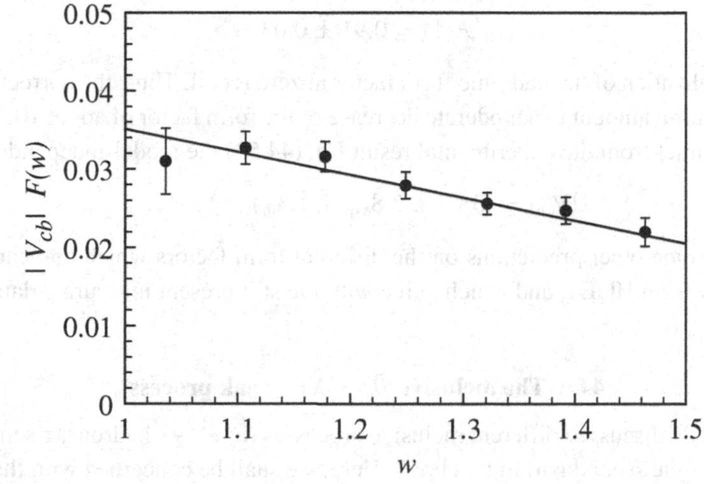


Fig. 44.6. CLEO data for the product $|V_{cb}|\,\mathcal{F}(w)$, as extracted from the recoil spectrum in $\bar{B} \to D^*\ell\,\bar{\nu}$ decays [575]. The line shows a linear fit to the data.

experimental results is [576]:

$$|V_{cb}|\,\mathcal{F}(1) = (35.2 \pm 2.6) \times 10^{-3}\,. \tag{44.53}$$

Heavy-quark symmetry implies that the general structure of the symmetry-breaking corrections to the form factor at zero recoil is [567]:

$$\mathcal{F}(1) = \eta_A \left(1 + 0 \times \frac{\Lambda_{\rm QCD}}{m_Q} + {\rm const} \times \frac{\Lambda^2_{\rm QCD}}{m_Q^2} + \cdots \right) \equiv \eta_A\,(1 + \delta_{1/m^2})\,, \tag{44.54}$$

where $\eta_A$ is a short-distance correction arising from the finite renormalization of the flavour-changing axial current at zero recoil, and $\delta_{1/m^2}$ parametrizes second-order (and higher) power corrections. The absence of first-order power corrections at zero recoil is a consequence of Luke's theorem [568]. The one-loop expression for $\eta_A$ has been known for a long time [577,565,578]:

$$\eta_A = 1 + \frac{\alpha_s(M)}{\pi}\left(\frac{m_b + m_c}{m_b - m_c}\ln\frac{m_b}{m_c} - \frac{8}{3}\right) \approx 0.96\,. \tag{44.55}$$

The scale $M \approx 0.51\sqrt{m_b m_c}$ in the running coupling constant can be fixed [579] by adopting the BLM prescription [173]. This lowest order value has been confirmed by the two-loop result [580]:

$$\eta_A|_{2-{\rm loop}} \simeq 0.960 \pm 0.007\,. \tag{44.56}$$

The different analysis of power corrections are more uncertain. The results are in the range:

$$\delta_{1/m^2} \simeq -(0.055 \pm 0.025)\,. \tag{44.57}$$

These different results lead to:

$$\mathcal{F}(1) = 0.91 \pm 0.03 \tag{44.58}$$

for the normalization of the hadronic form factor at zero recoil. Thus, the corrections to the heavy-quark limit amount to a moderate decrease of the form factor of about 10%. This can be used to extract from the experimental result Eq. (44.53) the model-independent value

$$|V_{cb}| = (38.7 \pm 2.8_{\rm exp} \pm 1.3_{\rm th}) \times 10^{-3}\,. \tag{44.59}$$

There are some other predictions on the different form factors which one can obtain in the same way from HQET, and which agree with the still present inaccurate data.

### 44.6 The inclusive $\bar{B} \to Xl\bar{\nu}$ weak process

We have already discussed different inclusive processes ($e^+e^- \to$ hadrons, $\tau$ semi-leptonic decays, ...) in the second part of this book. Here, we shall be concerned with the inclusive $\bar{B} \to Xl\bar{\nu}$ weak process involving a heavy quark. From a theoretical point of view such decays have two advantages: first, bound-state effects related to the initial state, such as the 'Fermi motion' of the heavy quark inside the hadron [581,582], can be accounted for in a systematic way using the heavy-quark expansion; secondly, the fact that the final state consists of a sum over many hadronic channels eliminates bound-state effects related to the properties of individual hadrons. This second feature is based on the hypothesis of quark-hadron duality, which is an important concept in QCD phenomenology. The assumption of duality is that cross-sections and decay rates, which are defined in the physical region (i.e. the region of time-like momenta), are calculable in QCD after a 'smearing' or 'averaging' procedure has been applied [583]. In semi-leptonic decays, it is the integration over the lepton and neutrino phase space that provides a smearing over the invariant hadronic mass of the final state (so-called global duality). For non-leptonic decays, on the other hand, the total hadronic mass is fixed, and it is only the fact that one sums over many hadronic states that provides an averaging (so-called local duality[2]). Clearly, local duality is a stronger assumption than global duality. It is important to stress that quark-hadron duality cannot yet be derived from first principles; still, it is a necessary assumption for many applications of QCD. The success of the QCD predictions for the hadronic $\tau$ widths is a strong test of the validity of global duality [325,328,346,345].

Using the optical theorem, the inclusive decay width of a hadron $H_b$ containing a $b$ quark can be written in the form:

$$\Gamma(H_b \to X) = \frac{1}{m_{H_b}} \,\mathrm{Im}\, \langle H_b | \,\mathbf{T}\, | H_b \rangle \,, \tag{44.60}$$

where the transition operator $\mathbf{T}$ is given by:

$$\mathbf{T} = i \int \mathrm{d}^4x \; T\{\mathcal{L}_{\mathrm{eff}}(x), \mathcal{L}_{\mathrm{eff}}(0)\} \,. \tag{44.61}$$

Inserting a complete set of states inside the time-ordered product, we recover the standard expression

$$\Gamma(H_b \to X) = \frac{1}{2m_{H_b}} \sum_X (2\pi)^4\, \delta^4(p_H - p_X)\, |\langle X | \,\mathcal{L}_{\mathrm{eff}}\, | H_b \rangle|^2 \tag{44.62}$$

for the decay rate. For the case of semi-leptonic and non-leptonic decays, $\mathcal{L}_{\mathrm{eff}}$ is the effective Fermi weak Lagrangian, which, in practice is corrected for short-distance effects [550,551,584–586] arising from the exchange of gluons with virtualities between $m_W$ and $m_b$. In the case of the inclusive semi-leptonic decay rate, for instance, the sum would include only those states $X$ containing a lepton-neutrino pair. In perturbation theory, some contributions to the transition operator are given by the two-loop diagrams shown on the left-hand side in Fig. 44.7. Because of the large mass of the $b$ quark, the momenta flowing through the internal propagator lines are large. It is thus possible to construct an OPE for the transition operator, in which $\mathbf{T}$ is represented as a series of local operators containing the heavy-quark fields. The operator with the lowest dimension, $d = 3$, is $\bar{b}b$. It arises by contracting the internal lines of the first diagram. In the usual OPE, the only gauge-invariant

[2] This terminology may differ with the local duality used in the QCD spectral sum rules analysis which will be discussed in a future part of this book.

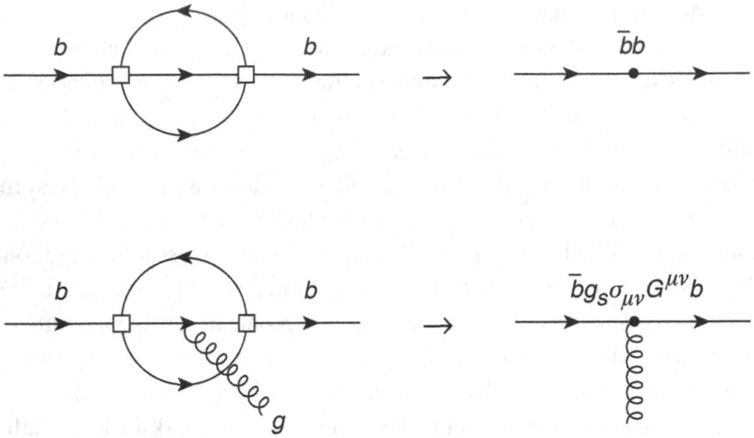


Fig. 44.7. Perturbative contributions to the transition operator **T** (left), and the corresponding operators in the OPE (right). The open squares represent a four-fermion interaction of the effective Lagrangian $\mathcal{L}_{\text{eff}}$, and the black circles represent local operators in the OPE.

operator with dimension four is $\bar{b}\, i\not{D}\, b$; however, the equations of motion imply that between physical states this operator can be replaced by $m_b\bar{b}b$. The first operator that is different from $\bar{b}b$ has dimension five and contains the gluon field. It is given by $\bar{b}\, g_s\sigma_{\mu\nu}G^{\mu\nu}b$. This operator arises from diagrams in which a gluon is emitted from one of the internal lines, such as the second diagram shown in Fig. 44.7. For dimensional reasons, the matrix elements of such higher-dimensional operators are suppressed by inverse powers of the heavy-quark mass. Thus, any inclusive decay rate of a hadron $H_b$ can be written as [587–589]:

$$\Gamma(H_b \to X_f) = \frac{G_F^2 m_b^5}{192\pi^3} \left\{ c_3^f\, \langle \bar{b}b\rangle_H + c_5^f\, \frac{\langle \bar{b}\, g_s\sigma_{\mu\nu}G^{\mu\nu}b\rangle_H}{m_b^2} + \cdots \right\}, \quad (44.63)$$

where the prefactor arises naturally from the loop integrations, $c_n^f$ are calculable coefficient functions (which also contain the relevant CKM matrix elements) depending on the quantum numbers $f$ of the final state, and $\langle O\rangle_H$ are the (normalized) forward matrix elements of local operators, for which we use the short-hand notation:

$$\langle O\rangle_H = \frac{1}{2m_{H_b}}\, \langle H_b|\, O\, |H_b\rangle\,. \quad (44.64)$$

In the next step, these matrix elements are systematically expanded in powers of $1/m_b$, using the technology of the HQET. The result is [562,587,589]:

$$\begin{aligned} \langle \bar{b}b\rangle_H &= 1 - \frac{\mu_\pi^2(H_b) - \mu_G^2(H_b)}{2m_b^2} + O\big(1/m_b^3\big)\,, \\ \langle \bar{b}\, g_s\sigma_{\mu\nu}G^{\mu\nu}b\rangle_H &= 2\mu_G^2(H_b) + O(1/m_b)\,, \end{aligned} \quad (44.65)$$

where we have defined the HQET matrix elements:

$$\mu_\pi^2(H_b) = \frac{1}{2m_{H_b}} \langle H_b(v)|\,\bar{b}_v\,(i\vec{D})^2\,b_v\,|H_b(v)\rangle\,,$$
$$\mu_G^2(H_b) = \frac{1}{2m_{H_b}} \langle H_b(v)|\,\bar{b}_v \frac{g_s}{2}\sigma_{\mu\nu}G^{\mu\nu}b_v\,|H_b(v)\rangle\,. \tag{44.66}$$

Here $(i\vec{D})^2 = (iv\cdot D)^2 - (iD)^2$; in the rest frame, this is the square of the operator for the spatial momentum of the heavy quark. Inserting these results into Eq. (44.63) yields:

$$\Gamma(H_b \to X_f) = \frac{G_F^2 m_b^5}{192\pi^3}\left\{c_3^f\left(1 - \frac{\mu_\pi^2(H_b) - \mu_G^2(H_b)}{2m_b^2}\right) + 2c_5^f\,\frac{\mu_G^2(H_b)}{m_b^2} + \cdots\right\}. \tag{44.67}$$

It is instructive to understand the appearance of the 'kinetic energy' contribution $\mu_\pi^2$, which is the gauge-covariant extension of the square of the $b$-quark momentum inside the heavy hadron. This contribution is the field-theory analogue of the Lorentz factor $(1 - \vec{v}_b^{\,2})^{1/2} \simeq 1 - \vec{k}^2/2m_b^2$, in accordance with the fact that the lifetime, $\tau = 1/\Gamma$, for a moving particle increases due to time dilatation.

The main result of the heavy-quark expansion for inclusive decay rates is the observation that the free quark decay (i.e. the parton model) provides the first term in a systematic $1/m_b$ expansion [590]. For dimensional reasons, the corresponding rate is proportional to the fifth power of the $b$-quark mass. The non-perturbative corrections, which arise from bound-state effects inside the $B$ meson, are suppressed by at least two powers of the heavy-quark mass; thus they are of relative order $(\Lambda_{\rm QCD}/m_b)^2$. Note that the absence of first-order power corrections is a consequence of the equations of motion, as there is no independent gauge-invariant operator of dimension four that could appear in the OPE. The fact that bound-state effects in inclusive decays are strongly suppressed explains a posteriori the success of the parton model in describing such processes [591,592].

The hadronic matrix elements appearing in the heavy-quark expansion in Eq. (44.67) can be determined to some extent from the known masses of heavy hadron states. For the $B$ meson, one finds that:

$$\mu_\pi^2(B) = -\lambda_1 = (0.3 \pm 0.2)\ \text{GeV}^2\,,$$
$$\mu_G^2(B) = 3\lambda_2 \approx 0.36\ \text{GeV}^2\,, \tag{44.68}$$

where $\lambda_1$ and $\lambda_2$ are the parameters appearing in the mass formula of Eq. (44.29). For the ground-state baryon $\Lambda_b$, in which the light constituents have total spin zero, it follows that:

$$\mu_G^2(\Lambda_b) = 0\,, \tag{44.69}$$

while the matrix element $\mu_\pi^2(\Lambda_b)$ obeys the relation:

$$(m_{\Lambda_b} - m_{\Lambda_c}) - (\bar{m}_B - \bar{m}_D) = \left[\mu_\pi^2(B) - \mu_\pi^2(\Lambda_b)\right]\left(\frac{1}{2m_c} - \frac{1}{2m_b}\right) + O(1/m_Q^2)\,, \tag{44.70}$$

where $\bar{m}_B$ and $\bar{m}_D$ denote the spin-averaged masses introduced in connection with Eq. (44.36). The above relation implies:

$$\mu_\pi^2(B) - \mu_\pi^2(\Lambda_b) = (0.01 \pm 0.03)\ \mathrm{GeV}^2\,. \tag{44.71}$$

What remains to be calculated, then, is the coefficient functions $c_n^f$ for a given inclusive decay channel.

To illustrate this general formalism, we discuss as an example the determination of $|V_{cb}|$ from inclusive semi-leptonic $B$ decays. In this case the short-distance coefficients in the general expression (44.67) are given by [587–589]

$$\begin{aligned} c_3^{\mathrm{SL}} &= |V_{cb}|^2[1 - 8x^2 + 8x^6 - x^8 - 12x^4 \ln x^2 + O(\alpha_s)]\,, \\ c_5^{\mathrm{SL}} &= -6|V_{cb}|^2(1 - x^2)^4\,. \end{aligned} \tag{44.72}$$

Here $x = m_c/m_b$, and $m_b$ and $m_c$ are the masses of the $b$ and $c$ quarks, defined to a given order in perturbation theory [133,147,148]. The $O(\alpha_s)$ terms in $c_3^{\mathrm{SL}}$ are known exactly [593], and reliable estimates exist for the $O(\alpha_s^2)$ corrections [594]. The theoretical uncertainties in this determination of $|V_{cb}|$ are quite different from those entering the analysis of exclusive decays. The main sources are the dependence on the heavy-quark masses, higher-order perturbative corrections, and above all the assumption of global quark-hadron duality. A conservative estimate of the total theoretical error on the extracted value of $|V_{cb}|$ yields [595]:

$$|V_{cb}| = (0.040 \pm 0.003)\left[\frac{\mathrm{B_{SL}}}{10.5\%}\right]^{1/2}\left[\frac{1.6\,\mathrm{ps}}{\tau_B}\right]^{1/2} = (40 \pm 1_{\mathrm{exp}} \pm 3_{\mathrm{th}}) \times 10^{-3}\,. \tag{44.73}$$

The value of $|V_{cb}|$ extracted from the inclusive semi-leptonic width is in excellent agreement with the value in Eq. (44.59) obtained from the analysis of the exclusive decay $\bar{B} \to D^*\ell\,\bar{\nu}$. This agreement is gratifying given the differences of the methods used, and it provides an indirect test of global quark-hadron duality.

Combining the two measurements gives the final result:

$$|V_{cb}| = 0.039 \pm 0.002\,. \tag{44.74}$$

After $V_{ud}$ and $V_{us}$, this is the third-best known entry in the CKM matrix.

## 44.7 Rare $B$ decays and $CP$-violation

One of the main objectives of $B$-factories is to test the CKM mechanism, which predicts that all CP violation results from a single complex phase in the quark mixing matrix.

Indeed, the determination of the sides and angles of the 'unitarity triangle' $V_{ub}^* V_{ud} + V_{cb}^* V_{cd} + V_{tb}^* V_{td} = 0$ depicted in Fig. 44.8 plays a central role in the $B$ factory program. Adopting the standard phase conventions for the CKM matrix, only the two smallest elements in this relation, $V_{ub}^*$ and $V_{td}$, have non-vanishing imaginary parts (to an excellent approximation). In the standard model the angle $\beta = -\arg(V_{td})$ can be determined in a theoretically clean way by measuring the mixing-induced CP asymmetry in the decays

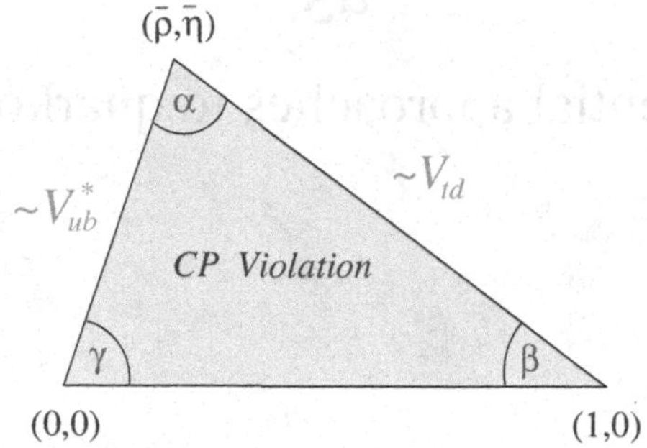


Fig. 44.8. The rescaled unitarity triangle representing the relation $1 + \frac{V_{ub}^* V_{ud}}{V_{cb}^* V_{cd}} + \frac{V_{tb}^* V_{td}}{V_{cb}^* V_{cd}} = 0$. The apex is determined by the Wolfenstein parameters $(\bar{\rho}, \bar{\eta})$. The area of the triangle is proportional to the strength of CP violation in the standard model.

$B \to J/\psi\, K_S$. Recents results from CDF [596] and especially from $B$-factories: Babar [597] and Belle [598] indicate a large value of $\beta$. The angle $\gamma = \arg(V_{ub}^*)$, or equivalently the combination $\alpha = 180° - \beta - \gamma$, is much harder to determine [595]. After the different announcements of evidence for a CP asymmetry in the decays $B \to J/\psi\, K_S$ and by direct CP violation in $K \to \pi\pi$ decays by the KTeV and NA48 groups [599], there are a lot of efforts for investigating theoretically these rare $B$ decay processes. Among others, two competing groups [600,601] work actively on these processes, but they have not yet reached any mutual agreements for their results.

# 45

# Potential approaches to quarkonia

## 45.1 The Schrödinger equation

As mentioned earlier, when the heavy quark mass $m_Q$ is much larger than the QCD scale $\Lambda_{\rm QCD}$, the running coupling $\alpha_s(m_Q)$ is small implying that at this scale of the order of the Compton wavelengh $\lambda \sim 1/m_Q$, one can safely use perturbative QCD for describing the hadrons. In this case the heavy $\bar{Q}Q$ bound states with the size $\lambda/\alpha_s(m_Q) \ll R_{\rm had} \sim 1$ fermi are hydrogen-like atoms to which ordinary quantum mechanics can be applied. In the *non-relativistic limit* (NR), it is possible to show that the interaction between the two $\bar{Q}$ and $Q$ states can be described by a local potential $V(\vec{r})$, where $\vec{r}$ is the relative coordinate between $Q$ and $\bar{Q}$ (spin is neglected for the moment). The energy levels and wave functions of the bound states can be found by solving the Schrödinger equation in three-dimensions:

$$E_{nl}\Psi_{nlm}(\vec{r}) = \left[-\frac{\hbar^2}{2\mu}\Delta + V(\vec{r})\right]\Psi_{nlm}(\vec{r}) \tag{45.1}$$

where $\mu \equiv m_Q/2$ is the reduced mass of the system; $\Psi_{nlm}(\vec{r})$ is the Schrödinger wave function; $V(\vec{r})$ is the interaction potential and $E_{nl}$ is the energy eigenvalue; $n, l$ and $m$ are respectively the principal quantum number, angular orbital, and eigenvalue of $l_z$ on the $z$-axis; $\hbar = 1$ in standard units:

$$\Delta \equiv \frac{\partial^2}{\partial x^2} + \frac{\partial^2}{\partial y^2} + \frac{\partial^2}{\partial z^2}\,, \tag{45.2}$$

is the Laplacian. In the case of central pontential, the wave function can be decomposed into its radial $R_{nl}(r)$ and spherical harmonic $Y_{lm}(\theta, \phi)$ components:

$$\Psi_{nlm}(\vec{r}) = R_{nl}(r)Y_{lm}(\theta, \phi)\,. \tag{45.3}$$

In terms of the reduced wave function:

$$u_{nl}(r) \equiv r R_{nl}(r)\,, \tag{45.4}$$

the Schrödinger equation becomes:

$$-\frac{d^2 u_{nl}}{dr^2} = -\frac{\hbar^2}{2\mu}\left[E_{nl} - V(r) - \frac{l(l+1)\hbar^2}{2\mu r^2}\right]u_{nl}(r)\,, \tag{45.5}$$

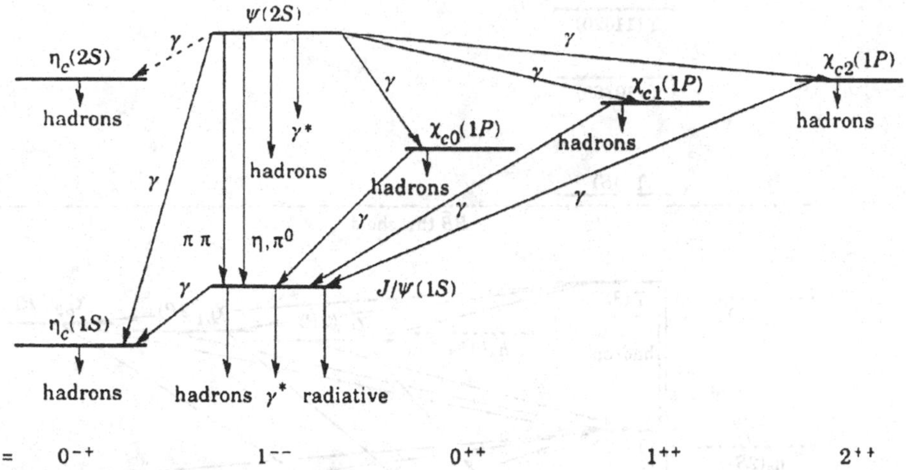


Fig. 45.1. Current situation of the charmonium system and transitions interpreted from some models. Dashed lines are uncertain states. $\gamma*$ refers to processes involving virtual photons, including decays into $e^+e^-$ and $\mu^+\mu^-$.

with the boundary conditions:

$$u_{nl}(0) = 0\,, \qquad \left.\frac{du_{nl}}{dr}\right|_{u=0} = R_{nl}(0)\,, \tag{45.6}$$

except that even parity solutions are inconsistent with the above boundary conditions. The wave function is normalized such that:

$$\int d^3\vec{r}\,|\Psi_{nlm}(\vec{r})|^2 = \int_0^\infty dr\,|u_{nl}(r)|^2 = 1\,. \tag{45.7}$$

It shows that the system is now described by the effective potential:

$$V_{\rm eff}(r) \equiv V(r) + \frac{l(l+1)\hbar^2}{2\mu r^2}\,, \tag{45.8}$$

Obviously, the main uncertainty for a quantitative spectroscopy is the choice of the correct $\bar{Q}Q$ potential $V(r)$ as far as its exact form is not yet known from first principles. We show in Fig. 45.1 the spectra of charmonium and in Fig. 45.2 those for the bottomium systems from [16]

## 45.2 The QCD static Coulomb potential

First, the model has to recover the short distance QCD *Coulomb static potential*. The expression of this potential can be derived from the tree-level scattering amplitude of the process:

$$\mathcal{A}[Q(p_1,\lambda_1,i_1) + \bar{Q}((p_2,\lambda_2,i_2) \to Q(p_1',\lambda_1',i_1') + \bar{Q}((p_2',\lambda_2',i_2')]\,, \tag{45.9}$$

shown at tree level in Fig. 45.3; $i$, $j$ and $\lambda_i$ are respectively colour and spinor indices.

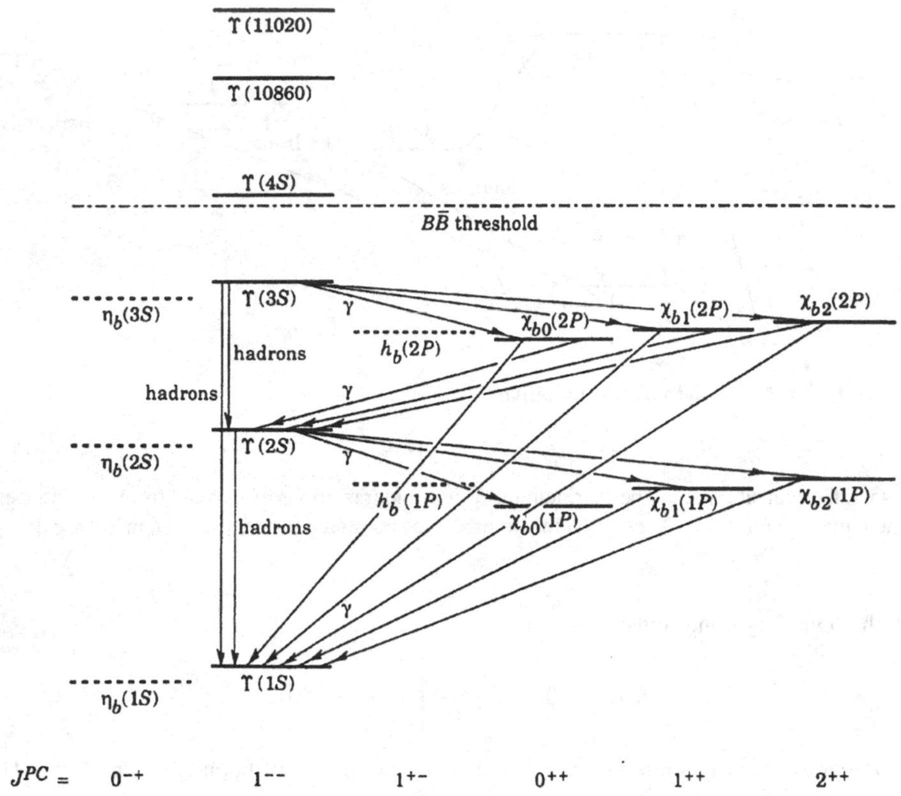


Fig. 45.2. Same as in Fig. 45.1 but for the Bottomium system.

In a covariant Feynman gauge, it is easy to obtain:

$$\mathcal{A} = \frac{1}{4}\sum_a \lambda^a_{i_2 i_2'}\lambda^a_{i_1 i_1'}\left(\frac{g^2}{4\pi^2}\right)\bar{u}(p_1', \lambda_1')\gamma^\mu u(p_1, \lambda_1)\frac{g_{\mu\nu}}{k^2}\bar{v}(p_2, \lambda_2)\gamma^\mu v(p_2', \lambda_2') , \quad (45.10)$$

where $\lambda^a$ are colour matrices. It can again be rearranged by using the relation:

$$\bar{v}(p_2, \lambda_2)\gamma_\mu v(p_2', \lambda_2') = -\bar{u}(p_2', \lambda_2')\gamma_\mu u(p_2, \lambda_2) . \quad (45.11)$$

The non-relativistic amplitude is related to $\mathcal{A}$ as:

$$\mathcal{T}_{\rm NR} = \frac{1}{4\sqrt{p_{10}p_{10}'p_{20}p_{20}'}}\mathcal{A} . \quad (45.12)$$

In the non-relativistic limit:

$$p_0 \equiv \sqrt{\vec{p}^2 + m_Q^2} \simeq m_Q + \frac{\vec{p}^2}{2m_Q} + \frac{\vec{p}^4}{8m_Q^2} ,$$

$$k^2 = (p_{10} - p_{20})^2 - \vec{k}^2 \simeq -\vec{k}^2 + \frac{\vec{p}^2 - \vec{p}'^2}{4m_Q^2} , \quad (45.13)$$

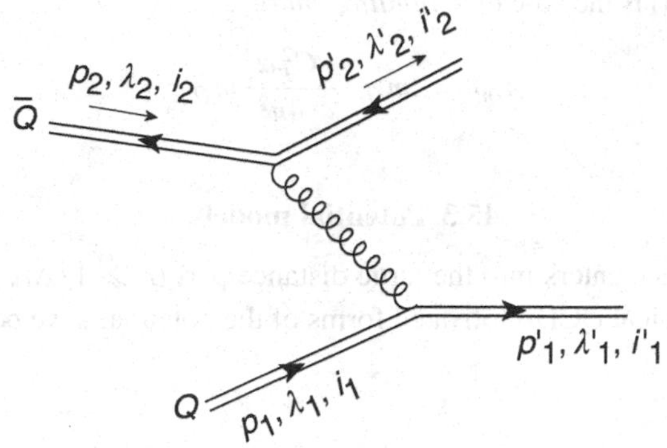


Fig. 45.3. Tree-level diagram for $\bar{Q}Q$ scattering.

and

$$\frac{1}{\sqrt{2p_0}}u(p,\lambda) \simeq \begin{pmatrix} (1-\vec{p}^2/4m_Q^2)\,\chi(\lambda) \\ (1/2m_Q)\vec{p}\cdot\vec{\sigma}\chi(\lambda) \end{pmatrix} \tag{45.14}$$

where the Pauli matrices $\vec{\sigma}$ act on the two-component spinor $\chi(\lambda_i)$. In the static limit, one can retain only the leading term in Eq. (45.13), and obtains:

$$\begin{aligned} \mathcal{T}_{\rm NR}(Born) &\simeq -\frac{1}{4}\sum_a \lambda^a_{i_2i'_2}\lambda^a_{i_1i'_1}\left(\frac{g^2}{4\pi^2}\right)\chi^\dagger(\lambda'_1)\chi^\dagger(\lambda'_2)\frac{1}{\vec{k}^2}\chi(\lambda_1)\chi(\lambda_2) \\ &= -\frac{1}{4}\sum_a \lambda^a_{i_2i'_2}\lambda^a_{i_1i'_1}\delta_{\lambda_1\lambda'_1}\delta_{\lambda_2\lambda'_2}\left(\frac{g^2}{4\pi^2}\right)\frac{1}{\vec{k}^2}\,. \end{aligned} \tag{45.15}$$

On the other hand, the non-relativistic amplitude can be related to the potential as:

$$\mathcal{T}_{\rm NR}(Born) = -\frac{1}{4\pi^2}\int d^3\vec{r}\; e^{i\vec{k}\vec{r}}\chi^\dagger(\lambda'_1)\chi^\dagger V(\vec{r})(\lambda'_2)\frac{1}{\vec{k}^2}\chi(\lambda_1)\chi\,. \tag{45.16}$$

By identification, taking the inverse Fourier transform, and using:

$$\frac{1}{4}\sum_a \lambda^a_{i_2i'_2}\lambda^a_{i_1i'_1} = \begin{cases} -C_F\ : & \text{Singlet} \\ \dfrac{1}{2N_c} = \dfrac{1}{6}\ : & \text{Octet} \end{cases} \tag{45.17}$$

one obtains the expression.[1]

$$\begin{aligned} V(r \ll 1/\Lambda_{\rm QCD}) &= -\left(C_F \equiv \frac{4}{3}\right)\frac{\alpha_s}{r}\ : \quad \text{Singlet} \\ &= \left(\frac{1}{2N_c}\right)\frac{\alpha_s}{r}\ : \quad \text{Octet}\,, \end{aligned} \tag{45.18}$$

where the running of the QCD coupling and the form of the potential have been verified on the lattice. Using this form of the potential, the eigenvalue of the previous Schrödinger

[1] We shall only consider the singlet case in the following.

equation in Eq. (45.5) is the so-called *binding energy*:

$$E_{nl} = 2m_Q - \frac{C_F^2 \alpha_s^2}{4n^2} m_Q \,. \tag{45.19}$$

## 45.3 Potential models

The model dependence enters into the large distance part ($r \gg 1/\Lambda_{\rm QCD}$) of the potential. Many phenomenological QCD-motivated forms of the potential have been proposed in the literature [12,81–94].

### 45.3.1 Cornell potential

The simplest phenomenological form is the Cornell linear potential [82]:

$$V(r \gg 1/\Lambda_{\rm QCD}) \simeq \sigma r \tag{45.20}$$

where $\sigma$ is the QCD string tension.

### 45.3.2 Richardson potential

In the Richardson potential [83], the QCD coupling in the Coulomb potential is allowed to run, and an interpolating formula for the Fourier transform of the potential has been proposed:

$$\tilde{V}(q) = -\frac{4}{3}\left(\frac{48\pi^2}{33 - 2n_f}\right)\frac{1}{q^2 \ln\left(1 + q^2/\Lambda_{\rm QCD}^2\right)} \,. \tag{45.21}$$

which behaves as:

$$\tilde{V}(q \gg \Lambda_{\rm QCD}) \sim \frac{1}{q^2 \ln\left(q^2/\Lambda_{\rm QCD}^2\right)} \,, \qquad V(q \ll \Lambda_{\rm QCD}) \sim \frac{1}{q^2} \,. \tag{45.22}$$

The charmonium and upsilon spectroscopy fix the parameters to be:

$$\Lambda_{\rm QCD} \approx 400 \text{ MeV} \,, \qquad \sigma \simeq (400 \text{ MeV})^2 \,. \tag{45.23}$$

### 45.3.3 Martin potential

Some more empirical models are the Martin potential [12,84–86]:

$$V(r) \sim A + Br^n \,, \tag{45.24}$$

where the different terms are fixed from the fit of the rich quarkonia families. The power of the potential is found to be:

$$n \simeq 0.1 \,. \tag{45.25}$$

Martin's potential is neither Coulombic at short distance nor linear at long distance, but is strongly constrained inside the region $0.1 \sim 1$ fermi. Its slight modification outside this region does not affect the results as the wave function vanishes rapidly.

From the concavity properties of the potential [12,84–86]:

$$\frac{dV}{dr} > 0 \,, \qquad \frac{d^2V}{dr^2} < 0 \quad \Longrightarrow \quad \frac{d}{dr}\left(\frac{1}{r}\frac{dV}{dr}\right) < 0 \,, \tag{45.26}$$

some impressive sets of inequalities can be derived. If $n$ is the number of nodes of the radial wave functions, and $l$ the orbital angular momentum, one has for $n \geq 0$:

$$E(n, l+1) > \frac{1}{2}[E(n+1, l) + E(n, l)] \,, \tag{45.27}$$

which is satisfied by the observed masses:

$$M_{\Upsilon'} - M_{\Upsilon} > M_{\Upsilon'} - M_{\Upsilon"} \ldots \tag{45.28}$$

The flavour independence assumption leads to the concavity relation:

$$2E(\bar{Q}q) > E(\bar{Q}Q) + E(\bar{Q}q) \,, \tag{45.29}$$

which is well satisfied by the observed masses. In particular, one expects to have:

$$M_{B_c} \geq \frac{1}{2}[M_{\psi} + M_{\Upsilon}] \,, \tag{45.30}$$

while the lower bound can also be obtained [86]. Analogous inequalities have also been derived among baryons and mesons.

However, despite the great phenomenological success of various types of potential models, some difficulties arise in attempting to relate them to field theory. Leutwyler and Voloshin criticize the locality of the potentials [90], whilst Bell and Bertlmann [91–93] do not see their flavour independence.

## 45.4 QCD corrections to the static Coulomb potential: Leutwyler–Voloshin model

In this section, we shall consider the Coulomb static potential given in Eq. (45.18) and we shall investigate the different QCD corrections to it.

### 45.4.1 Relativistic corrections

In this case, the interaction betwen the $Q$ and $\bar{Q}$ can be described by the Breit-fermi potential describing the positronium $e^+e^-$ bound state (see e.g. Schwinger [319], Bertstetski *et al.* [53]). It gives the relativistic corrections:

$$V^{(0)}(r) = V^{(0)}_{\text{stat}}(r) + V^{(0)}_{\text{rel}} \tag{45.31}$$

where:

$$V_{\rm rel}^{(0)} \equiv V_{\rm orb}^{(0)} + V_{\rm tens}^{(0)} + V_{LS}^{(0)} + V_{HF}^{(0)} , \tag{45.32}$$

which corresponds respectively to the purely orbital (spin independent + kinetic energy), tensor, spin-orbit and hyperfine potentials. They read:

$$\begin{aligned} V_{\rm orb}^{(0)} &= -\frac{1}{4m_Q^3}\Delta^2 + \frac{C_F\alpha_s}{m_Q^2}\frac{1}{r}\Delta \\ V_{\rm tens}^{(0)} &= \frac{C_F\alpha_s}{4m_Q^2}\frac{1}{r^3}S_{12} \\ V_{LS}^{(0)} &= \frac{3C_F\alpha_s}{2m_Q^2}\frac{1}{r^3}\vec{L}\cdot\vec{S} \\ V_{HF}^{(0)} &= \frac{4\pi C_F\alpha_s}{3m_Q^2}\vec{S}^2\delta(\vec{r}) . \end{aligned} \tag{45.33}$$

Here $\vec{L}$, $\vec{S}$ and $S_{12}$ are respectively the orbital angular momentum, total spin and tensor operators defined as:

$$\vec{L} = -i\vec{r}\times\Delta , \qquad \vec{S} = \frac{1}{2}(\sigma_1+\sigma_2) , \qquad S_{12} = 2\sum_{ij}\left(2\frac{r_ir_j}{r^2} - \delta_{ij}\right)S_iS_j . \tag{45.34}$$

In Eq. (45.33), one should notice that $r^{-1}$ and $\Delta$ do not commute, which is not important as one only considers diagonal matrix element of $r^{-1}\Delta$ between the $\Psi$ states. Another peculiar point is that one has to take the expectation values of terms containing $\vec{L}\vec{S}$ and $S_{12}$ to be zero between states with zero angular momentum as their angular average vanishes. This is despite the fact that the factor $1/r^3$ is singular at the origin.

Noting that, in the Coulombic approximation, from the average value:

$$\langle \vec{k}^2/m_Q^2\rangle_{nl} = \left(\frac{C_F\alpha_s}{2n}\right)^2 , \tag{45.35}$$

one can, for example, deduce the shift of the spin-independent energy levels:

$$E_{nl}^{(0)} \to E_{nl}^{(0)} + \left[\delta_{rel}E_{nl} \equiv \langle V_{\rm orb}^{(0)}\rangle_{nl}\right] , \tag{45.36}$$

with:

$$\delta_{rel}E_{nl} = \left(\frac{C_F\alpha_s}{m_Q^2a^3}\right)\left(\frac{2l+1-4n}{(2l+1)n^4}\right) - \frac{2}{m_Q^3a^4}\left[\frac{1}{(2l+1)n^4} - \frac{3}{8n^5}\right] , \tag{45.37}$$

where:

$$a \equiv \frac{2}{m_QC_F\alpha_s} , \tag{45.38}$$

is the Bohr radius. Analogously, the hyperfine splittings can be obtained from $\langle V_{HF}^{(0)}\rangle_{n0}$, which in the case $n = 1$ gives:

$$M_\Upsilon - M_{\eta_b} = \delta_{HF} E_{10} \equiv \langle V_{HF}^{(0)} \rangle_{10} = \frac{8 C_F \alpha_s}{3 m_Q^2 a^3} . \tag{45.39}$$

### 45.4.2 Radiative and non-perturbative corrections

Radiative corrections to the previous lowest order relativistic corrections are known. The readers can find a compilation of the results obtained within the $\overline{MS}$ scheme in, for example, [46].

A priori, one may expect that non-perturbative corrections to the static potential have complicated structure. However, as the heavy quarks move in a short distance region $\langle \vec{k}^2 \rangle \sim a \ll 1/\Lambda_{\rm QCD}$, one can be convinced that the first known leading non-perturbative correction in $1/m_Q$ is due to the gluon condensate. Treating the interaction Hamiltonian as a perturbation to the Coulomb potential and using a dipole approximation:

$$H_I = -\frac{g}{2} (\lambda_q^a - \lambda_{\bar{q}}^a) \vec{x} \vec{E}_a , \tag{45.40}$$

where $\vec{E}_a$ is the colour electric field which is related to the gluon condensate as:

$$\langle 0 | g^2 \vec{E}^a \vec{E}_a | 0 \rangle = -\pi \langle G^2 \rangle , \tag{45.41}$$

the energy levels are determined by the quadratic Stark-effect of the chromoelectric field:

$$\delta_{NP} E_{nl} = \langle \Psi_{nl} | 0 | H_I \frac{1}{E_n^{(0)} - H_{\rm Coul}^{(8)}} H_I | 0 \rangle | \Psi_{nl} \rangle = \frac{\pi}{18} \langle \alpha_s G^2 \rangle \langle \Psi_{nl} | \vec{x} \frac{1}{E_n^{(0)} - H_{\rm Coul}^{(8)}} \vec{x} | \Psi_{nl} \rangle , \tag{45.42}$$

which contributes to the level shift as [90]:

$$\delta_{NP} E_{nl} = m_Q \frac{\epsilon_{nl} n^6 \pi \langle \alpha_s G^2 \rangle}{(m_Q C_F \tilde{\alpha}_s)^4} , \tag{45.43}$$

where:

$$\epsilon_{nl} = \frac{2}{9} \frac{1}{n^3 (2l+1)} [(l+1)[F(n,l) - F(-n,l)] + l[F(n,-l-1) - F(-n,-l-1)]] , \tag{45.44}$$

with:

$$\begin{aligned} F(n,l) &= 2n[n^2 - (l+1)^2] + (n+l+2)(n+l+1) \\ &\quad \times \left[ \frac{(n-l)(n+l+3)}{9n+16} + 4 \frac{(2n-l)^2}{9n+8} \right] . \end{aligned} \tag{45.45}$$

Some particular values are:

$$\epsilon_{10} = \frac{624}{425} , \qquad \epsilon_{20} = \frac{1051}{663} , \qquad \epsilon_{21} = \frac{9929}{9945} . \tag{45.46}$$

The main feature of the result is that the level shift grows like $n^6$, showing that, even for heavy quarks, the non-perturbative corrections are important for excited states.

### 45.4.3 Validity range

The validity of the previous result can only be realized if the shift is much smaller than the Schrödinger binding energy in Eq. (45.19), which needs that $n^2/m_Q \ll 1$. Taking $n = 1$, this leads to the condition:

$$m_Q \gg 5 \text{ GeV} , \tag{45.47}$$

indicating that the model is quite inaccurate when applied to the bottomium system.

### 45.4.4 Some phenomenological applications

Collecting all different corrections, the vector-pseudoscalar mass-difference is [46]:

$$\begin{aligned} M_\Upsilon - M_{\eta_b} &= \frac{8C_F\alpha_s}{3m_Q^2a^3}[1+\delta_{\alpha_s}+\delta_{NP}]^2 \\ &\times \left(\frac{\alpha_s}{\pi}\right)\left\{1+\left[-\beta_1\left(\ln\frac{a\mu}{2}-1\right)+\frac{21}{4}(\ln C_F\tilde{\alpha}_s+1)+b_{HF}\right]\right. \\ &\left.+\frac{1161}{8704}\frac{\pi\langle\alpha_s G^2\rangle}{m_Q^4\tilde{\alpha}_s^6}\right\}, \end{aligned} \tag{45.48}$$

where:

$$\begin{aligned} b_{HF} &= \frac{11C_A-9C_F}{18}, \qquad \delta_{\alpha_s} = -\frac{3\beta_1}{2}\left(\frac{\alpha_s}{\pi}\right)\left(\ln\frac{a\mu}{2}-\gamma_E\right), \\ \delta_{NP} &= \frac{1}{2}\left[\frac{270\,459}{108\,800}+\frac{1\,838\,781}{2\,890\,000}\right]\frac{\pi\langle\alpha_s G^2\rangle}{m_Q^4\tilde{\alpha}_s^6}. \end{aligned} \tag{45.49}$$

The leptonic width of the $\Upsilon$ is:

$$\Gamma(\Upsilon\to e^+e^-) = \Gamma^{(0)}\times[1+\delta_{\alpha_s}+\delta_{NP}]^2\left(1-\frac{4C_F\alpha_s}{\pi}\right), \tag{45.50}$$

where:

$$\Gamma^{(0)} = 16\pi\left(\frac{Q_b\alpha}{M_\Upsilon}\right)^2|\psi(0)|^2 . \tag{45.51}$$

The wave function is found to be:

$$|\psi(0)|^2 = 2\left[m_Q C_F\tilde{\alpha}_s^3(\mu^2)\right], \tag{45.52}$$

Using:

$$\Lambda_{\text{QCD}}(n_f=4,\ 3\text{ loops}) = (0.23\pm0.06)\text{ GeV}, \qquad \langle\alpha_s G^2\rangle = (0.06\pm0.02)\text{ GeV}^4 , \tag{45.53}$$

they lead to the numerical predictions:

$$M_\Upsilon - M_{\eta_b} = (47 \pm 13)\ \text{MeV}\ , \qquad \Gamma(\Upsilon \to e^+e^-) \simeq (1.1 \pm 0.3)\ \text{keV}\ . \quad (45.54)$$

The electronic width is quite inaccurate but agrees within the errors with the data (1.32 ± 0.04) keV. The prediction for the mass splitting will be compared with the other QCD-based predictions in subsequent chapters. Some other predictions for the mass splittings are also available [46], which in general are in good agreement with the data. This method has been also used in [90,94,46] for extracting the values of the quark pole masses. We quote below the corresponding values of the $\overline{MS}$ running masses:

$$\bar{m}_b(m_b^2) = 4440^{+43}_{-28}\ \text{MeV}\ , \qquad \bar{m}_c(m_c^2) = 1531^{+132}_{-127}\ \text{MeV}\ , \qquad (45.55)$$

which are systematically higher than predictions from QCD spectral sum rules methods (see however, [602] and the next chapter on quark masses).

## 45.5 Bell–Bertlmann equivalent potentials

'Equivalent' potential reproducing the Leutwyler–Voloshin spectrum has been proposed in [93]. Using the form of the Stark effect in Eq. (45.42), in the static limit ($m_Q \to \infty$) where one can neglect the kinetic term $p^2/m_Q$, the energy denominator becomes a potential difference:

$$E_n^{(0)} - H^8_{\text{coul}} \overset{m_Q\to\infty}{\rightarrow} V^0_{\text{coul}} - V^8_{\text{coul}} = \frac{9\beta}{8}\frac{1}{r}\ , \qquad (45.56)$$

which leads to the cubic potential:

$$V_{\text{static}} = \frac{4\pi}{81\beta}\langle\alpha_s G^2\rangle r^3\ . \qquad (45.57)$$

This potential accounts for large quantum numbers, as the distance in a Coulombic state behaves as:

$$\langle r\rangle_n^{\text{Coul}} \sim n^2\ . \qquad (45.58)$$

Corrections of order $1/m_Q$ to this potential approximate low quantum numbers. Therefore, one arrives at the 'equivalent' potential [93]:

$$\delta V(r) = \frac{4\pi}{81\beta}\langle\alpha_s G^2\rangle\left[r^3 - \frac{304}{81}\frac{r^2}{m_Q\beta} + \frac{53}{10}\frac{r}{(m_Q\beta)^2} - \frac{113}{100}\frac{1}{(m_Q\beta)^3}\right]\ , \qquad (45.59)$$

which differs from the effective potential models as it is *flavour dependent*.

Another 'equivalent' potential has been proposed in [91–93] for interpreting the QCD spectral sum rule non-relativistic moments (see Part X, QCD spectral sum rules):

$$\mathcal{R}(\tau_N) = -\frac{d}{d\tau_N}\ln\left[\mathcal{M}(\tau_N) \equiv \int dE\, e^{-E\tau_N}\,\text{Im}\Pi(E)\right] \overset{\tau\to\infty}{\longrightarrow} E_0\ , \qquad (45.60)$$

in a potential theory, where $E_0$ is the energy of the ground state. Im$\Pi(E)$ is the spectral function which can be parametrized within the potential theory as:

$$\mathrm{Im}\Pi(E) = \frac{3}{8m_Q^2} \sum_n 4\pi |\psi_n(0)|^2 \delta(E - E_n) , \tag{45.61}$$

which shows that the moments $\mathcal{M}$ is nothing but the time-dependent Green function:

$$\mathcal{M}(\tau_N) = \frac{3}{8m_Q^2} 4\pi \langle \vec{x} | e^{-H\tau_N} | \vec{x} \rangle |_{\vec{x}=0} , \tag{45.62}$$

where $H \equiv p^2/m_Q + V$ is the total Hamiltonian of the system. Perturbing the kinetic term by the potential with respect to the time $\tau_N$:

$$e^{-H\tau_N} = e^{-\frac{p^2}{m_Q}\tau_N} - \int_0^{\tau_N} d\tau'_N e^{-\frac{p^2}{m_Q}(\tau_N - \tau'_N)} V e^{-\frac{p^2}{m_Q}\tau'_N} , \tag{45.63}$$

one obtains for a power-like potential: $V = \sum_n v_n r^n$:

$$\mathcal{M}(\tau_N) = \frac{3}{8m_Q^2} 4\pi \left( \frac{m_Q}{4\pi\tau_N} \right)^{3/2} \left[ 1 - \sum_n v_n \Gamma\left(\frac{n}{2} + 1\right) m_Q \left( \frac{\tau_N}{m_Q} \right)^{n/2+1} \right] . \tag{45.64}$$

An identification of this term with the QCD moments in Eq. (49.45) leads to the 'equivalent' potential:

$$V(r) = -\frac{4}{3}\frac{\alpha_s}{r} + \frac{\pi}{144} \langle \alpha_s G^2 \rangle m_Q r^4 , \tag{45.65}$$

which differs from the effective potential models and from the previous Leutwyler–Voloshin 'equivalent' potential. The main feature of the BB 'equivalent' potentials is that they are *flavour dependent* in contrast to the effective potential models.

### 45.6 Stochastic vacuum model

We have seen previously that for excited states the Voloshin–Leutwyler approach [90] cannnot be applied as $n^2/m_Q$ is no longer smaller than 1. It has been noted in [603], that this is due to the fact that the correlators $\langle G_{\mu\nu}(x) G_{\alpha\beta}(x) \rangle$ have been taken to be independent of $x$, although they should decrease exponentially for large spacelike $x$. Splitting the field strength $G^a_{\mu\nu}$ into a chromomagnetic piece $\mathbf{B}_a$ and a chromoelectric one $\mathbf{E}^i_a = G^{0i}$, one can show that in the non-relativistic limit, the spin-independent piece of the splitting will only involve $\mathbf{E}^i_a$. Therefore, the non-perturbative correlator reads:

$$\langle \mathbf{E}(x)\mathbf{E}(0) \rangle = \frac{1}{12} \left[ \delta_{ij} \Delta(x) + x_i x_j \frac{\partial}{x_\mu} \frac{\partial}{x_\mu} D_1(x^2) \right] \tag{45.66}$$

with:

$$\Delta(x) \equiv D(x^2) + D_1(x^2) + x^2 \frac{\partial}{x_\mu} \frac{\partial}{x_\mu} D_1(x^2) . \tag{45.67}$$

If one neglects the $x$ dependence of the correlator, the only surviving part is:

$$\Delta(0) = 2\pi \langle \alpha_s G^2 \rangle \, . \tag{45.68}$$

Therefore, Eq. (45.66) indicates that one can derive the Voloshin–Leutwyler formula for small $n$, but one can also obtain another potential for large $n$ if one takes into account the $x$ dependence of the correlator.

Defining the correlation time $T_Q$ for quarks as:

$$\langle x_i(\tau_1) x_j(\tau_2) \rangle_{nl} = \frac{\delta_{ij}}{3} \langle \vec{x}^2_{nl} \rangle \exp \left[ -\frac{|\tau_1 - \tau_2|}{T_Q} \right] , \tag{45.69}$$

and the one $T_G$ for gluons:

$$\left\langle \mathbf{E}^i_a(\tau_1) \mathbf{E}^j_b(\tau_2) \right\rangle_E = \frac{\delta^{ij}}{3} \frac{\delta_{ab}}{8} \langle \mathbf{E}^2 \rangle \exp \left[ -\frac{|\tau_1 - \tau_2|}{T_G} \right] , \tag{45.70}$$

one can also find that the sum rule approach within the Bell–Bertlmann 'equivalent' potential is applicable for $T_G \gg T_Q$ [604]. These features are the basis of the *stochastic model* discussed in details in [605,51].

### 45.6.1 The model

One assumes that the quarks and the gluons background fields fluctuate stochastically according to a Markov process. Let us consider the stochastic variable $\xi(t)$ depending on one or several variables $t$. It will be distributed according to some distribution which fixes the vacuum expectation values:

$$\langle \xi(t) \rangle \, , \qquad \langle \xi(t_1) \xi(t_2) \rangle \, , \ldots \tag{45.71}$$

The *cumulants* or *linked clusters* are defined by:

$$\left\langle \mathcal{P} \exp \int dt \, [\xi(t)] \right\rangle = \exp \left[ \int dt \, \langle\langle \xi(t) \rangle\rangle + \frac{1}{2!} \int \int dt_1 dt_2 \, \langle\langle \xi(t_1) \xi(t_2) \rangle\rangle + \cdots \right] , \tag{45.72}$$

where the path ordering prescription:

$$\Phi(t, t', C) \equiv \left\langle \mathcal{P} \exp \int_C dt \, [\xi(t)] \right\rangle = \lim_{t_{i+1} - t_i \to 0} \prod_{i=1}^{N+1} \exp \left[ \xi \left( \frac{t_{i+1} + t_i}{2} \right) (t_{i+1} - t_i) \right] , \tag{45.73}$$

with $t_i$ are ordered points on the path $C$ with $t_N = t'$ and $t_1 = t$, should be introduced if the stochastic variable $\xi(t)$ are non-commuting. In the case of commuting stochastic variables which we shall consider here, an expansion of Eq. (45.72) gives:

$$\begin{aligned} &\langle\langle\xi(t)\rangle\rangle = \langle\xi(t)\rangle\,, \\ &\langle\langle\xi(t_1)\xi(t_2)\rangle\rangle = \langle\xi(t_1)\xi(t_2)\rangle - \langle\langle\xi(t_1)\rangle\rangle\langle\langle\xi(t_2)\rangle\rangle\,, \\ &\ldots \end{aligned} \tag{45.74}$$

A *centered Gaussian process* is a process where only the $n = 2$ cumulants occur, that is, all expectation values can be determined by the correlators with $n = 2$:

$$\begin{aligned} &\langle\xi(t)\rangle = \langle\xi(t_1)\xi(t_2)\xi(t_3)\rangle = \cdots = 0\,, \\ &\langle\xi(t_1)\xi(t_2)\rangle = \langle\langle\xi(t_1)\xi(t_2)\rangle\rangle\,, \\ &\ldots \end{aligned} \tag{45.75}$$

It can be shown [605,51] that assumptions where the contributions of low frequency fields can be described by a functional integral (stochastic process) with converging clusters leads to an area law of the Wilson loop and then to linear confinement for static sources.

For that purpose, we consider the Wegner–Wilson loop in a pure gauge theory:

$$W[C] = \int \mathcal{D}A_\mu\, e^{-\frac{1}{4}G^2_{\mu\nu}(x)} \exp\left[ ig\int_C A^\mu(x)dx_\mu \right]\,. \tag{45.76}$$

Denoting:

$$\langle\ldots\rangle_A = \int\cdots\prod_{k<\mu} dA_\mu\, e^{-\frac{1}{4}G^2_{\mu\nu}(x)}\,, \tag{45.77}$$

these low frequency contributions to the Wilson loop are defined as:

$$\begin{aligned} W[C]_A &= \left\langle \exp\left[ ig\int_C A^\mu(x)dx_\mu \right]\right\rangle_A \\ &= \left\langle \exp\left[ ig\int_{\mathcal{F}} G^{\mu\nu}(x)d\sigma_{\mu\nu}(x) \right]\right\rangle_A\,. \end{aligned} \tag{45.78}$$

$\mathcal{F}$ is an area whose border is the loop $C$; $d\sigma_{\mu\nu}(x)$ $(\mu < \nu)$ is the surface element of $\mathcal{F}$ at point $x$, and $G_{\mu\nu}$ is the field strength. One has used the Stokes theorem which transforms the line into surface integral. These low frequency contributions are given by the cluster expansion:

$$W[C]_A = \exp\left[ -\frac{g^2}{2!}\int d\sigma_{\mu\nu}(x)d\sigma_{\alpha\beta}(x')\,\langle\langle G^{\mu\nu}(x)G^{\alpha\beta}(x')\rangle\rangle + \cdots \right], \tag{45.79}$$

$\langle G^{\mu\nu}\rangle = 0$ due to Lorentz invariance. Lorentz and translational invariances also yield the most general decomposition:

$$\begin{aligned} \langle G_{\mu\nu}(x_1)G_{\alpha\beta}(x_1')\rangle_A = \frac{1}{12}\langle G^2\rangle\Bigg\{ &(\delta_{\mu\alpha}\delta_{\nu\beta} - \delta_{\mu\beta}\delta_{\nu\alpha})D(x^2)\kappa + \Bigg[\Bigg(\frac{1}{2}\frac{\partial}{\partial x_\mu}(x_\alpha\delta_{\nu\beta} - x_\beta\delta_{\nu\alpha}) \\ &+ \frac{1}{2}\frac{\partial}{\partial x_\nu}(x_\beta\delta_{\mu\alpha} - x_\alpha\delta_{\mu\beta})\Bigg) D_1(x^2)(1-\kappa)\Bigg]\Bigg\}\,, \end{aligned} \tag{45.80}$$

where $\langle G^2 \rangle \equiv \langle G_{\mu\nu}(0) G^{\mu\nu}(0) \rangle$ and $\kappa$ a parameter. One can insert this expression into the cluster expansion in Eq. (45.79). Assuming that the correlator falls off for $|x - x'| > \lambda$ and using $|\mathcal{F}| \gg \lambda^2$, one obtains:

$$W[C]_A \simeq \exp[-\kappa \langle g^2 G^2 \rangle \lambda^2 |\mathcal{F}| K [1 + \mathcal{O}(1/\mathcal{F})]] , \tag{45.81}$$

where the constant $K$ depends on the shape of the scalar function $D(x^2)$. To leading order (two-cluster) of the cluster expansion, one can notice that the result is proportional to $\kappa$ as $D_1$ in Eq. (45.80) does not contribute to the term proportional to the area loop. The assumption of a convergent cluster expansion thus leads to the area law of the Wilson loop if the tensor structure with $D$ in Eq. (45.80) does not vanish. Thus leading apparently to a natural linear confinement for Abelian theory as well. However, one can show that the use of the Maxwell equations for QED:

$$\partial^\alpha \epsilon_{\alpha\mu\nu\beta} G^{\mu\nu} = 0 , \tag{45.82}$$

implies that:

$$\kappa = 0 , \tag{45.83}$$

and hence that we have no area law and then no confinement. For QCD, we have instead:

$$\partial_\alpha \epsilon_{\alpha\mu\nu\beta} G^a_{\mu\nu} = -ig \epsilon_{\alpha\mu\nu\beta} f_{abc} A^b_\mu A^c_\nu \neq 0 , \tag{45.84}$$

indicating that there is no reason why $\kappa$ should be equal to zero. A lattice measurement shows that, for QCD, the correlator $D(x^2)$ is dominant as one finds [606]:

$$\kappa \approx 0.74 . \tag{45.85}$$

### 45.6.2 Application to the static potential

We shall discuss here the application of the model to static potential.[2] Let us consider the gauge-invariant non-local operator:

$$\mathcal{O}(x, x') \equiv \bar{Q}(x) \phi(x, x') Q(x') , \tag{45.86}$$

where $Q(x)$ is the heavy quark field, and $\phi(x, x')$ is the string along the straightline:

$$\phi(x, x') = \mathcal{P} \exp \left[ ig \int_C A^\mu [x + \lambda(x' - x)] \left(x' - x\right)_\mu d\lambda \right] . \tag{45.87}$$

Applying the previous operator to the vacuum state of hadrons leads to a gauge- and Lorentz-invariant state composed of a quark at a position $x$ and an antiquark at a position $x'$. The evolution of this operator is given by the Green's function:

$$G(x, x'; y, y') = \int \mathcal{D}A \mathcal{D}Q \mathcal{D}\bar{Q} \; e^{-S} \mathcal{O}(x, x') \mathcal{O}^\dagger(y, y') . \tag{45.88}$$

[2] Some other applications of the model can be found in more specialized reviews (see e.g. [51]).

The QCD action is:

$$\mathcal{S} = \int dx\ \bar{Q}(x)[i\gamma_\mu(\partial^\mu + igA^\mu) - m_Q]Q(x) + \mathcal{S}_{YM}\ , \tag{45.89}$$

with:

$$\mathcal{S}_{YM} \equiv (1/4)G^a_{\mu\nu}G^{\mu\nu}_a\ . \tag{45.90}$$

Doing the integration over fermion fields, one obtains:

$$G(x, x'; y, y') = \int \mathcal{D}A\ e^{-\mathcal{S}_{YM}}\mathrm{Det[A]Tr}[S(x', y'; A)\phi(y', y)S(y, x; A)]\ , \tag{45.91}$$

where:

$$S(x, y, A) = \frac{\delta(x-y)}{[i\gamma_\mu(\partial^\mu + igA^\mu) - m_Q]}\ , \tag{45.92}$$

is the quark propagator in external field; Det[A] is the functional determinant from the quark field integration, and describes internal fermion loops as power series of $g$. Using $i\gamma_0 = \gamma_4$, one obtains to leading order in $1/m_Q$ and $g$:

$$S(x, y, A) \simeq \phi(x, y)\delta(\vec{x} - \vec{y}) \times \left[e^{-m_Q(x_4-y_4)}\theta(x_4 - y_4)\left(\frac{1+\gamma_0}{2}\right)\right.$$
$$\left. + e^{-m_Q(y_4-x_4)}\theta(y_4 - x_4)\left(\frac{1-\gamma_0}{2}\right)\right] + \mathcal{O}\left(\frac{1}{m_Q}\right) \tag{45.93}$$

Therefore, to this approximation, the Green's function becomes:

$$G[(\vec{x}, 0), (\vec{x}', 0); (\vec{y}, T), (\vec{y}', T)] \simeq \delta(\vec{x} - \vec{y})\delta(\vec{x}' - \vec{y}')e^{-2m_QT}$$
$$\times \int \mathcal{D}A\ e^{-\mathcal{S}_{YM}}\mathbf{Tr}[\phi(y, x)\phi(x, x')\phi(x', y')\phi(y', y)]$$
$$\equiv \delta(\vec{x} - \vec{y})\delta(\vec{x}' - \vec{y}')e^{-2m_QT}\,\mathbf{Tr}\ W[L]\ , \tag{45.94}$$

where the Wegner–Wilson loop has been defined in Eq. (45.76). The second loop integral entering in Eq. (45.76) is defined along the rectangle with corners $[(\vec{x}, 0), (\vec{x}', 0), (\vec{y}', T), (\vec{y}, T)]$. Using the fact that the Green's function scales like $e^{-E_nT}$, where $E_n$ is the energy of the system, one can deduce:

$$E_n = -\lim_{T\to\infty}\frac{1}{T}\ln \mathbf{Tr}\ W[L] + 2m_Q\ . \tag{45.95}$$

The term $2m_Q$ is the rest energy of the two quarks, while the first term can be identified with the potential $V(\vec{x} - \vec{x}')$ of the system. Evaluating the rectangular Wegner–Wilson loop using a strong coupling or lattice calculations, one finds in terms of the string tension $\sigma$ [491]:

$$V(\vec{x} - \vec{x}') = \sigma(\vec{x} - \vec{x}')\ . \tag{45.96}$$

Using the cluster decomposition in Eq. (45.79), one obtains, from the area law of the Wegner–Wilson loop, the spin-independent part of the potential:

$$V_0(r) = \frac{1}{24N_c}\langle g^2G^2\rangle \left[ r\int_0^r d\rho \int_{-\infty}^{+\infty} d\tau \; D(\tau^2+\rho^2) + \int_0^r \rho d\rho \int_{-\infty}^{+\infty} d\tau \left[ -D(\tau^2+\rho^2) + \frac{1}{2}D_1(\tau^2+\rho^2)\right]\right], \quad (45.97)$$

where:

$$\lim_{r\to\infty} V_0(r) \sim \sigma r\,, \quad (45.98)$$

corresponding to the standard linear potential. At short distance:

$$\lim_{r\to 0} V_0(r) \sim r^2\,, \quad (45.99)$$

which recovers the form expected from renormalon calculations (see previous chapter on renormalons).

However, one should notice that this result is only valid for:

$$T^{-1} \gg E_n\,, \quad (45.100)$$

which corresponds to the regime where $r$ is small but the state is located on average at a large distance from the centre of mass. Collecting the previous result, the full non-relativistic potential, from the stochastic model, is:

$$V(r) = -C_F\frac{\alpha_s}{r} + V_0(r)\,. \quad (45.101)$$

To this expression one can add spin-dependent corrections to order $1/m_Q^2$ (see e.g. [51]), which can also be expressed in terms of the correlators $D(x)$ and $D_1(x)$. One should notice that the spin-dependent part of the confining potential is known phenomenologically to be specifically different from the Coulomb potential, while good results are obtained if one assumes that the confining potential leads to the same spin-dependent force as a scalar exchange [607]. Radiative corrections can also be included into the Coulombic potential. Predictions for the higher excited mass splittings using the model are quite sucessful.

## 45.7 Non-relativistic effective theories for quarkonia

In a previous section, we have anticipated the different regimes (short and long distances) appearing in the $\bar{Q}Q$ system. In the present approach, it is convenient to introduce two UV scales $\Lambda_{1,2}$ which characterize such regimes, and which are ordered by the heavy quark velocity $v \ll 1$:

- The quark mass $m_Q$ is called the hard scale.
- The momentum $m_Q v$ is the soft scale (S).
- The binding energy $m_Q v^2$ is the ultrasoft scale (US).

Therefore, one can have the regime hierarchy:

$$m_Q v^2, \Lambda_{\rm QCD} \ll \Lambda_1 \ll m_Q v \ll \Lambda_2 \ll m_Q \,. \tag{45.102}$$

In this way, $\Lambda_1$ is the cut-off of the quark energy and of the gluon energy and momentum, whilst $\Lambda_2$ is the cut-off of the relative momentum $\vec{p}$ of the quark-antiquark system. For a Coulombic system, one has:

$$v \sim \alpha_s \,. \tag{45.103}$$

As the two scales are largely separated, one can (in principle) integrate out the UV scales step by step: after integrating out the heavy quark mass $m_Q$, one obtains the usual non-relativistic QCD (NRQCD) effective theory [608]. The Lagrangian of NRQCD is written in terms of an expansion in $1/m_Q$. Potential NRQCD (pNRQCD) is obtained by integrating out from NRQCD the soft scale S [609]. In this way, the Lagrangian of pNRQCD is expressed as an expansion in terms of $1/m_Q$ and of the relative coordinate $\vec{r}$ (*multipole expansion*) of the $\bar{Q}$ and $Q$.

The integration of the degrees of freedom is done using matching conditions (see e.g. [610,611,612,613] for details), namely by comparing the shell amplitudes order by order in QCD and NRQCD. The matching from QCD to NRQCD can always be done perturbatively since, by definition of the heavy quark, $m_Q \gg \Lambda_{\rm QCD}$. The matching from NRQCD to pNRQCD can only be carried out perturbatively when $m_Q v \gg \Lambda_{\rm QCD}$. This condition is assumed to be satisfied in the following discussion. Therefore, the matching coefficients in both NRQCD and pNRQCD can be computed order by order in $\alpha_s$. The non-analytical behaviour in $1/m_Q$ appears through logs in the matching coefficients of the NRQCD Lagrangian:

$$C_H \sim A\alpha_s \left( \ln \frac{m_Q}{\mu} + B \right), \tag{45.104}$$

where $\mu$ denotes the matching scale between QCD and NRQCD. In practice, one can choose:

$$\frac{\Lambda_2^2}{m_Q} \ll \Lambda_1 \,, \tag{45.105}$$

If one denotes by $\Lambda_{mp}$ any scale below $\Lambda_1$, the relevant small dimensionless scales involved in the analysis are:

$$\frac{p}{m_Q}\,, \quad \frac{1}{r m_Q}\,, \quad \text{and} \quad \Lambda_{mp} r \ll 1 \,. \tag{45.106}$$

Decomposing the $\bar{Q}Q$ state into a singlet $S(\vec{R},\vec{r},t)$ and an octet $O(\vec{R},\vec{r},t)$ states ($\vec{R} \equiv (\vec{x}_1 + \vec{x}_2)/2,\ r \equiv (\vec{x}_1 + \vec{x}_2)$), the minimal form in terms of the derivatives of the pNRQCD Lagrangian reads:

$$\begin{aligned}
\mathcal{L}_{\rm pNRQCD} &= -\frac{1}{4}G^a_{\mu\nu}G^{\mu\nu\,a} \\
&+ {\rm Tr}\left\{{\rm S}^\dagger\left(i\partial_0 - \frac{\mathbf{p}^2}{m_Q} + \frac{\mathbf{p}^4}{4m_Q^3} - V_s^{(0)}(r) - \frac{V_s^{(1)}}{m_Q} - \frac{V_s^{(2)}}{m_Q^2} + \cdots\right){\rm S}\right. \\
&\left. + {\rm O}^\dagger\left(iD_0 - \frac{\mathbf{p}^2}{m} - V_o^{(0)}(r) + \cdots\right){\rm O}\right\} \\
&+ gV_A(r){\rm Tr}\{{\rm O}^\dagger\mathbf{r}\cdot\mathbf{E}\,{\rm S} + {\rm S}^\dagger\mathbf{r}\cdot\mathbf{E}\,{\rm O}\} + g\frac{V_B(r)}{2}{\rm Tr}\{{\rm O}^\dagger\mathbf{r}\cdot\mathbf{E}\,{\rm O} + {\rm O}^\dagger{\rm O}\mathbf{r}\cdot\mathbf{E}\}\,,
\end{aligned} \tag{45.107}$$

where the dots indicate higher-order potentials in the $1/m_Q$ expansion; $\mathbf{p} \equiv \vec{p}$; E is the chromoelectric field. One has neglected the centre-of-mass variables $R$ and only kept $O(r)$ terms in the multipole expansion.

The structure of the potentials up to $O(1/m^2)$ are:

- To order $1/m_Q^0$, one has the Coulomb potential.

$$V_s^{(0)}(r) \equiv -C_F\frac{\tilde{\alpha}_s(r)}{r}\,. \tag{45.108}$$

- To order $1/m_Q$, and using dimensions plus time reversal $V_s^{(1)}(r)$, one can only have the following structure:

$$V_s^{(1)} \equiv -\frac{C_F C_A D_s^{(1)}}{2r^2}\,, \qquad C_A = N_c\,. \tag{45.109}$$

- To order $1/m_Q^2$, and using the present accuracy for the matching, one obtains the following potential:

$$\begin{aligned}
V_s^{(2)} &= -\frac{C_F D_{1,s}^{(2)}}{2}\left\{\frac{1}{r}, \mathbf{p}^2\right\} + \frac{C_F D_{2,s}^{(2)}}{2}\frac{1}{r^3}\mathbf{L}^2 + \pi C_F D_{d,s}^{(2)}\delta^{(3)}(\mathbf{r}) + \frac{4\pi C_F D_{S^2,s}^{(2)}}{3}\mathbf{S}^2\delta^{(3)}(\mathbf{r}) \\
&+ \frac{3C_F D_{LS,s}^{(2)}}{2}\frac{1}{r^3}\mathbf{L}\cdot\mathbf{S} + \frac{C_F D_{S_{12},s}^{(2)}}{4}\frac{1}{r^3}S_{12}(\hat{\mathbf{r}})\,.
\end{aligned} \tag{45.110}$$

Note that $\mathbf{p}$ appears analytically in the potentials, with a power ($\mathbf{p}^n$), which is constrained by the power in $1/m_Q$. The different matching coefficients : $\tilde{\alpha}$, $D^{(1)}$, $D^{(2)}$ ... in pNRQCD can be obtained by performing the matching between NRQCD and pNRQCD. A detailed description of the procedure can be found in [609,612,613]. They read:

$$\begin{aligned}
\tilde{\alpha}_s(r,\mu) = \alpha_s(r)\Bigg\{&1 + (a_1 + 2\gamma_E\beta_0)\frac{\alpha_s(r)}{4\pi} + \left[\gamma_E\left(a_1\beta_0 + \frac{\beta_1}{2}\right)\right. \\
&\left. + \left(\frac{\pi^2}{12} + \gamma_E^2\right)\beta_0^2 + \frac{a_2}{4}\right]\frac{\alpha_s^2(r)}{4\pi^2} + \frac{C_A^3}{12}\frac{\alpha_s^3(r)}{\pi}\ln r\mu\Bigg\}\,.
\end{aligned} \tag{45.111}$$

In terms of the $\beta$ function defined in the first part of the this book (Table 11.1), they read for $SU(n)_f$ flavours:

$$\beta_0 \equiv -2\beta_1 = 11 - \frac{2}{3}n_f\,, \qquad \beta_1 \equiv -8\beta_2 = 2\left(51 - \frac{19}{3}n_f\right)\,. \tag{45.112}$$

The one- and two-loop coefficients $a_1, a_2$ have been obtained in [614]. They read:

$$a_1 = \frac{31}{9}C_A - \frac{20}{9}T_F n_f\,, \tag{45.113}$$

$$\begin{aligned} a_2 = &\left(\frac{4343}{162} + 4\pi^2 - \frac{\pi^4}{4} + \frac{22}{3}\zeta_3\right) C_A^2 \\ &- \left(\frac{55}{3} - 16\zeta_3\right) C_F T_F n_f + \frac{400}{81} T_F^2 n_f^2 \\ &- \left(\frac{1798}{81} + \frac{56}{3}\zeta_3\right) C_A T_F n_f\,, \end{aligned} \tag{45.114}$$

respectively. For $SU(3)_c$, one has $T_F = 1/2$:

$$\begin{aligned} \tilde{\alpha}_s = \alpha_s^{\overline{MS}} \Bigg\{ 1 + &\left(\frac{\alpha_s^{\overline{MS}}}{\pi}\right)(2.6 - 0.3n_f) \\ + &\left(\frac{\alpha_s^{\overline{MS}}}{\pi}\right)^2 (53.4 - 7.2n_f + 0.2n_f^2) + \cdots \Bigg\}\,. \end{aligned} \tag{45.115}$$

which shows that the convergence of the QCD series is not good. The other coefficients are [615,611–613,616]:

$$\begin{aligned} D_s^{(1)} &= \alpha_s^2(r)\left\{1 + \frac{2}{3}(4C_F + 2C_A)\frac{\alpha_s}{\pi}\ln r\mu\right\}\,, \\ D_{1,s}^{(2)} &= \alpha_s(r)\left\{1 + \frac{4}{3}C_A\frac{\alpha_s}{\pi}\ln r\mu\right\}\,; \\ D_{2,s}^{(2)} &= \alpha_s(r)\,, \\ D_{d,s}^{(2)} &\simeq \alpha_s(r)\left\{1 + \frac{\alpha_s}{\pi}\left[\frac{2C_F}{3} + \frac{17C_A}{3}\right]\ln m_Q r + \frac{16}{3}\frac{\alpha_s}{\pi}\left(\frac{C_A}{2} - C_F\right)\ln r\mu\right\}\,, \\ D_{S^2,s}^{(2)} &\simeq \alpha_s(r)\left(1 - \frac{7C_A}{4}\frac{\alpha_s}{\pi}\ln m_Q r\right)\,, \\ D_{LS,s}^{(2)} &\simeq \alpha_s(r)\left(1 - \frac{2C_A}{3}\frac{\alpha_s}{\pi}\ln m_Q r\right)\,, \\ D_{S_{12},s}^{(2)} &\simeq \alpha_s(r)\left(1 - C_A\frac{\alpha_s}{\pi}\ln m_Q r\right)\,. \end{aligned} \tag{45.116}$$

The previous results can e.g. be used to compute the $\mathcal{O}(m_Q\alpha_s^5)$ corrections to the heavy quarkonium spectrum. The $N^3LL$ correction to the energy shift of the $\Upsilon(1S)$ is found to be [609]:

$$\delta E_{n=1;\, l=0;\, j=1} \simeq \frac{1730}{81\pi} m_b \alpha_s^4(\mu) \alpha_s(\mu') \ln\left[1/\alpha_s(\mu')\right] \simeq (80 \sim 100)\ \text{MeV}\,, \quad (45.117)$$

where we note that $\mu$ is the matching scale from QCD to NRQCD: $m_Q v < \mu < m_Q$, while $m_Q v^2 < \mu' < m_Q v$ is the one from NRQCD to pNRQCD. One should notice that the correction is relatively large, and the perturbative series has a bad convergence. This convergence might be improved by working with a renormalon-free quark mass definition other than the pole mass, which will then facilitate the main motivation of the approach for exploring the dynamics of the quark-antiquark bound states using the perturbative approach.

# 46

# On monopole and confinement

Though our knowledge of confinement is, at present, quite poor, it is necessary to discuss briefly the approach of monopoles where some activities have been investigated recently for understanding the mechanism of confinement.[1] One of the most favoured mechanism of confinement is the one due to monopole condensation [620], where one has to see if the monopole condensation occurs in the confined phase but not in the deconfined one. Dual-superconductivity mechanism of confinement assumes the formation of an Abrikosov-type tube between heavy quarks introduced into the vacuum via the Wilson loop,[2] while the tube itself is a classical solution of the equations of motion of the Higgs-type model Lagrangian of the action:

$$S_{\text{eff}} = \int d^4x \left[ |D_\mu \phi|^2 + \frac{1}{4} G_{\mu\nu}^2 + V(|\phi|^2) \right] , \tag{46.1}$$

where $\phi$ is a scalar field with a non-zero magnetic charge, $G_{\mu\nu}$ is the field strength tensor built from the dual-gluon field $B_\mu$, $D_\mu$ is the covariant derivative, and $V(|\phi|^2)$ is the potential energy:

$$V(|\phi|^2) = \frac{m^2}{2} |\phi|^2 - \frac{\lambda}{4} |\phi|^4 , \tag{46.2}$$

ensuring that $\langle \phi \rangle \neq 0$ in the vacuum. If $m^2 < 0$, the potential has the typical Mexican shape and $|\phi|^2 = m^2/\lambda$. However, the relation of these effective fields to the fundamental ones of QCD is not yet clear, which is the main limitations of the use of this effective theory. However, there is not, at present, any answer to this question, and the answer can only come from the data, which are, at present, lattice measurements. This lack of understanding concerns the nature of non-perturbative field configurations defined as monopoles in non-Abelian gauge theories. The few knowledge one has is that monopoles are intrinsically $U(1)$ configurations. However, it is not a priori clear which $U(1)$ subgroup of e.g. $SU(2)$ is to be taken for classifying the monopoles. If one takes the most successful *maximal Abelian projection* [617–619], and associates a conserved magnetic charge to any operator in the adjoint representation, we still have very little understanding of the field configurations describing

[1] For reviews, see e.g. [617–619].

[2] The energy of the flux tube is proportional to the length of the flux implying that an infinite energy is needed for dissociating at infinite distance a monopole–antimonopole pair.

monopoles in this projection, and in particular on the monopole size. Lattice measurements indicate that magnetic charges condense in the confined phase, and is independent of the specific choice of the Abelian projection [619]. On the other, a lattice measurement of the monopole size gives the radius [621]:

$$R_{\rm mono} \approx 0.06 \text{ fm} , \tag{46.3}$$

defined in terms of the full non-Abelian action associated with the monopole and not in terms of the projected action. It is relatively small compared with the temperature of the confinement–deconfinement phase transition:

$$T_c \approx 300 \text{ MeV} , \tag{46.4}$$

corresponding to a distance $d_{\rm mono} \sim 1/T_c \sim 0.5$ fm. An attempt to understand the origin of this scale hierarchy has been investigated in [617] using monopole cluster assuming that monopole condensation occurs when the monopole action is UV divergent. However, one expects that the onset of condensation in the standard field theoretical language corresponds to the zero mass of the magnetically charged field $\phi$. This apparent contradiction can be understood from the kinematical relation between the physical mass $m_{\rm phys}$ entering in the propagator of the scalar field and the mass $M \equiv S/L$ defined in terms of the Euclidian action, where $L$ is the length of the trajectory and $S$ the corresponding action on a cubic lattice with spacing $a$. To leading order in $ma$:

$$m^2_{\rm phys} \cdot a = M - \frac{\ln 7}{a} , \tag{46.5}$$

where $\ln 7$ originates from the fact that a trajectory of length $L$ can be realized on a cubic lattice in $N_L = 7^{L/a}$ various ways. At each step, the trajectory can be continued on an adjacent cube, where in four dimensions one has eight such cubes. Zakharov [617] argues that the data on monopole action imply a fine tuning:

$$M_{\rm mono}(a) - \frac{\ln 7}{a} \ll M_{\rm mono}(a) \sim \Lambda_{\rm QCD} , \tag{46.6}$$

where $\ln 7$ is of pure geometrical origin and $M_{\rm mono}$ is the monopole energy defined on a compact $U(1)$ group as:

$$M_{\rm mono}(a) = \frac{1}{8\pi} \int \vec{B}^2 d^3 r \sim \left(\frac{c}{e^2}\right)\left(\frac{1}{a}\right) , \tag{46.7}$$

where $c$ is a constant, $e$ is the electric charge and $g_m = 1/2e$ is the magnetic charge. Analysis of lattice data [618] suggests that the actual physical size $R_{\rm phys}$ of the monopole can be much smaller than that in Eq. (46.3). By $R_{\rm phys}$ one means the distance where the excess of the monopole action is parametrically smaller than the action associated with the zero-point fluctuations. Using the running of the QCD coupling and the condition due to the $U(1)$ critical coupling $e_c^2 \approx 1$ at which the monopole condenses, one obtains the scale:

$$M_{\rm phys} \approx 1 \text{ TeV} \tag{46.8}$$

giving the electroweak scale rather than the QCD one of the order of $\Lambda_{\rm QCD}$, therefore indicating that QCD projected onto the scalar-filed theory via monopoles corresponds to a fine-tuned theory. This result suggests a $SU(2)$ lattice measurements at $\beta = 4$ rather than the present results at $\beta = 2.6$, which is too low to see the dissolution of monopoles at short distance. This is a subject that deserves further investigations.

It is also often stated that the symmetry responsible for confinement is different in pure gauge theory and in the presence of quarks. In pure gauge theory, the order parameter is the vacuum expectation value of the Polyakov line, and the symmetry is $Z_N$, the centre of the group. Since in the presence of quarks, $Z_N$ is explicitly broken, one might expect that the order parameter is the chiral quark $\langle\bar\psi\psi\rangle$ condensate, which is responsible for spontaneous breaking of the chiral symmetry, although it is also known that the quark masses explicitly break chiral symmetry. However, the relation between confinement and chiral symmetry is not clear at all. Lattice simulations indicate that the two transitions take place at the same temperature, but there is no explanation of this numerical observation.

Another point is that if dual superconductivity in all Abelian projections is the symmetry behind confinement, then it should also work in full QCD.

Finally, there are the recent attempts [622,623] to tackle the confinement problem using QCD perturbation theory. The approach is based on a gluon chain model in the large $N_C$ QCD, which gives a *string-like* picture of hadrons although confinement is not built in. One can modify the Born approximation by introducing a non-local counterterm for an *IR renormalization* of the Coulomb potential, which now possesses a linear term proportional to the QCD string tension. The arbitrary IR subtraction point can be optimized by using a variational method. It reaches its optimal value at that of the string tension. The procedure induces a mass to the gluon which, in some sense, is similar to the tachyonic gluon mass introduced by [161] at short distance. Some further examples of the applications of the approach to confinement are discussed in [623].

From this short summary, we conclude that though there has been progress towards an understanding of confinement via monopole condensation, but there remain some unclarified points that still need further investigation. The perturbative approach to confinement looks promising.

# Part X

QCD spectral sum rules

# 47
# Introduction

We have discussed in the previous part several of the most popular QCD non-perturbative methods other than the QCD spectral sum rules (QSSR). Now, we shall dedicate this part of the book to the discussion of this non-perturbative approach, which has been used successfully for understanding the hadron properties and hadronic matrix elements, using those parameters (QCD coupling, quark masses and QCD condensates), derived from QCD first principles. This method was introduced by SVZ in 1979 [1] and reviewed in a book [3], numerous reviews and lecture notes [356–365]. Its basic concepts, based on the operator product expansion and dispersion relations, are well understood in quantum field theory, and is a fully relativistic approach in contrast to potential models, for example. Its applications are quite simple and transparent. However, on the one hand it has a limited accuracy (usually about 10–20% depending on the process), and some uses of the method in some QCD-like models show that its accuracy cannot be improved iteratively. On the other hand, confinement is not a result of the method but is put into it via the introduction of different QCD condensates. In practice, one has to introduce some assumptions, and the results are obtained from self-consistency.

However, in some cases, results obtained from the sum rules disagree with each other and have led to some polemics, which some people use to discredit the approach. We shall see how it is important to check that the results satisfy *some stability criteria* and that the matching between the low-energy hadronic region and the perturbative region described by QCD, which is called *global duality tests* in the literature, are obtained.

A further limitation of the method comes from the fact that the Green's function is computed in the Euclidian region from which observable quantities can be extracted by duality. Due to the approximate nature of the method, one can only extract the properties of the ground state of given quantum numbers but not those of its radial excitations, which are smeared by the perturbative QCD continuum used to parametrize these states of higher masses.

In this part, we shall follow closely the discussions in [3], not with the aim of reproducing this book, but to update the different discussions therein. However, the present book cannot replace the former as we shall not repeat the detailed derivations already included therein. We shall also not be able to give a complete presentation of the different existing QSSR results due to the large number of its applications. Instead, we will try to limit ourselves to some specific applications, which in my opinion, are representative of the QSSR results. This part of the book is organized as follows: in the first two chapters, we give an introduction to the method of QSSR and in the remaining chapters, we shall review the main developments and results from the method.

In the first part of this book, we have already discussed some current algebra sum rules prior to QCD, such as the Adler–Weisberger sum rules, the Weinberg and DMO sum rules, the electromagnetic $\pi^+$-$\pi^0$ mass difference. These current algebra sum rules are prototype QSSR. We shall discuss some of them in the context of QCD.

# 48

# Theoretical foundations

## 48.1 Generalities and dispersion relations

The fundamental concepts behind QCD spectral sum rules are the operator product expansion (OPE) and quark-hadron duality through the dispersion relation obeyed by the Green's functions due to their analytic properties. For illustrating the discussions, we shall consider, for definiteness, the generic two-point correlator:

$$\Pi_H(q^2) = i \int d^4x e^{iqx} \langle 0|\mathbf{T} J_H(x) J_H^\dagger(0)|0\rangle \ , \tag{48.1}$$

where $J_H(x)$ is a local gauge-invariant operators built from quark and/or gluon fields. In most applications, $J_H(x)$ are Noether currents associated to the global transformations of flavour degrees of freedom, such the vector $\bar\psi\gamma_\mu\psi$ or axial-vector $\bar\psi\gamma_\mu\gamma_5\psi$ current, but can also be the operators of gluon fields describing the gluonium $\mathbf{Tr}\, G_{\mu\nu}G^{\mu\nu}$, or operators describing the baryons $\psi\Gamma_1\psi\Gamma_2\psi$, the hybrids $\bar\psi\gamma_\mu\lambda_a G_a^{\mu\nu}\psi$ or weak matrix elements $\bar\psi\Gamma_1\psi\bar\psi\Gamma_2\psi \ldots$. Thanks to its analyticity property, it has been shown [624] that $\Pi(q^2)$ obeys the well-known Källen–Lehmann dispersion relation or *Hilbert representation*:

$$\Pi_H(q^2) = \int_{t_<}^{\infty} \frac{dt}{t - q^2 - i\epsilon} \frac{1}{\pi} \mathrm{Im}\Pi_H(t) + P(q^2) \tag{48.2}$$

where $P(q^2)$ represents subtraction terms, which are, in general, polynomial in $q^2$, with its degree depending on the convergence properties of the *spectral function* $\mathrm{Im}\Pi_H(t)$ for $t \to \infty$:

$$P(q^2) \equiv a + bq^2 + \cdots \ ; \tag{48.3}$$

$t_<$ is the hadronic threshold, which we shall take to be zero for simplifying the notation. The previous representation is a *QCD spectral sum rule*, which shows the *duality* between the LHS calculable theoretically in QCD, using the OPE, provided that $-q^2$ is much larger than $\Lambda^2$, with the RHS, where the *spectral function* $\mathrm{Im}\Pi_H(t)$ can be measured experimentally. In the case of the electromagnetic current:

$$J^\mu_{\mathrm{em}}(x) = \frac{2}{3}\bar u(x)\gamma^\mu u(x) - \frac{1}{3}\bar d(x)\gamma^\mu d(x) - \frac{1}{3}\bar s(x)\gamma^\mu s(x) + \cdots \tag{48.4}$$

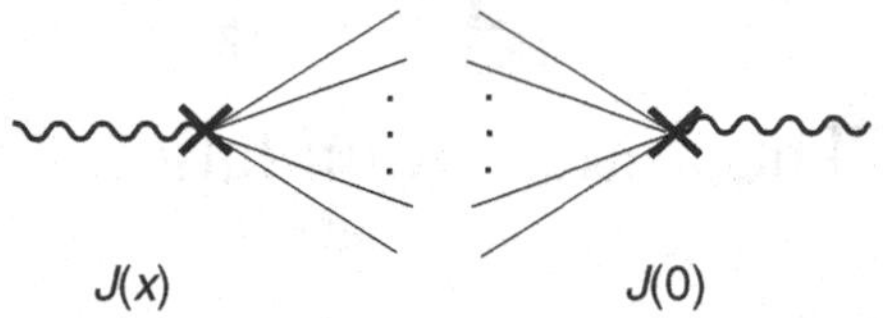


Fig. 48.1. Hadronic spectral function of Eq. (48.6).

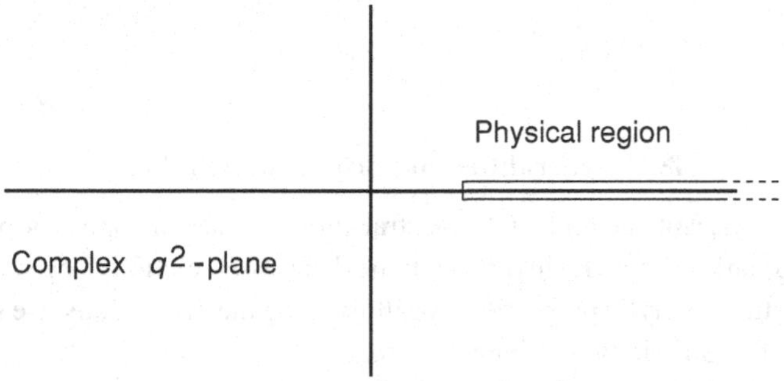


Fig. 48.2. The complex $q^2$-plane.

the spectral function is related to the $e^+e^- \to$ hadrons total cross-section $\sigma_H(t)$ through the optical theorem:

$$\sigma_H(t) = \frac{4\pi^2\alpha}{t}\frac{1}{\pi}\mathrm{Im}\Pi_{\mathrm{em}}(t)\,, \tag{48.5}$$

with:

$$-3\theta(q)\frac{t}{\pi}\mathrm{Im}\Pi_{\mathrm{em}}(t) = \sum_{\Gamma}\langle 0|J^{\mu}_{\mathrm{em}}(0)|\Gamma\rangle\langle\Gamma|J^{\mu}_{\mathrm{em}}(0)^{\dagger}|0\rangle(2\pi)^3\delta^{(4)}(q-p_{\Gamma})\,, \tag{48.6}$$

where the sum runs over all possible physical states, and the integration over the corresponding phase space is understood. This is represented in Fig. 48.1.

In this case, the lowest possible state is the two pions. Therefore, the function $\Pi(q^2)$ is analytic in the complex $q^2$-plane but for a cut near the real axis $4m_\pi^2 \le q^2 \le \infty$ shown in Fig. 48.2.

## 48.2 Explicit derivation of the dispersion relation

In so doing, we consider the lowest order two-point function:

$$\begin{aligned}\Pi^{\mu\nu}(q^2) &\equiv i\int d^4x e^{iqx}\langle 0|\mathbf{T}J^{\mu}_{V}(x)\left(J^{\nu}_{V}\right)^{\dagger}(0)|0\rangle \\ &= -(g^{\mu\nu}q^2 - q^{\mu}q^{\nu})\Pi(q^2)\,,\end{aligned} \tag{48.7}$$

shown in Fig. 8.31 but built from the electromagnetic current:

$$J^\mu_V = e\bar{\psi}\gamma^\mu\psi\,, \tag{48.8}$$

where $\psi$ is a massive quark field with mass m. We follow the same procedure as for the pseudoscalar current for evaluating this lowest-order diagram. It is easy to show that the renormalized two-point function subtracted at $q^2 = 0$ if we choose on-shell renormalization reads:

$$\Pi_R(q^2) \equiv \Pi(q^2) - \Pi(0) = -\frac{\alpha}{\pi}\int_0^1 dx\, 2x(1-x)\log\left(1 - \frac{q^2}{m^2 - i\epsilon}x(1-x)\right). \tag{48.9}$$

With the change of variables $y = 1 - 2x$, and using the fact that the resulting integral is symmetric when $y \to -y$, we get:

$$\Pi_R(q^2) = \frac{\alpha}{\pi}\int_0^1 dy(1-y^2)\log\left[1 - \frac{q^2}{4m^2 - i\epsilon}(1-y^2)\right]. \tag{48.10}$$

Integrating by parts this equation using the identity: $1 - y^2 = \frac{\partial}{\partial y}(y - \frac{1}{3}y^3)$, one obtains the integral:

$$\Pi_R(q^2) = \frac{\alpha}{\pi}\int_0^1 dy\, 2y\left(y - \frac{1}{3}y^3\right)\frac{q^2}{4m^2 - q^2(1-y^2) - i\epsilon}\,. \tag{48.11}$$

With a new change of variables: $t = 4m^2/(1-y^2)$, we finally obtains the representation of the renormalized two-point function:

$$\frac{\Pi_R(q^2)}{q^2} = \frac{\alpha}{\pi}\int_{4m^2}^\infty \frac{dt}{t}\frac{1}{t - q^2 - i\epsilon}\frac{1}{3}\left(1 + \frac{2m^2}{t}\right)\sqrt{1 - \frac{4m^2}{t}}\,. \tag{48.12}$$

The reason why this representation is interesting is that it is in fact *a dispersion relation*. We have succeeded in rewriting the initial Feynman parametric representation in Eq. (48.9) as a dispersion relation by simple changes of variables. Using the identity:

$$\frac{1}{t - q^2 - i\epsilon} = \mathrm{PP}\frac{1}{t - q^2} + i\pi\delta(t - q^2)\,, \tag{48.13}$$

we immediately see that:

$$\frac{1}{\pi}\mathrm{Im}\Pi(t) = \frac{\alpha}{\pi}\frac{1}{3}\left(1 + \frac{2m^2}{t}\right)\sqrt{1 - \frac{4m^2}{t}}\theta(t - 4m^2)\,. \tag{48.14}$$

Equation (48.12) is a particular case of the general dispersion relation written in Eq. (48.2), when the arbitrary polynomial is just a constant, and we have got rid of the constant because the on-shell renormalized $\Pi_R$ is defined as:

$$\Pi_R(q^2) = \Pi(q^2) - \Pi(0) = \int_0^\infty \frac{dt}{t}\frac{q^2}{t - q^2 - i\epsilon}\frac{1}{\pi}\mathrm{Im}\Pi(t)\,. \tag{48.15}$$

It is perhaps worth insisting on the fact that asymptotically:

$$\lim_{t\to\infty} \frac{1}{\pi}\text{Im}\Pi(t) \Longrightarrow \frac{\alpha}{\pi}\frac{1}{3}\,; \tag{48.16}$$

i.e. the electromagnetic spectral function goes to a constant. In fact, it is this constant which fixes the value of the lowest-order contribution to the $\beta$–function associated with the charge renormalization in QED.

### 48.3 General proof of the dispersion relation

We shall now sketch a proof of the dispersion relation property for two-point functions in full generality following [361]. The key of the proof lies in the definition of the time-ordered product implicit in Eq. (48.1):

$$\mathcal{T}(J_H(x)J_H(0)^\dagger = \theta(x)J_H(x)J_H(0)^\dagger + \theta(-x)J_H(0)^\dagger J_H(x)\,, \tag{48.17}$$

and the use of translation invariance. The function $\theta(x)$ is the Heaviside function: $\theta(x) = 1$ if $x_0 > 0$ and $\theta(x) = 0$ if $x_0 < 0$, which has the integral representation:

$$\theta(x) = \frac{1}{2\pi i}\int_{-\infty}^{+\infty} dw\, \frac{e^{iwx}}{w - i\epsilon}\,. \tag{48.18}$$

First we insert a complete set of states $\sum_\Gamma |\Gamma\rangle\langle\Gamma|$ between the two currents in the T–product definition. This leads to matrix elements of the type $\langle 0|J_H(x)|\Gamma\rangle$ to which we apply translation invariance:

$$\langle 0|J(x)|\Gamma\rangle = \langle 0|\mathcal{U}^{-1}\mathcal{U}J_H(x)\mathcal{U}^{-1}\mathcal{U}|\Gamma\rangle\,, \tag{48.19}$$

where $\mathcal{U}$ is the unitary operator induced by translations in space–time:

$$\mathcal{U}(a)J_H(x)\mathcal{U}^{-1}(a) = J_H(x+a) \quad \text{and} \quad \mathcal{U}(a)|\Gamma\rangle = e^{ip_\Gamma\cdot a}|\Gamma\rangle\,, \tag{48.20}$$

and $p_\Gamma$ denotes the sum of the energy–momenta of all the particles which define the state $|\Gamma\rangle$. The choice $a \equiv -x$ factors out the $x$-dependence of the matrix element into an exponential:

$$\langle 0|J_H(x)|\Gamma\rangle = e^{-ip_\Gamma\cdot x}\,\langle 0|J_H(0)|\Gamma\rangle\,. \tag{48.21}$$

All the particles in the state $|\Gamma\rangle$ are on-shell. This constrains the total energy–momentum $p_\Gamma$ to be a time–like vector: $p_\Gamma^2 = t$ with $t \geq 0$. With these constraints on $p_\Gamma$ we can insert the identity:

$$\int_0^\infty dt \int d^4p\,\theta(p)\,\delta(p^2 - t)\,\delta^{(4)}(p - p_\Gamma) = 1\,, \tag{48.22}$$

inside the sum $\sum_\Gamma$ over the complete set of states. Interchanging the order of sum over $\Gamma$ and integration over $t$ and $p$, there appears naturally the definition of the spectral function associated with the $J$-operator

$$\sum_{\Gamma} \langle 0|J_H(0)|\Gamma\rangle\langle\Gamma|J_H(0)^\dagger|0\rangle\,(2\pi)^4\delta^{(4)}(p-p_\Gamma) \equiv 2\pi\rho(p^2)\,. \tag{48.23}$$

The spectral function $\rho(p^2)$ is a scalar function of the Lorentz invariant $p^2$ and the masses of the particles in the states $|\Gamma\rangle$ only. By construction it is a real function and non-negative:

$$\rho(p^2)^* = \rho(p^2) \geq 0\,. \tag{48.24}$$

We can now rewrite the two-point function in Eq. (48.1) as follows:

$$\Pi_H(q^2) = \int d^4x\, e^{iq\cdot x} \int_0^\infty dt\, \rho(t)$$
$$\times \int \frac{d^4p}{(2\pi)^3}[i\theta(x)\,e^{-ip\cdot x}\theta(p)\delta(p^2-t) + i\theta(-x)\,e^{ip\cdot x}\theta(p)\delta(p^2-t)]. \tag{48.25}$$

Here, one can recognize the familiar functions of free field theory:

$$\Delta^+(x) = \int \frac{d^4p}{(2\pi)^3}\, e^{-ip\cdot x}\theta(p)\delta(p^2-t) \tag{48.26}$$

and:

$$\Delta^-(x) = \int \frac{d^4p}{(2\pi)^3}\, e^{ip\cdot x}\theta(p)\delta(p^2-t)$$
$$= \int \frac{d^4p}{(2\pi)^3}\, e^{-ip\cdot x}\theta(-p)\delta(p^2-t)\,; \tag{48.27}$$

and therefore the Feynman propagator function:

$$\Delta_F(x;t) = i\theta(x)\Delta^+(x;t) + i\theta(-x)\Delta^-(x;t)$$
$$= \int \frac{d^4p}{(2\pi)^4}\frac{e^{-ip\cdot x}}{t-p^2-i\epsilon}\,, \tag{48.28}$$

where the last expression can be obtained using the representation in Eq. (48.18) of the $\theta$–function (see e.g. ref. [625]). The two-point function $\Pi(q^2)$ appears then to be the Fourier transform of a scalar free-field propagating with an arbitrary mass squared $t$ weighted by the spectral function density $\rho(t)$ and integrated over all possible values of $t$:

$$\Pi_H(q^2) = \int d^4x\, e^{iq\cdot x} \int_0^\infty dt\, \rho(t)\, \Delta_F(x;t)\,. \tag{48.29}$$

Integrating over $x$ and $p$ results finally in the wanted representation:

$$\Pi_H(q^2) = \int_0^\infty dt\, \rho(t) \frac{1}{t-q^2-i\epsilon}\,. \tag{48.30}$$

With:

$$\Pi_H(q^2) = \mathrm{Re}\Pi_H(q^2) + i\mathrm{Im}\Pi_H(q^2)\,, \tag{48.31}$$

and the use of the identity in Eq. (48.13), it follows that:

$$\rho(t) \equiv \frac{1}{\pi}\mathrm{Im}\Pi_H(t) , \tag{48.32}$$

which identifies the spectral function with the imaginary part of the two-point function.

Notice that the formal manipulations above avoid the question of convergence of the principal value integral:

$$\mathrm{Re}\Pi_H(q^2) = \mathrm{PP}\int_0^\infty dt\,\frac{1}{t-q^2}\,\frac{1}{\pi}\mathrm{Im}\Pi_H(t) . \tag{48.33}$$

The convergence of the integral in the UV limit ($t \to \infty$) depends on the behaviour of the spectral function at large $t$-values. When doing above the exchange of sum over $\Gamma$ and integrations we have implicitly assumed good convergence properties; but in general the product of the distributions $\theta(x)$ and $\int_0^\infty dt\,\rho(t)\,\Delta^+(x;t)$ may not be a well-defined distribution. The ambiguity manifests by the presence of an arbitrary polynomial in $q^2$ in the RHS of the PP-integral:

$$\mathrm{Re}\Pi(q^2) = \mathrm{PP}\int_0^\infty dt\,\frac{1}{t-q^2}\,\frac{1}{\pi}\mathrm{Im}\Pi_H(t) + P(q^2) . \tag{48.34}$$

Notice that the coefficients of the arbitrary polynomial $P(q^2)$ have no discontinuities; in other words, the ambiguity of short-distance behaviour reflects only in the evaluation of the real part of the two-point function, not in the imaginary part. The physical meaning of these coefficients depends of course on the choice of the local operator $J_H(x)$ in the two-point function. In some cases the coefficients in question are fixed by low-energy theorems; e.g. if $\Pi(0)$ is known, we can trade the constant $a$ in Eq. (48.34) for $\Pi(0)$:

$$\mathrm{Re}\Pi_H(q^2) = \mathrm{Re}\Pi_H(0) + \mathrm{PP}\int_0^\infty \frac{dt}{t}\,\frac{q^2}{t-q^2}\,\frac{1}{\pi}\mathrm{Im}\Pi_H(t) + bq^2 + \cdots . \tag{48.35}$$

while the constant $b$ is related to its slope $\Pi'_H(0)$. In other cases the constants can be absorbed by renormalization constants. In general, it is always possible to get rid of the polynomial terms by taking an appropriate number of derivatives with respect to $q^2$. Various examples will be in the next chapter.

## 48.4 The QCD side of the sum rules

Using the SVZ expansion, one can express the two-point correlator in terms of the QCD condensates, where for large Euclidian $q^2$, one obtains:

$$\Pi_H = \sum_{D=0,2,4,\ldots} \frac{1}{(-s)^{D/2}} \sum_{\dim O=D} C^{(J)}(s,\mu)\,\langle O(\mu)\rangle , \tag{48.36}$$

where $\mu$ is an arbitrary scale that separates the long- and short-distance dynamics; $C^{(J)}$ are the Wilson coefficients calculable in perturbative QCD, while $\langle O\rangle$ are the non-perturbative quark and/or gluon condensates. The unit operator is the naïve perturbative contribution.

Table 48.1. *Values of perturbative QCD parameters used or obtained in the sum rules analysis (see chapter on $\alpha_s$ and on quark masses)*

| Perturbative QCD parameters | Values | Sources |
|---|---|---|
| QCD coupling | | |
| $\alpha_s(M_Z)$ | $0.118 \pm 0.002$ | [139,16] |
| Quark running masses to $\mathcal{O}\left(\alpha_s^2\right)$ | | see Section 11.11 |
| $\bar{m}_d(2\ \text{GeV})$ | $(3.6 \pm 0.6)$ MeV | average from different channels |
| $\bar{m}_d(2\ \text{GeV})$ | $(6.5 \pm 1.2)$ MeV | " |
| $\bar{m}_s(2\ \text{GeV})$ | $(117.4 \pm 23.4)$ MeV | " |
| $\bar{m}_c(m_c)$ | $(1.23 \pm 0.05)$ GeV | average from the $J/\psi$ and $D$, $D^*$ |
| $\bar{m}_b(m_b)$ | $(4.24 \pm 0.06)$ GeV | average from the $\Upsilon$ and $B$, $B^*$ |
| Perturbative pole masses $\mathcal{O}\left(\alpha_s^2\right)$ | | see Section 11.12 |
| $M_c$ | $(1.43 \pm 0.04)$ GeV | average from the $J/\psi$ and $D$, $D^*$ |
| $M_b$ | $(4.66 \pm 0.06)$ GeV | average from the $\Upsilon$ and $B$, $B^*$ |

One expects that, for enough large $q^2$ (usually of the order of 1–2 GeV$^2$), the first two-three lowest dimension condensates can give a good approximation of the QCD correlator. In practice, one usually truncates the series until the dimension-six condensates, which are already small corrections in the analysis. The well-known condensate is the quark $\langle\bar{\psi}\psi\rangle$ condensate responsible for the spontaneous breaking of chiral symmetry and is related to the pion and decay amplitude squared through the GMOR relation:

$$(m_u + m_d)\langle\bar{u}u + \bar{d}d\rangle = -2m_\pi^2 f_\pi^2 \qquad (48.37)$$

with $f_\pi = 93.2$ MeV. The other condensates are less known, and are not calculable from QCD first principles though one can determine them from phenomenological analysis. We summarize in Tables 48.1 and 48.2 the values of these QCD parameters which will be useful for the discussion in this part (Part X) of the book.

We have already anticipated the discussions of the theoretical input of the sum rules analysis in the previous chapters:

- In Part III, we discussed the different ingredients for treating and evaluating, within the $\overline{MS}$ scheme and using the renormalization group equation, the perturbative contributions to the unit operator. We have also given there and in Part VI the value of the running QCD coupling and the light and heavy quark masses in Sections 11.7, 11.11 and 11.12, entering the QCD Lagrangian and useful in the sum rules analysis.
- In Part VII, we have discussed the different non-perturbative contributions:
  - In Chapter 27, we have studied the operator product expansion (OPE) and classified the condensates versus their dimensions. We have also constructed renormalization group invariant condensates and given the values of some of the condensates which have been determined mainly from the sum rules.

Table 48.2. *Values of the non-perturbative QCD (NPQCD) parameters used or obtained in the sum rules analysis*

| Dimension | NPQCD parameters | Values | Sources |
|---|---|---|---|
| 2 | $(\alpha_s/\pi)\lambda^2$ | $-(0.06–0.07)$ GeV$^2$ | Chapter 30 |
| 3 | $\frac{1}{2}\langle\bar{u}u + \bar{d}d\rangle(2\text{ GeV})$ | $-[(254 \pm 15)\text{ MeV}]^3$ | Chapter 27 |
| | $\langle\bar{s}s\rangle/\langle\bar{u}u\rangle$ | $0.75 \pm 0.12$ | non-normal ordered |
| | | $0.66 \pm 0.10$ | normal ordered |
| 4 | $\langle\alpha_s G^2\rangle$ | $(7 \pm 1)10^{-2}$ GeV$^4$ | Chapter 27 |
| 5 | $g\langle\bar{\psi}\sigma_{\mu\nu}\frac{\lambda^a}{2}\psi G_a^{\mu\nu}\rangle \equiv M_0^2\alpha_s^{1/3\beta_1}\langle\bar{\psi}\psi\rangle$ | $M_0^2 = (0.80 \pm 0.02)$ GeV$^2$ | " |
| 6 | $g^3 f_{abc}\langle G^a G^b G^c\rangle$ | $(1.2\text{ GeV}^2)\langle\alpha_s G^2\rangle$ | " |
| | $\rho\alpha_s\langle\bar{\psi}\psi\rangle^2$ | $(5.8 \pm 0.9)10^{-4}$ GeV$^6$ | " |

- In Chapter 28, we discuss in details the evaluation of the Wilson coefficients in the OPE. In so doing, we give as an explicit example the evaluation of the light quark pseudoscalar two-point function including dimension-six condensates. We also discuss the evaluation of the heavy quark correlators.
- In Part VIII, we give a compilation of different QCD two-point functions including radiative corrections to the unit operator and the contributions of different condensates.
- In Chapter 29, we discuss the modifications of the OPE due to IR and UV renormalons. IR renormalons introduce perturbative contributions, which lead to some ambiguities for defining the condensates at higher order of perturbation theories, though such ambiguities can be absorbed by the Wilson coefficients when computing the Green's functions. In practice, the IR renormalon effects are so tiny such that they do not affect in a significant way the phenomenology of the sum rules. UV renormalons have also been discussed so far, and affect the uncertainties of the PT series. Again, within the sum rules uncertainties, these effects are not quantifiable in the sum rules analysis.
- In Chapter 30, we have discussed the different scenarios beyond the SVZ expansion. In the following, we shall only discuss the modification due to the tachyonic gluon mass which modifies the OPE due to the presence of the new $D = 2$ 'condensate', not present in the original SVZ-expansion owing to the fact that one cannot form a $D = 2$ local gauge-invariant operator in QCD. We shall not discuss the effects of (direct) instantons which act like high-dimension operators and should be suppressed like other high-dimension condensates in the sum rule working region. However, some other schools expect that this contribution is dominant for the (pseudo)scalar channels but surprisingly are not there if one works with the longitudinal part of the axial-vector correlator, though the two are related to each others by Ward identity. However, the inclusion of the large instanton effects leads to inconsistencies in the scalar channel. Another confusion for the sum rule practitioners is the fact that the instanton liquid model does not use a novel OPE but provides an alternative way of parametrizing the condensates. However, the fact that the analysis is done in the coordinate rather than in the momentum space may probe a new region not explored in the momentum space. Interested readers may consult [386] where this method is explored in detail.

# 49

# Survey of QCD spectral sum rules

QCD spectral sum rules are different versions and/or improvments of the previous Hilbert representation in Eq. (48.2). For the purposes of more general discussions, let us forget QCD for the moment, namely the theoretical side Re $\Pi(q^2)$, and we shall concentrate on the RHS spectral integral.

In some channels such as $e^+e^- \to$ hadrons or $\tau \to \nu_\tau +$ hadrons data, the spectral function $\text{Im}\Pi(t)$ is known from the data, and the sum rules can be used for determining the QCD parameters given in Tables 48.1 and 48.2. In other channels, the sum rules are used for determining the properties of the hadrons for a guide to their experimental searches. In this case, one has to introduce a model for parametrizing the spectral function. For this purpose, the most common model used in the sum rule analysis is the so-called *naïve duality ansatz*, where the spectral function reads:

$$\text{Im}\Pi(t) = f_H^2 M_H^{2d} \delta\big(t - M_H^2\big) + \theta(t - t_c)\text{Im}\Pi_{\text{QCD}}(t) . \tag{49.1}$$

$f_H$ is the coupling having the dimension of mass of the lowest hadron ground state $H$ to the hadronic current; $d$ is the power of $t$ in the asymptotic $t$-behaviour of the spectral function ($d = 0$ for the vector two-point function, ...); $t_c$ is the 'QCD continuum' threshold above which the spectral function is approximated by the discontinuity $\text{Im}\Pi_{\text{QCD}}(t)$ of the QCD diagram, which is expected to smear the contributions of the higher mass radial excitations. We shall test later on in some examples the accuracy of this simple duality ansatz for reproducing the measured spectral function.

An alternative parametrization can be provided by approximating the spectral function with an infinite sum of narrow resonances:

$$\text{Im}\Pi(t) = \sum_H f_H^2 M_H^2 \delta\big(t - M_H^2\big) , \tag{49.2}$$

where the model is supported by the large $N_c$-behaviour of QCD as discussed in the previous part of this book.

## 49.1 Moment sum rules in QCD

In QCD the number of derivatives required to obtain a well-defined two-point function is fixed by the asymptotic freedom property of the theory. For a gauge-invariant local operator

$J_H(x)$, the asymptotic behaviour of the associated two-point function is of the type:

$$\lim_{t\to\infty}\frac{1}{\pi}\text{Im}\Pi(t) \sim \text{A}t^d\left\{1+a_1\frac{\alpha_s(t)}{\pi}+\cdots\right\}, \tag{49.3}$$

with A and $a_1$ calculable coefficients, and $d$ a known integer $d = 0, 1, 2, \ldots$, depending on the dimensions of the operator $J_H(x)$. It is then sufficient to take $d+1$ derivatives with respect to $q^2$ to get rid of the arbitrary polynomial and obtain a convergent integral. The functions defined by the *moment integrals* ($Q^2 \equiv -q^2$):

$$\begin{aligned}\Pi^{(m)}(Q^2) &= \frac{(-1)^m}{(m-d-1)!}(Q^2)^{m-d}\frac{\partial^m}{(\partial Q^2)^m}\Pi(q^2)\\ &= \int_0^\infty dt\,\frac{m(m-1)\cdots(m-d)}{(t+Q^2)^{d+1}}\left(\frac{Q^2}{t+Q^2}\right)^{m-d}\frac{1}{\pi}\text{Im}\Pi(t)\,,\end{aligned} \tag{49.4}$$

for $m \geq d+1$ are then well-defined functions calculable in perturbative QCD at sufficiently large $Q^2$-values. To our knowledge, these sum rules were first discussed by Yndurain [631] in connection with the study of $e^+e^- \to$ hadrons data and used later on for heavy-quark systems [632,1,434]. One can notice that for high-derivative moments, the rôle of the ground state is enhanced in the sum rule. Therefore the sum rule in Eq. (49.4) is a good candidate for studying the low-energy properties of hadrons as we shall see later on.

A classical example of moments sum rules is the $D$-function defined in Eq. (33.20), which is superconvergent and therefore obeys an homogeneous RGE. From Eq. (33.25), one can deduce for three massless flavours:

$$\begin{aligned}D(Q^2) &\equiv -Q^2\frac{d}{dQ^2}\Pi_{\text{em}}(Q^2) = \int_0^\infty\frac{dt}{(t+Q^2)^2}\frac{1}{\pi}\text{Im}\Pi_{\text{em}}(t)\\ &= \frac{2}{16\pi^2}\left[1+\left(\frac{\bar\alpha_s}{\pi}\right)+1.64\left(\frac{\bar\alpha_s}{\pi}\right)^2+6.37\left(\frac{\bar\alpha_s}{\pi}\right)^3+\cdots+\text{non}-\text{perturbative}\right].\end{aligned} \tag{49.5}$$

However, when trying to confront this sum rule with experiment, there appears the problem that the integrand in the RHS is only known experimentally from the threshold up to finite values of $t$. This brings in a question of *matching* whatever is known about the low-energy hadronic spectral function with its asymptotic behaviour as predicted by pQCD.

## 49.2 Laplace sum rule (LSR)

This type of sum rule is derived from the previous dispersion relation in Eq. (48.2) by applying to both sides the *inverse Laplace operator* [1]:[1] ($Q^2 \equiv -q^2 \geq 0$):

$$\hat{\mathcal{L}} \equiv \lim_{n,\;Q^2\to\infty}(-1)^n\frac{(Q^2)^n}{(n-1)!}\frac{\partial^n}{\partial Q^2)^n}, \tag{49.6}$$

[1] This sum rule was originally called the Borel sum rule by SVZ [1].

where $n/Q^2 \equiv \tau$ is fixed, which is the Laplace sum rule variable. It has been found in the study of the radiative corrections that sum rule expression of these radiative terms naturally have the properties of the Laplace transform [626], whilst later on [405], it has also been noticed that the operator $\hat{\mathcal{L}}$ is an algebraic form of the Laplace inversion operator. These observations led to simplifications in the derivation of the QCD expressions of the sum rules once one knows the expression of the two-point correlator $\Pi(q^2)$. Useful expressions are collected in Appendix G. Therefore, one gets the exponential form of the sum rule:

$$\hat{\mathcal{L}}\Pi = \tau \int_0^\infty dt\, e^{-t\tau} \frac{1}{\pi} \mathrm{Im}\Pi(t) . \tag{49.7}$$

As can be seen in the derivation of the Laplace sum rule, one has to assume that various derivatives exist. For an approximate truncated series as in QCD improved by the renormalization group equation, this existence is satisfied as in the case of the moment sum rules. The advantages of $\hat{\mathcal{L}}\Pi$ are two-fold:

- First, the use of various derivatives helps to eliminate the subtraction terms in Eq. (48.2), which are often polynomials in $q^2$.
- Second, the exponential factor increases the role of the ground state into the spectral integral if the QSSR variable $\tau$ is not too small, but still not too large for the perturbative calculation to make sense. In practice $\tau$ is about the value of the hadronic scale. This fact is welcome for low-energy physics.

## 49.3 Ratio of moments

From Eq. (49.7), one can derive the ratio of moments [91–93]:

$$R(\tau) = -\frac{d}{d\tau} log \int_0^\infty dt\, e^{-t\tau} \frac{1}{\pi} \mathrm{Im}\Pi(t) , \tag{49.8}$$

or the finite energy-like [1]:

$$R_c(\tau) = \frac{\int_0^{t_c} dt\, t\, e^{-t\tau} \frac{1}{\pi} \mathrm{Im}\Pi(t)}{\int_0^{t_c} dt\, e^{-t\tau} \frac{1}{\pi} \mathrm{Im}\Pi(t)} . \tag{49.9}$$

Its non-relativistic version is obtained by transforming the variable $t$ into the non-relativistic energy $E$ and $\tau_N = 4m\tau$. In this way, the ratio becomes:

$$R(\tau_N) = -\frac{d}{d\tau_N} log \int_0^\infty dE\, e^{-E\tau_N} \frac{1}{\pi} \mathrm{Im}\Pi(t) , \tag{49.10}$$

where $\tau_N$ can be interpreted as the *imaginary time* variable. The advantage of the ratio of moments can be explicitly seen in the following way:

- If one uses the simple duality ansatz 'one resonance' plus 'QCD continuum' for parametrizing the spectral function, one can see that the two sum rules in Eqs. (49.8) and (49.9) give an expression of the mass squared of the ground state. More precisely, for large $\tau$ values, the RHS of the sum rule tends to the mass squared of the lowest resonance.

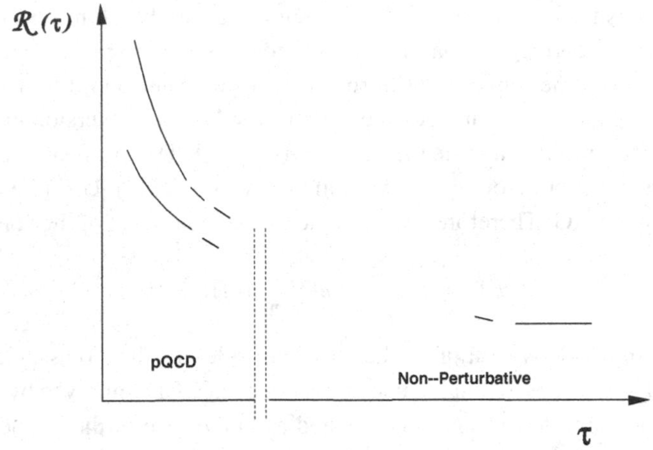


Fig. 49.1. Expected behaviour of $\mathcal{R}(\tau)$ at short and long distances.

- For small $\tau$-values, the ratio of moments has the parton model behaviour:

$$\mathcal{R}(\tau) = (d+1)\tau^{-1}[1 + \text{QCD corrections}] , \tag{49.11}$$

where $d$ is the only reminiscence left from the number of subtractions needed in the dispersion relation for the initial two-point function. For large $\tau$ values, the ratio of moments is dominated by the non-perturbative corrections. In the *sum rule window* compromise region, where the moments stabilize, these non-perturbative corrections are small though vital for stablizing the result. These features lead to the expected behaviour of $\mathcal{R}$ given in Fig. 49.1.

- Because of the positivity property of a spectral function $\text{Im}\Pi(t) \geq 0$, the function $-\log \mathcal{M}(\tau)$ is a concave function of $\tau$; or in other words, the slope of the function $\mathcal{R}(\tau)$ must always be negative. This of course implies severe restrictions on the way that the two asymptotic regimes illustrated in Fig. 49.1 can be joined. The proof of this property is rather straightforward. It can be understood very simply by making an analogy with statistical mechanics: $\mathcal{R}(\tau)$ can be viewed as the equilibrium 'energy' $\langle t \rangle$ of a system with variable 'energy' $t$ in thermal equilibrium with a second system at 'temperature' $1/\tau$. In this analogy, $\text{Im}\Pi(t)$ represents the 'density of states' with 'energy' $t$. Then the mean squared 'energy fluctuation' is given by:

$$-\frac{d}{d\tau}\mathcal{R}(\tau) \equiv \langle (t - \langle t \rangle)^2 \rangle = \langle t^2 \rangle - \langle t \rangle^2 \geq 0 , \tag{49.12}$$

which by definition is a positive quantity.

- In its non-relativistic version, the ratio of moments tends to the ground state energy $E_0$ for large imaginary time $\tau_N \to \infty$. In the corresponding theoretical perturbative expansion, the minimum of $\mathcal{R}$ gives an approximation of this ground state energy:

$$\min\mathcal{R}(\tau_N) = E_0 , \tag{49.13}$$

and the mass of the ground state is given by:

$$M = 2m + \mathcal{R}(\tau_N) . \tag{49.14}$$

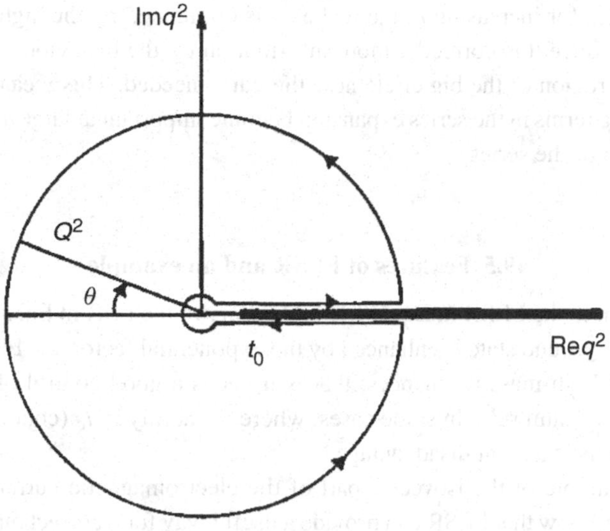


Fig. 49.2. Contour integral in the complex $q^2$-plane, with $q^2 = -Q^2 \exp(i\theta)$.

## 49.4 Finite energy sum rule (FESR)

Another version of QSSR is the FESR:

$$\mathcal{M}_n(Q^2) \equiv \int_0^{Q^2} dt\, t^n \frac{1}{\pi}\mathrm{Im}\Pi_{\mathrm{QCD}}(t) \simeq \int_0^{Q^2} dt\, t^n \frac{1}{\pi}\mathrm{Im}\Pi_{\mathrm{exp}}(t) \qquad : n = 0, 1, \ldots, \tag{49.15}$$

which was known a long time before QCD [627]. The previous FESR can be derived in many ways. One way to derive the FESR is the use of the Cauchy theorem on a finite radius contour in the complex $q^2$ plane (Fig. 49.2) à la Shankar [628].

Avoiding the cut along the real axis, it leads to [628,28,31]:

$$\frac{1}{2\pi i}\oint dz z^n \Pi(z) = 0. \tag{49.16}$$

If one neglects the contribution of the little circle around the origin which is safer if $\Pi(0) = 0$, one deduces the moments:

$$\mathcal{M}_n(Q^2) = \int_0^{Q^2} dt\, t^n \frac{1}{\pi}\mathrm{Im}\Pi(t) = (-1)^{n+1}\frac{(Q^2)^{n-1}}{2\pi}\cdot\int_{-\pi}^{+\pi} d\theta\, e^{i(n+1)\theta}\,\Pi(Q^2 e^{i\theta}), \tag{49.17}$$

where the LHS can be measured from the data and comes from the paths above and below the real axis which pick up the discontinuity of $\Pi(q^2)$ and then its imaginary part. The RHS comes from the big circle of radius $Q^2$, which can be computed in QCD provided it is large enough. The sum rule results from the matching of these two contributions. However, as the FESR diverges for increasing $n$, the real axis is dominated by the high $Q^2$ region. For the RHS to reproduce this correctly, more information on the behaviour of the two-point correlator in the region of the big circle near the cut is needed. This means that more and more non-leading terms in the series expansion become important at large $n$ and can destroy the convergences of the series.

### 49.5 Features of FESR and an example

Now, let us return to the FESR in Eq. (49.15). Contrary to the LSR in Eq. (49.7), where the role of the lowest ground state is enhanced by the exponential factor, the FESR is governed by the effects of high-mass resonances; that is it needs a good control of the continuum contributions to the sum rule. In some cases, where a stability in $t_c$ (continuum threshold) does not occur, this is a great disadvantage.

Taking the example of the isovector part of the electromagnetic current (the $\rho$-meson channel), one can show that FESR can provide a useful way for a correct matching between the low-energy hadronic spectral function and the onset of QCD perturbative continuum. In this sense, it complements the analysis from the LSR. Using the naïve duality ansatz for the hadronic spectral function, the spectral function reads:

$$\frac{1}{\pi}\mathrm{Im}\Pi(t)_{I=1} = \frac{M_\rho^2}{4\gamma_\rho^2}\delta\left(t - M_\rho^2\right) + \frac{N_c}{16\pi^2}\frac{2}{3}\theta(t-t_c)[1+\cdots] , \tag{49.18}$$

where the $\rho$-meson coupling $\gamma_\rho \simeq 2.55$ is normalized in Eq. (2.52). Using the $n = 0$ FESR moments, one can derive the constraint:

$$\frac{M_\rho^2}{4\gamma_\rho^2} \simeq \frac{N_c}{16\pi^2}\frac{2}{3}t_c[1+\cdots] . \tag{49.19}$$

Using the experimental values of the $\rho$-meson parameters, and adding QCD corrections, one obtains (see details in [405,3]):

$$t_c \simeq 1.7\ \mathrm{GeV}^2 , \tag{49.20}$$

which is reasonably high for pQCD calculation to make sense. It is worthwhile to notice that the *FESR fixes both the lowest ground state parameters and the correlated value of the QCD continuum threshold* $t_c$, as contrary to the LSR, the FESR is weighted by the high-energy region for positive values of the degree $n$ of the moment. However in some cases, this property become a great inconvenience of the method.

In general, this value of $t_c$ is slightly different from the phenomenological value of the first radial excitation mass. This might be not so surprising as the QCD model, which gives a smearing of the high-energy region, cannot take into account the complicated structure of the resonances in this region.

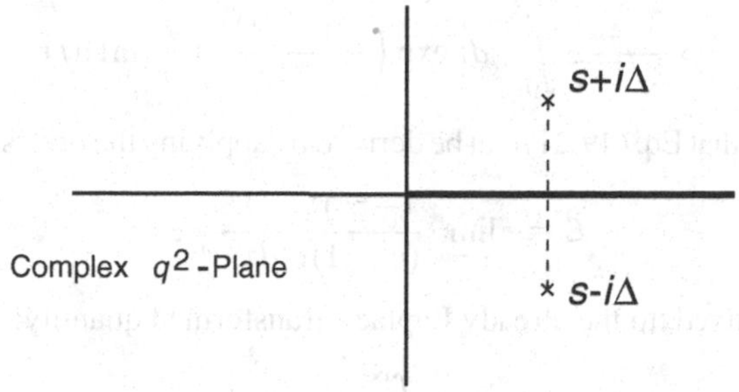


Fig. 49.3. Points in the complex $q^2$-plane where the two-point function in Eq. (49.22) is evaluated.

## 49.6 The Gaussian sum rules

Another way of deriving the FESR which casts light upon the meaning of local duality is the Gaussian sum rule which reads [405,406]:

$$G(s,\sigma) = \frac{1}{\sqrt{4\pi\sigma}} \int_0^\infty dt\, e^{-\frac{(t+s)^2}{4\sigma}} \frac{1}{\pi} \mathrm{Im}\Pi(t)\,, \tag{49.21}$$

for a Gaussian centred at $s$ with a finite width resolution $\sqrt{4\pi\sigma}$. Let us discuss how to get the Gaussian transform from a generic two-point function like $\Pi(q^2)$ in Eq. (48.1). First, one evaluates $\Pi(q^2)$ at a complex point $q^2 = s + i\Delta$ ($s$ and $\Delta$ are real positive variables) and at its complex conjugate $q^2 = s - i\Delta$ (see Fig. 49.3) and defines the combination, (one assumes for simplicity that the dispersion relation for $\Pi(q^2)$ requires at most one subtraction, but the argument can be easily generalized as in the case discussed for the Laplace transform):

$$\frac{\Pi(s+i\Delta)}{i\Delta} + \frac{\Pi(s-i\Delta)}{-i\Delta} = \int_0^\infty dt\, \frac{1}{(t-s)^2+\Delta^2} \frac{1}{\pi} \mathrm{Im}\Pi(t)\,. \tag{49.22}$$

The integral in the RHS brings in the convolution with a Lorentz-like kernel which we can write as a Laplace transform

$$\frac{1}{(t-s)^2+\Delta^2} = \int_0^\infty dx\, e^{-x\Delta^2} e^{-x(t-s)^2}\,. \tag{49.23}$$

Applying the techniques developed in the previous Section 49.2 to this integral representation allows us to construct the inverse Laplace transform operator which is needed to obtain the Gaussian transform in Eq. (49.21) from the Lorentz transform in Eq. (48.2). It is the operator:

$$\mathbf{L} \equiv \lim_{N,\Delta^2\to\infty} \Big|_{\frac{1}{N}\Delta^2=4\tau} \frac{(-1)^N}{(N-1)!} \left(\Delta^2\right)^N \frac{\partial^N}{\left(\partial\Delta^2\right)^N}\,. \tag{49.24}$$

We then have the desired relation:

$$\frac{1}{\sqrt{4\pi\tau}} 2\tau \mathbf{L} \left[ \frac{\Pi(s+i\Delta)}{i\Delta} + \frac{\Pi(s-i\Delta)}{-i\Delta} \right] \tag{49.25}$$

$$\Longrightarrow \frac{1}{\sqrt{4\pi\tau}} \int_0^\infty dt \exp\left(-\frac{(s-t)^2}{4\tau}\right) \frac{1}{\pi}\text{Im}\Pi(t) \,. \tag{49.26}$$

One can also note that Eq. (49.21) can be derived by applying the inverse Laplace operator:

$$\hat{\mathcal{L}} \equiv \lim_{n,\tau^2\to\infty} \frac{(-\tau^2)^n}{(n-1)!} \frac{d^n}{(d\tau^2)^n} \,, \tag{49.27}$$

where $n/\tau^2 \equiv \sigma$ is fixed, to the already Laplace-transformed quantity:

$$\mathcal{F}(\tau) = e^{-s\tau}\tau^{-1} \int_0^\infty dt\, e^{-t\tau} \frac{1}{\pi}\text{Im}\Pi(t) \,. \tag{49.28}$$

One can already note from Eq. (49.21) that in limit $\sigma = 0$, where the Gaussian kernel becomes a delta function, one has the *strict local duality*:

$$G(s, 0) = \frac{1}{\pi}\text{Im}\Pi(s) \,. \tag{49.29}$$

Also, Eq. (49.21) obeys the heat-evolution equation:

$$\left(\frac{\partial^2}{\partial s^2} - \frac{\partial}{\partial \sigma}\right) G(s, \sigma) = 0 \,, \tag{49.30}$$

with the initial condition in Eq. (49.29), where now $s$ is the position, $\sigma$ the time evolution and $\frac{1}{\pi}\text{Im}\Pi(t)$ the temperature distribution in the region $0 \le s \le \infty$. The two boundary conditions for $\sigma > 0$:

$$G(s = 0, \sigma) = 0 \,, \qquad \frac{\partial G}{\partial s}(s, \sigma)\Big|_{s=0} = 0 \,, \tag{49.31}$$

lead to two independent solutions $U^-(s, \sigma)$ and $U^+(s, \sigma)$ where $G(s, \sigma) = \frac{1}{2}(U^+ + U^-)(s, \sigma)$. These solutions can be expressed in terms of Hermite polynomials. The conservation of the total heat implies the duality relation:

$$\int_{-\infty}^{+\infty} ds\, G(s, \sigma) = \int_0^\infty ds\, \frac{1}{\pi}\text{Im}\Pi(s) = \int_0^\infty ds\, U^+(s, \sigma) \,, \tag{49.32}$$

where the last equality comes from the symmetry properties of $U^+(s, \sigma)$. A relation involving higher moments of the spectral function can also be deduced using the generating function of Hermite polynomials and leads to the sum rules:

$$\begin{aligned} \sigma^n \int_0^\infty ds\, H_{2n}\left(\frac{s}{2\sqrt{\sigma}}\right) U^+(s, \sigma) &= \int_0^\infty dt\, t^{2n} \frac{1}{\pi}\text{Im}\Pi(t) \,, \\ \sigma^{n+1/2} \int_0^\infty ds\, H_{2n+1}\left(\frac{s}{2\sqrt{\sigma}}\right) U^-(s, \sigma) &= \int_0^\infty dt\, t^{2n+1} \frac{1}{\pi}\text{Im}\Pi(t) \,, \end{aligned} \tag{49.33}$$

which only become useful once statements about the restriction to finite intervals can be made. In this case, Eq. (49.33) leads to the FESR in Eq. (49.15).

In [405,361], the example of the $\rho$ meson has been taken for illustrating the Gaussian sum rules and summarized in the following figures. Figure 49.4 shows the evolution in the pseudo-'time' variable $\sigma$ of the Gaussian transform of the spectral function *ansatz*

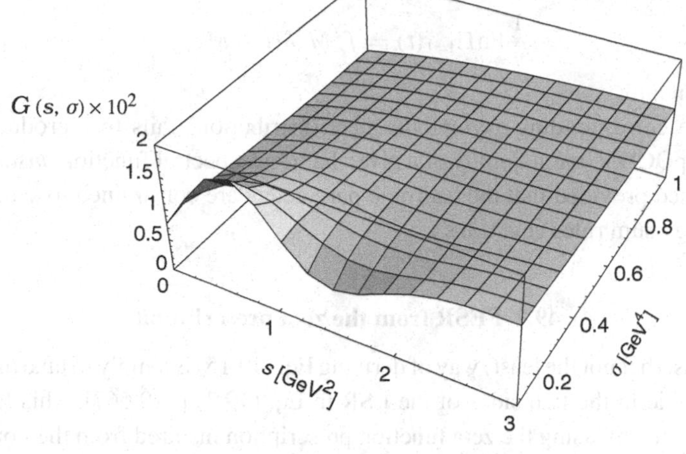


Fig. 49.4. The Gaussian transform of the spectral function in Eq. (49.18).

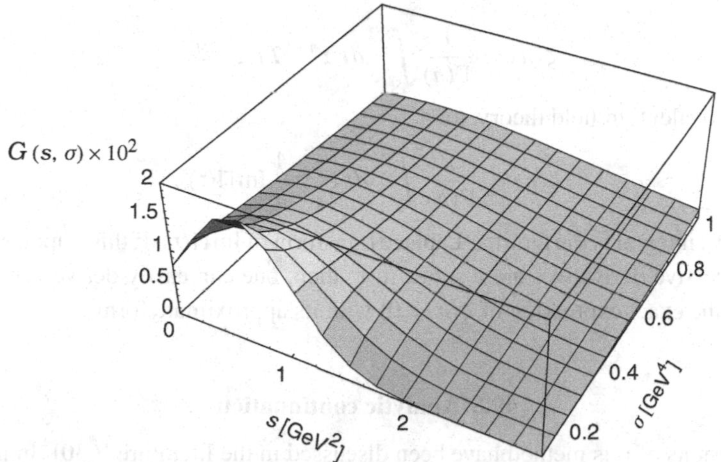


Fig. 49.5. The Gaussian transform of the spectral function in Eq. (49.35).

in Eq. (49.18) with the onset of the continuum $t_c$ fixed by the finite energy sum rule in Eq. (49.19).

In the 'heat evolution' analogy the spectral function in Eq. (49.18) corresponds to the initial 'heat distribution' in the $s$–axis. The picture shows the evolution in 'time' of this 'heat distribution' in the interval $0.1\,\text{GeV}^4 \leq \sigma \leq 1\,\text{GeV}^4$. We observe that asymptotically in 'time', i.e. for $\sigma$ large, the spectral function evolves very well to the asymptotic 'heat distribution' predicted by pQCD i.e.:

$$\lim_{\sigma\to\infty} G(s,\sigma) = \frac{1}{16\pi^2}\left(1 - \text{erf}\left(\frac{s}{2\sqrt{\sigma}}\right)\right)[1+\cdots]\,, \qquad (49.34)$$

where erf$(x)$ denotes the error function erf$(x) = \frac{2}{\sqrt{\pi}}\int_0^x dy\, e^{-y^2}$. By contrast, Fig. 49.5 shows the same evolution in the limit case of only a delta-function *ansatz* for the spectral function:

$$\frac{1}{\pi}\text{Im}\Pi_{\text{I}=1}(t) = f_\rho^2 M_\rho^2 \delta\left(t - M_\rho^2\right) , \tag{49.35}$$

with no continuum.

Clearly the corresponding asymptotic 'heat distribution' fails to reproduce the shape predicted by pQCD. *Global duality* of a given hadronic spectral function *ansatz* with QCD is only obtained provided that the hadronic parameters are constrained to satisfy a system of finite-energy sum rules equations.

## 49.7 FESR from the zeta prescription

Finally, the last (but not the least) way of deriving Eq. (49.15) is simply to take the coefficient of the $\tau$ variable in the two sides of the LSR in Eq. (49.7) [629,667]. This latter method can be formalized by using the zeta function prescription inspired from the non-relativistic approach [406]. In fact, if $H$ is a Hamilton operator, the associated zeta-function can be written as:

$$\zeta(n) = \frac{1}{\Gamma(n)} \int_0^\infty dt\ \tau^{n-1} Tr\ e^{-Ht} , \tag{49.36}$$

which is equivalent, in field theory, to:

$$\zeta(n) = \frac{1}{\Gamma(n)} \int_0^\infty dt\ e^{-t\tau} \frac{1}{\pi}\text{Im}\Pi(t) , \tag{49.37}$$

where the last integral is the familiar Laplace transform of $\text{Im}\Pi(t)$. If this Laplace-transform and its successive derivatives are a series in $\tau$, then, one can easily derive Eq. (49.15) by comparing the exact expression of $\zeta(n = 0)$ with its approximate form.

## 49.8 Analytic continuation

Various versions of this method have been discussed in the literature [630]. In most cases, the problem is formulated in terms of norm problems for the input errors and is quite similar to the standard $\chi^2$–minimization used in numerical analysis. More explicity let us take a simple example. A polynomial in $t$ is used for approximating the $1/(t - q^2)$ term of Eq. (48.2) in the real axis [630]. Then, applying the Cauchy theorem to the finite $Q^2$ contour in the complex $Q^2$ plane, one arrives at the sum rule:

$$\Pi(q^2) = \frac{1}{2i\pi} \oint_C dt \left( \frac{1}{t - q^2} - \sum_{a_n} t^n \right) \Pi(t)$$
$$+ \left[ \Delta_n \equiv \frac{1}{\pi} \int_0^{Q^2} dt \left( \frac{1}{t - q^2} - \sum_n a_n t^n \right) \right] \text{Im}\Pi(t) , \tag{49.38}$$

where $\Delta_n$ is the 'fit error' which should tend to zero, if the result is optimal. An important difference with the previous sum rules is that in the RHS the data enters only in $\Delta_n$ whilst the main part of $\Pi(q^2)$ is given by its theoretical side. However, it is difficult to appreciate the reliability of the results coming from the method due to:

- The *ad hoc* uses of the polynomial parametrization (or in general of the kernels in the integrals) and to the strong dependence of the results on the values of the input errors.
- Its form in Eq. (49.38) where the dependence of the sum rule on the arbitrary subtraction scale is unclear.
- The way of extrapolating the QCD information up to small $q^2$ which is model dependent.

Due to these weak points, all the beautiful mathematical forms used to formulate the sum rule might lose their efficiency in its physical applications. More refinements and more phenomenological tests of this approach are needed before a definite claim about its superiority can be made.

## 49.9 Summary

We have given a brief general survey of spectral function sum rule methods which we believe can be applied for a general class of QCD-like theories. As one can see all the methods presented here have their own advantages and disadvantages. For the particular case of QCD where the theory has not yet been solved exactly, some questions, though important, such as the existence of high derivatives at high $Q^2$ as well as a correct and convincing way of estimating the true theoretical systematic errors in the sum rules analysis remain academic. We have checked in a QCD-like model such as the non-linear $\sigma$ model in two dimensions, as suggested by Gabriele Veneziano, that the high derivatives for a two-point correlator exist unambiguously. Also, one can always test a posteriori whether the assumptions used for the analysis make sense.

In this review, we shall mainly concentrate on the uses of the LSR in Eq. (49.7) to Eq. (49.9) owing to their sensitivity with respect to the low-energy behaviour of the spectral functions. However, in most cases, we shall also discuss for a comparison, the constraints from FESR in Eq. (49.15) which complement the LSR results.

## 49.10 Optimization criteria

One can notice that the sum rule variables $\tau$ (LSR variable) or $n$ (finite number of derivatives) and the continuum threshold $t_c$ are, in general, free parameters in the sum rule analysis.

- In the original work of SVZ [1], the optimal result from the sum rule is obtained inside a window in $\tau$ or $n$, where one has a balance between the QCD continuum and the non-perturbative condensates contributions in the sum rule. In QSSR1 [3], one has shown that this feature corresponds to the existence of a minimum in $\tau$ or $n$, as can be illustrated by the example of three-dimensional harmonic oscillator in quantum mechanics and of the charmonium sum rules [91–93].

### 49.10.1 The harmonic oscillator

For this purpose, let consider the harmonic oscillator potential:

$$V(r) = \frac{1}{2} m\omega^2 r^2, \tag{49.39}$$

and the 'correlation function' for the $S$-wave states:

$$F(\tau) = \sum (R_n)^2 e^{-E_n \tau} \quad : \qquad n = 0, 2, 4, \ldots\,, \tag{49.40}$$

where $R_n$ is the radial wave function for zero angular momentum and $E_n$ the corresponding eigenvalue. $\tau$ is the parameter which regulates the energy resolution of the sum rule and plays the role of an 'imaginary time' variable. The exact solution of the LHS for the harmonic oscillator potential $V(r)$ reads:

$$F(\tau)_{\text{exact}} = \frac{2}{\sqrt{2\pi}} \left( \frac{m\omega}{\sinh \omega\tau} \right)^{3/2}, \tag{49.41}$$

where one can see that, in the limit $\tau \to \infty$, the exact expression:

$$\mathcal{R}(\tau)_{\text{exact}} \equiv -\frac{d}{d\tau} \log F(\tau) \tag{49.42}$$

tends to the lowest eigenvalue:

$$E_0 = \frac{3}{2}\omega\,. \tag{49.43}$$

At finite $\tau$ and for a truncated series in $\tau$, one can write the approximate solution:

$$\mathcal{R}(\tau)_{\text{approx}} \equiv E_0 \left[ \frac{1}{\omega\tau} + \frac{\omega\tau}{3} - \frac{(\omega\tau)^3}{45} + \frac{2(\omega\tau)^5}{945} + \cdots \right], \tag{49.44}$$

where the first term is the free motion, and the next ones are higher-order corrections in $\tau$ to this term. One can notice that in this approximate solution, one cannot take formally the limit $\tau \to \infty$, as the asymptotic series will blow up. Therefore, a comparison of the exact and approximate solution can only be done in a compromise region where the series converge and where the $S$ states contribution is dominant. This is exactly the situation which we shall encounter in the QCD sum rule analysis. The $\tau$ behaviour of $\mathcal{R}_{\text{approx}}(\tau)$ in shown in the Fig. 49.6, which one can compare with the eigenvalue $E_0$. One can notice that it stays above $E_0$ as a consequence of the positivity of $\mathcal{R}$. The agreement between $\mathcal{R}_{\text{approx}}$ and $\mathcal{R}_{\text{exact}}$ increases if one adds more and more terms in the $\tau$ expansion. The minimum of $\mathcal{R}_{\text{approx}}$ provides an upper bound to the value of $E_0$ while the distance between $\mathcal{R}_{\text{approx}}$ and $E_0$ controls the strength of the continuum contribution to the sum rule. One can notice that the optimal information from $\mathcal{R}_{\text{approx}}$ is obtained at the minimum, where there is a balance between the higher order terms in the expansion and the higher states contributions.

We shall see that this quantum mechanics example mimics quite well the case of QCD.

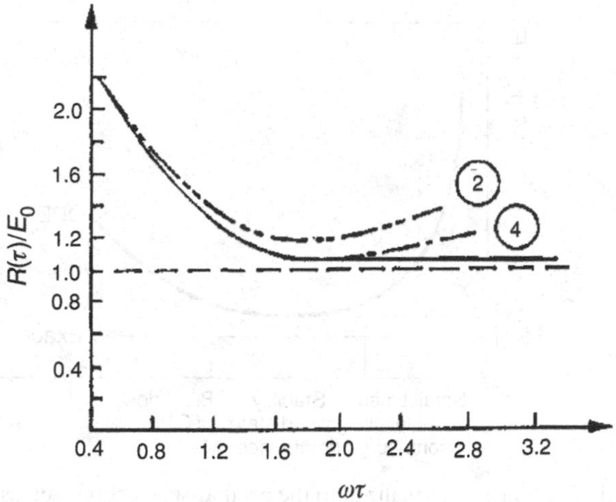


Fig. 49.6. The ratio of moments normalized to the ground-state energy versus the imaginary time for the case of the harmonic oscillator potential. (2) and (4): approximate series including the second and fourth order terms;– – – –exact solution.

### 49.10.2 Non-relativistic charmonium sum rules

Retaining the correction due to the gluon condensate, the QCD expression of the non-relativistic QCD moments is [91–93]:

$$\mathcal{M}(\tau_N) \equiv \int dE e^{-E\tau_N} \mathrm{Im}\Pi(E)$$

$$= \frac{3}{8m^2} 4\pi \left(\frac{m}{4\pi\tau_N}\right)^{3/2} \left[1 + \frac{4}{3}\alpha_s\sqrt{\pi m}\tau_N^{1/2} - \frac{4\pi}{288m}\langle\alpha_s G^2\rangle\tau_N^3\right], \tag{49.45}$$

where $\tau_N$ is the imaginary time variable, from which one can deduce the ratio of moments:

$$\mathcal{R}(\tau_N) = \frac{3}{2\tau_N} - \frac{2}{3}\alpha_s\sqrt{\pi m}\tau_N^{-1/2} + \frac{4\pi}{96m}\langle\alpha_s G^2\rangle\tau_N^2, \tag{49.46}$$

where $m$ is the charm quark (pole) mass. Using the QCD parameters given in Tables 48.1 and 48.2, one can show in Fig. 49.7 the $\tau_N$ behaviour of the ratio of moments. One can notice a strict ressemblence with the case of the harmonic oscillator.

The moments have the following features:

- The exact ratio reaches its limit $E_0$ very quickly as shown in Fig. 49.7.
- The theoretical curve which is a good approximation for small times stabilizes at medium time and blows up at large time indicating a breaking of the approximation for $\tau_N \geq \tau_N^c$. At the minimum, one has:

$$\min\mathcal{R}(\tau_N^c) = E_0^{\text{exact}} \tag{49.47}$$

within about 10% accuracy, indicating a good description of the ground state energy. This slight

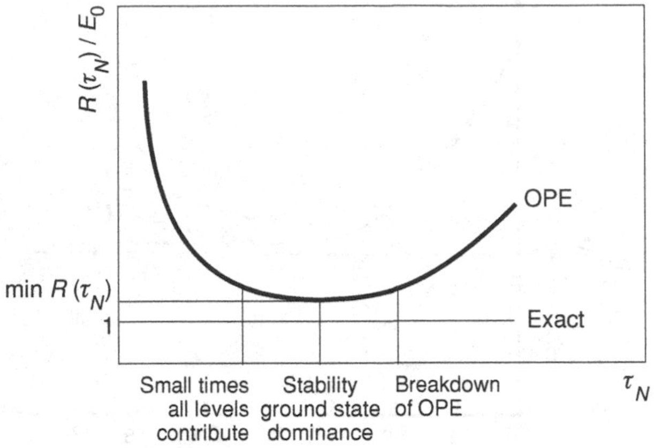


Fig. 49.7. The ratio of moments normalized to the ground-state energy versus the imaginary time in the case of the charmonium sum rules.

discrepancy can be reduced by including the contribution of the QCD continuum into the spectral function.

- However, it is quite surprising that, for a clearly emerging level where one would expect a dominance of the confinement force, while the moment shows that there exists a window where perturbation theory still works but the individual energy levels clearly emerge. For this reason, Bell–Bertlmann called it *magic moments*.

### 49.10.3 Implications for QCD

- However, by working with a truncated series as in QCD, we do not often have, in some other channels, a nice minimum for $\mathcal{R}_{\text{approx}}$. This minimum is replaced in some cases by an inflexion point where the optimal information on the resonance properties is obtained.
- Moreover, we need also a similar optimization for the value of the QCD contimuum threshold $t_c$, which, a priori, is also a free parameter. Optimal estimate can be obtained if the result presents stability in $t_c$. In various examples, this procedure can lead to an overestimate of the result, such that one can safely consider the result obtained in this way as an upper bound. On the contrary, a lower bound can be obtained at the value of $t_c$ where one starts to have a minimum or an infexion point with respect to the changes of the sum rule variables $\tau$ or $n$. A further test of the $t_c$ value is its comparison with the one obtained from FESR constraints.
- We conclude from the previous analysis that the optimal and most conservative results from the sum rule discussed in this book will obey the $\tau$ or $n$ optimization criterion (SVZ window), but in addition the corresponding $t_c$ values are in the range where we start to have these $\tau$ or $n$ minimum until the one where we have a stability in $t_c$. In many examples, the value of $t_c$ from a FESR constraint belongs to this range. In some cases, $t_c$ can be higher than the value intuitively expected around the mass of the radial excitation, which is not very surprising as the QCD continuum is an average of all the higher-state contributions. Finally, one can also test that at the optimal region, the OPE still makes sense as the QCD series converge quite well.

## 49.11 Modelling the $e^+e^- \to I = 1$ hadrons data using a QCD-duality ansatz

Due to the complexity and to the absence of the data in some channels, it appears necessary to introduce a simple model for parametrizing the spectral function. In the example of the $\rho$ meson, we have used the parametrization:

$$\frac{1}{\pi}\text{Im}\Pi_\rho(t) = \frac{M_\rho^2}{4\gamma_\rho^2}\delta(t - M_\rho^2) + \Theta(t - t_c) \text{ `QCD continuum'} , \tag{49.48}$$

where the first term is the lowest resonance contribution, whilst the second one takes into account *all* discontinuities coming from the QCD diagrams. $\gamma_\rho$ is the $\rho$-meson coupling to the vector current:

$$V_\mu = \frac{1}{2}(\bar{u}\gamma_\mu u - \bar{d}\gamma_\mu d) , \tag{49.49}$$

and is normalized as in Eq. (2.52). We have also seen that the lowest FESR moment leads to the constraint:

$$\frac{M_\rho^2}{4\gamma_\rho^2} \simeq \frac{t_c}{8\pi^2}\left[1 + \left(\frac{\alpha_s}{\pi}\right) + \mathcal{O}(\bar{\alpha}_s^2)\right] , \tag{49.50}$$

which, given the experimental value $\gamma_\rho \simeq 2.55$, leads to:

$$t_c \simeq 1.7 \text{ GeV}^2 . \tag{49.51}$$

As first noticed in [405], this FESR constraint shows that the properties of the lowest ground state is correlated to the value of the QCD continuum threshold, and permits one to check the (in)consistencies of various predictions done in the early literature on the sum rules.

We compare the prediction of this model with the available complete $e^+e^- \to I = 1$ hadrons data for the ratio of moments $\mathcal{R}(\tau)$ as shown in Fig. 49.8.[2]

We have used the $e^+e^-$ total cross-section shown in Fig. 49.9.

One can notice that the deviation of this *naïve and simple model* (dashed curve) from the data (black points) is at most 15%,[3] and that is very good. One can also notice that the QCD duality ansatz prediction is below the complete data one, which can be understood because the QCD continuum might give an underestimate of the radial excitation contributions as it only gives a smearing of the higher-state effects and does not account for the complex resonance structure between 1 and 2 GeV.

## 49.12 Test of the QCD-duality ansatz in the charmonium sum rules

Let us now test the validity of the QCD-duality ansatz in the heavy quark sector. In so doing, we consider the charmonium family ($J/\psi,\ \psi', \ldots$), which couples to the charm current

[2] More details discussions can be found in QSSR1 [3].

[3] The continuous curve corresponds to another set: $\gamma_\rho = 2.2$ and $t_c = 2.2$ GeV$^2$, which gives a worse prediction.

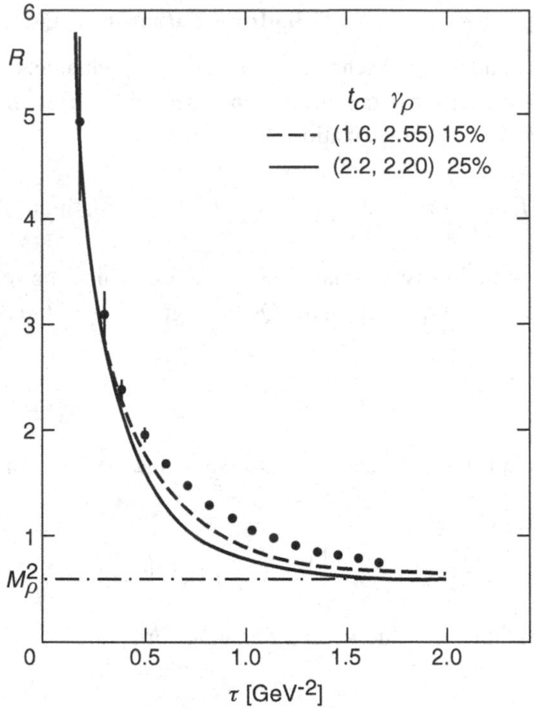


Fig. 49.8. Ratio of moments $\mathcal{R}(\tau)$ as function of the sum rule variable $\tau$ in the $\rho$-meson channel for two values of $t_c$ and $\gamma_\rho$. The data points are $e^+e^- \to I = 1$ hadrons.

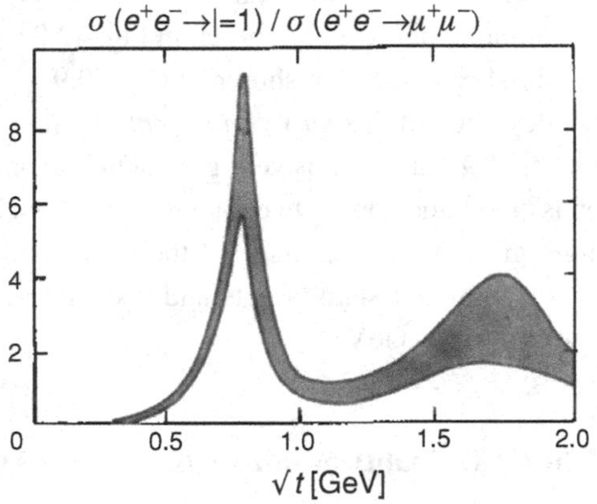


Fig. 49.9. Ratio of the $e^+e^- \to I = 1$ hadrons over the $e^+e^- \to \mu^+\mu^-$ total cross-section as function of the c.o.m energy $\sqrt{t}$.

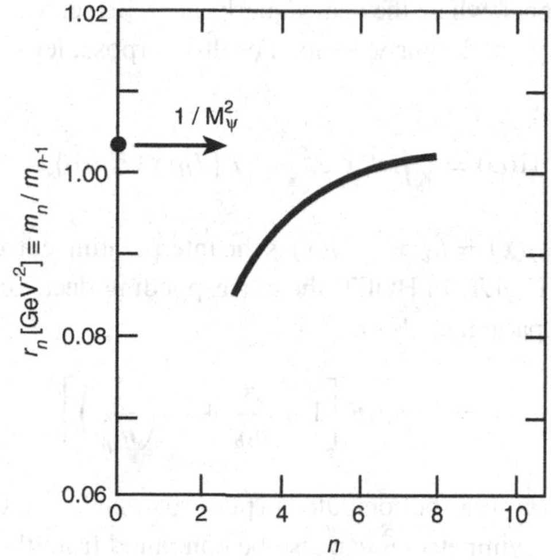


Fig. 49.10. Phenomenological side of the ratio of moments versus $n$.

via:

$$\langle 0|\bar{c}\gamma^\mu c|\psi\rangle = \sqrt{2}\frac{M_\psi^2}{2\gamma_\psi}\epsilon^\mu , \tag{49.52}$$

and the corresponding two-point correlator. The coupling $\gamma_\psi$ is normalized as in Eq. (1.51). The QCD continuum is simply approximated by the step function:

$$\frac{1}{\pi}\mathrm{Im}\Pi_\psi(t)_{\text{cont}} = \frac{1}{4\pi^2}\left[1+\left(\frac{\alpha_s}{\pi}\right)(t)\right]\Theta(t-t_c) , \tag{49.53}$$

which one can improve by including the available quark mass and higher-order radiative corrections. We show in Fig. 49.10 the ratio of the $Q^2 = 0$ moments:[4]

$$r_n \equiv \frac{\mathcal{M}_n}{\mathcal{M}_{n-1}} , \tag{49.54}$$

by using the data for the different leptonic widths of the $J/\psi$ family and by including the QCD continuum. One can notice that for larger value of $n \geq 6$, the ratio of moments is completely saturated by the lowest mass resonance, which shows that the QCD duality ansatz parametrization is a good approximation in the sum rule analysis of the heavy quark sector.

## 49.13 HQET sum rules

QCD spectral sum rules are often used in the Heavy Quark Effective Theory (HQET) for the estimate of meson masses and decay constants [164]. One considers the correlation

[4] More detaileds discussions can be found in QSSR1 [2].

functions of quark currents, where the heavy quarks are represented by their effective fields $h_v(x)$, $v$ being the heavy quark four-velocity. For this purpose, let's consider the two-point correlation function:

$$\Pi(\omega) = i \int d^4x \, e^{ik\cdot x} \langle 0|T\{J_H(x)J_H^\dagger(0)\}|0\rangle \tag{49.55}$$

where $\omega = 2v \cdot k$ and $J_H(x) = \bar{h}_v(x) i\gamma_5 q(x)$ is the interpolating current of the pseudoscalar heavy-light mesons in HQET. In HQET, the corresponding decay constant $f_B$ can be expressed in terms of the parameter $\hat{F}$ as:

$$f_B = \hat{C}(m_b)\hat{F}\left[1 - \frac{A}{m_b} + \mathcal{O}\left(\frac{1}{m_b^2}\right)\right], \tag{49.56}$$

where the coefficient $\hat{C}(m_b)$ can be computed in perturbation theory. One can notice that due to the heavy quark spin symmetry, $\hat{F}$ can also be computed from the two-point correlation function of the vector currents $J_V(x) = \bar{h}_v(x)(\gamma_\mu - v_\mu)q(x)$ interpolating heavy-light $1^-$ mesons. Isolating the ground state contribution from the integral over the excited states and the continuum, one can write the dispersion relation:

$$\Pi(\omega) = \frac{\hat{F}^2}{2\bar{\Lambda} - \omega} + \int_{E_0}^{\infty} dE \, \frac{\text{Im}\Pi(E)}{E - \omega} + \text{subtractions} \, . \tag{49.57}$$

The variable $E$ is related to the usual $t$ variable as:

$$t = (E + m_b)^2 \, . \tag{49.58}$$

The parameter $\bar{\Lambda} \simeq M_B - m_b$ represents the binding energy of the light degrees of freedom in the heavy meson. Here $m_b$ represents the heavy quark pole mass. The dispersion relation, Eq. (49.57), is then matched with the QCD expression, obtained for negative $\omega$ using the SVZ expansion:

$$\Pi(\omega) = \Pi_{\text{pert}}(\omega) + \sum_d C_d \frac{\langle O_d \rangle}{(-\omega)^d} \quad . \tag{49.59}$$

It is also convenient in the Laplace sum rule analysis to introduce the non-relativistic variable:

$$\tau_N = 4m_b\tau \, . \tag{49.60}$$

### *49.13.1 Decay constant, meson-quark mass gap, kinetic energy and chromomagnetic operator*

Different applications of this method to the two-point functions of heavy-light meson and baryon currents have been focused on the estimate of the decay constant, the hadron-quark mass gap $\bar{\Lambda}$, the kinetic energy $\lambda_1$ and the chromomagnetic interaction parameter $\lambda_2$.

The value of the $B$ meson decay constant obtained from the analysis is [164,166]:

$$\hat{F} = (0.4 \pm 0.06)\ \mathrm{GeV}^{3/2}\ , \qquad A = (0.9 \pm 0.2)\ \mathrm{GeV}\ , \tag{49.61}$$

which one can compare with the result obtained from the full theory discussed later on in the chapter of quark masses and decay constants.

The meson-quark mass gap $\bar{\Lambda}$ is in important input in HQET approach. Recall (see previous chapter on HQET) that it can be defined as [164,166]:[5]

$$M_{H_Q} = m_Q + \bar{\Lambda} + \frac{\Delta m^2}{2m_Q}\ , \tag{49.62}$$

with:

$$\Delta m^2 = -\lambda_1 + 2\left[J(J+1) - \frac{3}{2}\right]\lambda_2\ , \tag{49.63}$$

$J = j \pm 1/2$ being the total spin of the hadron states. Taking, for definiteness, the case of the $B$ meson, one has:

$$\lambda_1 \equiv \frac{1}{2M_B}\langle B(v)|\mathcal{O}_{\mathrm{kin}}|B(v)\rangle \quad \text{and} \quad \lambda_2 \equiv -\frac{1}{3M_B}\langle B(v)|\mathcal{O}_{\mathrm{mag}}|B(v)\rangle \tag{49.64}$$

which correspond respectively to the matrix elements of the kinetic and of the chromomagnetic operators:

$$\mathcal{O}_{\mathrm{kin}} \equiv \bar{h}(iD)^2 h \quad \text{and} \quad \mathcal{O}_{\mathrm{mag}} \equiv \frac{1}{4} g_s \bar{h}\sigma_{\mu\nu}G^{\mu\nu}h\ , \tag{49.65}$$

where $h$ is the heavy quark field and $G^{\mu\nu}$ the gluon field strength tensor.

The estimate of $\bar{\Lambda}$ from HQET-sum rules leads to [165]:

$$\bar{\Lambda} \simeq (0.52 - 0.70)\ \mathrm{GeV}\ , \tag{49.66}$$

in good agreement with the previous results [164,633], although less accurate as we have taken a larger range of variation for the continuum energy. An anologous sum rule in the full QCD theory leads to [634]:

$$\bar{\Lambda} \simeq (0.6 - 0.80)\ \mathrm{GeV}\ , \tag{49.67}$$

which combined together leads to the intersecting range of values [165]:

$$\bar{\Lambda} \simeq (0.65 \pm 0.05)\ \mathrm{GeV}\ . \tag{49.68}$$

The sum rule estimate of the kinetic energy gives [165]:

$$\lambda_1 \simeq -(0.5 \pm 0.2)\ \mathrm{GeV}^2 \tag{49.69}$$

where the large error, compared with the previous result of [166], is due to the absence of the stability point with respect to the variation of the continuum energy threshold. By

[5] We are aware of the fact that in the lattice calculations, $\bar{\Lambda}$ defined in this way can be affected by renormalons [798].

combining the previous estimates with the one of the chromomagnetic energy:

$$\lambda_2 \simeq \frac{1}{4}\left(M_{B^*}^2 - M_B^2\right) + \mathcal{O}(1/m_b) \simeq 0.49 \text{ GeV}^2 , \tag{49.70}$$

one deduces the value of the pole mass to two-loop accuracy:

$$M_b \equiv m_b = (4.61 \pm 0.05) \text{ GeV} , \tag{49.71}$$

in good agreement with the previous values from the sum rules in the full theory and (within the errors) with the HQET results in [164,633].

### *49.13.2 Isgur–Wise function*

This approach has been also extended to the three-point function for studying the Isgur–Wise function for the $B \to D^{(*)}$ semi-leptonic transition [164]. Compared with the sum rule in the full theory, the HQET sum rules have a much simpler QCD expression because it is a series in $1/M_b$. Therefore, the evaluation of radiative corrections like the one for the three-point function becomes feasible. We shall come back to this point in the chapter on $B$ and $D$ exclusive weak decays.

## 49.14 Vertex sum rules and form factors

The extension of QSSR two-point function sum rules into vertex or three-point function sum rules has been discussed by many authors [635–641] and in many reviews on sum rules [356–365], with the aim of estimating the three-hadron couplings and to study the $q^2$-dependence of the hadron form factors. In most of these applications, the vertex is saturated by the lowest hadronic state plus a QCD continuum, while the QCD expressions are evaluated in the Euclidian region using a configuration that is best suited to the processes considered. The mathematical validity of the spectral representation for the three-point function is not well established in general,[6] although one may expect that, in the case of narrow resonances, it simplifies, due to the disappearance of some anomalous thresholds.[7] Among other choices, the symmetric configuration:

$$p^2 = q^2 = (p+q)^2 = -(Q^2 \gg \Lambda^2) , \tag{49.72}$$

for the vertex depicted in Fig. 49.11 appears to be convenient for extracting the trilinear boson couplings, as we are only left with one variable in the sum rule analysis, while its QCD side is guaranteed to be safe from some eventual IR singularities. In this narrow-width approximation, one can write the duality diagrams between the two sides of the vertex sum rules (Fig. 49.12).

[6] To our knowledge, the most serious attempts to study such a problem is in [640] within perturbation theory.
[7] Related discussions will be done in the next chapter, taking the example of the $B_c$ meson semi-leptonic decays.

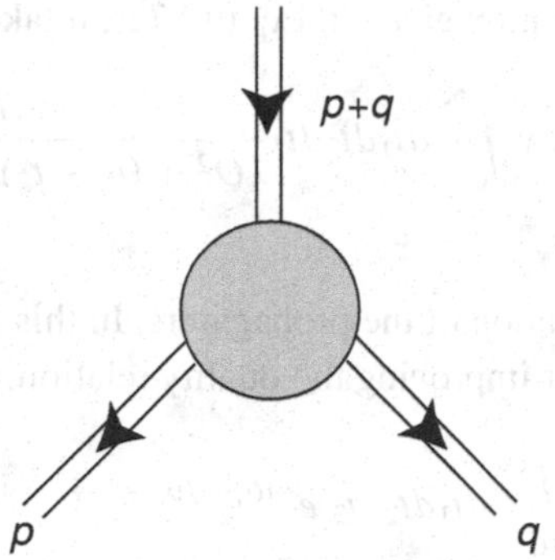


Fig. 49.11. Hadronic vertex.

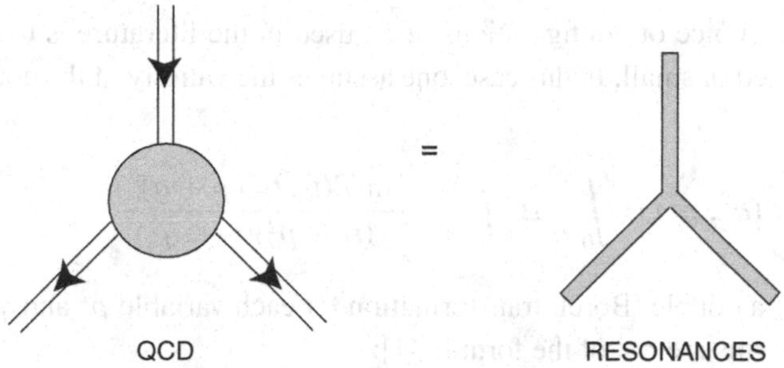


Fig. 49.12. Duality between a QCD vertex and a hadronic vertex in a narrow width approximation.

The discussions on the theoretical validity of the symmetric configuration method and its first phenomenological applications in QCD, for the case of trilinear mesons [637] and meson-baryon-baryon couplings, can be found in [636]. Some other applications of this method will be discussed later on in following chapters for the estimate of the decay widths of scalar mesons, gluonia and hybrids. The case of the heavy meson exclusive decays will be extensively discussed.

The uses of vertex sum rules for studying the $q^2$-dependence of different light and heavy hadron form factors have been also discussed extensively in the literature and will be discussed in later chapters. In connection to this, we shall also discuss light-cone sum rules that are an alternative to the vertex sum rules.

### 49.14.1 Spectral representation

The hadronic vertex in Fig. 49.11, can be represented by the spectral representation:

$$T(p^2, q^2, (p+q)^2) = \frac{1}{\pi} \int dt_1\, dt_2\, dt_3 \frac{\mathrm{Im}T(t_1, t_2, t_3)}{(t_1 - p^2)(t_2 - q^2)(t_3 - (p+q)^2)} . \quad (49.73)$$

In the symmetric representation given in Eq. (49.72), it takes the simple form:

$$T(Q^2) = \frac{2}{\pi}\int_0^1 xdx \int_0^1 dy \int_0^\infty dt_1 dt_2 dt_3 \frac{\mathrm{Im}T(t_1, t_2, t_3)}{[Q^2 + (t_1 - t_2)xy + (t_2 - t_3)x + t_3]^3}\,, \tag{49.74}$$

after a Feynmann parametrization of the propagators. In this form, it is trivial to apply the Laplace sum rule operator for improving the duality relation. One obtains:

$$\hat{\mathcal{L}}[T] = \frac{\tau^3}{\pi}\int_0^1 xdx \int_0^1 dy \int_0^\infty dt_1 dt_2 dt_3\, e^{-[(t_1-t_2)xy+(t_2-t_3)x+t_3]\tau}\, \mathrm{Im}T(t_1, t_2, t_3)\,, \tag{49.75}$$

where the rôle of the depressive factor is only manifest when the mass of the first excited state is much higher than any of the lowest ground states involved in the three channels.

An alternative choice of configurations often used in the literature is to take one of the three-moment fixed or small. In this case, one assumes the validity of the double-dispersion relation:

$$T(q^2, p^2) = \int_0^\infty dt_1 \int_0^\infty dt_2 \frac{\mathrm{Im}T(t_1, t_2, (p+q)^2)}{(t_1 - p^2)(t_2 - q^2)} + \cdots\,. \tag{49.76}$$

One can apply a double 'Borel' transformation for each variable $p^2$ and $q^2$ provided that the subtraction terms are not of the form [641]:

$$(p^2)^n \int_0^\infty \frac{\Delta(t)dt}{(t - q^2)} \quad \text{or} \quad (q^2)^n \int_0^\infty \frac{\Delta(t)dt}{(t - p^2)}\,, \tag{49.77}$$

which would induce non-controllable contributions to the sum rule. However, treating $(p^2, q^2)$ as independent variables may not be justified, as the spectral representation should be done in a $(q^2, p^2)$ plane along a straight-line which is a combination of the variables $p^2$ and $q^2$ [635].

### 49.14.2 Illustration from the evaluation of the $g_{\omega\rho\pi}$ coupling

The relevant three-point function is:

$$\begin{aligned} T_{\mu\nu}(p, q) &= i\int d^4x\, e^{-ip\cdot x} e^{i(p+q)\cdot y}\langle |\, \mathcal{T} J^\rho_\mu(x) J^\pi(y) J^\omega_\nu(0) \mid 0\rangle \\ &= \epsilon_{\mu\nu\alpha\beta}\, p^\alpha q^\beta T(Q^2, (p-q)^2)\,. \end{aligned} \tag{49.78}$$

where $T$ is the invariant amplitude. The quark interpolating currents are normalized as:

$$J^\rho_\mu =: \bar{u}\gamma_\mu d : \qquad J^\omega_\nu = \frac{1}{6} : \bar{u}\gamma_\nu u + \bar{d}\gamma_\nu d : \qquad J^\pi = (m_u + m_d)\, : \bar{d}(i\gamma_5)u :\,. \tag{49.79}$$

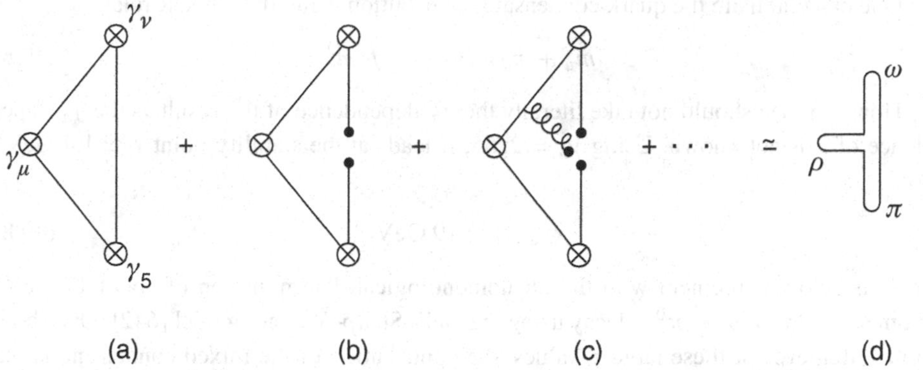


Fig. 49.13. Duality between the QCD and hadronic vertices for $g_{\omega\rho\pi}$: (a), (b) and (c) are respectively the QCD perturbative, quark condensate and mixed quark condensate contributions; (d) is the $g_{\omega\rho\pi}$ coupling.

The QCD expression of the vertex at the symmetric point can be evaluated from the diagrams depicted in Fig. 49.13 and gives [637]:

$$T(p,q) = \frac{1}{16\pi^2}\frac{(m_u+m_d)}{Q^2}\left\{(m_u+m_d)I_{x,y} - \frac{\langle\bar{u}u\rangle}{Q^2}\left(1+\frac{5}{36}\frac{M_0^2}{Q^2}\right)\right\}. \tag{49.80}$$

We have parametrized the mixed quark condensate effects by introducing the scale $M_0^2 \simeq 0.8\ \text{GeV}^2$. $I_{x,y}$ is a typical Feynman parameter integral:

$$I_{x,y} = \int_0^1 dx \int_0^{1-x} dy \frac{1}{x(1-x)+y(1-y)-xy} = 2.34\,. \tag{49.81}$$

Using a narrow width approximation (NWA) and retaining the lowest mass resonances, one obtains:

$$T_{\text{exp}} = \frac{\sqrt{2} f_\pi m_\pi^2}{Q^2+m_\pi^2}\frac{\sqrt{2}M_\rho^2}{2\gamma_\rho}\frac{\sqrt{2}M_\omega^2}{2\gamma_\omega}|g_{\omega\rho\pi}|\,. \tag{49.82}$$

We have used the usual normalization:

$$\langle 0|J^\pi|\pi\rangle = \sqrt{2}m_\pi^2 f_\pi\,, \qquad \langle 0|J_\mu^\rho|\rho\rangle = \sqrt{2}\frac{M_\rho^2}{2\gamma_\rho}\epsilon_\mu\,, \qquad \langle 0|J_\nu^\omega|\omega\rangle = \frac{M_\omega^2}{2\gamma_\omega}\epsilon_\nu\,. \tag{49.83}$$

The hadronic coupling is normalized as:

$$\langle\omega(p_1,\epsilon_1)|\rho(p_2,\epsilon_2)\pi(p_3)\rangle = |g_{\omega\rho\pi}|\epsilon_{\mu\nu\rho\sigma}\epsilon_1^\mu\epsilon_2^\nu p_1^\rho p_2^\sigma\,. \tag{49.84}$$

Invoking quark-hadron duality and taking the Laplace transform, one obtains in the chiral limit and for $\gamma_\omega \simeq 3\gamma_\rho$, $M_\rho \simeq M_\omega$:

$$|g_{\omega\rho\pi}| \simeq \frac{6\gamma_\rho^2}{M_\rho^4}\frac{(m_u+m_d)\langle\bar{u}u\rangle}{f_\pi m_\pi^2}\tau^{-1}e^{M_\rho^2\tau}\left(1+\frac{5}{36}M_0^2\tau\right). \tag{49.85}$$

One can eliminate the quark condensate contribution using the GMOR relation:

$$(m_u + m_d)\langle \bar{u}u\rangle = -f_\pi^2 m_\pi^2 . \tag{49.86}$$

However, one should not take literally the $f_\pi$ dependence of the result as the $f_\pi$ dependence of $\tau$ is not known. Using $\gamma_\rho = 2, 55$, it leads at the stability point $\tau \simeq 1.4$ GeV$^{-2}$ to:

$$|g_{\omega\rho\pi}| \approx 19 \text{ GeV}^{-1} \tag{49.87}$$

in satisfactory agreement with the phenomenological determination of about 17 GeV$^{-1}$ from $\omega \to 3\pi$ or $\omega \to \pi^0\gamma$ decay using the Gell–Sharp–Wagner model [642]. One should notice that even at these large $\tau$-values, the contribution of the mixed condensate is only about 15% indicating the convergence of the OPE. The effect of radial excitations have also been shown [637] to be negligible.

Similar approaches have been used in some other channels [636,3] (see also forthcoming chapters).

## 49.15 Light-cone sum rules

### *49.15.1 Basics and illustration by the $\pi^0 \to \gamma^*\gamma^*$ process*

The method of light-cone sum rules (LCSR) [643] is an alternative to the vertex sum rules for studying hadronic form factors.[8] It combines the SVZ technique and the theory of hard exclusive processes [644]. The basic idea is to expand the products of currents near the light cone. It can be illustrated by the analysis of the pion form factor in the process $\pi^0 \to \gamma\gamma$ for on-shell pion ($p^2 = m_\pi^2 = 0$) in the chiral limit. The corresponding amplitude, is:

$$\begin{aligned} T_{\mu\nu}(p,q) &= i\int d^4x\, e^{-iq\cdot x}\langle \pi^0(p)\mid T\, J_\mu^{\rm em}(x) J_\nu^{\rm em}(0)\mid 0\rangle \\ &= \epsilon_{\mu\nu\alpha\beta}\, p^\alpha q^\beta\, F(Q^2,(p-q)^2)\,, \end{aligned} \tag{49.88}$$

where $p$ is the pion momentum, $q$ and $(p-q)$ are the photon momenta, $Q^2 = -q^2$, $J_\mu^{em}$ is the quark electromagnetic current and $F$ is the invariant amplitude. To derive the LCSR, one has to calculate the correlation function of Eq. (49.88) in QCD, in the region of large $Q^2$ and $|(p-q)^2|$ and to use a dispersion relation to match the result of this calculation with hadronic matrix elements:

$$T_{\mu\nu}(p,q) = \frac{1}{\pi}\int_{s_0^h}^{\infty} ds\, \frac{Im T_{\mu\nu}(Q^2,s)}{s-(p-q)^2} . \tag{49.89}$$

The spectral function cn be saturated by the lowest masses $\rho$ and $\omega$ mesons via:

$$\langle V| J_\nu^{\rm em}(0)\mid 0\rangle = \epsilon_\nu \sqrt{2}\frac{M_V^2}{2\gamma_V} \quad : \qquad V \equiv \rho,\ \omega . \tag{49.90}$$

[8] Various recent applications of this method are reviewed in [360].

The correlation function in Eq. (49.88) can be calculated by expanding the $T$ product of quark currents near the light cone $x^2 = 0$, which, for large $Q^2$ and $|(p-q)^2|$, is expected to give the dominant contribution. This expansion is different from the local OPE as it no longer involves QCD vacuum condensates, but a summation of infinite series of local operators. It is convenient to introduce the DIS variables:

$$\nu \equiv p \cdot q\,, \qquad \xi = 2\nu/Q^2\,. \tag{49.91}$$

The leading contribution to the amplitude can be obtained by contracting the quark fields $\psi$ in Eq. (49.88), using the propagator of the free massless quark:

$$i\,S_0(x,0) = \langle 0 \mid T\{\psi(x)\bar{\psi}(0)\} \mid 0\rangle = \frac{i\, \not{x}}{2\pi^2 x^4}\,, \tag{49.92}$$

and transforming $\gamma_\mu\gamma_\alpha\gamma_\nu \to -i\epsilon_{\mu\alpha\nu\rho}\gamma^\rho\gamma_5 + \cdots$. Then, one obtains:

$$T_{\mu\nu}(p,q) = -i\epsilon_{\mu\nu\alpha\rho}\int d^4x \frac{x^\alpha}{\pi^2 x^4} e^{-iq\cdot x}\langle \pi^0(p)|\,\bar{\psi}(x)\gamma^\rho\gamma_5\psi(0)\,|0\rangle\,. \tag{49.93}$$

Expanding the local operators around $x = 0$:

$$\bar{\psi}(x)\gamma_\rho\gamma_5\psi(0) = \sum_n \frac{1}{n!}\bar{\psi}(0)(\overleftarrow{D}\cdot x)^n\gamma_\rho\gamma_5\psi(0)\,, \tag{49.94}$$

the matrix elements of these operators have the following general decomposition:

$$\begin{aligned}\langle \pi^0(p)|\bar{\psi}\,\overleftarrow{D}_{\alpha_1}\overleftarrow{D}_{\alpha_2}\cdots\overleftarrow{D}_{\alpha_r}\gamma_\rho\gamma_5\psi|0\rangle &= (-i)^n p_{\alpha_1}p_{\alpha_2}\cdots p_{\alpha_r}p_\rho M_n\\ &\quad + (-i)^n g_{\alpha_1\alpha_2}p_{\alpha_3}\cdots p_{\alpha_r}p_\rho M'_n + \cdots,\end{aligned} \tag{49.95}$$

where $M_n$, $M'_n$ are matrix elements coming respectively from twist-2 and twist-4 local operators. Substituting the decomposition Eq. (49.94) in Eq. (49.93), integrating over $x$ and using the definitions Eq. (49.95) and Eq. (49.91) one obtains:

$$F(Q^2,(p-q)^2) = \frac{1}{Q^2}\sum_{n=0}^{\infty}\xi^n M_n + \frac{4}{Q^4}\sum_{n=2}^{\infty}\frac{\xi^{n-2}}{n(n-1)}M'_n + \cdots\,. \tag{49.96}$$

Since the variable $\xi \sim 1$ in a generic exclusive kinematics with $p \neq 0$, all terms should be kept in each series in this expression. The second term containing $M'_n$ and further similar terms are suppressed by powers of a small parameter $1/Q^2$ as compared with the first term containing $M_n$. Keeping the lowest twist contribution, at $x^2 = 0$ (and $p^2 = 0$), the matrix element in Eq. (49.93) has the following parametrization ($f_\pi = 92.4$ MeV):

$$\langle \pi^0(p)|\bar{\psi}(x)\gamma_\mu\gamma_5\psi(0)|0\rangle_{x^2=0} = -ip_\mu f_\pi \int_0^1 du\, e^{iup\cdot x}\varphi_\pi(u,\mu)\,, \tag{49.97}$$

where the function $\varphi_\pi(u,\mu)$ is the pion *light-cone distribution amplitude* of twist 2, normalized to unity: $\int_0^1 \varphi_\pi(u,\mu)du = 1$. Furthermore, expanding both sides of Eq. (49.97) and comparing the LHS with the expansions, Eqs. (49.94) and (49.95), we find that the moments of $\varphi_\pi(u)$ are related to the matrix elements of local twist-2 operators:

$$M_n = -i f_\pi \int_0^1 du\, u^n \varphi_\pi(u,\mu)\,. \tag{49.98}$$

The function $\varphi_\pi(u)$, multiplied by $f_\pi$, is a universal non-perturbative object encoding the long-distance dynamics of the pion. Together with the corresponding higher-twist distribution amplitudes, $\varphi_\pi(u)$ plays a similar role as the vacuum condensates play in SVZ sum rules. However, to our opinion, a connection between the distribution amplitude and the vacuum condensates has not been yet clarified and needs further investigation. Substituting the definition Eq. (49.97) in Eq. (49.93), integrating over $x$, restoring the electromagnetic charge factor and summing the $u$ and $d$ quark contributions, one obtains the correlation function in the twist 2 approximation:

$$F^{(2)}(Q^2,(p-q)^2) = \frac{2}{3} f_\pi \int_0^1 \frac{du\, \varphi_\pi(u,\mu)}{\bar{u}Q^2 - u(p-q)^2}\,, \tag{49.99}$$

where $\bar{u} = 1-u$. We are now in a position to obtain a sum rule from the dispersion relation, Eq. (49.89), matching it with the result of the light-cone expansion. We define the matrix element

$$\langle \pi^0(p)| j_\mu^{\rm em} |\rho^0(p-q)\rangle = F^{\rho\pi}(Q^2) m_\rho^{-1} \epsilon_{\mu\nu\alpha\beta} \epsilon^{(\rho)\nu} q^\alpha p^\beta\,, \tag{49.100}$$

in terms of the transition form factor $F^{\rho\pi}(Q^2)$. We parametrize the higher state contributions by the QCD continuum ansatz from a threshold $t_c$ and write a dispersion relation for $F^{(2)}$. It is easy to obtain, the duality relation, to leading twist-2 accuracy:

$$\frac{\sqrt{2}M_\rho}{2\gamma_\rho} \frac{F^{\rho\pi}(Q^2)}{m_\rho^2 - (p-q)^2} + \frac{1}{\pi}\int_{t_c}^\infty ds \frac{{\rm Im}\, F(Q^2,s)}{s-(p-q)^2} = \frac{\sqrt{2}f_\pi}{3} \int_0^1 \frac{du\, \varphi_\pi(u)}{\bar{u}Q^2 - u(p-q)^2}\,, \tag{49.101}$$

where $\gamma_\rho$ is the coupling normalized as usual in this book:

$$\langle 0| \frac{1}{\sqrt{2}} (\bar{u}\gamma_\mu u - \bar{d}\gamma_\mu d)|\rho\rangle = \epsilon_\mu \frac{\sqrt{2}M_\rho^2}{2\gamma_\rho} \tag{49.102}$$

Introducing the continuum threshold:

$$u_c^\rho = Q^2/(t_c + Q^2) \tag{49.103}$$

and taking the Laplace transform, one obtains the LCSR for the form factor of the $\gamma^*\rho \to \pi$ transition to twist-2 accuracy [645,360]:

$$F^{\rho\pi}(Q^2) = \frac{2}{3}\frac{f_\pi}{M_\rho}\gamma_\rho \int_{u_c}^1 \frac{du}{u} \varphi_\pi(u,\mu) \exp\left(-\frac{\bar{u}Q^2}{uM^2} + \frac{m_\rho^2}{M^2}\right)\,. \tag{49.104}$$

### $\rho \to \pi\pi$ *form factor*

Another example of the application of LCSR is the calculation of the pion electromagnetic form factor defined as:

$$\langle \pi(p')|j_\mu^{\text{em}}|\pi(p)\rangle = F_\pi(q^2)(p+p')_\mu\,, \tag{49.105}$$

where $q = p' - p$ and $j_\mu^{\text{em}}$ is the electromagnetic current:

$$j_\mu^{\text{em}} = e_u \bar{u}\gamma_\mu u + e_d \bar{d}\gamma_\mu d\,. \tag{49.106}$$

The resulting LCSR, at zeroth order in $\alpha_s$ and in the twist 2 approximation, reads [649]:

$$F_\pi(Q^2) = \int_{u_c^\pi}^1 du\,\varphi_\pi(u,\mu_u)\exp\left(-\frac{\bar{u}Q^2}{uM^2}\right) \overset{Q^2\to\infty}{\longrightarrow} \frac{\varphi_\pi'(0,M^2)}{Q^4}\int_0^{t_c^\pi} ds\,s\,e^{-s/M^2}\,, \tag{49.107}$$

where $\varphi_\pi'(0) = -\varphi_\pi'(1)$, and $u_c^\pi = Q^2/(t_c^\pi + Q^2)$, $s_0^\pi$ is the duality threshold in the pion channel. The factorization scale $\mu_u^2 = \bar{u}Q^2 + uM^2$ corresponds to the average quark virtuality in the correlation function. At $O(\alpha_s)$, one recovers the leading $\sim 1/Q^2$ asymptotic behaviour corresponding to the hard scattering mechanism. Including this contribution in the LCSR and retaining the first two terms of the sum rule expansion in powers of $1/Q^2$ one obtains [650]:

$$F_\pi(Q^2) = \frac{2\alpha_s}{3\pi Q^2}\int_0^{s_0^\pi} ds\,e^{-s/M^2}\int_0^1 du\,\frac{\varphi_\pi(u)}{\bar{u}} + \varphi_\pi'(0)\int_0^{s_0}\frac{ds\,s\,e^{-s/M^2}}{Q^4} + O\left(\frac{\alpha_s}{Q^4}\right). \tag{49.108}$$

The $O(1/Q^2)$ term in Eq. (49.108) coincides with the well-known expression for the asymptotics of the pion form factor [644]:

$$F_\pi(Q^2) = \frac{8\pi\alpha_s f_\pi^2}{9Q^2}\left|\int_0^1 du\frac{\varphi_\pi(u)}{\bar{u}}\right|^2, \tag{49.109}$$

obtained by the convolution of two twist-2 distribution amplitudes $\varphi_\pi(u)$ of the initial and final pion with the $O(\alpha_s)$ quark hard-scattering kernel.

### 49.15.2 Distribution amplitudes

The model dependence and main uncertainties of the LCSR approach is in the parametrization of the distribution amplitude. It can be expanded using the conformal symmetry of massless QCD [360,646]. The conformal spin (partial wave) decomposition allows to represent each distribution amplitude as a sum of certain orthogonal polynomials in the variable $u$. The coefficients of these polynomials are multiplicatively renormalizable, and have growing anomalous dimensions, so that, at sufficiently large normalization scale $\mu$, only the first few terms in this expansion are relevant. The part of the distribution amplitude, which does not receive logarithmic renormalization is called *asymptotic*. Within this expansion, one can write:

$$\varphi_\pi(u,\mu) = 6u\bar{u}\left[1 + \sum_{n=2,4,\ldots} a_n(\mu)C_n^{3/2}(u-\bar{u})\right], \tag{49.110}$$

where $C_n^{3/2}$ are the Gegenbauer polynomials (for a derivation, see, e.g., [167]). The coefficients $a_n$ are multiplicatively renormalizable:

$$a_n(\mu) = a_n(\mu_0)\left(\frac{\alpha_s(\mu)}{\alpha_s(\mu_0)}\right)^{\gamma_n/\beta_0}, \tag{49.111}$$

and:

$$\gamma_n = C_F\left[-3 - \frac{2}{(n+1)(n+2)} + 4\left(\sum_{k=1}^{n+1}\frac{1}{k}\right)\right] \tag{49.112}$$

are the anomalous dimensions [647]. At $\mu \to \infty$, $a_n(\mu)$ vanish, and the limit $a_n = 0$ corresponds to the asymptotic distribution amplitude

$$\varphi_\pi^{(as)}(u) = 6u\bar{u}\,. \tag{49.113}$$

The values of the non-asymptotic coefficients $a_n$ at a certain intermediate scale $\mu_0$ can be estimated from two-point sum rules [647,648,3] for the moments $\int u^n\varphi_\pi(u,\mu)du$ at low $n$. This method is attractive because it employs non-perturbative information expressed in terms of quark and gluon condensates. However, in practice the two-point sum rule determination of $a_n$ is not very accurate, such that one should consider conservatively the large range spanned by $a_n$ from different analysis.

# 50

# Weinberg and DMO sum rules

As mentioned earlier in Subsection 2.2.7 of Part I, Weinberg and DMO sum rules are prototypes of QSSR, whilst their derivation is based on the asymptotic realization of chiral and flavour symmetries, or alternatively, in the world with massless quarks and without any interactions with external gluon fields. The convergence of these sum rules has been tested in QCD when the quark masses and non-perturbative power corrections are switched on [28,29,31,32]. The analysis has been reviewed in details in [30,3,34], where the QCD corrections to the WSR have been given explicitly.

We shall follow the notations and conventions in Subsection 2.2.7 of Part I. We shall be concerned here with the two-point correlator:

$$\Pi^{\mu\nu}_{LR}(q) \equiv i \int d^4x \, e^{iqx} \langle 0|\mathcal{T} J^{\mu}_{L}(x) \left(J^{\nu}_{R}(0)\right)^{\dagger}|0\rangle$$
$$= -(g^{\mu\nu}q^2 - q^{\mu}q^{\nu})\Pi^{(1)}_{LR}(q^2) + q^{\mu}q^{\nu}\Pi^{(0)}_{LR}(q^2) \,, \tag{50.1}$$

built from the left- and right-handed components of the local weak current:

$$J^{\mu}_{L} = \bar{u}\gamma^{\mu}(1-\gamma_5)d, \qquad J^{\mu}_{R} = \bar{u}\gamma^{\mu}(1+\gamma_5)d \,, \tag{50.2}$$

and/or using isospin rotation relating the neutral and charged weak currents. The indices (1) and (0) corresponds to the spins of the hadrons entering into the spectral function. In the chiral limit, the longitudinal part $\Pi^{(0)}_{LR}(q^2)$ of the two-point correlator vanishes, once the pion pole has been subtracted. The spectral function is normalized as:

$$\frac{1}{2\pi}\mathrm{Im}\Pi^{(1)}_{LR} \equiv \frac{1}{2\pi}\mathrm{Im}\Pi_{LR} \equiv \frac{1}{4\pi^2}(v-a) \,, \tag{50.3}$$

where the last term is the notation in [193,199].

### 50.1 Sacrosanct Weinberg sum rules (WSR) in the chiral limit

Here, we shall follow closely the discussions in [34].

### 50.1.1 The sum rules

The 'sacrosanct' Weinberg sum rules read in the chiral limit:

$$
\begin{aligned}
I_0 &\equiv \int_0^\infty ds\, \frac{1}{2\pi}\mathrm{Im}\Pi_{LR} = f_\pi^2\,, \\
I_1 &\equiv \int_0^\infty ds\, s\, \frac{1}{2\pi}\mathrm{Im}\Pi_{LR} = 0\,, \\
I_{-1} &\equiv \int_0^\infty \frac{ds}{s}\, \frac{1}{2\pi}\mathrm{Im}\Pi_{LR} = -4L_{10}\,, \\
I_{\rm em} &\equiv \int_0^\infty ds\, \left(s\, \log\frac{s}{\mu^2}\right) \frac{1}{2\pi}\mathrm{Im}\Pi_{LR} = -\frac{4\pi}{3\alpha} f_\pi^2 \left(m^2_{\pi^\pm} - m^2_{\pi^0}\right)\,,
\end{aligned}
\tag{50.4}
$$

where $f_\pi|_{\rm exp} = (92.4 \pm 0.26)$ MeV is the experimental pion decay constant that should be used here as we shall use data from $\tau$-decays involving physical pions; $m_{\pi^\pm} - m_{\pi^0}|_{\rm exp} \simeq 4.5936(5)$ MeV; $L_{10} \equiv f_\pi^2\langle r_\pi^2\rangle/3 - F_A$ [where $\langle r_\pi^2\rangle = (0.439 \pm 0.008) fm^2$ is the mean pion radius and $F_A = 0.0058 \pm 0.0008$ is the axial-vector pion form factor for $\pi \to e\nu\gamma$] is one of the low-energy constants of the effective chiral Lagrangian [498–502]. The last sum rule $I_{\rm em}$ is often called DMO sum rule and it governs the electro-magnetic mass shift of the pion.

It has been shown that in the case of massless quarks, the $SU(n)_L \times SU(n)_R$ chiral symmetry is not spontaneously broken by perturbative QCD radiative corrections in QCD to all orders of perturbation theory, in the framework where the Dirac matrix $\gamma_5$ anti-commutes with the remaining ones [118]. Therefore the WSR remains valid in this case.

Recent measurement of the difference between the vector and axial-vector spectral function has been performed by ALEPH/OPAL using hadronic $\tau$-decay data [33] as shown in Fig. 25.7. This has permitted us to have a detailed analysis of the spectral part of the WSR. In order to exploit these sum rules using the ALEPH/OPAL data, we shall work with their finite energy sum rule (FESR) versions (see e.g. [28,325] for such a derivation). In the chiral limit ($m_q = 0$ and $\langle\bar{u}u\rangle = \langle\bar{d}d\rangle = \langle\bar{s}s\rangle$), this is equivalent to truncate the LHS at $t_c$ until which the data are available, while the RHS of the integral remains valid to leading order in the $1/t_c$ expansion in the chiral limit, because, in this limit the breaking of these sum rules by higher dimension $D = 6$ condensates, which is of the order of $1/t_c^3$, is numerically negligible [29]. The analysis of these different sum rules using the $\tau$ decay data is shown in Fig. 50.1.

### 50.1.2 Matching between the low- and high-energy regions

In order to fix the $t_c$ values which separate the low and high energy parts of the spectral functions, we require that the second Weinberg sum rule (WSR) $I_1$ should be satisfied by the present data. As shown in Fig. 50.1, this is obtained for two values of $t_c$:

$$
t_c \simeq (1.4 \sim 1.5)\ \mathrm{GeV}^2 \quad \text{and} \quad (2.4 \sim 2.6)\ \mathrm{GeV}^2\,. \tag{50.5}
$$

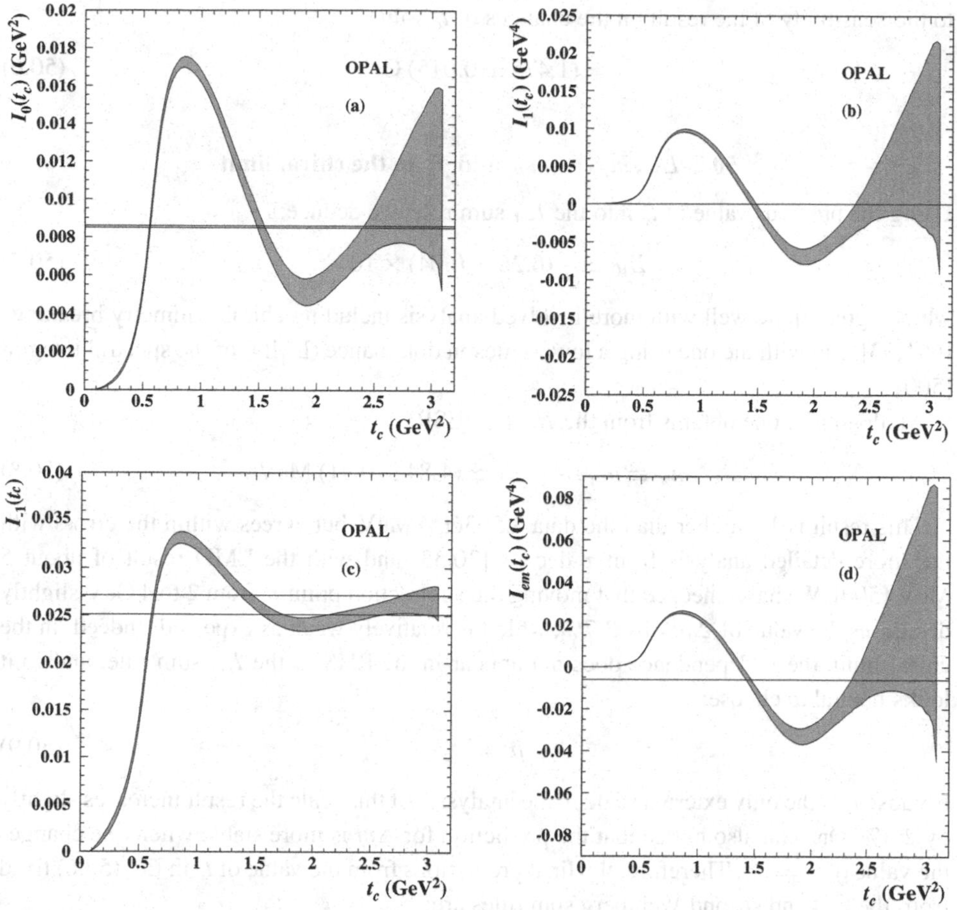


Fig. 50.1. Measurements of the different WSR until an energy cut $t_c$ from $\tau$-decay data by OPAL [33]. A similar result has been obtained by ALEPH. The RHS of the sum rules is given by the straight line ($\pm 1\sigma$) when two lines are present.

Although the second value is interesting from the point of view of the QCD perturbative calculations (better convergence of the QCD series), its exact value is strongly affected by the inaccuracy of the data near the $\tau$-mass (with the low values of the ALEPH/OPAL data points, the second Weinberg sum rule is only satisfied at the former value of $t_c$).

After having these $t_c$ solutions, we can improve the constraints by requiring that the first Weinberg sum rule $I_0$ reproduces the experimental value of $f_\pi$ [1] within an accuracy that is twice the experimental error. This condition allows us to fix $t_c$ in a very narrow margin due

[1] Although we are working here in the chiral limit, the data are obtained for physical pions, such that the corresponding value of $f_\pi$ should also correspond to the experimental one.

to the sensitivity of the result on the changes of $t_c$ values:[2]

$$t_c = (1.475 \pm 0.015)\ \text{GeV}^2\ . \tag{50.6}$$

## 50.2 $L_{10}$, $m_{\pi^\pm} - m_{\pi^0}$ and $f_\pi$ in the chiral limit

Using the previous value of $t_c$ into the $I_{-1}$ sum rule, we deduce:

$$L_{10} \simeq -(6.26 \pm 0.04) \times 10^{-3}\ , \tag{50.7}$$

which agrees quite well with more involved analysis including chiral symmetry breakings [651,33], and with the one using a lowest meson dominance (LMD) of the spectral integral [500].

Analogously, one obtains from the $I_{em}(t_c)$ FESR:

$$\Delta m_\pi \equiv m_{\pi^\pm} - m_{\pi^0} \simeq (4.84 \pm 0.21)\ \text{MeV}\ . \tag{50.8}$$

This result is $1\sigma$ higher than the data 4.5936(5) MeV, but agrees within the errors with the more detailed analysis from $\tau$-decays [30,33] and with the LMD result of about 5 MeV [500]. We have checked that moving the subtraction point $\mu$ from 2 to 4 GeV slightly decreases the value of $\Delta m_\pi$ by 3.7%, which is relatively weak as expected. Indeed, in the chiral limit, the $\mu$ dependence does not appear in the RHS of the $I_{em}$ sum rule, and so it looks natural to choose:

$$\mu^2 = t_c\ , \tag{50.9}$$

because $t_c$ is the only external scale in the analysis. At this scale the result increases slightly by 2.5%. One can also notice that the prediction for $\Delta m$ is more stable when one changes the value of $t_c = \mu^2$. Therefore, the final predictions from the value of $t_c$ in Eq. (50.6) fixed from the first and second Weinberg sum rules are:

$$\begin{aligned} \Delta m &\simeq (4.96 \pm 0.22)\ \text{MeV}\ , \\ L_{10} &\simeq -(6.42 \pm 0.04) \times 10^{-3}\ , \end{aligned} \tag{50.10}$$

which we consider as our 'best' predictions.

For some more conservative results, we also give the predictions obtained from the second $t_c$–value given in Eq. (50.5). In this way, one obtains:

$$\begin{aligned} f_\pi &= (87 \pm 4)\ \text{MeV}\ , \\ \Delta m &\simeq (3.4 \pm 0.3)\ \text{MeV}\ , \\ L_{10} &\simeq -(5.91 \pm 0.08) \times 10^{-3}\ , \end{aligned} \tag{50.11}$$

where one can notice that the results are systematically lower than those obtained in Eq. (50.10) from the first $t_c$ value given previously, which may disfavour a posteriori the second choice of $t_c$ values, although, in principle, we do not have a strong argument favouring one

[2] For the second set of $t_c$-values in Eq. (50.5), one obtains a slightly lower value: $f_\pi = (84.1 \pm 4.4)$ MeV.

with respect to the other. However, approach based on $1/N_c$ expansion and a saturation of the spectral function by the lowest state within a narrow width approximation (NWA) as discussed in Section 43.2 favours the former value of $t_c$ given in Eq. (50.6) [500]. A similar value of $t_c$ is also obtained from the FESR constraint using the naïve duality ansatz of the vector spectral function. Taking as a conservative value the largest range spanned by the two sets of results, one obtains:

$$\begin{aligned} f_\pi &= (86.8 \pm 7.1)\ \text{MeV}\ , \\ \Delta m &\simeq (4.1 \pm 0.9)\ \text{MeV}\ , \\ L_{10} &\simeq -(5.8 \pm 0.2) \times 10^{-3}\ , \end{aligned} \tag{50.12}$$

which we found to be quite satisfactory in the chiral limit. The previous tests are very useful, as they will allow us to gauge the confidence level of the sum rule predictions in the following chapters.

## 50.3 Masses and power corrections to the Weinberg sum rules

It has been shown [28,29,31,32] that:

- The $SU(n)_L \times SU(n)_R$ chiral symmetry is broken by massive quarks. The first WSR is broken to order $\alpha_s\, m_u m_d$ but is still convergent, whereas the second WSR is not convergent in its mathematical sense. However, this non-convergence does not affect the success of the $A_1$ mass prediction from the second WSR, as phenomenologically the light running quark mass effects are small.
- The $SU(n)_L \times SU(n)_R$ chiral symmetry is broken spontaneously by the dimension-six four-quark condensate, which affects the WSR. However, the effect is relatively small and vanishes as $1/q^4$, where $q^2$ is the typical scale of the sum rule.

Using the QCD expressions of the vector and axial-vector two-point correlators given in Part VIII from [325], it is easy to derive the different power corrections to the Weinberg sum rules. Introducing the running coupling $\bar{\alpha}_s$ and masses $\bar{m}_i$ evaluated at $Q^2$, one can deduce for the spin $1+0$ combination:

$$\begin{aligned} Q^2\big[\Pi^{(1+0)}_{ij,LR}\big]^{(D=4)} &= -\frac{1}{\pi^2}\left(\frac{\bar{\alpha}_s}{\pi}\right)\bar{m}_i\bar{m}_j \\ Q^4\big[\Pi^{(1+0)}_{ij,LR}\big]^{(D=4)} &= \frac{4}{3}\left(\frac{\bar{\alpha}_s}{\pi}\right)\langle m_j\bar{\psi}_i\psi_i + m_i\bar{\psi}_j\psi_j\rangle - \frac{8}{7\pi^2}\bar{m}_i\bar{m}_j\big[\bar{m}_i^2 + \bar{m}_j^2\big] \\ Q^6\big[\Pi^{(1+0)}_{ij,LR}\big]^{(D=6)} &= 8\pi\left(\frac{\bar{\alpha}_s}{\pi}\right)\big[\langle(\bar{\psi}_i\gamma_\mu T_a\psi_j)(\bar{\psi}_j\gamma_\mu T_a\psi_i)\rangle \\ &\quad -\langle(\bar{\psi}_i\gamma_\mu\gamma_5 T_a\psi_j)(\bar{\psi}_j\gamma_\mu\gamma_5 T_a\psi_i)\rangle\big] \\ &\simeq -\frac{64\pi}{9}\rho\alpha_s\langle\bar{u}u\rangle^2 \\ Q^8\big[\Pi^{(1+0)}_{ij,LR}\big]^{(D=8)} &\approx 8\pi\alpha_s M_0^2\langle\bar{u}u\rangle^2\ , \end{aligned} \tag{50.13}$$

where $\rho = 1$ in the large $N_c$-limit; $T_a \equiv \lambda_a/2$ is the $SU(3)_c$ matrix defined in Appendix B; $M_0^2 \simeq 0.8$ GeV$^2$ is the scale introduced in Chapter 27 in order to parametrize the mixed condensate. The $D = 8$ contribution comes from [652]. For the spin 0 component of the two-point function, one can also deduce from [325]:

$$\begin{aligned}
Q^2\left[\Pi^{(0)}_{ij,LR}\right]^{(D=2)} &= \frac{3}{2\pi^2} m_i m_j \left[\ln \frac{Q^2}{\nu^2} + \mathcal{O}(1)\right] \\
Q^4\left[\Pi^{(0)}_{ij,LR}\right]^{(D=4)} &= \langle (m_i - m_j)(\bar\psi_i\psi_i - \bar\psi_j\psi_j)\rangle - \langle (m_i + m_j)(\bar\psi_i\psi_i + \bar\psi_j\psi_j)\rangle \\
&\quad + \frac{1}{4\pi^2}\left[-\frac{12}{7}\left(\frac{\bar\alpha_s}{\pi}\right)^{-1} + \frac{11}{14}\right]\{[\bar m_i - \bar m_j][\bar m_i^3 - \bar m_j^3] \\
&\quad - [\bar m_i + \bar m_j][\bar m_i^3 + \bar m_j^3]\} \\
&\quad - \frac{3}{4\pi^2}\bar m_i \bar m_j[\bar m_i - \bar m_j]^2 + \frac{3}{4\pi^2}\bar m_i \bar m_j[\bar m_i + \bar m_j]^2 \,. \qquad (50.14)
\end{aligned}$$

With these expressions, it is easy to derive the QCD expressions of the different WSR. The phenomenology of these FESR sum rules and especially their Laplace transform have been explictly discussed in [3], which the readers may also consult.

## 50.4 DMO sum rules in QCD

The DMO sum rule which controls the $SU(n)$ flavour symmetry has been analyzed in QCD in [31], [32] and in [354]. Phenomenologically, it has been used to extract the value of the quark masses and to predict the splittings due to $SU(3)$ breakings among the mesons. In particular, its $\tau$-like version has been used to extract the value of the running strange quark mass, which has the advantage to be model-independent. We shall come back to this point in the chapter on light quark masses.

# 51
# The QCD coupling $\alpha_s$

In this chapter, we shall outline the different determinations of the QCD coupling $\alpha_s$ from $\tau$ decay, $e^+e^- \to I = 1$ hadrons processes and heavy quark systems. This discussion has been already anticipated in Sections 25.5, 25.6 and Part VI and will not be repeated in detail here This chapter will complete these discussions.

## 51.1 $\alpha_s$ from $e^+e^- \to I = 1$ hadrons and $\tau$-decays data

These channels have the great advantage that the spectral functions are measured in a region where pQCD is applicable and therefore the analysis does not suffer from any model dependence in the parametrization of the spectral functions. The $\tau$ decays and $\tau$-like decays in $e^+e^- \to I = 1$ hadron processes (see previous Sections 25.5 and 25.6) are *another prototype of QCD spectral sum rules* like the Weinberg sum rules

- The determinations of the QCD coupling from hadronic $\tau$-decays and $e^+e^- \to I = 1$ hadrons have been already discussed in details in Sections 25.5, 25.6 and are based on the approach proposed by BNP [325] for $\tau$-like decay processes. It relies on the fact that the non-perturbative contributions based on the 'standard SVZ expansion' give a tiny correction at the $\tau$ mass. In addition, recent analysis [161] including the new $D = 2$ dimension tachyonic gluon mass term not considered in the analysis of [325] shows that this effect is small and does not affect in a significant way the determination carried out without this term. It tends to reduce slightly (about 10%) the central value of $\alpha_s$ improving the agreement between the $\tau$-decay prediction and the world average at the $Z^0$ mass. However, this change is marginal as it is of similar size to the theoretical error in the $\alpha_s$ determination from $\tau$-decay.
- The previous original BNP approach [325] has been generalized to higher moments [333] that have been exploited by the different experimental groups [328,33] for extracting $\alpha_s$. The mean value coming from the different structure of pQCD series of the ALEPH/OPAL [33] measurements is:

$$\alpha_s(M_\tau) = 0.323 \pm 0.005(\text{exp}) \pm 0.030(\text{th}) \,. \tag{51.1}$$

This rather modest accuracy runned until the $Z^0$ mass leads to the precise determinations:

$$\alpha_s(M_{Z^0}) = 0.1181 \pm 0.0007\ (\text{exp})\ \pm 0.0030(\text{th}) \,, \tag{51.2}$$

which is in excellent agreement with the different determinations summarized in Part VI.

- In [346,338,391] and [329] (reprinted in Section 52.10) $e^+e^- \to I = 1$ hadrons below 2 GeV and the sum of exclusive $\Delta S = 0$ $\tau$-decay data have been also used to extract the value of $\alpha_s$ as a cross-check of the value obtained from inclusive $\tau$-decay data. This result has been given in Section 25.6. The most recent analysis in [329] gives:

$$\alpha_s(M_\tau) = 0.33 \pm 0.03 , \tag{51.3}$$

which is in excellent agreement with the $\tau$-decay data result.
- Stability of the previous determinations using lower $\tau$-mass values has been tested using $e^+e^- \to I = 1$ hadrons below 2 GeV [346] and the inclusive distributions of $\tau$-decay [345,33]. This test is reassuring as it indicates that for reasonable values of $M_\tau$ larger than 1 GeV, pQCD still applies, while the OPE parametrized by the few lowest dimension condensates describes the data quite well. It also indicates that the contribution near the real axis where QCD does not apply is negligible due to the phase space double suppression factor $(1 - s/M_\tau^2)^2$. It also indicates that *quark-hadron duality* which is the main idea behind this dispersion relation approach and then behind the QCD spectral sum rule approach is fulfilled by QCD.

## 51.2 $\alpha_s$ from heavy quarkonia mass-splittings

Examining the pioneering SVZ charmonium sum rule [1] (as one can also see in details in [3] and in the next chapter) the QCD side of this sum rule contains three relevant QCD parameters, namely the charm quark mass, the perturbative radiative correction $\alpha_s$ and the gluon condensate $\langle\alpha_s G^2\rangle$. In order to extract more reliably one of these parameters, one has to find appropriate sum rules or their combinations which can disentangle these parameters.

- In [313] (Section 51.3), it is observed that a double ratio $\mathcal{R}_{P_1^1}/\mathcal{R}_{P_1^3}$ of Laplace sum rules for the $P$ charmonium states can be used for extracting the value of $\alpha_s$:

$$\frac{\mathcal{R}_{P_1^1}}{\mathcal{R}_{P_1^3}} \simeq \frac{M^2_{P_1^1}}{M^2_{P_1^3}} \simeq 1 + \alpha_s\left[\Delta_\alpha^{13}(\text{exact}) = 0.014^{-0.004}_{+0.008}\right] , \tag{51.4}$$

where the gluon condensate effect vanishes to leading order. Using the experimental mass, one can deduce at the optimization point $\sigma \simeq (0.6 \pm 0.1)\ \text{GeV}^{-2}$:

$$\alpha_s(\sigma^{-1} = 1.3\ \text{GeV}) = 0.64^{+0.36}_{-0.18} \pm 0.02 \Longrightarrow \alpha_s(M_Z) = 0.127 \pm 0.009 \pm 0.002 . \tag{51.5}$$

- The non-relativistic $q^2 = 0$ moments sum rules for the $\Upsilon$ have also been used in [155] in order to extract $\alpha_s$. Recent improvements [156] of the previous analysis including the new $\alpha_s^2$ corrections of the two-point correlator [447] leads to the value:

$$\alpha_s(M_b) = 0.233^{+0.045}_{-0.030} , \tag{51.6}$$

which runned until $M_Z$ gives:

$$\alpha_s(M_Z) = 0.120^{+0.010}_{-0.008} . \tag{51.7}$$

These results are less accurate than the determination from $\tau$-decay, $e^+e^-$ and LEP data but are still interesting for an independent determination of $\alpha_s$.

## 51.3 Reprinted paper

# *Heavy quarkonia mass-splittings in QCD: gluon condensate, $\alpha_s$ and 1/m-expansion*

S. Narison



## 1. The double ratio of moments

QCD spectral sum rule (QSSR) after SVZ [1] (for a recent review, see, e.g. [2]) has shown since 15 years, its impressive ability for describing the complex phenomena of hadronic physics with the few universal "fundamental" parameters of the QCD Lagrangian (QCD coupling $\alpha_s$, quark masses and vacuum condensates built from the quarks and/or gluon fields), without waiting for a complete understanding of the confinement problem. In the example of the two-point correlator:

$$\Pi_Q(q^2) \equiv i \int d^4x\, e^{iqx} \langle 0|\mathcal{T} J_Q(x) (J_Q(o))^\dagger|0\rangle \,, \tag{1}$$

associated to the generic hadronic current: $J_Q(x) \equiv \bar{Q}\Gamma Q(x)$ of the heavy $Q$-quark ($\Gamma$ is a Dirac matrix which specifies the hadron quantum numbers), the SVZ-expansion reads:

$$\Pi_Q(q^2) \simeq \sum_{D=0,2,\ldots} \sum_{\dim 0=D} \frac{C^{(J)}(q^2, M_Q^2, \mu)\langle O(\mu)\rangle}{\left(M_Q^2 - q^2\right)^{D/2}} \,, \tag{2}$$

where $\mu$ is an arbitrary scale that separates the long- and short-distance dynamics; $C^{(J)}$ are the Wilson co-efficients calculable in perturbative QCD by means of Feynman diagrams techniques; $\langle O\rangle$ are the non-perturbative condensates of dimension $D$ built from the quarks or/and gluon fields ($D = 0$ corresponds to the case of the naïve perturbative contribution). Owing to gauge invariance, the lowest dimension condensates that can be formed are the $D = 4$ light quark $m_q\langle\bar{\psi}\psi\rangle$ and gluon $\langle\alpha_s G^2\rangle$ ones, where the former is fixed by the pion PCAC relation, whilst the latter is known to be $(0.07 \pm 0.01)\,\mathrm{GeV}^4$ from more recent analysis of the light [3] quark systems [2]. The validity of the SVZ-expansion has been understood formally, using renormalon techniques (absorption of the IR renormalon ambiguity into the definitions of the condensates and absence of some extra $1/q^2$-terms not included in the OPE) [4,5] and/or by building renormalization-invariant combinations of the condensates (Appendix of [6] and references therein). The SVZ expansion is phenomenologically confirmed from the unexpected accurate determination of the QCD coupling $\alpha_s$ and from a measurement of the condensates from semi-inclusive $\tau$-decays [6–8].

The previous QCD information is transmitted to the data through the spectral function Im $\Pi_Q(t)$ via the Källen–Lehmann dispersion relation (*global duality*) obeyed by the hadronic correlators, which can be improved from the uses of different versions of the sum rules [1,9–11]. In this paper, we shall use the simple duality ansatz parametrization: "*one narrow resonance*"+"*QCD continuum*", from a threshold $t_c$, which gives a good description of the spectral integral in the sum rule analysis, as has been tested successfully in the light-quark channel from the $e^+e^- \to I = 1$ hadron data and in the heavy-quark ones from the $e^+e^- \to \psi$ or $\Upsilon$ data. We shall work with the relativistic version of the Laplace or exponential sum rules where the QCD expression known to order $\alpha_s$ is given in terms of the pole mass $m(p^2 = m^2)$ [11–14]:[1]

$$
\begin{aligned}
\mathcal{L}(\sigma, m^2) &\equiv \int_{4m^2}^{\infty} dt \exp(-t\sigma)\frac{1}{\pi} \operatorname{Im} \Pi_Q(t) \\
&= 4m^2 A_H(\omega)\left[1 + \alpha_s a_H(\omega) + \frac{\pi}{36}\frac{\langle\alpha_s G^2\rangle}{m^4} b_H(\omega)\right], \\
\mathcal{R}_H(\sigma) &\equiv -\frac{d}{d\sigma} \log \mathcal{L}_H(\sigma, m^2) = 4m^2 F_H(\omega) \\
&\times \left[1 + \alpha_s P_H(\omega) + \frac{\pi}{36}\frac{\langle\alpha_s G^2\rangle}{m^4} Q_H(\omega)\right], \qquad (3)
\end{aligned}
$$

where

$$\omega \equiv 1/x = 4m^2\sigma \qquad (4)$$

is a dimensionless variable, while $\sigma \equiv \tau$ (notation used in the literature) is the exponential Laplace sum rule variable; $F_H$, $P_H$ and $Q_H$ are complete QCD Whittaker functions compiled in [12–14]; $H$ specifies the hadronic channel studied. In principle, the pair $(\sigma,\ t_c)$ are free external parameters in the analysis, so that the optimal result should be insensitive to their variations. *Stability criteria*, which are equivalent to the variational method, state that the best results should be obtained at the minimas or at the inflexion points in $n$ or $\sigma$, while stability in $t_c$ is useful to control the sensitivity of the result in the changes of $t_c$-values. These stability criteria are satisfied in the heavy quark channels studied here, as the continuum effect is negligible and does not exceed 1% of the ground state contribution [2,12], such that at the minimum in $\sigma$, one expects to a good approximation:

$$\min_\sigma \mathcal{R}(\sigma) \simeq M_H^2 . \qquad (5)$$

Moreover, one can a posteriori check that, at the stability point, where we have an equilibrium between the continuum and the non-perturbative contributions, which are both small, the OPE is still convergent such that the SVZ-expansion makes sense. The previous approximation can be improved by working with the double ratio of moments[2]:

$$
\begin{aligned}
\mathcal{R}_{HH'}(x) &\equiv \frac{\mathcal{R}_H}{\mathcal{R}_{H'}} \simeq \frac{M_H^2}{M_{H'}^2} \\
&= \Delta_0^{HH'}\left[1 + \alpha_s \Delta_{\alpha_s}^{HH'} + \frac{4\pi}{9}\langle\alpha_s G^2\rangle\sigma^2 x^2 \Delta_G^{HH'}\right], \qquad (6)
\end{aligned}
$$

[1] For consistency, we shall work with the two-loop order $\alpha_s$ expression of the pole mass [15].
[2] This method has also been used in [16] for studying the mass splittings of the heavy-light quark systems.

provided that each ratio of moments stabilizes at about the same value of $\sigma$, as in this case, there is a cancellation of the different leading terms such as the heavy quark mass (and its ambiguous definition used in some previous literatures), the negligible continuum effect (which is already small in the ratio of moments), and each leading QCD corrections. We shall limit ourselves here to the $\alpha_s$-correction for the perturbative contribution and to the leading order one in $\alpha_s$ for the gluon condensate effects. To the order we are working, the gluon condensate is well-defined as the ambiguity only comes from higher order terms in $\alpha_s$, which have, however, a smaller numerical effect than the one from the error of the phenomenological estimate of the condensate.

## 2. Test of the $1/m$-expansion

For this purpose, we use the complete *horrible* results expressed in terms of the pole mass to order $\alpha_s$ given by [12] and checked by various authors [2], which we expand with the help of the Mathematica program. We obtain for different channels the expressions given in Table 1. By comparing the complete and truncated series in $1/m$, one can notice that, at the $c$ and $b$ mass scales, the convergence of the $1/m$-expansion is quite bad due to the increases of the numerical coefficients with the power of $1/m$ and to the alternate signs of the $1/m$ series.

## 3. Balmer-mass formula from the ratio of moments

The Balmer formula derived from a non-relativistic approach ($m \to \infty$) of the Schrödinger levels reads [17] (see also [18–20]). for the $S_1^3$ vector meson:

$$M_{S_1^3} \simeq 2m\left[1 - \frac{2}{9}\alpha_s^2 + 0.23\frac{\pi}{(m\alpha_s)^4}\langle\alpha_s G^2\rangle\right] . \tag{7}$$

It is instructive to compare this result with the mass formula obtained from the ratio of moments within the $1/m$-expansion. Using the different QCD corrections in Table 1, one obtains the mass formula at the minimum in $\sigma$ of $\mathcal{R}$:

$$M_{S_1^3} \simeq \sqrt{\mathcal{R}(\sigma_{\min})} \simeq 2m\left(1 + \frac{3}{16m^2\sigma}\right) \times\left[1 - \frac{\sqrt{\pi}}{6m}\frac{\alpha_s(\sigma)}{\sqrt{\sigma}} + \frac{\pi}{12}\sigma^2\langle\alpha_s G^2\rangle\right] . \tag{8}$$

In the case of the $b$-quark, where we expect the *static approximation* to be more reliable, the minimum of this quantity is obtained to leading order at:

$$\sqrt{\sigma_{\rm coul}} \simeq \frac{9}{4m\alpha_s\sqrt{\pi}} \simeq 0.85\ \text{GeV}^{-1} , \tag{9}$$

Table 1. *Expanded expressions of different QCD corrections in the case of the pole mass $m(p^2 = m^2)$ known to order $\alpha_s$*

| | Vector $S_1^3$ |
|---|---|
| $\pi A_V$ | $\frac{3}{16\sqrt{\pi}}x^{3/2}\left(1-\frac{3}{4}x+\frac{45}{32}x^2-\frac{525}{128}x^3+\cdots\right)$ |
| $a_V$ | $\frac{4}{3\sqrt{\pi}}\left(\frac{\pi}{\sqrt{x}}+0.040+1.952\sqrt{x}-1.539x-\cdots\right).$ |
| $b_V$ | $-\frac{1}{2x^3}+\frac{3}{2x^2}+\frac{27}{8x}-\frac{21}{8}+\cdots$ |
| $F_V$ | $1+\frac{3}{2}x-\frac{5}{4}x^2+5x^3-\cdots$ |
| $P_V$ | $-\frac{2}{3}\sqrt{\pi x}+2.704x^{3/2}-10.093x^{5/2}+52.93x^{7/2}-\cdots$ |
| $Q_V$ | $\frac{3}{2x^2}-\frac{1}{4x}+\frac{13}{8}-\frac{41}{4}x+\cdots$ |
| | S-waves splitting |
| $\Delta_0^{VP}$ | $1-\frac{x^2}{2}+\frac{7}{2}x^3-\cdots$ |
| $\Delta_\alpha^{VP}$ | $\frac{\sqrt{\pi}}{9}x^{3/2}+1.539x^2-3.0258x^{5/2}-7.719x^3+26.307x^{7/2}+\cdots$ |
| $\Delta_G^{VP}$ | $\frac{5}{x}\left(1-\frac{4}{5}x+\frac{11}{10}x^2+\frac{17}{10}x^3-\cdots\right)$ |
| | P-waves splittings |
| $\Delta_0^{01}$ | $1$ |
| $\Delta_\alpha^{01}$ | $-3.18x^2(1-10.17x+102.1x^2+\cdots)$ |
| $\Delta_G^{01}$ | $-\frac{2}{x}+\frac{5}{2}-\frac{55}{4}x+\cdots$ |
| $\Delta_\alpha^{13}$ | $1.06x^2(1-9.5x+81.1x^2-\cdots)$ |
| $\Delta_0^{AT}$ | $1+x^2-\frac{23}{2}x^3+\cdots$ |
| $\Delta_\alpha^{AT}$ | $-0.1576x^{3/2}-2.545x^2+3.95x^{5/2}-\cdots$ |
| $\Delta_G^{AT}$ | $-\frac{6}{x}+\frac{31}{4}-\frac{89}{8}x-\frac{1715}{8}x^2+\cdots$ |
| | P-versus S-waves splittings |
| $\Delta_0^{VS}$ | $1-x+5x^2-30x^3+\cdots$ |
| $\Delta_\alpha^{VS}$ | $-\frac{2}{9}\sqrt{\pi x}-0.336x^{3/2}+4.244x^2+7.458x^{5/2}-42.017x^3-\cdots$ |
| $\Delta_G^{VS}$ | $-\frac{3}{x^2}-\frac{2}{x}-\frac{41}{4}+\frac{389}{4}x-\cdots$ |
| $\Delta_0^{VA}$ | $\Delta_0^{VS}$ |
| $\Delta_\alpha^{VA}$ | $-\frac{2}{9}\sqrt{\pi x}-0.336x^{3/2}+1.06x^2+7.458x^{5/2}-9.655x^3\cdots$ |
| $\Delta_G^{VA}$ | $-\frac{3}{x^2}-\frac{4}{x}-\frac{31}{4}+\frac{167}{2}x-\cdots$ |
| $\Delta_0^{VT}$ | $1-x+6x^2-\frac{85}{2}x^3+\cdots$ |
| $\Delta_\alpha^{VT}$ | $-\frac{2}{9}\sqrt{\pi x}-0.493x^{3/2}-1.484x^2+11.409x^{5/2}+18.248x^3-\cdots$ |
| $\Delta_G^{VT}$ | $-\frac{3}{x^2}-\frac{10}{x}+\frac{579}{8}x-\frac{16719}{16}x^2+\cdots$ |

where we have used for 5 flavours[3]: $\alpha_s(\sigma) \simeq 0.32 \pm 0.06$ after evolving the value $\alpha_s(M_z) = 0.118 \pm 0.006$ from LEP [21] and $\tau$-decay data [6–8]. The inclusion of the gluon

[3] In the approximation we are working, the effect of the number of flavours enters only through the $\beta$-function and therefore is not significant.

condensate correction shifts the value of $\sigma_{\min}$ to:

$$\sqrt{\sigma_{\min}} \simeq 0.74\sqrt{\sigma_{\text{coul}}} \,. \tag{10}$$

These previous values of $\sigma$ confirm the more involved numerical analysis in [12] and indicate the relevance of the gluon condensate in the analysis of the spectrum. By introducing the previous leading order expression of $\sigma_{\text{coul}}$ into the sum rule, one obtains:

$$M_\Upsilon \simeq 2m_b\left(1+\frac{\pi}{27}\alpha_s^2\right)\left[1-\frac{2}{9}\left(\frac{\pi}{3}\right)\alpha_s^2+\left(\frac{27}{128}\right)\left(\frac{3}{\pi}\right)^2\frac{\pi}{(m_b\alpha_s)^4}\langle\alpha_s G^2\rangle\right], \tag{11}$$

where one can deduce by identification in the *static limit* ($m_b \to \infty$) that the Coulombic effect is exactly the same in the two approaches. The apparent factor $\pi/3$ is due to the fact that we use here the approximate Schwinger interpolating formula for the two-point correlator. The gluon condensate coefficient is also about the same in the two approaches. This agreement indicates the consistency of the potential model and sum rule approach in the static limit, though a new extra $\alpha_s^2$ correction due to the $v^2$ (finite mass) terms in the free part appears here (for some derivations of the relativistic correction in the potential approach see [22,23]), and tends to reduce the Coulombic interactions. On the other hand, at the $b$-quark mass scale, the dominance of the gluon condensate contribution indicates that the $b$-quark is not enough heavy for this system to be Coulombic rendering the non-relativistic potential approach to be a crude approximation at this scale.

## 4. $S_1^3 - S_0^1$ hyperfine and $P-S$-wave splittings

In the non-relativistic approach used in [20], the hyperfine and $S$–$P$ wave splittings are given to leading order by:

$$M(S_1^3)-M(S_0^1) \simeq 2m\frac{(C_F\alpha_s)^4}{6}\left[1+3.255\frac{\pi}{2m^4\alpha_s^6}\langle\alpha_s G^2\rangle\right],$$
$$M(P_1^3)-M(S_3^1) \simeq 2m\left[\frac{3(C_F\alpha_s)^2}{32}+\frac{2^5\pi}{(C_F m\alpha_s)^4}\langle\alpha_s G^2\rangle\right], \tag{12}$$

where $C_F = 4/3$. Using the double ratio of moments and the QCD corrections given in Table 1, one obtains at $\sigma_{\text{coul}}$:

$$\frac{M(S_1^3)-M(S_0^1)}{M(S_1^3)} \approx -4\pi^2\left(\frac{\alpha_s}{9}\right)^4+\frac{8}{9}\left(\frac{\sqrt{\pi}\alpha_s}{9}\right)^3\alpha_s+\frac{45}{32m^4\alpha_s^2}\langle\alpha_s G^2\rangle+\cdots,$$
$$\frac{M(P_1^3)-M(S_1^3)}{M(S_1^3)} \approx \frac{4\pi}{81}\alpha_s^2+\frac{2\pi}{81}\alpha_s^2+\frac{27}{8m^4\alpha_s^2}\langle\alpha_s G^2\rangle+\cdots, \tag{13}$$

where the corrections are, respectively, relativistic, Coulombic and non-perturbative. By comparing the sum rules in Eqs. (8) and (13), one can realize that the leading $x$ or $1/\sigma$-terms cancel in the hyperfine splitting, while the $x$-expansion is slowly convergent for the

$\alpha_s$-term at the $b$-mass. Comparing now this result with the one from the non-relativistic approach, it is interesting to notice that both approaches lead to the same $\alpha_s$-behaviour of the Coulombic and gluon condensate contributions. A one to one correspondence of each QCD corrections is not very conclusive, and needs an evaluation of the correlator at the next-next-to-leading order for a better control of the $\alpha_s^2 x$ terms. However, as the Coulombic potential is a fundamental aspect of QCD, we shall, however, expect that, after the resummation of the higher order terms in $\alpha_s$, the coefficient of the $\alpha_s^4$-term in the hyperfine splitting will be the same in the two alternative approaches. In the case of the $S-P$ wave splitting, the sum of the $\alpha_s^2$ corrections agrees from the two methods, though one can also notice that the relativistic correction is larger than the Coulombic one. The discrepancy for the coefficients of the gluon condensate in the two approaches is more subtle and may reflect the difficulty of Bell-Bertlmann [19] to find a bridge between the field theory after SVZ (flavour-dependent confining potential) and the potential models (flavour-independence). Resolving the different puzzles encountered during this comparison is outside the scope of the present paper.

## 5. Leptonic width and quarkonia wave function

Using the sum rule $\mathcal{L}_H$ and saturating it by the vector $S_1^3$ state, we obtain, to a good approximation, the sum rule:

$$M_V \Gamma_{V\to e^+e^-} \simeq (\alpha e_Q)^2 \frac{e^{2\delta m M_V \sigma}}{72\sqrt{\pi}} \frac{\sigma^{-3/2}}{m} \times \left[1 + \frac{8}{3}\sqrt{\pi\sigma} m \alpha_s - \frac{4\pi}{9}\langle \alpha_s G^2\rangle m \sigma^{5/2}\right] , \tag{14}$$

where $e_Q$ is the quark charge in units of e; $\delta m \equiv M_V - 2m$ is the meson-quark mass gap. In the case of the $b$-quark, we use [15] $\delta m \simeq 0.26$ GeV, and the value of $\sigma_{\min}$ given in Eq. (10). Then:

$$\Gamma_{\Upsilon(S_1^3)\to e^+e^-} \simeq 1.2 \text{ keV} , \tag{15}$$

in agreement with the value found from the data, 1.3 keV. However, one should remark from Eq. (14) that the $\alpha_s$ correction is huge and needs an evaluation of the higher order terms (the gluon condensate effect is negligible), while the exponential factor effect is large, such that one can *reciprocally* use the data on the width to fix either $\alpha_s$ or/and the quark mass. Larger value of the heavy quark mass at the two-loop level (see, e.g. [26]) corresponding to a negative value of $\delta m$, would imply a smaller value of the leptonic width in disagreement with the data.

In the non-relativistic approach, one can express the quarkonia leptonic width in terms of its wave function $\Psi(0)_Q$:

$$\Gamma_{V\to e^+e^-} = \frac{16\pi\alpha^2}{M_V^2} e_Q^2 |\Psi(0)|_Q^2 \left(1 - 4C_F \frac{\alpha_s}{\pi}\right) , \tag{16}$$

where (see, e.g. [20]):

$$16\pi |\Psi(0)|_Q^2 \left(1 - 4C_F \frac{\alpha_s}{\pi}\right) \simeq 2(mC_F\alpha_s)^3 \approx 15 \text{ GeV}^3 . \tag{17}$$

In our approach, one can deduce:

$$16\pi|\Psi(0)|^2_Q\left(1-4C_F\frac{\alpha_s}{\pi}\right) \simeq \frac{1}{72\sqrt{\pi}}e^{2\delta m M_V\sigma}\sigma^{-3/2}\frac{M_V}{m}$$
$$\times\left[1+\frac{8}{3}\sqrt{\pi\sigma m}\alpha_s - \frac{4\pi}{9}\langle\alpha_s G^2\rangle m\sigma^{5/2}\right] \simeq 18.3 \text{ GeV}^3 . \tag{18}$$

Using the expression of $\sigma_{coul}$, one can find that, to leading order, the two approaches give a similar behaviour for $\Psi(0)_Q$ in $\alpha_s$ and in $m$ and about the same value of this quantity, though, one should notice that in the present approach, the QCD coupling $\alpha_s$ is evaluated at the scale $\sigma$ as dictated by the renormalization group equation obeyed by the Laplace sum rule [27] but *not* at the resonance mass!

## 6. Gluon condensate from $M_{\psi(S_1^3)} - M_{\eta_c(S_0^1)}$

The value of $\sigma$, at which, the $S$-wave charmonium ratio of sum rules stabilize is [12]:

$$\sigma \simeq (0.9 \pm 0.1)\ \text{GeV}^{-2} . \tag{19}$$

Using the range of the charm quark pole mass to order $\alpha_s$ accuracy [15][4]: $m_c \simeq 1.2$–$1.5$ GeV one can deduce the conservative value of $x$:

$$\omega \equiv 1/x \simeq 6.6 \pm 1.8 . \tag{20}$$

The ratio of the mass squared of the vector $V(S_1^3)$ and the pseudoscalar $P(S_0^1)$ is controlled by the double ratio of moments given generically in Eq. (6), where the exact expressions of the corrections read:

$$\Delta_0^{VP} \simeq 0.995^{+0.001}_{-0.004} , \qquad \Delta_\alpha^{VP} \simeq 0.0233^{-0.009}_{+0.011} ,$$
$$\Delta_G^{VP} \simeq 29.77^{+8.86}_{-10.23} , \tag{21}$$

where each terms lead to be about 0.5, 2 and 7% of the leading order one. One can understand from the approximate expressions in Table 1 that the leading $x$-corrections appearing in the ratio of moments cancel in the double ratio, while the remaining corrections are relatively small. However, the $x$-expansion is not convergent for the $\alpha_s$-term at the charm mass, which invalidates the use of the $1/m$-expansion done in [28] in this channel. Using for 4 flavours [15]: $\alpha_s(\sigma) \simeq 0.48^{+0.17}_{-0.10}$, and the experimental data [25]: $\mathcal{R}_{VP}^{exp} = 1.082$, one can deduce the value of the gluon condensate:

$$\langle\alpha_s G^2\rangle \simeq (0.10 \pm 0.04)\ \text{GeV}^4 . \tag{22}$$

We have estimated the error due to higher order effects by replacing the coefficient of $\alpha_s$ with the one obtained from the effective Coulombic potential, which tends to reduce the

[4]For a recent review on the heavy quark masses, see e.g. [24,25].

estimate to 0.07 GeV$^4$. We have tested the convergence of the QCD series in $\sigma$, by using the numerical estimate of the dimension-six gluon condensate $g\langle f_{abc}G^aG^bG^c\rangle$ contributions given in [14]. This effect is about 0.1% of the zeroth order term and does not influence the previous estimate in Eq. (22), which also indicates the good convergence of the ratio of exponential moments already at the charm mass scale in contrast with the $q^2 = 0$ moments studied in [1,29]. We also expect that in the double ratio of moments used here, the radiative corrections to the gluon condensate effects (their expression for the two-point correlator is however available in the literature [30]) are much smaller than in the individual moments, such that they will give a negligible effect in the estimate of the gluon condensate. This value obtained at the same level of $\alpha_s$-accuracy as previous sum rule results, confirm the ones of Bell-Bertlmann [11,12,14,30,2] from the ratio of exponential moments and from FESR-like sum rule for quarkonia [31,2] claiming that the SVZ value [1] has been underestimated by about a factor 2 (see also [29,32]). Our value is also in agreement with the more recent estimate $(0.07 \pm 0.01)$ GeV$^4$ from the $\tau$-like sum rules [3], and FESR [34] in $e^+e^- \to I = 1$ hadrons. A more complete comparison of different determinations is done in Table 2.

## 7. Charmonium $P$-wave splittings

The analysis of the different ratios of moments for the $P$-wave charmonium shows [11–14] that they are optimized for:

$$\sigma \simeq (0.6 \pm 0.1)\ \text{GeV}^{-2}\,, \qquad \Longrightarrow \alpha_s(\sigma) \simeq 0.41^{+0.11}_{-0.07}\,, \qquad 1/x = 4.5 \pm 1.5\,. \tag{23}$$

In the case of the *Scalar $P_0^3$-axial $P_1^3$ mass splitting*, the different exact QCD coefficient corrections of the corresponding double ratio of moments read:

$$\Delta_0^{01} = 1\,, \qquad \Delta_\alpha^{01} \simeq -\left(0.045^{-0.014}_{+0.028}\right)\,, \qquad \Delta_G^{01} \simeq -\left(7.75^{+2.84}_{-2.77}\right)\,. \tag{24}$$

Using the correlated values of the different parameters, one obtains the mass-splitting $\Delta M_{10}^3 \equiv M_{P_1^3} - M_{P_0^3} \simeq (60^{-16}_{+35})$ MeV, where we have used the experimental value $M_{P_1^3} =$ 3.51 GeV. Adding the $\langle gG^3\rangle$ dimension-six condensate effect, which is about $-1.6\%$ of the leading term in $\mathcal{R}_{01}$, one can finally deduce the prediction in Table 2, which is in excellent agreement with the data. One should remark that the previous predictions indicate that, for the method to reproduce correctly the mass-splittings of the $S$ and $P$-wave charmonium states, one needs *both* larger values of $\alpha_s$ and $\langle\alpha_s G^2\rangle$ than the ones favoured in the early days of the sum rules.

In the case of the *Tensor $P_2^3$-axial $P_1^3$ mass splitting*, the different exact QCD corrections for the double ratio of the tensor over the axial meson moments read:

$$\Delta_0^{TA} = \left(0.989^{+0.003}_{-0.006}\right)\,, \qquad \Delta_\alpha^{TA} = \left(0.029^{-0.004}_{+0.013}\right)\,, \qquad \Delta_G^{TA} = \left(22.1^{+8.5}_{-8.2}\right)\,, \tag{25}$$

from which, one can deduce the prediction in Table 2, which is slightly higher than the

Table 2. *Predictions for the gluon condensate, for the different mass-splittings (in units of MeV) and for the leptonic widths (in units of keV). We use $\alpha_s(M_Z) = 0.118 \pm 0.006$ from LEP and $\tau$-decay*

| Observables | Input | Predictions | Data/comments |
|---|---|---|---|
| $\langle\alpha_s G^2\rangle[\text{GeV}]^4 \times 10^2$ | $M_\Psi - M_{\eta c} = 108$ | $10 \pm 4$ | This work |
| | $M_{\chi b}^{\text{c.o.m.}} - M_\gamma = 440$ | $6.5 \pm 2.5$ | – |
| | Average | $7.5 \pm 2.5$ | Mass splittings |
| | Charmonium masses | $\approx 4$ | SVZ-value [1] |
| | | | $q^2 = 0$-mom. |
| | – | $5.3 \pm 1.2$ | $q^2$-mom. [9] |
| | – | $10 \pm 2$ | exp. mom. [11,12] |
| | – | $9.2 \pm 3.4$ | mom. [31] |
| | $e^+e^- \to I = 1$ hadrons | $4 \pm 1$ | ratio of mom. [33] |
| | – | $13^{+5}_{-7}$ | FESR [34] |
| | – | $7 \pm 1$ | $\tau$-like decay [3] |
| | $\tau$-decay (axial) | $6.9 \pm 2.6$ | [35] |
| | $\tau$-decay | | data |
| | ALEPH | | $7.5 \pm 3.1$ [8] |
| | CLEO | | $2.0 \pm 3.8$ [8] |
| $\alpha_s(1.3\ \text{GeV})$ | $M_{\chi c(P_1^1)} - M_{\chi c(P_1^3)}$ | $0.64^{+0.36}_{-0.18} \pm 0.02$ | $\alpha_s(M_Z) \simeq 0.127 \pm 0.011$ |
| $M_{\chi c(P_1^1)} - M_{\chi c(P_1^3)}$ | $\alpha_s$ from LEP/$\tau$-decay | $10.1^{-4.1}_{+9.9}$ | 11.1 (c.o.m.) |
| | | | 15.6 (data) |
| $M_{\chi c(P_1^3)} - M_{\chi c(P_0^3)}$ | $\langle\alpha_s G^2\rangle$average | $89^{-16}_{+35}$ | 95 |
| $M_{\chi c(P_2^3)} - M_{\chi c(P_1^3)}$ | – | $77^{+26}_{-11}$ | 50 |
| $M_\gamma - M_{\eta b}$ | $\langle\alpha_s G^2\rangle$average | $13^{-7}_{+10}$ | order $\alpha_s$ |
| | | $63^{-29}_{+51}$ | coeff.$\alpha_s$: Coul. pot. |
| $M_{\chi b(P_0^3)} - M_Y$ | – | $475^{+75}_{-64}$ | 400 |
| $M_{\chi b(P_1^3)} - M_Y$ | – | $485^{+25}_{-68}$ | 432 |
| $M_{\chi b(P_2^3)} - M_Y$ | – | $500 \pm 71$ | 453 |
| $M_{\chi b(P_1^1)} - M_Y$ | c.o.m. | $492^{+56}_{-69}$ | 440 |
| $M_\mathcal{T} - 2_{m_t}$ | $\langle\alpha_s G^2\rangle$average | $-906$ | two-loop pole mass |
| $M_\mathcal{T} - M_{\eta_t}$ | – | 1.8 | order $\alpha_s$ |
| | | 93 | coeff.$\alpha_s$: Coul. pot. |
| $M_{\chi_t} - M_\mathcal{T}$ | – | 1800 | – |
| $\Gamma_{\gamma \to e^+e^-}$ | – | 1.2 | 1.32 |
| $\Gamma_{\mathcal{T} \to e^+e^-}$ | – | 0.16 | – |

data of 50 MeV. This small discrepancy may be attributed to the unaccounted effects of the dimension-six condensate or/and to the (usual) increasing role of the continuum for state with higher spins. However, the prediction is quite satisfactory within our approximation.

## 8. $\alpha_s$ from the $P_1^1$-$P_1^3$ axial mass splitting

The corresponding double ratio of moments has the nice feature to be independent of the gluon condensate to leading order in $\alpha_s$ and reads:

$$\frac{M^2_{P_1^1}}{M^2_{P_1^3}} \simeq 1 + \alpha_s \left(\Delta_\alpha^{13}(\text{exact}) = 0.014^{-0.004}_{+0.008}\right) . \quad (26)$$

The recent experimental value $M_{P_1^1} = 3526.1$ MeV of the $P_1^1$ state denoted by $h_c(1P)$ in PDG [25] almost coincides with the one of the center of mass energy:

$$M_P^{\text{c.o.m}} = \frac{1}{9}\left[5M_{P_2^3} + 3M_{P_1^3} + M_{P_0^3}\right] \simeq 3521.6\,\text{MeV} \quad (27)$$

as expected from the short range nature of the spin-spin force [36]. Using this experimental value, one can deduce:

$$\alpha_s(\sigma^{-1} \simeq 1.3\,\text{GeV}) \simeq 0.64^{+0.36}_{-0.18} \pm 0.02 \Longrightarrow \alpha_s(M_Z) \simeq 0.127 \pm 0.009 \pm 0.002 . \quad (28)$$

The error is twice bigger than the one from LEP and $\tau$ decay data, but its value is perfectly consistent with the latter. The theoretical error is mainly due to the uncertainty in $\Delta_\alpha$, while a naïve exponential resummation of the higher order $\alpha_s$ terms leads to the second error much smaller than the previous ones. This value of $\alpha_s$ can be useful for an alternative derivation of this fundamental quantity at low energies and for testing its $q^2$-evolution. *Reciprocally*, using the value of $\alpha_s$ from LEP and $\tau$-decay data as input, one can deduce the prediction of the center of mass (c.o.m) of the $P_J^3$ states given in Table 2.

## 9. $\Upsilon - \eta_b$ mass splitting

For the bottomium, the analysis of the ratios of moments for the $S$ and $P$ waves shows that they are optimized at the same value of $\sigma$, namely [12]:

$$\sigma = (0.35 \pm 0.05)\,\text{GeV}^{-2} , \quad (29)$$

which implies for 5 flavours: $\alpha_s(\sigma) \simeq 0.32 \pm 0.06$. Using the conservative values of the two-loop $b$-quark pole mass: $m_b \simeq 4.2-4.7$ GeV, one can deduce:

$$1/x \simeq 28 \pm 7 , \quad (30)$$

where one might (a priori) expect a good convergence of the $1/m$ expansion.

The splitting between the vector $\Upsilon(S_1^3)$ and the pseudoscalar $\eta_b(S_0^1)$ can be done in a similar way than the charmonium one. The double ratio of moments reads numerically:

$$\mathcal{R}_{VP} \simeq \frac{M_V^2}{M_P^2} \simeq \left(0.9995^{+0.0002}_{-0.0003}\right) \times \left[1 + \alpha_s \left(2.4^{-0.7}_{+1.4}\right) \times 10^{-3} + (0.03 \pm 0.01)\,\text{GeV}^{-4} \langle \alpha_s G^2 \rangle\right] , \quad (31)$$

where we have used the exact expressions of the QCD corrections. This leads to the mass splitting in Table 2. To this order of perturbation theory, this result is in the range of the potential model estimates [36,37,20], with the exception of the one in [17,22], where in the latter it has been shown that the square of the quark velocity $v^2$ correction can cause a large value of about 100 MeV for the splitting. One should also notice that, to this approximation, the gluon condensate gives still the dominant effect at the $b$-mass scale (0.2% of the leading order) compared to the one 0.08% from the $\alpha_s$-term. However, the $1/m$ series of the QCD $\alpha_s$ correction is badly convergent, showing that the static limit approximation is quantitatively inaccurate in the $b$-channel. Therefore, one expects that the corresponding prediction of $(13^{-7}_{+10})$ MeV is a very crude estimate. In order to control the effect of the unknown higher order terms, it is legitimate to introduce into the sum rule, the coefficient of the Coulombic effect from the QCD potential as given by the $\alpha_s^2$-term in Eq. (12)[5]. Therefore, we deduce the "improved" final estimate in Table 2:

$$M_\Upsilon - M_{\eta_b} \approx (63^{-29}_{+51})\ \text{MeV}\ , \tag{32}$$

implying the possible observation of the $\eta_b$ from the $\gamma$ radiative decay.

## 10. $\Upsilon - \chi_b$ mass splittings and new estimate of the gluon condensate

As the $S$ and $P$ wave ratios of moments are optimized at the same value of $\sigma$, we can compare directly, with a good accuracy, the different $P$ states with the $\Upsilon_{(S_1^3)}$ one. As the coefficient of the $\alpha_s^2$ corrections, after inserting the expression of $\sigma_{\min}$, are comparable with the one from the Coulombic potential, we expect that the prediction of this splitting is more accurate than in the case of the hyperfine. The different double ratios of moments read numerically for the values in Eqs. (28)–(29):

$$\begin{aligned}
\mathcal{R}_{VS} &\simeq \frac{M_V^2}{M_S^2} \simeq (0.9696^{+0.0054}_{-0.0083}) \times \left[1 - \alpha_s(0.071^{-0.006}_{+0.011}) - (0.50^{+0.18}_{-0.11})\,\text{GeV}^{-4}\langle\alpha_s G^2\rangle\right]\ , \\
\mathcal{R}_{VA} &\simeq \frac{M_V^2}{M_A^2} \simeq (0.9696^{+0.0054}_{-0.0083}) \times \left[1 - \alpha_s(0.074^{-0.007}_{+0.012}) - (0.54^{+0.18}_{-0.12})\,\text{GeV}^{-4}\langle\alpha_s G^2\rangle\right]\ , \\
\mathcal{R}_{VT} &\simeq \frac{M_V^2}{M_T^2} \simeq (0.9704^{+0.0051}_{-0.0084}) \times \left[1 - \alpha_s(0.077^{-0.008}_{+0.006}) - (0.57^{+0.16}_{-0.13})\,\text{GeV}^{-4}\langle\alpha_s G^2\rangle\right]\ ,
\end{aligned} \tag{33}$$

where $V, S, A, T$ refer respectively to the $\gamma$ and to the different $\chi_b$ states $P_0^3$, $P_1^3$, $P_2^3$. Using the value of the gluon condensate obtained previously, these sum rules lead to the mass splittings in Table 2, which is in good agreement with the corresponding data, but definitely higher than the previous predictions of [39], where, among other effects, the values of $\alpha_s$ and of the gluon condensate used there are too low. Reciprocally, one can use the data for re-extracting *independently* the value of the gluon condensate. As usually observed in the

[5] In this case, the gluon condensate contribution is smaller than the Coulombic one as has been observed in [38].

literature, the prediction is more accurate for the c.o.m., than for the individual mass. The corresponding numerical sum rule is:

$$\frac{M_{\chi_b}^{\text{c.o.m.}} - M_\Upsilon}{M_\Upsilon} \simeq \left(1.53^{+0.26}_{-0.42}\right) \times 10^{-2} + \left(1.20^{+0.1}_{-0.2}\right) \times 10^{-2} + \left(0.28^{+0.08}_{-0.06}\right)\text{GeV}^{-4}\langle\alpha_s G^2\rangle\ , \tag{34}$$

which leads to:

$$\langle\alpha_s G^2\rangle \simeq (6.9 \pm 2.5) \times 10^{-2}\ \text{GeV}^4\ . \tag{35}$$

We expect that this result is more reliable than the one obtained from the $M_\psi - M_{\eta_c}$ as the latter can be more affected by the noncalculated next-next-to-leading perturbative radiative corrections than the former. An average of the two results from the $\psi - \eta_c$ and $\Upsilon - \chi_b$ mass splittings leads to:

$$\langle\alpha_s G^2\rangle \simeq (7.5 \pm 2.5) \times 10^{-2}\ \text{GeV}^4\ , \tag{36}$$

where we have retained the most precise error.

## 11. Update average value of $\langle\alpha_s G^2\rangle$

The previous result can be compared with different fits of the heavy and light quark channels given in Table 2, and which range from 4 (SVZ) to 14 in units of $10^{-2}$ GeV$^4$. The most recent estimate from $e^+e^- \to I = 1$ hadrons data using $\tau$-like decay is $(7 \pm 1) \times 10^{-2}$ GeV$^4$, where one should also notice that the different post-SVZ estimates favour higher values of the gluon condensate. If one considers this latter as the update of the light quark channel estimates and the former as an update of the heavy quark one, one can deduce the *update average* from the sum rule:

$$\langle\alpha_s G^2\rangle \simeq (7.1 \pm 0.9) \times 10^{-2}\ \text{GeV}^4\ . \tag{37}$$

More accurate measurements of this quantity than the already available results from $\tau$-decay data [8] are needed for testing the previous phenomenological estimates.

## 12. Toponium: illustration of the infinite mass limit

Since only in the case of the toponium, the $1/m$-expansion is ideal, we have extended the previous analysis in this channel, though, we are aware that this application can be purely academic because of the eventual non-existence of the corresponding bound states. We use the top mass: $m_t \simeq (173 \pm 14)$ GeV, obtained from the average of the CDF candidates $(174 \pm 16)$ GeV and of the electroweak data $(169 \pm 26)$ GeV as compiled by PDG [25]. We shall work with the ratio of moments in the vector channel for determining the mass of the $S_1^3$ state, and use with a good confidence the leading terms of the expressions given in

Table 1. Using the sum rules in Eqs. (8) and (11) and the value of the minimum $\sigma^{-1/2} \simeq$ 20 GeV from Eq. (9), we deduce the result for the meson-quark mass gap given in Table 2. For the splittings, we use the sum rules in Eqs. (12) and (13), while, for the leptonic width, we use the sum rule in Eq. (14). Our results are summarized in Table 2.

## 13. Conclusions

We have used *new double ratios* of exponential sum rules for directly extracting the mass-splittings of different heavy quarkonia states. Therefore, we have obtained from $M_\psi - M_{\eta_c}$ and $M_{\chi_b} - M_\Upsilon$ a more precise estimate of the value of the gluon condensate given in Eq. (36), which combined with the one from $\tau$-like decay in $e^+e^- \to I = 1$ hadrons data leads to the *update average* in Eq. (37) from the sum rule. We have also used $M_{\chi_c(P_1^1)} - M_{\chi_c(P_1^3)}$ for an alternative extraction of $\alpha_s$ at low energy (see Eq. (28)), with a value consistent with the one from LEP and $\tau$-decay. Our numerical results are summarized in Table 2, where a comparison with different estimates and experimental data is done.

We have also attempted to connect the sum rules and the potential model approaches, using a $1/m$-expansion. We found, that the Coulombic corrections, which are quite well understood in QCD, agree, in general, in the two approaches, except in the radiative corrections of the hyperfine splitting which requires the knowledge of the next-next-to-leading $\alpha_s$-corrections. Relativistic corrections due to finite value of the quark mass have been included in our analysis. However, the coefficients of the gluon condensate disagree in the two approaches, which may be related to the difficulty encountered by Bell-Bertlmann in finding a bridge between a field theory after SVZ and potential models.


## Acknowledgements

It is a pleasure to thank the French Foreign Ministry, the French embassy in Vienna and the Austrian Ministry of Research for a financial support within a bilateral scientific cooperation, the Institut für Theoretische Physik of Vienna for a hospitality, Reinhold Bertlmann for a partial collaboration in the early stage of this work, André Martin and Paco Yndurain for interesting discussions on the potential model results.


## References


[1] M.A. Shifman, A.I. Vainshtein and V.I. Zakharov, Nucl. Phys. B 147 (1979) 385, 448.
[2] For a review, see, e.g: S. Narison, QCD spectral sum rules, Lecture Notes in Physics 26 (World Scientific, 1989).
[3] S. Narison, Phys. Lett. B 361 (1995) 121.
[4] A. Mueller, QCD-20 Years Later Workshop, Aachen (1992) and references therein; F. David, Montpellier Workshop on Non-Perturbative Methods (World Scientific,

1985); M. Beneke and V.I. Zakharov, Phys. Rev. Lett. 69 (1992) 2472; G. Grunberg, QCD94, Montpellier (1994).
[5] P. Ball, M. Beneke and V. Braun, Nucl. Phys. B 452 (1995) 563; C.N. Lovett-Turner and C.J. Maxwell, Nucl. Phys. B 452 (1995) 188.
[6] E. Braaten, S. Narison and A. Pich, Nucl. Phys. B 373 (1992) 581.
[7] F. Le Diberder and A. Pich, Phys. Lett. B 286 (1992) 147; D. Buskulic et al., Phys. Lett. B 307 (1993) 209; A. Pich, QCD94 Workshop, Montpellier, Nucl. Phys. (Proc. Suppl) B 39 (1995); S. Narison, Tau94 Workshop, Montreux, Nucl. Phys. (Proc. Suppl) B 40 (1995).
[8] L. Duflot, Tau94 Workshop, Montreux, Nucl. Phys. (Proc. Suppl) B 40 (1995); R. Stroynowski, ibid.
[9] L.J. Reinders, H. Rubinstein and S. Yazaki, Nucl. Phys. B 186 (1981) 109; Phys. Rep. 127 (1985) I, and references therein.
[10] R.A. Bertlmann, G. Launer and E. de Rafael, Nucl. Phys. B 250 (1985) 61; N.V. Krasnikov, A.A. Pivovarov and A.N. Tavkhelidze, Z. Phys. C 19 (1983) 301.
[11] J.S. Bell and R.A. Bertlmann, Nucl. Phys. B 177 (1981) 218.
[12] R.A. Bertlmann, Nucl. Phys. B 204 (1982) 387.
[13] R.A. Bertlmann and H. Neufeld, Z. Phys. C 27 (1985) 437.
[14] J. Marrow and G. Shaw, Z. Phys. C 33 (1986) 237; J. Marrow, J. Parker and G. Shaw, Z. Phys. C 37 (1987) 103.
[15] S. Narison, Phys. Lett. B 197 (1987) 405, B 341 (1994) 73, and references therein.
[16] S. Narison, Phys. Lett. B 210 (1988) 238.
[17] H. Leutwyler, Phys. Lett. B 98 (1981) 304.
[18] H.G. Dosch and U. Marquard, Phys. Rev. D35 (1987) 2238.
[19] J.S. Bell and R.A. Bertlmann, Nucl. Phys. B 227 (1983) 435; R.A. Bertlmann, Acta Phys. Austriaca 53 (1981) 305; QCD90 Workshop, Montpellier, Nucl. Phys. (Proc. Suppl) B 23 (1991), and references therein.
[20] F.J. Yndurain and S. Titard, Phys. Rev. D (1994) 231.
[21] S. Bethke, QCD94 Workshop, Montpellier, Nucl. Phys. (Proc. Suppl) B 39 (1995).
[22] R. McLary and N. Byers, Phys. Rev. D 28 (1981) 1692.
[23] For a review, see, e.g. K. Zalewski, QCD96 Conf. Montpellier 5–12th July 1996.
[24] G. Rodrigo, Valencia preprint FTUV 95/30 (1995), talk given at the Int. Work. on Elem. Part. Phys: present and future, Valencia (1995).
[25] PDG94 by L. Montanet et al., Phys. Rev. D 50 (1994) 1173 and references therein.
[26] M.B. Voloshin, Minnesota preprint UMN-TH-1326-95 (1995).
[27] S. Narison and E. de Rafael, Phys. Lett. B 103 (1981) 87.
[28] C.A. Dominguez and N. Paver, Phys. Lett. B 293 (1992) 197; C.A. Dominguez, G.R. Gluckman and N. Paver, Phys. Lett. B 333 (1994) 184.
[29] S.N. Nikolaev and A.V. Radyushkin, Phys. Lett. B 124 (1983) 243; Nucl. Phys. B 213 (1983) 285.
[30] D.J. Broadhurst et al., Phys. Lett. B 329 (1994) 103.
[31] K.J. Miller and M.G. Olsson, Phys. Rev. D 25 (1982) 1247, 1253.
[32] A. Zalewska and K. Zalewski, Z. Phys. C 23 (1984) 233; K. Zalewski, Acta. Phys. Polonica B 16 (1985) 239 and references therein.
[33] G. Launer, S. Narison and R. Tarrach, Z. Phys. C 26 (1984) 433.
[34] R.A. Bertlmann, C.A. Dominguez, M. Loewe, M. Perrottet and E. de Rafael, Z. Phys. C 39 (1988) 231.
[35] C.A. Dominguez and J. Solà, Z. Phys. C 40 (1988) 63.

[36] A. Martin, 21st Int. Conf. on High-Energy Physics, Paris (1982) J. de Physique, Colloque C 3, supp 12, ed. Editions de Physique, Les Ulysses; J.M. Richard, Phys. Rep. 212 (1992) 1.
[37] W. Buchmüller, Erice Lectures (1984) and references therein.
[38] G. Curci, A. Di Giacomo and G. Paffuti, Z. Phys. C 18 (1983) 135; M. Campostrini, A. Di Giacomo and Š. Olejník, Pisa preprint, IFUP-TH 2/86 (1986).
[39] M.B. Voloshin, preprint ITEP-21 (1980).

# 52
# The QCD condensates

We anticipated this discussion in Chapter 27 when we discussed the anatomy of the SVZ expansion. Here, we shall review the different determinations of the QCD condensates from QSSR.

Indeed, a good control of the values of the QCD condensates is necessary in the phenomenological applications of QSSR. The non-vanishing value of the light quark condensate, which we shall discuss in the next section, is intimately related to the GMOR realization of chiral symmetry, as can be inferred from the PCAC relation. SVZ [1,654] have also postulated that QCD is spontaneously broken by the gluon condensate, which they confirm from their analysis of the charmonium sum rule. The non-vanishing value of the gluon condensate and the gluon correlation length has been also checked on the lattice [402]. Since the pioneering work of SVZ, [1] a lot of effort has been devoted to this issue as can be found in the long list of published papers in this subject (for reviews see e.g. [3],[51,46], [356–363]). The condensates have been extracted from the light mesons [403–409], [325], [33,328], [341,387], and in [329] (Section 52.10), from the baryons [424–430], from the heavy quarkonia [433,434], and [313] (Section 51.3), and from the heavy-light mesons [401].

The $e^+e^- \to$ hadrons and $\tau$-decays data have been always used as a laboratory for testing the perturbative and non-perturbative structure of QCD [1,3,325], [403–409], [346,338,341] and [329,161] (Sections 19.4 and 52.10). As already mentioned, these channels have the great advantage that the spectral functions are measured in a region where pQCD is applicable and therefore the analysis does not suffer from any model dependence in the parametrization of the spectral functions. Therefore, one expects that the determinations from these channels are model-independent.

### 52.1 Dimension-two tachyonic gluon mass

- $e^+e^- \to I = 1$ hadrons below 2 GeV has been also used for extracting the hypothetical dimension-two term beyond the SVZ expansion which has been interpreted in [161] as the effect due to a tachyonic gluon mass. In [341,329], ratios of Laplace sum rules and $\tau$-like sum rules which can disentangle the leading radiative perturbative corrections from the non-perturbative contributions have been used. As a result, one is able to extract the tiny contribution due to the dimension-two terms. The outcome of the analysis is:

$$d_2 \equiv -1.05 \left(\frac{\alpha_s}{\pi}\right) \lambda^2 \simeq (0.03 \sim 0.07)\ \text{GeV}^2\ . \tag{52.1}$$

This is not the case of other attempts (see e.g. [653]), where in these approaches the $\alpha_s$ contribution masks the one of the tachyonic gluon mass.

- In [161], an alternative estimate of this quantity has been produced in the pseudoscalar channel where one can notice that the size of this contribution is about four times the one in the $\rho$-meson channel, such that its effect is more sizeable. At the optimization scale of the sum rule, one obtains:

$$\left(\frac{\alpha_s}{\pi}\right) \lambda^2 \simeq -(0.12 \pm 0.06)\ \text{GeV}^2\ , \tag{52.2}$$

which is more precise than the etimate using $e^+e^-$ data but still inaccurate.

- The value quoted in Table 48.2 corresponds to the intersection of these two estimates which we take to be:

$$\left(\frac{\alpha_s}{\pi}\right) \lambda^2 \simeq -(0.06 \sim 0.07)\ \text{GeV}^2\ . \tag{52.3}$$

## 52.2 Dimension-three quark condensate

The derivation of the $\langle\bar\psi\psi\rangle$ condensate from the sum rules will be discussed in the chapter on light quark masses and light baryons.

## 52.3 Dimension-four gluon condensate

- The estimate of the gluon condensate from $\tau$-decays [328,33] is not very conclusive, which one can understand from [325] because its contribution has an extra $\alpha_s$ correction in the QCD expression of the $\tau$ width. In this case, the analysis of $e^+e^-$ data from the usual QSSR (in particular the Laplace sum rule) is superior. Most recent results using $e^+e^-$ data have been obtained in the Section 52.10. It reads:

$$\langle\alpha_s G^2\rangle \simeq (7.1 \pm 0.7) \times 10^{-2}\ \text{GeV}^4\ , \tag{52.4}$$

showing that the original SVZ result has been underestimated by a factor of about 2. An analogous result has been already obtained in the past by Bell–Bertlmann [91–93] using Laplace sum rules for heavy quark systems and adding a quantum mechanics argument for supporting their result. Similar conclusions have been reached in [655–658], while the validity of the SVZ value has been also questioned in [659]. Analogously [405,406] use high moments in $n$ of FESR in $e^+e^- \to I = 1$ hadrons and have found larger values of the different condensates but the results were not very convincing due to the large sensitivity of the moments on the high-energy tails of the spectral functions.

- In [3], we have reworked in detail the different estimates of the gluon condensate from charmonium systems and come to the conclusions that using the standard sum rules of SVZ, one cannot extract an accurate value of the gluon condensate from the charmonium sum rules because of the uncertainties induced by the correlated value and definition of the charm quark mass. The emerging value from different heavy quark sum rules analysis is [3]:

$$\langle\alpha_s G^2\rangle \simeq (4 \sim 6) \times 10^{-2}\ \text{GeV}^4\ . \tag{52.5}$$

- However, working with sum rules which can disentangle these different contributions, one can extract a more precise result. In Section 51.3, one has observed after examining the different QCD contributions in various quarkonia sum rules that the ones for the $J/\Psi - \eta_c$ and $\Upsilon - \eta_b$ mass splittings is quite sensitive to the value of the gluon condensate, which one can disentangle from the quark mass and leading $\alpha_s$ corrections. In the case of the $J/\psi$-$\eta_c$, the sum rule reads:

$$\mathcal{R}_{VP} \equiv \frac{M^2_{J/\psi}}{M^2_{\eta_c}} = \Delta_0^{VP}\left[1 + \alpha_s(\sigma)\Delta_{\alpha_s}^{VP} + \frac{4\pi}{9}\langle\alpha_s G^2\rangle\sigma^2 x^2 \Delta_G^{VP}\right], \tag{52.6}$$

where $\sigma \simeq (0.9 \pm 0.1)$ GeV$^{-2}$; $1/x \equiv 4M_c^2\sigma \simeq (6.6 \pm 1.8)$ if one uses the conservative value of the charm pole mass $M_c \simeq (1.2 \sim 1.5)$ GeV, while numerically, the complete non-expanded expressions in the quark mass read:

$$\Delta_0^{VP} = 0.995^{+0.001}_{-0.004} \quad \Delta_{\alpha_s}^{VP} = 0.0233^{-0.009}_{+0.011} \quad \Delta_G^{VP} = 29.77^{+8.86}_{-10.23}, \tag{52.7}$$

which leads, respectively, to a correction of about 0.5, 2 and 7% of the leading order term for a typical value of the QCD parameters. It is informative to give the expression of these terms in the limit where the quark mass is large. In this way, one obtains to leading order in $x$ from the table in Section 51.3

$$\Delta_0^{VP} = 1 - \frac{x^2}{2} \qquad \Delta_{\alpha_s}^{VP} = \frac{\sqrt{\pi}}{9}x^{3/2} + 1.539x^2 \qquad \Delta_G^{VP} = \frac{5}{x}\left(1 - \frac{4}{5}x\right), \tag{52.8}$$

which shows, in particular, that the $x$ dependence appearing in the gluon condensate correction is partially compensated by the $1/x$ behaviour of $\Delta_G^{VP}$. Using the experimental value $\mathcal{R}_{VP}^{exp} = 1.082$ and $\alpha_s(\sigma) = 0.48^{+0.17}_{-0.10}$ for four flavours, one obtains:

$$\langle\alpha_s G^2\rangle = (10 \pm 4)10^{-2}\ \text{GeV}^4. \tag{52.9}$$

A similar analysis for the $\Upsilon$-$\chi_b$ mass-splitting gives a much more accurate result as we work at higher scales. Numerically, the sum rule reads [313]:

$$\frac{M_{\chi_b}^{\text{c.o.m}} - M_\Upsilon}{M_\Upsilon} \simeq \left(1.53^{+0.26}_{-0.42}\right) \times 10^{-2} + \left(1.20^{+0.10}_{-0.20}\right) \times 10^{-2} + \left(0.28^{+0.08}_{-0.06}\right) \text{GeV}^{-4}\langle\alpha_s G^2\rangle, \tag{52.10}$$

where $M_{\chi_b}^{\text{c.o.m}}$ is the centre of mass energy:

$$M^{\text{c.o.m}} = \frac{1}{9}\left[5M_{P_2^3} + 3M_{P_0^3} + M_{P_0^3}\right], \tag{52.11}$$

with $P_0^3$, $P_1^3$ and $P_2^3$ refer respectively to the scalar, axial vector and tensor $\chi_b$ states. Using the experimental value of these mass-splittings, the analysis leads to:

$$\langle\alpha_s G^2\rangle = (6.9 \pm 2.5)10^{-2}\ \text{GeV}^4, \tag{52.12}$$

The two channels give the average:

$$\langle\alpha_s G^2\rangle = (7.5 \pm 2.5)10^{-2}\ \text{GeV}^4, \tag{52.13}$$

in good agreement with the previous value from $e^+e^- \to I = 1$ hadrons data, although less precise.

- The average of these results from different sources reads:

$$\langle\alpha_s G^2\rangle = (7.1 \pm 0.9)10^{-2}\ \text{GeV}^4, \tag{52.14}$$

which we consider as a final estimate from the sum rules compiled in Table 48.2. Lattice calculations support the above phenomenological estimate [402].

We then conclude from the previous analysis that *the gluon condensate* also breaks spontaneously QCD in a similar way that the quark condensate does for chiral symmetry. A more physical intuitive picture of the non-vanishing value of the gluon condensate is given in [654].

## 52.4 Dimension-five mixed quark-gluon condensate

- As discussed previously in Chapter 27, the mixed condensate can be parametrized as:

$$\left\langle \bar\psi \frac{\lambda_a}{2}\sigma_{\mu\nu}G_a^{\mu\nu}\psi\right\rangle \equiv M_0^2\alpha_s^{\gamma_M/-\beta_1}\langle\bar\psi\psi\rangle , \tag{52.15}$$

where $\gamma_M = 1/3$ is the anomalous dimension. Due to its important rôle in the baryon sum rules analysis of odd dimension $F_2$ part of the correlator, the size of the mixed condensate has been initially obtained from this channel [424–430], where the result also appears to be independent of the choice of the nucleon interpolating currents. It reads:

$$M_0^2 \simeq 0.8\ \text{GeV}^2 . \tag{52.16}$$

- Alternatively, the mixed condensate has been also obtained from the heavy-light quark systems [401], as it has been noticed that the $B$ and $B^*$ masses are quite sensitive to this quantity, which acts in opposing directions. A priori, this latter method is more reliable than the previous baryon sum rules, due to the smaller complication of this meson channel. It gives a result that is consistent with the one from the baryon sum rules, and extraordinarily accurate:

$$M_0^2 = (0.80 \pm 0.01)\ \text{GeV}^2 , \tag{52.17}$$

to the order we have used.
- From these two completely independent analyses, we deduce the value given in Table 48.2:

$$M_0^2 = (0.8 \pm 0.1)\ \text{GeV}^2 , \tag{52.18}$$

where we have estimated the error to be about 10% typical of the sum rule analysis. This result is in agreement, with the quenched lattice estimate [660], and with the one from an effective quark interaction model [661]. The result from an instanton liquid model [662] appears to be too high. The analysis of [663] also indicates that the mixed condensate shows a $SU(3)_F$ breaking analogous to the one of the quark condensate. However, this result is opposite with the one from baryon sum rules [426]. The discrepancy of these two results needs clarification.

## 52.5 Dimension-six four-quark condensates

In Section 52.10 the $e^+e^- \to I = 1$ hadrons data have been used for estimating the non-perturbative condensates. The result obtained in [329] after using different forms of the sum rules and the last iteration of different steps is quoted in Table 48.2. It reads:

$$\delta_V^{(6)} \simeq= (3.7 \pm 0.6) \times 10^{-2} \tag{52.19}$$

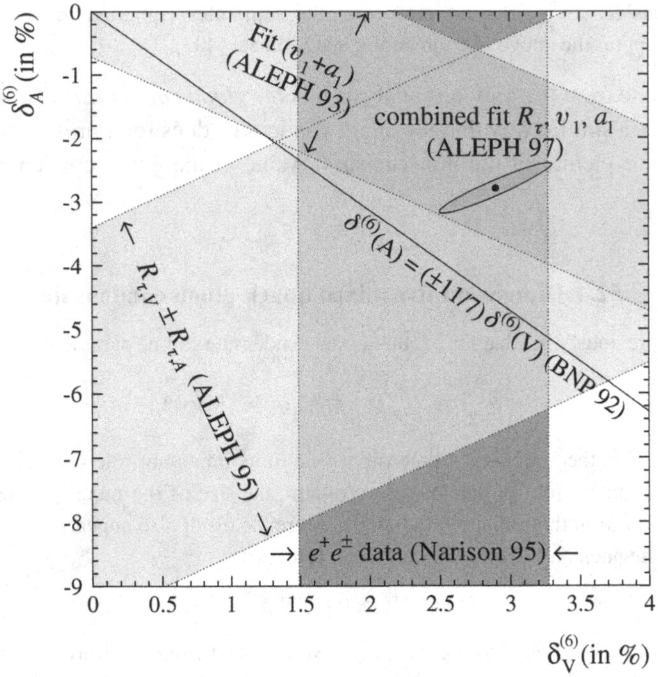


Fig. 52.1. Dimension-six condensate contributions to $R_{\tau,V/A}$.

which is normalized as in Eq. (25.49):

$$M_\tau^6 \delta_{V/A}^{(6)} = \begin{pmatrix} 7 \\ -11 \end{pmatrix} \frac{256\pi^3}{27} \rho\alpha_s \langle\bar{\psi}\psi\rangle^2 , \tag{52.20}$$

where $\rho = 1$ if one uses vacuum saturation for the estimate of the four-quark operators. For a comparison, Fig. 52.1 (figure taken from ALEPH) shows the different estimates from $\tau$-decay in the vector (V) and axial-vector (A) channels. The ALEPH result is [33]:

$$\delta_V^{(6)} \simeq -\delta_A^{(6)} = (2.9 \pm 0.4) \times 10^{-2} , \qquad \delta_{V+A}^{(6)} = -(0.1 \pm 0.4) \times 10^{-2} . \tag{52.21}$$

The BNP [325] result shown in Fig. 52.1 is based on the vacuum saturation assumption for the ratio of the axial-vector over the vector channel contributions. One should, however, notice that the result of Section 52.10 quoted by ALEPH in this figure corresponds to the first iteration result in the section. The improved result obtained in Section 52.10 and quoted in Table 48.2 has a central value slightly higher and more precise than the one quoted in Fig. 52.1. It corresponds to the value quoted in Eq. (52.19) which is more precise and in excellent agreement with the ALEPH result.

We conclude from the previous analysis that the vacuum saturation or equivalently the leading $1/N_c$ is a very crude estimate of the four-quark condensate values. In most cases, the real value is two to three times the vacuum saturation value. These recent results confirm earlier estimates from $e^+e^-$ data using the ratio of moments [404] and from baryon sum rules [424].

## 52.6 Dimension-six gluon condensates

These condensates have already been discussed in Chapter 27. To my knowledge, there are no sum rules estimates of these quantities. Therefore, we have nothing to add to the discussions in the previous chapter. We shall use the value given in Table 48.2 coming from a conservative range from lattice [402] and DIGA model.

## 52.7 Dimension-eight condensates

In Section 52.10, the analysis has been pursued for fixing the size of the dimension-eight condensates appearing in the OPE of the vector channel. The final result quoted there is:

$$d_{8,V} \equiv M_\tau^8 \delta_V^{(8)} \simeq -(1.5 \pm 0.6)\ \mathrm{GeV}^8\,, \tag{52.22}$$

normalized in the same way as in Eq. (25.49), although a value of $-(0.85 \pm 0.18)$ GeV$^8$ has also been obtained in the first stage of the iteration. This result is consistent with the one about $-0.95$ GeV$^8$ in [346] and the one from ALEPH [33]:

$$d_{8,V} = -(0.9 \pm 0.1)\ \mathrm{GeV}^8 \tag{52.23}$$

and of OPAL [33]:

$$d_{8,V} = -(0.8 \pm 0.1)\ \mathrm{GeV}^8 \tag{52.24}$$

from $\tau$-decay data. We can consider as a final result from the different estimates the (arithmetic) average value:

$$d_{8,V} \equiv M_\tau^8 \delta_V^{(8)} \simeq -(1.1 \pm 0.3)\ \mathrm{GeV}^8 \tag{52.25}$$

The axial-vector channel is only accessible from $\tau$-decay data. The results from ALEPH is:

$$d_{8,A} = (0.8 \pm 0.1)\ \mathrm{GeV}^8\,, \tag{52.26}$$

and from OPAL:

$$d_{8,A} = (0.4 \pm 0.2)\ \mathrm{GeV}^8\,, \tag{52.27}$$

where the two central values differ almost by a factor 2. We adopt the average of the two results as a final estimate:

$$d_{8,A} \equiv M_\tau^8 \delta_V^{(8)} \simeq (0.6 \pm 0.2)\ \mathrm{GeV}^8\,. \tag{52.28}$$

These estimates are about one order of magnitude bigger than the rough estimate coming from the vacuum saturation [325]:[1]

$$d_{8,V} \approx d_{8,A} \approx -\frac{39}{162}\pi^2\langle\alpha_s G^2\rangle^2 \approx 0.01\ \text{GeV}^8\ . \tag{52.29}$$

A fit of the (pseudo)scalar channel [394] also shows that the size of the $D = 8$ condensates needed to reproduce the lattice data [393] at large $x$ is also much bigger than the one from the vacuum saturation estimate of these operators.

## 52.8 Instanton like-contributions

In Section 52.10 an attempt has been made to estimate the contributions of the dimension-nine operators which can mimic the instanton-like effects to the vector correlator [382–387]. The result of the analysis is:

$$\delta_V^{(9)} \equiv \delta_V^{\text{inst}} = -(7.0 \pm 26.5)10^{-4}\left(\frac{1.78}{M_\tau}\right)^9\ , \tag{52.30}$$

which is completely negligible in the sum rule working region. This result confirms the alternative phenomenological estimate [387]:

$$\delta_V^{\text{inst}} \approx 3 \times 10^{-3}\ , \tag{52.31}$$

and theoretical estimate [385]:

$$\delta_V^{\text{inst}} \approx 2 \times 10^{-5} \tag{52.32}$$

at the $\tau$ mass. In [384], one also expects a further cancellation for the sum of the vector and axial-vector channels:

$$\delta_{V+A}^{\text{inst}} \approx \frac{1}{20}\delta_V^{\text{inst}}\ . \tag{52.33}$$

Although there is a concensus over the negligible effect of small size instantons in the $V/A$ channels, the situation in the (pseudo)scalar channel is more controversial. In [385], one also expects that the instanton effect is negligible at the sum rule working scale of about 1 GeV, while in [383] one expects that it breaks completely the OPE in this channel.

In the remaining part of the discussions of this book, we shall adopt the pragmatic attitude that the usual OPE describes the (pseudo)scalar channel and the inclusion of the quadratic term restores the discrepancy between the scales in the $\rho$ and $\pi$ meson sum rules [161]. This attitude is supported by the lattice result for the (pseudo)scalar two-point function [393], which can be fitted quite well until large $x$ by the OPE including quadratic $\lambda^2$ term and the dimension-eight condensate.

[1] A missprint of a factor $1/\pi^2$ has been corrected in the BNP formula.

### 52.9 Sum of non-perturbative contributions to $e^+e^- \to I = 1$ hadrons and $\tau$ decays

A fit of the sum of the different non-perturbative terms entering in the QCD expression of $\tau$ decays has been also done by ALEPH and OPAL [33,328]. Using the normalization in Section 25.5, ALEPH obtains:

$$\delta_{NP,V} = 0.020 \pm 0.017\,, \quad \delta_{NP,A} = -0.027 \pm 0.009\,, \quad \delta_{NP,V+A} = -0.003 \pm 0.005\,, \tag{52.34}$$

while OPAL obtains for different structures of the perturbative QCD series:

$$\delta_{NP,V} = 0.016 \pm 0.004\,, \quad \delta_{NP,A} = -0.023 \pm 0.004\,, \quad \delta_{NP,V+A} = -0.0035 \pm 0.0035\,. \tag{52.35}$$

In Section 52.10, the sum of the different non-perturbative contributions including the dimension-nine condensates from $e^+e^-$ data is found to be:

$$\delta_{NP,V} = 0.024 \pm 0.009 \tag{52.36}$$

in good agreement with the former results from $\tau$-decays.

### 52.10 Reprinted paper

## *QCD tests from $e^+e^- \to I = 1$ hadrons data and implication on the value of $\alpha_s$ from $\tau$-decays*

S. Narison

Reprinted from *Physics Letters B,* Volume 361. pp. 121–130, Copyright (1995) with permission from Elsevier Science.

#### 1. Introduction

Measurements of the QCD scale $\Lambda$ and of the $q^2$-evolution of the QCD coupling are one of the most important tests of perturbative QCD. At present LEP and $\tau$-decay data [1–7] indicate that the value of $\alpha_s$ is systematically higher than the one extracted from deep-inelastic low-energy data[1]. The existing estimate of $\alpha_s$ from QCD spectral sum rules [9] à la SVZ [10] in $e^+e^-$ data [11,12] apparently favours a low value of $\alpha_s$, a result, which is, however, in contradiction with the recent CVC-test performed by [13] using $e^+e^-$ data. It is therefore essential to test the reliability of the low-energy predictions before speculating on the phenomenological consequences implied by the previous discrepancy.

Deep-inelastic scattering processes need a better control of the parton distributions and of the power corrections in order to be competitive with the LEP and tau-decay measurements. In addition, perturbative corrections in these processes should be pushed so far such that the remaining uncertainties will only be due to the re-summation of the perturbative series at large order. Indeed, the $\tau$-decay rate has been calculated including the $\alpha_s^3$-term [3], while an

[1] However, new results of jet studies in deep-inelastic *ep*-scattering at HERA for photon momentum transfer $10 \le Q^2$ [GeV$^2$] $\le 4000$ give a value of $\alpha_s$ [8] compatible with the LEP-average.

estimate [14] and a measurement [15] of the $\alpha_s^4$ coefficient is done. Moreover, a resummation of the $(\beta_1\alpha_s)^n$ of the perturbative series is now available [16].

The QCD spectral sum rule (QSSR) [9] à la SVZ [10] applied to the $I = 1$ part of the $e^+e^- \to$ hadrons total cross-section has a QCD expression very similar to the $\tau$-decay inclusive width, such that on a theoretical basis, one can also have a good control of it.

In a previous paper [17], we have derived in a model-independent way the running mass of the strange quark from the difference between the $I = 1$ and $I = 0$ parts of the $e^+e^- \to$ hadrons total cross-section. In this paper, we pursue this analysis by re-examining the estimate of $\alpha_s$ and of the condensates including the instanton-like and the *marginal* $D = 2$-like operators obtained from the $I = 1$ channel of the $e^+e^-$ data. In so doing, we re-examine the exponential Laplace sum rule used by [11] in $e^+e^-$, which is a generalization of the $\rho$-meson sum rule studied originally by SVZ [10]. We also expect that the Laplace sum rule gives a more reliable result than the FESR due to the presence of the exponential weight factor which suppresses the effects of higher meson masses in the sum rule. This is important in the particular channel studied here as the data are very inaccurate above 1.4–1.8 GeV, where, at this energy, the optimal result from FESR satisfies the so-called heat evolution test [12,18,19]. That makes the FESR prediction strongly dependent on the way the data in this region are parametrized, a feature which we have examined [13,20] for criticizing the work of [21]. We also test the existing and controversial estimates [18,19] of the $D = 2$-type operator obtained from QSSR. Combining our different non-perturbative results with the recent resummed perturbative series [16], we re-estimate and confirm the value of $\alpha_s$ from $\tau$-decays.

## 2. $\alpha_s$ from $e^+e^- \to I = 1$ hadrons data

Existing estimates of $\alpha_s$ or $\Lambda$ from different aspects of QSSR for $e^+e^- \to I = 1$ hadrons data [11,12] lead to values much smaller than the present LEP and $\tau$-decay measurements [3–7]. However, such results contradict the stability-test on the extraction of $\alpha_s$ from $\tau$-like inclusive decay [13] obtained using CVC in $e^+e^-$ [22] for different values of the $\tau$-mass. In the following, we shall re-examine the reliability of these sum rule results.

We shall not reconsider the result from FESR [12] due to the drawbacks of this method mentioned previously, and also, because the FESR-analysis has been re-used recently [18,19] in the determination of the $D=2$-type operator, which we shall come back later on.

$\Lambda_3$ and the condensates have been extracted in [11] from the Laplace sum rule:

$$\mathcal{L}_1 \equiv \frac{2}{3}\tau \int_{4m_\pi^2}^{\infty} ds\, e^{-st} R^{I=1}(s) \tag{1}$$

and from its $\tau \equiv 1/M^2$ derivative:

$$\mathcal{L}_2 \equiv \frac{2}{3}\tau^2 \int_{4m_\pi^2}^{\infty} ds\, s\, e^{-s\tau} R^{I=1}(s)\,, \tag{2}$$

where:

$$R^I \equiv \frac{\sigma(e^+e^- \to I \text{ hadrons})}{\sigma(e^+e^- \to \mu^+\mu^-)}\,. \tag{3}$$

In the chiral limit $m_u = m_d = 0$, the QCD expressions of the sum rule can be written as:

$$\mathcal{L}_i = 1 + \sum_{D=0,2,4,\dots} \Delta_i^{(D)}. \tag{4}$$

The perturbative corrections can be deduced from the ones of $R^{I=1}$ obtained to order $\alpha_s^3$:

$$R^{I=1}(s) = \frac{3}{2}\left\{1 + a_s + F_3 a_s^2 + F_4 a_s^3 + \mathcal{O}\left(a_s^4\right)\right\}, \tag{5}$$

where, for 3 flavours: $F_3 = 1.623$ [23], $F_4 = 6.370$ [24]; the expression of the running coupling to three-loop accuracy is:

$$\begin{aligned} a_s(\nu) = a_s^{(0)} \Bigg\{ & 1 - a_s^{(0)} \frac{\beta_2}{\beta_1} \log\log\frac{\nu^2}{\Lambda^2} \\ & + \left(a_s^{(0)}\right)^2 \left[\frac{\beta_2^2}{\beta_1^2} \log^2 \log\frac{\nu^2}{\Lambda^2} - \frac{\beta_2^2}{\beta_1^2} \log\log\frac{\nu^2}{\Lambda^2} \right. \\ & \left. - \frac{\beta_2^2}{\beta_1^2} + \frac{\beta_3}{\beta_1}\right] + \mathcal{O}\left(a_s^3\right)\Bigg\}, \end{aligned} \tag{6}$$

with:

$$a_s^{(0)} \equiv \frac{1}{-\beta_1 \log(\nu/\Lambda)} \tag{7}$$

and $\beta_i$ are the $\mathcal{O}(a_s^i)$ coefficients of the $\beta$-function in the $\overline{\text{MS}}$ scheme for $n_f$ flavours:

$$\begin{aligned} \beta_1 &= -\frac{11}{2} + \frac{1}{3} n_f \\ \beta_2 &= -\frac{51}{4} + \frac{19}{12} n_f \\ \beta_3 &= \frac{1}{64}\left[-2857 + \frac{5033}{9} n_f - \frac{325}{27} n_f^2\right]. \end{aligned} \tag{8}$$

For three flavours, we have:

$$\beta_1 = -9/2, \qquad \beta_2 = -8, \qquad \beta_3 = -20.1198\,. \tag{9}$$

In the chiral limit, the $D = 2$-contribution vanishes. It has also been proved recently [16] that renormalon-type contributions induced by the resummation of the QCD series at large order cannot induce such a term.

In the chiral limit, the $D = 4$ non-perturbative corrections read [10,3]:

$$\begin{aligned} \Delta_1^{(4)} &= \frac{\pi}{3} \tau^2 \langle \alpha_s G^2 \rangle \left(1 - \frac{11}{18}\frac{\alpha_s}{\pi}\right) \\ \Delta_2^{(4)} &= -\Delta_1^{(4)}. \end{aligned} \tag{10}$$

The $D = 6$ non-perturbative corrections read [10]:

$$\Delta_1^{(6)} = -\frac{448\pi^3}{81}\tau^3\rho\langle\bar{u}u\rangle^2$$
$$\Delta_2^{(6)} = -2\Delta_1^{(6)} . \tag{11}$$

We shall use, in the first iteration of the analysis, the conservative values of the condensates [9,3]:

$$\langle\alpha_s G^2\rangle = (0.06 \pm 0.03)\,\text{GeV}^4,$$
$$\rho\langle\bar{u}u\rangle^2 = (3.8 \pm 2.0)10^{-4}\,\text{GeV}^6, \tag{12}$$

and *high* values of $\Lambda$ from LEP and tau-decay data [1–4] for 3 flavours:

$$\Lambda_3 = 375^{+105}_{-85}\,\text{MeV}, \tag{13}$$

corresponding to $\alpha_s(M_Z) = 0.118 \pm 0.06$.

The phenomenological side of the sum rule has been parametrized using analogous data as [11] and updated using the data used in [13]. The confrontation of the QCD and the phenomenological sides of the sum rules is done in Fig. 1a and in Fig. 2a for a giving value of $\Lambda_3 = 375$ MeV and varying the condensates in the range given previously. One can conclude that one has a good agreement between the two sides of $\mathcal{L}_1$ for $M \geq 0.8$ GeV and of $\mathcal{L}_2$ for $M \geq 1.0 \sim 1.2$ GeV. The effects of the condensates are important below 1 GeV for $\mathcal{L}_1$ and below 1.3 GeV for $\mathcal{L}_2$. In Fig. 1b and Fig. 2b, we fix the condensates at their central values and we vary $\Lambda_3$ in the range given above. One can notice that a value of $\Lambda_3$ as high as 525 MeV is still allowed by the data, while the shape of the QCD curve for $\mathcal{L}_2$ changes drastically for a high value of $\Lambda_3$. This phenomena is not informative as, below 1 GeV, higher dimension condensates can already show up and may break the Operator Product Expansion (OPE).

By comparing these results with the ones of [11], one can notice that our QCD prediction for $\mathcal{L}_1$ corresponding to the previous set of parameters is as good as the one of [11] obtained from a different set of the QCD parameters, while for that of $\mathcal{L}_2$, the agreement between the two sides of the sum rule is obtained here at a slightly larger value of $M$ for high values of $\Lambda_3$.

One can clearly conclude from our analysis is that the exponential Laplace sum rules applied to $e^+e^- \rightarrow I = 1$ hadrons data *do not exclude* values of $\Lambda_3$ obtained from LEP and $\tau$-decay data. Contrary to some claims in the literature, the sum rules cannot also give a *precise* information on the real value of $\Lambda_3$ if the condensates are allowed to move inside the conservative range of values given in Eq. (12). It is also important and reassuring, that our analysis supports the value of $\Lambda_3$ obtained from $\tau$-decay and used in $e^+e^-$ via CVC [22] for the stability-test of the prediction for different values of the $\tau$-mass [13] as we shall see also below.

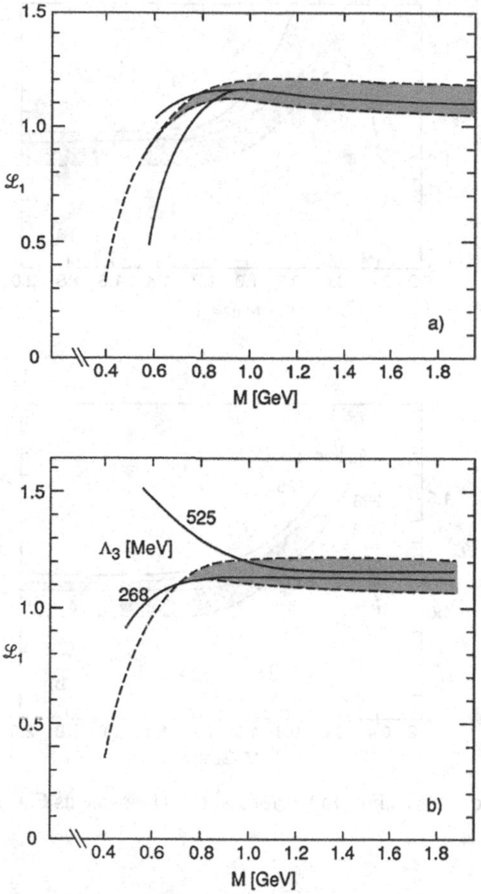


Fig. 1. (a) The Laplace sum rule $\mathcal{L}_1$ versus the sum rule parameter M. The dashed curves correspond to the experimental data. The full curves correspond to the QCD prediction for $\Lambda_3 = 375$ MeV, $\langle\alpha_s G^2\rangle = 0.06 \pm 0.03$ GeV$^4$ and $\rho\alpha_s\langle\bar{u}u\rangle^2 = (3.8 \pm 2.0)10^{-4}$ GeV$^6$. (b) The same as Fig. 1a but for different values of $\Lambda_3$ and for $\langle\alpha_s G^2\rangle = 0.06$ GeV$^4$ and $\rho\alpha_s\langle\bar{u}u\rangle^2 = 3.8\,10^{-4}$ GeV$^6$.

## 3. The condensates from $\tau$-like decays

In so doing, we shall work with the vector component of the $\tau$ decay-like quantity deduced from CVC [22]:

$$R_{\tau,1} \equiv \frac{3\cos^2\theta_c}{2\pi\alpha^2} S_{\rm EW} \times \int_0^{M_\tau^2} ds \left(1 - \frac{s}{M_\tau^2}\right)^2 \left(1 + \frac{2s}{M_\tau^2}\right) \frac{s}{M_\tau^2}\, \sigma_{e^+e^- \to I=1}\,, \tag{14}$$

where $S_{\rm EW} = 1.0194$ is the electroweak correction from the summation of the leading-log contributions [25].

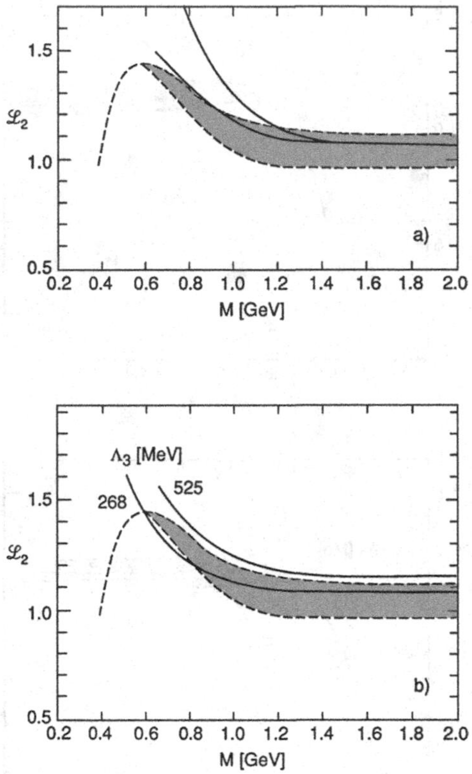


Fig. 2. (a) The same as Fig. 1a but for $\mathcal{L}_2$. (b) The same as Fig. 1b but for $\mathcal{L}_2$.

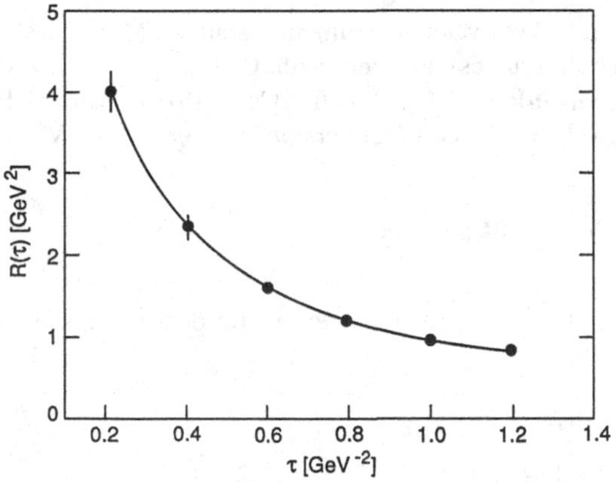


Fig. 3. Experimental value of the ratio of Laplace sum rules $\mathcal{R}(\tau)$ versus the sum rule variable $\tau = 1/M^2$.

This quantity has been used in [13] in order to test the stability of the $\alpha_s$-prediction obtained at the $\tau$-mass of 1.78 GeV. It has also been used to test CVC for different exclusive channels [13,26]. Here, we shall again exploit this quantity in order to deduce *model-independent* informations on the values of the QCD condensates. The QCD expression of $R_{\tau,1}$ reads:

$$R_{\tau,1} = \frac{3}{2}\cos^2\theta_c S_{\rm EW} \times \left(1 + \delta_{\rm EW} + \delta^{(0)} + \sum_{D=2,4,\dots} \delta_1^{(D)}\right), \tag{15}$$

where $\delta_{\rm EW} = 0.0010$ is the electroweak correction coming from the constant term [27]; the perturbative corrections read [3]:

$$\delta^{(0)}\left(a_s \equiv \frac{\alpha_s(M_\tau)}{\pi}\right) + 5.2023a_s^2 + 26.366a_s^3 + \cdots, \tag{16}$$

The $a_s^4$ coefficient has also been estimated to be about 103 [14,15], though we shall use $(78 \pm 25)\, a_s^4$ where the error reflects the uncalculated higher order terms of the $D$-function, while the first term is induced by the lower order coefficients after the use of the Cauchy integration.

In the chiral limit $m_i = 0$, the quadratic mass-corrections contributing to $\delta_1^{(2)}$ vanish. Moreover, it has been proved [16] that the summation of the perturbative series cannot induce such a term, while the one induced eventually by the freezing mechanism is safely negligible [28,18]. Therefore, we shall neglect this term in the first step of our analysis. We shall test, later on, the internal consistency of the approach if a such term is included into the OPE.

In the chiral limit $m_i = 0$, the $D = 4$ contributions read [3]:

$$\delta_1^{(4)} = \frac{11}{4}\pi a_s^2 \frac{\langle \alpha_s G^2 \rangle}{M_\tau^4}, \tag{17}$$

which, due to the Cauchy integral and to the particular $s$-structure of the inclusive rate, the gluon condensate starts at $\mathcal{O}(a_s^2)$. This is a great advantage compared with the ordinary sum rule discussed previously. The $D = 6$ contributions read [3]:

$$\delta_1^{(6)} \simeq 7\frac{256\pi^3}{27} \frac{\rho\alpha_s \langle \bar{\psi}_i \psi_i \rangle^2}{M_\tau^6}, \tag{18}$$

The contribution of the $D = 8$ operators in the chiral limit reads [3]:

$$\delta_1^{(8)} = -\frac{39\pi^2}{162} \frac{\langle \alpha_s G^2 \rangle^2}{M_\tau^8}. \tag{19}$$

The phenomenological parametrization of $R_{\tau,1}$ has been done using the same data input as in [18,13]. We give in Table 1 its value for different values of the tau-mass. Neglecting the $D = 4$-contribution which is of the order $\alpha_s^2$, we perform a two-parameter fit of the data for each value of $\Lambda_3$ corresponding to the world average value of $\alpha_s(M_Z) = 0.118 \pm 0.006$

Table 1. *Phenomenological estimate of $R_{\tau,1}$*

| $M_\tau$[GeV] | $R_{\tau,1}$ |
|---|---|
| 1.0 | $1.608 \pm 0.064$ |
| 1.2 | $1.900 \pm 0.075$ |
| 1.4 | $1.853 \pm 0.072$ |
| 1.6 | $1.793 \pm 0.070$ |
| 1.8 | $1.790 \pm 0.081$ |
| 2.0 | $1.818 \pm 0.097$ |

Table 2. *Estimates of $d_6$ and $d_8$ from $R_{\tau,1}$ for different values of $\Lambda_3$*

| $\Lambda_3$ [MeV] | $d_6$ [GeV$^6$] | $-d_8$ [GeV$^8$] |
|---|---|---|
| 480 | $-0.07 \pm 0.43$ | $1.15 \pm 0.40$ |
| 375 | $0.27 \pm 0.34$ | $0.69 \pm 0.31$ |
| 290 | $0.58 \pm 0.29$ | $0.83 \pm 0.27$ |

[1,2] and by letting the $D = 6$ and $D = 8$ condensates as free-parameters. We show the results of the fitting procedure in Table 2 for different values of $\Lambda_3$.

The errors take into account the effects of the $\tau$-mass moved from 1.6 to 2.0 GeV, which is a negligible effect, and the one due to the data. One can notice that the estimate of the $D = 8$ condensates is quite accurate, while the one of the $D = 6$ is not very conclusive for $\Lambda_3 \geq 350$ MeV. Indeed, only below this value, one sees that the $D = 6$ contribution is clearly positive as expected from the vacuum saturation estimate. This fact also explains the anomalous low value of $-d_8$ around this transition region. Using the average value of $\Lambda_3$ in Eq. (13), we can deduce the result:

$$d_8 \equiv M_\tau^8 \delta_1^{(8)} = -(0.85 \pm 0.18)\,\text{GeV}^8$$
$$d_6 \equiv M_\tau^6 \delta_1^{(6)} = (0.34 \pm 0.20)\,\text{GeV}^6\,, \qquad (20)$$

which we shall improve again later on once we succeed to fix the value of $d_6$.

## 4. The condensates from the ratio of the Laplace sum rules

Let us now improve the estimate of the $D = 6$ condensates. In so doing, one can remark that, though there are large discrepancies in the estimate of the absolute values of the condensates from different approaches, there is a consensus in the estimate of the ratio of the $D = 4$ over the $D = 6$ condensates:[2]

$$r_{46}\ [\text{GeV}^{-2}] \equiv \frac{\langle \alpha_s G^2 \rangle}{\rho \alpha_s \langle \bar{u}u \rangle^2} = 94.80 \pm 23\ [29]$$
$$96.20 \pm 35\ [12]$$
$$114.6 \pm 16\ [30]$$
$$92.50 \pm 50\ [32], \tag{21}$$

from which we deduce the *average*:

$$r_{46} = (105.9 \pm 11.9)\ \text{GeV}^{-2}\ . \tag{22}$$

We use the previous informations on $d_8$ and $r_{46}$ for fitting the value of the $D = 4$ condensates from the ratio of the Laplace sum rules:

$$\mathcal{R}(\tau) \equiv \tau^{-2} \frac{\mathcal{L}_2}{\mathcal{L}_1}\ , \tag{23}$$

used previously by [29] for a simultaneous estimate of the $D = 4$ and $D = 6$ condensates. We recall that the advantage of this quantity is its less sensitivity to the leading order perturbative corrections. The phenomenological value of $\mathcal{R}(\tau)$ is given in Fig. 2. Using a one-parameter fit, we deduce:

$$\langle \alpha_s G^2 \rangle = (6.1 \pm 0.7)\ 10^{-2}\ \text{GeV}^4\ . \tag{24}$$

Then, we re-inject this value of the gluon condensate into the tau-like width in Eq. (14), from which we re-deduce the value of the $D = 8$ condensate. After a re-iteration of this procedure, we deduce our *final* results:

$$\langle \alpha_s G^2 \rangle = (7.1 \pm 0.7)\ 10^{-2}\ \text{GeV}^4\ ,$$
$$d_8 = -(1.5 \pm 0.6)\ \text{GeV}^8\ . \tag{25}$$

Using the mean value of $r_{46}$, we also obtain:

$$\rho \alpha_s \langle \bar{u}u \rangle^2 = (5.8 \pm 0.9)\ 10^{-4}\ \text{GeV}^6\ . \tag{26}$$

We consider these results as an improvement and a confirmation of the previous result in Eq. (12). It is also informative to compare these results with the ALEPH and CLEO II measurements of these condensates from the moments distributions of the $\tau$-decay width. The most accurate measurement leads to [5]:

$$\langle \alpha_s G^2 \rangle = (7.8 \pm 3.1)\ 10^{-2}\ \text{GeV}^4\ , \tag{27}$$

while the one of $d_6$ has the same absolute value as previously but comes with the wrong sign. Our value of $d_8$ is in good agreement with the one $d_8 \simeq -0.95\ \text{GeV}^8$ in [13,6] obtained from

[2] We have multiplied the original error given by [30] by a factor 10. The constraint obtained in [31] is not very conclusive as it leads to $r_{46} \leq 110\ \text{GeV}^{-2}$ and does not exclude negative values of the condensates which are forbidden from positivity arguments.

the same quantity, but it is about one order of magnitude higher than the vacuum saturation estimate proposed by [33] and about a factor 5 higher than the CLEO II measurement. However, it is lower by a factor 2∼3 than the FESR result from the vector channel [32][3]. The discrepancy with the vacuum saturation indicates that this approximation is very crude, while the one with the FESR is not very surprising, as the FESR approach done in the vector and axial-vector channels [12,32] tends *always to overestimate* the values of the QCD condensates. The discrepancy with the CLEO II measurement can be understood from the wrong sign of the $D = 6$ condensate obtained there and to its correlation with the $D = 8$ one.

## 5. Instanton contribution

Let us now extract the size of the instanton-like contribution by assuming that it acts like a $D \geq 9$ operator. A good place for doing it is $\mathcal{R}_{\tau,1}$ as, in the Laplace sum rules, this contribution is suppressed by a 8! factor implying a weaker constraint. Using the previous values of the $D = 6$ and $D = 8$ condensates, we deduce:

$$\delta_1^{(9)} \equiv \delta_V^{\rm inst} = -(7.0 \pm 26.5)\ 10^{-4}(1.78/M_\tau)^9\ , \tag{28}$$

which, though inaccurate indicates that the instanton contribution is negligible for the vector current and has been overestimated in [34] ($\delta_V^{\rm inst} \approx 0.03 \sim 0.05$). Our result supports the negligible effects found from an alternative phenomenological [35] ($\delta_V^{\rm inst} \approx 3 \times 10^{-3}$) and theoretical [36] ($\delta_V^{\rm inst} \approx 2 \times 10^{-5}$) analysis. Further cancellations in the sum of the vector and axial-vector components of the tau widths are however expected [34,35] ($\delta^{\rm inst} \approx \frac{1}{20}\delta_V^{\rm inst}$).

## 6. Test of the size of the $1/M_\tau^2$-term

Let us now study the size of the $1/M_\tau^2$-term. From the QCD point of view, its possible existence from the resummation of the PTS due to renormalon contributions [28] has not been confirmed [16], while some other arguments [28,37] advocating its existence are not convincing and seems to be a pure speculation. Postulating its existence (whatever its origin!), [18] has estimated the strength of this term by using FESR and the ratio of moments $\mathcal{R}(\tau)$. As already mentioned earlier, the advantage in working with the ratio of moments is that the leading order perturbative corrections disappear such that in a *compromise* region where the high-dimension condensates are still negligible, there is a *possibility* to pick up the $1/M_\tau^2$-contribution. Indeed, using usual stability criteria and *allowing a large range of values around the optimal result*, [18] has obtained the *conservative* value:

$$d_2 \equiv C_2 \equiv \delta_1^2 M_\tau^2 \simeq (0.03 \sim 0.08)\ {\rm GeV}^2\ , \tag{29}$$

[3] In the normalization of [32], our value of $d_8$ translates into $C_8\langle O_8\rangle = (0.18 \pm 0.04)\ {\rm GeV}^8$.

Table 3. *Estimates of $\langle \alpha_s G^2 \rangle$ from $R(\tau)$ for different values of $d_2$*

| $d_2$ [GeV]$^2$ [18] | $-d_2$ [GeV]$^2$ [19] | $\langle \alpha_s G^2 \rangle$ $10^2$ [GeV$^4$] |
|---|---|---|
| 0.03 | | $7.8 \pm 0.5$ |
| 0.05 | | $8.1 \pm 0.5$ |
| 0.07 | | $8.6 \pm 0.5$ |
| 0.09 | | $9.1 \pm 0.5$ |
| | 0.2 | $3.2 \pm 0.29$ |
| | 0.3 | $1.2 \pm 0.29$ |
| | 0.4 | $-0.7 \pm 0.6$ |

while the estimate of [18] from FESR applied to the vector current has not been very conclusive, as it leads to the inaccurate value:

$$d_2 \simeq (0.02 \pm 0.12)\ \text{GeV}^2\ . \tag{30}$$

However, the recent FESR analysis from the axial-vector current [19] obtained at about the same value of the continuum threshold $t_c$ satisfying the so-called evolution test [12], disagrees in sign and magnitude with our previous estimate from the ratio of moments and is *surprisingly* very precise compared with the result in Eq. (30) obtained from the same method for the vector current. If one assumes like [19] a quadratic dependence in $\Lambda_3$, the result of [19] becomes for the value of $\Lambda_3$ in Eq. (13):

$$d_2 \simeq -(0.3 \pm 0.1)\ \text{GeV}^2\ , \tag{31}$$

We test the reliability of this result, by remarking that $d_2$ (if it exists!) is strongly correlated to $d_4$ in the analysis of the ratio of Laplace sum rules $R_{(\tau)}$, while this is not the case between $d_2$ or $d_4$ with $d_6$ and $d_8$. Using our previous values of $d_6$ and $d_8$, one can study the variation of $d_4$ given the value of $d_2$. The results given in Table 3 indicate that the present value of the gluon condensate excludes the value of $d_2$ in Eq. (31) and can only permit a negligible fluctuation around zero of this contribution, which should not exceed the value $0.03 \sim 0.05$. This result rules out the possibility to have a sizeable $1/M_\tau^2$-term [28,37] and justifies its neglect in the analysis of the $\tau$-width. More precise measurement of the gluon condensate or more statistics in the $\tau$-decay data will improve this constraint.

## 7. Sum of the non-perturbative corrections to $R_\tau$

Using our previous estimates, it is also informative to deduce the sum of the non-perturbative contributions to the decay widths of the observed heavy lepton of mass 1.78 GeV. In so doing, we add the contributions of operators of dimensions $D = 4$ to $D = 9$ and we neglect the expected small $\delta^{(2)}$-contribution.

Table 4. *QCD predictions for $R_\tau$ using the contour coupling-expansion*

| $\alpha_s(M_\tau)$ | $a_s^3$ | $a_s^4$ | $a_s^6$ | $a_s^8$ |
|---|---|---|---|---|
| 0.26 | $3.364 \pm 0.022$ | 3.370 | $3.380 \pm 0.019$ | 3.381 |
| 0.28 | $3.402 \pm 0.024$ | 3.411 | $3.426 \pm 0.019$ | 3.426 |
| 0.30 | $3.442 \pm 0.026$ | 3.453 | $3.474 \pm 0.021$ | 3.472 |
| 0.32 | $3.484 \pm 0.030$ | 3.498 | $3.526 \pm 0.023$ | 3.520 |
| 0.34 | $3.526 \pm 0.033$ | 3.546 | $3.582 \pm 0.031$ | 3.568 |
| 0.36 | $3.571 \pm 0.040$ | 3.594 | $3.640 \pm 0.045$ | 3.613 |
| 0.38 | $3.616 \pm 0.040$ | 3.645 | $3.706 \pm 0.069$ | 3.655 |
| 0.40 | $3.664 \pm 0.040$ | 3.700 | $3.775 \pm 0.108$ | 3.685 |

For the vector component of the tau hadronic width, we obtain:[4]

$$\delta_V^{\rm NP} \equiv \sum_{D=4}^{9} \delta_1^{(D)} = (2.38 \pm 0.89)10^{-2} , \tag{32}$$

while using the expression of the corrections for the axial-vector component given in [3], we deduce:

$$\delta_A^{\rm NP} = -(7.95 \pm 1.12)10^{-2} , \tag{33}$$

and then:

$$\delta^{\rm NP} \equiv \frac{1}{2}(\delta_V^{\rm NP} + \delta_A^{\rm NP}) = -(2.79 \pm 0.62)10^{-2} , \tag{34}$$

Our result confirms the smallness of the non-perturbative corrections measured by the ALEPH and CLEO II groups [5]:

$$\delta_{\rm exp}^{\rm NP} = (0.3 \pm 0.5)10^{-2} , \tag{35}$$

though the exact size of the experimental number is not yet very conclusive.

## 8. Implication on the value of $\alpha_s$ from $R_\tau$

Before combining the previous non-perturbative results with the perturbative correction to $R_\tau$, let us test the accuracy of the resummed $(\alpha_s \beta_1)^n$ perturbative result of [16]. In so doing, we fix $\alpha_s(M_\tau)$ to be equal to 0.32 and we compare the resummed value of $\delta^{(0)}$ including the $\delta_s^3$-corrections with the one where the coefficients have been calculated in the $\overline{\rm MS}$ scheme [23]. We consider the two cases where $R_\tau$ is expanded in terms of the usual coupling $\alpha_s$ or in terms of the *contour coupling* [4]. In both cases, one can notice that the approximation

[4] We have used, for $M_\tau = 1.78$ GeV, the conservative values: $\delta_V^{(9)} \approx \delta_A^{(9)} \simeq -(0.7 \pm 2.7)10^{-3}$ and $\delta^{(9)} \approx 1/20\delta_V^{(9)}$ [34].

used in the resummation technique tends to overestimate the perturbative correction by about 10%. Therefore, we shall reduce systematically by 10%, the prediction from this method from the $\alpha_s^5$ to $\alpha_s^9$ contributions. We shall use the coefficient 27.46 of $\alpha_s^4$ estimated in [14,15]. Noting that, to the order where the perturbative series (PTS) is estimated, one has alternate signs in the PTS, which is an indication for reaching the asymptotic regime. Therefore, we can consider, as the best estimate of the resummed PTS, its value at the minimum. That is reached, either for truncating the PTS by including the $\alpha_s^6$ or the $\alpha_s^8$ contributions. The corresponding value of $R_\tau$ including our non-perturbative contributions in Eq. (34) is given in Table 4. We show for comparison the value of $R_\tau$ including the $\alpha_s^3$-term, where we have used the perturbative estimate in [6] (the small difference with the previous papers [4,13,6,7,20] comes from the different non-perturbative term used here), while the error quoted there comes from the naïve estimate $\pm 50 a_s^4$. However, one can see that the estimate of this perturbative error has taken properly the inclusion of the higher order terms, while the truncation of the series at $\alpha_s^3$ already gives a quite good evaluation of the PTS. One can also notice that there is negligible difference between the PTS to order $\alpha_s^6$ and $\alpha_s^8$ for small values of $\alpha_s$, while the difference increases for larger values. We consider, as a final perturbative estimate of $R_\tau$, the one given by the PTS including the $\alpha_s^6$-term at which we encounter the first minimum. The error given in this column is the sum of the non-perturbative one from Eq. (34) with the perturbative conservative uncertainty, which we have estimated like the effect due to the last term i.e $\pm 34.53(-\beta_1 a_s/2)^6$ at which the minimum is reached, which is a legitimate procedure for asymptotic series [38]. We have also added to the latter the one due to the small fluctuation of the minimum of the PTS from the inclusion of the $\alpha_s^6$ or $\alpha_s^8$-terms. One can notice that for $\alpha_s \leq 0.32$, the error in $R_\tau$ is dominated by the non-perturbative one, while for larger value of $\alpha_s$, it is mainly due to the one from the PTS. Using the value of $R_\tau$ in Table 4, we deduce:

$$\alpha_s(M_\tau) = 0.33 \pm 0.030, \qquad (36)$$

where we have used the experimental average [2]:

$$R_\tau = 3.56 \pm 0.03. \qquad (37)$$

Our result from the optimized resummed PTS is in good agreement with the most recent estimate obtained to order $\alpha_s^3$ [6,5,7]:

$$\alpha_s(M_\tau) = 0.33 \pm 0.030. \qquad (38)$$

## 9. Conclusion

Our analysis of the isovector component of the $e^+e^- \rightarrow$ hadrons data has shown that there is a consistent picture on the extraction of $\alpha_s$ from high-energy LEP and low-energy $\tau$ and $e^+e^-$ data.

It has also been shown that the values of the condensates obtained from QCD spectral sum rules based on *stability criteria* are reproduced and improved by fitting the $\tau$-like decay widths and the ratio of the Laplace sum rules. Our estimates are in good agreement with the determination of the condensates from the $\tau$-hadronic width moment-distributions [5], which needs to be improved from accurate measurements of the $e^+e^-$ data or/and for more data sample of the $\tau$-decay widths which can be reached at the $\tau$-charm factory machine.

Finally, our consistency test of the effect of the $1/M_\tau^2$-term, whatever its origin, does not support the recent estimate of this quantity from FESR axial-vector channel [19] and only permits a small fluctuation around zero due to its strong correlation with the $D = 4$ condensate effects in the ratio of Laplace sum rules analysis, indicating that it cannot affect in a sensible way the accuracy of the determination of $\alpha_s$ from tau decays.

As a by-product, we have reconsidered the estimate of $\alpha_s(M_\tau)$ from the $\tau$-widths taking into account the recent resummed result of the perturbative series. Our result in Eq. (36) is a further support of the existing estimates.

**Acknowledgements**

It is a pleasure to thank A. Pich for exchanges and for carefully reading the manuscript.

**References**

[1] S. Bethke, talk given at the QCD94 Workshop, 7–13th July 1994, Montpellier, France and references therein; I. Hinchliffe, talk given at the 1994 Meeting of the American Physical Society, Albuquerque (1994); B. Weber, talk given at the IHEP-Conference, Glasgow (1994).
[2] PDG94, L. Montanet et al., Phys. Rev. D 50 (1994), Part 1, 1175.
[3] E. Braaten, S. Narison and A. Pich, Nucl. Phys. B 373 (1992) 581.
[4] F. Le Diberder and A. Pich, Phys. Lett. B 286 (1992) 147; B 289 (1992) 165.
[5] ALEPH Collaboration: D. Buskulic et al., Phys. Lett. B 307 (1993) 209; CLEOII Collaboration: R. Stroynowski, talk given at TAU94 Sept. 1994, Montreux, Switzerland; for a review of the different LEP data on tau decays, see e.g: L. Duflot, talks given at the QCD94 Workshop, 7–13th July 1994, Montpellier, France and TAU94 Sept. 1994, Montreux, Switzerland.
[6] A. Pich, talk given at the QCD94 Workshop, 7–13th July 1994, Montpellier, France.
[7] S. Narison, talk given at TAU94 Sept. 1994, Montreux, Switzerland.
[8] Hl Collaboration, Phys. Lett. B 346 (1995) 415.
[9] S. Narison, QCD spectral sum rules Lecture notes in physics, Vol 26 (1989).
[10] M.A. Shifman, A.I. Vainshtein and V.I. Zakharov, Nucl. Phys. B 147 (1979) 385, 448.
[11] S.I. Eidelman, L.M. Kurdadze and A.I. Vainshtein, Phys. Lett. B 82 (1979) 278.
[12] R.A. Bertlmann, G. Launer and E. de Rafael, Nucl. Phys. B 250 (1985) 61; R.A. Bertlmann, C.A. Dominguez, M. Loewe, M. Perrottet and E. de Rafael, Z. Phys. C 39 (1988) 231.
[13] S. Narison and A. Pich, Phys. Lett. B 304 (1993) 359.
[14] A.L. Kataev, talk given at the QCD94 Workshop, 7–13th July 1994, Montpellier, France.

[15] F. Le Diberder, talk given at the QCD94 Workshop, 7–13th July 1994, Montpellier, France.
[16] P. Ball, M. Beneke and V.M. Braun, CERN-TH/95-26 (1995) and references therein; C.N. Lovett-Turner and C.J. Maxwell, University of Durham preprint DTP/95/36 (1995).
[17] S. Narison, Montpellier preprint PM95/06 (1995).
[18] S. Narison, Phys. Lett. B 300 (1993) 293.
[19] C.A. Dominguez, Phys. Lett. B 345 (1995) 291.
[20] S. Narison, talk given at the QCD-LEP meeting on $\alpha_s$ (published by S. Bethke and W. Bernreuther as Aachen preprint PITHA 94/33).
[21] T.N. Truong, Ecole polytechnique preprint, EP-CPth.A266.1093 (1993); Phys. Rev. D 47 (1993) 3999; talk given at the QCD-LEP meeting on $\alpha_s$ (published by S. Bethke and W. Bernreuther as Aachen preprint PITHA 94/33).
[22] F.J. Gilman and S.H. Rhie, Phys. Rev. D 31 (1985) 1066; F.J. Gilman and D.H. Miller, Phys. Rev. D 17 (1978) 1846; F.J. Gilman, Phys. Rev. D 35 (1987) 3541.
[23] K.G. Chetyrkin, A.L. Kataev and F.V. Tkachov, Phys. Lett. B 85 (1979) 277; M. Dine and J. Sapirstein, Phys. Rev. Lett. 43 (1979) 668; W. Celmaster and R.J. Gonsalves, Phys. Rev. Lett. 44 (1980) 560.
[24] S.G. Gorishny, A.L. Kataev and S.A. Larin, Phys. Lett. B 259 (1991) 144; L.R. Surguladze and M.A. Samuel, Phys. Rev. Lett. 66 (1991) 560, 2416 (E).
[25] W. Marciano and A. Sirlin, Phys. Rev. Lett. 61 (1988) 1815; 56 (1986) 22.
[26] S.I. Eidelman and V.N. Ivanchenko, Phys. Lett. B 257 (1991) 437; talk given at the TAU94 Workshop, Sept. 1994, Montreux, Suisse.
[27] E. Braaten and C.S. Li, Phys. Rev. D 42 (1990) 3888.
[28] G. Altarelli, in: QCD-20 years later, eds. P.M. Zerwas and H.A. Kastrup, WSC (1994) 308; talk given at the TAU94 Workshop, Sept. 1994, Montreux, Suisse; G. Altarelli, P. Nason and G. Ridolfi, CERN-TH.7537/94 (1994).
[29] G. Launer, S. Narison and R. Tarrach, Z. Phys. C 26 (1984) 433.
[30] J. Bordes, V. Gimenez and J.A. Peñarrocha, Phys. Lett. B 201 (1988) 365.
[31] M.B. Causse and G. Menessier, Z. Phys. C 47 (1990) 611.
[32] C.A. Dominguez and J. Sola, Z. Phys. C 40 (1988) 63.
[33] E. Bagan, J.I. Latorre, P. Pascual and R. Tarrach, Nucl. Phys. B 254 (1985) 555.
[34] I.I. Balitsky, M. Beneke and V.M. Braun, Phys. Lett. B 318 (1993) 371.
[35] V. Kartvelishvili and M. Margvelashvili, Phys. Lett. B 345 (1995) 161.
[36] P. Nason and M. Porrati, Nucl. Phys. B 421 (1994) 518.
[37] M.A. Shifman, Minnesota preprint UMN-TH-1323-94 (1994); M. Neubert, CERN-TH.7524/94 (1994).
[38] G.N. Hardy, Divergent Series, Oxford University press (1949).

# 53

# Light and heavy quark masses, chiral condensates and weak leptonic decay constants

We review the present status for the determinations of the light and heavy quark masses, the light quark chiral condensate and the decay constants of light and heavy-light (pseudo)scalar mesons from QCD spectral sum rules (QSSR). Bounds on the light quark running masses at 2 GeV are found to be (see Tables 53.1 and 53.2): 6 MeV $< (\bar{m}_d + \bar{m}_u)(2) <$ 11 MeV and 71 MeV$< \bar{m}_s(2) <$ 148 MeV. The agreement of the ratio $m_s/(m_u + m_d) = 24.2$ in Eq. (53.45) from pseudoscalar sum rules with the one (24.4 $\pm$ 1.5) from ChPT indicates the consistency of the pseudoscalar sum rule approach. QSSR predictions from different channels for the light quark running masses lead to (see Section 53.9.3): $\bar{m}_s(2) = (117.4 \pm 23.4)$ MeV, $(\bar{m}_d + \bar{m}_u)(2) = (10.1 \pm 1.8)$ MeV, $(\bar{m}_d - \bar{m}_u)(2) = (2.8 \pm 0.6)$ MeV with the corresponding values of the RG invariant masses. The different QSSR predictions for the heavy quark masses lead to the running masss values: $\bar{m}_c(\bar{m}_c) = (1.23 \pm 0.05)$ GeV and $\bar{m}_b(\bar{m}_b) = (4.24 \pm 0.06)$ GeV (see Tables 53.5 and 53.6), from which one can extract the scale independent ratio $m_b/m_s = 48.8 \pm 9.8$. Runned until $M_Z$, the $b$-quark mass becomes: $\bar{m}_b(M_Z) = (2.83 \pm 0.04)$ GeV in good agreement with the average of direct measurements $(2.82 \pm 0.63)$ GeV from three-jet heavy quark production at LEP, and then supports the QCD running predictions based on the renormalization group equation. As a result, we have updated our old predictions of the weak decay constants $f_{\pi'(1.3)}$, $f_{K'(1.46)}$, $f_{a_0(0.98)}$ and $f_{K_0^*(1.43)}$ [see Eqs. (53.75) and (53.77)]. We obtain from a global fit of the light (pseudo)scalar and $B_s$ mesons, the flavour breakings of the *normal ordered* chiral condensate ratio: $\langle\bar{s}s\rangle/\langle\bar{u}u\rangle = 0.66 \pm 0.10$ [see Eq. (53.100)]. The last section is dedicated to the QSSR determinations of $f_{D_{(s)}}$ and $f_{B_{(s)}}$.

## 53.1 Introduction

One of the most important parameters of the standard model and chiral symmetry is the light and heavy quark masses. Light quark masses and chiral condensates are useful for a much better understanding of the realizations of chiral symmetry breaking [55–57] and for some eventual explanation of the origin of quark masses in unified models of interactions [664]. Within some popular parametrizations of the hadronic matrix elements [665], the strange quark mass can also largely influence the Standard Model prediction of the $CP$ violating parameters $\epsilon'/\epsilon$, which have been measured recently [599]. However, contrary to the QED case where leptons are observed, and then the physical masses can be identified with the pole of the propagator (on-shell mass value),[1] the quark masses are difficult to define because of confinement which does not allow us to observe free quarks. However, despite this difficulty, one can consistently treat the quark masses in perturbation theory like the QCD coupling constant.They obey a differential equation, where its boundary condition can be identified with the renormalized mass of the QCD Lagrangian. The corresponding solution is the running mass, which is gauge invariant but renormalization scheme and scale dependent, and the associated renormalization group-invariant mass. To our knowledge, these notions have been introduced for the first time in [28]. In practice, these masses are conveniently defined within the standard $\overline{MS}$ scheme discussed in previous chapters. In addition to the determination of the ratios of light quark masses (which are scale independent) from current algebra [55], and from chiral perturbation theory (ChPT), its modern version [498–502], a lot of effort reflected in the literature [16] has been put into extracting directly from the data the running quark masses using the SVZ [1] QCD spectral sum rules (QSSR) [3], LEP experiments and lattice simulations. The content of these notes is:

- a review of the light and heavy quark mass determinations from the different QCD approaches;
- a review of the direct determinations of the quark vacuum condensate using QSSR and an update of the analysis of its flavour breakings using a global fit of the meson systems;
- an update of the determinations of the light (pseudo)scalar decay constants, which, in particular, are useful for understanding the $\bar{q}q$ contents of the light scalar mesons; and
- a review of the determinations of the weak leptonic decay of the heavy-light pseudoscalar mesons $D_{(s)}$ and $B_{(s)}$.

This review develops and updates the review papers [54,364] and some parts of the book [3]. It also updates previous results from original works.

## 53.2 Quark mass definitions and ratios of light quark masses

Let us remind ourselves of the meaning of quark masses in QCD. One starts from the mass term of the QCD Lagrangian:

$$\mathcal{L}_m = m_i \bar{\psi}_i \psi_i \ , \tag{53.1}$$

where $m_i$ and $\psi_i$ are respectively the quark mass and field. The renormalized mass will be improved by the uses of the RGE leading to the running mass, for which a definition is given in Section 11.11. We shall also use the short-distance pole masses defined in Section 11.12 and the alternative definition in Section 11.13. Finally, we often use the value of the ratios of quark masses from ChPT given in Eq. (42.5.4).

[1] For a first explicit definition of the perturbative quark pole mass in the $\overline{MS}$ scheme, see [147,133] (renormalization-scheme invariance) and [148] (regularization-scheme invariance).

## 53.3 Bounds on the light quark masses

In QSSR, the estimate and lower bounds of the sum of the light quark masses from the pseudoscalar sum rule were first found in [167,626], while a bound on the quark mass difference was first derived in [666]. The literature in this subject of light quark masses increases with time.[2] However, it is in some sense quite disappointing that in most of the published papers no noticeable progress has been made since the early pioneering studies. The most impressive progress comes from the QCD side of the sum rules where new calculations have become available both on the perturbative radiative corrections known to order $\alpha_s^3$ [167,667,442] and on the non-perturbative corrections [1,3].[3] Another new contribution is due to the inclusion of the tachyonic gluon mass as a manifestation of the resummation of the pQCD series [162,161,394]. Alas, no sharp result is available on the exact size of direct instanton contributions advocated to be important in this channel [383], while [385] claims the opposite. Though the instanton situation remains controversial, recent analysis [668,669] using the results of [670] based on the Instanton Liquid Model (ILM) of [386] indicates that this effect is negligible justifying the neglect of this effect in different analysis of this channel. However, it might happen that adding together the effect of the tachyonic gluon to that of the direct instanton might also lead to a double counting in a sense that there can be two alternative ways for parametrizing the non-perturbative vacuum [394]. In the absence of precise control of the origin and size of these effects, we shall consider them as new sources of errors in the sum rule analysis.

### *53.3.1 Bounds on the sum of light quark masses from pseudoscalar channels*

Lower bounds for $(\bar{m}_u + \bar{m}_d)$ based on moments inequalities and the positivity of the spectral functions have been obtained, for the first time, in [167,626]. These bounds have been rederived recently in [671,672] to order $\alpha_s$. As checked in [54] for the lowest moment and redone in [669] for higher moments, the inclusion of the $\alpha_s^3$ term decrease by about 10 to 15% the strength of these bounds, which is within the expected accuracy of the result.

For definiteness, we shall discuss in details the pseudoscalar two-point function in the $\bar{u}s$ channel. The analysis in the $\bar{u}d$ channel is equivalent. It is convenient to start from the second derivative of the two-point function which is superficially convergent:

$$\Psi''(Q^2) = \int_0^\infty dt \frac{2}{(t+Q^2)^3} \frac{1}{\pi} \mathrm{Im}\Psi_5(t) . \tag{53.2}$$

The bounds follow from the restriction of the sum over all possible hadronic states which can contribute to the spectral function to the state(s) with the lowest invariant mass. The lowest hadronic state which contributes to the corresponding spectral function is the $K$–pole.

[2] Previous works are reviewed in [54,3].

[3] See also Part VIII on two-point functions where more references to original works are given.

From Eq. (53.2) we then have:

$$\Psi_5''(Q^2) = \frac{2}{\left(M_K^2 + Q^2\right)^3} 2 f_K^2 M_K^4 + \int_{t_0}^{\infty} dt \frac{2}{(t + Q^2)^3} \frac{1}{\pi} \mathrm{Im}\Psi_5(t)\,, \tag{53.3}$$

where $t_0 = (M_K + 2m_\pi)^2$ is the threshold of the hadronic continuum.

It is convenient to introduce the moments $\Sigma_N(Q^2)$ of the hadronic continuum integral:

$$\Sigma_N(Q^2) = \int_{t_0}^{\infty} dt \frac{2}{(t + Q^2)^3} \times \left(\frac{t_0 + Q^2}{t + Q^2}\right)^N \frac{1}{\pi} \mathrm{Im}\Psi_5(t)\,. \tag{53.4}$$

One is then confronted with a typical moment problem (see e.g. [673].) The positivity of the continuum spectral function $\frac{1}{\pi}\mathrm{Im}\Psi_5(t)$ constrains the moments $\Sigma_N(Q^2)$ and hence the LHS of Eq. (53.3) where the light quark masses appear. The most general constraints among the first three moments for $N = 0, 1, 2$ are:

$$\Sigma_0(Q^2) \geq 0, \quad \Sigma_1(Q^2) \geq 0, \quad \Sigma_2(Q^2) \geq 0\,; \tag{53.5}$$

$$\Sigma_0(Q^2) - \Sigma_1(Q^2) \geq 0, \quad \Sigma_1(Q^2) - \Sigma_2(Q^2) \geq 0\,; \tag{53.6}$$

$$\Sigma_0(Q^2)\Sigma_2(Q^2) - (\Sigma_1(Q^2))^2 \geq 0\,. \tag{53.7}$$

The inequalities in Eq. (53.6) are in fact trivial unless $2Q^2 < t_0$, which constrains the region in $Q^2$ to too small values for pQCD to be applicable. The other inequalities lead however to interesting bounds which we next discuss.

The inequality $\Sigma_0(Q^2) \geq 0$ results in a first bound on the running masses:

$$[m_s(Q^2) + m_u(Q^2)]^2 \geq \frac{16\pi^2}{N_c} \frac{2 f_K^2 M_K^4}{Q^4} \times \frac{1}{\left(1 + \frac{M_K^2}{Q^2}\right)^3} \frac{1}{\left[1 + \frac{11}{3}\frac{\alpha_s(Q^2)}{\pi} + \cdots\right]}\,, \tag{53.8}$$

where the dots represent higher order terms which have been calculated up to $\mathcal{O}(\alpha_s^3)$, as well as non-perturbative power corrections of $\mathcal{O}(1/Q^4)$ and strange quark mass corrections of $\mathcal{O}(m_s^2/Q^2)$ and $\mathcal{O}(m_s^4/Q^4)$ including $\mathcal{O}(\alpha_s)$ terms. Notice that this bound is non-trivial in the large–$N_c$ limit ($f_K^2 \sim \mathcal{O}(N_c)$) and in the chiral limit ($m_s \sim M_K^2$). The bound is of course a function of the choice of the Euclidean $Q$-value at which the RHS in Eq. (53.8) is evaluated. For the bound to be meaningful, the choice of $Q$ has to be made sufficiently large. In [671] it is shown that $Q \geq 1.4$ GeV is already a safe choice to trust the pQCD corrections as such. The lower bound which follows from Eq. (53.8) for $m_u + m_s$ at a renormalization scale $\mu^2 = 4$ GeV$^2$ results in the solid curves shown in Fig. 53.1.

The resulting value of the bound at $Q = 1.4$ GeV is:

$$(m_s + m_u)(2) \geq 80 \text{ MeV} \quad \Longrightarrow \quad (m_u + m_d)(2) \geq 6.6 \text{ MeV}\,, \tag{53.9}$$

if one uses either ChPT and the previous SR analysis for the mass ratios. Radiative corrections tend to decrease the strengths of these bounds. Their contributions to the second

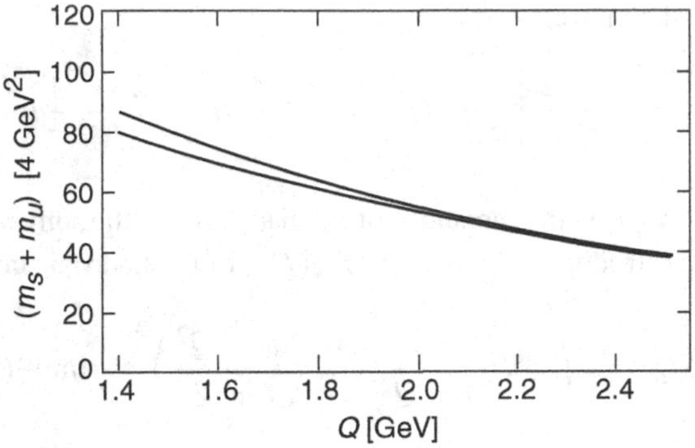


Fig. 53.1. Lower bound in MeV to order $\alpha_s$ for $(m_s + m_u)(2)$ versus $Q$ in GeV from Eq. (53.8) for $\Lambda_3 = 290$ MeV (upper curve) and 380 MeV (lower curve).

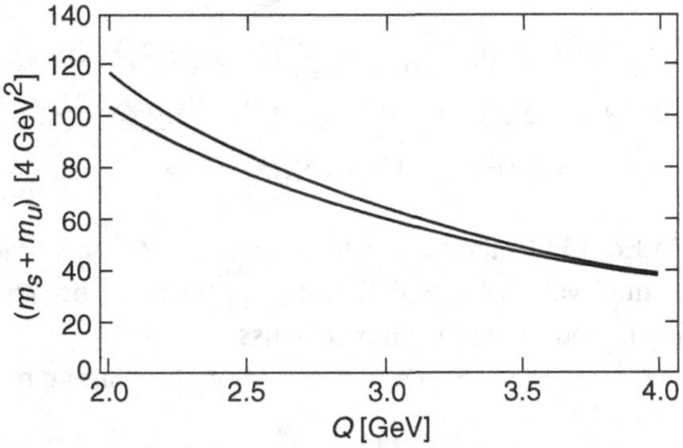


Fig. 53.2. The same curves as those in Fig. 53.1 but from the quadratic inequality to order $\alpha_s$.

moment of the two-point function are (see previous part of the book):

$$\Psi_5''(q^2) = \frac{3}{8\pi^2}\frac{(\bar{m}_u + \bar{m}_s)^2}{Q^2}\left[1 + \frac{11}{3}\left(\frac{\bar{\alpha}_s}{\pi}\right) + 14.179\left(\frac{\bar{\alpha}_s}{\pi}\right)^2 + 77.368\left(\frac{\bar{\alpha}_s}{\pi}\right)^3\right]. \quad (53.10)$$

At this scale, the PT series converges quite well and behaves as:

$$\text{Parton}[1 + 0.45 + 0.22 + 0.15]\,. \quad (53.11)$$

Including these higher order corrections, the bounds become:

$$(m_s + m_u)(2) > (71.4 \pm 3.7)\,\text{MeV} \quad \Longrightarrow \quad (m_u + m_d)(2) > (5.9 \pm 0.3)\,\text{MeV}\,, \quad (53.12)$$

The bound will be saturated in the extreme limit where the continuum contribution to the spectral function is neglected.

The quadratic inequality in Eq. (53.7) results in improved lower bounds for the quark masses which we show in Fig. 53.2.

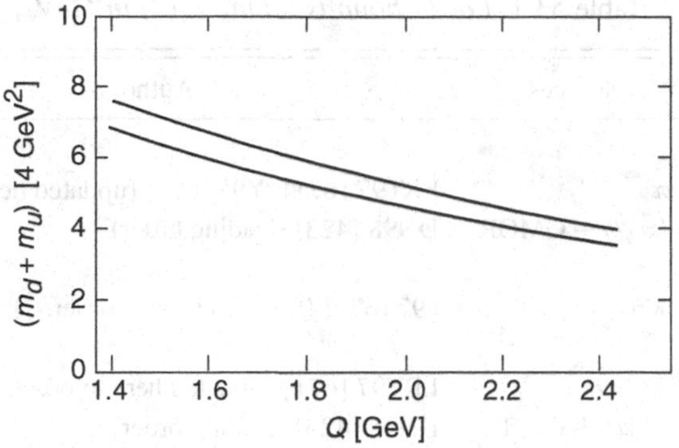


Fig. 53.3. Lower bound in MeV for $(m_d + m_u)(2)$ from the quadratic inequality to order $\alpha_s$.

The quadratic bound is saturated for a $\delta$–like spectral function representation of the hadronic continuum of states at an arbitrary position and with an arbitrary weight. This is certainly less restrictive than the extreme limit with the full hadronic continuum neglected, and it is therefore not surprising that the quadratic bound happens to be better than those from $\Sigma_N(Q^2)$ for $N = 0$, 1, and 2. Notice however that the quadratic bound in Fig. 53.2 is plotted at higher $Q$-values than the bound in Fig. 53.1. This is due to the fact that the coefficients of the perturbative series in $\alpha_s(Q^2)$ become larger for the higher moments. In [671] it is shown that for the evaluation of the quadratic bound $Q \geq 2$ GeV is already a safe choice.

Similar bounds can be obtained for $(m_u + m_d)$ when one considers the two–point function associated with the divergence of the axial current:

$$\partial_\mu A^\mu(x) = (m_d + m_u) :\bar{d}(x) i\gamma_5 u(x): . \tag{53.13}$$

The method to derive the bounds is exactly the same as the one discussed above and therefore we only show, in Fig. 53.3 below, the results for the corresponding lower bounds which one obtains from the quadratic inequality. At $Q = 2$ GeV, one can deduce the lower bounds from the quadratic inequality:

$$(m_s + m_u)(2) > 105 \text{ MeV}, \qquad (m_u + m_d)(2) > 7 \text{ MeV}. \tag{53.14}$$

The convergence of the QCD series is less good here than in the lowest moment. It behaves as [669]:

$$\text{Parton}\left[1 + \frac{25}{3}\left(\frac{\bar{\alpha}_s}{\pi}\right) + 61.79\left(\frac{\bar{\alpha}_s}{\pi}\right)^2 + 517.15\left(\frac{\bar{\alpha}_s}{\pi}\right)^3\right], \tag{53.15}$$

which numerically reads:

$$\text{Parton}[1 + 0.83 + 0.61 + 0.51]. \tag{53.16}$$

Table 53.1. *Lower bounds on $\bar{m}_{u,d,s}(2)$ in MeV*

| Observables | Sources | Authors |
|---|---|---|
| $\bar{m}_u + \bar{m}_d$ | | |
| 6 | $\pi$ | LRT97 [671], Y97 [672] (updated here to order$\alpha_s^3$) |
| 6.8 | $\langle\bar{\psi}\psi\rangle$ + GMOR | DN98 [423] (leading order) |
| $\bar{m}_d - \bar{m}_u$ | | |
| 1.1 | $K\pi$ | Y97 [672] (updated here to order$\alpha_s^3$) |
| $\bar{m}_s$ | | |
| 71.4 | $K$ | LRT97 [671] (updated here to order$\alpha_s^3$) |
| 90 | $\langle\bar{\psi}\psi\rangle$ + ChPT | DN98 [423] (leading order) |

This leads to the radiatively corrected lower bound to order $\alpha_s^3$:

$$(m_s + m_u)\,(2) > (82.7 \pm 13.3)\ \text{MeV}\ , \qquad (m_u + m_d)(2) > (6 \pm 1)\ \text{MeV}\ , \quad (53.17)$$

where the error is induced by the truncation of the QCD series which we have estimated to be about the contribution of the last known $\alpha_s^3$ term of the series.[4] From the previous analysis, and taking into account the uncertainties induced by the higher order QCD corrections, *the best lower bound* comes from the lowest inequality and is given in Eq. (53.12). The result is summarized in Table 53.1.

### 53.3.2 Lower bound on the light quark mass difference from the scalar sum rule

As in [666], one can extract lower bound on the light quark mass difference $(m_u - m_d)$ and $(m_u - m_s)$ working with the two-point function associated to the divergence of the vector current:

$$\partial_\mu V^\mu_{\bar{u}q} = (m_u - m_q) : \bar{\psi}_u(i)\psi_q : \ . \quad (53.18)$$

The most recent analysis has been done in [672]. We have updated the result by including the $\alpha_s^3$-term. It is given in Table 53.1.

### 53.3.3 Bounds on the sum of light quark masses from the quark condensate and $e^+e^- \to I = 0$ hadrons data.

Among the different results in [423], we shall use the range of the chiral $\langle\bar{\psi}\psi\rangle \equiv \langle\bar{u}u\rangle \simeq \langle\bar{d}d\rangle$ condensate from the vector form factor of $D \to K^* l\nu$. Using three-point function sum

[4] In [668], alternative bound has been derived using a Hölder type inequality. The lower bound obtained from this method, which is about 4.2 MeV is weaker than the one obtained previously.

rules, the form factor reads to leading order:

$$
\begin{aligned}
V(0) = & \frac{m_c(m_D + m_{K^*})}{4 f_D f_{K^*} m_D^2 m_{K^*}} \exp\left[(m_D^2 - m_c^2)\tau_1 + m_{K^*}^2 \tau_2\right] \\
& \times \langle\bar{\psi}\psi\rangle \Bigg\{ -1 + M_0^2 \left( -\frac{\tau_1}{3} + \frac{m_c^2}{4}\tau_1^2 + \frac{2m_c^2 - m_c m_s}{6}\tau_1\tau_2 \right) \\
& - \frac{16\pi}{9}\alpha_s \rho \langle\bar{\psi}\psi\rangle \left( \frac{2m_c}{9}\tau_1\tau_2 - \frac{m_c^3}{36}\tau_1^3 \right. \\
& \left. - \frac{2m_c^3 - m_c^2 m_s}{36}\tau_1^2\tau_2 + \frac{-m_c}{9}\tau_1^2 + \frac{2m_s}{9}\tau_2^2 + \frac{2}{9}m_s\tau_1\tau_2 + \frac{4}{9}\frac{\tau_2}{m_c} \right) \\
& + \frac{e^{m_c^2\tau_1}}{\langle\bar{\psi}\psi\rangle} \int_0^{s_{20}} ds_2 \int_{s_2+m_c^2}^{s_{10}} ds_1\, \rho_v(s_1, s_2) e^{-s_1\tau_1 - s_2\tau_2} \Bigg\}
\end{aligned}
\tag{53.19}
$$

$$
\begin{aligned}
\text{with}\,\rho_v(s_1, s_2) = & \frac{3}{4\pi^2 (s_1 - s_2)^3} \left\{ m_s \left( (s_1 + s_2)(s_1 - m_c^2) - 2s_1 s_2 \right) \right. \\
& \left. + m_c \left( (s_1 + s_2)s_2 - 2s_2(s_1 - m_c^2) \right) \right\} .
\end{aligned}
\tag{53.20}
$$

The factor $\rho \simeq 2 \sim 3$ expresses the uncertainty in the factorization of the four quark condensate. In our numerical analysis, we start from standard values of the QCD parameters and use $f_{K^*} = 0.15$ GeV($f_\pi = 93.3$ MeV). The value of $f_D \simeq (1.35 \pm 0.07) f_\pi$ is consistently determined by a two-point function sum rule including radiative corrections as we shall see in the next chapter, where the sum rule expression can, for example, be found in [3]. The following parameters enter only marginally: $m_s(1\ \text{GeV}) = (0.15 \sim 0.19)$ GeV, $s_{10} = (5 \sim 7)$ GeV$^2$, $s_{20} = (1.5 \sim 2)$ GeV$^2$. Using the conservative range of the charm quark mass: $m_c$(pole) between 1.29 and 1.55 GeV (the lower limit comes from the estimate in [3] and the upper limit is one-half of the $J/\Psi$ mass), one can deduce the running condensate value at 1 GeV [423]:

$$
0.6 \leq \langle\bar{\psi}\psi\rangle / [-225\ \text{MeV}]^3 \leq 1.5 . \tag{53.21}
$$

This result has been confirmed by the lattice [674]. Using the GMOR relation:

$$
2m_\pi^2 f_\pi^2 = -(m_u + m_d)\langle\bar{u}u + \bar{d}d\rangle + \mathcal{O}(m_q^2) . \tag{53.22}
$$

one can translate the upper bound into a lower bound on the sum of light quark masses. The lower bound on the chiral condensate can be used in conjunction with the positivity of the $m_q^2$ correction in order to give an upper bound to the quark mass value. In this way, one obtains:

$$
6.8\ \text{MeV} \leq (\bar{m}_u + \bar{m}_d)(2\ \text{GeV}) \leq 11.4\ \text{MeV} . \tag{53.23}
$$

The resulting values are quoted in Tables 53.1 and 53.2. We expect that these bounds are satisfied within the typical 10% accuracy of the sum rule approach.

We also show in Table 53.2 the upper bound obtained in [354] by using the positivity of the spectral function from the analysis of the $e^+e^- \to I = 0$ hadrons data where the determination will be discussed in the next section.

Table 53.2. *Upper bounds on $\bar{m}_{u,d,s}$ (2) in MeV*

| Observables | Sources | Authors |
|---|---|---|
| $\bar{m}_u+\bar{m}_d$ | | |
| 11.4 | $\langle\bar{\psi}\psi\rangle$ + GMOR | DN98 [423] (leading order) |
| $\bar{m}_s$ | | |
| 148 | $\langle\bar{\psi}\psi\rangle$ + ChPT | DN98 [423] (leading order) |
| $147\pm21$ | $e^+e^-+\tau$-decay | SN99 [354] (to order$\alpha_s^3$) |

## 53.4 Sum of light quark masses from pseudoscalar sum rules

### *53.4.1 The (pseudo)scalar Laplace sum rules*

The Laplace sum rule for the (pseudo)scalar two-point correlator reads (see [3,167,376, 400,582]:

$$\int_0^{t_c} dt\,\exp(-t\tau)\frac{1}{\pi}\mathrm{Im}\Psi_{(5)}(t)\simeq(\bar{m}_u\pm\bar{m}_d)^2\frac{3}{8\pi^2}\tau^{-2}\left[(1-\rho_1)\left(1+\delta_\pm^{(0)}\right)+\sum_{n=2}^{6}\delta_\pm^{(n)}\right], \tag{53.24}$$

where the indices 5 and + refer to the pseudoscalar current. Here, $\tau$ is the Laplace sum rule variable, $t_c$ is the QCD continuum threshold and $\bar{m}_i$ is the running mass to three loops:

$$\rho_1\equiv(1+t_c\tau)\exp(-t_c\tau)\,. \tag{53.25}$$

Using the results compiled in the previous chapter, the perturbative QCD corrections read for $n$ flavours:

$$\begin{aligned}\delta_\pm^{(0)}=&\left(\frac{\bar{\alpha}_s}{\pi}\right)\left[\frac{11}{3}-\gamma_1\gamma_E\right]\\&+\left(\frac{\bar{\alpha}_s}{\pi}\right)^2\left[\frac{10801}{144}-\frac{39}{2}\zeta(3)-\left(\frac{65}{24}-\frac{2}{3}\zeta(3)\right)n\right.\\&-\frac{1}{2}\left(1-\gamma_E^2\right)\left[\frac{17}{3}(2\gamma_1-\beta_1)+2\gamma_2\right]\\&\left.+\left(3\gamma_E^2-6\gamma_E-\frac{\pi^2}{2}\right)\frac{\gamma_1}{12}(2\gamma_1-\beta_1)\right]\\\delta_\pm^{(2)}=&-2\tau\left[\left[1+\left(\frac{\bar{\alpha}_s}{\pi}\right)C_F(4+3\gamma_E)\right]\left(\bar{m}_i^2+\bar{m}_j^2\right)\right.\\&\left.\mp\left[1+\left(\frac{\bar{\alpha}_s}{\pi}\right)C_F(7+3\gamma_E)\right]\bar{m}_i\bar{m}_j\right],\end{aligned} \tag{53.26}$$

where $C_F = 4/3$ and $\gamma_E = 0.5772\ldots$ is the Euler constant; $\gamma_1$, $\gamma_2$ and $\beta_1$, $\beta_2$ are respectively the mass-anomalous dimensions and $\beta$-function coefficients defined in a previous chapter. For three colours and three flavours, they read:

$$\gamma_1 = 2\,, \quad \gamma_2 = 91/12\,, \quad \beta_1 = -9/2\,, \quad \beta_2 = -8\,. \tag{53.27}$$

In practice, the perturbative correction to the sum rule simplifies as:

$$\delta_{\pm}^{(0)} = 4.82a_s + 21.98a_s^2 + 53.14a_s^3 + \mathcal{O}(a_s^4) \quad : \quad a_s \equiv \left(\frac{\bar{\alpha}_s}{\pi}\right). \tag{53.28}$$

Introducing the RGI condensates defined in the previous chapter, the non-perturbative contributions are [325]:

$$\begin{aligned}
\delta_{\pm}^{(4)} = \frac{4\pi^2}{3}\tau^2 \Bigg[ & \frac{1}{4}\left\langle\frac{\alpha_s}{\pi}G^2\right\rangle - \frac{\gamma_1}{\beta_1}\left(\frac{\bar{\alpha}_s}{\pi}\right)\sum_i \overline{\langle m_i\bar{\psi}_i\psi_i\rangle} - \frac{3}{8\pi^2}\frac{1}{(4\gamma_1+\beta_1)}\sum_i \bar{m}_i^4 \\
& + \left[1 + \left(\frac{\bar{\alpha}_s}{\pi}\right)C_F\left(\frac{11}{4} + \frac{3}{2}\gamma_E\right)\right]\left(\overline{\langle m_j\bar{\psi}_j\psi_j\rangle} + \overline{\langle m_i\bar{\psi}_i\psi_i\rangle}\right) \\
& \mp \left[2 + \left(\frac{\bar{\alpha}_s}{\pi}\right)C_F\,(7+3\gamma_E)\right]\left(\overline{\langle m_i\bar{\psi}_j\psi_j\rangle} + \overline{\langle m_j\bar{\psi}_i\psi_i\rangle}\right) \\
& - \frac{3}{2\pi^2}\left[\frac{1}{(4\gamma_1+\beta_1)}\left[\frac{\pi}{\bar{\alpha}_s} + C_F\left(\frac{11}{4} + \frac{3}{2}\gamma_E\right) + \frac{1}{6}(4\gamma_1+\beta_1)\right.\right. \\
& \left.\left. - \frac{1}{4\gamma_1}(4\gamma_2+\beta_2)\right] - \frac{1}{4}(1-2\gamma_E)\right]\left(\bar{m}_i^4 + \bar{m}_j^4\right) - \frac{3}{2\pi^2}\bar{m}_j^2\bar{m}_i^2 \\
& \pm \left[\frac{1}{(4\gamma_1+\beta_1)}\left[\frac{2\pi}{\bar{\alpha}_s} + \frac{1}{3}(4\gamma_1+\beta_1) - \frac{1}{2\gamma_1}(4\gamma_2+\beta_2) + C_F\,(7+3\gamma_E)\right]\right. \\
& \left. + \gamma_E\right]\left(\bar{m}_j^3\bar{m}_i + \bar{m}_i^3\bar{m}_j\right)\Bigg]\,, \\
\delta_{\pm}^{(6)} = \mp\frac{8\pi^2}{3}\tau^3 \Bigg[ & \frac{1}{2}\left[m_j\left\langle\bar{\psi}_i\sigma^{\mu\nu}\frac{\lambda_a}{2}G^a_{\mu\nu}\psi_i\right\rangle + m_i\left\langle\bar{\psi}_j\sigma^{\mu\nu}\frac{\lambda_a}{2}G^a_{\mu\nu}\psi_j\right\rangle\right] \\
& - \frac{16}{27}\pi\bar{\alpha}_s[\langle\bar{\psi}_j\psi_j\rangle^2 + \langle\bar{\psi}_i\psi_i\rangle^2 \mp 9\langle\bar{\psi}_j\psi_j\rangle\langle\bar{\psi}_i\psi_i\rangle]\Bigg]\,.
\end{aligned} \tag{53.29}$$

Beyond the SVZ expansion, one can have two contributions:

- The direct instanton contribution can be obtained from [386] and reads:

$$\delta_{+}^{\text{inst}} = \frac{\rho_c^2}{\tau^3}\exp(-r_c)\,[K_0(r_c) + K_1(r_c)] \tag{53.30}$$

with: $r_c \equiv \rho_c^2/(2\tau)$; $\rho_c \approx 1/600\ \text{MeV}^{-1}$ being the instanton radius; $K_i$ is the MacDonald function. However, one should notice that analogous contribution in the scalar channel leads to some contradictions ([386] and private communication from Valya Zakharov).

- The tachyon gluon mass contribution can be deduced from [161]:

$$\delta_{\pm}^{\text{tach}} = -4\left(\frac{\bar{\alpha}_s}{\pi}\right)\lambda^2 , \tag{53.31}$$

where $(\alpha_s/\pi)\lambda^2 \simeq -0.06$ GeV$^2$ [161],

which completes the different QCD contributions to the two-point correlator.

### *53.4.2 The $\bar{u}d$ channel*

From the experimental side, we do not still have a complete measurement of the pseudoscalar spectral function. In the past [3], one has introduced the radial excitation $\pi'$ of the pion using a NWA where the decay constant has been fixed from chiral symmetry argument [57] and from the pseudoscalar sum rule analysis itself [422,675,3], through the quantity:

$$r_\pi \equiv \frac{M_{\pi'}^4 f_{\pi'}^2}{m_\pi^4 f_\pi^2} . \tag{53.32}$$

Below the QCD continuum $t_c$, the spectral function is usually saturated by the pion pole and its first radial excitation and reads:

$$\int_0^{t_c} dt \exp(-t\tau)\frac{1}{\pi}\text{Im}\Psi_5(t) \simeq 2m_\pi^4 f_\pi^2 \exp\left(-m_\pi^2\tau\right)\left[1 + r_\pi \exp\left[\left(m_\pi^2 - M_{\pi'}^2\right)\tau\right]\right] . \tag{53.33}$$

The theoretical estimate of the spectral function enters through the not yet measured ratio $r_\pi$. Detailed discussions of the sum rule analysis can be found in [3,420,422]. However, this channel is quite peculiar due to the Goldstone nature of the pion, where the value of the sum rule scale ($1/\tau$ for Laplace and $t_c$ for FESR) is relatively large, being about 2 GeV$^2$, compared with the pion mass where the duality between QCD and the pion is lost. Hopefully, this paradox can be cured by the presence of the new $1/q^2$ [162,161,394] due to the tachyonic gluon mass, which enlarges the duality region to lower scale and then minimizes the role of the higher states into the sum rule. This naïve NWA parametrization has been improved in [676] by the introduction of threshold effect and finite width corrections. Within the advent of ChPT, one has been able to improve the previous parametrization by imposing constraints consistent with the chiral symmetry of QCD [677]. In this way, the spectral function reads:

$$\frac{1}{\pi}\text{Im}\Psi_5(t) \simeq 2m_\pi^4 f_\pi^2 \left[\delta\left(t - m_\pi^2\right) + \theta\left(t - 9m_\pi^2\right)\frac{1}{\left(16\pi^2 f_\pi^2\right)^2}\frac{t}{18}\rho^{3\pi}(t)\right] , \tag{53.34}$$

with:

$$\begin{aligned}
\rho^{3\pi}(t) = &\int_{4m_\pi^2}^{(\sqrt{t}-m_\pi)^2} \frac{du}{t}\sqrt{\lambda\left(1, \frac{u}{t}, \frac{m_\pi^2}{t}\right)}\sqrt{1 - \frac{4m_\pi^2}{u}}\left\{5 + \frac{1}{2\left(t - m_\pi^2\right)^2}\right. \\
&\times\left[\frac{4}{3}\left[t - 3\left(u - m_\pi^2\right)\right]^2 + \frac{8}{3}\lambda\left(t, u, m_\pi^2\right)\left(1 - \frac{4m_\pi^2}{u}\right) + 10m_\pi^4\right] \\
&\left. + \frac{1}{\left(t - m_\pi^2\right)}\left[3\left(u - m_\pi^2\right) - t + 10m_\pi^2\right]\right\} ,
\end{aligned} \tag{53.35}$$

where $\lambda(a, b, c) = a^2 + b^2 + c^2 - 2ab - 2bc - 2ca$ is the usual phase space factor. Based on this parametrization but including finite width corrections, a recent re-analysis of this sum rule has been given to order $\alpha_s^2$ [677]. Result from the LSR is, in general, expected to be more reliable than the one from the FESR due to the presence of the exponential factor which suppresses the high-energy tail of the spectral function, although the two analysis are complementary. In [677], FESR has been used for matching by duality the phenomenological and theoretical parts of the sum rule. This matching has been achieved in the energy region around 2 GeV$^2$, where the optimal value of $m_u + m_d$ has been extracted. In [54], the LSR analysis has been updated by including the $\alpha_s^3$ correction obtained in [442]. In this way, we get:

$$(\bar{m}_u + \bar{m}_d)(2\ \text{GeV}) = (9.3 \pm 1.8)\ \text{MeV}\ , \tag{53.36}$$

where we have converted the original result obtained at the *traditional* 1 GeV to the lattice choice of scale of 2 GeV through:

$$\bar{m}_i(1\ \text{GeV}) \simeq (1.38 \pm 0.06)\ \bar{m}_i(2\ \text{GeV})\ , \tag{53.37}$$

for running, to order $\alpha_s^3$, the results from 1 to 2 GeV. This number corresponds to the average value of the QCD scale $\Lambda_3 \simeq (375 \pm 50)$ MeV from PDG [16] and [139]. Analogous value of $(9.8 \pm 1.9)$ MeV for the quark mass has also been obtained in [678] to order $\alpha_s^3$ as an update of the [677] result. We take as a final result the average from [54] and [678]:

$$(\bar{m}_u + \bar{m}_d)(2\ \text{GeV}) = (9.6 \pm 1.8)\ \text{MeV}\ . \tag{53.38}$$

The inclusion of the tachyonic gluon mass term reduces this value to [161]:

$$\Delta^{\text{tach}}(\bar{m}_u + \bar{m}_d)(2\ \text{GeV}) \simeq -0.5\ \text{MeV}\ . \tag{53.39}$$

As already mentioned, adding to this effect the one of direct instanton might lead to a double counting in a sense that they can be alternative ways for parametrizing the non-perturbative QCD vacuum. Considering this contribution as another source of errors, it gives:

$$\Delta^{\text{inst}}(\bar{m}_u + \bar{m}_d)(2\ \text{GeV}) \simeq -0.5\ \text{MeV}\ . \tag{53.40}$$

Therefore, adding different sources of errors, we deduce from the analysis:

$$(\bar{m}_u + \bar{m}_d)(2\ \text{GeV}) = (9.6 \pm 1.8 \pm 0.4 \pm 0.5 \pm 0.5)\ \text{MeV}\ , \tag{53.41}$$

leading to the conservative result for the sum of light quark masses:

$$(\bar{m}_u + \bar{m}_d)(2\ \text{GeV}) = (9.6 \pm 2.0)\ \text{MeV}\ . \tag{53.42}$$

The first error comes from the SR analysis, the second one comes from the running mass evolution and the two last errors come, respectively, from the (eventual) tachyonic gluon and direct instanton contributions. This result is in agreement with previous determinations [3,419–423,679,680], although we expect that the errors given there have been underestimated. One can understand that the new result is lower than the old result [3,420] obtained

Table 53.3. *QSSR determinations of $\bar{m}_s$ (2) in MeV to order $\alpha_s^3$. Some older results have been updated by the inclusion of the higher-order terms. The error contains the evolution from 1 to 2 GeV. In addition, the errors in the (pseudo)scalar channels contain those due to the small size instanton and tachyonic gluon mass. Their quadratic sum increases the original errors by 8.9%. The estimated error in the average comes from an arithmetic average of the different errors*

| Channels | $\bar{m}_s$ (2) | Comments | Authors |
|---|---|---|---|
| Pion SR + ChPT | $117.1 \pm 25.4$ | $\mathcal{O}(\alpha_s^3)$ | SN99 [54]Eq. (53.43) |
| $\langle\bar{\psi}\psi\rangle$ + ChPT | $129.3 \pm 23.2$ | $N,\ B - B^*$ (l.o) | DN98 [423] Eq. (53.52) |
| | $117.1 \pm 49.0$ | $D \to K^* l\nu$ (l.o) | DN98 [423] Eq. (53.53) |
| Kaon SR | $119.6 \pm 18.4$ | updated to $\mathcal{O}(\alpha_s^3)$ | SN89 [420,3] |
| | $112.3 \pm 23.2$ | $\mathcal{O}(\alpha_s^3)$ | DPS99[681] |
| | $116 \pm 12.8$ | " | KM01 [669] |
| Scalar SR | $148.9 \pm 19.2$ | $\mathcal{O}(\alpha_s^3)$ | CPS97 [443] |
| | $103.6 \pm 15.4$ | " | CFNP97 [682] |
| | $115.9 \pm 24.0$ | " | J98 [683] |
| | $115.2 \pm 13.0$ | " | M99 [684] |
| | $99 \pm 18.3$ | " | JOP01 [685] |
| $\tau$-like $\phi$ SR: $e^+$-$e^-$ + $\tau$-decay | $129.2 \pm 25.6$ | average: $\mathcal{O}(\alpha_s^3)$ | SN99[354] |
| $\Delta S = -1$ part of $\tau$-decay | $169.5^{+46.7}_{-57}$ | $\mathcal{O}(\alpha_s^2)$ | ALEPH99* [348] |
| | $144.9 \pm 38.4$ | " | CKP98 [349] |
| | $114 \pm 23$ | " | PP99 [350] |
| | $125.7 \pm 25.4$ | " | KKP00 [351] |
| | $115 \pm 21$ | " | KM01 [352] |
| | $116^{+20}_{-25}$ | " | CDGHKK01[353] |
| **Average** | $117.4 \pm 23.4$ | | |

* Not included in the average.

without the $\alpha_s^2$ and $\alpha_s^3$ terms as both corrections enter with a positive sign in the LSR analysis. However, it is easy to check that the QCD perturbative series converge quite well in the region where the optimal result from LSR is obtained. Combining the previous value in Eq. (53.42) with the ChPT mass ratio, one can also deduce:

$$\bar{m}_s(2\ \text{GeV}) = (117.1 \pm 25.4)\ \text{MeV}\ . \tag{53.43}$$

### 53.4.3 The $\bar{u}s$ channel and QSSR prediction for the ratio $m_s/(m_u + m_d)$

Doing analogous analysis for the kaon channel, one can also derive the value of the sum $(m_u + m_s)$. The results obtained from [420] updated to order $\alpha_s^3$ and from [681] are shown in Table 53.3 given in [54] but updated. We add to the original errors the one from the tachyonic gluon (5.5%), from the direct instanton (5.5%) and the one due to the evolution from 1 to 2 GeV (4.4%), which altogether increases the original errors by 8.9%. Therefore, we deduce the (arithmetic) average from the kaon channel:

$$\bar{m}_s(2\ \text{GeV}) = (116.0 \pm 18.1)\ \text{MeV}\,, \qquad (53.44)$$

One should notice here that, unlike the case of the pion, the result is less sensitive to the contribution of the higher states continuum due to the relatively higher value of $M_K$, although the parametrization of the spectral function still gives larger errors than the QCD series. It is interesting to deduce from Eqs. (53.42) and (53.44), the sum rule prediction for the scale invariant quark mass ratios:

$$r_3 \equiv \frac{2m_s}{m_u + m_d} \simeq 24.2\,, \qquad (53.45)$$

where we expect that the ratio is more precise than the absolute values due to the cancellation of the systematics of the SR method. This ratio compares quite well with the ChPT ratio [57]:

$$r_3^{CA} = 24.4 \pm 1.5\,, \qquad (53.46)$$

and confirms the self-consistency of the pseudoscalar SR approach. This is a non-trivial test of the SR method used in this channel and may confirm a posteriori the neglect of less controlled contributions like direct instantons for example.

### 53.5 Direct extraction of the chiral condensate $\langle \bar{u}u \rangle$

As mentioned in previous section, the chiral $\bar{u}u$ condensate can be extracted directly from the nucleon, $B^*$-$B$ splitting and vector form factor of $D \to K^* l\nu$, which are particularly sensitive to it and to the mixed condensate $\langle \bar{\psi}\sigma^{\mu\nu}(\lambda_a/2)G^a_{\mu\nu}\psi \rangle \equiv M_0^2 \langle \bar{\psi}\psi \rangle$ [423]. We have already used the result from the $D \to K^* l\nu$ form factor in order to derive upper and lower bounds on $(m_u + m_d)$. Here, we shall use information from the nucleon and from the $B^*$-$B$ splitting in order to give a more accurate estimate. In the nucleon sum rules [424–430,3], which seem, at first sight, a very good place for determining $\langle \bar{\psi}\psi \rangle$, we have two form factors for which spectral sum rules can be constructed, namely the form factor $F_1$ which is proportional to the Dirac matrix $\gamma\, p$ and $F_2$ which is proportional to the unit matrix. In $F_1$ the four quark condensates play an important role, but these are not chiral symmetry breaking and are related to the condensate $\langle \bar{\psi}\psi \rangle$ only by the factorization hypothesis [1] which is known to be violated by a factor of two to three [424,404,3]. The form factor $F_2$ is dominated by the condensate $\langle \bar{\psi}\psi \rangle$ and the mixed condensate $\langle \bar{\psi}\sigma G\psi \rangle$, such that the baryon mass is essentially determined by the ratio $M_0^2$ of the two condensates:

$$M_0^2 = \langle \bar{\psi}\sigma G\psi \rangle / \langle \bar{\psi}\psi \rangle\,. \qquad (53.47)$$

Therefore, from the nucleon sum rules one gets quite a reliable determination of $M_0^2$ [430,426]:

$$M_0^2 = (.8 \pm .1)\ \text{GeV}^2. \qquad (53.48)$$

A sum rule based on the ratio $F_2/F_1$ would in principle be ideally suited for a determination of $\langle\bar\psi\psi\rangle$ but this sum rule is completely unstable [426] due to fact that odd parity baryonic excitations contribute with different signs to the spectral functions of $F_1$ and $F_2$. In the correlators of heavy mesons ($B$, $B^*$ and $D$, $D^*$) the chiral condensate gives a significant direct contribution in contrast to the light meson sum rules [3], since, here, it is multiplied by the heavy quark mass. However, the dominant contribution to the meson mass comes from the heavy quark mass and therefore a change of a factor two in the value of $\langle\bar\psi\psi\rangle$ leads only to a negligible shift of the mass. However, from the $B$-$B^*$ splitting one gets a precise determination of the mixed condensate $\langle\bar\psi\sigma G\psi\rangle$ with the value [401]:

$$\langle\bar\psi\sigma G\psi\rangle = -(9\pm 1)\times 10^{-3}\ \mathrm{GeV}^5\ , \tag{53.49}$$

which combined with the value of $M_0^2$ given in Eq. (53.48) gives our first result for the value of $\langle\bar\psi\psi\rangle$ at the nucleon scale:

$$\langle\bar\psi\psi\rangle(M_N) = -[(225\pm 9\pm 9)\ \mathrm{MeV}]^3\ , \tag{53.50}$$

where the last error is our estimate of the systematics and higher-order contributions. Using the GMOR relation, one can translate the previous result into a prediction on the sum of light quark masses. The resulting value is [423]:

$$(\bar m_u+\bar m_d)(2\ \mathrm{GeV}) = (10.6\pm 1.8\pm 0.5)\ \mathrm{MeV}\ , \tag{53.51}$$

where we have added the second error due to the quark mass evolution. Combining this value with the ChPT mass ratio, one obtains:

$$\bar m_s(2\ \mathrm{GeV}) \simeq 129.3\pm 23.2\ \mathrm{MeV}\ . \tag{53.52}$$

Alternatively, one can use the central value of the range given in Eq. (53.21) in order to deduce the estimate:

$$(\bar m_u+\bar m_d)(2\ \mathrm{GeV}) = (9.6\pm 4\pm 0.4)\ \mathrm{MeV} \quad\Longrightarrow\quad \bar m_s(2\ \mathrm{GeV}) \simeq (117.1\pm 49.0)\ \mathrm{MeV}\ . \tag{53.53}$$

The results for $m_s$ are shown in Table 53.3.

## 53.6 Final estimate of $(m_u+m_d)$ from QSSR and consequences on $m_u$, $m_d$ and $m_s$

One can also notice the impressive agreement of the previous results from pseudoscalar and from the other channels. As the two results in Eqs. (53.42), (53.51) and (53.53) come from completely independent analysis, we can take their geometric average and deduce *the final value from QSSR*:

$$(\bar m_u+\bar m_d)(2\ \mathrm{GeV}) = (10.1\pm 1.3\pm 1.3)\ \mathrm{MeV}\ , \tag{53.54}$$

where the last error is our estimate of the systematics. One can combine this result with the one for the light quark mass ratios from ChPT [57]:

$$r_2^{CA} \equiv \frac{m_u}{m_d} = 0.553 \pm 0.043\,, \qquad r_3^{CA} \equiv \frac{2m_s}{(m_d + m_u)} = 24.4 \pm 1.5\,. \tag{53.55}$$

Therefore, one can deduce the running masses at 2 GeV:

$$\bar{m}_u(2) = (3.6 \pm 0.6)\ \text{MeV}\,, \quad \bar{m}_d(2) = (6.5 \pm 1.2)\ \text{MeV}\,, \quad \bar{m}_s(2) = (123.2 \pm 23.2)\ \text{MeV}\,. \tag{53.56}$$

Alternatively, we can use the relation between the invariant mass $\hat{m}_q$ and running mass $\bar{m}_q(2)$ to order $\alpha_s^3$ in order to get:

$$\hat{m}_q = (1.14 \pm 0.05)\,\bar{m}_q(2)\,, \tag{53.57}$$

for $\Lambda_3 = (375 \pm 50)$ MeV. Therefore, one can deduce the invariant masses:

$$\hat{m}_u = (4.1 \pm 0.7)\ \text{MeV}\,, \qquad \hat{m}_d = (7.4 \pm 1.4)\ \text{MeV}\,, \qquad \hat{m}_s = (140.4 \pm 26.4)\ \text{MeV}\,. \tag{53.58}$$

## 53.7 Light quark mass from the scalar sum rules

As can be seen from Eq. (53.24), one can also (in principle) use the isovector-scalar sum rule for extracting the quark mass-differences $(m_d - m_u)$ and $(m_s - m_u)$, and the isoscalar-scalar sum rules for extracting the sum $(m_d + m_u)$.

### *53.7.1 The scalar $\bar{u}d$ channel*

In the isovector channel, the analysis relies heavily on the less controlled nature of the $a_0(980)$ [3,420,422,666], which has been speculated to be a four-quark state [73]. However, it appears that its $\bar{q}q$ nature is favoured by the present data [690], and further tests are needed for confirming its real $\bar{q}q$ assignment.

In the $I = 0$ channel, the situation of the $\pi$-$\pi$ continuum is much more involved due to the possible gluonium nature of the low mass and wide $\sigma$ meson [686,687,689,688,690], which couples strongly to $\pi$-$\pi$ and then can be missed in the quenched lattice calculation of scalar gluonia states.

Assuming that these previous states are quarkonia states, bounds on the quark mass difference and sum of quark masses have been derived in [666,671,672], while an estimate of the sum of the quark masses has been recently derived in [691]. However, in view of the hadronic uncertainties, we expect that the results from the pseudoscalar channels are much more reliable than the ones obtained from the scalar channel. Instead, we think that it is more useful to use these sum rules the other way around. Using the values of the quark masses from the pseudoscalar sum rules and their ratio from ChPT, one can extract their decay constants, which are useful for testing the $\bar{q}q$ nature of the scalar resonances [3,688] (we shall come back to this point in the next section). The agreement of the values of the quark masses from the isovector scalar channel with the ones from the pseudoscalar channel can be interpreted as a strong indication for the $\bar{q}q$ nature of the $a_0(980)$. In the isoscalar channel, the value of the sum of light quark masses obtained recently in [691], although slightly lower, agrees within the errors with that from the pseudoscalar channel. This result supports the maximal quarkonium-gluonium scheme for the broad low mass $\sigma$ and narrow $f_0(980)$ meson: the narrowness of the $f_0$ is due to a destructive interference, while the broad nature of the $\sigma$ is due to a contructive interference allowing its strong coupling with $\pi$-$\pi$. These features are very important for the scalar meson phenomenology, and need to be tested further.

### 53.7.2 The scalar $\bar{u}s$ channel

Here, the analysis is mostly affected by the parametrization of the $K\pi$ phase shift data, which strongly affects the resulting value of the strange quark mass as can be seen from the different determinations given in Table 53.3.

## 53.8 Light quark mass difference from $(M_{K^+} - M_{K^0})_{\rm QCD}$

The mass difference $(m_d - m_u)$ can be related to the QCD part of the kaon mass difference $(M_{K^+} - M_{K^0})_{\rm QCD}$ from the current algebra relation [57]:

$$r_2^{CA} \equiv \frac{(m_d - m_u)}{(m_d + m_u)} = \frac{m_\pi^2}{M_K^2} \frac{\left(M_{K^0}^2 - M_{K^+}^2\right)_{\rm QCD}}{M_K^2 - m_\pi^2} \frac{m_s^2 - \hat{m}^2}{(m_u + m_d)^2} = (0.52 \pm 0.05)10^{-3}\left(r_3^2 - 1\right), \tag{53.59}$$

where $2\hat{m} = m_u + m_d$; the QCD part of the $K^+ - K^0$ mass difference comes from the estimate of the electromagnetic term using the Dashen theorem including next-to-leading chiral corrections [677]. Using the sum rule prediction of $r_3$ from the ratio of $(m_u + m_d)$ in Eq. (53.54) with the average value of $m_s$ in Table 53.3 or the ChPT ratio given in the previous section, one can deduce to order $\alpha_s^3$:

$$(\bar{m}_d - \bar{m}_u)\,(2\ \text{GeV}) = (2.8 \pm 0.6)\ \text{MeV}\,. \tag{53.60}$$

An analogous result has been obtained from the heavy-light meson mass-differences [692]. We shall come back to the values of these masses at the end of this chapter.

## 53.9 The strange quark mass from $e^+e^-$ and $\tau$ decays

### 53.9.1 $e^+e^- \to I = 0$ hadrons data and the $\phi$-meson channel

Its extraction from the vector channel has been done in [693,3] and more recently in [354], while its estimate from an improved Gell-Mann–Okubo mass formula, including the quadratic mass corrections, has been done in [32,399,3]. More recently, the vector channel has been re-analysed in [354] using a $\tau$-like inclusive decay sum rule in a modern version of the Das–Mathur–Okubo (DMO) sum rule [27] discussed in a previous chapter. The analysis in this vector channel is interesting as we have complete data from $e^+e^-$ in this channel, which is not the case of (pseudo) scalar channels where some theoretical inputs related to the realization of chiral symmetry have to be used in the parametrization of spectral function. One can combine the $e^+e^- \to I = 0,\ 1$ hadrons and the rotated recent $\Delta S = 0$ component of the $\tau$-decay data in order to extract $m_s$. Unlike previous sum rules, one has the advantage to have a complete measurement of the spectral function in the region covered by the analysis. We shall work with:

$$R_{\tau,\phi} \equiv \frac{3|V_{ud}|^2}{2\pi\alpha^2} S_{EW} \int_0^{M_\tau^2} ds \left(1 - \frac{s}{M_\tau^2}\right)^2 \left(1 + \frac{2s}{M_\tau^2}\right) \frac{s}{M_\tau^2} \sigma_{e^+e^- \to \phi,\phi',\ldots} ,$$

and the $SU(3)$-breaking combinations [354]:

$$\Delta_{1\phi} \equiv R_{\tau,1} - R_{\tau,\phi}, \qquad \Delta_{10} \equiv R_{\tau,1} - 3R_{\tau,0} , \tag{53.61}$$

which vanish in the $SU(3)$ symmetry limit; $\Delta_{10}$ involves the difference of the isoscalar ($R_{\tau,0}$) and isovector ($R_{\tau,1}$) sum rules à la DMO. The PT series converges quite well at the optimization scale of about 1.6 GeV [354]. For example, normalized to $\bar{m}_s^2$, one has:

$$\Delta_{1\phi} \simeq -12 \frac{\bar{m}_s^2}{M_\tau^2} \left\{ 1 + \frac{13}{3} a_s + 30.4 a_s^2 + (173.4 \pm 109.2) a_s^3 \right\}$$
$$+ 36 \frac{\bar{m}_s^4}{M_\tau^2} - 36 \alpha_s^2 \frac{\langle m_s \bar{s}s - m_d \bar{d}d \rangle}{M_\tau^4} . \tag{53.62}$$

The different combinations $\Delta_{1\phi}$ and $\Delta_{10}$ have the advantage to be free (to leading order) from flavour-blind combinations like the tachyonic gluon mass and instanton contributions. We have checked using the result in [161] that, to non-leading in $m_s^2$, the tachyonic gluon contribution is also negligible. It has been argued in [355] that $\Delta_{10}$ can be affected by large $SU(2)$ breakings. This claim has been tested using some other sum rules not affected by these terms [354] but has not been confirmed. The average from different combinations is given in Table 53.3. An upper bound deduced from the positivity of $R_{\tau,\phi}$ is also given in Table 53.2.

### 53.9.2 Tau decays

As in the case of $e^+e^-$, one can use tau decays for extracting the value of $m_s$. However, data from $\tau$ decays are more accurate than those from $e^+e^-$. A suitable combination of sum rules that are sensitive to leading order to the $SU(3)$ breaking parameter is needed. It is easy to construct such a combination which is very similar to the one for $e^+e^-$. One can work with the DMO-like sum rule involving the difference between the $\Delta S = 0$ and $\Delta S = -1$ processes [348–353]:

$$\delta R_\tau^{kl} \equiv \frac{R_{\tau,V+A}^{kl}}{|V_{ud}|^2} - \frac{R_{\tau,S}^{kl}}{|V_{us}|^2} = 3 S_{EW} \sum_{D\geq 2} \left\{ \delta_{ud}^{kl(D)} - \delta_{us}^{kl(D)} \right\} , \tag{53.63}$$

where the moments are defined as:

$$R_\tau^{kl} \equiv \int_0^{M_\tau^2} ds \left(1 - \frac{s}{M_\tau^2}\right)^k \left(\frac{s}{M_\tau^2}\right)^l \frac{dR_\tau}{ds} , \tag{53.64}$$

with $R_\tau^{00} \equiv R_\tau$ is the usual $\tau$-hadronic width. The QCD expression reads:

$$\delta R_\tau^{kl} \simeq 24 S_{EW} \left\{ \frac{\bar{m}_s^2}{M_\tau^2} \Delta_{kl}^{(2)} - 2\pi^2 \frac{\langle m_s \bar{s}s - m_d \bar{d}d \rangle}{M_\tau^4} \Delta_{kl}^{(4)} \right\} , \tag{53.65}$$

where $\Delta_{kl}^{(D)}$ are perturbative coefficients known to order $\alpha_s^2$:

$$\Delta_{kl}^{(D)} \equiv \frac{1}{4} \{ 3\Delta_{kl}^{(D)}|_{\mathrm{L+T}} + \Delta_{kl}^{(D)}|_{\mathrm{L}} \} , \tag{53.66}$$

where the indices T and L refer to the tranverse and longitudinal parts of the spectral functions. For $D = 2$, the L part converges quite badly while the L + T converge quite well such that the combination can still have an acceptable convergence. For the lowest moments, one has:

$$\begin{aligned} \Delta_{00}^{(2)} &= 0.973 + 0.481 + 0.372 + 0.337 + \cdots \\ \Delta_{10}^{(2)} &= 1.039 + 0.558 + 0.482 + 0.477 + \cdots \\ \Delta_{20}^{(2)} &= 1.115 + 0.643 + 0.608 + 0.647 + \cdots \end{aligned} \tag{53.67}$$

The authors advocate that although the convergence is quite bad, the behaviour of the series is typical for an asymptotic series close to their point of minimum sensistivity. Therefore, the mathematical procedure for doing a reasonable estimate of the series is to truncate the expansion where the terms reach their minimum value. However, the estimate of the errors is still arbitrary. The authors assume that the error is given by the last term of the series. The result of the analysis is given in Table 53.3. The different numbers given in the table reflects the difference of methods used to get $m_s$ but the results are consistent with each other within the errors. As in the case of the $e^+e^-$ DMO-like sum rule, the combination used here is not affected to leading order by flavour-blind contribution like the tachyonic gluon and instanton contribution. We have checked [161] that the contribution of the tachyonic gluon to order $m_s^2 \alpha_s \lambda^2 / M_\tau^2$ gives a tiny correction and does not affect the estimate done without the inclusion of this term.

### 53.9.3 Summary for the estimate of light quark masses

Here, we summarize the results from the previous analysis:

- The sum $(\bar{m}_u + \bar{m}_d)$ of the running up and down quark masses from the pion sum rules is given in Eq. (53.42), while the one of the strange quark mass from the kaon channel is given in Eq. (53.44). Their values lead to the pseudoscalar sum rules prediction for the mass ratio in Eq. (53.45) which agrees nicely with the ChPT mass ratio.

- The sum $(\bar{m}_u + \bar{m}_d)$ of the running up and down quark masses averaged from the pseudoscalar sum rule and from a direct extraction of the chiral condensate $\langle \bar{u}u \rangle$ obtained from a global fit of the nucleon, $B^*$-$B$ mass-splitting and the vector part of the $D^* \to K^* l\nu$ form factor is given in Eq. (53.51) and reads for $\Lambda_3 = (375 \pm 50)$ MeV:

$$(\bar{m}_u + \bar{m}_d)(2\ \text{GeV}) = (10.1 \pm 1.8)\ \text{MeV}\ , \tag{53.68}$$

implying with the help of the ChPT mass ratio $m_u/m_d$, the value:

$$\bar{m}_u(2\ \text{GeV}) = (3.6 \pm 0.6)\ \text{MeV}\ , \qquad \bar{m}_d(2\ \text{GeV}) = (6.5 \pm 1.2)\ \text{MeV}\ , \tag{53.69}$$

which leads to the invariant mass in Eq. (53.58):

$$\hat{m}_u = (4.1 \pm 0.7)\ \text{MeV}\ , \qquad \hat{m}_d = (7.4 \pm 1.4)\ \text{MeV}\ , \tag{53.70}$$

- We have combined the result in Eq. (53.54) with the sum rule prediction for $m_s/(m_u + m_d)$ in order to deduce the quark mass difference $(m_d - m_u)$ from the QCD part of the $K^0 - K^+$ mass difference. We obtain the result in Eq. (53.60):

$$(\bar{m}_d - \bar{m}_u)(2\ \text{GeV}) = (2.8 \pm 0.6)\ \text{MeV}\ . \tag{53.71}$$

  This result indeed agrees with the one taking the difference of the mass given previously. The fact that $(m_u + m_d) \neq (m_d - m_u)$ does not favour the possibility of having $m_u = 0$.
- We give in Table 53.3 the different sum rules determinations of $m_s$. The results from the pion SR and $\langle \bar{\psi}\psi \rangle$ come from the determination of $(m_u + m_d)$ to which we have added the ChPT contraint on $m_s/(m_u + m_d)$. One can see from this table that different determinations are in good agreement with each others. Doing an average of these different results, we obtain:

$$\bar{m}_s(2\ \text{GeV}) = (117.4 \pm 23.4)\ \text{MeV} \quad \Longrightarrow \quad \hat{m}_s = (133.8 \pm 27.3)\ \text{MeV}\ . \tag{53.72}$$

  Aware on the possible correlations between these estimates, we have estimated the error as an arithmetic average which is about 10% as generally expected for the systematics of the SR approach.

It is informative to compare the above results with the average of different quenched and unquenched lattice values [694]:

$$\begin{aligned} \bar{m}_{ud}(2\ \text{GeV}) &\approx \frac{1}{2}(\bar{m}_u + \bar{m}_d)(2\ \text{GeV}) = (4.5 \pm 0.6 \pm 0.8)\ \text{MeV}\ , \\ \bar{m}_s(2\ \text{GeV}) &= (110 \pm 15 \pm 20)\ \text{MeV}\ , \end{aligned} \tag{53.73}$$

where the last error is an estimate of the quenching error. We show in Table 53.4 a compilation of the lattice unquenched results including comments on the lattice characterisitcs (action, lattice spacing $a$, $\beta$). Also shown is the ratio over $m_s/m_{ud}$ and quenched (quen) over unquenched (unq) results.

## 53.10 Decay constants of light (pseudo)scalar mesons

### 53.10.1 Pseudoscalar mesons

Due to the Goldstone nature of the pion and kaon, we have seen that their radial excitations play an essential rôle in the sum rule. This unusual property allows a determination of

Table 53.4. *Simulation details and physical results of unquenched lattice calculations of light-quark masses from [694], where original references are quoted*

| | Action | $a^{-1}$[GeV] | $\#_{(\beta,K_{sea})}$ | $Z_m$ | $\bar{m}_s(2)$ | | $\frac{m_s}{m_{ud}}$ | $\frac{\bar{m}_s^{\rm quen}}{\bar{m}_s^{\rm unq}}$ |
|---|---|---|---|---|---|---|---|---|
| SESAM 98 | Wilson | 2.3 | 4 | PT | 151(30) | $(m_{K,\phi})$ | 55(12) | 1.10(24) |
| MILC 99 | Fatlink | 1.9 | 1 | PT | 113(11)<br>125(9) | $(m_K)$<br>$(m_\phi)$ | 22(4) | 1.08(13) |
| APE 00 | Wilson | 2.6 | 2 | NP-RI | 112(15)<br>108(26) | $(m_K)$<br>$(m_\phi)$ | 26(2) | 1.09(20) |
| CP-PACS 00 | MF-Clover | $a \to 0$ | 12 | PT | $88^{+4}_{-6}$<br>$90^{+5}_{-11}$ | $(m_K)$<br>$(m_\phi)$ | 26(2) | 1.25(7) |
| JLQCD 00 | NP-Clover | 2.0 | 5 | PT | $94(2)^\dagger$<br>$88(3)^\ddagger$<br>$109(4)^\dagger$<br>$102(6)^\ddagger$ | $(m_K)$<br>$(m_\phi)$ | – | – |
| QCDSF + UKQCD 00 | NP-Clover | 2.0 | 6 | PT | 90(5) | $(m_K)$ | 26(2) | – |

$^\dagger$ From vector WI; $^\ddagger$ from axial WI. The errors on the ratios $m_s/m_{ud}$ and $\bar{m}_s^{\rm quen}/\bar{m}_s^{\rm unq}$ are estimates based on the original data.

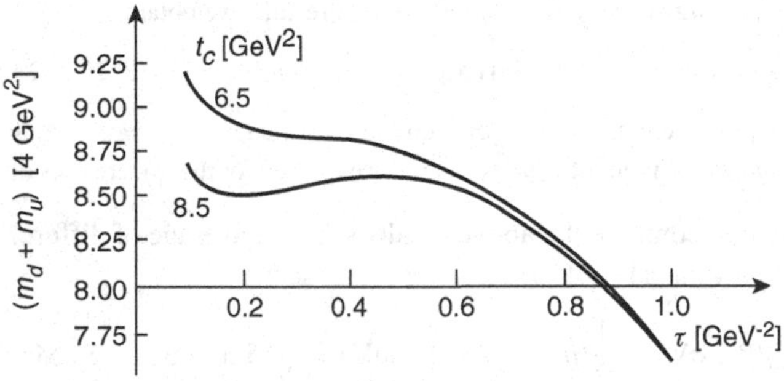


Fig. 53.4. LSR analysis of the ratio $r_\pi \equiv M^4_{\pi'} f^2_{\pi'}/M^4_\pi f^2_\pi$. For a given value $r_\pi = 9.5$, we show the value of $(\bar{m}_d + \bar{m}_u)(2)$ for two values of the QCD continuum $t_c$.

the radial excitation parameters. In the strange quark channels, an update of the results in [354,422,420,3,675] gives:

$$r_K \equiv M^4_{K'} f^2_{K'}/M^4_K f^2_K \simeq 9.5 \pm 2.5 \simeq r_\pi \ , \tag{53.74}$$

where $r_\pi$ has been defined previously. The optimal value has been obtained at the LSR scale $\tau \approx 0.5$ GeV$^{-2}$ and $t_c \simeq 4.5 - 6.5$ GeV$^2$ as shown in Fig. 53.4. This result implies for $\pi'(1.3)$ and $K'(1.46)$:

$$f_{\pi'} \simeq (3.3 \pm 0.6)\ \text{MeV} \ , \qquad f_{K'} \simeq (39.8 \pm 7.0)\ \text{MeV} \ . \tag{53.75}$$

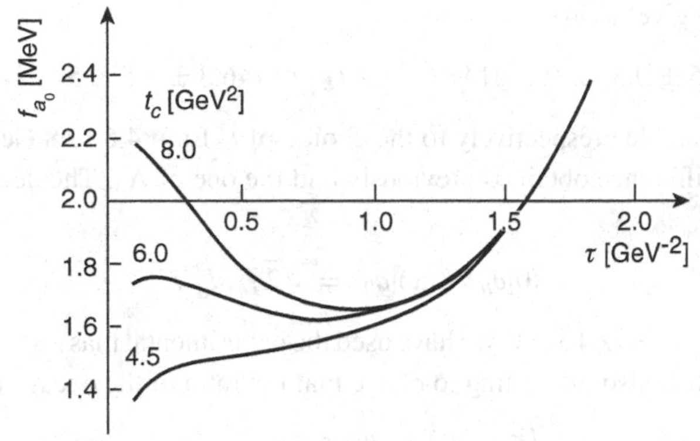


Fig. 53.5. LSR analysis of the decay constant $f_{a_0}$ of the $a_0(.98)$ meson normalized as $f_\pi = 92.4$ MeV. We use $(\bar{m}_d - \bar{m}_u)(2) = 2.8$ MeV.

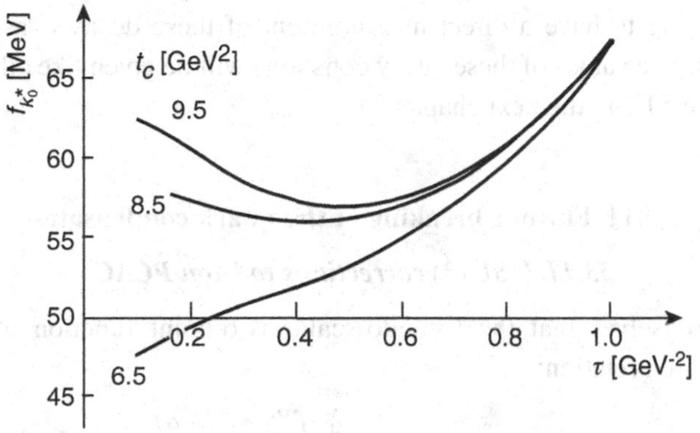


Fig. 53.6. LSR analysis of the decay constant $f_{K*0}$ of the $K_0^*(1.43)$ meson normalized as $f_\pi = 92.4$ MeV. We use $\bar{m}_s(2) = 117.4$ MeV.

It is easy to see that the result satisfies the relation:

$$\frac{f_{K'}}{f_{\pi'}} \approx \frac{M_K^2}{m_\pi^2} \approx \frac{m_s}{m_d}, \tag{53.76}$$

as expected from chiral symmetry arguments.

### 53.10.2 Scalar mesons

We expect that the scalar channel is more useful for giving the decay constants of the mesons which are not well known rather than predicting the value of the quark masses. Such a programme has been initiated in [420,422,3]. Since then, the estimate of the decay constants has not mainly changed. The analysis is shown in Figs. 53.5 and 53.6.

A recent estimate gives [354]:

$$f_{a_0} = (1.6 \pm 0.15 \pm 0.35 \pm 0.25)\,\mathrm{MeV}\,, \qquad f_{K_0^*} \simeq (46.3 \pm 2.5 \pm 5 \pm 5)\,\mathrm{MeV}\,, \quad (53.77)$$

where the errors are due respectively to the choice of $t_c$ from 4.5 to 8 GeV$^2$, the value of the quark mass difference obtained previously and the one of $\Lambda_3$. The decay constants are normalized as:

$$\langle 0|\partial_\mu V^\mu(x)|a_0\rangle = \sqrt{2} f_a M_a^2\,, \tag{53.78}$$

corresponding to $f_\pi = 92.4$ MeV. We have used the experimental masses 0.98 and 1.43 GeV in our analysis.[5] It is also interesting to notice that the ratio of the decay constants are:

$$\frac{f_{K_0^*}}{f_{a_0}} \simeq 29 \approx \frac{m_s - m_u}{m_d - m_u} \simeq 40\,, \tag{53.79}$$

as naïvely expected. We are aware that the values of these decay constants might have been overestimated due to the eventual proliferations of nearby radial excitations. Therefore, it will be interesting to have a direct measurement of these decay constants for testing these predictions. The values of these decay constants will be given like other meson decay constants in Table 54.1 in the next chapter.

## 53.11 Flavour breaking of the quark condensates

### *53.11.1* $SU(3)$ *corrections to kaon PCAC*

Let us remind ourselves that the (pseudo)scalar two-point function obeys the twice-subtracted dispersion relation:

$$\Psi_{(5)}(q^2) = \Psi_{(5)}(0) + q^2\Psi'_{(5)}(0) + q^4\int_0^\infty \frac{dt}{t^2(t-q^2-i\epsilon)}\mathrm{Im}\Psi_{(5)}(t)\,. \tag{53.80}$$

The deviation from kaon PCAC was first studied in [400] using the once-subtracted pseudoscalar sum rule of the quantity:

$$\frac{\Psi_{(5)}(q^2) - \Psi_{(5)}(0)}{q^2} \tag{53.81}$$

sensitive to the value of the value of the correlator at $q^2 = 0$.[6] The Ward identity obeyed by the (pseudo)scalar two-point function leads to the low-energy theorem:

$$\Psi_{(5)}(0) = -(m_i \pm m_j)\langle\bar\psi_i\psi_i \pm \bar\psi_j\psi_j\rangle\,, \tag{53.82}$$

in terms of the *normal ordered condensates*. However, as emphasized in different papers [167,399,444,442], $\Psi_{(5)}(0)$ contains a perturbative piece which cancels the mass singularities appearing in the OPE evaluation of $\Psi_{(5)}(q^2)$. This leads to the fact that the quark

[5] The masses of the $a_0$ and $K_0^*$ are also nicely reproduced by the ratio of moments [357,3].

[6] This sum rule has also been used in [260,261,265] for estimating the $U(1)_A$ topological suceptibility and its slope. The result has been confirmed on the lattice [266].

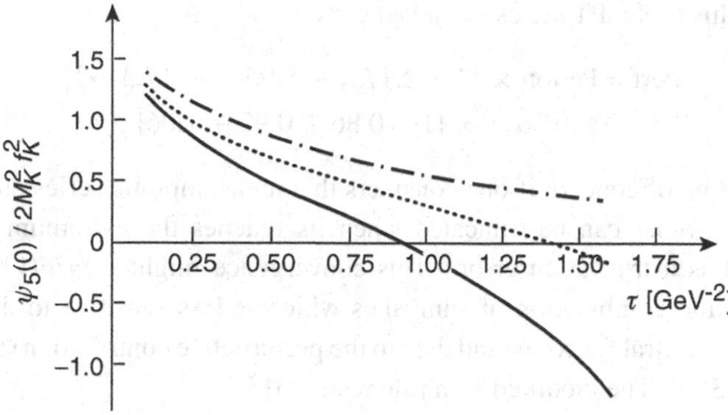


Fig. 53.7. LSR analysis of the subtraction constant $\Psi_5(0)$. We use $\bar{m}_s(2) = 117.4$ MeV, $r_K = 9.5$ and $t_c = 6\ \mathrm{GeV}^2$. The curves correspond to different truncations of the PT series: to $\mathcal{O}(\alpha_s)$: dotted-dashed; to $\mathcal{O}(\alpha_s^2)$: dashed; to $\mathcal{O}(\alpha_s^3)$: continuous.

condensate entering in Eq. 53.82 are defined as a *non-normal ordered condensate*, which has a slight dependence on the scale and renormalization scheme. This mass correction effect is only quantitatively relevant for the $\bar{u}s$ channel but not for the $\bar{u}d$ one. To order $\alpha_s^3$ for the perturbative term and to leading order for the condensates, the (pseudo)scalar sum rule for the $\bar{u}s$ channel reads, by neglecting the up quark mass:

$$\int_0^{t_c} \frac{dt}{t} \exp(-t\tau) \frac{1}{\pi} \mathrm{Im}\Psi_{(5)}(t) \simeq \Psi_{(5)}(0)$$
$$+ (\bar{m}_u \pm \bar{m}_s)^2 \frac{3}{8\pi^2}\tau^{-1} \left\{ (1-\rho_0) \left[ 1 + 6.82\left(\frac{\bar{\alpha}_s}{\pi}\right) + 58.55\left(\frac{\bar{\alpha}_s}{\pi}\right)^2 + 537.6\left(\frac{\bar{\alpha}_s}{\pi}\right)^3 \right] \right.$$
$$+ 3.15\bar{m}_s^2\tau \left[ 1 + 3.32\left(\frac{\bar{\alpha}_s}{\pi}\right) \right]$$
$$- \left[ \frac{\pi}{3}\langle\alpha_s G^2\rangle - \frac{8\pi^2}{3} \left[ \left( \bar{m}_s - \frac{\bar{m}_u}{2} \right) \langle\bar{u}u\rangle \pm (u \leftrightarrow s) \right] \right] \tau^2$$
$$\left. + \frac{1}{2}(2 \mp 9)\left(\frac{128}{81}\right) \pi^3 \rho \alpha_s \langle\bar{u}u\rangle^2 \tau^3 \right\}, \tag{53.83}$$

where we have neglected the $SU(3)$ breaking for the four-quark condensates. This assumption does not, however, affect the analysis due to the small contribution of this operator at the optimization scale. The analysis is shown in Fig. 53.7. Examining the different curves, on can notice that they deviate notably from the kaon PCAC prediction:

$$\Psi_5(0) \simeq 2M_K^2 f_K^2 , \tag{53.84}$$

therefore confirming the early findings in [400]. The LSR indicates a slight stability point at $\tau \approx (0.50 \sim 0.75)\ \mathrm{GeV}^{-2}$, where:

$$\Psi_5(0) \simeq (0.5 \pm 0.2) 2M_K^2 f_K^2 . \tag{53.85}$$

However, at this scale, PT series has a bad convergence:

$$\begin{aligned}\text{Pert} &= \text{Parton} \times \left\{1 + 2.17\alpha_s + 5.93\alpha_s^2 + 17.34\alpha_s^3\right\} \\ &\simeq \text{Parton} \times \{1 + 0.86 + 0.92 + 1.06\} ,\end{aligned} \tag{53.86}$$

which might not be of concern if one considers that an asymptotic series close to its point of *minimum sensitivity* can be truncated when its reaches the extremum value and the last term added as a truncation error.[7] This convergence might *a priori* be improved if one works with the combination of sum rules which is less sensitive to the high-energy behaviour of the spectral function (and then to the perturbative contribution) than the former [675,680,420,3,354]. The modified sum rule reads [3].[8]

$$\begin{aligned}&\int_0^\infty \frac{dt}{t}\exp(-t\tau)(1-t\tau)\frac{1}{\pi}\text{Im}\Psi_{(5)}(t) \simeq \Psi_{(5)}(0) + (\bar{m}_u \pm \bar{m}_s)^2\frac{3}{8\pi^2}\tau^{-1} \\ &\times\left\{2\left(\frac{\bar{\alpha}_s}{\pi}\right)\left[1 + 18.3\left(\frac{\bar{\alpha}_s}{\pi}\right) + 242.2\left(\frac{\bar{\alpha}_s}{\pi}\right)^2\right] + 5.15\bar{m}_s^2\tau\left[1 + 5.0\left(\frac{\bar{\alpha}_s}{\pi}\right)\right]\right. \\ &\left. + 2\left[\frac{\pi}{3}\langle\alpha_s G^2\rangle - \frac{8\pi^2}{3}\bar{m}_s\left[\langle\bar{u}u\rangle \mp \frac{1}{2}\langle\bar{s}s\rangle\right]\right]\tau^2 + \frac{3}{2}(2\mp 9)\left(\frac{128}{81}\right)\pi^3\rho\alpha_s\langle\bar{u}u\rangle^2\tau^3\right\} .\end{aligned} \tag{53.87}$$

The analysis also leads to a similar result. The LSR has been also studied recently in [695], by including threshold effects and higher mass resonances, which enlarge the region of stability in the LSR variable. Within the previous hadronic parametrization, one obtains:

$$\Psi_5(0) \simeq (0.56 \pm 0.04 \pm 0.15) 2M_K^2 f_K^2 , \tag{53.88}$$

where we have added the error due to our estimate of the truncation of the QCD PT series as deduced from Fig. 53.7. An alternative estimate is obtained with the use of FESR [679]. Parametrizing the subtraction constant as:

$$\Psi_5(0)_s^u = 2M_K^2 f_K^2 (1 - \delta_K) , \tag{53.89}$$

one has the sum rule [679]:

$$\delta_K \simeq \frac{3}{16\pi^2}\frac{\bar{m}_s^2 t_c}{f_K^2 M_K^2}\left\{1 + \frac{23}{3}a_s + \mathcal{O}\left(a_s^2\right)\right\} - r_K\left(\frac{M_K}{M_{K'}}\right)^2 , \tag{53.90}$$

which gives, after using the *correlated values* of the input parameters [420,3,354]:

$$\delta_K = 0.34^{+0.23}_{-0.17} , \tag{53.91}$$

[7] A similar argument has been used for the extraction of the strange quark mass from $\tau$-decay data discussed in the previous section, where the QCD series has also quite bad behaviour.

[8] Notice that we have not yet introduced the QCD continuum into the LHS of the sum rule.

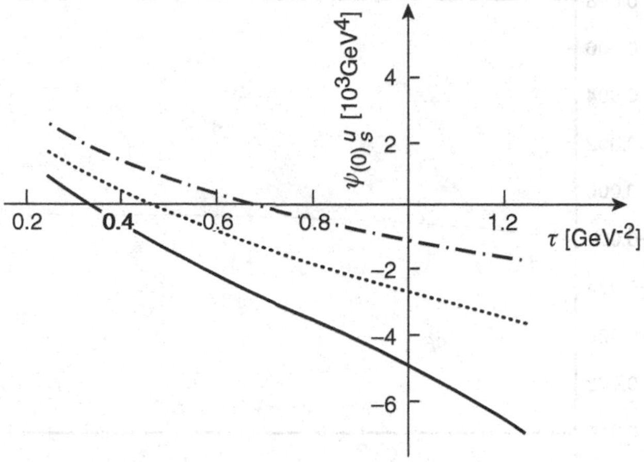


Fig. 53.8. LSR analysis of the subtraction constant $\Psi(0)$. We use $\bar{m}_s(2) = 117.4$ MeV, $f_{K*0} = 46$ MeV and $t_c = 6.5$ GeV$^2$. The curves correspond to different truncations of the PT series: to $\mathcal{O}(\alpha_s)$: dotted-dashed; to $\mathcal{O}(\alpha_s^2)$: dashed; to $\mathcal{O}(\alpha_s^3)$: continuous.

leading to:

$$\Psi_5(0) \simeq (0.66 \pm 0.20) 2M_K^2 f_K^2 \,, \tag{53.92}$$

confirming the large violation of kaon PCAC obtained from LSR.

### 53.11.2 Subtraction constant from the scalar sum rule

One can do a similar analysis for the scalar channel. The analysis from LSR is shown in Fig. 53.8. One can also see that there is a slight stability for $\tau \approx (0.50 \sim 0.75)$ GeV$^{-2}$, which gives:

$$\Psi(0) \approx -10^{-3}\ \text{GeV}^4 \,, \tag{53.93}$$

in agreement with previous results [3,420,422,675]. In [695], using LSR, a similar result but from a larger range of LSR stability, has been obtained within an Omnés representation for relating the scalar form factor to the $K\pi$ phase shift data:

$$\Psi(0) \simeq -(1.06 \pm 0.21 \pm 0.20)10^{-3}\ \text{GeV}^4 \,, \tag{53.94}$$

where the last term is our estimate of the error due to the truncation of the QCD series. We show the analysis in Fig. 53.9.

One can use an alternative approach by working with FESR:

$$\Psi(0)^u_s = 2M^2_{K_0^*} f^2_{K_0^*} - \frac{3}{16\pi^2}\bar{m}_s^2 t_c \left\{1 + \frac{23}{3}a_s + \mathcal{O}(a_s^2)\right\} , \tag{53.95}$$

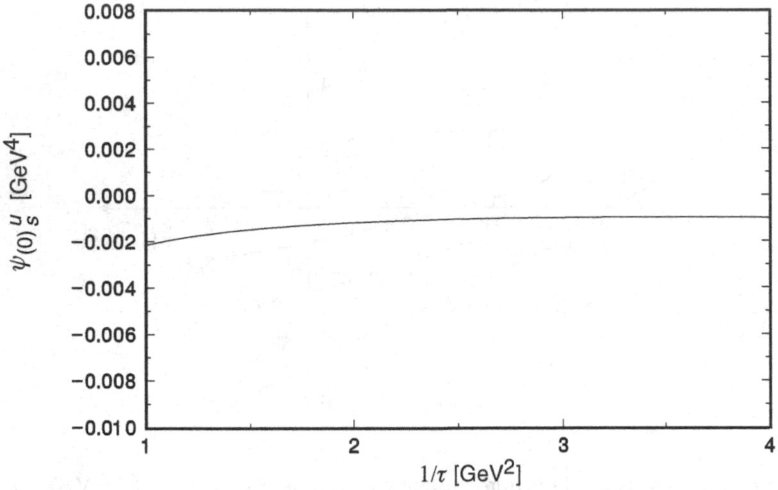


Fig. 53.9. LSR analysis of the subtraction constant $\Psi(0)$ versus the sum rule scale using $K\pi$ phase shift data, from [695].

which gives [354]:

$$\Psi(0)_s^u = -\left(7.8^{+5.5}_{-2.7}\right) 10^{-4}\ \mathrm{GeV}^4. \tag{53.96}$$

### 53.11.3 $\langle\bar{s}s\rangle/\langle\bar{u}u\rangle$ from the (pseudo)scalar sum rules

We take the arithmetic average of the previous determinations for our final estimate:

$$\Psi_5(0) \simeq (0.57 \pm 0.19) 2M_K^2 f_K^2\ , \qquad \Psi(0) \simeq -(0.92 \pm 0.35) 10^{-3}\ \mathrm{GeV}^4\ , \tag{53.97}$$

Taking the ratio of the scalar over the pseudoscalar subtraction constants expressed in terms of the *normal-ordered* condensates, one can deduce:

$$\langle\bar{s}s\rangle/\langle\bar{u}u\rangle = 0.57 \pm 0.12\ . \tag{53.98}$$

### 53.11.4 $\langle\bar{s}s\rangle/\langle\bar{u}u\rangle$ from the $B_s$ meson

One can also extract the flavour breakings of the condensates from a sum rule analysis of the $B_s$ and $B_s^*$ masses, which are sensitive to the chiral condensate as it enters like $m_b\langle\bar{s}s\rangle$ in the OPE of the heavy-light meson (see next section). The masses of the mesons are found to decrease linearly with the value of the chiral condensate. Using the observed value of the $B_s$ meson mass $M_{B_s} = 5.375$ GeV, one can deduce from Fig. 3 of [401]:

$$\langle\bar{s}s\rangle/\langle\bar{u}u\rangle \simeq 0.75 \pm 0.08\ , \tag{53.99}$$

where the error is the expected typical sum rule estimate. The effect of the strange quark mass is less important than the one here, such that the result given in [401] remains valid although obtained with slightly different values of $m_s$ and $\Lambda_5$. This estimate is expected to be more reliable than the one from the (pseudo)scalar light mesons, which are affected by the bad convergence of the PT QCD series. Using this value of ratio of the condensates in the (old) curve of the $B_s^*$ mass, leads to a value higher than the measured one, which needs to be clarified.

### 53.11.5 Final sum rule estimate of $\langle\bar{s}s\rangle/\langle\bar{u}u\rangle$

Using the previous results, one can deduce that the sum rules from the light (pseudo)scalar and from the $B_s$ meson predict for the *normal ordered* condensate ratio:

$$\langle\bar{s}s\rangle/\langle\bar{u}u\rangle \simeq 0.66 \pm 0.10\,, \tag{53.100}$$

confirming earlier findings [675,3,420,354] on the large flavour breaking of the chiral condensate. This number comes from the arithmetic average of the two values in Eqs. (53.98) and (53.99). If one instead works with the *non-normal ordered* condensate, one should add to the expression in Eq. (53.82) a small perturbative part first obtained by Becchi *et al.* [167] (see also [3,399,441]):

$$\langle\bar{s}s\rangle_{\overline{MS}} = \langle\bar{s}s\rangle - \frac{3}{2\pi^2}\frac{2}{7}\left(\frac{1}{a_s} - \frac{53}{24}\right)\bar{m}_s^3\,. \tag{53.101}$$

This leads to the ratio of the *non-normal ordered* condensates:

$$\langle\bar{s}s\rangle/\langle\bar{u}u\rangle|_{\overline{MS}} = 0.75 \pm 0.12. \tag{53.102}$$

The previous estimates are in good agreement with those from chiral perturbation theory [57] (see also [696]). They are also in fair agreement with the one from the baryonic sum rules [424–430], although we expect that the result from the latter is less accurate due to the complexity of the analysis in this channel (choice of the interpolating operators, eventual large effects of the continuum due to the nearby Roper resonances, . . . ).

### 53.11.6 $SU(2)$ breaking of the quark condensate

The $SU(2)$ breaking of the quark condensate has been studied for the first time in [680] and in [679,3]. Using similar approaches, the estimate is [3,420]:

$$\langle\bar{d}d\rangle/\langle\bar{u}u\rangle \simeq 1 - 9 \times 10^{-3}\,. \tag{53.103}$$

The previous estimate is in good agreement with the one from FESR [679].

## 53.12 Heavy quark masses

In the previous part of this book, we have already discussed the different definitions of the heavy quark masses and given their values. Contrary to the light quark masses, the definition of pole quark masses $p^2 = M_H^2$ can (in principle) be introduced perturbatively for heavy quarks [147,133,148], similarly to that of the electron, as here the quark mass is much heavier than the QCD scale $\Lambda$ such that the perturbative approach makes sense. However, a complication arises due to the resummation of the QCD series [154] such that the pole mass definition has an intrinsic ambiguity, which can be an obstacle to its improved accuracy determination, although the effect is relatively small. Alternative definitions that are free from such ambiguities have been proposed in the literature [157,159]. In this section, we shall discuss the determinations of the perturbative running quark masses which do not have such problems.

### 53.12.1 The quarkonia channel

Charmonium and bottomium are the standard channels for extracting the charm and bottom quark masses. Most of the sum rule analysis are based on the $Q^2 = 0$ moments (MOM) originally introduced by SVZ for the study of the charmonium systems:

$$\mathcal{M}_n \equiv \frac{1}{n!}\left(-\frac{d}{dQ^2}\right)^n \Pi\bigg|_{Q^2=0} = \int_{4m^2}^{\infty} \frac{dt}{t^{n+1}} \frac{1}{\pi} \text{Im}\Pi(t)\,, \tag{53.104}$$

but convenient for the bottomium systems due to a much better convergence of the OPE. In [357], the $Q^2 \neq 0$ moments have been introduced for improving the convergence of the QCD series:

$$\mathcal{M}_n(Q_0^2) \equiv \frac{1}{n!}\left(-\frac{d}{dQ^2}\right)^n \Pi\bigg|_{Q^2=Q_0^2} = \int_{4m^2}^{\infty} \frac{dt}{\left(t+Q_0^2\right)^{n+1}} \frac{1}{\pi} \text{Im}\Pi(t)\,, \tag{53.105}$$

The spectral function can be related to the $e^+e^- \to Q\bar{Q}$ total cross-section via the optical theorem:

$$\text{Im}\Pi(t+i\epsilon) = \frac{1}{12\pi Q_Q^2} \frac{\sigma(e^+e^- \to Q\bar{Q})}{\sigma(e^+e^- \to \mu^+\mu^-)}\,. \tag{53.106}$$

$Q_Q$ is the heavy quark charge in units of e. The contribution to the spectral function is as usual saturated by the lowest few resonances plus the QCD continuum above the threshold $t_c$:

$$\text{Im}\Pi_Q(t) = \frac{3}{4\alpha^2} \frac{1}{Q_Q^2} \sum_i \Gamma_i M_i \delta\left(t - M_i^2\right) + \theta(t - t_c)\text{Im}\Pi_Q^{\text{QCD}}(t), \tag{53.107}$$

where $\Gamma_i$ is the electronic width of the resonances with the value given in PDG [16]. Retaining the observed resonances, the value of $\sqrt{t_c}$ fixed from stability analysis is about $(11 \sim 12)$ GeV for the $\Upsilon$-and about 5 GeV for the $J/\Psi$-families. However, the result will be practically independent from this choice of $t_c$ due to the almost complete dominance of the lowest ground state to the spectral function at the stability point. An alternative approach that is used in [148,149] is the LSR:

$$\mathcal{L}(\tau) = \int_{4m^2}^{\infty} dt\, \exp^{-t\tau} \frac{1}{\pi} \text{Im}\Pi(t)\,. \tag{53.108}$$

This sum rule is particularly convenient for the analysis of the charmonium systems as the corresponding OPE converges faster than the moment sum rules. It has been noted in [149] that the ratios of sum rules (and their finite energy sum rule (FESR) variants) are more appropriate for the estimate of the quark mass as these ratios equate *directly* the mass squared of ground state to that of the quark:

$$\mathcal{R}_n \equiv \frac{\mathcal{M}^{(n)}}{\mathcal{M}^{(n+1)}} \quad \text{and} \quad \mathcal{R}_\tau \equiv -\frac{d}{d\tau}\log \mathcal{L}\,, \tag{53.109}$$

They also eliminate, to leading order, some artefact dependence due to the sum rules (exponential weight factor or number of derivatives) and some other systematic errors appearing in each of the individual moments. For the perturbative part, we shall use (without expanding in 1/M) the Schwinger interpolating formula to two loops:

$$\mathrm{Im}\Pi_Q^{pert}(t) \simeq \frac{3}{12\pi} v_Q \left(\frac{3 - v_Q^2}{2}\right)\left\{1 + \frac{4}{3}\alpha_s f(v_Q)\right\}, \tag{53.110}$$

where:

$$v_Q = \sqrt{1 - 4M_Q^2/t}\,, \qquad f(v_Q) = \frac{\pi}{2v_Q} - \frac{(3 + v_Q)}{4}\left(\frac{\pi}{2} - \frac{3}{4\pi}\right) \tag{53.111}$$

are respectively the quark velocity and the Schwinger function [319]. We express this spectral function in terms of the running mass by using the two-loops relation given in a previous chapter and including the $\alpha_s \log(t/M_Q^2)$-term appearing for off-shell quark. We shall add to this perturbative expression the lowest dimension $\langle\alpha_s G^2\rangle$ non-perturbative effect (it has been explained in a previous part of this book that, for a heavy-heavy quark correlator, the heavy-quark condensate contribution is already absorbed into the gluon one), which among the available higher-dimension condensate terms can only give a non-negligible contribution. The gluon condensate contribution to the moments $\mathcal{M}^{(n)}$ and so to $\mathcal{R}_n$ can be copied from the original work of SVZ [1] and reads:

$$\mathcal{M}_G^{(n)} = -\mathcal{M}_{pert}^{(n)} \frac{(n+3)!}{(n-1)!(2n+5)} \frac{4\pi}{9} \frac{\langle\alpha_s G^2\rangle}{\left(4M_Q^2\right)^2}\,, \tag{53.112}$$

where $\mathcal{M}_{pert}^{(n)}$ is the lowest perturbative expression of the moments. The one to the Laplace ratio $\mathcal{R}_\tau$ can be also copied from the original work of Bertlmann [93], which has been expanded recently in $1/M_Q$ by [697]. It reads:

$$\mathcal{R}_\tau^G \simeq \left(4M_Q^2\right)\frac{2\pi}{3}\langle\alpha_s G^2\rangle\tau^2\left(1 + \frac{4}{3\omega} - \frac{5}{12\omega^2}\right), \tag{53.113}$$

where $\omega \equiv 4M_Q^2\tau$. The results of the analysis from the ratios of moments and Laplace sum rules give the values of the running masses to order $\alpha_s$:[9]

$$\bar{m}_c(\bar{m}_c) = (1.23 \pm 0.03 \pm 0.03)\ \mathrm{GeV}\,, \qquad \bar{m}_b(\bar{m}_b) = (4.23 \pm 0.04 \pm 0.02)\ \mathrm{GeV}\,, \tag{53.114}$$

where the errors are respectively due to $\alpha_s(M_Z) = 0.118 \pm 0.006$ and $\langle\alpha_s G^2\rangle = (0.06 \pm 0.03)\ \mathrm{GeV}^4$ used in the original work. These running masses can be converted into the pole

[9] The inclusion of the $\alpha_s^2$ correction is under study.

Table 53.5. *QSSR direct determinations of $\bar{m}_c(\bar{m}_c)$ in $\overline{MS}$ scheme and of the pole mass $M_c$ from $J/\Psi$-family, $e^+e-$ data and D-meson and comparisons with lattice results. Determinations from some other sources are quoted in PDG [16]. The results are given in units of GeV. The estimated error in the SR average comes from an arithmetic average of the different errors. The average for the pole masses is given at NLO. The one of the running masses is almost unchanged from NLO to NNLO determinations. $\Longleftarrow$ means that the perturbative relations between the different mass definitions have been used to get the quoted values*

| Sources | $\bar{m}_c(\bar{m}_c)$ | $M_c$ | Comments | Authors |
|---|---|---|---|---|
| ***J*/Ψ-family** | | | | |
| MOM and LSR at NLO | $(1.27 \pm 0.02) \Longleftarrow$ | $(1.45 \pm 0.05)$ | $\Longleftarrow m\left(-m_c^2\right) = (1.26 \pm 0.02)$ | SN89 [148] |
| Ratio of LSR at NLO | $(1.23 \pm 0.04) \Longrightarrow$ | $(1.42 \pm 0.03)$ | | SN94 [149] |
| NRSR at NLO | $(1.23 \pm 0.04) \Longleftarrow$ | $(1.45 \pm 0.04)$ | | SN94 [149] |
| SR at NLO | $(1.22 \pm 0.06) \Longleftarrow$ | $(1.46 \pm 0.04)$ | | DGP94 [697] |
| NRSR at NNLO | $(1.23 \pm 0.09)$ | $(1.70 \pm 0.13)^*$ | | EJ01 [699] |
| **$e^+e^-$ data** | | | | |
| FESR at NLO | $(1.37 \pm 0.09)$ | | | PS01 [700] |
| MOM at NNLO | $(1.30 \pm 0.03)$ | | | KS01 [701] |
| NLO | $(1.04 \pm 0.04) \Longleftarrow$ | $1.33 \sim 1.4$ | | M01 [702] |
| ***D* meson** | | | | |
| Ratio of LSR at NNLO | $(1.1 \pm 0.04)$ | $(1.47 \pm 0.04)$ | | SN01 [150] |
| **SR average** | $(1.23 \pm 0.05)$ | $(1.43 \pm 0.04)$ | | |
| **Quenched lattice** | | | | |
| | $(1.33 \pm 0.08)$ | | | FNAL98 [703] |
| | $(1.20 \pm 0.23)$ | | | NRQCD99 [704] |
| | $(1.26 \pm 0.13)$ | | | APE01 [705] |

* Not included in the average.

masses at this order. Non-relativistic versions of these sum rules (NRSR) introduced by [155] have also been used in [148,149] for determining the $b$ quark mass. These NRSR approaches have been improved by the inclusion of higher-order QCD corrections and resummation of the Coulomb corrections from ladder gluonic exchanges. Some recent different determinations are given in Tables 53.5 and 53.6.

### *53.12.2 The heavy-light D and B meson channels*

Heavy quark masses can also be extracted from the heavy-light quark channels because the corresponding correlators are sensitive to leading order to the values of these masses

Table 53.6. *The same as in Table 53.5 but for the b-quark*

| Sources | $\bar{m}_b(\bar{m}_b)$ | | $M_b$ | Comments | Authors |
|---|---|---|---|---|---|
| **Υ-family** | | | | | |
| MOM and LSR at NLO | $(4.24 \pm 0.05)$ | $\Longleftarrow$ | $(4.67 \pm 0.10)$ | $\Longleftarrow m_b\left(-m_b^2\right) = (4.23 \pm 0.05)$ | SN89 [148] |
| Ratio of LSR at NLO | $(4.23 \pm 0.04)$ | $\Longrightarrow$ | $(4.62 \pm 0.02)$ | | SN94 [149] |
| NRSR at NLO | $(4.29 \pm 0.04)$ | $\Longleftarrow$ | $(4.69 \pm 0.03)$ | | SN94 [149] |
| FESR at NLO | $(4.22 \pm 0.05)$ | $\Longleftarrow$ | $(4.67 \pm 0.05)$ | | SN95 [149] |
| | $(4.14 \pm 0.04)$ | $\Longleftarrow$ | $(4.75 \pm 0.04)$ | | KPP98 [706] |
| NRSR at NNLO | $(4.20 \pm 0.10)$ | | | | PP99, MY99 [707] |
| MOM at NNLO | $(4.19 \pm 0.06)$ | | | | JP99 [708] |
| NR at NNNLO | $(4.45 \pm 0.04)$ | | | | PY00, LS00 [94,709] |
| NR at NNNLO | $(4.21 \pm 0.09)$ | | | | P01 [602] |
| NR at NNLO | $(4.25 \pm 0.08)$ | | | $\Longleftarrow$ Residual mass | BS99 [710] |
| NR at NNLO | $(4.20 \pm 0.06)$ | | | $\Longleftarrow$ 1$S$ mass | H00 [711] |
| MOM at NNNLO | $(4.21 \pm 0.05)$ | | | | KS01 [701] |
| ***B* and *B** mesons** | | | | | |
| Ratio of LSR at NLO | $(4.24 \pm 0.07)$ | $\Longleftarrow$ | $(4.63 \pm 0.08)$ | | SN94 [149] |
| Ratio of LSR at NNLO | $(4.05 \pm 0.06)$ | | $(4.69 \pm 0.06)$ | $B$-meson only | SN01 [150] |
| **SR average** | $(4.24 \pm 0.06)$ | | $(4.66 \pm 0.06)$ | $\Longrightarrow \bar{m}_b(M_Z) = (2.83 \pm 0.04)$ | |
| **Average LEP** | | | | | |
| Three-jets at $M_Z$ | $(4.23 \pm 0.94)$ | | | $\Longleftarrow \bar{m}_b(M_Z) = (2.82 \pm 0.63)$ | LEP [712] |
| **Unquenched lattice** | | | | | |
| | $(4.23 \pm 0.09)$ | | | | APE00 [713] |

[3,401,149,698,150]. Again, we shall be concerned here with the LSR $\mathcal{L}(\tau)$ and the ratio $\mathcal{R}(\tau)$. The latter sum rule, or its slight modification, is useful, as it is equal to the resonance mass squared, in the simple duality ansatz parametrization of the spectral function:

$$\frac{1}{\pi}\,\mathrm{Im}\psi_5(t) \simeq f_D^2 M_D^4 \delta\left(t - M_D^2\right) + \text{“QCD continuum”}\theta(t - t_c)\,, \tag{53.115}$$

where $f_D$ is the decay constant analogue to[10] $f_\pi = 130.56$ MeV. The QCD side of the sum rule reads:

$$\mathcal{L}_{\mathrm{QCD}}(\tau) = M_Q^2 \left\{ \int_{M_Q^2}^{\infty} dt\ \mathrm{e}^{-t\tau}\ \frac{1}{8\pi^2}\left[3t(1-x)^2\left(1 + \frac{4}{3}\left(\frac{\alpha_s}{\pi}\right)f(x)\right) + \left(\frac{\alpha_s}{\pi}\right)^2 R2s\right] \right.$$
$$\left. + \left[C_4\langle O_4\rangle + C_6\langle O_6\rangle\tau\right]\mathrm{e}^{-M_Q^2\tau} \right\}, \tag{53.116}$$

[10] Notice that we have adopted here the lattice normalization for avoiding confusion. We shall discuss its determination in the next chapter.

where $R2s$ is the new $\alpha_s^2$-term obtained semi-analytically in [448] and is available as a Mathematica package program Rvs.m. Neglecting $m_d$, the other terms are:

$$\begin{aligned}
x &\equiv M_Q^2/t\ ,\\
f(x) &= \frac{9}{4} + 2\mathrm{Li}_2(\mathrm{x}) + \log \mathrm{x} \log(1-\mathrm{x}) - \frac{3}{2}\log(1/\mathrm{x}-1)\\
&\quad - \log(1-x) + x\log(1/x-1) - (x/(1-x))\log x\,,\\
C_4\langle O_4\rangle &= - M_Q\langle \bar{d}d\rangle + \langle \alpha_s G^2\rangle/12\pi\\
C_6\langle O_6\rangle &= \frac{M_Q^3\tau}{2}\left(1-\frac{1}{2}M_Q^2\tau\right) g\left\langle \bar{d}\sigma_{\mu\nu}\frac{\lambda_a}{2}G_a^{\mu\nu}d\right\rangle\\
&\quad - \left(\frac{8\pi}{27}\right)\left(2-\frac{M_Q^2\tau}{2}-\frac{M_Q^4\tau^2}{6}\right)\rho\alpha_s\langle\bar{\psi}\psi\rangle^2\,. 
\end{aligned} \tag{53.117}$$

The previous sum rules can be expressed in terms of the running mass $\bar{m}_Q(\nu)$.[11] From this expression, one can easily deduce the expression of the ratio $\mathcal{R}(\tau)$, where the unknown decay constant disappears, and from which we obtain the running quark masses:

$$\bar{m}_c(m_c) = (1.10 \pm 0.04)\ \mathrm{GeV}\,. \tag{53.118}$$

The analysis is shown in Fig. 53.10, where a simultaneous fit of the decay constant from $\mathcal{L}$ and of $\bar{m}_c(\bar{m}_c)$ from $\mathcal{R}$ is shown.[12]

Our optimal results correspond to the case where both stability in $\tau$ and in $t_c$ are reached. However, for a more conservative estimate of the errors we allow deviations from the stability points, and we take:

$$t_c \simeq (6 \sim 9.5)\ \mathrm{GeV}^2\,, \qquad \tau \simeq (1.2 \pm 0.2)\ \mathrm{GeV}^{-2}\,, \tag{53.119}$$

and where the lowest value of $t_c$ corresponds to the beginning of the $\tau$-stability region. Values outside the above ranges are not consistent with the stability criteria. One can check that the dominant non-perturbative contribution is due to the dimension-four $M_c\langle\bar{d}d\rangle$ light quark condensate, and test that the OPE is not broken by high-dimension condensates at the optimization scale. However, the perturbative radiative corrections converge slowly, as the value of $f_D$ increases by 12% after the inclusion of the $\alpha_s$ correction and the sum of the lowest order plus $\alpha_s$-correction increases by 21% after the inclusion of the $\alpha_s^2$ term, indicating that the total amount of corrections of 21% is still a reasonnable correction despite the slow convergence of the perturbative series, which might be improved using a resummed series. However, as the radiative corrections are both positive, we expect that this slow convergence will not affect the final estimate in a significant way. A similar analysis is done for the pole mass. The discussions presented previously also apply here, including the one of the radiative corrections. We quote the final result:

$$M_c = (1.46 \pm 0.04)\ \mathrm{GeV}\,, \tag{53.120}$$

[11] It is clear that, for the non-perturbative terms which are known to leading order of perturbation theory, one can use either the running or the pole mass. However, we shall see that this distinction does not notably affect the present result.

[12] We shall discuss the decay constant in the next section.

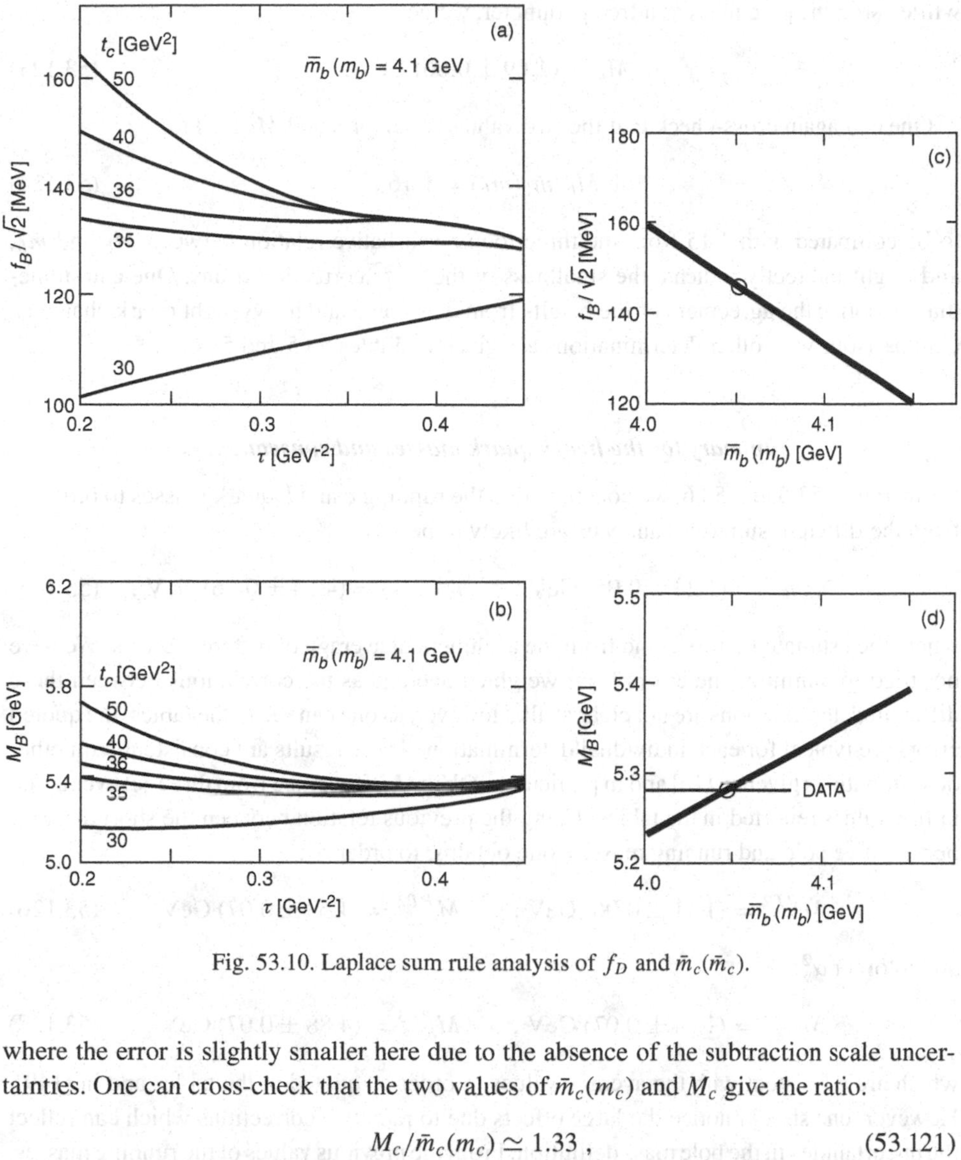


Fig. 53.10. Laplace sum rule analysis of $f_D$ and $\bar{m}_c(\bar{m}_c)$.

where the error is slightly smaller here due to the absence of the subtraction scale uncertainties. One can cross-check that the two values of $\bar{m}_c(m_c)$ and $M_c$ give the ratio:

$$M_c/\bar{m}_c(m_c) \simeq 1.33 \,, \tag{53.121}$$

which satisfies quite well the three-loop perturbative relation $M_c/\bar{m}_c(m_c) = 1.33$. This could be a non-trivial result if one has in mind that the quark pole mass definition can be affected by non-perturbative corrections not present in the standard SVZ-OPE. In particular, it may signal that $1/q^2$ correction of the type discussed in [162,161,394], if present, will only affect weakly the standard SVZ-phenomenology as observed explicitly in the light quark, gluonia and hybrid channels [161]. Using analogous analysis for the $B$ meson, we obtain at the optimization scale $\tau = 0.375$ GeV$^{-2}$ and $t_c = 38$ GeV$^2$:

$$\bar{m}_b(m_b) = (4.05 \pm 0.06) \text{ GeV} \,, \tag{53.122}$$

while using the pole mass as a free parameter, we get:

$$M_b = (4.69 \pm 0.06)\ \text{GeV}\ . \qquad (53.123)$$

One can again cross-check that the two values of $\bar{m}_b(m_b)$ and $M_b$ lead to:

$$M_b/\bar{m}_b(m_b) = 1.16\ , \qquad (53.124)$$

to be compared with 1.15 from the three-loop perturbative relation between $M_b$ and $\bar{m}_b$, and might indirectly indicate the smallness of the $1/q^2$ correction if any. One can immediately notice the agreement of the results from quarkonia and heavy-light quark channels. Comparisons with other determinations are given in Tables 53.5 and 53.6.

### *Summary for the heavy quark masses and consequences*

From Tables 53.5 and 53.6, we conclude that the running $c$ and $b$ quark masses to order $\alpha_s^2$ from the different sum rules analysis are likely to be:

$$\bar{m}_c(\bar{m}_c) = (1.23 \pm 0.05)\ \text{GeV}\ , \qquad \bar{m}_b(\bar{m}_b) = (4.24 \pm 0.06)\ \text{GeV}\ , \qquad (53.125)$$

where the estimated errors come from the arithmetical average of different errors. We have not tried to minimize the errors from weighted average as the correlations between these different determinations are not clear at all. However, as one can see in the tables, the quoted errors are typical for each individual determination. These results are consistent with other determinations given in [16] and in particular with the LEP average from three-jet events and lattice values reported in the tables. Using the previous relation between the short distance perturbative pole and running masses, one obtains, to order $\alpha_s$:

$$M_c^{PT2} = (1.41 \pm 0.06)\ \text{GeV}\ , \qquad M_b^{PT2} = (4.63 \pm 0.07)\ \text{GeV}\ , \qquad (53.126)$$

and to order $\alpha_s^2$:

$$M_c^{PT3} = (1.64 \pm 0.07)\ \text{GeV}\ , \qquad M_b^{PT3} = (4.88 \pm 0.07)\ \text{GeV}\ , \qquad (53.127)$$

which are consistent with the average values to order $\alpha_s$ quoted in the tables and in [149]. However, one should notice the large effects due to radiative corrections which can reflect the uncertainties in the pole mass definition. From the previous values of the running masses, one can also deduce the values of the RG invariant masses to order $\alpha_s^2$:

$$\hat{m}_c = (1.21 \pm 0.07)\ \text{GeV}\ , \qquad \hat{m}_b = (6.9 \pm 0.2)\ \text{GeV}\ . \qquad (53.128)$$

We have used $\Lambda_4 = 325 \pm 40$ MeV and $\Lambda_5 = 225 \pm 30$ MeV. Taking into account threshold effects and using matching conditions, we can also evaluate the running masses at the scale 2 GeV and obtain:

$$\bar{m}_c(2) = (1.23 \pm 0.05)\ \text{GeV}\ , \qquad \bar{m}_b(2) = (5.78 \pm 0.26)\ \text{GeV}\ . \qquad (53.129)$$

Combining the values of $m_b$ and $m_s$ obtained in the previous section, one can deduce the scale-independent mass ratio:

$$\frac{m_b}{m_s} = 48.8 \pm 9.8\,, \tag{53.130}$$

which is useful for model building.

One can also run the value of $m_b$ at the $Z$-mass, and obtains the value of $\bar{m}_b(M_Z)$ quoted in the table:

$$\bar{m}_b(M_Z) = (2.83 \pm 0.04)\ \text{GeV}. \tag{53.131}$$

This value compares quite well with the ones measured at $M_Z$ from three-jet heavy quark production at LEP where the average $(2.83 \pm 0.04)$ GeV of different measurements [712] is also given in the table. This is a *first indication* for the running of $m_b$ in favour of the QCD predictions based on the renormalization group equation.

## 53.13 The weak leptonic decay constants $f_{D_{(s)}}$ and $f_{B_{(s)}}$

In this section,[13] we summarize the different results obtained from the QCD spectral sum rules (QSSR) on the leptonic decay constants of the $B$ and $D$ mesons which are useful in the analysis of the leptonic decay and on the $B$-$\bar{B}$ mixings. Intensive studies have been carried out on this subject during the last few years using QSSR and lattice calculations.

The leptonic constant of the pseudoscalar $P \equiv D,\ B$ meson is defined as:

$$\langle 0|\partial_\mu A^\mu|P\rangle = f_P M_P^2 \vec{P}\,, \tag{53.132}$$

where $\vec{P}$ is the pseudoscalar meson field and $f_P$ is the pseudoscalar decay constant which controls the $P \to l\nu$ leptonic decay width, normalized as $f_\pi = 130.56$ MeV.[14] The current:

$$\partial_\mu A^\mu(x)^i_j = (m_i + M_j)\bar{\psi}_i(i\gamma_5)\psi_j \quad (i \equiv u,\ d,\ s;\ j \equiv c, b)\,, \tag{53.133}$$

is the divergence of the axial current. In the sum rule analysis, we shall be concerned with the pseudoscalar two-point correlator:

$$\Psi_5(q^2) = i\int d^4x e^{iqx}\langle 0|\mathbf{T}\partial_\mu A^\mu(x)^i_j\left(\partial_\mu A^\mu(0)^i_j\right)^\dagger|0\rangle\,. \tag{53.134}$$

In the case of the $B(\bar{u}b)$ meson, the decay width into $\tau\nu_\tau$ reads:

$$\Gamma(B\to\tau\nu_\tau\ +\ B\to\tau\nu_\tau\gamma) = \frac{G_F^2|V_{ub}|^2}{4\pi}M_B\left(1-\frac{M_\tau^2}{M_B^2}\right)^2 M_\tau^2 f_B^2\,, \tag{53.135}$$

where $M_\tau$ expresses the helicity suppression of the decay rate into light leptons $e$ and $\mu$.

[13] This is an extension and an update of the some parts of the reviews given in [364].

[14] In this chapter, we adopt this normalization used by the lattice and experimental groups. In the previous sections, we have used $f_\pi \equiv f_\pi/\sqrt{2} = 92.4$ MeV.

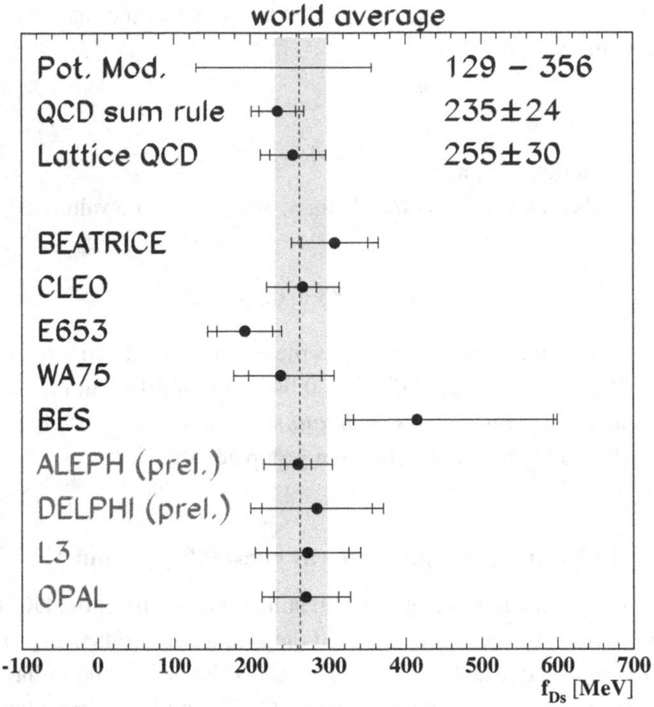


Fig. 53.11. Different measurements of $f_{D_s}$ compared with theoretical predictions from [714].

This expression shows that a good determination of $f_B$ will allow a precise extraction of the CKM mixing angle $V_{ub}$. One the other hand, $f_B$ and the so-called bag parameter $B_B$ also control the matrix element of the $\Delta B = 2$ $B^0$-$\bar{B}^0$ mixing matrix, which is of a non-perturbative origin, as we shall discuss in another chapter.

However, contrary to the case of the $\pi$ and $K$ mesons, the leptonic width of the heavy meson is small as the corresponding decay constant vanishes as $1/\sqrt{M_Q}$, while the presence of the neutrino in the final state renders the reconstruction of the signal and the rejection of background difficult. Moreover, the $B$ leptonic rate is Cabibbo suppressed, which makes it unreachable with present measurements. ($\sim |V_{ub}|^2$), while the $D_s$ leptonic rate is Cabbibbo favoured ($\sim |V_{cs}|^2$). Recent measurements of $f_{D_s}$ are given in Fig. 53.11, where the quoted average is [714]:

$$f_{D_s} \simeq (264 \pm 37) \text{ MeV} . \tag{53.136}$$

### 53.13.1 Upper bound on the value of $f_D$

Within the QSSR framework, the decay constants of the $B$ and $D$ mesons have been firstly estimated in [414], while the first upper bounds on their values have been derived in [29] and updated in the recent review [364]. Indeed, a *rigorous* upper bound on these couplings can be derived from the second-lowest superconvergent moment:

$$\mathcal{M}^{(2)} \equiv \frac{1}{2!} \frac{\partial^2 \Psi_5(q^2)}{(\partial q^2)^2}\bigg|_{q^2=0} , \tag{53.137}$$

where for this low moment, the OPE behaves well. Using the positivity of the higher-state contributions to the spectral function, one can deduce [29,399]:

$$f_P \leq \frac{M_P}{4\pi} \left\{ 1 + 3\frac{m_q}{M_Q} + 0.751\bar{\alpha}_s + \cdots \right\} , \tag{53.138}$$

where one should not misinterpret the mass dependence in this expression compared with that expected from heavy-quark symmetry. Applying this result to the $D$ meson, one obtains:

$$f_D \leq 2.14 f_\pi , \tag{53.139}$$

which is not dependent to leading order on the value of the charm quark mass. Although presumably quite weak, this bound, when combined with the recent determination of the $SU(3)_F$ breaking effects to two loops on the ratio of decay constants [716]:

$$\frac{f_{D_s}}{f_D} \simeq (1.15 \pm 0.04) , \tag{53.140}$$

implies

$$f_{D_s} \leq (2.46 \pm 0.09) f_\pi \simeq (321.2 \pm 11.8) \text{ MeV} , \tag{53.141}$$

which is useful for a comparison with the recent measurement of $f_{D_s}$, with the average value given previously. One cannot push, however, the uses of the moments to higher $n$ values in this $D$ channel, in order to minimize the continuum contribution to the sum rule with the aim of derive an estimate of the decay constant from this method, and to derive its 'correct' mass dependence, because the QCD series will not converge at higher $n$ values.

### 53.13.2 Estimate of the D decay constant $f_D$

The decay constant $f_D$ can be extracted from the pseudoscalar Laplace sum rules given in Eq. (53.116).[15] Prior to 1987, the different sum rules estimate of the decay constant $f_P$ were inconsistent among each other. To our knowledge, the first attempt to understand such discrepancies was reported in [717,718] (see also [719]), where it was shown, *for the first time* and a long time before the lattice results, that:

$$f_D \approx f_B \approx (1.2 \sim 1.5) f_\pi , \tag{53.142}$$

which differs from that expected from the heavy-quark symmetry scaling law [720]:

$$f_B \approx \sqrt{\frac{M_D}{M_B}} f_D \left( \frac{\alpha_s(M_c)}{\alpha_s(M_b)} \right)^{-1/\beta_1} , \tag{53.143}$$

[15] For reviews, see for example [715,360].

valid in the extreme case where the heavy-quark mass is infinite.[16] It has also been understood that the apparent disagreement among different existing QSSR numerical results in the literature is *not only* due to the choice of the continuum threshold $t_c$ [its effect is $(7 \sim 10)\%$ of the result when one moves $t_c$ from the one at the beginning of sum rule variable to the one where the $t_c$ stability is reached.][17] as misleadingly claimed in the literature. Indeed, the main effect is *also* due to the different values of the quark masses used,[18] which is shown explicitly in the table of [716].

In the $D$ channel, the most appropriate sum rule is the (relativistic) Laplace sum rule, as the OPE of the $q^2 = 0$ moments does not converge for larger values of the number of derivatives $n$, at which the $D$ meson contribution to the spectral integral is optimized. The results from different groups are consistent with each others for a given value of the $c$-quark mass. For the $D$ meson, the optimal result is obtained for:

$$6 \leq t_c \leq 9.5 \text{ GeV}^2 , \qquad \tau \simeq (1.2 \pm 0.2) \text{ GeV}^{-2} . \tag{53.144}$$

where the QCD corrections are still reasonnably small. The most recent estimate including $\alpha_s^2$ corrections from a simultaneous fit of the set either $(f_D, \bar{m}_c(m_c))$ or $(f_D, M_c^{pole})$ is given in Fig. 53.10. The obtained values of the quark masses have been quoted in Table 53.5. The resulting value of $f_D$ is [150]:

$$f_D \simeq (203 \pm 23) \text{ MeV} , \tag{53.145}$$

in agreement with the recent evaluation $(195 \pm 20)$ MeV at order $\alpha_s^2$ but using the pole mass as input [722].

### 53.13.3 Ratio of the decay constants $f_{D_s}/f_D$ and $f_{B_s}/f_B$

The $SU(3)$ breaking ratios $f_{D_s}/f_D$ and $f_{B_s}/f_B$ have been obtained semi-analytically in [716]. In order to have a qualitative understanding of the size of these corrections, we start from the global hadron-quark duality sum rule:

$$\int_0^{\omega_c} d\omega \ \text{Im}\Psi_5^{res}(\omega) \simeq \int_0^{\omega_c} d\omega \ \text{Im}\Psi_5^{\bar{q}Q}(\omega) , \tag{53.146}$$

where $\omega_c$ is the continuum energy defined as:

$$t = (E + M_Q)^2 \equiv M_Q^2 + \omega M_Q . \tag{53.147}$$

Keeping the leading order terms in $\alpha_s$ and in $1/M_Q$, it leads to:

$$R_P \simeq \rho_P \left\{ 1 + 3 \left( \frac{m_s}{\omega_c} \right) \left( 1 - \frac{m_s}{M_Q} \right) - 6 \left( \frac{m_s}{\omega_c}^2 \right) - \left( \frac{m_s}{M_Q} \right) \left( 1 - \frac{m_s}{M_Q} \right) \right\} , \tag{53.148}$$

[16] Finite mass corrections to this formula will be discussed later on.

[17] In some papers in the literature, the value of $t_c$ is taken smaller than the previous range. In this case, the $t_c$ effect is larger than the one given here.

[18] A critical review on the discrepancy between different existing estimates is given in [721].

Table 53.7. *Estimate of $f_{B_{(s)}}$ to order $\alpha_s^2$ and $f_{B_s}/f_B$ to order $\alpha_s$ from QSSR and comparison with the lattice*

| Sources | $f_B$ (MeV) | $f_{B_{(s)}}/f_B$ | $f_{B_s}$ (MeV) | Comments | Authors |
|---|---|---|---|---|---|
| **QSSR** | | | | | |
| LSR | $203 \pm 23$ | $1.16 \pm 0.04 \Longrightarrow$ | $236 \pm 30$ | $M_{\rm pole}, \bar{m}_b$: output | SN94,01 [716,150] |
| | $210 \pm 19$ | | $244 \pm 21$ | $\bar{m}_b$: input | JL01 [724] |
| HQETSR | $206 \pm 20$ | | | $M_{\rm pole}$: input | PS01 [722] |
| **SR average** | $207 \pm 21 \Longrightarrow$ | | $240 \pm 24$ | | |
| **Unq. lattice** | | | | | |
| | $200 \pm 30$ | $1.16 \pm 0.04 \Longrightarrow$ | $232 \pm 35$ | average | LAT01 [725] |

where:

$$\rho_P \equiv \left(\frac{M_P}{M_{P_s}}\right)^2 \left(1 + \frac{m_s}{M_Q}\right) . \tag{53.149}$$

The value of $\omega$ is fixed from stability criteria to be [717,721,634,165]:

$$\omega_c \simeq (3.1 \pm 0.1) \text{ GeV}. \tag{53.150}$$

The sum rule indicates that the $SU(3)$ breaking corrections are of two types, the one $m_s/M_Q$ and the other $m_s/\omega_c$. More quantitatively, we work with the Laplace sum rule:

$$\mathcal{L} = \int_0^{\omega_c} d\omega\, \mathrm{e}^{-\omega\tau}\, \mathrm{Im}\Psi_5^{res}(\omega). \tag{53.151}$$

Analogously the Laplace sum rule gives:

$$R_P^2 \simeq \rho_P^2 \left\{ 1 + 2(2.2 \pm 0.2)\left(\frac{m_s}{\omega_c}\right)\left(1 - \frac{m_s}{M_Q}\right) - 2\,(8.2 \pm 1.6)\left(\frac{m_s}{\omega_c}^2\right) \right\}, \tag{53.152}$$

where the numerical integration includes a slight $M_Q$ dependence. Including $m_s\alpha_s$ and $m_s^2\alpha_s$-corrections, the resulting values of the ratio are:

$$R_D \equiv \frac{f_{D_s}}{f_D} = 1.15 \pm 0.04\,, \quad R_B \equiv \frac{f_{B_s}}{f_B} = 1.16 \pm 0.05\,. \tag{53.153}$$

This result implies:

$$f_{D_s} \simeq (235 \pm 24) \text{ MeV}\,, \tag{53.154}$$

which agrees within the errors with the data [714] and lattice [723] averages both quoted in Fig. 53.11. This feature increases confidence in the use of the QSSR method for predicting the not yet measured decay constant of the $B$ meson.

### 53.13.4 Estimate of the $B$ decay constant $f_B$

For the estimate of $f_B$, one can either work with the Laplace, the moments or their non-relativistic variants because the $b$-quark mass is relatively heavy. The optimal result which we shall give here comes from the Laplace relativistic sum rules. They corresponds to the *conservative range* of parameters:

$$0.6 \leq E_c^{(b)} \equiv \sqrt{t_c} - M_B \leq 1.8 \text{ GeV} , \qquad \tau \simeq 0.38 \text{ GeV}^{-2} , \tag{53.155}$$

which have been used in the previous section for getting the $b$-quark mass. As shown in the figure given in [726,727], the dominant corrections come from the $\langle \bar{u}u \rangle$ quark condensate with the strengh (30~40)% of the lowest order term in $f_B$, while the higher condensate effects are smaller, which are respectively $-(20{\sim}30)\%$ , $+(5{\sim}8)\%$ for the $d = 5$ and 6 condensates. This shows, despite the large value of the quark condensate contribution, that the OPE is convergent. It has been noticed in [726,727], that the convergence of the OPE is faster for the relativistic LSR than for the moments, such that the most precise result should come from the LSR. In both cases the perturbative corrections are small. One obtains from the relativistic LSR , the results to order $\alpha_s^2$ [150]:

$$f_B \simeq (203 \pm 23) \text{ MeV} \simeq (1.55 \pm 0.18) f_\pi , \tag{53.156}$$

and to order $\alpha_s$ (see previous discussion) [716]:

$$\frac{f_{B_s}}{f_B} \simeq 1.16 \pm 0.04 . \tag{53.157}$$

These values of $f_B$ and $f_D$ agree quite well with the previous QSSR findings in [716], [3] and [719]. They also agree with the non-relativistic sum rules estimate in the full theory [717], in HQET [633,722] and in [634,165]. However, unlike the relativistic sum rules, the HQET sum rule is strongly affected by the huge perturbative radiative corrections of about 100%, which is important at the sum rule scale of about 1 GeV at which the HQET sum rule optimizes. These results are also in good agreement with the lattice average estimate given in Table 53.7.

### 53.13.5 Static limit and $1/M_b$-corrections to $f_B$

As noticed previously, the *first* result $f_B \simeq f_D$ in [716], which has been confirmed by recent estimates from different approaches, shows a large violation of the scaling law expected from heavy-quark symmetry. This result suggests that finite quark mass corrections are still huge at the $D$ and $B$ meson masses. The first attempt to understand this problem analytically is in [718] in terms of large corrections of the type $E_c/M_b$ if one expresses in this paper the continuum threshold $t_c$ in terms of the threshold energy $E_c$:

$$t_c \equiv (E_c + M_b)^2. \tag{53.158}$$

More recently different approaches have been investigated for the estimate of the size of these corrections.

In the lattice approach, these mass corrections have been estimated from a fit of the obtained value of the meson decay constant at finite and infinite (static limit) values of the heavy quark mass and by assuming that these corrections are polynomial in $1/M_Q$ up to log. corrections. A similar analysis has been done with the sum rule in the full theory [726,727], by studying numerically the quark mass dependence of the decay constant up to the quark mass value ($M_Q \leq 15$ GeV) until which one may expect that the sum rule analysis is still valid. In so doing, we use the parametrization:

$$f_B\sqrt{M_B} \simeq \tilde{f}_B \alpha_s^{1/\beta_1} \left\{1 - \frac{2}{3}\frac{\alpha_s}{\pi} - \frac{A}{M_b} + \frac{B}{M_b^2}\right\} , \tag{53.159}$$

by including the quadratic mass corrections. The analysis gives:[19]

$$\tilde{f}_B \equiv (f_B\sqrt{M_B})_\infty \simeq (0.65 \pm 0.06)\ \text{GeV}^{3/2} , \tag{53.160}$$

which one can compare with the results from the HQET Laplace sum rule [633] and [165,728]:

$$\tilde{f}_B \simeq (0.35 \pm 0.10)\ \text{GeV}^{3/2} , \tag{53.161}$$

and from FESR [165]:

$$\tilde{f}_B \simeq (0.57 \pm 0.10)\ \text{GeV}^{3/2} , \tag{53.162}$$

Taking the average of these three (independent) determinations, one can deduce:

$$\tilde{f}_B \simeq (0.58 \pm 0.09)\ \text{GeV}^{3/2} , \tag{53.163}$$

where we have evaluated an arithmetic average of the errors. This result is in good agreement with the lattice value given in [723,729] using non-perturbative clover fermions. One can translate this result into the value of $f_B$ in the static limit approximation:

$$f_B^{\text{stat}} \simeq (1.9 \pm 0.3) f_\pi . \tag{53.164}$$

We can also use the previous value of $f_B^{\text{stat}}$ together with the previous values of $f_B$ and $f_D$ at the 'physical' quark masses in order to determine numerically the coefficients $A$ and $B$ of the $1/M_b$ and $1/M_b^2$ corrections. In so doing, we use the values of the quark 'pole' masses given in Tables 53.5 and 53.6. Then, we obtain from a quadratic fit:

$$A \approx 0.98\ \text{GeV} \quad \text{and} \quad B \approx 0.35\ \text{GeV}^2 , \tag{53.165}$$

while a linear fit gives a large uncertainty:

$$A \approx (0.74 \sim 0.91)\ \text{GeV} . \tag{53.166}$$

One can notice that the fit of these coefficients depends strongly on the input values of $f_D$ and $f_B$. Indeed, using some other set of values as in [726,727,634], one obtains values about two-times smaller. Therefore, we consider as a conservative value of these coefficients an

[19] The numbers given in [726,727] correspond to the quark mass $M_b = 4.6$ GeV and should be rescaled until the meson mass $M_B$. In the following, we shall also work with $f_B$ normalized to be $\sqrt{2}$ bigger than in the original papers.

uncertainty of about 50%. The value of $\Lambda$ obtained is comparable with the one from HQET sum rules [166] and [726,634] of about $0.7 \sim 1.2$ GeV. A similar value of $\Lambda$ has been also obtained from the recent NR lattice calculations with dynamic fermions and using a linear fit [729]:

$$\Lambda \approx 0.7 \text{ GeV} . \tag{53.167}$$

One can *qualitatively* compare this result with the one obtained from the analytic expression of the moment in the full theory [3,718,728]:

$$f_B^2 \approx \frac{1}{\pi^2}\frac{E_c^3}{M_b}\left[1 - \frac{2}{3}\frac{\alpha_s}{\pi} - \frac{3(n+2)E_c}{M_b} + \cdots - \frac{\pi^2}{2}\frac{\langle \bar{u}u\rangle}{E_c^3} + \cdots\right] . \tag{53.168}$$

Here, one can notice that the size of the $1/M_b$ corrections depends on the number of moments, such that their estimate using literally the expression of the moments can be inconclusive. A qualitative estimate of these corrections can be done from the semi-local duality sum rule, which has more intuitive physical meaning due to its direct connection with the leptonic width and total cross-section through the optical theorem. It corresponds to $n = -2$, and leads to the *interpolating formula* [728]:

$$\sqrt{2} f_B \sqrt{M_B} \approx \frac{E_c^{3/2}}{\pi}\alpha_s^{1/\beta_1}\left(\frac{M_b}{M_B}\right)^{3/2}\left\{1 - \frac{2}{3}\frac{\alpha_s}{\pi} + \frac{3}{88}\frac{E_c^2}{M_b^2} - \frac{\pi^2}{2}\frac{\langle \bar{u}u\rangle}{E_c^3} + \cdots\right\} , \tag{53.169}$$

from which, one obtains:

$$\begin{aligned} \Lambda &\approx \frac{3}{2}(M_B - M_b) \simeq 1 \text{ GeV}, \\ B &\approx \frac{3}{88}E_c^2 + \frac{27}{32}(M_B - M_b)^2 \simeq 0.45 \text{ GeV}^2 , \end{aligned} \tag{53.170}$$

which is in good agreement with the previous numerical estimate.

## 53.14 Conclusions

We have reviewed the determinations of the quark masses and leptonic decay constants of (pseudo)scalar mesons which are useful in particle physics phenomenology. The impressive agreements of the QSSR results with the data when they are measured and/or with lattice calculations in different channels indicate the self-consistency of the QSSR approach, making it one of the most powerful semi-analytical QCD non-perturbative approaches available today. Applications of these results for studying the $B$-$\bar{B}$ mixings and $CP$-violation will be discussed in the next chapter.

# 54
# Hadron spectroscopy

## 54.1 Light $\bar{q}q$ mesons

The spectroscopy of the light mesons has been discussed extensively in the literature, using LSR [1,357,3,32,452], FESR [405,629], and Yndurain's moments within positivity [730]. Detailed derivations of the analysis in some channels can e.g. be found in QSSR1 [3]. In addition to the currents associated to the (axial)-vector and (pseudo)scalar channels, discussed in previous chapters, we shall also be concerned with the $2^{++}$ tensor current, where its renormalization has been discussed in Part VIII, while the corresponding sum rule has been re-analysed in [452]. The analysis is summarized in Table 54.1, where one can deduce from Table 54.1, that the predictions for the couplings are quite good compared with available data, whereas in some cases the meson masses are overestimated.

## 54.2 Light baryons

As mentioned earlier, the light baryon systems have been studied in [424–430]. The decuplet and the octet baryons can be respectively described by the operators:

$$\Delta^\mu = \frac{1}{\sqrt{2}} : \psi^T C \gamma_\lambda \psi \left( g^{\mu\lambda} - \frac{\gamma^\mu \gamma^\lambda}{4} \right) \psi \,, \tag{54.1}$$

and:

$$N = \frac{1}{\sqrt{2}} : [(\psi C \lambda_5 \psi)\, \psi + b\, (\psi C \psi)\, \lambda_5 \psi] \,, \tag{54.2}$$

where $C$ is the charge conjugation matrix, $b$ is an arbitrary mixing parameter and $\psi$ is the *valence* quark field. We have suppressed the colour indices. The corresponding two-point correlator:

$$S(q) = i \int d^4x \, e^{iqx} \langle 0 | \mathcal{T} \mathcal{O}(x) \mathcal{O}^\dagger(0) | 0 \rangle \,, \tag{54.3}$$

can be parametrized (without loss of generalities) in terms of two invariants:

$$S(q) = (\hat{q} F_1 + F_2) \Gamma + \cdots \,, \tag{54.4}$$

where for the decuplet ($\Delta(3/2)$) $\Gamma \equiv g^{\mu\nu}$, and for the octet (nucleon) $\Gamma = 1$. The expressions of the form factors are known including radiative corrections and non-perturbative

Table 54.1. *Light meson masses and couplings from QSSR. The coupling $f_P$ are in units of MeV and normalized as $f_\pi = 92.4$ MeV, while $\gamma_V$ has no dimension and is normalized as $\gamma_\rho = 2.55$ (Eq. 2.52). WSR refer to Weinberg sum rules in QCD (previous chapter)*

| $J^{PC}$ | Meson | Coupling | Mass (GeV) | $t_c$ (GeV$^2$) | $\tau$ (GeV$^{-2}$) | References |
|---|---|---|---|---|---|---|
| $1^{--}$ | $\rho$ | $\gamma_\rho \simeq 2.5 \sim 2.8$ | $0.80 \sim 0.85$ | $1.4 \sim 1.7$ | $0.4 \sim 1$ | [3] |
| | $K^*$ | $\gamma_{K^*} \simeq 2.0 \sim 2.5$ | | $1.4 \sim 2$ | $0.6 \sim 1$ | [3] |
| $1^{++}$ | $A_1$ | $\gamma_{A_1} \simeq (1.2 \sim 1.9)\gamma_\rho$ | 1.28 | $2.0 \sim 2.3$ | $0.6 \sim 1$ | WSR [29,3] |
| $0^{-+}$ | $\pi$ | $f_\pi \simeq (74 \sim 96)$ | $\pi^+ - \pi^0 \simeq 4.6 \times 10^{-3}$ | $1.8 \sim 2.3$ | $0.7 \sim 1.6$ | [30,29,3] |
| | $K'$ | $\frac{M_{K'}^4 f_{K'}^2}{M_K^4 f_K^2} \simeq 9.5 \pm 2.5$ | | $1.8 \sim 2.3$ | $0.7 \sim 1.6$ | [3] |
| $0^{++}$ | $a_0(\bar{u}d)$ | $f_a \simeq (1.6 \pm 0.5)$ | $1.0 \sim 1.05$ | $1.3 \sim 1.6$ | $0.4 \sim 0.8$ | [3,420,422] |
| | $f_0(\bar{u}u + \bar{d}d)$ | $f_{f_0} \simeq f_a$ | $M_{f_0} \simeq M_{a_0}$ | | | $SU(2)$ |
| | $K_0^*(\bar{u}s)$ | $f_{K_0^*} \simeq (46.3 \pm 7.5)$ | $1.3 \sim 1.4$ | $1.8 \sim 2.3$ | $0.7 \sim 1.6$ | [419,3,420,422] |
| | $f_0(\bar{s}s)$ | $f_3 \simeq (22 \sim 28)$ | $1.474 \pm 0.044$ | $\simeq t_c^{K^*0}$ | $0.3 \sim 0.5$ | [688,3] |
| $2^{++}$ | $f_2$ | $f_{f_2} \simeq (132 \sim 184)$ | $1.4 \sim 1.6$ | $2.5 \sim 3.5$ | $0.6 \sim 1.2$ | [452,3] |
| | $f_2'$ | $f_{f_2'} \simeq (112 \sim 152)$ | $\frac{M_{f_2'}}{M_{f_2}} \simeq 1.14 \sim 1.26$ | $3 \sim 4$ | $0.6 \sim 1.2$ | [452,3] |

terms. These terms are tabulated in [426] (see chapter on two-point function). The relevant quantities for the analysis are the LSR:

$$\mathcal{L}_i(\tau) = \int_0^{t_c} dt\, e^{-t\tau}\frac{1}{\pi}\text{Im}F_i(t) \;:\quad i = 1,2 \tag{54.5}$$

and their ratios:

$$R_{ii}(\tau) = \frac{\int_0^{t_c} dt\, t\, e^{-t\tau}\frac{1}{\pi}\text{Im}F_i(t)}{\int_0^{t_c} dt\, e^{-t\tau}\frac{1}{\pi}\text{Im}F_i(t)}\,, \tag{54.6}$$

and:

$$R_{12}(\tau) = \frac{\int_0^{t_c} dt\, e^{-t\tau}\frac{1}{\pi}\text{Im}F_2(t)}{\int_0^{t_c} dt\, e^{-t\tau}\frac{1}{\pi}\text{Im}F_1(t)}\,. \tag{54.7}$$

The baryon contribution to the spectral function can be introduced through its coupling and using a duality ansatz parametrization:

$$\begin{aligned} \frac{1}{\pi}\text{Im}F_2(t) &= M_B|Z_B|^2\delta\left(t - M_B^2\right) + \Theta(t - t_c)\ \text{`QCD continuum'}\,, \\ \frac{1}{\pi}\text{Im}F_1(t) &= |Z_B|^2\delta\left(t - M_B^2\right) + \Theta(t - t_c)\ \text{`QCD continuum'}\,. \end{aligned} \tag{54.8}$$

Qualitatively, $\mathcal{R}_{12}$ can provide a good explanation of the proton mass in terms of the chiral condensate:

$$M_N \approx 32\pi^2\langle\bar{\psi}\psi\rangle\tau\left(\frac{7 - 2b - 5b^2}{5 + 2b + 5b^2}\right)\,. \tag{54.9}$$

Table 54.2. *Light baryon masses and couplings from QSSR*

| Baryon | Coupling ($GeV^6$) | Mass (GeV) | Mass (GeV) (exp) | $t_c$ ($GeV^2$) |
|---|---|---|---|---|
| **Octet:** | $J^P = (1/2)^+$ | | | |
| $N$ | 0.14 | 1.05 | 0.94 | 1.58 |
| $\Sigma$ | 0.27 | 1.16 | 1.19 | 2.09 |
| $\Lambda$ | 0.23 | 1.24 | 1.1 | 2.15 |
| $\Xi$ | 0.31 | 1.33 | 1.31 | 2.42 |
| **Decuplet:** | $J^P = (3/2)^+$ | | | |
| $\Delta$ | 1.15 | 1.21 | 1.23 | 2.2 |
| $\Omega$ | 5.16 | 1.61 | 1.67 | 4.08 |
| $\Sigma*$ | 1.89 | 1.35 | 1.38 | 2.78 |
| $\Xi*$ | 3.07 | 1.48 | 1.51 | 3.39 |

One can optimize this relation in the change of the mixing parameter $b$ by requiring that its first derivative in $b$ is zero (Principle of Minimal Sensitivity), which gives:

$$b = -\frac{1}{5}, \tag{54.10}$$

which is the optimal choice of Chung *et al.* [424] obtained after an involved numerical analysis. The value of the sum rule at which the sum rule is optimized is approximately $\tau^{-2} \approx M_N^2$, from which one can deduce the interesting sum rule:

$$M_N \approx \left[-\pi^2 \frac{152}{3} \langle\bar{\psi}\psi\rangle\right]^{1/3} \approx 1.8 \text{ GeV}, \tag{54.11}$$

which is not too bad taking into account the crude approximation used to get this formula. However, it shows the rôle of the non-leading terms in correctly fixing the mass of the nucleon. A comparison of the numerical ability of these different sum rules has been discussed in [2], where it has been noted that $\mathcal{L}_2$ and $\mathcal{R}_{22}$ are the most advantageous sum rule (stability, small radiative correction, ...). The results from this analysis are given in [426] and discussed in details in [3]. We show them in Table 54.2.

We have only quoted in Table 54.2 the central value, where the error is typically about 10%. The different sum rules optimize for $\tau$ around $0.8 \sim 1.2$ GeV$^{-2}$, while the $t_c$ values quoted in the table come from the lowest FESR moment. The results for the octet corresponds to the optimal value $b = -1/5$ discussed earlier. A compromise value:

$$M_0^2 \simeq 0.8 \text{ GeV}^2, \tag{54.12}$$

of the scale parametrizing the mixed condensate is needed, as also required for fitting the heavy-light $B$ and $B^*$ meson masses [401]. The $SU(3)$ mass-splittings need large flavour breakings of the chiral condensates $\langle\bar{s}s\rangle$ and $\langle\bar{s}Gs\rangle$, which, at the order we are working to, seem to act in opposite directions. In particular, the $\Omega$ mass can be reproduced for a 40% increase of the mixed condensate value compared with its $SU(3)$ symmetric value.

## 54.3 Spectroscopy of the heavy-light hadrons

### 54.3.1 Beautiful mesons

The masses and mass-splittings of heavy-light mesons made with $\bar{q}b$ quarks ($q$ is the light quark $u$, $d$ and $s$) have been analysed in [401] using the ratio $r_n$ and double ratios $d_n$ of $q^2 = 0$ moments:

$$d_n \equiv \frac{r_n^H}{r_n^{H'}} , \tag{54.13}$$

where $r_n$ has been defined analogously to Eq. (54.6); $H$ and $H'$ are the indices of the corresponding meson. The analysis shows a good $n \simeq 7 \sim 9$ and $t_c \simeq (40 \sim 60)$ GeV$^2$ stabilities, which indicates that the sum rule can give a much better prediction for the ratio than for the absolute values of the meson masses. The observed masses of the $B$ and $B^*$ mesons have been used for fixing the $b$ quark mass and the value of the mixed condensate. The predictions for the mass splittings are given in the table of [313] (Section 51.3). The main features of the results are summarized below.

- **Splittings between the chiral partners**
  Typically, the mass splitting between the meson and its chiral partner is [313]:

$$B_\delta(0^{++}) - B(0^{-+}) \approx B_A^*(1^{++}) - B^*(1^{--}) \simeq (417 \pm 212) \text{ MeV} , \tag{54.14}$$

  which is mainly due to the chiral quark condensate as expected from general arguments.
- **Splittings due to $SU(3)$ breakings**
  The $SU(3)$ breaking mass-splitting is given in [313], as function of the ratio of the normal ordered condensate $\chi = \langle \bar{s}s \rangle / \langle \bar{u}u \rangle$. Using the experimental value $B_s = 5.37$ GeV, one can deduce:

$$\chi \approx 0.75 , \tag{54.15}$$

  in agreement with the result obtained from the light meson systems discussed previously. The mass of the $B_s^*$ meson is also given in [401] as a function of $\chi$. Using the previous value of $\chi$, into the prediction of the $B_s^*$ in [401] leads to a value slightly higher than the data [16], and needs to be reconsidered.
- **Decay constants and couplings**
  Besides the decay constants $f_{B_{(s)}}$ of the $B_{(s)}$ mesons which plays an important role in the $B^0_{(s)} - \bar{B}^0_{(s)}$ mixing matrix elements, which we shall discuss in more details in the next chapter, we give below the findings of [401] for the couplings and decay constants of the other mesons to order $\alpha_s$:

$$f_{B_\delta} \simeq (1.99 \pm 0.39) f_\pi , \qquad f_{B^*} \equiv \frac{M_{B^*}}{2\gamma_{B^*}} \simeq f_{B_A^*} \simeq (1.78 \pm 0.22) f_\pi \tag{54.16}$$

  compared with the values of $f_B$ and $f_{B_s}$ obtained to the same order in [717,698,716].

### 54.3.2 Baryons with one heavy quark

The masses and couplings of the baryons ($Quu$) where $Q \equiv b, c$ have been estimated [453] using $q^2 = 0$ moments and LSR. In the case of charmed baryons, the LSR stabilizes at

$\tau \simeq 0.4$ GeV$^{-2}$ and for $t_c$ in the range 8 to 16 GeV$^2$, where the first value corresponds to the beginning of $\tau$ stability, while the second one to the $t_c$ stability. Moreover, a study of the stability on the change of the mixing parameter $b$ for the $\Sigma_Q$ currents lead to the range:

$$-0.5 \leq b \leq 0.5\,, \tag{54.17}$$

in favour of the Chung *et al.* [424] choice $b = -1/5$ in the light baryons sector. In the case of beautiful baryons, the optimal results are obtained for $\tau \simeq 0.2$ GeV$^{-2}$ (LSR), $n \simeq 4 \sim 6$ (moments), and for $t_c \simeq 40 \sim 50$ GeV$^2$. The analysis leads, to a good accuracy, to the mass difference [453]:

$$\Sigma_b - \Sigma_c \simeq 3.4\,\text{GeV}\,, \qquad \Sigma_b^* - \Sigma_c^* \simeq 3.3\,\text{GeV}\,. \tag{54.18}$$

Using the experimental value of $\Sigma_c$, one can then predict [453]:

$$\Sigma_b \simeq 5.85\,\text{GeV}\,, \tag{54.19}$$

in agreement with the potential model estimate. The corresponding couplings are:

$$|Z_{\Sigma_c}|^2 \simeq (4.2 \sim 7.7) \times 10^{-4}\,\text{GeV}^6\,, \qquad |Z_{\Sigma_b}|^2 \simeq (0.10 \sim 0.45) \times 10^{-2}\,\text{GeV}^6\,, \tag{54.20}$$

where we have used the same normalization as in the light baryon systems. To a lesser accuracy, one has also obtained [453]:

$$\Sigma_c^* \simeq (2.15 \sim 2.92)\,\text{GeV}\,, \qquad \Sigma_b^* \simeq (5.4 \sim 6.2)\,\text{GeV}\,, \tag{54.21}$$

in agreement with potential model estimates. The corresponding couplings are:

$$|Z_{\Sigma_c^*}|^2 \simeq (1.1 \sim 2.2) \times 10^{-3}\,\text{GeV}^6\,, \qquad |Z_{\Sigma_b}|^2 \simeq (2.0 \sim 5.4) \times 10^{-3}\,\text{GeV}^6\,. \tag{54.22}$$

The analysis has been also applied to the $\Lambda_Q$ baryon. One has obtained [453]:

$$\Sigma_c - \Lambda_c \leq 207\,\text{MeV}\,, \qquad \Sigma_b - \Lambda_b \leq 163\,\text{MeV}\,, \tag{54.23}$$

where the bounds should be understood as 'practical' though not 'rigorous'. One can also notice that the value of the $\Lambda_Q$ mass decreases with the value of the gluon condensate. Finally, the previous analysis for the baryons has been extended in the case where the $b$ quark mass tends to infinity (HQET sum rule) [454]. In so doing one has considered the combination of form factors:

$$\mathcal{S}_B(\mathcal{D}_B) = M_Q \text{Im}\, F_1^B(t) \pm \text{Im}\, F_2^B(t)\,, \tag{54.24}$$

corresponding, respectively, to the positive $B_Q^+$ and negative $B_Q^-$ parity states. Doing the analysis for the $\Sigma_Q$, and taking a conservative range of the QCD continuum energy $E_c \approx (1.5 \sim 3)$ GeV, one obtains the mass gap:

$$\delta M_{\Sigma^+} \equiv \Sigma^+ - M_b \approx (1.1 \sim 2.1)\,\text{GeV}\,, \qquad \delta M_{\Sigma^-} \equiv \Sigma^- - M_b \approx (1.8 \sim 2.5)\,\text{GeV}\,, \tag{54.25}$$

respectively for the positive and negative parity states. This result shows that the baryon mass gap is systematically higher than the meson mass one which is about 0.65 GeV. Analogous analysis for the $\Lambda^{\pm}$ baryons shows that, to the approximation we are working, the $\delta M_{\Lambda^+}$ sum rule does not present any stability, while one finds that the $\Sigma^-$ and $\Lambda^-$ are almost degenerate.

### 54.4 Hadrons with charm and beauty

From the point of view of quark-gluon interactions, the $B_c(\bar{b}c)$ meson is intermediate between the $\bar{c}c$ and $\bar{b}b$ systems, and it shares with the two heavy-quarkonia common dynamic properties. It is possible to consider the heavy quark and anti-quark as non-relativistic particles, and describe the bound state, adding then the relativistic corrections. On the other hand, $B_c$, being the lightest hadron with open beauty and charm, decays weakly. Therefore, it provides us with a rather unique possibility of investigating weak decay form factors in a quarkonium system.

The spectroscopy of the $(\bar{c}b)$ mesons and of the $(bcq)$, $(ccq)$ and $(bbq)$ baryons ($q \equiv d$ or $s$), the decay constant and the (semi)leptonic decay modes of the $B_c$ meson have been extensively discussed in [731] using combined informations from potential models and QSSR. As a result, one obtains [731]:

- **Spectra**
  The spectra of the $B_c$-like hadrons from potential models are:

$$\begin{aligned} M_{B_c(\bar{b}c)} &= (6.26 \pm 0.02)\ \text{GeV}\,, & M_{B_c^*(\bar{b}c)} &= (6.33 \pm 0.02)\ \text{GeV}, \\ M_{\Lambda(bcu)} &= (6.93 \pm 0.05)\ \text{GeV}, & M_{\Omega(bcs)} &= (7.00 \pm 0.05)\ \text{GeV}, \\ M_{\Xi^*(ccu)} &= (3.63 \pm 0.05)\ \text{GeV}, & M_{\Xi^*(bbu)} &= (10.21 \pm 0.05)\ \text{GeV}, \end{aligned} \tag{54.26}$$

  which are consistent with, but more precise than, the sum-rule results given in [453,454].
- **The decay constant $f_{B_c}$ and other residues**
  The decay constant of the $B_c$ meson is better determined from QSSR than from potential models. The average of the LSR and $q^2 = 0$ moments sum rule gives the result [731]:

$$f_{B_c} \simeq (2.94 \pm 0.12) f_\pi\,, \tag{54.27}$$

  which leads to the leptonic decay rate into $\tau\nu_\tau$ of about $(3.0 \pm 0.4) \times (V_{cb}/0.037)^2 \times 10^{10}\ \text{s}^{-1}$. This result has been obtained for $\tau \simeq (0.04 \sim 0.12)\ \text{GeV}^{-2}$, for $n \simeq 2 \sim 3$, and for $t_c \simeq 50 \sim 67\ \text{GeV}^2$, or equivalently for $E_c \simeq (1.0 \sim 2.1)$ GeV, where $t_c \equiv (M_b + M_c + E_c)^2$. By comparing it with $f_B$, one can notice that their difference is about $M_c$ as intuitively expected.
  Residues of the different baryons have been also estimated. Their values with the corresponding normalizations can be found in [731,453,454].
- **Semi-leptonic decays of the $B_c$**
  We have also studied the semi-leptonic decay of the $B_c$ mesons and the $q^2$-dependence of the form factors, which differs from the usual VDM expectation. We shall come back to this point in the next chapter.

Detection of these particles in the next $B$-factory machine will then serve as a stringent test of the results from combined potential models and QSSR analysis obtained in [731].

## 54.5 Mass splittings of heavy quarkonia

The mass splittings of heavy quarkonia have been studied recently using double ratios of exponential sum rules [313] (Section 51.3). One can notice that the sum rule analysis of the mass splittings is insensitive to the change of the continuum threshold $t_c$, whilst it optimizes at the sum rule scale $\sigma \equiv \tau \simeq 0.9$ (respectively 0.35) GeV$^{-2}$ for the charmonium (respectively bottomium) systems. As emphasized earlier, some observed mass splittings can be used for fixing the QCD parameters $\alpha_s$ and gluon condensate (see the table of Section 51.3). In this section, we give the different predictions obtained once we know these QCD parameters. These predictions are given in the table of Section 51.3. One can notice that there is a fair agreement between the theoretical predictions and the data when available. For a particular interest is the prediction on the $\Upsilon - \eta_b$ mass-splitting in the range $30 \sim 110$ MeV, which can imply the observation of the $\eta_b$ through the $\Upsilon$ radiative decay. The (non) observation of the $\eta_b$ through this process is a test of the validity of the resummed Coulombic correction, which differs from the correction obtained from the truncation of the QCD series. In this latter case, the predicted mass splitting is only of the order of 3 to 20 MeV.

## 54.6 Gluonia spectra

The properties of gluonia from QSSR and some low-energy theorems have been discussed recently in the update work of [688] where complete references to the original works can also be found), which we shall summarize. The most relevant papers for our discussions will be those in [3,734] and [455–457] for the sum rules analysis, those in [686] for the low-energy theorems and vertex sum rules and those in [687,458,450] for the mixings.[1]

We shall consider the lowest dimension gluonic currents that can be built from the gluon fields and which are gauge invariant:

$$\begin{aligned} J_s &= \beta(\alpha_s) G_{\alpha\beta} G^{\alpha\beta} , \\ \theta^g_{\mu\nu} &= -G^\alpha_\mu G_{\nu\alpha} + \frac{1}{4} g_{\mu\nu} G_{\alpha\beta} G^{\alpha\beta} \\ Q(x) &= \left(\frac{\alpha_s}{8\pi}\right) \text{tr}\, G_{\alpha\beta} \tilde{G}^{\alpha\beta} , \end{aligned} \tag{54.28}$$

where the sum over colour is understood. The $\beta$ function has been defined in Chapter 2, while

$$\tilde{G}_{\mu\nu} = \frac{1}{2} \epsilon_{\mu\nu\alpha\beta} G^{\alpha\beta} , \tag{54.29}$$

is the dual of the gluon field strengths. These currents have respectively the quantum numbers of the $J^{PC} = 0^{++}$, $2^{++}$ and $0^{-+}$ gluonia,[2] which are familiar in QCD. The former two

[1] Some theoretical reviews and experimental results on the status of gluonia can be found in [688,365,690,735–738].
[2] The pseudotensor $2^{-+}$ will not be considered here.

enter into the QCD energy-momentum tensor $\theta_{\mu\nu}$, while the later is the $U(1)_A$ axial-anomaly current. We shall also consider the scalar three-gluon local current:

$$J_{3G} = g^3 f_{abc} G^a G^b G^c . \tag{54.30}$$

The spectra of the gluonia has been obtained from a QSSR analysis of the generic two-point correlator:

$$\psi_G(q^2) \equiv i \int d^4x \, e^{iqx} \langle 0|\mathcal{T} J_G(x) \left(J_G(0)\right)^\dagger |0\rangle , \tag{54.31}$$

built from the previous gluonic currents $J_G(x)$, using the LSR and its ratio $\mathcal{R}(\tau)$. The gluonium contribution to the spectral function enters through its coupling:

$$\langle 0| J_G(x)|G \rangle = \sqrt{2} f_G M_G^2 . \tag{54.32}$$

The results of the analysis are summarized in Table 54.3.

The mass of the three-gluonic bound state as well as its small mixing with the two-gluonic state have been obtained in [457]. The predictions obtained in the table show that the $0^{++}$ gluonium is the lightest gluonia, which is in good agreement with lattice calculations [739] and QCD-like inequalities [740] results. This agreement is an a posteriori confirmation of the fact that the neglect of the instanton contributions qualitatively discussed in [382] is a good approximation. Moreover, different analysis in the literature also shows that using current models, instanton effects are negligible in the mass calculations.

One should notice that, in addition to the $G(1.5)$ also found from lattice quenched simulation, the low-mass $\sigma_B$ coupled to the gluonic current is needed in the QSSR analysis for a consistency of the subtracted and unsubtracted sum rules which optimize in different energy regimes. Moreover, a low-mass $\sigma_B$ is also found in different experiments [690] and is needed in the linear $\sigma$ model [741,742,743].

## 54.7 Unmixed scalar gluonia

### 54.7.1 Masses and decay constants

The mass spectrum of the scalar gluonium has been given in Table 54.3, where, as previously mentioned, a low-mass scalar $\sigma_B$ below 1 GeV (we have chosen 1 GeV for definiteness in our discussion but a lower value is not also excluded) is needed in addition to the one $G$ of 1.5 GeV, in order to have consistency between the results from the substracted and unsubtracted sum rules.

### 54.7.2 $\sigma_B$ and $\sigma'_B$ couplings to $\pi\pi$

The decays of pure gluonia states have been estimated from the vertex function shown in Fig. 54.1 constrained by a low-energy theorem (LET):

$$V(q^2) \equiv \langle P|\theta_\mu^\mu|P\rangle = q^2 + 2m_P^2 , \qquad P \equiv \pi, K, \eta , \tag{54.33}$$

Table 54.3. *Gluonia masses and couplings from QSSR*

| $J^{PC}$ | Name | Mass (GeV) Estimate | Mass (GeV) Upper bound | $f_G$ (MeV) | $\sqrt{t_c}$ (GeV) |
|---|---|---|---|---|---|
| $0^{++}$ | $G$ | $1.5 \pm 0.2$ | $2.16 \pm 0.22$ | $390 \pm 145$ | $2.0 \sim 2.1$ |
| | $\sigma_B$ | $\leq 1.00$ (input) | | $\geq 1000$ | |
| | $\sigma'_B$ | 1.37 (input) | | 600 | |
| | $3G$ | 3.1 | | 62 | |
| $2^{++}$ | $T$ | $2.0 \pm 0.1$ | $2.7 \pm 0.4$ | $80 \pm 14$ | 2.2 |
| $0^{-+}$ | $P$ | $2.05 \pm 0.19$ | $2.34 \pm 0.42$ | $8 \sim 17$ | 2.2 |
| | $E/\iota$ | 1.44 (input) | | $7 : J/\psi \to \gamma\iota$ | |

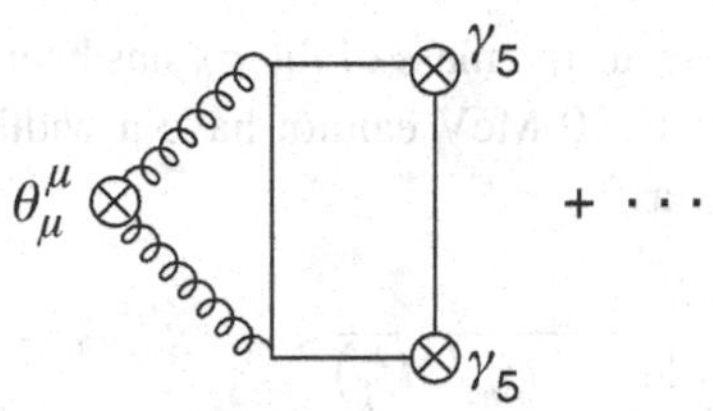


Fig. 54.1. QCD diagram contributing to the scalar gluonium decay into two pseudoscalar Goldstone bosons.

where $q$ is the scalar meson momentum. In the chiral limit, it obeys the dispersion relation:

$$V(q^2) = q^2 \int \frac{dt}{t - q^2 - i\epsilon} \frac{1}{\pi} \text{Im} V(t) , \tag{54.34}$$

which gives the first Narison–Veneziano (NV) sum rule [686]:

$$\frac{1}{4} \sum_{S \equiv \sigma_B, \sigma'_B, G} g_{S\pi\pi} \sqrt{2} f_S \simeq 0 . \tag{54.35}$$

Using $V'(0) = 1$, and a $\sigma_B$ meson dominance,[3] it leads to the second NV sum rule:

$$\frac{1}{4} \sum_{S \equiv \sigma_B, \sigma'_B, G} g_{SPP} \frac{\sqrt{2} f_S}{M_S^2} = 1 . \tag{54.36}$$

These two sum rules are a generalization of the Goldberger–Treiman relation. It gives the value of the $\sigma_B \pi^+ \pi^-$ and $\sigma'_B \pi^+ \pi^-$ couplings given in Table 54.4 leading to the unexpected large width of the lowest mass scalar gluonium into pairs of Goldstone bosons.[4]

[3] We could identify the $G(1.5 \sim 1.6)$ with the one observed at GAMS [744] which coupling to $\pi\pi$ is negligible, while for definiteness, we use as input $M_{\sigma'_B} \approx 1.37$ GeV.

[4] Similar results have been obtained in [689].

Table 54.4. *Unmixed scalar gluonia and quarkonia decays*

| Name | Mass (GeV) | $\pi\pi$ (GeV) | $KK$ (MeV) | $\eta\eta$ (MeV) | $\eta\eta'$ (MeV) | $(4\pi)_S$ (MeV) | $\gamma\gamma$ (keV) |
|---|---|---|---|---|---|---|---|
| $\sigma_B$ | $0.75 \sim 1.0$ (input) | $0.2 \sim 0.8$ | $SU(3)$ | $SU(3)$ | | | $0.2 \sim 0.3$ |
| $\sigma'_B$ | 1.37 (input) | $0.8 \sim 2.0$ | $SU(3)$ | $SU(3)$ | | $43 \sim 316$ (exp) | 0.7 |
| $G$ | 1.5 | $\approx 0$ | $\approx 0$ | $1.1 \sim 2.2$ | $5 \sim 10$ | $60 \sim 138$ | 0.5 |
| $S_2$ | 1. | 0.12 | $SU(3)$ | $SU(3)$ | | | 0.67 |
| $S'_2$ | $1.3 \approx \pi'$ | 0.3 | $SU(3)$ | $SU(3)$ | | 4.4 | |
| $S_3$ | $1.474 \pm 0.044$ | | $73 \pm 27$ | $15 \pm 6$ | | 0.4 | |
| $S'_3$ | $\approx 1.7$ | | 112 | $SU(3)$ | | 1.1 | |

The behaviour of this width versus the mass of the $\sigma_B$ has been given in [688], where it has been shown that a $\sigma_B$ of about 600 MeV cannot have a width larger than 150 MeV. The couplings behave analytically as:[5]

$$g_{\sigma_B\pi\pi} \approx \frac{4}{\sqrt{2}f_{\sigma_B}}\frac{1}{\left(1 - M^2_{\sigma_B}/M^2_{\sigma'_B}\right)}\ , \qquad g_{\sigma'_B\pi\pi} \approx g_{\sigma_B\pi\pi}\left(\frac{f_{\sigma_B}}{f_{\sigma'_B}}\right)\ . \tag{54.37}$$

The unexpected relatively large width of the $\sigma_B$ indicates a *large violation of the OZI rule* in the $U(1)$ scalar sector like in the case of the $\eta'$ decay. As a result from these sum rules, one expects that the scalar gluonium $\sigma_B$ has an universal coupling (up to $SU(3)$ symmetry breaking terms) to pairs of Goldstone bosons, which is a characteristic feature that should be tested experimentally.

### *54.7.3 G(1.5) coupling to $\eta\eta'$*

The coupling of the scalar gluonium to the pairs of $U(1)_A$ mesons ($\eta'\eta'$, $\eta\eta'$) is gouverned by a three-point function made with a glue line shown in Fig. 54.2.

Analogous low-energy theorem [686] gives at $q^2 = 0$:

$$\langle\eta_1|\theta^\mu_\mu|\eta_1\rangle = 2M^2_{\eta_1}\ , \tag{54.38}$$

where $\eta_1$ is the unmixed $U(1)$ singlet state of mass $M_{\eta_1} \simeq 0.76$ GeV. Writing the dispersion relation for the vertex, one obtains the NV sum rule:

$$\frac{1}{4}\sum_{S\equiv\sigma_B,\sigma'_B,G} g_{S\eta_1\eta_1}\sqrt{2}f_S = 2M^2_{\eta_1}\ . \tag{54.39}$$

[5] In some L$\sigma$M approaches, the coupling vanishes in the chiral limit as a reflection of the arbitrariness of the scalar potential term of the effective Lagrangian.

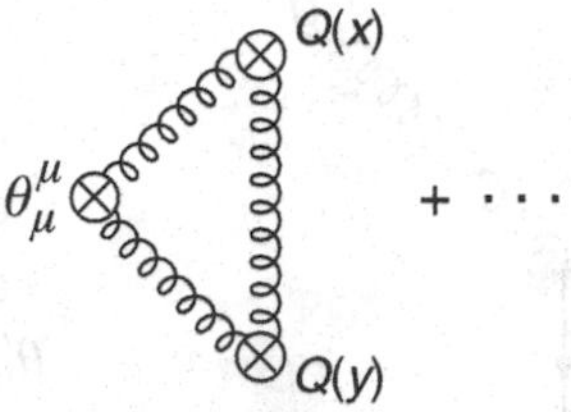


Fig. 54.2. QCD diagram contributing to the scalar gluonium decay into 2 $U(1)_A$ pseudoscalar mesons. $Q(x)$ represents the gluon part of the axial-anomaly.

Assuming a $G(1.5)$ dominance of the vertex, the sum rule leads to:

$$g_{G\eta_1\eta_1} \approx (1.2 \sim 1.7) \text{ GeV} . \tag{54.40}$$

Introducing the 'physical' $\eta'$ and $\eta$ through:

$$\eta' \equiv \cos\theta_P \eta_1 - \sin\theta_P \eta_8 , \qquad \eta \equiv \sin\theta_P \eta_1 + \cos\theta_P \eta_8 , \tag{54.41}$$

where [16,200] $\theta_P \simeq -(18 \pm 2)^\circ$ is the pseudoscalar mixing angle, one obtains the width given in Table 54.4. The previous scheme is also known to predict (see NV and [746]):

$$r \equiv \frac{\Gamma_{G\eta\eta}}{\Gamma_{G\eta\eta'}} \simeq 0.22 , \qquad g_{G\eta\eta} \simeq \sin\theta_P g_{G\eta\eta'} , \tag{54.42}$$

typical for the $U(1)_A$ anomaly dominance in the decays into $\eta'$ and $\eta$. It can be compared with the GAMS data $r \simeq 0.34 \pm 0.13$, and implies the width $\Gamma_{G\eta\eta}$ in Table 54.4. This result can then suggest that the $G(1.6)$ seen by the GAMS group is a pure gluonium, which is not the case of the particle seen by Crystal Barrel [735] which corresponds to $r \approx 1$. It also shows that the G(1.6) is a relatively narrow state, which may justify the validity of the lattice quenched approximation in the evaluation of its mass.

### 54.7.4 $\sigma'_B(1.37)$ and $G(1.5)$ couplings to $4\pi$

Within our scheme, we expect that the $4\pi$ are mainly $S$-waves initiated from the decay of pairs of $\sigma_B$. Using:

$$\langle\sigma_B|\theta^\mu_\mu|\sigma_B\rangle = 2M^2_{\sigma_B} , \tag{54.43}$$

and writing the dispersion relation for the vertex, one obtains the sum rule:

$$\frac{1}{4} \sum_{i=\sigma_B,\sigma'_B,G} g_{S\sigma_B\sigma_B}\sqrt{2}f_S = 2M^2_{\sigma_B} . \tag{54.44}$$

We identify the $\sigma'_B$ with the observed $f_0(1.37)$, and use its observed width into $4\pi$, which is about (46 ~ 316) MeV [16] ($S$-wave part). Neglecting, to a first approximation, the $\sigma_B$ contribution to the sum rule, we can deduce:

$$g_{G\sigma_B\sigma_B} \approx (2.7 \sim 4.3) \text{ GeV} , \tag{54.45}$$

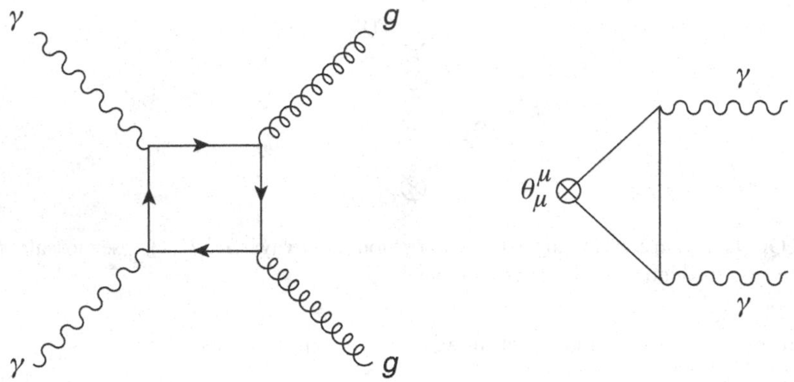


Fig. 54.3. Vertex controlling the gluonium couplings to $J/\psi(\gamma)\gamma$: (a) box diagram; (b) anomaly diagram.

which leads to the width of 60–138 MeV, much larger than the one into $\eta\eta$ and $\eta\eta'$ in Table 54.4. This feature seems to be satisfied by the states seen by GAMS [744] and Crystal Barrel [735]. Our previous approaches show the consistency in interpreting the $G(1.6)$ seen at GAMS as an 'almost' pure gluonium state (ratio of the $\eta\eta'$ versus the $\eta\eta$ widths), while the state seen by the Crystal Barrel, though having a gluon component in its wave function, cannot be a pure gluonium because of its prominent decays into $\eta\eta$ and $\pi^+\pi^-$. We shall see later on that *the Crystal Barrel state can be better explained from a mixing of the GAMS gluonium with the $S_3(\bar{s}s)$ and $\sigma'_B$ states.*

### 54.7.5 $\sigma_B$, $\sigma'_B$ and $G$ couplings to $\gamma\gamma$

The two-photon widths of the $\sigma_B$, $\sigma'_B$ and $G$ can be obtained by identifying the Euler–Heisenberg effective Lagrangian (see Fig. 54.3) [747]:[6]

$$\mathcal{L}_{\gamma g} = \frac{\alpha\alpha_s Q_q^2}{180 m_q^2}[28 F_{\mu\nu}F_{\nu\lambda}G_{\lambda\sigma}G_{\sigma\mu} + 14 F_{\mu\nu}G_{\nu\lambda}F_{\lambda\sigma}G_{\sigma\mu} - F_{\mu\nu}G_{\mu\nu}F_{\alpha\beta}G_{\alpha\beta} - F_{\mu\nu}F_{\mu\nu}G_{\alpha\beta}G_{\alpha\beta}] \,, \tag{54.46}$$

with the scalar-$\gamma\gamma$ Lagrangian

$$\mathcal{L}_{S\gamma\gamma} = g_{S\gamma\gamma}\sigma_B(x)F^{(1)}_{\mu\nu}F^{(2)}_{\mu\nu} \,. \tag{54.47}$$

This leads to the sum rule:

$$g_{S\gamma\gamma} \simeq \frac{\alpha}{60}\sqrt{2} f_S M_S^2 \left(\frac{\pi}{-\beta_1}\right) \sum_{q\equiv u,d,s} \frac{Q_q^2}{m_q^4} \,, \tag{54.48}$$

[6] $F^{\mu\nu}$ is the photon field strength, $Q_q$ is the quark charge in units of $e$, $-\beta_1 = 9/2$ for three flavours, and $m_q$ is the 'constituent' quark mass, which we shall take to be $m_u \simeq m_d \simeq M_\rho/2$, $m_s \simeq M_\phi/2$.

from which we deduce the couplings:[7]

$$g_{S\gamma\gamma} \approx (0.4 \sim 0.8)\alpha \text{ GeV}^{-1} , \tag{54.49}$$

$(S \equiv \sigma_B,\ \sigma'_B,\ G)$ and the widths in Table 54.3, smaller (as expected) than the well-known quarkonia width: $\Gamma(f_2 \to \gamma\gamma) \simeq 2.6$ keV. Alternatively, one can use the trace anomaly: $\langle 0|\theta^\mu_\mu|\gamma_1\gamma_2\rangle$ and the fact that its RHS is $\mathcal{O}(k^2)$, in order to get the sum rule [748,382] $(R \equiv 3\sum Q_i^2)$:

$$\langle 0|\frac{1}{4}\beta(\alpha_s)G^2|\gamma_1\gamma_2\rangle \simeq -\langle 0|\frac{\alpha R}{3\pi}F_1^{\mu\nu}F_2^{\mu\nu}|\gamma_1\gamma_2\rangle , \tag{54.50}$$

from which one can deduce the couplings:

$$\frac{\sqrt{2}}{4}\sum_{S\equiv\sigma_B,\sigma'_B,G} f_S g_{S\gamma\gamma} \simeq \frac{\alpha R}{3\pi} . \tag{54.51}$$

It is easy to check that the previous values of the couplings also satisfy the trace anomaly sum rule.

### 54.7.6 $J/\psi \to \gamma S$ radiative decays

As stated in [747], one can estimate the width of this process, using dispersion relation techniques, by saturating the spectral function by the $J/\psi$ plus a continuum. The glue part of the amplitude can be converted into a physical non-perturbative matrix element $\langle 0|\alpha_s G^2|S\rangle$ known through the decay constant $f_S$ estimated from QSSR. By assuming that the continuum is small, one obtains:[8]

$$\Gamma(J/\psi \to \gamma S) \simeq \frac{\alpha^3\pi}{\beta_1^2 656100}\left(\frac{M_{J/\psi}}{M_c}\right)^4\left(\frac{M_\sigma}{M_c}\right)^4\frac{\left(1-M_S^2/M_{J/\psi}^2\right)^3}{\Gamma(J/\psi \to e^+e^-)}f_\sigma^2 . \tag{54.52}$$

This leads to:[9]

$$B(J/\psi \to \gamma S) \times B(S \to \text{ all}) \approx (0.4 \sim 1) \times 10^{-3} . \tag{54.53}$$

for $S \equiv \sigma_B,\ \sigma'_B,\ G$. These branching ratios can be compared with the observed $B(J/\psi \to \gamma f_2) \simeq 1.6 \times 10^{-3}$. The $\sigma_B$ could already have been produced, but might have been confused with the $\pi\pi$ background. The 'pure gluonium' $G$ production rate is relatively small, contrary to the naïve expectation for a glueball production. In our approach, this is due to the relatively small value of its decay constant, which controls the non-perturbative dynamics. Its observation from this process should wait for the $\tau$CF machine. However, we do not exclude the possibility that a state resulting from a quarkonium-gluonium mixing may be produced at higher rates.

7 Here and in the following, we shall use $M_{\sigma_B} \approx (0.75 \sim 1.0)$ GeV.

8 We use $M_c \simeq 1.5$ GeV for the charm constituent quark mass and $-\beta_1 = 7/2$ for six flavours.

9 From the previous results, one can also deduce the corresponding stickiness which is proportional to $\Gamma(J/\psi \to X\gamma)/\Gamma(X \to \gamma\gamma)$ in [736].

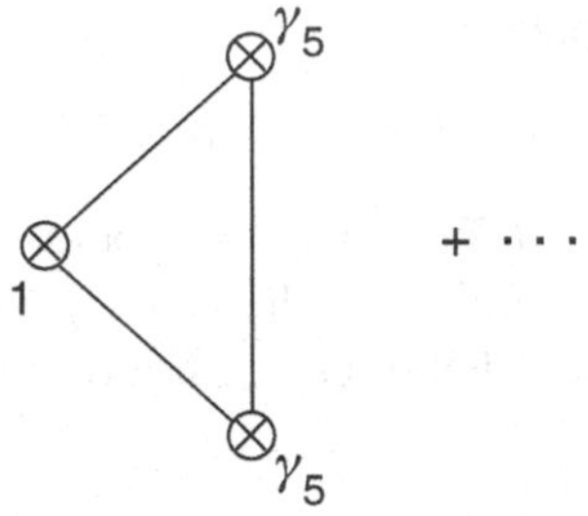


Fig. 54.4. QCD diagram contributing to the scalar $\bar{q}q$ decay into two pseudoscalar mesons Goldstone bosons.

### 54.7.7 $\phi \to \sigma_B \gamma$ and $D_s \to \sigma_B l \nu$ decays

From the previous approaches, one also expects to produce the $\sigma_B$ from $\phi$ radiative decays [749,688]. Similar analysis can also be done for the $D_s \to \sigma_B l \nu$ semi-leptonic decays [750], where the $\pi\pi$ final state is uniquely produced from a glue rich channel. One should note that due to the large OZI-violation of the $\sigma_B \to \pi\pi$ process, one expects that the $\phi \to \sigma_B \gamma$ and $D_s \to \sigma_B l \nu$ decay rates are sizeable.

## 54.8 Unmixed scalar quarkonia

The masses of pure $\bar{q}q$ states are given in Table 54.1 and have been obtained using QSSR [3] and [688]. Here $S_2$ and $S_3$ denote isoscalar ($I = 0$) scalar states $\bar{u}u + \bar{d}d$ and $\bar{s}s$ and their radial excitations $S'_2$ and $S'_3$. QSSR predicts a degeneracy between the isovector $a_0$[10] and isoscalar mass $S_2$ of about 1 GeV in the absence of mixing, while due to $SU(3)$ breaking the mass of the $S_3$ is predicted to be in the range of $1.3 \sim 1.4$ GeV. Their widths have been estimated using QSSR for the vertex shown in Fig. 54.4 [751,3] and [688]. These results are compiled in Table 54.4 and indicate that the *$\bar{q}q$ interpretation of the wide $\sigma(0.6)$ and $\kappa(0.9)$ indicated by some recent data is incompatible with these QSSR results*. The discrepancy with the $\kappa$ mass is more intriguing, where in this isovector channel, one expects no possible gluon component. The existence of this particle should be further tested from independent and cleaner processes. In the table we quote the updated values from [688], where we have used the fact that the $G$ couplings to $\pi\pi$ and $KK$ are negligible as indicated by the GAMS data [744].

## 54.9 Mixing schemes for scalar mesons

Previous results for the decay widths of unmixed scalar states can be used using phenomenological mixing schemes in order to explain or to predict the widths of the observed scalar mesons.

[10] One should notice (see Chapter 53), that the $a_0$ is the particle naturally associated to the divergence of the $SU(2)$ vector current, and which satifies the different chiral symmetry breaking constraints: $m_d - m_u, \ldots$

### 54.9.1 Nature of the $\sigma$ and $f_0(0.98)$

Below or around 1 GeV, one can use a two-component mixing as in [687,688] between the gluonium $\sigma_B$ and quarkonium $S_2$, and fixes the mixing angle from the predicted $\Gamma(\sigma_B \to \gamma\gamma)$ in Table 54.4 and the observed $\Gamma(f_0 \to \gamma\gamma) \approx 0.3$ keV. In this way one obtains a maximal mixing angle:

$$|\theta_S| \simeq (40 \sim 45)^\circ , \qquad (54.54)$$

indicating that the $f_0$ and $\sigma$ have a large amount of glue and quarks in their wave functions, which is a situation quite similar to the case of the $\eta'$ in the pseudoscalar channel (mass given by its gluon component but strong coupling to quarkonia).

By recapitulating, our scheme suggests that around 1 GeV, there are two mesons that have 1/2 gluon and 1/2 quark in their wave functions resulting from a maximal destructive mixing of a quarkonium ($S_2$) and gluonium ($\sigma_B$) states:

- The $f_0(0.98)$ is narrow, with a width $\leq$ 134 MeV, and couples strongly to $\bar{K}K$, with the strength $g_{f_0K^+K^-}/g_{f_0\pi^+\pi^-} \simeq 2$, a property seen in $\pi\pi$ and $\gamma\gamma$ scatterings [742] and in $\bar{p}p$ [737] experiments. Its production from $\phi$ radiative decays is [749]:

$$\Gamma(\phi \to f_0(980)\gamma) \simeq 1.3 \times 10^{-4} , \qquad (54.55)$$

  in good agreement with recent data from Novosibirsk and Daphne of about $1.1 \times 10^{-4}$.
- The $\sigma$, with a mass around $(0.75 \sim 1)$ GeV, is large, with a width of about $(400 \sim 900)$ MeV, and has universal couplings to $\pi\pi$ and $KK$. However, our analysis shows that a $\sigma$ with lower mass cannot be large.

### 54.9.2 Nature of the $f_0(1.37)$, $f_0(1.5)$ and $f_J(1.7)$

We use a three-component mixing scheme between the $\sigma'(1.37)$, $S_3(1.47)$ and $G(1.5)$ bare states in order to explain the nature of the above-mentioned three observed states. In order to fix the different angles of the CKM-like $3 \times 3$ mixing matrices, we use the following input deduced from the present Crystal Barrel data [752]:

$$\Gamma(f_0(1.50) \to \pi\pi) \simeq (20 \sim 31)\ \text{MeV} \qquad \Gamma(f_0(1.50) \to \bar{K}K) \simeq (3.6 \sim 5.6)\ \text{MeV}$$
$$\Gamma(f_0(1.50) \to \eta\eta) \simeq (2.6 \sim 3.3)\ \text{MeV} \qquad \Gamma(f_0(1.50) \to 4\pi^0) \simeq (68 \sim 105)\ \text{MeV}, \qquad (54.56)$$

and:

$$\Gamma(f_0(1.50) \to \eta\eta') \simeq 1.3\ \text{MeV} . \qquad (54.57)$$

We shall also use the widths of the $\sigma'$ given in the previous table. The resulting values of the mixing angles read:

$$\begin{pmatrix} f_0(1.37) \\ f_0(1.50) \\ f_0(1.60) \end{pmatrix} \approx \begin{pmatrix} 0.01 \sim 0.22 & -(0.44 \sim 0.7) & 0.89 \sim 0.67 \\ 0.11 \sim 0.16 & 0.89 \sim 0.71 & 0.43 \sim 0.69 \\ -(0.99 \sim 0.96) & -(0.47 \sim 0.52) & 0.14 \sim 0.27 \end{pmatrix} \begin{pmatrix} \sigma'(1.37) \\ S_3(1.47) \\ G(1.5) \end{pmatrix} , \tag{54.58}$$

where the first (respectively second) numbers correspond to the case of large (respectively narrow) $\sigma'$ widths. From the previous schemes, we deduce the predictions in units of MeV:[11]

$$\Gamma(f_0(1.37) \to \pi\pi) \approx (22 \sim 48) \quad \Gamma(f_0(1.37) \to \eta\eta) \leq 1. \quad \Gamma(f_0(1.37) \to \eta\eta') \leq 2.5 , \tag{54.59}$$

and

$$\Gamma(f_0(1.5) \to \bar{K}K) \approx (3 \sim 12) \quad \Gamma(f_0(1.5) \to \eta\eta) \approx (1 \sim 2) \quad \Gamma(f_0(1.5) \to \eta\eta') \leq 1 , \tag{54.60}$$

despite the crude approximation used, these are in good agreement with the data. These results suggest that the observed $f_0(1.37)$ and $f_0(1.5)$ comes from a maximal mixing between gluonia ($\sigma'$ and $G$) and quarkonia $S_3$. The mixing of the $S_3$ and $G$ with the quarkonium $S_2'$, which we have neglected compared with the $\sigma'$, can restore the small discrepancy with the data. One should notice that the state seen by GAMS [744] is more likely to be similar to the unmixed gluonium state $G$ (dominance of the $4\pi$ and $\eta\eta'$ decays), as already emphasized in [686], which can be due to some specific features of the central production for the GAMS experiment.[12]

*Nature of the $f_J(1.71)$*

For the $f_0(1.6)$, we obtain in units of GeV:

$$\begin{aligned} \Gamma(f_0(1.6) \to \bar{K}K) &\approx (0.5 \sim 1.6) & \Gamma(f_0(1.6) \to \pi\pi) &\approx (0.9 \sim 2.) \\ \Gamma(f_0(1.6) \to \eta\eta) &\approx (0.04 \sim 0.6) & \Gamma(f_0(1.6) \to \eta\eta') &\approx (0.03 \sim 0.07) , \end{aligned} \tag{54.61}$$

and

$$\Gamma(f_0(1.6) \to (4\pi)_S) \approx (0.02 \sim 0.2) , \tag{54.62}$$

which suggests that the $f_0(1.6)$ is very broad and can again be confused with the continuum. Therefore, the $f_J(1.7)$ observed to decay into $\bar{K}K$ with a width of the order $(100 \sim 180)$ MeV, can be essentially composed by the radial excitation $S_3'(1.7 \sim 2.4)$ GeV of the $S_3(\bar{s}s)$, as they have about the same width into $\bar{K}K$ (see Section 6.4). This can also explain the smallness of the $f_J(1.7)$ width into $\pi\pi$ and $4\pi$. Our predictions of the $f_J(1.71)$ width can agree with the result of the OBELIX collaboration [737], while its small decay width into $4\pi$

[11] Recall that we have used as inputs: $\Gamma(f_0(1.37) \to \bar{K}K) \approx 0$, $\Gamma(f_0(1.5) \to \pi\pi) \approx 25$ MeV, while our best prediction for $\Gamma(f_0(1.5) \to (4\pi)_S)$ is about 150 MeV. The present data also favour negative values of the $f_0\eta\eta$, $f_0\eta'\eta$ and $f_0 KK$ couplings.

[12] Some alternative scenarios are discussed in [754,755].

is consistent with the best fit of the Crystal Barrel collaboration (see Abele *et al.* in [752]), which is consistent with the fact that the $f_0(1.37)$ likes to decay into $4\pi$. However, the $f_0(1.6)$ and the $f_J(1.71)$ can presumably interfere destructively for giving the dip around $1.5 \sim 1.6$ GeV seen in the $\bar{K}K$ mass distribution from the Crystal Barrel and $\bar{p}p$ annihilations at rest [753,737].

### 54.10 Mixing and decays of the tensor gluonium

The mass mixing between the tensor gluonium and quarkonium states has been estimated to be small:

$$\theta_T \simeq -10^\circ\ , \tag{54.63}$$

in [450] and [2] using the off-diagonal two-point correlator:

$$\begin{aligned}\psi^{gq}_{\mu\nu\rho\sigma} &\equiv i\int d^4x\ e^{iqx}\langle 0|T\theta^g_{\mu\nu}(x)\theta^q_{\rho\sigma}(0)^\dagger|0\rangle \\ &= \frac{1}{2}\left(\eta_{\mu\rho}\eta_{\nu\sigma}+\eta_{\mu\sigma}\eta_{\nu\rho}-\frac{2}{3}\eta_{\mu\nu}\eta_{\rho\sigma}\right)\psi_{gq}(q^2)\ ,\end{aligned} \tag{54.64}$$

where

$$\theta^q_{\mu\nu}(x) = i\bar{q}(x)(\gamma_\mu\bar{D}_\nu+\gamma_\nu\bar{D}_\mu)q(x)\ . \tag{54.65}$$

Here, $\bar{D}_\mu \equiv \vec{D}_\mu - \overleftarrow{D}_\mu$ is the covariant derivative, and the other quantities have been defined earlier.

The hadronic width of the tensor gluonium has been constrained to be [452,450]:

$$\Gamma(T \to \pi\pi + KK + \eta\eta) \leq (119 \pm 36)\ \text{MeV}\ , \tag{54.66}$$

from the $f_2 \to \pi\pi$ data, and assuming an universal coupling of the $T$ to pairs of Goldstone bosons. A vertex sum rule analogous to the case of the scalar gluonia assumed to be saturated by the $f_2$ and $T$ leads to:

$$\Gamma(T \to \pi\pi) \leq 70\ \text{MeV}\ . \tag{54.67}$$

These results show that the tensor gluonium cannot be wide. Its width into $\gamma\gamma$ can be related to that of the scalar gluonium $G$ using a non-relativistic relation. In this way one obtains:

$$\Gamma(T \to \gamma\gamma) \simeq \frac{4}{15}\left(\frac{M_T}{M_{0^+}}\right)^3 \Gamma(0^{++} \to \gamma\gamma) \simeq 0.06\ \text{keV}, \tag{54.68}$$

which shows again a small value typical of a gluonium state.

### 54.11 Mixing and decays of the pseudoscalar gluonium

The mass mixing angle between the pseudoscalar gluonium and quarkonium states has also been estimated from the off-diagonal two-point correlator, with the value [458,3,734]:

$$\theta_P \simeq 12^\circ, \tag{54.69}$$

from which one can deduce:

$$\begin{aligned}\Gamma(P \to \gamma\gamma) &\simeq \tan^2\theta_P \left(\frac{M_P}{M_{\eta'}}\right)^3 \Gamma(\eta' \to \gamma\gamma) \approx 1.3 \text{ keV} \\ \Gamma(P \to \rho\gamma) &\simeq \tan^2\theta_P \left(\frac{k_P}{k_{\eta'}}\right)^3 \Gamma(\eta' \to \rho\gamma) \approx 0.3 \text{ MeV}\,,\end{aligned} \tag{54.70}$$

where $k_i$ is the momentum of the particle $i$. We have used $\Gamma(\eta' \to \gamma\gamma) \simeq 4.3$ keV and $\Gamma(\eta' \to \rho\gamma) \simeq 72$ keV. Measurements of the $P$ widths can test the amount of glue inside the $P$-meson.

Some other aplications of the sum rules in the pseudoscalar channel will be discussed later on. These concern the estimate of the topological charge and its implications to the proton spin.

### 54.12 Test of the four-quark nature of the $a_0(980)$

The four-quark nature of the $a_0(980)$ and $f_0(975)$ has been conjectured from the bag model approach [73,757] in order to explain their degeneracy and their large couplings to $K^+K^-$ pairs, where in this scheme, their quark content would be:

$$\begin{aligned}|a_0\rangle &= \frac{1}{\sqrt{2}}\bar{s}s(\bar{u}u - \bar{d}d) \\ |f_0\rangle &= \frac{1}{\sqrt{2}}\bar{s}s(\bar{u}u + \bar{d}d)\,.\end{aligned} \tag{54.71}$$

On the other hand, this scheme is unlikely as it does not explain why the usual $\bar{q}q$ scalar states are absent, whilst in addition, it leads to a proliferation of states (many cryptoexotics, ...). Moreover, the relation of this model with the usual chiral and flavour symmetries is not obvious. Within the framework of QSSR, the two-point correlator associated to the colour singlet operators:

$$\mathcal{O}_1^{\pm} = \frac{1}{\sqrt{2}} \sum_{\Gamma=1,\gamma_5} \bar{s}\Gamma s(\bar{u}\Gamma u - \bar{d}\Gamma d)\,, \tag{54.72}$$

has been firstly studied in [465], where the LSR analysis leads to:

$$M_E \simeq 1 \text{ GeV}\,, \qquad f_E \simeq 2.5 \text{ MeV}\,, \tag{54.73}$$

where $E$ corresponds to exotic and $f_E$ is normalized as $f_\pi = 92.4$ MeV. In [756], one has

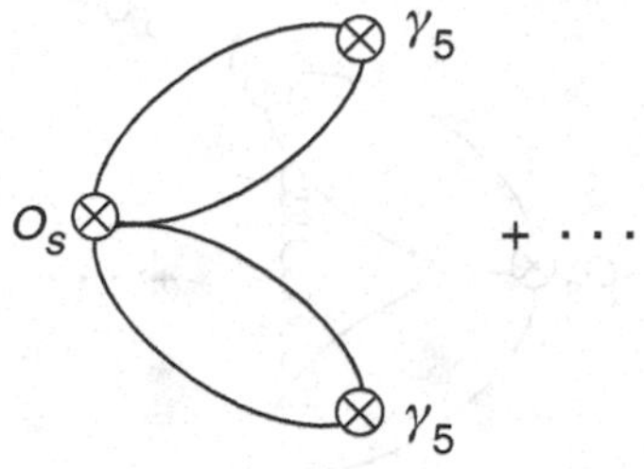


Fig. 54.5. QCD diagram contributing to the $a_0$ decay into two pseudoscalar bosons in the assumption of a four-quark state.

also introduced the operator:

$$\mathcal{O}_2^{\pm} = \frac{1}{\sqrt{2}} \sum_{\Gamma=1,\gamma_5} \bar{s}\Gamma\lambda_a s(\bar{u}\Gamma u - \bar{d}\Gamma\lambda^a d) , \tag{54.74}$$

and studied the more general combination:

$$\mathcal{O}^{\pm} = \mathcal{O}_1^{\pm} + t\mathcal{O}_2^{\pm} . \tag{54.75}$$

In this case the decay constant $f_E$ becomes function of $t$ as:

$$f_E \simeq \left(1 + \frac{32}{9}t^2\right)^{1/2} . \tag{54.76}$$

A sum rule analysis of the vertex shown in Fig. 54.5 leads to the $a_0 \to \eta\pi, \bar{K}K$ widths [756,3]:

$$\Gamma(a_0 \to \eta\pi) \simeq (52 \sim 88)\ \text{MeV} \left(1 + \frac{32}{9}t^2\right)^{-1} , \tag{54.77}$$

and:

$$g_{a_0\bar{K}K} \simeq \sqrt{\frac{3}{2}} g_{a_0\eta\pi} \times \left(1 + \frac{16}{3}t\right) \simeq (5 \sim 8)\ \text{GeV} . \tag{54.78}$$

These results indicate that the $a_0\bar{K}K$ coupling can strongly deviate from the $SU(3)$ expectation owing to the extra $(1 + \frac{16}{3}t)$ factor, which is a result expected in this scenario. However, an estimate of the $\gamma\gamma$ width within the same framework, as shown in Fig. 54.6, gives:

$$\Gamma(E \to \gamma\gamma) \approx (2 \sim 5) \times 10^{-4}\ \text{keV} , \tag{54.79}$$

which is too small compared with the data for the $a_0$ of about 0.3 keV. The smallness of this quantity can be better understood if one compares the ratio of the $\gamma\gamma$ couplings obtained

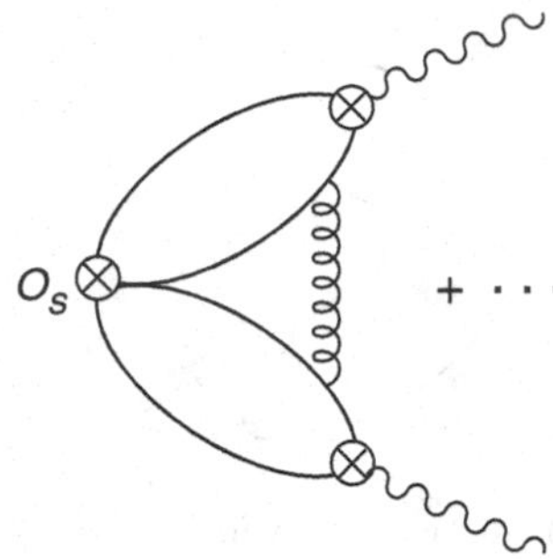


Fig. 54.6. QCD diagram contributing to the $a_0$ decay into two photons in the assumption of a four-quark state.

from QSSR for the $\bar{q}q$ and four-quark representation of the $a_0$. At the stability point, one obtains:

$$\frac{g_{E\gamma\gamma}}{g_{a_0\gamma\gamma}} \simeq -\left(\frac{\alpha_s}{\pi}\right)\frac{4t}{27}\frac{\langle\bar{s}s\rangle}{M_E^2 f_E}\frac{f_{a_0}}{(m_d - m_u)}, \tag{54.80}$$

where $f_{a_0}$ is proportional to the running quark mass difference $m_d - m_u$. This relation shows the relative suppression of $g_{E\gamma\gamma}$ with respect to the one in the $\bar{q}q$ scheme reproduced correctly by the QSSR method. The coefficients in the previous equation does not support the *rough* estimate [758]:

$$\Gamma(E \to \gamma\gamma)_{(\bar{q}q)^2} \approx 0.24\alpha_s^2\Gamma(a_0 \to \gamma\gamma)_{\bar{q}q}, \tag{54.81}$$

which overestimates the width by about two orders of magnitude ($\approx 1/16\pi^2$). This negative result does not support the four-quark nature of the $a_0$. Further experimental tests are needed for checking this result. We plan to come back to this point in a future work.

## 54.13 Light hybrids

Hybrid mesons are interesting due to the *exotic* quantum numbers, such that they are not expected to mix with ordinary mesons. A lot of studies have investigated their activities within QSSR [459,460] and the final correct QCD result has been obtained in [461]. In this approach, one can consider the colourless, local and gauge-invariant operators:

$$\mathcal{O}_V^\mu(x) =: g\bar{\psi}\lambda_a\gamma_\nu\psi G_a^{\mu\nu} : \qquad \mathcal{O}_A^\mu(x) =: g\bar{\psi}\lambda_a\gamma_\nu\gamma_5\psi G_a^{\mu\nu} : \tag{54.82}$$

corresponding to the vector and axial-vector channels. They are the only lowest dimension operators that can be used to study the quantum numbers of the exotic mesons $1^{-+}$ an $0^{--}$. The corresponding two-point correlator can be decomposed into its transverse spin 1 and

Table 54.5. *Light hybrid masses and couplings from QSSR*

| $J^{PC}$ | Name | Mass (GeV) | $f_H$ (MeV) | $\sqrt{t_c}$ (GeV) |
|---|---|---|---|---|
| $1^{-+}$ | $\tilde{\rho}(\bar{u}gd)$ | $1.6 \sim 1.7$ | $25 \sim 50$ | $M_{\tilde{\rho}} + 0.2$ |
| | $\tilde{\phi}(\bar{s}gs)$ | $M_{\tilde{\rho}} + 0.6$ | | |
| $0^{--}$ | $\tilde{\eta}(\bar{u}gu + \bar{d}gd)$ | 3.8 | – | 4.1 |

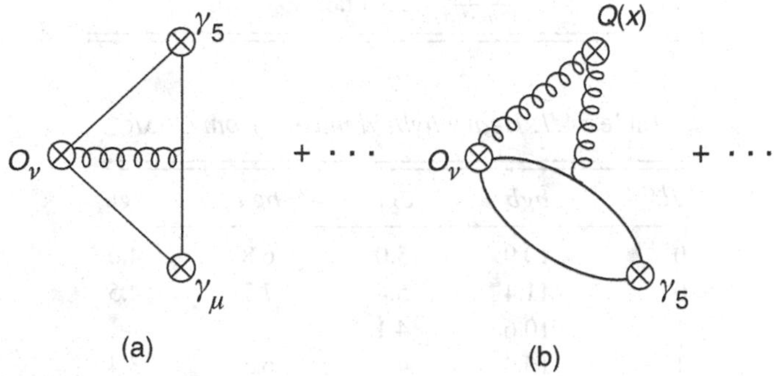


Fig. 54.7. QCD diagram contributing to the $\tilde{\rho}$ decay into (a) $\pi\rho$ and (b) $\pi\eta'$. $Q(x)$ represents the gluon component of the axial-anomaly.

longitudinal spin 0 parts:

$$\begin{aligned}\Pi^{\mu\nu}_{V/A}(q^2) &\equiv i \int d^4x\, e^{iqx} \langle\, 0|\mathcal{T}\mathcal{O}^{\mu}_{V/A}(x) \left(\mathcal{O}^{\nu}_{V/A}(0)\right)^{\dagger} |0\,\rangle \\ &= -(g^{\mu\nu}q^2 - q^{\mu}q^{\nu})\Pi^{(1)}_{V/A}(q^2) + q^{\mu}q^{\nu}\Pi^{(0)}_{V/A}(q^2)\,. \end{aligned} \tag{54.83}$$

### *54.13.1 Spectra*

The masses of the light hybrids have been obtained in [461,3,462] using LSR within stability criteria and a two-parameter fit of the ratio of moments. The inclusion of the contribution of a new $D = 2$ operator due to tachyonic gluon mass has been considered in [462], but the effect on the mass and coupling predictions is negligible. The updated results [462] are given in Table 54.5.

### *54.13.2 Decay widths of the $\tilde{\rho}$*

The different exclusive decays of the $\tilde{\rho}$ have been studied in [461,3,462] using vertex sum rules shown in Fig. 54.7.

We give the results in Table 54.6. These results show that the $\rho\pi$ is the dominant mode, while $\eta'\pi$ is the most characteristic signal for detecting the $\tilde{\rho}$.

Table 54.6. *Decay modes of the $\tilde{\rho}$ from QSSR*

| Decay modes | Width (MeV) | Comments |
|---|---|---|
| $\rho\pi$ | 274 | |
| $K^*K$ | 8 | |
| $\gamma\pi$ | 3 | |
| $\pi\pi, KK$ | $\approx 0$ | $\mathcal{O}(m_q^2)$ |
| $\eta'\pi$ | 3 | $U(1)$ anomaly |
| $\eta\pi$ | $\frac{\Gamma(\tilde{\rho}\to\eta\pi)}{\Gamma(\tilde{\rho}\to\eta'\pi)} \simeq 3.1\tan^2\theta_P$ | |

Table 54.7. *Heavy hybrid masses from QSSR*

| $J^{PC}$ | $\bar{b}gb$ | $\bar{c}gc$ | $\bar{b}gu$ | $\bar{c}gu$ |
|---|---|---|---|---|
| $0^{++}$ | 10.9 | 5.0 | 6.8 | 4.0 |
| $0^{--}$ | 11.4 | 5.4 | 7.7 | 4.5 |
| $1^{+-}$ | 10.6 | 4.1 | – | – |
| $1^{++}$ | 10.9 | 4.7 | 6.5 | 3.4 |

## 54.14 Heavy hybrids

The heavy hybrids have been studied within QSSR in [463]. Unfortunately, no independent group has checked their calculations.[13] The results presenting $\tau$ stability are given in Table 54.7.

One should notice that the results obtained in [463] have no $t_c$ stability as they increase with $t_c$. The results in the table correspond to the beginning of $\tau$ stability. They correspond to the value of $\sqrt{t_c}$ of about (0.3–0.4) and (0.6–0.7) GeV above the meson masses respectively for the $\bar{c}gc$ and $\bar{b}gb$. For the $\bar{c}gu$ and $\bar{b}gu$, the $\sqrt{t_c}$ values are respectively 0.2 and (0.3–0.4) GeV above the meson masses. One can see from this table that the splitting between two opposite $C$-parity states is typically (300 $\sim$ 500) MeV, while the spin zero state is much heavier than the spin one, which is similar to the case of light hybrids. These results are in general in agreement with lattice values [759].

### *54.14.1 Conclusions*

There are some progress in the long-term study and experimental search for the exotics. Before some definite conclusions, one still needs improvments of the present data, and some improved unquenched lattice estimates which should complement the QCD spectral sum rule (QSSR) and low-energy theorem (LET) results. Our results cannot absolutely exclude the existence of the $1^{-+}$ state at 1.4 GeV seen recently in hadronic machines (BNL and

[13] Recently, we have checked some of these results [464].

Crystal Barrel) [760], but at the same time predict the existence of a $1^{--}$ hybrid almost degenerate with the $1^{-+}$, and which could manifest in $e^+e^- \to$ hadrons by mixing with the radial excitations of the $\rho$ mesons. The relatively low value of $t_c$ might also indicate a rich population of (axial-)vector hybrid states in the region aroud 1.8 GeV. In our analysis, the lightest $0^-$ states are in the range of the charmonium states, such that they could mix with these states as well. A search for heavy hybrid mesons at LHCb or some other $B$-factories should be useful for testing the theoretical predictions.

# 55

# $D$, $B$ and $B_c$ exclusive weak decays

This chapter[1] contains a discussion of the different results obtained from the vertex sum rules for the weak semi-leptonic decays of the $D$, $B$ and $B_c$. Intensive activities have been devoted to these processes during the last few years using different non-perturbative QCD approaches (QSSR [3,364,761–771]; light-cone sum rules [360,780–782]; lattice calculations [723]; heavy quark symmetry [164], and a perturbative factorization treatment within a heavy quark approach [784,600,601]). Here, we shall concentrate on the study of the previous decays from the point of view of QSSR from which we can extract the values of the form factors and some CKM mixing angles.

## 55.1 Heavy to light exclusive decays of the $B$ and $D$ mesons

### *55.1.1 Introduction and notations*

One can extend the analysis carried out for the two-point correlator to the more complicated case of three-point function, in order to study the form factors related to the heavy to light transitions: $B \to K^*\gamma$ and $B \to \rho/\pi$ semi-leptonic decays. In contrast to the heavy to heavy transitions, where the symmetry of the heavy quarks can be exploited and which considerably simplifies the analysis, the heavy to light processes need non-perturbative approaches such as lattice or/and QSSR. For the QSSR approach, which we shall discuss here, we can consider the generic process:

$$B(D) \to L + \gamma(l\bar{\nu}) , \tag{55.1}$$

and the corresponding three-point function:

$$V(p, p', q^2) \equiv -\int d^4x \, d^4y \, e^{i(p'x-py)} \langle 0|\mathcal{T} J_L(x)\mathcal{O}(0)J_B^\dagger(y)|0\rangle , \tag{55.2}$$

where $J_L$, $J_B$ are the currents of the light and $B$ mesons; $\mathcal{O}$ is the weak operator specific for each process (penguin for the $K^*\gamma$, weak current for the semi-leptonic); $q \equiv p - p'$

[1] This is an extension and an update of the part of book [3] and the review given in [364].

and $q^2 \equiv t$.[2] The vertex obeys the double dispersion relation:

$$V(p^2, p'^2) \simeq \int_{M_b^2}^{\infty} \frac{ds}{s - p^2 - i\epsilon} \int_{m_L^2}^{\infty} \frac{ds'}{s' - p'^2 - i\epsilon} \frac{1}{\pi^2} \operatorname{Im} V(s, s') . \qquad (55.3)$$

As usual, the QCD part enters the LHS of the sum rule, while the experimental observables can be introduced through the spectral function after the introduction of the intermediate states:

$$V(p^2, p'^2) \simeq \frac{\langle 0|J_L(x)|L\rangle\langle L|\mathcal{O}(0)|B\rangle\langle B|J_B(y)|0\rangle}{(M_L^2 - p'^2)(M_B^2 - p^2)} + \text{higher states} . \qquad (55.4)$$

$M_L$ and $M_B$ are respectively the masses of the final $L$ and $B$ mesons. The matrix elements are:

$$\langle 0|J_{P,B}(x)|P, B\rangle = \sqrt{2} f_{P,B} M_{P,B}^2 , \qquad \langle 0|J_V^{\mu}(x)|V\rangle = \frac{\sqrt{2} M_V^2}{2\gamma_V} \epsilon^{\mu} , \qquad (55.5)$$

respectively, for pseudoscalar and vector states, where $f_\pi = 92.4$ MeV and $\gamma_\rho = 2.55$. The meson decay constants have been obtained either from the meson leptonic width or from the analysis of the two-point function discussed in the previous chapter of this book. Here, we shall be interested on the evaluation of the matrix element:

$$\langle L|\mathcal{O}(0)|B\rangle . \qquad (55.6)$$

The improvement of the dispersion relation can be done in the way discussed previously for the two-point function. In the case of the heavy to light transition, where the two sum rule scales are quite different, the only possible improvement with a good $M_b$ behaviour at large $M_b$ (convergence of the QCD series[3]) is the so-called hybrid sum rule (HSR) corresponding to the uses of the moments for the heavy-quark channel and to the Laplace for the light one [721,761]:

$$\mathcal{H}(n, \tau') = \frac{1}{\pi^2} \int_{M_b^2}^{\infty} \frac{ds}{s^{n+1}} \int_0^{\infty} ds' \, e^{-\tau' s'} \operatorname{Im} V(s, s') . \qquad (55.7)$$

Assuming that the higher state contributions to the spectral function are approximated by those of the QCD continuum from a threshold $t_c$ and $t'_c$, and assuming that the QCD contribution also obeys a double dispersion relation, one obtains the FESR:

$$\mathcal{H}(n, \tau') = \frac{1}{\pi^2} \int_{M_b^2}^{t_c} \frac{ds}{s^{n+1}} \int_0^{t'_c} ds' \, e^{-\tau' s'} \operatorname{Im} V(s, s') . \qquad (55.8)$$

[2] It has to be noticed that we shall use here, like in [761–764], the pseudoscalar current $J_P = (m_u + m_d)\bar{u}(i\gamma^5)d$ for describing the pion, where the QCD expression of the form factor can be deduced from the one in [767] by taking $m_c = 0$ and by remarking that the additional effect due to the light quark condensate for $B \to \pi$ relative to $B \to D$ vanishes in the sum rule analysis. In the literature [768,771], the axial-vector current has been used. However, as it is already well known in the case of the two-point correlator of the axial-vector current, by keeping its $q_\mu q_\nu$ part, (which is similarly done in the case of the three-point function) one obtains the contribution from the $\pi$ plus the $A_1$ mesons but *not* the $\pi$ contribution alone. Although, the $A_1$ effect can be numerically small in the sum rule analysis due to its higher mass, the mass behaviour of the form factor obtained in this way differs significantly from the one where the pseudoscalar current has been used due to the different QCD expressions of the form factor in the two cases.

[3] One should notice here that contrary to the case of the double exponential (Borel) sum rule where the mixed condensate explodes, the OPE behaves quite well, at least to leading order in $\alpha_s$.

Finally, in order to minimize the effects of the sum rule parameters into the analysis, it is usual to introduce the two-point sum rule expression of the decay constants and use a suitable relation between the three-point and two-point sum rules variables. These relations are [773] (obvious index notations):

$$\tau_3 = \frac{\tau_2}{2} \tag{55.9}$$

for the double Laplace sum rule (DLSR), and [763]:

$$n_3 = \frac{1}{2}\left(n_2 - \frac{1}{2}\right), \tag{55.10}$$

for the hybrid sum rule (HSR) and lead to a cancellation of the $\tau$ or $n$ dependences in the sum rule analysis [768,731,761–766]. The different form factors entering the previous semi-leptonic and radiative processes are defined as:

$$\begin{aligned}\langle\rho(p')|\bar{u}\gamma_\mu(1-\gamma_5)b|B(p)\rangle &= (M_B+M_\rho)A_1\epsilon^*_\mu - \frac{A_2}{M_B+M_\rho}\epsilon^* p'(p+p')_\mu \\ &\quad + \frac{2V}{M_B+M_\rho}\epsilon_{\mu\nu\rho\sigma}p^\rho p'^\sigma, \\ \langle\pi(p')|\bar{u}\gamma_\mu b|B(p)\rangle &= f_+(p+p')_\mu + f_-(p-p')_\mu,\end{aligned} \tag{55.11}$$

and:

$$\begin{aligned}\langle K^*(p')|\bar{s}\sigma_{\mu\nu}\left(\frac{1+\gamma_5}{2}\right)q^\nu b|B(p)\rangle &= i\epsilon_{\mu\nu\rho\sigma}\epsilon^{*\nu}p^\rho p'^\sigma F_1^{B\to K^*} \\ &\quad + \left\{\epsilon^*_\mu(M_B^2-M_\rho^2) - \epsilon^* q(p+p')_\mu\right\}\frac{F_1^{B\to K^*}}{2}.\end{aligned} \tag{55.12}$$

For completeness, we give the relations of these form factors to the decay rates of the $B$ meson. In the case of the pseudoscalar final state, we have:

$$\frac{d\Gamma_+}{dt} = \frac{G_F^2|V_{bq}|^2}{192\pi^3 M_B^3}\lambda^{3/2}(M_B^2, M_L^2, t)F_+^2(t), \tag{55.13}$$

while for the vector final state:

$$\begin{aligned}\frac{d\Gamma_+}{dt} &= \frac{G_F^2|V_{bu}|^2}{192\pi^3 M_B^3}\langle^{1/2}(M_B^2, M_L^2, t) \\ &\quad\times\left[2(F_0^A)^2 + \langle F_V^2 + \frac{1}{4M_F^2}\left((M_B^2 - M_L^2 - t)F_0^A + \langle F_+^A\right)^2\right], \\ \lambda &= \lambda(M_B^2, M_L^2, t).\end{aligned} \tag{55.14}$$

We have introduced the notation (often used in the literature):

$$F_+ = f_+ \quad ; F_0^A = (M_B + M_L)A_1 ;$$
$$F_+^A = \frac{-A_2}{M_B + M_L} ; F_V = 0 \frac{V}{M_B + M_L} . \qquad (55.15)$$

### 55.1.2 Estimate of the form factors and of $V_{ub}$

In the numerical analysis, we obtain at $q^2 = 0$, the value of the $B \to K^*\gamma$ form factor [762]:

$$F_1^{B\to\rho} \simeq 0.27 \pm 0.03 , \qquad \frac{F_1^{B\to K^*}}{F_1^{B\to\rho}} \simeq 1.14 \pm 0.02 , \qquad (55.16)$$

which leads to the branching ratio $(4.5 \pm 1.1) \times 10^{-5}$, in perfect agreement with the CLEO data [16] $Br(B_0 \to K^{*0}\gamma = (4.55 \pm 0.7 \pm 0.34) \times 10^{-5}$, and with the estimate in [781,820] and in [723]. for the $D$ meson, one obtains:

$$F_1^{D\to\rho} \simeq 0.62 \pm 0.10 , \qquad \frac{F_1^{D\to K^*}}{F_1^{D\to\rho}} \simeq 1.22 \pm 0.04 . \qquad (55.17)$$

One should also notice that, in this case, the coefficient of the $1/M_b^2$ correction is very large, which makes the extrapolation of the $c$-quark results to higher values of the quark mass dangerous. This extrapolation is often done in some lattice calculations.

For the semi-leptonic decays $B \to \rho, \pi + l\nu$, QSSR gives a good determination of the ratios of the form factors with the values for the B-decays [763]:

$$\frac{A_2(0)}{A_1(0)} \simeq \frac{V(0)}{A_1(0)} \simeq 1.11 \pm 0.01 , \qquad \frac{A_1(0)}{F_1^{B\to\rho}(0)} \simeq 1.18 \pm 0.06 , \qquad \frac{A_1(0)}{f_+(0)} \simeq 1.40 \pm 0.06 , \qquad (55.18)$$

although their absolute values are quite inaccurate [761] and [771]. The direct determinations of the absolute values are given in Table 55.1, showing that different results are consistent with each others. The precise measurement of the ratios is due to the cancellation of systematic errors.

Combining these results with the 'world average' value of $f_+(0) = 0.25 \pm 0.02$ and the one of $F_1^{B\to\rho}(0)$, one can deduce the rates:

$$\Gamma_\pi \simeq (4.3 \pm 0.7)|V_{ub}|^2 \times 10^{12}\ \mathrm{s}^{-1} , \qquad \Gamma_\rho/\Gamma_\pi \simeq 0.9 \pm 0.2 . \qquad (55.19)$$

These results indicate:

- The possibility to reach $V_{ub}$ with a good accuracy from the exclusive modes. Using the accurate $B$ lifetime $\tau_{B^+} = (1.655 \pm 0.027) \times 10^{-12}\ s$, and the measured branching ratio into $\pi$ [16], one can deduce:

$$V_{ub} = (3.6 \pm 0.3) \times 10^{-3} , \qquad (55.20)$$

inside the range $(2 - 5) \times 10^{-3}$ given by PDG.

Table 55.1. *Values of the different form factors in the D, B semi-leptonic processes at zero momentum from hybrid (HSR), double Laplace (DLSR)* and light cone (LCSR) sum rules*

| Process | $f_+(0)$ | $A_1(0)$ | $A_2(0)$ | $V(0)$ | Ref. |
|---|---|---|---|---|---|
| $D^0 \to \pi^- l\bar{\nu}$ | $0.7 \pm 0.2$ | | | | [769] |
| | $0.5 \pm 0.2$ | | | | [771] |
| | $0.65 \pm 0.10$ | | | | [772] |
| | $0.80^{+0.21}_{-0.14}$ | | | | [16] Data |
| $D^0 \to K^- l\bar{\nu}$ | $0.8 \pm 0.2$ | | | | [769] |
| | $0.6 \pm 0.13$ | | | | [773] |
| | $0.75 \pm 0.12$ | | | | [772] |
| | $0.6 \pm 0.1$ | | | | [775] |
| | $0.76 \pm 0.03$ | | | | [16] Data |
| $D^+ \to$ scalar $(\bar{u}u, \bar{s}d)\, l\bar{\nu}$ | $0.42 - 0.57$ | | | | [776] |
| $D_s^+ \to \eta\, l\bar{\nu}$ | $0.50 \pm 0.15$ | | | | [777] |
| $D^0 \to \rho^- l\bar{\nu}$ | | $0.5 \pm 0.2$ | $0.4 \pm 0.1$ | $1.0 \pm 0.2$ | [771] |
| | | $0.34 \pm 0.08$ | $0.57 \pm 0.08$ | $0.98 \pm 0.11$ | [772] |
| $D^0 \to K^{*-} l\bar{\nu}$ | | $0.50 \pm 0.15$ | $0.60 \pm 0.15$ | $1.1 \pm 0.25$ | [773] |
| | | $0.54 \pm 0.04$ | $0.67 \pm 0.08$ | $1.1 \pm 0.1$ | [772] |
| | | $0.58 \pm 0.03$ | $0.41 \pm 0.06$ | $1.06 \pm 0.09$ | [16] Data |
| $\bar{B}^0 \to \pi^+ l\bar{\nu}$ | $0.23 \pm 0.02$ | | | | [761] (DLSR + HSR) |
| | $0.26 \pm 0.02$ | | | | [771] |
| | $0.24 \pm 0.03$ | | | | [774] |
| | $0.29 \pm 0.04$ | | | | [772] |
| | $0.24 - 0.29$ | | | | [778] (LCSR) |
| $\bar{B}^0 \to D l\bar{\nu}$ | $1.0 \pm 0.2$ | | | | [767] |
| | $0.62 \pm 0.06$ | | | | [761] |
| $\bar{B}^0 \to \rho^+\, l\bar{\nu}$ | | $0.35 \pm 0.16$ | $0.42 \pm 0.12$ | $0.47 \pm 0.14$ | [761] (DLSR + HSR) |
| | | $0.5 \pm 0.1$ | $0.4 \pm 0.2$ | $0.6 \pm 0.2$ | [771] |
| $\bar{B}^0 \to K^{*+} \bar{\nu}\nu$ | | $0.37 \pm 0.03$ | $0.40 \pm 0.03$ | $0.47 \pm 0.03$ | [779] |
| $\bar{B}^0 \to D^* l\bar{\nu}$ | | $0.46 \pm 0.02$ | $0.53 \pm 0.09$ | $0.58 \pm 0.03$ | [761] (DLSR + HSR) |

* If not mentioned DLSR have been used.

- One should also notice that the ratio between the widths into $\rho$ and into $\pi$ is about 1 due to the non-pole behaviour of $A_1^B$, while in different pole models, it ranges from 3 to 10. This result is in agreement within $1\sigma$ with the data $(1.5 \pm 0.5)$. Data on $B \to K(K^*) + \psi(\psi')$ decays [785] also favour this non-pole behaviour, while LCSR and lattice calculations indicate a slight increase of $A_1$ for increasing $q^2$. However, the arguments given in [782] for explaining the failure of the standard QSSR approach is unclear to us and should deserve a further investigation. In the case of the non-pole behaviour, one obtains a large negative value of the asymmetry $\alpha$, contrary to the case of the pole models.

### 55.1.3 $SU(3)_F$ breaking in $\bar{B}/D \to Kl\bar{\nu}$ and determination of $V_{cd}/V_{cs}$ and $V_{cs}$

We extend the previous analysis for the estimate of the $SU(3)_F$ breaking in the ratio of the form factors:

$$R_P \equiv f_+^{P\to K}(0)/f_+^{P\to\pi}(0) , \tag{55.21}$$

where $P \equiv \bar{B}, D$. As mentioned before, we use the hybrid moments for the $B$ and the double exponential sum rules for the D. The analytic expression of $R_P$ is given in [764], which leads to the numerical result:

$$R_B = 1.007 \pm 0.020 , \qquad R_D = 1.102 \pm 0.007 , \tag{55.22}$$

where one should notice that for $M_b \to \infty$, the SU(3) breaking vanishes, while its size at finite mass is typically of the same sign and magnitude as the one of $f_{D_s}$ or of the $B \to K^*\gamma$ discussed before. The previous value of $R_D$ can be used with the data [16]:

$$\frac{Br(D^+ \to \pi^0 l\nu)}{Br(D^+ \to \bar{K}^0 l\nu)} = (8.5 \pm 3.4)\% , \tag{55.23}$$

for deducing the value of $|V_{cd}|/|V_{cs}|$

We can also determine directly from QSSR the absolute value of the $D \to K$ form factor. We obtain [764]:

$$f_+^{D\to K}(0) \simeq 0.80 \pm 0.16 . \tag{55.24}$$

The data from $D$ lifetime and $De_3$ [16] gives:

$$\left|f_+^{D\to K}(0)\right|^2 \left|V_{cs}\right|^2 \simeq 0.531 \pm 0.027 . \tag{55.25}$$

Using the previous prediction for $f_+^{D\to K}(0)$ leads to:

$$V_{cs} = 0.91 \pm 0.18 , \tag{55.26}$$

which differs slightly from the PDG prediction as the value of the form factor used there was $0.7 \pm 0.1$. It is also expected that the most reliable result is the lower bound derived from Eq. (55.25) and from $f_+^{D\to K}(0) \le 1$, which is:

$$V_{cs} \ge 0.73 . \tag{55.27}$$

### 55.1.4 Large $M_b$-limit of the form factors

We have studied analytically the large $M_b$ limit of some of the previous form factors [762–764]. We found that, within the approximation at which we are working, and to leading order in $M_b$, they are dominated, for $M_b \to \infty$, by the effect of the light-quark condensate, which dictates (to leading order) the $M_b$ behaviour of the form factors to be typically of the form:

$$F(0) \sim \frac{\langle \bar{d}d \rangle}{f_B} \left\{ 1 + \frac{\mathcal{I}_F}{M_b^2} \right\} , \tag{55.28}$$

where $\mathcal{I}_F$ is the integral from the perturbative triangle graph, which is constant as $t_c'^2 E_c / \langle \bar{d}d \rangle$ ($t_c'$ and $E_c$ are the continuum thresholds of the light and $b$ quarks) for large values of $M_b$. It indicates that at $q^2 = 0$ and to leading order in $1/M_b$, all form factors behave like $\sqrt{M_b}$, although, in most cases, the coefficient of the $1/M_b^2$ term is large, which explains the numerical dominance of the perturbative contribution at finite $M_b$. It should be finally noticed that owing to the overall $1/f_B$ factor, all form factors for the heavy to light transitions have a large $1/M_b$ correction.

### *55.1.5 $q^2$-behaviour of the form factors*

Although the sum rules give a reliable prediction for the value of the form factors at zero momentum transfer, the analysis of their $q^2$ behaviour is more delicate due to the eventual presence of non-Landau singularities [774] above a critical value:

$$t_{cr} = (M_Q + m_{q_2})^2 , \tag{55.29}$$

for a $\bar{Q}q_1$ meson decaying semi-leptonically into a $\bar{q}_1 q_2$ meson, where the weak current is $\bar{q}_2 \gamma_\mu q_1$; $M_Q$ and $m_q$ are constituent quark masses. Much below $t_{cr}$, the sum rule result is expected to provide the right $q^2$-behaviour of the form factor. The study of the $q^2$ behaviours of the $B$ semi-leptonic form factors shows that, with the exception of the $A_1$ form factor, their $q^2$ dependence is only due to the non-leading (in $1/M_b$) perturbative graph, so that for $M_b \to \infty$, these form factors remain almost constant from $q^2 = 0$ to $q^2_{\rm max}$, with a cautious for the accuracy of the result at $q^2_{\rm max}$. The resulting $M_b$ behaviour at $q^2_{\rm max}$ is the one expected from the heavy quark symmetry. The numerical effect of this $q^2$-dependence at finite values of $M_b$ is a polynomial in $q^2$ (which can be resummed), and mimics the pole parametrization quite well for a pole mass of about 6–7 GeV. The situation for the $A_1$ is drastically different from the other ones, as here the Wilson coefficient of the $\langle \bar{d}d \rangle$ condensate contains a $q^2$ dependence with a *wrong* sign and reads [763]:

$$A_1(q^2) \sim \frac{\langle \bar{d}d \rangle}{f_B} \left\{ 1 - \frac{q^2}{M_b^2} \right\} , \tag{55.30}$$

which, for $q^2_{\rm max} \equiv (M_B - M_\rho)^2$, gives the expected behaviour:

$$A_1 \left( q^2_{\rm max} \right) \sim \frac{1}{\sqrt{M_b}} . \tag{55.31}$$

One can notice that the $q^2$ dependence of $A_1$ is in complete contradiction with the pole behaviour due to its wrong sign. This result may explain the numerical analysis of [771].

One should notice that a recent phenomenological analysis of the data [783] on the large longitudinal polarization observed in $B \to K^* + \psi$ and a relatively small ratio of the rates $B \to K^* + \psi$ over $B \to K + \psi$ [785] can only be simultaneously explained if the $A_1(q^2)$ form factor decreases as indicated by our previous result, while larger choices of increasing or/and monotonically form factors fail to explain the data [786]. It is important to test this *anomalous* feature of the $A_1$-form factor from some other data. One should notice that the $q^2$ behaviour of $A_1$ has also been studied from lattice calculations [723] and light-cone sum rule (LCSR) [780,782]. The latter result shows a slower increase of $A_1$ for increasing $q^2$. Contrary to the interpretation given in [782] where the arguments given there are not clear to us (in particular the connection between the LCSR and the SVZ sum rule), the increase of the form factor may indicate that non-leading contributions at finite $M_b$, not accounted for in our approximation, can be numerically important, and competes with the leading-order contribution presented here. We plan to come back to this point in the near future. Finally, as a complement of the heavy quark symmetry which will be discussed in the next section, we have also presented in Table 55.1 the results for the $B \to D, D^* \, l\bar{\nu}$ form factors.

## 55.2 Slope of the Isgur–Wise function and value of $V_{cb}$

Using heavy quark symmetry in the infinite quark mass limit, the different form factors of the semi-leptonic $B \to D^*$ and $B \to D$ can be related to each others and expressed in terms of a single form factor:

$$f_+(q^2) = V(q^2) = A_0(q^2) = A_2(q^2) = \left(1 - \frac{q^2}{(M_B + M_D)^2}\right)^{-1} A_1(q^2)\,, \tag{55.32}$$

where:

$$A_1(q^2) = \frac{M_B + M_D}{2\sqrt{M_B M_D}} \zeta(y)\,. \tag{55.33}$$

$\zeta(y \equiv v.v')$ is the so-called Isgur–Wise (IW) function and contains all non-perturbative QCD effects ($v$ and $v'$ are respectively the $B$ and $D$ meson velocity). At zero recoil $y = 1$, or at $q^2_{\rm max}$, it is normalized as $\zeta(1) = 1$ from the conservation of the vector current. In this limit, the $B \to D^*$ decay distribution can be written as:

$$\frac{d\Gamma}{dy} = \frac{G_F^2}{48\pi^3} (M_B - M_{D^*})^2 M_{D^*}^3 \sqrt{y^2 - 1}(y+1)^2 \times \left[1 + \frac{4y}{y+1} \frac{M_B^2 - 2yM_B M_{D^*} + M_{D^*}^2}{(M_B - M_{D^*})^2}\right] |V_{cb}|^2 \mathcal{F}^2(y)\,, \tag{55.34}$$

where $\mathcal{F}(y)$ is the IW function including perturbative and power corrections. Near zero recoil, one can write the expansion:

$$\mathcal{F}(1) = \eta_A \left[ 1 + \frac{C_2}{M_Q^2} + \cdots \right] , \tag{55.35}$$

where $\eta_A$ includes the perturbative corrections, which to two-loop accuracy reads [580]:

$$\eta_a = 0.960 \pm 0.007 , \tag{55.36}$$

while, there is no $1/M$ correction in virtue of the Luke's theorem [568]. There are some attempts to estimate the size of the $1/M^2$ terms in the literature [164,562,587], which remain not under good control. The resuting compromise value is:

$$\mathcal{F}(1) \simeq 0.91(6) , \tag{55.37}$$

where we have multiplied the quoted error by a factor 2 in order to be more conservative. Let me now discuss the slope of the IW function. De Rafael and Taron [789] have exploited the analyticity of the elastic $b$-number form factor $F$ defined as:

$$\langle B(p')|\bar{b}\gamma^\mu b|B(p)\rangle = (p + p')^\mu F(q^2) , \tag{55.38}$$

which is normalized as $F(0) = 1$ in the large mass limit $M_B \simeq M_D$. Using the positivity of the vector spectral function and a mapping in order to get a bound on the slope of $F$ outside the physical cut, they obtained a rigorous though weak bound:

$$F'(vv' = 1) \geq -6 . \tag{55.39}$$

Including the effects of the $\Upsilon$ states below $\bar{B}B$ thresholds by assuming that the $\Upsilon\bar{B}B$ couplings are of the order of 1, the bound becomes stronger:

$$F'(vv' = 1) \geq -1.5 . \tag{55.40}$$

Using QSSR, we can estimate the part of these couplings entering in the elastic form factor. We obtain the value of their sum [765]:

$$\sum g_{\Upsilon\bar{B}B} \simeq 0.34 \pm 0.02 . \tag{55.41}$$

In order to be conservative, we have multiplied the previous estimate by a factor 3 larger. We thus obtain the improved bound:

$$F'(vv' = 1) \geq -1.34 , \tag{55.42}$$

but the gain over the previous one is not much. Using the relation of the form factor with the slope of the IW function, which differs by $-16/75 \log \alpha_s(M_b)$ [790], one can deduce the final bound [765]:[4]

$$\zeta'(1) \geq -1.04 . \tag{55.43}$$

[4] Voloshin in [791] derives also the upper bound $\rho^2 \leq 1/4 + \bar{\Lambda}/[2(M_{B'} - M_B)]$, which, however, depends crucially on the less controlled value of $\bar{\Lambda}$ and the mass of the radial excitation $M_{B'}$.

The previous bound combined with the Bjorken lower bound [792] leads to the allowed domain:

$$\frac{1}{4} \leq \rho^2 \equiv -\zeta'(1) \leq 1.04 \,. \tag{55.44}$$

However, one can also use the QSSR expression of the IW function from vertex sum rules [655] in order to extract the slope *analytically*. To leading order in $1/M$, the *physical* IW function reads [Rep. 18.5]:

$$\zeta_{\text{phys}}(y \equiv vv') = \left(\frac{2}{1+y}\right)^2 \left\{1 + \frac{\alpha_s}{\pi} f(y) - \langle \bar{d}d\rangle \tau^3 g(y) \right. \\ \left. + \langle \alpha_s G^2\rangle \tau^4 h(y) + g\langle \bar{d}Gd\rangle \tau^5 k(y)\right\}, \tag{55.45}$$

where $\tau$ is the Laplace sum rule variable and $f$, $h$ and $k$ are analytic functions of $y$. From this expression, one can derive the analytic form of the slope [765]:

$$\zeta'_{\text{phys}}(y=1) \simeq -1 + \delta_{\text{pert}} + \delta_{NP} \,, \tag{55.46}$$

where at the $\tau$-stability region:

$$\delta_{\text{pert}} \simeq -\delta_{NP} \simeq -0.04 \,, \tag{55.47}$$

which shows the near-cancellation of the non-leading non-perturbative corrections at this *leading order in* $1/M$. Adding a generous 50% error of 0.02 for the correction terms, we finally deduce the *leading order result in* $1/M$:

$$\zeta'_{\text{phys}}(y=1) \simeq -1 \pm 0.02 \,. \tag{55.48}$$

Using this result in different existing model parametrizations, we deduce the value of the mixing angle, *to leading order in* $1/M$:

$$V_{cb} \simeq \left(\frac{1.48 \text{ ps}}{\tau_b}\right)^{1/2} \times (37.3 \pm 1.2 \pm 1.4) \times 10^{-3}, \tag{55.49}$$

where the first error comes from the data and the second one from the model-dependence.[5]

In order to discuss the effects due to the $1/M$ corrections, we proceed in the following phenomenological way:

- We use the predicted value of the form factor $0.91 \pm 0.03$ at $y=1$,
- We also use the value $0.53 \pm 0.09$ at $q^2 = 0$ [761][6] from the sum rule in the full theory (i.e without using a $1/M$-expansion).
- We join the two results, where the model dependence of the analysis enters through the concavity of the form factor between these two extreme boundaries.

[5] A recent analysis [574] relates the curvature with the slope, such that in this case, the model dependence of the result disappears.
[6] This value is just on top of the CLEO data [793].

Table 55.2. *Different QSSR estimates of the slope of the IW-function compared with the lattice result*

| $-\zeta' \equiv \rho^2$ | References | Comments |
|---|---|---|
| $0.84 \pm 0.02$ | [795] | Numerical fit |
| $0.70 \pm 0.1$ | [796] | |
| $0.70 \pm 0.25$ | [797] | |
| $1.00 \pm 0.02$ | [765] | Analytic expression |
| $0.91 \pm 0.04$ | | QSSR average |
| $(0.9^{+0.2}_{-0.3}{}^{+0.4}_{-0.4})$ | [798] | Lattice |

If we use the value of the concavity given by [574], the form factor can be parametrized as:

$$F(y) = F(1)\{1 + \hat{\xi}'(y-1) + \hat{c}(y-1)^2\}\,, \tag{55.50}$$

where:

$$\hat{\xi}' = \zeta' - (0.16 \pm 0.02)\,, \qquad \hat{c} \approx -0.66\hat{\xi}' - 0.11\,. \tag{55.51}$$

Therefore, we can deduce the slope:

$$\zeta' \simeq -(0.75 \pm 0.1)\,, \tag{55.52}$$

which can indicate that the $1/M$ correction also tends to decrease the value of $|\zeta'|$. This leads to the *final* estimate:

$$V_{cb} \simeq \left(\frac{1.48\ \text{ps}}{\tau_b}\right)^{1/2} \times (38.8 \pm 1.2 \pm 1.5 \pm 1.5) \times 10^{-3}\,, \tag{55.53}$$

where the new last error is induced by the error from the slope, while the model dependence only brings a relatively small error. Using the measured $B^0$-lifetime $\tau_B = 1.548 \pm 0.032$ ps, one obtains:

$$V_{cb} \simeq (37.9 \pm 2.4) \times 10^{-3}\,, \tag{55.54}$$

compared with the value $0.0402 \pm 0.0019$ from LEP measurements of exclusive and inclusive decays [16] and CLEO data [793]. Our result for the slope is also in good agreement with the data. Finally, we compare the different results from the sum rules in Table 55.2, from which we deduce the *weighted average* from the sum rules given in Table 55.2, where we have taken the error of the most precise determinations which we have multiplied by a factor 2 in order to be conservative. This average is in good agreement with the lattice value, which is also given in this table.

### 55.3 $B^*(D^*) \to B(D)\,\pi(\gamma)$ couplings and decays

In [766], we have further applied the vertex sum rules in order to study the decays and couplings of the $B^*(D^*) \to B(D)\,\pi(\gamma)$, using systematically a $1/M_b$ expansion in the full theory. The couplings are defined as:

$$\langle B^*(p)B(p')\pi(q)\rangle = g_{B^*B\pi}q_\mu\epsilon^\mu\,,$$
$$\langle B^*(p)B(p')\gamma(q)\rangle = -e g_{B^*B\gamma}\,p_\alpha p'_\beta\epsilon^{\mu\nu\alpha\beta}\epsilon_\mu\epsilon'_\nu\,, \tag{55.55}$$

where $q \equiv p' - p$ and $-Q^2 \equiv q^2 \le 0$, while $\epsilon_\mu$ are the polarization of the vector particles. Our numerical predictions for the couplings are:[7]

$$g_{B^*B\pi} \simeq 14 \pm 4\,, \qquad g_{D^*D\pi} \simeq 6.3 \pm 1.9\,, \tag{55.56}$$

in good agreement with the results in [800] obtained by combining QSSR with soft pion techniques. These results lead to the prediction:

$$\Gamma_{D^{*-}\to D^0\pi^-} \simeq 1.54\Gamma_{D^{*0}\to D^0\pi^0} \simeq (8 \pm 5)\text{ keV}\,, \tag{55.57}$$

where we have assumed an isospin invariance for the couplings.

We notice that in the large $M_b$ limit, the perturbative graph gives the leading contribution like some other heavy-to-heavy processes studied within the same approach. In this limit, we obtain:

$$g_{B*B\pi} \simeq \frac{2M_B}{\sqrt{2}f_\pi}g^\infty\left\{1 + \frac{E_c^B}{M_b} + \frac{\pi^2}{2}\frac{\langle\bar{u}u\rangle}{(E_c^B)^3}\right\}, \tag{55.58}$$

where $g^\infty$ is the static coupling:

$$g^\infty \equiv \frac{N_C}{2}\left(\frac{m_u + m_d}{m_\pi^2}\right)(0.12E_c^\infty) \simeq (0.15 \pm 0.03)\,, \tag{55.59}$$

for $E_c^\infty \simeq (1.6 \pm 0.1)$ GeV.

In the same way, we have also estimated the $B^*B\gamma$ and $D^*D\gamma$ couplings. We obtain:

$$\Gamma_{B^{*-}\to B^-\gamma} \simeq 2.5\Gamma_{B^{*0}\to B^0\gamma} \simeq (0.10 \pm 0.03)\text{ keV}\,. \tag{55.60}$$

For the $D^*$ meson, one obtains:

$$\Gamma_{D^{*0}\to D^0\gamma} \simeq (7.3 \pm 2.7)\text{ keV}\,, \quad \Gamma_{D^{*-}\to D^-\gamma} \simeq (0.03 \pm 0.08)\text{ keV}\,, \tag{55.61}$$

which despite the large errors, shows in the analysis that the heavy quark contribution acts in the right direction for explaining the large charge dependence of the observed rates:

$$\Gamma_{D^{*0}\to D^0\pi^0}/\Gamma_{D^{*0}\to D^0\gamma}, \qquad \Gamma_{D^{*-}\to D^0\pi^-}/\Gamma_{D^{*-}\to D^-\gamma}\,. \tag{55.62}$$

[7] The application of the $1/M$ expansion to the $D$ and $D^*$ mesons might be a very crude approximation. A comparison of the result with the recent CLEO data [799] $g_{D^*D\pi} = 17.9 \pm 0.3 \pm 1.9$ needs further investigation using a complete QCD expression of the vertex.

Table 55.3. *Comparison of semi-leptonic form factors for different decays. We compare the dimensionless quantities $f_+$, $A_1$, $A_2$, $V$ related to $F_0^A$, $F_+^A$ and $F_V$ through Eq. (55.15)*

| Channels | Reference | $f_+$ | $V$ | $A_2$ | $A_1$ |
|---|---|---|---|---|---|
| $c\bar{c}$ | [731] | $0.55 \pm 0.10$ | $0.48 \pm 0.07$ | $0.30 \pm 0.05$ | $0.30 \pm 0.05$ |
| | [801] | $0.20 \pm 0.01$ | $0.37 \pm 0.1$ | $0.27 \pm 0.03$ | $0.28 \pm 0.01$ |
| $b\bar{s}$ | [731] | $0.60 \pm 0.12$ | $1.6 \pm 0.3$ | $0.06 \pm 0.06$ | $0.40 \pm 0.10$ |
| | [801] | $0.30 \pm 0.05$ | $2.1 \pm 0.25$ | $0.39 \pm 0.05$ | $0.35 \pm 0.20$ |
| $B \to D^{(*)}$ | [771] | $0.75 \pm 0.05$ | $0.8 \pm 0.1$ | $0.68 \pm 0.08$ | $0.65 \pm 0.10$ |
| | [803] | 0.69 | 0.71 | 0.69 | 0.65 |
| | [48] | $0.62 \pm 0.06$ | $0.58 \pm 0.03$ | $0.53 \pm 0.09$ | $0.46 \pm 0.02$ |

| | $B_c \to \eta_c$ <br> $B_c \to J/\psi$ | $B_c \to B_s$ <br> $B_c \to B_s^*$ | $B_c \to B$ <br> $B_c \to B^*$ | $B_c \to D$ <br> $B_c \to D^*$ |
|---|---|---|---|---|
| $F_+(0)$ | $0.55 \pm 0.10$ | $0.60 \pm 0.12$ | $0.48 \pm 0.14$ | $0.18 \pm 0.08$ |
| $F_V(0)$ [GeV$^{-1}$] | $0.048 \pm 0.007$ | $0.15 \pm 0.02$ | $0.11 \pm 0.02$ | $0.02 \pm 0.01$ |
| $F_+^A(0)$ [GeV$^{-1}$] | $-0.030 \pm 0.003$ | $-0.005 \pm 0.005$ | $0.005 \pm 0.005$ | $0.010 \pm 0.010$ |
| $F_0^A(0)$ [GeV] | $3.0 \pm 0.5$ | $3.3 \pm 0.7$ | $1.7 \pm 0.7$ | $0.8 \pm 0.4$ |

One should also notice that the non-leading $1/M_b$ corrections are large in the two channels. For the $PP^*\pi$ coupling, these corrections coming mainly from the perturbative graph tend to cancel, which imply the validity of the HQET result:

$$g_{B^*B\pi} f_{B^*}\sqrt{M_B} \simeq g_{D^*D\pi} f_{D^*}\sqrt{M_D} \,. \tag{55.63}$$

For the electromagnetic, these large corrections are necessary to explain the large charge dependence of the ratio of the $D^{*0} \to D^0\gamma$ over the $D^{*-} \to D^-\gamma$ observed widths. However, the new CLEO data give $\Gamma_{D^{*-}\to D^-\gamma} \simeq (96 \pm 22)$ keV indicating that the $1/M$ approach for the absolute width can be a bad approximation.

## 55.4 Weak semi-leptonic decays of the $B_c$ mesons

The analysis of the semi-leptonic decays of the $B_c$ meson has been performed in [731] using QSSR methods. The procedure is very similar to the one used in the previous sections. The principal results of the sum rules evaluation of the form factors in Eq (55.11) are collected in Table 55.3. The value with the lower (resp. larger) modulus corresponds to the value of the continuum energy $E_c = 1$ GeV (resp. 2 GeV). In Fig. 55.1, we show the $q^2$ behaviour of the $B_c \to \eta_c\, l\bar{\nu}$ process, which shows a net deviation of the QSSR prediction from the monopole fit:

$$F_+(t) = \frac{F_+(0)}{1 - t/M_{\rm pole}^2} \,, \tag{55.64}$$

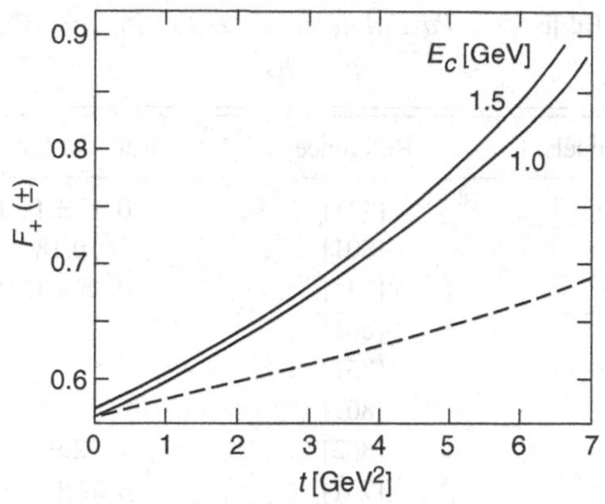


Fig. 55.1. $q^2$ behaviour of the $B_c \to \eta_c l \bar{\nu}$ form factor: QSSR predictions with polynomial fit (continuous line) for two values of $E_c$; monopole parametrization with $M_{\text{pole}} = M_{B_c^*} = 6.33$ GeV.

with $M_{\text{pole}} = M_{B_c^*}$, which can be tested experimentally. The same feature is observed in other channels. For the $B_c \to J/\psi\, l \bar{\nu}$ process, one obtains the fitted QSSR pole mass:

$$F_V: \quad M_{\text{pole}} \simeq 4.08 \text{ GeV}, \quad F_+^A: \quad M_{\text{pole}} \simeq 4.44 \text{ GeV}, \quad F_0^A: \quad M_{\text{pole}} \simeq 4.62 \text{ GeV}. \tag{55.65}$$

However, the question is, whether it is known if the previous $q^2$-behaviour deviation from the monopole model is an artifact of sum rule or something more fundamental. In VDM, the vector current couples to the hadrons with appropriate flavour content, where the intermediate vector mesons leads to the pole of the form factor $F_+(t)$ at $t \equiv q^2 = M_V^2$ giving the $t$-behaviour in Eq. (55.64), while an intuitive quark model form factor determined by the Fourier transform of the hadron wave function gives:

$$F_+(t) = 1 + \frac{\langle r^2 \rangle}{6} t + \mathcal{O}(t^2), \tag{55.66}$$

where $\langle r^2 \rangle$ is the hadron mean-squared radius in the quark model. For light hadrons, the vector meson is the $\rho$ meson. Expanding Eq. (55.64) in $t$, and identifying with Eq. (55.66), one obtains:

$$\sqrt{\langle r^2 \rangle_\pi} = \frac{\sqrt{6}}{M_\rho} \simeq 0.6 \text{ fm}, \tag{55.67}$$

which is a reasonable value for the quark model, while for the case of the $B_c$ meson, the vector meson is the $B_c^*$ with a mass of 6.4 GeV leading to a mean radius of about 0.08 fm, which is too small for a reliable validity of the non-relativistic quark model. This feature might indicate that the non-relativistic picture is not reasonable, such that, we have, instead, to discuss the problem within a relativistic field theory approach such as QSSR. We compare in Table 55.4 different theoretical predictions based on QCD-like models.

Table 55.4. *Partial decay rates for $B_c$ and $B_c^*$ mesons*

| Channels | Reference | Rates in $10^{10}$s$^{-1}$ |
|---|---|---|
| $B_s l\nu$ | [731] | $0.35 \pm 0.10$ |
| | [801] | 0.18 |
| $B_s^* l\nu$ | [731] | $0.35 \pm 0.10$ |
| | [801] | 0.87 |
| $b\bar{s}l\nu$ | [731] | |
| | [801] | |
| | [802] | 2.91 |
| $\eta_c l\nu$ | [731] | $0.27 \pm 0.07$ |
| | [801] | 0.03 |
| $J/\psi l\nu$ | [731] | $0.32 \pm 0.08$ |
| | [801] | 0.21 |
| $c\bar{c}l\nu$ | [731] | |
| | [801] | |
| | [802] | 6.90 |

### 55.4.1 Anomalous thresholds

Another subtle point in the vertex sum rule approach is the eventual existence of anomalous thresholds in the study of the $q^2$ behaviour of the form factors [804–806]. Let's illustrate the analysis from the study of the process $B_c(\bar{b}c) \to \Psi(\bar{c}c)l\bar{\nu}$. The interaction between $\bar{b}$ and $c$ leads to the formation of the vector meson $B_c^*(\bar{c}b)$ allowing us to approximate the singularity by the pole of the vector meson mass near the *normal threshold* $t_n = (M_b + M_c)^2$, where $t$ should be large enough for ensuring the $\bar{b}$ and $c$ quark on-mass shell. *Anomalous threshold* occurs, if under a certain condition, for smaller value $t_a$ of $t$, one is able to give on-shell mass $b$ a gentle kick in order to transform it into an on-shell mass $c$. The derivation of the existence of anomalous thresholds can be simply done using dual diagrams [804] shown in Fig. 55.2. The dual (b) of the QCD three-point function (a) can be obtained by transforming the plane segments $A,\ B,\ C,\ O$ of the planar diagram (a) into points of the dual diagram connected by lines with lengths given by the masses of the particles dividing the segments. If the point $O$ of the dual diagram is inside the triangle $ABC$, then there exists an anomalous threshold and its value is given by the square of the distance $AB$. For the $B_c$ of mass 6.25 GeV, anomalous thresholds do exist for the decay $B_c \to \eta_c,\ J/\Psi$ provided the quark mass fulfill the conditions: $m_c < 2.1$ GeV and $m_b < 5.9$ GeV, which are satisfied by any quark models. The exact position of the anomalous threshold depends strongly on the value of the quark masses. Using the constituent (pole) masses $m_b = 4.9$ GeV and $m_c = 1.57$ GeV, one would obtain $\sqrt{t_a} = 4.6$ GeV, while the minimum possible value is $\sqrt{t_a} \geq m_b - m_c = 3.3$ GeV. These values are consistent with the effective pole mass of about 4.2 GeV found in the sum rule analysis. Everything looks consistent except

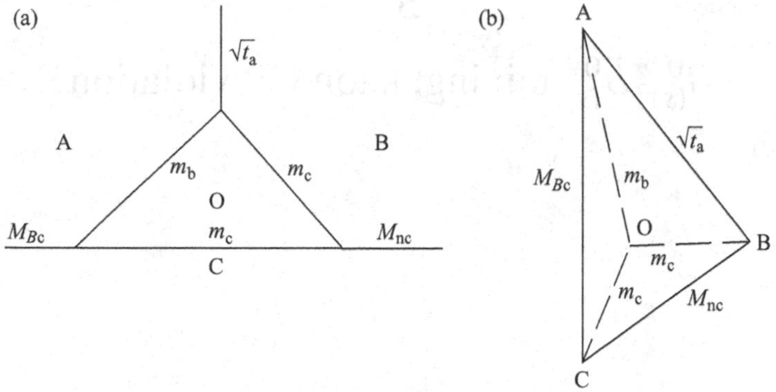


Fig. 55.2. QCD three-point function (a) and its dual diagram (b).

that quarks are not free particles while perturbation theory breaks down if one tries to approach the pole of the quark propagators. Therefore, an anomalous threshold should not be present in the $t$-dependence of the (hadronic) form factors. The same conclusion holds for the normal thresholds, as in the $e^+e^-$ process, we do not observe a quark-antiquark state but hadrons. Quark-hadron duality tells us that the discontinuity across the quark-antiquark cut, if viewed from a certain distance or smeared over some energy interval describes the hadronic production in $e^+e^-$ quite well. In a QCD-like model, [805,806] found that although quark anomalous thresholds are absent, at some distance from the calculated threshold, the amplitude can be quite well approximated by anomalous singularities at the quark level.

# 56

# $B^0_{(s)}$-$\bar{B}^0_{(s)}$ mixing, kaon CP violation

In this chapter, we provide the basics of the phenomenological description of the $B^0$-$\bar{B}^0$ and $K^0$-$\bar{K}^0$ systems, and summarize the different results obtained from QCD spectral sum rules (QSSR), on the bag constant parameters entering in the analysis of the $B^0_{(s)}$-$\bar{B}^0_{(s)}$ mass differences, and on different operators entering in the analysis of kaon CP violation. There is practically no theory behind this description. It is only based on first principles: the superposition principle, Lorentz invariance, and general invariance properties under the P, C and T symmetries. The basic idea is to reduce the description of this system to a minimum of phenomenological parameters which, eventually, an underlying theory, like the Standard Model (SM) should be able to predict.

## 56.1 Standard formalism

This section has been inspired from the lectures given in [500].

### *56.1.1 Phenomenology of $B^0$-$\bar{B}^0$ and $K^0$-$\bar{K}^0$ mixings*

In the absence of the weak interactions, the $K^0$ and $\bar{K}^0$[1] particles produced by the strong interactions are stable eigenstates of strangeness with eigenvalues $\pm 1$. In the presence of the weak interaction they become unstable. The states with an exponential time dependence law ($\tau$ is the proper time):

$$|K_L\rangle \to e^{-iM_L\tau}\,|K_L\rangle \quad \text{and} \quad |K_S\rangle \to e^{-iM_S\tau}\,|K_S\rangle\,, \tag{56.1}$$

are linear superpositions of the eigenstates of strangeness:

$$|K_L\rangle = \frac{1}{\sqrt{|\,p\,|^2 + |\,q\,|^2}}(p \mid K^0\rangle + q\,|\bar{K}^0\rangle) \tag{56.2}$$

$$|K_S\rangle = \frac{1}{\sqrt{|\,p\,|^2 + |\,q\,|^2}}(p \mid K^0\rangle - q\,|\bar{K}^0\rangle)\,, \tag{56.3}$$

where $p$ and $q$ are complex numbers and CPT invariance, which is a property of the SM in

[1] Discussions for the $B^0$ and $\bar{B}^0$ particles are very similar.

any case, has been assumed. The parameters $M_{L,S}$ in Eq. (56.1) are also complex:

$$M_{L,S} = m_{L,S} - \frac{i}{2}\Gamma_{L,S}\,, \tag{56.4}$$

with $m_{L,S}$ the masses and $\Gamma_{L,S}$ the decay widths of the long-lived and short-lived neutral kaon states.

As we shall see, experimentally, the $|K_S\rangle$ and $|K_L\rangle$ states are very close to the CP eigenstates:

$$|K_1^0\rangle = \frac{1}{\sqrt{2}}(|K^0\rangle - |\bar{K}^0\rangle) \quad \text{and} \quad |K_2^0\rangle = \frac{1}{\sqrt{2}}(|K^0\rangle + |\bar{K}^0\rangle) \tag{56.5}$$

with:

$$\mathrm{CP}|K_1^0\rangle = +|K_1^0\rangle \quad \text{and} \quad \mathrm{CP}|K_2^0\rangle = -|K_2^0\rangle\,. \tag{56.6}$$

This is characterized by the small complex parameter $\tilde{\epsilon}$:

$$\tilde{\epsilon} = \frac{p-q}{p+q}\,; \tag{56.7}$$

in terms of which:

$$|K_{L,S}\rangle = \frac{1}{\sqrt{1+|\tilde{\epsilon}\,|^2}}\left(|K_{2,1}^0\rangle + \tilde{\epsilon}|K_{1,2}^0\rangle\right). \tag{56.8}$$

According to Eqs. (56.1) and (56.2), a state initially pure $|K^0\rangle$ evolves, in a period of time $\tau$ to a state which is a superposition of $|K^0\rangle$ and $|\bar{K}^0\rangle$:

$$|K^0\rangle \to \frac{1}{2}[e^{-iM_L\tau} + e^{-iM_S\tau}]\,|K^0\rangle + \frac{1}{2}\frac{p}{q}[e^{-iM_L\tau} - e^{-iM_S\tau}]\,|\bar{K}^0\rangle\,; \tag{56.9}$$

and, likewise:

$$|\,\bar{K}^0\rangle \to \frac{1}{2}[e^{-iM_L\tau} + e^{-iM_S\tau}]\,|\,\bar{K}^0\rangle + \frac{1}{2}\frac{q}{p}[e^{-iM_L\tau} - e^{-iM_S\tau}]\,|\,K^0\rangle\,. \tag{56.10}$$

For a small period of time $\delta\tau$ we then have:

$$|\,K^0\rangle \to |\,K^0\rangle - i\delta\tau(\mathcal{M}_{11}\,|\,K^0\rangle + \mathcal{M}_{12}\,|\,\bar{K}^0\rangle)\,; \tag{56.11}$$

$$|\,\bar{K}^0\rangle \to |\,\bar{K}^0\rangle - i\delta\tau(\mathcal{M}_{21}\,|\,K^0\rangle + \mathcal{M}_{22}\,|\,\bar{K}^0\rangle), \tag{56.12}$$

where:

$$\mathcal{M}_{ij} = \frac{1}{2}\begin{pmatrix} M_L + M_S & \frac{p}{q}(M_L - M_S) \\ \frac{q}{p}(M_L - M_S) & M_L + M_S \end{pmatrix}. \tag{56.13}$$

This is the complex mass matrix of the $K^0 - \bar{K}^0$ system.

In full generality, the mass matrix $\mathcal{M}_{ij}$ admits a decomposition, similar to the one of the complex parameters $M_{L,S}$ in Eq. (56.4), in terms of an absorptive part $\Gamma_{ij}$ and a dispersive part $M_{ij}$:

$$\mathcal{M}_{ij} = M_{ij} - \frac{i}{2}\Gamma_{ij} \, . \tag{56.14}$$

In a given quantum field theory, like, for example, the Standard Electroweak Model, the complex $K^0 - \bar{K}^0$ mass matrix is defined via the transition matrix $T$ which characterizes $S$-matrix elements. More precisely, the off-diagonal absorptive matrix element $\Gamma_{12}$ for example, is given by the sum of products of on-shell matrix elements:

$$\Gamma_{12} = \sum_{\Gamma} \int d\Gamma (\langle \Gamma \mid T \mid \bar{K}^0 \rangle)^* \langle \Gamma \mid T \mid K^0 \rangle \, , \tag{56.15}$$

where the sum is extended to all possible states $\mid \Gamma \rangle$ to which the states $\mid K^0 \rangle$ and $\mid \bar{K}^0 \rangle$ can decay. The symbol $d\Gamma$ denotes the phase space measure appropriate to the particle content of the state $\Gamma$. The corresponding matrix element $M_{12}$ is defined by the dispersive principal part integral:

$$M_{12} = \frac{1}{\pi} \wp \int ds \frac{1}{m_K^2 - s} \Gamma_{12}(s) + \text{`local -- terms'} \, . \tag{56.16}$$

The fact that $\mathcal{M}_{11} = \mathcal{M}_{22}$ in Eq. (56.13) is a consequence of CPT invariance. In general, if we have a transition between an initial state $\mid IN \rangle$ and a final state $\mid FN \rangle$, CPT invariance relates the matrix elements of this transition to the one between the corresponding CPT-transformed states $\mid \overline{FN}' \rangle$ and $\mid \overline{IN}' \rangle$, where $\mid \overline{IN}' \rangle$ denotes the state obtained from $\mid IN \rangle$ by interchanging all particles into antiparticles (this is the meaning of the bar symbol in $\overline{IN}$), and taking the mirror image of the kinematic variables: $[(E, \vec{p}) \to (E, -\vec{p}); (\sigma^0, \vec{\sigma}) \to (-\sigma^0, \vec{\sigma})]$, as well as their motion reversal image: $[(E, \vec{p}) \to (E, -\vec{p}); (\sigma^0, \vec{\sigma}) \to (\sigma^0, -\vec{\sigma})]$. (These kinematic changes are the meaning of the prime symbol in $IN' \rangle$.)

Altogether, CPT invariance implies then:

$$\langle FN \mid T \mid IN \rangle = \langle \overline{IN}' \mid T \mid \overline{FN}' \rangle \, . \tag{56.17}$$

Since, for the $K^0$-states: $\mid (\bar{K}^0)' \rangle = \mid K^0 \rangle$, the CPT invariance relation implies:

$$\mathcal{M}_{11} = \mathcal{M}_{22} \, . \tag{56.18}$$

The off–diagonal matrix elements in Eq. (56.13) are also related by CPT invariance, plus the hermiticity property of the $T$-matrix in the absence of strong final-state interactions; certainly the case when the $\mid IN \rangle$ and $\mid FN \rangle$ states are $\mid K^0 \rangle$ and $\mid \bar{K}^0 \rangle$. In general, in the absence of strong final-state interactions, we have:

$$\langle \overline{IN}' \mid T \mid \overline{FN}' \rangle = (\langle \overline{FN}' \mid T \mid \overline{IN}' \rangle)^* \, . \tag{56.19}$$

This relation, together with the CPT invariance relation in Eq. (56.17) implies then:

$$\mathcal{M}_{12} = (\mathcal{M}_{21})^* \, . \tag{56.20}$$

There are a number of interesting constraints between the various phenomenological parameters we have introduced. With $M_{12}$ and $\Gamma_{12}$ defined in Eqs. (56.16) and (56.15) and using Eqs. (56.13), (56.4) and (56.7), we have:

$$\frac{q}{p} = \frac{1-\tilde{\epsilon}}{1+\tilde{\epsilon}} = \frac{1}{2}\frac{\Delta m + i\frac{1}{2}\Delta\Gamma}{M_{12} - \frac{i}{2}\Gamma_{12}} = \frac{M_{21} - \frac{i}{2}\Gamma_{21}}{\frac{1}{2}(\Delta m + \frac{i}{2}\Delta\Gamma)}, \tag{56.21}$$

where

$$\Delta m \equiv m_L - m_S \quad \text{and} \quad \Delta\Gamma \equiv \Gamma_S - \Gamma_L. \tag{56.22}$$

As already discussed, CPT invariance implies:

$$M_{21} = (M_{12})^* \quad \text{and} \quad \Gamma_{21} = (\Gamma_{12})^* . \tag{56.23}$$

Experimentally, the masses $m_{L,S}$ and widths $\Gamma_{L,S}$ are well measured, and in what follows they will be used as known parameters. (There is no way for theory at present to do better than experiments in the determination of these parameters…) The precise values for the masses and widths can be found in PDG [16]. Nevertheless, it is important to keep in mind some orders of magnitude:

$$\Gamma_S^{-1} \simeq 0.9 \times 10^{-10}\mathrm{s}; \tag{56.24}$$
$$\Gamma_L \simeq 1.7 \times 10^{-3}\Gamma_S; \tag{56.25}$$
$$\Delta m \simeq 0.5\Gamma_S. \tag{56.26}$$

#### 56.1.2 The Bell–Steinberger unitarity constraint

Let us consider a state $\mid \Psi\rangle$ to be an arbitrary superposition of the short-lived and long-lived kaon states:

$$\mid \Psi\rangle = \alpha \mid K_S\rangle + \beta \mid K_L\rangle . \tag{56.27}$$

The total decay rate of this state must be compensated by a decrease of its norm:

$$\sum_\Gamma \mid \langle \Gamma \mid T \mid \Psi\rangle \mid^2 = -\frac{d}{d\tau} \mid \Psi \mid^2 . \tag{56.28}$$

The change in rate is governed by the mass matrix defined by Eq. (56.11). Equating terms proportional to $\mid \alpha \mid^2$ and $\mid \beta \mid^2$ in both sides of Eq. (56.28) results in the trivial relations:

$$\Gamma_L = \sum_\Gamma \int d\Gamma \mid \langle \Gamma \mid T \mid K_L\rangle \mid^2 , \tag{56.29}$$

$$\Gamma_S = \sum_\Gamma \int d\Gamma \mid \langle \Gamma \mid T \mid K_S\rangle \mid^2 . \tag{56.30}$$

The mixed terms, proportional to $\alpha\beta^*$ and $\alpha^*\beta$, lead however to a highly non-trivial relation, first derived by Bell and Steinberger [807]:

$$-i(M_L^* - M_S)\langle K_L \mid K_S\rangle = \sum_\Gamma \int d\Gamma\, (\langle \Gamma \mid T \mid K_L\rangle)^* \langle \Gamma \mid T \mid K_S\rangle . \tag{56.31}$$

Notice that

$$\langle K_L \mid K_S \rangle = \frac{\mid p \mid^2 - \mid q \mid^2}{\mid p \mid^2 + \mid q \mid^2} = \frac{2\mathrm{Re}\tilde{\epsilon}}{1+ \mid \tilde{\epsilon} \mid^2} . \tag{56.32}$$

The LHS of Eq. (56.31) can be expressed in terms of measurable physical parameters with the result:

$$\left( \frac{\Gamma_S + \Gamma_L}{2} - i\Delta m \right) \frac{2\mathrm{Re}\tilde{\epsilon}}{1+ \mid \tilde{\epsilon} \mid^2} = \sum_\Gamma \int d\Gamma \left( \langle \Gamma \mid T \mid K_L \rangle \right)^* \langle \Gamma \mid T \mid K_S \rangle . \tag{56.33}$$

The RHS of this equation can be bounded, using the Schwartz inequality, with the result:

$$\left| \frac{\Gamma_S + \Gamma_L}{2} - i\Delta m \right| \frac{2\mathrm{Re}\tilde{\epsilon}}{1+ \mid \tilde{\epsilon} \mid^2} \leq \sqrt{\Gamma_L \Gamma_S} . \tag{56.34}$$

Inserting the experimental values for $\Gamma_{S,L}$ and $\Delta m$, results in an interesting bound for the non-orthogonality of the $K_L$ and $K_S$ states [see Eq. (56.32)]:

$$\frac{2\mathrm{Re}\tilde{\epsilon}}{1+ \mid \tilde{\epsilon} \mid^2} \leq 2.9 \times 10^{-2} , \tag{56.35}$$

indicating also that the admixture of $K_1^0(K_2^0)$ in $K_L(K_S)$ has to be rather small.

It is possible to obtain further information from the unitarity constraint in Eq. (56.33), if one uses the experimental fact that the $2\pi$ states are by far the dominant terms in the sum over hadronic states $\Gamma$.

One can then write the RHS of Eq. (56.33) in the form:

$$\sum_{\pi\pi} \int d(\pi\pi) \left( \langle \pi\pi \mid T \mid K_L \rangle \right)^* \langle \pi\pi \mid T \mid K_S \rangle + \gamma \Gamma_S . \tag{56.36}$$

It is possible to obtain a bound for $\gamma$, by considering other states than $2\pi$ in the sum of the RHS in Eq. (56.33) and applying the Schwartz inequality to individual sets of states separated by selection rules.

The contribution from the various semi-leptonic modes, for example, is known to be smaller than:

$$\left| \sum_{\text{lep.modes}} \int \cdots \right| \ll 10^{-3} \Gamma_S ; \tag{56.37}$$

and the contribution from the $3\pi$-states:

$$\left| \sum_{3\pi} \int \cdots \right| \ll 10^{-3} \Gamma_S . \tag{56.38}$$

We conclude that, to a good approximation, we can restrict the Bell–Steinberger relation to $2\pi$-states. We shall later come back to this inequality, but first we have to discuss the phenomenology of the dominant $K \to \pi\pi$ transitions.

### 56.1.3 $K \to 2\pi$ amplitudes

In the limit where CP is conserved the states $K_S(K_L)$ become eigenstates of CP; namely, the states $K_1^0(K_2^0)$ introduced in Eq. (56.5) with eigenvalues CP $= +1$(CP $= -1$). On the other hand a state of two-pions with total angular momentum $J = 0$ has CP $= +1$. Therefore, the observation of a transition from the long-lived component of the neutral kaon system to a two-pion final state is evidence for CP violation. The first observation of such a transition to the $\pi^+\pi^-$ mode was made by Christenson *et al.* [808] in 1964, with the result:

$$\frac{\Gamma_L(+,-)}{\Gamma_L(\mathrm{all})} = (2 \pm 0.4) \times 10^{-3} \,. \tag{56.39}$$

Since then the transition to the $\pi^0\pi^0$ mode has also been observed, as well as the phases of the amplitude ratios:

$$\eta_{+-} = \frac{\langle \pi^+\pi^- \mid T \mid K_L \rangle}{\langle \pi^+\pi^- \mid T \mid K_S \rangle} \quad \text{and} \quad \eta_{00} = \frac{\langle \pi^0\pi^0 \mid T \mid K_L \rangle}{\langle \pi^0\pi^0 \mid T \mid K_S \rangle} \,, \tag{56.40}$$

with the results [16]:

$$\eta_{+-} = (2.269 \pm 0.023) \times 10^{-3} e^{i(44.3 \pm 0.8)^\circ} \,; \tag{56.41}$$

$$\eta_{00} = (2.259 \pm 0.023) \times 10^{-3} e^{i(43.3 \pm 1.3)^\circ} \,. \tag{56.42}$$

In order to make a phenomenological analysis of $K \to \pi\pi$ transitions, it is convenient to express the states $\mid \pi^+\pi^- \rangle$ and $\mid \pi^0\pi^0 \rangle$ in terms of well defined isospin $I = 0$, and $I = 2$ states. (The $I = 1$ state in this case is forbidden by Bose statistics.):

$$\mid +- \rangle = \sqrt{\frac{2}{3}} \mid 0 \rangle + \sqrt{\frac{1}{3}} \mid 2 \rangle \,; \tag{56.43}$$

$$\mid 00 \rangle = \sqrt{\frac{2}{3}} \mid 2 \rangle - \sqrt{\frac{1}{3}} \mid 0 \rangle \,. \tag{56.44}$$

The reason for introducing pure isospin states, is that the matrix elements of transitions from $K^0$ and the $\bar{K}^0$ states to the same $(\pi\pi)_I$-state can be related by *CPT* invariance plus Watson's theorem on final-state interactions. The relation in question is the following:

$$e^{-2i\delta_I} \langle I \mid T \mid K^0 \rangle = (\langle I \mid T \mid \bar{K}^0 \rangle)^* \,, \tag{56.45}$$

where $\delta_I$ denotes the appropriate $J = 0$, isospin $I$ $\pi\pi$ phase-shift at the energy of the neutral kaon mass.

The proof of this relation is rather simple. With $S = 1 + iT$, the unitarity of the $S$ matrix, $SS^\dagger = 1$, implies:

$$T^\dagger T = i(T^\dagger - T) \,. \tag{56.46}$$

If one takes matrix elements of this operator relation between an initial state $K^0$, and a final $2\pi$-state with isospin $I$, we then have:

$$\sum_F \langle I \mid T^\dagger \mid F \rangle \langle F \mid T \mid K^0 \rangle = i \langle I \mid T^\dagger \mid K^0 \rangle - i \langle I \mid T \mid K^0 \rangle \,, \tag{56.47}$$

where we have inserted a complete set of states $\sum | F\rangle\langle F |= 1$ between $T$ and $T^\dagger$. The crucial observation is that, in the strong interaction sector of the $S$ matrix, only the state $F = I$ can contribute to the $T^\dagger$-matrix element. All the other states are suppressed by selection rules; for example, the $3\pi$-states have opposite $G$-parity than the $2\pi$-states; the $\pi l\nu$-states are not related to $2\pi$-states by the strong interactions alone; etc. Then, introducing the $\pi\pi$ phase-shift definition:

$$\langle I \mid S \mid I\rangle = e^{2i\delta_I} , \tag{56.48}$$

results in the relation:

$$\begin{aligned} i(e^{-2i\delta_I} - 1)\langle I \mid T \mid K^0\rangle &= i\langle I \mid T^\dagger \mid K^0\rangle - i\langle I \mid T \mid K^0\rangle , \\ &= i(\langle K^0 \mid T \mid I\rangle)^* - i\langle I \mid T \mid K^0\rangle . \end{aligned} \tag{56.49}$$

We can next use CPT invariance [recall Eq. (56.17), which in our case implies the relation: $\langle K^0 \mid T \mid I\rangle^* = (\langle I \mid T \mid \bar{K}^0\rangle)^*$.] The result in Eq. (56.45) then follows.

As a consequence of the relation we have proved, we can use in full generality the following parametrization for $K^0(\bar{K}^0) \to (\pi\pi)_I$ amplitudes:

$$\langle I \mid T \mid K^0\rangle = i A_I e^{i\delta_I} ; \tag{56.50}$$

$$\langle I \mid T \mid \bar{K}^0\rangle = -i A_I^* e^{i\delta_I} . \tag{56.51}$$

One possible quantity we can introduce to characterize the amount of CP violation in $K \to 2\pi$ transitions is the parameter:

$$\epsilon = \frac{A[K_L \to (\pi\pi)_{I=0}]}{A[K_S \to (\pi\pi)_{I=0}]} . \tag{56.52}$$

This parameter is related to the $\tilde{\epsilon}$–parameter introduced in Eq. (56.7); as well as to the complex $A_0$-amplitude defined in Eqs. (56.50) and (56.51), in the following way:

$$\epsilon = \frac{(1+\tilde{\epsilon})A_0 - (1-\tilde{\epsilon})A_0^*}{(1+\tilde{\epsilon})A_0 + (1-\tilde{\epsilon})A_0^*} . \tag{56.53}$$

namely:

$$\epsilon = \frac{\tilde{\epsilon} + i\frac{\mathrm{Im}A_0}{\mathrm{Re}A_0}}{1 + i\tilde{\epsilon}\frac{\mathrm{Im}A_0}{\mathrm{Re}A_0}} . \tag{56.54}$$

This is a good place to comment on the history of phase conventions in neutral $K$-decays. In their pioneering paper on the phenomenology of the $K - \bar{K}$ system, Wu and Yang [809] chose to freeze the arbitrary relative phase between the $K^0$ and $\bar{K}^0$ states, with the choice $\mathrm{Im}A_0 = 0$. With this convention, $\epsilon = \tilde{\epsilon}$. In fact, the parameter $\epsilon$ is phase-convention independent; while neither $\tilde{\epsilon}$, nor $A_I$ are. Indeed, under a small arbitrary phase change of the $K^0$-state:

$$| K^0\rangle \to e^{-i\varphi} | K^0\rangle , \tag{56.55}$$

the parameters $A_I$, $M_{12}$, and $\tilde{\epsilon}$ change as follows:

$$\mathrm{Im}A_I \to \mathrm{Im}A_I - \varphi \mathrm{Re}A_I \; ; \tag{56.56}$$
$$\mathrm{Im}M_{12} \to \mathrm{Im}M_{12} + \varphi \Delta m \; ; \tag{56.57}$$
$$\tilde{\epsilon} \to \tilde{\epsilon} + i\varphi \; ; \tag{56.58}$$

while $\epsilon$ remains invariant. The Wu–Yang phase convention was made prior to the development of the electroweak theory. In the standard model, the conventional way by which the freedom in the choice of relative phases of the quark fields has been frozen, is not compatible with the Wu–Yang convention. Since $\epsilon$ is convention independent, we shall keep it as one of the fundamental parameters. Then, however, we need a second parameter which characterizes the amount of *intrinsic* CP violation specific to the $K \to 2\pi$ decay, by contrast to the CP violation in the $K^0 - \bar{K}^0$ mass matrix. The parameter we are looking for has to be sensitive then to the lack of relative reality of the the two isospin amplitudes $A_0$ and $A_2$. This is the origin of the famous $\epsilon'$–parameter, which we shall next discuss.

In general, we can define three independent ratios of the $K_{L,S} \to (2\pi)_{I=0,2}$ transition amplitudes. One is the $\epsilon$ parameter in Eq. (56.52). Two other natural ratios are

$$\frac{A[K_L \to (\pi\pi)_{I=2}]}{A[K_S \to (\pi\pi)_{I=0}]} \quad \text{and} \quad \omega \equiv \frac{A[K_S \to (\pi\pi)_{I=2}]}{A[K_S \to (\pi\pi)_{I=0}]} \, . \tag{56.59}$$

Both ratios can be expressed in terms of the $\tilde{\epsilon}$ parameter introduced in Eq. (56.7), and the complex $A_I$ amplitudes defined in Eqs. (56.50) and (56.51):

$$\begin{aligned}\frac{A[K_L \to (\pi\pi)_{I=2}]}{A[K_S \to (\pi\pi)_{I=0}]} &= \frac{(1+\tilde{\epsilon})A_2 - (1-\tilde{\epsilon})A_2^*}{(1+\tilde{\epsilon})A_0 + (1-\tilde{\epsilon})A_0^*} e^{i(\delta_2-\delta_0)} \\ &= \frac{i\frac{\mathrm{Im}A_2}{\mathrm{Re}A_0} + \tilde{\epsilon}\frac{\mathrm{Re}A_2}{\mathrm{Re}A_0}}{1 + i\tilde{\epsilon}\frac{\mathrm{Im}A_0}{\mathrm{Re}A_0}} e^{i(\delta_2-\delta_0)} \; ;\end{aligned} \tag{56.60}$$

and:

$$\begin{aligned}\omega \equiv \frac{A[K_S \to (\pi\pi)_{I=2}]}{A[K_S \to (\pi\pi)_{I=0}]} &= \frac{(1+\tilde{\epsilon})A_2 + (1-\tilde{\epsilon})A_2^*}{(1+\tilde{\epsilon})A_0 + (1-\tilde{\epsilon})A_0^*} e^{i(\delta_2-\delta_0)} \\ &= \frac{\frac{\mathrm{Re}A_2}{\mathrm{Re}A_0} + \tilde{\epsilon}\frac{\mathrm{Im}A_2}{\mathrm{Re}A_0}}{1 + i\tilde{\epsilon}\frac{\mathrm{Im}A_0}{\mathrm{Re}A_0}} e^{i(\delta_2-\delta_0)} \, .\end{aligned} \tag{56.61}$$

The $\epsilon'$ parameter is then defined as the following combination of these ratios:

$$\epsilon' = \frac{1}{\sqrt{2}} \left( \frac{A[K_L \to (\pi\pi)_{I=2}]}{A[K_S \to (\pi\pi)_{I=0}]} - \epsilon \times \omega \right) . \tag{56.62}$$

From these results, and using the expression for $\epsilon$ we obtained in Eq. (56.54), we finally get:

$$\epsilon' = \frac{i}{\sqrt{2}} \frac{(1-\tilde{\epsilon}^2)e^{i(\delta_2-\delta_0)}}{(\mathrm{Re}A_0 + i\tilde{\epsilon}\mathrm{Im}A_0)^2} (\mathrm{Im}A_2 \mathrm{Re}A_0 - \mathrm{Im}A_0 \mathrm{Re}A_2) \, , \tag{56.63}$$

an expression which clearly shows the proportionality to the lack of relative reality between the $A_0$ and $A_2$ amplitudes.

We shall next establish contact with the parameters $\eta_{+-}$ and $\eta_{00}$, which were introduced in Eq. (56.40), and which are directly accessible to experiment. Using Eqs. (56.43), (56.44), as well as the definitions of $\epsilon$, $\epsilon'$, and $\omega$ above, one finds:

$$\eta_{+-} = \epsilon + \epsilon' \frac{1}{1 + \frac{1}{\sqrt{2}}\omega}\ ; \tag{56.64}$$

$$\eta_{00} = \epsilon - 2\epsilon' \frac{1}{1 - \sqrt{2}\omega}\ . \tag{56.65}$$

So far, we have made no approximations in our phenomenological analysis of the $K^0 - \bar{K}^0$ mass matrix and $K \to 2\pi$ decays. It is however useful to try to thin down in some way the exact expressions we have derived, by taking into account the relative size of the various phenomenological parameters which appear in the expressions above. The strategy will be to neglect first, terms which are products of CP violation parameters. For example, in Eq. (56.61), we have introduced the parameter $\omega$, which a priori we can reasonably expect to be dominated by the term:

$$\omega \simeq \frac{\mathrm{Re}A_2}{\mathrm{Re}A_0} e^{i(\delta_2 - \delta_0)}, \tag{56.66}$$

where experimentally [16]:

$$\delta_2 - \delta_0 = -(42 \pm 4)^0\ . \tag{56.67}$$

We can justify this approximation by the fact that non-leptonic $\Delta I = \frac{3}{2}$ transitions, although suppressed with respect to the $\Delta I = \frac{1}{2}$ transitions, are nevertheless larger than the observed CP violation effects. Notice that the amplitude $A_2$ is responsible for the deviation from an exact $\Delta I = \frac{1}{2}$ rule. The ratio $\frac{\mathrm{Re}A_2}{\mathrm{Re}A_0}$ can be obtained from the experimentally known branching ratios $\Gamma(K_S \to \pi^+\pi^-)$ and $\Gamma(K_S \to \pi^0\pi^0)$.

More precisely, correcting for the phase–space effects, one must compare the normalized decay rates:

$$\gamma(1,2) \equiv \frac{\Gamma(K \to \pi_1\pi_2)}{\frac{1}{16\pi M}\sqrt{1 - \frac{(m_1+m_2)^2}{M^2}}\sqrt{1 - \frac{(m_1-m_2)^2}{M^2}}}\ , \tag{56.68}$$

where the denominator here is the two-body phase space factor for the mode $K \to \pi_1\pi_2$, ($M$ is the mass of the $K$-particle and $m_{1,2}$ the pion masses.) Then, we have:

$$\frac{\gamma_S(+-)}{2\gamma_S(00)} = 1 + 3\sqrt{2}\frac{\mathrm{Re}A_2}{\mathrm{Re}A_0}\cos(\delta_2 - \delta_1) + \mathcal{O}\left(\frac{\alpha}{\pi}\right)\ . \tag{56.69}$$

Experimentally, from the PDG [16], one finds:

$$\frac{\gamma_S(+-)}{2\gamma_S(00)} = 1.109 \pm 0.012\ , \tag{56.70}$$

and using the present experimental information on $(\delta_2 - \delta_1)$, we find, with neglect of radiative corrections:

$$\frac{\mathrm{Re}A_2}{\mathrm{Re}A_0} = (+22.2)^{-1}\,. \tag{56.71}$$

We shall discuss later some of the qualitative dynamical explanations, within the standard model, of how this small number appears. It is fair to say however, that a reliable calculation of this ratio is still lacking at present. Using the approximations:

$$\tilde{\epsilon}\,\mathrm{Im}A_0 \ll \mathrm{Re}A_0 \quad \text{and} \quad \tilde{\epsilon}^2 \ll 1\,, \tag{56.72}$$

we can rewrite $\epsilon'$ in a simpler form:

$$\epsilon' \simeq \frac{1}{\sqrt{2}} e^{i(\delta_2-\delta_0+\frac{\pi}{2})} \frac{\mathrm{Re}A_2}{\mathrm{Re}A_0}\left(\frac{\mathrm{Im}A_2}{\mathrm{Re}A_2} - \frac{\mathrm{Im}A_0}{\mathrm{Re}A_0}\right), \tag{56.73}$$

clearly showing the fact that $\epsilon'$ *is proportional to direct CP–violation in* $K \to 2\pi$ *transitions and is also suppressed by the* $\Delta I = \frac{1}{2}$ *selection rule.*

The same approximations in Eq. (56.72), when applied to $\epsilon$, lead to:

$$\epsilon \simeq \tilde{\epsilon} + i\frac{\mathrm{Im}A_0}{\mathrm{Re}A_0}\,. \tag{56.74}$$

Let us next go back to the mass matrix equations in Eq. (56.21) which, expanding in powers of $\tilde{\epsilon}$, we can rewrite as follows:

$$1 - 2\tilde{\epsilon} \simeq \frac{\mathrm{Re}M_{12} - \frac{i}{2}\mathrm{Re}\Gamma_{12}}{\frac{1}{2}(\Delta m + \frac{i}{2}\Delta\Gamma)} - i\frac{\mathrm{Im}M_{12} - \frac{i}{2}\mathrm{Im}\Gamma_{12}}{\frac{1}{2}(\Delta m + \frac{i}{2}\Delta\Gamma)}\,. \tag{56.75}$$

To a first approximation, neglecting CP violation effects altogether, we find that:

$$\mathrm{Re}M_{12} \simeq \frac{\Delta m}{2} \quad \text{and} \quad \mathrm{Re}\Gamma_{12} \simeq -\frac{\Delta\Gamma}{2}\,. \tag{56.76}$$

If furthermore, we restrict the sum over intermediate states in $\Gamma_{12}$ [see Eq. (56.15)] to $2\pi$ states, an approximation which we have already seen to be rather good [see Eqs. (56.37) and (56.38)] we can write

$$\Gamma_{12} \simeq (-iA_0^* e^{i\delta_0})^* iA_0 e^{i\delta_0} = -(\mathrm{Re}A_0 + i\mathrm{Im}A_0)^2\,, \tag{56.77}$$

from where it follows that:

$$\frac{\mathrm{Im}\Gamma_{12}}{\mathrm{Re}\Gamma_{12}} \simeq \frac{2\mathrm{Re}A_0\mathrm{Im}A_0}{\mathrm{Re}A_0^2 + \mathrm{Im}A_0^2} \simeq 2\frac{\mathrm{Im}A_0}{\mathrm{Re}A_0}\,. \tag{56.78}$$

Then, using the empirical fact that $\Delta m \simeq \frac{\Gamma_S}{2}$, and $\Gamma_L \ll \Gamma_S$, we finally arrive at the simplified expression:

$$\tilde{\epsilon} \simeq \frac{1}{1+i}\left(i\frac{\mathrm{Im}M_{12}}{\Delta m} + \frac{\mathrm{Im}A_0}{\mathrm{Re}A_0}\right), \tag{56.79}$$

and, using Eq. (56.74):

$$\epsilon \simeq \frac{1}{\sqrt{2}} e^{i\frac{\pi}{4}} \left( \frac{\mathrm{Im} M_{12}}{\Delta m} + \frac{\mathrm{Im} A_0}{\mathrm{Re} A_0} \right) . \tag{56.80}$$

This is as much as one can do, within a strict phenomenological analysis of the CP violation in $K$ decays. We have reduced the problem to the knowledge of two parameters: $\epsilon$ in Eq. (56.80), and $\epsilon'$ in Eq. (56.73). Their present experimental values are [16,599]:

$$\epsilon \simeq (2.280 \pm 0.013) \times 10^{-3} e^{i(43.5\pm 0.1)^0} , \qquad \mathrm{Re}(\epsilon'/\epsilon) \simeq (17.2 \pm 1.8) \times 10^{-4} . \tag{56.81}$$

We shall come back to these parameters in the next section. There, we shall discuss what predictions for these fundamental parameters can be made at present within the framework of the Standard Model. As we shall see, the main difficulty comes from the lack of quantitative understanding of the low-energy sector of the strong interactions. In terms of QCD, the sector in question is the one of the interactions between the states with lowest masses: the octet of the pseudoscalar particles ($\pi$, $K$, $\eta$) and presumably the singlet ($\sigma$, $\eta'$).

## 56.2 $B^0_{(s)}$-$\bar{B}^0_{(s)}$ mixing

### *56.2.1 Introduction*

$B^0_{(s)}$ and $\bar{B}^0_{(s)}$ are not eigenstates of the weak Hamiltonian, such that their oscillation frequency is governed by their mass difference $\Delta M_q$. The measurement by the UA1 collaboration [810] of a large value of $\Delta M_d$ was the *first* indication of an heavy top quark mass. In the SM, the mass difference is approximately given by [665,475]:

$$\Delta M_q \simeq \frac{G_F^2}{4\pi^2} M_W^2 |V_{tq} V_{tb}^*|^2 S_0 \left( \frac{m_t^2}{M_W^2} \right) \eta_B C_B(\nu) \frac{1}{2M_{B_q}} \langle \bar{B}^0_q | \mathcal{O}_q(\nu) | B^0_q \rangle , \tag{56.82}$$

where the $\Delta B = 2$ local operator $\mathcal{O}_q$ is defined as:

$$\mathcal{O}_q(x) \equiv (\bar{b}\gamma_\mu L q)(\bar{b}\gamma_\mu L q) , \tag{56.83}$$

with: $L \equiv (1-\gamma_5)/2$ and $q \equiv d, s,$; $S_0$, $\eta_B$ and $C_B(\nu)$ are short-distance quantities and Wilson coefficients which are calculable perturbatively [811,475,665,812], while the matrix element $\langle \bar{B}^0_q | \mathcal{O}_q | B^0_q \rangle$ requires non-perturbative QCD calculations, and is usually parametrized for $SU(N)_c$ colours as:

$$\langle \bar{B}^0_q | \mathcal{O}_q | B^0_q \rangle = N_c \left( 1 + \frac{1}{N_c} \right) f_{B_q}^2 M_{B_q}^2 B_{B_q} . \tag{56.84}$$

$f_{B_q}$ is the $B_q$ decay constant normalized as $f_\pi = 92.4$ MeV, and $B_{B_q}$ is the so-called bag parameter which is $B_{B_q} \simeq 1$ if one uses a vacuum saturation of the matrix element. From Eq. (56.82), it is clear that the measurement of $\Delta M_d$ provides a measurement of the CKM mixing angle $|V_{td}|$ if one uses $|V_{tb}| \simeq 1$. One can also extract this quantity from the ratio:

$$\frac{\Delta M_s}{\Delta M_d} = \left|\frac{V_{ts}}{V_{td}}\right|^2 \frac{M_{B_d}}{M_{B_s}} \frac{\langle \bar{B}^0_s|\mathcal{O}_s|B^0_s\rangle}{\langle \bar{B}^0_d|\mathcal{O}_d|B^0_d\rangle} \equiv \left|\frac{V_{ts}}{V_{td}}\right|^2 \frac{M_{B_d}}{M_{B_s}} \xi^2 , \qquad (56.85)$$

since in the SM with three generations and unitarity constraints, $|V_{ts}| \simeq |V_{cb}|$. Here:

$$\xi \equiv \sqrt{\frac{g_s}{g_d}} \equiv \frac{f_{B_s}\sqrt{B_{B_s}}}{f_B\sqrt{B_B}} . \qquad (56.86)$$

The great advantage of Eq. (56.85) compared with the former relation in Eq. (56.82) is that in the ratio, different systematics in the evaluation of the matrix element tends to cancel out, thus providing a more accurate prediction. However, unlike $\Delta M_d = 0.473(17)$ ps$^{-1}$, which is measured with a good precision [16], the determination of $\Delta M_s$ is an experimental challenge due to the rapid oscillation of the $B^0_s$-$\bar{B}^0_s$ system. At present, only a lower bound of 13.1 ps$^{-1}$ is available at the 95% confidence level from experiments [16], but this bound already provides a strong constraint on $|V_{td}|$.

### 56.2.2 Two-point function sum rule

Pich [813] has extended the analysis of the $K^0$-$\bar{K}^0$ systems of [814], using a two-point correlator of the four-quark operators in the analysis of the quantity $f_B\sqrt{B_B}$ which governs the $B^0$-$\bar{B}^0$ mass difference. The two-point correlator defined as:

$$\psi_H(q^2) \equiv i\int d^4x\, e^{iqx}\, \langle 0|\mathcal{T}\mathcal{O}_q(x)(\mathcal{O}_q(0))^\dagger|0\rangle , \qquad (56.87)$$

is built from the $\Delta B = 2$ weak operator defined previously. Its QCD expression is given in the chapter on the two-point function. The hadronic part of the spectral function can be conveniently parametrized using the effective realization [813]:

$$\mathcal{O}^{\rm eff}_q = \frac{2}{3}\left(g_B \equiv f^2_{B-q} B_{B_q}\right) \partial_\mu B^0_q \partial^\mu B^0_q + \cdots , \qquad (56.88)$$

where $\cdots$ corresponds to higher mass hadronic states. It leads to the general form [814]:

$$\begin{aligned}
\frac{1}{\pi}{\rm Im}\hat{\Psi}^{\rm had}(t) = {}& \theta(t-4M_B^2)\frac{2}{9}\left(\frac{g_B}{4\pi}\right)^2 t^2 \cdot \int_{t_{10}}^{(\sqrt{t}-\sqrt{t_{20}})^2} dt_1 \int_{t_{20}}^{(\sqrt{t}-\sqrt{t_1})^2} dt_2\, \lambda^{1/2}\left(1,\frac{t_1}{t},\frac{t_2}{t}\right)\\
&\cdot\left\{\left(\frac{t_1}{t}+\frac{t_2}{t}-1\right)^2 \frac{1}{\pi}{\rm Im}\Pi^{(0)}(t_1)\frac{1}{\pi}{\rm Im}\Pi^{(0)}(t_2)\right.\\
&+2\lambda\left(1,\frac{t_1}{t},\frac{t_2}{t}\right)\frac{1}{\pi}{\rm Im}\Pi^{(1)}(t_1)\frac{1}{\pi}{\rm Im}\Pi^{(0)}(t_2)\\
&\left.+\left[\left(\frac{t_1}{t}+\frac{t_2}{t}-1\right)^2+8\frac{t_1t_2}{t^2}\right]\frac{1}{\pi}{\rm Im}\Pi^{(1)}(t_1)\frac{1}{\pi}{\rm Im}\Pi^{(1)}(t_2)\right\}\\
&+\Theta(t-t_c)\frac{1}{\pi}{\rm Im}\Psi^{QCD}(t),
\end{aligned} \qquad (56.89)$$

where the index $i = 0, 1$ refers to the hadronic states with spin 0, 1, and:

$$\frac{1}{\pi}\mathrm{Im}\Pi^{(i)}(t) \equiv \frac{1}{\pi}\mathrm{Im}\Pi_V^{(i)}(t) + \frac{1}{\pi}\mathrm{Im}\Pi_A^{(i)}(t) , \tag{56.90}$$

are the correlators associated to the vector (index $V$) and axial-vector (index $A$) currents; $\lambda^{1/2}$ is the usual phase space factor. In the following, we shall retain the contributions from the $B$-$\bar{B}$ and $B^*$-$\bar{B}^*$ states, and we (reasonnably) assume that:

$$g_B \simeq g_{B^*} , \tag{56.91}$$

which is supported by the HQET and QSSR results ($f_B \approx f_{B^*}$) and the vacuum saturation assumption ($B_B \approx B_{B^*} \approx 1$) a posteriori recovered from our analysis. The corresponding Laplace (resp. moment) sum rules are:

$$\mathcal{L}(\tau) = \int_{4M_B^2}^{\infty} dt\, \mathrm{e}^{-t\tau}\mathrm{Im}\psi_H(t) , \qquad \mathcal{M}_n = \int_{4M_B^2}^{\infty} dt\, t^n\, \mathrm{Im}\psi_H(t) , \tag{56.92}$$

The two-point function approach is very convenient due to its simple analytic properties which is not the case for the approach based on three-point functions.[2] However, it involves non-trivial QCD calculations which become technically complicated when one includes the contributions of radiative corrections due to non-factorizable diagrams. These perturbative radiative corrections due to *factorizable and non-factorizable* diagrams have been already computed in [816] (referred as NP), where it has been found that the factorizable corrections are large while the non-factorizable ones are negligibly small. NP analysis has confirmed the estimate in [323] from lowest order calculations, where under some assumptions on the contributions of higher mass resonances to the spectral function, the value of the bag constant $B_B$ has been found to be:

$$B_{B_d}\left(4m_b^2\right) \simeq (1 \pm 0.15) . \tag{56.93}$$

This value is comparable with the value $B_{B_d} = 1$ from the vacuum saturation estimate, which is expected to be quite a good approximation due to the relative high scale of the $B$-meson mass. Equivalently, the corresponding RGI quantity is:

$$\hat{B}_{B_d} \simeq (1.5 \pm 0.2) , \tag{56.94}$$

where we have used the relation:

$$B_{B_q}(\nu) = \hat{B}_{B_q}\alpha_s^{-\frac{\gamma_0}{\beta_1}}\left\{1 - \left(\frac{5165}{12696}\right)\left(\frac{\alpha_s}{\pi}\right)\right\} , \tag{56.95}$$

with $\gamma_0 = 1$ as the anomalous dimension of the operator $\mathcal{O}_q$ and $\beta_1 = -23/6$ for five flavours. The NLO corrections have been obtained in the $\overline{MS}$ scheme [665]. We have also used, to this order, the value [148,149,3]:

$$\bar{m}_b(m_b) = (4.24 \pm 0.06)\ \mathrm{GeV} , \tag{56.96}$$

and $\Lambda_5 = (250 \pm 50)$ MeV [139]. In a forthcoming paper [817], we study (*for the first time*), from the QSSR method, the $SU(3)$ breaking effects on the ratio: $\xi$ defined previously

[2] For detailed criticisms, see [3].

in Eq. (56.86), where a similar analysis of the ratios of the decay constants has given the values [716]:

$$\frac{f_{D_s}}{f_D} \simeq 1.15 \pm 0.04\,, \qquad \frac{f_{B_s}}{f_B} \simeq 1.16 \pm 0.04\,. \tag{56.97}$$

### 56.2.3 Results and implications on $|V_{ts}|^2/|V_{td}|^2$ and $\Delta M_s$

We deduce by taking the average from the moments and Laplace sum rules results [817]:

$$\xi \equiv \frac{f_{B_s}\sqrt{B_{B_s}}}{f_B\sqrt{B_B}} \simeq 1.18 \pm 0.03\,, \qquad f_B\sqrt{\hat{B}_B} \simeq (247 \pm 59)\ \text{GeV}, \tag{56.98}$$

in units where $f_\pi = 130.7$ MeV. For the ratio, one expects small errors due to the cancellation of the systematics, while for $f_B\sqrt{\hat{B}_B}$, the error estimate comes mainly from the one of $m_b$ and the estimate of higher-order terms of the QCD series. These results can be compared with different lattice determinations compiled in [823,723]. By comparing the ratio with the one of $f_{B_s}/f_{B_d}$ in Eq. (56.97),[3] one can conclude (to a good approximation) that:

$$\hat{B}_{B_s} \approx \hat{B}_{B_d} \simeq (1.65 \pm 0.38) \Longrightarrow B_{B_{d,s}}(4m_b^2) \simeq (1.10 \pm 0.25)\,, \tag{56.99}$$

indicating a negligible $SU(3)$ breaking for the bag parameter. For a consistency, we have used the estimate to order $\alpha_s$ [698]:

$$f_B \simeq (1.47 \pm 0.10) f_\pi\,, \tag{56.100}$$

and we have assumed that the error from $f_B$ compensates the one in Eq. (56.98). The result is in excellent agreement with the previous result of [816] in Eqs. (56.93) and (56.94). Using the experimental values:

$$\Delta M_d = 0.472(17)\ ps^{-1}\,, \qquad \Delta M_s \geq 13.1\ ps^{-1}\ \ (95\%\ \text{CL})\,, \tag{56.101}$$

one can deduce from Eq. (56.85):

$$\rho_{sd} \equiv \left|\frac{V_{ts}}{V_{td}}\right|^2 \geq 20.0(1.1)\,. \tag{56.102}$$

Alternatively, using:

$$\rho_{sd} \simeq \frac{1}{\lambda^2[(1-\bar{\rho})^2 + \bar{\eta}^2]} \simeq 28.4(2.9)\,, \tag{56.103}$$

with [723]:

$$\lambda \simeq 0.2237(33)\,, \qquad \bar{\rho} \equiv \rho\left(1 - \frac{\lambda^2}{2}\right) \simeq 0.223(38)\,,$$
$$\bar{\eta} \equiv \eta\left(1 - \frac{\lambda^2}{2}\right) \simeq 0.316(40)\,, \tag{56.104}$$

[3] One can notice that similar strengths of the $SU(3)$ breakings are obtained for the $B \to K^*\gamma$ and $B \to Kl\nu$ form factors [818].

$\lambda,\ \rho,\ \eta$ being the Wolfenstein parameters, we deduce:

$$\Delta M_s \simeq 18.6(2.2)\mathrm{ps}^{-1}\ , \tag{56.105}$$

in good agreement with the present experimental lower bound.

### 56.2.4 Conclusions

We have applied QCD spectral sum rules for extracting *(for the first time)* the $SU(3)$ breaking parameter in Eq. (56.98). The phenomenological consequences of our results for the $B^0_{d,s}$-$\bar{B}^0_{d,s}$ mass differences and CKM mixing angle have been discussed. An extension of this work to the study of the $B^0_{s,d}$-$\bar{B}^0_{s,d}$ width difference is in progress.

## 56.3 The $\Delta S = 2$ transition of the $K^0$-$\bar{K}^0$ mixing

### 56.3.1 Estimate of the bag constant $B_K$

This parameter plays an important rôle for the $CP$ violation parameter in connection with the previous quantities $f_B$ and $B_B$. The $B_K$-parameter is associated to the $K^0$-$\bar{K}^0$ mixing matrix as:

$$\langle \bar{K}^0 | \bar{b}\gamma^L_\mu d \bar{d}\gamma^L_\mu b | K^0 \rangle = \frac{4}{3} f_K^2 M_K^2 B_K(\nu)\ , \tag{56.106}$$

where as before, one has also introduced the RGI parameter $\hat{B}_K$. We estimate this quantity using the four-quark two-point correlator as in [814,815]. Using the Laplace sum rule (LSR) and adopting the parametrization of the spectral function in [814], we have obtained the *conservative* estimate [815]:

$$\hat{B}_K \simeq (0.58 \pm 0.22)\ , \tag{56.107}$$

where the central value is slightly higher than the one from FESR [814]: $\hat{B}_K \simeq (0.39 \pm 0.10)$. This difference might be attributed to the fact that FESR is strongly affected by the higher radial excitation contributions that are not under good control. LSR has the advantage is less sensitive to these effects due to the exponential factor which suppresses their relative contributions. One can also notice that this result from the two-point function sum rule is more accurate than the one from the three-point function [3], which ranges from 0.2 to 1.3, although the result of [3] is in good agreement with ours. This inaccuracy can be intuitively understood from the relative complexity of the three-point function sum rule analysis for parametrizing the higher-states contributions to the spectral function.

### 56.3.2 Estimate of the CP violation parameters $(\bar{\rho}, \bar{\eta})$

We are now ready to discuss the implications of the previous results for the estimate of the CKM parameters $(\bar{\rho}, \bar{\eta})$ defined in the standard way within the Wolfenstein parametrization [16,665,500,820].

Within this parametrization, one can express the CP violation of the kaon system as:

$$|\epsilon| = C_\epsilon A^2\lambda^6\bar{\eta}[-\eta_1 S(x_c) + \eta_2 S(x_t)(A^2\lambda^4(1-\bar{\rho})) + \eta_3 S(x_c, x_t)]\hat{B}_K \,, \quad (56.108)$$

where:

$$C_\epsilon = \frac{G_F^2 f_K^2 M_K M_W^2}{3\sqrt{2}\pi^2 \Delta M_K} \,; \quad (56.109)$$

$S(x_i)$, $S(x_i, x_j)$, $\eta_1 \simeq 1.38$, $\eta_2 \simeq 0.574$, $\eta_3 \simeq 0.47$ are short-distance functions calculable perturbatively [811,475,665,812] with $x_q \equiv m_q^2/M_W^2$; $(A, \lambda, \bar{\rho}, \bar{\eta})$ are set of CKM parameters within the Wolfenstein parametrization. For a self-consistent analysis, it is essential to use the previous values of $f_B$, $B_B$ and $B_K$, which are all obtained from a unique method. Using the phenomenological analysis in [723,820], one can approximately obtain:

$$|\epsilon| \simeq \frac{4}{3}\hat{B}_K \mathrm{Im}(V_{ts}^* V_{td})(18.9 - 14.4\bar{\rho}) \,, \quad (56.110)$$

where $\mathrm{Im}(V_{ts}^* V_{td}) \simeq (1.2 \pm 0.2) \times 10^{-4}$ and $\bar{\rho} \simeq 0.2 \pm 0.1$. With such values, one can, for example, deduce:

$$|\epsilon| \simeq (14.8 \pm 5.6) \times 10^{-4} \,, \quad (56.111)$$

which agrees within about $1\sigma$ with the experimental value in Eq. (56.81).

## 56.4 Kaon penguin matrix elements and $\epsilon'/\epsilon$

### *56.4.1 SM theory of $\epsilon'/\epsilon$*

In the SM, it is customary to study the $\Delta S = 1$ process from the weak Hamiltonian:

$$\mathcal{H}_{\mathrm{eff}} = \frac{G_F}{\sqrt{2}} V_{ud} V_{us}^* \sum_{i=1}^{10} C_i(\mu) Q_i(\mu) \,, \quad (56.112)$$

where $C_i(\mu)$ are known perturbative Wilson coefficients including complete NLO QCD corrections [665], which read in the notation of [665]:

$$C_i(\mu) \equiv z_i(\mu) - \frac{V_{td} V_{ts}^*}{V_{ud} V_{us}^*} y_i(\mu) \,, \quad (56.113)$$

where $V_{ij}$ are elements of the CKM-matrix; $Q_i(\mu)$ are non-perturbative hadronic matrix elements which need to be estimated from different non-perturbative methods of QCD (chiral perturbation theory, lattice, QCD spectral sum rules, ...). In the choice of basis of [665], the dominant contributions come from the four-quark operators which are classified as:

- **Current-current:**

$$\mathcal{Q}_1 \equiv (\bar{s}_\alpha u_\beta)_{V-A} (\bar{u}_\beta d_\alpha)_{V-A} \,, \qquad \mathcal{Q}_2 \equiv (\bar{s}u)_{V-A} (\bar{u}d)_{V-A} \,. \quad (56.114)$$

- **QCD penguins:**

$$\mathcal{Q}_3 \equiv (\bar{s}d)_{V-A}\sum_{u,d,s}(\bar{\psi}\psi)_{V-A}\,, \qquad \mathcal{Q}_4 \equiv (\bar{s}_\alpha d_\beta)_{V-A}\sum_{u,d,s}(\bar{\psi}_\beta\psi_\alpha)_{V-A}\,,$$
$$\mathcal{Q}_5 \equiv (\bar{s}d)_{V-A}\sum_{u,d,s}(\bar{\psi}\psi)_{V+A}\,, \qquad \mathcal{Q}_6 \equiv (\bar{s}_\alpha d_\beta)_{V-A}\sum_{u,d,s}(\bar{\psi}_\beta\psi_\alpha)_{V+A}\,. \tag{56.115}$$

- **Electroweak penguins:**

$$\mathcal{Q}_7 \equiv \frac{3}{2}(\bar{s}d)_{V-A}\sum_{u,d,s}e_\psi(\bar{\psi}\psi)_{V+A}\,, \qquad \mathcal{Q}_8 \equiv \frac{3}{2}(\bar{s}_\alpha d_\beta)_{V-A}\sum_{u,d,s}e_\psi(\bar{\psi}_\beta\psi_\alpha)_{V+A}\,,$$
$$\mathcal{Q}_9 \equiv \frac{3}{2}(\bar{s}d)_{V-A}\sum_{u,d,s}e_\psi(\bar{\psi}\psi)_{V-A}\,, \qquad \mathcal{Q}_{10} \equiv \frac{3}{2}(\bar{s}_\alpha d_\beta)_{V-A}\sum_{u,d,s}e_\psi(\bar{\psi}_\beta\psi_\alpha)_{V-A}, \tag{56.116}$$

where $\alpha$, $\beta$ are colour indices; $e_\psi$ denotes the electric charges[4] reflecting the electroweak nature of $\mathcal{Q}_{7,\ldots 10}$, while $V-(+)A \equiv (1-(+)\gamma_5)\,\gamma_\mu$. Using an OPE of the amplitudes, one obtains:

$$\frac{\epsilon'}{\epsilon} \simeq \mathrm{Im}\lambda_t\big[P^{(1/2)} - P^{(3/2)}\big]e^{i\Phi}\,, \tag{56.117}$$

where $\Phi \equiv \Phi_{\epsilon'} - \Phi_\epsilon \approx 0$ (see previous section); $\lambda_t \equiv V_{td}V_{ts}^*$ can be expressed in terms of the CKM matrix elements as ($\delta$ being the CKM phase) [665,820]:

$$\mathrm{Im}\lambda_t \approx |V_{ub}||V_{cb}|\sin\delta \simeq (1.33 \pm 0.14)\times 10^{-4}\,, \tag{56.118}$$

from $B$-decays and $\epsilon$. The QCD quantities $P^{(I)}$ read:

$$P^{(1/2)} = \frac{G_F|\omega|}{2|\epsilon|\mathrm{Re}A_0}\sum_i C_i(\mu)\langle(\pi\pi)_{I=0}|\mathcal{Q}_i|K^0\rangle_0\,(1-\Omega_{IB})\,,$$
$$P^{(3/2)} = \frac{G_F}{2|\epsilon|\mathrm{Re}A_2}\sum_i C_i(\mu)\langle(\pi\pi)_{I=2}|\mathcal{Q}_i|K^0\rangle_2\,. \tag{56.119}$$

$\Omega_{IB} \simeq (0.16 \pm 0.03)$ quantifies the $SU(2)$-isospin breaking effect, which includes the one of the $\pi^0$-$\eta$ mixing [821], and which reduces the usual value of $(0.25 \pm 0.08)$ [665] due to $\eta'$-$\eta$ mixing. It is also expected that the QCD- and electroweak-penguin operators:

$$\mathcal{Q}_8^{3/2} \approx B_8^{3/2}/m_s^2 + \mathcal{O}(1/N_c)\,, \qquad \mathcal{Q}_6^{1/2} \approx B_6^{1/2}/m_s^2 + \mathcal{O}(1/N_c)\,, \tag{56.120}$$

give the dominant contributions to the ratio $\epsilon'/\epsilon$ [822]; $B$ are the bag factors which are expected to be 1 in the large $N_c$-limit. Therefore, a simplified approximate but very informative

[4] Though apparently suppressed, the effect of the electroweak penguins are enhanced by $1/\omega$ as we shall see later on in Eq. (56.119).

expression of the theoretical predictions can be derived [665]:

$$\frac{\epsilon'}{\epsilon} \approx 13\ \mathrm{Im}\lambda_t \left(\frac{110}{\bar{m}_s(2)\ [\mathrm{MeV}]}\right)^2 \times \left[B_6^{1/2}(1-\Omega_{IB}) - 0.4 B_8^{3/2}\left(\frac{m_t}{165\ \mathrm{GeV}}\right)\right]\left(\frac{\Lambda^{(4)}_{\overline{MS}}}{340\ \mathrm{MeV}}\right), \quad (56.121)$$

where the average value $\hat{B}_K = 0.80 \pm 0.15$ of the $\Delta S = 2$ process has been used. This value includes the conservative value $0.58 \pm 0.22$ from Laplace sum rules [815]. The values of the top quark mass and the QCD scale $\Lambda^{(4)}_{\overline{MS}}$ [16,139] are under quite good control and have small effects. A recent review of the light quark mass determinations [54] also indicates that the strange quark mass is also under control and a low value advocated in the previous literature to explain the present data on $\epsilon'/\epsilon$ is unlikely to be due to the lower bound constraints from the positivity of the QCD spectral function or from the positivity of the $m^2$ corrections to the GMOR PCAC relation. For a consistency with the approach used in this paper, we shall use the average value of the light quark masses from QCD spectral sum rules(QSSR), $e^+e^-$ and $\tau$-decays given in [54] (previous chapter):

$$\bar{m}_s(2) \simeq (117 \pm 23)\ \mathrm{MeV}\,, \quad \bar{m}_d(2) \simeq (6.5 \pm 1.2)\ \mathrm{MeV}\,, \quad \bar{m}_u(2) \simeq (3.6 \pm 0.6)\ \mathrm{MeV}\,. \quad (56.122)$$

Using the previous experimental values, one can deduce the constraint in [54] updated:

$$\mathcal{B}_{68} \equiv B_6^{1/2} - 0.48 B_8^{3/2} \simeq 1.4 \pm 0.6\ (\mathrm{resp.} \geq 0.5)\,, \quad (56.123)$$

if one uses the value of $m_s$ in Eq. (56.122) (resp. the lower bound of 71 MeV reported in [54]). This result shows a possible violation of more than $2\sigma$ for the leading $1/N_c$ vacuum saturation prediction $\approx 0.52$ corresponding to $B_6^{1/2} \approx B_8^{3/2} \approx 1$. Consulting the available predictions reviewed in [665], which we will summarize and update in Table 56.1, one can notice that the values of the $B$ parameters have large errors. One can also see that results from QCD first principles (lattice and $1/N_c$) fail to explain the data, which however can be accomodated by various QCD-like models. We shall come back to this discussion when we shall compare our results with presently available predictions. It is, therefore, clear that the present estimate of the four-quark operators, and in particular the estimates of the dominant penguin ones given previously in Eq. (56.120), need to be re-investigated. Due to the complex structures and large size of these operators, they should be difficult to extract unambiguously from different approaches. In this paper, we present alternative theoretical approaches based also on first principles of QCD ($\tau$–decay data, analyticity), for predicting the size of the QCD- and electroweak-penguin operators given in Eq. (56.120). In performing this analysis, we shall also encounter the electroweak penguin operator:

$$\mathcal{Q}_7^{3/2} \approx B_7^{3/2}/m_s^2 + \mathcal{O}(1/N_c)\,. \quad (56.124)$$

Table 56.1. *Penguin B parameters for the $\Delta S = 1$ process from different approaches at $\mu = 2$ GeV. We use the value $m_s(2) = (117 \pm 23)$ MeV from [54], and predictions based on dispersion relations [833,832] have been rescaled according to it. We also use for our results $f_\pi = 92.4$ MeV [16], but we give in the text their $m_s$ and $f_\pi$ dependences. Results without any comments on the scheme have been obtained in the $\overline{MS}-NDR$–scheme (see discussions on $\gamma_5$ in Appendix D). However, at the present accuracy, one cannot differentiate these results from the ones of $\overline{MS}-HV$–scheme. More recent results can also found in [838].*

| Methods | $B_6^{1/2}$ | $B_8^{3/2}$ | $B_7^{3/2}$ | Comments |
|---|---|---|---|---|
| Lattice [823,824,825] | $0.6 \sim 0.8$<br>unreliable | $0.7 \sim 1.1$ | $0.5 \sim 0.8$ | Huge NLO<br>at matching [826] |
| Large $N_c$ [827] | $0.7 \sim 1.3$ | $0.4 \sim 0.7$ | $-0.10 \sim 0.04$ | $\mathcal{O}(p^0/N_c,\ p^2)$<br>scheme? |
| | $1.5 \sim 1.7$ | – | – | $\mathcal{O}(p^2/N_c)$; $m_q = 0$<br>scheme? |
| **Models** | | | | |
| Chiral QM [828] | $1.2 \sim 1.7$ | $\sim 0.9$ | $\approx B_8^{3/2}$ | $\mu = 0.8$ GeV<br>rel. with $\overline{MS}$? |
| ENJL + IVB [829] | $2.5 \pm 0.4$ | $1.4 \pm 0.2$ | $0.8 \pm 0.1$ | $NLO$ in $1/N_c$<br>$m_q = 0$ |
| L$\sigma$-model [830] | $\sim 2$ | $\sim 1.2$ | – | Not unique<br>$\mu \approx 1$ GeV; scheme? |
| NL $\sigma$-model [831] | $1.6 \sim 3.0$ | $0.7 \sim 0.9$ | – | $M_\sigma$: free; $SU(3)_F$ trunc.<br>$\mu \approx 1$ GeV; scheme? |
| **Dispersive** | | | | |
| Large $N_c$+ LMD<br>+ LSD–match. [832] | – | – | 0.9<br>strong $\mu$-dep. | $NLO$ in $1/N_c$, |
| DMO-like SR [833] | – | $1.6 \pm 0.4$<br>huge NLO | $0.8 \pm 0.2$ | $m_q = 0$<br>Strong $s$, $\mu$–dep. |
| FSI [834] | $1.4 \pm 0.3$ | $0.7 \pm 0.2$ | – | Debate for fixing<br>the Slope [835] |
| **This work [836,34]** | | | | |
| DMO-like SR:<br>[833] revisited | – | $2.2 \pm 1.5$<br>inaccurate | $0.7 \pm 0.2$ | $m_q = 0$<br>Strong $s$, $\mu$–dep. |
| $\tau$-like SR | – | – | inaccurate | $t_c$–changes |
| $R_\tau^{V-A}$ | – | $1.7 \pm 0.4$ | – | $m_q = 0$ |
| $S_2 \equiv (\bar{u}u + \bar{d}d)$<br>from QSSR | $1.0 \pm 0.4$<br>$\leq 1.5 \pm 0.4$ | – | – | $\overline{MS}$ scheme<br>$m_s(2) \geq 90$ MeV |

### 56.4.2 Soft pion and kaon reductions of $\langle(\pi\pi)_{I=2}|\mathcal{Q}^{3/2}_{7,8}|K^0\rangle$ to vacuum condensates

We shall consider here the kaon electroweak penguin matrix elements:

$$\langle\mathcal{Q}^{3/2}_{7,8}\rangle_{2\pi} \equiv \langle(\pi\pi)_{I=2}|\mathcal{Q}^{3/2}_{7,8}|K^0\rangle\,, \tag{56.125}$$

defined as:

$$\mathcal{Q}^{3/2}_7 \equiv \frac{3}{2}(\bar{s}d)_{V-A}\sum_{u,d,s} e_\psi(\bar\psi\psi)_{V+A}\,,$$
$$\mathcal{Q}^{3/2}_8 \equiv \frac{3}{2}(\bar{s}_\alpha d_\beta)_{V-A}\sum_{u,d,s} e_\psi(\bar\psi_\beta\psi_\alpha)_{V+A}\,, \tag{56.126}$$

where $\alpha, \beta$ are colour indices; $e_\psi$ denotes the electric charges. In the chiral limit $m_{u,d,s} \sim m_\pi^2 \simeq m_K^2 = 0$, one can use soft pion and kaon techniques in order to relate the previous amplitude to the four-quark vacuum condensates [833] (see also [832]):[5]

$$\langle\mathcal{Q}^{3/2}_7\rangle_{2\pi} \simeq -\frac{2}{f_\pi^3}\langle\mathcal{O}^{3/2}_7\rangle\,,$$
$$\langle\mathcal{Q}^{3/2}_8\rangle_{2\pi} \simeq -\frac{2}{f_\pi^3}\left\{\frac{1}{3}\langle\mathcal{O}^{3/2}_7\rangle + \frac{1}{2}\langle\mathcal{O}^{3/2}_8\rangle\right\}\,, \tag{56.127}$$

where we use the shorthand notations: $\langle 0|\mathcal{O}^{3/2}_{7,8}|0\rangle \equiv \langle\mathcal{O}^{3/2}_{7,8}\rangle$, and $f_\pi = (92.42 \pm 0.26)$ MeV.[6] Here:

$$\mathcal{O}^{3/2}_7 = \sum_{u,d,s} \bar\psi\gamma_\mu\frac{\tau_3}{2}\psi\bar\psi\gamma_\mu\frac{\tau_3}{2}\psi - \bar\psi\gamma_\mu\gamma_5\frac{\tau_3}{2}\psi\bar\psi\gamma_\mu\gamma_5\frac{\tau_3}{2}\psi\,,$$
$$\mathcal{O}^{3/2}_8 = \sum_{u,d,s} \bar\psi\gamma_\mu\lambda_a\frac{\tau_3}{2}\psi\bar\psi\gamma_\mu\lambda_a\frac{\tau_3}{2}\psi - \bar\psi\gamma_\mu\gamma_5\lambda_a\frac{\tau_3}{2}\psi\bar\psi\gamma_\mu\gamma_5\lambda_a\frac{\tau_3}{2}\psi\,, \tag{56.128}$$

where $\tau_3$ and $\lambda_a$ are flavour and colour matrices. Using further pion and kaon reductions in the chiral limit, one can relate this matrix element to the $B$-parameters [833]:

$$B^{3/2}_7 \simeq \frac{3}{2}\frac{(m_u+m_d)}{m_\pi^2}\frac{(m_u+m_s)}{m_K^2}\frac{1}{f_\pi}\langle\mathcal{Q}^{3/2}_7\rangle_{2\pi}$$
$$B^{3/2}_8 \simeq \frac{1}{2}\frac{(m_u+m_d)}{m_\pi^2}\frac{(m_u+m_s)}{m_K^2}\frac{1}{f_\pi}\langle\mathcal{Q}^{3/2}_8\rangle_{2\pi} \tag{56.129}$$

where all QCD quantities will be evaluated in the $NDR$-$\overline{MS}$ scheme and at the scale $M_\tau$.

[5] In the following discussion, we shall use a normalization of the matrix elements which differ by a factor 2 from the one used in [833,836]. This is due to the uses of the operator $\mathcal{Q}^{3/2}_8$ in Eq. 56.126 currently used in the literature rather the one: $(\bar{s}_\alpha d_\beta)_{V-A}[(\bar{u}_\beta u_\alpha)_{V+A} - (\bar{d}_\beta d_\alpha)_{V+A} + (\bar{s}_\beta s_\alpha)_{V+A}]$ used in [833] and [836].

[6] In the chiral limit $f_\pi$ would be about 87 MeV. However, it is not clear to us what value of $f_\pi$ should be used here because we shall use real data from $\tau$-decay. Therefore, we shall leave it as a free parameter which the reader can fix at his convenience.

### 56.4.3 The $\langle \mathcal{O}_{7,8}^{3/2} \rangle$ condensates from DMO-like sum rules in the chiral limit

In previous papers [833,832], the vacuum condensates $\langle \mathcal{O}_{7,8}^{3/2} \rangle$ have been extracted using Das–Mathur–Okubo(DMO)- and Weinberg-like sum rules based on the difference of the vector and axial-vector spectral functions $\rho_{V,A}$ of the $I = 1$ component of the neutral current:

$$2\pi\langle \alpha_s \mathcal{O}_8^{3/2} \rangle = \int_0^\infty ds\, s^2 \frac{\mu^2}{s+\mu^2} (\rho_V - \rho_A) ,$$
$$\frac{16\pi^2}{3}\langle \mathcal{O}_7^{3/2} \rangle = \int_0^\infty ds\, s^2 log\left(\frac{s+\mu^2}{s}\right) (\rho_V - \rho_A) , \quad (56.130)$$

where $\mu$ is the subtraction point. In this normalization, the first Weinberg sum rule gives in the chiral limit:

$$\int_0^\infty ds\, (\rho_V - \rho_A) = f_\pi^2 . \quad (56.131)$$

Due to the quadratic divergence of the integrand, the previous sum rules are expected to be sensitive to the high energy tails of the spectral functions where the present ALEPH/OPAL data from $\tau$-decay [193,199] are inaccurate. This inaccuracy can a priori affect the estimate of the four-quark vacuum condensates. On the other hand, the explicit $\mu$–dependence of the analysis can also induce another uncertainty. En passant, we check below the effects of these two parameters $t_c$ and $\mu$. After evaluating the spectral integrals, we obtain at $\mu = 2$ GeV and for our previous values of $t_c \simeq (1.48 \pm 0.02)$ GeV$^2$ (see Chapter on Weinberg sum rules), the values (in units of $10^{-3}$ GeV$^6$) using the cut-off momentum scheme (c.o):

$$\alpha_s\langle \mathcal{O}_8^{3/2} \rangle_{c.o} \simeq -(0.69 \pm 0.06) , \qquad \langle \mathcal{O}_7^{3/2} \rangle_{c.o} \simeq -(0.11 \pm 0.01) , \quad (56.132)$$

where the errors come mainly from the small changes of $t_c$-values. If instead, we use the second set of values of $t_c \simeq (2.4 \sim 2.6)$ GeV$^2$ (see Chapter on Weinberg sum rules), we obtain by setting $\mu = 2$ GeV:

$$\alpha_s\langle \mathcal{O}_8^{3/2} \rangle_{c.o} \simeq -(0.6 \pm 0.3) , \qquad \langle \mathcal{O}_7^{3/2} \rangle_{c.o} \simeq -(0.10 \pm 0.03) , \quad (56.133)$$

which is consistent with the one in Eq. (56.132), but with larger errors as expected. We have also checked that both $\langle \mathcal{O}_8^{3/2} \rangle$ and $\langle \mathcal{O}_7^{3/2} \rangle$ increase in absolute value when $\mu$ increases where a stronger change is obtained for $\langle \mathcal{O}_7^{3/2} \rangle$, a feature which has been already noticed in [832]. In order to give a more conservative estimate, we consider as our final value the largest range spanned by our results from the two different sets of $t_c$-values. This corresponds to the one in Eq. (56.133) which is the less accurate prediction. We shall use the relation between the momentum cut-off (c.o) and $\overline{MS}$ schemes given in [833]:

$$\langle \mathcal{O}_7^{3/2} \rangle_{\overline{MS}} \simeq \langle \mathcal{O}_7^{3/2} \rangle_{c.o} + \frac{3}{8} a_s \left(\frac{3}{2} + 2d_s\right) \langle \mathcal{O}_8^{3/2} \rangle$$
$$\langle \mathcal{O}_8^{3/2} \rangle_{\overline{MS}} \simeq \left(1 - \frac{119}{24} a_s \pm \left(\frac{119}{24} a_s\right)^2\right) \langle \mathcal{O}_8^{3/2} \rangle_{c.o} - a_s \langle \mathcal{O}_7^{3/2} \rangle , \quad (56.134)$$

where $d_s = -5/6$ (resp $1/6$) in the so-called Naïve Dimensional Regularization NDR (resp. t'Hooft-Veltmann HV) schemes;[7] $a_s \equiv \alpha_s/\pi$. One can notice that the $a_s$ coefficient is large in the second relation (50% correction), and the situation is worse because of the relative minus sign between the two contributions. Therefore, we have added a rough estimate of the $a_s^2$ corrections based on the naïve growth of the PT series, which here gives 50% corrections of the sum of the two first terms. For a consistency of the whole approach, we shall use the value of $\alpha_s$ obtained from $\tau$-decay, which is [193,199]:

$$\alpha_s(M_\tau)|_{\rm exp} = 0.341 \pm 0.05 \quad \Longrightarrow \quad \alpha_s(2~{\rm GeV}) \simeq 0.321 \pm 0.05\,. \qquad (56.135)$$

Then, we deduce (in units of $10^{-4}$ GeV$^6$) at 2 GeV:

$$\langle \mathcal{O}_7^{3/2} \rangle_{\overline{MS}} \simeq -(0.7 \pm 0.2)\,, \qquad \langle \mathcal{O}_8^{3/2} \rangle_{\overline{MS}} \simeq -(9.1 \pm 6.4)\,, \qquad (56.136)$$

where the large error in $\langle \mathcal{O}_8^{3/2} \rangle$ comes from the estimate of the $a_s^2$ corrections appearing in Eq. (56.134). In terms of the $B$ factor and with the mean value of the light quark masses quoted in [54], this result, at $\mu = 2$ GeV, can be translated into:

$$B_7^{3/2} \simeq (0.7 \pm 0.2) \left( \frac{m_s(2)~[{\rm MeV}]}{119} \right)^2 k^4\,,$$
$$B_8^{3/2} \simeq (2.5 \pm 1.3) \left( \frac{m_s(2)~[{\rm MeV}]}{119} \right)^2 k^4\,, \qquad (56.137)$$

where:

$$k \equiv \frac{92.4}{f_\pi~[{\rm MeV}]}\,. \qquad (56.138)$$

- Our results in Eqs. (56.136) compare quite well with the ones obtained by [833] in the $\overline{MS}$ scheme (in units of $10^{-4}$ GeV$^6$) at 2 GeV:

$$\langle \mathcal{O}_8^{3/2} \rangle_{\overline{MS}} \simeq -(6.7 \pm 0.9)\,, \qquad \langle \mathcal{O}_7^{3/2} \rangle_{\overline{MS}} \simeq -(0.70 \pm 0.10)\,, \qquad (56.139)$$

  using the same sum rules but presumably a slightly different method for the uses of the data and for the choice of the cut-off in the evaluation of the spectral integral.
- Our errors in the evaluation of the spectral integrals, leading to the values in Eqs. (56.132) and (56.133), are mainly due to the slight change of the cut-off value $t_c$.[8]
- The error due to the passage into the $\overline{MS}$ scheme is due mainly to the truncation of the QCD series, and is important (50%) for $\langle \mathcal{O}_8^{3/2} \rangle$ and $B_8^{3/2}$, which is the main source of errors in our estimate.
- As noticed earlier, in the analysis of the pion mass difference, it looks more natural to do the subtraction at $t_c$. We also found that moving the value of $\mu$ can affects the value of $B_{7,8}^{3/2}$.

For the above reasons, we expect that the results given in [833] for $\langle \mathcal{O}_8^{3/2} \rangle$ although interesting are quite fragile, while the errors quoted there have been presumably underestimated.

[7] The two schemes differ by the treatment of the $\gamma_5$ matrix (see Section 8.2).

[8] A slight deviation from such a value affects notably previous predictions as the $t_c$-stability of the results ($t_c \approx 2$ GeV$^2$) does not coincide with the one required by the second Weinberg sum rules. At the stability point the predictions are about a factor 3 higher than the one obtained previously.

Therefore, we think that a reconsideration of these results using alternative methods are mandatory.[9]

### 56.4.4 The $\langle \mathcal{O}_{7,8}^{3/2} \rangle$ condensates from hadronic tau inclusive decays

In the following, we shall not introduce any new sum rule, but, instead, we shall exploit known informations from the total $\tau$-decay rate and available results from it, which have not the previous drawbacks. The $V$-$A$ total $\tau$-decay rate, for the $I = 1$ hadronic component, can be deduced from BNP [325], and reads:[10]

$$R_{\tau,V-A} = \frac{3}{2}|V_{ud}|^2 S_{EW} \sum_{D=2,4,\ldots} \delta_{V-A}^{(D)} \,. \tag{56.140}$$

$|V_{ud}| = 0.9753 \pm 0.0006$ is the CKM-mixing angle, while $S_{EW} = 1.0194$ is the electroweak corrections [326]. In the following, we shall use the BNP results for $\mathcal{R}_{\tau,V/A}$ in order to deduce $R_{\tau,V-A}$:

- The chiral invariant $D = 2$ term due to a short distance tachyonic gluon mass [162,161] cancels in the $V$-$A$ combination. Therefore, the $D = 2$ contributions come only from the quark mass terms:

$$M_\tau^2 \delta_{V-A}^{(2)} \simeq 8\left[1 + \frac{25}{3} a_s(M_\tau)\right] m_u m_d \,, \tag{56.141}$$

  as can be obtained from the first calculation [28], where $m_u \equiv m_u(M_\tau) \simeq (3.6 \pm 0.6)$ MeV, $m_d \equiv m_d(M_\tau) \simeq (6.5 \pm 1.2)$ MeV [54] (previous chapter) are respectively the running coupling and quark masses evaluated at the scale $M_\tau$.
- The dimension-four condensate contribution reads:

$$M_\tau^4 \delta_{V-A}^{(4)} \simeq 32\pi^2 \left(1 + \frac{9}{2} a_s^2\right) m_\pi^2 f_\pi^2 + \mathcal{O}\left(m_{u,d}^4\right) \,, \tag{56.142}$$

  where we have used the $SU(2)$ relation $\langle \bar{u}u \rangle = \langle \bar{d}d \rangle$ and the Gell-Mann–Oakes–Renner PCAC relation:

$$(m_u + m_d)\langle \bar{u}u + \bar{d}d \rangle = -2m_\pi^2 f_\pi^2 \,. \tag{56.143}$$

- By inspecting the structure of the combination of dimension-six condensates entering in $R_{\tau,V/A}$ given by BNP [325], which are renormalizaton group invariants, and using a $SU(2)$ isospin rotation which relates the charged and neutral (axial)-vector currents, the $D = 6$ contribution reads:

$$M_\tau^6 \delta_{V-A}^{(6)} = -2 \times 48\pi^4 a_s \left[\left[1 + \frac{235}{48} a_s \pm \left(\frac{235}{48} a_s\right)^2 - \frac{\lambda^2}{M_\tau^2}\right] \langle \mathcal{O}_8^{3/2} \rangle + a_s \langle \mathcal{O}_7^{3/2} \rangle\right], \tag{56.144}$$

  where the overall factor 2 in front expresses the different normalization between the neutral isovector and charged currents used respectively in [833] and [325], whilst all quantities are evaluated at the scale $\mu = M_\tau$. The last two terms in the Wilson coefficients of $\langle \mathcal{O}_8^{3/2} \rangle$ are new: the first term is an estimate of the NNLO term by assuming a naïve geometric growth of the $a_s$ series; the second one is the effect of a tachyonic gluon mass introduced in [161], which takes into account

[9] In recent works [838], these results have been also reconsidered.
[10] Hereafter we shall work in the $\overline{MS}-$ $NDR$ scheme.

the re-summation of the QCD asymptotic series, with: $a_s\lambda^2 \simeq -0.06$ GeV$^2$.[11] Using the values of $\alpha_s(M_\tau)$ given previously, the corresponding QCD series behaves quite well as:

$$\text{Coef. } \langle \mathcal{O}_8^{3/2} \rangle \simeq 1 + (0.53 \pm 0.08) \pm 0.28 + 0.18 , \tag{56.145}$$

where the first error comes from the one of $\alpha_s$, while the second one is due to the unknown $a_s^2$-term, which introduces an uncertainty of 16% for the whole series. The last term is due to the tachyonic gluon mass. This leads to the numerical value:

$$M_\tau^6 \delta^{(6)}_{V-A} \simeq -(1.015 \pm 0.149) \times 10^3 \left[(1.71 \pm 0.29)\langle \mathcal{O}_8^{3/2} \rangle + a_s \langle \mathcal{O}_7^{3/2} \rangle\right] , \tag{56.146}$$

- If, one estimates the $D = 8$ contribution using a vacuum saturation assumption, the relevant $V$-$A$ combination vanishes to leading order of the chiral symmetry breaking terms. Instead, we shall use the combined ALEPH/OPAL [193,199] fit for $\delta^{(8)}_{V/A}$, and deduce:

$$\delta^{(8)}_{V-A}\Big|_{\text{exp}} = -(1.58 \pm 0.12) \times 10^{-2} . \tag{56.147}$$

We shall also use the combined ALEPH/OPAL data for $R_{\tau,V/A}$, in order to obtain:

$$R_{\tau,V-A}|_{\text{exp}} = (5.0 \pm 1.7) \times 10^{-2} , \tag{56.148}$$

Using the previous information in the expression of the rate given in Eq. (56.140), one can deduce:

$$\delta^{(6)}_{V-A} \simeq (4.49 \pm 1.18) \times 10^{-2} . \tag{56.149}$$

This result is in good agreement with the result obtained by using the ALEPH/OPAL fitted mean value for $\delta^{(6)}_{V/A}$:

$$\delta^{(6)}_{V-A}|_{\text{fit}} \simeq (4.80 \pm 0.29) \times 10^{-2} . \tag{56.150}$$

We shall use as a final result the average of these two determinations, which coincides with the most precise one in Eq. (56.150). We shall also use the result:

$$\frac{\langle \mathcal{O}_7^{3/2} \rangle}{\langle \mathcal{O}_8^{3/2} \rangle} \simeq \frac{1}{8.3} \left( \text{resp. } \frac{3}{16} \right) , \tag{56.151}$$

where, for the first number we use the value of the ratio of $B_7^{3/2}/B_8^{3/2}$ which is about $0.7 \sim 0.8$ from, for example, lattice calculations quoted in Table 56.1, and the formulae in Eqs. (56.127) to (56.129); for the second number we use the vacuum saturation for the four-quark vacuum condensates [1]. The result in Eq. (56.151) is also comparable with the estimate of [833] from the sum rules given in Eq. (56.130). Therefore, at the scale $\mu = M_\tau$, Eqs. (56.144), (56.150) and (56.151) lead, in the $\overline{MS}$ scheme, to:

$$\langle \mathcal{O}_8^{3/2} \rangle (M_\tau) \simeq -(0.94 \pm 0.21) \times 10^{-3} \text{ GeV}^6 , \tag{56.152}$$

where the main errors come from the estimate of the unknown higher-order radiative corrections. It is instructive to compare this result with the one using the vacuum saturation

[11] This contribution may compete with the dimension-eight operators discussed in [837].

assumption for the four-quark condensate (see e.g. BNP):

$$\langle\mathcal{O}_8^{3/2}\rangle|_{v.s} \simeq -\frac{32}{18}\langle\bar{u}u\rangle^2(M_\tau) \simeq -0.65\times 10^{-3}\ \text{GeV}^6\ , \tag{56.153}$$

which shows about $1\sigma$ violation of this assumption. We have used for the estimate of $\langle\bar{\psi}\psi\rangle$ the value of $(m_u+m_d)(M_\tau)\simeq 10$ MeV [54] and the GMOR pion PCAC relation. However, this violation of the vacuum saturation is not quite surprising, as a similar fact has also been observed in other channels [3,193,199], although it also appears that the vacuum saturation gives a quite good approximate value of the ratio of the condensates [3,193,199]. The result in Eq. (56.152) is comparable with the value $-(0.98\pm 0.26)\times 10^{-3}$ GeV$^6$ at $\mu=2$ GeV $\approx M_\tau$ obtained by [833] using a DMO-like sum rule, but, as discussed previously, the DMO-like sum rule result is very sensitive to the value of $\mu$ if one fixes $t_c$ as 1.48 GeV$^2$ (see chapter on Weinberg sum rules) according to the criterion discussed above. Here, the choice $\mu=M_\tau$ is well-defined, and then the result becomes more accurate (as mentioned previously our errors come mainly from the estimated unknown $\alpha_s^3$ term of the QCD series). Using Eqs. (56.127) and (56.151), our previous results in Eq. (56.136) for $\mathcal{O}_7^{3/2}$ and in Eq. (56.152) for $\mathcal{O}_8^{3/2}$ can be translated into the prediction on the weak matrix elements in the chiral limit and at the scale 2 GeV for the NDR scheme ($k\equiv 92.4/f_\pi$[MeV] is defined in Eq. (56.138)):[12]

$$\begin{aligned}\langle(\pi\pi)_{I=2}|\mathcal{Q}_7^{3/2}|K^0\rangle(2) &\simeq (0.18\pm 0.05)\ \text{GeV}^3\ k^3\\ \langle(\pi\pi)_{I=2}|\mathcal{Q}_8^{3/2}|K^0\rangle(2) &\simeq (1.35\pm 0.30)\ \text{GeV}^3\ k^3\ ,\end{aligned} \tag{56.154}$$

normalized to $f_\pi$, which avoids the ambiguity on the real value of $f_\pi$ to be used in such an expression. Our result is in agreement with different determinations from dispersive approaches [832,833,838]. Our result is higher by about a factor of 2 than the quenched lattice result [823]. A resolution of this discrepancy can only be found after the inclusion of chiral corrections in Eqs. (56.127) to (56.129), and after the use of dynamic fermions on the lattice. However, some parts of the chiral corrections in the estimate of the vacuum condensates are already included into the QCD expression of the $\tau$-decay rate and these corrections are negligibly small. We might expect that chiral corrections, which are smooth functions of $m_\pi^2$ will not strongly affect the relation in Eqs. (56.127) to (56.129), although an evaluation of their exact size is mandatory. Using the previous mean values of the light quark running masses [54], we deduce in the chiral limit and at the scale $M_\tau$:

$$B_8^{3/2}\simeq(1.70\pm 0.39)\left(\frac{m_s(M_\tau)\ [\text{MeV}]}{119}\right)^2 k^4\ , \tag{56.155}$$

where $k$ is defined in Eq. (56.138). One should notice that, contrary to the $B$-factor, the result in Eq. (56.154) is independent to leading order of the value of the light quark masses.

[12] As already mentioned, this normalization differs by a factor 2 than the one used in [833,836].

### 56.4.5 Impact of the results on the CP violation parameter $\epsilon'/\epsilon$

One can combine the previous result of $B_8$ with the value of the $B_6$ parameter of the QCD penguin diagram [665]:

$$\begin{aligned}\langle \mathcal{Q}_6^{1/2}\rangle_{2\pi} &\equiv \langle (\pi^+\pi^-)_{I=0}|\mathcal{Q}_6^{1/2}|K^0\rangle \\ &\simeq -[2\langle\pi^+|\bar{u}\gamma_5 d|0\rangle\langle\pi^-|\bar{s}u|K^0\rangle \\ &\quad + \langle\pi^+\pi^-|\bar{d}d+\bar{u}u|0\rangle\langle 0|\bar{s}\gamma_5 d|K^0\rangle\,] \\ &\simeq -4\sqrt{\frac{3}{2}}\left(\frac{m_K^2}{m_s+m_d}\right)^2 \\ &\quad \times\sqrt{2}(f_K-f_\pi)\,B_6^{1/2}(m_c)\,. \end{aligned} \tag{56.156}$$

We have estimated the $\langle \mathcal{Q}_6^{1/2}\rangle_{2\pi}$ matrix element by relating its first term to the $K\to\pi\, l\nu_l$ semi-leptonic form factors as usually done (see e.g. [822]), while the second term has been obtained from the contribution of the $S_2\equiv(\bar{u}u+\bar{d}d)$ scalar meson having its mass and coupling fixed by QCD spectral sum rules [3,688] and in the scheme where the observed low mass $\sigma$ meson results from a maximal mixing between the $S_2$ and the $\sigma_B$ associated to the gluon component of the trace of the anomaly [686,680,688]:[13]

$$\theta_\mu^\mu = \frac{1}{4}\beta(\alpha_s)G^2 + (1+\gamma_m(\alpha_s))\sum_{u,d,s} m_i\bar{\psi}_i\psi_i\,, \tag{56.157}$$

where $\beta$ and $\gamma_m$ are the $\beta$ function and mass anomalous dimension. In this way, one obtains at the scale $m_c$:

$$B_6^{1/2}(m_c)\simeq 3.7\left(\frac{m_s+m_d}{m_s-m_u}\right)^2\times\left[(0.65\pm0.09)-(0.53\pm0.13)\left(\frac{(m_s-m_u)\,[\text{MeV}]}{142.6}\right)\right], \tag{56.158}$$

which satisfies the double chiral constraint. We have used the running charm quark mass $m_c(m_c)=1.2\pm0.05$ GeV [54]. Evaluating the running quark masses at 2 GeV, with the values given in [54], one deduces:

$$\begin{aligned} B_6^{1/2}(2) &\simeq (1.1\pm0.4) \quad \text{for } m_s(2)=117\ \text{MeV}\,, \\ &\le (2.1\pm0.4) \quad \text{for } m_s(2)\ge 71\text{MeV}\,. \end{aligned} \tag{56.159}$$

The errors added quadratically have been relatively enhanced by the partial cancellations of the two contributions. Therefore, we deduce the combination:

$$\begin{aligned} \mathcal{B}_{68} &\equiv B_6^{3/2}-0.48B_8^{3/2} \\ &\simeq (0.3\pm0.4) \quad \text{for } m_s(2)=117\ \text{MeV}\,, \\ &\le (1.3\pm0.4) \quad \text{for } m_s(2)\ge 71\ \text{MeV}\,, \end{aligned} \tag{56.160}$$

[13] Present data appear to favour this scheme [690].

where we have added the errors quadratically. Using the approximate simplified expresssion [665]:

$$\frac{\epsilon'}{\epsilon} \approx 14.5 \times 10^{-4} \left( \frac{110}{\bar{m}_s(2)\ [\mathrm{MeV}]} \right)^2 \mathcal{B}_{68} \ , \tag{56.161}$$

one can deduce the result in units of $10^{-4}$:

$$\begin{aligned} \frac{\epsilon'}{\epsilon} &\simeq (4 \pm 5) \quad \text{for } m_s(2) = 117 \text{ MeV} \ , \\ &\leq (45 \pm 14) \ , \quad \text{for } m_s(2) \geq 71 \text{ MeV} \ , \end{aligned} \tag{56.162}$$

where the errors come mainly from $\mathcal{B}_{68}$ (40%). The upper bound, though rather weak, agrees quite well with the experimental world average data [599]:

$$\frac{\epsilon'}{\epsilon} \simeq (17.2 \pm 1.8) \times 10^{-4} \ . \tag{56.163}$$

We expect that the failure of the inaccurate estimate for reproducing the data is not naïvely due to the value of the quark mass, but may indicate the need for other important contributions than the single $\bar{q}q$ scalar meson $S_2$ (not the observed $\sigma$)-meson which have not been considered so far in the analysis. Among others, a much better understanding of the effects of the gluonium (expected large component of the $\sigma$-meson [686,688,687]) in the amplitude, through presumably a new operator, needs to be studied. This effect might be signalled by the success of the final state interaction approach within an effective approach (quark and gluon content blind) for reproducing the previous data [835].

### 56.4.6 Summary and conclusions

We have explored the $V$-$A$ component of the hadronic tau decays for predicting non-perturbative QCD parameters. Our main results are summarized as:

- Electroweak penguins:
  - Eq. (56.137): $B_7^{3/2}$ ,
  - Eq. (56.155): $B_8^{3/2}$
  - Eq. (56.154): $\langle(\pi\pi)_{I=2}|\mathcal{Q}_8^{3/2}|K^0\rangle$ .
- QCD penguin: Eq. (56.159).
- $\epsilon'/\epsilon$: Eq. (56.162) .

Our results are compared with some other predictions in Table 56.1 (see also [838]). However, as mentioned in the table caption, a direct comparison of these results is not straightforward due to the different schemes and values of the scale where the results have been obtained. In most of the approaches, the values of $B_7^{3/2}$ are in agreement within the errors and are safely in the range $0.5 \sim 1.0$. For $B_8^{3/2}$ the predictions can differ by a factor 2 and cover the range $0.7 \sim 2.1$. There are strong disagreements by a factor 4 for the values of $B_6^{1/2}$ which range from $0.6 \sim 3.0$. We are still far from having good control of these non-perturbative parameters. This weak point does not permit us to give a reliable prediction of the $CP$ violation parameter $\epsilon'/\epsilon$. Therefore, no definite bound for new physics effects can be derived at present, before improvements of these SM predictions.

# 57

# Thermal behaviour of QCD

## 57.1 The QCD phases

We study here the uses of QCD spectral sum rules (QSSR) in a matter with non-zero temperature $T$ and non-zero chemical potential $\mu$ (so-called Quark-Gluon Plasma (QGP)). Since the corresponding critical temperature for the *colour deconfinement* is expected to be rather small ($T_c \leq 1$ GeV), these new states of matter can be investigated in high-energy hadron collisions. At high enough temperature $T \geq T_c \approx 150 - 200$ MeV corresponding to a vacuum pressure of about 500 MeV/fm$^3$, QGP phase occurs and can be understood without confinement. In this phase, one also expects that chiral symmetry is restored (*chiral symmetry restoration*). However, it is a priori unclear, if the QGP phase and the chiral symmetry restoration occurs at the same temperature or not. Untuitively, one can expect that the deconfinement phase occurs before the chiral symmetry restoration. An attempt to show that the two phases are reached at the same temperature has been made in [839] using the FESR version of the Weinberg sum rules, which we shall discuss later on, where the constraint has been obtained by assuming that in the QGP phase, the continuum threshold starts from zero. In the QGP phase, the thermodynamics of the plasma is governed by the Stefan–Boltzmann law, as in an ordinary black body transition. This feature has been confirmed by a large number of lattice simulations. The RHIC-BNL heavy ion program is dedicated to the study of transitions to this phase. Actually, the AGS ($2 + 2$ GeV per nucleon in the c.o.m.), CERN-SPS ($10 + 10$ GeV) and RHIC ($100 + 100$ GeV) are expected to reach this phase transition. The phase diagram of QCD is shown in Fig. 57.1 in the $T$ plane versus the baryonic chemical potential $\mu$ normalized per quark (not per baryon).

CS2 (two flavours) and CS3 (three flavours) are colour superconducting phases corresponding to large density and low $T$ regions, which are not crossed by the heavy ion collisions but belong to the neutron star physics. The small drop of CS matter is expected to be due to one Cooper pair composed with a (ud scalar diquark) and one massive quark [840]. Some new crystalline phases due to oscillating $\langle \bar{q}q(x) \rangle$ condensates may also appear [841], which may compete with CS2 at $\mu \approx 400$ MeV. Hadronic phase at small value of $(T,\ \mu)$ and the QGP phase at large value of $(T,\ \mu)$ have been known for a long time. However, it is not quite clear if the phase transition line separating them starts at $(T = T_c,\ \mu = 0)$ but at an endpoint $E$, a remnant of the so-called QCD tricritical point which QCD possesses in the massless quark limit.

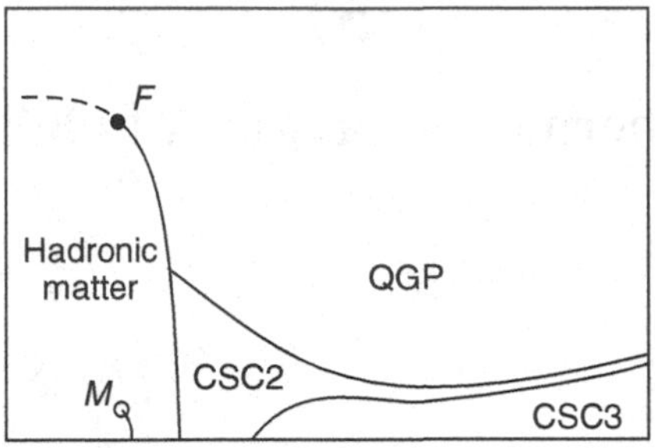


Fig. 57.1. QCD phase diagram.

## 57.2 Big-bang versus heavy ion collisions

- In both cases, the expansion law is roughly the Hubble law $v(r) \sim r$, but strongly isotropic in the case of the heavy ion collisions.
- The acceleration history is not well measured (for Big Bang, one uses distant supernovae) but both show small dipole components.
- In both cases, the major puzzle is how the large entropy was actually produced, and why it happened so early, with a subsequent expansion close to the adiabatic expansion of equilibrated hot medium.

## 57.3 Hadronic correlations at finite temperature

The analysis of hadronic correlations at finite temperature is of great interest in connection with the modifications of hadronic properties in hot hadronic matter. The structure of the QCD correlations at intermediate distances also reflects the changes in the interaction between quarks and gluons. In the context of QSSR, this analysis has been extensively studied [839,842–851]. However, further assumptions on the $T$-behaviour of the condensates or information from some other approaches [852–854] are needed in the analysis as well as some assumptions on the shape of the hadron spectrum, which limit the accuracy of the results. Neverthless, interesting quantitative results can be extracted. We shall be concerned here with the retarted (advanced) correlator (the causal one does not have useful analytic properties [855]) of the hadronic current $J(x)$ (generic notation):

$$\Pi_R(\omega, \mathbf{q}) = i \int d^4x \, e^{iqx} \mathrm{theta(x^0)} \langle\langle [\mathrm{J(x), J^\dagger(0)}] \rangle\rangle \,, \tag{57.1}$$

where $q \equiv (\omega, \mathbf{q})$ and $\langle\langle \ldots \rangle\rangle$ stands for the Gibbs average:

$$\langle\langle \ldots \rangle\rangle = \sum_n W_n \langle n| \ldots |n\rangle \,, \quad W_n \equiv \exp[(\Omega - E'_n)/T] \,. \tag{57.2}$$

$|n\rangle$ is a complete set of the eigenstates of the QCD effective Hamiltonian:

$$H'|n\rangle = E'|n\rangle \,, \quad H = H - \mu N \,, \tag{57.3}$$

where $H$ is the usual Hamiltonian, $N$ is some conserved additive quantum number (baryonic charge, strangeness,...), $\mu$ is the corresponding chemical potential, and $\Omega = -T \ln Tr \exp(-H'/T)$. The two-point correlator is analytic in the upper half-plane of the complex variable $\omega$ and obeys the dispersion relation:

$$\Pi_R(\omega,\ \mathbf{q}) = \int_{-\infty}^{+\infty} \frac{du}{u-\omega-i\epsilon}\rho(u,\ \mathbf{q})\ . \tag{57.4}$$

The spectral density $\rho$ is:

$$\rho(\omega,\ \mathbf{q}) = (2\pi)^3 \sum_{n,m} \langle n|J|m\rangle\langle m|J|n\rangle \\ \times W_n[1-\exp(-\omega/T)]\delta(\omega-\omega_{mn})\delta^{(3)}(\mathbf{q}-\mathbf{k}_{mn})\ , \tag{57.5}$$

where:

$$\omega_{mn} = E'_m - E_n\ , \qquad \mathbf{k}_{mn} = \mathbf{k}_m - \mathbf{k}_n\ . \tag{57.6}$$

Noting that in the set of discrete points of the imaginary half-axis, the retarded correlator in Eq. (57.1) coïncides with the corresponding Matsubara Green functions [856]:

$$\Pi_R(i\omega_n) = \mathcal{G}(\omega_n - i\mu)\ , \tag{57.7}$$

where:

$$\omega_n = 2\pi n T\ , \qquad n = 0, 1, 2, \ldots, \tag{57.8}$$

one can then calculate it using Feynmann diagram techniques in the Matsubara representation [856] (imaginary time formalism). [1] For $\omega \to \infty$, the correlator in Eq. (57.1) tends to its perturbative QCD value. Therefore, it is reasonable to use quarks and gluons in Eq. (57.5), in order to get its asymptotic behaviour. Using the OPE, one can add, to the perturbative part, the NP contributions due to the quark and gluon condensates:

$$\Pi_R(q) = \sum_n C_n(q)\langle\langle\mathcal{O}_n\rangle\rangle\ , \tag{57.9}$$

where $C_n$ is the Wilson coefficient obtained at zero temperature. The temperature dependence appears when one takes the Gibbs averages $\langle\langle\ldots\rangle\rangle$ of the local operators $\mathcal{O}_n$. Summing up all contributions from different $\mathcal{O}_n$, one expects to recover the Matsubara results. The radiative corrections appearing in the evaluation of the condensates should be of the form $\alpha_s(\nu)\ln(T/\nu)$ as only $T$ is the dimensional scale in the calculation of condensates at finite $T$, such that it is important to determine the $T$ value above which one can rely on the calculation. Formally, $T$ should be much bigger than the QCD scale $\Lambda$. For the time being, we shall ignore radiative corrections and assume that perturbation theory works for $T$ around 150 MeV.

[1] Real time formalism as discussed in [857] is not convenient for the problem discussed here, where one evaluates the spectral function.

## 57.4 Asymptotic behaviour of the correlator in hot hadronic matter

Let us consider the correlator associated with the light quark vector current:

$$J_\mu(x) = \bar{\psi}\gamma_\mu\psi \tag{57.10}$$

which has the quantum number of the photon $\gamma$. The imaginary part of the correlator defined as in Eq. (57.5) is then the probability of the absorption of a virtual $\gamma$-quanta of time-like momenta $\omega^2 - \mathbf{q}^2 > 4m_q^2(m_q \to 0)$ by the matter. The virtual quanta consisting of free fermions $n_F(E)$ are converted into a quark-antiquark pair at a rate proportional to $[1 - n_F(E_1)][1 - n_F(E_2)]$ according to Pauli's principle where $E_{1,2}$ are the quark energies. At the same time, the $\gamma$-quanta are produced with the rate $n_F(E_1)n_F(E_2)$. Therefore, the rate of the disappearance of time-like $\gamma$-quanta in the fermionic medium $\rho^g_{\mu\nu}$ is:

$$\rho^g_{\mu\nu}(\omega,\ \mathbf{q}) = \sum_s \int \mathrm{LIPS}(E_1, \mathbf{k_1}, E_2, \mathbf{k_2}) \\ \times \langle 0|J_\mu|\bar{\psi}\psi\rangle\langle\bar{\psi}\psi|J_\nu|0\rangle \left\{[1 - n_F(E_1)][1 - n_F(E_2)] - n_F(E_1)n_F(E_2)\right\}\ , \tag{57.11}$$

where:

$$\mathrm{LIPS}(E_1, \mathbf{k_1}, E_2, \mathbf{k_2}) \equiv \frac{d^3k_1}{2E_1(2\pi)^3}\frac{d^3k_2}{2E_2(2\pi)^3}\delta(\omega - E_1 - E_2)\delta^{(3)}(\mathbf{q} - \mathbf{k_1} - \mathbf{k_2})\ . \tag{57.12}$$

The numbers of fermions and bosons inside the plasma are:

$$n_F(z) = 1/(1 + e^z)\ , \qquad n_B = 1/(1 - e^z)\ . \tag{57.13}$$

The $\gamma$-virtual space-like quanta ($\omega^2 - \mathbf{q}^2 < 0$) can be absorbed by the (anti) quarks of the medium at a rate, proportional to $n_F(E_1)[1 - n_F(E_2)]$ and emitted at the rate $n_F(E_2)[1 - n_F(E_1)]\ :\ E_1 = \omega + E_2$. Thus, it disappears at a scattering rate:

$$\rho^s_{\mu\nu}(\omega,\ \mathbf{q}) = \sum_g \int \mathrm{LIPS}(E_1, \mathbf{k_1}, -E_2, -\mathbf{k_2}) \\ \times \langle\psi|J_\mu|\bar{\psi}\rangle\langle\bar{\psi}|J_\nu|\psi\rangle \left\{n_F(E_1)[1 - n_F(E_2)] - n_F(E_2)[1 - n_F(E_1)]\right\}\ , \tag{57.14}$$

where one can notice that the location of the singularities at $T \neq 0$ differs qualitatively from that at $T = 0$ as shown in Fig. 57.2.

Therefore the spectral function reads:

$$\rho(\omega,\ \mathbf{q}) = \theta(\omega^2 - \mathbf{q}^2 - t_c)\rho^g + \theta(\omega^2 - \mathbf{q}^2)\rho^s\ , \tag{57.15}$$

where $t_c$ is some threshold and the spectral function does not vanish for both time-like and space-like momenta. $\rho^s$ corresponds to the scattering term which increases with the particle density and appears as we use a mixed state of matter instead of the vacuum, when

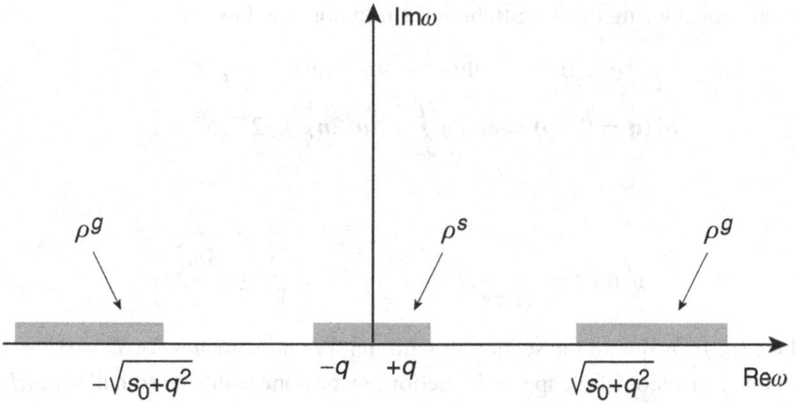


Fig. 57.2. Location of singularities in the complex $\omega$ plane.

averaging the initial commutator. At $T \neq 0$, the correlator of the vector isovector current:

$$J_\mu = \frac{1}{2}(\bar{u}\gamma_\mu u - \bar{d}\gamma_\mu d)\,, \tag{57.16}$$

contains two invariants $\Pi_T$ and $\Pi_L$, which, in the rest frame of matter, can be defined as follows:

$$\Pi_{00} = \mathbf{q}^2\Pi_L\,, \qquad \Pi_{ij} = \left(\delta_{ij} - \frac{q_i q_j}{\mathbf{q}^2}\right)\Pi_T + \frac{q_i q_j}{\mathbf{q}^2}\omega^2\Pi_L\,. \tag{57.17}$$

Using Eqs. (57.11) to (57.14), one can deduce Im $\Pi_{L,T}$. To lowest order in $\alpha_s$, the Euclidian asymptotic of the form factors reads [842]:

$$\begin{aligned}
\rho_L^g &= \frac{3}{32\pi^2}\int_{-v}^{+v} dx\,(1-x^2)\,\mathrm{th}(q^+/2T)\,,\\
\rho_L^s &= \frac{3}{32\pi^2}\int_{-v}^{+v} (1-x^2)[2n_F(q^+/2T) - 2n_F(q^-/2T)]\,,\\
\rho_T^g &= \frac{3}{64\pi^2}(\omega^2-q^2)\int_{-v}^{+v} dx\,(2+x^2-v^2)\,\mathrm{th}(q^+/4T)\,,\\
\rho_T^s &= \frac{3}{64\pi^2}(\omega^2-q^2)\int_{-v}^{+v} dx\,(2+x^2-v^2)\,[2n_F(q^+/2T) - 2n_F(q^-/2T)]\,,
\end{aligned} \tag{57.18}$$

where: $q^+ \equiv qx + \omega$; $q^- \equiv qx - \omega$ and:

$$v = \sqrt{1 - \frac{4m_q^2}{(\omega^2 - \mathbf{q}^2)}}\,. \tag{57.19}$$

In the case of a resonance at rest with respect to the medium ($\mathbf{q} \to 0$), one knows that the two previous form factors are proportional to each other [858]:

$$\Pi_T(\mathbf{q}=0) = (\omega^2 - \mathbf{q}^2)\Pi_L|_{\mathbf{q}=0}\,. \tag{57.20}$$

Therefore, considering the longitudinal form factor, one has:

$$\rho_L^g(\mathbf{q}=0,\ \omega) = \text{th}\big(\omega^2 - 4m_q^2\big)\text{th}(\omega/4T)\rho^0(\omega^2)\,,$$
$$\rho_L^s(\mathbf{q}=0,\ \omega) = \delta(\omega^2)\int_{4m_q^2}^{\infty} du^2 2n_F(u/2T)\rho^0(u^2)\,, \tag{57.21}$$

with:

$$\rho^0(\omega^2) = \frac{3}{16\pi^2}v(3-v^2)\,,\quad v = \sqrt{1-\frac{4m_q^2}{\omega^2}}\,, \tag{57.22}$$

where the term $\rho^s$ is due to the scattering with matter and vanishes for $T=0$.

Systematic expansion of the spectral function can be done using the so-called *hard thermal loop expansion* [859], where the order parameters are the 'hard scale' of the order of $T$ or larger and the 'soft scale' of the order of $gT$ ($g$ being the QCD coupling, which is assumed to be small), where, at this soft scale, collective effects in the plasma lead to effective (resummed) propagators and vertices parametrizing the modification of the physics at this scale, which one can formulate using an effective Lagrangian [860].

To one loop, the vector spectral function behaves as [861]:

$$\rho^g(\omega,\mathbf{q}) \sim \alpha_s T^2\left[\ln\left(\frac{\omega T}{m^2}\right) + \text{constant}\right] \tag{57.23}$$

where $m^2 \sim \alpha_s T^2$ is related to the quark thermal mass. Therefore, the large $\ln(1/\alpha_s)$ dominates over the constant term. However, the two-loop contribution is of the same order as the one-loop graph due to the enhancement factor $1/m^2 \sim 1/\alpha_s$ which compensates the $\alpha_s$ factor associated to the quark-gluon coupling from the gluon exchange. This $1/m^2$ factor originates from the presence of collinear singularities. This result questions the validity of the QCD perturbative calculation in this regime. Phenomenological applications of these results to the the lepton pair production in the quark-gluon plasma are discussed in [861].

### 57.5 Quark condensate at finite $T$

The temperature dependence of the $\langle\bar\psi\psi\rangle$ quark condensate can be expressed as:

$$\langle\langle\bar\psi\psi\rangle\rangle = \frac{Tr\left(\bar\psi\psi\exp -H/T\right)}{Tr\left(\exp -H/T\right)}\,. \tag{57.24}$$

It has been studied in [852] using an effective Lagrangian approach, where one has exploited the fact that, at low temperatures, the behaviour of the partition function is dominated by the contributions from the lightest particles occuring in the spectrum. In QCD, this lightest particle is the pion, which is massless in the chiral limit. Interaction among the pions generates power corrections controlled by the expansion parameter $T^2/8f_\pi^2$, while the contribution due to a massive state $i$ is suppressed as $\exp(-M_i/T)$, where $M_i$ is the corresponding hadron mass. At low temperature, one can express the pressure as the temperature-dependent part of the free energy density:

$$P = E_0 - Z\,, \tag{57.25}$$

where $E_0$ is the vacuum energy density and $Z$ is the partition function:

$$Z = -T \lim_{L\to\infty} L^{-3} \ln[Tr\,\exp(-H/T)]\,. \tag{57.26}$$

The quark condensate is given by the *log*-derivative of the partition function with respect to the quark mass:

$$\langle\langle\bar\psi\psi\rangle\rangle = \frac{\partial Z}{\partial m}\,, \tag{57.27}$$

where at zero temperature:

$$\langle\bar\psi\psi\rangle = \frac{\partial E_0}{\partial m}\,. \tag{57.28}$$

Therefore, at finite temperature:

$$\langle\langle\bar\psi\psi\rangle\rangle = \langle\bar\psi\psi\rangle\left(1 + \frac{c}{F^2}\frac{\partial P}{\partial m_\pi^2}\right), \tag{57.29}$$

where:

$$c \simeq 0.90 \pm 0.15 \tag{57.30}$$

is a constant ($c = 1$ in the massless case) fixed from $\pi$-$\pi$ scattering data, while $F =$ 88 MeV is the value of the pion decay constant in the chiral limit ($f_\pi = 92.4$ MeV). In the massless limit:

$$P = \frac{\pi^2}{30}T^2\left(1 + \frac{16}{9}x_T^2 \ln\frac{\Lambda_q}{T} + \mathcal{O}(T^6)\right). \tag{57.31}$$

The log-scale is related to the $p^4$-term of the effective Lagrangian and is fixed from $\pi$-$\pi$ scattering analysis to be:

$$\Lambda_q \simeq (470 \pm 110)\,\text{MeV}\,. \tag{57.32}$$

To order $T^8$, the temperature dependence of the condensate in the chiral limit reads:

$$\langle\langle\bar\psi\psi\rangle\rangle = \langle\bar\psi\psi\rangle\left(1 - x_T - \frac{1}{6}x_T^2 - \frac{16}{9}x_T^3 \ln\frac{\Lambda_q}{T} + \mathcal{O}(T^8)\right), \tag{57.33}$$

where:

$$x_T = \frac{T^2}{8F^2}\,, \tag{57.34}$$

indicating that the temperature scale is set by $\sqrt{8}F \simeq 250$ MeV. The behaviour of the condensate versus $T$ is given in Fig. 57.3 for massless quarks. One can deduce that the condensate gradually melts for increasing $T$, and vanishes for massless quarks at $T_c \approx$ 190 MeV, indicating the occurence of a phase transition at this temperature.

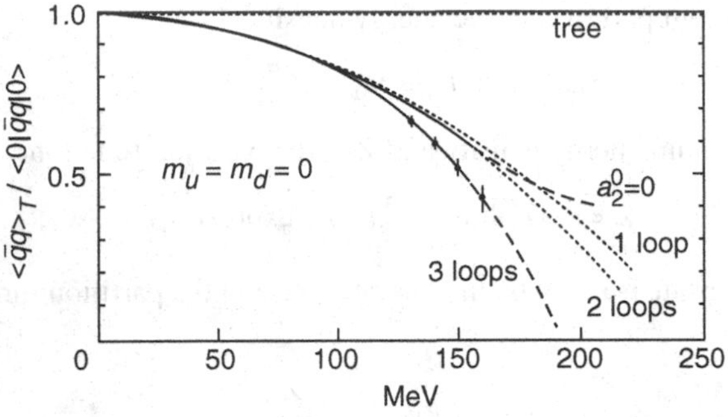


Fig. 57.3. $T$-behaviour of the quark condensate in the chiral limit from [852]. Here and in the following $\langle\bar{q}q\rangle \equiv \langle\bar{\psi}\psi\rangle$.

The inclusion of the quark masses can be obtained on the basis of Eq. 57.29. It is shown that the effect shifts the value of $T_c$ to higher value of about 240 MeV.

Using a dilute gas approximation (i.e. neglecting the self-interactions of hadrons, as they manifest as the product of their density $n_i n_j \sim \exp[-(M_i + M_j)/T]$) in the low-temperature region, the effect of massive states induces a positive correction of order $T^2$ to the condensate:

$$\Delta\langle\bar{\psi}\psi\rangle \simeq -\langle\bar{\psi}\psi\rangle\frac{T^2}{F^2}\sum_i h_1\left(\frac{T}{M_i}\right)\frac{m}{m_\pi^2}M_i\frac{\partial M_i}{\partial m}\;, \tag{57.35}$$

with:

$$h_1(\tau) \equiv \frac{1}{2\pi^2\tau^2}\int_0^\infty dx\ \mathrm{sh}^2x[\exp(\mathrm{ch}x/\tau) - 1]^{-1}\;, \tag{57.36}$$

and:

$$\frac{\partial M_i}{\partial m} = \frac{1}{2M_i}\langle p_i|\bar{\psi}\psi|p_i\rangle\;, \tag{57.37}$$

where $|p_i\rangle$ denotes a one-particle state of momentum $p$. Therefore, Eq. (57.37) counts the number of valence quarks of type $u$ and $d$ inside the hadrons. One can estimate the uncertainties in this approximation by considering the kaon mass formula:

$$M_K^2 = (m + m_s)B\;, \tag{57.38}$$

from which one can deduce:

$$m\frac{\partial M_K}{\partial m} = \frac{m}{2(m + m_s)}M_K[1 + \text{corrections}]\;. \tag{57.39}$$

A comparison of the numerical value from Eqs. (57.37) and (57.39) show that the naïve relation in Eq. (57.37) underestimates the real value by a factor about 1.4. The effect of massive states is such that it accelerates the melting of the condensate and implies a fast drop until the phase transition of about 200 MeV. We show in Figs. 57.4 and 57.5, the sum

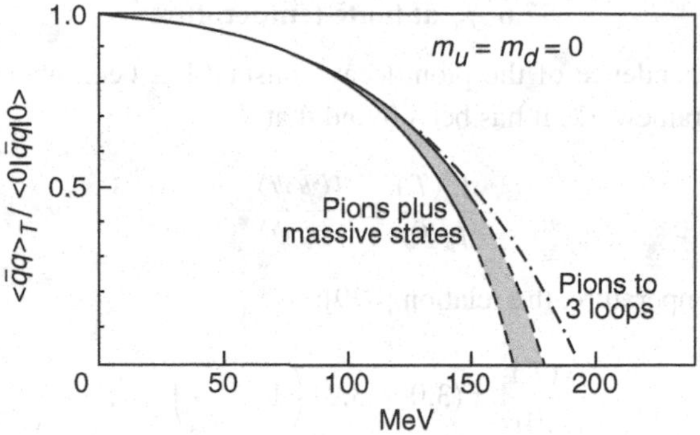


Fig. 57.4. $T$-behaviour of the quark condensate in the chiral limit including massive states.

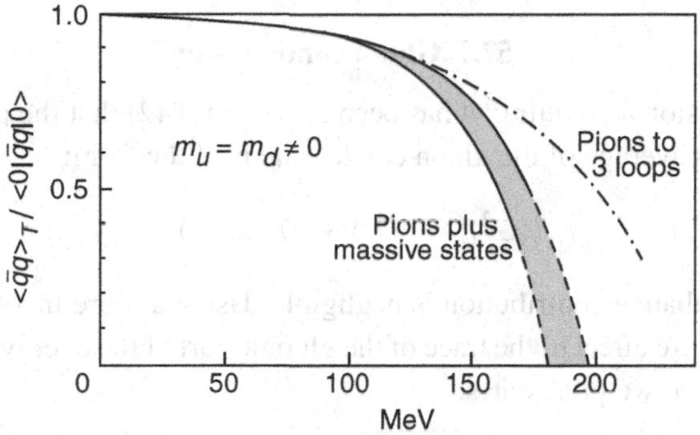


Fig. 57.5. $T$-behaviour of the quark condensate in the massive quark limit including massive states.

of the different effects in the case of massless and massive quarks. The uncertainties in evaluating the quantity in Eq. (57.37) are given by the shaded region in Figs. 57.4 and 57.5.

One can deduce that the critical temperature is:

$$T_c \simeq 170 \text{ MeV} \qquad (m_u = m_d = 0)$$
$$190 \text{ MeV} \qquad (m_u, m_d \neq 0) . \tag{57.40}$$

Such a value of $T_c$ has been confirmed by lattice simulation with two dynamical Kogut–Susskind fermions [862]:

$$T_c(n_f = 2) \simeq 150 \text{ MeV} , \tag{57.41}$$

and by instanton liquid model [386]. In the case of the $\langle \bar{s}s \rangle$, the $m_s$ corrections shifts $T_c$ to higher values of about 250 MeV [863]. Similar analysis can be done respectively for the entropy density $S$, energy density $U$ and heat capacity $C_v$:

$$S = \frac{\partial P}{\partial T} , \qquad U = TS - P , \qquad C_v = \frac{\partial U}{\partial T} = T\frac{\partial S}{\partial T} , \tag{57.42}$$

where for massless quarks, $P$ is given in Eq. (57.31).

### 57.6 $f_\pi$ at finite temperature

In [853], the $T$ dependence of the pion decay constant has been also studied within the composite model framework. It has been found that:

$$\frac{f_\pi^2(T)}{f_\pi^2(0)} \simeq \frac{\langle\langle\bar\psi\psi\rangle\rangle}{\langle\bar\psi\psi\rangle} . \tag{57.43}$$

Near the critical temperature, the relation [839]:

$$\frac{f_\pi(T)}{f_\pi(0)} \simeq (3.0 \sim 3.2)\left(1-\frac{T}{T_c}\right)^{1/2} , \tag{57.44}$$

might be more accurate than the one derived from Eq. (57.33) in the chiral limit.

### 57.7 Gluon condensate

Using naïve dimensional counting, it has been argued in [842] that the perturbative contribution to the Gibbs average of the gluon condensate is of the form:

$$\langle\langle G^2\rangle\rangle \sim \langle G^2\rangle + cT^4\alpha_s(T) , \tag{57.45}$$

such that this pertubative contribution is negligible. Using a more involved estimate, they relate this temperature effect to the trace of the gluonic part of the energy momentum tensor sandwiched between two pion states:

$$\langle\langle G^2\rangle\rangle \sim \langle G^2\rangle + F(m_\pi,\ T)\langle\pi|G^2|\pi\rangle , \tag{57.46}$$

where $F$ is a function that is dependent on $m_\pi$ and $T$. An explicit evaluation gives [854]:

$$\langle\langle G^2\rangle\rangle \sim \langle G^2\rangle - \pi^3\frac{16384}{3465}n_f^2(n_f^2-1)x_T^4\left(\ln\frac{\Lambda_q}{T}-\frac{1}{4}\right) , \tag{57.47}$$

where the temperature-dependent term is a very small correction, although showing that the gluon condensate melts very smoothly for increasing temperature.

### 57.8 Four-quark condensate

The temperature dependence of the four-quark condensate has been often estimated using the vacuum saturation assumption:

$$\langle\langle \mathcal{O}_4\rangle\rangle \equiv \langle\langle\bar\psi\Gamma_1\psi\bar\psi\Gamma_2\psi\rangle\rangle \sim \langle\langle\psi\rangle\rangle^2 , \tag{57.48}$$

which should only be taken very qualitatively, as there are evidences that, already at zero temperature, the vacuum saturation assumption is violated by a factor of 2 to 3 (see previous sections). In [848], the temperature dependence of the four-quark operators in different channels has been studied within the sum rule itself, where the approximate behaviour:

$$\langle\langle \mathcal{O}_4\rangle\rangle \approx \langle \mathcal{O}_4\rangle \left[1-\frac{T^2}{(330\ \text{MeV})^2}\right], \tag{57.49}$$

has been obtained to be compared with that:

$$\langle\langle \mathcal{O}_4\rangle\rangle \sim \langle\bar\psi\psi\rangle^2 \left[1-\frac{T^2}{(177\ \text{MeV})^2}\right], \tag{57.50}$$

which one would have obtained if the vacuum saturation for the four-quark condensate has been used together with the previous $T$-dependence of $\langle\bar\psi\psi\rangle$. The smooth dependence of the four-quark dependence can be qualitatevely understood when using the fact that at the optimization scale of the LSR, one has the relation:

$$M_\rho \sim \langle\langle \alpha_s \mathcal{O}_4\rangle\rangle^{1/6}, \tag{57.51}$$

and as one will see later on, the $\rho$-meson mass varies smoothly with temperature below $T_c$.

## 57.9 The $\rho$-meson spectrum in hot hadronic matter

The $\rho$ meson mass spectrum was originally studied in [842] using Laplace transform sum rules. The $\rho$ coupling to the longitudinal part of the spectral function is introduced as:

$$\rho_l = \frac{M_\rho^2}{2\tilde\gamma_\rho^2}\delta\left(\omega^2-\mathbf{q}^2-m_l^2\right). \tag{57.52}$$

The LSR reads:

$$\begin{aligned}\mathcal{F}(\tau) &\simeq 4\pi^2\frac{M_\rho^2}{2\tilde\gamma_\rho^2}\exp\left(-M_\rho^2\tau\right)\\ &=\int_0^{t_c} dt\ \exp(-t\tau)\text{th}(\sqrt{t}/4T)\\ &+2\int_0^\infty dt\ [n_f(\sqrt{t}/2T)-\frac{1}{3}n_f(\sqrt{t}/2T)]\\ &+\frac{\pi}{3}\langle\alpha_s G^2\rangle\tau-2\pi^3\alpha_s\langle\bar\psi\psi\rangle^2\tau^2.\end{aligned} \tag{57.53}$$

Instead, we work with moments:

$$\mathcal{R}(\tau)\equiv -\frac{d}{d\tau}\ln\mathcal{F}(\tau), \tag{57.54}$$

which is convenient for studying the spectrum as it free from the unknown longitudinal coupling $\tilde\gamma_\rho$. In addition, width corrections which might be large here partially cancel in the ratio. Stability of the LSR moments versus the usual parameters $(\tau, t_c)$ has been studied in

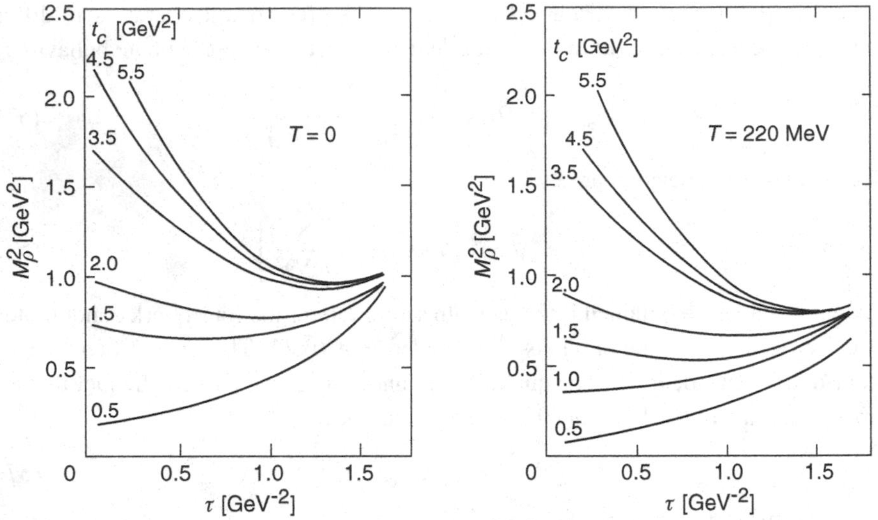


Fig. 57.6. $\tau$ and $t_c$ stabilities analysis of the rho meson mass.

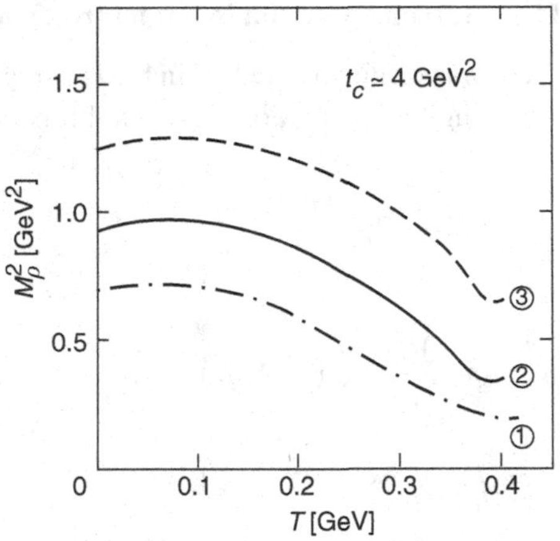


Fig. 57.7. $T$-behaviour of the rho meson spectrum.

detail in [843] using different values of the QCD input parameters, where, in the analysis, the small variations of the gluon and four-quark condensates with temperature have been neglected. This analysis is shown in Fig. 57.6.

The optimal result of the $\rho$-meson mass from [843] is shown in Fig. 57.7, where the smooth $T$-dependence decrease of the spectrum differs from the sharp change at $T_c$ obtained from [842]. This difference is understood to be due to the effect of the change of the continuum threshold from 0.8 to 1.5 GeV$^2$ with $T$ in [842], where stability in the variable $\tau$ of the LSR has not yet been reached. This result also differs from the expectation [864] that $M_\rho(T) > M_\rho(0)$. We conclude from this figure that the hot fermi gas does not lead to a drastic change of the spectra in the region below 200 MeV. Such behaviour just indicates that the $\rho$-meson mass at zero and at $T \leq T_c$ has qualitatively the same feature, and a posteriori indicates that the quark hadron duality used at zero temperature is also applicable here. This feature just indicates that some criticisms [847] on the non-validity of the quark-gluon basis at small $T$ in Eq. (57.2) are obviously wrong, as in this region the quark-hadron duality is expected to work better. In the same way, some other arguments raised by [847] like the non-consideration of non-peturbative effects (confinement) within the approach of [842] reviewed here are not at all justified, as has been demonstrated in the analysis of the exactly solvable QCD-like two-dimensional sigma $O(N)$ and Schwinger models [846]. The analysis of [843] has been completed by giving sets of FESR:

$$4\pi^2 \frac{M_\rho^2}{\bar{\gamma}_\rho^2} \simeq \int_0^{t_c} dt \; \mathrm{th}(\sqrt{\mathrm{t}}/4\mathrm{T}) + 2\int_0^{\infty} \mathrm{dt} \left( \mathrm{n_F} - \frac{1}{3}\mathrm{n_B} \right)$$

$$4\pi^2 \frac{M_\rho^4}{\bar{\gamma}_\rho^2} \simeq \int_0^{t_c} dt \; t \; \mathrm{th}(\sqrt{\mathrm{t}}/4\mathrm{T}) - \frac{1}{3}\pi \langle \alpha_s \mathrm{G}^2 \rangle$$

$$4\pi^2 \frac{M_\rho^6}{\bar{\gamma}_\rho^2} \simeq \int_0^{t_c} dt \; t^2 \; \mathrm{th}(\sqrt{\mathrm{t}}/4\mathrm{T}) - \frac{896}{81}\pi^3 \alpha_s \langle \bar{\psi}\psi \rangle^2 \,, \tag{57.55}$$

that are necessary for checking the consistency of the parameters obtained from the LSR. However, one should understand that the accuracy of the constraints decreases with increasing dimensions. These FESR have been used in [845] for studying the $T$ dependence of the continuum threshold $t_c$, a result which has been confirmed by [839] within the framework of composite models. Namely a smooth dependence with $T$ has been found:

$$\frac{t_c(T)}{t_c(0)} \simeq \frac{f_\pi(T)}{f_\pi(0)} \,. \tag{57.56}$$

## 57.10 $\rho$-meson coupling and width

The $T$ dependence of the $\rho$ meson coupling has been also studied in [849] in which a smooth dependence was found as well:

$$\frac{1}{\bar{\gamma}_\rho(T)} \simeq \frac{1}{\bar{\gamma}_\rho(0)} \left( 1 - \frac{2}{3} x_T + \cdots \right) . \tag{57.57}$$

These smooth dependences of the *rho*-meson mass and coupling were confirmed later in [850].

The $T$-dependence of the $\rho$-meson width has been proposed in [865] by using the relation between the $\rho\pi\pi$ coupling and $f_\pi$ through the KSFR relation (see Part I):

$$g_{\rho\pi\pi} = \frac{M_\rho}{\sqrt{2} f_\pi} \,, \tag{57.58}$$

and its relation to the $\rho \to \pi\pi$ width:

$$\Gamma_{\rho\to\pi\pi} = \frac{g^2_{\rho\pi\pi}}{4\pi}\frac{M_\rho}{12} , \tag{57.59}$$

from which, it is easy to deduce the desired $T$-dependence:

$$\Gamma_\rho(T) \simeq \Gamma_\rho(0)\left(1 - \frac{T^2}{4f_\pi^2}\right)^{-1} , \tag{57.60}$$

showing that for increasing $T$-values, the $\rho$-meson becomes broader and difficult to assign as a *true* resonance. This result has been confirmed from a recent study of the $T$-dependence of the imaginary part of the $\rho$-meson propagator [866].

Therefore, one can conclude from our previous analysis that the $\rho$-meson mass is almost insensitive to the $T$-variation, while its non-identification as a resonance is mainly due to the large increase of its hadronic $\pi\pi$ width at non-zero temperature. As a result, the study of dilepton-invariant mass in nuclear collisions through Drell–Yan processes at the rho-meson mass energy will not reveal a clear resonance structure.

## 57.11 Deconfinement phase and chiral symmetry restoration

We mentioned in the introduction that the Weinberg sum rules have been used in [839] for studying the relation between the deconfinement temperature $T_d$ and the chiral symmetry restoration temperature $T_c$. At finite temperature the first Weinberg sum rule reads [845]:

$$8\pi^2 f_\pi^2(T) \simeq \int_0^{t_c(T)} d\omega^2\ \mathrm{th}(\omega/4T) + 2\int_0^\infty d\omega^2\ n_F(\omega/2T) , \tag{57.61}$$

where $t_c$ is the continuum threshold separating the hadron from the continuum. Assuming that in the QGP, $t_c(T_d) = 0$, one can derive the constraint [839]:

$$f_\pi(T_d) = \frac{T_c}{\sqrt{6}}\left(\frac{T_d}{T_c}\right) , \tag{57.62}$$

relating the chiral restoration temperature $T_c$ with the deconfinement one $T_d$. The graphical resolution of the previous equality is obtained for:

$$T_d \approx T_c , \tag{57.63}$$

showing that the chiral restoration and the deconfinement phase are obtained at about the same temperature.

## 57.12 Hadronic couplings

The extension of the present framework for evaluating the meson trilinear couplings has been done in [845] using the symmetric point configuration of the vertex firstly proposed in [636,637]. It has been found that the couplings vanish like $f_\pi^n$, where $n$ is a positive model-dependent number, showing that hadrons decouple in the deconfinement phase. It is also remarkable to notice that the $\pi^0\gamma\gamma$ coupling vanishes at high temperature indicating the absence of the QED anomaly in the QGP phase. This result has been confirmed by some field theory model calculation [867].

## 57.13 Nucleon sum rules and neutron electric dipole moment

The extension of the previous application for studying the nucleon sum rule is straightforward. Noting (see previous section) that:

$$M_N \sim \langle\bar{\psi}\psi\rangle\tau \tag{57.64}$$

where $\tau$ is the sum rule scale variable, one also expects that its $T$-dependence is similar to that of $\langle\bar{\psi}\psi\rangle$. Finally, nucleon sum rules have been also used in [851] for studying the $T$-dependence of the ratio of the neutron electric dipole moment over the QCD-$\theta$ angle responsible for the strong $CP$ problem, where a smooth variation with temperature has been obtained.

# 58

# More on spectral sum rules

QSSR have wider applications and can also be applied to theories other than QCD as it is based on first principles of analyticity and duality.

## 58.1 Some other applications in QCD

Some other applications of QSSR have been already reviewed in details in the book [3], which we list below:

- Baryons at large $N_c$.
- String tension from Wilson loops.
- Relation between lattice correlators and chiral symmetry breaking in the continuum limit.
- Two-dimensional QCD.

## 58.2 Electroweak models with dynamic symmetry breaking

QSSR has been also extended in order to give dynamic constraints on fermions and $W$, $Z$ bosons assumed to be bound states of preons (haplons) where the structures may manifest at the TeV scale. The analysis has been done by assuming that at that scale, one can have a strong interaction theory of preons that is closely analogous to QCD describing the electroweak interactions, with the exception that one has to be careful on the chirality of the theories:

- In [868] and [869],[1] (reviewed in [870]), QSSR (Laplace and FESR) have been used in order to test the consistency of the compositeness assumption for the $W$ and $Z$ bosons and its spin zero partners, in the haplon model proposed by Fritzsch and Mandelbaum [871],[2] leading to a duality constraint between the boson masses and couplings with the continuum threshold (compositeness scale).
- In [873], QSSR has been used for an alternative derivation of the Dashen formula and some constraints among the Goldstone parameters with the condensate in supersymmetric QCD.
- In [874], analogous constraints have been derived between the 'composite' fermions and the vacuum structure of the (non)-supersymmetric theory, where it is found that for supersymmetric theories, there are unlikely to be composite fermions below the TeV scale.

[1] Some phenomenological implications of the scenario are also discussed in this paper.
[2] However, the obtained constraints are more general and can be applied to some other classes of composite models discussed in [872].

# Epilogue

We have tried to put together an elementary introduction to QCD and its modern developments in order to have as far as possible a *complete QCD handbook*, which I hope will be useful for a large spectrum of readers. After reading this book, I hope that the reader have learned more on the developments and aspects of perturbative and non-perturbative QCD, which, owing to the gluonic self-interactions, are based on asymptotic freedom, dynamical symmetry breaking and confinement. The former enables us to apply perturbation theory at large momenta where the coupling is small, and this has given successful predictions for various hard processes in terms of the single-order parameter $\alpha_s(Q)$. The running of the QCD coupling (and of the quark masses) as predicted by QCD and the renormalization group equation has been verified experimentally at different energy scales. On the other hand, the growth of the coupling at low energies indicates that QCD dynamics are governed by the confinement of quarks and gluons into colour-singlet hadronic states. However, a rigorous proof of this property is still lacking and remains one of the challenging problems in QCD. At present, non-perturbative approaches such as QCD spectral sum rules, lattice calculations provide an indirect evidence that QCD also provides the proper pattern of chiral symmetry breaking, where the quark and gluon condensates breaks dynamically the symmetry of the QCD Lagrangian. Thus, we have at present an overwhelming experimental and theoretical evidence that QCD gauge theory is the most robust theory of hadrons, though we have still to tackle the longstanding problem of confinement.

Some remarks concerning the presentation of this book have to be made:

- The readers may have noticed that, some specialized topics, like, for example, monopoles and more generally confinement, have not been discussed in details due to space–time limitations, but, mainly, due to the fact that our understanding on these subjects is not yet mature.
- Unlike, the case of the book and review in [3,2], we have not discussed in detail the derivation of each results from QCD spectral sum rules, but we have only summarized the different results after discussing some particular examples. The readers who wish to look into more details in the derivation of the QSSR results can then consult the previous references and the original papers.

I hope that, after reading this book, the readers have acquired the necessary information and technology for tackling new research projects in this field, which, after looking at the large number of QCD publications and conferences, has been and still remains one the most active fields in high-energy physics.

*'Physics will change even more ... If it radical and unfamiliar ... we think that the future will be only more radical and not less, only more strange and not more familiar, and that it will have its own new insights for inquiring human spirit.'*

*(Oppenheimer, Reith Lectures, BBC, 1953)*

# Part XI

Appendices

# Appendix A
## Physical constants and units

### A.1 High-energy physics conversion constants and units

Table A.1. *High-energy physics conversion constants and units*

| Quantity | Name | Value |
|---|---|---|
| Speed of light | $c$ | $299\,792\,458$ m s$^{-1}$ |
| Reduced Planck constant | $\hbar \equiv h/2\pi$ | $1.054\,572\,66(63) \times 10^{-34}$ J s = |
| | | $6.582\,122\,0(20) \times 10^{-23}$ MeV s |
| Conversion constants | $\hbar c$ | $197.327\,053(59)$ MeV fm |
| | $(\hbar c)^2$ | $0.389\,379\,66(23)$ GeV$^2$ mbarn |
| Units where $\hbar = c = 1$ | Mass, energy | 1 eV = $1.602\,177\,33(49) \times 10^{-19}$ J |
| | | 1 GeV = $10^3$ MeV = $10^6$ keV = $10^9$ eV |
| | | 1 erg = $10^{-7}$ J |
| | | 1 eV/$c^2$ = $1.782\,662\,70(54) \times 10^{-36}$ kg |
| | Length | 1 GeV$^{-1}$ = 0.197 327 053 fm = |
| | | $0.197\,327\ldots \times 10^{-13}$ cm |
| | | 1 in = 0.0254 m    1 Å = 0.1 nm |
| | Lifetime | 1 GeV$^{-1}$ = $6\,582\,122\,0 \times 10^{-25}$ s |
| | Decay rate | 1 GeV = $(1/6\,582\,122\,0) \times 10^{25}$ s$^{-1}$ |
| | Cross-section | 1 GeV$^{-2}$ = $0.389\,379\,66(23) \times 10^6$ barn |
| | | 1 barn = $10^{-28}$ m$^2$    1 nb = $10^{-9}$ barn |
| | Others | 0 °C = 273.15 K    1 G = $10^{-4}$ T |
| | | $kT$ at 300 K = $[38.681\,49(33)]^{-1}$ eV |
| | | 1 atmosphere = 760 torr = 101 325 Pa |
| | | 1 dyne = $10^{-5}$ N |

### A.2 High-energy physical constants

A complete list of physical constants is given in PDG [16]. Among them, we have:

Table A.2. *Some high-energy physical constants*

| Observable | Symbol | Value |
|---|---|---|
| Electron mass | $m_e$ | 0.510 999 06(15) MeV/$c^2$ <br> = 9.109 389 7(54) × $10^{-31}$ kg |
| Muon mass | $m_\mu$ | 105.658357(5) MeV/$c^2$ |
| Tau mass | $m_\tau$ | $1777.03^{+30}_{-26}$MeV/$c^2$ |
| Proton mass | $m_p$ | 938.272 31(28) MeV/$c^2$ <br> = 1836.152 701(37) $m_e$ |
| Electron charge | $e$ | 1.602 177 33(49) × $10^{-19}$ C <br> = 4.803 206 8(15) × $10^{-10}$ esu |
| Permittivity of free space | $\epsilon_0$ | 8.854 187 817... × $10^{-12}$ F m$^{-1}$ |
| Fine structure constant | $\alpha = e^2/4\pi\epsilon_0\hbar c$ | 1/137.035 999 58(52) at $q^2 = m_e^2$ <br> 1/128 at $q^2 = M_Z^2$ |
| Electron anomaly | $a_e \equiv \frac{1}{2}(g_e - 2)$ | 115 965 218 84(43) × $10^{-13}$ |
| Muon anomaly | $a_\mu \equiv \frac{1}{2}(g_\mu - 2)$ | 116 592 023(151) × $10^{-11}$ |
| Tau anomaly | $a_\tau \equiv \frac{1}{2}(g_\tau - 2)$ | 0.004 ± 0.027 ± 0.023 |
| Electron radius | $r_e = e^2/4\pi\epsilon_0 m_e c^2$ | 2.817 940 92(38) × $10^{-15}$ m |
| Bohr radius ($m_{nucleus} = \infty$) | $a_\infty = 4\pi\epsilon_0\hbar^2/m_e c^2$ <br> $= r_e\alpha^{-2}$ | 0.529 177 249(24) × $10^{-10}$ m |
| Electron Compton wavelength | $\lambda_e/2\pi = \hbar/m_e c$ <br> $= r_e/\alpha$ | 3.861 593 23(35) × $10^{-13}$ m |
| Rydberg energy | $hcR_\infty = m_e c^2\alpha^2/2$ | 13.605 698 1(40) eV |
| Thomson cross-section | $\sigma_T = 8\pi r_e^2/3$ | 0.665 246 16(18) barn |
| Bohr magneton | $\mu_B = e\hbar/2m_e$ | 5.788 382 63(52) × $10^{-11}$ MeV T$^{-1}$ |
| Nuclear magneton | $\mu_B = e\hbar/2m_P$ | 3.152 451 66(28) × $10^{-14}$ MeV T$^{-1}$ |
| Electron cyclotron freq./field | $\omega^e_{cycl}/B = e/m_e$ | 1.758 819 62(53) × $10^{11}$ rad s$^{-1}$ T$^{-1}$ |
| Fermi coupling constant | $G_F/(\hbar c)^2$ | 1.166 39(2) × $10^{-5}$ GeV$^{-2}$ |
| Weak mixing angle | $\sin^2\theta_W(M_Z)\ \overline{MS}$ | 0.2315(4) |
| $W^\pm$ boson mass | $M_W$ | 80.33(15) GeV/$c^2$ |
| $Z^0$ boson mass | $M_Z$ | 91.187(7) GeV/$c^2$ |
| Strong coupling constant | $\alpha_s(M_Z)$ | 0.118(3) |

## A.3 CKM weak mixing matrix

In the electroweak standard model $SU(2)_L \times U(1)$, where both quarks and leptons left-handed doublets and right-handed singlets, the quark mixing matrix can be represented as:

$$\begin{pmatrix} d' \\ s' \\ b' \end{pmatrix} = \begin{pmatrix} V_{ud} & V_{us} & V_{ub} \\ V_{cd} & V_{cs} & V_{cb} \\ V_{td} & V_{ts} & V_{tb} \end{pmatrix} \begin{pmatrix} d \\ s \\ b \end{pmatrix}, \tag{A.1}$$

where from weak decays, the mixing matrix has the value:

$$\begin{pmatrix} 0.9745 - 0.9757 & 0.219 - 0.224 & 0.002 - 0.005 \\ 0.218 - 0.224 & 0.9736 - 0.9750 & 0.036 - 0.046 \\ 0.004 - 0.014 & 0.034 - 0.046 & 0.9989 - 0.9993 \end{pmatrix} \tag{A.2}$$

In the Wolfenstein parametrization:

$$\begin{aligned} V_{us} &\simeq \lambda\,, & V_{ub} &\simeq \lambda^3 A(\rho - i\eta) \\ V_{cb} &\simeq \lambda^2 A\,, & V_{td} &\simeq \lambda^3 A(1 - \rho - i\eta) \end{aligned} \tag{A.3}$$

## A.4 Some astrophysical constants

Table A.3. *Some astrophysical constants*

| Observable | Symbol | Value |
|---|---|---|
| Newton gravitation constant | $G_N$ | $6\,672\,59(85) \times 10^{-11}$ m$^3$kg$^{-1}$s$^{-2}$ |
| Astronomical unit | AU | $1.495\,978\,706\,6(2) \times 10^{11}$ m |
| Tropical year(equinox to equinox) | yr | 31 556 925.2 s |
| Age of the universe | $t_0$ | 15(5) Gyr |
| Planck mass | $\sqrt{\hbar c/G_N}$ | $1.221\,047(79) \times 10^{19}$ GeV/c$^2$ |
| parsec(1AU/1 arc sec) | pc | $3.085\,677\,580\,7(4) \times 10^{16} m = 3.262\ldots$ ly |
| light year | ly | $0.306\,6\ldots$ pc $= 0.9461\ldots \times 10^{16}$ m |
| Solar mass | $M_\odot$ | $1.968\,92(25) \times 10^{30}$ kg |
| Solar luminosity | $L_\odot$ | $3.846 \times 10^{26}$ W |
| Solar equatorial radius | $R_\odot$ | $76.96 \times 10^8$ m |
| Earth mass | $M_\oplus$ | $5.973\,70(76) \times 10^{24}$ kg |
| Earth equatorial radius | $R_\oplus$ | $6.378\,140 \times 10^6$ m |
| Hubble constant | $H_0$ | $100\, h_0$ km s$^{-1}$Mpc$^{-1}$ = $h_0 \times (9.778\,13\text{ Gyr})^{-1}$ |
| Normalized Hubble constant | $h_0$ | $0.5 \le h_0 \le 0.85$ |
| Critical density of the universe | $\rho_c = 3H_0^2/8\pi G_N$ | $2.775\,366\,27 \times 10^{11} h_0^2 M_\odot$Mpc$^{-3}$ |
| Local halo density | $\rho_{\text{halo}}$ | $(2 - 13)10^{-25}$ g cm$^{-3}$ ≈ $(0.1 - 0.7)$ GeV/$c^2$ cm$^{-3}$ |
| Scaled cosmological constant | $\lambda_0 = \Lambda c^2/3H_0^2$ | $-1 < \lambda_0 < 2$ |
| Scale factor for cosmological constant | $c^2/3H_0^2$ | $2.853 \times 10^{51} \mathrm{h}_0^2$ m$^2$ |

# Appendix B

## Weight factors for $SU(N)_c$

### B.1 Definition

The generators $T_a$ of the $SU(N)_c$ Lie algebra obey the commutation relation:

$$[T_a, T_b] = i f_{abc} T_c \tag{B.1}$$

and the trace properties:

$$Tr\ T_a = 0\,. \tag{B.2}$$

$f_{abc}$ are constants which are *real* and totally antisymmetric and normalized as:

$$f_{abc} f_{dbc} = N \delta_{ad}\,. \tag{B.3}$$

### B.2 Adjoint representation of the gluon fields

In this representation, one has:

$$(T_a)_{bc} = -i f_{abc}\,, \tag{B.4}$$

with the properties:

$$\begin{aligned} f_{abe} f_{cde} &= \frac{2}{N}[\delta_{ac}\delta_{bd} - \delta_{ad}\delta_{bc}] + d_{ace}d_{dbe} - d_{ade}d_{bce}\,, \\ & f_{abe}d_{cde} + f_{ace}d_{dbe} + f_{ade}d_{bce} = 0\,, \end{aligned} \tag{B.5}$$

where $d_{abc}$ is a real and totally symmetric tensor:

$$\begin{aligned} d_{abb} &= 0\,, \\ d_{abc}d_{dbc} &= (N - 4/N)\,\delta_{ad}\,. \end{aligned} \tag{B.6}$$

In this representation, the trace properties are:

$$\begin{aligned} Tr\ T_a T_b &= N\delta_{ab}\,, \\ Tr\ T_a T_b T_c &= \frac{i}{2} N \delta_{ab}\,, \\ Tr\ T_a T_b T_c T_d &= \delta_{ab}\delta_{cd} + \delta_{ad}\delta_{bc} + \frac{N}{4}\left(d_{abe}d_{cde} - d_{ace}d_{dbe} + d_{ade}d_{bce}\right)\,. \end{aligned} \tag{B.7}$$

### B.3 Fundamental representation of the quark fields

In this case:

$$T_a = \frac{1}{2}\lambda_a \,, \tag{B.8}$$

with the properties:

$$\begin{aligned}
[\lambda_a, \lambda_b] &= 2i f_{abc}\lambda_c \,, \\
\{\lambda_a, \lambda_b\} &= \frac{4}{n}\delta_{ab} + 2d_{abc}\lambda_c \,, \\
\lambda_a\lambda_b &= \frac{2}{N}\delta_{ab} + d_{abc}\lambda_c + i f_{abc}\lambda_c \,.
\end{aligned} \tag{B.9}$$

The trace properties are:

$$\begin{aligned}
Tr\ \lambda_a &= 0 \\
Tr\ \lambda_a\lambda_b &= 2\delta_{ab} \\
Tr\ \lambda_a\lambda_b\lambda_c &= 2(d_{abc} + i f_{abc}) \\
Tr\ \lambda_a\lambda_b\lambda_c\lambda_d &= \frac{4}{N}(\delta_{ab}\delta_{cd} - \delta_{ac}\delta_{bd} + \delta_{ad}\delta_{bc}) \\
&\quad + 2(d_{abe}d_{cde} - d_{ace}d_{abe} + d_{ade}d_{bce}) \\
&\quad + 2i(d_{abe}f_{cde} - d_{ace}f_{abe} + d_{ade}f_{bce}) \,.
\end{aligned} \tag{B.10}$$

Some other useful relations are:

$$\begin{aligned}
(\lambda_a)_{\alpha\beta}(\lambda_a)_{\gamma\delta} &= 2\left(\delta_{\alpha\delta}\delta_{\beta\gamma} - \frac{1}{N}\delta_{\alpha\beta}\delta_{\gamma\delta}\right) \\
&= \frac{2(N^2-1)}{N^2}\delta_{\alpha\delta}\delta_{\beta\gamma} - \frac{1}{N}(\lambda_a)_{\alpha\beta}(\lambda_a)_{\gamma\delta} \,, \\
(\lambda_a)_{\alpha\beta}(\lambda_a)_{\beta\gamma} &= 4\left(C_2(R) \equiv \frac{N^2-1}{2N}\right)\delta_{\alpha\gamma} \,, \\
(\lambda_b\lambda_a\lambda_b)_{\alpha\beta} &= -\frac{2}{N}(\lambda_a)_{\alpha\beta} \,, \\
(\lambda_a\lambda_b)_{\alpha\beta}(T_b)_{ca} &= N(\lambda_c)_{\alpha\beta} \,.
\end{aligned} \tag{B.11}$$

In the adjoint representation:

$$(T_a)_{bc}(T_a)_{cd} = (C_2(G) \equiv N)\,\delta_{bd} \,. \tag{B.12}$$

### B.4 The case of $SU(3)_c$

In this case, one can write explicitly:

$$\lambda_1 = \begin{pmatrix} 0 & 1 & 0 \\ 1 & 0 & 0 \\ 0 & 0 & 0 \end{pmatrix} \quad \lambda_2 = \begin{pmatrix} 0 & -i & 0 \\ i & 0 & 0 \\ 0 & 0 & 0 \end{pmatrix} \quad \lambda_3 = \begin{pmatrix} 1 & 0 & 0 \\ 0 & -1 & 0 \\ 0 & 0 & 0 \end{pmatrix}$$

$$\lambda_4 = \begin{pmatrix} 0 & 0 & 1 \\ 0 & 0 & 0 \\ 1 & 0 & 0 \end{pmatrix} \quad \lambda_5 = \begin{pmatrix} 0 & 0 & -i \\ 0 & 0 & 0 \\ i & 0 & 0 \end{pmatrix} \quad \lambda_6 = \begin{pmatrix} 0 & 0 & 0 \\ 0 & 0 & 1 \\ 0 & 1 & 0 \end{pmatrix}$$

$$\lambda_7 = \begin{pmatrix} 0 & 0 & 0 \\ 0 & 0 & -i \\ 0 & i & 0 \end{pmatrix} \quad \lambda_8 = \frac{1}{\sqrt{3}} \begin{pmatrix} 1 & 0 & 0 \\ 0 & 1 & 0 \\ 0 & 0 & -2 \end{pmatrix} . \tag{B.13}$$

Therefore:

$$\begin{aligned} &f_{123} = +1 \\ &f_{147} = f_{156} = f_{246} = f_{257} = f_{345} = -f_{367} = \frac{1}{2} , \\ &f_{458} = f_{678} = \frac{\sqrt{3}}{2} , \end{aligned} \tag{B.14}$$

and:

$$\begin{aligned} &d_{118} = d_{228} = d_{338} = d_{888} = \frac{1}{\sqrt{3}} \\ &d_{146} = d_{157} = -d_{247} = d_{256} = d_{344} = d_{355} = -d_{366} = -d_{377} = \frac{1}{2} , \\ &d_{448} = d_{558} = d_{668} = d_{778} = -\frac{1}{2\sqrt{3}} . \end{aligned} \tag{B.15}$$

The other components which cannot be obtained by permutation of indices of the above ones are zero.

# Appendix C

## Coordinates and momenta

The space–time coordinates $(t, x, y, z) \equiv (t, \vec{x})$ are denoted by the *contravariant* four-vector $x$, which is defined as:[1]

$$x^{\mu} \equiv (t, x, y, z) \equiv (x^0, x^1, x^2, x^3) . \tag{C.1}$$

The *covariant* four-vector is defined as:

$$x^{\mu} \equiv (t, -x, -y, -z) \equiv (x_0, x_1, x_2, x_3) = g_{\mu\nu} z^{\nu} , \tag{C.2}$$

where:

$$g_{\mu\nu} = \begin{pmatrix} 1 & 0 & 0 & 0 \\ 0 & -1 & 0 & 0 \\ 0 & 0 & -1 & 0 \\ 0 & 0 & 0 & -1 \end{pmatrix} . \tag{C.3}$$

The three-vector is also often denoted as:

$$\vec{x} \equiv \mathbf{x} \tag{C.4}$$

The momentum vector is defined in the same way:

$$p^{\mu} = (E, p_x, p_y, p_z) \tag{C.5}$$

The scalar products are:

$$\begin{aligned} x^2 &= x_{\mu} x^{\mu} = t^2 - \vec{x}^2 , \\ p_1 \cdot p_2 &= p_1^{\mu} p_{2,\mu} = E_1 E_2 - \vec{p}_1 \vec{p}_2 , \\ x \cdot p &= tE - \vec{x} \cdot \vec{p} \end{aligned} \tag{C.6}$$

The derivative operator is:

$$\partial_{\mu} \equiv \frac{\partial}{\partial x_{\mu}} \equiv \left( \frac{\partial}{\partial t}, -\vec{\nabla} \right) \equiv \left( \frac{\partial}{\partial t}, -\frac{\partial}{\partial x}, -\frac{\partial}{\partial y}, -\frac{\partial}{\partial z} \right) . \tag{C.7}$$

The Dalembertian operator is:

$$\nabla^2 \equiv \partial x_{\mu} \partial x^{\mu} = \frac{\partial^2}{\partial t^2} - \vec{\nabla}^2 . \tag{C.8}$$

[1] We shall follow the notations of Bjorken–Drell and Landau–Lifchitz.

The electromagnetic four-vector potential is:

$$A_\mu = (\Phi, \vec{A}) \tag{C.9}$$

The electromagnetic field strength is:

$$F_{\mu\nu} = \frac{\partial}{\partial x_\nu} A_\mu - \frac{\partial}{\partial x_\mu} A_\nu \tag{C.10}$$

The electromagnetic and magnetic fields are:

$$\mathbf{E} = (F^{01}, F^{02}, F^{03}) \qquad \mathbf{B} = (F^{23}, F^{31}, F^{12}) \tag{C.11}$$

The gluon field tensor is:

$$G^a_{\mu\nu} = \frac{\partial}{\partial x_\nu} A^a_\mu - \frac{\partial}{\partial x_\mu} A^a_\nu + g f_{abc} A^b_\mu A^c_\nu \tag{C.12}$$

where $A^a_\mu$ is the guon fields and $a = 1, 2, \ldots 8$ are the colour indices. The electromagnetic covariant derivative is:

$$D_\mu = \partial_\mu + ieA_\mu \tag{C.13}$$

The gluon covariant derivative acting on the quark colour componet $\alpha, \beta =$ red, blue, yellow is:

$$(D_\mu)_{\alpha\beta} \equiv \delta_{\alpha\beta}\partial_\mu - ig \sum_a \frac{1}{2} \lambda^a_{\alpha\beta} A^a_\mu \,, \tag{C.14}$$

where $\lambda^a_{\alpha\beta}$ are eight $3 \times 3$ colour matrices.

# Appendix D

## Dirac equation and matrices

### D.1 Definition and notations

If $\psi$ is a generic notation of a fermion field, it can be expressed in terms of the usual annihilation and creation operators as:

$$\psi(x) = \int \frac{d^3p}{(2\pi)^3 2E} \sum_{\lambda} [u(\vec{p},\lambda)a(\vec{p},\lambda)e^{-ipx} v(\vec{p},\lambda)b^{\dagger}(\vec{p},\lambda)e^{ipx}] \tag{D.1}$$

where the integration is over the mass hyperboloid with $p^2 = m^2$ and $p^0 > 0$. $\lambda$ is the two possible fermion helicities. The annihilation and creation operators satisfy the commutation relations:

$$\begin{aligned}
[a(p), a^{\dagger}(p')] &= [b(p), b^{\dagger}(p')] = (2\pi)^3 2E\delta^3(p' - p)\,, \\
[a(p), a(p')] &= 0 = [b(p), b(p')]\,.
\end{aligned} \tag{D.2}$$

The fermion spinors $u(p)$ (particle) and $v(p)$ (anti-particle) of mass m obey the Dirac equation:

$$\begin{aligned}
(\hat{p} - m)u(p) &= 0 = \bar{u}(p)(\hat{p} - m)\,, \\
(\hat{p} + m)v(p) &= 0 = \bar{v}(p)(\hat{p} - m)\,,
\end{aligned} \tag{D.3}$$

and normalized as:

$$\bar{u}(\vec{p},\lambda)u(\vec{p},\lambda) = 2m = -\bar{v}(\vec{p},\lambda)v(\vec{p},\lambda) \tag{D.4}$$

with:

$$\begin{aligned}
\bar{u} &= u^{\dagger}\gamma^0 \\
\bar{v} &= v^{\dagger}\gamma^0\,, \\
\hat{p} &= \gamma_{\mu}p^{\mu} = \gamma_0 p^0 - \gamma \cdot \mathbf{p}\,,
\end{aligned} \tag{D.5}$$

where $\gamma_{\mu}$ are the Dirac matrices. In four dimensions, these matrices can be defined as:

$$\gamma_5 = \begin{pmatrix} 0 & \mathbf{1} \\ \mathbf{1} & 0 \end{pmatrix} \quad \gamma_{\mu} = \begin{pmatrix} 0 & \sigma_{\mu} \\ -\sigma_{\mu} & 0 \end{pmatrix} \text{ for } \mu = 1, 2, 3 \quad \gamma_0 = \begin{pmatrix} \mathbf{-1} & 0 \\ 0 & \mathbf{1} \end{pmatrix} \tag{D.6}$$

in terms of the Pauli matrices $\sigma$:

$$\sigma_1 = \begin{pmatrix} 0 & 1 \\ 1 & 0 \end{pmatrix} \quad \sigma_2 = \begin{pmatrix} 0 & -i \\ i & 0 \end{pmatrix} \quad \sigma_3 = \begin{pmatrix} 1 & 0 \\ 0 & -1 \end{pmatrix} \tag{D.7}$$

They obey the properties:

$$\{\gamma_\mu, \gamma_\nu\} = 2g_{\mu\nu}\,, \qquad \sigma_{\mu\nu} \equiv \frac{i}{2}[\gamma_\mu, \gamma_\nu]\,, \tag{D.8}$$

and:

$$(\gamma_5)^2 = \mathbf{1}\,, \quad \text{and} \quad \gamma_5\gamma_\mu = -\gamma_\mu\gamma_5\,, \tag{D.9}$$

with the definition:

$$\gamma_5 = i\gamma_0\gamma_1\gamma_2\gamma_3 \tag{D.10}$$

or:

$$\gamma_5 = \frac{1}{4!}\epsilon_{\mu\nu\lambda\rho}\gamma^\mu\gamma^\nu\gamma^\rho\gamma^\sigma\,. \tag{D.11}$$

The Dirac matrices are (anti)hermitians:

$$\gamma_\mu = -\gamma_\mu^+\,, \qquad \mu = 1, 2, 3\,, \qquad \gamma_0^+ = \gamma_0 \quad \text{and} \quad \gamma_5^+ = \gamma_5\,. \tag{D.12}$$

## D.2 CPT transformations

The action of the operators:
C ≡ charge conjugation, P ≡ parity transformation, T ≡ time reversal ,
on the fermion field $\psi(t, \vec{r})$ are:

$$\begin{aligned}
\mathrm{C}\,\psi(t,\vec{r}) &= \gamma_2\psi^\dagger(t,\vec{r}) \\
\mathrm{T}\,\psi(t,\vec{r}) &= -i\gamma_1\gamma_3\psi^\dagger(-t,\vec{r}) \\
\mathrm{PT}\,\psi(t,\vec{r}) &= \gamma_0\gamma_1\gamma_3\psi^\dagger(-t,-\vec{r}) \\
\mathrm{CPT}\,\psi(t,\vec{r}) &= \gamma_2\gamma_0\gamma_1\gamma_3\psi(-t,-\vec{r}) \\
&= i\gamma_5\psi(-t,-\vec{r})\,,
\end{aligned} \tag{D.13}$$

where:

$$\psi^\dagger = \bar{\psi}\gamma_0\,. \tag{D.14}$$

## D.3 Polarizations

In the evaluation of unpolarized cross-section, one has to sum over polarizations of, for example, fermions:

$$\sum_\lambda u(p,\lambda)\bar{u}(p,\lambda) = \hat{p} + m\,, \qquad \sum_\lambda v(p,\lambda)\bar{v}(p,\lambda) = \hat{p} - m\,, \tag{D.15}$$

while for polarized cross-section, one inserts the projection matrices:

$$\begin{aligned}
u\left(p, \lambda = \pm\frac{1}{2}\right)\bar{u}\left(p, \lambda = \pm\frac{1}{2}\right) &= \frac{1}{2}(\hat{p} + m)\left(\frac{1 \pm \gamma_5\hat{s}}{2}\right)\,, \\
v\left(p, \lambda = \pm\frac{1}{2}\right)\bar{v}\left(p, \lambda = \pm\frac{1}{2}\right) &= \frac{1}{2}(\hat{p} - m)\left(\frac{1 \pm \gamma_5\hat{s}}{2}\right)\,,
\end{aligned} \tag{D.16}$$

where: $s$ is the polarization four-vector of the (anti-)particle with energy-momnetum $p$:

$$s \cdot p = 0 \quad \text{and} \quad s^2 = -1 \,. \tag{D.17}$$

For a photon or massless vector boson, the polarization is transverse:

$$\epsilon^{\mu} = (0, \vec{\epsilon}) \quad \text{with} \quad \vec{p} \cdot \vec{\epsilon} = 0 \,. \tag{D.18}$$

For unpolarized cross-section involving (massless) photons, one has to sum over polarizations:

$$\sum_{polar.} \epsilon_{\mu}^{*} \epsilon^{\mu} = -g_{\mu\nu} \,. \tag{D.19}$$

## D.4 Fierz identities

In some calculations, it is useful to arrange products of fermion bilinears using Fierz identities. Denoting by $\psi_i$ the field of a fermion $i$, one has in four dimensions:

$$(\bar{\psi}_1 \psi_4)(\bar{\psi}_3 \psi_2) = \frac{1}{4} \sum_{\mu} (\bar{\psi}_1 \gamma_{\mu} \psi_2)(\bar{\psi}_3 \gamma^{\mu} \psi_4) \,. \tag{D.20}$$

Similar relation can be obtained by the substitution:

$$\psi_4 \to \gamma_{\nu} \psi_4 \,, \qquad \psi_2 \to \gamma_{\rho} \psi_2 \,, \tag{D.21}$$

and by using the decomposition:

$$\gamma_{\mu} \gamma_{\nu} = \frac{1}{4} \sum_{\sigma} (Tr\; \gamma_{\mu} \gamma_{\nu} \gamma_{\sigma}) \gamma^{\sigma} \,. \tag{D.22}$$

A typical Fierz rearrangement is the one of weak four-fermion operator:

$$(\bar{\psi}_{1L} \gamma^{\mu} \psi_{2L})(\bar{\psi}_{3L} \gamma_{\mu} \psi_{4L}) = -(\bar{\psi}_{1L} \gamma^{\mu} \psi_{4L})(\bar{\psi}_{3L} \gamma_{\mu} \psi_{2L}) \tag{D.23}$$

where:

$$\psi_{iL} \equiv \frac{1}{2}(1 - \gamma_5) \psi_i \,. \tag{D.24}$$

Additional relations can be obtained by using:

$$(\sigma_{\mu})_{\alpha\beta} (\sigma^{\mu})_{\gamma\delta} = 2 \epsilon_{\alpha\gamma} \epsilon_{\beta\delta} \,. \tag{D.25}$$

## D.5 Dirac algebra in $n$-dimensions

The (anti)-commutation properties of the Dirac matrices in four dimensions given in Eq. (D.8) are maintained, but the algebra becomes:[1]

$$\begin{aligned}
\gamma_{\mu} \gamma^{\mu} &= n\mathbf{1} = g_{\mu\nu} g^{\mu\nu} \,, \\
\gamma_{\mu} \gamma_{\alpha} \gamma^{\mu} &= (2 - n) \gamma_{\alpha} \,, \\
\gamma_{\mu} \gamma_{\alpha} \gamma_{\beta} \gamma^{\mu} &= 4 g_{\alpha\beta} \mathbf{1} + (n - 4) \gamma_{\alpha} \gamma_{\beta} \,, \\
\gamma_{\mu} \gamma_{\alpha} \gamma_{\beta} \gamma_{\gamma} \gamma^{\mu} &= -2 \gamma_{\gamma} \gamma_{\beta} \gamma_{\alpha} - (n - 4) \gamma_{\alpha} \gamma_{\beta} \gamma_{\gamma} \,.
\end{aligned} \tag{D.26}$$

[1] See also the discussions in Section 8.2 for different aspects of dimensional regularization.

The traces in $n$ dimensions can be chosen to be the same as in four dimensions. The usual properties are:

$$Tr\,\mathbf{1} = 4 \tag{D.27}$$

and:

$$\begin{aligned} Tr\ (\gamma^{\mu_1} \ldots \gamma^{\mu_m}) &= 0 \quad \text{for } m \text{ odd}\,, \\ Tr\ (\gamma^{\mu_1} \ldots \gamma^{\mu_m}) &= -Tr\ (\gamma^{\mu_m} \gamma^{\mu_1} \ldots \gamma^{\mu_{m-1}}) \\ &\quad + 2\sum_{i=1}^{m-1} (-1)^{i+1} Tr\ (\gamma^{\mu_1} \ldots \gamma^{\mu_{i-1}} \gamma^{\mu_{i+1}} \gamma^{\mu_{m-1}})\, g_{\mu_i \mu_m}\,. \end{aligned} \tag{D.28}$$

Therefore, one can deduce:

$$\begin{aligned} Tr\ \gamma_\mu \gamma_\nu &= 4 g_{\mu\nu}\,, \\ Tr\ \gamma_\mu \gamma_\nu \gamma_\rho \gamma_\sigma &= 4(g_{\mu\nu} g_{\rho\sigma} - g_{\mu\rho} g_{\nu\sigma} + g_{\mu\sigma} g_{\nu\rho})\,, \\ Tr\ \gamma_{\lambda\mu\nu\rho\sigma\tau} &= g_{\lambda\mu} T_{\nu\rho\sigma\tau} - g_{\nu\lambda} T_{\mu\rho\sigma\tau} + g_{\lambda\rho} T_{\mu\nu\sigma\tau} - g_{\lambda\sigma} T_{\mu\nu\rho\tau} + g_{\lambda\tau} T_{\mu\nu\rho\sigma}\,, \end{aligned} \tag{D.29}$$

with:

$$\gamma_{\lambda\mu\nu\rho\sigma\tau} \equiv \gamma_\lambda \gamma_\mu \gamma_\nu \gamma_\rho \gamma_\sigma \gamma_\tau\,, \qquad T_{\mu\nu\rho\sigma} \equiv Tr\ \gamma_\mu \gamma_\nu \gamma_\rho \gamma_\sigma\,. \tag{D.30}$$

The definition of $\gamma_5$ is more delicate in $n$-dimensions. There are many definitions in the literature (see e.g. [116] and the review in [2]). These definitions are good if the corresponding Green's functions satisfy constraints imposed by the Ward identities, and do not induce *unphysical* anomalous term [116,119], which cannot be absorbed in the Lagrangian counterterms. The most convenient and unambiguous definition is the one encountered in four dimensions, which is either the one in Eq. (D.10) or the one in Eq. (D.11), although the one in Eq. (D.10) does not exist for $n < 4$. In both cases, the most important properties are:

$$(\gamma_5)^2 = \mathbf{1}, \qquad \gamma_5^\dagger = \gamma_5, \quad \text{and} \quad \gamma_5 \gamma_\mu = -\gamma_\mu \gamma_5\,. \tag{D.31}$$

and:

$$[\gamma_5, \sigma_{\mu\nu}] = 0 \quad \text{and} \quad \gamma_5 \sigma_{\mu,\nu} = \frac{i}{2} \epsilon_{\mu\nu\rho\sigma} \sigma^{\rho\sigma}\,. \tag{D.32}$$

The traces involving $\gamma_5$ are:

$$\begin{aligned} Tr\ \gamma_5 &= 0\,, \\ Tr\ \gamma_5 \gamma_\mu \gamma_\nu &= 0\,, \\ Tr\ \gamma_5 \gamma_\mu \gamma_\nu \gamma_\rho \gamma_\sigma &= 4i \epsilon_{\mu\nu\rho\sigma}\,, \\ Tr\ \gamma_5 \gamma_\lambda \gamma_\mu \gamma_\nu \gamma_\rho \gamma_\sigma \gamma_\tau &= 4i [g_{\mu\lambda} \epsilon_{\nu\rho\sigma\tau} - g_{\lambda\nu} \epsilon_{\mu\rho\sigma\tau} + g_{\mu\nu} \epsilon_{\lambda\rho\sigma\tau} + g_{\sigma\tau} \epsilon_{\lambda\mu\nu\rho} \\ &\quad - g_{\rho\tau} \epsilon_{\lambda\mu\nu\sigma} + g_{\rho\sigma} \epsilon_{\lambda\mu\nu\tau}] \\ Tr\ \gamma_5 \gamma_{\mu_1} \ldots \gamma_{\mu_m} &= \text{ for } m \text{ odd}\,. \end{aligned} \tag{D.33}$$

Finally, in order to complete the presentation of the Dirac algebra in $n$ dimensions, it is also useful to remind the hermiticity:

$$\gamma^0 \gamma^\mu \gamma^0 = (\gamma^\mu)^\dagger\,, \qquad \gamma^0 \gamma_5 \gamma^0 = -\gamma_5^\dagger = -\gamma_5\,, \tag{D.34}$$

and the parity properties:

$$C\gamma_\mu C^{-1} = -\gamma_\mu^T \qquad C\gamma_5 C^{-1} = \gamma_5^T ,$$
$$C\sigma_{\mu\nu}C^{-1} = -\sigma_{\mu\nu}^T \qquad C(\gamma_5\gamma_\mu)C^{-1} = (\gamma_5\gamma_\mu)^T , \tag{D.35}$$

where $C$ is the charge conjugate operator normalized as:

$$C^2 = -1 . \tag{D.36}$$

## D.6 The totally anti-symmetric tensor

The totally anti-symmetric tensor has the same definition as in four dimensions:

$$\epsilon_{\mu\nu\rho\sigma} = \begin{cases} 0, & \text{if two indices are equal} \\ -1, & \text{if } \mu\nu\rho\sigma = 0123 \\ +1, & \text{if } \mu\nu\rho\sigma = 1230 , \end{cases} \tag{D.37}$$

while one can choose its properties as:

$$\epsilon_{\mu\nu\alpha\beta}\epsilon^{\rho\nu\alpha\beta} = -(n-3)(n-2)(n-1)g_\mu^\rho ,$$
$$\epsilon_{\mu\nu\alpha\beta}\epsilon^{\rho\sigma\alpha\beta} = -(n-3)(n-2)\left(g_\mu^\rho g_\nu^\sigma - g_\nu^\rho g_\mu^\sigma\right) ,$$
$$\epsilon_{\mu\nu\alpha\beta}\epsilon^{\rho\sigma\tau\beta} = -(n-3)\begin{vmatrix} g_\mu^\rho & g_\mu^\sigma & g_\mu^\tau \\ g_\nu^\rho & g_\nu^\sigma & g_\nu^\tau \\ g_\alpha^\rho & g_\alpha^\sigma & g_\alpha^\tau \end{vmatrix} \tag{D.38}$$

# Appendix E

## Feynman rules

### E.1 Factors induced by external or internal lines

ingoing quark: 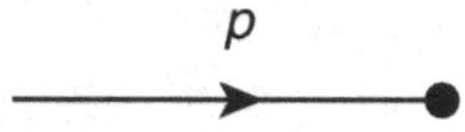

$(2\pi)^{-3/2}u(p, \lambda)$

ingoing antiquark: 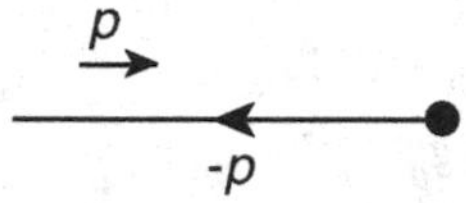

$(2\pi)^{-3/2}\bar{v}(p, \lambda)$

outgoing quark: 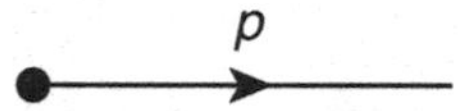

$(2\pi)^{-3/2}\bar{u}(p, \lambda)$

outgoing antiquark: 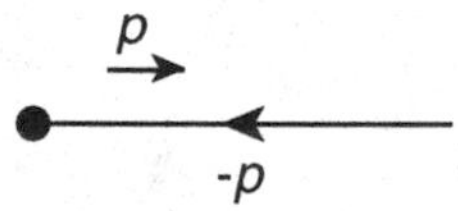

$(2\pi)^{-3/2}v(p, \lambda)$

ingoing gluon: 

$(2\pi)^{-3/2}\epsilon^{\mu}(k, \eta)$

outgoing gluon: 

$(2\pi)^{-3/2}\epsilon^{*}_{\mu}(k, \eta)$

### E.2 Factors induced by closed loops

$\int \frac{d^n p}{(2\pi)^n}$ for each loop integration

$(-1)$ for each closed fermion or ghost loop

$\frac{1}{2!}$ for 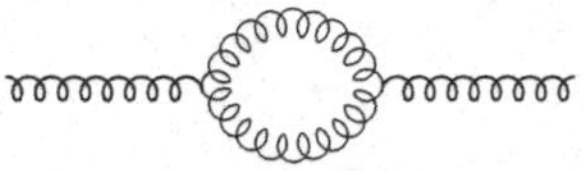 or 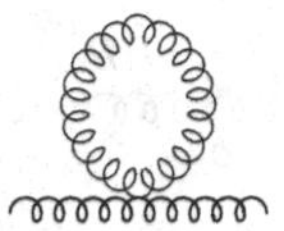 or 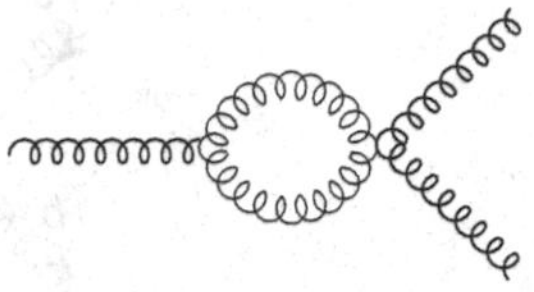

$\frac{1}{3!}$ for 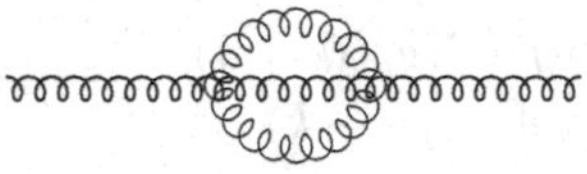

## E.3 Propagators and vertices

### Propagators

quark: 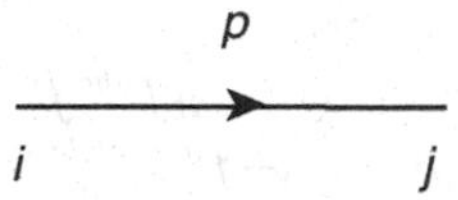

$\frac{i}{\hat{p}-m+i\epsilon'}\delta_{ij}$

gluon: 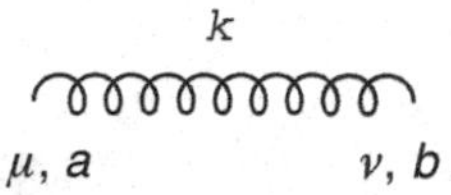

$(-i)\frac{\delta_{ab}}{k^2+i\epsilon'}\left[g_{\mu\nu}-(1-\alpha_G)\frac{k_\mu k_\nu}{k^2}\right]$

ghost: 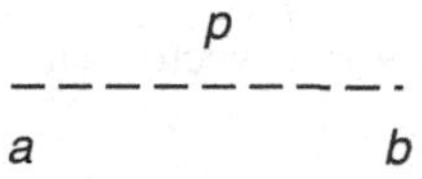

$(-i)\frac{\delta_{ab}}{k^2+i\epsilon'}$

### Vertices

quark-gluon-quark: 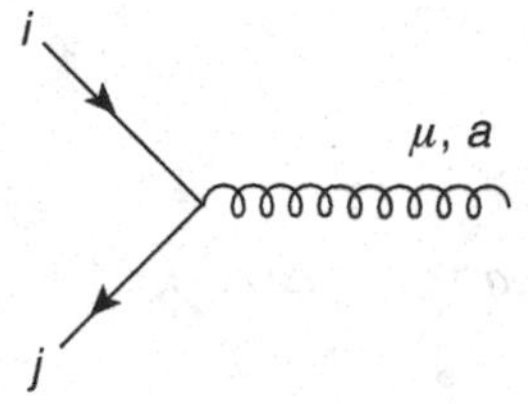

$(ig)\gamma_\mu T^a_{ij}$

ghost-gluon-ghost: 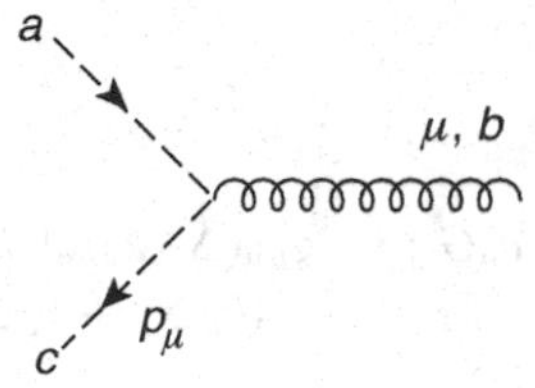

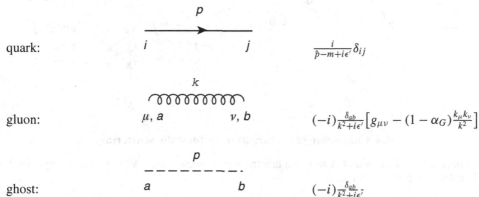

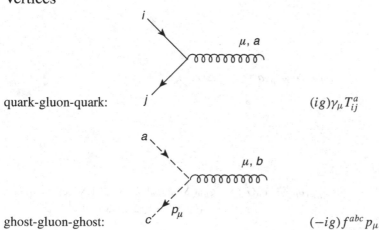

$(-ig)f^{abc}p_\mu$

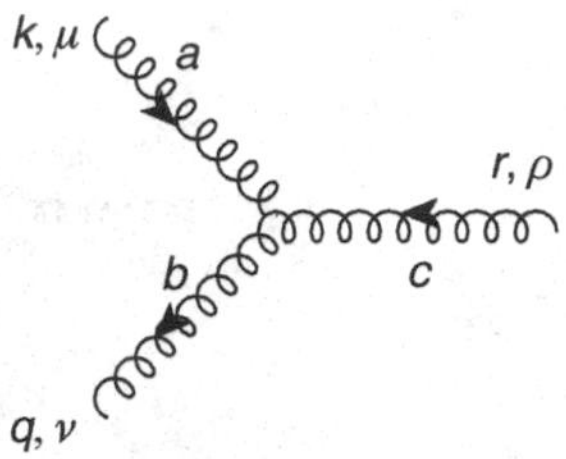


3-gluon:

$$(gf^{abc})[g_{\mu\nu}(k+q)_\rho - g_{\nu\rho}(q+r)_\mu + g_{\rho\mu}(r-k)_\nu]$$

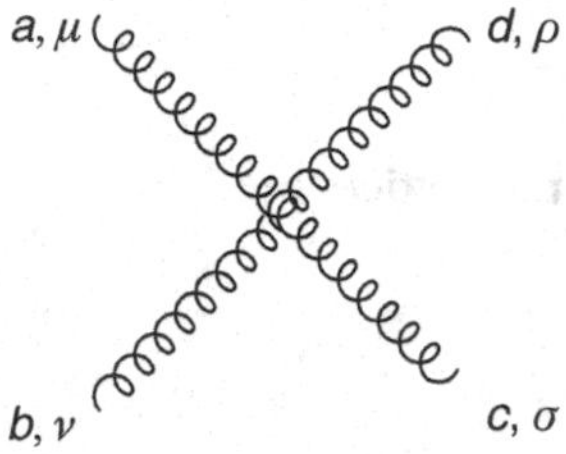


4-gluon:

$$(-ig^2)[f^{abe}f^{cde}(g^{\mu\sigma}g^{\nu\rho} - g^{\mu\rho}g^{\nu\sigma}) + f^{ace}f^{bde} \times (g^{\mu\nu}g^{\sigma\rho} - g^{\mu\rho}g^{\nu\sigma}) + f^{ade}f^{cbe} \times (g^{\mu\sigma}g^{\nu\rho} - g^{\mu\nu}g^{\sigma\rho})]$$

## E.4 Composite operators in deep-inelastic scattering

We define $\Gamma \equiv 1$ or $\gamma_5$ and $\Delta$ to be an arbitrary four-vector with $\Delta^2 = 0$. The composite operators are defined at $x = 0$.

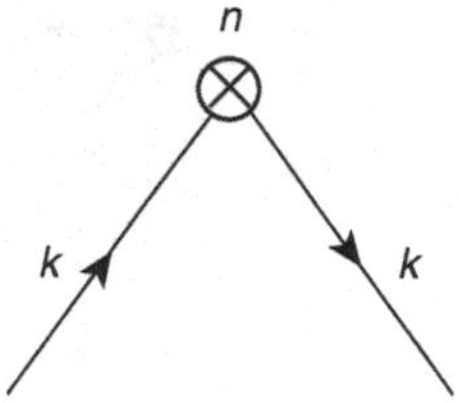


$: \bar{q}\gamma_{\mu_1} \cdots \partial_{\mu_n} q :$ $\qquad \hat{\Delta}(\Delta \cdot k)^{n-1}\Gamma$

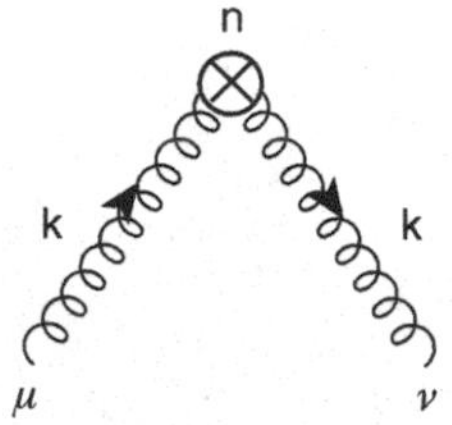


$: G_{\mu\mu_1}\partial_{\mu_2} \cdots \partial_\nu G :$ $\qquad g_{\mu\nu}(\Delta \cdot k)^n + k^2 \Delta_\mu \Delta_\nu (\Delta \cdot k)^{n-2} - (k_\mu \Delta_\nu + k_\nu \Delta_\mu)(\Delta \cdot k)^{n-1}$

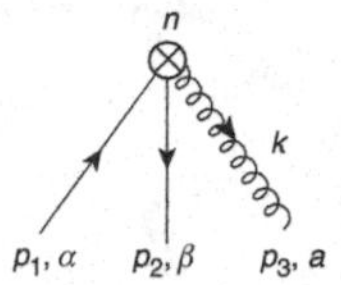


$$: \bar{q}_\alpha \gamma_{\mu_1} \cdots g B_a^\mu T_{ij}^a \cdots \gamma_{\mu_n} \Gamma q_\beta : \qquad g T^a_{\alpha\beta} \Delta^\mu \hat{\Delta} \sum_{j=0}^{n-2} (\Delta \cdots p_1)^j \times (\Delta \cdots p_2)^{n-j-2} \Gamma$$

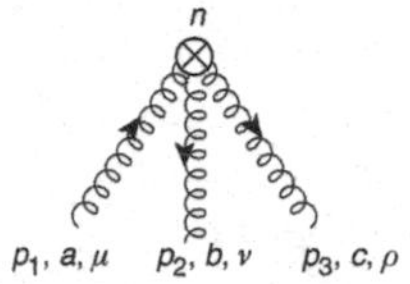


$$: G_{\mu\mu_1} \partial_{\mu_2} \cdots g B_{\mu_i} \cdots G_{\mu_n \nu} : \qquad \frac{ig}{3!} f^{abc} \{ \Delta_n u [\Delta_\rho p_{3,\mu} (\Delta \cdots p_1) + p_{1,\rho} \Delta_\mu (\Delta \cdot p_3) - g_{\mu\rho} \times (\Delta \cdot p_1)(\Delta \cdot p_3) - \Delta_\mu \Delta_\rho \times (p_3 \cdot p_1)] + \sum_{j=1}^{n-2} (-1)^j \times (\Delta \cdot p_1)^{j-1} (\Delta \cdot p_3)^{n-j-2} + g_{\mu\rho} \Delta_\nu - g_{\nu\rho} \Delta_\mu)(\Delta \cdot p_3) + \Delta_\rho (\Delta_\mu p_{3,\nu} - p_{3,\mu} \Delta_\nu)] \times (\Delta \cdot p_3)^{n-2} + \text{perm.} \}$$

## E.5 Rules in the background field approach

The background field is represented by $A$. The combinations of gauge fields not shown below vanish. For instance, there is no quadrilinear vertices with three or four background fields. We use the conventions in [127].

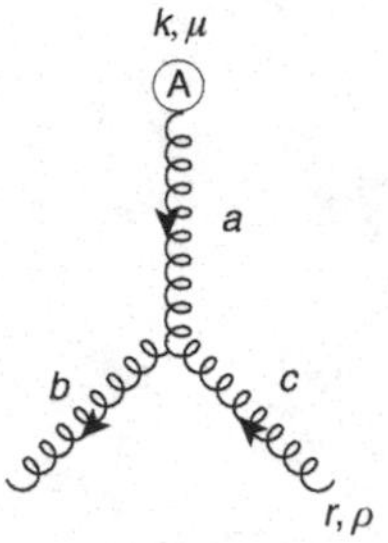


$$(g f^{abc}) \Big[ g_{\mu\nu} \Big( k + q - \frac{r}{\alpha_G} \Big)_\rho - g_{\nu\rho} (q + r)_\mu + g_{\rho\mu} \Big( r - k - \frac{q}{\alpha_G} \Big)_\nu \Big]$$

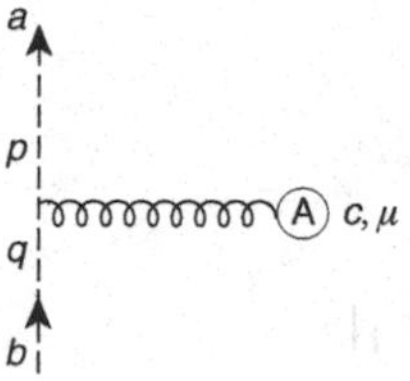


$$(-g) f^{bca} (p + q)_\mu$$

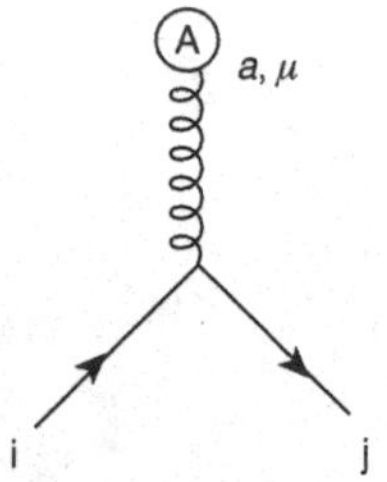


$$(ig)T_{ij}^{(a)}\gamma_\mu$$

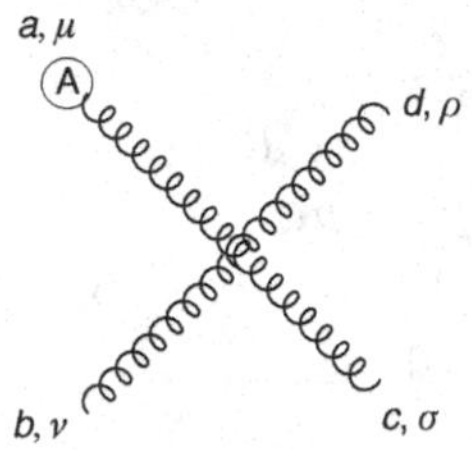


$\equiv$

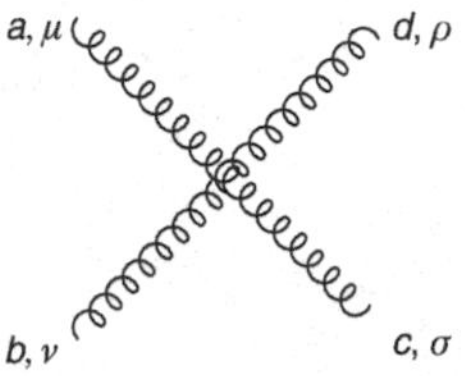


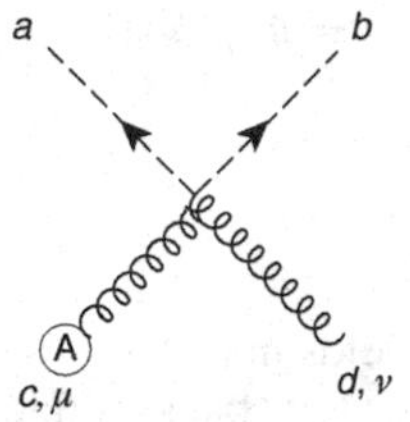


$$(-ig^2)f^{ace}f^{edb}g_{\mu\nu}$$

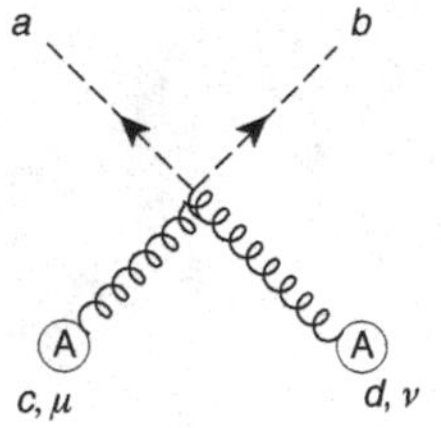


$$(-ig^2)\left(f^{ace}f^{edb}+f^{adx}f^{xcb}\right)g_{\mu\nu}$$

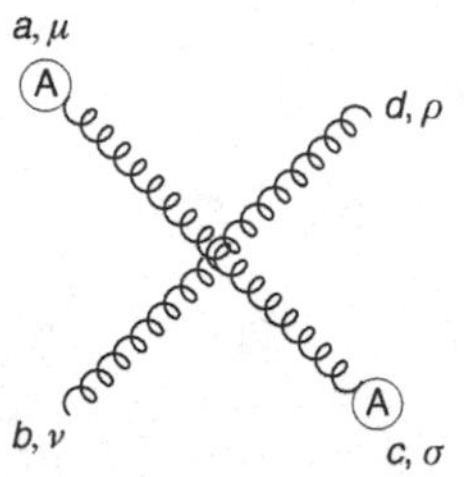


$$\begin{aligned}&(-ig^2)\Big[f^{abe}f^{cde}\Big(g^{\mu\sigma}g^{\nu\rho}-g^{\mu\rho}g^{\nu\sigma}\\&\quad+\tfrac{1}{\alpha_G}g^{\mu\nu}g^{\sigma\rho}\Big)+f^{ace}f^{bde}\\&\quad\times(g^{\mu\nu}g^{\sigma\rho}-g^{\mu\rho}g^{\nu\sigma})+f^{ade}f^{cbe}\\&\quad\times\Big(g^{\mu\sigma}g^{\nu\rho}-g^{\mu\nu}g^{\sigma\rho}+\tfrac{1}{\alpha_G}g^{\mu\rho}g^{\nu\sigma}\Big)\Big]\end{aligned}$$

# Appendix F
## Feynman integrals

### F.1 Feynman parametrization

The Feynman parametrization is needed to recombine the product of denominators appearing in the momentum integral. We shall discuss the most usual ways of parametrization.

#### *F.1.1 Schwinger representation*

The first one consists of an exponentiation of the propagator denominators and leads to:

$$\frac{1}{a_1 \cdots a_n} = \int_0^\infty dz_1 \cdots \int_0^\infty dz_n \exp\left(-\sum_{i=1}^{n} a_i z_i\right) . \tag{F.1}$$

In connection to this, the following Gaussian integral is useful:

$$\int \frac{d^n k}{(2\pi)^n} e^{-\alpha k^2} = \frac{1}{(4\pi\alpha)^{n/2}} . \tag{F.2}$$

#### *F.1.2 Original Feynman parametrization*

The second alternative is obtained from the original Feynman parametrization:

$$\frac{1}{a_1 \cdots a_n} = (n-1)! \int_0^1 dz_1 \cdots \int_0^1 dz_n\, \delta\left(1 - \sum_i z_i\right) \frac{1}{\left(\sum_i a_i z_i\right)^n} . \tag{F.3}$$

After a suitable change of variables, one can eliminate the $\delta$-function and one obtains:

$$\begin{aligned}\frac{1}{a_1 \cdots a_n} &= (n-1)! \int_0^1 u_1^{n-2} du_1 \int_0^1 u_2^{n-3} du_2 \cdots \int_0^1 du_{n-1} \\ &\quad \times [(a_1 - a_2)u_1 \cdots u_{n-1} + (a_2 - a_3)u_1 \cdots u_{n-2} + \cdots + a_n]^{-n} .\end{aligned} \tag{F.4}$$

This parametrization is quite convenient as it allows possible cancellations among terms of two propagators, and has the advantage to provide finite bounds of integration, which is convenient in various numerical integration calculations encountered e.g. in QED calculations (g-2, ...). A particularly useful case of Eq. (F.4) are:

$$\frac{1}{a^\alpha b^\beta} = \frac{\Gamma(\alpha+\beta)}{\Gamma(\alpha)\Gamma(\beta)} \int_0^1 dx \frac{x^{\alpha-1}(1-x)^{\beta-1}}{[(a-b)x+b]^{\alpha+\beta}} , \tag{F.5}$$

and:

$$\frac{1}{a^n b^m c^r} = \frac{\Gamma(n+m+r)}{\Gamma(n)\Gamma(m)\Gamma(r)} \int_0^1 dx\, x^{m+n-1}(1-x)^{r-1} \times \int_0^1 dy \frac{(1-y)^{m-1} y^{n-1}}{[(a-b)xy+(b-c)x+c]^{m+n+r}} , \tag{F.6}$$

entering in a one-loop calculation. In the case where $a_1$ is $\ln k^2$, the following representation integral is useful:

$$\frac{1}{(\ln k^2)^{n+1}} = \frac{1}{\Gamma(n+1)} \int_0^\infty dx\, x^n (k^2)^{-x} . \tag{F.7}$$

## F.2 The Γ function

It is defined for complex $z$ by the Euler integral:

$$\Gamma(z) = \int_0^\infty dt\, t^{z-1} e^{-t} , \tag{F.8}$$

If the previous integral does not exist, it can be defined, using an analytic continuation, by:

$$\Gamma(z) = \sum_{n=0}^\infty \frac{(-1)^n}{n!(z+n)} + \int_1^\infty dt\, t^{z-1} e^{-t} , \tag{F.9}$$

which expresses that $\Gamma(z)$ is analytic in the entire $z$-plane but contains simple poles at $z = 0, -1, -2, \dots$. It has the properties:

$$\Gamma(1+z) = z\Gamma(z) , \tag{F.10}$$

and:

$$\Gamma(1+z) = \exp\left\{-z\gamma_E + \sum_{n=2}^\infty (-1)^n \frac{z^n}{n} \zeta(n)\right\} , \tag{F.11}$$

with:

$$\gamma_E \equiv \gamma = \lim_{n\to\infty} \left\{1 + \frac{1}{2} + \cdots + \frac{1}{n}\right\} = 0.577\,215\,664\,9\dots \tag{F.12}$$

and:

$$\zeta(n) = \sum_{k=1}^\infty \frac{1}{k^n} , \tag{F.13}$$

is the Riemann function: with:

$$\zeta(2) = \frac{\pi^2}{6} , \qquad \zeta(3) = 1.202\,056\,903\,1\dots , \qquad \zeta(4) = \frac{\pi^4}{90} . \tag{F.14}$$

The following expansion is particularly useful in dimensional regularization:

$$\lim_{\epsilon\to 0}\Gamma(1+\epsilon) = 1 - \epsilon\gamma_E + \frac{\epsilon^2}{2}\left(\gamma_E^2 + \frac{\pi^2}{6}\right) - \frac{\epsilon^3}{3}\left(\frac{\gamma_E^3}{2} + \frac{\pi^2}{4}\gamma_E + \zeta(3)\right) + \mathcal{O}(\epsilon^4), \tag{F.15}$$

from which one can deduce $\Gamma(\epsilon)$ with the help of Eq. (F.10). For integer $n$, one has:

$$\Gamma(n) = (n-1)!, \tag{F.16}$$

while one also has the following properties:

$$\Gamma\left(\frac{1}{2}\right) = \sqrt{\pi},$$
$$\Gamma(x)\Gamma(1-x) = \frac{\pi}{\sin \pi x} \tag{F.17}$$

## F.3 The beta function $B(x, y)$

It is defined as:

$$B(x, y) = \int_0^1 dt\, t^{x-1}(1-t)^{y-1} = \frac{\Gamma(x)\Gamma(y)}{\Gamma(x+y)}, \tag{F.18}$$

and has the useful properties:

$$B(x+1, y) = \left(\frac{x}{x+y}\right) B(x, y),$$
$$B(x, 1+y) = \left(\frac{y}{x+y}\right) B(x, y). \tag{F.19}$$

Therefore, one can deduce:

$$B(1+az, 1+bz) = \frac{1}{1+(a+b)z}\exp\left\{-z\gamma_E + \sum_{n=2}^{\infty}(-1)^n \frac{z^n}{n}\zeta(n)[a^n + b^n - (a+b)^n]\right\}. \tag{F.20}$$

In the limit $\epsilon \to 0$, it has the Taylor expansion:

$$\lim_{\epsilon\to 0} B(1-a\epsilon, 1-b\epsilon) = 1 - \epsilon(a+b) + \epsilon^2\left[(a+b)^2 - ab\frac{\pi^2}{6}\right] + \epsilon^3(a+b)[-(a+b)^2 + ab\zeta(2) + ab\zeta(3)] + \cdots,$$
$$\lim_{\epsilon\to 0} B\left(n - \frac{\epsilon}{2}, 1 - \frac{\epsilon}{2}\right) = \frac{1}{n}\left\{1 + \frac{\epsilon}{2}\left[\frac{2}{n} + \sum_{j=1}^{n-1}\frac{1}{j}\right]\right\} + \mathcal{O}(\epsilon^2),$$
$$\lim_{\epsilon\to 0} B\left(n - \frac{\epsilon}{2}, 2 - \frac{\epsilon}{2}\right) = \frac{1}{n(n+1)}\left\{1 - \frac{\epsilon}{2}\left[1 - \frac{2}{n} - \frac{2}{n+1} - \sum_{j=1}^{n-1}\frac{1}{j}\right]\right\} + \mathcal{O}(\epsilon^2), \tag{F.21}$$

### F.4 The incomplete beta function $B_a(x, y)$

It is defined as:

$$B_a(x, y) = \int_0^a dt\, t^{x-1}(1-t)^{y-1}\,. \tag{F.22}$$

Defining the function:

$$I_a(x, y) = \frac{B_a(x, y)}{B_1(x, y)}\,, \tag{F.23}$$

one has the properties:

$$\begin{aligned} &aI_a(x, y) - I_a(x+1, y) + (1-a)I_a(x+1, y-1) = 0\,, \\ &(x+y-xa)I_a(x, y) - yI_a(x, y+1) - x(1-a)I_a(x+1, y-1) = 0\,, \\ &yI_a(x, y+1) + xI_a(x+1, y) - (x+y)I_a(x, y) = 0\,. \end{aligned} \tag{F.24}$$

### F.5 The hypergeometric function ${}_2F_1(a, b, c; z)$

It is defined as:

$${}_2F_1(a, b, c; z) = \frac{\Gamma(c)}{\Gamma(b)\Gamma(c-b)} \int_0^1 dt\, t^{b-1}(1-t)^{c-b-1}(1-tz)^{-a} \tag{F.25}$$

for Re $c$ and Re $b > 0$, and arg(1 − z)$< \pi$. It has the properties:

$$\begin{aligned} {}_2F_1(a, b, c; z) &= 1 + \frac{ab}{1.c}z + \frac{a(a+1)b(b+1)}{1.2.c(c+1)}z^2 + \cdots \\ &= \frac{\Gamma(c)}{\Gamma(b)\Gamma(c-b)} \sum_{n=0}^{\infty} \frac{\Gamma(a+n)\Gamma(b+n)}{\Gamma(c+n)} \frac{z^n}{n!}\,. \end{aligned} \tag{F.26}$$

The hypergeometric function enters frequently in the calculation of multiloop Feynman integrals when the Gegenbauer polynomial techniques are used.

### F.6 One-loop massless integrals

The most useful integral is:

$$\begin{aligned} I(\alpha, \beta) &\equiv \int \frac{d^n k}{(2\pi)^n} \frac{1}{(k^2 + i\epsilon')^\alpha} \frac{1}{[(k-q)^2 + i\epsilon']^\beta} \\ &= \frac{i}{(16\pi^2)^{n/4}}(-1)^{-\alpha-\beta}(-q^2)^{-\alpha-\beta+n/2} \frac{\Gamma(\alpha+\beta-n/2)}{\Gamma(\alpha)\Gamma(\beta)} B\left(\frac{n}{2}-\beta, \frac{n}{2}-\alpha\right)\,. \end{aligned} \tag{F.27}$$

Combining this result with the one in Eq. (8.24), one can derive in $n = 4 - \epsilon$ dimensions:

$$I^\mu(\alpha, \beta) \equiv \int \frac{d^n k}{(2\pi)^n} \frac{k^\mu}{(k^2 + i\epsilon')^\alpha((k-q)^2 + i\epsilon')^\beta}$$

Table F.1. *Some values of $I(\alpha, \beta)$*

| $\alpha$ | $\beta$ | $I(\alpha,\beta)\nu^{\epsilon}\left(\frac{16\pi^2}{i}\right)(q^2)^{\alpha+\beta-2}$ |
|---|---|---|
| 1 | 1 | $\frac{2}{\tilde{\epsilon}}+2$ |
| 2 | 1 | $-\frac{2}{\tilde{\epsilon}}+0$ |
| 3 | 1 | $-1$ |
| 2 | 2 | $-\frac{4}{\tilde{\epsilon}}-2$ |
| 4 | 1 | $-1/3$ |
| 3 | 2 | $-\frac{4}{\tilde{\epsilon}}-5$ |

Table F.2. *Some values of $I^{\mu}(\alpha, \beta)$*

| $\alpha$ | $\beta$ | $I^{\mu}(\alpha,\beta)\nu^{\epsilon}\left(\frac{16\pi^2}{i}\right)(q^{\mu})^{-1}(q^2)^{\alpha+\beta-2}$ |
|---|---|---|
| 1 | 1 | $\frac{1}{\tilde{\epsilon}}+1$ |
| 2 | 1 | $+1$ |
| 1 | 2 | $-\frac{2}{\tilde{\epsilon}}-1$ |
| 3 | 1 | $-\frac{1}{\tilde{\epsilon}}-\frac{1}{2}$ |
| 2 | 2 | $-\frac{2}{\tilde{\epsilon}}-1$ |
| 1 | 3 | $\frac{1}{\tilde{\epsilon}}-\frac{1}{2}$ |

$$= \nu^{-\epsilon}\frac{i}{16\pi^2}\left(\frac{-q^2}{4\pi\nu^2}\right)^{-\epsilon/2}(q^2)^{2-\alpha-\beta}q^{\mu} \times\frac{\Gamma(3-\alpha-\epsilon/2)\Gamma(2-\beta-\epsilon/2)\Gamma(\alpha+\beta-2+\epsilon/2)}{\Gamma(\alpha)\Gamma(\beta)\Gamma(5-\alpha-\beta-\epsilon)}, \tag{F.28}$$

$$\begin{aligned} I^{\mu\nu}(\alpha,\beta) &\equiv \int\frac{d^nk}{(2\pi)^n}\frac{k^{\mu}k^{\nu}}{(k^2+i\epsilon')^{\alpha}(k-q)^2+i\epsilon')^{\beta}} \\ &= \nu^{-\epsilon}\frac{i}{16\pi^2}\left(\frac{-q^2}{4\pi\nu^2}\right)^{-\epsilon/2}(q^2)^{2-\alpha-\beta} \\ &\times\left\{g^{\mu\nu}q^2\frac{\Gamma(3-\alpha-\epsilon/2)\Gamma(3-\beta-\epsilon/2)\Gamma(\alpha+\beta+\epsilon/2)}{2\Gamma(\alpha)\Gamma(\beta)\Gamma(6-\alpha-\beta-\epsilon)}\right. \\ &\left.+q^{\mu}q^{\nu}\frac{\Gamma(4-\alpha-\epsilon/2)\Gamma(2-\beta-\epsilon/2)\Gamma(\alpha+\beta-2+\epsilon/2)}{2\Gamma(\alpha)\Gamma(\beta)\Gamma(6-\alpha-\beta-\epsilon)}\right\}. \end{aligned} \tag{F.29}$$

We give (Tables F.1–F.3 in values of the integrals for some values of $\alpha$ and $\beta$, where:

$$\frac{2}{\tilde{\epsilon}} \equiv \frac{2}{\epsilon} - \gamma_E - \ln\frac{-q^2}{4\pi\nu^2}. \tag{F.30}$$

Table F.3. *Some values of*
$I^{\mu\nu}(\alpha,\beta) \equiv \nu^{-\epsilon}\left(\frac{i}{16\pi^2}\right)(q^2)^{2-\alpha-\beta}[Aq^2 g^{\mu\nu} + Bq^\mu q^\nu]$

| $\alpha$ | $\beta$ | A | B |
|---|---|---|---|
| 1 | 1 | $-\frac{1}{6\bar{\epsilon}} - \frac{2}{9}$ | $\frac{2}{3\bar{\epsilon}} + \frac{13}{18}$ |
| 2 | 1 | $\frac{1}{2\bar{\epsilon}} + \frac{1}{2}$ | $\frac{1}{2}$ |
| 1 | 2 | $\frac{1}{2\bar{\epsilon}} + \frac{1}{2}$ | $-\frac{2}{\bar{\epsilon}} - \frac{3}{2}$ |
| 3 | 1 | $-\frac{1}{2\bar{\epsilon}} - \frac{1}{4}$ | $\frac{1}{2}$ |
| 2 | 2 | $\frac{1}{2}$ | $-\frac{2}{\bar{\epsilon}} - 2$ |
| 1 | 3 | $-\frac{1}{2\bar{\epsilon}} - \frac{1}{4}$ | $\frac{2}{\bar{\epsilon}} + \frac{1}{2}$ |

## F.7 Two- and three-loop massless integrals

Most of the integral encountered in the evaluation of loop diagrams can be reduced to the following integrals by means of the formula:

$$kq = \frac{1}{2}[k^2 + q^2 - (k-q)^2] \,. \tag{F.31}$$

These integral come from [2] and reads:

$$\begin{aligned}
I_2 &\equiv \left\{\int \frac{d^n k_1}{(2\pi)^n} \frac{1}{k_1^2(k_1-q)^2}\right\}^2 , \\
&= \frac{(-1)}{(16\pi^2)^{n/2}}(-1)^{-4}(-q^2)^{-4+n}\left\{\Gamma\left(2-\frac{n}{2}\right)B\left(\frac{n}{2}-1,\frac{n}{2}-1\right)\right\}^2 \\
&= \frac{(-1)}{(4\pi)^{4-\epsilon}}(-q^2)^{-\epsilon}4\left\{\frac{1}{\epsilon^2}+\frac{1}{\epsilon}(2-\gamma)+3-2\gamma+\frac{\gamma^2}{2}-\frac{\pi^2}{24}\right\} \quad \text{for } n = 4-\epsilon \,, \\
I_3 &\equiv \int \frac{d^n k_1}{(2\pi)^n}\int \frac{d^n k_2}{(2\pi)^n}\frac{1}{k_1^2(k_1-k_2)^2(k_1-q)^2} \\
&= \frac{(-1)}{(16\pi^2)^{n/2}}(-1)^{-3+n}(-q^2)^{-3+n}\Gamma(3-n)\,B\left(\frac{n}{2}-1,\frac{n}{2}-1\right)B\left(\frac{n}{2}-1,n-2\right) \\
&= \frac{(-1)^{-\epsilon}}{(4\pi)^{4-\epsilon}}(q^2)(-q^2)^{-\epsilon}\frac{1}{2}\left\{\frac{1}{\epsilon}+\frac{13}{4}-\gamma\right\} \quad \text{for } n = 4-\epsilon \,, \\
I_4 &\equiv \int \frac{d^n k_1}{(2\pi)^n}\int \frac{d^n k_2}{(2\pi)^n}\frac{1}{k_1^2(k_1-q)^2(k_1-k_2)^2(k_2-q)^2} \\
&= \frac{(-1)}{(16\pi^2)^{n/2}}(-1)^{-4+n}(-q^2)^{-4+n}\frac{\Gamma(2-n/2)\,\Gamma(4-n)}{\Gamma(3-n/2)} \\
&\quad \times B\left(\frac{n}{2}-1,\frac{n}{2}-1\right)B\left(\frac{n}{2}-1,n-3\right) \\
&= \frac{(-1)^{1-\epsilon}}{(4\pi)^{4-\epsilon}}(-q^2)^{-\epsilon}\left\{\frac{2}{\epsilon^2}+\frac{5-2\gamma}{\epsilon}+\frac{19}{2}-5\gamma\gamma^2-\frac{\pi^2}{12}\right\} \quad \text{for } n = 4-\epsilon \,, \\
I_5 &\equiv \int \frac{d^n k_1}{(2\pi)^n}\int \frac{d^n k_2}{(2\pi)^n}\frac{k_1^2}{k_1^2(k_1-q)^2(k_1-k_2)^2(k_2-q)^2}
\end{aligned}$$

$$= \frac{4}{(4\pi)^{4-\epsilon}}(-q^2)^{1-\epsilon}\frac{\Gamma(\epsilon)}{\epsilon}B\left(1-\frac{\epsilon}{2}, 2-\frac{\epsilon}{2}\right)B\left(2-\frac{\epsilon}{2}, 1-\epsilon\right)$$

$$= \frac{1}{(4\pi)^{4-\epsilon}}(-q^2)^{1-\epsilon}\left\{\frac{1}{\epsilon^2}+\frac{1}{\epsilon}\left(\frac{11}{4}-\gamma\right)\right.$$
$$\left.+\frac{1}{2}\left(\frac{89}{8}-\frac{11}{2}\gamma+\gamma^2-\frac{\pi^2}{12}\right)\right\} \quad \text{for } n = 4-\epsilon\,. \tag{F.32}$$

Some more complicated integrals entering in the three-loop calculation has been done analytically in [875] using Gegenbauer techniques. Within our notations and conventions, these integrals are generally of the form:

$$I(\alpha,\beta,\lambda,\delta) \equiv \int\frac{d^nk_1}{(2\pi)^n}\int\frac{d^nk_2}{(2\pi)^n}\frac{1}{k_1^{2\alpha}(k_1-q)^{2\beta}(k_1-k_2)^{2\gamma}(k_2-q)^{2\delta}}$$
$$= \frac{(-1)}{(4\pi)^{4-\epsilon}}(-q^2)^{-\epsilon}(q^2)^{4-\alpha-\beta-\gamma-\delta}\frac{\Gamma(\gamma+\delta-2+\epsilon/2)}{\Gamma(\gamma)\Gamma(\delta)}$$
$$\times\frac{\Gamma(\alpha+\beta+\gamma+\delta-4+\epsilon)}{\Gamma(\alpha)\Gamma(\beta+\gamma+\delta-2+\epsilon/2)}B\left(2-\gamma-\frac{\epsilon}{2}, 2-\delta-\frac{\epsilon}{2}\right)$$
$$\times B\left(2-\alpha-\frac{\epsilon}{2}, 4-\beta-\gamma-\delta-\epsilon\right) \quad \text{for } n=4-\epsilon\,. \tag{F.33}$$

One also has:

$$I_\lambda(\alpha,\beta) \equiv \int\frac{d^nk_1}{(2\pi)^n}\int\frac{d^nk_2}{(2\pi)^n}\frac{1}{k_1^{2\alpha}(k_1-q)^{2\beta}(k_1-k_2)^2k_2^2(k_2-q)^2}$$
$$= \frac{(-1)}{(16\pi^2)^{\lambda+1}}(-1)^{2\lambda-1-\alpha-\beta}(-q^2)^{2\lambda-1-\alpha-\beta}F_\lambda(\alpha,\beta) \tag{F.34}$$

for $\alpha, \beta \geq 1$ and where:

$$\lambda \equiv n/2-1\,,$$
$$F_1(1,1) = 6\sum_{k=o}^{\infty}\frac{1}{(k+1)^3} = 6\zeta(3)\,,$$
$$F_\lambda(\alpha>1,\beta>1) = \frac{\Gamma(1-2\lambda)\Gamma(\lambda-\alpha)\Gamma(\lambda-\beta)\Gamma(\lambda)\Gamma(\alpha+\beta-2\lambda)}{\Gamma(\alpha)\Gamma(\beta)\Gamma(3\lambda-\alpha-\beta)}$$
$$\times\left\{\frac{\Gamma(3\lambda-\alpha-\beta)}{\Gamma(1+\lambda-\alpha-\beta)}-\frac{\Gamma(\alpha+\beta-\lambda)}{\Gamma(\alpha+\beta+1-3\lambda)}+\frac{\Gamma(\alpha)}{\Gamma(1+\alpha-2\lambda)}\right.$$
$$\left.-\frac{\Gamma(2\lambda-\alpha)}{\Gamma(1-\alpha)}+\frac{\Gamma(\beta)}{\Gamma(1+\beta-2\lambda)}-\frac{\Gamma(2\lambda-\beta)}{\Gamma(1-\beta)}\right\}. \tag{F.35}$$

A final type of integral is:

$$I_\lambda(\alpha,\beta,\gamma) \equiv \int\frac{d^nk_1}{(2\pi)^n}\int\frac{d^nk_2}{(2\pi)^n}\frac{1}{k_1^{2\alpha}(k_1-q)^2(k_1-k_2)^{2\beta}k_2^{2\gamma}(k_2-q)^2}$$
$$= \frac{(-1)}{(4\pi^{4+\epsilon}}(q^2)^{2-\alpha-\beta-\gamma}2\lambda-1-\alpha-\beta F_\lambda(\alpha,\beta,\gamma)\,. \tag{F.36}$$

where:

$$F_\lambda(\alpha,\beta,\gamma) = \frac{\Gamma(\lambda+1-\alpha)\Gamma(\lambda+1-\beta)\Gamma(\lambda+1-\gamma)\Gamma(\lambda)}{\Gamma(\alpha)\Gamma(\beta)\Gamma(\gamma)\Gamma(2\lambda)}$$
$$\times \sum_{m,n=0}^{\infty} \frac{(-1)^m}{m!n!(n+\lambda)} \frac{\Gamma(n+2\lambda)\Gamma(m+n+\alpha+\beta+\gamma-2\lambda)}{\Gamma(1+3\lambda-m-\alpha-\beta-\gamma)\Gamma(m+n+\lambda+1)}$$
$$\times \left\{ \frac{1}{(n+\beta)(m+n+\alpha+\beta-\lambda)} + \frac{1}{(m+n+\alpha)(m+n+\alpha+\beta-\lambda)} \right.$$
$$\left. + \frac{1}{(m+n+\alpha)(n+2\lambda-\beta)} + (\alpha \leftrightarrow \gamma) \right\}, \qquad \text{(F.37)}$$

where the series is convergent for ($A \equiv \alpha+\beta+\gamma$):

$$A < 3\lambda+1\,, \qquad A < 2\lambda+2\,, \qquad A < \lambda+4\,. \qquad \text{(F.38)}$$

## F.8 One-loop massive integrals

$$I(\alpha,\beta,m^2) \equiv \int \frac{d^nk}{(2\pi)^n} \frac{1}{(k^2+i\epsilon')^\alpha} \frac{1}{[k^2-m^2+i\epsilon']^\beta}$$
$$= \nu^{-\epsilon} \frac{i}{16\pi^2} (-m^2)^{2-\alpha-\beta} \left(\frac{m^2}{4\pi\nu^2}\right)^{-\epsilon/2}$$
$$\times \frac{\Gamma(2-\alpha-\epsilon/2)\,\Gamma(\alpha+\beta-2+\epsilon/2)}{\Gamma(2-\epsilon/2)\Gamma(\beta)}$$

$$I(\alpha,\beta,q^2,m^2) \equiv \int \frac{d^nk}{(2\pi)^n} \frac{1}{[(k-q)^2-m^2+i\epsilon']^\alpha} \frac{1}{(k^2+i\epsilon')^\beta}$$
$$= \nu^{-\epsilon} \frac{i}{16\pi^2} (q^2)^{2-\alpha-\beta} \left(\frac{-q^2}{4\pi\nu^2}\right)^{-\epsilon/2}$$
$$\times \frac{\Gamma(2-\beta-\epsilon/2)\,\Gamma(\alpha+\beta-2+\epsilon/2)}{\Gamma(2-\epsilon/2)\Gamma(\alpha)} \left(1-\frac{m^2}{q^2}\right)^{2-\alpha-\beta-\epsilon/2}$$
$$\times {}_2F_1\left(\alpha+\beta-2+\epsilon/2, 2-\beta-\epsilon/2, 2-\epsilon/2; \frac{1}{1-\frac{m^2}{q^2}}\right)$$

$$I^\mu(\alpha,\beta,q^2,m^2) \equiv \int \frac{d^nk}{(2\pi)^n} \frac{k^\mu}{[(k-q)^2-m^2+i\epsilon']^\alpha} \frac{1}{(k^2+i\epsilon')^\beta}$$
$$= \nu^{-\epsilon} \frac{i}{16\pi^2} (q^2)^{2-\alpha-\beta} \left(\frac{-q^2}{4\pi\nu^2}\right)^{-\epsilon/2}$$
$$\times q^\mu \frac{\Gamma(2-\beta-\epsilon/2)\,\Gamma(\alpha+\beta-2+\epsilon/2)}{\Gamma(2-\epsilon/2)\Gamma(\alpha)} \left(1-\frac{m^2}{q^2}\right)^{2-\alpha-\beta-\epsilon/2}$$
$$\times {}_2F_1\left(\alpha+\beta-2+\epsilon/2, 3-\beta-\epsilon/2, 3-\epsilon/2; \frac{1}{1-\frac{m^2}{q^2}}\right) \qquad \text{(F.39)}$$

One also has:

$$\tilde{I}(\alpha,\beta,q^2,m^2) \equiv \int \frac{d^n k}{(2\pi)^n} \frac{1}{[(k-q)^2 - m^2 + i\epsilon']^\alpha} \frac{1}{(k^2 - m^2 + i\epsilon')^\beta}$$
$$= \nu^{-\epsilon} \frac{i}{16\pi^2} (q^2)^{2-\alpha-\beta} \left(\frac{-q^2}{4\pi\nu^2}\right)^{-\epsilon/2} \frac{\Gamma(\alpha+\beta-2+\epsilon/2)}{\Gamma(\alpha)\Gamma(\beta)}$$
$$\times \int_0^1 dx\, x^{\alpha-1}(1-x)^{\beta-1} \left[x(1-x) - \frac{m^2}{q^2}\right]^{2-\alpha-\beta-\epsilon/2}, \tag{F.40}$$

with:

$$\tilde{I}(\alpha,\beta,q^2,m^2) = \tilde{I}(\beta,\alpha,q^2,m^2). \tag{F.41}$$

For $\alpha+\beta > 2$, one can rewrite the $x$-integral by letting $\epsilon \to 0$. For some particular values of $\alpha$ and $\beta$, one has, by using the definition of $\tilde{\epsilon}$ in Eq. (F.30):

$$I(1,1,q^2,m^2) = \nu^{-\epsilon} \frac{i}{16\pi^2} \left\{ \frac{2}{\tilde{\epsilon}} - \frac{m^2}{q^2} \ln \frac{m^2}{-q^2} - \left(1 - \frac{m^2}{q^2}\right) \ln\left(1 - \frac{m^2}{q^2}\right) + 2 \right\},$$
$$I(1,2,q^2,m^2) = \nu^{-\epsilon} \frac{i}{16\pi^2} \frac{1}{q^2 - m^2} \left\{ -\frac{2}{\tilde{\epsilon}} - \frac{m^2}{q^2} \ln \frac{m^2}{-q^2} + \left(1 + \frac{m^2}{q^2}\right) \ln\left(1 - \frac{m^2}{q^2}\right) \right\},$$
$$I^\mu(1,1,q^2,m^2) = \nu^{-\epsilon} \frac{i}{16\pi^2} \frac{q^\mu}{2} \left\{ \frac{2}{\tilde{\epsilon}} - \frac{m^2}{q^2}\left(2 - \frac{m^2}{q^2}\right) \ln \frac{m^2}{q^2} - \left(1 - \frac{m^2}{q^2}\right)^2 \right.$$
$$\left. \times \ln\left(1 - \frac{m^2}{q^2}\right) - \frac{m^2}{q^2} + 2 \right\},$$
$$I^\mu(1,2,q^2,m^2) = \nu^{-\epsilon} \frac{i}{16\pi^2} \frac{q^\mu}{q^2} \left\{ -\frac{m^2}{q^2} \ln \frac{m^2}{-q^2} + \frac{m^2}{q^2} \ln\left(1 - \frac{m^2}{q^2}\right) + 1 \right\},$$
$$\tilde{I}(1,1,q^2,m^2) = \nu^{-\epsilon} \frac{i}{16\pi^2} \left\{ \frac{2}{\tilde{\epsilon}} - \sqrt{1 - 4m^2/q^2} \ln \frac{\sqrt{1-4m^2/q^2}+1}{\sqrt{1-4m^2/q^2}-1} + 2 \right\},$$
$$\tilde{I}(1,2,q^2,m^2) = \nu^{-\epsilon} \frac{i}{16\pi^2} \left\{ \frac{1}{q^2\sqrt{1-4m^2/q^2}} \ln \frac{\sqrt{1-4m^2/q^2}+1}{\sqrt{1-4m^2/q^2}-1} \right\}, \tag{F.42}$$

## F.9 A two-loop massive integral

$$I_\lambda(\alpha,\beta,m^2) \equiv \int \frac{d^n k_1}{(2\pi)^n} \int \frac{d^n k_2}{(2\pi)^n} \frac{1}{(k_1^2 + m^2)^\alpha (k_2^2 + m^2)^\beta (k_1 - k_2)^2}$$
$$= \frac{(-1)}{(4\pi)^{4+\epsilon}} (m^2)^{3-\alpha-\beta+\epsilon} \frac{B(\alpha-\lambda,\beta-\lambda)\Gamma(\alpha+\beta-2\lambda-1)}{(1+\epsilon/2)\Gamma(\alpha)\Gamma(\beta)}. \tag{F.43}$$

### F.10 The dilogarithm function

For complex $z$, it is defined as:

$$\mathrm{Li}_2(z) \equiv -\int_0^z \frac{dt}{t}\ln(1-t). \tag{F.44}$$

When $z$ is real and bigger than one, the log. is complex, and there is a branch cut from $z = 1$ to $\infty$. Therefore, the function develops an imaginary part:

$$\mathrm{Li}_2(x + i0) \to -i\pi \ln x\ . \tag{F.45}$$

For $z \leq 1$, one can write as:

$$\mathrm{Li}_2(z) = \sum_{n=1}^{\infty} \frac{z^n}{n^2}\ , \tag{F.46}$$

which is convenient for numerical calculations. In particular, its values are:

$$\begin{aligned} \mathrm{Li}_2(1) &= \frac{\pi^2}{6}\ , & \mathrm{Li}_2(-1) &= -\frac{\pi^2}{12}\ , \\ \mathrm{Li}_2\left(\frac{1}{2}\right) &= \frac{\pi^2}{12} - \frac{1}{2}\ln^2 2\ , & \mathrm{Li}_2(2 - i0) &= \frac{\pi^2}{4} - i\pi \ln 2\ . \end{aligned} \tag{F.47}$$

Some of its useful properties are:

$$\begin{aligned}
&\mathrm{Li}_2\,(x) + \mathrm{Li}_2(1-x) + \ln x \ln(1-x) = \frac{\pi^2}{6}\ , \\
&\mathrm{Li}_2\left(-\frac{1}{x}\right) + \mathrm{Li}_2(-x) + \frac{1}{2}\ln^2 x = -\frac{\pi^2}{6} \quad : x > 0\ , \\
&\mathrm{Li}_2\left(\frac{1}{x}\right) + \mathrm{Li}_2(x) + \frac{1}{2}\ln^2 x = \frac{\pi^2}{3} - i\pi \ln x \quad : x > 1\ , \\
&\mathrm{Li}_2(x) + \mathrm{Li}_2\left(\frac{x}{1-x}\right) = \frac{\pi^2}{2} - 2i\pi \ln x + i\pi \ln(x-1) - \frac{1}{2}\ln^2(1-x) \quad : x > 1\ , \\
&\mathrm{Li}_2(x) + \mathrm{Li}_2\left(-\frac{x}{1-x}\right) = -\frac{1}{2}\ln^2(1-x) \quad : x < 1\ , \\
&\mathrm{Li}_2\left(\frac{1}{1+x}\right) - \mathrm{Li}_2(-x) = \frac{\pi^2}{6} - \frac{1}{2}\ln(1+x)\ln\left(\frac{1+x}{x^2}\right)\ , \\
&\mathrm{Li}_2(1-x) - \mathrm{Li}_2\left(\frac{1}{x}\right) = -\frac{\pi^2}{6} + \frac{1}{2}\ln x \ln\frac{x}{(x-1)^2}\ , \\
&\mathrm{Li}_2(x) + \mathrm{Li}_2(-x) = \frac{1}{2}\mathrm{Li}_2(x^2)\ , \\
&\mathrm{Li}_2(z) = \frac{1}{1+z}\left[2(1-z)\ln(1-z) + z\left[3 + \sum_{n=1}^{\infty}\frac{z^n}{n^2(n+1)^2}\right]\right].
\end{aligned} \tag{F.48}$$

## F.11 Some useful logarithmic integrals

$$\int dx\, x^n \ln(ax-b) = \frac{1}{n+1}\left\{x^{n+1} - \left(\frac{b}{a}\right)^{n+1} \ln(ax-b) - \sum_{i=1}^{n+1}\left(\frac{b}{a}\right)^{n+1-i}\frac{x^i}{i}\right\}$$

$$\int_0^1 dx\, x^n \ln x = -\frac{1}{(n+1)^2} \tag{F.49}$$

Defining:

$$I_n = \int_0^1 dx\, x^n \ln[a - x(1-x)]\,, \tag{F.50}$$

one has:

$$\begin{aligned} I_0 &= -2 + \ln a + \sqrt{1-4a}\ln\frac{\sqrt{1-4a}+1}{\sqrt{1-4a}-1}\,,\\ I_1 &= \frac{1}{2}I_0\,,\\ I_2 &= \frac{1}{3}\left[-\frac{13}{6} + 2a + \ln a + (1-a)\sqrt{1-4a}\ln\frac{\sqrt{1-4a}+1}{\sqrt{1-4a}-1}\right],\\ I_3 &= \frac{1}{2}[I_0 - 3I_1 + 3I_2]\,, \end{aligned} \tag{F.51}$$

where one has exploited the invariance under the change $x \leftrightarrow (1-x)$ allowing to deduce the integrals odd in $n$ from the even ones. In addition to the properties of dilogarithm functions [876], one also needs [877,45,3]:

$$\begin{aligned} \int_0^x \frac{dt}{t}\ln(1+t^2) &= -\frac{1}{2}\mathrm{Li}_2(-x^2)\,,\\ \int_0^{x>0} \frac{dt}{t}\ln(1-t+t^2) &= -\frac{1}{3}\mathrm{Li}_2(-x^3) + \mathrm{Li}_2(-x)\,,\\ \int_0^{x>0} \frac{dt}{t}\ln(1+t+t^2) &= -\frac{1}{3}\mathrm{Li}_2(x^3) + \mathrm{Li}_2(x)\,,\\ \int_0^1 \frac{dt}{1+t}\ln(1+t)\ln^2 t &= -\frac{3}{2}\zeta^2(2) + 4S_4 + \frac{7}{2}\zeta(3)\ln 2 - \zeta(2)\ln^2 2 + \frac{1}{6}\ln^4 2\\ \int_0^1 \frac{dt}{1+t}\ln^2(1+t)\ln t &= -\frac{4}{5}\zeta^2(2) + 2S_4 + \frac{7}{4}\zeta(3)\ln 2 - \frac{1}{2}\zeta(2)\ln^2 2 + \frac{1}{12}\ln^4 2\,,\\ \int_0^1 \frac{dt}{t}\ln t\ln^2(1+t) &= \frac{7}{10}\zeta^2(2) = -\frac{1}{3}\int_0^1\frac{dt}{1+t}\ln^2 t\,,\\ \int_0^1 \frac{dt}{t}\ln(1+t)\mathrm{Li}_2(-t) &= \frac{1}{8}\zeta^2(2)\,,\\ \int_0^1 \frac{dt}{1+t}\mathrm{Li}_2(-t)\ln(1+t) &= -\frac{6}{5}\zeta^2(2) - 3S_4 - \frac{21}{8}\zeta(3)\ln 2 + \frac{1}{2}\zeta(2)\ln^2 2 - \frac{1}{8}\ln^4 2\,,\\ \int_0^1 \frac{dt}{1+t}\mathrm{Li}_2(-t)\ln t &= \frac{13}{8}\zeta^2(2) - 4S_4 - \frac{7}{2}\zeta(3)\ln 2 + \zeta(2)\ln^2 2 - \frac{1}{6}\ln^4 2\,, \end{aligned}$$

$$\int_0^1 \frac{dt}{1+t}\ln^2(1+t)\ln(1-t) = -\frac{4}{5}\zeta^2(2) + 2S_4 + 2\zeta(3)\ln 2 - \zeta(2)\ln^2 2 + \frac{1}{3}\ln^4 2\,, \tag{F.52}$$

with:

$$S_4 \equiv \sum_{n=1}^{\infty} \frac{1}{2^n n^4} = 0.571\,479\,061\,6\ldots \tag{F.53}$$

Some other useful integrals are:

$$\begin{aligned}\int_0^1 dt\, t^{\alpha-1}(1-t)^{\beta-1}\ln t &= \frac{\Gamma(\alpha)\Gamma(\beta)}{\Gamma(\alpha+\beta)}[S_1(\alpha-1) - S_1(\alpha+\beta-1)]\,,\\ \int_0^1 dt\, t^{\alpha-1}(1-t)^{\beta-1}\ln^2 t &= \frac{\Gamma(\alpha)\Gamma(\beta)}{\Gamma(\alpha+\beta)}[[S_1(\alpha-1) - S_1(\alpha+\beta-1)]^2\\ &\quad + S_2(\alpha+\beta-1) - S_2(\alpha-1)]\,,\end{aligned} \tag{F.54}$$

where:

$$S_l(n) \equiv \sum_{k=1}^{n} \frac{1}{k^l} \quad : l = 1, 2, 3, \ldots \tag{F.55}$$

Differentiating with respect to $\beta$, one gets:

$$\begin{aligned}\int_0^1 dt\, t^{\alpha-1}(1-t)^{\beta-1}\ln t\ln(1-t) &= \frac{\Gamma(\alpha)\Gamma(\beta)}{\Gamma(\alpha+\beta)}[S_2(\alpha+\beta-1) - \zeta(2)\\ &\quad + \left[S_1(\alpha-1) - S_1(\alpha+\beta-1)\right]\\ &\quad \times [S_1(\beta-1) - S_1(\alpha+\beta-1)]].\\ &\quad \times \int_0^1 dx\, \frac{1-x^\alpha}{1-x} = S_1(\alpha)\,.\end{aligned} \tag{F.56}$$

Integrals of the form:

$$\int_0^1 dt\, t^{\alpha-1}(1-t)^{\beta}\ln^m t, \tag{F.57}$$

for integer or half-integer values of $\alpha$ and $\beta$ are also useful and are given in [876]. For some particular values, one has:

$$\begin{aligned}\int_0^1 \frac{dt}{(1-t)^2}\ln^2 t &= -4\zeta(3) + \frac{\pi^2}{3} + 2\,,\\ \int_0^1 \frac{dt}{(1-t)^2} t^{3/2}\ln^2 t &= -21\zeta(3) + \pi^2 + 16,\end{aligned} \tag{F.58}$$

and:

$$\int_0^1 \frac{dt}{(1-t)} t^{\alpha-1}\ln^m t = (-1)^m m![\zeta(m+1) - S_{m+1}(\alpha-1)]\,. \tag{F.59}$$

## F.12 Further useful functions

The functions:

$$S_1(z) = \sum_{k=1}^{\infty} \frac{z}{k(z+k)}\,, \qquad S_l(z) = \zeta(l) - \sum_{k=1}^{\infty} \frac{1}{(z+k)^l}\,, \tag{F.60}$$

appear often in the evaluation of the parametric integrals. For $z = n$ integer, it can be reduced to the one in Eq. (F.55). They have the properties:

$$\begin{aligned} S_1(z+1) &= S_1(z) + \frac{1}{z+1}\,, \\ S_l(\infty) &= \zeta(l)\,, \\ \frac{dS_l(z)}{dz} &= l[\zeta(l+1) - S_{l+1}(z)], \end{aligned} \tag{F.61}$$

for $l = 1, 2, 3, \ldots$ They are related to the psi-function defined as:

$$\psi(z) \equiv \frac{d}{dz} \ln \Gamma(z)\,, \tag{F.62}$$

with the properties:

$$\begin{aligned} \psi(z) &= -\gamma_E - \frac{1}{z} + S_1(z)\,, \qquad \psi(1) = -\gamma_E\,, \\ \frac{d^l}{dz^l}\psi(z) &= l!(-1)^{l+1}[\zeta(l+1) - S_{l+1}(z-1)] \quad \text{for } l \geq 1, \end{aligned} \tag{F.63}$$

and therefore:

$$\psi'(1) = \zeta(2)\,, \quad \psi'(2) = \zeta(2) - 1\,, \quad \psi''(1) = -2\zeta(3)\,. \tag{F.64}$$

# Appendix G

## Useful formulae for the sum rules

### G.1 Laplace sum rule

Let the Laplace transform operator ($Q^2 \equiv -q^2 > 0$):

$$\mathcal{L} \equiv \lim_{\substack{n,\, Q^2 \to \infty \\ n/Q^2 \equiv \tau \text{ fixed}}} (-1)^n \frac{(Q^2)^n}{(n-1)!} \frac{\partial^n}{(\partial Q^2)^n} . \tag{G.1}$$

Then, one has the properties:

$$\begin{aligned}
\mathcal{L}\left[\frac{1}{(Q^2+m^2)^\alpha}\right] &= \frac{1}{\Gamma(\alpha)}\tau^\alpha e^{-m^2\tau} ,\\
\mathcal{L}\left[\frac{1}{(Q^2)^\alpha}\ln\frac{Q^2}{\nu^2}\right] &= \frac{1}{\Gamma(\alpha)}\tau^\alpha[-\ln\tau\nu^2+\psi(\alpha)] ,\\
\mathcal{L}\left[\frac{1}{(Q^2)^\alpha}\ln^2\frac{Q^2}{\nu^2}\right] &= \frac{1}{\Gamma(\alpha)}\tau^\alpha[\ln^2\tau\nu^2-2\psi(\alpha)\ln\tau\nu^2+\psi^2(\alpha)-\psi'(\alpha)] ,\\
\mathcal{L}\left[\frac{1}{(Q^2)^\alpha}\ln^3\frac{Q^2}{\nu^2}\right] &= \frac{1}{\Gamma(\alpha)}\tau^\alpha[-\ln^3\tau\nu^2+3\psi(\alpha)\ln^2\tau\nu^2\\
&\quad -(3\psi^2(\alpha)-\psi'(\alpha))\ln\tau\nu^2+\psi^3(\alpha)-3\psi(\alpha)\psi'(\alpha)+\psi''(\alpha)] ,\\
\mathcal{L}\left[\frac{1}{x^\alpha}\frac{1}{(\ln x)^\beta}\right] &= y\,\mu(y,\beta-1,\alpha-1),\\
&\underset{y\to 0}{\simeq} \frac{1}{\Gamma(\alpha)}y^\alpha\frac{1}{(-\ln y)^\beta}\left[1+(\beta)\psi(\alpha)\frac{1}{\ln y}+\mathcal{O}\left(\frac{1}{\ln^2 y}\right)\right] ,\\
\mathcal{L}\left[\frac{\ln\ln x}{x^\alpha(\ln x)^\beta}\right] &\underset{y\to 0}{\simeq} \frac{1}{\Gamma(\alpha)}y^\alpha\frac{\ln\ln y}{(-\ln y)^\beta}\left[1+\beta\psi(\alpha)\frac{1}{\ln y}+\mathcal{O}\left(\frac{1}{\ln^2 y}\right)\right] ,
\end{aligned} \tag{G.2}$$

where:

$$\begin{aligned}
\mu(y,\beta,\alpha) &= \int_0^\infty dx \frac{x^\beta}{\Gamma(\beta+1)}\frac{y^{\alpha+x}}{\Gamma(\alpha+x+1)} ,\\
\mu(y,-m,\alpha) &= (-1)^{m-1}\frac{d^{m-1}}{(dx)^{m-1}}\left(\frac{y^{\alpha-x}}{\Gamma(\alpha+x+1)}\right)_{x=0} \qquad m=1,2,\ldots ,
\end{aligned} \tag{G.3}$$

with the properties:

$$\mu(y,-1,\alpha)=\frac{y^\alpha}{\Gamma(\alpha+1)},$$
$$\mu(y,-2,\alpha)=\frac{y^\alpha}{\Gamma(\alpha+1)}[-\ln y+\psi(\alpha+1)],$$
$$\mu(y,-3,\alpha)=\frac{y^\alpha}{\Gamma(\alpha+1)}[\ln^2 y-2\psi(\alpha+1)\ln y+\psi^2(\alpha+1)-\psi'(\alpha+1)]. \quad (G.4)$$

For the treatment of the QCD continuum, we need the integral:

$$\int_0^{t_c} dt\, t^n\, e^{-t\tau}=(n-1)!\tau^{-n}\,(1-\rho_n)\,, \quad (G.5)$$

where:

$$\rho_n=e^{-t_c\tau}\left(1+t_c\tau+\cdots+\frac{(t_c\tau)^n}{n!}\right). \quad (G.6)$$

## G.2 Finite energy sum rule

For the FESR, the integral:

$$\int_0^{t_c} dt\, t^n \ln\frac{t}{\nu^2}\,, \quad (G.7)$$

induces the extra-term:

$$\frac{t_c^{n+1}}{n+1}\left(-\frac{1}{n}\right), \quad (G.8)$$

after a renormalization group improvement of the QCD series.

## G.3 Coordinate space integrals

In some applications, one works in the $x$-space instead of the usual momentum one. Using the Fourier transform:

$$f(x)=\int\frac{d^4q}{(2\pi)^4}\,e^{iqx}\,f(q)\,, \quad (G.9)$$

one has the correspondence ($Q^2\equiv -q^2>0$) for $x\to 0$ [394]:

## G.4 Cauchy contour integrals

We shall be concerned with the integral entering e.g. into the $\tau$-like decay processes (see Section 25.5), and which can be evaluated using the Cauchy contour integral along the circle of radius $M_\tau$.

$$I_{ij}=\oint_{|s|=M_\tau^2} dt(-t)^i\left(\ln\frac{\nu^2}{-t}\right)^j. \quad (G.10)$$

Results are given for some particular values of $i$ and $j$ [878]. $L\equiv\ln\nu^2/M_\tau^2$.

Table G.1. *Some useful Fourier transforms*

| $Q$-space | $x$-space |
|---|---|
| $Q^2 \ln Q^2$ | $\frac{8}{\pi^2}\frac{1}{x^6}$ |
| $\ln Q^2$ | $-\frac{1}{\pi^2}\frac{1}{x^4}$ |
| $\frac{1}{Q^2}$ | $\frac{1}{4\pi^2}\frac{1}{x^2}$ |
| $\frac{1}{Q^2}\ln Q^2$ | $-\frac{1}{4\pi^2}\frac{1}{x^2}\ln^2 x^2$ |
| $\frac{1}{Q^4}$ | $-\frac{1}{16\pi^2}\ln x^2$ |
| $\frac{1}{Q^4}\ln Q^2$ | $\frac{1}{64\pi^2}\ln^2 x^2$ |
| $\frac{1}{Q^6}$ | $\frac{1}{8\times 16\pi^2}x^2\ln x^2$ |
| $\frac{1}{Q^6}\ln Q^2$ | $-\frac{1}{496\pi^2}x^2\ln^2 x^2$ |
| $\frac{1}{Q^8}$ | $-\frac{1}{8\times 16\times 24\pi^2}x^4\ln x^2$ |
| $\frac{1}{Q^8}\ln Q^2$ | $\frac{1}{496\times 24\pi^2}x^4\ln^2 x^2$ |

Table G.2. *Some useful Cauchy integrals*

| $i$ | $j$ | $I_{ij}/2i\pi$ | $i$ | $j$ | $I_{ij}/2i\pi$ |
|---|---|---|---|---|---|
| $-3$ | 0 | 0 | $-2$ | 0 | 0 |
| | 1 | $-\frac{1}{2}$ | | 1 | 1 |
| | 2 | $\frac{1}{2}-L$ | | 2 | $-2+2L$ |
| | 3 | $-\frac{3}{4}+\frac{3}{2}L-\frac{3}{2}\pi L^2+\frac{1}{2}\pi^2$ | | 3 | $6-6L+3L^2-\pi^2$ |
| $-1$ | 0 | $-1$ | 0 | 0 | 0 |
| | 1 | $-L$ | | 1 | $-1$ |
| | 2 | $-L^2+\frac{\pi^2}{3}$ | | 2 | $-2-2L$ |
| | 3 | $-L^3+\pi^2 L$ | | 3 | $-6-6L-3L^2+\pi^2$ |
| 1 | 0 | 0 | 2 | 0 | 0 |
| | 1 | $\frac{1}{2}$ | | 1 | $-\frac{1}{3}$ |
| | 2 | $\frac{1}{2}+L$ | | 2 | $-\frac{2}{9}-\frac{2}{3}L$ |
| | 3 | $\frac{3}{4}+\frac{3}{2}L+\frac{3}{2}L^2-\frac{\pi^2}{2}$ | | 3 | $-\frac{2}{9}-\frac{2}{3}L-L^2+\frac{\pi^2}{3}$ |
| 3 | 0 | 0 | | | |
| | 1 | $\frac{1}{4}$ | | | |
| | 2 | $\frac{1}{8}+\frac{1}{2}L$ | | | |
| | 3 | $\frac{3}{32}+\frac{3}{8}L+\frac{3}{4}L^2-\frac{\pi^2}{4}$ | | | |

# Bibliography


### I General introduction

*Introductory references*

1. M.A. Shifman, A.I. Vainshtein and V.I. Zakharov, (hereafter referred as SVZ), *Nucl. Phys.* **B 147** (1979) 385, 448.
2. S. Narison, *Phys. Rep.* **84** (1982) 263.
3. S. Narison, QCD spectral sum rules, *World Sci. Lect. Notes Phys.* **26** (1989) 1.
4. S. Narison, *Role and Present Status of Particle Physics* (in French), Montpellier preprint PM 96/20 (1996), Talks given at the Malagasy Academy of Sciences and Universities.
5. Scientific information service, CERN, Geneva.
6. J. Ellis, talk given at HEP-MAD'01 International Conference, 27 Sept–5 Oct. 2001, Antananarivo (Madagascar).

*The pre-QCD era*

*Quark models*

7. M. Gell-Mann, *Phys. Rev.* **125** (1961) 1067;
   Y. Ne'eman, *Nucl. Phys.* **26** (1961) 222;
   M. Gell-Mann and Y. Ne'eman, *The Eightfold Way*, W.A. Benjamin, New York, (1964).
8. M. Gell-Mann, *Phys. Lett.* **8** (1964) 214.
9. G. Zweig, *Cern-reports* TH-401, TH-412 (1964). In: *Developments in Quark Theory of Hadrons*, a reprint collection, vol. 1, 1964/1978, eds. D.B. Lichtenberg and S.P. Rosen, Hadronic Press, Nonamtum, MA, (1980).
10. S. Sakata *Prog. Theor. Phys.* **16** (1956) 636.
11. M. Gell-Mann, *The Eightfoldway* (1961), printed in [7]; *Phys. Rev.* **125** (1962) 1067;
    S. Okubo, *Prog. Theor. Phys.* **27** (1962) 949.
12. For a review see e.g. J.M. Richard, *Phys. Rep.* **212** (1992) 1.

*Current algebra*

13. For reviews on current algebra, see e.g. S.L. Adler and R.F. Dashen, *Current Algebra and Applications to Particle Physics*, Benjamin, New York, (1968);

H. Pagels, *Phys. Rep.* **16** (1975) 221;
V.D. Alfaro, S. Fubini, G. Furlan and C. Rosseti, *Currents in Hadron Physics*, North Holland, Amsterdam, (1973);
E. Reya, *Rev. Mod. Phys.* **46** (1974) 545.
14. Y. Nambu and J.J. Sakurai, *Phys. Rev. Lett.* **8** (1962) 79;
J.J. Sakurai, *Current and Mesons*, University Chicago Press, Chicago, IL, (1969).
15. N. Cabibbo, *Phys. Rev. Lett.* **10** (1963) 531.
16. PDG96, R.M. Barnett *et al.*, *Phys. Rev.* **D 54** (1996) 1; PDG97, L. Montanet *et al.*, *Phys. Rev.* **D 54** (1997) 1;
PDG98, C. Caso *et al.*, *Eur. Phys. J.* **C 3** (1998) 1;
PDG2000, D.E. Groom *et al.*, *Eur. Phys. J.* **C 15** (2000) 1.

*PCAC*

17. Y. Nambu, *Phys. Rev.* **117** (1960) 648;
Y. Nambu and G. Iona-Lasinio, *Phys. Rev.* **122** (1961) 345;
J. Goldstone, *Nuov. Cimento.* **19** (1961) 154.
18. E. Wigner, *Group Theory*, Princeton University Press, Princeton, NJ, (1939);
H. Weyl, *The Classical Group*, Princeton University Press, Princeton, NJ, (1939).
19. See e.g., T. Das *et al.*, *Phys. Rev. Lett.* **18** (1967) 759.
20. J.C. Ward, *Phys. Rev.* **78** (1950) 1824.
21. M. Gell-Mann, R.J. Oakes and B. Renner, *Phys. Rev.* **175** (1968) 2195.
22. M. L. Goldberger and S. B. Treiman, *Phys. Rev.* **110** (1958) 1178.
23. S.L. Adler, *Phys. Rev.* (1964) **139**, B1638.
24. S.L. Adler, *Phys. Rev.* **140** (1965) B736;
W.L. Weisberger, *Phys. Rev.* **143** (1966) 1302.

*Weinberg and DMO sum rules*

25. K. Kawarabayashi and M. Suzuki, *Phys. Rev. Lett.* **16** (1966) 255;
N. Ryazuddin and Fayazuddin, *Phys. Rev.* **147** (1966) 1071.
26. S. Weinberg, *Phys. Rev. Lett.* **18** (1967) 507.
27. T. Das, V.S. Mathur and S. Okubo, *Phys. Rev. Lett.* **19** (1967) 470.
28. E.G. Floratos, S. Narison and E. de Rafael, *Nucl. Phys.* **B 155** (1979) 155.
29. S. Narison, *Z. Phys.* **C 14** (1982) 263.
30. R.D Peccei and J. Sola, *Nucl. Phys.* **B 281** (1987) 1.
31. S. Narison and E. de Rafael, *Nucl. Phys.* **B 169** (1980) 253.
32. S. Narison, *Z. Phys.* **C 22** (1984) 161.
33. The ALEPH collaboration, R. Barate *et al.*, *Eur. Phys. J.* **C 4** (1998) 409;
The OPAL collaboration, K. Ackerstaff *et al.*, *Eur. Phys. J.* **C 7** (1999) 571.
34. S. Narison, *Nucl. Phys. Proc. Suppl.* **96** (2001) 364 [HEP-PH 0012019].

*Parton models*

35. R.P. Feynman, *Photon-hadron Interactions*, Benjamin, Reading, MA, (1972).
36. J.D. Bjorken, *Phys. Rev.* **179** (1969) 1547.
37. C.J. Callan and D.J. Gross, *Phys. Rev. Lett.* **22** (1969) 156.
38. G. Altarelli and G. Parisi, *Nucl. Phys.* **B 126** (1977) 298.

*S-matrix*

39. T. Regge, *Nuovo Cim.* **14** (1959) 951;
A. Bottino, A.M. Longoni, and T. Regge, *Nuovo Cim.* **23** (1962) 954.
40. G. Veneziano, *Nuovo Cim.* **57 A** (1968) 190.
41. H. Harari, *Phys. Rev. Lett.* **22** (1969) 562;
J.L. Rosner, *Phys. Rev. Lett.* **22** (1969) 689;
T. Matsuoka, K. Ninomiga and S. Savada, Nagoya preprint 1969.

***The QCD story***

*For QCD books, see:*

42. H. Fritzsch, *Quarks*. Allen Lane Penquin Books Ltd (1983).
43. T. Muta, *Foundations of Quantum Chromodynamics*. Lecture Notes in Physics-Vol. 5, World Scientific Publishing Co., Singapore, (1987).
44. R.K. Ellis, W.J. Stirling and B.R. Weber, *QCD and Collider Physics*. Cambridge University Press, Cambridge, (1996).
45. P. Pascual and R. Tarrach, *QCD: Renormalization for Pratitioners*. Springer, Berlin, (1984).
46. F.J. Yndurain, *The Theory of Quark and Gluon Interactions*, Texts and Monographs in Physics, Springer-Verlag, New York, (1999).

*For QCD reviews, see:*

47. H.D. Politzer, *Phys. Rep.* **14 C** (1974) 129;
W. Marciano and H. Pagels, *Phys. Rep.* **36 C** (1978) 137;
E. de Rafael, *Quantum Chromodynamics*. Gif summer school institute (1980);
A. Peterman, *Phys. Rep.* **53 C** (1979) 157;
J. Ellis and C. Sachrajda, *QCD and its Applications*. Quarks and Leptons, Cargese, ed. M. Levy *et al.*, Plenum Press, New York, (1979);
Yu. L. Dokshitzer, D. L. Dyakonov and S.I. Troyan, *Phys. Rep.* **58 C** (1980) 269;
C.H. Llewellyn Smith, Topics in QCD, In quantum flavourdynamics, quantum chromodynamics and unified theories (1980), ed. K.T. Mahanthappa and J. Randa, Plenum Press, New-York;
A.H. Mueller, *Phys. Rep.* **73** (1981) 237;
E. Reya, *Phys. Rep.* **69 C** (1981) 195;
F. Wilczek, *Ann. Rev. Nucl. Part. Sci.* **32** (1982) 177;
F.E. Close, *Phys. Scripta* **25** (1982) 86.
48. G. Altarelli, *Phys. Rep.* **81** (1982) 1.
49. A.J. Buras, *Rev. Mod. Phys.* **52** (1980) 199
50. A. Pich, *Aspects of Quantum Chromodynamics*, ICTP summer school in particle physics, Trieste (1999).
51. H.G. Dosch, Non-perturbative methods in QCD Prog. *Part. Nucl. Phys.* **33** (1994) 121.
52. M. Jacob, *HEP International Conference*, Singapore (1990).

*For QFT books, see:*

53. J.D. Bjorken and S.D. Drell, *Relativistic Quantum Fields*. McGraw-Hill, New York, (1964);

V. Berestetski, E. Lifchitz and L. Pitayevki, *Relativistic Quantum Theory*, Part 1, Editions Mir, Moscow, (1972);
E.M. Lifshitz and L.P. Pitaevskii, *Relativistic Quantum Theory*, Part 2, Pergamon Press, New York, (1971);
E. de Rafael, *Lectures on Quantum ElectroDynamics*. Univ. Autonoma di Barcelona, UAB-FT-D1 (1977);
N.N. Bogoliubov and D.V. Shirkov, *Introduction to Theories of Quantized Fields*, 3rd edition. Wiley-Interscience, New York, (1980);
C. Itzykson and B. Zuber *Introduction to Quantum Field Theory*. McGraw-Hill, New York, (1980);
T-P. Cheng and L-F. Li, *Gauge Theory of Elementary Particle Physics*. Oxford University Press, Oxford, (1984);
G. Sterman, *Introduction to Quantum Field Theory*. Cambridge University Press, Cambridge, (1993);
M.E. Peskin and D.V. Schroeder, *Introduction to Quantum Field Theory*. Addison Wesley, (1995);
S. Weinberg, *The Quantum Theory of Fields*, Vol.I and II, Cambridge University Press, Cambridge, (1996).

*Review on light quark masses*

54. S. Narison, *Nucl. Phys. (Proc. Suppl.)* **B 86** (2000) 242 [HEP-PH 9911454];
S. Narison, hep-ph/0202200.

*Quark masses from current algebras*

55. S. Weinberg, *Festschrift for I.I. Rabi*, ed. L. Uotz, Academic Sciences, New York. (1977) p. 185.
56. P. Langacker and H. Pagels, *Phys. Rev.* **D 19** (1979) 2070;
P. Langacker, *Phys. Rev.* **D 20** (1979) 2983;
J. Gasser, *Ann. Phys. (N.Y.)* **136** (1981) 62;
J. Gasser and H. Leutwyler, *Phys. Reports* **87** (1982) 79.
57. For a recent review, see: H. Leutwyler, BERN preprint, BUTP-94/8 (1994) and hep-ph/0011049.

*c,b and t Quarks*

58. J.D. Bjorken and S.L. Glashow, *Phys. Lett.* **11** (1964) 255;
Z. Maki, *Prog. Theor. Phys* **31** (1964) 331, 333;
Y. Hara, *Phys. Rev.* **B 134** (1964) 701.
59. J.J. Aubert *et al.*, *Phys. Rev. Lett.* **33** (1974) 1404;
J.E. Augustin *et al.*, *Phys. Rev. Lett.* **33** (1974) 1406.
60. S.L. Glashow, J. Illiopoulos and L. Maiani, *Phys. Rev.* **D 2** (1970) 1285.
61. S.L. Glashow, *Nucl. Phys.* **22** (1961) 579;
S. Weinberg, *Phys. Rev. Lett.* **19** (1967) 1264;
A. Salam, *Elementary Particle Physics*, ed. N. Svartholm, Almqvist and Wiksells, Stockholm, (1968), p. 367.
62. M.L. Perl *et al.*, *Phys. Rev. Lett.* **35** (1975) 1489.
63. S.W. Herb *et al.*, *Phys. Rev. Lett.* **39** (1977) 252.
64. F. Abe *et al.*, *Phys. Rev. Lett.* **74** (1995) 2626;
S. Abachi *et al.*, *Phys. Rev. Lett.* **74** (1995) 2632.

65. M. Kobayashi and K. Maskawa, *Prog. Theor. Phys.* **49** (1973) 652.
66. H. Harari, *Phys. Lett.* **B 57** (1975) 265; *proc. Summer Inst. on Particle Phys.*, SLAC report 191 (1975) 159.
67. M. Kobayashi and K. Maskawa, *Prog. Theor. Phys.* **49** (1972) 282.
68. LEP Electroweak Working Group, www.cern.ch/lepewwg/; CERN-EP/2000-016.
69. M. Gell-Mann, *Acta Phys. Aust. Suppl.* **9** (1972) 733;
H. Fritzsch and M. Gell-Mann, *Proc of the XVI Intern. Conf. on High Energy Phys*, Chicago **2** (1972) 135;
H. Fritzsch, M. Gell-Mann and H. Leutwyler, *Phys. Lett.* **B 47** (1973) 365;
M. Han and Y. Nambu, *Phys. Rev.* **139** (1965) 1006.
70. O.W. Greenberg, *Phys. Rev. Lett.* **13** (1964) 598.

*QED anomaly*

71. J. Steinberger, *Phys. Rev.* **76** (1949) 1180;
S.L. Adler, *Phys. Rev.* **177** (1969) 2426;
J.S. Bell and R. Jackiw, *Nuov. Cim.* **19** (1961) 154.
72. S.L. Adler and W.A. Bardeen, *Phys. Rev.* **182** (1969) 156.

*The four-quark states*

73. R.L. Jaffe, *Phys. Rev* **D 15** (1977) 267;
D. Wong and K.F. Liu, *Phys. Rev.* **D 21** (1980) 2039;
J. Weinstein and N. Isgur, *Phys. Rev.* **D 27** (1983) 588.

*Asymptotic freedom*

74. G. 't Hooft, unpublished remarks at the Aix-en Provence (Marseille) Meeting (1972).
75. E.C.G. Stueckelberg and A. Peterman, *Helv. Phs. Acta* **26** (1953) 499.
76. M. Gell-Mann and F.E. Low, *Phys. Rev.* **95** (1954) 1300.
77. H.D. Politzer, *Phys. Rev. Lett.* **30** (1973) 157;
D.J. Gross and F. Wilczek, *Phys. Rev. Lett.* **30** (1973) 1343.
78. G. 't Hooft, *Nucl. Phys (Proc. Suppl.)* **74** (1999) 413;
D.J. Gross, *Nucl. Phys (Proc. Suppl.)* **74** (1999) 426.
79. L.D. Landau, *Niels Bohr and the Development of Physics*, Pergamon Press, New York, (1955), p. 52.
80. S. Coleman and D.J. Gross, *Phys. Rev. Lett.* **31** (1973) 851;
A. Zee, *Phys. Rev.* **D 7** (1973) 3630.

*Potential models*

81. A. De Rujula, H. Georgi and S.L. Glashow, *Phys. Rev.* **D 12** (1975) 147.
82. E. Eichten, K. Gottfried, T. Kinoshita, K.D. Lane and T.M. Yan, *Phys. Rev.* **D 17** (1978) 3090; **D 21** (1980) 203.
83. J.L. Richardson, *Phys. Lett.* **B 82** (1979) 272.
84. H. Grosse and A. Martin, *Phys. Rep.* **60** (1979) 341.
85. A. Martin, *Phys. Lett.* **B 100** (1981) 511.
86. J.M. Richard, *QCD 94 Conference*, Montpellier, *Nucl. Phys. Proc. Suppl.* **B 39** (1995) 441 and references therein.
87. W. Buchmüller, *Erice lectures* (1984) and references therein.

88. K. Zalewski, *QCD 96 Conference*, Montpellier, *Nucl. Phys. Proc. Suppl.* **B, A 54** (1997).
89. C. Quigg and J.L. Rosner, *Phys. Rep.* **56** (1979) 167.
90. H. Leutwyler,*Phys. Lett.* **B 98** (1981) 447;
M.B. Voloshin, *Nucl. Phys.* **B 187** (1981) 365.
91. J.S. Bell and R.A. Bertlmann, *Nucl. Phys.* **B 177** (1981) 218.
92. R.A. Bertlmann, *Nucl. Phys.* **B 204** (1982) 387.
93. J.S. Bell and R.A. Bertlmann, *Nucl. Phys.* **B 227** (1983) 435;
R.A. Bertlmann, *Acta. Phys. Austriaca.* **53** (1981) 305; *QCD 90 Conference*, Montpellier, *Nucl. Phys. (Proc. Suppl.)* **B 23** (1991) and references therein.
94. S. Titard and F.J. Yndurain, *Phys. Rev.* **D 49** (1994) 6007; **D 51** (1995) 6348;
A. Pineda and F.J. Yndurain, *Phys. Rev.* **D 58** (1998) 094022; **D 61** (2000) 077505.

### *Field Theory*

95. G.C. Wick, *Phys. Rev.* **80** (1950) 268.
96. R.P. Feynman, *The Principle of Least Action in Quantum Mechanics*, Thesis, Princeton University, (1942).
97. L.D. Faddeev and Y.N Popov, *Phys. Lett.* **B 25** (1967) 29;
B. De Wit, *Phys. Rev. Lett.* **12** (1964) 742.
98. G.'t. Hooft, *Nucl. Phys.* **B 35** (1971) 167.

## II QCD gauge theory

### *Lagrangian and quantization*

99. V. Fock, *Z. Phys.* **39** (1927) 226.
100. H. Weyl, *Z. Phys.* **56** (1929) 330.
101. C.N. Yang and R.L. Mills, *Phys. Rev.* **96** (1954) 191.
102. P. Higgs, *Phys. Lett.* **12** (1964) 132; *Phys. Rev. Lett.* **13** (1964) 508; *Phys. Rev.* **145** (1964) 1156;
F. Englert and R. Brout, *Phys. Rev. Lett.* **13** (1964) 321;
G.S. Guralnik, C.R. Hagen and T.W. Kibble, *Phys. Rev. Lett.* **13** (1964) 585.
103. C. Becchi, A. Rouet and R. Stora, *Phys. Lett.* **B 52** (1974) 344; *Commun. Math. Phys.* **42** (1975) 127;
I.V. Tyutin, Lebedev Institute preprint N39 (1975).
104. A.A. Slanov, *Sov. Journ. Part. Nucl.* **5** (1975) 303;
J.C. Taylor, *Nucl. Phys.* **B 33** (1971) 436.
105. A.A. Belavin *et al.*, *Phys. Lett.* **B 59** (1975) 15;
G.'t. Hooft, *Phys. Rev. Lett.* **37** (1976) 8; *Phys. Rev.* **D 14** (1976) 3432.

## III $\overline{MS}$ scheme for QCD and QED

### *The $\overline{MS}$ scheme*

106. R.P. Feynman, *Rev. Mod. Phys.* **20** (1948) 367; *Phys. Rev.* **74** (1948) 939, 1430; **76** (1949) 749, 769; **80** (1950) 440;

J. Schwinger, *Phys. Rev.* **73** (1948) 146; **74** (1948) 1439; **75** (1949) 651; **76** (1949) 790; **82** (1951) 664, 914; **91** (1953) 713; *Proc. Nat. Acad. Sci. USA.* **37** (1951) 452;
S. Tomonaga, *Prog. Theor. Phys.* **1** (1946) 27;
Z. Koba, T. Tati and S. Tomonoga, *ibid* **2** (1947) 101;
F.J. Dyson, *Phys. Rev.* **75** (1949) 486, 1736.

107. N. Pauli and F. Villars, *Rev. Mod. Phys.* **21** (1949) 434.
108. G.'t. Hooft and M. Veltman, *Nucl. Phys.* **B 44** (1972) 189.
109. C.G. Bollini and J.J. Giambiagi, *Nuov. Cimento* **B 12** (1972) 20; J. Ashmore, *Nuov. Cim. Lett.* **4** (1972) 289.
110. E. de Rafael, *Gif Summer Institute* (1980).
111. S. Narison, *Phys. Rep.* **84** (1982) 263.
112. G. Leibbrandt, *Rev. Mod. Phys.* **47** (1975) 849; G.'t. Hooft and M. Veltmann, *Diagrammar CERN Yellow report* (1973).
113. G.'t. Hooft, *Nucl. Phys.* **B 33** (1971) 173.
114. E.R. Speer, *Generalized Feynman Amplitudes*, Princeton University Press, Princeton, NJ, (1969).
115. M. Creutz, *Quarks, Gluons and Lattices*, Cambridge University Press, Cambridge (1983).
116. D.A. Akyeampong and R. Delbourgo, *Nuov. Cim.* **19 A** (1974) 219; **17 A** (1973) 578.
117. P. Breitenlohner and D. Maison, *Comm. Math. Phys.* **52** (1977) 39.
118. T.L. Trueman, *Phys. Lett.* **B 88** (1979) 331; I. Antoniadis, *Phys. Lett.* **B 88** (1979) 223.
119. G. Bonneau, *Phys. Lett.* **B 96** (1980) 147.
120. W. Seigel, *Phys. Lett.* **B 84** (1979) 193; B 94 (1980) 37.
121. For reviews on supersymmetry, see e.g.: P. Fayet and S, Ferrara, *Phys. Rep.* **C 32** (1977) 250;
A. and J. Strathdee, *Fortschr. Phys.* **26** (1978) 57;
H.P. Nilles, *Phys. Rep.* (1982):
For an encyclopedia, see e.g. S. Duplij, http:// www-home.univer.kharkov.ua/duplij.
122. R. Delbourgo and V.B. Prasad, *J. Phys.* **G 1** (1975) 377.
123. G.'t. Hooft, *Nucl. Phys.* **B 61** (1973) 455.
124. W.A. Bardeen, A.J. Buras, D.W. Duke and T. Muta, *Phys. Rev.* **D 18** (1978) 3998.
125. B.W. Lee and J. Zinn-Justin, *Phys. Rev.* **D 5** (1972) 3121, 3137; H. Kluberg-Stern and J.B. Zuber, *Phys. Rev.* **D 12** (1975) 467, 482.

### ***Renormalizations of operators***

126. B.S. DeWitt, *Phys. Rev.* **162** (1967) 1195, 1239;
G.'t. Hooft, *Nucl. Phys.* **B 62** (1973) 444;
H. Kluberg-Stern and J.B. Zuber, *Phys. Rev.* **D 12** (1975) 3159.
127. L.F. Abbott, *Nucl. Phys.* **B 185** (1981) 189.
128. R. Tarrach, *Nucl. Phys.* **B 196** (1982) 45.
129. D. Espriu and R. Tarrach, *Z. Phys.* **C 16**, (1982) 77.
130. S. Narison and R. Tarrach, *Phys. Lett.* **B 125** (1983) 217.
131. M. Jamin and M. Kremer, *Nucl. Phys.* **B 277** (1986) 349;
V.P. Spiridonov and K.G. Chetyrkin, *Sov. J. Nucl. Phys.* **47** (1988) 522;
L.-E. Adam and K.G. Chetyrkin, Univ. Karlsruhe preprint TTP93-26 (1993).

### The renormalization group

132. C.G. Callan, *Phys. Rev.* **D 2** (1970) 1541;
K. Symanzik, *Commun. Math. Phys.* **18** (1970) 227.
133. R. Tarrach, *Nucl. Phys.* **B 183** (1981) 384.
134. W. Caswell, *Phys. Rev. Lett.* **33** (1974) 244;
D.R.T. Jones, *Nucl. Phys.* **B 75** (1974) 531.
135. O.V. Tarasov, Dubna preprint JINR P2-82-900 (1982);
O.V. Tarasov, A.A. Vladimirov and A.Yu. Zharkov, *Phys. Lett.* **B 93** (1980) 429.
136. T. van Ritbergen, J.A.M. Vermaseren and S.A. Larin, hep-ph/9701390 (1997); hep-ph/9703284 (1997).
137. S. Weinberg, *Phys. Rev.* **118** (1960) 838.
138. A.J. Buras *et al.*, *Nucl. Phys.* **B 131** (1977) 308.
139. S. Bethke, *QCD 96 Conference*, Montpellier, *Nucl. Phys. (Proc. Suppl.)* **B, A 54** (1997); *J. Phys.* **G 26** (2000) R27 [hep-ex/0004021]; *QCD 02 Conference*, Montpellier (to appear) and private communication.
140. T. Appelquist and J. Carazzone, *Phys. Rev.* **D 11** (1975) 2856.
141. P. Binetruy and T. Sucker, *Nucl. Phys.* **B 178** (1981) 293.
142. S. Weinberg, *Phys. Lett.* **B 91** (1980) 51.
143. B. Ovrut and H. Schnitzer, Brandeis Univ. preprint (1980).
144. W. Bernreuther and W. Wetzel, *Nucl. Phys.* **B 197** (1982) 228;
W. Bernreuther, *Ann. Phys.* **151** (1983) 127; talk given at the *QCD-LEP meeting on* $\alpha_s$ (published by S. Bethke and W. Bernreuther as Aachen preprint PITHA 94/33) and private communication.
145. G. Rodrigo and A. Santamaria, *Phys. Lett.* **B 313** (1993) 441;
A. Peterman, CERN-TH.6487/92 (1992) (unpublished).
146. K.G. Chetyrkin, B.A. Kniehl and M. Steinhauser, *Nucl. Phys.* **B 510** (1998) 61.

### Heavy quark pole masses

147. R. Coquereaux, *Annals of Physics* **125** (1980) 401.
148. S. Narison, *Phys. Lett.* **B 197** (1987) 405; **B 216** (1989) 191.
149. S. Narison, *Phys. Lett.* **B 341** (1994) 73.
150. S. Narison, *Phys. Lett.* **B 520** (2001) 115.
151. N. Gray, D.J. Broadhurst, W. Grafe and K. Schilcher, *Z. Phys.* **C 48** (1990) 673;
N. Gray, Open University Thesis *OUT*- 4102-35 (1991) (unpublished).
152. M. Steinhauser and K. Chetyrkin, *Nucl. Phys.* **B 573** (2000) 617;
K. Melnikov and T. van Ritbergen, hep-ph/9912391.
153. S.A. Larin, T. Van Ritbergen and J.A.M. Vermaseren, *Nucl. Phys.* **B 438** (1995) 278.
154. P. Ball, M. Beneke and V. Braun, *Nucl. Phys.* **B 452** (1995) 563;
M. Beneke, *Phys. Rep.* **317** (1999) 1 and references therein;
M. Beneke and V. Braun, hep-ph/0010208 and references therein.
155. M.B. Voloshin, *Int. J. Mod. Phys.* **A 10** (1995) 2865; *Sov. J. Nucl. Phys.* **29** (1979) 703;
M.B. Voloshin and M. Zaitsev, *Sov. Phys. Usp* **30** (1987) 553.
156. M. Jamin and A. Pich, *Nucl. Phys.* **B 507** (1997) 334.
157. A.A. Penin and A.A. Pivovarov, *Nucl. Phys.* **B 549** (1999) 217;
M. Beneke, *Nucl. Phys. (Proc. Suppl.)* **B 86** (2000) 547 and references therein.
K. Melnikov and A. Yelkhovsky, *Phys. Rev.* **D 59** (1999) 114009.

158. M. Beneke, *Phys. Lett.* **B 434** (1998) 115.
159. A. Hoang, *Nucl. Phys. (Proc. Suppl.)* **B 86** (2000) 512 and references therein.
160. I.I. Bigi, M.A. Shifman and N.G. Uraltsev, *Ann. Rev. Nucl. part. Sci.* **47** (1997) 591.
161. K.G. Chetyrkin, S. Narison and V.I. Zakharov, *Nucl. Phys.* **B 550** (1999) 353.
162. F.V. Gubarev, M.I. Polikarpov and V.I. Zakharov, *Nucl. Phys. (Proc. Suppl.)* **B 86** (2000) 457;
V.I. Zakharov, *Nucl. Phys. (Proc. Suppl.)* **B 74** (1999) 392;
R. Akhoury and V.I. Zakharov, *Nucl. Phys. (Proc. Suppl.)* **A 54** (1997) 217.
163. V. Zakharov, private communication.
164. For a review, see e.g.: M. Neubert, *Phys. Rep.* **245** (1994) 259; *Int. J. Mod. Phys.* **A 11** (1996) 4173; hep-ph/9702375 (1997) and references therein;
T. Mannel, *QCD 94 Conference*, Montpellier, *Nucl. Phys. Proc. Suppl.* **B, C 39** (1995) 426.
165. S. Narison, *Phys. Lett.* **B 352** (1995) 122.
166. V. Eletsky and E. Shuryak, *Phys. Lett.* **B 276** (1992) 191;
P. Ball and V.M. Braun, *Phys. Lett.* **B 389** (1996); *Phys. Rev.* **D 49** (1994) 2472;
M. Neubert, *Phys. Lett.* **B 389** (1996) 727; **322** (1994) 419.

*Light quark pseudoscalar channel*

167. C. Becchi, S. Narison, E. de Rafael and F.J. Ynduràin, *Z. Phys.* **C 8** (1981) 335.

***Other renormalization schemes***

168. H. Georgi and D. Politzer, *Phys. Rev.* **D 9** (1974) 416.
169. R. Barbieri, L. Caneschi, G. Gurci and E. d'Emilio, *Phys. Lett.* **B 81** (1979) 207.
170. W. Celmaster and R.J. Gonsalves, *Phys. Rev. Lett.* **42** (1979) 1435; *Phys. Rev.* **D 20** (1979) 1420.
171. S. Weinberg, *Phys. Rev.* **D 8** (1973) 3497.
172. P. Pascual and R. Tarrach, *Nucl. Phys.* **B 174** (1980) 123.
173. S.J. Brodsky, G.P. Lepage and P.B. Mackenzie, *Phys. Rev.* **D 28** (1983) 228.
174. W. Celmaster and P.M. Stevenson, *Phys. Lett.* **B 125** (1983) 493.
175. G. Grunberg and A.L. Kataev, *Phys. Lett.* **B 279** (1992) 352.
176. P.M. Stevenson, *Phys. Rev.* **D 23** (1981) 2916.
177. G. Grunberg, *Phys. Lett.* **B 95** (1980) 70; *Phys. Rev.* **D 29** (1984) 2315.
178. A.L. Kataev and V.V. Starshenko, *QCD 94 Conference*, Montpellier, *Nucl. Phys. (Proc. Suppl.)* **B 39** (1995) 312; *Mod. Phys. Lett.* **A 10** (1995) 235.

***$\overline{MS}$ scheme for QED and high-precision tests of QED***

179. R. Jost and J.M. Luttinger, *Hel. Phys. Acta.* **23** (1950) 201;
E. de Rafael and J. Rosner, *Ann. Phys.* **82** (1974) 369.
180. S. Narison, *Phys. Lett.* **B 167** (1986) 214.
181. R.S. Van Dyck Jr., P.B. Schwinberg and H.G. Dehmelt, *Phys. Rev. Lett.* **59** (1987) 26.
182. P.J. Mohr and B.N. Taylor, *Rev. Mod. Phys.* **72** (2000) 351.
183. S. Laporta and S. Remiddi, *Acta. Phys. Pol.* **B 28** (1997) 959;
A. Czarnecki and W.J. Marciano, *Nucl. Phys. (Proc. Suppl.)* **B 76** (1999) 245.

184. V.W. Hughes and T. Kinoshita, *Rev. Mod. Phys.* **71, 2** (1999) S133.
185. T. Kinoshita, *Rep. Prog. Phys.* **59** (1996) 1459.
186. A. Czarnecki and W.J. Marciano, hep-ph/0102122.
187. C. Bouchiat and L. Michel, *J. Phys. Radium* **22** (1961) 121.
188. L. Durand III, *Phys. Rev.* **128** (1962) 441; *erratum* **129** (1963) 2835.
189. T. Kinoshita and R. Oakes, *Phys. Lett.* **B 25** (1967) 143.
190. J.E. Bowcock, *Z. Phys.* **211** (1968) 400.
191. M. Gourdin and E. de Rafael, *Nucl. Phys.* **B 10** (1969) 667.
192. B. Lautrup and E. de Rafael, *Phys. Rev.* **174** (1968) 1835.
193. The ALEPH collaboration, R. Barate *et al.*, *Eur. Phys. J.* **C 76** (1997) 15; **C 4** (1998) 409; A. Hocker, hep-ex/9703004.
194. S. Dolinsky *et al.*, *Phys. Rep.* **C 202** (1991) 99.
195. (ADH98) R. Alemany, M. Davier and A. Höcker, *Eur. Phys. J.* **C 2** (1998) 123.
196. The DM2 collaboration, A. Antonelli *et al.*, *Z. Phys.* **C 56** (1992) 15; D. Bisello *et al.*, *Z. Phys.* **C 39** (1988) 13.
197. The DM1 collaboration, F. Mane *et al.*, *Phys. Lett.* **B 112** (1982) 178; A. Cordier *et al.*, *Phys. Lett.* **B 110** (1982) 335.
198. S. Eidelman and F. Jegerlehner, *Z. Phys.* **C 67** (1995) 585.
199. The OPAL collaboration, K. Ackerstaff *et al.*, *Eur. Phys. J.* **C 7** (1999) 571.
200. F.J. Gilman and S.H. Rhie, *Phys. Rev.* **D 31** (1985) 1066.
201. S. Narison, *Phys. Lett.* **B 513** (2001) 53, Erratum-ibid **526** (2002) 414. [HEP-PH 0103199].
202. S. Narison, hep-ph/0108065;
S. Narison, hep-ph/0203053, talk given at 1st Madagascar International High-Energy Physics conference, HEP-MAD'01, Antananarivo (Madagascar).
203. M. Davies *et al.*, hep-ph/0208177; K. Illagiwara *et al.*, hep-ph/0209187.
204. The CMD collaboration, R.R. Akhmetshin *et al.*, *Nucl. Phys.* **A 675** (2000) 424c;
S.I. Serednyakov, *Nucl. Phys. (Proc. Suppl.* **B 96** (2001) 197;
The BES collaboration, J. Z. Bai *et al.*, *Phys. Rev. Lett.* **84** (2000) 594; hep-ex/0102003.
205. (DH98) M. Davier and D. Höcker, *Phys. Lett.* **B 435** (1998) 427;
(YT01) J.F. De Troconiz and F.J. Yndurain, hep-ph/0106025, hep-ph/0111258, hep-ph/0107318.
206. E821 Muon g-2 collaboration, D.W. Hertzog, hep-ex/0202024.
207. H.N. Brown *et al.*, hep-ex/0009029; and *Phys. Rev. Lett.* **86** (2000) 2227.
208. G.W. Bennett *et al.*, *Phys. Rev. Lett.* **89** (2002) 1001804; C.S. Ogben, hep-ex/0211044.
209. J. Calmet, S. Narison, M. Perrottet and E. de Rafael, *Phys. Lett.* **B 61** (1976) 283; *Rev. Mod. Phys.* **49** (1977) 21.
210. B. Krause, *Phys. Lett.* **B 390** (1997) 392.
211. M. Hayakawa and T. Kinoshita, *Phys. Rev.* **D 57** (1998) 465;
M. Hayakawa and T. Kinoshita and A.I. Sanda, *Phys. Rev.* **D 54** (1996) 3137.
212. J. Bijnens, E. Pallante and J. Prades, *Nucl. Phys.* **B 474** (1996) 379;
E. de Rafael, *Phys. Lett.* **B 322** (1994) 239.
213. M. Knecht and A. Nyffeler, hep-ph/0111058;
M. Knecht, A. Nyffeler, M. Perrottet and E. de Rafael, hep-ph/0111059;
M. Hayakawa and T. Kinoshita, hep-ph/0112102;
I. Blokland, A. Czarnecki and K. Melnikov, hep-ph/0112117;
J. Bijnens, E. Pallante and J. Prades, hep-ph/0112255.

214. T. Kinoshita, B. Nizić and Y. Okamoto, *Phys. Rev.* **D 31** (1985) 2108;
S. Laporta and E. Remiddi, *Phys. Lett.* **B 301** (1993) 440.
215. S. Narison, *J. Phys.* **G 4** (1978) 1849.
216. J. Bailey *et al.*, *Nucl. Phys.* **B 150** (1979) 1.
217. L. Taylor, *Nucl. Phys. (Proc. Suppl.)* **B 76** (1999) 273.
218. (J 01) F. Jegerlehner, hep-ph/0104304 and references therein; W.J. Marciano and B.L. Roberts, hep-ph/0105056.
219. (MOR 01) A.D. Martin, J. Outhwaite and M.G. Ryskin, hep-ph/0012231; (BP 01) H. Burkhardt and B. Pietrzyk, LAPP-EXP 2001–03 (2001).
220. (CEK 01) A.Czarnecki, S. Eidelman and S.G. Karshenbiom, hep-ph/0107327 and references therein.
221. (FKM 99) R.N. Faustov, A. Karimkhodzhaev and A.P. Martynenko, *Phys. Rev.* **A 59** (1999) 2498.

## IV Deep inelastic scatterings...

### *The operator product expansion and unpolarized DIS*

222. K.G. Wilson, *Phys. Rev.* **179** (1969) 1499; **D 10** (1974) 2445.
223. W. Zimmermann, *Ann. Phys.* **77** (1973) 570;
R.A. Brandt, *Phys. Rev. Lett.* **23** (1969) 1260.
224. H. Fritzsch and M. Gell-Mann, *Proc. Coral Gables Conf. on Fund. Int. at High Energy, Gordon and Breach* **Vol 2** (1971) 1, eds. M.D. Clin *et al.*
225. D. Gross and S.B. Treiman, *Phys. Rev.* **D 4** (1971) 1059.

### *Unpolarized DIS processes*

226. N. Christ, B. Hasslacher and A.H. Mueller, *Phys. Rev.* **D 6** (1972) 3543.
227. E.C. Titchmarsh, *Theory of Functions*, Oxford University Press, Oxford, (1939).
228. E.G. Floratos, D. Ross and C.T. Sachrajda, *Nucl. Phys.* **B 129** (1977) 66; E. **B 139** (1978) 545.
229. O. Nachtmann, *Nucl. Phys.* **B 63** (1973) 237.
230. A. De Rujula, H. Georgi and D. Politzer, *Ann. Phys.* **103** (1977) 387; *Phys. Rev.* **D 15** (1977) 2495.
231. A. Gonzalez-Arroyo, C. López and F.J. Yndurain, *Nucl. Phys.* **B 153** (1979) 161.
232. W. Furmanski and R. Petronzio, *Phys. Lett.* **B 97** (1980) 437.
233. W.A. Bardeen *et al.*, *Phys. Rev.* **D 18** (1978) 3998.
234. D.J. Gross and F. Wilczek, *Phys. Rev.* **D 9** (1974) 980.
235. J. Sanchez-Guillen *et al.*, *Nucl. Phys.* **B 353** (1991) 337.
236. W.L. van Neerven and E.B. Zijlstra, *Phys. Lett.* **B 272** (1991) 127, 476.
237. S.A. Larin and J.A.M. Vermaseren, *Z. Phys.* **C 57** (1993) 93.
238. S.A. Larin *et al.*, *Nucl. Phys.* **B 492** (1997) 338.
239. G. Altarelli and G. Parisi, *Nucl. Phys.* **B 126** (1977) 298.
240. V.N. Gribov and L.N. Lipatov, *Sov. J. Nucl. Phys.* **15** (1972) 438, 675.
241. D. Amati *et al.*, *Nucl. Phys.* **B 173** (1980) 429;
M. Ciafaloni and G. Curci, *Phys. Lett.* **B 102** (1981) 352;
G. Sterman, *Nucl. Phys.* **B 281** (1987) 310;
S. Catani and L. Trentadue, *Nucl. Phys.* **B 327** (1989) 323.

242. A. Donnachie, H.G. Dosch, P. Landshoff and O. Nachtmann, Cambridge University Press Monograph series (to appear).
243. A.A. Migdal, A.M. Polyakov and K.A. Ter-Martirosian, *Phys. Lett.* **B 48** (1974) 239; H.D. Abarbanel and J.B. Bronzan, *Phys. Rev.* **D 8** (1974) 2397; L.N. Lipatov, *Sov. J. Nucl. Phys.* **23** (1976) 338.
244. E.A. Kuraev, L.N. Lipatov and V.S. Fadin, *Sov. Phys. JETP* **44** (1976) 443; Y.Y. Balitskii and L.N. Lipatov, *Sov. J. Nucl. Phys.* **28** (1978) 822; M. Ciafaloni, *Nucl. Phys.* **B 296** (1988) 49; S. Catani, S. Fiorini and M. Ciafaloni, *Nucl. Phys.* **B 336** (1990) 18.
245. K. Adel, F. Barreiro and F.J. Yndurain, *Nucl. Phys.* **B 495** (1997) 221.
246. A. Bodek *et al.*, *Phys. Rev.* **D 20** (1979) 1471.
247. S.L. Adler, *Phys. Rev.* **143** (1966) 1144.
248. D.J. Gross and C.H. Llewellyn Smith, *Nucl. Phys.* **B 14** (1969) 337.
249. J. Chyla and A.L. Kataev, *Phys. Lett.* **B 297** (1992) 385; S.A. Larin and J.A.M. Vermaseren, *Phys. Lett.* **B 259** (1991) 345.
250. J. Santiago and F.J. Yndurain, *Nucl. Phys.* **B 563** (1999) 45; S.I. Alekhin, *Phys. Rev.* **D 59** (1999) 114016.
251. M. Virchaux and A. Milsztajn, *Phys. Lett.* **B 274** (1992) 221.
252. S. Albino *et al.* CERN-EP/2002–029.


### *Polarized DIS processes and the proton spin crisis*


253. J. Ashman *et al.*, *Phys. Lett.* **B 206** (1988) 364; *Nucl. Phys.* **B 328** (1990) 1; G. Baum *et al.*, *Phys. Rev. Lett.* **51** (1983) 1135.
254. S. Okubo, *Phys. Lett.* **5** (1963) 5; J. Izuki, K. Okada and D. Shito, *Prog. Theor. Phys.* **35** (1966) 1061.
255. G. Veneziano, Lecture at the Okubofest, May 1990, CERN-TH.5840/90; Mod. Phys. Lett. A4 (1989) 1605.
256. G.M. Shore and G. Veneziano, *Phys. Lett.* **B 244** (1990) 75; *Nucl. Phys.* **B 381** (1992) 23.
257. G.M. Shore, *Nucl. Phys. (Proc. Suppl.)* **B, C 39** (1995) 303; **B, A 54** (1997) 122; **B 64** (1998) 167.
258. G. Altarelli, *Lectures at the International School of Subnuclear Physics, Erice*; G. Altarelli *et al.*, hep-ph/9803237; R.G. Roberts, *The Structure of the Proton*, Cambridge University Press, Cambridge, (1990); J. Kodaira and K. Tanaka, *Nucl. Phys. (Proc. Suppl.)* **B 86** (1999) 134.
259. J. Ellis and R.L. Jaffe, *Phys. Rev.* **D 9** (1974) 1444, **D 10** (1974) 1669.
260. S. Narison, G.M. Shore and G. Veneziano, *Nucl. Phys.* **B 433** (1995) 209.
261. S. Narison, G.M. Shore and G. Veneziano, *Nucl. Phys.* **B 546** (1999) 235.
262. E. Witten, *Nucl. Phys.* **B 156** (1979) 269.
263. G. Veneziano, *Nucl. Phys.* **B 159** (1979) 213.
264. G.'t Hooft, *Phys. Rev. Lett.* **37** (1976) 8; *Phys. Rev.* **D 14** (1976) 3432; Erratum **D 18** (1978) 2199.
265. S. Narison, *Phys. Lett.* **B 255** (1991) 101.
266. G. Briganti, A. Di Giacomo and H. Panagopoulos, *Phys. Lett.* **B 253** (1991) 427; A. Di Giacomo, *QCD 90 Conference*, Montpellier, *Nucl. Phys. (Proc. Suppl.)* **B, 23** (1991) 303.

267. D. de Florian, G.M. Shore and G. Veneziano, hep-ph/9711353.
268. L.Trentadue and G. Veneziano, *Phys. Lett.* **B 323** (1994) 201;
L. Trentadue, *QCD 96 Conference*, Montpellier, *Nucl. Phys. (Proc. Suppl.)* **B, A 54** (1997) 176.


### *Drell–Yan processes*


269. S.D. Drell and T.M. Yan, *Ann. Phys.* **66** (1971) 578.
270. G. Altarelli, R.K. Ellis and G. Martinelli, *Nucl. Phys.* **B 143** (1978) 521; (E) **B 146** (1978) 544; **B 157** (1979) 461;
J. Kubar-Andrè and F.E. Paige, *Phys. Rev.* **D 19** (1979) 221;
B. Humpert and W.L. van Neerven, *Nucl. Phys.* **B 184** (1981) 225;
K. Harada and T. Muta, *Phys. Rev.* **D 22** (1980) 663.
271. W.L. van Neerven and E.B. Zijlstra, *Nucl. Phys.* **B 382** (1992) 11.
272. G. Parisi, *Phys. Lett.* **B 90** (1980) 295;
G. Curci and M. Greco, *Phys. Lett.* **B 92** (1980) 175.


### *Prompt photon processes*


273. P. Aurenche *et al.*, hep-ph/9811382.
274. P. Aurenche *et al.*, *Nucl. Phys.* **B 297** (1988) 661;
L.E. Gordon and W. Vogelsang, *Phys. Rev.* **D 50** (1994) 1901;
F. Aversa *et al.*, *Nucl. Phys.* **B 327** (1989) 104.
275. UA6 collaboration, M. Werlen *et al.*, *Phys. Lett.* **B 452** (1999) 201.


## V Hard processes in $e^+e^-$

### *Book on* $e + e-$


276. F.M. Renard, *Basics of Electron Positron Collisions* (1981) Editions Frontières, Gif-sur-Yvette, France.


### *Fragmentation functions*


277. P. Chiappetta, hep-ph/9301254 and references therein.
278. ALEPH collaboration, D. Buskulic *et al.*, *Phys. Lett.* **B 357** (1995) 487; **E B 364** (1995) 247.
279. DELPHI collaboration, P. Abreu *et al.*, *Phys. Lett.* **B 398** (1997) 194.


### $\gamma\gamma$ *Processes*


280. S.J. Brodsky, T. Kinoshita and H. Terezawa, *Phys. Rev. Lett.* **27** (1971) 280;
H. Terezawa, *Rev. Mod. Phys.* **45** (1973) 615.
281. S, Narison, G. Shore and G. Veneziano, *Nucl. Phys.* **B 391** (1993) 69.
282. E. Witten, *Nucl. Phys.* **B 120** (1977) 189.

283. W.A. Bardeen and A.J. Buras, *Phys. Rev.* **D 20** (1979) 166.
284. G.M. Shore, hep-ph/0111165.

### *QCD jets*

285. F. Bloch and A. Nordsieck, *Phys. Rev.* **52** (1937) 54.
286. T. Kinoshita, *J. Math. Phys.* **3** (1962) 650;
T.D. Lee and M. Nauenberg, *Phys. Rev.* **133** (1964) **B 1549**.
287. G. Sterman and S. Weinberg, *Phys. Rev. Lett.* **39** (1977) 1436.
288. P. Binetruy and G. Girardi, *Phys. Lett.* **B 83** (1979) 382.
289. E. Farhi, *Phys. Rev. Lett.* **39** (1977) 1587.
290. H. Georgi and M. Machacek, *Phys. Rev. Lett.* **39** (1977) 1237.
291. A. De Rujula *et al.*, *Nucl. Phys.* **B 138** (1978) 387.
292. G.C. Fox and S. Wolfram, *Nucl. Phys.* **B 149** (1978) 387.
293. C.L. Basham *et al.*, *Phys. Rev. Lett.* **41** (1979) 413.
294. A. Ali and F. Barriero, *Phys. Lett.* **B 118** (1982) 155;
S.D. Ellis, D. Richards and J.W. Stirling, *Phys. Lett.* **B 119** (1982) 193.
295. Jade collaboration, W. Bartel *et al.*, *Z. Phys.* **C 33** (1986) 23; *Phys. Lett.* **B 213** (1988) 235.
296. Yu. Dokshitzer, *J. Phys.* **G17** (1991) 1572;
S. Bethke *et al.*, *Nucl. Phys.* **B 370** (1992) 310.
297. S. Catani *et al.*, *Phys. Lett.* **B 263** (1991) 491; **B 269** (1991) 432; **B 272** (1991) 368;
N. Brown and J. Stirling, *Z. Phys.* **C 53** (1992) 629.
298. ALEPH collaboration, CERN-EP/2002–029 (2002).
299. DELPHI collaboration, K. Hamacher *et al.*,
P. Abreu *et al.*, *Phys. Lett.* **B 449** (1999) 383.
300. ZEUS collaboration, M. Derrick *et al.*, *Phys. Lett.* **B 363** (1995) 201;
H1 collaboration, C. Adolff *et al.*, *Eur. Phys. J.* **6 C** (1999) 575 and references therein.
301. J. Huth *et al.*, *Research Directions for the Decade*, ed. E.L. Berger, World Scientific, Singapore, (1992).
302. CDF collaboration, F. Abe *et al.*, *Phys. Rev. Lett.* **68** (1992) 1104; ibid. **88** (2002) 042001.
303. M. Cacciari, *QCD'02 International Conference*, Montpellier, (July 2002).

### ***Total inclusive hadron productions***

#### *Heavy quarkonia decays*

304. G. Veneziano (private communication).
305. T. Appelquist and H.D. Politzer, *Phys. Rev. Lett.* **34** (1975) 43;
A. De Rújula and S.L. Glashow, *Phys. Rev. Lett.* **34** (1975) 46.
306. R. Barbieri *et al.*, *Phys. Lett.* **B 57** (1975) 455.
307. P.B. Mackenzie and G.P. Lepage, *Phys. Rev. Lett.* **47** (1981) 1244.
308. P.B. Mackenzie and G.P. Lepage, In *Proc. Conf. Pert, QCD*, Tallahassee, (1981).
309. R. Barbieri *et al.*, *Nucl. Phys.* **B 154** (1979) 535.

310. N.V. Krasnikov and A.A. Pivovarov, *Phys. Lett.* **B 116** (1982) 168;
M.R. Pennington, R.G. Roberts and G.G. Ross, *Nucl. Phys.* **B 242** (1984) 69.
311. M. Kobel *et al.*, *Z. Phys.* **C 53** (1992) 193.
312. M. Jamin and A. Pich, *Nucl. Phys.* **B 507** (1997) 334;
A. Penin and A.A. Pivovarov, *Phys. Lett.* **B 435** (1998) 413;
A.H. Hoang, *Phys. Rev.* **D 61** (2000) 034005.
313. S. Narison *Phys. Lett.* **B 387** (1996) 162;
S. Narison, *Nucl. Phys. Proc. Suppl.* **A 54** (1997) 238 [HEP-PH 9609258].
314. A.X. El-Khadra *et al.*, *Phys. Rev. Lett.* **69** (1992) 729.
315. C. Davies *et al.*, *Phys. Rev.* **D 56** (1997) 2755.
316. A. Spitz *et al.*, *Phys. Rev.* **D 60** (1999) 074502.

*$e + e-$ into hadrons*

317. K.G. Chetyrkin, A.L. Kataev and F.V. Tkachov, *Phys. Lett.* **B 85** (1979) 277.
318. G. Källen and A. Sabry, *Dan. Mat. Phys. Medd.* **29** (1955) 17.
319. J. Schwinger, *Phys. Rev.* **82** (1951) 664; *Particles, Sources and Fields*, Vol. II, Addison Wesley, Reading, MA, (1973) p. 407.
320. M. Dine and J. Sapirstein, *Phys. Rev. Lett.* **43** (1979) 668;
W. Celsmaster and R. Gonsalves, *Phys. Rev. Lett.* **44** (1980) 560.
321. S.G. Gorishny, A.L. Kataev and S.A. Larin, *Phys. Lett.* **B 259** (1991) 144; **B 212** (1988) 238;
See also: L.R. Surguladze and M.A. Samuel, *Phys. Rev. Lett.* **66** (1991) 560 and Erratum 2416.
322. D. Haidt, In *Directions in High Energy Physics*, Vol. 14, ed. P. Langacker, World Scientific, Singapore, (1995).

*Tau-decays and QCD perturbative series*

323. A. Pich, *QCD 94 Conference*, Montpellier, *Nucl. Phys. Proc. Suppl.* **B, C 39** (1995) 303; hep-ph/9701305 (1997) and references therein.
324. E. Braaten, *Phys. Rev. Lett.* **60** (1988) 1606; *Phys. Rev.* **D 39** (1989) 1458; *Proc. τCf Workshop*, Stanford (1989), ed. L.V. Beers;
S. Narison and A. Pich, *Phys. Lett.* **B 211** (1988) 183;
A. Pich, *Proc. τCf Workshop*, Stanford (1989), ed. L.V. Beers; *Proc. Workshop on τ-physics*, Orsay (1990), eds. M. Davier and B. Jean-Marie (Ed. Frontières).
325. E. Braaten, S. Narison and A. Pich, *Nucl. Phys.* **B 373** (1992) 581.
326. W. Marciano and A. Sirlin, *Phys. Rev, Lett.* **61** (1988) 1815; **56** (1986) 22.
327. E. Braaten and C.S. Li, *Phys. Rev.* **D 42** (1990) 3888.
328. ALEPH collaboration, R. Barate *et al.*, *Phys. Lett.* **B 307** (1993) 209;
CLEOII collaboration : T. Coan *et al.*, *Phys. Lett.* **B 356** (1995) 580;
R. Stroynowski, *τ94 Workshop*, Montreux, *Nucl. Phys. (Proc. Suppl.)* **B 40** (1995) 569 ed. L. Rolandi;
L. Duflot, *QCD 94 Conference*, Montpellier, *Nucl. Phys. Proc. Suppl.* **B 39** (1995) 322 and references therein; *τ94 Workshop*, Montreux, *Nucl. Phys. (Proc. Suppl.)* **B 40** (1995) 37 ed. L. Rolandi;
A. Höcker, *τ96 Workshop*, Colorado (1996).

### Renormalons 1


329. S. Narison, *Phys. Lett.* **B 361** (1995) 121.
330. D.J. Broadhurst, *Z. Phys.* **C 58** (1993) 179;
D.J. Broadhurst and A.L. Kataev, *Phys. Lett,* **B 315** (1993) 179.
331. M. Beneke and V.I. Zakharov, *Phys. Rev. Lett.* **69** (1992) 2472;
M. Beneke, V. Braun and V.I. Zakharov, *Phys. Rev. Lett.* **73** (1994) 3058;
G. Grunberg,*QCD 94 Conference*, Montpellier, *Nucl. Phys. (Proc. Suppl.)* **B, C 39** (1995) 303;
C.N. Lovett-Turner and C.J. Maxwell, *Nucl. Phys.* **B 452** (1995) 188; **B 432** (1994) 147;
L.S. Brown and L.G. Yaffe, *Phys. Rev.* **D 45** (1992) 398;
M. Neubert, *Nucl. Phys.* **B 463** (1996) 511;
S. Groote, J.G. Körner and A.A. Pivovarov, hep-ph/9703268 (1997).
332. F.le Diberder, *QCD 94 Conference*, Montpellier, *Nucl. Phys. (Proc. Suppl.)* **B 39** (1995) 318.
333. F. Le Diberder and A. Pich, *Phys. Lett.* **B 286** (1992) 147 and **B 289** (1992) 165.
334. J. Chýla, A.L. Kataev and S.A. Larin, *Phys. Lett.* **B 267** (1991) 269.
335. A.A. Pivovarov, *Z. Phys.* **C 53** (1992) 461.
336. P.A. Raczka and A. Szymacha, *Z. Phys.* **C 70** (1996) 125.
337. G.N. Hardy, *Divergent Series*, Oxford University Press, Oxford, (1949).
338. S. Narison, *Nucl. Phys. Proc. Suppl.* **40** (1995) 47;
S. Narison, hep-ph/9404211.
339. C.J. Maxwell and D.G. Tonge, Durham preprint DTP/96/52 (1996), hep-ph/96066392.
340. E. Braaten, *τ96 Workshop*, Colorado (1996).
341. S. Narison, *Phys. Lett.* **B 300** (1993) 293.
342. A.I. Vainshtein and V.I. Zakharov, *Phys. Rev. Lett.* **73** (1994) 1207, and private communications from V.I. Zakharov.
343. S. Peris and E. de Rafael, hep-ph/9701418 (1997).
344. G. Altarelli, *τ94 Workshop*, Montreux, *Nucl. Phys. (Proc. Suppl.)* **B 40** (1995) 40 ed. L. Rolandi;
G. Altarelli, G. Ridolfi and P. Nason, *Z. Phys.* **C 68** (1995) 257.
345. M. Girone and M. Neubert, *Phys. Rev. Lett.* **76** (1996) 359.
346. S. Narison and A. Pich, *Phys. Lett.* **B 304** (1993) 359.
347. F.J. Gilman and S.H. Rhie, *Phys. Rev.* **D 31** (1985) 1066;
F.J. Gilman and D.H. Miller, *Phys. Rev.* **D 17** (1978) 1846;
F.J. Gilman, *Phys. Rev.* **D 35** (1987) 3541.


### Strange quark mass


348. S. Chen, *Nucl. Phys. (Proc. Suppl.)* **B 64** (1998) 256;
E. Tournefier, *Nucl. Phys. (Proc. Suppl.)* **B 86** (2000) 228 and references therein.
349. K.G. Chetyrkin, J.H. Kühn and A. Pivovarov, *Nucl. Phys.* **B 533** (1998) 473.
350. A. Pich and J. Prades, *Nucl. Phys. (Proc. Suppl.)* **B 86** (2000) 236.
351. J.G. Körner, F. Krajewski and A.A. Pivovarov, hep-ph/0003165.
352. J. Kambor and K. Maltman, *Nucl. Phys. (Proc. Suppl.)* **B 98** (2001) 314.
353. S. Chen *et al.*, hep-ph/0105253 (2001).
354. S. Narison, *Phys. Lett.* **B 466** (1999) 345; **B 358** (1995) 113.
355. K. Maltman, *Phys. Lett.* **B 428** (1998) 179.

## VI Summary of QCD tests

...

## VII Power corrections in QCD

### *Introduction*

*Reviews on QCD spectral sum rules*


356. V.A. Novikov *et al.*, *Phys. Rep.* **41** (1978) 1;
Different contributions in the *Conference on Non-perturbative Methods*, Montpellier, (1985) ed. S. Narison, World Scientific Company, Singapore, (1985);
R.A. Bertlmann, *Acta Phys. Austriaca.* **53** (1981) 305;
E.V. Shuryak, *Phys. Reports* **115** (1984) 158; *Rev. Mod. Phys.* **65** (1993) 1;
B.L. Ioffe, *International Conference in High-Energy Physics*, Leipzig (1984);
M.A. Shifman, *Vacuum Structure and QCD Vacuum, Nucl. Phys. (Proc. Suppl.* **B 10** (1992); *Prog. Theor. Phys. Suppl.* **131** (1998) 1;
A. Radyushkin, *Introduction to the QCD Sum Rule Approach*, hep-ph/0101227.
357. L.J. Reinders, H.R. Rubinstein and S. Yazaki, *Phys. Rep.* 127 (1985) 1.
358. H.G. Dosch, *Conference on Non-perturbative Methods*, Montpellier (1985) ed. S. Narison, World Scientific Company, Singapore, (1985).
359. D.B. Leinweber, *QCD Sum Rules for Skeptics*, hep-ph/9510051 (1995).
360. P. Colangelo and A. Khodjamirian, *QCD Sum Rules, a Modern Prespectives*, hep-ph/0010175.
361. E. de Rafael, *An Introduction to Sum Rules in QCD*, Les Houches Summer Institute (1998), hep-ph/9802448;
Some comments on QCD sum rules, *Proc. Topical Conf. on Phenomenology of unified theories*, Dubrovnik L(1983) University of Marseille preprint CPT-84/P.1587 (1984); CERN lectures on QCD sum rules 1986 (unpublished); *Current algebra quark masses in QCD, Proc. NSF-CNRS Joint Seminar on Recent Developments in QCD,* University of Marseille preprint CPT-81/P.1344 (1981).
362. S. Narison, *QCD Duality Sum Rules: Introduction and Some Recent Developments*, Gif Summer Institute, Paris, CERN-TH 4624/86 (1986).
363. S. Narison, *Low Mass Hadrons of QCD, LEAR Workshop*, Les Arcs, University of Montpellier preprint PM/85-02 (1985); Chiral symmetry breaking and the light meson systems, *Nuovo Cim.* **10** (1987) 1; *A short note on QCD sum rules, Conference on Non-perturbative Methods*, Montpellier, (1985) ed. S. Narison, World Scientific Company, Singapore, (1985); *QCD parameters from the light mesons systems, International Conference in High-Energy Physics*, Leipzig, (1984).
364. S. Narison, *QSSR for Heavy Flavours*, Epiphany Conference, Cracow, *Acta Phys. Polon.* **B 26** (1995) 687 [HEP-PH 9503234]; *Heavy Flavours from QCD spectral sum rules*, hep-ph/9510270; *Heavy quarks from QSSR, QCD 94 Conference,* Montpellier, *Nucl. Phys. (Proc. Suppl.)* **BC 39** (1995) 446; Heavy quarks from QSSR, hep-ph/9406206; *Beautiful Beauty Mesons from QSSR, QCD 90 Conference,* Montpellier, *Nucl. Phys. Proc. Suppl.* **B 23** (1991) 336;
J.M. Brom, S.J. Brodsky, L. Montanet, S. Narison and M. Poulet, *Nucl. Phys. Proc. Suppl.* **8** (1989) 73.
365. S. Narison, *On the Quark and Gluon Substructure of the Sigma and other Scalar Mesons*, hep-ph/0009108; Gluonia in QCD with massless quarks, hep-ph/9711495;

*Scalar Mesons and the Phenomenology of the Trace of the Energy-momentum Tensor, QCD 90 Conference,* Montpellier, *Nucl. Phys. Proc. Suppl.* **B 23** (1991) 280; Gluonia, scalar mesons and the light Higgs boson, In *QCD, Hadron 89 Workshop*, Ajaccio (1989) Univ. Montpellier preprint PM-89/31 (1989); Some properties of gluonia In *QCD, Moriond Conference,* Les Arcs (1989), CERN-TH 5353/89 (1989); Tensor mesons and gluonium of QCD, Munich High Energy Phys. (1988); Exotic mesons of QCD, LEAR Workshop (1987); *Light Exotic Mesons from QCD Duality Sum Rules, International Conference in High-Energy Physics,* Berkley (1986), Univ. Montpellier preprint PM/86-27 (1986).

*Renormalons 2*

366. B.R. Webber, *Phys. Lett.* **B 339** (1994);
Yu. L. Dokshitzer, G. Marchesini and B.R. Webber, *Nucl. Phys.* **B 444** (1995) 602.
367. Analysis including renormalization effects is in Novikov, V.A. *et al.*, *Nucl. Phys.* **B 249** (1985) 445; *Phys. Report* **116** (1984) 105.
368. H.R Quinn and S. Gupta, *Phys. Rev.* **D 26** (1982) 499;
C.Taylor and B.Mc Clain, *Phys. Rev.* **D 28** (1983) 1364.
369. For a review, see eg.: F. David, *Proc. Non-perturbative Methods*, Montpellier, ed. S. Narison, World Scientific Company, Singapore, (1985).
370. M. Soldate, *Ann. Phys.* **158** (1984) 433;
A.N. Tavkhelidze and V.F. Tokarev, *Sov. J. Part. and Nucl.* **16** (1985) 431;
G.S. Danilov and I.T. Dyatolov, *Yad Fiz* **14** (1985) 1298.
371. F. David, *Phys. Lett.* **B 138** (1984) 139.
372. D. Gross and A. Neveu, *Phys. Rev.* **D 10** (1974) 3235.
373. F. David and H. Hamber, *Nucl. Phys.* **B 248** (1984) 381.
374. F. David, *Nucl. Phys.* **B 209** (1982) 433; **B 234** (1984) 237.
375. G.'t. Hooft, *'The Why's of Subnuclear Physics'*, Erice 1977, ed. A. Zichichi, Plenum, New York;
B. Lautrup, *Phys. Lett.* **B 69** (1977) 109;
G. Parisi, *Nucl. Phys.* **B 150** (1979) 163;
J.D. Bjorken, SLAC preprint SLAC-PUB-5103 (1989);
G.B. West, *Phys. Rev. Lett.* **67** (1991) 1388 [Erratum : **67** (1991)].
376. A.H. Mueller, *Nucl. Phys.* **B 250** (1985) 327.
377. A. Mueller, *QCD-20 Years Later Workshop*, Aachen (1992) and references therein.
378. V.I. Zakharov, *Nucl. Phys.* **B 385** (1992) 452.
379. F. Wilczek, *Lepton-Photon Conference*, Cornell, Ithaca, NY, (1993).
380. G. Martinelli and C.T. Sachrajda, *Phys. Lett.* **B 354** (1995) 423.

*Instantons*

381. A.A. Belavin, A.M. Polyakov, A.S. Schwartz and Y.S. Tyupkin, *Phys. Lett.* **B 59** (1975) 85;
R. Jackiw and C. Rebbi, *Phys. Rev. Lett.* **37** (1976) 172;
C.G. Callan, R. Dashen and D. Gross, *Phys. Lett.* **B 63** (1976) 334.
382. V.A. Novikov *et al.*, *Nucl. Phys.* **B 191** (1981) 301.
383. B.V. Geshkeinben and B.L. Ioffe, *Nucl. Phys.* **B 166** (1980) 340.
384. I.I. Balitsky, M. Beneke and V.M. Braun, *Phys. Lett.* **B 318** (1993) 371.
385. E. Gabrielli and P. Nason, *Phys. Lett.* **B 313** (1993) 430;
P. Nason and M. Porrati, *Nucl. Phys.* **B 421** (1994) 518.

386. E.V. Shuryak, *Phys. Reports* **115** (1984) 158; World Scientific book (1988); *Rev. Mod. Phys.* **65** (1993) 1;
T. Schafer, E.V. Shuryak, *Rev. Mod. Phys.* **70** (1998) 323.
387. V. Kartvelishvili and M. Margvelashvili, *Phys. Lett.* **B 345** (1995) 161.
388. For a review, see e.g.: A. Ringwald, *QCD'02 International Conference*, July 2002, Montpellier (to appear) and references therein.
389. M.A. Shifman, A.I. Vainshtein and V.I. Zakharov, *Phys. Lett.* **B 76** (1978) 971.
390. E. Ilgenfritz and AM. Müller-Preussker, *Nucl. Phys.* **B 184** (1981) 443;
D. Diakonov and V.Y. Petrov, *Nucl. Phys.* **B 245** (1984) 259.
391. M.A. Shifman, A.I. Vainshtein and V.I. Zakharov, *Nucl. Phys.* **B 165** (1980) 45.
392. E.V. Shuryak, *Nucl. Phys.* **B 198** (1982) 83.
393. T. DeGrand, hep-lat/0106001.
394. S. Narison and V.I. Zakharov, *Phys. Lett.* **B 522** (2001) 266.
395. M.N. Chernodub, F.V. Gubarev, M.I. Polikarpov, V.I. Zakharov, *Nucl. Phys.* **B 592** (2001) 107 (hep-th/0003138); *Nucl. Phys.* **B 600** (2001) 163 (hep-th/0010265).
396. F.V. Gubarev, M.I. Polikarpov, V.I. Zakharov, *Mod. Phys. Lett.* **A 14** (1999) 2039;
M.N. Chernodub, F.V. Gubarev, M.I. Polikarpov, V.I. Zakharov, *Phys. Lett.* **B 475** (2000) 303;
F.V. Gubarev , V.I. Zakharov, *Phys. Lett.* **B 501** (2001) 28.
397. V. Shevchenko, Yu. Simonov, hep-ph/0109051.
398. E.V. Shuryak, hep-ph/9909458.

## *SVZ Expansion*

*Quark condensates*

399. D.J. Broadhurst, *Phys. Lett.* **B 101** (1981) 423;
D.J. Broadhurst and S.C. Generalis, Open University preprint *OUT* **4102–8** (1982);
D.J. Broadhurst and S.C. Generalis, Open University preprint *OUT* **4102–12** (1982);
S.C. Generalis, *Ph. D. Thesis* **4102–13** (1983). (unpublished);
S.C. Generalis, *Journal of Phys.* **G 15** (1989) L225; **G 16** (1990) 785;
A. Djouadi and P. Gambino, *Phys. Rev.* **D 49** (1994) 3499; *Erratum-ibid* **D 53** (1996) 4111.
400. S. Narison, *Phys. Lett.* **B 104** (1981) 485.

*Mixed condensate from $B - B^*$*

401. S. Narison, *Phys. Lett.* **B 210** (1988) 238.

*Gluon condensate from lattice*

402. A. Di Giacomo and G.C. Rossi, *Phys. Lett.* **B 100** (1981) 481;
G. Curci, A. Di Giacomo and G. Paffuti, *Z. Phys.* **C 18** (1983) 135;
M. Campostrini, A. Di Giacomo and Š. Olejník, Pisa preprint, IFUP-TH 2/86 (1986).

*Gluon condensate from $e^+e^-$*

403. S.I. Eidelman, L.M. Kurdadze and A.I. Vainshtein, *Phys. Lett.* **B 82** (1979) 278.
404. G. Launer, S. Narison and R. Tarrach, *Z. Phys.* **C 26** (1984) 433.

405. R.A. Bertlmann, G. Launer and E. de Rafael, *Nucl. Phys.* **B 250** (1985) 61.
406. R.A. Bertlmann, C.A. Dominguez, M. Loewe, M. Perrottet and E. de Rafael, *Z. Phys.* **C 39** (1988) 231.
407. C.A. Dominguez and J. Solà, *Z. Phys.* **C 40** (1988) 63.
408. G. Bordes, V. Gimenez and J.A. Peñarrocha, *Phys. Lett.* **B 201** (1988) 365.
409. M.B. Causse and G. Mennessier, *Z. Phys.* **C 47** (1990) 611.

*High-dimension gluon condensates*

410. D.J. Broadhurst and S.C. Generalis, *Phys. Lett.* **B 139** (1984) 85; **B 165** (1985) 175.
411. E. Bagan *et al.*, *Nucl. Phys.* **B 254** (1985) 555; *Z. Phys.* **C 32** (1986) 43.
412. S.N. Nikolaev and A.V. Radyushkin, *Phys. Lett.* **B 124** (1983) 243; *Nucl. Phys.* **B 213** (1983) 285.
413. D. Espriu, E. de Rafael and J. Taron, *Nucl. Phys.* **B 345** (1990) 22.
414. V.A. Novikov *et al.*, Neutrino 78, Purdue Univ. Lafayette, (1978).
415. D. Espriu, M. Gross and J.F. Wheater, *Phys. Lett.* **B 146** (1984) 67.

*Quadratic corrections*

416. H.B.G. Casimir and D. Polder, *Phys. Rev.* **73** (1948) 360.
417. Y.Y. Balitskii, *Nucl. Phys.* **B 254** (1985) 166;
H.G. Dosch and Y.A. Simonov, *Nucl. Phys.* **B 254** (1985) 166.
418. M.B. Voloshin, *Nucl. Phys.* **B 154** (1979) 365;
H. Leutwyler, *Phys. Lett.* **B 98** (1981) 447.

*Quark condensates 2*

419. S. Narison, *Phys. Lett.* **B 358** (1995) 113.
420. S. Narison, *Riv. Nuov. Cim.* **10 N2** (1987) 1.
421. S. Narison, N. Paver and D. Treleani, *Nuovo. Cim.* **A 74** (1983) 347.
422. S. Narison, *Phys. Lett.* **B 216** (1989) 191.
423. H.G. Dosch and S. Narison, *Phys. Lett.* **B 417** (1998) 173.

*Baryon sum rules*

424. Y. Chung *et al.*, *Phys. Lett.* **B 102** (1981) 175; *Nucl. Phys.* **B 197** (1982) 55; *Z. Phys.* **C 25** (1984) 151;
For a review, see e.g. [358].
425. M. Jamin, *Z. Phys.* **C 37** (1988) 635 and private communication.
426. H.G. Dosch, M. Jamin and S. Narison, *Phys. Lett.* **B 220** (1989) 251.
427. D. Espriu, P. Pascual and R. Tarrach, *Nucl. Phys.* **B 214** (1983) 285;
L.J. Reinders, H. Rubinstein and S. Yazaki, *Phys. Lett.* **B 120** (1983) 209.
428. A.A. Ovhinnikov, A.A. Pivovarov and L.R. Surguladze, *Int. J. Mod. Phys.* **A 6** (1991) n. 11, 2025.
429. For a review, see e.g. [359].
430. B.L. Ioffe, *Nucl. Phys.* **B 188** (1981) 317; **191** (1981) 591;
For a review, see e.g.: B.L. Ioffe in [356].

### *Technologies for evaluating the Wilson coefficients*

*Fock–Schwinger fixed point*


431. V.A Novikov *et al.*, *Forsch. Phys.* **32** (1984) 585.
432. W. Hubschmid and A. Mallik, *Nucl. Phys.* **B 207** (1982) 29.
433. S.N. Nikolaev and A.V. Radyushkin, *Phys. Lett.* **B 124** (1983) 243; *Nucl. Phys.* **B 213** (1983) 285.
434. L.J. Reinders, H. Rubinstein and S. Yazaki, *Nucl. Phys.* **B 186** (1981) 109 and 357.
435. V.A. Fock, *Sov. J. Phys.* **12** (1937) 404 and *Works on Quantum Field Theory*, Leningrad University Press, Leningrad (1957).
436. M.S. Dubovikov and A.V. Smilga, *Nucl. Phys.* **B 185** (1981) 109.
437. D.J. Broadhurst and S.C. Generalis, *Phys. Lett.* **B 142** (1984) 75; E. Bagan, J.I. Lattore and P. Pascual, *Z. Phys.* **C 32** (1986) 43; S.G. Gorishny, S.A. Larin and F.V. Tkachov, *Phys. Lett.* **B 124** (1983) 217.


## VIII QCD two-point functions

### *(Axial-)vector and (pseudo)scalar correlators*


438. P.A. Baikov, K.G. Chetyrkin and J.H. Kühn, hep-ph/01008197.
439. P. Pascual and E. de Rafael, *Z. Phys.* **C 12** (1982) 127.
440. K.G. Chetyrkin, Private communication; K.G. Chetyrkin, S.G. Gorishny and V.P. Spiridonov, *Phys. Lett.* **B 160** (1985) 149; K.G. Chetyrkin, J.H. Kühn and A. Kwiatkowski, Univ. Karlsruhe preprint TTP94-32 (1994) and references therein.
441. M. Jamin and M. Münz, *Z Phys.* **C 66** (1995) 633.
442. K.G. Chetyrkin, C.A. Dominguez, D. Pirjol and K. Schilcher, *Phys. Rev.* **D 51** (1995) 5090.
443. K.G. Chetyrkin, D. Pirjol and K. Schilcher, *Phys. Lett.* **B 404** (1997) 337.
444. M. Jamin and M. Münz, *Z. Phys.* **C 60** (1993) 569.
445. S.G. Gorishny, A.L. Kataev, S.A. Larin and L.R. Surguladze, *Mod. Phys. Lett.* **A 5** (1990) 2703.
446. K.G. Chetyrkin, *Phys. Lett.* **B 390** (1997) 309.
447. K.G. Chetyrkin, R.V. Harlander, J.H. Kuehn, *Nucl. Phys.* **B 586** (2000) 56.
448. K.G. Chetyrkin and M. Steinhauser, *Phys. Lett.* **B 502** (2001) 104; hep-ph/0108017.
449. K.G. Chetyrkin, R. Harlander, J.H. Kuehn, M. Steinhauser, *Nucl. Phys.* **B 503** (1997) 339.


### *Tensors*


450. E. Bagan, A. Bramon and S. Narison, *Phys. Lett.* **B 196** (1987) 203.
451. T.M. Aliev and M.A. Shifman, *Phys. Lett.* **B 112** (1982) 401.
452. E. Bagan and S. Narison, *Phys. Lett.* **B 214** (1988) 451.


### *Heavy baryons*


453. E. Bagan, M. Chabab, H.G. Dosch and S. Narison, *Phys. Lett.* **B 278** (1992) 367; **B 287** (1992) 176; E. Bagan, M. Chabab and S. Narison, *Phys. Lett.* **B 306** (1993) 350.

454. E. Bagan, M. Chabab, H.G. Dosch and S. Narison, *Phys. Lett.* **B 301** (1993) 243.

### *Gluonia*

455. A.L. Kataev, N.V. Krasnikov and A.A. Pivovarov, *Nucl. Phys.* **B 198** (1982) 508;
A.L. Kataev, N.V. Krasnikov and A.A. Pivovarov, erratum, hep-ph/9612326 (1996);
K. Chetyrkin (private communication).
456. E. Bagan and T.G. Steele, *Phys. Lett.* **B 243** (1990) 413;
D. Asner *et al.*, *Phys. Lett.* **B 296** (1992) 171.
457. J.I. Latorre, S. Paban and S. Narison, *Phys. Lett.* **B 191** (1987) 437.
458. N. Pak, S. Narison and N. Paver, *Phys. Lett.* **B 147** (1984) 162.

### *Hybrids*

459. I.I. Balitsky, D.I. D'Yakonov and A.V. Yung, *Phys. Lett.* **B 112** (1982) 71;
*Sov. J. Nucl. Phys.* **35** (1982) 781.
460. J. Govaerts, F. de Viron, D. Gusbin and J. Weyers, *Phys. Lett.* **B 128** (1983) 262 and *Nucl. Phys.* **B 248** (1984) 1;
J.I. Latorre, S. Narison, P. Pascual and R. Tarrach, *Phys. Lett.* **B 147** (1984) 169.
461. J.I. Latorre, S. Narison and P. Pascual, *Z. Phys.* **C 34** (1987) 347.
462. K. Chetyrkin and S. Narison, *Phys. Lett.* **B 485** (2000) 145.
463. J. Govaerts, L.J. Reinders, H. Rubinstein and J. Weyers, *Nucl. Phys.* **B 258** (1985) 215;
J. Govaerts, L.J. Reinders and J. Weyers, *Nucl. Phys.* **B 262** (1985) 575.
464. K. Chetyrkin, S. Narison and A.A. Pivovarov (unpublished).

### *Four-quarks*

465. J.I. Latorre and P. Pascual, *J. Phys.* **G 11** (1985) 231.
466. S. Narison, *Phys. Lett.* **B 175** (1986) 88.
467. B. Guberina, B. Machet and E. de Rafael, *Phys. Lett.* **B 128** (1983) 269.
468. A. Pich and E. de Rafael, *Phys. Lett.* **B 158** (1985) 477; *Nucl. Phys.* **B 358** (1991) 311.
469. B. Guberina, A. Pich and E. de Rafael, *Phys. Lett.* **B 163** (1985) 198.
470. B. Guberina, A. Pich and E. de Rafael, *Nucl. Phys.* **B 277** (1986) 197.
471. M. Jamin and A. Pich, *Nucl. Phys.* **B 425** (1994) 15.
472. A. Pich, *Phys. Lett.* **B 206** (1988) 322.
473. S. Narison and A.A. Pivovarov, *Phys. Lett.* **B 327** (1994) 341.
474. K. Hagiwara, S. Narison and D. Nomura, hep-ph/0205092.
475. A.J. Buras, M. Jamin and P.H. Weisz, *Nucl. Phys.* **B 347** (1990) 491.
476. A.J. Buras and P.H. Weisz, *Nucl. Phys.* **B 333** (1990) 66.
477. A. Djouadi and P. Gambino, *Phys. Rev.* **D 49** (1994) 3499.

### *Renormalons 3*

478. G. Parisi and R. Petronzio, *Phys. Lett.* **B 94** (1980) 51;
J.M. Cornwall, *Phys. Rev.* **D 26** (1982) 1453.

479. V.P. Spiridonov and K.G. Chetyrkin, *Sov. J. Nucl. Phys.* **47** (1988) 522.
480. P.A. Movilla Fernández, *Nucl. Phys. (Proc. Suppl.)* **B 74** (1999) 384.
481. S. Luiti, *Nucl. Phys. (Proc. Suppl.)* **B 74** (1999) 380.
482. E. Stein *et al.*, *Phys. Lett.* **B 376** (1996) 177; hep-ph/9803342.
483. V.M. Braun and A.V. Kolesnichenko, *Nucl. Phys.* **B 283** (1987) 723.
484. X. Ji, *Nucl. Phys.* **B 448** (1995) 51.
485. A.H. Hoang *et al.*, hep-ph/9804227.
486. U. Aglietti and Z. Ligeti, *Phys. Lett.* **B 364** (1995) 75.
487. G. Burgio *et al.*, hep-ph/9808258; hep-ph/9809450.
488. G. Grunberg, hep-ph/9705290; hep-ph/9711481.

## *IX QCD non-perturbative methods*

### *Lattice gauge theory*

489. M. Creutz, *Quarks, Gluons and Lattices*, Cambridge University Press, Cambridge, (1983);
C. Rebbi, *Lattice Gauge Theories and Monte-Carlo Simulations*, World Scientific Company, Singapore, (1983);
H.J. Rothe, *Lattice Gauge Theories*, World Scientific Company, Singapore (1992).
490. F.J. Wegner, *J. Math. Phys.* **12** (1971) 2259.
491. see the second reference in [222].
492. K.G. Wilson, *in New Phenomena in Subnuclear Physics*, ed. A. Zichichi, Plenum, New York, (1977).
493. H. B. Nielsen and M. Ninomiya, *Nucl. Phys.* **B 185** (1981) 20; **B 193** (1981) 173.
494. H. Kawai, R. Nakayama and K. Seo, *Nucl. Phys.* **B 189** (1981) 40.
495. A. Hasenfratz and P Hasenfratz, *Phys. Lett.* **B 93** (1980) 165.
496. P. Goddard, J. Goldstone, C. Rebbi and C.B. Thorn, *Nucl. Phys.* **B 56** (1973) 109.

### *Chiral perturbation theory*

497. S. Weinberg, *Phys. Rev. Lett.* **18** (1967) 188;
J. Schwinger, *Phys. Lett.* **B 24** (1967) 473;
J.A. Cronin, *Phys. Rev.* **161** (1967) 1483;
J. Wess and B. Zumino, *Phys. Rev.* **163** (1967) 1727;
R. Dashen and M. Weinstein, *Phys. Rev.* **183** (1969) 1261;
S. Gasiorowicz and D.A. Geffen, *Rev. Mod. Phys.* **41** (1969) 571.
498. S. Weinberg, *Physica.* **A 96** (1979) 327.
499. J. Gasser and H. Leutwyler, *Ann. Phys., NY* **158** (1984) 142; *Nucl. Phys.* **B 250** (1985) 465.
500. E. de Rafael, *Lectures given the 1994th TASI-School on CP-violation and limits of the Standard Model*, Univ. Colorado at Boulder, hep-ph/9502254; *Nucl. Phys. (Proc. Suppl.)* **B 96** (2001) 316 and references therein.
501. A. Pich, hep-ph/9502366 and references therein.
502. J.F. Donoghue, Chiral symmetry as an experimental science, in *Medium Energy Antiprotons and the Quark-gluon Structure of Hadrons*, eds. R. Landua, J.M. Richard and R. Klapisch, Plenum Press, New York, (1991), p. 39;
H. Leutwyler, Chiral effective Langrangians, In *Recent Aspects of Quantum Fields*,

eds. H. Mitter and M. Gausterer, Lecture Notes in Physics, Vol. 396, Springer Verlag, Berlin, (1991);
U.G. Meissner, *Proc. of the Workshop on Effective Field Theories of the Standard Model*, Dobogókö, Hungary, 1991, World Scientific, Singapore, (1992);
G. Ecker, Chiral perturbation theory, In *Quantitative Particle Physics*, eds. M. Levy *et al.*, Plenum Publ. Co., New York, (1993); Chiral dynamics, In *Proc. Workshop on Physics and Detectors for DAΦNE*, ed. G. Pancheri, Frascati, (1991), p. 291;
H. Georgi, *Weak Interactions and Modern Particle Theory*, Benjamin/Cummings, Menlo Park, (1984);
J.F. Donoghue, E. Golowich and B.R. Holstein, *Dynamics of Standard Model*, Cambridge University Press, Cambridge, (1992);
A. Manohar, *Effective Field Theories*, hep-ph/9607484 (1996);
J. Bijnens, *Int. J. Mod. Phys.* **A 8** (1993) 3045;
T. Hatsuda and T. Kunihiro, *QCD phenomenology based on ChPT*, University of Tsukuba preprint UTHEP-270 (1994) (pub. in *Phys. Rep.* (1994).

503. E. Euler, *Ann. Phys., Lpz* **26** (1936) 398;
E. Euler and W. Heisenberg, *Z. Phys.* **98** (1936) 714.
504. S. Weinberg, *Phys. Rev. Lett.* **17** (1966) 616.
505. J.M. Knecht and J. Stern, *QCD 94 Conference*, Montpellier, *Nucl. Phys. B (Proc. Suppl.)* **B, C 39** (1995) 249 and references therein.
506. M. Faber *et al.*, *QCD 94 Conference Montpellier*, *Nucl. Phys. B (Proc. Suppl.)* **B, C 39** (1995) 228.
507. R. Dashen, *Phys. Rev.* **183** (1969) 1245.
508. J. Wess and B. Zumino, *Phys. Lett.* **B 37** (1971) 95.
509. E. Witten, *Nucl. Phys.* **B 223** (1983) 422.
510. W.A. Bardeen, *Phys. Rev.* **184** (1969) 1848.
511. M. Soldate and R. Sundrum, *Nucl. Phys.* **B 340** (1990) 1;
R.S. Chivukula, M.J. Dugan and M. Golden, *Phys. Rev.*, **D 47** (1993) 2930.
512. H. Georgi and A. Manohar, *Nucl. Phys.* **B 234** (1984) 189.
513. H. Leutwyler and M. Roos, *Z. Phys.* **C 25** (1984) 91.
514. S.R. Amendolia *et al.*, *Nucl. Phys.* **B 277** (1986) 168.
515. H. Leutwyler, *Nucl. Phys. B (Proc. Suppl.)* **A 7** (1989) 42.
516. W.R. Molzon *et al.*, *Phys. Rev. Lett.* **41** (1978) 1213.
517. E.B. Dally *et al.*, *Phys. Rev. Lett.* **48** (1982) 375; **45** (1980) 235.
518. J. Bijnens, *Phys. Lett.* **B 306** (1993) 343;
J.F. Donoghue, B.R. Holstein and D. Wyler, *Phys. Rev.*, **D 47** (1993) 2089.
519. D.B. Kaplan and A.V. Manohar, *Phys. Rev. Lett.* **56** (1986) 2004;
H. Georgi and I. Mc Arthur, Harvard Univ. Report HUTP-81/A011, (1981).

### *Models of the effective QCD action*

#### *Large $N_c$*

520. G.'t. Hooft, *Nucl. Phys.* **B 72** (1974) 461.
521. G. Veneziano, *Nucl. Phys.* **B 117** (1976) 519.
522. E. Witten, *Nucl. Phys.* **B 156** (1979) 269; **B 160** (1979) 57.
523. E. Witten, *Ann. Phys.* **128** (1980) 363.
524. W.A. Bardeen, A.J. Buras and J.-M. Gérard, *Nucl. Phys.* **B 293** (1987) 787; *Phys. Lett.* **B 192** (1987) 138; **B 211** (1988) 343.

525. S. Coleman and E. Witten, *Phys. Rev. Lett.* **45** (1980) 100.
526. M. Knecht and E. de Rafael, it Phys. Lett. **B 424** (1998) 335.
527. M. Davier, A. Höcker, L. Girlanda and J. Stern, *Phys. Rev.* **D 58** (1998) 096014.
528. C.A. Dominguez and K. Schilcher, *Phys. Lett.* **B 448** (1999) 93.
529. E. de Rafael, *Nucl. Phys. (Proc. Suppl.)* **B 74** (1999) 399.
530. S. Peris, M. Perrottet and E. de Rafael, *JHEP* **05** (1998) 011;
M. Golterman and S. Peris, *Phys. Rev.* **D 61** (2000) 034018.
531. B.L. Ioffe and M.A. Shifman, *Nucl. Phys.* **B 202** (1982)...
532. H.G. Dosch and S. Narison, *Phys. Lett.* **B 180** (1986) 390; **184 B** (1987) 78.

*Effective models*

533. G. Ecker, J. Gasser, A. Pich and E. de Rafael, *Nucl. Phys.* **B 321** (1989) 311.
534. J.F. Donoghue, L. Ramirez and G. Valencia, *Phys. Rev.* **D 39** (1989) 1947.
535. G. Ecker, J. Gasser, H. Leutwyler, A. Pich and E. de Rafael, *Phys. Lett.* **B 223** (1989) 425.
536. D. Espriu, E. de Rafael and J. Taron, *Nucl. Phys.* **B 345** (1990) 22; [Erratum: **B 355** (1991) 278.]
537. A. Pich and E. de Rafael, *Nucl. Phys.* **B 358** (1991) 311.
538. J.L. Kneur and D. Reynaud, *Nucl. Phys. (Proc. Suppl.)* **B 96** (2001) 255;
C. Arvanitis *et al.*, *Int. J. Mod. Phys.* **A 12** (1997) 3307.

*The extended Nambu–Jona–Lasinio model*

539. Y. Nambu and G. Jona-Lasinio, *Phys. Rev.* **122** (1961) 345; *ibid.* **124** (1961) 246.
540. J. Bijnens, Ch. Bruno and E. de Rafael, *Nucl. Phys.* **B 390** (1993) 501.
541. J. Bijnens, E. de Rafael and H. Zheng, *Z. Phys.* **C 62**(1994) 437.
542. S. Peris and E. de Rafael, *Phys. Lett.* **B 309** (1993) 389.
543. [263,522], P. di Vecchia and G. Veneziano, *Nucl. Phys.* **B 171** (1980) 253;
C. Rosenzweig, J. Schechter and C.G. Trahern, *Phys. Rev.* **D 21** (1980) 253.
544. J. Bijnens and J. Prades, *Phys. Lett.* **B 320** (1994) 130.

## ***Heavy quark effective theory***

*Heavy quark symmetry and HQET*

545. M. Neubert, *Phys. Rep.* **245** (1994) 259; hep-ph/9702375; hep-ph/0001334 and references therein;
N. Isgur and M.B. Wise, In *Heavy Flavours*, eds. A.J. Buras and M. Lindner, World Scientific Company, Singapore, (1992).
546. H. Georgi, *Phys. Lett.* **B 240** (1990) 447 .
547. M.A. Shifman, A.I. Vainshtein, and V.I. Zakharov, *Nucl. Phys.* **B 120** (1977) 316.
548. E. Witten, *Nucl. Phys.* **B 122** (1977) 109.
549. J. Polchinski, *Nucl. Phys.* **B 231** (1984) 269.
550. G. Altarelli and L. Maiani, *Phys. Lett.* **B 52** (1974) 351.
551. M.K. Gaillard and B.W. Lee, *Phys. Rev. Lett.* **33** (1974) 108.
552. F.J. Gilman and M.B. Wise, *Phys. Rev.* **D 27** (1983) 1128.
553. T. Mannel, W. Roberts, and Z. Ryzak, *Nucl. Phys.* **B 368** (1992) 204.
554. J. Soto and R. Tzani, *Phys. Lett.* **B 297** (1992) 358.

555. E. Eichten and B. Hill, *Phys. Lett.* **B 234** (1990) 511; **243** (1990) 427.
556. A.F. Falk, B. Grinstein, and M.E. Luke, *Nucl. Phys.* **B 357** (1991) 185.
557. N. Isgur and M.B. Wise, *Phys. Lett.* **B 232** (1989) 113; **237** (1990) 527.
558. A.F. Falk, M. Neubert and M.E. Luke, *Nucl. Phys.* **B 388** (1992) 363.
559. N. Isgur and M.B. Wise, *Phys. Rev. Lett.* **66** (1991) 1130.
560. U. Aglietti, *Phys. Lett.* **B 281** (1992) 341; *Int. J. Mod. Phys.* **A 10** (1995) 801.
561. A.F. Falk, *Nucl. Phys.* **B 378** (1992) 79.
562. A.F. Falk and M. Neubert, *Phys. Rev.* **D 47** (1993) 2965 and 2982.

*Semi-leptonic form factors: the Isgur–Wise function*

563. S. Nussinov and W. Wetzel, *Phys. Rev.* **D 36** (1987) 130.
564. M.B. Voloshin and M.A. Shifman, *Yad. Fiz.* **45** (1987) 463 [*Sov. J. Nucl. Phys.* **45** (1987) 292].
565. M.B. Voloshin and M.A. Shifman, *Yad. Fiz.* **47** (1988) 801 [*Sov. J. Nucl. Phys.* **47** (1988) 511].
566. M. Neubert and V. Rieckert, *Nucl. Phys.* **B 382** (1992) 97.
567. M. Neubert, *Phys. Lett.* **B 264** (1991) 455.
568. M.E. Luke, *Phys. Lett.* **B 252** (1990) 447.
569. M. Ademollo and R. Gatto, *Phys. Rev. Lett.* **13** (1964) 264.
570. M. Neubert, *Nucl. Phys.* **B 416** (1994) 786.
571. P. Cho and B. Grinstein, *Phys. Lett.* **B 285** (1992) 153.
572. H.D. Politzer and M.B. Wise, *Phys. Lett.* **B 206** (1988) 681; **208** (1988) 504.
573. C.G. Boyd, B. Grinstein, and R.F. Lebed, *Phys. Lett.* **B 353** (1995) 306; *Nucl. Phys.* **B 461** (1996) 493; *Phys. Rev.* **D 56** (1997) 6895;
C.G. Boyd and R.F. Lebed, *Nucl. Phys.* **B 485** (1997) 275.
574. I. Caprini and M. Neubert, *Phys. Lett.* **B 380** (1996) 376;
I. Caprini, L. Lellouch, and M. Neubert, *Nucl. Phys.* **B 530**, (1998) 153.
575. CLEO Collaboration (B. Barish *et al.*), *Phys. Rev.* **D 51** (1995) 1041.
576. P.S. Drell, In: *Proceedings of the 18th International Symposium on Lepton-Photon Interactions*, Hamburg, Germany, July 1997, eds. A. De Roeck and A. Wagner, World Scientific, Singapore, (1998), p. 347.
577. J.E. Paschalis and G.J. Gounaris *Nucl. Phys.* **B 222** (1983) 473;
F.E. Close, G.J. Gounaris and J.E. Paschalis, *Phys. Lett.* **B 149** (1984) 209.
578. M. Neubert, *Nucl. Phys.* **B 371** (1992) 149.
579. M. Neubert, *Phys. Lett.* **B 341** (1995) 367.
580. A. Czarnecki, *Phys. Rev. Lett.* **76** (1996) 4124.

*Inclusive weak processes*

581. M. Neubert, *Phys. Rev.* **D 49** (1994) 3392 and 4623.
582. I.I. Bigi, M.A. Shifman, N.G. Uraltsev, and A.I. Vainshtein, *Int. J. Mod. Phys.* **A 9** (1994) 2467.
583. E.C. Poggio, H.R. Quinn, and S. Weinberg, *Phys. Rev.* **D 13** (1976) 1958.
584. F.G. Gilman and M.B. Wise, *Phys. Rev.* **D 20** (1979) 2392.
585. G. Altarelli, G. Curci, G. Martinelli, and S. Petrarca, *Phys. Lett.* **B 99** (1981) 141; *Nucl. Phys.* **B 187** (1981) 461.
586. A.J. Buras and P.H. Weisz, *Nucl. Phys.* **B 333** (1990) 66.
587. I.I. Bigi, N.G. Uraltsev, and A.I. Vainshtein, *Phys. Lett.* **B 293** (1992) 430 [E: **B 297** (1993) 477];

I.I. Bigi, M.A. Shifman, N.G. Uraltsev, and A.I. Vainshtein, *Phys. Rev. Lett.* **71** (1993) 496;
I.I. Bigi *et al.*, In: *Proceedings of the Annual Meeting of the Division of Particles and Fields of the APS*, Batavia, Illinois, 1992, eds. C. Albright *et al.*, World Scientific, Singapore, (1993), p. 610.

588. B. Blok, L. Koyrakh, M.A. Shifman, and A.I. Vainshtein, *Phys. Rev.* **D 49** (1994) 3356 [E: **D 50** (1994) 3572].
589. A.V. Manohar and M.B. Wise, *Phys. Rev.* **D 49** (1994) 1310.
590. J. Chay, H. Georgi, and B. Grinstein, *Phys. Lett.* **B 247** (1990) 399.
591. G. Altarelli *et al.*, *Nucl. Phys.* **B 208** (1982) 365.
592. C.H. Jin, W.F. Palmer, and E.A. Paschos, *Phys. Lett.* **B 329** (1994) 364;
A. Bareiss and E.A. Paschos, *Nucl. Phys.* **B 327** (1989) 353.
593. Y. Nir, *Phys. Lett.* **B 221** (1989) 184.
594. A. Czarnecki and K. Melnikov, *Nucl. Phys.* **B 505** (1997) 65; *Phys. Rev. Lett.* **78** (1997) 3630; *Phys. Rev.* **D 59** (1999) 014036.
595. For a review, see: *The BaBar Physics Book*, eds. P.F. Harison and H.R. Quinn, SLAC Report No. SLAC-R-504 (1998).
596. A documentation of this measurement can be found at `http://www-cdf.fnal.gov/physics/new/bottom/cdf4855.`
597. The Babar collaboration, for a recent review, see e.g. F. Anulli, *HEP-MAD'01 International Conference*, World Scientific, Singapore, (to appear)
598. The Belle collaboration, for a recent review, see e.g. S. Schrenk, *HEP-MAD'01 International Conference*, World Scientific, Singapore, (to appear)
599. For a recent review, see e.g. M. Pepe, *HEP-MAD'01 International Conference*, World Scientific, Singapore, (to appear)
600. H.N. Li, hep-ph/0110365 and references therein.
601. M. Beneke, G. Buchalla, M. Neubert, and C.T. Sachrajda, *Phys. Rev. Lett.* **83**, 1914 (1999).

## *Potential approaches for quarkonia*

### *Stochastic vacuum model*

602. A. Pineda, *JHEP* **0106** (2001) 022.
603. M. Campostrini, A. Di Giacomo and S. Olejnik, *Z. Phys.* **C 31** (1986) 577.
604. H.G. Dosch and U. Marquard, *Phys. Rev.* **D 37** (1987) 2238;
R.A. Bertlmann, H.G. Dosch and A. Krämer, *Phys. Lett.* **B 223** (1989) 105.
605. H.G. Dosch, *Phys. Lett.* **B 190** (1987) 555;
H.G. Dosch and Yu.A. Simonov, *Phys. Lett.* **B 205** (1988) 339;
Yu. A. Simonov, *A. Yad. Fiz* **48** (1988) 1381; **50** (1988) 213.
606. A. Di Giacomo and H. Panagopoulos, *Phys. Lett.* **B 285** (1992) 133.
607. H. Schnitzer, *Phys. Rev.* **D 18**, (1978) 3481;
D. Gromes, *Z. Phys.* **C 57** (1984) 401.

### *Non-relativistic effective theories for quarkonia*

608. W.E. Caswell and G.P. Lepage, *Phys. Lett.* **B 167** (1986) 437;
G.T. Bodwin, E. Braaten and G.P. Lepage, *Phys. Rev.* **D 51** (1995) 1125, Erratum *ibid.* **D 55** (1997) 5853.

609. A. Pineda and J. Soto, *Nucl. Phys. (Proc. Suppl.)* **B 64**, (1998) 428;
A. Pineda, *Nucl. Phys. (Proc. Suppl.)* **B 86** (2000) 517;
A. Vairo, *Nucl. Phys. (Proc. Suppl.)* **B 86** (2000) 521;
N. Brambilla, *Nucl. Phys. (Proc. Suppl.)* **B 96** (2001) 410.
610. A.V. Manohar, Phys. Rev. **D 56** (1997) 230.
611. A. Pineda and J. Soto, *Phys. Rev.* **D 58** (1998) 114011.
612. A. Pineda and J. Soto, *Phys. Rev.* **D 59** (1999) 016005.
613. N. Brambilla, A. Pineda, J. Soto and A. Vairo, *Nucl. Phys.* **B 566** (2000) 275.
614. Y. Schröder, Phys. Lett. **B 447** (1999) 321; *Nucl. Phys. (Proc. Suppl.)* **B 86** (2000) 525;
M. Peter, *Phys. Rev. Lett.* **78** (1997) 602.
615. A. Duncan, *Phys. Rev.* **D 13** (1976) 2866.
616. B.A. Kniehl and A.A. Penin, hep-ph/9907489.

### *Monopoles and confinement*

617. V.I. Zakharov, hep-ph/0202040.
618. M.N. Chernodub and M.I.Polikarpov, hep-th/9710205;
T. Suzuki, *Prog. Theor. Phys. Suppl.* **131** (1998) 633.
619. A. Digiacomo, talk given at HEP-MAD'01, 27 Sept–5Oct, Antanarivo-Madagascar; *Prog. Theor. Phys. Suppl.* **131** (1998) 161.
620. Y. Nambu, *Phys. Rev.* **D 10** (1974) 4262;
G.'t. Hooft, *in High Energy Physics* (1975), Ed. Compositori, Bologna;
S. Mandelstam, *Phys. Rep.* **C 23** (1976) 516.
621. V.G. Bornyakov *et al.*, hep-lat/0103032.
622. J. Greensite and B. Thorn, hep-ph/0112326.
623. G.'t. Hooft, *QCD 02 Conference*, Montpellier.

## X QCD spectral sum rules

### *Theoretical foundations*

624. G. Källen, *Helv. Phys. Acta.* **25** (1952) 417;
H. Lehmann, *Nuovo. Cimento.* **11** (1954) 342.
625. M. Veltman, *Diagrammatica: the Path of Feynman Diagrams*, Cambridge Lecture Notes in Physics (1994) 50, 51.

### *Survey of QCD spectral sum rules*

*Laplace, FESR sum rules,...*

626. S. Narison and E. de Rafael, *Phys. Lett.* **B 103** (1981) 57.
627. A.A. Lugunov, L.D. Soloviev and A.N. Tavkhelidze, *Phys. Lett.* **B 24** (1967) 181;
J.J. Sakurai, *Phys. Lett.* **B 46** (1973) 207;
A. Bramon, E. Etim and M. Greco, *Phys. Lett.* **B 41** (1972) 609.
628. R. Shankar, *Phys. Rev.* **D 15** (1977) 755.
629. K. Chetyrkin, N.V. Krasnikov and A.N. Tavkhelidze, *Phys. Lett.* **B 76** (1978) 83;
N.V. Krasnikov, A.A. Pivovarov and A.N. Tavkhelidze, *Z. Phys.* **C 19** (1983) 301.

630. M. Kremer *et al.*, *Phys. Rev.* **D 34** (1986) 2127;
I. Caprini and C. Verzegnassi, *Nuovo Cimento* **A 75** (1983) 275;
S. Ciulli *et al.*, *J. Math. Phys.* **25** (1984) 3194;
G. Auberson and G. Mennessier, Montpellier Preprint PM 87/29 (1987);
B. Causse, Montpellier University Thesis, (1989) unpublished.
631. F.J. Yndurain, *Phys. Lett.* **B 63** (1976) 211.
632. V.A. Novikov *et al.*, *Phys. Reports.* **41** (1978) 3.

*HQET sum rules*

633. D.J. Broadhurst and A.G. Grozin, *Phys. Lett.* **B 274** (1992) 421;
E. Bagan, P. Ball, V.M. Braun and H.G. Dosch, *Phys. Lett.* **B 278** (1992) 457;
M. Neubert, *Phys. Rev.* **D 45** (1992) 2451;
V. Eletsky and A.V. Shuryak, *Phys. Lett.* **B 276** (1992) 191;
H.G. Dosch, Heidelberg preprint HD-THEP-93-52 (1993) and references therein.
634. S. Narison, *Phys. Lett.* **B 308** (1993) 365.

*Vertex sum rules, form factors*

635. N.S. Craigie and J. Stern, *Nucl. Phys.* **B 216** (1983) 204.
636. S. Narison and N. Paver, *Phys. Lett.* **B 135** (1984) 159.
637. S. Narison and N. Paver, *Z. Phys.* **C 22** (1984) 69.
638. R. Koniuk and R. Tarrach, *Z. Phys.* **C 18** (1983) 179.
639. A. Radyushkin, *QCD 96 Conference*, Montpellier, *Nucl. Phys. (Proc. Suppl.)* **B, A 54** (1997) ed. S. Narison.
640. N. Nakanishi, *Progr. Theor. Phys.* **25** (1961) 361; *Theor. Phys. Suppl.* **18** (1961) 1;
G. Källen and S. Wightman, *Kgl. Danske Vidensk. Selsk. Mat. Fys. Skr.* **1** n. 6 (1958).
641. V.A. Novikov *et al.*, *Nucl. Phys.* **B 237** (1984) 525.
642. M. Gell-Mann, D. Sharp and W.G. Wagner, *Phys. Rev. Lett.* **8** (1962) 261; H.M. Pilkuhn, *Relatvistic Particle Physics*, Springer Verlag, Berlin, (1979) 176.

*Light cone sum rules*

643. I.I. Balitsky, V.M. Braun and A.V. Kolesnichenko, *Nucl. Phys.* **B 312** (1989) 509;
V.L. Chernyak and I.R. Zhitnitsky, *Nucl. Phys.* **B 345** (1990) 137.
644. G.P. Lepage, *Phys. Lett.* **B 87** (1979) 359; *Phys. Rev.* **D 22** (1980) 2157; **D 24** (1981) 1808;
A. V. Efremov and A. V. Radyushkin, *Phys. Lett.* **B 94** (1980) 245; *Theor. Math. Phys.* **42** (1980) 97; V. L. Chernyak and A. R. Zhitnitsky, *JETP Lett.* **25** (1977) 510; *Sov. J. Nucl. Phys.* **31** (1980) 544.
645. A. Khodjamirian, *Eur. Phys. J.* **C 6** (1999) 477.
646. P. Ball *et al.*, *Nucl. Phys.* **B 529** (1998) 323;
P. Ball and V. Braun, *Nucl. Phys.* **B 543** (1999) 2001.
647. V.L. Chernyak and A.R. Zhitnitsky, *Phys. Rep.* **112** (1984) 173.
648. S. Narison, *Phys. Lett.* **B 224** (1989) 184.
649. V.M. Braun and I. Halperin, *Phys. Lett.* **B 328** (1994) 457.
650. V.M. Braun, A. Khodjamirian and M. Maul, *Phys. Rev.* **D 61** (2000) 073004.

### *Weinberg sum rules 2*

651. M. Davier, L. Girlanda, A. Höcker and J. Stern, hep-ph/9802447.
652. B.L. Ioffe and K.N. Zyablyuk, hep-ph/0010089.

### *QCD condensates 2*

*Tachyonic gluon*

653. C.A. Dominguez and K. Schilcher, *Phys. Rev.* **D 61** (2000) 114020; B.V. Geshkenbein, B.L. Ioffe and K.N. Zyablyuk, hep-ph/0104048.
654. V.I. Zakharov, *The 1999 Sakurai Prize lecture* hep-ph/9906264.

*Anti-SVZ gluon condensate*

655. R.A. Bertlmann and H. Neufeld, *Z. Phys.* **C 27** (1985) 437.
656. J. Marrow and G. Shaw, *Z. Phys.* **C 33** (1986) 237; J. Marrow, J. Parker and G. Shaw, *Z. Phys.* **C 37** (1987) 103.
657. D.J. Broadhurst *et al. Phys. Lett.* **B 329** (1994) 103.
658. K.J. Miller and M.G. Olsson, *Phys. Rev.* **D 25** (1982) 1247, 1253.
659. A. Zalewska and K. Zalewski, *Z. Phys.* **C 23** (1984) 233; K. Zalewski, *Acta. Phys. Polonica.* **B 16** (1985) 239 and references therein.

*Mixed condensate 2*

660. M. Kremer and G. Schierholz, *Phys. Lett.* **B 194** (1987) 283.
661. T. Meissner, hep-ph/9702293 (1997).
662. M.V. Polyakov and C. Weiss, *Phys. Lett.* **B 387** (1996) 841.
663. M. Beneke and H.G. Dosch, *Phys. Lett.* **B 284** (1992) 116.

### *Light and heavy quark masses,...*

*Light quark masses*

664. See e.g. H. Fritzsch, *Nucl. Phys. (Proc. Sup.)* **B64** (1997) 271 .
665. S. Bosch *et al.*, hep-ph/9904408 and references therein.
666. S. Narison, N. Paver, E. de Rafael and D. Treleani, *Nucl. Phys.* **B 212** (1983) 365.
667. S. Gorishny, A.L. Kataev and S.A. Larin, *Phys. Lett.* **B 135** (1984) 457.
668. T.G. Steele and D. Harnett, hep-ph/0108232.
669. J. Kambor and K. Maltman, hep-ph/0108227.
670. A.E. Dorkhov *et al.*, *J. Phys.* **G 23** (1997) 643.
671. L. Lellouch, E. de Rafael and J. Taron, *Phys. Lett.* **B 414** (1997) 195; E. de Rafael, *Nucl. Phys. (Proc. Sup.)* **B 64** 258 (1998).
672. F.J. Yndurain, *Nucl. Phys.* **B 517** (1998) 324.
673. N.I. Ahiezer and M. Krein, 'Some Questions in the Theory of Moments', In: *Translations of Math. Monographs*, Vol. 2, Amer. Math. Soc., Providence, Rhode Island, (1962).
674. L.Giusti *et al.*, hep-lat/9807014.
675. S.Narison, N. Paver and D. Treleani, *Nuov. Cim.* **A 74** (1983) 347.

676. C.A. Dominguez and M. Loewe, *Phys. Rev.* **D 31** (1985) 2930;
C.Ayala, E. Bagan and A. Bramon, *Phys. Lett.* **B 189** (1987) 347.
677. J.Bijnens, J. Prades and E. de Rafael, *Phys. Lett.* **B 348** (1995) 226.
678. J.Prades, *Nucl. Phys. (Proc. Sup.)* **B 64** (1998) 253.
679. C.A. Dominguez and E. de Rafael, *Ann. Phys.* **174** (1987) 372.
680. E.Bagan, A. Bramon, S. Narison and N. Paver, *Phys. Lett.* **B 135** (1984) 463.
681. C.A. Dominguez, L. Pirovano and K. Schilcher, *Nucl. Phys. (Proc. Sup.)* **B 74** (1999) 313 and references therein.
682. P. Colangelo, F. De Fazio, G. Nardulli and N. Paver, hep-ph/9704249 (1997).
683. M. Jamin, *Nucl. Phys. (Proc. Sup.)* **B 64** (1998) 250.
684. K. Maltman, *Phys. Lett.* **B 462** (1999) 195.
685. M. Jamin, A. Pich and J. Oller, hep-ph/0110194.
686. S. Narison and G. Veneziano, *Int. J. Mod. Phys.* **A 4, 11** (1989) 2751.
687. A. Bramon and S. Narison, *Mod. Phys. Lett.* **A 4** (1989) 1113.
688. S. Narison, *Nucl. Phys.* **B 509** (1998) 312; *Nucl. Phys. Proc. Suppl.* **64** (1998) 210 [HEP-PH 9710281]; *Nucl. Phys. Proc. Suppl.* **96** (2001) 244 [HEP-PH 0012235]; *Nucl. Phys.* **A 675** (2000) 54c.
689. P. Minkowski and W. Ochs, *Eur. Phys. J.* **C 9** (1999) 283; W. Ochs, Hadron 2001, Novosibirsk, hep-ph/0111309.
690. L. Montanet, *Nucl. Phys. (Proc. Sup.)* **B 86** (2000) 381.
691. S.N. Cherry and M.R. Pennington, hep-ph/0111158.
692. V.L. Eletsky and B.L. Ioffe, Bern preprint BUTP-93/2 (1993).
693. L.J. Reinders and H. Rubinstein, *Phys. Lett.* **B 145** (1984) 108.
694. V. Lubicz, hep-lat/0012003.
695. C.A. Dominguez, A. Ramlakan and K. Schilcher, *Phys. Lett.* **B 511** (2001) 59.
696. M. Jamin, hep-ph/0201174.

### *Heavy quark masses*

697. C.A. Dominguez and N. Paver, *Phys. Lett.* **B 293** (1992) 197; C.A. Dominguez, G.R. Gluckman and N. Paver, *Phys. Lett.* **B 333** (1994) 184.
698. S. Narison, *Nucl. Phys. Proc. Suppl.* **B 74** (1999) 304;
S. Narison, hep-ph/9712386.
699. M. Eidemueller and M. Jamin, *Phys. Lett.* **B 498** (2001) 203.
700. A. Penarrocha and K. Schilcher, *Phys. Lett.* **B 515** (2001) 291.
701. J.H. Kuhn and M. Steinhauser, hep-ph/0109084.
702. A.D. Martin, J. Outhwaite and M.G. Ryskin, *Eur. Phys. Jour.* **C 19** (2001) 681.
703. A.S. Kronfeld, *Nucl. Phys. (Proc. Suppl.)* **B 63** (1998) 311.
704. K.Hornbostel (NRQCD Coll.), *Nucl. Phys. (Proc. Suppl.)* **B 73** (1999) 339.
705. D. Becirevic, V. Lubicz and G. Martinelli, hep-ph/0107124.
706. J.H. Kuhn, A.A. Penin and A.A. Pivovarov, *Nucl. Phys.* **B 534** (1998) 356.
707. A.A. Penin and A.A. Pivovarov, *Nucl. Phys.* **B 549** (1999) 217;
K. Melnikov and A. Yelkhovsky, *Phys. Rev.* **D 59** (1999) 114009.
708. M. Jamin and A. Pich, *Nucl. Phys. (Proc. Suppl.)* **B 74** (1999) 300.
709. W. Lucha and F.F. Shoeberl, *Phys. Rev.* **D 62** (2000) 097501.
710. M. Beneke and A. Signer, *Phys. Lett.* **B 471** (1999) 233.
711. A.H. Hoang, *Phys. Rev.* **D 61** (2000) 034005; **D 59** (1999) 014039.
712. G. Rodrigo, A. Santamaria and M.S. Bilenky, *Phys. Rev. Lett.* **79** (1997) 193;
DELPHI collaboration, P. Abreu *et al.*, *Phys. Lett.* **B 418** (1998) 430;

A. Brandenburg *et al.*, *Phys. Lett.* **B 468** (1999) 168;
ALEPH collaboration, R. Barate *et al.*, *EPJ.* **C 18** (2000) 1 and F. Palla, *Nucl. Phys. (Proc. Suppl.)* **B 96** (2001) 396.;
OPAL collaboration, G. Abbiendi *et al.*, *EPJ.* **C 21** (2001) 441; R. Seuster, Ph.D. thesis Aachen (2001) and private communication from S. Bethke.
713. V. Gimenez *et al.*, *JHEP.* **0003** (2000) 018.

*D and B decay constants*

714. S. Söldner-Rembold, *ICHEP Conference*, hep-ex/0109023 (2001);
J.D. Richman, *ICHEP Conference*, Warsaw, hep-ex/9701014 (1996).
715. For reviews, see e.g.: C.A. Dominguez, *3rd Workshop on Tau-Charm Factory* Marbella-Spain (1993), ed. J. and R. Kirkby; ibid S. Narison; The BABAR Collaboration, P.F. Harrison and H.R. Quinn, Slac-R-0504 (1997).
716. S. Narison, *Phys. Lett.* **B 322** (1994) 247.
717. S. Narison, *Phys. Lett.* **B 198** (1987) 104.
718. S. Narison, *Phys. Lett.* **B 218** (1989) 238.
719. C.A. Dominguez and N. Paver, *Phys. Lett.* **B 197** (1987) 423; **B 246** (1990) 493.
720. M.A. Shifman and M.B. Voloshin, *Sov. Nucl. Phys.* **45 (2)** (1987) 292;
H.D. Politzer and M.B. Wise, *Phys. Lett.* **B 208** (1988) 504.
721. S. Narison, *Z. Phys.* **C 55** (1992) 671.
722. A. Penin and M. Steinhauser, hep-ph/0108110.
723. For recent reviews, see e.g.: M. Cuichini *et al.*, hep-ph/001308;
C.W. Bernard, *Nucl. Phys. (Proc. Suppl.)* **B 94** (2001) 159.
724. M. Jamin and B.O. Lange, hep-ph/0108135.
725. The MILC collaboration, C. Bernard *et al.*, hep-lat/0011029;
The CP-PAC collaboration, A. Ali Khan *et al.*, hep-lat/0103020.
726. S. Narison, *Phys. Lett.* **B 279** (1992) 137.
727. S. Narison, *Phys. Lett.* **B 285** (1992) 141.
728. S. Narison and K. Zalewski, *Phys. Lett.* **B 320** (1994) 369.
729. S. Collins *et al.*, *Phys. Rev.* **D 63** (2001) 034505.

## *Hadron spectroscopy*

*Mesons*

730. S. Narison, *Nucl. Phys.* **B 182** (1981) 59.
731. E. Bagan, H.G. Dosch, P.Godzinsky, S. Narison and J.M. Richard, *Z. Phys.* **C 64** (1994) 57.
732. M. Lusignoli and M. Masetti, *Z. Phys.* **C 51** (1991) 549.
733. P. Colangelo, G. Nardulli and N. Paver, *Z. Phys.* **C 57** (1993) 43.

*Gluonia*

734. S. Narison, *Z. Phys.* **C 26** (1984) 209; *Phys. Lett.* **B 125** (1983) 501.
735. L. Montanet in [690];
R. Landua, *Proc. of the Int. High-Energy Phys. Conf.*, Varsaw, 1996;
S. Spanier, *QCD 96 Conference*, Montpellier, *Nucl. Phys. (Proc. Suppl.)* **B, A 54** (1997);

A. Palano, *QCD 94 Conference*, Montpellier, *Nucl. Phys. (Proc. Suppl.)* **B, C 39** (1995) 303;
A. Bediaga, *Nucl. Phys. (Proc. Suppl.)* **B 96** (2001) 225.

736. M.S. Chanowitz, *Proc. of the VI Int. Workshop on $\gamma\gamma$ collisions,* ed. R. Lander, World Scientific Company, Singapore, (1984).
737. V. Ableev *et al.*, Legnaro preprint LNL-INFN (Rep) 105/96 (1996); U. Gastaldi, Legnaro preprint LNL-INFN (Rep) 99/95 (1995);
738. D. Bugg, *Nucl. Phys. (Proc. Suppl.)* **B 96** (1999) 218.
739. A. Patel *et al.*, *Phys. Rev. Lett.* **57** (1986) 1288; T.H. Burnett and S.R. Sharpe, *Annu. Rev. Nucl. and Par. Science* **40** (1990) 327 and references therein; M. Teper, University of Oxford preprint OUTP-95-06P (1994); G. Bali *et al.*, *Phys. Lett.* **B 309** (1993) 378; F. Butler *et al.*, *Nucl. Phys.* **B 430** (1994) 179; **B 421** (1994) 217; J. Sexton, A. Vaccarino and D. Weingarten, *Nucl. Phys. (Proc. Suppl.)* **B 42** (1995) 279; C. Morningstar and M. Peardon, *Phys. Rev,* **D 60** (1999) 034509.
740. G. West, *QCD 96 Conference*, Montpellier, *Nucl. Phys. (Proc. Suppl.)* **B, A 54** (1997) and private communication.
741. N.A. Törnquist, hep-ph/9510256 *Proc. HADRON95* Manchester, 1995 and references therein; V.V. Anisovich, *QCD 96 Conference*, Montpellier, *Nucl. Phys. Proc. Suppl.* **B, A 54** (1997); A. Bodyulkov and V. Novozhilov, Trieste preprint IC/89/141 (1989); P. Jain, R. Johnson and J. Schechter, *Phys. Rev,* **D 35** (1987) 2230; M.D. Scadron, *HADRON95*, Manchester (1995).
742. K.L. Au, Thesis RALTO32 (1986) unpublished; K.L. Au, D. Morgan and M.R. Pennington, *Phys. Lett.* **B 167** (1988) 229; G. Mennessier, *Z. Phys.* **C 16** (1983) 241.
743. Different contributions at the workshop on sigma meson, Kyoto, (2000).
744. F. Binon *et al.*, *Nuovo Cimento* **A 78** (1983) 13; D. Alde *et al.*, *Nucl. Phys.* **B 269** (1988) 485; D. Alde *et al.*, *Phys. Lett.* **B 201** (1988) 160.
745. F. Gilman and R. Kauffman, *Phys. Rev.* **D 36** (1987) 2761; L. Montanet, *Proc. Non-Perturbative Methods*, Montpellier, 1985, ed. S. Narison, World Scientific Company, Singapore, (1985); P. Ball, J.M. Frère and M. Tytgat, hep-ph/9508359 (1995).
746. S.S. Gershtein, A.A. Likhoded and Y.D. Prokoshkin, *Z. Phys.* **C 24** (1984) 305.
747. V.A. Novikov *et al.*, *Nucl. Phys.* **B 165** (1980) 67.
748. R.J. Crewther, *Phys. Rev. Lett.* **28** (1972) 1421; J. Ellis and M.S. Chanowitz, *Phys. Lett.* **B 40** (1972) 397; *Phys. Rev.* **D 7** (1973) 2490.
749. S. Narison (unpublished); *Nucl. Phys.* **A 675** (2000) 54c.
750. H.G. Dosch and S. Narison Nucl. Phys. Proc. Suppl. **B 121** (2003) 114 (hep-ph/0208271).
751. G. Mennessier, S. Narison and N. Paver, *Phys. Lett.* **B 158** (1985) 153.
752. C. Amsler *et al.*, *Phys. Lett.* **B 342** (1995) 433; **B 355** (1995) 425; **B 353** (1995) 571; **B 358** (1995) 389; **B 322** (1994) 431; A. Abele *et al.*, *Phys. Lett.* **B 380** (1996) 453.
753. A. Abele *et al.*, *Phys. Lett.* **B 385** (1996) 425.

754. C. Amsler and F.E. Close, *Phys. Lett.* **B 353** (1995) 385; *Phys. Rev.* **D 53** (1996) 295; F.E. Close, *Proc. LEAP96*, Dinkelsbul (1996) (hep-ph/9610426) and references therein.
755. A. Lahiri and B. Bagchi, *J. Phys.* **G 11** (1985) L147.

*Exotics*

756. S. Narison, *Phys. Lett.* **B 175** (1986) 88.
757. For reviews, see e.g.: T.H. Hansson, *Proc. of Non-Perturbative Methods*, Montpellier, ed. S. Narison, World Scientific Company, Singapore, (1985);
A.W. Thomas, *Advances in Nucl. Phys.* **13** (1983) 1;
T.H. Burnett and S.R. Sharpe, in [739];
M.S. Chanowitz, in [735];
T. Barnes and F.E. Close, *τCf Workshop*, Marbella (1993).
758. N.N. Achasov, S.A. Devyanin and G.N. Shestakov, *Phys. Lett.* **B 96** (1980) 168; *Z. Phys.* **C 16** (1982) 55.
759. S. Menke, *Nucl. Phys. (Proc. Suppl.)* **B 86** (2000) 397.
760. S.U. Chung, QCD 99 (Montpellier), *Nucl. Phys. (Proc. Suppl.)* **B 86** (2000) 341.

***B and D exclusive weak decays***

*Heavy to light exclusive decays*

761. S. Narison, *Phys. Lett.* **B 283** (1992) 384.
762. S. Narison, *Phys. Lett.* **B 327** (1994) 354.
763. S. Narison, *Phys. Lett.* **B 345** (1995) 166.
764. S. Narison, *Phys. Lett.* **B 337** (1994) 163.
765. S. Narison, *Phys. Lett.* **B 325** (1994) 197.
766. H.G. Dosch and S. Narison, *Phys. Lett.* **B 368** (1996) 163;
H.G. Dosch, *QCD 96 Conference*, Montpellier, *Nucl. Phys. Proc. Suppl.* **B, A 54** (1997) 253.
767. A.A. Ovchinnikov and V.A. Slobodenyuk, *Z. Phys.* **C 44** (1989) 433;
V.N. Baier and D. Grozin, *Z. Phys.* **C 47** (1992) 669.
768. P. Colangelo and P. Santorelli, *Phys. Lett.* **B 327** (1994) 123; A.A. Ovchinnikov, *Phys. Lett.* **B 229** (1989) 127.
769. T.M. Aliev, A.A. Ovchinnikov and V.A. Slobodenyuk, Trieste preprint IC/89/382 (1989) (unpublished).
770. M. Nielsen and F.S. Navarra, HEP-MAD'01 International Conference, 27Sept–5Oct, Antananarivo-Madagascar.
771. P. Ball, *Phys. Rev.* **D 48** (1993) 3190.
772. K-C. Yang and W-Y. P. Hwang, *Z. Phys.* **C 73** (1997) 275.
773. P. Ball, V.M. Braun and H.G. Dosch, *Phys. Rev.* **D 44** (1991) 3567.
774. P. Ball, V.M. Braun and H.G. Dosch, *Phys. Lett.* **B 273** (1991) 316.
775. T.M. Aliev, V.I. Eletsky and Ya.I. Kogan, *Sov. J. Nucl. Phys.* **40** (1984) 527.
776. H.G. Dosch, E.M. Ferreira, F.S. Navarra and M. Nielsen, hep-ph/0203225.
777. P. Colangelo and F. De Fazio, *Phys. Lett.* **B 520** (2001) 78.
778. V. Belyaev, A. Khodjamirian and R. Rückl, *Z. Phys.* **C 60** (1993) 349.
779. P. Colangelo *et al.*, *Phys. Rev.* **D 53** (1996) 3672.

780. V. Belyaev, V.M. Braun, A. Khodjamirian and R. Rückl, *Phys. Rev.* **D 51** (1995) 6177; T.M. Aliev, A. Ozpineci and M. Savci, hep-ph/9612480 (1996).
781. A. Ali, V. Braun and H. Simma, CERN-TH.7118/93 (1993);
782. P. Ball and V.M. Braun, hep-ph/9701238 (1997).
783. M. Gourdin, Y.Y. Keum and X.Y. Pham, hep-ph/9501257.
784. R. Akhoury, G. Sterman and Y.-P. Yao, *Phys. Rev.* **D 50** (1994) 358.
785. CLEO collaboration, M.S. Alam *et al. Phys. Rev.* **D 50** (1994) 43; CDF collaboration, FERMILAB Conf-94/127-E (1994); ARGUS collaboration, *Phys. Lett.* **B 340** (1994) 217.
786. M. Gourdin, A.N. Kamal and X.Y. Pham, *Phys. Rev. Lett.* **73** (1994) 3355; R. Aleksan, A. Le Yaounac, L. Oliver, O. Pène and J.C. Raynal, hep-ph/9408215; B. Stech, Heidelberg preprint HD-THEP-95-4 (1995).
787. CLEO II collaboration, M.S. Alan *et al.*, *Phys. Rev. Lett.* **71** (1993) 1311.
788. MARK III collaboration, J.S. Adler *et al.*, *Phys. Rev. Lett.* **62** (1989) 1821.

*Heavy to heavy transitions*

789. E. de Rafael and J. Taron, Marseille preprint CPT-93/P.2908 (1993).
790. A.F. Falk, H. Georgi and B. Grinstein, *Nucl. Phys.* **B 343** (1990) 1.
791. M.B. Voloshin, *Phys. Rev.* **D 46** (1992) 3062.
792. J.D. Bjorken, *Rencontre de Physique de la vallée d'Aoste*, la Thuile *Editions Frontières*, ed. M. Greco, (1990).
793. CLEO II collaboration, W. Ross, *QCD 94 Conference*, Montpellier, *Nucl. Phys. (Proc. Suppl.)* **B, C 39** (1995) 303 ed. S. Narison and references therein.
794. M. Neubert, *Phys. Lett.* **B 338** (1994) 84.
795. E. Bagan, P. Ball and P. Gosdzinsky, *Phys. Lett.* **B 301** (1993) 249.
796. M. Neubert, *Phys. Rev.* **D 47** (1993) 4063.
797. B. Blok and M.A. Shifman, *Phys. Rev.* **D 47** (1993) 2949.
798. For reviews, see e.g.: C.T. Sachrajda, *Acta Phys. Polonica.* **B 26** (1995) 731; G. Martinelli, hep-ph/9610455; L. Lellouch, *Acta Phys. Polonica.* **B 25** (1994) 1679.
799. The CLEO collaboration: S. Ahmed *et al.*, *Phys. Rev. Lett.* **87** (2001) 251801; A. Anastassov *et al.*, hep-ex/0108043.
800. P. Colangelo, G. Nardulli, A. Deandrea, N. Di Bartolomeo, R. Gatto and F. Feruglio, *Phys. Lett.* **B 339** (1994) 151;
F.S. Navarra *et al.*, hep-ph/0005026;
For more complete references see e.g.: D. Guetta and P. Singer, hep-ph/0009057.

*$B_c$ decays*

801. P. Colangelo, G. Nardulli and N. Paver, *Z. Phys.* **C 57** (1993) 43.
802. E. Eichten and C. Quigg, *Workshop on Beauty Physics at Colliders*, Fermilab-pub-91/032-T (1994); C. Quigg, Fermilab-Conf-93/265-T (1993).
803. M. Bauer, B. Stech and M. Wirbel, *Z. Phys.* **C 29** (1985) 637.
804. H.G. Dosch, *Nucl. Phys. (proc. Suppl.)* **B 50** (1996) 50.
805. R.L. Jaffe, *Phys. Lett.* **B 245** (1990) 221.
806. R.L. Jaffe and P.F. Mende, *Nucl. Phys.* **B 369** (1992) 189.

### $B^0_{(s)}$-$\bar{B}^0_{(s)}$ mixing, kaon CP-violation

*Generalities*

807. J.S. Bell and J. Steinberger, *Weak Interactions of Kaons*, Proc. Oxford Int. Conf. on Elementary Particles (1965) 195.
808. J.M. Christenson, J. Cronin, W.L. Fitch and R. Turlay, *Phys. Rev. Lett.* **13** (1964) 138.
809. T.T. Wu and C.N. Yang, *Phys. Rev. Lett.* **13** (1964) 380.

*$B^0$-$\bar{B}^0$ mixing*

810. C. Albajar *et al.*, UA1 collaboration, *Phys. Lett.* **B 186** (1987) 237, 247.
811. T. Inami and C.S. Lim, *Prog. Theor. Phys.* **65** (1981) 297; **65** (1981) 1772.
812. S. Herrlich and U. Nierste, *Nucl. Phys.* **B 419** (1994) 192.
813. A. Pich, *Phys. Lett.* **B 206** (1988) 322.
814. B. Guberina *et al.*, *Phys. Lett.* **B 128** (1983) 269;
A. Pich and E. de Rafael, *Phys. Lett.* **B 158** (1985) 477; *Nucl. Phys.* **B 358** (1991) 311;
J. Prades *et al.*, *Z. Phys.* **C 51** (1991) 287.
815. S. Narison, *Phys. Lett.* **B 351** (1995) 369.
816. S. Narison and A. Pivovarov, *Phys. Lett.* **B 327** (1994) 341.
817. K. Hagiwara, S. Narison and D. Nomura, hep-ph/0205092;
S. Narison, hep-ph/0203171, HEP-MAD'01 International High-Energy Physics Conference, World Scientific, Singapore.
818. S. Narison, *Phys. Lett.* **B 327** (1994) 354; **B 337** (1994) 163; **B 283** (1992) 384.
819. Work in preparation.
820. A. Ali and D. London, *Nucl. Phys.* **B, A 54** (1997) 297; *Eur. Phys. J.* **C 9** (1999) 687;
A. Hocker *et al.*, hep-ph/0104062.

*Theory of $\epsilon'/\epsilon$*

821. G. Ecker *et al.*, *Phys. Lett.* **B 477** (2000) 88.
822. A.I. Vainshtein, V.I. Zakharov and M.A. Shifman, *Sov. Phys. JETP* **45**(4) (1977) 670; V.I. Zakharov (private communication).
823. For a review, see.e.g., G. Martinelli, hep-ph/9910237.
824. Rome collaboration, A. Donini *et al.*, hep-lat/9910017;
L. Giusti, QCD99 Euroconference (Montpelllier), *Nucl. Phys. (Proc. Suppl.)* **B 86** (2000) 299.
825. For a review, see e.g., R. Gupta, *Nucl. Phys. (Proc. Suppl.)* **B 63** (1998) 278.
826. D. Pekurovsky and G. Kilcup, hep-lat/9812019.
827. T. Hambye *et al.*, *Phys. Rev.* **D 58** (1998) 014017; hep-ph/9906434;
T Hambye and P. Soldan, hep-ph/9908232.
828. S. Bertolini, J.O. Eeg and M. Fabbrichesi, hep-ph/0002234; *Rev. Mod. Phys.* **72** (2000) 65.
829. J. Bijnens and J. Prades, *JHEP.* **01** (1999) 23;
J. Prades, QCD 00 Euroconference (Montpellier), *Nucl. Phys. (Proc. Suppl.)* **B 96** (2001) 354.
830. T. Morozumi, C.S. Lim and A.I. Sanda, *Phys. Rev. Lett.* **65** (1990) 404;
T. Morozumi, A.I. Sanda and A. Soni, *Phys. Rev.* **D 46** (1992) 2240;
Y.Y. Keum, U. Nierste and A.I. Sanda, *Phys. Lett.* **B 457** (1999) 157;
E.P. Shabalin, *Sov. J. Nucl. Phys.* **48** (1) (1988) 172.

831. M. Harada *et al.*, hep-ph/9910201.
832. M. Knecht, S. Peris and E. de Rafael, *Phys. Lett.* **B 457**(1999) 227; ibid, QCD 99 Euroconference (Montpellier), *Nucl. Phys. (Proc. Suppl.)* **B 86** (2000) 279 and references therein; S. Peris and E. de Rafael (private communications).
833. J.F. Donoghue and E. Golowich, hep-ph/9911309.
834. E. Pallante and A. Pich, hep-ph/9911233.
835. A.J. Buras *et al.*, hep-ph/0002116; see however: E. Pallante, A. Pich and I. Scimemi, hep-ph/0105011; G. Mennessier (private communication).
836. S. Narison, *Nucl. Phys.* **B 593** (2001) 3; S. Narison, *Nucl. Phys. Proc. Suppl.* **B 96** (2001) 364.
837. J.F. Donoghue, *Nucl. Phys. Proc. Suppl.* **B 96** (2001) 329; V. Cirigliano, J.F. Donoghue and E. Golowich, hep-ph/0007196.
838. For references to recent works, see e.g., J. Prades, E. Gamiz and J. Prades, *JHEP*, **0110** (2001) 009; E. de Rafael, hep-ph/0110195; S. Peris, hep-ph/0204181; J. Bijnens, hep-ph/0207082; J. Prades (private communication).


## *Plasma*


839. A. Barducci *et al.*, *Phys. Lett.* **B 244** (1990) 311.
840. M. Alford, K. Rajagopal and F. Wilczek, *Phys. Lett.* **B 422** (1998) 247; R. Rapp *et al.*, *Phys. Rev. Lett.* **81** (1998) 53.
841. R. Rapp, T. Schäfer and I. Zahed, *Phys. Rev.* **D 63** (2001) 034008; M. Alford, J. Bowers and K. Rajagopal, *Phys. Rev.* **D 63** (2001) 074016.
842. A.I. Bochkarev and M.E.Shaposhnikov, *Nucl. Phys.* **B 268** (1986) 220.
843. H.G. Dosch and S. Narison, *Phys. Lett.* **B 203** (1988) 155.
844. R.J. Furnstahl, T. Hatsuda and S.H. Lee, *Phys. Rev.* **D 42** (1990) 1744.
845. C.A. Dominguez and M. Loewe, hep-ph/0008250 and references therein, *Phys. Lett.* **B 233** (1989) 201.
846. C.A. Dominguez and M. Loewe, hep-ph/9406213.
847. V.L. Eletsky and B.L. Ioffe, *Phys. Rev.* **D 47** (1993) 3083.
848. M.B. Johnson and S. Kisslinger, hep-ph/9908322.
849. S.H. Lee, C. Song and H. Yabu, *Phys. Lett.* **B 341** (1995) 407.
850. S. Mallik and K. Mukherjee, *Phys. Rev.* **D 58** (1998) 096011.
851. M. Chabab, N.E. Biaze and R. Markazi, hep-ph/001324.
852. J. Gasser and H. Leutwyler, *Phys. Lett.* **B 184** (1987) 83; P. Gerber and H. Leutwyler, *Nucl. Phys.* **B 321** (1989) 387 and references therein.
853. A. Barducci *et al.*, *Phys. Rev.* **D 46** (1992) 2203.
854. V.L. Eletsky, P.J. Ellis and J.L. Kapusta, hep-ph/9302300.
855. L.D. Landau, *ZhETF* **37** (1958) 805.
856. T. Matsubara, *Prog. Theor. Phys.* **14** (1955) 351.
857. L. Dolan and R. Jackiw, *Phys. Rev.* **D 9** (1974) 3320.
858. E.S. Fradkin, *Proc. P.N. Lebedev Phys. Inst.* **29** (1965) 3.
859. E. Braaten and R.D. Pisarski, *Nucl. Phys.* **B 337** (1990) 569; **B 339** (1990) 310; J. Frenkel and J.C. Taylor, *Nucl. Phys.* **B 334** (1990) 199; **B 374** (1992) 156.
860. J.C. Taylor and S.M.H. Wong, *Nucl. Phys.* **B 346** (1990) 115; E. Braaten and R.D. Pisarski, *Phys. Rev.* **D 45** (1992) 1827.

861. For a review, see e.g. P. Aurenche, *Nucl. Phys. (Proc. Suppl.)* **B 96** (2001) 179 and references therein.
862. T.L. Blum *et al.*, *Phys. Rev.* **D 51** (1995) 5133;
C. Boyd *et al.*, hep-lat/9501029.
863. J.R. Pelaez, QCD 02 Montpellier Conference; hep-ph/0202265.
864. R.D. Pisarski, *Phys. Rev.* **D 52** (1995) 3773.
865. C.A. Dominguez and M. Loewe, *Z. Phys.* **C 49** (1991) 423.
866. V.L. Eletsky *et al.*, *Phys. Rev.* **C 64** (2001) 035202.
867. R.D. Pisarski, T.L. Trueman and M.H.G. Tytgat, *Phys. Rev.* **D 56** (1997) 7077.

### *More on spectral sum rules*

868. S. Narison, *Phys. Lett.* **B 122** (1983) 171.
869. G. Girardi, S. Narison and M. Perrottet, *Phys. Lett.* **B 133** (1983) 234;
S. Narison and M. Perrottet, *Nuov. Cim.* **A 90** (1985) 49.
870. S. Narison, *Z. Phys.* **C 28** (1985) 591; *HEP Conference*, Bari, (1985) 152.
871. H. Fritzsch and G. Mandelbaum, *Phys. Lett.* **B 102** (1981) 319;
L. Abbott and E. Farhi, *Nucl. Phys.* **B 189** (1981) 547.
872. R.D. Peccei, Munich preprint MPI-PAE/PTh 35/84 (1984);
W. Buchmuller, R.D. Peccei and T. Yanagida, *Phys. Lett.* **B 124** (1983) 67;
W. Buchmuller, CERN preprint Th 4189/85 (1985);
H. Harari, *SLAC Summer Institute*, Stanford, SLAC report 259 (1982);
H. Fritzsch, Munich preprint MPI-PAE/PTh 31/84 (1984);
L. Lyons, Oxford preprint 58/82 (1982);
M. Peskins, SLAC-pub 3852 (1985);
D. Amati, K. Konishi, Y. Meurice, G.C. Rossi and G. Veneziano, *Phys. Rep.* **C 162** (1988) 169.
873. S. Narison, *Phys. Lett.* **B 142** (1984) 168.
874. S. Narison and P. Pascual, *Phys. Lett.* **B 150** (1985) 363.

### XI Appendices

875. K.G. Chetyrkin and F.V. Tkachov, *A New Approach to Multiloop Feynman Integrals*, INR preprint (1979).
876. L. Lewin, *Dilogarithms and Associated Functions*, MacDonald, London, (1958);
H. Bateman, *Higher Transcendental Functions*, Vol. 1, MacGraw-Hill, New-York, Toronto, London, (1953).
877. H.B. Dwight, *Tables of Integrals and Other Mathematical Data*, Macmillian, New York, (1961);
I.S. Gradshtein and I.M. Ryzhik, *Tables of Integrals, Series and Products*, Academic Press, New York, (1980).
878. K. Chetyrkin (private communication).

# Index

Printed in the United States
by Baker & Taylor Publisher Services

Printed in the United States
by Baker & Taylor Publisher Services